Environmental Forensics

Environmental Forensics
Contaminant Specific Guide

Editors
Robert D. Morrison, Ph.D.
Brian L. Murphy, Ph.D.

AMSTERDAM • BOSTON • HEIDELBERG • LONDON
NEW YORK • OXFORD • PARIS • SAN DIEGO
SAN FRANCISCO • SINGAPORE • SYDNEY • TOKYO
Academic Press is an imprint of Elsevier

Acquisitions Editor	Christine Minihane
Project Manager	A. B. McGee
Editorial Assistant	Andrea Sherman
Marketing Managers	Steve Brewster, Tara Isaacs
Cover Design	Eric DeCicco
Composition	Integra Software Services
Cover Printer	Phoenix Color
Interior printer	Maple-Vail Book Manufacturing Group

Academic Press is an imprint of Elsevier
30 Corporate Drive, Suite 400, Burlington, MA 01803, USA
525 B Street, Suite 1900, San Diego, California 92101-4495, USA
84 Theobald's Road, London WC1X 8RR, UK

This book is printed on acid-free paper. ∞

Library of Congress Cataloging-in-Publication Data
Application Submitted

British Library Cataloguing-in-Publication Data
A catalogue record for this book is available from the British Library

ISBN 13: 978-0-12-507751-4
ISBN 10: 0-12-507751-3

For information on all Elsevier Academic Press publications
visit our Web site at www.books.elsevier.com

Printed and bound by CPI Group (UK) Ltd, Croydon, CR0 4YY

Transferred to Digital Print 2011

Contents

CHAPTER 9   PERCHLORATE 167

Robert D. Morrison, Emily A. Vavricka, and P. Brent Duncan

CHAPTER 10   POLYCHLORINATED BIPHENYLS 187

Glenn W. Johnson, John F. Quensen, III,
Jeffrey R. Chiarenzelli, and M. Coreen Hamilton

Scott A. Stout, Gregory S. Douglas, and Allen D. Uhler

Foreword

Environmental forensics is the systematic and scientific evaluation of physical, chemical and historical information for the purpose of developing defensible scientific and legal conclusions regarding the source or age of a contaminant release into the environment. Within this general definition, the science of environmental forensics has evolved into a global scientific discipline with numerous applications.

The purpose of this book is to provide the student or scientist of environmental forensics with a contaminant-specific resource for investigating and solving the questions of when a contaminant release occurred, the origin of the release, and a basis for apportioning liability among multiple responsible parties. The impetus for structuring a book on a contaminant-specific approach rather than a methodological or analytical perspective is to provide the reader with the ability to quickly review contaminant-specific information that is most germane to their particular contaminant-based inquiry. While the media for these investigations vary widely (air, soil, sediments, groundwater, surface water, etc.), the available forensic techniques are remarkably similar.

The basic structure of each contaminant-specific chapter was so devised that the reader is provided with an overview of the chemistry of the contaminant, current analytical methods used for detection, identification of natural and anthropogenic sources, and finally a presentation of forensic techniques for consideration. As with many sciences, certain contaminants are more studied and forensically evolved than others. Petroleum hydrocarbons, for example, have decades of forensic technique development, as contrasted with the emerging contaminants, such as dioxins or perchlorate. The sophistication of the forensic technique is also often a function of the global interest of the scientific community in a particular contaminant or forensic technique. As a result, many forensic techniques are rapidly changing not only as a function of their global interest but also due to their migration from an existing scientific field or study. An example of this migration is the development of microbiological forensics. While this scientific discipline is extensively employed in criminal and bioterrorism applications, it is only recently being employed in environmental forensic investigations.

Another reason for the evolution of environmental forensics is rapid advances in analytical methods, such as comprehensive gas chromatography, gas chromatography with isotopic ratio mass spectrophometry, and laser ablation inductively coupled plasma mass spectrometry, that provide greater analytical precision, lower detection limits, and a greater number of contaminants to examine forensically. When used in concert, these techniques as well as emerging statistical tools provide the forensic investigator with a multitude of possible techniques for answering the forensic questions of contaminant age dating and origin.

It is our hope that the material in this book will be a useful contaminant-specific reference and will nurture ideas for developing additional techniques for use by the environmental forensic community. We wish you the greatest success in your environmental investigation and in your contributions to this emerging science.

Robert D. Morrison, Ph.D.
Brian L. Murphy, Ph.D.

Contributors

EDITORS **Robert D. Morrison**, Ph.D. is a soil physicist with over 35 years of experience as an environmental consultant. Dr. Morrison has a B.S. in Geology, an M.S. in Environmental Studies, an M.S. in Environmental Engineering, and a Ph.D. in Soil Physics from the University of Wisconsin at Madison. Dr. Morrison published the first book on environmental forensics in 1999, is the editor of the *Environmental Forensics Journal* and is Director of the International Society of Environmental Forensics (ISEF). Dr. Morrison specializes in the use of environmental forensics for contaminant age dating and source identification.

Brian L. Murphy, Ph.D. is a principal scientist at Exponent. He is the author of more than 30 journal publications and is on the editorial board of the journal *Environmental Forensics*. Dr. Murphy has an Sc.B. from Brown University and M.S. and Ph.D. degrees from Yale University in physics. Dr. Murphy has had Visiting Instructor positions at Harvard School of Public Health and the University of South Florida. His practice focuses on reconstructing past contaminating events, either for purpose of remedial cost allocation or in order to determine doses in toxic torts.

CHAPTER AUTHORS **Andrew S. Ball**, Ph.D. is Foundation Chair in Environmental Biotechnology at Flinders University of South Australia in Adelaide. Dr. Ball previously spent 16 years working at the University of Essex in the United Kingdom. Environmental issues such as contaminated land and water, climate change, and loss of biodiversity are key issues at both local and international scales. His research is focused on the response, in terms of activity and diversity of the microbial community, to environmental perturbations. One of his major interests is the treatment of contaminated land and water.

Donna M. Beals attended the University of South Carolina as a marine science major, and then attended graduate school at the University of California, Santa Cruz. Her research involved the use of naturally occurring radionuclides to study ocean processes. Prior to joining the Savannah River National Laboratory, she worked as a research chemist at a commercial laboratory which provided analytical services to several Department of Energy sites, thereby learning about the man-made radionuclides in the environment. Since joining the Savannah River National Lab in 1991, Ms. Beals has been involved with developing very sensitive analytical techniques for environmental monitoring and for nuclear forensic analyses. Her publications cover the periodic table, from tritium to plutonium, from analytical method to instrument development, with sampling sites from the central Pacific gyre (near Hawaii) to the Arctic Ocean.

Philip B. Bedient, Ph.D. is a recognized expert in surface-water and groundwater hydrology, simulation modeling, and the fate and transport of contaminants in the environment. Dr. Bedient is Herman Brown Professor of Engineering and has authored four books and more than 125 professional articles in the past 30 years. Dr. Bedient holds B.S., M.S., and Ph.D. degrees from the University of Florida. He serves as an expert witness in toxic tort and environmental litigation cases.

Laurie Benton, Ph.D. received a Ph.D. in Geosciences from the University of Tulsa in 1997 and an M.S. in Geochemistry from New Mexico Institute of Mining and Technology in 1991. She has been an environmental consultant for 7 years at Exponent, Geraghty & Miller, and the law firm of Gardere & Wynne. Prior to joining Exponent, Dr. Benton conducted isotopic geochemistry research as a National Science Foundation postdoctoral fellow at the Carnegie Institution of Washington, Department of Terrestrial Magnetism in Washington, DC.

Brad Bessinger, Ph.D. is a senior geochemist with Exponent in Lake Oswego, OR, and specializes in environmental chemistry and the processes affecting the fate and transport of organic compounds and metals in the environment. He received his Ph.D. from the University of California at Berkeley in 2000, with a dissertation focused on the speciation and mobility of metals such as mercury under hydrothermal conditions. His previous professional activities include the development of conceptual and numerical models of mercury cycling and bioaccumulation in the San Francisco Bay estuary, the evaluation of treatment alternatives for mercury-contaminated soils and sediments, and the behavior of mercury spilled from gas pressure regulators and mercury vapor in indoor air. Dr. Bessinger's analyses in the field of environmental forensics have been used in a number of lawsuits concerned with source allocation and exposure.

Gary N. Bigham, LG, is a principal with Exponent in Bellevue, WA, and specializes in the evaluation of contaminant and sediment transport and fate in the environment. He received his B.S. in Geology from Oregon State and his M.S. in Geophysical Sciences from Georgia Tech. Gary has undertaken several investigations of mercury in the environment and in indoor air over the past 15 years. The largest has been the comprehensive remedial investigation of mercury cycling and bioaccumulation in Onondaga Lake, NY. He also recently participated in a natural resource damage assessment of the Guadalupe River, CA, which drains the New Almaden Mining District, the largest mercury mining area in the United States. Over the past ten years, he has been involved with litigation regarding the influence of nutrients on mercury cycling and bioaccumulation in the Florida Everglades. Gary has participated in investigations at many other mercury-contaminated sites and published numerous papers and presentation abstracts. He also led an extensive evaluation of the behavior of mercury spilled from gas pressure regulators and mercury vapor in indoor air and served as an expert witness in litigation involving mercury spilled in buildings and homes.

Paul D. Boehm, Ph.D. is Group Vice President and Principal Scientist of Exponent's Environmental business. Dr. Boehm has devoted 28 years to environmental consulting experience on chemical aspects of aquatic and terrestrial contamination. Dr. Boehm is a recognized environmental chemist and marine scientist; an originator of advanced techniques for petroleum and PAH fingerprinting; a developer of the US "Mussel Watch" national monitoring program; and an authority on fate/transport/effects of oil and chemical spills. His expertise includes extensive knowledge of the strategic application and practice of environmental forensics (chemical fingerprinting, fate/transport, source attribution, and allocation) using polycyclic aromatic hydrocarbons (PAHs), polychlorinated biphenyls (PCBs), dioxins, and other tracers. Dr. Boehm has been extensively involved in the *Exxon Valdez* oil spill science programs for over 16 years. He has published more than 100 articles in peer-reviewed journals and has been appointed to serve on several National Academy of Sciences panels.

Teresa S. Bowers, Ph.D. is environmental scientist with a Ph.D. in Geochemistry, with experience in exposure, mathematical and geochemical modeling, and the application of this information to risk-based environmental strategies. Dr. Bowers has worked as a consultant at Gradient Corporation for 15 years, where she has developed blood lead and urine arsenic models that relate human exposure to environmental sources of lead and arsenic. She has also published unique statistical approaches to calculating soil cleanup levels for a variety of contaminants. Prior to her work at Gradient, Dr. Bowers held research and visiting faculty positions at the Massachusetts Institute of Technology and Harvard University, where she taught courses in resource geology and applied thermodynamics.

Leigh A. Burgoyne, Ph.D. is a biochemist/molecular biologist at the School of Biological Sciences, Flinders University in Adelaide, South Australia. Dr. Burgoyne has interests in the degradation of DNA, chromatin structure, genome structures, and practical applications of these fields. Dr. Burgoyne has extensively studied issues regarding forensic DNA issues and the problems of long-term DNA storage. Dr. Burgoyne co-authored one of the first articles on the use of forensic DNA issues in 1987 titled "Hypervariable lengths of human DNA associated with a human satellite III sequence found in the 3.4kb Y-specific fragment" (*Nucleic Acids Research* 15: 3929). Dr. Burgoyne also designed one of the earliest DNA storage materials specifically designed to protect DNA and co-edited one of the earliest books on the topic, *DNA in Forensic Science* (ISBN 0-13-217506-1), in 1990.

David E.A. Catcheside, Ph.D. is Professor of Genetics and Head of the School of Biological Sciences at Flinders University in Adelaide, South Australia. Dr. Catcheside's primary research interest is the understanding of how information is exchanged between chromosomes during meiotic recombination. In recent years, he has collaborated with Leigh Burgoyne in developing novel DNA-based forensic tools.

Jeffrey R. Chiarenzelli, Ph.D. is a geologist and environmental scientist living and working in northern New York. His research interests include fate, transport, and transformation of polychlorinated biphenyls in the environment. Dr. Chiarenzelli's current projects include the relationship between atmospheric deposition of contaminants and lake-effect precipitation, and differentiating locally derived from far-traveled contaminants in the Arctic.

Jan H. Christensen, Ph.D. is assistant professor at the Department of Natural Sciences, The Royal Veterinary and Agricultural University, Copenhagen, Denmark. His specialties include oil spills in the marine and terrestrial environment; oil spill identification; transport and transformation of petroleum hydrocarbon mixtures (e.g., crude oil and refined petroleum products); chemical analysis of complex mixtures of contaminants in the environment (environmental profiling) using modern analytical techniques; development of novel tools for automated chromatographic preprocessing; chemometric data analysis of preprocessed analytical data (e.g., chromatographic data and fluorescence landscapes); statistical evaluation of sample similarities; fate of complex chemical mixtures in environmental samples; and risk assessment of pollutant mixtures. Dr. Christensen manages a laboratory at the Royal Veterinary and

Agricultural University, Denmark, and has authored 36 publications on analytical chemistry subjects.

Winnie Dejonghe, Ph.D. is a project manager within the Environmental and Process Technology Centre of the Flemish Institute for Technological Research (VITO) in Belgium. Dr. Dejonghe is involved in research concerning the remediation of contaminated soils and aquifers through natural attenuation (aerobic and anaerobic removal of BTEX, CAH, heavy metals, and pesticides) and the study of the microbial community structure by the use of molecular techniques (RT-PCR-DGGE).

Richard E. Doherty, PE, LSP, is a PE in LSP and he holds an M.S. in Civil/Environmental Engineering from the Massachusetts Institute of Technology and a B.S. from the University of Lowell. Mr. Doherty has worked in the environmental investigation and remediation field since 1987, and is the founder and President of Engineering & Consulting Resources, Inc. (www.ecr-consulting.com). Mr. Doherty has extensively researched the history of use and manufacturing of industrial/commercial chemicals, particularly chlorinated solvents. Articles by Mr. Doherty on the subject of chlorinated solvents are recognized classic articles on this subject. He is a Registered Professional Engineer in four states, and is a Massachusetts Licensed Site Professional.

Gregory S. Douglas, Ph.D. has over 25 years of experience in the field of environmental chemistry. He has designed, implemented, managed, audited, and defended a wide range of environmental forensic chemistry studies for government and industry concerning complex petroleum and fuel contamination issues in marine and soil/groundwater systems. Dr. Douglas has performed extensive research concerning the fate of gasoline NAPL and gasoline additives in groundwater. He has written interpretive reports on more than 100 site or incident investigations involving gasoline source and age dating issues, and has published and presented on the development and application of environmental forensic analytical methods, and source identification and allocation within complex contaminated environments.

P. Brent Duncan, Ph.D. has a B.A. in Geology (Earth Science) and Environmental Studies from Baylor University, an M.S. in Environmental Science from Baylor University, and a Ph.D. in Environmental Science from the University of North Texas. Recent emphasis has been in freshwater systems, geographic information systems, and hydrologic modeling.

Arthur F. Eidson, Ph.D. is a chemist with 30 years of diversified experience studying the interactions of metals with biological systems and assessing risks associated with environmental contaminants. He received a Batchelor of Science in Chemistry degree from the University of Michigan and a Ph.D. from the University of Illinois, Chicago. While at the Inhalation Toxicology Research Institute (now named the Lovelace Respiratory Research Institute), Dr. Eidson conducted field investigations and laboratory experiments on the biological fate of inhaled industrial uranium and mixed uranium-plutonium aerosols by their physical and chemical properties. Since 1991, he has managed risk assessments of radioactive, hazardous, and mixed-waste contamination for commercial chemical and oil company clients, state environmental agencies, and governmental agencies, including the Federal Aviation Administration, the US Department of Defense, and the US Department of Energy. He is a member of the Society for Risk Analysis and the Health Physics Society, and served as a member of the National Council on Radiation Protection and Measurements (NCRP), Task Group-15 on Uranium.

Melanie R. Edwards received her M.S. in Statistics from the University of Wisconsin at Madison in 1998 and a B.S. in mathematics from the University of Washington in 1991. She has 12 years of experience in environmental consulting both as a statistician and as a data manager. Her work ranges from preparing summary statistics for risk calculations to predicting contaminant trends in biological tissues to evaluating results of relative bioavailability studies.

Merv Fingas, Ph.D. is Chief of the Emergencies, Science and Technology Division of Environment Canada. Dr. Fingas has a Ph.D. in Environmental Physics from McGill University and three master degrees. His specialties include spill dynamics and behavior, spill-treating agent studies, remote sensing and detection, in-situ burning, and the technology of personal protection equipment. He has devoted the last 30 years of his life to spill research and has over 600 papers and publications in the field. Dr. Fingas is a member of several editorial boards including being editor-in-chief of the *Journal of Hazardous Materials*, the leading scientific journal covering chemical fate, behavior, and countermeasures.

A. Mohamad Ghazi, Ph.D. is an analytical geochemist with ERS Consulting, Atlanta, GA. He has a bachelor's degree in Geology and Chemistry and a Ph.D. in Geology. Previously he was Associate Professor of Geology and Director of the Laser Ablation Plasma Mass Spectrometry (LA-ICPMS) at Georgia Sate University. Dr. Ghazi's areas of research are in analytical and environmental geochemistry, with special focus in the field of forensic chemistry. With more than 15 years of laboratory and classroom experience, in addition to classic analytical chemistry techniques, Dr. Ghazi has made significant contributions in developing an application of laser ablation for in-situ analysis of materials. Dr. Ghazi is the author of 65 peer reviewed articles and principal investigator on 15 successful research proposals to National Science Foundations and other private agencies. He is an honorary member of Sigma Xi International Scientific Society and he serves on the editorial board for the *International Journal of Environmental Forensics*.

Duane Graves, Ph.D. received a Ph.D. in Life Sciences from the University of Tennessee and is currently a principal with GeoSyntec Consultants with over 20 years of research and consulting experience. His professional practice areas include environmental biotechnology; environmental forensics; in situ groundwater, soil, and sediment remediation; and detection and remediation of biological contaminants. Dr. Graves organized and led the first large-scale commercial application of real-time on-site polymerase chain reaction technology for the detection of *Bacillus anthracis* in response to the 2001 bioterrorism attack launched through the United States Postal Service. He is also actively involved in the development of new bioremediation technologies and innovative sampling and analytical techniques for environmental applications.

M. Coreen Hamilton, Ph.D. is an analytical chemist specializing in method development for monitoring organic contaminants in the environment and in biological tissue samples. Dr. Hamilton has been the senior scientist at AXYS Analytical for over 20 years, where she has directed the research leading to the development of methods for the analysis for perfluorinated chemicals, polybrominated diphenylethers, PCB congeners, brominated dioxins and furans, and a variety of other compounds. In conjunction with the EPA Office of Water and her AXYS colleagues, Dr. Hamilton assisted in the development of Method 1668A, for the analysis of all 209 PCB congeners by high resolution GC/MS and EPA Method 1614 for brominated diphenyl ethers. Dr. Hamilton is also an adjunct professor at the University of Victoria.

Betsy Henry, Ph.D. is a managing scientist at Exponent and has been working in the field of mercury transport and fate for 15 years. She earned a Ph.D. from Harvard University in 1992 with a dissertation on bacterial methylation of mercury in the environment. Her main focus at Exponent has been complex mercury-contaminated sites including Onondaga Lake in Syracuse, New York, and the Ventron/Velsicol Superfund Site on Berry's Creek in New Jersey. In addition to project management, she has designed and implemented studies of mercury methylation and remineralization in lakes and mercury volatilization from soils. She has also co-authored several published papers and abstracts on mercury cycling and bioaccumulation in the environment and served on an expert review panel for investigations at Lavaca Bay, another mercury-contaminated site.

Randy D. Horsak, PE, is a principal and President of 3TM International, Inc., an environmental consulting firm in Houston, Texas, that specializes in environmental science, engineering, and forensic investigations. Mr. Horsak has more than 30 years of professional experience in environmental project management, multi-media sampling and analysis, engineering feasibility studies, and remediation. Mr. Horsak often serves as a consultant or testifying expert in toxic tort and environmental litigation cases.

Jacqui Horswell, Ph.D. graduated from the University of Aberdeen in soil microbiology and is currently a senior research scientist with ESR, Ltd in Poriru, New Zealand. Dr. Horswell specializes in the investigation of microbial ecology of soils, soil biotechnology, and environmental contamination. Her primary research interest is the use of cultural and molecular techniques to characterize soil microbial community structure and dynamics. Dr. Horswell has published pivotal articles regarding the identification of microbial communities in soil and their application in criminal and environmental forensic investigations.

Glenn W. Johnson, Ph.D. is Research Associate Professor at the Energy and Geoscience Institute (EGI) at the University of Utah. His work at EGI includes research and application of chemometric methods in environmental forensics; the study of sources, fate, and transport of persistent organic pollutants; and teaching within the Department of Civil and Environmental Engineering. Dr. Johnson is also President and Chief Scientist of GeoChem Metrix, a small consulting company specializing in environmental forensic investigations for litigation support projects.

Natalie Leys, Ph.D. is a project manager at the laboratory for microbiology at the Belgian Nuclear Research Centre (SCK/CEN). Dr. Leys is the principal investigator of several research projects studying the physiological and genetic response of heavy metal resistant *Cupriavidus metallidurans* strains in different stress situations. Gene transfer, gene modifications, and gene expression are analyzed in cells stressed by heavy metals, space flight, or related parameters such as microgravity, cosmic radiation, UV-radiation, and vibrations.

Paul D. Lundegard, Ph.D. is a principal scientist with Unocal Corporation. Dr. Lundegard received his Bachelor and Masters degrees in geology and his Ph.D. in Geochemistry from the University of Texas at Austin. Paul is a geochemist with 28 years of experience in the environmental and petroleum business. He has worked on a wide variety of contaminant fate and forensic issues at sites ranging from upstream exploration and production facilities to downstream refining and marketing facilities. His litigation experience includes claims for natural resource damages, cross-contamination, property value diminution, and human health impacts.

James R. Millette, Ph.D. is a scientist with MVA Scientific Consultants in Atlanta, Georgia. Since 1972, Dr. Millette's research focus is the investigation of environmental toxicology/particles using electron microscopy techniques and has over 60 peer-reviewed articles published on this subject. Dr. Millette's previous work included 11 years as a research scientist at the United States Environmental Protection Agency Research Center in Cincinnati, Ohio, and 5 years at McCrone Environmental Services performing and supervising analysis of particulates and product constituent analysis by microscopic techniques. Dr. Millette has testified in court on several occasions concerning environmental lead-containing particles.

Stephen M. Mudge, Ph.D. is a senior lecturer in marine chemistry and environmental forensics in the School of Ocean Sciences, University of Wales—Bangor, United Kingdom. Dr. Mudge is responsible for developing the world's first undergraduate degree program in Environmental Forensics and has worked for many years on the application of forensic techniques to track sewage and other organic matter in the environment, including the use of biomarkers, and interpreting and linking these results with multivariate statistical methods. Dr. Mudge has published extensively and is considered a world leader in the field of environmental forensics.

Rachel A. Parkinson, Ph.D. is a molecular microbiologist with a B.Sc. in Microbiology and Biochemistry from the University of Otago in New Zealand and received an M.Sc. (Honors) in Cell and Molecular Biology from Victoria University of Wellington in New Zealand. Ms. Parkinson's primary research focus is the use of non-human DNA forensic applications.

Priyabrata Pattnaik, Ph.D. received his B.Sc. in Zoology with distinction from Utkal University in Bhubaneswar, India, an M.Sc. (microbiology) from Orissa University of Agriculture and Technology, in Bhubaneswar, and a Ph.D. in microbiology from the National Dairy Research Institute in Karnal, India. Dr. Pattnaik served at the International Centre for Genetic Engineering and Biotechnology in New Delhi, where he worked on the development of a malaria vaccine. Dr. Pattnaik is also the recipient of the young scientist award-2000 from the Association of Microbiologists of India and has more than 50 publications to his credit. Currently Dr. Pattnaik is working as staff scientist at Defense Research and Development Establishment in Gwalior, India. His research interest is recombinant sub-unit vaccines, bioprocess scale-up and receptor–ligand interaction involved in biology and pathogenesis.

Ioana G. Petrisor, Ph.D. has a Ph.D. in Biology/Environmental Biotechnology from Romanian Academy of Sciences (awarded in 2000) and a Bachelor degree in Chemistry with a major in Biochemistry from Bucharest University in Romania, Faculty of Chemistry (awarded in 1992). In December 1999, she completed a UNESCO training program on Plant Molecular Genetics at the University of Queensland, Department of Botany in Brisbane, Australia. She is co-author of 63 scientific papers published in peer-review journals and proceedings. Dr. Petrisor is the managing editor for the *Environmental Forensics Journal* and a member of the editorial board of several other peer-review journals. She was recently elected as vice-chairman of the newly formed ASTM sub-committee on Forensic Environmental Investigations. She is a member of ACE (American Chemical Society), AEHS (Association for Environmental Health and Science), ITRC (Interstate Technology Regulatory Council), and other professional organizations. Her work experience includes conducting innovative research (at lab, field, and pilot scales) for the United States Department of Energy and the European Community on bioremediation, phytoremediation, environmental characterization, and risk assessment, as well as forensic investigations on petroleum products, chlorinated solvents, and heavy metal releases.

John F. Quensen, III, Ph.D., is a research professor in the Department of Crop and Soil Sciences at Michigan State University. His research interests include both aerobic and anaerobic microbial transformations of normally recalcitrant environmental contaminants including polychlorinated biphenyls, dioxins, and pesticides, and the biological effects of microbial metabolites of these compounds. He was the first scientist to demonstrate the microbially mediated reductive dechlorination of PCBs and the DDT metabolite DDE.

Kim Reynolds Reid is an environmental chemist with 13 years of experience in evaluating and interpreting analytical data. As a consultant at Gradient Corporation, she applies her knowledge of analytical techniques to the usability and integrity of inorganic, wet chemistry, and organic data generated in support of human health and ecological risk assessments, environmental forensic studies, and litigation matters. Ms. Reid also designs and provides quality assurance oversight for a wide variety of sampling and laboratory analysis programs. Prior to joining Gradient, Ms. Reid was an inorganic analyst at Enseco, Inc.

Jennifer K. Saxe is a consulting environmental engineer at Gradient Corporation whose work focuses on evaluating and modeling the environmental transport, transformation, and ecological bioavailability of trace elements and hydrophobic organic compounds in soil and water, as well as their exposure to organisms. She also has experience in designing and evaluating chemical characterization studies for environmental media containing these analytes. Prior to becoming a consultant, Dr. Saxe worked in the United States Environmental Protection Agency Office of Research and Development, National Risk Management Research Laboratory. She works and resides near Boston, Massachusetts.

Walter J. Shields, Ph.D. received his Ph.D. in soil science from the University of Wisconsin at Madison in 1979. He conducted research on land applications of biosolids and herbicide impacts to watersheds as an environmental scientist with Crown Zellerbach Corporation until 1985. He has been an environmental consultant since then, first with CH2M Hill and then with Exponent, where he is currently Director of the Environmental Sciences. Dr. Shields has testified as an expert in source identification related to toxic tort and cost allocation cases, and in the timing of releases in insurance coverage cases.

Scott A. Stout, Ph.D. is an organic geochemist with 19 years of petroleum and coal industry experience. He has extensive knowledge of the chemical compositions and chemical fingerprinting applications of coal, petroleum, gasoline, and other fuel-derived sources of contamination in terrestrial and marine environments. Dr. Stout has written interpretive reports on more than 250 site or incident investigations and has authored or co-authored nearly 100 papers published in scientific journals and books. He has conducted environmental research while employed at Unocal Corporation, Battelle Memorial Institute, and is currently a partner at NewFields Environmental Forensics Practice, Rockland, Massachusetts.

Julie K. Sueker, Ph.D., P.H., PE, is a hydrogeochemist with more than 13 years of professional experience in physical hydrology, isotope hydrology, hydrogeology, and environmental geochemistry. Dr. Sueker's hydrology experience includes determining sources of solutes in surface- and groundwater-environments using chemical and isotopic tracers, assessing and modeling the transport and fate of solutes in surface- and groundwater-environments, and predicting flow paths of water by using chemical and isotopic hydrograph separation techniques. Dr. Sueker's forensics focus is the use of isotopes to distinguish sources of constituents of concern (COCs) in the environment and to demonstrate microbial degradation of COCs in both regulatory and litigation contexts.

F. Ben Thomas, Ph.D. is a principal and Vice President at Risk Assessment and Management Group, Inc. (RAM Group), an environmental and health consulting firm in Houston, Texas. Dr. Thomas's professional training is in pathology and toxicology, and he has over 25 years of experience in the adverse effects associated with chemical and physical toxicants. Other focuses of his consulting practice include risk assessment, decision analysis, strategic planning, program development, regulatory negotiation, and program management. Dr. Thomas often serves as a consulting and/or testifying expert in toxic tort and environmental litigation cases.

Yves Tondeur, Ph.D. received his Ph.D. in Chemistry from the Free University of Brussels, Belgium. Although his formal education related to the field of physical organic chemistry, Dr. Yondeur did postdoctoral research work in environmental trace analyses involving polychlorinated dibenzo-*p*-dioxins and polychlorinated dibenzofurans (PCDD/Fs) and other persistent organic pollutants at Florida State University and then with the National Institute of Environmental Health Sciences. Dr. Tondeur developed the United States Environmental Protection Agency Method 8290 for PCDD/Fs in 1987. He is founder of Alta Analytical Perspectives located in Wilmington, North Carolina, where he serves as president and CEO.

Allen D. Uhler, Ph.D. has over 25 years experience in environmental chemistry. Dr. Uhler has developed advanced analytical methods for petroleum, coal-derived, and anthropogenic hydrocarbons, and other man-made organic compounds in waters, soils, sediments, vapor, and air. He has conducted assessments of the occurrence, sources, and fate of fugitive petroleum at refineries, offshore oil and gas production platforms, bulk petroleum storage facilities, along petroleum pipelines, at varied industrial facilities, and in sedimentary environments. Dr. Uhler has studied coal-derived wastes at former gas plants, wood-treating facilities, and in nearby sedimentary environments. His experience includes expertise in the measurement and environmental chemistry of man-made industrial chemicals including PCB congeners and Aroclors, persistent pesticides, dioxins and furans, metals, and organometallic compounds.

Karolien Vanbroekhoven, Ph.D. is a researcher and project manager within the Environmental and Process Technology Centre of Expertise of the Flemish Institute for Technological Research (VITO) in Belgium. Dr. Vanbrokehoven specializes in microbiology, molecular ecology, and chemistry. Dr. Vanbroekhoven has 6 years of experience in soil and groundwater remediation and is involved in the development of technologies for the removal of heavy metals from groundwater and contaminated soils (both by in situ bioprecipitation and abiotic techniques).

Drew R. Van Orden, PE, is a senior scientist at RJ Lee Group, Inc., where he has been involved in asbestos research for the past 18 years. Mr. Van Orden was educated at Millersville State College, where he earned a B.A. in Earth Sciences; at the Pennsylvania State University, where he earned an MS in Mineral Process Engineering; and at the University of Pittsburgh, where he earned an M.A. in Applied Statistics. He is a Professional Engineer (Mineral Engineering) in the Commonwealth of Pennsylvania. Mr. Van Orden has authored or co-authored numerous articles on asbestos analyses and is an active participant in asbestos analytical methodology development.

Emily A. Vavricka is a geologist with DPRA. Ms. Vavricka specializes in the investigation and forensic determination of perchlorate in soil and groundwater. Ms Vavricka has investigated the sources of naturally occurring perchlorate in southern California and has examined forensic techniques for distinguishing between naturally occurring and anthropogenic sources of perchlorate. Ms. Vavricka has a B.S. in Environmental Studies from the University of California at Riverside.

Zhendi Wang, Ph.D. is a senior research scientist and Head of Oil Spill Research of Environment Canada, working in the oil and toxic chemical spill research field. His specialties and research interests include development of oil spill fingerprinting and tracing technology; environmental forensics of oil spill; oil properties, fate, and behavior of oil and other hazardous organics in the environment; oil burn emission and products study; oil bioremediation; identification and characterization of oil hydrocarbons; spill treatment studies; and applications of modern analytical techniques to oil spill studies and other environmental science and technology. Dr. Wang has authored 270 academic publications including 76 peer-reviewed articles and 4 invited reviews in highly respected journals within environmental science and analytical chemistry, and 8 books and book chapters. Dr. Wang has received numerous national and international scientific honors. He is editorial board member of *Environmental Forensics* and had been guest editor for several journals. He has been invited by many international agencies (such as UNDP, UNEP, UNIAEA, IOC/UNISCO, APEC, US EPA, US MMS, ASTM, CCME, and Canadian International Development Agency) and many countries to conduct numerous workshops and give seminars.

James M. Waters, Ph.D. graduated from Flinders University in Adelaide, South Australia. His dissertation examined ways in which soil microbial communities could be compared by DNA profiling methods. He is currently a postdoctoral researcher at the University of Durham investigating epithelial stem cells.

Chun Yang, Ph.D. is a scientist with the Environmental Technology Center at Environment Canada. He received Ph.D. from Nanyang Technological University in Singapore and currently specializes in the use of forensic analysis using a variety of analytical techniques. His primary scientific interest is in the use of chemical fingerprinting for petroleum hydrocarbons to identify the source of a release into the environment. He has authored or co-authored over 25 peer-reviewed journals.

1

Mercury

Gary N. Bigham, Betsy Henry, and Brad Bessinger

Contents

1.1 INTRODUCTION

Mercury is a naturally occurring element that exists in three oxidation states: Hg(0) (Hg^0, elemental mercury), Hg(I) (Hg^+, mercurous mercury), and Hg(II) (Hg^{2+}, mercuric mercury). Of the three forms, only elemental mercury and mercuric mercury contribute to the global mass balance (the rarity of Hg(I) compounds can be explained by the instability of Hg–Hg bonds that characterize the Hg(I) forms). Although mercury is nondegradable, it readily converts between the oxidation states and chemical species listed in Table 1.1.1. Chemical transformations between forms govern the fate and transport of elemental mercury in the atmosphere. The production of mercuric methylmercury (CH_3Hg^+) in aqueous environments is similarly dependent.

1.1.1 Elemental Mercury

Zero-valent elemental mercury (Hg(0)) occurs as a liquid ($Hg^0(l)$), gas ($Hg^0(g)$), and dissolved constituent in water ($Hg^0(aq)$) at ambient temperatures (Table 1.1.1). Owing to its relatively high (for a metal) saturation vapor pressure, the gaseous concentration in equilibrium with the liquid is predicted to be $13,000\,\mu g/m^3$ (Winter, 2003). This concentration is three orders of magnitude higher than the time-weighted average threshold limit value of $25\,\mu g/m^3$ used for occupational exposure (ACGIH, 2000). It is also six orders of magnitude higher than the reported background concentration of 0.002–$0.005\,\mu g/m^3$, reported in Table 1.1.1 (Seigneur et al., 1994). Although slow evaporation kinetics and ventilation will necessarily reduce concentrations in indoor environments where mercury spills have occurred (Winter, 2003), measurements between 0.5 (Carpi and Chen, 2001) and $140\,\mu g/m^3$ (Smart, 1986) have been observed.

The fate and transport of elemental mercury released to the atmosphere are conceptualized in Figure 1.1.1. Consistent with Table 1.1.1, Figure 1.1.1 shows that gaseous elemental mercury is the predominant species in outdoor air, accounting for as much as 98 percent of the total elemental mass. This speciation is partially attributed to mercury's high saturated vapor pressure. (Because gaseous elemental mercury concentrations are significantly below saturation with respect to the liquid form, atmospheric releases are thermodynamically restricted from converting to $Hg^0(l)$.) Other chemical properties that contribute to the stability of gaseous elemental mercury in the atmosphere include its low water solubility (which prevents it from being quickly removed by precipitation via the chemical species $Hg^0(aq)$) and its high ionization potential (which inhibits its oxidation to reactive gaseous mercury (Hg(II)), a species with a higher water solubility and rate of sequestration, by cloud droplets and precipitating aerosols) (Mason et al., 1994).

From a forensic science perspective, the most important transformation of elemental mercury depicted in Figure 1.1.1 is its conversion to mercuric mercury, a species that is readily deposited and potentially transported to

Table 1.1.1 Typical Mercury Species and Concentrations of Mercury Species in Gas and Water Phases in the Atmosphere

Oxidation State	Phase	Atmospheric Concentration[a]	Species[b,c]
Hg(0)	Liquid		$Hg^0(l)$
	Gas	$0.002 - 0.005\,\mu g/m^3$	$Hg^0(g)$
	Water	$6 - 27 \times 10^{-6}\,\mu g/L$	$Hg^0(aq)$
Hg(I)	Water		Hg_2^{+2}
Hg(II)[d,e]	Solid[f]	—[g]	$HgS(s), HgO(s), Hg(OH)_2(s),$ $HgCl_2(s), HgSe(s), HgSO_4(s),$ $(CH_3)HgCl(s),$ $>OHg^+, >SHg^+, (>S)_2Hg^0$ $>SCH_3Hg^0$
	Liquid[h]		$(CH_3)_2Hg(l), (C_2H_5)_2Hg(l)$
	Gas	$0.9 - 1.9 \times 10^{-4}\,\mu g/m^3$	$Hg(OH)_2(g), HgCl_2(g), CH_3HgCl(g),$ $CH_3HgCH_3(g)$
	Water	$0.4 - 1.3 \times 10^{-2}\,\mu g/L^i$	$Hg^{+2},$ $Hg(OH)^+, Hg(OH)_2(aq),$ $HgOHCl(aq),$ $HgCl^+, HgCl_2\,(aq), HgCl_3^-, HgCl_4^{-2},$ $HgSO_3(aq), Hg(SO_3)_2^{-2},$ $HgC_2O_4(aq),$ $HgHS^+, Hg(HS)_2(aq), HgHS_2^-, HgS\,(aq),$ $ROHg^+, RSHg^+, (RS)_2Hg(aq)$ $CH_3Hg^+, (CH_3)_2Hg(aq)$

[a] Atmospheric concentrations in gas and water phases from Seigneur et al. (1994) (where μg represents micrograms).
[b] Typical species reported in air from Lin and Pehkonen (1999), water from Hintelmann et al. (1997), Benoit et al. (1999, 2001), and Haitzer et al. (2002, 2003) (polysulfides in groundwater not included [Paquette and Helz, 1997; Jay et al., 2000]), and soil/sediment from API (2004).
[c] Note that air, water, and sediment each contain multiple phases (gas, water, and solid).
[d] ">" refers to surface adsorption sites for mercury (e.g., carboxyl and sulfide sites on soot and organic material).
[e] "R" designates a natural organic ligand such as a humic and fulvic acid.
[f] HgS(s) is the most abundant naturally occurring form.
[g] HgO may be present in the solid particulate phase (Seigneur et al., 1994).
[h] Not significant in concentration.
[i] A large fraction of the mercury concentration dissolved in water droplets in the atmosphere is associated with soot particles (Petersen et al., 1995).

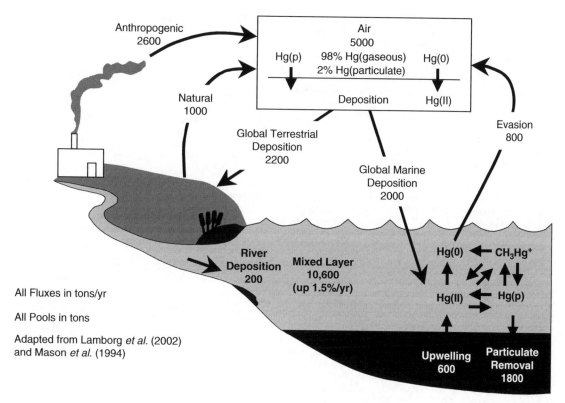

Figure 1.1.1 Global mercury budget and fluxes.

Notes: Lamborg et al. (2002) calculated a net increase in mercury in the oceanic mixed layer on the order of 1.5% per year. Also, Mason et al. (1994) calculated a rever source to the ocean of 200 tons/yr (most of which is deposited near shore).

shallow-sediment aquatic environments that are conducive to methylation. The oxidation of elemental mercury shown in the figure occurs primarily at the solid/liquid interface in fog and cloud droplets (Munthe, 1992), but some gas-phase oxidation also occurs (Hall, 1995). The primary oxidants involved in the conversion include ozone (Lin and Pehkonen, 1999), hydroxyl radicals (Bergan and Rodhe, 2001), and/or halogens such as chlorine and bromine (Hedgecock and Pirrone, 2001; Ariya et al., 2002; Temme et al., 2003). Although mercuric mercury can also be reduced back to elemental mercury by sulfite (Munthe et al., 1991) or by photoreduction (Xiao et al., 1994), oxidation typically exceeds reduction. The relatively high mobility of elemental mercury is reflected in atmospheric residence times that are estimated to be in the order of several months (Sommar et al., 2001) to one year (Fitzgerald and Mason, 1997). This longevity allows atmospheric emissions to contribute to terrestrial and oceanic mercury loads on both regional and global scales.

Although not shown in the figure, liquid elemental mercury (Hg(*l*)) that is released directly to soil can also be converted to methylmercury. Once spilled onto soil, the dense liquid travels downward in the soil column, especially through cracks and fissures (Turner, 1992). Counteracting this process is mercury's tendency to form balls or beads and become trapped in soil pore spaces due to its high surface tension (Turner, 1992). Elemental mercury can be oxidized to more soluble forms of mercury (i.e., Hg(II) species) in soil, which can then be transported by leaching; however, the process of oxidation is usually slow, and elemental mercury is long-lived at contaminated sites.

1.1.2 Methylmercury

The predominant oxidation state of mercury in terrestrial and aquatic environments is Hg(II), of which methylmercury (composed of ionic Hg(II) [Hg^{+2}] and a single methyl group [CH_3^-] to form CH_3Hg^+) is a subset (Table 1.1.1). Although methylmercury is created in aqueous environments, it rarely accounts for more than a few percent of the total mercury concentration (Krabbenhoft et al., 2003). Methylmercury bioaccumulates because it is a cation, is lipophilic, and has a high affinity for sulfur-containing organic matter. These properties cause it to readily sorb to phytoplankton and particulate humic substances, which are subsequently consumed by fish or benthic macroinvertebrates. Within higher trophic levels, methylmercury tends to concentrate in muscle tissue (muscle tissue is where sulfhydryl groups of protein are located).

The production of methylmercury involves both a methylation (i.e., addition of a methyl group to a mercuric ion) and demethylation (i.e., removal of a methyl group from a methylmercury ion) aqueous reaction (Figure 1.1.1). Methylation is mediated primarily by sulfate-reducing bacteria, which are present in anoxic sediment of freshwater and estuarine environments (Compeau and Bartha, 1985; Winfrey and Rudd, 1990; Gilmour et al., 1992). Demethylation is dominated by abiotic processes in many waters, effectively removing methylmercury from the system by transforming it into elemental mercury, which volatilizes to the atmosphere (Figure 1.1.1). (Gardfeldt [2003] discussed evidence that Hg(II) is the more probable product of photolysis.) In aquatic sediment, demethylation is apparently accomplished by sulfidogens (sulfate-reducing bacteria) and methanogens (methane-producing bacteria) (Oremland *et al.*, 1991, 1995; Pak and Bartha, 1998). Because of demethylation, methylmercury is relatively short-lived in surface water in the presence of oxygen and light; however, because methylmercury is formed continually, concentrations remain high enough to be of concern.

Net rates of methylmercury production in sediments depend on a number of factors that mainly affect the activity of mercury-methylating bacteria and/or the availability of inorganic mercury for methylation. These factors include temperature, dissolved oxygen, and sulfate concentration. In littoral (i.e., located in shallow water) sediments of lakes, temperature is positively correlated with net methylmercury production (Matilainen *et al.*, 1991). High rates of mercury methylation in profundal surficial sediments during late summer may also result from elevated temperatures (Callister and Winfrey, 1986; Korthals and Winfrey, 1987). Dissolved oxygen is another critical factor. Numerous studies of both freshwater (Callister and Winfrey, 1986; Korthals and Winfrey, 1987; Regnell, 1990; Matilainen *et al.*, 1991) and estuarine (Olson and Cooper, 1976; Compeau and Bartha, 1984) sediments have shown higher rates of net methylmercury production under anoxic, rather than oxic, conditions. Finally, sulfate additions have been found to stimulate mercury methylation in freshwater sediments (Gilmour *et al.*, 1992).

The factors that primarily affect the availability of inorganic mercury for methylation include concentrations of sulfate/sulfide, dissolved and particulate organic matter, and total mercury. High sulfate concentrations may limit mercury methylation in estuarine sediments (Blum and Bartha, 1980; Compeau and Bartha, 1984, 1987; Gilmour and Capone, 1987), either due to complexation and precipitation of mercury as mercuric sulfide or because of a change in aqueous speciation that affects the supply of neutral, bioavailable mercury sulfide species to methylating bacteria (Benoit *et al.*, 1999). Elevated concentrations of total organic matter may also reduce the net methylation potential of sediment, due to adsorption of Hg^{+2} on the organics and a subsequent reduction in dissolved ionic mercury available for methylation (Hammerschmidt and Fitzgerald, 2004). Finally, an increase in the total mercury concentration may be related to an increase in methylmercury (Krabbenhoft *et al.*, 2003), although the strength of this correlation likely varies among sites. Bloom *et al.* (2003) found that the relative degree of mercury solubility and methylation depends on its chemical form, with sorbed or organo-chelated compounds exhibiting higher solubility than mercuric sulfide ($HgS(s)$).

1.2 NATURALLY OCCURRING SOURCES OF ELEMENTAL MERCURY

Terrestrial Emissions—Sources of mercury in the environment are both natural and anthropogenic. A third category (or sub-category), termed "re-emission", refers to mercury that was originally deposited in water or soil by natural or anthropogenic processes but is re-emitted to the environment (Ebinghaus *et al.*, 1999). Natural mercury emission and re-emission cannot be quantified separately, and together account for a substantial portion of the global mercury cycle.

Natural terrestrial sources of elemental mercury to the atmosphere include volcanic emissions, evasion from the subsurface crust via geothermal activity, and volatilization of mercury from soil (Table 1.2.1). Geologic regions, specifically those associated with volcanically active areas, that contain elevated concentrations of mercury (termed "mercuriferous belts") contribute measurably more mercury to the atmosphere through degassing (Lindqvist *et al.*, 1991;

Table 1.2.1 Global Budget Estimates for Atmospheric Mercury (tons/yr)

Source	Seigneur et al. (2004)[a]	Lamborg et al. (2002)[b]
Natural sources		
Volcanic eruptions	130[c]	1000[f]
Mercuriferous soils	500[d]	
Subsurface crust emissions	—[e]	
Re-emitted from land	560	
Direct from oceans	440	400[g]
Re-emitted from oceans	510	
Total	2100	1400
Anthropogenic sources		
Direct from land	2100	2600[h]
Re-emitted from land	1100	
Re-emitted from oceans	1000	400
Total	4200	3000

[a] Used reported base value for Year 1998 of Seigneur *et al.* (2004).
[b] Used data from best-fit solution of global mass balance model of Lamborg *et al.* (2002).
[c] Estimate does not include passive, degassing volcanoes with estimated emissions of 30 tons/yr or geothermal sources with emissions of 60 tons/yr (Varekamp and Buseck, 1986).
[d] Estimate does not include 200 tons/yr from unmineralized soil.
[e] The emission from this source category is estimated to be between 3000 and 6000 tons/yr (Rasmussen, 1994).
[f] Total natural terrestrial emissions.
[g] Total natural ocean emissions.
[h] Total anthropogenic land emissions.

Ferrara *et al.*, 2000; Coolbaugh *et al.*, 2002), compared to areas that are not enriched in mercury (Gustin *et al.*, 2000). This volatilization is most likely a combination of elemental mercury volatilization and reduction of Hg(II) in soil water followed by volatilization (Lindberg *et al.*, 1998), and is positively correlated with soil and air temperatures (Gustin *et al.*, 1995; Lindberg *et al.*, 1995), solar radiation and soil moisture (Carpi and Lindberg, 1997), and weather conditions such as wind speed, relative humidity, and turbulence (Kim, 1995).

According to Seigneur *et al.* (2004) and Lamborg *et al.* (2002) (Table 1.2.1), total natural emissions and re-emissions account for approximately one-third of the total annual mercury load to the atmosphere. Although the natural terrestrial component is estimated to constitute one-half to two-thirds of that value, there are large uncertainties in the estimates. For example, Rasmussen (1994) discussed evidence that natural emissions from terrestrial bedrock are between 3000 and 6000 tons/year. If this subsurface crustal emission were included in the global budgets of Seigneur *et al.* (2004) and Lamborg *et al.* (2002), natural terrestrial sources of mercury would be predicted to be equal to, or greater than, anthropogenic sources.

Oceanic Evasions—The reduction of mercuric mercury to elemental mercury and its subsequent volatilization is an important pathway by which mercury enters or re-enters the atmosphere (Table 1.2.1). The concentration of elemental mercury in surface waters tends to be supersaturated relative to the atmosphere (Vandal *et al.*, 1991), suggesting that elemental mercury is formed in surface water. The reduction reaction itself is apparently photo-induced (Amyot *et al.*, 1997a,b; Krabbenhoft *et al.*, 1998). In temperate lakes, the flux of mercury from water to the atmosphere by volatilization is estimated to be 10 percent of the annual input of mercury to the lakes from atmospheric deposition (Vandal *et al.*, 1991). Seigneur *et al.* (1994) estimated that natural oceanic evasions account for approximately one-sixth of the total current global air emissions, with the fraction that is remobilized mercury accounting for more than half of that value (Table 1.2.1). These estimates are likely to be revised downward in the future due to evidence that mercury is cycled more rapidly in the marine boundary layer relative to the continents (Hedgecock and Pirrone, 2004).

1.3 SOURCES OF MERCURIC MERCURY AND METHYLMERCURY

Natural atmospheric sources of mercury to aquatic environments include the following: (1) mercury originally introduced as the gaseous elemental form but transformed to Hg(II) and deposited in wet precipitation or adsorbed to aerosols and (2) particulate-bound Hg(II) derived from the erosion of soil and rocks by wind. Mercury deposited from the atmosphere is subject to methylation in acquatic environments. Air Deposition—Methylmercury forms predominantly in anoxic sediment and waters of estuarine, lake, and riverine environments where sulfate-reducing bacteria actively metabolize inorganic mercuric mercury to methylated forms. (Methylmercury formation is usually negligible in surface soil, because the process requires anoxia and moisture, conditions rarely found in upland surface soil.)

Most mercuric mercury that is deposited to the earth's surface is dissolved in wet precipitation or adsorbed to aerosols such as soot (Mason et al., 1994; Ebinghaus et al., 1999). Annual rates of atmospheric deposition of mercury (primarily from rain and snow) are a function of both the concentration of mercury in deposition (which is affected by geologic and geographic variables) and the annual rate of deposition. Estimates of the total mercury flux from wet deposition include $15 \mu g/m^2$-year in northeastern Minnesota (Glass et al., 1991), $6.8 \pm 2 \mu g/m^2$-year over Little Rock Lake in Wisconsin (Fitzgerald et al., 1991), $14 \mu g/m^2$-year in Scandinavia (Iverfeldt, 1991; Bindler, 2003), and $10–20 \mu g/m^2$-year around Chesapeake Bay and western Maryland (Mason et al., 1997). The fraction of the total deposition rates attributable to natural sources is purported to be small. For example, Bindler (2003) measured mercury concentrations in 500- to 4000-year-old peat cores in south-central Sweden and determined that preindustrial natural deposition rates were in the range $0.5–1 \mu g/m^2$-year. These values are at least an order of magnitude lower than air measurements in Sweden over the past 30 years ($5–30 \mu g/m^2$-year) (Munthe et al., 2001). Recent modeling studies have similarly concluded that

deposition rates have increased by a factor of 2–10 during the past 200 years (Bergan et al., 1999).

Of the mercury that is deposited from wet and dry precipitation, approximately 10 to 25 percent is believed to be transported from drainage basins to lakes, either through direct deposition or run-off (Lorey and Driscoll, 1999; Swain et al., 1992). This 75–90 percent estimated retention of mercury by terrestrial soils is consistent with loads for riverine (Quemerais et al., 1999) and oceanic environments (Mason and Sheu, 2002).

Upstream Discharges—In addition to direct atmospheric deposition and run-off, natural sources of mercury to aquatic ecosystems include erosion of bedrock, leaching of soil, and discharge from wetlands. For example, mercury occurs as a trace element in geologic formations and can be eroded with fine particles by surface waters (Plouffe, 1995; Quémerais et al., 1999). In addition, although Hg(II) is strongly adsorbed to soil particles, soil organic matter, and sulfides (Schuster, 1991), a small fraction can be leached over time by groundwater (Hultberg et al., 1995; Krabbenhoft et al., 1995). Finally, wetlands discharge is of particular importance, because wetlands have been found to be natural locations of methylmercury production (St. Louis et al., 1994; Hurley et al., 1995; Brianfireun et al., 1996). Driscoll et al. (2002), for example, investigated watershed ecosystems in the Adirondack region of New York and concluded that riparian wetlands and lakes were sources of methylmercury to downstream surface waters.

The fate of upstream mercury discharges in a receiving water body is shown in Figure 1.3.1 and includes sedimentation and burial, methylation, volatilization, and downstream transport/loading. Because mercuric mercury (Hg(II)) is a weak Lewis acid, it preferentially forms stable complexes with Lewis bases, such as sulfur functional groups, that are present on the surfaces of particulate humic material (Haitzer et al., 2002, 2003). Consequently, mercury partitions strongly to organic-rich particles. Because mercury also tends to adsorb to mineral colloids (Schuster, 1991), the primary mass transport process controlling mercury transport from upstream sources is believed to be sedimentation and sediment resuspension/burial (Le Roux et al., 2001).

In moving waters, such as rivers, creeks, and marshes, surface sediments are often resuspended and transported, resulting in transport of mercury downstream. In quiescent waters, sediment remains in place, and the ultimate fate of mercury associated with particles in these systems is burial. Mercury in buried sediments is not remobilized by

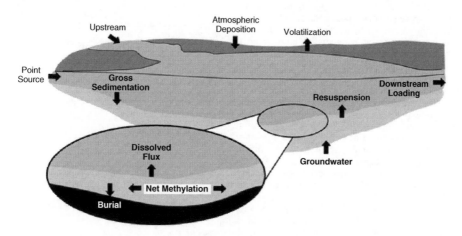

Figure 1.3.1 *Mercury cycling in aquatic systems.*

processes such as cation exchange or redox reactions that can mobilize other metals such as iron (Hurley *et al.*, 1991; Watras *et al.*, 1994). However, mercury in surface sediments in shallow water may be involved with the redox cycling of iron, resulting in flux into overlying water (Gill *et al.*, 1999).

Although mass transfer mechanisms such as dissolved fluxes from sediment and volatilization likely contribute to the cycling of mercury and methylmercury (Figure 1.3.1), measurements of dissolved fluxes of mercury from sediment are highly variable, and their relevance to the overall transport and fate of mercury in aquatic systems is still unclear. Fluxes of methylmercury from Lavaca Bay, Texas, sediments, for example, varied across a range of three orders of magnitude, and fluxes at a single site varied seasonally (maximum fluxes occurred in late winter to early spring) and diurnally (Gill *et al.*, 1999). Methylmercury flux from sediment was six times higher during dark periods (at night) than during light periods (Gill *et al.*, 1999). The authors suggest that chemical or biological processes at the sediment/water interface, such as migration of the oxic/anoxic boundary downward during the day as oxygen is produced by photosynthesis, control methylmercury flux. Lavaca Bay is a shallow estuary (average water depth of 1 m), so sediments there would be more responsive to changing light conditions than sediments in deeper waters would be. As regards the production and volatilization of dissolved elemental mercury in fresh water, its transport is also variable, depending on photolysis rates (Gardfeldt *et al.*, 2001) and microbial mercury oxidase activity (which additionally varies inversely with hydrogen peroxide [H_2O_2] diurnal patterns) (Siciliano *et al.*, 2002).

Total mercury concentrations in upstream rivers tend to be highly variable, because suspended particles can contribute to the total mercury concentration. A survey of 60 streams in Maine that were unaffected by any known mercury sources found a median total mercury concentration of 1.5 ng/L and a 95th-percentile concentration of 4.5 ng/L (Maine Department of Environmental Protection, 1999). In Wisconsin, analysis of mercury in water samples from seven rivers found median total mercury concentrations ranging from 1.13 to 8.70 ng/L (Babiarz *et al.*, 1998). In some cases, however, single high-flow events (Babiarz *et al.*, 1998) or storm events in general (Mason and Sullivan, 1998; Scherbatskoy *et al.*, 1998) can carry a major fraction of the annual load of mercury from a watershed. High flows tend to transport particulate-bound mercury. As with lakes, methylmercury is generally a small fraction of total mercury in rivers, except in watersheds with high proportions of wetlands (i.e., sites of methylmercury production) and high dissolved organic carbon (DOC) concentrations.

Background concentrations of mercury in receiving water and sediment can also vary widely depending on the type of watershed, the type of water body, water quality characteristics, and sediment quality characteristics. Total mercury concentrations in lake water not affected by point sources of mercury range from less than 1 ng/L in surface waters to greater than 40 ng/L in anoxic bottom waters (Watras *et al.*, 1994; Jacobs *et al.*, 1995). Total mercury concentrations in surface waters of lakes in Wisconsin range from 0.3 to 2.9 ng/L (Babiarz and Andren, 1995). Methylmercury usually constitutes a small fraction of the total mercury in surface waters (e.g., 5–15 percent in Wisconsin lake surface waters [Watras *et al.*, 1994]), but the portion can increase to as much as 50 percent in anoxic bottom waters (Watras *et al.*, 1994; Jacobs *et al.*, 1995).

Background mercury concentrations in sediment depend a great deal on sediment characteristics. Given similar loading and depositional patterns, the strongest factor controlling sediment mercury concentration appears to be the organic content of the sediment (Langston, 1986; Benoit *et al.*, 1998; Mason and Lawrence, 1999). Mercury concentration is also positively correlated with silt content in estuarine sediments (Bartlett and Craig, 1981). In a study of 34 lakes in Newfoundland that were unaffected by industrial discharge, the mean total mercury concentration was 0.039 mg/kg, and the range was 0.003 to 0.156 mg/kg (French *et al.*, 1999). In 24 of Massachusetts' least affected water bodies, mercury concentrations in sediment ranged from 0.008 to 0.425 mg/kg (Rose *et al.*, 1999). Methylmercury concentrations are generally low (i.e., from less than 0.01 to 2 μg/kg) in sediment and make up less than 0.1 to 16 percent of total mercury (Gilmour and Henry, 1991). Methylmercury concentrations tend to be highest near the sediment surface, in the zone of active mercury methylation.

1.4 ANTHROPOGENIC SOURCES

1.4.1 Combustion

Anthropogenic sources of mercury in the atmosphere are varied; however, most mercury emissions are attributed to combustion. For example, combustion sources accounted for 87 percent of emissions in the United States (US EPA, 1997) and 77 percent of emissions in the world (Pirrone *et al.*, 2001; UNEP, 2002) in 1994–1995. The principal sources of combustion include the following: coal-fired utility boilers, municipal waste combustors, coal- and oil-fired commercial/industrial boilers, and medical waste incinerators. Table 1.4.1 (Schierow, 2004) summarizes the estimated annual contributions of anthropogenic sources to the atmosphere in the United States between 1995 and 1999. As shown in the table, electric utilities are currently the largest combustion source. By contrast, state-mandated emission reductions by solid waste incinerators have dramatically reduced the contribution from this source category.

Pacyna *et al.* (2004) provide estimates of global total anthropogenic mercury emissions over the period 1990 to 2000 by continent (Table 1.4.2). Global mercury emissions have increased over this period while the contributions from Europe and North America have decreased. As of 2000, Asian countries contributed about 53 percent of the global mercury emissions to the atmosphere.

Table 1.4.1 *Major Anthropogenic Mercury Emission Sources in the United States (1995–1999) (Schierow, 2004)*

Source	Year 1995[a] Tons/yr	Year 1998[b] Tons/yr	Year 1999[c] Tons/yr
Major combustion sources			
Electric utilities	51	46	48
Waste incineration/ combustion	77	32	15
Industrial boilers	12	14	12
Major manufacturing sources			
Chlor-alkali manufacturing	8	7	7
Major mining/refining sources Gold mining	12	7	12
Major area sources			
Mobile sources	—[d]	27[c]	—[d]
Other sources	37	34	25
Total	197	167	118

[a] US EPA (1997).
[b] Seigneur *et al.* (2004).
[c] US EPA (2004).
[d] Mobile sources not included.

Table 1.4.2 *Anthropogenic Mercury Emissions by Continent (1990–2000) (Pacyna et al., 2004)*

Continent	Year 1990 Tons/yr	Year 1995 Tons/yr	Year 2000 Tons/yr
Africa	178	389	407
Asia	705	1121	1204
Australia	48	113	125
Europe	627	338	239
North America	261	215	202
South America	62	84	92
Total	1881	2260	2269

Table 1.4.3 *Mercury Emission Profiles from Anthropogenic Sources in 1995, in Fractional Abundance (Pacyna and Pacyna, 2002)*

Source	Hg^0 (gas)	Hg^{2+}	Hg (particulate)
Combustion sources			
Coal	0.5	0.4	0.1
Oil	0.5	0.4	0.1
Waste incineration	0.2	0.6	0.2
Major manufacturing sources			
Chlor-alkali manufacturing	0.7	0.3	0.0
Cement	0.8	0.15	0.05
Non-ferrous metals	0.8	0.15	0.05
Pig iron	0.8	0.15	0.05
Other Sources	0.8	0.15	0.05
Average	0.6	0.3	0.1

Coal-Fired Power Plants—Coal contains only trace concentrations of mercury, but the high volume combusted annually results in emissions of mercury into the atmosphere that are significant relative to other US combustion sources. Of the 48 tons/year of mercury emitted from utilities, 45 tons/year is attributable to coal-fired power plants (EPRI, 2000).

The factors that affect the speciation and concentration of mercury emitted from power plants include the chemical characteristics of flue gases (CO, hydrocarbons, H_2O, O_2, NO, SO_x), the physical characteristics of unburned carbon fly ash (abundance, size, surface area) (Niksa et al., 2001 and 2002), and numerous operational considerations (temperature, residence time, and air pollution control devices) (Senior, 2001; Kilgore and Senior, 2003). The primary release of mercury from coal combustion is $Hg^0(g)$; however, it can be oxidized through homogeneous (i.e., gas phase) and heterogeneous reactions (i.e., on activated carbon substrates) to produce more water-soluble species such as $HgCl_2(g)$ and solids like $HgO(s)$ (Johansen, 2003) prior to emission. Elemental mercury and Hg(II) can also sorb onto fly ash to produce particulate-bound phases ($Hg(p)$). Finally, particulate control devices such as fabric filters and flue gas desulfurization systems (i.e., scrubbers) can remove up to 90 percent of mercury from bituminous coals before it enters the atmosphere (Kilgore and Senior, 2003).

Table 1.4.3 reports typical speciation profiles for coal-fired power plant emissions relative to other sources. The fraction of total emissions that is gaseous elemental mercury is important because particulate-bound and Hg(II) are more likely deposited on a local or regional scale (within 100 km of the source) (Judson, 2002), whereas elemental mercury is transported globally (Constantinou et al., 1995). Although Pacyna and Pacyna (2002) estimated that gaseous elemental mercury accounts for approximately one-half of the total atmospheric mercury releases from power plants, the United States Environmental Protection Agency's (EPA's) Information Collection Request has shown that elemental mercury may constitute anywhere between 0 and 95 percent of the emissions for a particular facility (Senior, 2001). In fact, the relatively low total mercury concentrations in Lake Velenje, Slovenia, have been attributed to a low abundance of oxidized mercury (5–20 percent) in the emissions from the adjacent coal-fired power plant (Kotnik et al., 2000, 2002).

In addition to direct air emissions, coal-fired power plants also generate wastewater and fly ash that require disposal. Mercury in these sources can potentially be converted to methylmercury after being transported and deposited in methylating environments. An estimate of the relative global mass of mercury released as solid and aqueous waste is shown in Figure 1.4.1. Although the data are more than 15 years old (and deserve greater quantification [Trip et al.,

2004]), power plant releases to soil and water likely exceed air emissions. Also, because the releases are necessarily more localized than atmospheric emissions, they merit consideration in evaluating methylmercury production in nearby water bodies. For example, without distinguishing between anthropogenic and natural sources, Rolfhus et al. (2003) concluded that methylmercury in Lake Superior was more strongly influenced by watershed sources than wet deposition.

Other Combustion Sources—The second-largest combustion source in the United States is waste incineration in 1998 and 1999. The information in presented in Table 1.4.1, with emissions predominantly characterized by reactive gaseous mercury (Hg(II)) (Table 1.4.3). The relative importance of incinerator emissions to local air deposition has been demonstrated by Dvonch et al. (1999), who used modeling of multi-element tracers and analysis of wet deposition samples in southern Florida to conclude that 71–73 percent of the mercury in wet deposition at 17 sites could be accounted for by local anthropogenic sources such as municipal waste incineration and oil combustion. The majority of aerosols in the study area were similarly attributable to municipal waste incinerators (Graney et al., 2004).

Fate and Transport—The most common approach to tracking the fate of combustion emissions in the atmosphere is the implementation of numerical atmospheric mercury models that include anthropogenic sources, chemical speciation, and meteorological inputs (US EPA, 1997; Bergan et al., 1999; Seigneur et al., 2001; Ryaboshapko et al., 2002; Lin and Tao, 2003; Munthe et al., 2003; Dastoor and Larocque, 2004; Seigneur et al., 2004). For example, the Northeast Mercury Study (NESCAUM, 1998) performed model simulations with and without local, regional, and global anthropogenic sources, and concluded that approximately one-third of the total mercury depositing in New Jersey originated within the state. More recently, Cohen et al. (2004) used a computer model to conclude that coal combustion is the most important source of atmospheric mercury deposition to the Great Lakes.

Whereas methods have been developed to understand atmospheric releases, the fate of mercury introduced from combustion sources to aquatic ecosystems is more problematic. In addition to being subjected to the numerous transport and transformation processes, the methylation potential of the mercury depends on its chemical form within the methylating environment. For example, Bloom (2001) investigated the methylation of various dissolved and

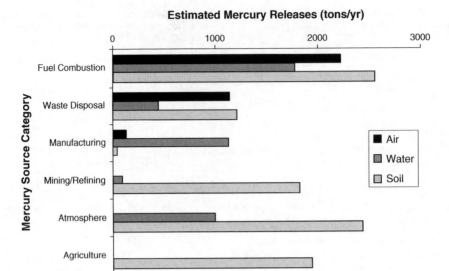

¹ Data from Nriagu and Pacyna (1988)

² Waste disposal includes incineration, wastewater, sludge, and landfilling

³ Atmosphere is re-deposited air emissions (assumed 70% is on land)

⁴ Agriculture includes fertilizer, peat, food, plant, and animal wastes

Figure 1.4.1 *Global anthropogenic inputs of mercury to the environment.*

sediment-bound mercury species added to sediment slurries. It was found that the smallest net change in methylation occurred for mercuric sulfide, and the largest occurred for dissolved mercury spikes.

1.4.2 Manufacturing

The second-largest atmospheric mercury emission source category in the United States is manufacturing, which includes chlor-alkali (at 7 tons/yr), Portland cement (at 5 tons/yr), and pulp and paper production (at 2 tons/yr) (US EPA, 1997). In chlor-alkali plants, elemental mercury is used as a cathode in the electrolysis of sodium chloride to produce chlorine and caustic soda. In cement manufacturing, mercury is introduced as a trace element in raw materials such as limestone and fuels for cement kilns (Johansen and Hawkins, 2003). As with combustion sources, the load contribution and speciation data presented in Tables 1.4.1 and 1.4.2 should be interpreted cautiously. For example, fugitive air emissions from chlor-alkali plants likely vary with factory operating conditions (Southworth *et al.*, 2004). Also, the 30 percent reactive gaseous mercury concentration (Hg(II)) in Table 1.4.3 may instead be closer to 2 percent (Landis *et al.*, 2004). Finally, emissions from manufacturing sources are likely to change dramatically in the future, because the US industry has largely switched to a mercury-free membrane-cell process in chlor-alkali plants, and the European Union is poised to either decommission existing plants or convert them to mercury-free facilities by 2020 (European Commission 2004).

In contrast to fuel combustion, the largest potential releases of mercury from manufacturing sites are estimated to occur through direct or indirect discharges into waterways (Table 1.4.2). Numerous industries such as mercury-cell chlor-alkali and electroplating facilities may discharge mercury to surface water bodies under National Pollutant Discharge Elimination System (NPDES) regulations. Also, municipal treatment plants are recipients of mercury from common sources such as hospitals, dental laboratories, and university laboratories.

Typical effluent concentrations of mercury in treated wastewater are not well characterized, but current information shows that they can vary from plant to plant and within a plant. Analysis of total mercury in effluent grab samples collected at 75 municipal sewage treatment plants in Maine showed that concentrations in secondary treated effluent averaged 11.3 ng/L and ranged from 0.74 to 99.23 ng/L (Maine Department of Environmental Protection 1999). Monthly sampling (24-hour, flow-weighted, composite samples) of Metropolitan Syracuse (New York) sewage treatment plant discharge showed total mercury concentrations ranging from 17.7 to 104 ng/L and methylmercury concentrations ranging from 0.60 to 3.1 ng/L (Bigham and Vandal, 1996). In general, sewage treatment removes most of the mercury that enters the plant (Glass *et al.*, 1990; Balogh and Liang, 1995; Mugan, 1996), primarily by sedimentation of solids. However, certain configurations such as combined sewer overflows and outside sewage lagoons may become anoxic, resulting in increased discharge of mercury and methylmercury (Bodaly *et al.*, 1998).

Wastewater releases to waterways are particularly important, because the mercury is predominantly dissolved and/or complexed with organic material (Hsu and Sedlak, 2003), and these forms can be more readily converted to methylmercury than sediment-bound forms. For example, the higher methylation potential of dissolved species may explain mercury concentrations in sunfish from Reality Lake (Oak Ridge, Tennessee), a site highly contaminated by mercury in the sediment. After the primary source of dissolved mercury to the lake was eliminated in 1998, fish-tissue mercury concentrations decreased dramatically (Southworth *et al.*, 2002).

Although leaching and transport of contaminated soils is a potentially important source of mercury to aquatic environments, load estimates are inherently uncertain and are likely variable among sites. Mercury is widely used in many industries, and anthropogenic sources, such as former battery recycling facilities, mercury retorting facilities, manufactured gas plants, mercury-cell chlor-alkali facilities, metering stations on natural gas pipelines, and sewage sludge applications to soil, can produce locally elevated mercury concentrations in environmental media surrounding the facilities. At mercury-contaminated sites, concentrations as high as 3000 mg/kg in soils have been reported (Turner and Southworth, 1999). This is significantly higher than the background concentration (<1 mg/kg).

1.4.3 Mining and Refining

Another important mercury emission source category is mining and refining. The specific source identified in Table 1.4.1 for the United States is gold mining, which uses mercury amalgam to recover gold during ore processing. Other mining- and refining-related atmospheric emission sources include smelting of nonferrous metals (in which mercury occurs as a trace element) (US EPA, 1997), direct mercury mining (or emissions from historical tailings containing cinnabar (HgS(s)) (Gustin et al., 2003), and petroleum refining (with an estimated contribution of 1.5 tons/yr in the United States) (Wilhelm, 2001). Although Table 1.4.1 indicates that combustion is the largest anthropogenic emission in the United States, nonferrous metal smelting operations were the largest source category in Mexico and Canada for the year 1990 (31 and 24 tons/yr, respectively) (Pai et al., 2000).

Similar to manufacturing sources, Figure 1.4.1 indicates that the largest releases of mercury from mining sites are to the soil or water. Examples include mercury mines at Mt. Amiata in central Italy (Ferrara, 1999), Almaden in Spain (Ferrara, 1999), Idrija in Slovenia (Miklavcic, 1999), Pinchi Lake in British Columbia (Turner and Southworth, 1999), and Clear Lake in California, United States (Turner and Southworth, 1999). Mercury contamination of soil has also resulted from gold and silver mining, with about 500,000 tons of mercury used from 1540 to 1900 for silver and gold production in the Americas (Nriagu, 1994). Soil contamination associated with these areas resulted from tailing deposits and erosional transport of fine particles containing mercury (Lacerda and Salomons, 1999). Not only has mercury amalgamation resulted in elevated mercury and methylmercury levels in the Sacramento River Basin, California (Domagalski, 1998), but continued use in gold mining in South America has resulted in widespread mercury contamination in the Amazon region of Brazil (Barbosa et al., 1994).

1.4.4 Miscellaneous Sources of Mercury

The primary sources of elemental mercury to indoor air are occupation-related contamination and consumer products that contain either mercury salts or liquid elemental mercury (Hg^0). The clothing and personal gear of employees in mercury-related industries can be a source of contamination to indoor air in the home (ATSDR, 1990). In a study conducted in homes of thermometer plant workers, 10 percent of 39 randomly selected homes had indoor mercury vapor concentrations that exceeded the EPA reference concentration (RfC) (Hudson et al., 1987). More commonly, spills from the following household devices that contain liquid elemental mercury are a source of mercury to indoor air (ATSDR, 1999):

• Fluorescent light bulbs
• Thermostats
• Glass thermometers
• Barometers
• Switches in large appliances
• Natural gas pressure regulators.

Consumer products such as latex paint and chlorine-based detergents and cleansers that can decompose and volatilize to produce gaseous elemental mercury (Beusterien et al., 1991) are also potential sources. In addition to these sources, elemental mercury is sometimes released intentionally during cultural and religious practices (Riley et al., 2001).

The level of exposure to mercury vapor via inhalation is the critical health issue related to elemental mercury in indoor environments (ATSDR, 1999). Potential toxic effects include damage to the respiratory tract, heart, gastrointestinal tract, reproductive system, brain, and kidneys (ATSDR, 1999); however, the primary adverse effects are usually limited to the brain and kidneys, which have been identified as the most sensitive target organs. Either acute or chronic exposure apparently elicits a similar toxicological syndrome. The lowest-observed-adverse-effect level (LOAEL), established on the basis of central nervous system (CNS) effects during human occupational exposure, is estimated to be $26 \mu g/m^3$ (Ashe et al., 1953; Fawer et al., 1983).

Existing regulatory guidance on mercury vapor air concentrations and corresponding exposure durations that are protective of human health in occupational and residential environments is presented in Table 1.4.4. As shown in the table, the Occupational Safety and Health Administration (OSHA) has set a legally enforceable concentration ceiling for workplace exposure at $100 \mu g/m^3$. In addition, two recommended time-weighted average concentrations have been established by the National Institute for Occupational Safety and Health (NIOSH) and the American Conference of Governmental Industrial Hygienists (ACGIH). The corresponding values of 50 and $25 \mu g/m^3$, respectively, are designed to be protective of occupational exposure during an 8- or 10-hour workday (and a 40-hour work week). The $25-\mu g/m^3$ concentration is based on the LOAEL noted above.

As illustrated in Table 1.4.4, permissible concentrations in residential environments are lower than occupational settings, owing to the longer expected exposure durations. In response to concerns about mercury vapor exposure in homes, schools, or businesses due to accidental releases during the removal of mercury-seal gas-pressure regulators, the Agency for Toxic Substances and Disease Registry (ATSDR) established residential relocation and occupancy levels of 10 and $1 \mu g/m^3$, respectively (ATSDR, 2005). The upper value is used to determine whether residents can remain in a house or apartment during a spill. The lower value is used as an acceptable re-occupancy level after cleanup, and is deemed protective of pregnant women and children under the age of six. The last two values in Table 1.4.4 are the EPA and ATSDR reference minimal risk levels, which were calculated using the LOAEL studies as a basis. ATSDR considers the RfC and chronic minimum risk level to equivalently represent the daily human exposure concentration that can occur without appreciable risk of adverse, non-cancer health effects.

In 1999 and 2000, ATSDR's Hazardous Substances Emergency Events Surveillance system reported more than 300 incidents of mercury spills per year in the northeastern United States. Evidence shows that very small quantities of elemental mercury spilled from consumer products such as thermometers can result in indoor air mercury concentrations that exceed the standards and action levels presented in Table 1.4.4 (Smart, 1986). In addition, a random sampling of indoor sites in the New York metropolitan area revealed that 11 of the 12 investigated residences had mercury vapor concentrations that were higher than outdoor levels (Carpi and Chen, 2001). In one case, a concentration of $0.045 \mu g/m^3$ was observed in a bathroom following a spill that had occurred 16 years prior.

Table 1.4.4 *Summary of Mercury Vapor Guidance Values*

Guidance Value	Air Concentration ($\mu g/m^3$)	Basis for Standard or Action Level	ATSDR Recommended Method of Analysis[a]
OSHA ceiling (occupational)	100	Maximum concentration not to be exceeded in workplace	Real-time air monitoring (Jerome® or Lumex® equivalent)
NIOSH Recommended Exposure Limit (REL) (occupational)	50	Time-weighted average (TWA) concentration	NIOSH 6009 or passive dosimeters[b]
ACGIH Threshold Limit Value (TLV) (occupational)	25	Human LOAEL in occupational studies	Real-time air monitoring (Jerome or Lumex eq.)
ASTDR Action Level	10	Residential relocation based on subtle health effects	
ASTDR Action Level	3	Re-occupancy level in occupational or commercial setting	NIOSH 6009 or equivalent
ASTDR Action Level	1	Re-occupancy level in residential setting (assumes 24 hrs/day)	NIOSH 6009 or real-time monitoring (Lumex equiv.)
EPA RfC	0.3	LOAEL adjusted and uncertainty factor of 30 (assumes 24 hrs/day)	
ASTDR Minimum Risk Level	0.2	LOAEL adjusted and uncertainty factor of 30 (assumes 24 hrs/day)	

[a] Agency for Toxic Substances and Disease Registry (ATSDR, 2005) unless otherwise indicated.
[b] OSHA recommendation.

Examples of litigation involving indoor mercury releases include the NICOR Power Company in Chicago, which was ordered to inspect more than 200,000 households following the discovery that liquid elemental mercury was accidentally spilled during the routine replacement of natural gas pressure regulators (NY Times, 2000). No systematic environmental forensic technique was employed in the related investigations. Instead, breathing-zone mercury vapor levels were measured, without direct assessment of all possible sources.

1.5 DETECTING MERCURY IN INDOOR AIR

Several methods are available for assessing mercury vapor concentrations in indoor air. Some of the more common techniques are summarized in Table 1.5.1. According to procedures developed by ATSDR for assessing indoor air mercury contamination from natural gas mercury pressure regulators (ATSDR, 2005), air monitoring should occur during initial characterization, confirmation, cleanup, and post-remediation. The methods described below each offer advantages in terms of portability, detection limits, and response time during the investigation.

Jerome®—The Jerome® 431-X mercury vapor analyzer is a portable, hand-held device that allows rapid, screening-level assessments. The instrument uses a gold film sensor to collect mercury from a known volume of indoor air. The measured change in electrical resistance in the sensor is compared to a reference sensor to produce a quantifiable mercury vapor concentration. The circuit in the instrument is automatically zeroed after each reading, and the gold film is periodically cleaned of mercury by electric heating. Although the Jerome® can be used to rapidly characterize indoor air, its detection limit is just below the threshold concentration for residential occupancy (Table 1.4.4). Because the Jerome® is also subject to interferences from volatile gases that can adsorb to the gold film, Bigham (2004) used a Lumex® for confirmation when the Jerome® recorded

values equal to or greater than $3 \mu g/m^3$ in samples collected 5 in. above the floor.

Lumex®—The Lumex® RA-915+ mercury analyzer is a portable device, with many advantages over the Jerome® 431-X. Among its advantages, the instrument is an atomic absorption spectrometer with a Zeeman background correction for removing interferences. It offers a lower detection limit than the Jerome® and also generates real-time data.

NIOSH 6009—This method involves the collection of an 8-hour air sample by passing room air drawn by a pump through a solid sorbent material in a glass collection tube. NIOSH recommends that the sampling height be approximately 3 ft (corresponding to the height of a small child), and that the sample pump be located in the area of the home where the maximum exposure to mercury contamination is expected to occur. Analysis of the sorbent in the laboratory involves digesting the sorbent in concentrated acids (HCl and HNO_3), reducing the mercury using stannous chloride, and finally quantifying the mercury vapor concentration using cold-vapor atomic absorption spectroscopy.

Dosimeters—Both NIOSH Method 6009 and passive dosimeters are OSHA-approved methods for measuring mercury vapor levels in workplace environments (OSHA, 2005). The collection medium is typically hopcalite (oxides of manganese and copper). Although no interferences are listed, disadvantages of passive dosimeters include a sampling rate dependence on face velocity. Consequently, dosimeters are not recommended where the air velocity is greater than 229 m/min.

Tekran®—Ultra-low detection limits are possible with the Tekran® Model 2537A mercury vapor analyzer by first preconcentrating mercury from the air using a pure gold adsorbent that amalgamates the mercury. Cold-vapor atomic fluorescence spectroscopy is then used to quantify the desorbed vapor using a method developed by Temmerman et al. (1990). Tekran® instruments are potentially useful tools in mercury source evaluations. This is particularly true

Table 1.5.1 *Summary of Residential Mercury Vapor Measurement Techniques*

Method	Description	Proposed Uses

Jerome® 431-X	Type: Portable hand-held
	   Range: $1\,\mu g/m^3$–$999\,\mu g/m^3$
	   Accuracy: $\pm 5\%$ at $100\,\mu g/m^3$
	   Sample Frequency: 13 seconds
	   Conditions: 0–40°C
	   Interferences: Cl, NO_2, H_2S, thiols (e.g., cat litter, ammonia, smoke, liquid bleach, spoiled food)
		Screening Level Investigation (ATSDR, 2005)

Lumex®	Type: Portable (7.5 kg)
	   Range: $0.002\,\mu g/m^3$–$100\,\mu g/m^3$
	   Accuracy: $\pm 5\%$ at 5–$40\,\mu g/m^3$
	   Sample Frequency: 1 second
	   Conditions: 5–40°C
	   Interferences: Ambient environment during calibration should have mercury $<0.1\,\mu g/m^3$
		Confirmation and Characterization Sampling (Bigham, 2004)

NIOSH 6009	Type: Solid sorbent tube
	   Range: $0.1\,\mu g/m^3$–$500\,\mu g/m^3$
	   Accuracy: Not determined
	   Sample Frequency: 8 hours
	   Interferences: Inorganic and organic mercury compounds
		Verification Sampling and Workplace Sampling (ATSDR, 2005; Bigham, 2004)

Passive Dosimeter	Type: Mobile
	   Range: 2–$4\,\mu g/m^3$ lower limit
	   Accuracy: $\pm 9\%$
	   Sample Frequency: 8 hours
	   Interferences: None listed
		Workplace Sampling (OSHA, 2005)

Tekran®	Type: Stationary Sampler (23 kg)
	   Range: $0.0001\,\mu g/m^3$–$10\,\mu g/m^3$
	   Accuracy: Linearity at $\pm 2\%$
	   Sample Frequency: 2.5–360 minutes
	   Interferences: None
		Low-level Ambient Air Monitoring

in outdoor air, where background mercury vapor concentrations are typically in the range of $0.002 - 0.005\,\mu g/m^3$ (Seigneur *et al.*, 1994).

Biological Sampling Methods—An additional method for identifying exposure is measuring mercury concentrations in biological samples. Of the various biological media, the concentration of mercury in urine is believed to be the most accurate indicator of chronic exposure to mercury vapor (blood better approximates short-term exposure and is more sensitive to dietary pathways) (ATSDR, 1999). Tsuji *et al.* (2003) found that mercury levels in air and urine are correlated below $0.050\,mg/m^3$; however, chronic exposure to mercury vapor concentrations below $0.010\,mg/m^3$ corresponds to urinary mercury concentrations that are indistinguishable from background urinary mercury levels.

1.6 MERCURY FORENSICS

Investigation of mercury in an environmental forensic context has had limited application. Perhaps the most well-developed methods are the use of radioisotopes to track the fate of mercury in terrestrial and aquatic environments. Mercury has seven naturally occurring (nonradioactive), stable isotopes, listed in Table 1.6.1. Different stable isotopes can be added to soils, sediment, or water to follow its movement, including chemical transformation. A major experiment, known as METAALICUS (Mercury Experiment To Assess Atmospheric Loading In Canada and the United States), is underway in Ontario, Canada, to evaluate the behavior of mercury contributed by atmospheric deposition. The study is being conducted in two phases. Background studies (Phase 1) began in 1999. Phase 2, the artificial loading of mercury isotopes to the whole ecosystem, began in 2001

and will continue through at least 2005 (EPRI, 2005). Different mercury isotopes (upland, ^{200}Hg; wetland, ^{198}Hg; and lake, ^{202}Hg) have been applied at a rate of about three times background level ($25\,\mu g/m^2$-yr) (METAALICUS, 2003). The different isotopes can be detected in sediments and as methylmercury in fish tissue to distinguish it from natural background and to determine the original source of the mercury. The objective of the drainage-basin-scale experiment is to determine the response of the ecosystem to a change in atmospheric mercury loading.

A second application of environmental forensics to mercury is to evaluate the rate of volatilization of spilled elemental mercury. Spillage of elemental mercury in homes, schools, or the workplace is relatively common, and most state health departments provide advice on actions to take in response to a spill. Spillage of mercury has been a particular problem during removal of natural gas regulators that

Table 1.6.1 *Naturally Occurring Mercury Isotopes[a]*

Isotope	Atomic Mass (m_a/u)	Natural Abundance (percent)
196Hg	195.965807	0.15
198Hg	197.966743	9.97
199Hg	198.968254	16.87
200Hg	199.968300	23.10
201Hg	200.970277	13.18
202Hg	210.970617	29.86
204Hg	203.973467	6.87

[a]IUPAC (1998).

contain mercury as part of a pressure relief mechanism. Originally installed, typically in basements, in the 1940s–1960s, mercury-containing regulators have been replaced by regulators with a spring pressure-release mechanism. As noted earlier, the issue of mercury spilled from regulators became a much-publicized issue for Nicor Gas, EPA, and the State of Illinois in 2000. About the same time, similar problems developed for other gas utilities in the Chicago area and in Detroit. Most recently, a class action lawsuit related to spillage of mercury in homes was filed against a gas utility in Philadelphia in 2005 and was subsequently settled.

While the action levels for mercury in air have been set (Table 1.4.4), the relationship between the mass of mercury spilled and the resulting concentration of mercury vapor in air has not been well established. The typical problem related to mercury spilled from gas pressure regulators is to hindcast the historical mercury vapor concentrations in the basement or entire home for a spill that occurred up to ten years ago. Commonly, an effort would have been made to recover the spilled mercury, but small amounts of mercury remained in floor cracks, under carpet, and under or absorbed into wood molding. While the evaporation of mercury vapor from a spherical drop or bead can be readily calculated, it is unclear how applicable these calculations are to the typical form of spilled mercury. Winter (2003) evaluated the evaporation of mercury from two beads, one about 0.2 g and 3 mm in diameter, and the other a larger, approximately 2 g, nonspherical drop. Based on theory, Winter (2003) calculated the evaporation rate of the smaller drop to be about 11 μg/hr at 20°C and estimated that the drop would evaporate completely in about 3 years. Such rapid evaporation seems inconsistent with observations in homes where spills occurred (Bigham, 2004). Through repeated weighing of the two mercury drops, Winter (2003) found that the rate of evaporation decreased significantly, from an initial rate of 7 μg/hr for the small drop to negligible weight loss after 2.5 months. Evaporation of the larger drop decreased from an initial rate of 6.2 μg/hr to 0.5 μg/hr six months later.

The explanation for the decrease in the evaporation rate comes from an unusual source—astronomy. Telescopes with mirrors of elemental mercury have been in use for more than 20 years (Borra, 1982). Known as liquid mirror telescopes (LMTs), these instruments take advantage of the fact that spinning liquid takes on the form of a parabola, ideal for telescope mirrors. Current LMTs are about 2–4 m in diameter. An LMT constructed by the National Aeronautics and Space Administration (NASA) is 3 m in diameter and contains 14 L of elemental mercury (http://www.astro.ubc.ca/LMT/Nodo/index.html). The mercury spreads over the parabolic-shaped support to a thickness of 1.6 mm as the support spins at a rate of 10 rpm. LMTs are limited to pointing straight up or tilting only a few degrees from vertical, but are much less expensive to build and provide superior reflectivity than conventional telescopes.

Extensive evaluation of the stability of the reflective surface has shown that elemental mercury oxidizes in the presence of water vapor and impurities on the time scale of a few hours. The mercuric oxide surface layer reduces the evaporation rate by five orders of magnitude, so that after two weeks, evaporation is negligible (Mulrooney, 2000).

Based on the work of Mulrooney (2000) and Winter (2003), it is reasonable to expect that mercury spilled in homes or the workplace will remain indefinitely if not cleaned up or periodically disturbed. Because of oxidation, the evaporation rate of elemental mercury hidden in floor cracks or elsewhere will decline to negligible levels over the course of a few months. If disturbed by foot traffic, for example, the oxidized crust can be destroyed and the oxidation process must begin again.

A further consequence of mercury oxidation and the resulting negligible long-term evaporation is that forensic techniques based on isotope fractionation are unlikely to be useful for determining the timing of a mercury spill. The rates of evaporation of mercury's stable isotopes (Table 1.6.1) are slightly different because of the slight difference in atomic mass. If the isotopic ratio of the mercury originally spilled was known, it might be possible to determine how long the mercury had been evaporating by determining the degree of isotopic fractionation in the spilled sample. Unfortunately, this technique cannot be used when the rate of evaporation (fractionation) is variable or negligible.

REFERENCES

ACGIH, 2000. *Mercury, elemental and inorganic. Threshold limit values for chemical substances and physical agents and biological exposure indices.* American Council of Governmental and Industrial Hygienists, Cincinnati, OH.

Amyot, M., D.R.S. Lean, and G. Mierle. 1997a. Photochemical formation of volatile mercury in high arctic lakes. *Environ. Toxicol. Chem.* 16:2054–2063.

Amyot, M., G. Mierle, D.R.S. Lean, and D.J. McQueen. 1997b. Effect of solar radiation on the formation of dissolved gaseous mercury in temperate lakes. *Geochim. Cosmochim. Acta* 61:975–988.

API. 2004. *Mercury: Chemistry, fate, toxicity, and wastewater treatment options.* Unpublished API Publication. American Petroleum Institute, Washington, DC.

Ariya, P., A. Khalizov, and A. Gidas. 2002. Reactions of gaseous mercury with atomic and molecular halogens: Kinetics, product studies, and atmospheric implications. *J. Phys. Chem. A.* 106:7310–7320.

Ashe, W.F., E.J. Largent, F.R. Dutra, D.M. Hubbard, and M. Blackstone. 1953. Behavior of mercury in the animal organism following inhalation. *Ind. Hyg. Occup. Med.* 17:19–43.

ATSDR. 1990. Final report. Technical assistance to the Tennessee Department of Health and Environment. Mercury exposure study Charleston, Tennessee. Agency for Toxic Substances and Disease Registry, US Department of Health and Human Services, Public Health Service, Centers for Disease Control, Atlanta, GA.

ATSDR. 1999. *Toxicological profile for mercury.* Agency for Toxic Substances and Disease Registry, US Department of Health and Human Services, Public Health Service.

ATSDR. 2005. Suggested action levels for indoor mercury vapors in homes. Available at www.publichealth. columbus.gov/Asset/iu_files/Indoor_ Air_Table.pdf. Accessed May 11, 2005. Agency for Toxic Substances and Disease Registry.

Babiarz, C.L. and A.W. Andren. 1995. Total concentrations of mercury in Wisconsin (USA) lakes and rivers. *Water Air Soil Pollut.* 83:173–183.

Babiarz, C.L., J.P. Hurley, J.M. Benoit, M.M. Shafer, A.W. Andren, and D.A. Webb. 1998. Seasonal influences on partitioning and transport of total and methylmercury in rivers from contrasting watersheds. *Biogeochemistry* 41:237–257.

Balogh, S. and L. Liang. 1995. Mercury pathways in municipal wastewater treatment plants. *Water Air Soil Pollut.* 83:173–183.

Barbosa, A.C., A.A. Boischio, G.A. East, I. Ferrari, A. Goncalves, P.R.M. Silva, and T.M.E. da Cruz. 1994. Mercury contamination in the Brazilian Amazon: Environmental and occupational aspects. *Water Air Soil Pollut.* 80:109–121.

Bartlett, P.D. and P.J. Craig. 1981. Total mercury and methyl mercury levels in British estuarine sediments-II. *Water Res.* 15:37–47.

Benoit, J.M., C.C. Gilmour, R.P. Mason, G.S. Reidel, and G.F. Reidel. 1998. Behavior of mercury in the Patuxent River estuary. *Biogeochemistry* 40:249–265.

Benoit, J.M., C.C. Gilmour, and R.P. Mason. 1999. Sulfide controls on mercury speciation and bioavailability to methylating bacteria in sediment pore water. *Environ. Sci. Technol.* 33:951–957.

Benoit, J.M., R.P. Mason, C.C. Gilmour, and G.R. Aiken. 2001. Constants for mercury binding by dissolved organic matter isolates from the Florida Everglades. *Geochim. Cosmochim. Acta.* 65:4445–4451.

Bergan, T. and H. Rodhe. 2001. Oxidation of elemental mercury in the atmosphere: Constraints imposed by global scale modeling. *J. Atmos. Chem.* 40:191–212.

Bergan, T., L. Gallardo, and H. Rodhe. 1999. Mercury in the global troposphere: A three-dimensional model study. *Atmos. Environ.* 33:1575–1585.

Beusterien, K.M., R.A. Etzel, M.M. Agocs, G.M. Egeland, E.M. Socie, M.A. Rouse, and B.K. Mortensen. 1991. Indoor air mercury concentrations following application of interior latex paint. *Arch. Environ. Contam. Toxicol.* 21:62–64.

Bigham, G.N. 2004. Assessment of exposure to mercury vapor in indoor air from spilled elemental mercury. Abstract: Proc. 7th International Conference on Mercury as a Global Pollutant. Ljubljana, Slovenia, June 27–July 2, 2004.

Bigham, G.N. and G.M. Vandal. 1996. A drainage basin perspective of mercury transport and bioaccumulation: Onondaga Lake, New York. *Neurotoxicology* 17(1):279–290.

Bindler, R. 2003. Estimating the natural background atmospheric deposition rate of mercury utilizing ombrotrophic bogs in Southern Sweden. *Environ. Sci. Technol.* 37:40–46.

Bloom, N.S. 2001. Assessment of ecological and human health impacts of mercury in the Bay-Delta watershed. CALFED Bay-Delta Mercury Project, Final Report.

Bloom, N.S., E. Preus, J. Katon, and M. Hiltner. 2003. Selective extractions to assess the biogeochemically relevant fractionation of inorganic mercury in sediments and soil. *Anal. Chim. Acta.* 479:233–248.

Blum, J.E. and R. Bartha. 1980. Effect of salinity on methylation of mercury. *Bull. Environ. Contam. Toxicol.* 25:404–408.

Bodaly, R.A., J.W.M. Rudd, and R.J. Flett. 1998. Effect of urban sewage treatment on total and methyl mercury concentrations in effluents. *Biogeochemistry* 40:279–291.

Borra, E.F. 1982. The liquid-mirror telescope as a viable astronomical tool. *Royal Astron. Soc. Can. J.* 76:245–256.

Brianfireun, B.A., A. Heyes, and N.T. Roulet. 1996. The hydrology and methylmercury dynamics of a Precambrian Shield headwater peatland. *Water Resources Res.* 32:1785–1794.

Callister, S.M. and M.R. Winfrey. 1986. Microbial methylation of mercury in upper Wisconsin River sediments. *Water Air Soil Pollut.* 29:453–465.

Carpi, A. and Y-F. Chen. 2001. Gaseous elemental mercury as an indoor air pollutant. *Environ. Sci. Technol.* 35:4170–4173.

Carpi, A. and S.E. Lindberg. 1997. Sunlight-mediated emission of elemental mercury from soil amended with municipal sewage sludge. *Environ. Sci. Technol.* 31:2085–2091.

Cohen, M., R. Artz, R. Draxler, P. Miller, L. Poissant, D. Niemi, D. Ratte, M. Deslauriers, R. Duval, R. Laurin,

J. Slotnick, T. Nettesheim, and J. McDonald. 2004. Modeling the atmospheric transport and deposition of mercury to the Great Lakes. *Environ. Res.* 95:247–265.

Compeau, G., and R. Bartha. 1984. Methylation and demethylation of mercury under controlled redox, pH, and salinity conditions. *Appl. Environ. Microbiol.* 48:1203–1207.

Compeau, G. and R. Bartha. 1985. Sulfate-reducing bacteria: Principal methylators of mercury in anoxic estuarine sediment. *Appl. Environ. Microbiol.* 50:498–502.

Compeau, G. and R. Bartha. 1987. Effect of salinity of mercury-methylating activity of sulfate-reducing bacteria in estuarine sediments. *Appl. Environ. Microbiol.* 53:261–265.

Coolbaugh, M.F., M.S. Gustin, and J.J. Rytuba. 2002. Annual emissions of mercury to the atmosphere from natural sources in Nevada and California. *Environ. Geol.* 42:338–349.

Constantinou, E., X.A. Wu, and C. Seigneur. 1995. Development and application of a reactive plume model for mercury emissions. *Water Air Soil Pollut.* 80:325–335.

Dastoor, A.P. and Y. Larocque. 2004. Global circulation of atmospheric mercury: A modeling study. *Atmos. Environ.* 38:147–161.

Domagalski, J. 1998. Occurrence and transport of total mercury and methylmercury in the Sacramento River Basin, California. *J. Geochem. Explor.* 64:277–291.

Driscoll, C.T., M. Kalicin, E. McLaughlin, C. Liussi, R. Newton, R. Munson, and J. Yavitt. 2002. Chemical and biological control of mercury cycling in upland, wetland, and lake ecosystems in the Adirondack region of New York. In: *Mercury in the Environment: Assessing and Managing the Multimedia Risks.* Preprints of Extended Abstracts, American Chemical Society, Orlando, FL 42:809–812.

Dvonch, J.T., J.R. Graney, G.J. Keeler, and R.K. Stevens. 1999. Use of elemental tracers to source apportion mercury in south Florida precipitation. *Environ. Sci. Technol.* 33:4522–4527.

Ebinghaus, R., R.M. Tripathi, D. Wallschlager, and S.E. Lindberg. 1999. Natural and anthropogenic mercury sources and their impact on the air-surface exchange of mercury on regional and global scales. pp. 3–50. In: *Mercury Contaminated Sites: Characterization, Risk Assessment, and Remediation.* R. Ebinghaus, R.R. Turner, L.D. de Lacerda, O. Vasiliev, and W. Salomons (eds). Springer-Verlag, Berlin.

EPRI. 2000. *An Assessment of Mercury Emissions from US Coal-fired Power Plants.* EPRI, Palo Alto, CA.

EPRI. 2005. Mercury experiment to assess atmospheric loading in Canada and the United States (METAALICUS): Phase II, evaluating the effects of loadings. Available at www.epriweb.com/public/000000000001011964.pdf. Accessed May 11, 2005.

European Commission. 2004. Mercury flows in Europe and the world: The impacts of decommissioned chlor-alkali plants. Final Report. 88pp. Directorate General for Environment, Brussels.

Fawer, R.F., Y. De Ribaupierre, M.P. Guillemin, M. Berode, and M. Lob. 1983. Measurement of hand tremor induced by industrial exposure to metallic mercury. *Br. J. Ind. Med.* 40:204–208.

Ferrara, R. 1999. Mercury mines in Europe: Assessment of emissions and environmental contamination. pp. 51–72. In: *Mercury Contaminated Sites: Characterization, Risk Assessment, and Remediation.* R. Ebinghaus, R.R. Turner, L.D. de Lacerda, O. Vasiliev, and W. Salomons (eds). Springer-Verlag, Berlin.

Ferrara, R., B. Mazzolai, E. Lanzillotta, E. Nucaro, and N. Pirrone. 2000. Volcanoes as emission sources of atmospheric mercury in the Mediterranean basin. *Sci. Tot. Environ.* 259:115–121.

Fitzgerald, W.F. and R.P. Mason. 1997. Biogeochemical cycling of mercury in the marine environment. *Metal Ions Biol. Syst.* 34:53–111.

Fitzgerald, W.F., R.P. Mason, and G.M. Vandal. 1991. Atmospheric cycling and air-water exchange of mercury over mid-continental lacustrine regions. *Water Air Soil Pollut.* 56:745–767.

French, K.J., D.A. Scruton, M.R. Anderson, and D.C. Schnieder. 1999. Influence of physical and chemical characteristics on mercury in aquatic systems. *Water Air Soil Pollut.* 110:347–362.

Gardfeldt, K. 2003. Transformation of mercury species in the aqueous phase. Thesis, Goteborgs Universitet.

Gardfeldt, K., J. Sommar, D. Stromberg, and X. Feng. 2001. Oxidation of atomic mercury by hydroxyl radicals and photoinduced decomposition of methylmercury in the aqueous phase. *Atmos. Environ.* 35:3039–3047.

Gill, G.A., N.S. Bloom, S. Cappellino, C.T. Driscoll, C. Dobbs, L. McShea, R. Mason, and J.W.M. Rudd. 1999. Sediment-water fluxes of mercury in Lavaca Bay, Texas. *Environ. Sci. Technol.* 33:663–669.

Gilmour, C.C. and D.G. Capone. 1987. Relationship between Hg methylation and the sulfur cycle in estuarine sediments. *EOS, Trans. Amer. Geo. Union* 68:1718.

Gilmour, C.C. and E.A. Henry. 1991. Mercury methylation in aquatic systems affected by acid deposition. *Environ. Pollut.* 71:131–169.

Gilmour, C.C., E.A. Henry, and R. Mitchell. 1992. Sulfate stimulation of mercury methylation in freshwater sediments. *Environ. Sci. Technol.* 26:2281–2287.

Glass, G.A., J.A. Sorensen, K.W. Schmidt, and G.R. Rapp. 1990. New source identification of mercury contamination in the Great Lakes. *Environ. Sci. Technol.* 24:1059–1069.

Glass, G.E., J.A. Sorenson, K.W. Schmidt, G.R. Rapp, D. Yap, and D. Fraser. 1991. Mercury deposition and sources for the Upper Great Lakes region. *Water Air Soil Pollut.* 56:235–249.

Graney, J.R., J.T. Dvonch, and G.J. Keeler. 2004. Use of multi-element tracers to source apportion mercury in south Florida aerosols. *Atmos. Environ.* 38:1715–1726.

Gustin, M.S., G.E. Taylor, T.L. Leonard, and R.E. Keislar. 1995. Atmospheric mercury concentrations associated with geologically and anthropogenically enriched sites in central western Nevada. *Environ. Sci. Technol.* 30:2572–2579.

Gustin, M.S., S.E. Linberg, K. Austin, M. Coolbaugh, A. Vette, and H. Zhang. 2000. Assessing the contribution of natural sources to regional atmospheric mercury budgets. *Sci. Tot. Environ.* 259:61–71.

Gustin, M.S., M.F. Coolbaugh, M.A. Engle, B.C. Fitzgerald, R.E. Keislar, S.E. Lindberg, D.M. Nacht, J. Quashnick, J.J. Rytuba, C. Sladek, H. Zhang, and R.E. Zehner. 2003. Atmospheric mercury emissions from mine wastes and surrounding geologically enriched terrains. *Environ. Geol.* 43:339–351.

Haitzer, M., G.R. Aiken, and J.N. Ryan. 2002. Binding of mercury(II) to dissolved organic matter: The role of the mercury-to-DOM concentration ratio. *Environ. Sci. Technol.* 36:3564–3570.

Haitzer, M., G.R. Aiken, and J.N. Ryan. 2003. Binding of mercury(II) to aquatic humic substances: Influence of pH and source of humic substances. *Environ. Sci. Technol.* 37:2436–2441.

Hall, B. 1995. The gas phase oxidation of elemental mercury by ozone. *Water Air Soil Pollut.* 85:301–315.

Hammerschmidt, C.R. and W.F. Fitzgerald. 2004. Geochemical controls on the production and distribution of methylmercury in near-shore marine sediments. *Environ. Sci. Technol.* 38:1487–1495.

Hedgecock, I.M. and N. Pirrone. 2001. Mercury and photochemistry in the marine boundary layer—modeling studies suggest the *in situ* production of reactive gaseous phase mercury. *Atmos. Environ.* 35:3055–3062.

Hedgecock, I.M. and N. Pirrone. 2004. Chasing quicksilver: Modeling the atmospheric lifetime of $Hg^0(g)$ in the marine boundary layer at various latitudes. *Environ. Sci. Technol.* 38:69–76.

Hintelmann, H., P.M. Welbourn, and R.D. Evans. 1997. Measurement of complexation of methylmercury(II) compounds by freshwater humic substances using equilibrium dialysis. *Environ. Sci. Technol.* 31:489–495.

Hsu, H. and D. Sedlak. 2003. Strong Hg(II) complexation in municipal wastewater effluent and surface waters. *Environ. Sci. Technol.* 37:2743–2749.

Hudson, P.J., R.L. Vogt, J. Brondum, L. Witherell, G. Myers, and D.C. Paschal. 1987. Elemental mercury exposure among children of thermometer plant workers. *Pediatrics* 79:935–938.

Hultberg, H., J. Munthe, and A. Iverfeldt. 1995. Cycling of methylmercury and mercury–responses in the forest roof catchment to three years of decreased atmospheric deposition. *Water Air Soil Pollut.* 80:415–424.

Hurley, J.P., C.J. Watras, and N.S. Bloom. 1991. Distribution and flux of particulate mercury in temperate seepage lake. *Water Air Soil Pollut.* 56:543–551.

Hurley, J.P., J.M. Benoit, C.L. Babiarz, M.M. Shafer, A.W. Andren, J.R. Sullivan, R. Hammond, and D.A. Webb. 1995. Influences of watershed characteristics on mercury levels in Wisconsin rivers. *Eviron. Sci. Technol.* 29:1867–1875.

IUPAC. 1998. Commission on Atomic Weights and Isotopic Abundances report for the International Union of Pure and Applied Chemistry. In: *Isotopic Compositions of the Elements. Pure and Applied Chemistry* 70:217.

Iverfeldt, A. 1991. Mercury in forest canopy throughfall water and its relation to atmospheric deposition. *Water Air Soil Pollut.* 56:553–564.

Jacobs, L.A., S.M. Klein, and E.A. Henry. 1995. Mercury cycling in the water column of a seasonally anoxic urban lake (Onondaga Lake, New York). *Water Air Soil Pollut.* 80:553–562.

Jay, J.A., F.M.M. Morel, and H.F. Hemond. 2000. Mercury speciation in the presence of polysulfides. *Environ. Sci. Technol.* 34:2196–2200.

Johansen, V.C. 2003. Mercury speciation in other combustion sources: A literature review. R&D Serial No. 2578. Portland Cement Association, Skokie IL, USA. 20pp.

Johansen, V.C. and G.J. Hawkins. 2003. Mercury emission and speciation from Portland cement kilns. R&D Serial No. 2567a. Portland Cement Association, Skokie IL, USA. 19pp.

Judson, G. 2002. Analysis of mercury speciation profiles currently used for atmospheric chemistry modeling. Wisconsin Department of Natural Resources. Available at www.dnr.state.wi.us/org/aw/air/staff/hganalysisteam/docs/hgspeciation.pdf. Accessed May 11, 2005.

Kilgore, J. and C. Senior. 2003. Fundamental science and engineering of mercury control in coal-fired power plants. In: *Air Quality IV Conference*, September 22–24, 2003. Arlington, VA. 15pp.

Kim, J.P. 1995. Methylmercury in rainbow trout (*Oncorhynchus mykiss*) from lakes Okareka, Okaro, Rotomahana, Rotorua and Tarawera, North Island, New Zealand. *Sci. Tot. Environ.* 164:209–219.

Korthals, E.T. and M.R. Winfrey. 1987. Seasonal and spatial variations in mercury methylation and demethylation in an oligotrophic lake. *Appl. Environ. Microbiol.* 53:2397–2404.

Kotnik, J., M. Horvat, V. Mandic, and M. Logar. 2000. Influence of the Sostanj coal-fired thermal power plant on mercury and methylmercury concentrations in Lake Velenje, Slovenia. *Sci. Tot. Environ.* 259:85–95.

Kotnik, J., M. Horvat, V. Fajon, and M. Logar. 2002. Mercury in small freshwater lakes: A case study: Lake Velenje, Slovenia. *Water Air Soil Pollut.* 134:319–339.

Krabbenhoft, D.P., J.M. Benoit, C.L. Babiarz, J.P. Hurley, and A.W. Andren. 1995. Mercury cycling in the Allequash Creek watershed, northern Wisconsin. *Water Air Soil Pollut.* 80:425–433.

Krabbenhoft, D.P., J.P. Hurley, M.L. Olson, and L.B. Cleckner. 1998. Diel variability of mercury phase and species distributions in the Florida Everglades. *Biogeochemistry* 40:311–325.

Krabbenhoft, D.P., J.G. Wiener, W.G. Brumbaugh, M.L. Olson, J.F. DeWild, and T.J. Sabin. 2003. *A national pilot study of mercury contamination of aquatic ecosystems along multiple gradients.* US Geological Survey.

Lacerda, L.D. and W. Salomons. 1999. Mercury contamination from New World gold and silver mine tailings. pp. 73–87. In: *Mercury Contaminated Site: Characterization, Risk Assessment, and Remediation.* R. Ebinghaus (ed.). Springer, Berlin.

Lamborg, C.H., W.F. Fitzgerald, J. O'Donnell, and T. Torgersen. 2002. A non-steady state compartmental model of global-scale biogeochemistry with interhemispheric atmospheric gradients. *Geochim. Cosmochim. Acta* 66:1105–1118.

Landis, M.S., G.J. Keeler, K.I. Al-Wali, and R.K. Stevens. 2004. Divalent inorganic reactive gaseous mercury emissions from a mercury cell chlor-alkali plant and its impact on near-field atmospheric dry deposition. *Atmos. Environ.* 38:613–622.

Langston, W.J. 1986. Metals in sediments and benthic organisms in the Mersey estuary. *Estuar. Coast. Shelf Sci.* 23:239–261.

Le Roux, S., A. Turner, G.E. Millward, L. Ebdon, and P. Appriou. 2001. Partitioning of mercury onto suspended sediments in estuaries. *J. Environ. Monit.* 3:37–42.

Lin, C.J. and S.O. Pehkonen. 1999. The chemistry of atmospheric mercury: A review. *Atmos. Environ.* 33:2067–2079.

Lin, X. and Y. Tao. 2003. A numerical modeling study on regional mercury budget for eastern North America. *Atmos. Chem. Phys. Discuss.* 3:983–1015.

Lindberg, S.E., K.-H. Kim, T.P. Meyers, and J.G. Owens. 1995. Micrometeorological gradient approach for quantifying air/surface exchange of mercury vapor: Tests over contaminated sites. *Environ. Sci. Technol.* 29:126–135.

Lindberg, S.E., P.J. Hanson, T.P. Meyers, and K.-H. Kim. 1998. Air/surface exchange of mercury vapor over forests: The need for a reassessment of continental biogenic emissions. *Atmos. Environ.* 32:895–908.

Lindqvist, O., K. Johansson, M. Aastrup, A. Andersson, L. Bringmark, G. Hovsenius, L. Hakanson, A. Iverfeldt, M. Meili, and B. Timm. 1991. Mercury in the Swedish environment–recent research on causes, consequences, and corrective methods. *Water Air Soil Pollut.* 55:1–261.

Lorey, P. and C.T. Driscoll. 1999. Historical trends of mercury deposition in Adirondack lakes. *Environ. Sci. Technol.* 33:718–722.

Maine Department of Environmental Protection. 1999. Mercury in wastewater: Discharges to the waters of the state 1999. Maine Department of Environmental Protection, Augusta, ME.

Mason, R.P. and A.L. Lawrence. 1999. Concentration, distribution, and bioavailability of mercury and

methylmercury in sediments of Baltimore Harbor and Chesapeake Bay, Maryland, USA. *Environ. Toxicol. Chem.* 18(11):2438–2447.

Mason, R.P. and G.R. Sheu. 2002. Role of the ocean in the global mercury cycle. *Global Biogeochem. Cycles* 16:1093.

Mason, R.P. and K.A. Sullivan. 1998. Mercury and methylmercury transport through an urban watershed. *Water Res.* 32:321–330.

Mason, R.P., W.F. Fitzgerald, and F.M.M. Morel. 1994. The biogeochemical cycling of mercury: Anthropogenic influences. *Geochim. Cosmochim. Acta* 58:3191–3198.

Mason, R.P., N.M. Lawson, and K.A. Sullivan. 1997. Atmospheric deposition to the Chesapeake Bay watershed–regional and local sources. *Atmos. Environ.* 31:3531–3540.

Matilainen, T., M. Verta, M. Niemi, and A. Uusi-Rauva. 1991. Specific rates of net methylmercury production in lake sediments. *Water Air Soil Pollut.* 56:595–605.

METAALICUS. 2003. Available at www.umanitoba.ca/institutes/fisheries/ METAALICUS1.html. Accessed May 11, 2005.

Miklavcic, V. 1999. Mercury in the town of Idrija (Slovenia) after 500 years of mining and smelting. In: *Mercury Contaminated Sites: Characterization, Risk Assessment, and Remediation.* R. Ebinghaus, R.R. Turner, L.D. de Lacerda, O. Vasiliev, and W. Salomons (eds), pp. 259–269. Springer-Verlag, Berlin.

Mugan, T.J. 1996. Quantification of total mercury discharges from publicly owned treatment works to Wisconsin surface waters. *Water Environ. Res.* 68:229–234.

Mulrooney, M. 2000. A 3.0-meter liquid mirror telescope. Thesis, Rice University. 393pp.

Munthe, J. 1992. The aqueous oxidation of elemental mercury by ozone. *Atmos. Environ.* 26A:1461–1468.

Munthe, J., Z.F. Xiao, and O. Lindqvist. 1991. The aqueous reduction of divalent mercury by sulfite. *Water Air Soil Pollut.* 56:621–630.

Munthe, J., K. Kindbom, O. Kruger, G. Petersen, J. Pacyna, and A. Iverfeldt. 2001. Examining source-receptor relationships for mercury in Scandinavia—modeled and empirical evidence. *Water Air Soil Pollut. Focus.* 1:299–310.

Munthe, J., I. Wangberg, A. Iverfeldt, O. Lindqvist, D. Stromberg, J. Sommar, K. Gardfeldt, G. Petersen, R. Ebinghaus, E. Prestbo, K. Larjavi, and V. Siemens. 2003. Distribution of atmospheric mercury species in Northern Europe: Final results from the MOE project. *Atmos. Environ.* Supp. No. 1 S9–S20.

NESCAUM. 1998. Northeast states and eastern Canadian provinces mercury study: A framework for action. NESCAUM/NEWMOA/NEIWPCC/EMAN.

Niksa, S., J.J. Helble, and N. Fujiwara. 2001. Kinetic modeling of homogenous mercury oxidation: the importance of NO and H_2O in predicting oxidation in coal-derived systems. *Environ. Sci. Technol.* 35:3701–3706.

Niksa, S., N. Fujiwara, Y. Fujita, K. Tomura, H. Moritomi, T. Tuji, and S. Takasu. 2002. A mechanism for mercury oxidation in coal-derived exhaust. *J. Air Waste Manage. Assoc.* 52:894–901.

Nriagu, J.O. 1994. Mercury pollution from past mining of silver and gold in the Americas. *Sci. Tot. Environ.* 149:167–181.

NY Times. 2000. 200,000 homes in Illinois to be Searched for Mercury. *New York Times*, August 31, 2000.

Olson, B.H. and R.C. Cooper. 1976. Comparison of aerobic and anaerobic methylation of mercuric chloride by San Francisco Bay sediments. *Water Res.* 10:113–116.

Oremland, R.S., C.W. Culbertson, and M.R. Winfrey. 1991. Methylmercury decomposition in sediments and bacterial cultures: involvement of methanogens and sulfate reducers in oxidative demethylation. *Appl. Environ. Microbiol.* 57:130–137.

Oremland, R.S., L.G. Miller, P. Dowdle, T. Connel, and T. Barkay. 1995. Methylmercury oxidative degradation potentials in contaminated and pristine sediments of the Carson River, Nevada. *Appl. Environ. Microbiol.* 61:2745–2753.

OSHA. 2005. Mercury vapor in workplace atmospheres. Available at www.osha.gov/dts/sltc/methods/inorganic/id140/id140.html. Accessed May 11, 2005.

Pacyna, E.G. and J.M. Pacyna. 2002. Global emission of mercury from anthropogenic sources in 1995. *Water Air Soil Pollut.* 137: 149–165.

Pacyna, J.M., E.G. Pacyna, F. Steenhuisen, and S. Wilson. 2004. Global anthropogenic emissions of mercury in 2000. Abstract: Proc. 7th International Conference on Mercury as a Global Pollutant. Ljubljana, Slovenia, June 27–July 2, 2004.

Pai, P., D. Niemi, and B. Powers. 2000. A North American inventory of anthropogenic mercury emissions. *Fuel Process. Technol.* 65–66:101–115.

Pak, K. and R. Bartha. 1998. Products of mercury demethylation by sulfidogens and methanogens. *Bull. Environ. Contam. Toxicol.* 61:690–694.

Paquette, K.E. and G.R. Helz. 1997. Inorganic speciation of mercury in sulfidic waters: The importance of zero-valent sulfur. *Environ. Sci. Technol.* 31:2148–2153.

Petersen, G., A. Iverfeldt, and J. Munthe. 1995. Atmospheric mercury species over central and northern Europe—model calculations and comparison with observations from the Nordic air and precipitation network for 1987 and 1988. *Atmos. Environ.* 29:46–67.

Pirrone, N., J. Munthe, L. Barregård, H.C. Ehrlich, G. Petersen, R. Fernandez, J.C. Hansen, P. Grandjean, M. Horvat, E. Steinnes, R. Ahrens, J.M. Pacyna, A. Borowiak, P. Boffetta, and M. Wichmann-Fiebig. 2001. EU ambient air pollution by mercury (Hg) position paper. Available at europa.eu.int/comm/environment/air/background.htm#mercury. Accessed May 11, 2005. Office for Official Publications of the European Communities.

Plouffe, A. 1995. Glacial dispersal of mercury from bedrock mineralization along Pinchi Fault, North Central British Columbia. *Water Air Soil Pollut.* 80: 1109–1112.

Quémerais, B., D. Cossa, B. Rondeau, T. Pham, P. Gagnon, and B. Fortin. 1999. Sources and fluxes of mercury in the St. Lawrence River. *Environ. Sci. Technol.* 33:840–849.

Rasmussen, P.E. 1994. Current methods of estimating atmospheric mercury fluxes in remote areas. *Environ. Sci. Technol.* 28:2233–2241.

Regnell, O. 1990. Conversion and partitioning of radiolabeled mercury chloride in aquatic model systems. *Can. J. Fish. Aquat. Sci.* 47:548–553.

Riley, D.M., C.A. Newby, T.O. Leal-Almarez, and V.M. Thomas. 2001. Assessing elemental mercury vapor exposure from cultural and religious practices. *Environ. Health. Perspect.* 109:779–784.

Rolfhus, K.R., H.E. Sakamoto, L.B. Cleckner, R.W. Stoor, C.L., Babiarz, R.C., Back, H. Manolopoulos, and J.P. Hurley. 2003. Distribution and fluxes of total and methylmercury in Lake Superior. *Environ. Sci. Technol.* 37:865–872.

Rose, J., M.S. Hutchinson, C.R. West, O. Pancorbo, K. Hulme, A. Cooperman, G. DeCesare, R. Isaac, and A. Screpetis. 1999. Fish mercury distribution in Massachusetts, USA lakes. *Environ. Toxicol. Chem.* 18:1370–1379.

Ryaboshapko, A., R. Bullock, R. Ebinghaus, I. Ilyan, K. Lohman, J. Munthe, G. Petersen, C. Seigneur, and I. Wangberg. 2002. Comparison of mercury chemistry models. *Atmos. Environ.* 36:3881–3898.

Scherbatskoy, T., J.B. Shanley, and G.J. Keeler. 1998. Factors controlling mercury transport in an upland forested catchment. *Water Air Soil Pollut.* 105:427–438.

Schierow, L.-J. 2004. Mercury in the environment: Sources and health risks. CRS Report to Congress. 26pp.

Schuster, E. 1991. The behavior of mercury in the soil with special emphasis on complexation and adsorption processes: A review of the literature. *Water Air Soil Pollut.* 56:667–680.

Seigneur, C., J. Wrobel, and E. Constantinou. 1994. A chemical kinetic mechanism for atmospheric inorganic mercury. *Environ. Sci. Technol.* 28:1589–1597.

Seigneur, C., P. Karamchandani, K. Lohman, and K. Vijayaraghavan. 2001. Multiscale modeling of the atmospheric fate and transport of mercury. *J. Geophys. Res.* 106:27795–27809.

Seigneur, C., K. Vijayaraghavan, K. Lohman, P. Karamchandani, and C. Scott. 2004. Global source attribution for mercury deposition in the United States. *Environ. Sci. Technol.* 38:555–569.

Senior, C.L. 2001. Behavior of mercury in air pollution control devices on coal-fired utility boilers. In: *21st Century: Impacts of Fuel Quality and Operations*, Engineering Foundation Conference. Snowbird, UT, October 28–November 2, 2001. 17pp.

Siciliano, S.D., N.J. O'Driscoll, and D.R.S. Lean. 2002. Microbial reduction and oxidation of mercury in freshwater lakes. *Environ. Sci. Technol.* 36:3064–3068.

Smart, E.R. 1986. Mercury vapor levels in a domestic environment following breakage of a clinical thermometer. *Sci. Tot. Environ.* 57:99–103.

Sommar, J., K. Gardfeldt, D. Stromberg, and X. Feng. 2001. A kinetic study of the gas-phase reaction between the hydroxyl radical and atomic mercury. *Atmos. Environ.* 35:3049–3054.

Southworth, G.R., M.J. Peterson, and M.A. Bogle. 2002. Effect of point source removal on mercury bioaccumulation in an industrial pond. *Chemosphere* 49:455–460.

Southworth, G.R., S.E. Lindberg, H. Zhang, and F.R. Anscombe. 2004. Fugitive mercury emissions from a chlor-alkali factory: Sources and fluxes to the atmosphere. *Atmos. Environ.* 38:597–611.

St. Louis, V.L., J.W.M. Rudd, C.A. Kelly, K.G. Beaty, N.S. Bloom, and R.J. Flett. 1994. Importance of wetlands as sources of methyl mercury to boreal forest ecosystems. *Can J. Fish. Aquat. Sci.* 51:1065–1076.

Swain, E.B., D.R. Engstrom, M.E. Brigham, T.A. Henning, and P.L. Brezonik. 1992. Increasing rates of atmospheric deposition of atmospheric mercury deposition in midcontinental North America. *Science* 257:784–787.

Temme, C., J.W. Einax, R. Ebinghaus, and W.H. Schroder. 2003. Measurements of atmospheric mercury species at a coastal site in the Antarctic and over the South Atlantic Ocean during polar summer. *Environ. Sci. Technol.* 37:22–31.

Temmerman, E., C. Vandecasteele, G. Vermeir, *et al.* 1990. Sensitive determination of gaseous mercury in air by cold-vapour atomic-fluorescence spectrometry after amalgamation. *Anal. Chim. Acta.* 236(2):371–376.

Trip, L., T. Bender, and D. Niemi. 2004. Assessing Canadian inventories to understand the environmental impacts of mercury releases to the Great Lakes region. *Environ. Res.* 95:266–271.

Tsuji, J., P. Williams, M. Edwards, K. Allamneni, M. Kelsh, D. Paustenbach, and P. Sheehan. 2003. Evaluation of mercury in urine as an indicator of exposure to low levels of mercury vapors. *Environ. Health Perspect.* 111(4):623–630.

Turner, R.R. 1992. Elemental mercury in soil and the subsurface: transformations and environmental transport. p. 69. In: *Arsenic and Mercury: Workshop on Removal, Recovery, Treatment, and Disposal, Abstract Proceedings*. Alexandria

VA, August 17–20, 1992. EPA/600/R-92/105. US Environmental Protection Agency, Office of Research and Development, Cincinnati, Ohio, and Office of Solid Waste and Emergency Response, Washington, DC.

Turner, R.R. and G.R. Southworth. 1999. Mercury-contaminated industrial and mining sites in North America: An overview with selected case studies. pp. 89–112. In: *Mercury Contaminated Sites: Characterization, Risk Assessment, and Remediation*. R. Ebinghaus, R.R. Turner, L.D. de Lacerda, O. Vasiliev, and W. Salomons (eds). Springer-Verlag, Berlin.

UNEP. 2002. Global mercury assessment. Available at www.chem.unep.ch/mercury/Report/Final%20Assessment%20report.htm. Accessed on May 11, 2005. United Nations Environment Program.

US EPA. 1997. Mercury study report to Congress. EPA/452/R-97/003. US Environmental Protection Agency, Office of Air Quality Planning and Standards, Research Triangle Park, NC, and Office of Research and Development, Washington, DC.

US EPA. 2004. National emissions inventory. Available at www.epa.gov/ttn/chief/net/1999inventory.html#final3haps. Accessed May 11, 2005.

Vandal, G.M., R.P. Mason, and W.F. Fitzgerald. 1991. Cycling of volatile mercury in temperate lakes. *Water Air Soil Pollut.* 56:791–803.

Varekamp, J.C. and P.R. Buseck. 1986. Global mercury fluxes from volcanic and geothermal sources. *Appl. Geochem.* 1:65–73.

Watras, C.J., N.S. Bloom, R.J.M. Hudson, S. Gherini, R. Munson, S.A. Claas, K.A. Morrison, J. Hurley, J.G. Wiener, W.F. Fitzgerald, R. Mason, G. Vandal, D. Powell, R. Rada, L. Rislov, M. Winfrey, J. Elder, D. Krabbenhoft, A.W. Andren, C. Babiarz, D.B. Porcella, and J.W. Huckabee. 1994. Sources and fates of mercury and methylmercury in Wisconsin lakes. pp. 153–180. In: *Mercury Pollution: Integration and Synthesis*. C.J. Watras and J.W. Huckabee (eds). Lewis Publishers, Boca Raton, FL.

Wilhelm, S.M. 2001. Estimate of mercury emissions to the atmosphere from petroleum. *Environ. Sci. Technol.* 35:4704–4710.

Winfrey, M.R. and J.W.M. Rudd. 1990. Environmental factors affecting the formation of methylmercury in low pH lakes: A review. *Environ. Toxicol. Chem.* 9:853–869.

Winter, T.G. 2003. The evaporation of a drop of mercury. *Am. J. Phys.* 71:783–786.

Xiao, Z.F., J. Munthe, D. Stromberg, and O. Lindqvist. 1994. Photochemical behavior of inorganic mercury compounds in aqueous solution. pp. 581–592. In: *Mercury Pollution–Integration and Synthesis*. C.J. Watras and J.W. Huckabee (eds). CRC Press, Boca Raton, FL.

2

Asbestos

Drew R. Van Orden

Contents

2.1 INTRODUCTION

"Asbestos" is a commercial term applied to a group of naturally occurring minerals that have grown in a specific form and that exhibit characteristics of flexibility (tensile strength), large surface area, and resistance to heat and chemical degradation. With some "asbestos", the minerals can be woven into fabrics, ropes, braids, etc.; other "asbestos" can be used as filler in molded products. "Asbestos" is usually defined as: "(1) A collective mineralogical term encompassing the asbestiform varieties of various minerals; (2) an industrial product obtained by mining and processing primarily asbestiform minerals" (Campbell *et al.*, 1977). As used in this chapter, "asbestos" refers to the asbestiform varieties of the minerals. "Asbestiform" indicates the minerals have a fibrous nature, similar to cotton.

The term "asbestos" includes six regulated minerals: chrysotile, crocidolite (riebeckite asbestos), amosite (grunerite asbestos), anthophyllite asbestos, tremolite asbestos, and actinolite asbestos (OSHA, 1994). These minerals, a serpentine (chrysotile) and five amphiboles, are a few of the many minerals that can grow in an asbestiform habit. They are, however, the fibrous minerals that were commercially exploited as "asbestos".

Asbestos is identified as a causative agent in three diseases (asbestosis, lung cancer, and mesothelioma) (Ross, 1981; Cossette, 1984; Crump, 1991; Ross and Nolan, 2003) and is defined as a Group 1 carcinogen (IARC, 1987). The non-asbestos forms of the minerals are not defined as carcinogens.

The purpose of this chapter is to provide an overview of the mineralogy and sources of asbestos, appropriate analytical techniques for its identification and potential biases that a forensic investigation should consider when evaluating information from an asbestos investigation. The primary forensic issue associated with asbestos is determining its presence and distinguishing asbestos from non-asbestos fibers.

2.2 MINERALOGY AND SOURCES

The common regulated asbestos minerals fall into two mineral classifications: serpentine (chrysotile) and amphiboles (riebeckite asbestos, grunerite asbestos, anthophyllite asbestos, tremolite asbestos, and actinolite asbestos). Serpentine is a sheet silicate; the amphiboles are double chain silicate minerals. Commercial deposits of these minerals occur worldwide, but large-scale exploitation has been limited to a few countries. Of the regulated asbestos minerals, chrysotile (serpentine) has been produced in the largest quantities.

2.2.1 Mineralogy

The theoretical formula for these regulated minerals are shown in Table 2.2.1 (Virta and Mann, 1994). These formula are written to show the minerals chemically as hydrated silicate minerals with varying amounts of sodium, calcium, magnesium, and iron. Photographs of the minerals are shown in Figure 2.2.1.

2.2.1.1 Serpentine Minerals

Chrysotile, also known as white asbestos, is the most exploited asbestos mineral and is the only regulated (as asbestos) serpentine mineral. Other serpentine minerals that can occur in a fibrous habit are lizardite and antigorite. These minerals, while similar chemically, have slightly different structural forms: chrysotile a cylindrical form, lizardite with a planar structure, and antigorite with a corrugated structure (Wicks, 1979).

Table 2.2.1 *Types of Asbestos*

Type	Theoretical Formula
Chrysotile (serpentine)	$Mg_6[(OH)_4Si_2O_5]_2$
Riebeckite asbestos (crocidolite, amphibole)	$Na_2Fe_5[(OH)Si_4O_{11}]_2$
Grunerite asbestos (amosite, amphibole)	$MgFe_6[(OH)Si_4O_{11}]_2$
Anthophyllite asbestos (amphibole)	$(Mg, Fe)_7[(OH)Si_4O_{11}]_2$
Tremolite asbestos (amphibole)	$Ca_2(Mg, Fe)_5[(OH)Si_4O_{11}]_2$
Actinolite asbestos (amphibole)	$Ca_2(Mg, Fe)_5[(OH)Si_4O_{11}]_2$

Chrysotile is composed of a sheet of silica tetrahedra (SiO_4) layered with an octahedral sheet composed of magnesium octahedra with triagonal symmetry (similar to brucite) (Wicks, 1979). This layer lattice is described as similar to that of kaolinite clay (McCrone, 1980). As the brucite layer has a slightly larger lattice dimension than the silica tetrahedral, the layers roll into scrolls or tubular structures, with the brucite on the "outside" of the scroll.

The crystal parameters of the serpentine minerals are shown in Table 2.2.2 (Deer *et al.*, 1962). Chrysotile fibrils (a single fiber that cannot be split longitudinally into other fibers) are generally $0.02\,\mu m$ in width; the non-asbestiform minerals can occur in nearly any width. Within any particular deposit of chrysotile, the unit width of the fibril is fairly constant.

The chemical formula of chrysotile is shown in Table 2.2.1, with possible minor substitution of aluminum for silicon, or iron or nickel for magnesium. Table 2.2.3 shows reported chemical compositions of selected serpentine minerals.

Chrysotile is a hydrated mineral that decomposes, with heat and time, into an amorphous silicate and/or forsterite. The rate of conversion varies with heating temperature, Figure 2.2.2. (Virta and Mann, 1994). Langer (Langer, 2003) reports that the conversion from chrysotile to a form of amorphous silica can occur at temperatures as low as $150\,°C$. Complete dehydroxilation occurs at temperatures around $650\,°C$.

The effect of heat is also reflected in the tensile strength of chrysotile. Virta and Mann (1994) show that heating Canadian chrysotile for 3 minutes can reduce the tensile strength of the mineral by 9–68%, depending upon the heating temperature, Figure 2.2.3. Many investigators believe the biologically active properties of chrysotile are lost after chrysotile is heated to $810–820\,°C$ (the olivine transformation temperature); however, the above information indicates the physical properties of chrysotile begin to change at temperatures as low as $150\,°C$, suggesting the properties that make chrysotile biologically active may also be affected at these low temperatures.

2.2.1.2 Amphibole Minerals

The regulated amphiboles (riebeckite asbestos, grunerite asbestos, anthophyllite asbestos, tremolite asbestos, and actinolite asbestos) are a complex assemblage of minerals that have variable composition and extensive elemental substitutions. These minerals occur in a variety of forms, ranging from fibrous to acicular to massive particles (Campbell *et al.*, 1977). The nomenclature for these minerals has been standardized and can be found in Leake *et al.* (1997).

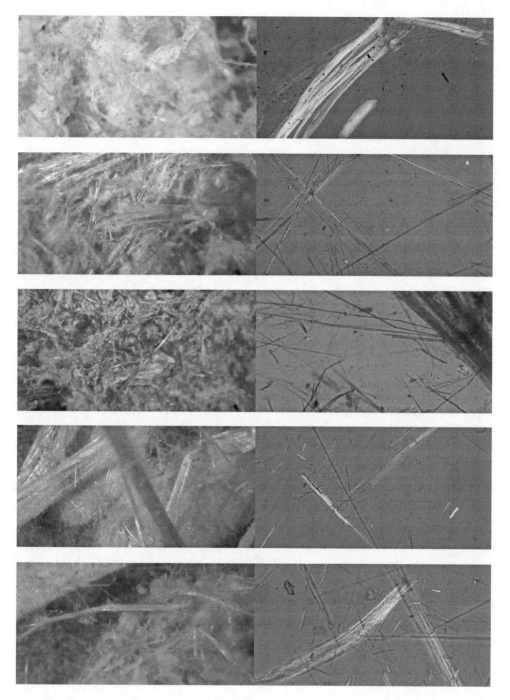

Figure 2.2.1 *Photographs of the regulated asbestos minerals. From top to bottom: chrysotile, amosite, crocidolite, anthophyllite, and tremolite/actinolite. The photographs on the left are taken at a magnification of 3×. The photographs on the right are taken in a polarized light microscope at a magnification of approximately 100×, with crossed polars, inserted wave retardation plate, and the mineral immersed in a matching refractive index oil (see color insert).*

The regulated amphiboles are often referred to by generic names. Riebeckite asbestos is also known as crocidolite or blue asbestos while grunerite asbestos is known as amosite or brown asbestos. "Amosite" is derived from the term "AMOSA", a shortened form of Asbestos Mines of South Africa (Ross *et al.*, 1984). Tremolite and actinolite form a single solid solution series with one end member (tremolite) containing no iron and the other end member (actinolite) containing a nearly even amount of iron and magnesium.

Table 2.2.2 Summary of Crystal Information for the Serpentine Minerals

	Chrysotile			Lizardite	Antigorite
	Clino-	Ortho-	Para-		
a, Å	5.34	5.34	5.3	5.31	43.3
b, Å	9.25	9.2	9.24	9.20	9.23
c, Å	14.65	14.63	14.7	7.31	7.27
β,°	93.25	90	90	90	91.1
Z	2	2	2	2	2
Space group				Cm	Cm
Crystal system	Monoclinic	Monoclinic	Monoclinic	Monoclinic	Monoclinic

Table 2.2.3 Chemical Composition of Selected Serpentine Minerals

Oxide	1[a]	2[a]	3[b]	4[b]	5[b]	6[b]
SiO_2	41.97	43.45	38.75	39.00	39.70	39.93
TiO_2		0.02				
Al_2O_3	0.10	0.81	3.09	4.66	3.17	3.92
Fe_2O_3	0.38	0.88	1.59	0.54	0.27	0.10
FeO	1.57	0.69	2.03	1.53	0.70	0.45
MnO			0.08	0.11	0.26	0.05
MgO	42.50	41.90	39.78	38.22	40.30	40.25
CaO		0.04	0.89	2.03	1.08	1.02
K_2O	0.08	0.02	0.18	0.07	0.05	0.09
Na_2O		0.05	0.10	0.07	0.04	0.09
H_2O	13.56	12.29	12.82	12.14	12.81	13.28
Total	100.16	100.19	99.79	100.20	100.51	100.22

[a] From Wicks (1979).
[b] From Ross *et al.* (1984).

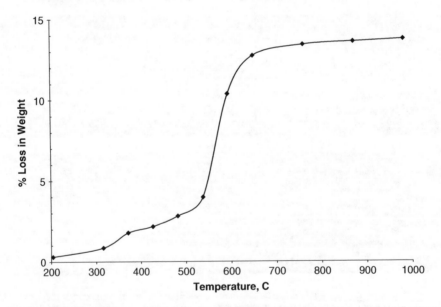

Figure 2.2.2 Observed weight loss of chrysotile heated for 2 hours at various temperatures (Langer, 2003).

Table 2.2.4 shows the range of chemical compositions for the regulated minerals. More detailed information on chemical compositions of specific locales are found in Deer, Howie, and Zussman (Deer *et al.*, 1997a,b,c,d).

The amphiboles crystallize into two general crystal forms: orthorhombic and monoclinic. Both are double chain silicate minerals (Deer *et al.*, 1997a). Table 2.2.5 summarizes the unit cell dimensions for the regulated amphiboles. Amphibole asbestos fibers are usually 0.2 to 0.3 µm in diameter, but each deposit should be evaluated for the distribution of fiber widths.

The solubility of the amphiboles in acids range from insoluble (anthophyllite) to slightly soluble (actinolite) (Virta and Mann, 1994).

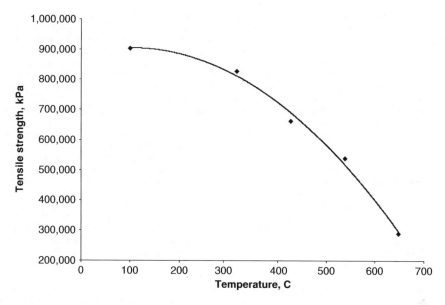

Figure 2.2.3 *Effect of heating chrysotile for three minutes at various temperatures on tensile strength (Virta and Mann, 1994).*

Table 2.2.4 *Approximate Chemical Composition of Regulated Amphiboles, (%)*
(after Virta and Mann, 1994)

	Riebeckite	Grunerite	Anthophyllite	Tremolite	Actinolite
SiO_2	49–53	49–53	56–58	53–62	53–62
MgO	0–3	1–7	28–34	0–30	10–30
FeO	13–20	34–44	3–12	1.5–5	5–16
Fe_2O_3	17–20				0–2
Al_2O_3		2–9	0.5–1.5	1–4	1–3
H_2O	2.5–4.5	2–5	1–6	0–5	0–5
CaO				10–18	10–18
Na_2O	4–6.5			0–9	0–1.5
$CaO + Na_2O$		0.5–2.5			

Table 2.2.5 *Summary of Crystal Information for the Regulated Amphiboles*

	Riebeckite	Grunerite	Anthophyllite	Tremolite—Actinolite
a, Å	9.64–9.82	9.45–9.62	18.5–18.6	~9.85
b, Å	17.88–18.07	17.95–18.45	17.7–18.1	~18.1
c, Å	5.31–5.33	5.27–5.34	5.27–5.32	~5.3
β,°	103.5	102–103		~105
Z	2	2	4	2
Space group	C2/m	C2/m or $P2_1/m$	Pnma	C2/m
Crystal	Monoclinic	Monoclinic	Orthorhombic	Monoclinic

Other amphibole minerals are also of interest, particularly those associated with vermiculite minerals. Though these amphiboles have historically been referred to as tremolite (or soda-tremolite) (Boetcher, 1966), recent studies using the current preferred nomenclature have shown these minerals to be winchite and richterite (Wylie and Verkouteren, 2000; Bandi, 2003; Gunter, 2003; Meeker, 2003). Microprobe characterization of the material from Libby, MT, indicates the mineral has an average percent composition of 56.8 SiO_2, 1.9 FeO, 4.5 Fe_2O_3, 20.4 MgO, 7.5 CaO, 3.8 Na_2O, 1.9 H_2O, with the remainder being minor amounts of Al_2O_3, TiO_2, MnO, K_2O, and F.

2.2.2 Sources of Asbetos

The regulated asbestos minerals are all ferro- or magnesiosilicates, and members of the amphibole or serpentine mineral families. The asbestos forms of these minerals occur as secondary or accessory minerals in metamorphic and igneous deposits where stresses within the earth have fractured the rocks. Chrysotile is primarily found as veins in host ultra-basic rocks, but commercial deposits have also occurred in serpentinized dolomitic limestone. Amosite and crocidolite are usually mined from metamorphosed ferruginous sedimentary formations, while tremolite/actinolite and anthophyllite are associated with highly metamorphosed ultra-basic rocks. Folding, faulting, or shearing has evidently played a major role in the formation of asbestos deposits. Deposits of asbestos of commercial significance are infrequent. Trace to minute quantities of asbestos can be found locally in almost any metamorphic or igneous mineral deposit. Thus, it is virtually impossible to certify a quarry or mining operation (in these rock types) as asbestos-free. Review of the geological maps of the US illustrates that metamorphic and igneous rock formations underlie the majority of the densely populated areas in the country.

Asbestos minerals occur in veins and lenses located in ultra-basic rock, metamorphosed ultra-basic rock, serpentinized dolomitic limestone, and metamorphosed ferruginous sedimentary formations. The asbestos fibers are found as cross fibers (fibers transverse to the vein), slip fibers (fibers in the plane of the vein), or as mass fibers (short fiber in lenses). Figure 2.2.4 shows the locations of asbestos mineralization in the United States (based on US Geological Survey data).

Asbestos has been mined worldwide, with current production occurring primarily in Canada, Russia, Kazakhstan, Zimbabwe, Brazil, and China (Virta, 2003). Historical production of asbestos (compiled from US Geological Survey records) for selected countries is shown in Figure 2.2.5. Production peaked in the mid-1980s and has declined substantially in the developed countries. Current production is almost exclusively chrysotile.

2.2.2.1 Historical Products
Asbestos are used in a wide variety of products ranging from cloth to friction materials, depending on the length of the fiber. Chrysotile was traditionally produced into various grades, as defined by the Quebec Asbestos Mining Association (Cossette and Delvaux, 1979), with grades 1 and 2 used for spinning and textile manufacturing. Table 2.2.6 shows the various grades of chrysotile asbestos and a listing of the types of products made with each grade. A similar system of grading the length of the chrysotile was also used in British Columbia, the former USSR, and in Africa.

Amosite and crocidolite were also graded according to fiber length, with product usage depending on fiber size. Within sizes that are equivalent to the Quebec sizes, Table 2.2.6 also shows the possible uses of crocidolite and amosite fibers. Long-fiber crocidolite may be used as textiles, ropes, and gaskets. Long amosite fibers may be used in insulation blankets, but spinning and weaving amosite fibers is difficult.

2.2.2.2 Current Products
In 1986, the US Environmental Protection Agency (EPA) issued regulations banning the production of asbestos-containing products. The ban was subsequently overturned in court, but asbestos products are currently limited to friction products, packing and gaskets, roofing products, and minor amounts of coatings and plastics. Asbestos has been banned in the European Union (Directive 76/769/EEC) since 1999 (Benjelloun, 2000). Other countries (including Australia and Japan) have also passed asbestos bans. Most current consumption is occurring in developing countries, with asbestos-cement products comprising the majority of produced goods.

2.2.2.3 Naturally Occurring Materials
Asbestos is a naturally occurring contaminant of many minerals. As noted earlier, vermiculite mined in Libby, MT,

has been found to contain trace levels of an amphibole mineral that occurs both in asbestiform and non-asbestos forms. Within the United States, greenstone building materials (such as from the Catoctin formation in the Appalachian Mountains) may occasionally be found to contain asbestos fibers (Dietrich, 1990). Marble in northwest New Jersey contains veinlets of tremolite in a non-asbestos form, with trace occurrences of asbestos fibers (Berman, 2003).

In recent years, allegations were made that suggested that asbestos was found in the talc used to make crayons, that asbestos occurred in play sand, and that asbestos was found in cosmetic talc. Subsequent investigations into these allegations have shown them to be false. Each of these materials was found to contain non-asbestiform amphibole particles, but no asbestos (McCrone, 1980; Langer and Nolan, 1987; Saltzman and Hatlied, 2000). The misinterpretation of these non-asbestiform minerals as "asbestos" occurred due to improper implementation of analytical protocols.

Naturally occurring asbestos (primarily amphibole minerals) can be found throughout the United States. In California, housing developments in the foothills of the Sierra Nevada Mountains, especially in El Dorado County (Anonymous, 2000; Churchill et al., 2000), have led to widespread public concern about possible health hazards from the minerals found in the ground at the housing sites. Tremolite/actinolite asbestos is reported in these locations in varying amounts, depending on the presence of veinlets in the host rock.

2.3 ANALYTICAL METHODS

Various analytical procedures have been developed for determining the presence and concentration of chrysotile and amphibole minerals in samples. Some methods have been promulgated by various governmental agencies for use in determining whether a material contains asbestos (usually above a specified concentration) or whether airborne particulate meet clearance levels following abatement of asbestos-containing materials.

The methods have been divided into groups on the basis of the material (bulk, airborne particulate, or water) to be analyzed for asbestos content.

2.3.1 Standard Methods
The methods discussed below are considered to be standard methods – methods that are published either by an agency of a government or by a standard methods organization. Analytical methods for some matrices, such as tissue samples, have been described in the literature but have not been published as standard methods.

2.3.1.1 Bulk Materials
Bulk samples are those that are used to represent either building materials, raw materials used in various manufacturing processes, or the ore/host rock samples. All of the methods assume that the sample analyzed has been properly collected, documented, and is representative of some larger population of material. Most regulations specify a material as "asbestos-containing" if the concentration of asbestos exceeds 1% by weight. The current trend in bulk asbestos analyses is toward lower levels of quantitation, such as 0.1% wt by weight or lower.

2.3.1.1.1 Microscopy
Optical microscopy has been used to analyze rocks and minerals for over 100 years. Tables of the optical properties of minerals, as determined by oil immersion techniques, were first published in 1900 (USGS, 1934). Optical microscopy (and polarized light microscopy, PLM, in particular) is a well-known analytical procedure (McCrone, 1980;

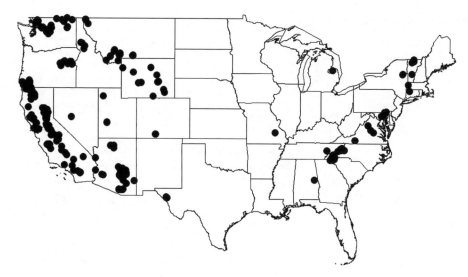

Figure 2.2.4 Map of the continental United States showing reported locations of asbestos mineralization (after McFaul, 2000).

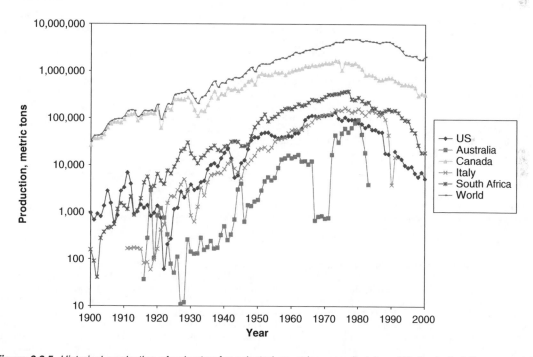

Figure 2.2.5 Historical production of asbestos for selected countries, compiled from US Geological Survey records (Virta, 2003) (see color insert).

Bloss, 1999). Originally used for rock and ore examination, optical microscopy is now used for determining the mineral content of numerous materials, including bulk building materials and soils.

In 1982, the EPA issued an Interim Method (EPA, 1982) that created a uniform procedure for mineral identification and quantification of asbestos in bulk building materials. Though it contains several typographic errors, the method was published in the Code of Federal Regulations (currently at Appendix E to Subpart E, 40 CFR Part 763). The original version of the method required the use of "point counting" to quantify the asbestos content of the material. In its current version in the regulations, the method permits either point counting or a visual estimation technique for quantitation.

Both the Occupational Safety & Health Administration (OSHA) and the National Institute for Occupational Safety and Health (NIOSH) have issued similar procedures. (NIOSH, 1989; Crane, 1992) The NIOSH method, however, relies on visual estimation of the sample to determine the amount of asbestos in the sample. This can lead to errors in quantitation, especially of samples with low levels of asbestos.

Table 2.2.6 *Asbestos Grades and Product Usage[a]*

QAMA[b] Chrysotile Grade		Product Uses	
Specification[c]	Fiber Grade	Chrysotile	Amphibole
66 - 24 - 8 - 2	3F	Textiles, packings, brake linings, clutch	Crocidolite—use similar to
44 - 44 - 9 - 3	3K	facings, electrical, high pressure and	chrysotile, provided color is
25 - 44 - 25 - 6	3R	marine insulation, electrolytic	acceptable
12 - 50 - 25 - 13	3T	diaphragms, pipe coverings, insulating	
6 - 56 - 25 - 13	3Z	blocks, resin laminates	
0 - 50 - 38 - 12	4A	Asbestos cement products, heavy duty	Crocidolite—asbestos-cement
0 - 44 - 38 - 18	4D	brake blocks, gaskets, plastics, paper,	pipes, packing, gaskets
0 - 31 - 50 - 19	4H	pipe covering, millboards	
0 - 25 - 56 - 19	4K		
0 - 25 - 50 - 25	4M		Amosite—magnesia block, pipe
0 - 19 - 56 - 25	4R		covering, insulation
0 - 12 - 62 - 26	4T		
0 - 9 - 59 - 32	4Z		
0 - 3 - 66 - 31	5D	Disk brakes, heavy duty brake blocks,	Crocidolite—asbestos-cement pipe
0 - 0 - 75 - 25	5K	gaskets, asbestos cement products,	
0 - 0 - 69 - 31	5M	molded products, paper products,	
0 - 0 - 62 - 38	5R	corrugated or flat boards, electrical	
0 - 0 - 54 - 46	5Z	panels, millboard, pipe covering	
0 - 0 - 44 - 56	6D	Asbestos cement products, heavy duty	Crocidolite—asbestos-cement
		brake blocks, paper, gaskets, plaster,	products, filler
		backing for vinyl tile, shingles,	Amosite—filler in
		millboards	asbestos-cement slurries
0 - 0 - 31 - 69	7D	drum brakes, clutch facings, adhesives,	Crocidolite and amosite were only
0 - 0 - 25 - 75	7F	roof coatings, vinyl asbestos tile,	used as a filler where color was
0 - 0 - 19 - 81	7H	paints, adhesives, thermal insulation,	not an issue
0 - 0 - 12 - 88	7K	caulking compounds, drilling mud	
0 - 0 - 6 - 94	7M	additive, plastics, asphalt compounds,	
0 - 0 - 0 - 100	7R	gaskets, welding rods	
0 - 0 - 0 - 100	7T		
0 - 0 - 0 - 100	7RF		

[a] After Crump (1991).
[b] Quebec Asbestos Mining Association.
[c] The grades are presented in terms of the percentage which would be retained on 1/2-inch, 4-mesh, and 10-mesh screens. A grade of 44 - 44 - 9 - 3 (grade 3K) indicates 44% longer than 1/2-inch, 44% between 1/2-inch and 4-mesh, 9% between 4- and 10-mesh, and 3% finer than 10-mesh.

Recognizing problems with established protocols for the analysis of raw materials, several states in the United States have issued their own PLM methods, the use of which is required by state regulation. California issued method CARB 435 (CARB, 1991) for use in determining the asbestos content in road aggregate. Samples of aggregate are collected, pulverized to a size finer than 75 μm, and analyzed using PLM. New York issued a PLM method (NYELAP, 1990) that utilized a stratified method for estimating the asbestos content of the material. For samples with "substantial amounts of asbestos", a visual estimation of the content is acceptable; other samples require a point-count quantitation.

In 1993, the EPA published an improved PLM method (Perkins and Harvey, 1993) that corrected some of the errors in the interim method. The new method, never promulgated though recommended for use (EPA, 1994), includes additional information for reducing the amount of interfering materials and for evaluating the bulk sample by electron microscopy.

International methods have also been issued (HSE, 1994) or are under consideration (ISO, undated). These methods utilize the same basic procedures of polarized light microscopy.

The most recent procedure was published by the European Union (Schneider *et al.*, 1997) and was specifically designed for the determination of low concentrations of asbestos in bulk materials. The procedure is similar to that of OSHA in that it uses a combination of polarized light microscopy with phase contrast microscopy. The method specifies a number of procedures to use in removing the matrix material, thus improving the precision and accuracy of the asbestos determination.

Electron microscopy has been used for the determination of the asbestos content of some materials. New York has issued a procedure (NYELAP, 1997) that uses a visual estimation technique in the transmission electron microscope for determining the asbestos content of non-friable organically bound materials (such as floor tile). The improved EPA method (Perkins and Harvey, 1993) also incorporates an electron microscopy method, though the procedure is not clearly defined. Some laboratories have used a variation of a dust procedure (see Section 2.3.1.3) for determination of low levels of asbestos.

Recently a method (EPA, 1997) has been designed to determine the amount of releasable fibers from a sample. A sample (sieved to pass a 1 cm screen) is tumbled with air passed through the tumbling material. The entrained particles then enter an expansion chamber where the larger particles fall out and only fine particles remain entrained in the air. These particles are collected on a filter and analyzed using an electron microscopy method (see Section 2.3.1.4.2). The final result is presented as a number of releasable fibers per gram. The procedure requires about a day to generate the filter for analysis. Proponents (Berman, 2000) suggest the method offers improved precision over other methods. However, for some samples, in order to generate the required dust loading on the filter, the sample must be tumbled for such long times that autogenous grinding takes place, not suspension of releasable particles. Because of this possibility, care must be exercised in the interpretation of these results.

2.3.1.1.2 X-Ray Powder Diffraction
X-ray powder diffraction (XRD) is a well-established procedure and is included in the initial EPA bulk procedure (EPA, 1982). The method is capable of determining the concentration of various minerals to about 0.5% by weight, but is not capable of discriminating between the asbestos and non-asbestos forms of the same mineral. For most building materials, any error caused by including non-asbestos minerals as asbestos will be minimal as any deliberately included "asbestos" minerals will be asbestiform. X-ray powder diffraction should not be used to quantify the asbestos content of raw materials due to the inability of discriminating between the asbestiform and non-asbestiform forms of the same mineral.

2.3.1.1.3 Remote Sensing
The US Geological Survey (USGS) has promoted a remote sensing technique for use as a rapid survey procedure (Clark, 1999) and is working to develop the procedure for asbestos minerals (USGS, undated). As noted by the USGS, the procedure is still developmental and requires the use of other techniques (PLM, electron microscopy) to fully understand the true nature of the minerals. The procedure cannot differentiate between the asbestiform and non-asbestiform varieties of the same mineral and has some difficulty in distinguishing tremolite-actinolite from talc. However, the technique is used to map the occurrences of ultramafic rocks, serpentine, and tremolite-actinolite in El Dorado County and does show promise for use in large-area mapping for possible occurrences of asbestos minerals (Swayze et al., 2004).

2.3.1.2 Water Samples
Transmission electron microscopy (TEM) is used to evaluate the suspended particulate in water for asbestos content. The US EPA regulates drinking water at a level of 7 MFL (million fibers per liter) for fibers longer than $10 \mu m$ (40 CFR Part 141.23) as determined using EPA Methods 100.1 or 100.2 (Chatfield and Dillon, 1983; Brackett et al., 1994). As implemented by various laboratories, the primary difference between the two methods is the use of a cellulose ester filter in Method 100.2 as opposed to a polycarbonate filter in Method 100.1. Both methods require that an aliquot of water be collected, agitated in an ultrasonic bath, and deposited onto the filter substrate. The filter is then prepared for examination in the TEM using direct preparation procedures.

There are local rules and regulations limiting the disposal of waste water that may contain asbestos fibers. Neither method is specifically written for a waste water matrix, but can be used for such material provided any organic materials can be removed prior to analysis. One technique that can be used to remove the organic material is to filter an aliquot of suspension, dry and ash the filter in a furnace, and analyze the ash for asbestos. An ozone generator (Chatfield and Dillon, 1983) may also be used to remove some of the organic material.

2.3.1.3 Surface Dust Samples
In recent years, the presence of asbestos in surface dust has been used by some investigators as an indicator of past and possible future exposure to airborne asbestos. Surface dust techniques have also been used to determine the cleanliness of a surface prior to or following cleaning. (Millette and Hays, 1994) Three published methods are available for the evaluation of surface dusts: ASTM D5755, D5756, and D6480 (ASTM 1999, 2002a,b). Other procedures have been described in the literature, but have not been formalized into a standard procedure (Millette and Hays, 1994).

The first two dust methods (D5755 and D5756) use an air cassette, connected to a portable sampling pump, to vacuum the dust off of surfaces in such a manner as to avoid scraping the surface while avoiding large (>1 mm) particles. The collected dust is washed out of the cassettes and suspended in an aqueous solution. For D5756, this initial suspension is filtered, the filter and particles ashed in a muffle furnace, and the ash resuspended. These suspensions are agitated using an ultrasonic bath and an aliquot of the suspension is withdrawn and deposited onto a new filter. The redeposit filter is prepared for analysis in the TEM using a direct preparation procedure.

A wipe material is used to sample the settled dust in ASTM D6480. The wetted wipe is rubbed across the surface to collect the dust, thus eliminating the sampling restrictions contained in D5755 or D5756. The collected wipe is placed in a beaker, submerged in water, and agitated in an ultrasonic bath to dislodge the collected particulate from the wipe. The wipe is removed and preparation of the suspension continues as in D5755. The recovery of the collected dust from the wipe can be improved by ashing the wipe and then suspending the ash in water.

The three methods are useful for determining the presence of asbestos on a surface, but controversy exists over the interpretation of the data (Chatfield, 1999; Lee et al., 1999). Indirect preparation procedures are known to disaggregate matrices, thus liberating entrapped asbestos fibers from the surrounding matrix materials. The use of the ultrasonic energy can fracture particles, thus changing the discrete particle count when compared with the material originally deposited onto the sampled surface (Lee et al., 1995; Lee et al., 1996; Chatfield, 1999).

2.3.1.4 Airborne Fibers
Samples of airborne fibers are collected and analyzed to determine exposure risks, doses, or for comparison with regulatory levels. Past methods used konimeters, impingers, or various impactors to collect particles. Optical microscopy was used to count all particles above a specified size. The 1960s saw the development of the membrane filter technique, which has replaced the earlier particle counters. These filters are now analyzed using optical or electron microscopy.

2.3.1.4.1 Optical Microscopy Techniques
In the early 1960s (Ayer et al., 1965), air filters began to achieve acceptance for the collection of the airborne particulate. These early studies, conducted first in the United Kingdom and later in the United States (Edwards and Lynch, 1968), showed the method to be an improvement in air sampling. Protocols for the method were established by the Public Health Service and NIOSH (NIOSH, 1977) for the use of phase contrast microscopy (PCM). Following studies that showed variability in results due, in part, to varying qualities

of the microscopes, NIOSH published the 7400 (NIOSH, 1994) method in 1984. This method specified sample collection procedures, material (filter and microscope) qualities, and counting protocols. Phase contrast microscopy, under NIOSH 7400, does not differentiate between asbestos and non-asbestos fibers, but counts all visible fibers that are longer than 5 μm and have a minimum aspect ratio of 3:1. This constraint is an artificial one since experienced analysts, with knowledge of the sampling environment, can differentiate between asbestos and non-asbestos fibers in many instances. The purpose of the constraint is to maximize precision in the analysis while minimizing operator training.

Various other organizations have similar analytical procedures. The OSHA Salt Lake City laboratory issued its own version of PCM method (OSHA ID-160) (Crane, 1988). The World Health Organization (WHO, 1997), ISO (ISO, 1993), and other groups use similar methods.

2.3.1.4.2 Transmission Electron Microscopy

Electron microscopes have been used to examine airborne particulates since at least 1953 (Fraser, 1953). Early transmission electron microscopy (TEM) studies of asbestos were principally directed at unraveling the crystal structure of the asbestos minerals, particularly that of chrysotile. By the mid- to late 1960s, the TEM was being used, notably by Pooley and Henderson in the United Kingdom (Stewart, 1988), for the study of mineral particles in lung tissue. Similar work was pursued by Mt Sinai School of Medicine and by the late 1960s, the TEM had been used for the study of environmental samples.

The first generally accepted TEM method was developed by IIT Research Institute (Yamate *et al.*, 1984). The draft method was never officially published by the EPA, but is still widely used. Among the improvements over earlier procedures was the specification of counting rules—defining what constitutes a fiber and how it should be counted. This draft method also specified the use of direct preparation procedures.

The first fully reviewed and promulgated TEM protocol in the United States was a method for testing the cleanliness of air in schools, following abatement actions. Under the authority of the Asbestos Hazard Emergency Response Act (AHERA), the EPA developed a TEM method (EPA, 1987) for use in clearance testing at abatement sites. The method specified sample collection procedures and required a direct transfer preparation method. To reduce the analysis time, the method did not require recording of fiber dimensions, but did require listing the fibers as either greater than 5 μm or less than 5 μm in length. One significant change over the Draft Method (Yamate *et al.*, 1984) was the increase in minimum aspect ratio from 3:1 to 5:1. A minimum length for asbestos fibers (0.5 μm) was specified for the first time to improve the reproducibility of fiber counts. This analytical method is widely used in the United States.

Not all airborne fibers are asbestos and, in 1989, NIOSH issued its first TEM asbestos method (NIOSH, 1989). The method, NIOSH 7402, was designed for use in conjunction with PCM (NIOSH, 7400) to allow determination of the proportion of countable fibers in mixed fiber environments that were asbestos. The method used a magnification comparable to the magnification used in the optical microscope, counted only fibers longer than 5 μm, wider than 0.25 μm, and that had an aspect ratio of at least 3:1. OSHA (OSHA, 1991) later issued a similar protocol with a minimum diameter of 0.2 μm. Both methods produce similar results. The OSHA permits the use of either TEM method when analyzing air samples for OSHA compliance purposes (in conjunction with PCM).

A second TEM air sample protocol for use at Superfund sites was published by EPA (EPA, 1990). The method retained some of the improvements of AHERA and added a general recording protocol from an earlier water method. (Chatfield and Dillon, 1983) The method utilizes both direct and indirect preparation procedures for the analysis.

In 1995, the International Standards Organization (ISO) adopted a standard method for the analysis of airborne asbestos, ISO 10312 (ISO, 1995). The method is an outgrowth of the Superfund method. The ISO method is a direct preparation method developed for measuring the concentrations in buildings or in ambient air. Because the method is complex is not promulgated by the government, it has gained little acceptance in the United States; many countries have used the ISO method as the basis upon which to further develop their own asbestos method (AFNOR, 1996). A second ISO method (ISO 13794 (ISO, 1999)) was adopted that permits indirect preparation of the air filter.

As in AHERA, the ISO methods count asbestos fibers 0.5 μm and longer that have a minimum aspect ratio of 5:1. Unlike AHERA, complex structures (such as matrices or clusters) can be counted as up to five asbestos structures, depending upon the actual composition. For example, in AHERA three fibers touching each other defines a single cluster. It is counted as "1" toward the final concentration. Under ISO, the overall cluster is considered a primary structure, but the three individual fibers (provided they meet the counting rules) are listed as substructures to the primary structure. The count used to calculate the final concentration would be "3" under the ISO counting rules.

The ISO 10312 method was further refined by the American Society for Testing and Materials (ASTM) with the adoption of ASTM D-6281 in 1998 (ASTM, 1998). Other than some editorial changes and improvements to definitions in the statistics section, this method revised the procedure for counting complex clusters and matrices. As with ISO 10312 and ISO 13794, substructures of the cluster or matrix are listed; however, only the primary structure is counted toward the final concentration calculation. Thus, when ISO counts a cluster of three fibers as a "3", ASTM counts it as a "1", as do AHERA and other TEM methods. All information is provided to the end user as required by the method.

2.3.1.4.3 Scanning Electron Microscopy

In addition to the TEM, a scanning electron microscope (SEM) can be used for asbestos analysis. Like a TEM, which uses an electron beam to view the particles, the SEM has the capability of reliably viewing fibers as thin as 0.2 μm and, when properly equipped, can also determine the chemistry of the particle using energy dispersive X-ray analyses (EDXA). Unlike the TEM, the SEM cannot determine the crystal structure through SAED. Newer field emission SEMs are capable of resolving thinner fibers than are conventional SEMs.

ISO has published a method, ISO 14966 (ISO, 2002), for the analysis of airborne fibers using the SEM. The method is based on an earlier German method (VDI 3492) that has been widely used in Europe. Fibers longer than 5 μm with an aspect ratio of at least 3:1 are counted at a magnification of about $2,000\times$.

Because the diameters of amphibole asbestos fibers are about 0.2 to 0.3 μm, the SEM method is a reliable technique for these particles. Chrysotile fibers thinner than 0.2 μm may not be resolved—but the TEM can be used for the thin chrysotile fibers.

2.3.1.4.4 Automated Fiber Counters

The Fibrous Aerosol Monitor (FAM) was developed in the late 1970s as a technique for real-time determination of airborne fiber concentrations. The method utilizes the scattering of a laser beam to detect and count fibers. The instrument is not specific for asbestos, but counts all elongated particles. Testing conducted through the mid-1980s showed comparable results to standard PCM data for amphibole fibers, but were not acceptable for chrysotile fibers at airborne concentrations below 2 f/cc (Baron, 1982).

The FAM continues to be used industrially to monitor abatement operations, primarily to detect leaks in the containment barrier.

2.3.2 Interferences and Biases

Various interferences occur in every analytical technique. In part, interferences exist due to the close similarity of the many amphibole minerals that exist (Leake, 1997). For example, winchite asbestos can be confused with the regulated tremolite asbestos. Other minerals may also posses a similar appearance to a regulated asbestos mineral, such as sepiolite for chrysotile or vermiculite scrolls for chrysotile.

Inherent in microscopic techniques is the use of the length and width dimensions of suspect particles in determining whether the particles are to be counted or not counted. The basis for this is a ruling in 1974 (Berndt and Brice, 2003) that defined asbestos particles as having an aspect ratio (length/width) of 3:1 or greater. Many analytical laboratories have used this definition (or other aspect ratio definition contained in a method) as the single characteristic defining whether a particle is asbestos or nonasbestos. They have not considered other distinguishing

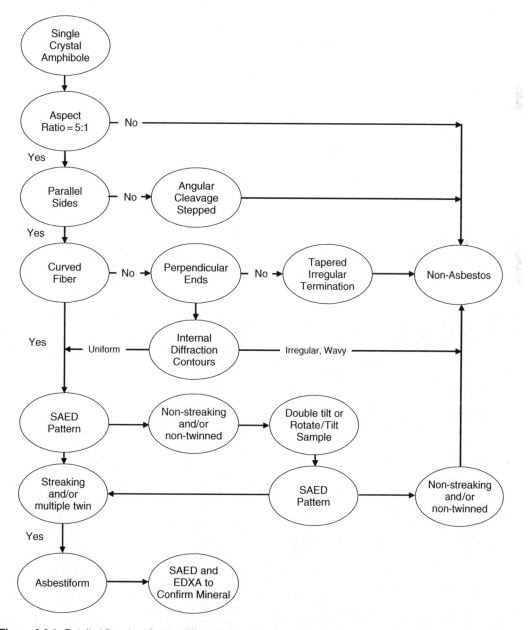

Figure 2.3.1 *Detailed flowchart for the differentiation of asbestos from non-asbestos analogs of the same mineral.*

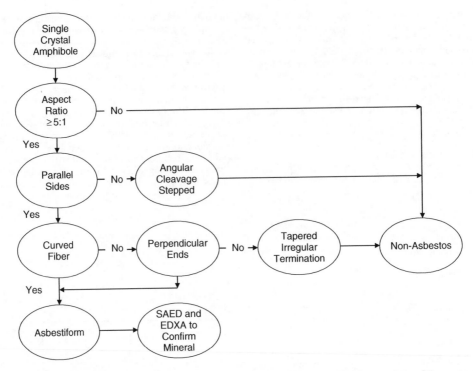

Figure 2.3.2 *Simplified flowchart for the differentiation of asbestos from non-asbestos analogs of the same mineral.*

characteristics (such as parallel extinction in the PLM or the presence of internal diffraction contours in the TEM) in helping to determine whether a particle is asbestos before deciding if it meets the counting rules of a particular method.

The definition of asbestos is one of the most widely published descriptions of any toxic material. Yet it is perhaps also the most poorly understood. The definition of asbestos suffers from a lack of precision in definition at the microscopic level. The problem occurs for several reasons. The major issue is the incorporation of an aspect ratio in the operational definition of asbestos in regulatory methods. While most regulations clearly specify that asbestos is being regulated and recognize that minerals occur in both asbestos and non-asbestos habits, they specify a minimum aspect ratio or length of the asbestos fiber to be counted. The asbestos habit requirement is generally lost in the practice of asbestos counting. Rather, the general practice has become that asbestos is identified as any particle that meets the aspect ratio specified in the method and is consistent with the mineralogy or chemistry of the regulated mineral. The aspect ratio thus has become the de facto definition of what constitutes asbestos. Many laboratories and industrial hygienists also employ a non-scientific theory—if in doubt, count it, under the misguided assumption that false positives are less significant than false negatives.

Each asbestos analytical method has a definition of a "fiber" and of "asbestos", but no clear mineralogical definition of what is truly an asbestos fiber and what is an elongated non-asbestos particle (Campbell *et al.*, 1977; Wylie, 2000). A recent risk model (Berman and Crump, 1989) makes an attempt to incorporate a width characteristic into the definition of asbestos by specifying only fibers thinner than 0.5 μm be considered for risk estimation. Amphibole asbestos fibers should be defined on a combination of fiber width and diffraction characteristics, as well as aspect ratios.

As part of the investigation into the amphibole mineral found in marble located at Southdown, NJ, a procedure was used to differentiate between asbestos and non-asbestos fibers (Lioy, 2001). The procedure was based on well-known physical characteristics of single fiber amphibole minerals: (1) the width of amphibole asbestos fibers is generally 0.2 to 0.3 μm; (2) the aspect ratio of asbestos fibers is > 20:1; (3) asbestos fibers have parallel sides; (4) the ends of asbestos fibers show regular termination; and (5) asbestos fibers show internal diffraction contours. This procedure, accepted by EPA Region 2 for the Southdown project, is shown in Figure 2.3.1. A simplified version is shown in Figure 2.3.2.

2.4 SUMMARY

Asbestos analytical procedures have been developed for a variety of matrices, including bulk materials, surface dusts, water, and airborne particles. These methods are based on well-understood physical properties of asbestos, such as refractive indices or electron diffraction properties. The data derived from the analysis of these materials is generally well accepted, though some (particularly for surface dust) can be controversial. Problems associated the implementation of the methods, particularly the differentiation between asbestos and the non-asbestos forms of the same minerals, have led to incorrect analyses in many cases. Due to the implementation of asbestos bans in various countries, standardized procedures are needed for determining low levels of asbestos in raw materials as well as soils and sediments. These procedures are needed to certify materials for import or use in which asbestos fibers may be present as a result of naturally occurring contamination of a material in the product.

REFERENCES

AFNOR (1996). Qualité de l'air, Détermination de la concentratiob en fibres d'amiante par microscopie électronique à transmission, norme française, NF X 43-050, l'Association Française de Normalisation, January 1996.

Anonymous (2000). "A General Location Guide fro Ultramafic Rocks in California—Areas More Likely to Contain Naturally Occurring Asbestos", Open File Report 2000-19, Department of Conservation, Division of Mines and Geology, Sacramento, California.

ASTM (1998). "Standard Test Method for Airborne Asbestos Concentration in Ambient and Indoor Atmospheres as Determined by Transmission Electron Microscopy Direct Transfer (TEM)", American Society for Testing and Materials, Method D-6281, adopted July 10, 1998, published October 1998.

ASTM (1999). "Standard Test Method for Wipe Sampling of Surfaces, Indirect Preparation, and Analysis for Asbestos Structure Number Concentration by Transmission Electron Microscopy", D6480, American Society for Testing and Materials, Conshohocken, PA.

ASTM (2002a). "Standard Test Method for Microvacuum Sampling and Indirect Analysis of Dust by Transmission Electron Microscopy for Asbestos Structure Number Surface Loading", ASTM D5755, American Society for Testing and Materials, Conshohocken, PA.

ASTM (2002b). "Standard Test Method for Microvacuum Sampling and Indirect Analysis of Dust by Transmission Electron Microscopy for Asbestos Mass Concentration", ASTM D5756, American Society for Testing and Materials, Conshohocken, PA.

Ayer, H. E., J. R. Lynch, and J. H. Fanney (1965). "A Comparison of Impinger and Membrane Filter Techniques for Evaluating Air Samples in Asbestos Plants", *Annals of New York Academy of Sciences, 132*, pp. 274–287.

Bandli, B. R., M. E. Gunter, B. Twamley, F. F. Foit, Jr., and S. B. Cornelius (2003). "Optical, Compositional, Morphological, and X-Ray Data on Eleven Particles of Amphibole form Libby, Montana, USA.", *Canadian Mineralogist, 41*, pp. 1241–1253.

Baron, P. A. (1982). "Review of Fibrous Aerosol Monitor (FAM-1) Evaluations", National Institute for Occupational Safety and Health, Cincinnati, Ohio.

Benjelloun, H. (2000). "The European Union's Ban on Asbestos: A Battle May be Over, But the War Continues", *Synergist, 11* (8).

Berman, D. W. (2000). "Asbestos Measurement in Soils and Bulk Materials: Sensitivity, Precision, and Interpretation—You can have it all", *Advances in Environmental Measurement Methods for Asbestos*, ASTM STP 1342, M. Beard and H. Rook, eds, American Society for Testing and Materials, West Conshohocken, PA, pp. 70–89.

Berman, D. W. (2003). "Analysis and Interpretation of Measurements for the Determination of Asbestos in Core Samples Collected at the Southdown Quarry in Sparta, New Jersey", Aeolus, Inc., November 12, 2003.

Berman, D. W., and K. Crump (1989). "Proposed Interim Methodology for Conducting Risk Assessment at Asbestos Superfund Sites, Part 2: Technical Background Document", ICF Technology Inc., May 18, 1989.

Berndt, M. E. and W. C. Brice (2003). "Reserve Mining and the Asbestos Case", presented at The International Symposium on the Health Hazard Evaluation of Fibrous Particles Associated with Taconite and the Adjacent Duluth Complex, St. Paul, MN, March 30–April 1, 2003.

Bloss, F. D. (1999). Optical Crystallography, MSA Monograph, Publication #5, Mineralogical Society of America, Washington, DC.

Boetcher, A. L. (1966). The Rainy Creek Igneous Complex Near Libby, Montana, PhD Thesis, The Pennsylvania State University, University Park, PA.

Brackett, K. A., P. J. Clark, and J. R. Millette (1994). "Determination of Asbestos Structures Over $10 \mu m$ in Length in Drinking Water", EPA Method 100.2, EPA/600/R-94/134.

CARB (1991). "Determination of Asbestos Content in Serpentine Aggregate", California Air Resources Board, Method 435, June 6, 1991.

Campbell, W. J., R. L. Blake, L. L. Brown, E. E. Cather, and J. J. Sjoberg (1977). "Selected Silicate Minerals and Their Asbestiform Varieties: Mineralogical Definitions and Identification-Characterization", Bureau of Mines Information Circular IC-8751.

Chatfield, E. J. (1999). "Correlated Measurements of Airborne Asbestos-Containing Particles and Surface Dust", *Advances in Environmental Measurement Methods for Asbestos*, ASTM STP 1342, M. Beard and H. Rook, eds, American Society for Testing and Materials, West Conshohocken, PA, pp. 378–402.

Chatfield, E. and M. J. Dillon (1983). "Analytical Method for Determination of Asbestos Fibers in Water", EPA Method 100.1, EPA-600/4-83-043.

Churchill, R. K., C. T. Higgins, and B. Hill (2000). "Areas More Likely to Contain Natural Occurrences of Asbestos in Western El Dorado County, California", Open File Report 2000-002, Department of Conservation, Division of Mines and Geology, Sacramento, California.

Clark, R. N. (1999). "Chapter 1: Spectroscopy of Rocks and Minerals, and Principles of Spectroscopy", *Manual of Remote Sensing*, Vol. 3, *Remote Sensing for Earth Sciences*, A. N. Rencz, ed., John Wiley & Sons, New York, pp. 3–58.

Cossette, M. (1984). "Defining Asbestos Particulate for Monitoring Purposes", *Definitions of Asbestos and Other Health-Related Silicates*, ASTM STP 834, B. Levadie, ed., American Society for Testing and Materials, Philadelphia, pp. 5–50.

Cossette, M. and P. Delvaux (1979). "Technical Evaluation of Chrysotile Asbestos Ore Bodies", *Short Course in Mineralogical Techniques of Asbestos Determination*, R. L. Ledoux, ed., Mineralogical Association of Canada, pp. 79–110.

Crane, D. T. (1988). "Asbestos in Air", *OSHA Analytical Methods Manual*, Method ID-160, July 1988.

Crane, D. T. (1992). "Polarized Light Microscopy of Asbestos", *OSHA Analytical Methods Manual*, Method ID-191, October 21, 1992.

Crump, K. S. (1991). Testimony for OSHA Hearing on Occupational Exposure to Asbestos, Tremolite, Anthophyllite and Actinolite; Proposed Rule, April 26, 1991.

Deer, W. A., R. A. Howie, and J. Zussman (1962). "Serpentines", *Rock-Forming Minerals: Sheet Silicates*, Vol. 3, John Wiley & Sons, New York, pp. 170–190.

Deer, W. A., R. A. Howie, and J. Zussman (1997a). "Anthophyllite Gedrite", *Rock-Forming Minerals: Double-chain silicates*, Vol. 2B, second edition, The Geological Society, London, pp. 21–78.

Deer, W. A., R. A. Howie, and J. Zussman (1997b). "Cummingtonite-Grunerite", *Rock-Forming Minerals: Double-chain silicates*, Vol. 2B, second edition, The Geological Society, London, pp. 86–133.

Deer, W. A., R. A. Howie, and J. Zussman (1997c). "Magnesio-riebeckite, Riebeckite", *Rock-Forming Minerals: Double-chain silicates*, Vol. 2B, second edition, The Geological Society, London, pp. 681–703.

Deer, W. A., R. A. Howie, and J. Zussman (1997d). "Tremolite-Actinolite-Ferro-actinolite", *Rock-Forming Minerals: Double-chain silicates*, Vol. 2B, second edition, The Geological Society, London, pp. 137–231.

Dietrich, R. V. (1990). Minerals of Virginia, Department of Mines, Minerals and Energy, Division of Mineral Resources, Charlottesville, VA.

Edwards, G. H. and J. R. Lynch (1968). "The Method Used by the US Public Health Service for Enumeration of Asbestos Dust on Membrane Filters", *Annals of Occupational Hygiene, 11*, pp. 1–6.

Fraser, D. A. (1953). "Absolute Method of Sampling and Measurement of Solid Airborne Particulates—Combined Use of the Molecular Filter Membrane and Electron Microscopy", *Archives of Industrial Hygiene and Occupational Medicine, 8*, p. 412.

Gunter, M. E., M. D. Dyar, B. Twamley, F. F. Foit, Jr., and S. Cornelius (2003). "Composition, $Fe^{3+}/\sum Fe$, and Crystal Structure of Non-Asbestiform and Asbestiform Amphiboles from Libby, Montana, USA.", *American Mineralogist, 88*, pp. 1970–1978.

Health & Safety Executive (1994). "Asbestos in Bulk Materials: Sampling and Identification by Polarized Light Microscopy (PLM)", Methods for the Determination of Hazardous Substances, MDHS 77, Occupational Medicine and Hygiene Laboratory, Health & Safety Executive (HSE).

IARC (1987). "Asbestos", Supplement 7, http://www-cie.iarc.fr/htdocs/monographs/suppl7/asbestos.html, p. 106. Accessed January 3, 2005.

International Organization for Standardization (1993). "Air quality—Determination of the Number Concentration of Airborne Inorganic Fibres by Phase Contrast Microscopy—Membrane Filter Method", ISO 8672, International Organization for Standardization (ISO), Geneva, Switzerland.

International Organization for Standardization (1995). "Ambient Air—Determination of Asbestos Fibres—Direct-Transfer Transmission Electron Microscopy Method", International Organization for Standardization, Geneva, Switzerland, Method ISO 10312, first edition, May 1, 1995.

International Organization for Standardization (1999). "Ambient Air—Determination of Asbestos Fibres—Indirect-Transfer Transmission Electron Microscopy Method", International Organization for Standardization, Geneva, Switzerland, Method ISO 13794, first edition, July 15, 1999.

International Organization for Standardization (2002). "Ambient Air—Determination of Numerical Concentration of Inorganic Fibrous Particles—Scanning Electron Microscopy Method", ISO 14966, International Organization for Standardization, Geneva, Switzerland.

International Organization for Standardization Draft PLM Method, personal communication from E. Chatfield.

Langer, A. M. (2003). "Reduction of the Biological Potential of Chrysotile Asbestos Arising from Conditions of Service on Brake Pads", Regulatory Toxicology and Pharmacology, *38*, pp. 71–77.

Langer, A. M. and R. P. Nolan (1987). "Asbestos in Play Sand", correspondence, *New England Journal of Medicine, 316*, p. 882.

Leake, B. E., *et al.* (1997). "Nomenclature of Amphiboles: Report of the Subcommittee on Amphiboles of the International Mineralogical Association, Commission on New Minerals and Mineral Names", Canadian Mineralogist, *35*, pp. 219–246.

Lee, R. J., D. R. Van Orden, and I. M. Stewart (1999). "Dust and Airborne Concentrations—Is There a Correlation?", *Advances in Environmental Measurement Methods for Asbestos*, ASTM STP 1342, M. Beard and H. Rook, eds, American Society for Testing and Materials, West Conshohocken, PA, pp. 313–322.

Lee, R. J., T. V. Dagenhart, G. R. Dunmyre, I. M. Stewart, and D. R. Van Orden (1995). "Effect of Indirect Sample Preparation Procedures on the Apparent Concentration of Asbestos in Settled Dusts", *Environmental Science & Technology, 29*, pp. 1728–1736.

Lee, R. J., D. R. Van Orden, G. R. Dunmyre, and I. M. Stewart (1996). "Interlaboratory Evaluation of the Breakup of Asbestos-Containing Dust Particles by Ultrasonic Agitation", *Environmental Science & Technology, 30*, pp. 3010–3015.

Lioy, P., *et al.* (2001). "Quality Assurance Project Plan: Assessment of Population Exposure and Risks to Emissions of Protocol Structures and Other Biologically Relevant Structures from the Southdown Quarry", January 24, 2001.

McCrone, W. C. (1980). *The Asbestos Particle Atlas*, Ann Arbor Science Publishers, Inc., pp. 1–66.

McFaul, E. J., G. T. Mason, Jr., W. B. Ferguson, and B. R. Lipin (2000). US Geological Survey Mineral Databases—MRDS and MAS/MILS, US Geological Survey Digital Data Series DDS-52, US Geological Survey, Washington, DC.

Meeker, G. P., A. M. Bern, I. K. Brownfield, H. A. Lowers, S. J. Sutley, T. M. Hoefen, and J. S. Vance (2003). "The Composition and Morphology of Amphiboles from the Rainey Creek Complex, Near Libby, Montana", *American Mineralogist, 88*, pp. 1955–1969.

Millette, J. R., and S. M. Hays (1994). *Settled Asbestos Dust Sampling and Analysis*, Lewis Publishers, Boca Raton, FL.

NIOSH (1989). "Asbestos (bulk) by PLM)", *NIOSH Manual of Analytical Methods*, Method 9002, May 15, 1989.

NIOSH (1977). *NIOSH Manual of Analytical Methods*, second edition, Vol. 1, P&CAM 239, US Department of Health, Education, and Welfare, publication (NIOSH) 77-157-A.

NIOSH (1994). *NIOSH Manual of Analytical Methods*, NIOSH 7400, "Fibers", The current version as listed as "Asbestos and Other Fibers by PCM", issue 2, August 15, 1994.

NIOSH (1989). NIOSH Manual of Analytical Methods, "Asbestos by TEM", Method 7402, May 15, 1989.

New York ELAP (1990). "Polarized-Light Microscope Methods for Identifying and Quantitating Asbestos in Bulk Samples", New York Environmental Laboratory Approval Program, Certification Manual, Item 198.1. Current version dated March 1, 1997.

New York ELAP (1997). "Transmission Electron Microscope Method for Identifying and Quantitating Asbestos in Non-Friable Organically Bound Samples", NY ELAP 198.4, Environmental Laboratory Approval Program Certification Manual, New York Department of Health, Albany, New York.

Perkins, R. L. and B. W. Harvey (1993). "Method for the Determination of Asbestos in Bulk Building Materials", US Environmental Protection Agency, EPA/600/R-93/116, July, 1993.

Ross, M. (1981). "The Geologic Occurrences and Health Hazards of Amphibole and Serpentine Asbestos", *Reviews in Mineralogy: Amphiboles and Other Hydrous Pyriboles—Mineralogy*, D. R. Veblen, ed., Mineralogical Society of America, Washington, DC, pp. 279–324.

Ross, M., R. A. Kuntze, and R. A. Clifton (1984). "A Definition for Asbestos", Definitions of Asbestos and Other Health-Related Silicates, ASTM STP 834, B. Levadie, ed., American Society for Testing and Materials, Philadelphia, pp. 139–147.

Ross, M. and R. P. Nolan (2003). "History of Asbestos Discovery and Use and Asbestos-Related Disease in Context with the Occurrence of Asbestos within Ophiolite Complexes", *Ophiolite Concept and the Evolution of Geological*

Thought: Boulder, Colorado, Y. Dilek and S. Newcomb, eds, Geological Society of America Special Paper 373, pp. 447–470.

Saltzman, L. E. and K. M. Hatlelid (2000). "CPSC Staff Report on Asbestos Fibers in Children's Crayons", US Consumer Product Safety Commission, Washington, DC.

Schneider, T., *et al.* (1997). Development of a method for the determination of low contents of fibres in bulk materials: Final Report, European Community Contract No. MAT1-CT93-0003, November, 1997.

Stewart, I. M. (1988). "Asbestos Analytical Techniques", *Applied Industrial Hygiene, 3*, pp. F24–F27.

Swayze, G. A., *et al.* (2004). "Preliminary Report on Using Imaging Spectroscopy to Map Ultramafic Rocks, Sepentinites, and Tremolite-Actinolite-Bearing Rocks in California", US Geological Survey Open-file Report 2004-1304.

US Environmental Protection Agency (1982). Interim Method for the Determination of Asbestos in Bulk Building Insulation Samples, EPA Report 600/M-82-020.

US Environmental Protection Agency (1987). "Interim Transmission Electron Microscopy Analytical Methods—Mandatory and Nonmandatory—And Mandatory Section to Determine Completion of Response", *Federal Register, 52*, pp. 41857–41897, October 30, 1987.

US Environmental Protection Agency (1990). "Environmental Asbestos Assessment Manual; Superfund Method for the Determination of Asbestos in Ambient Air; Part 1: Method", EPA/540/2-90-005a, May 1990.

US Environmental Protection Agency (1994). "Advisory Regarding Availability of an Improved Asbestos Bulk Sample Analysis Test Method; Supplementary Information on Bulk Sample Collection and Analysis", *Federal Register, 59*, pp. 38970–38971, August 1, 1994.

US Environmental Protection Agency (1997). "Superfund Method for the Determination of Releasable Asbestos in Soils and Bulk Materials", EPA 540-R-97-028, US Environmental Protection Agency. Unpublished modification to method by Berman and Kolk, May 23, 2000.

US Geological Survey (1934). The Microscopic Determination of the Nonopaque Minerals, second edition,

US Department of the Interior, Geological Survey, Bulletin 848.

US Geological Survey (undated). "Remote Sensing of Fibrous and Non-Fibrous Asbestos Forming Minerals", http://minerals.cr.usgs.gov/projects/dusts/task2.html, accessed January 3, 2005.

US Occupational Safety and Health Administration (1994). "Occupational Exposure to Asbestos: Final Rule", *Federal Register, 59*, pp. 40964–41162.

US Occupational Safety and Health Administration (1991). "TEM Methodology", OSHA SLC Analytical Laboratory, December 20, 1991.

Virta, R. L. and E. L. Mann (1994). "Asbestos", *Industrial Minerals and Rocks*, sixth edition, D. D. Carr, Sr. ed., Society for Mining, Metallurgy, and Exploration, Inc., Littleton, Co., pp. 97–124.

Virta, R. L. (2003). "Worldwide Asbestos Supply and Consumption Trends from 1900 to 2000", Open File Report 03-83, US Geological Survey, Reston, VA.

Wicks, F. J. (1979). "Mineralogy, Chemistry, and Crystallography of Chrysotile Asbestos", *Short Course in Mineralogical Techniques of Asbestos Determination*, R. L. Ledoux, ed., Mineralogical Association of Canada, pp. 35–78.

World Health Organization (1997). *Determination of Airborne Fibre Number Concentrations*, World Health Organization, Geneva, Switzerland.

Wylie, A. G. (2000). "The Habit of Asbestiform Amphiboles: Implications for the Analysis of Bulk Samples", Advances in Environmental Measurement Methods for Asbestos, ASTM STP 1342, M. Beard and H. Rook, eds, American Society for Testing and Materials, West Conshohocken, PA, pp. 53–69.

Wylie, A. G. and J. R. Verkouteren (2000). "Amphibole asbestos from Libby, Montana: Aspects of nomenclature", *American Mineralogist, 85*, pp. 1540–1542.

Yamate, G., S. C. Agarwal, and R. D. Gibbons (1984). "Methodology for the Measurement of Airborne Asbestos by Electron Microscopy", IIT Research Institute, Contract No. 68-02-3266, July, 1984. The method is referred to as the "Yamate Method" and also as "EPA Level II".

3

Sewage

Stephen M. Mudge and Andrew S. Ball

Contents

3.1 INTRODUCTION

"Sewage" is a generic term for the fecal wastes from animals although it is usually only applied to human-derived materials. This highlights one to the principal problems with assessing sewage inputs–is it derived from human or other, predominantly agricultural, sources? Notwithstanding this issue, which will be addressed below, the term "sewage" used here will predominantly apply to human wastes. Unlike most chemicals and compounds presented in this book, it is not a single compound, element or even class of compounds. Sewage is primarily a mixture of organic and inorganic components along with intact biological entities (bacteria and viruses); together, this makes a very complex cocktail. This cocktail changes during the day and season and in response to the treatment process that it receives. In this chapter, the major components of sewage and how these might be measured are presented, so that the extent to which sewage has dispersed in the environment can be assessed. Some of the more water-soluble components will be moved with the liquid phase while others are principally associated with the solid or particulate phase. Therefore, tracers may exist in a range of environments and at a range of distances from potential sources. Chemicals that are intrinsic to sewage, such as the stanols and sterols associated with human fecal matter; additives like detergents; microbiological communities present in the wastewaters; and effects caused by sewage to benthic communities will be examined. In specific cases, other chemicals may be present in discharges related to industries or activities in the wastewater receiving area from which the sewage originates.

A workshop consisting of proponents of a range of different biological and chemical tracers for identifying sewage sources was conducted in 2004 and concluded that at present there is no common methodology for the identification of fecal matter arising from sewage discharges (see Pond et al., 2004 for a review). However, most of the effort was directed toward bacterial identification and it may be that a combined chemical and biological approach is necessary.

The ability to quantify fecal material within the marine environment has long been a requirement under several portions of legislation including the European Shellfish Waters Directive (79/923/EEC), European Bathing Waters Directive (76/160/EEC) and, more recently, the European Water Framework Directive (2000/60/EC). Within the United States, the Clean Water Act (1972 as amended in 1977) and the Water Quality Act of 1987 have been key drivers in this field. Fecal material can enter such waters from a range of potential sources including sewage treatment facilities, domestic drainage from either septic tank systems or misconnections to surface water drains as well as non-point sources including agricultural drainage and surface water run-off. Most water quality guidelines focus on the fecal bacteria, especially *Escherichia coli* and fecal streptococci. Although these organisms are not directly harmful to man, they are considered good indicators of pathogenic organisms that are associated with sewage. Therefore, most water quality standards state the concentration of these fecal bacteria that are allowed to be present in water.

Although these organisms are of human origin, fecal bacteria also occur in numerous other animals from birds through to herbivores in agriculture systems. Challenges have arisen due to the commonality of the sources for these organisms and significant effort is being directed toward identifying the different bacterial strains from the different animal sources (Pond et al., 2004). These methods are based on DNA signatures and can be conducted on small volumes with relatively high efficiency (see Chapter xx). The environmental hazard associated with sewage can be related to the animal (human) pathogens, the organic matter itself (e.g., BOD_5), the nutrients that may contribute to eutrophic events and biologically active compounds especially the environmental estrogens.

Composition

The composition of sewage varies considerably, especially with regard to the trace organic components. However, it is generally considered that sewage has a gross composition as summarized in Table 3.1.1.

3.1.1 Physical Composition

The major component of untreated or treated sewage is water. The influent of a sewage treatment plant (STP) is typically 95%+ water by volume. The other major physical components include grit and sediment; the concentration of these varies in response to nature of the sewage infrastructure. Sewerage systems may be exclusively foul water drains from domestic and industrial premises. However, there are many systems that include surface water drains as well as foul sewers. In some extreme cases, entire streams may also form part of the system. In the latter two cases, rainfall is a major influence on the volume flow. It is possible to determine the Dry Weather Flow (DWF, the amount of liquid flow produced daily by the total population and industry in the wastewater receiving area) for a system by quantifying the number of premises and people in the catchment. In the United Kingdom, each person produces ~200 litres/day to the sewer while this value is nearly 300 litres/day in the United States of America. In other less developed parts of the world, the DWF may be substantially less.

Other physical components of sewage are sanitary products including plastics and rags. The nature and quantity of these materials is also dependent on the culture of the people in the catchment and may vary from location to location. It is also possible to find objects such as branches, leaves, and even animals in the influent sewer.

3.1.2 Chemical

The chemical nature of sewage varies widely according to the catchment. In regions with sparse industrial activity, domestic wastes comprise the majority of the matter. These materials are rich in proteins, carbohydrate, lipids,

Table 3.1.1 *The Typical Gross Composition of Sewage (Bungay, unpub.)*

Component		Concentration
Solids	Total	$700 \ mg \cdot l^{-1}$
	Dissolved total	500
	Volatile	200
	Suspended total	200
	Volatile	150
	Settleable solids	$10 \ mg \cdot l^{-1}$
BOD_5		200 ppm
TOC		200
COD		500
Nitrogen	Total as N	40
	Organic	15
	Ammonia (free)	25
	Nitrate or nitrite	0
Phosphorus	Total as P	10
	Organic	3
	Inorganic	7
Chloride		50
Alkalinity	as $CaCO_3$	100
Grease		100

Table 3.1.2 *Gross Chemical Composition of Sewage Waters (after Sophonsiri and Morgenroth, 2004)*

COD fractions (%)				$mg \cdot l^{-1}$
Protien	Carbohydrate	Lipid	Unidentified	Total COD
31	16	45	8	203
30	10	nd	60	813
15	7	nd	78	394
8	12	10	70	530
12	6	19	63	259
28	18	31	22	nd
18	16	7	59	967
12	6	82	0	309
38	11	44	7	35
Means				
21	11	34	41	439

nd = not determined.

Table 3.1.3 *The Major Bacterial Pathogens and Associated Diseases Associated with Sewage*

Bacterial Pathogens	Related Disease
Salmonella	*Salmonellosis*
S. typhimurium	Typhoid fever
Shigella	Shigellosis
Enterococcus	Diarrhea
E. coli	Diarrhea
Vibro cholerae	Cholera
Camplyobacter jejuni	Gastroenteritis

Table 3.1.4 *The Major Viral Pathogens Associated with Sewage*

Viral Pathogens	Related Disease
Hepatitis A	Hepatitis
Norwalk-like agents	Gastroenteritis
Virus-like 27 nanometer particles	Gastroenteritis
Rotavirus	Gastroenteritis and polio

and non-digestible matter (Table 3.1.2). There are many organic chemicals present within the influent waters and several of these are discussed above.

3.1.3 Biological
The heterogeneous nature of sewage provides an excellent growth medium for a multitude of microorganisms. Many of these microbes are necessary for the degradation and stabilization of organic matter and thus are beneficial. The diversity of participating microorganisms ensures a complex ecosystem will exist during sewage treatment. Bacteria represent the most abundant form of microorganisms in sewage, with in excess of 10^{12} cells per litre. Microbial species such as *Bacillus* and *Clostridium* are always present in large numbers in sewage; these species can resist severe changes in environmental conditions by forming spores. Other bacteria that appear to consistently contribute to sewage are the enteric bacteria (*Salmonella*, *Shigella*, *Escherichia*), *Staphylococcus*, *Streptococcus*, *Pseudomonas*, and *Vibrio*. Bacteria of two chemolithotrophic genera, *Nitrosomonas* and *Nitrobacter*, play a central role in the fate of nitrogen in sewage. The removal of nitrogen from sewage is carried out through the conversion of dissolved nitrate to gaseous nitrogen (denitrification). Bacteria which undertake this process include *Alcaligenes*, *Micrococcus*, and *Pseudomonas*. Their growth occurs under conditions where oxygen is absent because denitrification is an anaerobic process. Fungal cells are eukaryotic cells and although found in sewage treatment plants their numbers are usually low with no single characteristic species routinely detected. Where present, they are often found as an external biofilm of flocs.

Three classes of protozoa are found in sewage:

- Flagellates (subphylum mastigophora) are single-celled protozoa that move using flagella. The protozoa found in this classification are the most abundant type found in sewage.
- The ciliates (subphylum cilophora) represent the most diverse group found in sewage, although they are present in low numbers. One genus, *Vorticella*, is often found to adhere to flocs.
- The last of the three main types of protozoa found in wastewater are the amoebas (subphylum sarcodina).

Sewage contains pathogenic or potentially pathogenic microorganisms which pose a threat to public health. By definition, a pathogen is an organism capable of inflicting damage on its host. Waterborne diseases whose pathogens are spread by the fecal–oral route (with water as the intermediate medium) can be caused by bacteria, viruses, and parasites (including protozoa, worms, and rotifers). The major bacterial pathogens and their associated diseases are shown in Table 3.1.3.

Viruses are defined as genetic elements, containing either DNA or RNA and a protein capsid membrane, which are able to alternate between intracellular and extracellular states, the latter being the infectious state. Over 100,000 different viral types have been identified in human feces. Some of the major viral pathogens are shown in Table 3.1.4. A person infected with a disease-causing virus may excrete up to 10^6 (1,000,000) infectious particles per gram of feces.

Two protozoan disease causing organisms commonly associated with sewage are *Giardia lamblia* (responsible for the most widespread protozoan-caused disease in the world) and members of the genus *Cryptosporidium*.

3.2 TREATMENT PROCESSES

The treatment of sewage (Arundel, 1995) varies from completely untouched through to full secondary treatment with additional tertiary processes such as UV (ultra-violet) irradiation. The chemical nature of the effluent discharged to receiving waters also varies in response to both the catchment composition and the degree of treatment. The BOD_5 (biological oxygen demand with 5 days incubation) will be particularly decreased by treatment. However, the chemical composition varies as well and key lipid or bacterial biomarkers change with treatment. Thus, it may be appropriate to determine the "signature" of any effluent directly at the discharge point. Non-point sources such as agricultural runoff will have no treatment and should exhibit a different chemical profile than treated materials.

3.2.1 Screening
In most cases, influent material to an STP undergoes primary screening. This process removes the solid component that may cause damage to pumps or parts of the infrastructure. As well as degradation products of the sewerage pipe works, debris from streams and rivers, grit from road runoff and industrial discharges are removed. Sanitary material is also removed at this time. The most common system is

a coarse screen (wire mesh) which may be static or may rotate. The type of system used is determined by the nature of the waste in the sewer and typical mesh sizes range from 1 mm to 20 mm. The filterable material is then removed to a land disposal site.

3.2.2 Primary

Settlement of the solids in the influent has the largest effect on the BOD_5 of sewage. Typical reductions are in the order of 50% but may range up to 90% depending on the nature of the influent. Most primary treatment systems comprise a cone shaped tank with a series of concentric weirs; the influent is introduced into the middle of the tank, the sludge settles to the bottom and is removed *via* a separate pipe. The cleared liquid overflows the weir and proceeds to the next stage of treatment. Any floating materials can be skimmed off during this process using a rotating scoop.

3.2.3 Secondary

Cleared liquid can be further treated (secondary treatment) using a biological step. In general, the two major approaches are percolating bed filters or activated sludge treatment. In a percolating bed filter arrangement, the clarified liquid is introduced to the top of a 1.2–2 m thick bed of gravel, shale, coke, coal, or rigid plastic shapes. The material has a typical surface area of $220 \, m^2 \cdot m^{-3}$. The bed is constructed to allow a free flow of air to pass upwards through the system. The bed supports a wide variety of microorganisms that actively utilise the organic matter and inorganic nutrients in the production of further biomass. Some of this material escapes with the effluent and must be removed in a humus tank, a structure similar to a primary settlement tank. The effects of these first two steps in sewage treatment are summarised in Table 3.2.1.

The other major biological treatment process is through the use of an activated sludge tank. In this system, the liquor is aerated and stirred by either mechanical mixers or air diffusers as it passes through a long channel. A feed stock of activated sludge is added and the microbial community degrade the sewage aerobically. There are several advantages of this process over that of percolating bed filters:

- It uses less land.
- The odor is less and certain pests (e.g., *Psychoda* fly) are reduced or eliminated.
- As the process is temperature independent, the final effluent is of a better, more consistent quality. Typical values for BOD_5 removal, NH_3-N reduction, and bacterial removal exceed 90%.

However, this treatment method does require an energy input but this may be offset by the generation of power from methane produced during the anaerobic sludge digestion process.

In many cases, secondary treatment and its subsequent settlement is the final stage in an STP. In which case, the effluent is discharged directly to the receiving waters. In some cases, however, it is necessary to improve the quality of the final effluent to meet a regulatory standard. Therefore, tertiary treatment may be required.

3.2.4 Tertiary

Tertiary treatment may take several different forms depending on the quality of the final effluent required. Near bathing waters (European Bathing Waters Directive 76/160/EEC), UV irradiation is the most common method and has become a more favoured option in reducing the number of viable bacteria. Previously, chlorine had been used although this may lead to additional environmental problems such as bleaching and direct toxicity to the flora and fauna in the receiving waters.

In some instances, the concentration of phosphorus may have to be reduced, especially if the receiving waters are lakes or rivers (EU Urban Wastewater Directive 91/271/EEC). In these environments, increased PO_4 availability might lead to eutrophic conditions. A typical method of removing dissolved phosphorus is through precipitation as $FePO_4$ or $Fe_3(PO_4)_2$ after the addition of $FeCl_3$ or $Fe_2(SO_4)_3$.

If the solids need to be reduced, sand filters or other clarifiers may be used. The collected materials are then usually bulked with the other sludges on site for further treatment and disposal.

3.3 SLUDGE TREATMENT AND DISPOSAL

The nature of the sludge depends upon the materials disposed to the sewer in the catchment. In some cases, they may be principally inorganic from extraction or process industries (e.g., coal, chalk, dusts etc.). Organic sludges are a valuable resource as they can improve soils and add nutrients. The most common treatment route is to reduce the volume by settling or through mechanical dewatering. Thickening agents could be added to give the right consistency for transport and pumping. As most countries can no longer dispose of sewage sludge to the sea, on-land treatment and disposal is the norm. Digestion, either aerobically or more commonly, anaerobically, significantly reduces the BOD_5 and many microorganism pathogens.

The final waste may be pressed into pellets or cake and sold/given away to farmers or gardeners. In the United States, these are called bio-solids. Some sludge may be incinerated, applied directly to agricultural land, or go to landfill.

3.4 DISCHARGES

The planned discharge of sewage to receiving waters is regulated in most countries. In Europe, the European Union essentially dictates the standards although each country authorizes discharges under national legislation.

Table 3.2.1 *Summary of Typical Parameters Associated with Sewage Treatment Stages (Adapted from Arundel, 1995). Removal Percentages are Relative to the Feed Stock*

Primary Treatment Property	Value	Secondary Treatment Property	Value
Suspended solids removal	45–75%	NH_3-N removal	50–90%
BOD_5 removal	20–80%	BOD_5 removal	60–90%
Bacterial removal	10%	Bacterial removal	70+%
Solids' half-life in tank	2–8 h	Liquid retention time	20–35min for 80–90% reduction

3.4.1 European Union

The *Mandatory (or Imperative) Standards* in the EU Bathing Waters Directive (76/160/EEC), which should not be exceeded, are

1. 10000 total coliforms per 100 ml of water
2. 2000 fecal coliforms per 100 ml of water

In order for a bathing water to comply with the Directive, 95% of the samples must meet these bacterial standards.

The *Guideline Standards*, which should be achieved where possible, are

1. not more than 500 total coliforms per 100 ml of water in at least 80% of the samples,
2. not more than 100 fecal coliforms per 100 ml of water in at least 80% of the samples,
3. not more than 100 fecal streptococci per 100 ml of water in at least 90% of the samples.

There are other aspects of the Directive regarding beach facilities etc. but the majority of bathing water only fail on the bacterial standards. The European Commission has proposed new minimum bathing water quality standards of

1. 200 intestinal enterococci per 100 ml at 95% compliance, and
2. 500 *E. coli* per 100 ml at 95% compliance.

This is roughly equivalent to the Guideline standard in the existing Bathing Water Directive. The decrease in fecal bacterial numbers allowable under the proposed changes will lead to a reduction in the number of beaches complying although with on-going improvements to STPs and sewage infrastructure; it is anticipated that compliance will improve over a 10-year period.

The implementation of the EU Water Framework Directive (WFD) (2000/60/EC) will eventually subsume many of the water quality measures currently spread across a number of directives. It is envisaged that several directives or parts of directives will be repealed by 2013 once the WFD has been fully implemented and its first cycle of control completed.

3.4.2 United States of America

Section 303(d) of the federal Clean Water Act requires each State (together with the United States Environmental Protection Agency) establish the total maximum daily load (TMDL) of each pollutant that may cause a water body to not meet water quality standards. A TMDL is the pollution that a water body can assimilate before beneficial uses are affected (Barreca and Seiders, 2001). Typical values for fecal bacteria are (e.g., Santa Barbara County Environmental Health Services)

1. Total coliform exceeds 10000 MPN (Most Probable Number)
2. Fecal coliform exceeds 400 MPN
3. Enterococcus exceeds 104 MPN

3.5 TRACERS

3.5.1 Sterols

Sterols are structural components of cells and are synthesized by most organisms. In humans and other higher animals, cholesterol is biohydrogenated by the intestinal micro-flora to form 5β-coprostanol (Grimalt *et al.*, 1990). The conversion process includes ketone intermediaries of both cholesterol and 5β-coprostanol. In recent years, 5β-coprostanol has been identified in non-fecal derived material from other organisms including algae and bacteria (Green *et al.*, 1992).

Notwithstanding these possible extra sources, 5β-coprostanol has been widely used as an indicator for human sewage in marine and freshwaters. Typical sterol concentrations in human fecal matter are $\sim$5.6 mg $\cdot$ g^{-1}, of which about 60% can be 5β-coprostanol (Leeming and Nichols, 1998). Several key sterol structures are shown in Figure 3.5.1.

Most other higher animals and birds produce 5β-coprostanol as well but it is present in lower concentrations. This multiple sourcing of the marker can make interpretation difficult in regions where other point and non-point sources of agricultural slurries occur. However, due to the herbivorous diet of these animals, they produce the 24-ethyl derivative of coprostanol in a greater concentration than in humans. The ratio of 5β-coprostanol to 24-ethyl coprostanol is $\sim$2.8 in humans compared to 0–1.2 in animals (Table 3.5.1 and Leeming *et al.*, 1997). In some cases, this might be sufficient to distinguish between the different sources. However, in complex environments more sophisticated techniques are needed. Recently, multivariate statistical methods including principal component analysis and partial least squares have been employed in signature analyses. Discussion of these methods is beyond the scope of this chapter but further details may be found in Mudge and Duce (2005).

Due to grain surface area effects, muddy sediments will have greater organic matter content than sandy sediments. Comparative studies, therefore, have normalized the 5β-coprostanol concentration to remove these effects. Several measures have been used, including

- 5β-coprostanol/total sterols
- 5β-coprostanol/cholesterol
- 5β-coprostanol/(5β-coprostanol + 5α-cholestanol)

Critical values have been proposed for these indicator ratios; Grimalt *et al.*, 1990 have found that values of 0.7 of the 5β-coprostanol/(5β-coprostanol + 5α-cholestanol) ratio are indicative of contaminated areas. Grimalt and Albaiges (1990) have proposed that values of the 5β-coprostanol/cholesterol ratio greater than 0.2 are indicative of faecal contamination (Table 3.5.1).

It is possible to determine the degree of treatment or age of domestic sewage through use of another isomer of 5β-coprostanol. In the environment or during sewage treatment, some of the 5β-coprostanol is converted to epi-coprostanol (5β-cholestan 3α-ol) by bacterial action (McCalley *et al.*, 1981). Combined use of the 5β-coprostanol/cholesterol ratio, a measure of sewage contamination, with the epi-coprostanol/5β-coprostanol ratio, a measure of age or treatment, can help interpret data. An example is shown in Figure 3.5.2 (Mudge *et al.*, 1999).

3.5.1.1 Extraction Methodologies

Due to the lipophilic nature of the compounds, non-polar solvents are used in the extraction and these should be of good quality (distilled grade). Avoid all contact with plastics throughout the extractions to prevent contamination with plasticizers. Glass, metal or Polytetrafluoroethylene (PTFE) are the preferred materials. Glassware should be pre-cleaned with detergents prior to use, rinsed thoroughly in distilled or deionised water, dried, and finally rinsed with an organic solvent (e.g. dichloromethane (DCM) or hexane).

3.5.2 Total Sterols

This method is based on the *in situ* saponification of the organic matter. Due to the low water solubility of these compounds, they are essentially present in or on the solid phase in aqueous systems. Water samples should be collected in glass, PTFE, or high density polycarbonate if possible (in preference order). Since the sterols are associated with the

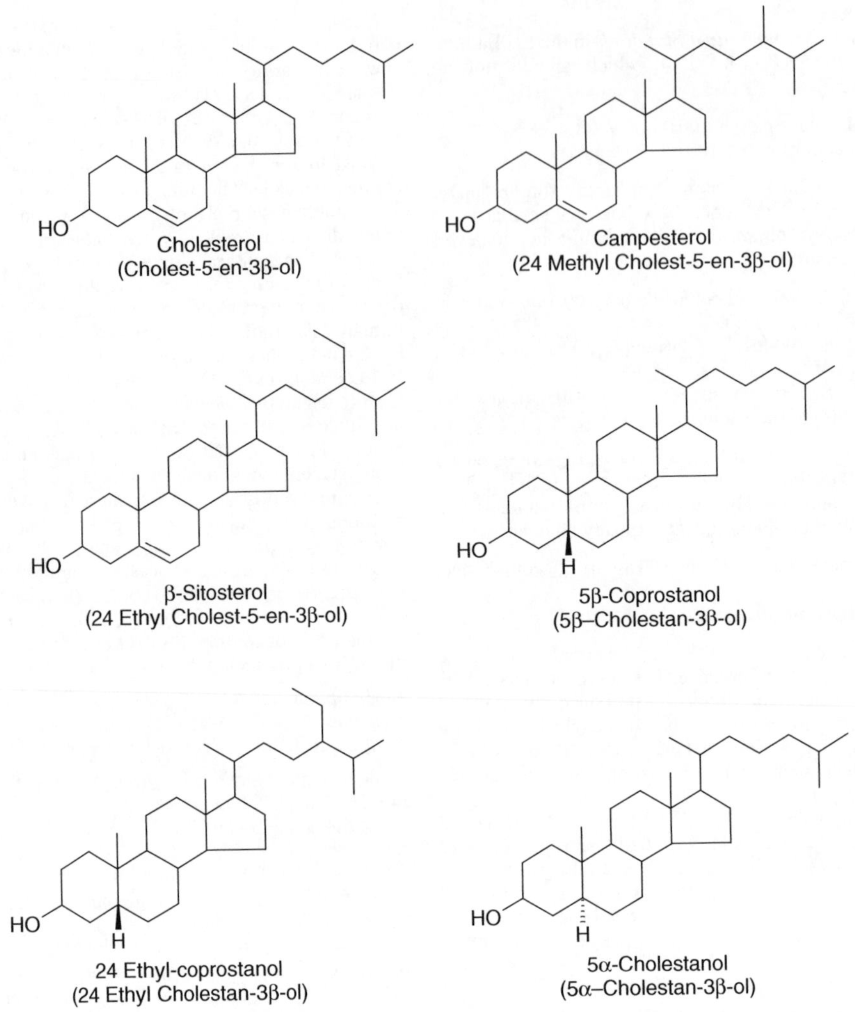

Figure 3.5.1 *The structure of several key environmental sterols.*

Table 3.5.1 *Typical Concentrations and Ratios of Key Sterol Markers in Potential Sewage/Faecal Matter (Gilpin et al., 2002)*

Source	5β-coprostanol mg · l^{-1}	5β-coprostanol/24-ethyl coprostanol
Septic tanks	34–170	2.9–3.7
Community wastewater	3–31	2.6–4.1
Meatworks-sheep, beef	1–16	0.5–0.9
Dairy shed wash-down	9–15	0.2
Urban lake	0.1–0.3	0.8

suspended particulate matter, these should be filtered from the bulk water sample as soon as possible. If this cannot be done on site, water samples should be cooled and returned to the laboratory as quickly as possible. Refrigeration is the main preservation method. At the time of sampling, a known amount of a sterol standard could be added. This will need to be one that does not occur in the natural samples. The suspended solids should be filtered out through a glass fiber filter (e.g., GF/F from Whatman, www.whatman.com) with

minimal vacuum. The filter paper or papers should then be transferred to a round bottom flask for solvent extraction.

Settled sediments should be collected as a surface scrape taking as thin a slice as possible. This will ensure that only recent sediments are collected. Alternatively, filter papers may be deployed on the surface of the sediments to collect the settling organic and inorganic matter. Experiments have shown that about four tides are needed to collect sufficient material for analysis (Mudge and Duce, 2005). These

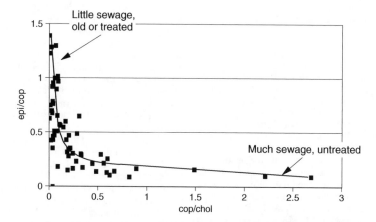

Figure 3.5.2 *A bi-plot of the epi-coprostanol/5β-coprostanol with the 5β-coprostanol/cholesterol. Different regions of the plot indicate those sites with untreated sewage or old or treated waste matter (redrawn from Mudge et al., 1999).*

samples should be returned to the laboratory in pre-cleaned glass jars, preferably while being refrigerated.

If core samples are needed to recreate the depositional history, a range of corers can be used. If the barrels or liners are made of plastic, ensure sub-samples for each depth are obtained from the center of the core. Due to the problems of core shortening (a function of core barrel diameter and sediment type), independent means of dating the samples should be considered (e.g., Pb deposition, PAHs (identifying the beginning of the industrial revolution, 1750 in UK), ^{14}C in older samples, ^{137}Cs in more recent samples, etc.).

For biological samples, specimens can be extracted wet or dry. Due to the lipophilic nature of the sterols, care and judgement should be used in the choice of tissues or organisms to analyze. Wet samples should be cut up as finely as possible to increase the surface area in the extraction vessel; dry samples can be obtained by freeze drying and light grinding afterwards.

Typical sample weights can range from a few grams to 60 grams in the case of sandy sediments with low TOC values. An internal standard can be added to sediment or biological samples while in the round bottom flask. In all cases, they should be extracted in 6% (w/v) KOH in methanol. In general, 50 ml is usually sufficient although more could be added to ensure complete sample coverage. The sample should then be refluxed for four hours; experiments have shown that this is sufficient for >90% recovery.

After allowing the extract to cool, the solids should be separated from the liquor by either centrifugation or filtration. If using centrifugation, glass centrifuge tubes of nominally 40 ml capacity should be balanced within 1 g and spun at 2500 rpm for ~5 min or until the solids have separated from the liquor. For improved recovery, the solids can be re-suspended in methanol, re-centrifuged, and the liquid combined with the initial extract.

If using filtration, GF/F filters should be pre-washed with methanol before use. The lowest vacuum necessary for extraction should be used to lessen the flow of fines through the filter. To speed up the separation, allow the solid–liquid mixture to settle for 10–20 min before carefully pouring off the upper layers. This approach reduces the clogging of the filters.

The clear liquor ranging in color from pale yellow to dark brown can be poured into a glass separating funnel using a glass funnel to aid in the transfer. The tap should be either glass or PTFE: no grease should be applied as this will

be extracted by the organic solvents and contaminate the sample. Add 20–30 ml of hexane to the liquor. This phase is less dense than the methanol and forms the upper layer. Shake and vent the funnel repeatedly. Care should be taken due to the accumulation of pressure in the funnel from the warm methanol and hexane; the funnel should be vented frequently initially. The two phases should be vigorously shaken and allowed to separate. The non-polar compounds in the methanol phase will partition into the hexane phase leaving the polar compounds behind. As well as the sterols, the fatty alcohols and PAHs will preferentially partition into the hexane.

The lower methanolic phase should be run out of the funnel and retained. The hexane phase should be collected in a florentine flask ready for rotary evaporation. The methanolic phase can be returned to the separating funnel and the process repeated. This improves the extraction efficiency. The lower polar phase can be runoff and either retained or discarded. The hexane fraction should be combined with the previous hexane component.

The fatty acids remain in the alkaline methanol as potassium salts. If these are required, the methanolic phase can be titrated back to an acid pH by the addition of HCl. If the sediments were initially anaerobic, hydrogen sulfide may be liberated at this point. The fatty acids can now be extracted in a similar manner to the sterols using 20% DCM in hexane as the extracting solvent (see below).

The hexane phase containing the non-polar compounds can now be rotary evaporated to a small volume (~2 ml). The sample can be transferred to a small vial by addition of a small volume of hexane using a Pasteur pipette. The solvent should be removed by nitrogen blow down so that dry lipids remain. This could be in a small (~1 ml) pre-weighed autosampler vial such that the data can be presented either on a dry weight (see below) or lipid weight basis.

Since the sterols have a hydroxyl group in the 3 position, they are best analyzed as their trimethyl silyl (TMS) derivatives. These TMS ethers are readily formed by the addition of a few drops of BSTFA (bis (trimethylsilyl) trifluoroacetamide), back flushing the vial with nitrogen to remove any moist air which would also react with the BSTFA, sealing with a PTFE-lined cap and heating for 10 min at 60 °C. After the allotted time, the remaining BSTFA can be removed by reducing to dryness under a stream of nitrogen. The now derivatized sterols can be dissolved in an exact volume (1.00 ml) of hexane prior to injection into a gas chromatograph–mass spectrometer (GC-MS).

3.5.3 Free Sterols

The concentration of the free sterol component is significantly less than that of the bound compounds and so greater sample sizes may be required. The profile of sterols extracted will be different from that of the bound sterols and so different interpretations may be placed on the results. Inter-conversion between free and bound phases occurs in the sediments (.Sun and Wakeham, 1998) and total analysis (free and bound) might be the most appropriate.

Sterols that are not bound to other materials may be extracted through Soxhlet extraction instead of simple reflux. If this is the case, samples should be dried before use (free drying in the case of biological samples and simple air or cool oven drying for sediments). Typical extraction solvents are DCM and mixtures of DCM with hexane. Reflux should be continued for ~8 hours to ensure high extraction efficiencies. Once in the liquor, samples may be cleaned up by solvent exchange, TLC or column chromatography. The compounds need to be derivatized in the same way as above before GC analysis.

3.5.4 Chromatographic Analysis

The derivatized sterols can be injected into a GC-MS preferably with an autosampler to improve consistency between repeat injections. With typical environmental concentrations dissolved in 1.0 ml, 1.0 µl injected provides adequate responses for analysis. The analytical column should be of the DB-5, HP-5, BPX-5 variety although better baselines have been seen using a high temperature column such as SGE's (Scientific Glass Engineering, Australia) HT-5. The temperature program needs to go to about 360 °C to ensure removal of all compounds from the column and at these elevated temperatures increased column bleed can become troublesome with some columns. Typical column lengths

are 30 to 60 m and the best separations can be seen with narrow bores and thin films (0.25 mm and 0.1 µm).

The temperature program should start at 60 °C, increasing at 15 °C min^{-1} to 300 °C, then at 5 °C min^{-1} to a maximum of 360 °C. Other gradients are possible and may be recommended if the fatty alcohols and polyaromatic hydrocarbons are to be quantified as well. The mass spectrometer (MS) is best configured for electron impact ionization at 70 eV and a mass scan range of 45–545 m/z per second. Alternatively, a GC–flame ionization detector (FID) could be employed although it is harder to identify unambiguously which peaks are which. An example gas chromatogram trace is shown in Figure 3.5.3.

Mass Spectrometer identification is usually through a series of diagnostic ions. The key ions are listed in Table 3.5.2. On an XX-5 column, these sterols elute in this order. One of the key advantages to tracking sewage matter by chemical means is that additional information regarding source and environmental conditions may be obtained through use of the other sterol and fatty alcohol biomarkers present in the sample. For example, β-sitosterol is the primary sterol in vascular plants and can be used as an indicator of terrestrial runoff. This may be of particular use when linking sewage discharge to non-piped sources and may point to rivers or diffuse pollution from agricultural areas.

Quantification can take three forms: use of the internal standard, through use of an external calibration curve, or by combination of the two. In the first case, the area under each of the sterol peaks of interest and the internal standard (IS) can be quantified and the concentration calculated by the following equation:

$$\text{Sterol Concentration} = \frac{\text{Area of sterol peak}}{\text{Area of IS peak}} \times \text{Concentration of added IS}$$

While this is sufficient to quantify the sterols present, it gives no direct information about the yield. The second

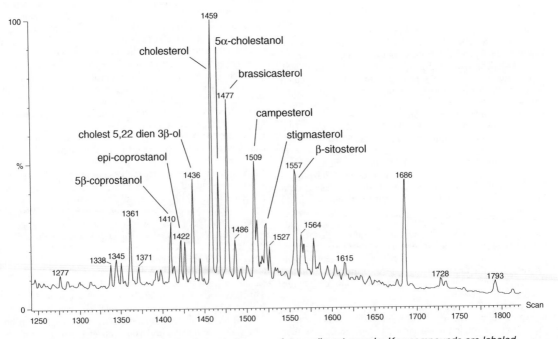

Figure 3.5.3 An example trace of sterols from a surface sediment sample. Key compounds are labeled.

Table 3.5.2 *Diagnostic Ions for Key Sterol Markers (as their TMS Ethers) of Sewage*

Compound	Molecular weight	Diagnostic ion (m/z)	Potential marker for
5β-coprostanol	460	370	Higher animal fecal matter
Epi-coprostanol	460	370	Treated sewage or old fecal matter
cholesterol	458	329	Fauna (natural and anthropogenic inputs)
5α-cholestanol	460	370	*In situ* reduction of cholesterol
24-ethyl coprostanol	488		Herbivore fecal matter
β-sitosterol	486	396	Terrestrial plants

method makes an assumption of 100% (or some other pre-determined extraction efficiency) and uses an external cali-bration of a known sterol. The usual practice is to weigh an exact amount of cholesterol out, derivatize it with BSTFA, and make a dilution series. This series can then be used to generate an external calibration curve. The area of any related sterol peak can then be applied to this curve and the compound quantified. The problem with this approach is that no real information is present about the extraction efficiency.

The third method is a combination of the two and does provide both sterol quantification and yield information. If an IS is added and used as above and an external calibration curve is constructed, the amount of the IS recovered gives a measure of the extraction efficiency. No correction to the compounds is required as the IS already takes account of any less than 100% yield.

3.6 QUALITY CONTROL ISSUES

The most efficient way of tracking yield and blank contam-ination is to use a sequence of samples interspersed with blanks and standards. When an autosampler is used, this becomes a simple process. Any sequence should begin with a blank, followed by a series of external calibration stan-dards or a single reference solution. This should be followed by another blank to ensure there is no carry over from one sample to another. Five samples can then be run fol-lowed by a standard and blank. This sequence can then be repeated and if there is any change in the response of the instrument between pairs of standards, interpolated values can be used as samples in between provided no more than 10% exists between the values. The blanks can be used to check for cross contamination and this allows removal of any background in the region of interest.

If accredited reference materials are available, these sam-ples can be extracted in parallel with the samples and used to determine if the method is consistent and appropriate. Sta-tistical process control methods are possible using Shewhart Plots to determine whether the method has deviated from a statistically significant range.

3.7 DRY WEIGHT DETERMINATION

The results are most often expressed on a dry weight basis ($\mu g \cdot g^{-1}$ or $ng \cdot g^{-1}$ DW of sediment). Since the sediment samples were extracted wet, it is necessary to take a rep-resentative sub-sample and dry it at 60 °C. From the water

loss, it is possible to correct the sterol results and express the concentrations on the basis of the dry solids component. This can also be done using the suspended materials, which usually have a much higher concentration when expressed on a weight basis. Alternatively, the results can be expressed on a volume basis if trying to calculate sterol loads from STP sources where the volume flow rate is known even if the suspended solid load is not.

In a series of samples across a lagoon receiving settled or untreated sewage, concentrations of total sterols ranged from 0.6 to 681 $\mu g \cdot g^{-1}$ in the suspended sediments and 0.1 to 16.3 $\mu g \cdot g^{-1}$ in the settled (surface) sediments (Mudge and Duce, 2005).

3.8 BILE ACIDS

Two series of bile acids are found in nature–those contain-ing 24 carbon atoms and those containing 27 or 28 carbon atoms. The former group are derivatives of cholanic acid, whereas the latter are derivatives of cholestanoic acid (Elh-mmali *et al.*, 1997). The structure of key bile acids can be seen in Figure 3.8.1. Source identification of terrestrial runoff has been conducted with these compounds since hyocholic (VIII) and hyodeoxycholic (V) are produced in substantial amounts by pigs compared to humans, while the latter produce deoxycholic acid (II), which is virtually absent in pigs (Elhmmali *et al.*, 1997). Ruminant animals such as cows produce predominantly deoxycholic acid (II) whereas omnivores (e.g., dogs and humans) also produce significant quantities of lithocholic acid (I) (Bull *et al.*, 2002). The absence of deoxycholic acid (II) and the presence of hyocholic acids (VIII) in pigs, therefore, allow those samples to be distinguished from human and canine contamination. In the case of dogs (Bull *et al.* (2002)) state that bile acids are a particularly useful biomarker as these animals do not produce 5β-stanols, in contrast with other mammals. Chaler *et al.* (2001) have also suggested that these compounds may be potential tracers for urban wastewaters.

3.9 FATTY ACIDS

Fatty acids are present in all living organisms and often com-prise the largest lipid component in environmental samples. The use of these compounds in tracing sewage may occur in several different ways.

3.9.1 Bacterial Biomarkers

Sewage contains a wide diversity and concentration of bacte-ria. Bacteria have a different suite of fatty acids compared to most animals and plants. Most organisms higher in the evo-lutionary tree than bacteria produce even chain fatty acids through the fatty acid synthase system. Here, two carbon sub-units in the form of acetyl Co-A are added sequentially to form long-chain molecules. As the cycle progresses, the chain length grows in acetyl (C_2) increments leading to a series of even carbon numbered compounds. Bacteria can also use other precursors (e.g. valine) and this gives rise to a series of odd carbon chain lengths. The mechanism of addition can generate three different acids: straight chain, and *iso* or *anteiso* branched chains (Figure 3.9.1). The pro-prionyl component may be added in three different ways giving rise to these compounds.

3.9.2 Cooking Oils

Domestic wastewaters contain fatty acids derived from cook-ing as well as the bacteria in the intestine of animals. Typi-cal vegetable cooking oils contain fatty acids with 16 or 18

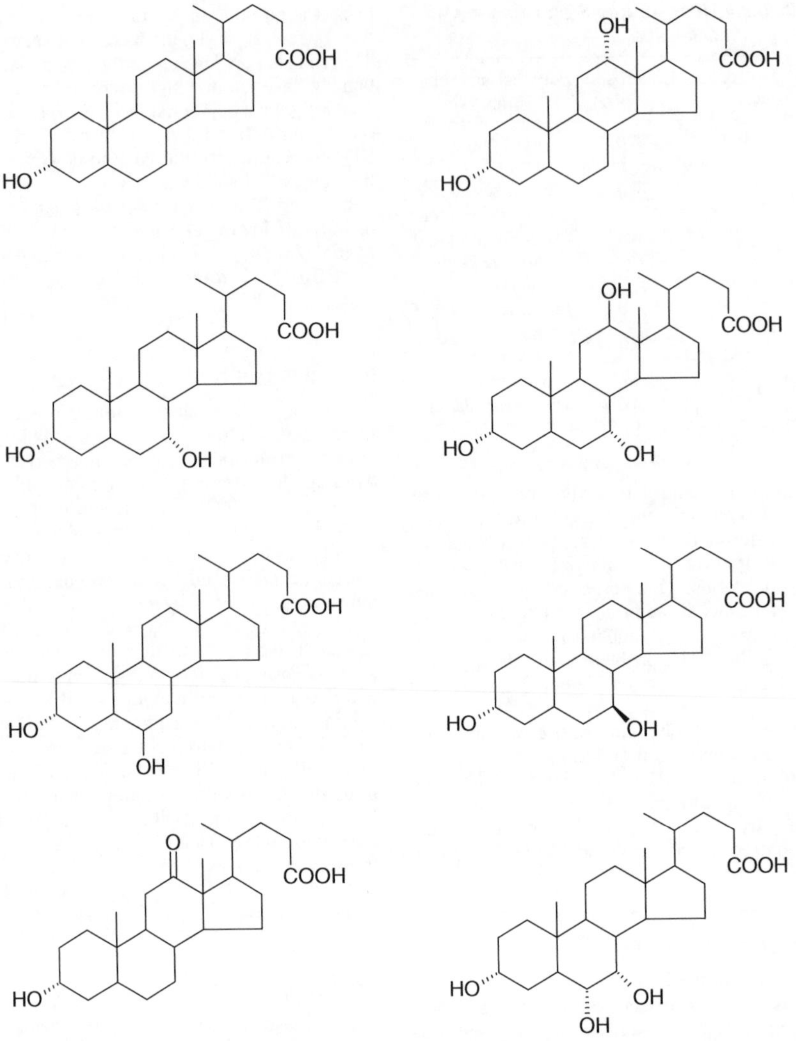

Figure 3.8.1 *Bile acid structures. I = Lithocholic acid (3α-hydroxy-5β-cholanoic acid), II = deoxycholic acid (3α, 12α-dihydroxy-5β-cholanoic acid), III = chenodeoxycholic acid (3α,7α-dihydroxy-5β-cholanoic acid), IV = cholic acid (3α,7α,12α-trihydroxy-5β-cholanoic acid), V = hyodeoxycholic acid (3α,6α-dihydroxy-5β-cholanoic acid), VI = ursodeoxycholic acid (3α,7β-dihydroxy-5β-cholanoic acid), VII = 3α-hydroxy-12-oxo-5β-cholanoic acid and VIII = 3α,6α,7α-trihydroxy-5β-cholanoic acid.*

carbons. Together with the saturated components, several unsaturated fatty acids are present depending on the nature of the oil. The fatty acid composition of typical cooking oils can be seen in Table 3.9.1. Analysis of these components can sometimes identify the type or source of urban waste-waters depending on the domestic or restaurant effluent (e.g. Mudge and Lintern, 1999).

The extraction and analysis of these compounds is partially described above in the sterols method. After *in situ* saponification, the fatty acids are present in the methanolic phase, which needs to be titrated back to an acid pH to convert the potassium salts into free acids. This may subsequently be extracted into 20% DCM in hexane by liquid–liquid separation. Once reduced in volume by rotary evaporation and nitrogen blow down, the fatty acids need to be derivatized to form either their methyl esters (preferred) or

TMS ethers. Most MS libraries have more entries for the Fatty Acid Methyl Esters (FAMEs) than the TMS derivatives.

3.9.3 Effect of UV treatment

Some analyses (Mudge, *unpub.*) have shown that the quantity of free fatty acids in sewage samples increases after the UV treatment stage in an STP. This may lead to a more rapid degradation rate in the environment and potentially increase the BOD_5 at the expense of the COD.

3.9.4 Fluorescent Whitening Agents

In order to improve the optical (brightness) properties of textiles, fabrics are treated with a group of chemicals which absorb in the UV and emit in the blue end of the visible

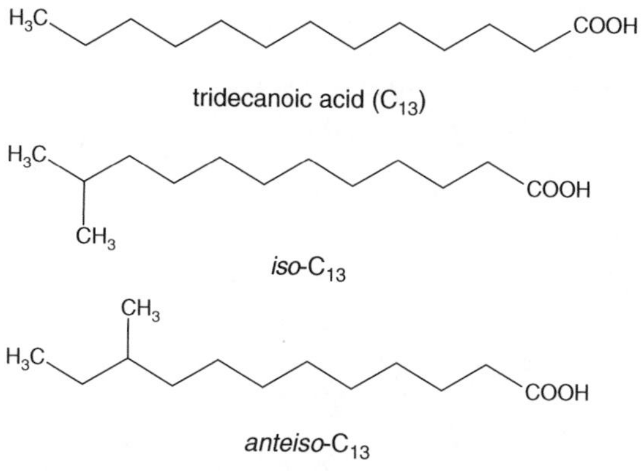

tridecanoic acid (C$_{13}$)

iso-C$_{13}$

anteiso-C$_{13}$

Figure 3.9.1 *Branched chain fatty acids indicative of bacteria.*

spectrum (Gilpin *et al.*, 2003). After repeated washing, these chemicals gradually wash out of the fabric but they can be replaced by compounds added to washing powders. The major compounds are shown in Figure 3.9.2. As well as being present in washing powders, they are used in improving the optical properties of papers.

The quantities in use are (Poiger *et al.*, 1999)

- Whitening paper – FWA8 (7, 600 t · a^{-1}, tonnes per year), FWA5 (100 t · a^{-1})
- Textiles FWA8 (1200 t · a^{-1})
- Detergents FWA1 (3000 t · a^{-1}) & FWA5 (500 t · a^{-1})

Washing powders contain between 0.03 and 0.3% of FWAs, notably FWA1 and FWA5. Between 5 and 80% of these compounds are discharged to the sewage treatment system. Therefore, these compounds will be present in the influent of domestic wastewater systems. Table 3.9.2 indicates the concentrations of these compounds as they pass through a typical STP.

Several studies have used FWAs as indicators of fecal contamination (Boving *et al.*, 2004, Gilpin *et al.*, 2002, Hayashi *et al.*, 2002) and with other measures were successful in differentiating sources.

3.9.4.1 Extraction Methodologies

Due to the relatively polar nature of these compounds, they are most likely to be found in the liquid phase and the most appropriate extraction will be using solid phase extraction (SPE) with C$_{18}$ as the stationary phase in either columns or on disks (e.g. Poiger *et al.*, 1996). Filtered water samples (~100 ml) should be passed through conditioned SPE columns and eluted with tetrabutylammonium hydrogen sulfate in methanol and finally water: dimethylformamide (1:1) mix. The eluted FWA agents can be quantified by adding a standard (an FWA not present in the system naturally) and determined by HPLC with fluorescence detection λ_{ex} 350 nm and λ_{em} 430 nm.

3.10 CAFFEINE AND TRICLOSAN

Some studies have been performed where caffeine (Figure 3.10.1) has been used as a marker for human derived sewage and 5β-coprostanol as an indicator of human and other herbivorous animal sources. This pre-supposes that cattle and sheep do not drink coffee or caffeinated

Table 3.9.1 *The Fatty Acid Profiles Typical of Cooking Fats and oils. Totals may not Reach 100% due to Other Compounds Present in the Oils (e.g., Cod Liver Oil has Considerable Amounts of 20:5ω3 and 22:6ω3 Fatty Acids)*

Oil or Fat	Unsat./Sat. ratio	Saturated				Mono unsaturated		Poly unsaturated	
		Capric Acid 10:0	Lauric Acid 12:0	Myristic Acid 14:0	Palmitic Acid 16:0	Stearic Acid 18:0	Oleic Acid 18:1ω9	Linoleic Acid (ω6) 18:2ω6	Linolenic Acid 18:3ω3
Beef Tallow	0.9	–	–	3	24	19	43	3	1
Butterfat (cow)	0.5	3	3	11	27	12	29	2	1
Butterfat (human)	1.0	2	5	8	25	8	35	9	1
Canola (rapeseed) oil	15.7	–	–	–	4	2	62	22	10
Cocoa Butter	0.6	–	–	–	25	38	32	3	–
Cod liver oil	2.9	–	–	8	17	–	22	5	–
Coconut oil	0.1	6	47	18	9	3	6	2	–
Corn oil	6.7	–	–	–	11	2	28	58	1
Cottonseed oil	2.8	–	–	1	22	3	19	54	1
Flaxseed oil	9.0	–	–	–	3	7	21	16	53
Grape seed oil	7.3	–	–	–	8	4	15	73	–
Lard (Pork fat)	1.2	–	–	2	26	14	44	10	–
Olive oil	4.6	–	–	–	13	3	71	10	1
Palm oil	1.0	–	–	1	45	4	40	10	–
Palm Kernel oil	0.2	4	48	16	8	3	15	2	–
Peanut oil	4.0	–	–	–	11	2	48	32	–
Safflower oil	10.1	–	–	–	7	2	13	78	–
Sesame oil	6.6	–	–	–	9	4	41	45	–
Soybean oil	5.7	–	–	–	11	4	24	54	7
Sunflower oil	7.3	–	–	–	7	5	19	68	1
Walnut oil	5.3	–	–	–	11	5	28	51	5

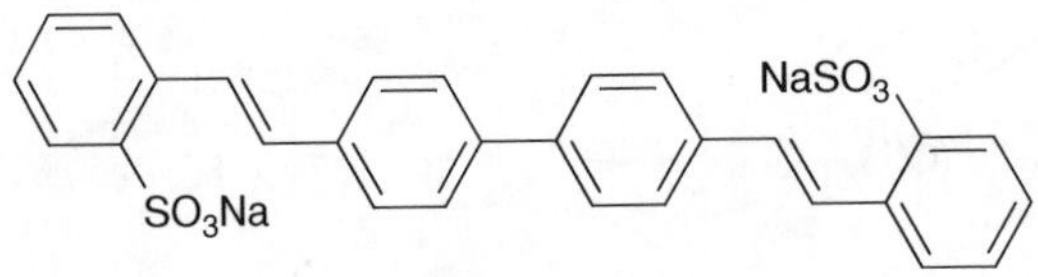

Figure 3.9.2 Structure of key fluorescent whitening agents (FWAs).

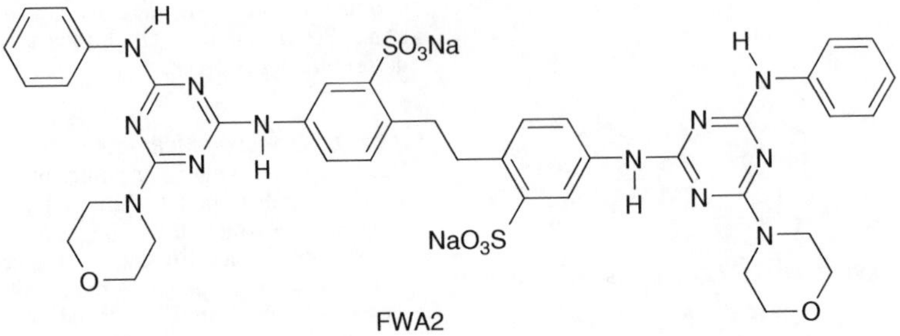

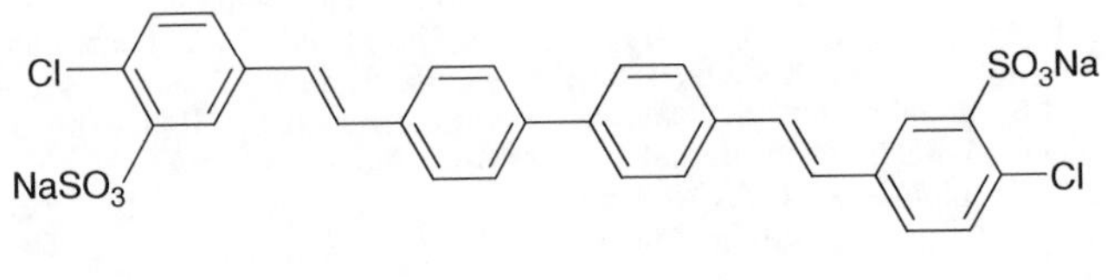

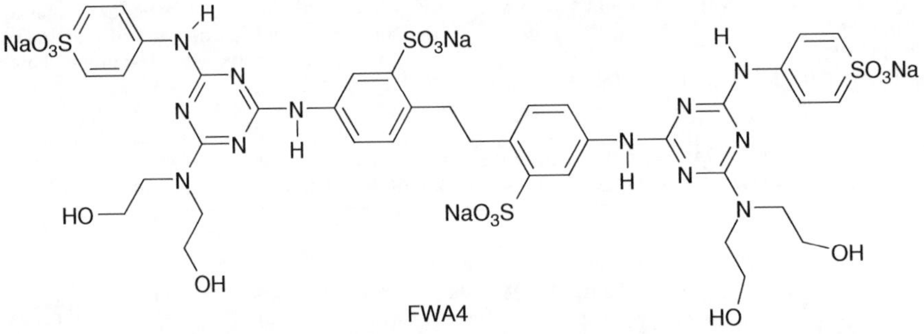

Table 3.9.2 The Concentration of Selected Fluorescent Whiteners in Domestic Wastewaters (after Poiger, 1998)

	FWA1 ($\mu g \cdot l^{-1}$)	FWA5 ($\mu g \cdot l^{-1}$)	FWA8 ($\mu g \cdot l^{-1}$)
Raw sewage	10	14	0.5
Primary effluent	6.9	10.6	0.018
Secondary effluent	2.4	6.4	0.024

drinks whereas humans do. This supposition may be valid; however, the solubility of caffeine is significantly greater than that of the sterols (1 g dissolves in 46 ml of water, Merck Index). In natural systems, caffeine will be transported in the dissolved phase while the sterols and other non-polar compounds will partition onto the solid phase. These solids will settle out at different places in the environment, making any use of ratios between the compounds very difficult to interpret.

Triclosan (Figure 3.10.1) is an antibacterial agent added to such preparations as toothpaste. It is a resistant chemical and can be found in both raw sewage and, due to its lower solubility in water, in the sludge. It may be possible to track domestic discharges by using this chemical together with other known markers.

Weigel et al., 2004a analyzed caffeine and a number of other synthetic chemicals in sewage from Norway. The caffeine concentration ranged between 0.06 and 293 $\mu g \cdot l^{-1}$ in a

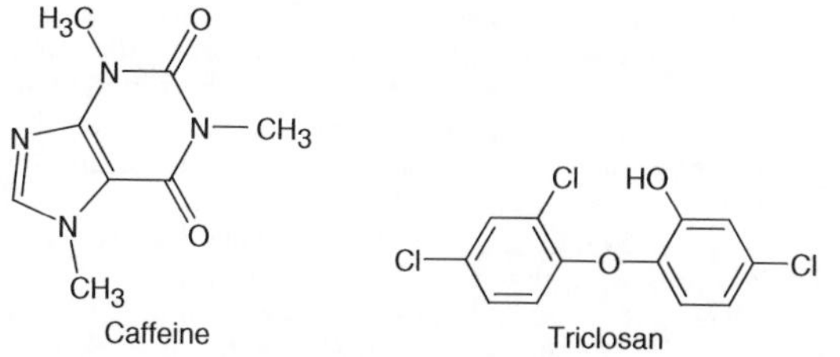

Figure 3.10.1 Chemical structure of caffeine and triclosan

range of sewage samples; the mean was $89\,\mu g \cdot l^{-1}$. For Triclosan, concentrations ranged between 0.16 and $2.38\,\mu g \cdot l^{-1}$, with a mean of $0.73\,\mu g \cdot l^{-1}$. In seawater, the concentrations of the relatively water soluble caffeine were in the nanogram range ($7–87\,ng \cdot l^{-1}$) with a discernable gradient away from the potential sources.

3.10.1 Extraction Methodologies

Due to the more polar nature of caffeine, extraction is usually conducted from the liquid phase. Solid phase extraction tubes are commonly used for this type of work. Any liquid is first filtered through either a 0.22 or $0.45\,\mu m$ membrane filter to remove particulates, which may block the columns. Experiments by Weigel et al. (2004b) found the greatest recoveries of spikes samples were obtained with styrene–methacrylate and styrene–N-vinyl pyrrolidone co-polymers. In their experiments, the column contents were first washed with a sequence of solvents from non-polar through to polar to remove any impurities (5 ml n-hexane, 5 ml ethyl acetate, 10 ml methanol and 10 ml water). The collected samples were then passed through the column at a medium flow rate ($10–20\,ml \cdot min^{-1}$), rinsed with distilled/deionised water, and eluted in 30 ml of methanol.

Analysis is traditionally by means of reverse phase HPLC (RP-18, ODS-2 column or alike) and a gradient of methanol/water (10 mM ammonium acetate, 0.1% triethylamine, acetic acid to pH 5). Compounds can be detected at 230 nm in a UV detector.

3.11 LAS AND OTHER SURFACTANTS

Surfactants and detergents are very widely used both in the home and in industry; they are used as household detergents, in personal care products such as cosmetics and pharmaceuticals and in different industrial formulations used in textile and fibre processing; mining, flotation and petroleum production; paint, plastics and lacquers production; food industry; pulp and paper industry; leather and fur industry; and agriculture (Gonzalez et al., 2004). They pass into the STP with all the other materials in sewage, where they may be degraded. The resistant compounds and/or their degradation products may then be released with the effluent or be concentrated into the sludge.

The most often studied compounds are the linear alkylbenzene sulfonate (LAS) and nonyl phenol polyethoxylate (Figure 3.11.1). Linear Alkylbenzene Sulphonates comprise 50% of the anionic detergents in use and recent estimates suggest the production of 4×10^6 tonnes annually (Fernandez et al., 2004).

As befitting detergents, they have both hydrophilic and hydrophobic properties and are typically extracted by SPE (C_{18}) cartridges. These should be pre-conditioned by washing with methanol and then water. Careful control of the pH may be required depending on the nature of the individual detergents and experiments should be conducted to

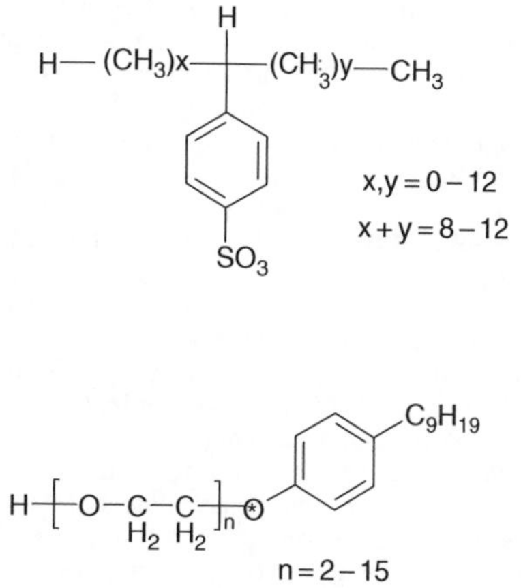

Figure 3.11.1 Linear alkylbenzene sulfonate (LAS) and nonyl phenol polyethoxylate (NPEO). For Nonyl Phenol, $n = 0$.

determine the most appropriate values for the compounds of interest. Elution is usually by methanol and analysis by HPLC using a reverse phase column (ODS-2 or alike) in a methanol–water gradient. LC-MS methods can also be applied with the 183 m/z ion diagnostic of LAS (Fernandez et al., 2004).

3.12 BIOLOGICAL ORGANISMS

Various micro-organisms present in fecal discharges offer the possibility of distinguishing between sources (Sinton et al., 1998; Scott et al., 2002). Investigations into the use of microbes for fecal source identification have involved three approaches (Sinton et al., 1998):

1. speciation, based on findings that particular species may be indicative of human or animal sources,
2. phenotyping,
3. genotyping, which is considered by some authors to be more reliable than phenotypic biochemical reactions.

The source-specific microbial indicators include assays for Bacteroides fragilis and phages of Bacteroides, F-specific RNA coliphages, species of Bifidobacterium, Rhodococcus coprophilus and human enteric viruses. Phenotypic source-determination methods include multiple antibiotic resistance (MAR) profiles of both E. coli and Fecal Streptococci (FS), and O-serotyping. Genotypic methods have included ribotyping (RT), pulsed-field gel electrophoresis (PFGE), repetitive-PCR profiles, and finally detection of host-specific molecular markers (e.g., Bacteroides-Prevotella PCR marker) (Sinton et al., 1998; Gilpin et al., 2002; Scott et al., 2002; Fogarty et al., 2003). Bacterial indicators as well as some of the phenotypic and genotypic methods used in fecal source identification are discussed below.

3.12.1 Escherichia coli and Fecal Streptococci

Escherichia coli is consistently and exclusively found in the feces of warm-blooded animals (Cabelli et al., 1982), where it may attain concentrations of 10^9 per g. It is found in sewage, treated effluents, all natural waters and soils that are subject to recent fecal contamination, whether from humans, agriculture, or wild animals, and birds. However, even in the remotest areas, fecal contamination by wild animals, including birds, can never be excluded. Recently, new methods based on molecular biology have been developed that directly detect and enumerate *E. coli* without requiring an additional confirmation test (George *et al.*, 2001). An alternative method is serotyping of *E. coli*: this is an immunological technique based on the presence of different somatic (O) antigenic determinants and it has been used by several investigators to differentiate *E. coli* from various sources (Scott *et al.*, 2002). Different serotypes of *E. coli* are characteristic of different animal sources, but there are also many serotypes that are shared between humans and animals (Scott *et al.*, 2002). However, serotyping may be useful in differentiating between *E. coli* from human and animal sources. Parveen *et al.* (2001) suggested that serotyping can be used in combination with other techniques, such as RT, in order to allow the testing of a more limited number of serotypes. Since one of the disadvantages of serotyping is the necessity for a large bank of antisera (Scott *et al.*, 2002), the possibility of testing only certain serotypes makes this a potential method for source-discrimination of fecal pollution in selected cases only.

The term "fecal streptococci" is a functional definition used to describe streptococci associated with the gastrointestinal tracts of man and animals (Godfree *et al.*, 1997). The taxonomy of streptococci has undergone major changes and has been divided to give birth to three new genera, *Lactococcus, Enterococcus*, and *Streptococcus*. The intestinal or fecal species mainly belong to either the genera *Enterococcus* or *Streptococcus* and all possess the Lancefiels group D antigen (Godfree *et al.*, 1997).

Fecal streptococci, in conjunction with total and fecal coliforms (FC), are generally regarded as the most useful fecal indicators. The usefulness of FS as microbiological indicators is generally accepted because of their inability to multiply in sewage effluents (Volterra *et al.*, 1986). In sewage, the numbers of FS are usually 10–100 times lower than those of FC and the presence of FS in a sample of treated water in the absence of coliforms would not be expected. In contrast, FS are present in greater numbers in animal feces than coliforms (up to 10^7 per g in sheep feces and 10^6 in cow feces, Sinton *et al.*, 1998). Human sewage are characterized by a predominance of enterococci (e.g., *E. faecalis* and *E. faecium*). *Streptococcus bovis* and *S. equinus* are the predominant streptococci in the feces of cows and horses respectively (Khalaf and Muhammad, 1989). However, it is not possible to differentiate the source of fecal contamination based on speciation of FS (Sinton *et al.*, 1998). Obiri-Danso and Jones (1999) concluded from their study that, at least at the phenotypic level, FS are not discriminating enough to distinguish different species as sources of pollution.

Another approach to the use of FS as fecal source indicators has been the fecal coliform: fecal streptococci (FC/FS) ratio. The rationale behind the use of this ratio was the observation that human feces contain higher numbers of FC, while animal feces contain higher levels of FS (Scott *et al.*, 2002). According to Geldreich and Kenner (1969), a ratio greater than 4 suggests a sewage effluent/human source of contamination while a ratio lower than 0.7, an animal one. However, the FC/FS ratio is dependent on the differential die-away rate of *E. coli* and FS and also on the different survival rates of the species within the FS group. According to Geldreich and Kenner (1969), the ratio would be valid only in the 24 h after the discharge of bacteria into the water. Even in the first hours after the collection of fresh feces, the FC/FS ratio is not constant for samples of the same origin (Pourcher *et al.*, 1991). For the above reasons, the ratio is no longer recommended as a means of differentiating between sources of fecal contamination.

Multiple antibiotic resistance (MAR) analysis belongs to the phenotypic group of approaches for fecal source identification. This method has been most widely used to differentiate bacteria (usually *E. coli* or FS) from different sources using antibiotics that are commonly associated with human and animal therapy, as well as animal feed (Scott *et al.*, 2002). This method is based on the underlying principle that the bacterial flora present in the gut of humans and various types of animals are subjected to different types, concentrations, and frequencies of antibiotics (Scott *et al.*, 2002). The technique is easy and rapid and involves the isolation and cultivation of a target organism and then the isolates are replica plated on media containing different antibiotics at various concentrations. The plates are then incubated and the organisms are scored according to their susceptibilities to various antibiotics. An antibiotic resistance profile can then be generated and the resulting fingerprints are characterized, analyzed by discriminate or cluster analysis, and finally compared to a reference database in order to identify an isolate as being either human or animal derived (Scott *et al.*, 2002).

The MAR technique has been used successfully in differentiating *E. coli* or fecal streptococci isolated from specific animal species including livestock, wildlife, and humans (Harwood *et al.*, 2000). Hagedorn *et al.* (1999) used MAR for fecal streptococci and identified cattle as the predominant source of fecal pollution in the Page Brook watershed in rural Virginia. In addition, previous studies by Wiggins (1996) showed that patterns of antibiotic resistance could be used to differentiate between human and animal sources of fecal pollution in natural waters. The disadvantages of MAR include the fact that antibiotic resistance is often carried on plasmids that can be lost from cells via cultivation and storage or by changes in environmental conditions. Secondly, strains from different locations may show variations in specific sensitivities due to variable antibiotic use between humans and livestock species (Scott *et al.*, 2002; Fogarty *et al.*, 2003). For these reasons, large reference databases are required that must contain antibiotic resistance profiles from multiple organisms from a large geographical area.

3.12.2 Bacteroides and Bifidobacteria

Detection of host-specific markers in raw water samples seems promising as an effective method for characterizing a microbial population without first culturing the organisms of interest (Scott *et al.*, 2002). Bernhard and Field (2000) used length heterogeneity polymerase chain reaction (PCR) and terminal restriction fragment length polymorphism to characterize members of the *Bacteroides-Prevotella* group as an attempt to discriminate human from ruminant fecal pollution. This approach is rapid and does not require a culturing step. In addition, the use of *Bacteroides* spp. is desirable, as anaerobic bacteria are unlikely to reproduce in the environment. However, not much is known about the survival and persistence of *Bacteroides* spp. in the environment and this issue raises questions concerning its utility as an indicator organism (Scott *et al.*, 2002). Gilpin *et al.* (2003) used *Bacteroides-Prevotella* markers specific to humans in an attempt to evaluate a range of alternative indicators of fecal pollution in river water samples. They detected those markers in only two of the seven samples tested, suggesting that these markers deserve further investigation.

Bifidobacteria are among the most numerous inhabitants of human and animal guts (Gilpin *et al.*, 2002). The advantage of using bifidobacteria in tracing sources is that their strict growth requirements make them unlikely to grow in the environment (Scott *et al.*, 2002). Bifidobacteria are also source specific; certain species (e.g., *B. adolescentis*) are only found in humans, while others such as *B. thermophilum* are specific to animals (Gilpin *et al.*, 2002). Two simple tests show promise as providing some human versus animal source differentiation: sorbitol fermentation and growth at 45°C in trypticase phytone yeast broth (TPYB) (Sinton et al., 1998). Sorbitol-fermenting bifidobacteria strains are only present in human feces and in water contaminated by human feces (Sinton et al., 1998; Scott et al., 2002). Recently, Lynch et al. (2002) enumerated B. adolescentis in effluent samples via growth on selective Bifidobacter media and identified B. adolescentis using an oligonucleotide probe specific for this species. Gilpin et al. (2003) applied the above molecular technique and successfully detected and enumerated B. adolescentis in river water samples collected downstream from human fecal inputs. Bifidobacteria, therefore, have potential as fecal indicators; however, their survival is highly variable suggesting that they can only be used as indicators of recent fecal contamination (Scott et al., 2002).

3.12.3 Rhodococcus coprophilus and Clostridium perfringens

Rhodococcus coprophilus is a bacterium which forms a fungus-like mycelium which breaks up into bacteria-like elements that contaminate the grass eaten by herbivores. They are able to survive passage through the digestive system and, therefore, contaminate fresh dung (Jagals et al., 1995). The organism has never been found in human feces. Instead, it is a natural inhabitant of the feces of herbivores such as cows, donkeys, goats, horses and sheep (Sinton et al., 1998; Gilpin et al., 2002). Jagals et al. (1995) detected high numbers of R. coprophilus in river and stream samples exposed to fecal pollution of domestic animal origin (upstream of human settlement) but counts dropped in the downstream samples evidently due to the dilution by effluents from human settlement which contained low numbers of these organisms. Those observations confirmed the findings of Oragui and Mara (1983), according to which R. coprophilus bacteria are specific tracers which may be used for detecting fecal pollution of animal origin (Jagals et al., 1995). However, the major disadvantage of using this micro-organism as a fecal indicator is that it grows slowly so that conventional detection methods require up to 21 days in order to obtain a result (Gilpin et al., 2002). To overcome this, Savill et al. (2001) developed PCR-based methods for the detection and enumeration of R. coprophilus. Successful amplification was achieved using DNA extracted from cow, sheep, horse, and deer feces but it was negative for fecal samples from humans, pigs, possums, ducks, dogs, and rabbits (Savill et al., 2001).

Clostridium perfringens has been suggested as an alternative or accompaniment to the traditional water quality indicators. This alternative indicator has been used in several studies to trace sewage wastes in the marine environment. Although there has been some controversy regarding the use of C. perfringens as a water quality indicator, because of its persistence in the environment and its greater resistance to toxic pollution than E. coli, a number of scientists continue to recommend its use, especially in situations where the prediction of the presence of viruses is desirable or when remote or old pollution is being examined (Scott et al., 2002). Finally, Leeming (1996) suggested the use of C. perfringens for distinguishing fecal pollution from birds and domestic animals. This was based on findings that dog and cat feces contained roughly equal and comparatively high numbers ($10^6 \cdot 10^8$ cfu $\cdot$ g^{-1}, colony forming units per gram) of both FC and C. perfringens spores, while the feces of native birds contained $10^6 \cdot 10^8$ cfu $\cdot$ g^{-1} of FC but generally less than 10^2 cfu $\cdot$ g^{-1} of C. perfringens spores.

3.12.4 Bacteriophage and human viruses

Coliphages are viruses that infect E. coli and have been proposed as potential tracers of sewage pollution in aquatic environments because they are ubiquitous in sewage and easier to enumerate than enteric viruses (Ferguson et al., 1996). There are two main groups of coliphages: somatic coliphages and male-specific (F-specific, or F$^+$) coliphages (Embrey, 2001; Scott et al., 2002). F$^+$ RNA coliphages may be classified into four main subgroups: group I, group II, group III, and group IV (Scott et al., 2002; Sinton et al., 2002). Furuse (1987), using serotyping, concluded that

- group I coliphages are isolated from animals,
- groups II and III tend to be isolated from human faeces, and
- group IV phages are of mixed origin (sewage, human and animals).

Havelaar et al. (1990) confirmed the above findings, but demonstrated that phages from groups II and III were found in sewage rather than feces (Sinton et al., 1998; Embrey, 2001). F$^+$ RNA coliphages can be enumerated by a variety of methods and identification of the phage subgroups can be achieved by either serotyping or genotyping (Sinton et al., 1998; Scott et al., 2002).

Serotyping of phages is expensive and time consuming and has been shown to produce ambiguous results (Sinton et al., 1998; Scott et al., 2002). For this reason, genotyping of F$^+$ RNA phages has been utilised using a nucleic acid hybridization approach. This technique has been shown to be successful in identifying the four subgroups of F$^+$ RNA phages and subsequently for use in differentiating between sources of fecal pollution. There are currently some limitations associated with the use of F$^+$ RNA coliphages as tracers; this is due to variable survival and difficulty with the assay, making it difficult to use routinely (Gilpin et al., 2002).

Bacteroides fragilis is one of the most numerous bacteria found in the human gut (Sinton et al., 1998). Because B. fragilis has a short survival in the environment and bacteriophages tend to be more persistent in aquatic environments than their hosts, phages that specifically infected this strain were recommended as human source indicators instead of B. fragilis itself (Sinton et al., 1998). The use of B. fragilis bacteriophage as a fecal indicator has the advantage of being highly specific for human fecal pollution (Sinton et al., 1998; Scott et al., 2002). In addition, these phages do not replicate in the environment, and their presence in the environment correlates well with the presence of human enteric viruses (Scott et al., 2002). However, the absence or low levels of B. fragilis phage in highly polluted waters and human sewage in some parts of the world has created some doubts as far as the validity of this approach in environmental samples is concerned. This problem, together with the inherent difficulty in performing the assay, limits the usefulness of this method in identifying human fecal pollution in the environment (Gilpin et al., 2002; Scott et al., 2002).

Over 100 different enteric viruses are specifically associated with the human gastrointestinal tract and these include enteroviruses, adenoviruses, and round structured viruses (Sinton et al., 1998; Scott et al., 2002). They are, therefore, potentially useful as tracers of human sewage. Direct monitoring for human pathogens, such as enteric viruses, circumvents the need to assay for fecal indicator organisms. However, many of these viruses are not easily cultivated in environmental samples (Scott et al., 2002) and are intermittently present in human effluents (with the possible exception of enteroviruses) (Sinton et al., 1998). In addition, the methods involved in their detection and enumeration tend to be time consuming and expensive (Sinton et al.,

1998). More recently, new molecular methods based around the use of the PCR have been applied for the detection of adenoviruses in environmental samples and as a result, routine monitoring for adenoviruses as an index of human fecal contamination has been suggested. Molecular methods (reverse transcription-PCR) have the ability to detect non-cultivable viruses that cannot be detected by simple cultivation methods; however, non-viable viruses are also detected by this procedure, and this provides no information about the associated human health risk. In addition, the low numbers of enteric viruses in the environment limit their use as fecal indicators.

3.13   OTHER BACTERIAL METHODS

It is possible to add components that are not normally present in sewage and track the occurrence of these materials in the environment (e.g., *Bacillus globigii* spores, Hodgson *et al.*, 2003, Hodgson *et al.*, 2004)

3.14   BIOLOGICAL MACROFAUNAL COMMUNITIES

It has long been known that different species react in different ways to anthropogenic stressors, including sewage. However, chemical analyses have traditionally been used in preference to biological monitoring after accidents or spills (Hewitt and Mudge, 2004). It is possible that evidence of sewage spills or discharges may have been lost or diluted to such an extent that chemical analyses do not identify effects. The faunal community in an affected area may respond to the discharge and record such events.

Recent work (Glemarec and Hily, 1981, Grall and Glemarec, 1997, Borja *et al.*, 2000, Borja *et al.*, 2003) has developed an index of biological community along a pollution gradient. This process involves identifying the species present in the community, ascribing them a value from 1 to 5 according to a table of European species and their response to contaminants in the environment. The groups are (from Borja *et al.*, 2000)

- *Group I.* Species very sensitive to organic enrichment and present under unpolluted conditions (initial state). They include the specialist carnivores and some deposit-feeding tubicolous polychaetes.
- *Group II.* Species indifferent to enrichment, always present in low densities with non-significant variations with time (from initial state to slight unbalance). These include suspension feeders, less selective carnivores, and scavengers.
- *Group III.* Species tolerant to excess organic matter enrichment. These species may occur under normal conditions, but their populations are stimulated by organic enrichment (slight unbalance situations). They are surface deposit-feeding species, such as tubicolous spionids.
- *Group IV.* Second-order opportunistic species (slight to pronounced unbalanced situations). Mainly small sized polychaetes: subsurface deposit-feeders, such as *cirratulids.*
- *Group V.* First-order opportunistic species (pronounced unbalanced situations). These are deposit-feeders, which proliferate in reduced sediments.

The percentage of the total community in each group is calculated and a final index can be devised from the following equation:

Biotic Coefficient =

$$\frac{(0 \times \%G_I) + (1.5 \times \%G_{II}) + (3 \times \%G_{III}) + (4.5 \times \%G_{IV}) + (6 \times \%G_V)}{100}$$

Higher values of this coefficient indicate the environment is under pressure, usually a pollution gradient. Borja *et al.*, 2003, have extended the original work to more Northern European locations and the index appears to work well for these locations as well. However, the community will respond to a range of contaminants and other physico-chemical factors and from these data alone it may not be possible to identify sewage-affected sites specifically.

A newer approach using multivariate statistics including signature analysis has been able to improve the specificity of contaminant identification. For example, Hewitt and Mudge (2004) identified meiofauna in lagoon sediments and developed a signature based on the community structure adjacent to known sewage discharges. Fitting of this signature to the remaining environmental data highlighted several sites that had a "sewage component". Further investigation of these sites identified contaminating surface water drains. Work with macrofauna (Hopkins and Mudge, 2004) in the same gradients was unable to show the same degree of sensitivity possibly due to mechanical disturbance of the site. More recent work (Chenery and Mudge, in press) using the biotic index in the same region was able to highlight the known sewage discharge points using the macrofauna.

3.15   TRANSFORMATIONS IN THE ENVIRONMENT
3.15.1   Chemical degradation processes

The inorganic components present in sewage, especially the nitrogen and phosphorus nutrients, may be readily assimilated into new biological matter. This may lead to problems such as eutrophication (e.g., Newton *et al.*, 2003) and is outside the scope of this chapter. The organic components suggested as tracers above may undergo diagenetic changes depending on the nature of the receiving waters. Many compounds are readily utilised as a food resource and removed from the system as biomass or ultimately CO_2. This would not be so much of an issue if all compounds behaved the same, but evidence suggests they do not (e.g., Hudson *et al.*, 2001; Jeng and Huh, 2001; Sun and Wakeham, 1994).

In general, smaller compounds will degrade more rapidly than larger ones. A good example are fatty acids–short chain ones are essentially produced in the marine environment while long chain ones are derived from terrestrial plants (Mudge *et al.*, 1998). When apportioning sources in core samples, the lack of short chain fatty acids could be interpreted as a lack of marine input when it may be due to diagenesis. In these cases, resistant markers such as the sterols might prove more reliable although they also do degrade with time, especially in oxic environments (Jeng and Huh, 2001).

3.15.2   Bacterial die-off

A consistent problem with the use of bacterial enumeration techniques in tracing is the fact that different bacteria die off at different rates in environmental waters. For example, it is thought that fecal streptococci are more resistant to environmental stress than coliforms and *E. coli* and therefore survive longer (Table 3.15.1).

However, the persistence of both bacterial indicators following the discharge of feces in the water column is not clear. Some investigators have found that FC die off more rapidly than FS while others have noticed the reverse. Recent studies have indicated that *E. coli* is more sensitive than fecal streptococci in natural water (Pourcher *et al.*, 1991). In addition, differences in the survival rates of the different species within the fecal streptococci group of organisms may also change the indicator density relationships (i.e., the relationship with fecal coliforms). *Streptococcus bovis* and *Strep. equinus*, for example, die off more rapidly in

Table 3.15.1 *Survival Rates of Microbes when Discharged into Coastal Waters (after Rees, 1993)*

Type	Cool seawater	Settled sediment
Coliforms	1–a few hours	A few days
E. coli	Hours–1 day	Days–weeks
Fecal streptococci	1–a few days	Weeks
Human pathogenic viruses	A few days	Weeks–months

the aquatic environment than *Strep. faecalis* and *Strep. faecium*. This differential decay is exaggerated when wastewater is disinfected prior to discharge in the water column (American Public Health Association, 1995; Godfree *et al.*, 1997). This is not a problem only with these two organisms; Jagals *et al.* (1995) has suggested that bifidobacteria are less resistant to conditions in a river environment than fecal coliforms since their numbers are observed to decrease more rapidly along the river than those of fecal coliforms. Further, bacteriophages are also prone to die away; F⁺ RNA phages have been found to be unreliable for use in marine and tropical waters due to variable survival rates (Scott *et al.*, 2002).

3.16 SUMMARY

This chapter illustrates the diverse chemical and biological components that make up sewage. Any program to trace the origin or track known sources in the environment should incorporate the nature of the environment such as sedimentation régime, redox potential, natural insolation, etc. This also needs to be complemented by an understanding of the catchment and the likely materials that will be present. For example, fluorescent whitening agents (FWAs) may be of great use in one area where domestic uses of detergents are prevalent but useless in a region where industrial discharges dominate.

Care must also be exercised in regions where the nature of the environment leads to rapid changes in the viability of biological components (e.g., high ultra violet insolation) as false negatives may be found. In a chemical sense, harsh redox conditions may also alter the signature of traditional sewage discharges (Seguel *et al.*, 2001). In these cases, the heterogenous nature of the sewage may become beneficial, with approaches using combined chemical and/or biological investigation being most likely to succeed.

The cost of analyses varies considerably depending on the analytes being measured and it may be most economically sound to have a tiered approach to sewage analysis. The analytes to be investigated should depend on the question being asked. The biological methods have advantages in being relatively cheap and related to the legislative structure. However, the presence or absence of selected bacterial strains does not always give sufficient information to allow the source to be identified. Chemical analyses, as well as giving direct measurements of fecal materials, can provide supplementary information on other biomarkers. For example, the presence of sterols such as β-sitosterol in samples may lead investigators to look for terrestrial sources (agricultural runoff) rather than piped marine discharges.

REFERENCES

American Public Health Association (APHA), *Standard Methods for the Examination of Water and Wastewater*, 1995, American Public Health Association, Washington D.C.

Arundel, J. *Sewage and Industrial Effluent Treatment*, Blackwell Science Ltd, 1995, P.242.

Barreca, J. and K. Seiders, Skokomish River Basin Fecal Coliform Total Maximum Daily Load (Water Cleanup Plan), Washington State Department of Ecology, 2001, 20.

Bernhard, A. E. and K. G. Field, A PCR assay to discriminate human and ruminant feces on the basis of host differences in Bacteroides-Prevotella genes encoding 16S rRNA, *Applied and Environmental Microbiology*, 2000, 66, 4571–4574.

Borja, A., I. Muxika and J. Franco, The application of a Marine Biotic Index to different impact sources affecting soft-bottom benthic communities along European coasts, *Marine Pollution Bulletin*, 2003, 46, 835–845.

Borja, A., J. Franco and V. Perez, A marine Biotic Index to establish the ecological quality of soft-bottom benthos within European estuarine and coastal environments, *Marine Pollution Bulletin*, 2000, 40, 1100–1114.

Boving, T. B., D. L. Meritt and J. C. Boothroyd, Fingerprinting sources of bacterial input into small residential watersheds: Fate of fluorescent whitening agents, *Environmental Geology*, 2004, 46, 228–232.

Bull, I. D., M. J. Lockheart, M. M. Elhmmali, D. J. Roberts and R. P. Evershed, The origin of faeces by means of biomarker detection, *Environment International*, 2002, 27, 647–654.

Cabelli, V. J., A. P. Dufour, M. A. Levin and L. J. McCage., Swimming associated gastroenteritis and water quality, *American Journal of Epidemiology*, 1982, 115, 606–616.

Chaler, R., B. R. T. Simoneit and J. O. Grimalt, Bile acids and sterols in urban sewage treatment plants, *Journal of Chromatography A*, 2001, 927, 155–160.

Chenery, A. M. and Mudge, S. M. Detecting anthropogenic stress in an ecosystem: 3. mesoscale variability and biotic indices. *Environmental Forensics* (in press).

Elhmmali, M. M., D. J. Roberts and R. P. Evershed, Bile acids as a new class of sewage pollution indicator, *Environmental Science & Technology*, 1997, 31, 3663–3668.

Embrey, S. S., Microbiological quality of Puget Sound Basin streams and identification of contaminant sources, *Journal of the American Water Resources Association*, 2001, 37, 407–421.

Ferguson, C. M., B. G. Coote, N. J. Ashbolt and I. M. Stevenson, Relationships between indicators, pathogens and water quality in an estuarine system, *Water Research*, 1996, 30, 2045–2054.

Fernandez, J., J. Riu, E. Garcia-Calvo, A. Rodriguez, A. R. Fernandez-Alba and D. Barcelo, Determination of photodegradation and ozonation by products of linear alkylbenzene sulfonates by liquid chromatography and ion chromatography under controlled laboratory experiments, *Talanta*, 2004, 64, 69–79.

Fogarty, L. R., S. K. Haack, M. J. Wolcott and R. L. Whitman, Abundance and characteristics of the recreational water quality indicator bacteria Escherichia coli and enterococci in gull faeces, *Journal of Applied Microbiology*, 2003, 94, 865–878.

Furuse, K., Distribution of coliphages in the environment: General considerations, In: *Phage Ecology*, S.M. Goyal, C.P. Gerba and G. Bitton, 1987, John Wiley & Sons Inc, New York, pp. 87–124.

Geldreich, E. E. and B. A. Kenner, Concepts of fecal streptococci in stream pollution, *J. Water Pollut. Control Fed.*, 1969, 41, R336–R352.

George, I., M. Petit, C. Theate and P. Servais, Use of rapid enzymatic assays to study the distribution of faecal coliforms in the Seine river (France), *Water Science and Technology*, 2001, 43, 77–80.

Gilpin, B. J., J. E. Gregor and M. G. Savill, Identification of the source of faecal pollution in contaminated rivers, *Water Science and Technology*, 2002, 46, 9–15.

Gilpin, B., T. James, F. Nourozi, D. Saunders, P. Scholes and M. Savill, The use of chemical and molecular microbial indicators for faecal source identification, *Water Science and Technology*, 2003, 47, 39–43.

Glemarec, M. and C. Hily, Effects of Urban and Industrial Discharges on the Benthic Macrofauna in the Bay of Concarneau, *Acta Oecologica-Oecologia Applicata*, 1981, 2, 139–150.

Godfree, A. F., D. Kay and M. D. Wyer, Faecal streptococci as indicators of faecal contamination in water, *Journal of Applied Microbiology*, 1997, 83, S110–S119.

Gonzalez, S., M. Petrovic and D. Barcelo, Simultaneous extraction and fate of linear alkylbenzene sulfonates, coconut diethanol amides, nonylphenol ethoxylates and their degradation products in wastewater treatment plants, receiving coastal waters and sediments in the Catalonian area (NE Spain), *Journal of Chromatography A*, 2004, 1052, 111–120.

Grall, J. and M. Glemarec, Using biotic indices to estimate macrobenthic community perturbations in the Bay of Brest, *Estuarine Coastal and Shelf Science*, 1997, 44, 43–53.

Green, G., J. H. Skerratt, R. Leeming and P. D. Nichols, Hydrocarbon and Coprostanol Levels in Seawater, Sea-Ice Algae and Sediments near Davis-Station in Eastern Antarctica—a Regional Survey and Preliminary-Results for a Field Fuel Spill Experiment, *Marine Pollution Bulletin*, 1992, 25, 293–302.

Grimalt, J. O. and J. Albaiges, Characterization of the depositional-environments of the Ebro Delta (Western Mediterranean) by the study of sedimentary lipid markers, *Marine Geology*, 1990, 95, 207–224.

Grimalt, J. O., P. Fernandez, J. M. Bayona and J. Albaiges, Assessment of fecal sterols and ketones as indicators of urban sewage inputs to coastal waters, *Environmental Science & Technology*, 1990, 24, 357–363.

Hagedorn, C., S. L. Robinson, J. R. Filtz, S. M. Grubbs, T. A. Angier and R. B. Reneau, Determining sources of fecal pollution in a rural virginia watershed with antibiotic resistance patterns in fecal streptococci, *Applied and Environmental Microbiology*, 1999, 65, 5522–5531.

Harwood, V. J., J. Whitlock and V. Withington, Classification of antibiotic resistance patterns of indicator bacteria by discriminant analysis: Use in predicting the source of fecal contamination in subtropical waters, *Applied and Environmental Microbiology*, 2000, 66, 3698–3704.

Havelaar, A. H., W. M. Pothogeboom, K. Furuse, R. Pot and M. P. Hormann, F-Specific Rna bacteriophages and sensitive host strains in feces and waste-water of human and animal origin, *Journal of Applied Bacteriology*, 1990, 69, 30–37.

Hayashi, Y., S. Managaki and H. Takada, Fluorescent whitening agents in Tokyo Bay and adjacent rivers: Their application as anthropogenic molecular markers in coastal environments, *Environmental Science & Technology*, 2002, 36, 3556–3563.

Hewitt, E. J. and S. M. Mudge, Detecting anthropogenic stress in an ecosystem: 1. Meiofauna in a sewage gradient, *Environmental Forensics*, 2004, 5, 155–170.

Hodgson, C. J., J. Perkins and J. C. Labadz, Evaluation of biotracers to monitor effluent retention time in constructed wetlands, *Letters in Applied Microbiology*, 2003, 36, 362–371.

Hodgson, C. J., J. Perkins and J. C. Labadz, The use of microbial tracers to monitor seasonal variations in effluent retention in a constructed wetland, *Water Research*, 2004, 38, 3833–3844.

Hopkins, F. E. and S. M. Mudge, Detecting anthropogenic stress in an ecosystem: 2. Macrofauna in a sewage gradient, *Environmental Forensics*, 2004, 5, 213–223.

Hudson, E. D., C. C. Parrish and R. J. Helleur, Biogeochemistry of sterols in plankton, settling particles and recent sediments in a cold ocean ecosystem (Trinity Bay, Newfoundland), *Marine Chemistry*, 2001, 76, 253–270.

Jagals, P., W. O. K. Grabow and J. C. Devilliers, Evaluation of indicators for assessment of human and animal fecal pollution of surface run-off, *Water Science and Technology*, 1995, 31, 235–241.

Jeng, W. L. and C. A. Huh, Comparative study of sterols in shelf and slope sediments off northeastern Taiwan, *Applied Geochemistry*, 2001, 16, 95–108.

Khalaf, S. H. and A. M. Muhammad, Studies on fecal streptococci in the River Tigris, *Microbios*, 1989, 57, 99–103.

Leeming, R. *Coprostanol and Related Sterols as Tracers for Faecal Contamination in Australian Aquatic Environments*, 1996 Canberra: Australia, Canberra University.

Leeming, R. and P. D. Nichols, Determination of the sources and distribution of sewage and pulp-fibre-derived pollution in the Derwent Estuary, Tasmania, using sterol biomarkers, *Marine and Freshwater Research*, 1998, 49, 7–17.

Leeming, R., V. Latham, M. Rayner and P. Nichols, Detecting and distinguishing sources of sewage pollution in Australian inland and coastal waters and sediments, in: *Molecular Markers in Environmental Geochemistry*, 1997, 306–319.

Lynch, P. A., B. J. Gilpin, L. W. Sinton and M. G. Savill, The detection of Bifidobacterium adolescentis by colony hybridization as an indicator of human faecal pollution, *Journal of Applied Microbiology*, 2002, 92, 526–533.

McCalley, D. V., M. Cooke and G. Nickless, Effect of sewage-treatment on fecal sterols, *Water Research*, 1981, 15, 1019–1025.

Mudge, S. M. and C. Duce, Identifying the source, transport path and sinks of sewage derived organic matter, *Environmental Pollution*, 2005, 136,2,209–220.

Mudge, S. M. and D. G. Lintern, Comparison of sterol biomarkers for sewage with other measures in Victoria Harbour, BC, Canada, *Estuarine Coastal and Shelf Science*, 1999, 48, 27–38.

Mudge, S. M., J. A. East, M. J. Bebianno and L. A. Barreira, fatty acids in the Ria Formosa lagoon, Portugal, *Organic Geochemistry*, 1998, 29, 963–977.

Mudge, S. M., M. Bebianno, J. A. East and L. A. Barreira, Sterols in the Ria Formosa lagoon, Portugal, *Water Research*, 1999, 33, 1038–1048.

Newton, A., J. D. Icely, M. Falcao, A. Nobre, J. P. Nunes, J. G. Ferreira and C. Vale, Evaluation of eutrophication in the Ria Formosa coastal lagoon, Portugal, *Continental Shelf Research*, 2003, 23, 1945–1961.

Obiri-Danso, K. and K. Jones, The effect of a new sewage treatment plant on faecal indicator numbers, campylobacters and bathing water compliance in Morecambe Bay, *Journal of Applied Microbiology*, 1999, 86, 603–614.

Oragui, J. I. and D. D. Mara, Investigation of the survival characteristics of Rhodococcus-Coprophilus and certain fecal indicator bacteria, *Applied and Environmental Microbiology*, 1983, 46, 356–360.

Parveen, S., N. C. Hodge, R. E. Stall, S. R. Farrah and M. L. Tamplin, Phenotypic and genotypic characterization of human and nonhuman Escherichia coli, *Water Research*, 2001, 35, 379–386.

Poiger, T., F. G. Kari and W. Giger, Fate of fluorescent whitening agents in the River Glatt, *Environmental Science & Technology*, 1999, 33, 533–539.

Poiger, T., J. A. Field, T. M. Field and W. Giger, Occurrence of fluorescent whitening agents in sewage and river water determined by solid-phase extraction and high-performance liquid chromatography, *Environmental Science & Technology*, 1996, 30, 2220–2226.

Poiger, T., J. A. Field, T. M. Field, H. Siegrist and W. Giger, Behavior of fluorescent whitening agents during sewage treatment, *Water Research*, 1998, 32, 1939–1947.

Pond, K. R., R. Rangdale, W. G. Meijer, J. Brandao, L. Falcao, A. Rince, B. Masterson, J. Greaves, A. Gawler, E. McDonnell, A. A. Cronin and S. Pedley, Workshop report: Developing pollution source tracking for recreational and shellfish waters, *Environmental Forensics*, 2004, 5, 237–247.

Pourcher, A. M., L. A. Devriese, J. F. Hernandez and J. M. Delattre, Enumeration by a miniaturized method of Escherichia-Coli, Streptococcus-Bovis and Enterococci as indicators of the origin of fecal pollution of waters, *Journal of Applied Bacteriology*, 1991, 70, 525–530.

Rees, G., Health Implications of Sewage in Coastal Waters—the British Case, *Marine Pollution Bulletin*, 1993, 26, 14–19.

Savill, M. G., J. A. Hudson, A. Ball, J. D. Klena, P. Scholes, R. J. Whyte, R. E. McCormick and D. Jankovic, Enumeration of campylobacter in New Zealand recreational and drinking waters, *Journal of Applied Microbiology*, 2001, 91, 38–46.

Scott, T. M., J. B. Rose, T. M. Jenkins, S. R. Farrah and J. Lukasik, Microbial source tracking: Current methodology and future directions, *Applied and Environmental Microbiology*, 2002, 68, 5796–5803.

Seguel, C. G., S. M. Mudge, C. Salgado and M. Toledo, Tracing sewage in the marine environment: Altered signatures in Conception Bay, Chile, *Water Research*, 2001, 35, 4166–4174.

Sinton, L. W., C. H. Hall, P. A. Lynch and R. J. Davies-Colley, Sunlight inactivation of fecal indicator bacteria and bacteriophages from waste stabilization pond effluent in fresh and saline waters, *Applied and Environmental Microbiology*, 2002, 68, 1122–1131.

Sinton, L. W., R. K. Finlay and D. J. Hannah, Distinguishing human from animal faecal contamination in water: A review, *New Zealand Journal of Marine and Freshwater Research*, 1998, 32, 323–348.

Sophonsiri, C. and E. Morgenroth, Chemical composition associated with different particle size fractions in municipal, industrial, and agricultural wastewaters, *Chemosphere*, 2004, 55, 691–703.

Sun, M. Y. and S. G. Wakeham, Molecular evidence for degradation and preservation of organic-matter in the anoxic Black-Sea Basin, *Geochimica Et Cosmochimica Acta*, 1994, 58, 3395–3406.

Sun, M.-Y. and S. G. Wakeham, A study of oxic/anoxic effects on degradation of sterols at the simulated sediment-water interface of coastal sediments, *Organic Geochemistry*, 1998, 28, 773–784.

Volterra, L., L. Bonadonna and F. A. Aulicino, Fecal streptococci recoveries in different marine areas, *Water Air and Soil Pollution*, 1986, 29, 403–413.

Weigel, S., U. Berger, E. Jensen, R. Kallenborn, H. Thoresen and H. Huhnerfuss, Determination of selected pharmaceuticals and caffeine in sewage and seawater from Tromso/Norway with emphasis on ibuprofen and its metabolites, *Chemosphere*, 2004a, 56, 583–592.

Weigel, S., R. Kallenborn and H. Huhnerfuss, Simultaneous solid-phase extraction of acidic, neutral and basic pharmaceuticals from aqueous samples at ambient (neutral) pH and their determination by gas chromatography-mass spectrometry, *Journal of Chromatography A*, 2004b, 1023, 183–195.

Wiggins, B. A., Discriminant analysis of antibiotic resistance patterns in fecal streptococci, a method to differentiate human and animal sources of fecal pollution in natural waters, *Applied and Environmental Microbiology*, 1996, 62, 3997–4002.

4

Lead

A. Mohamad Ghazi and James R. Millette

Contents

4.1 INTRODUCTION

The element lead (Pb), *plumbum* in Latin, is a dense, bluish-gray metallic element that was one of the first known metals. The atomic number of lead is 82; the element is in group 14 (or IVa) of the periodic table. Metallic lead is soft, malleable, and ductile. It has low tensile strength and is a poor conductor of electricity. A freshly cut surface has a bright silvery luster, which quickly turns to the dull, bluish-gray color characteristic of the metal. Lead melts at 328 °C (662 °F), boils at 1740 °C (3164 °F), and has a specific gravity of 11.34; the atomic weight of lead is 207.20. Lead exists mostly in the oxidation state +2. Inorganic compounds with valance +4 are unstable and strong oxidizing agents. Lead is very resistant to corrosion. Even in powder form, lead has a diminished reactivity by the formation of a thin protective layer of insoluble compounds, such as oxides, sulfates, and oxycarbonates. Lead is not biodegradable, it never disappears and only accumulates where it is deposited. Lead provides no known biological benefit to humans. Young children absorb lead more readily than adults (42–48% versus 8–10%). Lead may accumulate in the body over decades, and it is stored in the bones and teeth (half-life is 19 years). There is no acceptable normal or safe levels of lead. More than 95% of retained lead is in bone, acting as a reservoir, where it is in continuous exchange with the soft tissue pools. The half-life of circulating lead in blood is about one month.

In terms of lead detection and analysis, significant developments in analytical instrumentations have occurred in recent years, in particular in mass spectrometry (e.g., single and multicollector magnetic sector double focusing ICP-MS) and some of the associated auxiliary devices such as laser ablation.

This chapter provides a background review of the origin, geochemistry, formation, production, and usage of lead. The pathways and the entrance of lead into the environment and a brief review of some of the guidelines related to lead in the environment and lead poisoning are discussed. The later sections of the chapter are devoted to a review and discussion of different techniques (e.g., TIMS, ICP-MS, MC-ICP-MS, LA-ICP-MS) for lead analysis, with special focus on lead isotope ratio measurements. Finally, some perspectives are presented in terms of the role of new instrument development in establishing better connections between natural and environmental sciences in the context of forensic studies.

4.2 GEOCHEMISTRY

The element lead is composed of four naturally occurring stable isotopes: ^{204}Pb (1.0%), ^{206}Pb (25%), ^{207}Pb (23%), and ^{208}Pb (53%). These isotopes, with the exception of ^{204}Pb, are products of the radioactive decay of either uranium or thorium. In addition, two radioactive isotopes ^{210}Pb ($t_{1/2} =$ 22 years) and ^{212}Pb ($t_{1/2} = 10$ hours) are used for isotope tracers. Uranium has three naturally occurring isotopes: ^{234}U (0.01%), ^{235}U (0.72%), and ^{238}U (99.3%) with half-lives of 2.5×10^5 years, 0.7×10^9 years, and 4.5×10^9 years respectively. Thorium has only one naturally occurring isotope: ^{232}Th with a half life of 14×10^9 years. Radioactive decay of uranium and thorium takes place through a series of transformations rather than in a single step (Table 4.2.1). Until the last step, these radionuclides emit energy or particle with each transformation and become another radionuclide. Different lead daughter products arise from each isotope of uranium and thorium. Uranium and thorium coexist in nature because of their similar chemical properties and

the fact that both have the same oxidation state valances of +4. However, uranium is more mobile than thorium as a result of its capacity to form the uranyl ion (UO_2^{+2}). Under oxidizing conditions, the uranium has a valance of +6, allowing formation of (UO_2^{+2}), which is water-soluble. As a result of this water solubility, the uranium is able to leave a system (Russell and Farquhar, 1960; Faure, 1986). Consequently such mobility increases the thorium concentration in some systems as uranium is drawn away, while increasing the proportion of uranium in other systems as uranium is deposited. This results in differing concentrations of their respective lead daughter products at different

Table 4.2.1 *Radioactive Decay of Uranium and Thorium*

Radionuclides	Emitted Radiation(s)	Half-life
(a) Uranium-238 Decay Series		
^{238}U	Alpha	4.46×10^9 Years
^{234}Th	Beta, Gamma	24.1 Days
^{234}Pa	Beta, Gamma	114 Minutes
^{234}U	Alpha	235,000 Years
^{230}Th	Alpha	80,000 Years
^{226}Ra	Alpha	1,622 Years
^{222}Rn (gas)	Alpha	3.85 Days
^{218}Po	Alpha	3.05 Minutes
^{214}Pb	Beta, Gamma	26.8 Minutes
^{214}Bi	Beta, Gamma	19.7 Minutes
^{214}Po	Alpha	0.00015 Seconds
^{210}Pb	Beta, Gamma	22.2 Years
^{210}Bi	Beta, Gamma	4.97 Days
^{210}Po	Alpha	138 Days
^{206}Pb		Stable
(b) Uranium-235 Decay Series		
^{235}U	Alpha	0.703×10^9 Years
^{231}Th	Beta, Gamma	25.64 hours
^{231}Pa	Alpha	3.43×10 Years
^{227}Ac	Beta, Gamma	21.8 Years
^{227}Th	Alpha	18.4 Days
^{223}Fr	Beta, Gamma	21 Minutes
^{223}Ra	Alpha	11.68 Days
^{219}Rn (gas)	Alpha	3.92 Seconds
^{215}Po	Alpha	0.00183 Seconds
^{211}Pb	Beta, Gamma	36.1 Minutes
^{211}Bi	Alpha	2.16 Minutes
^{211}Po	Alpha	0.52 Seconds
^{207}Tl		4.78 Minutes
^{207}Pb		Stable
(c) Th-232 Decay Series		
^{232}Th	Alpha	1.4×10^{10} Years
^{228}Ra	Beta, Gamma	6.7 Years
^{228}Ac	Beta, Gamma	6.13 Hours
^{228}Th	Alpha	1.91 Years
^{224}Ra	Alpha	3.64 Days
^{220}Rn (gas)	Alpha	52 Seconds
^{216}Po	Alpha	0.158 Seconds
^{212}Pb	Beta, Gamma	10.64 Hours
^{212}Bi	Beta, Gamma	60.5 Minutes
^{212}Po	Alpha	3.04×10.7
^{208}Tl	Beta, Gamma	3.1 Minutes
^{208}Pb		Stable

Table 4.2.2 *Average Concentrations of U, Th, and Pb in Common Igneous, Sedimentary, and Metamorphic Rocks (after Faure, 1986, Table 18.1, p. 283)*

Rock Type	U ppm	Th ppm	Pb ppm
Ultramafic rocks	0.014	0.05	0.3
Gabbro	0.84	3.8	2.7
Basalt	0.43	1.6	3.7
Andesite	~2.4	~8	5.8
Nepheline syenite	8.2	17.0	14.4
Granitic rocks	4.8	21.5	23.0
Shale	3.2	11.7	22.8
Sandstone	1.4	3.9	13.7
Carbonate rocks	1.9	1.2	5.6
Granitic gneiss	3.5	12.9	19.6
Granulite	1.6	7.2	18.7

sites. Table 4.2.1 shows the production of lead from uranium and thorium.

Thorium is generally more abundant in nature than uranium; the estimated crustal abundance for Th and U are 9.6 mg/kg and 2.7 mg/kg, respectively. As a result, thorium is the more common element in most systems, thus contributing to the greatest proportion of lead as the ^{208}Pb isotope. The abundances of thorium and uranium as the parent isotopes of lead vary depending upon the rock type and geological age of the rock. Table 4.2.2 obtained from the Faure (1986) book on isotope geology shows the varying concentrations of thorium and uranium in different rock types. These variations combined with (1) the decay rate of the parent isotopes; (2) the initial parent-to-daughter ratios (i.e., ^{238}U/^{206}Pb, ^{232}Th/^{208}Pb) in the source reservoir; (3) the initial isotopic composition of reservoir lead; and (4) the length of time the reservoir evolved before fractionation of lead from the system by geochemical processes are responsible for variation in lead isotopic composition in worldwide lead ore deposits (Sangster *et al.*, 2000). A detailed account on the principle of isotope geochemistry of U-Th-Pb system is given in Faure (1986) and Dickin (1995).

4.3 PRINCIPLE OCCURANCE OF LEAD

Lead is widely distributed all over the world, and has been known since ancient times. Elemental lead is rarely found in nature. Lead is present in ores such as galena (cubic lead sulfide, PbS), anglesite (rhombic lead sulfate, $PbSO_4$), cerussite (rhombic lead carbonate, $PbCO_3$), minim (a form of lead oxide with formula Pb_3O_4), and other minerals. Galena is the most important source. Lead also accumulates with Zn and Cd, Fe and other metals in ore deposits. Lead occurs in rocks as a discrete mineral, or as the major portion of the metal in the earth's crust. It replaces K, Sr, Ba and even Ca and Na in the mineral lattice of silicate minerals. Among silicate minerals, potassium feldspars and micas are notable for their affinity to accumulate Pb; therefore, granitic rocks tend to have higher levels than basaltic ones (Table 4.2.2). Lead ranks about 36th in natural abundance among elements in the earth's crust with an average crustal abundance of 16 ppm, and is the most abundant of the heavy elements with an atomic number >60 (Marshal and Fairbridge, 1999).

Lead in surface water run-off originates from chemical weathering, municipal and industrial water discharges, and from atmospheric deposition. The concentration of lead in natural waters is much lower than would be expected from the inputs because of adsorption of the element onto particulate matter (clay minerals, oxides and hydroxides of aluminum, iron and manganese). The adsorption decreases with lowering pH of the water. Under reducing conditions lead precipitates as highly insoluble sulfide.

During production (mining, roasting, refining), use (antiknock agents, tetramethyllead, batteries, pigments, ceramics, plastics, glasses), recycling, and disposal, lead enters the environment. Estimates of the emissions of individual sources of lead indicate that atmosphere is the major initial recipient and that contributions from the anthropogenic sources are at least 1–2 orders of magnitude greater than natural sources. Lead occurs in the atmosphere as fine particulates (<1 um), generated mainly by anthropogenic high temperature sources. Atmospheric transport and deposition of airborne lead increases lead levels in soils, surface waters, and the food chain. Lead is a non-essential and toxic metal, the biogeochemical cycle which is affected by human activity to a greater degree than those of any other elements.

4.3.1 Type of Lead Ore Deposits

Lead deposits are members of a class of ore deposits known as the "hydrothermal" deposits, in which metals were initially leached by hot highly saline aqueous fluids in the subsurface of the Earth's crust and transported to the site of deposition. Lead and zinc are often found together in ore deposits; however, less so with copper and iron. Lead compounds are formed when the metals are precipitated from the ore fluids by various processes depending on specific local conditions; the most common processes are cooling, mixing with other fluids, and pH change. Metals are precipitated as the sulfide minerals galena (PbS; 87 wt. % Pb), sphalerite (ZnS), chalcopyrite ($CuFeS_2$), pyrite (FeS_2), and pyrrhotite (FeS). Lead is most commonly obtained from the ore mineral galena and occurs in several distinctly different types of base metal ore bodies. They are found almost entirely in hydrothermal deposits which formed in three major geologic environments (Sangster, 1990; Kirkham *et al.*, 1993; Kesler, 1994; Eckstrand, 1995; Sangster *et al.*, 2000). Basinal hydrothermal systems formed the most important lead deposits, including Mississippi Valley-type (MVT) and sedimentary exhalative (SEDEX) deposits. The MVT deposits are so named because they were first found in the midcontinent United States in the valley of the Mississippi River. One of the most prominent MVT lead deposit is located in the tri-state district (Oklahoma–Kansas–Missouri) near the city of Joplin, Missouri; as a result, a particular type of MVT deposits with distinct lead isotope signatures are also known as the J-type deposits. The MVT deposits consist of galena, sphalerite, barite, fluorite, and other minerals that fill secondary porosity in limestone and dolomite. These deposits originally formed due to the tectonically related expulsion of high salinity metal sulfide-rich hydrothermal fluids from sedimentary basins into surrounding carbonate platforms (Kesler, 1994; Sangster *et al.*, 2000). Fluid inclusion studies show that the hydrothermal fluid had salinities of 15% or more and temperatures of 90 to 180 °C (Kesler, 1994).

Sedimentary exhalative (SEDEX) deposits consist of layers of lead–zinc–iron sulfides which, as with MVT deposits, are the products of sedimentary processes and are found within large ancient sedimentary basins. The SEDEX were precipitated by submarine hot springs that flowed into basins filled with fine-grained, clastic sediments. In many cases, these solutions did not actually reach the sea floor and instead replaced lenses of sediment just below the water surface, usually shortly after the sediments were deposited (Kesler, 1994; Sagster *et al.*, 2000). The SEDEX are recognized largely by their spectacular internal layering; SEDEX deposits are almost invariably bimetallic Pb and Zn; the Ag content may range from essentially zero in some deposits to several hundred grams per ton in others (Sangster *et al.*, 2000). The largest SEDEX deposits, including Broken Hill, Mt Isa, and McArthur River in Australia, and Sullivan in British Columbia, are in a system of 1.8 to 1.3 Ga-old rifts that cut the early continents (Young, 1992).

Volcanogenic massive sulfide (VMS) deposits are also related to seawater hydrothermal systems. These deposits are among the most abundant base metal deposits on earth, and lead and zinc deposits of this type are found largely in continents and island arcs. Although several types of VMS deposits have been recognized according to their geological environment, all are considered to be developmentally related to the volcanic rocks in which they are found, with metals being leached from the surrounding sub-seafloor rocks by circulating seawater. The hot buoyant metal-bearing fluids then find their way back to the seafloor along sub-sea fractures. Sulfide precipitation takes place on the seafloor in response to abrupt changes in temperature, pressure, and pH when the debouching fluids are brought into contact with seawater. The most important deposits of this type are found in the Kuroko district of Japan (Kesler, 1994). They are found in rocks as old as 3.5 billion years and are still forming today on the seafloor in several parts of the world (Sangster *et al.*, 2000). In addition, lead and zinc of this type are also found in large vein deposits at Keno Hill in the Yukon, Coeur d' Alene in Idaho, Hidalgo del Parral in Mexico, and Casapalca in Peru, all of which have important amounts of co-product silver (Beaudoin and Sangster, 1992).

Skarn and chimney-manto deposits are related to vein deposits and appear to form where the country rock is cabonate rather than clastic sediment. These deposits are spatially and developmentally associated with some kind of intrusive activities of magmatic origin, such as granitic plutons where initially the temperature is relatively high. Some skarn deposits form at the contact between the limestone and igneous rock. Others form chimneys (vertical) or mantos (horizontal) that extend for hundreds of meters into the limestone away from igneous intrusions (Kesler, 1994). Precipitation of the metal-containing sulfides was brought about by reaction with, and replacement of, the limestone and dolomite rocks by fluids expelled during cooling of the nearby igneous bodies. In general, skarn and manto deposits are similar, except that manto deposits are formed farther away from the intrusive bodies, where the temperatures of the ore fluids are somewhat lower than those for the formation of the skarn deposits. Both deposit types form large, "massive" deposits containing mainly Pb and Zn sulfides (i.e., high sulfide, normally >60%) (Sangster *et al.*, 2000). These deposits are best developed in Mexico, Honduras, and Peru.

4.3.2 Mining and Production of Lead

Lead and zinc ores often occur together, and there are a large number of countries that mine and process lead ore, making it among the most widespread metals in terms of primary production (Kesler, 1994). Most of lead mining is done by highly mechanized underground methods, although some deposits are mined from open pits. Deposits of SEDEX and VMS types, which were typically deposited as result of rapid cooling of submarine hot-spring fluids entering cold ocean water, are too fine-grained for ore and gangue minerals to be separated by conventional beneficiation methods. On the contrary, those deposits which have gone through slower rate of cooling, or some kind of metamorphism and recrystallization typically can produce larger

crystals that are easily separated. Broken Hill, Australia, and Aggeneys, South Africa, are two outstanding examples of this process. Some of the galena crystals can measure up to tens of centimeters in width (Kesler, 1994).

Total refined or metallic lead production is of two types: (1) primary, which originates directly from ore concentrate and (2) secondary, which refers to recycled or recovered lead. The principal method of extracting lead from galena, which is its most prominent ore, is done by pyrometallurgical method of roasting to melting the ore. In other words, the ore is converted to the oxide, and the oxide reduced with coke in a blast furnace. Conventional smelting is carried out in a blast furnace, producing lead bullion, which must be refined to remove other metals such as zinc and silver that were also in the concentrate. When a mixed galena–sphalerite concentrate must be smelted, it is roasted and then put through the imperial smelting process, in which zinc is driven off as a vapor (to be condensed elsewhere), leaving less volatile lead in the residue (Kesler, 1994). Another primary method of lead production is to roast the ore in a reverberatory furnace until part of the lead sulfide is converted to lead oxide and lead sulfate. The air supply to the furnace is then cut off and the temperature rises; then the original lead sulfide combines with the lead sulfate and lead oxide to form metallic lead and SO_2.

The secondary production of lead is done basically from recycling waste material, such as battery scrap, recovered from various industrial processes, which is also smelted, and constitutes an important source of lead. Because galena often has other minerals associated with it, the crude lead, or pig lead, that is obtained from the smelting processes contains metals such as copper, zinc, silver, and gold as impurities. The recovery of precious metals from lead ores is often as important economically as the production of lead itself. Silver and gold are recovered by Parkes process, whereby a small amount of zinc stirred into molten lead dissolves the precious metals. This molten alloy then rises to the surface of the lead as an easily removed scum, and the zinc is removed from the silver or gold by distillation. Pig lead is often purified by stirring molten lead in the presence of air. The oxides of the metallic impurities rise to the top and are skimmed off. The purest grades of lead are refined electrolytically.

4.4 USAGE OF LEAD

Lead is used in many different ways, both as the metal and as its chemical compounds. The use of lead as the metal predominates; however, when alloyed with other compounds it has significant industrial applications.

The usage of lead is known to date from 5000 to 7000 years ago, when it was used in glazes on pottery and metal objects and medicines in ancient Egypt and China (Blaskett and Boxall, 1990 and Kesler, 1994). Usage of lead also has a special place in the history of the Roman Empire, where it was used to make pipes, eating and cooking utensils, jewelries, coins, weaponry, and writing equipment (e.g., lead pencils instead of present-day graphite pencils). There is historical evidence that suggests lead was used to sweeten wine and other foods. It has been also suggested that lead caused the decline of the Roman Empire, whose later rulers displayed symptoms of lead poisoning. The hypothesis is supported by the elevated concentrations of lead found in the bones of Roman nobility (Nriagu, 1983; Patterson *et al.*, 1987; Kesler, 1994). Similarly, but historically more recently, extensive lead-made artifacts such cooking utensils, weaponry (e.g., musket balls), jewelries, as well as lead-based pigments (i.e., as coloring make-up on human skulls) have been discovered in archeological sites of Native Americans. For example, in a study of the Omaha Indians, Reinhard and Ghazi (1992) and Ghazi (1994) discovered extensive use of lead as cooking utensils, jewelries, weaponry, as well as cosmetic pigment.

Currently, the most significant usage of lead and lead alloys is lead-acid batteries (in the grid plates, posts, and connector straps) used in cars, electric vehicles, telecom, ammunition, cable sheathing, and building construction materials (such as sheets, pipes, solder, and wool for caulking). Although lead pipes have not been used for the past three decades in domestic water applications, due of lead's corrosion resistant properties, in chemical plants they are widely used for carrying corrosive compounds. For example, lead is used to line tanks that store corrosive liquids, such as sulfuric acid (H_2SO_4). Lead's high density makes it useful as a shield against x-ray and gamma-ray radiation and it is used in x-ray machines and nuclear reactors. Lead is also used as a covering on some wires and cables to protect them from corrosion, as a material to absorb vibration and sound, and in the manufacture of weaponry. Other important applications include counterweights, battery clamps and other cast products such as bearings, ballast, gaskets, type metal, and foil.

4.5 LEAD GRADES AND ALLOYS OF LEAD

In general there are three grades of lead that are manufactured and used. Corroding lead, which is the most widely produced lead in the United States, has a purity of 99.94% and exhibits outstanding corrosion resistance. Corroding lead is used in making pigments, lead oxides, and a wide variety of other lead chemicals. Chemical lead is refined lead with a residual copper content of 0.04 to 0.08% and a residual silver content of 0.002 to 0.02%. This type of lead is particularly desirable in the chemical industries and thus is called chemical lead. Copper-bearing lead provides corrosion protection comparable to that of chemical lead in most applications that require high corrosion resistance. Because lead is a soft and malleable element, it is normally used commercially as alloying compounds. For example, common lead, which contains higher amounts of silver and bismuth than does corroding lead, is used for battery oxide and general alloying. Antimony, tin, arsenic, and calcium are the most common alloying elements. Solder, an alloy that is nearly half lead and half tin, is a material with a relatively low melting point that is used to join electrical components. Type-metal, an alloy of lead, tin and antimony, is a material used to make the type used in printing industry as presses and plates. Antimony alone produces an alloy with greater hardness and strength, as in storage battery grids, sheet, pipe, and castings. Arsenic is often used to harden lead–antimony alloys and is essential to the production of round dropped shot. Babbitt metal, another lead alloy, is used to reduce friction in bearings.

4.6 COMPOUNDS OF LEAD

There are many useful compounds of lead, including oxides, carbonates, arsenates, sulfates, chromates, nitrates, and silicates. Lead is also used as additives. Lead monoxide (PbO), also known as litharge, is a yellowish orange crystalline solid used in fire assaying since the early 1900s. Litharge is produced by contacting molten lead with air. Important uses of litharge are in the manufacturing of batteries, ceramics, specialized types of glass, such as lead crystal and flint glass, in the vulcanizing of rubber, and as a paint pigment.

Lead dioxide (PbO_2) is a brown material that is used in lead-acid storage batteries. Trilead tetraoxide (Pb_3O_4), also known as red lead, is used to make a reddish-brown paint that prevents rust on outdoor steel structures. Lead arsenate ($Pb_3(AsO_4)_2$) is a poisonous white solid used as an insecticide, a chemical used to kill insects. Lead arsenate contains 22% arsenic and is very slightly soluble in cold water. Lead carbonate ($PbCO_3$), also known as mineral cerussite, is a white, poisonous substance that was once widely used as a pigment for white paint. Use of lead carbonate in paints has largely been stopped in favor of titanium oxide (TiO_2). Lead sulfate ($PbSO_4$), also known as mineral anglesite, is used in a paint pigment known as sublimed white lead. Lead chromate ($PbCrO_4$), also known as crocoite, is used to produce chrome yellow paint. Lead nitrate ($Pb(NO_3)_2$) is used to make fireworks and other pyrotechnics. Finally, lead silicate ($PbSiO_3$) is used to make some types of glass and in the production of rubber and paints.

As an additive, up to about couple of decades ago lead was an important ingredient of three other important compounds. The most important was the use of tetraethyl and tetramethyl lead (TEL and TML) and related compounds as anti-knock additives in gasoline (e.g., Morison, 2000). Unfortunately this type of gasoline is still available in some of the third-world countries. TEL and TML are added in petrol as the most economic method of improving the "octane rating" to provide the proper grade of petrol for efficient operation of engines. The use of lead oxides as an additive in interior paint, glass, ceramics, and other chemicals has also declined because of its hazardous characteristics (Kesler, 1994). Lead compounds are also used in the production of colored plastic (in which lead chromate is used as pigment), also in the manufacture of both rigid and plasticized polyvinyl chloride (PVC). The purpose for using lead compound in this application is as a stabilizer. Typical of the compounds used in the manufacture of PVC plastics are lead salts including lead oxides, phthalate, sulfate, or carbonate, depending on the desired quality of the final product. More recent novel usage of lead include nuclear waste disposal applications, additives to asphalt, shields against radon gas and electromagnetic fields, and as an earthquake vibration damper in buildings (Keating, 1997).

4.6.1 Lead-Based Paint

From the turn of the century through the 1940s, paint manufacturers frequently used lead as a primary ingredient—as a pigment and drying agent in oil-based paint to extend the protective properties of paint. Usage of lead in paint gradually decreased throughout the 1950s and 1960s as lead-free latex paint became more widespread. The Consumer Product Safety Commission (CPSC) banned lead-based paints from residential use in 1978. Even though it is no longer used in this application, millions of homes remain painted with lead-based paint. Lead-based paint chips, as well as soil and household dust contaminated with lead are the primary sources of childhood lead poisoning.

Lead from exterior house paint can flake off or leach into the soil around the outside of a home, contaminating children's playing areas. Dust caused during normal lead-based paint wear (especially around windows and doors) can create an invisible film over surfaces in a house. In some cases, cleaning and renovation activities can actually increase the threat of lead-based paint exposure by dispersing fine lead dust particles back into the air and over accessible household surfaces. Both adults and children can receive hazardous exposures by inhaling the fine dust or by ingesting paint-dust during hand-to-mouth activities. At the heights of application of lead-based paint, the following compounds were commonly used as protective or cosmetic coating of materials. Lead carbonate (white lead) was used in house paint; lead acetate was commonly used in paint, varnish and other coatings; lead oxide (red lead) was commonly used as primer on steel to prevent rusting; gray or blue lead was commonly used on ships for corrosion control and lead chromate was commonly used on highways and parking structures.

4.6.2 Lead Arsenate

Lead arsenate was the most extensively used of the arsenical insecticides. It was first prepared as an insecticide in 1892 for use against gypsy moth (*Lymantria dispar*) in Massachusetts, USA (Peryea, 1998). Lead arsenate applied in foliar sprays adhered well to the surfaces of plants, so its effects as a pesticide was longer lasting. Home-made versions of lead arsenate were initially prepared by farmers by reacting soluble lead salts with sodium arsenate, a practice that continued in some countries through the 1930s and likely 1940s. Their formulations became more refined over time and eventually two principal forms were marketed: basic lead arsenate [$Pb_5OH(AsO_4)_3$] and acid lead arsenate [$PbHAsO_4$] for all other locations. Frequent applications of lead arsenate at increasing rates over time eventually caused lead and arsenic to accumulate in topsoil, and studies showed that lead acetate is not stable in soil environments and will convert over time to less soluble mineral forms (Millette *et al.*, 1995; Peryea, 1998). The simultaneous presence of lead and arsenic in contaminated soils greatly complicates remediation possibilities. There have been substantial studies on remediation of lead arsenate contaminated soil (e.g., Ma and Rao, 1997; Peryea, 1998; Ruby *et al.*, 1999). Lead arsenate use in USA effectively terminated in 1948, when DDT became widely available to the public. By that time, the principal target pest, codling moth, had developed resistance to the arsenate compound, and DDT was found to be a much more effective control agent (Peryea, 1998).

4.6.3 Lead Acetate

Lead acetate is a white crystalline compound of lead with a sweetish taste. Known as "sugar of lead", it is water-soluble and one of the most bioavailable forms of lead. Similar to other lead compounds, it is very poisonous and soluble in water. In the presence of water, lead acetate forms the trihydrate, $Pb(CH_3COO)_2 \cdot 3H_2O$, a colorless or white monoclinic crystalline substance that is commonly known as sugar of lead. The commercial form of lead acetate, lead acetate trihydrate, is used as a mordant in textile printing and dyeing, as a lead coating for metals, as a drier in paints, varnishes, and pigment inks, and as a colorant in hair dyes (IARC, 1980; Sittig, 1985; Sax, 1987). It is also used in antifouling paints, waterproofing, insecticides, and the gold cyanidation process (Sax, 1987).

The primary routes of potential human exposure to lead acetate are ingestion, inhalation, and dermal contact. Lead acetate is absorbed approximately 1.5 times faster than other lead compounds (Sittig, 1985). The National Occupational Hazard Survey, conducted by NIOSH from 1972 to 1974, estimated that 132,000 workers were possibly exposed to lead acetate in the workplace (NIOSH, 1976).

4.7 LEAD IN THE ENVIRONMENT AND LEAD POISONING

Lead has been one of the most important metals in human history with a wide range of applications. As result of such widespread usage, the annual anthropogenic emissions of lead is about 300,000 tons in comparison to its natural emissions, which is less than 20,000 tons for years (Nriagu,

1978; Jaworski, 1987; Nriagu and Pacyna, 1988; Nriagu, 1990; Kesler, 1994; George, 1999). Historically, lead from combustion of gasoline and burning coal have been the primary contributors to such high level of anthropogenic lead in our environment. The other important anthropogenic source is the lead from mining processes such as smelting. Older smelters, including those that were made up to the 1970s, emitted galena particles from the handling of ore concentrates, as well as an aerosol of lead oxides and sulfates that condensed in smelter exhaust gas. In developed countries, lead exposure is on the rapid decline, largely due to implementation of environmental and occupational regulations, which include the significant reduction or the complete banning of the use of leaded gasoline and installation of better recovery systems on coal-burning electric power plants and smelters. Nevertheless, in developing countries exposure to high doses of lead continues to be one of the most important problems of environmental and occupational origin, and studies have shown that even low levels of lead exposure can decrease cognitive ability in children (Bellinger *et al.*, 1997; Dietrich, 1987; Romieu *et al.*, 1994, 1997; Banks *et al.*, 1997; Hernandez-Avila *et al.*, 1999). Even in US, the director of the US Centers for Disease Control (CDC) has called lead poisoning "the number one environmental problem facing America's children" (Hilts, 1991). According to United States Environmental Protection Agency's Superfund Chemical Data Matrix (EPA-SCDM), lead is one of the compounds with highest toxicity (i.e., 10,000), environmental bioaccumulation (i.e., 50,000) indices, and it has an ecotoxicity of 1000 (SCDM, 2004).

Human lead exposure occurs when dust and fumes are inhaled and when lead is ingested via food, water, cigarettes, or lead-contaminated hands and clothing. Currently, excess lead uptake due to overexposure to lead-bearing materials found in the environment is one of the major causes of workplace illness and in terms of the presence of lead as environmental contaminant, lead-based paint is one of the major sources of lead poisoning. In most cases, the suspected real pathway is soil that has been enriched with lead from old paint flaking from houses or lead that has been transported and distributed as aerosol from natural sources via air. Lead entering the respiratory and digestive systems is released into the blood and distributed throughout the body. More than 90% of the total body burden of lead is accumulated in the bones, where it is stored indefinitely. Lead in bones may be released into the blood and re-expose organ systems long after the original environmental exposure. This process can also expose the fetus to lead in pregnant women. There are several biological indices of lead exposure, including increasing lead concentrations in blood, urine, teeth, and hair.

Lead poisoning is the leading environmentally induced illness in children. Children under the age of six are at the greatest risk because they are undergoing rapid neurological and physical development. Lead poisoning can retard mental and physical development and reduce attention span. Even extremely low levels of lead can retard fetal development. In adults, overexposure to lead causes irritability, poor muscle coordination, and nerve damage to the sense organs and nerves controlling the body. Lead poisoning may also cause other problems such as affecting the reproduction system by decreasing sperm counts. Overexposure to lead may also increase blood pressure. Therefore, young children, fetuses, infants on one hand, and adults with high blood pressure on the other hand, are the most vulnerable human groups to the effects of lead. Because the early symptoms of lead poisoning are easy to confuse with other illnesses, it is difficult to diagnose lead poisoning without medical testing. Early symptoms may include persistent tiredness, irritability, and loss of appetite, stomach discomfort, reduced attention span, insomnia, and constipation. Failure to treat children in the early stages can cause long-term or permanent health damage.

4.8 ENVIRONMENTAL TRANSPORT, DISTRIBUTION, AND TRANSFORMATION

As mentioned above, historically, the primary source of transport of lead into environment has been anthropogenic emissions into atmosphere. Most lead emissions are deposited near the source, although some particulate matter (<2 mm in diameter) is transported over long distances and results in the contamination of remote sites such as arctic glaciers (Parsons and Chisolm, 1999). These long transports usually take place within approximately 10 days, which is the average residence time of lead in the atmosphere (ATSDR, 1999). The distribution and transformation of lead in the environment takes place through the earth system processes; specifically, through a complete cycle, so-called "lead cycle" where lead is continuously transferred between the earth's four major "spheres," i.e., air (atmosphere), water (hydrosphere), soil and sediments (lithosphere), and biota (biosphere). These natural transformations are possible through a set of complicated and well-balanced chemical and physical processes such as weathering, runoff, precipitation, dry deposition of dust, and by way of stream/river flow. However, soil and sediments appear to be important sinks for lead (ATSDR, 1999). Although combustion of leaded gasoline was once the primary source of anthropogenic atmospheric releases of lead, industrial releases to soil from nonferrous smelters, battery plants, chemical plants, and deterioration or removal of older structures containing lead-based paints are now major contributors to total lead releases. However, in terms of lead-based paint, releases are frequently confined to the area in the immediate vicinity of painted surfaces, and can result in highly localized concentrations of lead in indoor air (e.g., from sanding and sandblasting) and on exposed surfaces.

From the large amounts of lead that are discharged to soil and water, a significant portion of such material tends to remain localized because of the extreme persistence of lead and the poor solubility of lead compounds in water. The speciation of lead in the four major media varies depending upon several important factors such as temperature, pH, and the presence of humic materials. Very little lead deposited on soil is transported to surface or ground water except through erosion or geochemical weathering; it is normally quite tightly bound to organic matter. Airborne lead can be transferred to biota directly or through uptake from soil. Animals can be exposed to lead directly through grazing and soil ingestion or by inhalation. There is little biomagnification of inorganic lead through the food chain (Parsons and Chisolm, 1999).

4.9 REGULATIONS AND GUIDELINES

Lead is regulated by several federal statutes in the United States and is a priority water pollutant and a hazardous air pollutant. Similarly, there are a number of international guidelines and regulations that are being administered by major organizations such as the World Health Organization (WHO) and the International Agency for Research on Cancer (IARC). The IARC's regulations are largely based on the impacts of the organolead compounds and their carcinogenicity to humans (e.g., IARC, 1987, 1989), whereas the WHO regulations are more comprehensive and consider all aspects of lead exposure (e.g., WHO, 2002, 2004).

In the United States, the EPA regulates lead and certain lead compounds under the Clean Water Act (CWA), Comprehensive Environmental Response, Compensation, and Liability Act (CERCLA), Resource Conservation and Recovery Act (RCRA), Superfund Amendments and Reauthorization Act (SARA), and Safe Drinking Water Act (SDWA). The Department of Housing and Urban Development (HUD) through the Lead-Based Paint Poisoning Prevention Act, as amended by the National Consumer Information and Health Promotion Act of 1976, mandates that the use of lead-based paint in residential structures constructed or rehabilitated by any federal agency or with federal assistance in any form be prohibited (HUD, 1998). Lead also appears on the list of poisonous and deleterious substances of the Federal Drug Administration (FDA), which was established to control levels of contaminants in human food and animal feed. The action levels established for these substances represent limits at or above which the FDA will take legal action to remove the affected consumer products from the market (FDA, 1994). In addition, FDA also regulates the use of lead acetate in hair dyes under the Food, Drug, and Cosmetic Act (FDA, 1994). Other agencies that have participated in setting guidelines for lead exposure, particularly in work environments, include the National Institute of Occupational Health (NIOSH), which in its report to Congress, summarizes occupational exposure information and provides recommendations for workers (NIOSH, 1997), and Occupational Safety and Health Administration (OSHA), which has established a permissible exposure limit (PEL) for an 8-hour time weighted average (TWA) for lead. The OSHA also regulates lead on the basis of acute and chronic toxicity for several organ systems, but not on the basis of carcinogenicity. A detailed summary of the above documents is available in Toxicological Profile for Lead document, which is published by the US Department of Health and Human Services (ATSDR, 1999).

4.10 ANALYTICAL METHODS: ELEMENTAL VS ISOTOPE RATIO

Sampling and elemental analysis of lead (e.g., biological, environmental, agricultural, food, fish and other games, archaeological, crime) is well established; thus, the intent of the remaining of the section is not to reproduce and provide an extensive list of analytical methods. Rather, the intention is to identify and provide a more detailed description and application of those techniques that are uniquely applicable for forensic and environmental forensic investigations (i.e., "source" identification and fingerprinting), more specifically, isotope ratio determination of lead.

Regardless of the type of samples, in general, there are three methods for the analysis of lead. These are bulk or total lead analysis (also known as elemental analysis), isotope ratio analysis, and scanning electron microscope—x-ray analysis used on individual particles. Elemental analysis of lead, similar to that of all other elements, is very well established and is considered being a trace element technique. These types of analysis are used for routine concentration measurements of elemental lead in all the above matrices and are performed according to approved methods, regulated by federal organizations such as USEPA, NIOSH, Association of Official Analytical Chemists (AOAC), and the American Public Health Association (APHA), which have approved a number of analytical methods. There are also other groups or organizations, for example, the NIOSH Manual of Analytical Methods (NMAM) DHHS Publication No. 94–113 (NIOSH, 1994) which provides a collection of methods for sampling and analysis of lead in workplace air, and in the blood and urine of workers who are occupationally exposed to lead. The US Department of Health and Human Services (USDHHR) provides a detailed account on toxicological profile for lead, which includes a very thorough section on analytical methods on lead analysis (ATSDR, 1999). Although data from the elemental analysis of lead is most critical for human health and environmental studies, in terms of forensic studies, it is only applicable when it is used in the context of some form of statistical analysis (e.g., Principal Component Analysis (PCA) or cluster analysis).

On the other hand, data from isotopic ratio analysis of lead, though not formally approved by any regulatory agencies, provides some of the most powerful tools for precise identification of the "sources" of lead exposed to environment and humans and for providing scientifically supported evidence for forensic studies. Isotopic ratio studies have gained huge popularity in human health, environmental, and forensic studies, particularly for building strong defensible cases that will be acceptable in a court of law. This popularity is in large part fueled by significant advances in instrumentation and the development of new generations of instruments, specifically different varieties of ICP-MS [e.g., quadruple: Q-ICP-MS, time of flight: TOF-ICP-MS, sector field: SF-HR-ICP-MS, and multicollector: MC-ICP-MS)] and a wide range of associated auxiliary equipment that have been developed largely for sample introduction in order to meet the demand of a specific application. It should be mentioned that in the literature the latter two sector instruments have often been referred to as "high resolution machines", i.e., SF-HR-ICP-MS and MC-HR-ICP-MS, or in some cases they are known as double focusing magnetic sector ICP-MS.

4.10.1 Elemental Analysis

In terms of instrumentation for elemental analysis, depending on the sample media (e.g., solid, solution, air, biological, environmental, food, tissues), analysis of lead has been accomplished by a variety of instruments. In general, most of the instruments have been used for a number of applications, though some of the instruments may be more suitable for specific sample type. Depending on the sample type, each instrument is associated with a certain set of advantages and disadvantages (e.g., instrumental capabilities, price, ease of operation, and cost per sample analysis). The literature on various applications of the elemental analysis of lead is large; however, a thorough summary together with appropriate references are given in Tables 6.1 and 6.2 of the Toxicological Profile for Lead document (ATSDR, 1999). The following is a brief list of the most commonly used instruments for analysis of elemental lead.

The primary methods of analyzing for elemental lead in solution are atomic absorption spectroscopy (AAS), graphite furnace AAS (GFAAS), inductively coupled plasma atomic emission spectroscopy (ICP-AES), and anode stripping voltametry (ASV). Less commonly used techniques include inductively coupled plasma mass spectrometry (ICP-MS), gas chromatography/photoionization detector (GC-PID), isotope dilution mass spectrometry (IDMS), and differential pulse anode stripping voltametry (DPASV). The use of ICP-MS has become more routine because of the lower detection limits, and higher sensitivity and specificity of the technique. The new generation of ICP-MSs are generally several orders of magnitude more sensitive than ICP-AES. Similarly, some chromatography techniques such as gas or high performance liquid chromatography (GC and HPLC) in conjunction with ICP-MS have also been used for the separation and quantification of organometallic and inorganic forms of lead. Solution samples are typically prepared by various acid digestion methods according to the guidelines established by regulatory agencies. In most extreme cases (e.g., very low concentration levels or complex matrices) where a very precise analysis is required, lead is pre-concentrated or separated from interfering elements, such as Fe and Zn, by anion-exchange chromatography in a hydrobromic acid medium.

Quantitative lead analysis in solid samples is not as routine as the solution analysis. With the exception of x-ray fluorescence (XRF), which is readily available (even as portable units), other techniques are more sophisticated and less available, typically housed in research academic or large national and international laboratories. For example, a large number of academic laboratories on a routine basis use electron microprobe (EMP) for lead analysis as part of the package for mineralogical and petrological studies. At more advanced and less available levels lead analyses are performed by using synchrotron XRF (SXRF) (e.g., Minder et al., 1994); photon-induced x-ray emission (PIXE) (e.g., Winchester, 1983; Uribe-Hernández et al., 1996); and instrumental neutron activation analysis (INAA). These types of instrumentation are only available in large national and international laboratories such National Synchrotron Light Source (NSLS, Brookhaven National Laboratory), Advanced Light Source (ALS, Lawrence Livermore National Laboratory), Synchrotron Ultraviolet Radiation Facility (SURF II at NIST, Gaithersburg, MD, USA), the European Synchrotron Radiation Facility (ESRF), and Advanced Photon Source (APS, Argonne National Laboratory). There are a number of other major laboratories in other countries (e.g., China, Korea, Canada, Germany, Switzerland, and Japan) around the world which can provide similar type analyses, but one of the disadvantages of such techniques is the lack of availability.

More recently, laser ablation microanalysis in conjunction with ICP-AES or a variety of ICP-MS has received substantial attention, particularly in the area of forensics, and environmental forensics, which will be discussed in greater detail in the following sections. The aforementioned solid sample techniques can provide very accurate data on lead concentrations; however, if a project involves analysis of particles including contaminated soil, the information relating to influence of size, shape, and internal particle structure may be overlooked in the process (Kennedy et al., 2002). This type of analysis is typically very specialized and information has shown to influence the distribution and bioavailability of lead or any heavy element (e.g., Davis et al., 1993; Ma et al., 1995; Gasser et al., 1996; Laperche et al., 1996; Eighmy et al., 1997; Ruby et al., 1999). Using scanning electron microscopy (SEM) coupled with energy dispersive (x-ray) spectroscopy analysis capabilities (SEM-EDS), it is possible to obtain both chemical (compositional) and physical (shape and morphology) information from contaminated particle samples (Vanderwood and Brown, 1992; VanderWood, 1992; Johnson and Hunt, 1995; Kennedy et al., 2002). For example, during an environmental forensic investigation of the cause for a high level of lead in residential dust, SEM-EDS showed that lead-containing flyash was responsible (Brown et al., 1995) and during another investigation of lead in residential dust, SEM-EDS showed that lead arsenate (Figures 4.10.1–4.10.3) was present (Millette et al., 1995). In a recent investigation of elevated lead levels in the computer and communications cable runways in a commercial building, SEM-EDS showed that the cable insulation coating contained a high number of lead-chlorine particles used as a plastic stabilizer (Millette, unpublished data, 2005). Figure 4.10.4 shows a piece of the cable with PVC insulating coating. Figures 4.10.5 and 4.10.6 show the SEM image in the backscatter mode of lead particles in the insulation coating and a representative x-ray spectrum of a lead-chlorine particle. For example, in a forensic study of plastic cable insulation that was collected from a construction site, using energy dispersive scanning electron, samples were found to contain lead particles in their plastic PVC insulation coating (Millette, unpublished data). Figures 4.10.1 and 4.10.2 show scanning electron microscope image of lead and the accompanying x-ray spectrum.

4.10.2 Lead Isotope Ratio Analysis

Lead isotope ratio analysis is a very powerful technique which has long been used in geology as one of the more important tools in isotope geochemistry and geochronology for absolute age determination of rocks

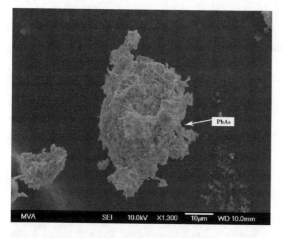

Figure 4.10.1 *Lead Arsenate Particle as Imaged by Scanning Electron Microscopy (Secondary Electron Mode).*

Figure 4.10.2 *Close-up view of the Lead Arsenate crystals in a particle of Lead Arsenate pesticide as Imaged by Scanning Electron Microscopy (Secondary Electron Mode).*

and minerals (e.g., Faure, 1986; Dickin, 1995). It has also been used in mineral exploration as isotope tracer to obtain detailed information about the source of lead (Gulson, 1984, 1985, 1996). Similarly, in more recent years the application of lead isotope ratios has found its way into other areas of applied sciences (e.g., environmental, human health, archaeological). Isotopic abundance is expressed as ratios such as $^{206}Pb/^{204}Pb$, $^{207}Pb/^{206}Pb$, and $^{208}Pb/^{206}Pb$, or more commonly as the 204-based ratios $^{206}Pb/^{204}Pb$, $^{207}Pb/^{204}Pb$, and $^{208}Pb/^{204}Pb$ (e.g., Gulson, 1999). These isotope ratios vary systematically with respect to concentration of the original parent elements and ^{204}Pb; the initial $^{206}Pb/^{204}Pb$, $^{207}Pb/^{204}Pb$, and $^{208}Pb/^{204}Pb$; age of the host rock; and with respect to the source rock environment of the ore forming fluids. Therefore, they can uniquely define the temporal and spatial relationship of the formation of an ore deposit. Lead isotope ratios are not influenced by chemical or physical weathering or even, in many circumstances, by regional or contact metamorphism. As a result, they are able to provide isotopic ratio signatures—'fingerprints' that are unique for ore forming events in a particular environment.

Prior to development of ICP-MS, thermal ionization mass spectrometry (TIMS) was the only method for measuring very precise and accurate lead isotope ratio analysis and "fingerprinting" of lead in environmental samples (e.g., Chow and Johnstone, 1965; Rabinowitz and Wetherill, 1972; Maning *et al.*, 1987; Rabinowitz, 1987; Gulson *et al.*, 1989; Rosman *et al.*, 1993, 1994a,b; Erel and Patterson, 1994; Gulson *et al.*, 1994; Gulson, *et al.*, 1995a; Chiaradia *et al.*, 1997; Manton, 1997). Similarly, TIMS also was extensively used for lead isotope ratio determination and source identification of anthropogenic lead found in the human body and clinical studies (e.g., Manton, 1977; Yaffe *et al.*, 1983; Rabinowitz, 1987; Gulson *et al.*, 1994; Angle *et al.*, 1995; Gulson *et al.*, 1995b; Gulson *et al.*, 2004). Isotope ratio measurements using TIMS on Pb-bearing artifacts have also been used in archo-anthropological studies to determine the possible geographic location of the lead ore (e.g., Brill and Wampler, 1967; Brill, 1970; Gale, 1989; Gale and Stos-Gale, 1992). Furthermore, the clinical effect of lead exposure and measurement of Pb concentration and isotope ratio in human skeletal remains has been the subject of research of several groups (e.g., Ericson *et al.*, 1979; Elias *et al.*, 1982; Patterson *et al.*, 1987).

Although isotopic analyses by TIMS are inherently labor intensive and require rigorous and extensive sample preparation, some workers still argue that TIMS produces the most accurate lead isotope ratio results of all mass spectrometric techniques (Walczyk, 2004). This is despite the fact that since the later half of 1980s there have been major advances in overall mass spectrometric development, including production of new generations of double-focusing magnetic sector field ICP-MS and multicollector version of these instruments (MC-ICP-MS).

4.10.2.1 ICP-MS Analytical Technique
4.10.2.1.1 Quadropole ICP-MS
Since the late 1980s, among the high-end analytical instrumentation, inductively coupled plasma instruments (ICPs) arguably have received the largest share of attention in instrument development and application. This is particularly true since ICP has been coupled with mass spectrometry, and now ICP-MS has matured into one of the most sophisticated and successful methods in atomic spectrometry (e.g., Houk *et al.*, 1980; Date and Gray, 1981, 1983). This is due to a number of advantages this technique offers over all other atomic spectrometry methods, such as a superior sensitivity, lower detection limits and the ability to make multielement and isotopic ratio measurements (Date and Gray, 1989;

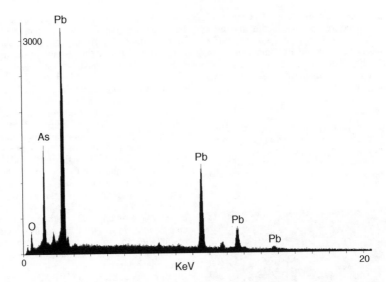

Figure 4.10.3 X-ray Spectrum from a Lead Arsenate Particle as determined by SEM-EDS.

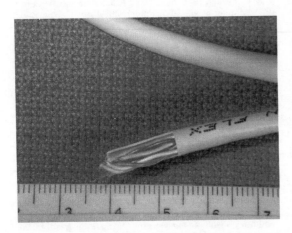

Figure 4.10.4 Computer Cable Insulated with PVC Coating that contains Lead Particles in the Coating.

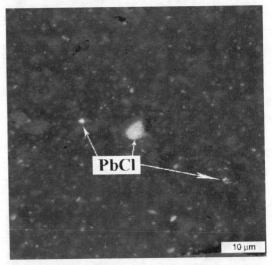

Figure 4.10.5 Scanning Electron Microscope Image (Electron Backscatter Mode) of the Surface of the PVC Insulating Coating shown in Figure 4.10.4. The bright particles are Lead-Chlorine.

Russ, 1989; Ketterrer *et al.*, 1991, Jarvis and Jarvis, 1992; Jarvis *et al.*, 1992; Koppenaal, 1992; Reinhard and Ghazi, 1992; Ghazi, 1994; Monna *et al.*, 1997; Hinners *et al.*, 1998; Chaudhary-Webb *et al.*, 1999, 2003). However, perhaps the most important advantage of ICP-MS techniques is the versatility in sample introduction. Regardless of the type of introduction method (i.e., solution, solid, slurry) all samples are introduced at atmospheric condition. This gives ICP-MS the potential of being a true application-specific instrument with perhaps the widest range of inorganic analytical applications and there is a vast literature on various aspect of this technique (e.g., Gray and Jarvis, 1989; Gray *et al.*, 1992; Hieftje, 1992; Montaser and Golightly, 1992; Ghazi *et al.*, 1993; Longerich *et al.*, 1993; Sargent and Webb, 1993; Ghazi *et al.*, 1996; Newman, 1996; Montaser, 1998).

The traditional ICP-MS instruments were based on using Q-ICP-MS, which provides a resolution of one atomic mass unit. The newer sector field magnetic sector ICP mass spectrometers (SF-HR-ICP-MS) offer yet greater advantages, such as significantly greater resolving power, lower detec-

tion limits [routinely in parts per trillion (ppt) and under the right conditions, as low as in parts per quadrillion (ppq)], and analyses of elements that are traditionally difficult to analyze by quadropole instruments, such as phosphorous, sulfur, and halogens. Nevertheless, in routine elemental and isotopic analysis, the Q-ICP-MS instruments are by far the most commonly used and the work horse for multielemental analysis in most analytical laboratories.

In terms of lead isotope ratio analysis in environmental studies the bulk of Pb isotope ratio determinations have been largely carried out by Q-ICP-MS and because of the lower sensitivity and narrow linear dynamic range, [204]Pb would not often be reported (Gulson *et al.*, 1989; Monna

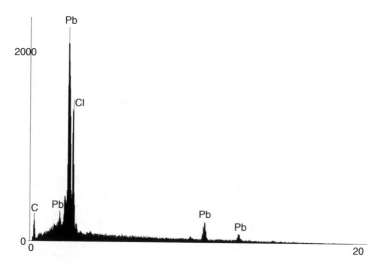

Figure 4.10.6 X-ray Spectrum from a Lead-Chlorine Particle shown in Figure 4.10.5.

et al., 1997; Hinners *et al.*, 1998). Instead, the ratios were typically reported on the basis of more abundant ^{206}Pb and ^{207}Pb isotopes (Hinners *et al.*, 1998; Monna *et al.*, 1997). In general, for Pb isotope ratio measurements, Q-ICP-MS does not possess the high precision (0.01% RSD or better) of thermal ionization mass spectrometry. The best precision for Pb isotope ratio is expected to be in the range of 0.1% RSD (Russ, 1989; Reinhard and Ghazi, 1992; Ghazi, 1994). This level of precision has been sufficient for first-order characterization of sources of lead, particularly when the differences between ratios are large (e.g., anomalous or radiogenic lead versus common lead, Faure (1986)).

For example, Ghazi (1994) and Reinhard and Ghazi (1992) using ^{204}Pb-based ratio were able to identify two different sources for lead that was used by the Omaha Indian tribe (AD

Table 4.10.1 *Short-term Precision of Q-ICP-MS for Lead Isotope Ratios Analysis of Pb(NO$_3$)$_2$ Laboratory Standard Solution. Each Analysis is a Mean of Seven 130-second Scans*

Analysis Number	^{206}Pb/^{204}Pb	^{207}Pb/^{204}Pb	^{208}Pb/^{204}Pb
A	16.034	15.418	35.794
RSD%	0.287	0.358	0.316
B	16.06	15.446	35.983
RSD%	0.287	0.365	0.436
C	16.062	15.44	35.821
SD%	0.111	0.137	0.105
D	16.028	15.414	35.934
RSD%	0.325	0.209	0.248
E	16.073	15.445	35.876
RSD%	0.328	0.208	0.262
F	15.988	15.402	35.797
RSD%	0.455	0.572	0.448
G	15.985	15.483	35.483
RSD%	0.47	0.37	0.491
TIMS Reference	16.023 ± 0.008	15.399 ± 0.012	35.703 ± 0.036

1780–1820) who lived in present-day Nebraska-Iowa, along the Missouri River. Table 4.10.1 shows the short-term precision of the isotope ratio values determined by a Q-ICP-MS for a Pb(NO$_3$)$_2$ reference solution which was determined by a TIMS. Values of short-term precision are expressed as percent (RSD%) for ^{206}Pb/^{204}Pb, ^{207}Pb/^{204}Pb, and ^{208}Pb/^{204}Pb for seven repeats and varies between (0.10% and 0.45%), (0.13% and 0.57%), and (0.10% and 0.49%), respectively (Table 4.10.1). The accuracy of isotope ratio measurements was given in terms of percent deviation from the values obtained by a TIMS, (i.e., ±TIMS). These values are (+0.02%) to (+0.31%) for ^{206}Pb/^{204}Pb, (+0.02%) to (+0.55%) for ^{207}Pb/^{204}Pb, and (−0.38%) to (+0.55%) for ^{208}Pb/^{204}Pb (Table 4.10.2). The long-term precision in RSD% is 0.221%, 0.175%, and 0.416% for ^{206}Pb/^{204}Pb, ^{207}Pb/^{204}Pb, and ^{208}Pb/^{204}Pb, respectively, and indicates excellent instrumental stability over a period of approximately one year (Table 4.10.3).

A lead geochron diagram was used to identify the source of lead found in artifacts and skeletal remains (Figure 4.10.7). Three of the artifact samples gave lead isotope signatures that were characteristic of Mississippi Valley-type deposits (J-Type or radiogenic lead), also known and anomalous lead signature (Table 4.10.4). One of the artifacts which was an alloy of Pb-Sn gave an intermediate isotope ratio signature (Table 4.10.6). On the other hand, lead found in cosmetic pigment gave signatures that were characteristic of ore deposits from New Jersey and Pennsylvania, also known as common lead (Faure, 1986) (Table 4.10.6). However, because of low concentrations of lead in bones, low sensitivity Q-ICP-MS, and low precision of analysis, it was not possible to decipher the source of the lead that was found in human remains. It was clear that these people exposed to lead were of two difference origins (Table 4.10.5).

In a similar study, Chaudhary-Webb *et al.* (1999, 2003) using Q-ICP-MS were able to identify the source for lead poisoning among women residing in Mexico City. Their study was based on tracing a specific route or source of exposure of the individual's blood lead isotope ratio. They compared blood, ceramic, and gasoline lead isotope ratios and concluded that determining lead isotope ratios can be an efficient tool to identify a major source of lead exposure and to support the implementation of public health prevention and control measures. Finally, in another study, using

Table 4.10.2 *Accuracy of Q-ICP-MS Analysis for Pb(NO₃)₂ Laboratory Standard Solution, Compared with TIMS Values for the Same Solution*

Pb (ppb)	$^{206}Pb/^{204}Pb$ ($\pm$TIMS)	$^{207}Pb/^{204}Pb$ ($\pm$TIMS)	$^{208}Pb/^{204}Pb$ ($\pm$TIMS)
10	14.873 (−) 7.1%	14.146 (−) 8.1%	32.128 (−) 10%
RSD%	4.536	3.682	3.736
75	16.376 (+) 0.2%	15.384 (−) 0.09%	35.304 (−) 1.1%
RSD%	0.682	0.43	0.823
150	16.144 (+) 0.7%	15.303 (−) 1.2%	35.284 (−) 1.1%
RSD%	0.588	0.967	0.831
500	16.276 (+) 1.5%	15.417 (+) 0.1%	35.284 (+) 0.2%
RSD%	0.502	0.576	0.571
750	16.329 (+) 1.9%	15.417 (+) 0.1%	35.793 (+) 0.2%
RSD%	0.678	0.675	0.733
1000	16.336 (+) 1.9%	15.465 (+) 0.4%	35.041 (−) 0.9%
RSD%	0.595	0.529	0.555
TIMS Values	16.023 ± 0.008	15.399 ± 0.012	35.703 ± 0.036

Table 4.10.3 *Long-term Precision of Q-ICP-MS (RSD%) from Seven Sets of Isotope Ratio Analyses. Data Obtained Within Approximately One Year. True Means are Calculated at 95% Confidence Interval*

Analysis Number	$^{206}Pb/^{204}Pb$ ($\pm$) TIMS %	$^{207}Pb/^{204}Pb$ ($\pm$) TIMS %	$^{208}Pb/^{204}Pb$ ($\pm$) TIMS %
A	16.034 (+) 0.07	15.418 (+) 0.12	35.794 (+) 0.25
B	16.060 (+) 0.23	15.446 (+) 0.31	35.983 (+) 0.53
C	16.062 (+) 0.02	15.440 (+) 0.27	35.821 (+) 0.33
D	16.028 (+) 0.03	15.414 (+) 0.10	35.934 (+) 0.65
E	16.073 (+) 0.31	15.445 (+) 0.30	35.876 (+) 0.48
F	15.988 (−) 0.16	15.402 (+) 0.02	35.797 (+) 0.26
G	15.985 (−) 0.24	15.483 (+) 0.55	35.483 (−) 0.61
Mean	16.032 (+) 0.033	15.435 (+) 0.025	35.799 (+) 0.138
RSD%	0.221	0.175	0.416
TIMS Reference	16.023 (±) 0.008	15.399 (±) 0.012	35.703 (±) 0.036

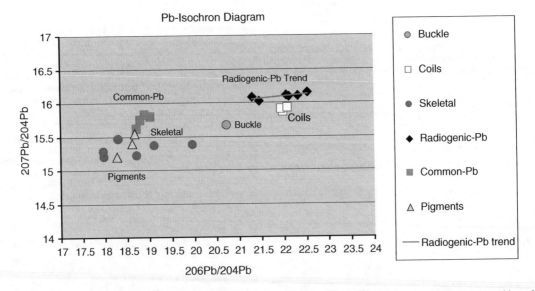

Figure 4.10.7 *Pb isotope ratio correlation diagram for $^{206}Pb/^{204}Pb$ vs $^{207}Pb/^{204}Pb$, showing isotope compositions for Pb from a number of Mississippi Valley-type deposits and non-Mississippi Valley-type deposits. Plotted in this diagram are also the isotope compositions for Pb from bones, pigments, and artifacts. The solid line is the anomalous-Pb line. This is the least-square regression through all the Pb isotopic data point with Mississippi Valley-type signature (modified after Ghazi, 1994).*

Table 4.10.4 Lead Isotope Ratio Data for Artifacts and Pigments. Each Analysis is a Mean of Seven 130-Second Scans

Sample Number	Type Sample	$^{206}Pb/^{204}Pb$	$^{207}Pb/^{204}Pb$	$^{208}Pb/^{204}Pb$
DK10-F20	Coil-1	21.981	15.861	41.388
RSD%		0.52	0.491	0.525
DK10-F18	Coil-2	21.943	15.9	41.294
RSD%		0.658	0.628	0.541
DK10-F14	Musket	22.091	15.92	41.58
RSD%	Ball	0.423	0.414	0.564
DK10-F30	Buckle	20.718	15.666	38.853
RSD%		0.658	0.628	0.541
CS50	Pigmant-1	18.603	15.399	38.264
RSD%		0.52	0.491	0.525
CS52	Pigmant-2	18.666	15.663	38.748
RSD%		0.075	0.071	0.285
CS53	Pigmant-3	18.122	15.202	38.875
RSD%		0.151	0.15	0.626

Table 4.10.5 Lead Isotope Ratio Data for Skeletal Remains with Highest Lead Concentration. Each Analysis is a Mean of Seven 130-Second Scans

Sample	$^{206}Pb/^{204}Pb$	$^{207}Pb/^{204}Pb$	$^{208}Pb/^{204}Pb$
DK10-F20	19.958	15.379	38.638
RSD%	1.63	1.514	1.524
DK 10-F19	18.301	15.469	37.864
RSD%	0.801	0.822	0.639
DK 10-F31	19.109	15.364	37.699
RSD%	1.098	1.224	1.078
DK 10-F33	17.987	15.198	36.689
RSD%	0.859	0.972	0.879
DK10-F26	18.711	15.213	36.902
SD%	1.827	1.679	1.818
DK2A-F9	17.976	15.219	36.765
RSD%	0.897	1.201	0.932

$^{205}Tl/^{203}Tl$ correction, Ketterrer et al. (1991) was able use ^{204}Pb-based isotope ratio data from Q-ICP-MS to identify the sources of lead in environmental samples.

4.10.2.1.2 Magnatic Sector ICP-MS

Magnetic sector ICP-MSs (also known as high resolution ICP-MSs), both single and multicollector ICP-MSs, are the most recent instrumental developments in analytical inorganic chemistry for elemental and isotope ratio analysis. Both of these instruments are double focusing sector field magnetic ICP-MSs in reverse Nier-Johnson geometry (e.g., Walder et al., 1993a,b; Walder and Furuta, 1993; Shuttleworth and Kemser, 1996; Becker and Dietze, 1997; Jakubowski et al., 1998; Becker, 2002). The major difference between the above two instruments is that one is equipped (or has the option) with a multicollector detector system for simultaneous detection of isotopes for more precision than isotope ratio analysis.

The operational philosophy of sector field instruments is significantly different to that of quadrupole instruments.

The ions are resolved by double focusing with the magnetic sector and the electrostatic analyzer (ESA). When the slit apertures at the entrance and exit of the mass spectrometer are small, the double focusing of ions results in high resolution. Three fixed resolutions—300, 3000, and 10,000—low, medium and high, respectively (dependent on the slit aperture widths) are possible with the instrument by ThermoQuest (Finnigan MAT) the "Element and Element2".

These instruments scan over different masses by changing the field strength of the magnetic sector or by changing the applied accelerating voltage. Three models of data acquisition are available: magnet scanning, acceleration voltage scanning, and a combination of the two. Magnet scanning is a relatively slow way of scanning the mass range because each magnet scan requires a settling time (1–300 ms) to reach stability with the magnetic field strength of the magnet. This mode of operation is of limited use with laser ablation acquisitions because temporal differences in the laser signal can result in poor accuracy and precision when settling times are required to be long. In contrast, acceleration voltage scanning is a relatively fast way of scanning the mass range. Jumps from peak to peak can be achieved in less than one millisecond but unfortunately, this method is limited to only thirty percent of the mass range relative to the magnet peak mass. Therefore, this method of acquisition is well suited to analysis of laser signals but is limited to the collection of subsets of elements. When high and low mass elements are to be analyzed by laser ablation in one acquisition, then we employ a combination of magnet and acceleration voltage scanning. Wide analyzer slit apertures produce peaks that have large flat tops enabling scanning of only a small portion of the peak and thus reducing the magnet settling time (1–10 ms) without compromising analytical performance. In this way the instrument can quickly peak hop across the mass range with a series of magnet and acceleration voltage jumps and acceleration voltage scans.

4.10.2.1.3 Lead Isotope Ratio Analysis by Magnetic Sector-ICP-MS

Although the lead isotope technique using thermal ionization mass spectrometry (TIMS) produces the most accurate isotope ratio data, the use of magnetic-sector ICP-MS, particularly the multicollector instrument, may be a very viable alternative for routine application of lead isotopic ratio measurements (Gwiazda et al., 1998; Gwiazda and Smith, 2000). In comparison with thermal ionization mass spectrometry, MC-ICP-MS certainly has several advantages for lead isotope ratio measurements, particularly in terms of analysis of environmental samples. These include (1) greater speed of analysis; (2) higher throughput of samples; and (3) relatively simple sample preparation (i.e., in most environmental cases lead isotope ratio analysis does not require column chromatography for purification of lead). A major disadvantage of magnetic-sector ICP-MS is that TIMS has greater precision and reproducibility. Another important factor that is making magnetic sector ICP-MS the instrument of choice for lead isotope analysis is that they are becoming more available in laboratories. However, despite the rapidly growing popularity of single and multicollector sector ICP-MS, there are groups of workers who strongly believe TIMS is still the more powerful instrument in producing superior quality lead isotope data (e.g., Woolard et al., 1998; Walczyk, 2004). Furthermore, these workers have shown that even in working with MC-ICP-MS, in order to obtain the best quality lead isotope data, sample preparation is critical and it is essential to follow the well-established anion exchange chromatography procedure (Walczyk, 2004).

In terms of sample introduction in ICP-MS, after proper acid digestion, evaporation, and dilution to proper concentration, samples are introduced by nebulization into the plasma for ion generation (Gray *et al.*, 1992). Sample preparation for TIMS is more labor intensive, time consuming, and requires purification by using anion exchange chromatography. Samples are loaded onto single rhenium filaments using the conventional silica gel-phosphoric acid technique (e.g. Tatsumoto and Unruh, 1976; Fey *et al.*, 1999; Unruh, 2002).

4.10.2.1.4 Lead Isotope Ratio by Laser Ablation ICP-MS

A significant breakthrough associated with ICP mass spectrometry is the development and the use of lasers as microsampling mechanisms for high sensitivity ICP-MSs (Shuttleworth, 1996; Shuttleworth and Kemser, 1996; Jeffries *et al.*, 1998; Pearce *et al.*, 1992; Perkins and Pearce, 1995; Günther *et al.*, 1998). Among the spectrum of in situ microanalytical methods, laser ablation ICP-MS analysis is increasingly recognized for its strengths and versatility, including major-to-trace element analyses, isotopic analyses, and application to a wide variety of matrices (e.g., metals, oxides, organics, and fluids), short analysis time, and low costs.

In comparison to other solid phase microanalytical sampling (e.g., SXRF, PIXE, ion probe), the LA-ICP-MS method has been flourishing in technological innovations and applications, particularly for UV laser ablation. UV laser systems use either solid-state lasers (e.g., quadrupled wavelength ($\lambda = 266\,nm$) and quintupled wave length ($\lambda = 213\,nm$) Nd-YAG lasers) or excimer lasers (e.g., Kr-F, ArF, and F2, ($\lambda = 193\,nm$)) for ablation sampling.

Laser ablation ICP mass spectrometers have produced trace element and isotopic ratio analyses of solid samples such as minerals and glasses (e.g., Arrowsmith, 1987; Jarvis *et al.*, 1992; Perkins *et al.*, 1992; Pearce *et al.*, 1992, 1997; Lichte, 1995; Günther *et al.*, 1997a; Ghazi *et al.*, 2000a,b, 2002a), soft tissues (Wataha *et al.*, 2001; Ghazi *et al.*, 2002b), liquids such as fluid inclusions (Shepherd and Chenery, 1995; Ghazi *et al.*, 1996; Günther *et al.*, 1997b; McCandless *et al.*, 1997; Audétat *et al.*, 1998; Ulrich *et al.*, 1999; Ghazi and Shuttleworth, 2000).

Calibration strategy is typically achieved by using one of the well-established solid standard reference materials

and comparing the response for a suitable element in a reference standard material and the unknown. This element response also allows correction for all elements included in the element menu. For example, Standard Reference Material 612 (SRM-612) from National Institute of Standards and Technology, which has a nominal concentration of each trace element of 50 mg/kg, is a popular standard that is commonly used in geological studies. SRM-612 contains 61 trace elements in a glass matrix. The concentration of Pb in SRM-612 is 35.88∀2 ppm (Pearce *et al.*, 1997). Figure 4.10.8 shows an example of time-resolved spectrum of SRM-612 to demonstrate the instrumental stability during the course of 25-second analyses.

In terms of forensic and environmental forensic studies, LA-ICP-MS has been successfully applied in a number of applications. This is becoming the instrument of choice because it offers high sample throughput, and rapid, accurate, and relatively inexpensive analysis with minimal sample preparation time. For example, McGill *et al.* (2003) used a 266 nm UV laser ablation with a multicollector ICP-MS to analyze press pellets (similar to those used for XRF analysis) of well-homogenized highly contaminated soil samples from landfill and brown-field sites around Nottingham and Wolverhampton. They concluded it is possible to use LA-MC-ICP-MS to establish the extent of lead source variability in this type of environmental samples.

Ghazi and Millette (2004), used a 213 nm UV laser ablation with a single collector magnetic sector ICP-MS to analyze elemental contents and lead isotope ratios in a layered lead-based paint sample (i.e., six layers, each 50–100 μm thick) (Figure 4.10.9). The beam size used for this study was about 20 μm (Figure 4.10.10), and the results show that only four of the six layers of the paint sample contained lead. Figure 4.10.11 shows cross sectional views of ablation holes in the paint sample. Analysis of each layer yielded a set of signal intensity values for two isotopes of mercury (^{202}Hg and ^{204}Hg) and four isotopes of lead (^{204}Pb, ^{206}Pb, ^{207}Pb, ^{208}Pb). After correction for ^{204}Hg isotope, intensity data (ion counts) was used to construct a set of time-resolved intensity profiles for laser ablation analysis across each layer of the sample (Figure 4.10.12). Finally the intensity data was used to calculate Pb isotopic ratios for individual layers (Figures 4.10.13 and 4.10.14 and Table 4.10.6).

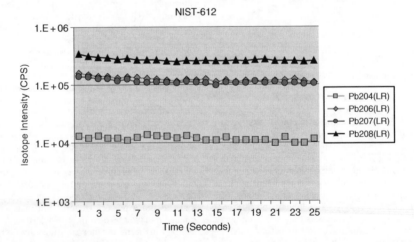

Figure 4.10.8 *A twenty-five second time-series spectrum in counts per seconds (cps) for laser ablation sector field ICP-MS analysis for four isotopes of lead (^{204}Pb, ^{206}Pb, ^{207}Pb, ^{208}Pb) in NIST-SRM-612 glass standard (modified after Ghazi and Millette, 2004).*

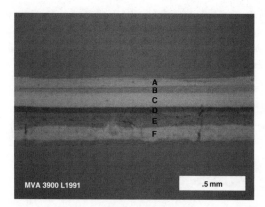

Figure 4.10.9 Light microscope photomicrograph of cross-section of layered paint sample, standing up, showing all six visible layers of different paint. Alphabets are used to identify each layer and correspond with text. The sample was prepared for analysis by scanning electron microscopy and laser ablation ICP-MS, by cutting a section of paint chip approximately 10 mm wide with a scalpel (modified after Ghazi and Millette, 2004).

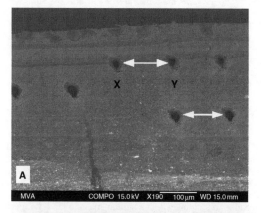

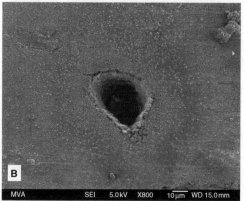

Figure 4.10.10 (A) Scanning electron microscope (SEM) photomicrograph of several equally spaced laser ablation pits along straight lines in the center of each layer of paint. Ablation holes are 20 μm in diameter and the distance between holes is 100 μm. (B) A close up of two of the ablation pits showing more detail of the morphology and texture of the paint as well as the shape of the ablation hole (modified after Ghazi and Millette, 2004).

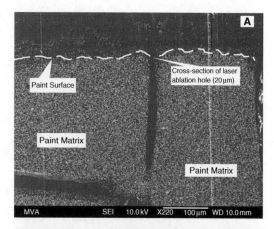

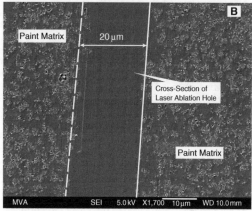

Figure 4.10.11 (A) Scanning electron microscope (SEM) photomicrograph showing the depth of a laser ablation hole in cross sectional view. As a result of interaction of laser beam with paint which is relatively soft it appeared that holes were slightly curved, thus the actual depth (~500 μm) of holes was greater than what it is shown in this diagram. (B) Close-up SEM photomicrograph of the above ablation pit (20 μm diameter). Also note the porous nature of the paint matrix (modified after Ghazi and Millette, 2004).

A lead geochron diagram was used to identify the source of lead (Figure 4.10.15). Lead isotope ratio data from one of the layers showed radiogenic signatures, very similar to Pb-Zn mines of the central United States region (Mississippi Valley-type deposit), most likely from the eastern Missouri mining district of Fredricktown/Avon area (Figure 4.10.14). The isotope ratio data for another layer also indicated radiogenic signature; however, this signature was more similar to that of the Pb-Zn-Ag mining districts of the Sierra Madre of eastern Mexico. Of the remaining two layers, one layer showed a common lead signature whereas the last layer was difficult to identify, because of presence of high concentrations of mercury in the matrix, thus presenting a significant interference problem at m/z 204.

Corrections For Lead Isotope Ratio

One of the potential problems that exist in measuring Pb isotope ratio by ICP-MS is the elemental interference at $m/z = 204$ that involves ^{204}Hg and ^{204}Pb isotopes with 6.85%

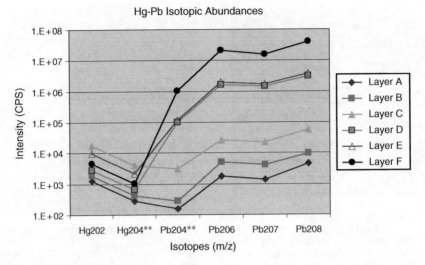

Figure 4.10.12 *A plot of isotope (m/z) intensity profile in count per seconds (CPS) for lead (^{204}Pb, ^{206}Pb, ^{207}Pb, ^{208}Pb) and two isotopes of mercury (^{202}Hg and ^{204}Hg). The intensity (CPS) is proportional to concentration of Pb and Hg in each layer. Symbol (**) indicates calculated values for intensity of ^{204}Hg and ^{204}Pb, by calculating ^{204}Hg intensity, using natural abundance ratio of ^{204}Hg/^{202}Hg = 6.85/29.8 = 0.229. ^{204}Pb isotope corrections were made by subtracting calculated ^{204}Hg values from total ions at m/z = 204. Layers A and B are virtually Pb- and Hg-free. The concentration of Hg is highest in layer C, followed by the layer E, and layer D also has virtually no Hg. The concentration profile of Pb in all six layers is as follow: A < B < C < D < E < F. However, concentration of Pb in layers D and E are basically identical. Note the crossing-cutting of the patterns for layers D, E, and F by that of layer C is a significant indication of different origin (i.e., mixture) for layer C (modified after Ghazi and Millette, 2004).*

and 1.4% isotopic composition, respectively (Figure 4.10.13). This interference is particularly problematic when the concentration of Hg is relatively high and the Pb isotope ratios are being measured with respect to ^{204}Pb (i.e., ^{206}Pb/^{204}Pb; ^{207}Pb/^{204}Pb; ^{208}Pb/^{204}Pb). In order to calculate the correct ^{204}Pb values, it is a common practice to measure the intensity of another mercury isotope (e.g., ^{202}Hg = 29.80%) and apply the following procedure to make the appropriate isotopic corrections (Ghazi, 1994; Ghazi and Millette, 2004):

$$(^{204}\text{Pb})_{\text{Corrected}} = (^{204}\text{Pb})_{\text{ICPMS}} - (^{204}\text{Hg})_{\text{Elemental}}$$

where $(^{204}\text{Pb})_{\text{ICPMS}}$ represents the total ion counts at m/z = 204, which includes both lead and mercury isotopes, and $(^{204}\text{Hg})_{\text{Elemental}}$ is that portion of the ion count that represents mercury, which can be calculated by the following equation:

$$(^{204}\text{Hg})_{\text{Elemental}} = (^{202}\text{Hg})_{\text{ICPMS}}/4.35$$

where 4.35 is the ratio of naturally occurring ^{202}Hg/^{204}Hg (i.e., 29.8/6.85 = 4.35), and $(^{202}\text{Hg})_{\text{ICPMS}}$ represents the actual concentration of Hg in the sample obtained by ICPMS.

An additional correction that needs to be made during isotopic ratio measurements is mass bias correction. The mass bias derives from the differential transmission of ions of different mass as from the point of entry from sampling device until they are detected by the detectors. The mass bias effects occur in the interface region in magnetic sector ICP. Several processes are considered to contribute to the mass bias effect, including the space-charge effect in the plasma or vacuum interface regions. The space-charge effect results in the preferential transmission of the heavier ions because the lighter ions tend to migrate to the exterior of

the plasma and are focused less efficiently into the mass analyzer.

In terms of lead isotope ratio measurements, application of ^{205}Tl/^{203}Tl ratios has become nearly a routine procedure for lead isotope (Ketterer *et al.*, 1991; Walder and Furuta, 1993; Walder, 1997; White *et al.*, 2000; Niederschlag *et al.*, 2003). This measurement is basically an internal isotopic standardization, and the reason for using thallium isotope is because this element has a similar mass to lead, Longerich *et al.* (1987).

4.11   LEAD FORENSICS

Although in principle geological and environmental sciences are closely related (e.g., a large number of geology departments around the world are also home to environmental sciences), there are significant gaps in basic knowledge and lack of communication between these two disciplines (Sangster *et al.*, 2000). One of the major gaps is the extremely limited amount of shared language between the two fields, particularly in terms of isotopic signature of lead ore deposits. For example, when the word "source" is used by a geologist, it is invariably used to identify the geographic location of the original ore deposit from which lead was mined.

When the same word is used by an environmental scientist of non-geology origin, the term is used to identify the type of material (e.g., paint, gasoline, contaminated soil) or the pathway (e.g., soil, air, water) that lead comes in contact with humans or the environment. Sangster *et al.* (2000) have identified at least two other major inconsistencies between geological and environmental knowledge, and have argued that in order to be able to correctly identify the true geological "source" for lead, if indeed that is the true objective, it is essential to establish a broader commonality

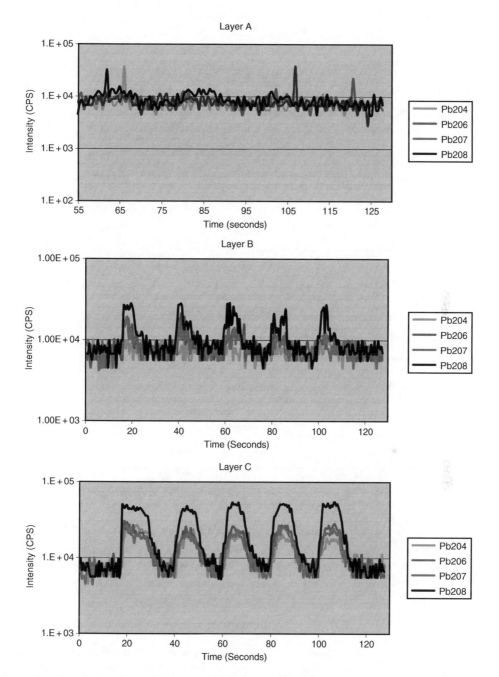

Figure 4.10.13 *Time-series Pb isotopic profile of distribution of ^{204}Pb, ^{206}Pb, ^{207}Pb, and ^{208}Pb in layered paint sample (Figure 4.10.5). This figure shows the spectra for layers A, B, and C. Spectrum A is for layer A and shows the presence of very minute amount of lead, analyzed through five ablation holes. Spectrum B is for layer B and shows five relatively small peaks representing Pb isotopic analysis of this layer through five ablation holes within a distance of approximately 500 μm. Spectrum C shows the Pb isotopic profile for layer C, which has higher concentrations of Pb. However, note the isotopic composition of Pb does not follow that of natural abundance (i.e., ^{204}Pb (1.4%) $<^{207}$Pb (22.1%) $<^{206}$Pb (24.1%) $<^{208}$Pb (52.4%)), and ^{204}Pb (1.4%) is nearly at the same level as ^{207}Pb (22.1%) and ^{206}Pb (24.1%) (modified after Ghazi and Millette, 2004).*

between these two fields of natural sciences. This should include broader common vocabulary, larger shared literature, particularly in terms of isotope ratio data sets from lead mines around the world, and reduction in instrumen-

tal use and differences in reporting isotope ratio data. For example, in geological science, most often the ratios have been traditionally expressed as ^{204}Pb-based ratios, knowing that ^{204}Pb (1.4%) is the only naturally formed stable

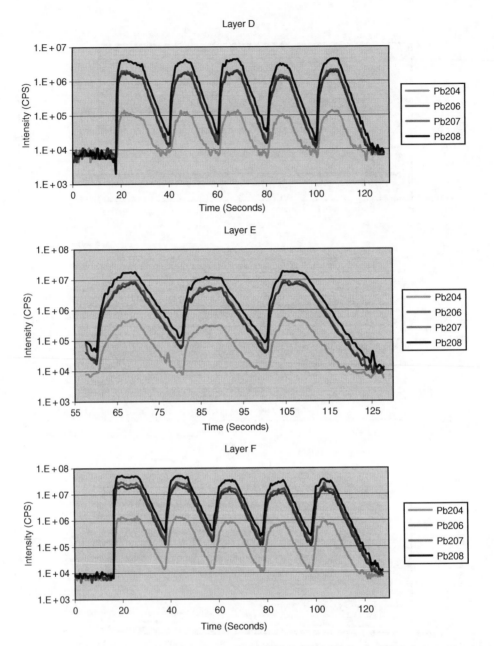

Figure 4.10.14 *Time-series Pb isotopic profile of distribution of [204]Pb,[206] Pb,[207] Pb and [208]Pb in layered paint sample (Figure 4.10.5). This figure shows the spectra for layers D, E, and F. These spectra show presence of significantly higher concentrations of Pb than layers A, B, and C (Figure 4.10.9). Spectra D and E show the data taken during 125 seconds, analyzing five ablation holes. Spectrum E is for layer E and shows the section of scan that was collected between 55 and 125 seconds during which three holes were drilled (modified after Ghazi and Millette, 2004).*

isotope of lead (Dickin, 1995; Faure, 1986). On the contrary, one could argue that environmental sciences, in particular, environmental forensic and forensic sciences, as relatively new and rapidly growing applied fields, are in the process of establishing their own terminology and methodology in employing lead isotope ratio data.

For example, for the purpose of lead-related environmental problems, it has been a tradition to use changes in $^{206}\text{Pb}/^{207}\text{Pb}(\text{R}_{\text{Pb}})$ ratios to demonstrate the origin of the lead

found in environmental samples (Kaplan, 2003). This ratio has been used successfully by Sturges and Barrie (1987) to show variability in sources of lead additives in gasoline used by Canadian and US automobiles. They found that the lowest $^{206}\text{Pb}/^{207}\text{Pb}$ ratios were from older deposits such as the Canadian and Australian deposits with R_{Pb} values of 0.92–1.04, and the highest ratios were from Mississippi Valley-type deposits (i.e., J-type deposits) with R_{Pb} of 1.3–1.39. Similarly, the same isotope ratio has also been

Table 4.10.6 Lead Isotope Ratio Data for Paint Chip. N Represents Total Number of Analysis in a
120-second Scan. Each Set of Isotopic Value is the Mean on N Analyses for Each Ablation Hole.
Mean Represents the Average Value for all Analyses in Each Layer

	N		^{206}Pb/^{204}Pb	±	^{207}Pb/^{204}Pb	±	^{208}Pb/^{204}Pb	±
Layer C								
C1	21		2.271	0.482	2.069	0.471	4.345	0.667
C2	15		2.567	0.429	2.334	0.39	5.093	0.921
C3	23		2.767	0.455	2.651	0.745	5.14	1.699
C4	22		3.196	0.576	2.829	0.488	5.76	1.237
C5	15		3.313	0.517	2.869	0.378	6.454	1.548
Sum	96	Mean	2.823	0.492	2.55	0.494	5.358	1.214
Layer D								
D1	21		16.205	0.677	14.877	1.281	35.597	3.504
D2	15		15.917	1.680	14.392	1.555	35.215	4.161
D3	23		15.722	1.032	14.393	1.642	35.101	5.319
D4	22		15.135	0.977	13.349	0.941	32.139	3.097
D5	15		16.326	1.519	14.148	1.056	34.096	2.131
Sum	96	Mean	15.861	1.177	14.232	1.295	34.430	3.643
Layer E								
E3	25		18.333	1.087	15.696	0.953	39.994	2.996
E4	25		18.119	1.293	15.876	1.363	40.158	3.609
E5	25		18.404	0.977	16.066	1.058	39.354	4.008
Sum	75	Mean	18.285	1.119	15.879	1.125	39.835	3.538
Layer F								
F1	26		21.248	1.142	15.852	1.421	40.013	2.890
F2	25		21.397	0.954	16.292	1.212	40.003	3.554
F3	25		20.971	1.222	16.250	1.680	40.026	2.132
F4	16		20.866	0.764	16.205	1.189	40.804	2.549
F5	16		21.164	2.651	16.550	2.764	39.653	3.410
Sum	108	Mean	21.129	1.347	16.230	1.653	40.100	2.907

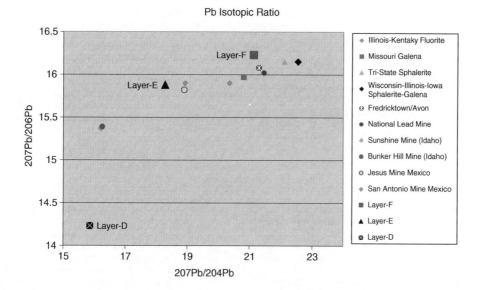

Figure 4.10.15 Pb isotope ratio correlation diagram for ^{206}Pb/^{204}Pb (vs) ^{207}Pb/^{204}Pb, showing isotope compositions for
Pb from a number of Mississippi Valley-type deposits and non-Mississippi Valley-type deposits. Plotted in this diagram
are also the isotope compositions for Pb (larger symbols) found in layers D, E, and F of Figure 4.10.5 (modified after
Ghazi and Millette, 2004).

used by a number of other environmental scientists as a monitoring tool to identify the origin of lead and to reconstruct the pathway of transport of trace amounts of lead found in samples in Europe, North America, Greenland, and Atlantic and Pacific Oceans (e.g., Chow and Johnstone, 1965; Chow et al., 1975; Petit et al., 1984; Sturges and Barrie, 1987; Flegal et al., 1987, 1989; Sherrell et al., 1992; Sturges et al., 1993; Rosman et al., 1993, 1994a,b; Grousset et al., 1994; Ritson et al., 1994; Watmough et al., 1999; Hansmann and Köppel, 2000; Bollhofer and Rossman, 2001). In another example, Kaplan (2003) as part of an invited paper published in the *Environmental Forensics* provides a thorough review of global use of $^{206}Pb/^{207}Pb$ ratio tracer and presented a cumulative database for $^{206}Pb/^{207}Pb$ versus time on North American samples. Similarly in another paper published on-line, as part of the Conference on Lead Poisoning Prevention and Treatment, Banaglore, India, Gulson (1999) provides an extensive list of references on all aspects of lead isotope ratio, including applications of $^{206}Pb/^{207}Pb$ ratio.

Another unique example of applications of $^{206}Pb/^{207}Pb$ isotope ratio in environmental sciences is the development of the so-called ALAS Model (Applications of Anthropogenic Lead ArchaeoStratigraphy) (Hurst et al., 1996). In this model, it has been suggested that $^{206}Pb/^{207}Pb$ in combination with $\delta^{206}Pb_{ALAS}$ ‰ (delta notation, similar to notations for the light element stable isotopes, C, N, O) can be used for estimating the year of release of hydrocarbon into the atmosphere (i.e., "age dating" of gasoline releases). Since the development and publication of this technique, there has been a series of articles published in *Environmental Forensic*, debating the applicability of $^{206}Pb/^{207}Pb$ for the purpose of "age dating" of leaded hydrocarbon and the true meaning the "age dating" (Hurst et al., 1996; Hurst, 2003; Kaplan, 2003).

As it can be seen from these examples and suggested by Sangster et al. (2000), there are many areas of overlap (e.g., terminology, instrumental techniques, and basic geological concepts) among geological, environmental, and forensic sciences. However, the approaches taken to arrive at certain conclusions, particularly in terms of "age dating" and "source identification", are not quite consistent. Perhaps with recent advances that have occurred in mass spectrometric techniques such as the development of new generations of multicollector ICP-MS and laser ablation MC-ICP-MS, which can produce fast, definitive, cost-effective elemental and lead isotopic ratio data (with nearly the same quality as TIMS data), gaps among these fields can be narrowed. Thus, better use of data and information can be established. As a result, in environmental forensic investigations, it should be expected that such robust and tightly connected geological-environmental data set can be most effectively used for building strong defensible cases that will be acceptable in a court of law.

REFERENCES

Angle, C.R., W.I. Manton, and Stanek, K.L. Stable isotope identification of lead sources in preschool children—the Omaha Study. *Clinical Toxicology*, 1995, 33, 657–662.

Arrowsmith, P. Laser ablation of solids for elemental analysis by inductively coupled plasma mass spectrometry. *Analytical Chemistry*, 1987, 59, 1437–1444.

(ATSDR) Agency for Toxic Substances and Disease Registry. Toxicological Profile For Lead. US Department Of Health and Human Services, Public Health Service, Agency for Toxic Substances and Disease Registry, 1999, http://www.atsdr.cdc.gov/toxprofiles.

Audétat, A., D. Günther, and Heinrich, C.A. Formation of a mgmatic-hydrothermal ore deposit-insights with LA-ICP-MS analysis of fluid inclusions. *Science*, 1998, 279, 2091–2094.

Banks, E.C., L.E., Ferreti, and Shucard, D.W., Effects of low level lead exposure on cognitive function in children: A review of behavioral, neuropsychological and biological evidence. *Neurotoxicology*, 1997, 18, 237–282.

Beaudoin, G., and Sangster, D.F., A descriptive model for silver-lead-zinc veins in clastic metasedimentary terranes. *Economic Geology*, 1992, 87, 1005–1021.

Becker, J.S. State-of-the-art and progress in precise and accurate isotope ratio measurements by ICP-MS and LA-ICP-MS. *Journal of Analytical Atomic Spectroscopy*, 2002, 17, 1172–1185.

Becker, J.S. and H.-J. Dietze. Double-focusing sector field inductively coupled plasma mass specrometry for highly sensitive multi-element and isotopic analysis. *Journal of Analytical Atomic Spectrometry*, 1997, 12, 881–889.

Bellinger, D, A. Leviton, C. Waternauz, H. Needleman, and Rabinowitz, M. Longitudinal analyses of prenatal and postnatal lead exposure and early cognitive development. *New England Journal of medicine*, 1997, 316, 1037–1043.

Bellinger, D., A. Leviton, M. Rabinowitz, E. Allred, H. Needleman, and Schoenbaum, S. Weight gain and maturity in fetuses exposed to low levels of lead. *Environmental Ressearch*, 1999, 54, 151–158.

Blaskett, D.R., and Boxall, D. *Lead and its alloys*. Horwood, New York, 1990, p. 161.

Bollhofer, A., and Rossman, K.J.R. Isotopic source signatures for atmospheric lead: The Northern Hemisphere. *Geochimica et Cosmochimica Acta*, 2001, 65, 1727–1740.

Brill, R.H. Lead and oxygen isotopes in ancient objects. Philosophical Transaction Royal Society of Loudon, 1970, A269, 143–164.

Brill, R.H. and Wampler, J.M. Isotope studies of ancient lead. *American Journal of Archaeology*, 1967, 71, 63–77.

Brown, R.S., J.R. Millette, and Mount, M.D. Application of Scanning Electron Microscopy for Pollution Particle Source Determination in Residential Dust and Soil. *Scanning*, 1995, 17, 302–305.

Chaudhary-Webb, M., D. Paschal, C.W. Elliott, H. Hopkins, A.M. Ghazi, B. Ting, and Romieu, I. Determination of lead isotope ratios in whole blood and pottery-lead sources in Mexico City. *Atomic Spectroscopy*, 1999, 19, 156–163.

Chaudhury-Webb, M., D. Paschal, I. Romieu, B.Ting, W.C. Elliott, H.P. Hopkins, L.E. Sanin, and Ghazi, A.M. Determining lead sources in Mexico using lead isotopes. *Revista Salud Pública de México*, 2003, 45, S183–S188.

Chiaradia, M., B. Chenhall, A.M. Depers, B.L. Gulson, and Jones, B.G. Identification of historical lead sources in roof dusts and recent lake sediments from an industrialized area: Indications from lead isotopes. *Science Total Environment*, 1997, 205, 107–128.

Chow, T.J. and Johnstone, M.S. Lead isotopes in gasoline and aerosols of Los Angeles basin, California. *Science*, 1965, 147, 502–503.

Chow, T.J., C.B. Snyder, and Earl, J.L. Isotope ratios of lead as pollutant source indicators. Proceedings IAEA-SM 191/4, International Atomic Energy Commission, Vienna, 1975, 95–108.

Date, A.R., and Gray, A.L. Plasma source mass spectrometryr using an inductively coupled plasma and high resolution quadropole mass filter. *Analyst*, 1981, 106, 1255–1267.

Date, A.R., and Gray, A.L. Development progress in plasma source mass spectrometry. *Analyst*, 1983, 108, 159–165.

Date, A.R., and Gray, A.L. *Applications of inductively coupled plasma mass spectrometry*, Blackie, Glasgow, 1989, p. 254.

Davis, A., J.W. Drexler, M.V. Ruby, and Nicholson, A. Micromineralogy of mine wastes in relation to lead bioavailability, Butte, Montana. *Environmental Science Technology*, 1993, 27, 1415–1425.

Dickin, A.P. *Radiogenic isotope geology*, Cambridge University Press, Cambridge, 1995, p. 506.

Dietrich, K.N., K.M., Krafft, R.L. Bornschein, P.B., Hammond, O. Berger, P.A. Succop, and Bier, M. Low-level fetal lead exposure effect on neurobehavioral development in early infancy. *Pediatrics*, 1987, 80, 721–730.

Eckstrand, O.R., W.D. Sinclair, and Thorpe, R.I. Geology of Canadian mineral deposit types. Geological Survey of Canada, Geology of Canada, 1995, No. 8. p. 633.

Eighmy, T.T., B.S. Crannell, L.G. Butler, F.K. Cartledge, E.F. Emery, D. Oblas, I.E. Krzanowski, J.D. Eusden, E.L. Shaw, and Francis, C.A. Heavy metal stabilization in municipal solid waste combustion dry scrubber residue using soluble phosphate. *Environmental Science and Teehnology*, 1997, 31, 3334–3338.

Elias, R.W., Y. Hirao, and Patterson, C.C. The circumvention of the biopurification of calcium along nutrient pathways by atmospheric inputs of industrial lead. *Geochimca et Cosmochimca Acta*, 1982, 46, 2561–2580.

Erel, Y., and Patterson, C.C. Leakage of industrial lead into the hydrocycle. *Geochimica et Cosmochimica Acta*, 1994, 58, 3289–3296.

Ericson, J.E., H. Shirahata, and Patterson, C.C. Skeletal concentration of lead in ancient pruvian. *New England Journal of Medicine*, 1979, 300, 946–951.

Faure, G. *Principles of Isotope Geology* (2nd Edition). John Wiley & Sons, New York, 1986, p. 589.

FDA. Direct food substances affirmed as generally recognized as safe. US Food and Drug Administration. Code of Federal Regulations. 1994, 21 CFR 184.

Fey, D.L., D.M. Unruh, and Church, S.E. Chemical data and lead isotopic compositions in stream-sediment samples from the Boulder River Watershed, Jefferson County, Montana: US Geological Survey Open File Report 99-575, 1999, p. 147.

Flegal, A.R., K.J.R Rosman, and Stephenson, M.D. Isotope systematic of contaminant leads in Monterey Bay. *Environmental Science and Technology*, 1987, 21, 1075–1079.

Flegal, A.R., A.R. Duda, and Niemeyer, S. High gradients of lead isotopic composition in north-east Pacific up-welling filaments. *Nature*, 1989, 339, 458–460.

Gale, N.H. Lead isotope studies applied to provenance studies—a brief review. In *Archaeometry*, Proceeding of the 25th international symposium, Y. Maniatis ed., Elsevier, 1989, 469–502.

Gale, N.H. and Stos-Gale, Z.A. Bronze Age archaeometallurgy of the Mediterranean: The impact of lead isotope studies. In: *Archaeological Chemistry IV*, R.O. Allen ed., American Chemical Society, Advances in Chemistry, Series, 1992, 220, 165–169.

Gasser, U.G., W.J. Walker, R.A. Dahlgren, R.S. Borch, and Burau, R.G. Lead release from smelter and mine waste impacted materials under simulated gastric conditions and relation to speciation. *Environmental Science Technology*, 1996, 30, 761–769.

George, A.M. Introductory remarks by organizer & host Conference on Lead Poisoning Prevention & Treatment, Banglore, India. February 8–10, 1999. http://www.leadpoison.net/general/intro.htm.

Ghazi, A.M. Laser ablation ICP-MS: A new elemental and isotopic ratio technique in environmental forensic investigation, *Environmental Forensic*, 6, 7–8.

Ghazi, A.M. Lead in Archaeological samples: An isotopic study by ICP-MS. *Applied Geochemistry*, 1994, 9, 627–636.

Ghazi, A. Mohamad, and Millette, J.R. Pb-Isotope Fingerprinting for Origin of Lead-Based Paint: A New Environmental Forensic Application for Laser Ablation High Res-

olution ICPMS (LA-HR-ICPMS). *Environmental Forensics*, 2004, 5, 97–108.

Ghazi, A.M. and Shuttleworth, S. Quantitative Trace elemental analysis of single fluid inclusions by laser ablation ICPMS (LA-ICPMS), *Analyst*, 2000, 125, 205–210.

Ghazi, A.M., K.J., Reinhard, and Durrance, E.M. Brief communication, further evidence of lead contamination of Omaha skeleton: An application of ICP-MS. *Journal of American Physical Anthropologists*, 1994, 95, 427–434.

Ghazi, A.M., D.A. Vanko, E. Roedder, and Seeley, R.C. Determination of rare earth elements in fluid inclusions by inductively coupled plasma mass spectrometry (ICP-MS). *Geochimica et Cosmochimica Acta*, 1993, 57, 4513–4516.

Ghazi, A.M., T. McCandless, J. Ruiz, and Vanko, D.A. Quantitative elemental determination in fluid inclusions by laser ablation ICPMS: Applications to Sr and Rb measurements in fluid inclusions in halite. *Journal of Analytical Atomic Spectrometry*, 1996, 11, 667–674.

Ghazi, A.M., S. Shuttleworth, S.J. Angula, and Pashley, D.H. Gallium Diffusion in Human Root Dentin: Quantitative Measurements by Pulsed Nd: YAG Laser Ablation ICPMS (LA-ICPMS). *Journal of Clinical Laser Medicine and Surgery*, 2000a, 18, 173–183.

Ghazi, A.M., S. Shuttleworth, S.J. Angula, and Pashley, D.H. New applications for laser ablation high resolution ICPMS (LA-HR-ICPMS): Quantitative measurements of gallium diffusion across human root dentin. *Journal of Analytical Atomic Spectrometry*, 2000b, 15, 335–341.

Ghazi, A.M., J.W. Wataha, S. Shuttleworth, R.B. Simmons, N.L. O'dell, Singh, B.B. Quantitative concentration profiling of nickel in tissues around metal implants: A new biomedical application of laser ablation sector field ICP-MS. *Journal of Analytical Atomic Spectrometry*, 2002a, 17, 1295–1299.

Ghazi, A.M., S. Shuttleworth, R.B. Simmons, S.J. Paul, and Pashley, D.A. Nanoleakage at the dentin adhesive interface vs μ-tensile bond strength: A new application for laser ablation high resolution ICPMS. *Journal of Analytical Atomic Spectrometry*, 2002b, 17, 682–687.

Gray, A.L. and Jarvis, K.E. *Applications of Inductively Coupled Plasma Mass Spectrometry*. Chapman & Hall, 1989, p. 324.

Gray, A.L., S. Houk, and Jarvis, K.E. *Handbook of Inductively Coupled Plasma Mass Spectrometry*. Chapman & Hall, 1992, p. 380.

Grousset, F.E., C.R. Quetel, B. Thomas, B., P. Buat-Menard, O.F.X. Donard, and Bucher, A. Transient Pb isotopic signatures in the western European atmosphere. *Environmental Science and Technology*, 1994, 28, 1605–1608.

Gulson, B.L. Uranium–lead and lead–lead investigations of minerals from the Broken Hill lodes and mine sequence rocks. *Economic Geology*, 1984, 79, 476–490.

Gulson B.L. Shale-hosted lead–zinc deposits in northern Australia; lead isotope variations. *Economic Geology*, 1985, 80, 2001–2012.

Gulson, B.L. *Lead isotopes in mineral exploration*. Elsevier, Amsterdam, 1986, p. 245.

Gulson, B.L. Tooth analyses of sources and intensity of lead exposure in children. *Environmental Health Perspective*, 1996, 104, 306–312.

Gulson, B.L. A brief review of the lead isotope fingerprinting method. Conference on Lead Poisoning Prevention & Treatment, Banaglore, India. February 8–10, 1999, http://www.leadpoison.net/screen/a-brief.htm.

Gulson, B.L., J.J. Davis, K.J. Mizon, M.J. Korsch, and Bawden-Smith, J. Sources of soil and dust and the use of dust fallout as a sampling medium. *Science of Total Environment*, 1995a, 166, 245–262.

Gulson, B.L., J.J. Davis, and Bawden-Smith, J. Paint as a source of recontamination of houses in urban environments and its role in maintaining elevated blood leads in children. *Science of Total Environment*, 1995b, 164, 221–235.

Gulson B.L., K.J. Mizon, A.J. Law, M.J. Korsch, and Davis, J.J. Source and pathway of lead in humans from the Broken Hill mining community—an alternative use of exploration methods. *Economic Geology*, 1994, 89, 889–908.

Gulson, B.L., K.J. Mizon, M.J. Korsch, and Noller, B.N. Lead isotopes as seepage indicators around a uranium tailings dam. *Environmental Science Technology*, 1989, 23, 290–294.

Gulson, B.L., K.J. Mizon, J.D. Davis, J.M. Palmer, and Vimpani, G. Identification of sources of lead in children in a primary zinc-lead smelter environment. *Environmental Heath Prespective*, 2004, 112, 52–60.

Günther, D., A. Audetat, R. Frischknecht, and Heinrich, C.A. Quantitative analysis of major, minor and trace elements using laser ablation inductively coupled plasma mass spectrometry. *Journal of Analytical Atomic Spectrometry*, 1998, 13, 263–270.

Günther, D., R. Frischknecht, C.A. Heinrich, and Kahlert, H.J. Capabilities of an argon fluoride 193 nm excimer laser for laser ablation inductively coupled plasma mass spectrometry microanalysis of geological materials. *Journal of Analytical Atomic Spectrometry*, 1997a, 12, 939–944.

Günther, D., R. Frischknecht, H.J. Müschenborn, and Heinrich, C.A. Direct liquid ablation: A new calibration strategy for laser ablation-ICP-MS microanalysis of solids and liquids. *Fresenius Journal of Analytical Chemistry*, 1997b, 359, 390–393.

Gwiazda, R.H., and Smith, D.R. Lead isotopes as a supplementary tool in the routine evaluation of household lead hazards. *Environmental Health Perspective*, 2000, 108, 1091–1097.

Gwiazda, R., D. Woolard, and Smith, D. Improved lead isotope ratio measurements in environmental and biological samples with a double focusing magnetic sector inductively coupled mass spectrometer (ICP-MS). *Journal of Analytical atomic Spectrometry*, 1998, 13, 1233–1238.

Hansmann, W., and Köppel, V. Lead-isotopes as tracers of pollutants in soils. *Chemical Geology*, 2000, 171, 123–144.

Hernandez-Avila, M., M. Cortez-Lugo, I. Munoz, and Rojo-Soliz, M.M.T. Lead exposure in developing countries. Conference on Lead Poisoning Prevention & Treatment, Banaglore, India. February 8–10, 1999, http://www.leadpoison.net/studies/exposure.htm.

Hieftje, G.M. Towards the next generation of plasma source mass spectrometers. *Journal of Analytical Atomic Spectrometry*, 1992, 7, 783–790

Hilts, P.J. US opens a drive on lead poisoning in nation's young. The New York Times, December, 20, 1991.

Hinners, T.A., R. Hughes, P.M. Outridge, W.J., Davis, K. Simon, and Woolard, D.R. Interlaboratory comparison of mass spectrometric methods for lead isotopes and trace elements in NIST SRM 1400 Bone Ash. *Journal of Analytical Atomic Spectrometry*, 1998, 13, 963–970.

Houk, R.S., V.A. Fassle, G.D. Flesch, H.J. Svec, A.L. Gary, and Taylor, C.E. Inductively coupled plasma as an ion source for mass spectrometric determination of trace elements. *Analytical Chemistry*, 1980, 52, 2238–2289.

HUD. Lead-based paint poisoning prevention in certain residential structures. US Department of Housing and Urban Development. Code of Federal Regulations 1998, 24 CFR 35.

Hurst, R.W. Application of anthropogenic lead archaeostratigraphy (ALAS Model) to hydrocarbon remediation, *Environmental Forensic*, 2002, 1, 11–23.

Hurst, R.W. Invited Commentary on Dr. Isaac Kaplan's Paper "Age Dating of Environmental Organic Residues". *Environmental Forensic*, 2003, 4, 145–152.

Hurst, R.W., T.E. David, and Chinn, B.D. The lead fingerprint of gasoline contamination. *Environmental Scince and Technology*, 1996, 30, 304A–307A.

IARC (International Agency for Research on Cancer). IARC monographs on the evaluation of the carcinogenic risk of chemicals to humans. Some metals and metallic compounds. Lyon, France, 1980, 23, p. 438.

IARC (International Agency for Research on Cancer). IARC monographs on the evaluation of the carcinogenic risk of chemicals to humans. Vol. 23: Some metals and metallic compounds. Lyons France: World Health Organization, International Agency for Research on Cancer, 1980, 352–415.

IARC (International Agency for Research on Cancer). IARC monographs on the evaluation of the carcinogenic risk of chemicals to humans: Overall evaluations of carcinogenicity. Supplement 7: An updating of the IARC monographs volumes 1 to 42. Lyon, France: World Health Organization, International Agency for Research on Cancer, 1987, 230–232.

IARC (International Agency for Research on Cancer). Directory of on-going research in cancer epidemiology: 1989/90. Lyon, France: World Health Organization, International Agency for Research on Cancer, 1989, 54, 453–454. IARC No. 101.

Jakubowski, N., L. Moens, and Vanhaecke, F. Sector field mass spectrometry in ICP-MS. *Spectrochimica Acta B*, 1998, 53, 1739–1763.

Jarvis, I., and Jarvis, K.E. Plasma spectrometry in the Earth Sciences: Techniques, applications and future trends. *Chemical Geology*, 1992, 95, 1–33.

Jarvis, K.E., A.L. Gray, and Houk, R.S. Handbook of Inductively Coupled Plasma Mass Spectrometry. Blackie, Glasgow, 1992, p. 380.

Jaworski, J. Lead. In: *Lead. Mercury. Cadmium and Arsenic in the Environment, Scope 31*, Te. Hutchinson and K.M. Meema eds, New York: Wiley & Sons, 1987, pp. 3–17.

Jeffries, T.E., S.E. Jackson, and Longerich, H.P. Application of a frequency quintupled Nd:YAG source (=213 nm) for laser ablation inductively coupled plasma mass spectrometric analysis of minerals. *Journal of Analytical Atomic Spectrometry*, 1998, 13, 935–940.

Johnson, D.L., and Hunt, A. Analysis of Lead in Urban Soils by Computer Assisted SEM/EDX—Method Development and Early Results. In: *Lead in Paint, Soil, and Dust: Health Risks, Exposure Studies, Control Measures, Measurement Methods, and Quality Assurance, ASTM STP 1226*, M.E. Beard and S.D.A. Iske, eds, American Society for Testing and Materials, Philadelphia, 1995, 283–299.

Kaplan, I.R., Age dating of environmental organic residue. *Environmental Forensic*, 2003, 4, 95–141.

Keating, J. Lead. In: *Canadian minerals yearbook, 1997*. Natural Resources Canada, Ottawa, Ont. 1997, pp. 28.1–28.14.

Kennedy, S.K., W. William Walker, and Forslund, B. Speciation and Characterization of Heavy Metal-Contaminated Soils Using Computer-Controlled Scanning Electron Microscopy. *Environmental Forensics*, 2002, 3, 131–143.

Kesler, S. *Mineral Resources, Economics and the Environment*. MacMillan College Publishing Company, Inc. New York, 1994, p. 391.

Ketterer, M.E., M.J. Peters, and Tisdale, P.J. Verification of correction procedure for measurements of isotope ratio by inductively coupled plasma mass spectrometry. *Journal of Analytical Atomic Spectrometry*, 1991, 6, 439–443.

Kirkham, R.V., W.D. Sinclair, R.I. Thorpe, and Duke, J.M. Mineral deposit modeling. Geological Association of Canada Special Paper, 1993, No. 40, p. 798.

Koppenaal, D. W. Atomic Mass Spectrometry. *Analytical Chemistry*, 1992, 64, 320R–340R.

Laperche, V., S.J. Trains, P. Gaddam, and Logan, T.J. Chemical and mineralogical characterizations of Pb in a contaminated soil: Reactions with synthetic apatite. *Environmental Science and Technology*, 1996, 30, 3321–3326.

Lichte, F.C. Determination of elemental content of rocks by laser ablation inductively coupled plasma mass spectrometry. *Analytical Chemistry*, 1995, 67, 2479–2485.

Longerich, H.P., B.J. Fryer, and Strong, D.F. Determination of lead isotope ratios by inductively coupled plasma-mass spectrometry (ICP-MS), *Spectrochimica Acta*, 1987, 42B, 39–48.

Longerich, H.P., S.E. Jackson, B.J. Fryer, and Strong, D.F. The laser ablation microprobe-inductively coupled plasma-mass spectrometer. *Geoscience Canada*, 1993, 20, 21–27.

Ma, L.Q., and Rao, G.N. *Journal of Environmental Quality*, 1997, 26, 788–794.

Ma, Q.Y., T.J. Logan, and Trains, S.J. Lead immobilization from aqueous solutions using phosphate rocks. *Environmental Science and Teehnology*, 1995, 29, 1118–1126.

McCandless, T.E., D.J. Lajack, Ruiz, J. and Ghazi, A.M. Trace element determination of single fluid inclusions in quatz by laser ablation ICP-MS, *Gesotandard Newsletter*, 1997, 21, 271–278.

McGill, R.A.R, J.M. Pearce, N.J. Fortey, J. Watt, L. Ault, L. and Parrish, R.R. Contaminant Source Apportionment by PIMMS Lead Isotope Analysis and SEM-Image Analysis. *Environmental Geochemistry and Health*, 2003, 25, 25–32.

Maning, H., D.M. Settle, P. Buat-Menard, F. Dulac, and Patterson, C.C. Stable lead isotope tracers of air mass trajectories in the Mediterranean region. *Nature*, 1987, 300, 154–156.

Manton, W.I. Sources of lead in blood. Identification by stable isotopes. *Archives of Environmental Health*, 1997, 32, 149–159.

Marshal, C.P., and Fairbridge, R.W. *Encyclopedia of Geochemistry*, Kluwer Academic Publisher, 1999, p. 712.

Millette, J.R., R.S. Brown, and Mount, M.D. Lead Arsenate. *Microscope*, 1995, 43(4), 187–191.

Minder, B., E.A. Das-Smaal, E.F. Brand, and Orlebeke, J.F. Exposure to lead and specific attentional problems in schoolchildren. *Journal of Learning Disability*, 1994, 6, 393–399.

Monna, F., J. Lancelot, I.W. Croudace, A.B. Cundy, and Lewis, J.T. Pb isotopic composition of airborne particulate material from France and the southern United Kingdom: Implications for lead pollution sources in urban areas. *Environmental Science and Technology*, 1997, 31, 2277–2286.

Montaser, A., and Golightly, D.W. *Inductively Coupled Plasmas in Analytical Atomic Spectrometry*, John Wiley & Sons, 2nd edition, 1992, p. 1017.

Montaser, A. 1998. *Inductively Coupled Plasmas Mass Spectrometry*. John Wiley & Sons, 2000, p. 1000.

Morrison, R. Application of forensic techniques for age dating and source identification in environmental litigation. *Environmental Forensics*, 2000, 1, 131–153.

Newman, A. Elements of ICP-MS. *Analytical Chemistry*, 1996, 65, 46A–51A.

Niederschlag, E., E. Pernicka, T.H., Seifert, and Bartelheim, M. The determination of lead isotope ratios by multiple collector ICP-MS: A case study of early bronze age artefacts and their possible relation with ore deposits of the Erzgebirge. *Achaeometry*, 2003, 45, 61–100.

NIOSH (National Institute for Occupational Safety and Health). National Occupational Hazard Survey (1972–74). Cincinnati, OH: Department of Health, Education, and Welfare, 1976.

NIOSH. NIOSH Manual of Analytical Methods, 4th edition. Methods 7082 (Lead by Flame AAS), 7105 (Lead by HGAAS), 7505 (Lead Sulfide), 8003 (Lead in blood and urine), 9100 (Lead in Surface Wipe Samples), US Department of Health and Human Services, Centers for Disease Control, National Institute for Occupational Safety and Health, 1994.

NIOSH. Protecting workers exposed to lead-based paint hazards. A report to congress. DHHS (NIOSH) Publication No. 98–112. January 1997. US Department of Health and Human Services, Center for Disease Control and Prevention, and National Institute for Occupational Safety and Health, 1997, pp. 1–74.

Nriagu, J.O. Lead in soils, sediments and major rock types. In: *The Geochemistry of Lead in the Environment*, J.O. Nriagu, ed., Amsterdam: Elsevier, 1978, pp. 15–72.

Nriagu, J.O. *Lead and lead poisoning in antiquity*, Wiley Interscience, New York, 1983, p. 437.

Nriagu, J.O. The rise and fall of leaded gasoline. *The Science of Total Environment*, 1990, 92, 13–28.

Nriagu, J.O., and Pacyna, J.M. Quantitative assessment of worldwide contamination of air, water and soils by trace metals. *Nature*, 1988, 333, 134–139.

Parsons, P.J. and Chisolm, J.J., Jr. Screening & Diagnosis the lead laboratory. Conference on Lead Poisoning Prevention & Treatment, Banaglore, India. February 8–10, 1999. http://www.leadpoison.net/screen/the-lead-lab.htm.

Patterson, C.C., H. Shirahata, and Ericson, J.E. Lead in ancient human bones and its relevance to historical developments of social problems with lead. *Science of Total Environment*, 1987, 61, 167–200.

Perkins, W.T., and Pearce, N.J.G. Microprobe techniques in Earth Sciences, P.J. Potts *et al.* eds, Chapman and Hall, London, 1995, pp. 291–325.

Perkins, W.T., N.J.G. Pearce, and Fuge, R. Analysis of zircon by laser ablation and solution inductively coupled plasma mass spectrometry. *Journal of Analytical Atomic Spectrometry*, 1992, 7, 612–616.

Pearce, N.J.G., W.T. Perkins, I. Abell, G.A.T Duller, and Fuge, R. Mineral microanalysis by laser ablation inductively coupled plasma mass spectrometry. *Journal of Analytical Atomic Spectrometry*, 1992, 7, 53–57.

Pearce, N.J.G., W.T. Perkins, J.A. Westgate, M.P. Gorton, S.E. Jackson, C.R. Neal, and Chenery, S.P. A compilation of new and published major and trace element data for NIST SRM 610 and NIST SRM 612 glass reference materials, *Geostandards Newsletter*, 1997, 21, 115–144.

Peryea, F.J. Historical use of lead arsenate insecticides, resulting soil contamination and implications for soil remediation. Proceedings, 16th World Congress of Soil Science, Montpellier, France, 1998. http://soils.tfrec.wsu.edu/leadhistory.htm.

Petit, D., J.P. Mennessier, and Lamberts, L. Stable lead isotopes in pond sediments as tracer of past and present atmospheric lead pollution in Belgium. *Atmospheric Environment*, 1984, 18, 1189–1193.

Rabinowitz, M.B. Stable isotope mass spectrometry in childhood lead poisoning. *Biological Trace Element Research*, 1987, 12, 223–229.

Rabinowitz, M.B., and Wetherill, G.W. Identifying sources of lead contamination by stable isotope techniques. *Environmental Science and Technology*, 1972, 6, 705–709.

Reinhard, K.J., and Ghazi, A.M. Environmental lead contamination of a 17th century Omaha Indian population as evidenced by Inductively Coupled Plasma Spectrometric (ICP-MS) analysis of skeletal remains and artifacts. *Journal of American Physical Anthropologist*, 1992, 89, 183–195.

Ritson, P.I., B.K. Esser, S. Niemeyer, and Flegal, A.R. Lead isotopic determination of historical sources of lead to

Lake Erie, North America. *Geochimca et Cosmochimica Acta*, 1994, 58, 3297–3305.

Romieu, I., M. Lacasana, and McConnell, R. Lead exposure in Latin America and the Caribbean. *Environmental Health Perspectives*, 1997, 105, 398–405.

Romieu, I., E. Palazuelos, A.M. Hernandez, C.M.I. Rios, C. Jimenez, and Cahero, G. Sources of lead exposure in Mexico City. *Environmental Health Perspectives*, 1994, 102, 384–389.

Rosman, K.J.R., W. Chisolm, C.F. Boutron, J.P. Candelone, and Gorlach, U. Isotopic evidence for the source of lead in Greenland snows since the late 1960s. *Nature*, 1993, 362, 333–334.

Rosman, K.J.R., W. Chisolm, C.F. Boutron, J.P. Candelone, and Hong, S. Isotopic evidence to account for changes in the concentration of Pb in Greenland between 1960 and 1988. *Geochimica et Cosmochimica Acta*, 1994a, 58, 3265–3269.

Rosman, K.J.R., W. Chisolm, C.F. Boutron, J.P. Candelone, and Patterson, C.C. Anthropogenic lead isotopes in Antarctica. *Geophysical Research Letters*, 1994b, 21, 2269–2672.

Ruby, M.V., R. Schoof, W. Brattin, M. Goldade, G. Post, M. Harnois, D.E. Mosby, S.W. Casteel, W. Berti, M. Carpenter, D. Edwards, D. Cragin, and Chappell, W. Advances in evaluating the oral bioavailability of inorganics in soil for use in human health risk assessment. *Environmental Science and Technology*, 1999, 33, 3697–3705.

Russ, G.P. Isotope measurements using ICP-MS. In: *Application of Inductively Coupled Plasma Mass Spectrometry*, A.R. Date and A.L. Gray, Blackie, Glagow, 1989, pp. 90–114

Russell, R.D., and Farquhar, R.M. Dating galenas by means of their isotopic constitutions, II. *Geochimica et Cosmochimica. Acta*, 1960, 19, 41–52.

Sangster, D.F. Mississippi Valley-type and SEDEX lead–zinc deposits: A comparative examination. Transaction Institution of Mining and Metallurgy, Section B, *Applied Earth Sciences*, 1990, 99, B21–B42.

Sangster, D.F., P.M. Outridge, and Davis, W.J. Stable lead isotope characteristics of lead ore deposits of environmental significance. *Environmental Review*, NRC Research, Canada, 2000, 8, 115–147.

Sargent, M., and Webb, K. Instrumental aspects of inductively coupled plasma-mass spectrometry. Spectroscopy Europe, 1993, 5, 21–28.

Sax, N.I. Hawley's Condensed Chemical Dictionary, 11th edition. New York: Van Nostrand Reinhold Corporation, 1987, pp. 276, 490, 633, 635, and 732.

Sheppherd, T.J. and Chenery, S.R. Laser ablation ICP-MS elemental analysis of individual fluid inclusions: An evaluation study. *Geochimica Cosmochimica Acta*, 1995, 59, 3997–4007.

Sherrell, R.M., E.A. Boyle, and Hamelin, B. Isotopic equilibration between dissolved and suspended particulate lead in the Atlantic Ocean: Evidence for 210Pb and stable Pb isotopes. *Journal of Geophysical Research*, 1992, 97, 11257–11268.

Shuttleworth, S. Optimization of laser wavelength in the ablation sampling of glass materials. *Surface Science*, 1996, 96–98, 513–517.

Shuttleworth, S. and Kemser, D.T. An assessment of laser ablation and sector field Inductively coupled plasma mass spectrometry for elemental analysis of solid samples. *Journal of Analytical Atomic Spectrometry*, 1996, 12, 412–422.

Sittig, M. *Handbook of Toxic and Hazardous Chemicals and Carcinogens*, 2nd edition, Park Ridge, N.J., Noyes Publications, 1985, p. 950.

Sturges, W.T., and Barrie, L.A. Lead 206/207 isotope ratios in the atmosphere of North America as tracers of US and Canadian emissions. *Nature*, 1987, 329, 144–146.

Sturges, W.T., J.F. Hopper, L.A. Barrie, and Schnell, R.C. Stable lead isotope ratios in Alaskan Arctic aerosols. *Atmospheric Environment*, 1993, 27A, 2865–2871.

Superfund Chemical Data Matrix (SCDM), US-EPA, January 2004 edition.

Tatsumoto, M. and Unruh, D.M. KREEP basalt age: Grain by grain U-Th-Pb systematics study of the quartz monzodiorite clast 15405, 88, Proceedings of the 7th Lunar Science Conference, 1976, 2107–2129.

Ulrich, T., D. Günther, and Heinrich, C.A. Gold concentrations of magmatic brines and the metal budget of porphyry copper deposits. *Nature*, 1999, 399, 676–679.

Unruh1, D.M. Lead isotopic analyses of selected soil samples from the USEPA Vasquez Blvd.-I-70 study area Denver, Colorado. Open-File Report 02-321, 2002.

Uribe-Hernández, R., A.J. Pérez-Zapata, M.J. Flores, F. Aldape, and Hernández-Méndez, B. Lead contents in blood samples of a children population of Mexico City related to levels of airborne lead determined by PIXE. *International Journal of PIXE*, 1996, 6, Nos 1–2, 255–262.

VanderWood, T.B. The Characterization of Lead Smelter Dusts by Automated Scanning Electron Microscopy. 1492–1493 Proceeding 50th Annual Meeting of the Electron Microscopy Society of America, G.W. Bailey, J. Bentley and J.A. Small, eds, San Francisco Press, 1992, 1492–1493.

Vander Wood, T.B., and Brown, R.S. The Application of Automated Scanning Electron Microscopy/Energy Dispersive X-Ray Spectrometry to the Identification of Lead-Rich Particles in Soil and Dust. *Environmental Choices Technical Supplement*, 1992, 1–1, 26–32.

Walczyk, T. TIMS versus multicollector-ICP-MS: Coexistence or struggle for survival? *Analytical Bioanalytical Chemistry*, 2004, 378(2), 229–231.

Walder, A.J. Advanced isotope ratio mass spectrometry II: Isotope ratio measurement by multiple collector inductively coupled plasma mass spectrometry. In: *Modern Isotope Mass Spectrometry*, I.T. Platzner ed., John Wiley, Chichester, 1997, 83–108,

Walder, A.J., and Furuta, N. High-precision lead isotope ratio measurement by inductively coupled plasma multiple collector mass spectrometry. *Analytical Sciences*, 1993, 9, 675–680.

Walder, A.J., I.D. Abell, I. Platzner, and Freedman, P.A. Lead isotope ratio easurements of NIST 610 glass by laser ablation inductively coupled mass spectrometry. *Spectrochimica Acta*, 1993a, 48B, 397–402.

Walder, A.J., D. Koller, N.M. Reed, R.C. Hutton, and Freedman, P.A. Isotope ratio measurement by inductively coupled plasma multiple collector mass spectrometry incorporating a high efficiency nebulization system. *Journal of Analytical Atomic Spectrometry*, 1993b, 8, 1037–1041.

Wataha, J.C., N.L. O'Dell, B.B. Singh, A.M. Ghazi, G.M. Whiteford, and Lockwood, P.E. Relating Nickle-induced tissue inflammation to nickle release in vivo. *Journal of Biomedical Material Research*, 2001, 49, 537–544.

Watmough, S.A., R.J. Hughes, and Hutchinson, T.C. 206Pb/207Pb ratios in tree rings as monitors of environmental changes. *Environmental Science Technology*, 1999, 33, 670–673.

White, W.M., F. Albarède, and Télouk, P. High-precision analysis of Pb isotope ratios by multi-collector ICP-MS, *Chemical Geology*, 2000, 167, 257–270.

WHO (World Health Organization). *Regional Office for Europe: Air quality guidelines*, Geneva, Switzerland, 2002, 6, p. 17.

WHO (World Health Organization). *Guidelines for Drinking-Water Quality*. Geneva, Switzerland, Volume I: Recommendations, 2004, pp. 392–394.

Winchester, J.W. Aerosol composition in remote and contaminated atmospheres: Application of PIXE analysis. *Neurotoxicology*, 1983, 4, 369–390.

Woolard, D., R. Franks, and Smith, D. An inductively coupled plasma-magnetic sector mass spectrometry method for stable lead isotope tracer studies. *Journal of Analytical atomic Spectrometry*, 1998, 13, 1015–1019.

Yaffe, Y., C.P. Flessel, J.J. Wesolowski, A. del Rosario, G.N. Guirguis, V. Matias, J. Gramlich, W.R. Kelly, T.E. Degarmo, and Coleman, G.C. Identification of lead sources in Califonia children using the stable isotope ratio technique. *Archives of Environmental Health*, 1983, 38, 237–245

Young, G.M. Late Proterozoic stratigraphy and the Canada–Australia connection. *Geology*, 1992, 20, 215–218.

5

Chromium

Julie K. Sueker

Contents

5.1 INTRODUCTION

Chromium (Cr) is a hard, steel-grey metallic element that is listed by the US Environmental Protection Agency (USEPA) as one of 129 priority pollutants (Keith and Telliard, 1979). Chromium is also listed among the 25 hazardous substances thought to pose the most significant potential threat to human health at priority superfund sites (USDHHS, 1987). Chromium has an atomic weight of 51.9961, a density of 7.14, a high melting point of 1900 °C, moderate thermal expansion, and a stable cubic crystalline structure (Merck, 1983). The name "chromium" is derived from the Greek word "chroma" meaning "color" and refers to the many colored compounds that contain chromium.

Chromium is the sixth most abundant element in the earth's crust, where it is found in small quantities associated with other metals, particularly with iron as chromite ($FeCr_2O_4$) and with lead as crocoite ($PbCrO_4$). The average crustal abundance of chromium is 100 mg/kg (Klein and Hurlbut, 1993). Average concentrations of chromium in granite, shale, and basalt are 10, 90, and 170 mg/kg, respectively (Drever, 1997), and natural chromium concentrations in western US soils range from 1 to 2000 mg/kg (Schacklette and Boerngen, 1984). Soils derived from serpentine strata may contain 500 to 62,000 mg/kg chromium (Irwin, 1997).

Chromium is used in three basic industries: (1) *metallurgical*, for the making of stainless steel and various metal alloys; (2) *chemical*, in dyes and pigments, chrome plating, leather tanning, and wood preservation; and (3) *refractory*, including magnesite-chrome firebrick for metallurgical furnace linings and granular chromite for various other heat-resistant applications (ATSDR, 2000). Smaller amounts of chromium are used in drilling muds, water treatment, catalysts, safety matches, copy-machine toner, corrosion inhibitors, photographic chemicals, and magnetic tapes (ATSDR, 2000).

Chromium is a common groundwater and soil contaminant, particularly in industrial areas, and is the second most abundant inorganic groundwater contaminant at hazardous waste sites (NAS, 1994). Typical concentrations of chromium in natural water range from 0.5 to 10 μg/L (Lillebo *et al.*, 1986; Hem, 1989). Contamination of groundwater by chromium may occur due to leaching of chromium from land disposal of solid wastes including mining wastes, seepage from industrial lagoons, and spills and leaks from industrial metal processing or wood preserving facilities. Releases from chrome plating operations at aircraft manufacturing facilities during World War II were the cause of the first known instances of groundwater contamination by chromium, and many of these plumes moved thousands of feet over 50 years, with little or no natural attenuation (USEPA, 2000).

Chromium exists in oxidation states ranging from −2 to +6; however, the most stable forms are elemental chromium (Cr^0), trivalent chromium (Cr(III) or Cr^{3+}), and hexavalent chromium (Cr(VI) or Cr^{6+}) (Palmer and Puls, 1994). Cr(III) is the most common form of naturally occurring chromium. Most of the Cr(VI) found in the environment is a result of domestic and industrial emissions (ATSDR, 2000). Although rare, natural Cr(VI) does occur in the mineral crocoite ($PbCrO_4$) (ATSDR, 2000). In addition, Cr(III) can be oxidized to Cr(VI) under certain geochemical conditions (USEPA, 2000)

Geochemical controls determine the speciation, transport, and fate of chromium in the environment. In both freshwater and marine environments, hydrolysis and precipitation are the most important processes that determine the environmental fate of chromium, whereas sorption and bioaccumulation are relatively minor (Eisler, 1986). Cr(III) is largely immobile in the environment. However, Cr(III) can be mobilized by forming complexes with dissolved organic carbon (DOC). Cr(VI) compounds are reduced to Cr(III) in the presence of certain iron species and oxidizable organic matter (ATSDR, 2000). However, Cr(VI) is generally stable and mobile in natural waters where there are low concentrations of reducing materials (ATSDR, 2000).

In addition to understanding chromium mobility in the environment, chromium speciation is important for understanding potential human and ecological health risks. Cr(III) is an essential dietary nutrient that is required for maintaining efficient glucose, lipid, and protein metabolism (ATSDR, 2000; Eisler, 1986) and for vascular integrity (Irwin, 1997; ATSDR, 2000). Chromium deficiency has been associated with a number of health problems including impaired glucose tolerance, elevated percent body fat, cardiovascular disease, and impaired fertility (ATSDR, 2000). Although Cr(III) is an essential nutrient, exposure to high levels via inhalation, ingestion, or dermal contact may cause adverse health effects (ATSDR, 2000). Cr(VI) is a Class A (known) carcinogen by inhalation of high doses of sparingly soluble and soluble chromate salts (ATSDR, 2000), possibly inducing mutations in living cells by damaging DNA–protein cross-linkages via its role as a strong oxidizing agent (ATSDR, 2000) and high membrane permeability (Eisler, 1986).

This chapter provides information regarding the principal occurrence of chromium in the natural environment, the history of chromium usage, the geochemistry of chromium, health concerns from chromium exposure, analytical procedures for determining chromium concentrations in environmental samples, and forensic techniques for chromium.

5.2 PRINCIPAL OCCURRENCE OF CHROMIUM

High amounts of chromium are found naturally in two minerals: chromite ($FeCr_2O_4$) and crocoite ($PbCrO_4$), also known as lead chromate or "red lead." Crocoite is a brilliant reddish-orange colored, four-sided crystal mineral that was discovered in 1765 at the Beresof mine near Ekaterinburg in the Ural Mountains of Siberia. Crocoite is too rare to be useful commercially, but is prized by mineral collectors for its brilliant color. Notable occurrences of crocoite are located in the Dundas District of Tasmania, Australia; the Ural Mountains, Russia; and Inyo and Riverside Counties in California, USA. In 1797, the French chemist Louis Nicholas Vauquelin produced chromium oxide (CrO_3) by mixing crocoite with hydrochloric acid. In 1978, Vequelin isolated elemental chromium by heating chromium oxide in a charcoal oven. Vauquelin discovered that chromium compounds can form reds, brilliant yellows, and deep greens, and that traces of chromium were responsible for the green color of a Peruvian emerald and the red of rubies.

Chromite is a dark dull mineral that forms in igneous environments and is the primary commercial chromium ore. In 1798, Lowitz and Klaproth independently discovered chromium in samples of heavy black rock that is now called chromite, found in a deposit north of the Beresof Mines. In 1798 a German chemist named Tassaert discovered chromium in chromite ore in a small deposit in the Var region of South-Eastern France. Chromite forms in deep ultra-mafic magmas and is one of the first minerals to crystallize from cooling magma. Because of their density, chromite crystals fall to the bottom of the magma body and concentrate there. Chromite is also found in metamorphic rocks such as serpentines. As is indicated by its early crystallization, chromite is resistant to the altering affects of high temperatures and pressures, thus it is generally unaffected by the metamorphic processes. This characteristic explains the use of chromites as a refractory component in the bricks and linings of blast furnaces.

The discovery of chromite ore deposits in the Ural Mountains greatly increased the supplies of chromium to the growing paint industry and resulted in a chromium chemicals factory being set up in Manchester, England around 1808. In 1827, Isaac Tyson identified deposits of chromite ore on the Maryland–Pennsylvania border and the USA became the monopoly supplier of chromium for a number of years. High-grade chromite deposits were found near Bursa in Turkey in 1848 and with the exhaustion of the Maryland deposits around 1860, it was Turkey that then became the main source of supply. The mining of chromium ore started in India and Southern Africa around 1906. Today, roughly one third to one half of the chromite ore in the world is produced from South Africa; Kazakhstan, India, and Turkey are also substantial producers (ATSDR, 2002). Approximately 15 million tons of marketable chromite ore were produced in 2003 (ICDA, 2005). Untapped chromite deposits are plentiful, but are geographically concentrated in Kazakhstan and southern Africa. Chromite mining in the United States ceased in 1961 (ATSDR, 2000).

Native chromium deposits are rare; however, some native chromium metal has been discovered. The Udachnaya Mine in Russia is a kimberlite pipe rich in diamonds and produces samples of the native metal. The reducing environment within the kimberlite pipe helped produce both elemental chromium and diamond.

5.3 HISTORY OF USAGE

Chromium metal is obtained commercially by heating chromite ore in the presence of carbon, aluminum, or silicon, and subsequent purification (ATSDR, 2000). Sodium chromate and dichromate are produced by roasting chromite ore with soda ash and are the primary chromium salts from which most other chromium compounds are derived (ATSDR, 2000). Historical uses of chromium by major industry classifications are presented on Table 5.3.1 (ATSDR, 2000).

The first known commercial use for chromium began in the late 1700s as a paint pigment, chrome yellow, derived from crocoite directly or from chromate produced from chromite. As early as 1820 the textile industry was using large amounts of chromium compounds, such as potassium dichromate, as mordants (chemical agents) to fix or stabilize dyes. Mordants bind with the dye and the fibers of a material and prevent bleeding and fading of the colored dye. Other uses of chromium as a colorant were developed. Chromium salts and chromium oxide (Cr_2O_3) are used to

color glass and enamel paint an emerald green. Chromium is used in producing synthetic rubies.

Beginning in the mid-1800s, iron manufactures discovered that adding chromium to steel produced a harder, more useful metal by delaying the transformation that occurs as steel is cooled. Steels with three to five percent chromium were produced beginning in 1865. The high strength and corrosion resistant properties of "stainless" steel containing more than five percent chromium were discovered in the early 1900s. Chromium steel resists warping and melting under conditions of extreme heat and is ideal for high-temperature applications such as jet-engine components. Chromium compounds are also used to anodize aluminum, a process which coats aluminum with a thick, protective layer of oxide (literally changing the surface into ruby). Stainless and other chrome-containing steels have many applications and the use of chromium in the production of stainless steel and other metal alloys currently accounts for approximately 85% of chromium consumption (ICDA, 2005).

Investigations into chromium plating techniques using chromium chloride and sulfate salts began in the mid-1800s; however, the fundamental principals of chromium electroplating were not discovered until 1924. Most metals plate from chloride or sulfate salts but chromium plates best from chromic acids. This technique was discovered by chance when a chromic acid solution was electrolyzed and a chromium deposit was noted. Chromic acid has the hypothetical structure H_2CrO_4. However, chromium trioxide, CrO_3, the acid anhydride of chromic acid, is sold industrially as "chromic acid". Chromium plating, or "chrome" plating, was first used in the production of jewelry, then for the plating of plumbing fixtures and household appliances, and car manufacturers soon began making chrome bumpers and molding. Chrome-plated articles were esthetically pleasing and functionally desirable due to their shiny surface and resistance to corrosion. Because chrome is a very hard metal and has a low coefficient of friction, chrome plating was also used for extending the life of parts that receive heavy wear, such as automobile cylinders, and was useful in boiler pipes to prevent build-up of scale (mineral deposits).

Refractory materials are highly resistant to heat and are chemically stable. These materials are used as insulation to line the inside of blast furnaces and crucibles used in metal manufacturing. Chromite was initially used as a refractory in France, and bricks of solid chromite cut straight from the mine were used without further refinement or processing up until the 1890s. As a cost-saving measure, manufacturers developed refractory bricks made of crushed chromite and resins that were shaped into bricks. The use of chromium as a refractory material declined in the late 1900s as other refractory materials gained popularity.

Most woods are susceptible to attack by fungus and insects. To make wood more resistant to attack, insecticides and anti-fungal agents can be added. The use of chromium as a wood preservative began in the early 1930s. The most common chromium-containing wood preservative is chromated copper arsenate, or CCA, that is applied to the wood under pressure (also known as the "Wolmanizing" process). Due to operating procedures that were standard practices at the time, nearly all wood preserving plants 30 years or older present some degree of soil and groundwater contamination (USEPA, 2000). Hexavalent chromium is the most significant soil and groundwater contaminant associated with the use of CCA at these older wood preserving plants (USEPA, 2000). Within the wood itself, Cr(VI) typically is reduced to Cr(III) through a complex series of "fixation" reactions (Lebow *et al.*, 2003). Potential health concerns associated with direct contact with CCA lumber

Table 5.3.1 *Historical Use of Chromium in the United States and the Western World (in %)*

Use	Western World 1996	United States 1996	United States 1951
Wood Preservation	15	52	2
Leather tanning	40	13	20
Metals finishing	17	13	25
Pigments	15	12	35
Refractory	3	3	1
Other	10	7	17

Source: USEPA, 2000.

(primarily due to risks associated with arsenic exposure) led to an agreement between USEPA and CCA producers in the early 2000s to phase out use of CCA-treated wood for most residential applications (USEPA, 2003).

Some additional chromium applications include the use of chromium salts in leather tanning, which was adopted commercially in 1884, and the use of chromium(IV) oxide (CrO_2) to manufacture magnetic tape, where its higher coercivity than iron oxide tapes gives better performance. Potassium dichromate is a powerful oxidizing agent and is the preferred compound for cleaning laboratory glassware of any possible organics. Dichromates are also used as oxidizing agents in laboratories for quantitative analysis. As can be seen, chromium is a commercially useful element that is used in a wide variety of industrial, chemical, and refractory applications.

5.4 GEOCHEMISTRY

Geochemical controls determine the speciation, transport, and fate of chromium in the environment. Chromium exists in oxidation states ranging from -2 to $+6$ but the most stable states are elemental chromium (Cr^0), trivalent chromium (Cr(III) or Cr^{3+}), and hexavalent chromium (Cr(VI) or Cr^{6+}) (Palmer and Puls, 1994). Cr(III) is the most common form of naturally occurring chromium. Most of the Cr(VI) found in the environment is a result of domestic and industrial emissions (ATSDR, 2000); however, Cr(III) can be oxidized to Cr(VI) under certain geochemical conditions (USEPA, 2000) and natural occurrence of Cr(VI) in groundwater has been discovered (Robertson, 1975; Ball and Izbicki, 2002; Farías et al., 2003).

Chemical processes responsible for inter-converting chromium between the two main oxidation states have been the subject of many investigations. Most studies were conducted in laboratories to identify important redox reactions of chromium and to determine reaction kinetics. As cited by Lin (2002), many investigators have identified reactants that are capable of transforming chromium between Cr(III) and Cr(VI) in water. These reactants include ferrous iron [Fe(II)], sulfite [S(IV)], manganese [Mn(III) and Mn(IV)], hydroxyl radical ($\cdot$OH), vanadium [V(II), V(III), and V(IV)], and arsenic [As(III)] (Lin, 2002). Expected reactions and estimated reaction rates for transformations between Cr(III) and Cr(VI) have been compiled by Ball and Nordstrom (1998) and Lin (2002).

The equilibrium oxidation states and chemical forms of chromium within specified reduction–oxidation (redox) potential (Eh) and pH ranges are described by Eh–pH diagrams. Figure 5.4.1 shows an Eh–pH diagram for chromium for 'typical' aqueous conditions (Palmer and Wittbrodt, 1991). As shown, hexavalent chromium (i.e., $HCrO_4^-$ and CrO_4^{2-}) is generally stable under oxidative (high Eh) and/or high pH conditions and trivalent chromium (i.e., Cr^{3+}, $CrOH^{2+}$, $Cr(OH)_2^+$, $Cr(OH)_3^0$, and $Cr(OH)_4^-$) is stable under most Eh conditions at lower pH and generally more reducing conditions at higher pH (Palmer and Wittbrodt, 1991). Chromium is unusual in that both Cr(VI) and Cr(III) can be thermodynamically stable in the presence of atmospheric oxygen, depending on pH (Hering and Harmon, 2004).

The Eh–pH diagram (Figure 5.4.1) implies that boundaries separating one species from another are distinct; however, the transformations are not always so clear (USEPA, 2000). Chromium speciation is affected by concentration, pressure, temperature, and the absence or presence of other aqueous ions. Cr(III) has been observed to remain stable outside of theoretical solubility limits (Ahern et al., 1985).

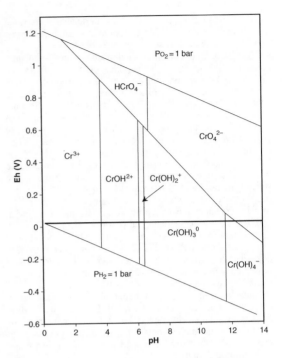

Figure 5.4.1 Eh–pH diagram for chromium (Source: Palmer and Wittbrodt, 1991).

Unlike the relatively non-reactive Cr(VI), Cr(III) is readily hydrolyzed and binds strongly to particles and organic material (Eary and Rai, 1987). If the binding materials are present as dissolved species, these reactions may lower the availability of Cr(III) for oxidation and thus cause an apparent redox disequilibrium (Johnson et al., 1992).

Kinetic controls on redox processes may also affect redox disequilibrium for chromium species (Figure 5.4.2). For example, Cr(VI) can be reduced by oxidation of reduced sulfur or iron species or by organic reducing agents on a time scale of minutes to hours. However, the presence of strong reductants is required for the reduction of Cr(VI) to Cr(III) to occur (Schroeder and Lee, 1975). On the other hand, oxidation of Cr(III) by manganese oxides is relatively slow – on the order of days to months (Schroeder and Lee, 1975; Eary and Rai, 1987), potentially allowing Cr(III) to remain in solution under oxidizing conditions. Therefore, a measure of caution must be exercised when using the Eh–pH diagram as site-specific conditions can significantly alter actual Eh–pH boundaries (USEPA, 2000).

Because Cr(III) and Cr(VI) have different solubilities and affinities for mineral surfaces, the mobility of chromium will depend strongly on its redox speciation (Hering and Harmon, 2004). Cr(III) is largely immobile in the environment (Figure 5.4.2). Cr(III) is strongly sorbed as Cr(III) or hydroxyl ionic species onto hydrated iron and manganese oxides or clay mineral surfaces (USEPA, 2000; Hering and Harmon, 2004), and as with most cationic species, sorption of Cr(III) increases with pH (Bartlett and Kimble, 1976a; Dzomback and Morel, 1990). Cr(III) oxidation appears to be largely governed by manganese oxides that occur mainly as surface coatings, crack deposits, or finely disseminated grains (Bartlett and James, 1979). Manganese oxides have a high affinity for sorption of cations and may provide a local surface environment in which the coupled processes

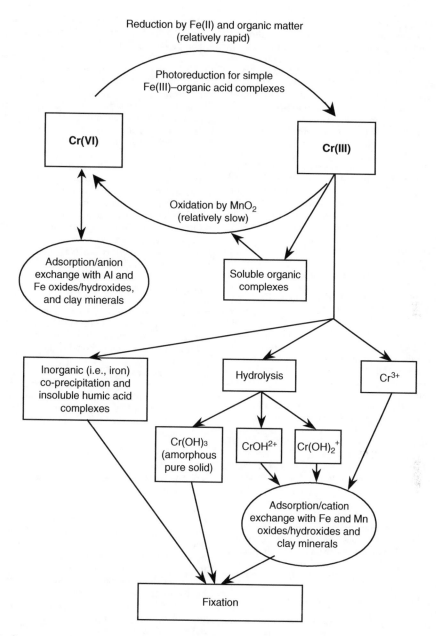

Figure 5.4.2 *Conceptual model depicting potential processes affecting inorganic and organic chromium in the aqeous environment (Adapted from USEPA, 2000).*

of aqueous Cr(III) oxidation and manganese reduction may take place (Eary and Rai, 1987). The proposed reaction involves the sequential adsorption of Cr(III) onto MnO_2 surface sites, followed by oxidation by surface Mn^{4+} and subsequent release or desorption of Cr(VI) (Puls *et al.*, 1994). Eary and Rai (1987) determined that Cr(III) is not oxidized by oxygen to any appreciable extent, and that the rate of Cr(III) oxidation by β-MnO_2 decreases with increasing pH and decreasing surface area to solution volume.

Cr(III) is only sparingly soluble in the neutral pH range. The maximum concentration of dissolved, inorganic Cr(III) in equilibrium with amorphous $Cr(OH)_3(s)$ is $7.5\,\mu g/L$ (Ball and Nordstrom, 1998). Chromium hydroxide solid solutions

may also form, particularly as amorphous $Cr_xFe_{1-x}(OH)_3$ (Puls *et al.*, 1994). This low solubility typically causes Cr(III) to be retained in the solid phase as colloids or precipitates (Johnson *et al.*, 1992). In a mixture of soil solution and $CrCl_3$, Bartlett and Kimble (1976a) observed a decrease in inorganic Cr(III) solubility with increased pH above 4, and apparent complete loss of Cr(III) from solution at pH 5.5. However, Cr(III)–organic acid complexes were observed to form under acidic conditions and these complexes remained stable after pH was raised to neutrality (Bartlett and Kimble, 1976a). Therefore, Cr(III)–organic acid complexes may remain in solutions at much greater concentrations than predicted by Cr(III) solubility or Eh–pH conditions.

Kaczynski and Kieber (1994) demonstrated that organic Cr concentration was correlated with DOC concentration and that for high-DOC surface water, concentrations of organic Cr exceeded that of total inorganic Cr. Humic substances were found to be important complexing agents in all water studied (Figure 5.4.2) (Kaczynski and Kieber, 1994). Photolysis experiments by Kaczynski and Kieber (1994) indicated that the organo-chromium species were photodegradable even after short-term exposure to ambient sunlight. The assumption of Cr(VI) reduction to Cr(III) leading to the immobilization of Cr may be unrealistic in water with high DOC or manganese concentrations. Reduced Cr(III) may form complexes with organic carbon, resulting in increased leaching and mobility of dissolved organically-bound Cr(III) (Walsh and O'Halloran, 1996a,b); and re-oxidation of the dissolved organically-bound Cr(III) to Cr(VI) by transport of the organically-bound Cr to regions of more oxidizing conditions may occur (Masscheleyn et al., 1992).

Some Cr(VI) salts are generally soluble, leading to potentially high concentrations of Cr(VI) in aqueous environments (ATSDR, 2000). Important mineral phases governing the solubility of Cr(VI) include $PbCrO_4$ (crocoite), $PbCrO_4 \cdot H_2O$ (iranite), K_2CrO_4 (tarapacite), and $BaCrO_4$ (hashemite) (Puls et al., 1994). In contrast to Cr(III), groundwater concentrations of Cr(VI) are not expected to be limited by the solubility of solid chromate-containing phases (Hering and Harmon, 2004). Although chromate minerals, such as crocoite, do occur in nature, they are generally found in evaporite deposits and would not be predicted to precipitate from dilute solution (Hering and Harmon, 2004).

Cr(VI) anions [bichromate ($HCrO_4^-$) and chromate (CrO_4^{2-})] can be sorbed through anion exchange by positively charged aluminum and iron oxides and hydroxides, and by clay minerals (Figure 5.4.2) (Bartlett, 1991). Cr(VI) adsorption is strongly pH dependent and decreases with increasing pH (Bartlett and James, 1979). Cr(VI) anion species can exchange with chloride (Cl^-), nitrate (NO_3^-), sulfate (SO_4^{2-}), and phosphate (PO_4^{3-}) (USEPA, 2000). Sorption of Cr(VI) is strong compared with sorption of chloride and nitrate (Bartlett and Kimble, 1976b). However, suppression of Cr(VI) sorption can occur in the presence of competing anions, such as sulfate and phosphate, especially at circumneutral pH (Bartlett and Kimble, 1976b; Zachara et al., 1987). Cr(VI) adsorption can also decrease due to increased ionic strength (Puls et al., 1994). The sorption behavior of Cr(VI) suggests an outer-sphere surface complexation reaction with various mineral surfaces (Puls et al., 1994). Interestingly, phosphate extraction of sorbed Cr(VI) from acidic soils (dominated by amorphous aluminum and iron oxides and organic complexes) indicated that adsorption of Cr(VI) protected the sorbed Cr(VI) from reduction (Bartlett, 1991).

Cr(VI) is generally stable and mobile in natural waters. Cr(VI) compounds are reduced to Cr(III) in the presence of certain iron and sulfur species and oxidizable organic matter (Figure 5.4.2) (Schroeder and Lee, 1975; Bartlett and Kimble, 1976b). Mineral phases capable of providing ferrous iron through dissolution or through surface-mediated reactions include biotite, magnetite, iron sulfides, chlorite, and nontronite (Puls et al., 1994). Chromate is a strong oxidizing agent in acid solutions, meaning that Cr(VI) can be easily reduced under acidic conditions (Bartlett, 1991), as is demonstrated by the instability of Cr(VI) at low pH (Figure 5.4.1). However, Cr(VI) can be stable in neutral water (Bartlett, 1991) or in oxic waters where there are low concentrations of reducing materials (Bartlett and Kimble, 1976b; ATSDR, 2000).

Photoreduction of Cr(VI) to Cr(III) has been demonstrated for simple Fe(III)–organic acid complexes (Figure 5.4.2) (Gaberell et al., 2003). Gaberell et al. (2003) investigated the role of the composition of dissolved organic matter (DOM) on the Cr(VI) photoreduction capability of organic matter. Photoreduction rates were found to increase with increasing iron concentrations, but the type of DOM influenced the rate of photoreduction, with faster reduction rates for DOM derived from terrestrial sources compared with autochthonous materials (Gabarell et al., 2003). Larger more aromatic substrates were observed to induce faster reduction.

Lin (2002) developed a hybrid chemical kinetic and equilibrium model to quantitatively assess the redox chemistry of chromium in the natural water system. The model incorporated identified chromium redox reactions with known kinetics, interactions among the chromium oxidants/reductants, and the aqueous-phase chemical equilibria important in chromium chemistry (Lin, 2002). Ferrous iron was predicted to be the predominant reductant for Cr(VI) at neutral pH (e.g., 6.5–7.5) due to rapid reduction rate of Cr(VI) by Fe(II). At pH greater than 8.0, dissolved oxygen was predicted to rapidly oxidize Fe(II), which greatly decreased the Cr(III) formation. At low pH values (e.g., 4.0–5.0), S(IV) was the predominant reductant for Cr(VI). Since Fe(II) does not form highly stable complexes with common organic ligands, the inhibition of Cr(VI) reduction caused by the complexation of Fe(II) with natural chelating agents should be limited in natural waters (Lin, 2002).

In summary, as Bartlett and Kimble (1976b) state, "Cr(VI) will remain mobile only if its concentration exceeds both the adsorbing and the reducing capacities of the soil." In addition, Bartlett (1991) states, "the marvel of the chromium cycle in soil is that oxidation and reduction can take place at the same time."

5.5 REGULATORY STANDARDS AND HEALTH EFFECTS

Irwin (1997) and ATSDR (2000) provide excellent overviews of potential beneficial and adverse effects of chromium in the environment. Potential beneficial effects and potential adverse effects due to chromium deficiency were briefly described in the introduction. Potential adverse effects due to excessive exposure to chromium in the environment and regulatory standards considered to be protective of health are the focus of this section. The USEPA regards all chromium compounds as toxic or potentially toxic, although the biological effects of chromium depend on chemical form, solubility, valence, and dose (Eisler, 1986). Chromium (VI) is much more toxic than Cr(III) for both acute and chronic exposures (IRIS, 2000a and b).

Routes of human exposure to chromium include inhalation, ingestion, skin and/or eye contact. Chromium is released to air primarily by combustion processes and metallurgical industries. Chromium release to water and soil is primarily from permitted or accidental releases from industrial and chemical industries that utilize chromium in their operations.

The USEPA has classified Cr(VI) as a Group A, known human carcinogen by the inhalation route of exposure (IRIS, 2000b). Cr(VI) causes cellular damage via its role as a strong oxidizing agent and ability to penetrate biological membranes, and is associated with cancer risk as well as kidney and liver damage (Eisler, 1986; Meyers, 1990; ATSDR, 2000). Chromic acid, dichromates, and other Cr(VI) compounds are powerful skin irritants and can be corrosive to skin (ATSDR, 2000). Inhalation of certain Cr(VI) compounds can cause airway irritation, airway obstruction, and possibly induce lung tumors in humans (ATSDR, 2000). Adverse renal effects have been reported in humans after inhalation, ingestion, and dermal exposure to chromium and acute chromium exposure can result in hepatic necrosis (ATSDR, 2000). However, the human body does have some capacity to neutralize potential adverse effects due to chromium exposure. Unless the dose of Cr(VI) is sufficient to overwhelm the body's reductive capacity, Cr(VI) can be reduced to less toxic Cr(III) by the epithelial lining fluid of the respiratory tract, and normal stomach acid pH can convert Cr(VI) to Cr(III) (ATSDR, 2000) although some absorption of both chromium species can occur through the lungs, skin, and intestinal tract.

Cr(III) is an essential nutrient and exhibits low acute and chronic toxicity (ATSDR, 2000). The National Academy of Science has established a safe and adequate daily intake for chromium in adults of 50–200 micrograms per day (ATSDR, 2000). No evidence exists to indicate that Cr(III) can cause cancer in humans or animals and the USEPA (IRIS, 2000a) has classified Cr(III) in Group D, not classifiable as carcinogenic in humans. However, sensitization of lung tissue resulting in asthmatic response can occur due to exposure to Cr(III) compounds, although it is more common from exposure to Cr(VI) (ATSDR, 2000).

Various United States regulatory agencies have established guidelines or criteria to reduce adverse health effects due to exposure to chromium (Table 5.5.1). The United States Occupational Safety and Health Administration (OSHA) mandated a permissible exposure level of $0.1 mg CrO_3/m^3$ for chromic acid and chromates (OSHA, 2005). The time-weighted average (TWA) PEL is $1 mg Cr/m^3$ for chromium metal and all insoluble chromium salts and $0.5 mg Cr/m^3$ for Cr(II) and Cr(III) salts (OSHA, 2005). The National Institute of Occupational Safety and Health (NIOSH) considers all Cr(VI) compounds potentially carcinogenic and has set a recommended exposure limit (REL) of $0.001 mg Cr/m^3$ and an imminently dangerous to life and health (IDLH) PEL of $15 mg Cr/m^3$ (NIOSH, 2005).

The USEPA has set a maximum contaminant level (MCL) for water of $0.1 mg/L$ total chromium (USEPA, 2002a). The USEPA national generic soil screening levels (SSLs) for Cr(VI) in soils range from 2 to $38 mg/kg$ for protection from migration to groundwater to $390 mg/kg$ for the ingestion pathway (USEPA, 1996).

Chromium toxicity to aquatic biota is significantly influenced by geochemical variables—such as water hardness or alkalinity, temperature, pH, and salinity—and biological factors—such as species, life stage, and potential differences in sensitivities of local populations (Eisler, 1986). High levels of Cr(VI) are toxic to both higher plants and to microorganisms (Bartlett, 1991). Chromium toxicity risk to plants is primarily related to acidic sandy soil with low organic content (Irwin, 1997). In plants, chromium interferes with uptake translocation and accumulation by plant tops of several essential elements and aggravates iron deficiency chlorosis by interfering with iron metabolism (Irwin, 1997). For Cr(VI), the water quality criteria maximum concentration (CMC) and criteria continuous concentrations (CCC) are $16 \mu g/L$ and $11 \mu g/L$, respectively, for freshwater, and $1100 \mu g/L$ and $50 \mu g/L$, respectively, for saltwater (USEPA, 2002b).

Table 5.5.1 *Regulations and Guidelines Applicable to Chromium*

Agency	Description	Information	Reference
International			
Guidelines			
IARC	Cancer Classification		IARC (1990)
	Chromium(0)	Group 3[a]	
	Chromium(III)	Group 3[a]	
	Chromium(VI)	Group 1[b]	
	European standards for drinking water-Cr(VI)	$0.05 \mu g/L$	WHO (1988)
United States			
(a) Air			
ACGIH	TLV-TWA—Cr, metal and inorganic compounds as Cr		ACGIH (1999)
	Metal and Cr(III) compounds	$0.5 mg/m^3$	
	Water soluble Cr(VI)	$0.05 mg/m^3$	
	Insoluble Cr(VI)	$0.01 mg/m^3$	
EPA	Cr(III) RfC	Not available	IRIS (2000a)
	Cr(VI) carcinogenic risk from inhalation exposure	$1.2 \times 10^{-2} mg/m^3$	IRIS (2000b)
	Chromic acid mists and dissolved Cr(VI) aerosols RfC	$8 \times 10^{-6} mg/m^3$	IRIS (2000b)
	Cr(VI) particulates RfC	$1 \times 10^{-4} mg/m^3$	IRIS (2000b)
NIOSH	REL 8-hour TWA		
	Chromium metal	$0.5 mg/m^3$	NIOSH (1999)
	Chromium(II)	$0.5 mg/m^3$	NIOSH (1999)
	Chromium(III)	$0.5 mg/m^3$	NIOSH (1999)
	Chromium(VI) carcinogenic	$0.001 mg/m^3$	NIOSH (1999)
	Chromyl chloride (carcinogenic)	$0.001 mg Cr(VI)/m^3$	NIOSH (1999)

(Continued)

Table 5.5.1 *(Continued)*

Agency	Description	Information	Reference
OSHA	8-Hour TWA		29 CFR 1910.1
	Chromium(II)	$0.5\,mg/m^3$	
	Chromium(III)	$0.5\,mg/m^3$	
	Cr metal and insoluble salts	$1.0\,mg/m^3$	
	Chromic acid and chromates	$1.0\,mg\ CrO^3/10m^3$	
(b) Water EPA			
	MCL—Chromium	$0.1\,mg/L$	40 CFR 141.62
	MCLG—Chromium	$0.1\,mg/L$	40 CFR 141.51
	MCL groundwater	$0.05\,mg/L$	40 CFR 264.94
	Drinking water standards—Cr		40 CFR 141.32
	Ambient water quality criteria (water and fish consumption)		EPA (1980, 1987)
	Chromium(III)	$170\,mg/L$	
	Chromium(VI)	$0.05\,mg/L$	
	Water quality criteria		40 CFR 264.94
	Chromium(II)		
	Freshwater	$74\,\mu g/L$	
	Saltwater	No Value	
	Chromium(III)		
	Freshwater	$11\,\mu g/L$	
	Saltwater	$50\,\mu g/L$	
	Chromium(VI)		
	Freshwater	$11\,\mu g/L$	
	Saltwater	$50\,\mu g/L$	
	Health advisories for Cr(III+VI) total		EPA 1996c
	10-kg child		
	1 day	$1.0\,mg/L$	
	10 day	$1.0\,mg/L$	
	Longer term	$0.2\,mg/L$	
	70-kg adult		
	Longer term	$0.8\,mg/L$	
	Lifetime	$0.1\,mg/L$	
	DWEL[c]	$0.2\,mg/L$	
(c) Food			
FDA	Reference daily intake for vitamins and minerals—chromium	$120\,\mu g$	21 CFR 101.9
(d) Other			
ACGIH	Cancer classification		ACGIH 1999
	Metal and Cr(III) compounds	A4[d]	
	Water soluble Cr(VI) compounds	A1[e]	
	Insoluble Cr(VI) compounds	A1[e]	
EPA	Chromium(III)		IRIS 2000a
	RfD (oral)	$1.5\,mg/kg/day$	
	RfC	Not available	
	Cancer classification	D—not classified	
	Chromium(VI)		IRIS 2000b
	RfD (oral)		
	Oral cancer classification	$1.2 \times 10^{-2}\,\mu g/m^3$	
	Carcinogenic risk-inhalation exposure		
	Inhalation cancer classification	A—known human carcinogen	

[a] Group 3: Not classifiable as to have carcinogenic potential.

[b] Group 1: Carcinogenic in humans.

[c] DWEL: Drinking water equivalent level. A lifetime exposure concentration protective of adverse, non-cancer health effects, which assumes all of the exposure to a contaminant is from a drinking water source.

[d] A4: Not classifiable as a human carcinogen.

[e] A1: Confirmed human carcinogen.

ACGIH—American Conference of Governmental Industrial hygienists; EPA—Environmental Protection Agency; FDA—Food and Drug Administration; IARC—International Agency for Research on Cancer; MCL—maximum contaminant level; MCLG—maximum contaminant level goal; NIOSH—National Institute for Occupational Safety and Health; OSHA—Occupational Safety and Health Administration; PEL—permissible exposure limit; REL—recommended exposure limit; RfC—inhalation reference concentration; RfD—reference dose; TLV—threshold limit value; TWA—time weighted average; WHO—World Health Organization.

Source: ATSDR (2000).

5.6 ANALYTICAL TECHNIQUES

There are many analytical techniques available for determination of chromium in the environment. Water, soil, sediment, and sludge samples are generally analyzed for total chromium and Cr(VI); Cr(III) is then determined by subtracting the results of the Cr(VI) analysis from the total chromium values (USEPA, 2000). However, due to increasing concern regarding stability of Cr(VI) during sample handling and storage and the need for better characterizing chromium speciation, several techniques have been developed to separate different chromium species in the field prior to sample transport or in the laboratory prior to sample analysis. The following sections describe standard and developing analytical techniques for characterizing chromium concentrations and speciation in environmental samples (Table 5.6.1). In addition, analyses that may be useful for characterizing site-specific chromium geochemistry are presented.

5.6.1 Analysis of Chromium in Water

Suggested analytical techniques for determination of total chromium in water include acid digestion (SW-846 3005A)

followed by analysis using inductively-coupled plasma-atomic emission spectroscopy (ICP-AES) [SW-846 6010B] or inductively-coupled plasma-mass spectroscopy (ICP-MS) [SW-846 6020] (Table 5.6.1). Dissolved chromium (total) can be determined by 0.45 micron filtration, followed by acid digestion [SW-846 3020A] and analysis using atomic absorption (AA) spectroscopy [SW-846 7191].

A common direct analytical method for determining concentrations of Cr(VI) in water is colorimetry by reaction with diphenylcarbazide in acid solution [SW-846 7196A]. A red-violet color of unknown composition is produced when Cr(VI) is present. The absorbance of the solution at a wavelength of 530 nm is then measured to determine the concentration of Cr(VI) in the sample. Cr(VI) can also be pre-concentrated using ion chromatography (IC) followed by the diphenylcarbazide determination technique [SW-846 7199]. Other EPA-approved analytical techniques for determining concentrations of Cr(VI) in water include co-precipitation of Cr(VI) as lead chromate using lead sulfate, followed by resolubilization and analysis by AA [SW-846 7195]; chelation of Cr(VI) using ammonium pyrrolidine dithiocarbamate (APDC) extraction with methyl isobutyl ketone (MIBK) and aspiration into flame of AA

Table 5.6.1 *Analytical Methods for Determining Chromium in Environmental Samples*

Sample Matrix	Preparation method	Analytical method	Sample detection limit	Reference/Method
Dissolved Cr(VI): drinking water, surface water, and certain domestic and industrial effluents	Complex Cr(VI) in water with APDC at pH 2.4 and extract with MIBK	Furnace AAS	2.3 μg/L	Method 218.5
Cr(VI): drinking water, groundwater, and water effluents	Buffer solution introduced into ion chromat. Derivitized with diphenylcarbazide	Ion chromatography at 530 nm	0.3 μg/L	Method 7199
Total Cr: water	Calcium nitrate added to water and Cr is converted to Cr(III) by acidified hydrogen peroxide	GFAAS or ICP/AES	1.3 μg/L ICP/AES, 7.0 μg/L GFAAS	Method 218.2 and 7191
Total Cr: industrial wastes, soils, sludges, sediments, and other solid wastes	Digest with nitric acid/hydrogen peroxide	ICP-AES	4.7 μg/L	Method 6010
Soluble Cr: oil wastes, oils, greases, waxes, crude oil	Dissolve in xylene or MIBK	AAS or GFAAS	0.05 mg/L	Method 7190
Cr(VI): groundwater, domestic and industrial waste	Cr(VI) is co-precipitated with lead sulfate, reduced, and resolubilized in nitric acid	AAS or GFAAS	0.05 mg/L AAS, 2.3 μg/L GFAAS	Method 7195
Cr(VI): groundwater-EP extract, domestic, and industrial waste	Chelation with ammonium pyrrolidine dithiocarbonate and extraction with MIBK	AAS	No data	Method 218.4 and 7197
Cr(VI): water, wastewater, and EP extracts	Direct	DPPA	10 μg/L	Method 7198
Cr(VI): soil, sediment, and sludges	Alkaline digestion extraction using Na_2CO_3 and NaOH	UV-VIS	No data	Method 3060A and 7196A

AAS—atomic absorption spectrophotometry; APDC—ammonium pyrrolidine dithiocarbonate; DPPA—differential pulse polarographic analysis; EP—extraction procedure; GFAAS—graphite furnace atomic absorption spectrometry; ICP-AES—inductively coupled plasma-atomic emission spectrometry; MIBK—methyl isobutyl ketone; NaOH—sodium hydroxide; Na_2CO_3—sodium carbonate. Source: ATSDR (2000).

[SW-846 7197]; and differential pulse polarography (DPP) [SW-846 7198]. A greater risk of sample contamination is posed by co-precipitation and extraction methods due to the generally greater sample handling requirements (Kotaś and Stasicka, 2000).

Hering and Harmon (2004) provide an overview of potential shortcomings of standard EPA-approved protocols for analysis of chromium in water. Typically, determinations of total Cr (unfiltered) or dissolved Cr (filtered) are used to assess compliance with the MCL for Cr and determinations of Cr(VI) are used to assess health risks presumed to be associated with exposure to inorganic Cr(VI). Cr(III) is then determined by subtracting Cr(VI) from total Cr. However, this method of determination for Cr(III) includes both colloidal and organic Cr as shown in Figure 5.6.1 (Hering and Harmon, 2004). Inclusion of the colloidal and organic fractions may lead to an overestimation of true Cr(III) concentration in water and may result in an apparent exceedance of solubility limits for Cr(III) (Hering and Harmon, 2004).

Reducing environments, such as wetlands, are thought to act as sinks for chromium; however, little is known about chromium cycling and mobility in organic-rich environments (Icopini and Long, 2002). The fraction of chromium associated with organic matter can be determined by collection of Cr onto a hydrophobic resin with subsequent leaching and analysis by ICP-AES or ICP-MS (Kaczynski and Kieber, 1994; Icopini and Long, 2002). Kaczynski and Kieber (1994) demonstrated that organic Cr concentration was correlated with DOC concentration and that for high-DOC surface water, concentrations of organic Cr exceeded that of total inorganic Cr. As described above, the assumption of Cr(VI) reduction to Cr(III) leading to the immobilization of Cr may be unrealistic in high-DOC water because reduced Cr(III) may form complexes with DOC, result-ing in increased mobility of Cr(III) (Walsh and O'Halloran, 1996a,b), and re-oxidation of the organically-bound Cr(III) to Cr(VI) by transport of the organically-bound Cr to regions of more oxidizing conditions may occur (Masscheleyn et al., 1992). In addition, the oxidation state of organically-bound Cr is generally not identified (Icopini and Long, 2002) and DOC could potentially aid in the mobility of the more toxic Cr(VI) form of chromium.

The importance of organically bound chromium is likely to be less in low-DOC groundwater than in high-DOC surface waters (Hering and Harmon, 2004) and the assumption of little or no organic Cr may be valid for low-DOC groundwater. However, if the difference between reported total "dissolved" Cr and Cr(VI) exceeds the solubility limit for Cr(III), neglect of colloidal Cr species may be more problematic (Hering and Harmon, 2002). As cited in Hering and Harmon (2004), contributions of the colloidal size fraction to "dissolved" metals concentrations can be rigorously assessed by ultrafiltration (Wen et al., 1996; Benoit and Rozan, 1999), field-flow filtration (Taylor et al., 1992; Murphy et al., 1993), or photon correlation spectroscopy (Ledin et al., 1995). However, such an assessment has not been made for chromium in groundwater (Hering and Harmon, 2004).

As described by Hering and Harmon (2004), sample handling and storage protocols for determination of Cr(VI) pose some potential problems. For example, significant interconversion of Cr(III) and Cr(VI) during storage of spiked water and wastewater samples has been observed. At low pH, loss of Cr(VI) was observed within days, but could be prevented by adjusting the sample to a pH of 9. In samples with an excess of Cr(III) over Cr(VI), the concentration of Cr(VI) was observed to increase with time in samples buffered to pH 9 unless addition of the strong organic complexing agent EDTA was used to bind the Cr(III).

Due to the potential interconversion of Cr(VI) and Cr(III), there is a 24-hour holding time for liquid sample Cr(VI) analytical procedures, starting at time of sample extraction. This 24-hour hold time requirement may be quite onerous under certain sampling conditions such as sampling for Cr(VI) at remote site location or for certain applications such as effluent water quality monitoring for Cr(VI). To overcome potential problems with interconversion of Cr(VI) and Cr(III) species during sample handling and storage, several chromatographic techniques have been developed to separate Cr(VI) and Cr(III) prior to analysis (e.g., Johnson, 1990; Cox and McLeod, 1992). Some of these separation procedures can be applied in the field (e.g., Cox and McLeod, 1992; Icopini and Long, 2002; Ball and McClesky, 2003). The advantages of these techniques include accurate determination of Cr(VI) at concentrations as low as $0.05\,\mu g/L$, sample storage at least several weeks prior to analysis, and use of readily available instrumentation (Ball and McClesky, 2003).

Additional analytical techniques have been developed and continue to be developed for trace and ultra-trace determination of hexavalent chromium in water and for separation of chromium species in the field. Kotaś and Stasicka (2000) provide an in depth compilation and discussion of chromium speciation and measurement methods. As mentioned, many of these techniques offer lower detection limits than the standard methods, which typically are above $1\,\mu g/L$ (Table 5.6.1), and may provide accurate values at low concentration levels rather than non-detect values at higher detection limits. Hexavalent chromium concentrations generated by these lower detection limit methods may provide information for risk assessment, treatment, and remediation investigation that is more valuable than a series of non-detect values at a higher detection limit.

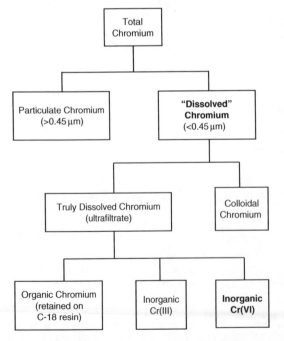

Figure 5.6.1 *Schematic diagram showing distribution of Cr species. Bold text indicates Cr species determined in standard analyses of potable water (Source: Hering and Harmon, 2004).*

5.6.2 Analysis of Chromium in Soil

Suggested analytical techniques for determination of total chromium in soil, sediment, and sludges are similar to those for water and include digestion by addition of nitric acid or nitric acid and hydrogen peroxide [SW-846 3050B, 3051, or 3052] followed by analysis using AA, ICP-AES, or ICP-MS [SW-846 7190, 6010B, or 6020] (Table 5.6.1). Other techniques for measuring chromium in soils include total chromium by neutron activation analysis (Helmke, 1996), x-ray absorption near-edge structures (XANES) spectroscopy for determination of Cr(VI) (Szulczewski et al., 1997, 2001); and isotope dilution analysis for exchangeable Cr(VI), i.e., chromium that is not permanently fixed in soil [SW-846 6800].

5.6.3 Other Analyses for Characterizing Chromium Geochemistry

Several additional analyses that may be useful for characterizing the potential mobility or remediation of chromium at a site are recommended by USEPA (2000) (USEPA, 2000). Determination of TOC in groundwater and soil and DOC in groundwater will indicate the availability of soluble organic ligands for Cr(III) complexing, which may provide a mobilization vehicle for potential oxidation to Cr(VI), and also the availability of more complex organic matter that has the potential for reduction of Cr(VI) to Cr(III). The total Cr(VI) reducing capacity of the soil can be determined using the Walkley–Black method to measure the portion of organic matter in the soil that is oxidizable by Cr(VI) (USEPA, 2000). Cation exchange capacity (CEC) of soil can be measured to determine whether sites are available for Cr(III)-hydroxy cation complexes to adsorb onto soil particles. Other tests that can be performed are porosity, grain size, soil moisture, and total manganese (USEPA, 2000). This list of potential analyses for understanding chromium dynamics at a particular site is by no means exhaustive, but should provide an understanding of the ongoing geochemical processes at a site as they relate to chromium characterization and remediation.

5.7 NATURAL CHROMIUM IN THE ENVIRONMENT

Chromium is a naturally occurring element found in rocks, animals, plants, and volcanic dust and gases. Natural chromium is widely disseminated in the environment and typically is found associated with iron as chromite in soils and rocks in concentrations that average about $100\,mg/kg$ (Drever, 1997).

Concentrations of chromium in natural waters generally are less than $10\,\mu g/L$ (e.g., Lillebo et al., 1986; Hem, 1989). Factors controlling chromium concentrations in natural waters can be quite complex. For example, Abu-Saba and Flegal (1997) determined that concentrations of dissolved chromium showed marked spatial and seasonal gradients throughout the San Francisco Bay Estuary during wet and dry season flows. Relatively large concentrations of dissolved chromium ($>1.0\,\mu g/L$) were observed during high flow periods while summer low flow concentrations (0.1 to $0.2\,\mu g/L$) were similar to oceanic surface water concentrations. Essentially all of the dissolved chromium present at concentrations greater than $0.15\,\mu g/L$ was present as Cr(III) and Cr(VI) concentrations were consistently about $0.1\,\mu g/L$. High flow chromium sources in the northern portion of the estuary were attributed to transport of chromium leached from alluvial sediments. Episodic increases in dissolved chromium concentrations in the southern portion of the estuary were attributed to sediment diagenesis and suboxic conditions linked to

decay of organic matter following spring diatom bloom (Abu-Saba and Flegal, 1997). Dissolved chromium sorbed onto particles, demonstrating that sediments serve as both a source and a sink in the chromium geochemical cycle.

Naturally occurring chromium in water typically occurs as the Cr(III) oxidation state. However, natural occurrences of elevated Cr(VI) in groundwater have been observed, primarily in arid or semi-arid regions (e.g., Robertson, 1975; Ball and Izbicki, 2002; Fantoni et al., 2002; Farías et al., 2003). In Argentina, concentrations of total Cr ranging from 20 to $230\,\mu g/L$ were found in Pampean Plain aquifers dominated by aeolean deposits with substantial contribution of volcanic materials (Farías et al., 2003).

Robertson (1975) measured concentrations of Cr(VI) up to $220\,\mu g/L$ in groundwater in Paradise Valley, Arizona, USA. Pardise Valley is located in the semi-arid Basin and Range physiographic province. The highest Cr(VI) concentrations were measured in the approximate center of the valley and concentrations declined toward the valley margins. The highest Cr(VI) concentrations were associated with fine grained sediments deposited in a playa environment and elevated pH. However, total Cr concentrations in the clay-silt fraction were not elevated. Thermodynamic calculations and Eh and pH measurements of the groundwater indicated that groundwater containing Cr(VI) was sufficiently oxidizing and alkaline to account for the observed Cr(VI) concentrations. Robertson (1975) determined that the alkaline environment, largely a result of primary silicate hydrolysis in the fine-grained sediments, allowed chemical oxidation of solid-phase Cr(III) to soluble Cr(VI).

In the western Mojave Desert, California, USA, elevated concentrations of Cr(VI) up to $60\,\mu g/L$ were detected in groundwater in fan alluvial deposits of mafic rock and concentrations of Cr(VI) up to $38\,\mu g/L$ were detected in groundwater from alluvial deposits of weathered volcanic and metamorphic rocks (Ball and Izbicki, 2002). Near recharge areas along the mountain front, Cr(VI) concentrations were below the $0.1\,\mu g/L$ detection limit; Cr(VI) concentrations and pH values increased downgradient. In groundwater underlying the desert floor, decreasing Cr(VI) concentrations were positively correlated with decreasing dissolved oxygen (O_2) concentrations along flow paths, indicating that reduction of Cr(VI) was occurring. Measurements of the isotopic composition of Cr(VI) in groundwater (described in Section 5.9) support this interpretation (Ball et al., 2001). Chromium(III) was detected at a maximum concentration of $2.6\,\mu g/L$ and was the predominant form of chromium only in areas where dissolved O_2 concentrations were below $1\,mg/L$.

In groundwater samples from the La Spezia Province, Italy, Fantoni et al. (2002) measured Cr(VI) concentrations ranging from 5 to $73\,\mu g/L$. Local ophiolites, especially serpentinites and ultramafites, are Cr-rich and represent a Cr source for groundwater. However, oxidation of the Cr(III) in these minerals is required for release of Cr(VI) to groundwater. Higher Cr(VI) concentrations were measured in Mg-HCO_3 type waters than in Ca-HCO_3 type waters. Fantoni et al. (2002) attributed this pattern to weathering of different minerals—periodites for Mg-rich waters and basalts for Ca-rich waters.

5.8 ANTHROPOGENIC CHROMIUM IN THE ENVIRONMENT

Anthropogenically-derived Cr(III) and Cr(VI) are released to the environment primarily from stationary point sources. Total atmospheric chromium emissions in the United States are about 2840 tons per year (ATSDR, 2000). Of these atmospheric chromium emissions, approximately 64% are comprised of Cr(III) from fuel combustion (residential, commercial, and industrial) and from steel production and approximately 32% are comprised of Cr(VI) from chemical manufacturing, chrome plating, and industrial cooling towers that used chromate chemicals as rust inhibitors (ATSDR, 2000). Relatively large amounts of chromium are released to surface waters from electroplating, leather tanning, wood preservation, and textile industries. Other environmental sources of chromium include road dust contaminated by emissions from chromium-based catalytic converters or erosion products of asbestos brake linings, cement dust, tobacco smoke, and foodstuffs. Cigarettes contain 0.24 to 14.6 milligrams per kilogram and it is plausible that cigarette smoking might constitute a significant source of human exposure to chromium (ATSDR, 2000).

From 1905 to the early 1970s, three companies located in Hudson County, New Jersey, USA, processed chromite ore to produce chromium (NJDEP, 1997). Starting in the 1950s over two million tons of chromite ore processing residue (COPR) were sold or given away and used as fill material in residential, commercial, and recreational settings (NJDEP, 1997). COPR was used as fill in preparation for building foundations, construction of tank berms, roadway construction, filling of wetlands, sewerline construction, and other construction and development projects. This COPR fill material contained total chromium at concentrations that exceeded 10,000 mg/kg, with Cr(VI) content ranging from about 1 to 50% (Burke et al., 1991). Continued leaching of COPR resulted in yellow-colored surface water runoff and yellow deposits on soil surfaces and inside basement walls (Burke et al., 1991) as well as elevated levels of Cr(VI) in alkaline soils and underlying groundwater (James, 1998). Chromate contamination has been found in a variety of places including interior and exterior building surfaces, surfaces of driveways and parking lots, and in the surface and subsurface of unpaved material (NJDEP, 1997). The New Jersey Department of Environmental Protection (NJDEP, 1998) based risk to health from exposure to CR(VI) in COPR on dermal contact, soil ingestion, inhalation of soil particles, and impact of soil contamination on groundwater quality. Known chromate waste contamination at most residential sites has been remediated while investigation and remediation of some non-residential properties continues (NJDEP, 1997).

In 1987, hexavalent chromium was detected at a concentration of 580 μg/L in groundwater sampled from a monitoring well in Hinkley, CA, a small town in the Mojave Desert (Pellerin and Booker, 2000). Cr(VI) was being used as an anticorrosive in the cooling towers of a gas compressor station. Spent cooling water, disposed of in unlined ponds from 1951 to 1982, leached to groundwater and impacted the drinking water supply for the town of Hinkley (Daugherty, 2001). People who lived in Hinkley had experienced an array of health problems including liver, heart, respiratory, and reproductive failure; cancer of the brain, kidney, breast, uterus, and gastrointestinal systems; Hodgkin disease; and frequent miscarriages (Pellerin and Booker, 2000). In 1993, seventy seven Hinkley plaintiffs filed a law suit against the owner of the gas compressor station. Eventually, a total of 648 plaintiffs recovered for injury claims and settled for $333 (US) million dollars (Pellerin and Booker, 2000). The station owner agreed to stop using Cr(VI) and to clean up the Cr(VI) in groundwater. The case remains controversial among chromium experts because most of the Hinkley exposures involved consuming drinking water impacted by Cr(VI). Ingestion of Cr(VI) is considered to be less toxic than inhalation of Cr(VI) since Cr(VI) can be reduced to Cr(III) in the stomach. In addition, some experts considered Cr(VI) exposure levels too low to cause health effects and

there are few data linking Cr(VI) exposures to the Hinkley residents' symptoms. However others believe there are too many data gaps in chromium toxicity studies to dismiss the Hinkley residents' case (Pellerin and Booker, 2000). Until more data about the effects of different doses and routes of exposure to Cr(VI) are available, it is too soon to rule out drinking water exposure as a health risk (Pellerin and Booker, 2000).

5.9 ISOTOPE APPLICATIONS FOR CHROMIUM INVESTIGATIONS

Stable chromium isotope abundances can be used to evaluate reduction of Cr(VI) to Cr(III) in natural environments and in remedial systems (Ball et al., 2001; Ellis et al., 2002, 2004). Stable isotope abundances of nitrogen and sulfur are well established as indicators of nitrate and sulfate reduction (e.g., Clark and Fritz, 1997; Kendall and McDonnell, 1998; Cook and Herczeg, 2000). Chromium isotope fractionation has been shown to occur during Cr(VI) reduction as well (Ball et al., 2001; Ellis et al., 2002).

Isotopes are atoms of the same element that have different masses due to differing numbers of neutrons in the nucleus. For example, hydrogen atoms can contain nuclei with no neutrons (hydrogen, ^1H or H), one neutron (deuterium or ^2H or D), or two neutrons (tritium, ^3H or T). Stable isotopes remain unchanged while radioactive isotopes decay to stable isotopes.

Stable isotope concentrations typically are measured against a known standard and are reported in δ (delta) notation in units of ‰ (per mil):

$$\delta^{53}Cr(‰) = ([(^{53}Cr/^{52}Cr)_{sam}/(^{53}Cr/^{52}Cr)_{std}] - 1) \times 1000$$

Where "sam" and "std" refer to sample and standard, respectively. Negative δ values indicate that there is less of the rare isotope (typically the heavier isotope) in the sample relative to the standard, i.e., the sample is "depleted" relative to the standard. Positive δ values indicate that there is more of the rare isotope in the sample relative to the standard, i.e., the sample is "enriched" relative to the standard.

Although isotopes of the same element behave the same physically and chemically, reaction rates differ due to the mass difference between the isotopes. This mass difference causes a preferential partitioning, or fractionation, that results in varying isotopic compositions between the reactant and product. Chemical isotopic fractionation can be described by equilibrium and non-equilibrium reactions. In equilibrium isotopic reactions, the reversible reactions occur at equivalent rates for a particular isotope. However, the reaction rates are different for the rare and the abundant isotope species. An example of an equilibrium isotope reaction is the exchange of liquid and vapor water in a closed system.

Non-equilibrium, or kinetic, isotopic reactions are irreversible biotic (e.g., microbial degradation of chlorinated solvents) or abiotic reactions (e.g., treatment of limestone with acid to liberate CO_2). Rates for kinetic isotope reactions are mass dependent. Lighter isotopes form weaker bonds than do heavier isotopes, and these weaker bonds are energetically favorable and are more easily broken during biotic and abiotic reactions (Clark and Fritz, 1997). Therefore, during kinetic reactions, the remaining reactant will be isotopically enriched and the product will be isotopically depleted relative to the starting reactant isotopic composition. For example, when sulfate is reduced to sulfide, the sulfide is depleted in ^{34}S relative to the sulfate, and the sulfate becomes increasingly enriched in ^{34}S as the reaction proceeds. It is this fractionation process that is most widely exploited in environmental isotope applications (Sueker, 2002, 2003).

Chromium has four stable isotopes of masses 50 (4.35%), 52 (83.8%), 53 (0.50%), and 54 (2.37%) (Rotaru et al., 1992). Ellis et al. (2002) developed a double isotope spike method, with a measurement precision of ±0.2‰, for measuring mass-dependent fractionation of chromium isotopes. This method was tested in laboratory experiments to evaluate fractionation of chromium isotopes during Cr(VI) reduction by magnetite, an estuarine sediment, and a freshwater sediment.

In all three experiments, decreases in Cr(VI) concentrations were accompanied by increases of the $\delta^{53}Cr$ in the remaining pool of reactant Cr(VI) (Ellis et al., 2002). Autoclaved duplicates of the two sediment experiments reduced Cr(VI) as quickly as the non-autoclaved experiments, indicating that abiotic reduction dominated over microbial reduction. Ellis et al. (2002) used a Rayleigh fractionation model to calculate the size of the kinetic isotope effect. The instantaneous isotope fractionation, ε, is the difference between the $\delta^{53}Cr$ value of the pool of reactant Cr(VI) and that of the Cr reduced to Cr(III) at any instant in time. Ellis et al. (2002) calculated ε values of -3.3 to -3.5‰ for the three experiments using the following relation:

$$\delta^{53}Cr = \delta^{53}Cr_{ini} + e \ln(f)$$

Where $\delta^{53}Cr$ and $\delta^{53}Cr_{ini}$ apply to the unreacted Cr(VI) pool at the time of sampling and at the start of the experiment, respectively, and f is the fraction of remaining Cr(VI). The similarity of the ε values determined for three contrasting experiments suggests that the reduction mechanisms were similar (Ellis et al., 2002).

Ball et al. (2001) measured isotopic compositions of Cr(VI) in natural groundwater in the Mojave Desert. The $\delta^{53}Cr$ values ranged from -0.3 to $+3.2$‰, with lighter $\delta^{53}Cr$ values observed in the younger waters of the upper aquifer and heavier $\delta^{53}Cr$ values observed in the lower aquifer. Ball et al. (2001) attributed this difference in isotopic compositions to the occurrence of Cr(VI) reduction along the groundwater flow path.

Ellis et al. (2002) measured isotopic compositions of chromium in basalt samples, chromium in reagents, and Cr(VI) in plating baths and in groundwater impacted by plating operations. The rocks, reagents, and plating baths exhibited a narrow range of $\delta^{53}Cr$ values from -0.07 to 0.36‰. In contrast, the Cr(VI) in groundwater ranged from 1.1 to 5.8‰. All of the groundwater samples showed Cr(VI) enrichment of $\delta^{53}Cr$ relative to the plating baths, indicating the reduction of Cr(VI) was likely occurring (Ellis et al., 2002). Ellis et al. (2004) determined that adsorption of Cr(VI) was non-fractionating, therefore, sorption of Cr(VI) to aquifer materials likely would not affect the isotopic composition of Cr(VI) in groundwater. The extent of Cr(VI) reduction for two groundwater samples was estimated to be 31 and 68%, based on chromium fractionation estimated from the laboratory experiments and on differences in Cr(VI) isotopic compositions between the plating bath and groundwater samples (Ellis et al., 2002).

As Ellis et al. (2002) contend, constraining rates of reduction is essential in many Cr(VI) contamination cases. For example, sufficient rates of natural reduction at a given site allow use of the "monitored natural attenuation" approach, which is much less expensive and disruptive than active remediation. In real systems, determining reduction rates can be difficult. Multiple rounds of groundwater sampling and analysis, with years of waiting time in between them, may be required to detect the decrease of Cr(VI) mass in a contaminant plume. Chromium isotope techniques allow a rapid means to demonstrate reduction of Cr(VI) and to estimate the extent of reduction of Cr(VI). Combined with a groundwater flow and solute transport model, estimates of the extent of Cr(VI) reduction can be used to constrain Cr(VI) reduction rates.

REFERENCES

Abu-Saba, K.E. and A.R. Flegal, 1997. Temporally variable freshwater sources of dissolved chromium to the San Francisco Bay Estuary. *Environ. Sci. Technol.*, 311(12): 3455–3460.

Agency for Toxic Substances and Disease Registry (ATSDR), 2000. Toxicological Profile for Chromium. TP-7. US Department of Health and Human Services, Atlanta, GA.

Ahern, F., J.M. Eckert, N.C. Payne, and K.L. Williams, 1985. Speciation of chromium in sea water. *Anal. Chim. Acta*, 175: 147–151.

American Conference of Governmental Industrial Hygienists (ACGIH), 1999. Chromium. Documentation of the threshold limit values and biological exposure indices. ACGIH, Cincinnati, OH.

Ball, J.W. and D.K. Nordstrom, 1998. Critical evaluation and selection of standard state thermodynamic properties for chromium metal and its aqueous ions, hydrolysis species, oxides, and hydroxides. *J. Chem. Eng. Data*, 43(6): 895–918.

Ball, J.W., T.D. Bullen, J.A. Izbicki, and T.M. Johnson, 2001. Stable isotope variations of hexavalent chromium in groundwaters of the Mojave Desert, California, USA. Abstract, Geological Society of America Annual Meeting, November 5–8, Boston, MA.

Ball, J.W. and J.A. Izbicki, 2002. Occurrence of hexavalent chromium in ground water in the western part of the Mojave Desert, California. Abstract, Geological Society of America Annual Meeting, Denver, CO.

Ball, J.W. and R.B. McClesky, 2003. A new cation-exchange method for accurate field speciation of hexavalent chromium. WRI 03-4018, US Geological survey Water Resources Investigations. Boulder, CO.

Bartlett, R.J., 1991. Chromium cycling in soils and water: Links, gaps, and methods. *Environ. Health Perspect.*, 92: 17–24.

Bartlett, R.J. and J.M. James, 1979. Behavior of chromium in soils: III. Oxidation. *J. Environ. Qual.*, 8: 31–35.

Bartlett, R.J. and J.M. Kimble, 1976a. Behavior of chromium in soils: I. Trivalent forms. *J. Environ. Qual.*, 5(4): 379–383.

Bartlett, R.J. and J.M. Kimble, 1976b. Behavior of chromium in soils: II. Hexavalent forms. *J. Environ. Qual.*, 5(4): 383–386.

Benoit, G. and T.F. Rozan, 1999. The influence of size distribution on the particle concentration effect and trace metal partitioning in rivers. *Geochim. et Cosmochim. Acta*, 63(1): 113–127.

Burke, T., J. Fagliano, M. Goldoft, R.E. Hazen, R. Iglewicz, and T. McKee, 1991. Chromite ore processing residue in Hudson County, New Jersey. *Environ. Health Perspect.*, 92: 131–137.

Clark, I. and P. Fritz, 1997. *Environmental Isotopes in Hydrogeology*. Lewis Publishers, Boca Raton.

Cook, P. and A.L. Herczeg, 2000. *Environmental Tracers in Subsurface Hydrology*. Kluwer Academic Publishers, Boston.

Cox, A.G. and C.W. McLeod, 1992. Field sampling technique for speciation of inorganic chromium in rivers. *Mikrochim Acta*, 109(1–4): 161–164.

Daugherty, R., 2001. *Hinkley chromium 6 problems continue.* Desert Dispatch, Barstow, CA, July 3.

Drever, James I. 1997. *Geochemistry of Natural Waters,* Third Edition. Prentice Hall, Englewood Cliffs, New Jersey.

Dzombak, L.E. and F.M. Morel, 1990. *Surface Complexation Modeling: Hydrous Ferric Oxide.* Wiley-Interscience, New York.

Eary, L.E. and D. Rai, 1987. Kinetics of chromium(III) oxidation to chromium(VI) by reaction with manganese dioxide. *Environ. Sci. Technol.,* 21(12): 1187–1193.

Eisler, R., 1986. Chromium hazards to fish, wildlife, and invertebrates: A synoptic review. US Fish and Wildlife Service Biological Report 85(1.6), p. 60.

Ellis, A.S., T.M. Johnson, and T.D. Bullen, 2002. Mass-dependent fractionation of Cr isotopes and the fate of hexavalent Cr in the environment. *Science,* 295: 2060–2062.

Ellis, A.S., T.M. Johnson, and T.D. Bullen, 2004. Using chromium stable isotope ratios to quantify Cr(VI) reduction: Lack of sorption effects. *Environ. Sci. Technol.,* 38(13): 3604–3607.

Fantoni, D., G. Brozzo, M. Canepa, F. Cipolli, L. Marini, G. Ottonello, and M.V. Zuccolini, 2002. Natural hexavalent chromium in groundwaters interacting with ophiolitic rocks. *Environ. Geology,* 42(8): 871–882.

Farías, S.S., V.A. Casa, C. Vázquez, L. Ferpozzi, G.N. Pucci, and I.M. Cohen, 2003. Natural contamination with arsenic and other trace elements in ground waters of Argentine Pampean Plain. *Sci. Tot. Environ.,* 309(1–3): 187–199.

Gaberell, M., Y. Chin, S.J. Hug, and B. Sulzberger, 2003. Role of dissolved organic matter composition on the photoreduction of Cr(VI) to Cr(III) in the presence of iron. *Environ. Sci. Technol.,* 37(19): 4403–4409.

Helmke. P.A.,1996 *In Methods of Soil Analysis: Part 3. Chemical Methods*; Sparks, D.L., A.L. Page, P.A. Helmke, R.H. Loeppert, P.N. Soltanpour, M.A. Tabatabai, C.T. Johnson, and M.E. Sumner, Eds. Soil Science Society of America, Inc., Madison, WI, pp. 141–159.

Hem, J.D., 1989. *Study and interpretation of the chemical characteristics of natural water,* Third Edition, US geological Survey Water-Supply Paper 2253. Government Printing Office, Alexandria, VA.

Hering, J. and T. Harmon, 2004. Geochemical Controls on Chromium Occurrence, Speciation and Treatability. American Water Works Association Research Foundation Report 91043F.

Icopini, G.A. and D.T. Long, 2002. Speciation of aqueous chromium by use of solid-phase extractions in the field. *Environ. Sci. Technol.,* 36(13): 2994–2999.

Integrated Risk Information System (IRIS), 2000a. Chromium III. USEPA, Office of Health and Environmental Assessment, Environmental Criteria and Assessment Office, Cincinnati, OH.

Integrated Risk Information System (IRIS), 2000b. Chromium VI. USEPA, Office of Health and Environmental Assessment, Environmental Criteria and Assessment Office, Cincinnati, OH.

International Agency for Research on Cancer (IARC), 1990. IARC monographs on the evaluation of carcinogenic risks to humans. *Chromium, Nickel, and Welding.* Vol. 49. World Health Organization, Lyons, France, pp. 49–256.

International Chromium Development Association (ICDA). 2005. Chromium. http://www.chromium-asoc.com, accessed May 1, 2005.

Irwin, R.J., 1997. *Environmental Contaminants Encyclopedia, Chromium.* National Park Service, Water Resources Division, Water Operations Branch, Fort Collins, CO, July 1.

James, B.R., 1998. Chromium oxidation and reduction chemistry in soils: Relevance to chromate contamination of groundwater if the northeastern United States. Maryland

Water Resources Research Center Annual Report 1997–1998, University of Maryland, College Park, MD.

Johnson, C.A., 1990. Rapid ion-exchange techniques for the separation and preconcentration of chromium(VI) and chromium(III) in fresh waters. *Analytica Chimica Acta,* 238(2): 273–278.

Johnson, C.A., L. Sigg, and U. Lindauer, 1992. The chromium cycle in a seasonally anoxic lake. *Limnol. Oceanogr.,* 37(2): 315–321.

Kaczynski, S.E. and R.J. Kieber, 1994. Hydrophobic C-18 bound organic complexes of chromium and their potential impact on the geochemistry of chromium in natural waters. *Environ. Sci. Technol.,* 28(5): 799–804.

Kendall, C. and J.J. McDonnell, 1998. *Isotope Tracers in Catchment Hydrology.* Elsevier, Amsterdam.

Klein, C. and J. Hurlbut, 1993. *Manual of Mineralogy.* John Wiley & Sons, New York, p. 681

Keith, L.H. and W.A. Telliard, 1979. Priority Pollutants: I – a perspective view. *Environ. Sci. Technol.,* 13(4): 416–423.

Kotaś, J. and Z. Stasicka, 2000. Chromium occurrence in the environment and methods of its speciation. *Environ. Pollution,* 107(3): 263–283.

Lebow, S., R.S. Williams, and P. Lebow, 2003. Effect of simulated rainfall and weathering on release of preservative elements from CCA treated wood. *Environ. Sci. Technol.,* 37(18): 4077–4082.

Ledin, A., S. Karlsson, A. Duker and B. Allard, 1995. Characterization of the submicrometer phase in surface waters – a review. *Analyst,* 120(3): 603–608.

Lillebo, H.P., S. Shaner, D. Carlson, N. Richard, and P. Dubowy, 1986. Report to the "85-1" committee: Water quality criteria for selenium and other trace elements for protection of aquatic life and its uses in the San Joaquin Valley. Division of Water Quality, State Water Resources Control Board, State of California, Sacramento, CA.

Lin, C., 2002. The chemical transformation of chromium in natural waters – a model study. *Water, Air, Soil Poll.,* 139:137–158.

Masscheleyn, P.H., J.H. Pardue, R.D., DeLaune, and W.H. Patrick, 1992. Chromium redox chemistry in a Lower Mississippi Valley bottomland hardwood wetland. *Environ, Sci. Technol.,* 26(6): 1217–1226.

Merck Co. Inc. 1983. The Merck Index, Tenth Edition. Rahway, N.J., p. 317.

Meyers, E. 1990. *Chemistry of Hazardous Materials.* Prentice Hall Career and Technology, Prentice Hall, Englewood Cliffs, New Jersey, p. 509.

Murphy, D.M., J.R. Garbarino, H.E. Taylor, B.T. Hart, and R. Beckett, 1993. Determination of size and element composition distributions of complex colloids by sedimentation field-flow fractionation inductively-coupled plasma-mass spectrometry. *J. Chromatography,* 642(1–2): 459–467.

National Academy of Sciences, 1994. *Committee on Ground Water Cleanup Alternatives, Alternatives for Ground Water Cleanup.* National Academy Press, Washington, D.C.

National Institute for Occupational Safety and Health (NIOSH), 2005. Online pocket guide to chemical hazards. May 5, 2005. http://www.cdc.gov/niosh/npg/npg.html.

New Jersey Department of Environmental Protection (NJDEP), 1997. Hudson County chromate chemical production waste sites, Background. Site Remediation & Waste Management, Site Information Program, Hudson Chromate Project. http://www.state.nj.us/dep/srp/ siteinfo/chrome/bkgrnd.htm, accessed May 8, 2005.

New Jersey Department of Environmental Protection (NJDEP), 1998. Summary of the Basis and Background of the Soil Cleanup Criteria for Trivalent and Hexavalent

Chromium (September, 1998). Site Remediation & Waste Management, Site Information Program, Hudson Chromate Project. http://www.state.nj.us/dep/srp/siteinfo/chrome/b_b_sum_1.htm, accessed May 8, 2005.

Palmer, C.D. and P.R. Wittbrodt, 1991. Processes affecting the remediation of chromium-contaminated sites. *Environ. Health Perspect.*, 92: 24–40.

Palmer, C.D. and R.W. Puls, 1994. Natural Attenuation of Hexavalent Chromium in Groundwater and Soils. EPA Ground Water Issue. US EPA, Technology Innovation Office, Office of Solid Waste and Emergency Response, Washington, D.C.

Pellerin, C. and S.M. Booker, 2000. Reflections on hexavalent chromium health hazards of an industrial heavyweight. *Environ. Health Perspect.*, 108(9): A402–A407.

Puls, R.W., D.A. Clark, C.J. Paul, and J. Vardy, 1994. Transport and transformation of hexavalent chromium through soils and into ground water. *J. Soil Contam.*, 3(2): 203–224.

Robertson, F.N., 1975. Hexavalent chromium in the ground water in Paradise Valley, Arizona. *Ground Water*, 13(6): 516–527.

Rotaru, M., J.L. Brick and C.J. Allegre, 1992. *Nature*, 358: 465–470.

Schacklette, H.T. and J.G. Boerngen, 1984. Element concentration in soils and other surficial materials. US Geological Survey Professional Paper 1710.

Schroeder, D.C., and G.F. Lee, 1975. Potential transformations of chromium in natural waters. *Water Air Soil Pollut.*, 4: 355–365.

Sueker, J.K., 2002. Isotope applications in environmental investigations: Theory and use in chlorinated solvent and petroleum hydrocarbon studies. *Remediation*, Winter: 5–21.

Sueker, J.K., 2003. Isotope applications in environmental investigations Part II: Groundwater age dating and recharge processes, and provenance of sulfur and methane. *Remediation*, Spring: 71–90.

Szulczewski, M.D., P.A. Helmke, and W.F. Bleam,1997. Comparison of XANES analyses and extractions to determine chromium speciation in contaminated soils. *Environ. Sci. Technol.*, 31: 2954–2959.

Szulczewski, M.D., P.A. Helmke, and W.F. Bleam, 2001. XANES spectroscopy studies of Cr(VI) reduction by thiols in organosulfur compounds and humic substances. *Environ. Sci. Technol.*, 35(6): 1134–1141.

Taylor, H.E., J.R. Garbarino, D.M. Murphy, and R. Beckett, 1992. Inductively coupled plasma mass-spectrometry as an element-specific detector for field-flow fractionation particle separation. *Analytical Chemistry*, 64(18): 2036–2041.

United States Department of Health and Human Services and United States Environmental Protection Agency, 1987. Notice of the first priority list of hazardous substances that will be the subject of toxicological profiles. Federal Register, 52: 12866–12874.

United States Environmental Protection Agency, 1980. Ambient water quality criteria for chromium. Office of Water Regulations and Standards, Criteria Standards Division, Washington, D.C. EPA-440/5-80-035.

United States Environmental Protection Agency, 1987. Quality criteria for water 1986. Office of Water Regulations and Standards, Criteria Standards Division, Washington, D.C. EPA 440/5-86-001.

United States Environmental Protection Agency, 1996. Soil Screening Guidance: Technical Background. EPA/540/R-95/128.

United States Environmental Protection Agency, 2000. In *Situ Treatment of Soil and Groundwater Contaminated with Chromium Technical Resource Guide*. Office of Research and Development, Washington, D.C. EPA/625/R-00/005.

United States Environmental Protection Agency, 2002a. List of Drinking Water Contaminants and MCLs. EPA 816-F-02-013. www.epa.gov/safewater/contaminants/dw_contamfs/chromium.html, accessed May 7, 2005.

United States Environmental Protection Agency, 2002b. National Recommended Water Quality Criteria: 2002. Office of Water, Office of Science and Technology, EPA-822-R-02-047.

United States Environmental Protection Agency, 2003. Manufacturers to use new wood preservatives replacing most residential uses of CCA. http://www.epa.gov/pesticides/citizens/cca_transition.htm.

United States Occupational Health and Safety Administration (OSHA), 2005. 29 CFR, 1910.1000, Table Z-1 OSHA Limits for Air Contamination. www.osha.gov, accessed May 7, 2005.

United Stated National Institute of Occupation Safety and Health (NIOSH), 2005. NIOSH Pocket Guide to Chemical Hazards. www.cdc.gov/niosh/npg, accessed May 7, 2005.

Walsh, A.R. and J. O'Halloran, 1996a. Chromium speciation in tannery effluent – I. an assessment of techniques and the role of organic Cr(III) complexes. *J. Water. Res.*, 30(10): 2393–2400.

Walsh, A.R. and J. O'Halloran, 1996b. Chromium speciation in tannery effluent – II. Speciation in the effluent and in a receiving estuary. *J. Water. Res.*, 30(10): 2401–2412.

Wen, L.S., M.C. Stordal, D.G. Tang, G.A. Gill, and P.H. Santschi, 1996. An ultraclean cross-flow ultrafiltration technique for the study of trace metals phase speciation in seawater. *Marine Chemistry*, 55(1–2): 129–152.

World Health Organization (WHO), 1988. Environmental health criteria 61: Chromium. Geneva: WHO, 197.

Zachara, J.M., D.C. Girvin, R.L. Schmidt, and C.T. Resch, 1987. Chromate adsorption on amorphous iron oxyhydroxide in the presence of major groundwater ions. *Environ. Sci. Technol.*, 21(6): 589–594.

6

Methane

Paul D. Lundegard

Contents

6.1 INTRODUCTION

Methane (CH_4) is a colorless, odorless gas, and the simplest of all hydrocarbon molecules. In the earth's atmosphere, CH_4 is ubiquitously present at approximately 2 parts per million by volume (ppmv) and is of considerable environmental concern because it is a greenhouse gas and is increasing in concentration at a rate of about 1% per year (Khalil and Rasmussen, 1990). Even though CH_4 is non-toxic, its presence in soil gas creates environmental concern because it can act as an asphyxiant and is an explosion hazard when present at concentrations between 5 and 15 percent by volume in air. Recognizing the potential hazards associated with near-surface CH_4, some municipalities have included assessment and mitigation procedures in their building codes (e.g., City of Los Angeles, 1996; City of Huntington Beach, 1997). The discovery of methane in shallow soil gas can have major financial implications for building construction projects, possibly leading to delays, denial of permits, additional costs associated with engineered mitigation measures, and litigation (e.g. Groves, 2003; Wilson and Sauerwein, 2003). Determining the source and origin of CH_4 is important to the determination of environmental liability and to the selection of appropriate mitigation measures.

6.2 METHANE IN THE ENVIRONMENT

Fires and explosions related to occurrences of methane (and associated hydrocarbon gases) have been attributed to several causes. Underground mine explosions are among the most familiar and deadly events (e.g. Los Angeles Times, 2001). Leaking gas well casings have also been linked to explosions (e.g. Baldassare and Laughrey, 1997). An explosion and fire in Hutchinson, Kansas, was apparently caused by leaks in an underground gas storage facility (Allison, 2001). Leaking well casings and unrecognized geologic conduits allowed gas from the storage facility to migrate under pressure for approximately 12 kilometers. A fatal explosion in Quesnel, Canada, was attributed to a break in a pressurized natural gas pipeline (Quesnel-Cariboo Observer, 1997). Alternatively, the risk associated with non-pressurized sources of biogenic methane is much lower.

Methane is a major constituent of natural gas. On the other hand, methane is generally a minor constituent of natural petroleum liquids, with the exception of thermally mature oils and gas condensates. While liquid petroleum products seldom contain significant amounts of CH_4, accounts of high CH_4 content in soil gas near petroleum spills are common. In addition, several well-documented soil gas studies have shown elevated concentrations of carbon dioxide (CO_2) and CH_4 in association with subsurface hydrocarbon contamination (Marrin, 1987, 1991; Kerfoot et al., 1988; Robbins et al., 1990; Deyo et al., 1993; Ririe and Sweeney, 1995; Lundegard et al., 1998 and 2000). In some investigations, CO_2 and CH_4 concentrations have been shown to correlate with the distribution of soil contamination, for both volatile and semi-volatile hydrocarbons (Marrin, 1987, 1989, 1991; Conrad et al., 1999a; Gandoy-Bernasconi et al., 2004).

Methane in the environment can have a variety of sources and origins (Schoell, 1988; Kaplan, 1994). Naturally occurring CH_4 can be classified, based on the predominant process by which it formed, as either thermogenic (produced by abiotic processes) or biogenic (produced by biological processes). Thermogenic CH_4 is produced at depth within sedimentary basins by the thermal degradation of sedimentary organic matter, and is commonly associated with coal or accumulations of oil and natural gas. Roughly 80% of the natural gas in geologic reservoirs has a thermogenic

origin (Whiticar, 1996). Thermogenic CH_4, while typically generated at depths of several thousand feet within sedimentary basins, commonly migrates upward to escape to the land surface or accumulate in shallow geologic structures bounded by low-permeability strata. The detection of near-surface gas seeps been a geochemical prospecting method for locating possible oil and gas accumulation in the deeper subsurface for many years.

Biogenic CH_4 is produced under anaerobic, near-surface conditions by microbial degradation of organic matter. Such microbially-produced CH_4 occurs widely in association with organic-rich sediments and materials, including marine, lake, and river sediments; marshes and swamps; glacial drift; and in landfills and sewers (Schoell, 1988; Coleman et al., 1995). Roughly 20% of the natural gas in geologic reservoirs has a biogenic origin (Whiticar, 1996), while at least 80% of the CH_4 emissions to the atmosphere are of biogenic origin (Khalil and Rasmussen, 1983, Table 4). Landfills are thought to be the single largest anthropogenic source of CH_4 in the atmosphere (Chanton and Liptay, 2000). There are two principal enzymatic pathways by which biogenic CH_4 is produced (Schoell, 1980): (1) acetate fermentation and (2) carbon dioxide reduction.

(1) acetate fermentation—$CH_3COOH \rightarrow CH_4 + CO_2$
(2) CO_2 reduction—$CO_2 + 4H_2 \rightarrow CH_4 + 2H_2O$

The fermentation pathway involves production of short-chain methylated precursors (commonly acetate but also including formate, methanol, and methylated amines) from the source organic matter. Subsequently, methanogenic bacteria disproportionate the precursor into CO_2 and CH_4. The CO_2 reduction pathway involves reduction of CO_2 by molecular hydrogen. Biogenic CH_4 formed in near-surface, non-marine environments (e.g., marshes, swamps, and landfills) is primarily formed by acetic acid fermentation (Coleman et al., 1995). Methanogens are capable of metabolizing a wide variety of substrates.

From the standpoint of environmental liability, it is useful to distinguish between near-surface CH_4 occurrences into those with natural and anthropogenic origins. Natural occurrences include (1) seeps originating from natural gas accumulations, (2) coal-bed gas, and (3) gas produced by biodegradation of indigenous sedimentary organic matter (e.g., organic-rich shales or marsh deposits). Anthropogenic CH_4 occurrences are caused by human activities and therefore are associated with potential environmental liability. Anthropogenic CH_4 can be derived from: (1) gas pipelines, (2) oil and gas wells, (3) underground gas storage facilities, (4) sewer pipes and septic systems, (5) buried compost, (6) buried animal waste, (7) landfills, and (8) spilled petroleum. In addition, hazards associated with natural occurrences of CH_4 can be increased by certain types of human activities. For example, the underground mining of coal seams creates opportunities for the de-gassing of coal into mine shafts, and thus increased risk of explosion. Construction of buildings, roads, and parking lots over natural gas seeps may cause up-ward moving gases to accumulate in places where they would otherwise escape to the atmosphere at low concentration or biodegrade to safe levels.

While correlations between CH_4 concentration in soil gas and petroleum-contaminated soil have led some environmental scientists to infer a direct causal relationship, the process by which carbon in petroleum would be converted to CH_4 is not clear. Methanogenesis requires (1) strict anaerobic conditions (i.e., an absence of oxygen), and (2) either molecular hydrogen and CO_2, or certain oxygen-bearing precursor compounds (e.g., simple organic acids and alcohols). Most natural and refined petroleum liquids do not contain significant amounts of the short-chain precursor compounds necessary for direct fermentation reactions. However, under certain conditions, petroleum degradation can generate the necessary precursor compounds for methanogenesis by fermentation (Eganhouse *et al.*, 1993; Revesz *et al.*, 1995). Though less well understood, some groups of bacteria produce H_2 under anaerobic conditions (Claypool and Kaplan, 1974). The molecular H_2 is then available for methanogenesis by CO_2 reduction (Whiticar *et al.*, 1986). Elevated H_2 concentrations have been found in contaminated ground water (Lovely *et al.*, 1994), and soil gas associated with petroleum contamination (Ririe and Sweeney, 1995; Lundegard *et al.*, 2000). While these studies indicate that generation of CH_4 from petroleum by indirect processes can occur, the abundance of other anthropogenic and natural sources of CH_4 requires that the possible causal relationship between petroleum and elevated CH_4 concentrations be evaluated on a site-specific basis.

6.3 COLLECTION OF SOIL GAS SAMPLES

Most forensic investigations of CH_4 occurrences involve the collection and analysis of soil gas samples from the vadose zone. The collection of valid, high quality soil gas samples is not unduly difficult but it does require care. There is generally nothing in the appearance of a gas sample to tell us at the time of collection whether it is a good or bad sample. Consequently, rigorous, proven sampling protocols should be followed, and procedures should be validated at appropriate intervals. Analysis of samples in the field at the time of collection is highly beneficial for assuring the quality of the final data. The gas sample that is analyzed must represent the chemical composition of the soil pore gas at a known location. The sample must be collected in such a manner that it is not diluted with air, and all the materials that contact a soil gas sample during collection and storage must not adversely alter its chemical composition. Probably the biggest cause of bad soil gas data is dilution of the sample with air as a result of leaky sampling devices and collection probes that are not adequately sealed to the native soil matrix. Several soil gas sampling guidance documents are available that address this and other sampling issues and should be consulted prior to sample collection (USEPA, 1996; CSDDEH, 2004; Wilson *et al.*, 2004).

6.4 TYPES OF FORENSIC DATA

Rarely is a single type of data sufficient to resolve an environmental forensics question. More often, it is the integration of several types of information that leads to an effective resolution. In the case studies presented here, it will be evident that it is the synthesis of geological, geochemical, and historical land use data that leads to a credible interpretation of the origin and source of the soil gas CH_4. Spatial relationships and trends in the data are particularly important. Below, several types of geochemical data useful in investigations of soil gas CH_4 are discussed.

6.4.1 Molecular Composition Data

Soil gas compositional analysis is generally the starting point for most investigations, since one first needs to determine the abundance of CH_4 and its spatial variation. Real-time, gas chromatographic analysis in the field is strongly preferred so that the quality of soil gas samples can be assured and useful sampling locations can be properly selected. Sometimes the composition of the trace constituents in the CH_4 gas mixture is very revealing. In some instances, gross gas compositional data alone allow different potential gas

sources to be distinguished (Schoell, 1980). For example, the content of higher molecular weight hydrocarbon gases (e.g., ethane, propane, butane, and pentane) relative to CH_4 may be used. Exploration geochemists have traditionally described the bulk composition of natural gases in terms of their "dryness" or "wetness". A dry gas is one with a high concentration of methane. Conversely, a wet gas is one with a lower concentration of methane and higher concentration of higher molecular weight compounds. A variety of gas dryness and gas wetness indices have been used, depending on the purpose of the investigation and the data that are available. An example of a gas dryness index might be the ratio of CH_4 to the sum of C_1 to C_5 hydrocarbons ($C_1/[C_1–C_5]$). Biogenic gases (e.g., landfill gas) are very dry, consisting predominantly of CH_4 and CO_2 and do not typically contain significant concentrations of C_2 through C_5 hydrocarbons ($C_1/[C_1–C_5] > 0.98$) (Rice and Claypool, 1981). In contrast, most thermogenic gases (e.g., from natural gas pipelines) do contain significant concentrations of C_2 through C_5 hydrocarbons ($C_1/[C_1–C_5] = 0.6$ to 1.0). Using a gas dryness index, thermogenic and biogenic gases can often be distinguished (Figure 6.4.1).

Natural gas at different depths or reservoir zones within a geologic basin typically varies in bulk composition or dryness. Consequently, produced natural gas, natural gas seepage, and gas leaking from damaged or improperly abandoned wells tends to vary in composition. From shallow to deep locations within a basin, natural gas tends to vary from dry biogenic gas, to wet thermogenic gas, to dry thermogenic gas (Figure 6.4.2).

Landfill gas is a consideration in the forensic investigation of many CH_4 occurrences. Landfills generate and emit large quantities of CH_4 produced by the anaerobic decomposition of the wastes they contain. Emissions from landfills account for an estimated 5–15% of the global anthropogenic sources of atmospheric CH_4 (Doorn and Barlaz, 1995). Undiluted landfill gas typically contains 50–60% CH_4, 40–50% CO_2, and negligible oxygen. This composition does not vary greatly, apparently reflecting similarity in the composition of landfill waste and the processes of methanogenesis. The consistently high content of CH_4 and CO_2 (in undiluted gas) is a useful preliminary indication of the possible contribution by landfill gas. However, gas of other origins can have similar concentrations of CH_4 and CO_2. Landfill gas commonly contains trace concentrations of other constituents, including

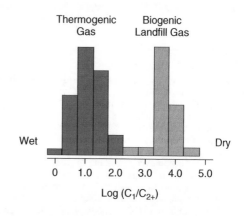

Figure 6.4.1 *Frequency histogram of a gas dryness index of thermogenic gases from the Appalachian and Turim Basins (Laughrey and Baldassare, 1998; Chen* et al.*, 2000) and biogenic landfill gases from Pennsylvania (Baldassare and Laughrey, 1997).*

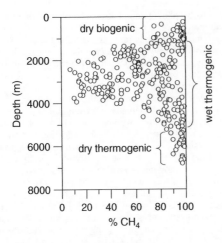

Figure 6.4.2 *Plot of subsurface depth versus percent* CH_4 *in gas from the Sverdrup Basin (after Snowdon and Roy, 1975).*

halogenated hydrocarbons. The presence of halogenated volatile organic compounds in a gas sample is strong evidence of an anthropogenic component in the gas mixture. Such compounds can also be derived directly from spills of products such as chlorinated solvents into soil and groundwater.

Gas transmission companies typically add low concentrations of odorants (e.g., mercaptans), or other tracers, to their gas to aid in the detection of leaks and to enhance consumer safety. The detection of these compounds is strong evidence that pipeline gas is present. Local gas companies should be contacted to determine what compounds are used in a particular market area.

6.4.2 Stable Isotope Data

Carbon and hydrogen isotopic data have been successfully used to discriminate natural gases (including CH_4) of various origins (Figure 6.4.3; Schoell, 1988; Suchomel *et al.*, 1990; Coleman *et al.*, 1995; Baldassare and Laughrey, 1997; Hackley *et al.*, 1999). Carbon isotopic data are typically reported using the delta (δ) notation relative to the PDB international standard (Hoefs, 1980). Similarly, hydrogen isotopic data are reported using the delta (δ) notation relative to the SMOW international standard.

$$\delta \times (\text{per mil}) = \frac{(R_{\text{sample}} - R_{\text{standard}})}{R_{\text{standard}}} \times 10^3$$

where X refers to either ^{13}C or D (D represents deuterium, ^2H) and R refers to either the $^{13}C/^{12}C$ or D/H isotopic ratio. Analytical laboratories typically request a gas sample that contains a few milliliters of the compound of interest (e.g., $\sim 100\,\text{ml}$ of a gas that is 2% CH_4). Analytical precision is typically about 0.2 per mil (‰) for $\delta\,^{13}C$ and about 1 per mil (‰) for $\delta\,D$ on standard size samples.

The carbon isotopic composition of CH_4 depends on the composition of the organic source, fractionations that depend on the process by which it is formed, and post-generation alteration. Microbial methanogenesis yields CH_4 that is highly depleted in $\delta^{13}C$ of relative to the organic matter from which it is formed. Depending on environmental conditions and the microbial pathway, $\delta^{13}C$ of biogenic CH_4 can be 25 to 90 per mil more depleted (i.e., negative) than the source organic matter. Once formed, transport processes in shallow subsurface settings do not significantly

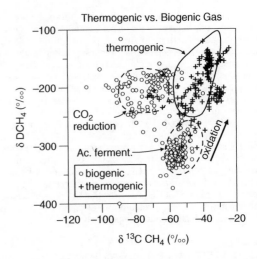

Figure 6.4.3 *Plot of hydrogen and carbon isotopic composition of* CH_4 *in gas samples of different origins and from different sources. Crosses are thermogenic gas samples. Open circles are biogenic gas samples. Arrow indicates general direction of isotopic shift resulting from partial oxidation of* CH_4. *Genetic fields after Coleman* et al. *(1995). Data from various sources (Schoell, 1980; Rigby and Smith, 1981; Smith* et al., *1985; Whiticar* et al., *1986; Coleman* et al., *1988; Jenden* et al., *1988; Coleman* et al., *1993; Jenden* et al., *1993; Kaplan, 1994; Baldassare and Laughrey, 1997; Laughrey and Baldassare, 1998; Lundegard* et al., *1998; Conrad* et al., *1999a,b; Chen* et al., *2000; Pierce and LaFountain, 2000)*

alter the stable isotopic composition of CH_4, but partial oxidation by methanotrophs can impart large isotopic changes in the residual CH_4 (see arrow in Figure 6.4.3; Coleman *et al.*, 1995).

Together, carbon and hydrogen isotopic data can often be used to distinguish gas samples of different origins (e.g. Schoell, 1980; Coleman *et al.*, 1995; Whiticar, 1996). Compositional ranges have been suggested for CH_4 produced by acetate fermentation, CO_2 reduction, and by thermogenic processes (Figure 6.4.3). These suggested ranges are useful guides but should not be used as absolute indications of gas origin because exceptions do occur. Methane in natural gas produced from oil and gas reservoirs and gas associated with coal beds is predominantly thermogenic in origin and generally has a carbon and hydrogen isotopic composition that distinguishes it from biogenic gas. Such CH_4 generally is more enriched in ^{13}C than biogenic CH_4 produced by CO_2 reduction, and more enriched in deuterium than biogenic CH_4 produced by acetate fermentation. Biogenic CH_4 produced by acetate fermentation is generally more enriched in ^{13}C and more depleted in deuterium than biogenic CH_4 produced by CO_2 reduction (Whiticar *et al.*, 1986). Isotope studies have shown that methanogenesis by acetate fermentation predominates in near-surface freshwater environments such as lakes and swamps (Whiticar *et al.*, 1986), yet it has been shown that in some systems the fraction of CH_4 produced by carbon dioxide reduction and acetate fermentation varies with time (Martens *et al.*, 1986; Sugimoto and Wada, 1995). Biogenic CH_4 in marine sediments and glacial drift is predominantly produced by the CO_2 reduction pathway (Claypool and Kaplan, 1974; Coleman *et al.*, 1988).

The vast majority of samples of CH_4 produced in landfills plot within the acetate fermentation field (Figure 6.4.4). Those samples that plot within the thermogenic gas field likely represent the effects of partial oxidation of landfill CH_4 or mixing of landfill and thermogenic CH_4. The isotopic composition of CH_4 produced from biodegradation of petroleum contamination has not yet been studied in a wide range of settings. Available data, however, suggest some interesting characteristics (Figure 6.4.5). While data plot

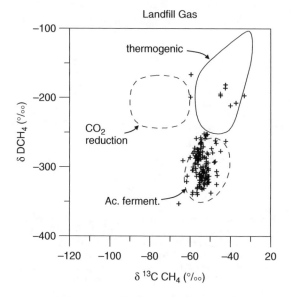

Figure 6.4.4 *Plot of hydrogen and carbon isotopic composition of CH_4 in gas samples collected from landfills. Data from various sources (Coleman* et al.*, 1993; Kaplan, 1994; Baldassare and Laughrey, 1997; Pierce and LaFountain, 2000).*

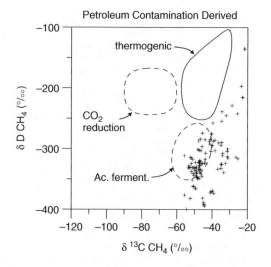

Figure 6.4.5 *Plot of hydrogen and carbon isotopic composition of CH_4 thought to be derived from biodegradation of spilled petroleum in near-surface environments. Data from various sources (author's unpublished data; Conrad* et al.*, 1999a,b).*

within or near the acetate fermentation field, most values are shifted toward higher δ ^{13}C and lower δ D values than landfill CH_4. In fact, a significant number of the samples of CH_4 produced through the process of petroleum biodegradation have isotopic compositions outside the previously known range for biogenic CH_4 of any origin. Careful examination of data sets from individual sites also shows that within the vadose zone, partial or complete oxidation of CH_4 produced from petroleum contamination seems to be common (e.g., Conrad *et al.* 1999b) and is sometimes reflected in its isotopic composition.

The hydrogen isotopic composition of biogenic CH_4 is a function of the isotopic composition of the ambient water and pathway-dependent fractionation. In the case of CO_2 reduction, all four H atoms in CH_4 come from ambient water (Daniels *et al.*, 1980). In the case of acetate fermentation, three of the H atoms originate from the methyl group of acetate, and one comes from ambient water (Pine and Barker, 1956). As a result, CH_4 produced by CO_2 reduction is more sensitive to the hydrogen isotopic composition of ambient pore water than CH_4 produced by acetic acid fermentation (Figure 6.4.6).

Carbon isotope ratios are used to estimate end member abundances in two-component mixtures, provided the concentration and isotopic composition of the CH_4 (or other gas species) in each end member is known. This is made possible by the degree of natural variation in isotope ratios and the high precision with which isotope ratios can be measured. The applicable mathematical expression is (Schoell *et al.*, 1993)

$$f_1 = \frac{C_{i2}{}^{*}(\delta_{iM} - \delta_{i2})}{[C_{i2}{}^{*}(\delta_{iM} - \delta_{i2}) - C_{i1}{}^{*}(\delta_{iM} - \delta_{i1})]}$$

In the above expression, f_1 is the fraction of gas 1 in the mixture. C_{i1} and C_{i2} are the concentrations of species i (e.g., CH_4) in gases 1 and 2, respectively. δ_{i1} and δ_{i2} are the isotopic compositions of species i in each end member. δ_{iM} is the isotopic composition of species i in the gas mixture.

Partial oxidation of CH_4 can have a substantial effect on the isotopic composition of residual CH_4 and associated CO_2. The potential for oxidation effects must be carefully considered before making final interpretations of stable isotopic data. Using a Rayleigh distillation model, the isotopic shifts caused by partial CH_4 oxidation can be predicted for

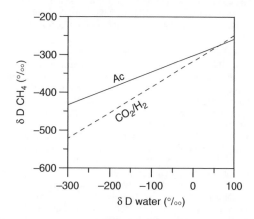

Figure 6.4.6 *Predicted relationship between the hydrogen isotopic composition of CH_4 and H_2O for the acetate fermentation pathway (solid line) and the CO_2 reduction pathway (dashed line) (after Sugimoto and Wada, 1995).*

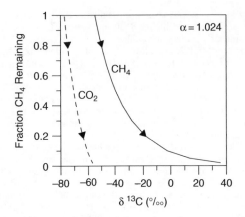

Figure 6.4.7 Plot showing the isotopic shift in residual CH_4 and produced CO_2 during oxidation. An isotopic fractionation factor (α) of 1.024 was used in the calculations.

assumed isotopic fractionation factors and initial isotopic compositions (Hoefs, 1980). For example, −55 per mil CH_4 might shift to −34.5 per mil after 60% oxidation and to −2.6 per mil after 90% oxidation, using a fractionation factor of 1.024 (Figure 6.4.7). Deuterium is also enriched in residual CH_4 undergoing oxidation (Coleman et al., 1981; Whiticar, 1996).

Rates of CH_4 oxidation by methanotrophs in natural environments can be appreciable. Using measured emission rates and isotopic fractionation of residual CH_4, CH_4 oxidation in landfill cover soil has been estimated. Schuetz et al. (2003) estimated average CH_4 oxidation rates of 1.3–1.5 g m^{-2} d^{-1} in soil cover over a landfill in France. Methane oxidation rates up to 166 g m^{-2} d^{-1} have been measured in soil microcosms (Kightley et al., 1995). Using a method based on measured diffusion rates for CH_4 and oxygen, the author estimated an oxidation rate of 0.6 g m^{-2} d^{-1} for CH_4 diffusing upward from a zone of free product at a petroleum release site.

Although rarely used in environmental forensic studies, there may be useful supplementary information in the carbon isotope ratio of individual C_{2+} gaseous hydrocarbons (James, 1983; Schoell et al., 1993). Such data are likely to be most applicable to the recognition and allocation of thermogenic gas sources where CH_4 isotopic compositions are very similar or microbial alteration has affected isotope ratios.

6.4.3 Radiogenic Isotopic Data

Carbon-14 analysis has forensic value because of the information it provides about the age of the organic carbon source. Carbon-14 is present in all living things. It is a naturally occurring isotope of carbon that is formed in the upper atmosphere by the reaction of cosmic-ray neutrons with nitrogen (Faure, 1977). Plants extract CO_2 from the atmosphere, thereby incorporating a ^{14}C concentration in their cells that reflects the ^{14}C concentration in atmospheric CO_2. This ^{14}C concentration is passed up the food chain. In most environmental investigations, the ^{14}C concentration in a sample is reported in terms of percent modern carbon (pMC), where 100 pMC is defined as the "normal" ^{14}C concentration in atmospheric CO_2 prior to anthropogenic disturbances such as nuclear bomb testing and extensive burning of fossil fuels (Stuiver and Polach, 1977).

The time significance of ^{14}C concentrations comes from two factors. The first is the natural decay of this radioactive isotope (Figure 6.4.8a). When a plant or animal dies,

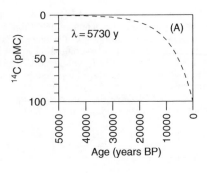

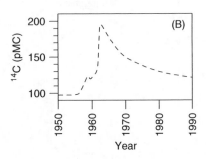

Figure 6.4.8 Temporal changes in ^{14}C concentrations. (A) Change related to radioactive decay of ^{14}C. (B) Change in ^{14}C concentration in atmospheric CO_2 related to nuclear bomb testing in the 1950s and 1960s. Note that CH_4 derived from very old (>50,000 years) carbon sources like petroleum will have a ^{14}C concentration of 0 pMC.

the ^{14}C concentration in its tissue gradually decreases as the ^{14}C undergoes radioactive decay with a half-life of 5730 years. After approximately 50,000 years, the remaining ^{14}C is too dilute to detect by standard methods. Organic matter greater than 50,000 years old, and CH_4 derived from such organic matter, will have a ^{14}C concentration of 0 pMC. This would include CH_4 formed from coal, oil, kerogen, and petroleum products made from these materials (e.g., gasoline, kerosene, and diesel fuel). Methane formed from organic matter less than 50,000 years old will have detectable ^{14}C (greater than 0 pMC) and can be "dated." A ^{14}C "date" obtained on CH_4 reflects not the time of methanogenesis, but the age of the carbon source from which CH_4 formed.

The second factor giving time significance to ^{14}C concentrations is the enormous increase in ^{14}C resulting from atmospheric testing of nuclear bombs during the 1950s and 1960s (Figure 6.4.8b; Levin et al., 1980). Consequently, organic material formed since this time has an elevated ^{14}C concentration. Methane produced by bacterial degradation of such young organic matter will also contain the so-called "bomb carbon", and have a ^{14}C concentration greater than 100 pMC. This is typically the case for landfill and sewage gases since they are largely produced from organic material less than a few decades in age (Coleman et al., 1995).

By considering mathematical mixing relationships, ^{14}C results can be used to investigate source contributions to a methane sample of mixed origin. In a two component mixture, the fraction of CH_4 derived from each source can be calculated from the measured ^{14}C concentration of the CH_4, and the anticipated ^{14}C concentration of CH_4 from two suspected sources,

$$f_1 = \frac{\left(^{14}C_{SPL} - {}^{14}C_2\right)}{\left(^{14}C_1 - {}^{14}C_2\right)}$$

where f_1 is the fraction of CH_4 from source 1 in the sample, and $^{14}C_{SPL}$, $^{14}C_1$, and $^{14}C_2$ are the ^{14}C concentrations in CH_4 in the sample, source 1, and source 2, respectively.

Much like ^{14}C, tritium has forensic value in identifying CH_4 produced from modern carbon sources. Tritium is a naturally occurring radioactive isotope of hydrogen (^3H) and has a half-life of 12.3 years. Its content in the atmosphere and hydrosphere was also markedly increased by atomic bomb testing in the 1950s and 1960s. Consequently, elevated tritium content in soil gas CH_4 could indicate the CH_4 was produced from organic matter no more than a few decades in age. Methane in some landfill gases is very high in tritium (Coleman et al., 1995).

6.4.4 Microbial Data

While still developing, methods for the analysis of microbial DNA have promise as a way to identify soil zones where

active methanogenesis (and CH_4 oxidation) is taking place (Inagaki et al., 2004; Kleikemper et al., 2005). By combining information on the distribution of active methanogens, methanotrophs, CH_4, oxygen, and organic carbon types, a fully integrated picture of biogenic CH_4 generation and oxidation might emerge. In the mean time, established methods for soil microcosms can be used for demonstrating the CH_4 generation potential of different soil/organic matter samples, and perhaps the relative generation rates from different samples (Figure 6.4.9). In a set of anaerobic microcosm experiments, several soil/organic matter samples were shown to naturally generate CH_4. Methane generation was most rapid for samples containing modern plant debris in landscaped soil. Soil containing tarry, weathered crude oil also generated CH_4, but at a lower rate.

6.5 CASE STUDIES

Several case studies, based on the author's personal investigations and others in the literature, are presented to illustrate application of the forensic tools discussed above. Note that in each case the spatial relationships of data and land use information are key to the interpretations.

6.5.1 Case Study #1

A service station, located in an urban area near a lake, was the site of a large accidental release of gasoline in 1980. An estimated $303 \, m^3$ (80,000 gallons) of leaded, premium gasoline was released. Light non-aqueous phase liquid (LNAPL) gasoline eventually covered an area roughly the size of a city block (~2 acres; Figure 6.5.1). During subsequent monitoring of a soil vapor extraction system, it was noticed that CH_4 comprised a substantial percentage of the influent vapor. The concentration of CH_4 in the SVE influent was greater on average than the concentration of total volatile hydrocarbons, and reached values greater than 10% (by volume). The CH_4 was initially attributed to bacterial degradation of the spilled gasoline. However, the persistence of high CH_4 concentrations over time raised questions about its origin and led to further investigation (Lundegard et al., 1998, 2000).

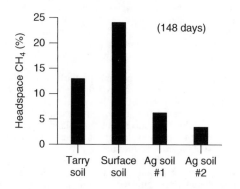

Figure 6.4.9 Methane concentration in the headspace of canned soil samples after 148 days of incubation. The tarry soil contains weathered crude oil from a former oilfield sump. The surface soil is a sandy soil that contains 4 weight percent organic carbon from modern plant debris. Ag Soil #1 and Ag Soil #2 are clay-rich agricultural soils that contain plant debris.

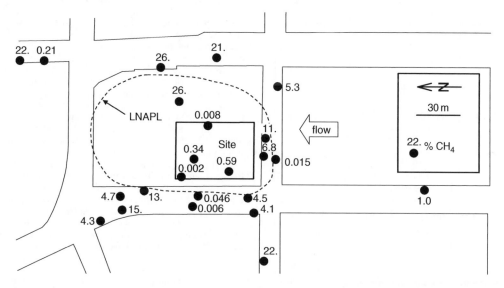

Figure 6.5.1 Site map for case study #1 showing original gasoline LNAPL plume and CH_4 concentrations measured in monitoring wells.

Exploratory borings and review of historical land use records revealed that the site vicinity was underlain by other carbon sources, including fill material with abundant zones of sawdust and wood, as well as historic lake sediments. Forensic investigations focused on whether gasoline or these other carbon sources could be the source of the CH_4.

Soil gas was collected from 23 monitoring wells within the limits of the original gasoline spill and beyond. Gas compositional analysis was conducted in the field with a portable gas chromatograph. Samples were also collected for subsequent isotopic analysis in the laboratory.

Methane content of soil gas samples varied by several orders of magnitude, from 0.002 to 26% (by volume) (Figure 6.5.1). Carbon dioxide concentrations ranged from 0.265 to 27% (by volume). High CH_4 and CO_2 concentrations were not restricted to the original area of contaminated soil and LNAPL, as is the case at other contaminated sites (Marrin, 1987, 1989, 1991; Suchomel *et al.*, 1990). Elevated CH_4 concentrations were encountered upgradient, downgradient, and cross-gradient of the original LNAPL plume (Figure 6.5.1). The occurrence of high CH_4 concentrations well beyond the limits of the original gasoline spill suggested that, perhaps, gasoline was not the sole source of the CH_4.

Within the limits of the gasoline spill some C_2–C_6 gasoline constituents were observed using gas chromatographic techniques. However, in high-CH_4 samples well beyond the limits of the original gasoline spill, less than 0.01% C_{2+} hydrocarbons (ethane and heavier) were observed. Low concentrations of C_2–C_6 alkanes are characteristic of biogenic gas accumulations.

Stable isotope data on the CH_4 showed that it was produced bacterially by the acetate fermentation pathway (Figure 6.5.2). Together with the gas composition the isotope data rule out a thermogenic source of the CH_4. Sources of thermogenic gas would include natural seepage from subsurface oil and gas deposits, and leaking gas pipelines.

However, stable isotope data alone did not resolve whether the source of the CH_4 was the spilled gasoline or the other carbon sources in the environment (e.g., woody fill debris or lake sediments). Carbon isotopic analysis of wood from the fill beneath the site and of gasoline from a monitoring well indicated that the wood and gasoline were essentially indistinguishable ($\delta^{13}C$ wood = −25.95‰; $\delta^{13}C$ gasoline = −25.83‰). Therefore, stable isotope analyses, while identifying the predominant mechanism for CH_4 formation (acetic acid fermentation), were unable to uniquely identify the carbon source for the CH_4.

When potential carbon sources have very similar stable isotopic compositions, or when bacterial oxidation has occurred, the source and origin of near-surface CH_4 is more difficult to determine. Under these conditions, ^{14}C (or tritium) data can be a powerful supplement to stable isotope data.

Five high-CH_4 samples were selected for ^{14}C analysis. Included were two samples more than 49 m beyond the limits of the original spill, on both the upgradient and downgradient sides (Figure 6.5.3). The ^{14}C concentrations in the CH_4 in the five samples were 60, 87, 92, 96, and 97 pMC, respectively. The two samples farthest from the original gasoline spill had the highest ^{14}C concentrations. If CH_4 were derived wholly from the spilled gasoline its ^{14}C concentration would be 0.0 pMC. It is clear, therefore, that the CH_4 in these samples is not derived predominantly from gasoline.

Using the ^{14}C data, the possible proportions of CH_4 from different sources can be explored through a simple mixing model. The sample with a CH_4 ^{14}C concentration of 60 pMC came from a well within the original extent of the gasoline plume. The ^{14}C concentration of this sample could be explained by formation solely from organic matter approximately 4000 years old. Alternatively, it could represent a mixture of two or more carbon sources. For example, if a two-component mixture involved contributions from the woody fill material (estimated ^{14}C concentration of 97 pMC) and the gasoline (^{14}C concentration of 0 pMC), 38% of the CH_4 could theoretically have been produced from gasoline. Any contribution of CH_4 from organic matter older than the wood fill, such as the underlying lake sediments, would reduce the calculated percentage of gasoline-derived CH_4. The ^{14}C concentration in the other four samples suggests that the actual contribution of CH_4 from gasoline is probably much lower than 38%.

The CH_4 in the other four samples, with ^{14}C concentrations of 87, 92, 96, and 97 pMC, could be produced entirely from organic matter with an average age of between 200 and 1100 years. While the precise age of lake sediments underlying the site is not known, their age could plausibly lie within this range. Derivation of CH_4 exclusively from the wood fill beneath the site (estimated age of 150–250 years) would yield a CH_4 ^{14}C concentration of approximately 97 pMC. One of these four CH_4 samples, therefore, could be completely derived from the wood fill. A two-component mixture calculation for these four samples, assuming the two sources are the gasoline (0 pMC) and the wood fill debris (97 pMC), indicates that the maximum contribution from gasoline ranges from 0 to 10%. Contributions of CH_4 from organic matter in the lake sediments would reduce the calculated contribution from gasoline. With the knowledge that two of these four samples came from well beyond the limits of the original gasoline spill, it is very unlikely that more than a few percent of the CH_4 in these samples was derived from the gasoline spilled at the site. The ^{14}C data indicate that the primary sources of the CH_4 are the woody debris within the fill material and the underlying lake sediments.

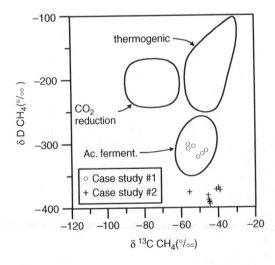

Figure 6.5.2 Hydrogen and carbon isotopic data for CH_4 samples in case study #1 (open circles) and case study #2 (crosses).

6.5.2 Case Study #2

At a large industrial facility where crude oil was stored, a number of accidental releases of the crude oil were known

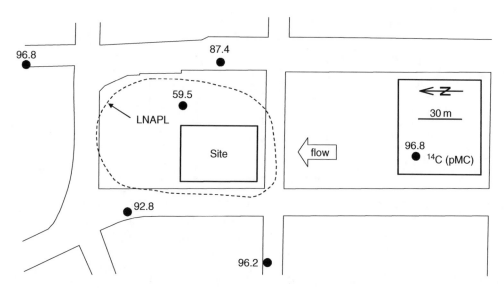

Figure 6.5.3 *Site map for case study #1 showing CH_4 ^{14}C results. Note that CH_4 derived from gasoline would have a ^{14}C concentration of 0 pMC. High values indicate little input from gasoline.*

to have occurred. Natural seeps of liquid and gaseous hydrocarbons from underlying petroleum reservoirs are also well known in the region surrounding the facility. Numerous pipelines and utility corridors exist in the area. In an area of the site where the soils were substantially impacted by spilled crude oil, high concentrations of combustible hydrocarbons were detected in soil gas. As a result of a property transaction, the source of the vapors was a liability issue and an important factor in determining the suitability of the area for construction. There was also interest in the fate of the CH_4 and whether there was significant upward transport toward the ground surface.

Shallow soil gas samples were collected in the area of the spilled crude oil. Driven soil probes were used to obtain samples at 0.6 m depth below grade. Gas chromatographic analysis revealed that the soil gas contained up to 24% CH_4 (by volume) as the predominant hydrocarbon. C_2 to C_5 hydrocarbons were present in low concentrations, from a few ppmv to over 1000 ppmv ($C_1/[C_1–C_5] > 0.98$). Having identified areas of very high CH_4 concentrations, vertical soil gas profiles to a depth of 6.1 m were conducted in order to understand better the occurrence and source of the CH_4. The soils throughout the profiled depth range contained substantial amounts of crude oil.

Carbon and hydrogen isotopic data on CH_4 in selected samples confirmed that the CH_4 is biogenic rather than thermogenic in origin (Figure 6.5.2). This origin is also consistent with the low concentrations of C_2 to C_5 hydrocarbons. Carbon-14 analysis indicated that the CH_4 and associated carbon dioxide contained no detectable ^{14}C, indicating that these components were derived from a carbon source greater than 50,000 years in age. Together, this information rules out several possible sources for the CH_4. The ^{14}C data rules out "modern" carbon sources such as buried refuse, compost, plant debris, and sewer gas. The stable isotope data excludes sources of thermogenic gas such as leaking gas pipelines and natural seepage from thermogenic gas in the underlying sedimentary basin. Based on these geochemical data, the CH_4 could be produced in the near-surface soil environment from biogenic degradation of the spilled crude oil, or produced biogenically at greater depth from

naturally occurring, old (>50,000 years) organic matter and transported to the near-surface soil environment.

A detailed vertical profile of soil gas composition was developed by careful sampling at different depths within the zone of contaminated soil from 0.15 to 6.1 m below ground surface. Soil gas profiles shed light on both the source and fate of the CH_4.

In the upper 1.5 m the CH_4 concentration increased dramatically with increasing depth, leveling off at between 70 and 75% (by volume) at 1.5 m and greater (Figure 6.5.4). Oxygen concentration showed the opposite depth trend, decreasing sharply in the upper 0.6 m and reaching zero at 0.9 m and greater, indicating the occurrence of anaerobic conditions. Carbon dioxide concentrations varied in a manner similar to CH_4, increasing rapidly with increasing depth to about 23% at 2.1 m and greater. However, in the aerobic zone of the soil column (upper 0.9 m) the CO_2/CH_4 ratio is significantly greater than it is below. This observation, together with the fact that CH_4 in the upper 1.5 m has a heavier (more positive) carbon isotopic composition, and CO_2 has a lighter (more negative) carbon isotopic composition than deeper samples, is evidence that CH_4 oxidation is occurring near the ground surface (Figure 6.5.5). In general, the presence of oxygen in soil gas will indicate that CH_4 oxidation is likely occurring, provided that the oxygen was not derived from an air leak in the sampling equipment.

The vertical profile of H_2 concentration was found to be important to understanding the source and origin of the CH_4 (Lundegard *et al.*, 2000). From a depth of 0.15 to 0.9 m, H_2 concentration increased more than 2 orders of magnitude from 41 ppm to 15,378 ppm (1.5%). Below 0.9 m, the H_2 concentration decreases sharply to less than 100 ppm by 2.1 m depth, indicating that H_2 utilization is occurring within the anaerobic zone of the soil column. In addition, the heavy carbon isotopic composition of CO_2 in the anaerobic zone (greater than 10‰) is evidence of methanogenesis by CO_2 reduction (Figure 6.5.4). These geochemical depth trends and isotopic data strongly suggest that CH_4 is being generated from petroleum contamination in the interval from 0.9 to 2.1 m by the reduction of CO_2 with molecular H_2. Thus, while the compositional and isotopic data indicate a biogenic origin for the CH_4 at this industrial facility, it is the detailed

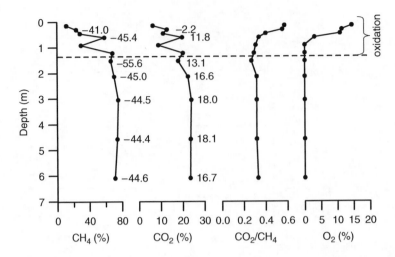

Figure 6.5.4 *Detailed vertical profile of soil gas composition at industrial facility (case study #2). Carbon isotope results for CH_4 and CO_2 are posted next to data points. Entire soil column is contaminated with crude oil. Upper 1.5 m is a zone of inferred CH_4 oxidation.*

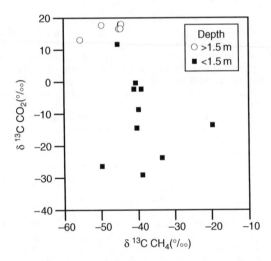

Figure 6.5.5 *Carbon isotopic composition of CH_4 and associated CO_2 for soil gas samples collected from depths greater than (open circles) and less than 1.5 m (solid squares) at the site described in case study #2. The isotopic enrichment of CH_4 and depletion of CO_2 in the shallow samples is evidence of CH_4 oxidation.*

depth profile that suggests the crude oil contamination as the source of the CH_4.

6.5.3 Case Study #3

Federal and state regulations in the US require that landfill gas containing greater than 5% CH_4 be prevented from migrating beyond the facility boundary. Generally, if CH_4 is found near a landfill, landfill gas is presumed to be the source. Such occurrences lead to compliance and sometimes litigation problems for the owner of the landfill. Advective escape of landfill gas can occur because positive pressure develops within the pores of the landfill waste. Pressures develop for two reasons. First there is a volume

increase associated with the methanogenic degradation of solid, ligno-cellulose into the gases, CH_4 and CO_2 (approximated by equation below).

$$2CH_2O \text{ (s)} \rightarrow CH_4 \text{ (g)} + CO_2 \text{ (g)}$$

Second, the waste within the landfill compacts under the weight of overburden, which reduces the porosity of the waste.

Near a municipal waste landfill, elevated concentrations of CH_4 in soil gas samples from site periphery and off-site locations raised questions about the effectiveness of the facility's gas recovery system (Pierce and LaFountain, 2000). Landfill gas recovered by the vapor extraction system is burned at a flare station. Effectiveness of the recovery system and regulatory compliance is monitored by analyzing samples from numerous gas probes along the periphery of the facility (Figure 6.5.6). Unexpectedly high gas pressures (over 10 inches of water) and CH_4 concentrations (over 60%) in some locations raised suspicions about the source of the CH_4.

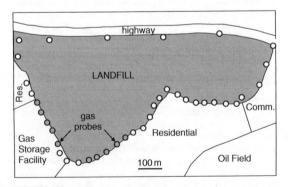

Figure 6.5.6 *Map of landfill site described in case study #3. Surrounding features include residential neighborhoods, commercial property, an oil field, and an underground gas storage facility. Gray circles indicate those peripheral gas probes that showed anomalous methane concentrations or gas pressures.*

The landfill is adjacent to residential and commercial properties, as well as an oil and gas field, and an underground gas storage facility (Figure 6.5.6). In the investigation of the CH_4 anomalies, gas samples were collected from site-periphery gas probes, offsite gas probes, the landfill flare station, the oil field, and the gas storage reservoir. Stable isotope data showed that most of the samples from the site-periphery gas probes plot within the field generally accepted for biogenic landfill gas (Figure 6.5.7). Gas samples from the oil field and the gas storage reservoir plot within the thermogenic field. And, several of the gas probe samples also plot within the thermogenic field, suggesting that some of the anomalous CH_4 localities are affected by non-landfill thermogenic gas. A smaller number of samples were also analyzed for Carbon-14 activity (Figure 6.5.8). As would be expected for thermogenic gas, CH_4 from the gas storage reservoir and the nearby oil field had ^{14}C activities of less then 1 pMC. A sample of landfill flare gas had a ^{14}C activity of 119 pMC, a value consistent with an origin from post-1950 organic refuse. Methane from some of the peripheral gas probes had ^{14}C activities consistent with a landfill gas origin (>119 pMC), while at other peripheral and offsite locations lower ^{14}C activities were evidence of gas of mixed origin containing variable amounts of non-landfill thermogenic gases. Methane from one of the peripheral gas probes had a ^{14}C activity of 2.5 pMC, indicating that this gas was about 98% thermogenic. These results confirmed that several of the site-periphery and offsite locations are affected by non-landfill thermogenic gas.

6.5.4 Case Study #4

Owners of a house built on remediated petroleum-impacted soil had trees that were not growing well. Methane derived from residual soil contamination was alleged to be the cause of the stressed vegetation. The sandy soil originally contained tarry crude oil contamination. The soil had been remediated to less than 100 mg/kg total petroleum hydrocarbons, then re-placed and re-compacted before housing construction. An assessment of indoor air failed to detect elevated concentrations of CH_4. A soil gas survey found

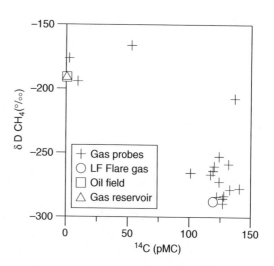

Figure 6.5.8 *Plot of hydrogen isotopic composition and ^{14}C activity of methane samples collected in case study #3. Note that several gas probe samples do not plot in the vicinity of the landfill flare gas, and two gas probe samples plot very close to samples from the oil field and gas storage reservoir. Data from Pierce and LaFountain (2000).*

highly variable concentrations of CH_4 in soil, from 0.15 to 3 m depth. Concentrations varied from less than 10 ppmv to 12%. Carbon-14 analysis of the soil gas sample with the highest CH_4 concentration yielded a result of 109 pMC, a value which indicates the CH_4 was predominantly derived from post-1950 organic matter. The low concentration of petroleum hydrocarbons in the remediated soil, therefore, did not significantly contribute to the CH_4. Microcosm experiments also demonstrated that local soils with modern organic debris readily generated methane when they became sufficiently wet (Figure 6.4.9). Agricultural experts attributed the stressed vegetation at the site to the unfavorable physical character of the densely compacted sand.

6.5.5 Case Study #5

Methane was encountered during a soil gas assessment of a small petroleum spill at a 2900-acre residential development site. Follow-up investigations revealed that shallow CH_4 anomalies were more widely distributed than the petroleum contaminated soil and multiple theories for the source of the CH_4 emerged. Sources that were initially considered included leaking pipelines, spilled petroleum, organic matter in lowland or wetland soils, animal manure, sedimentary organic matter in underlying geologic strata, and landfill waste. Site-wide reconnaissance, review of historic air photos, and the molecular composition of soil gas helped to narrow the list of likely sources. Stable and radiogenic isotopic data confirmed that the CH_4 was biogenic and derived from recent organic matter (Golightly, personal communication 2004).

After soil gas sampling at every graded lot, it was noted that the occurrence of elevated CH_4 correlated strongly with the location of re-compacted fill material (Figure 6.5.9; AMEC Earth & Environmental, 2001a).

The re-compacted fill material consisted of on-site native soil that was removed during mass grading of the site and replaced in areas where soil was needed to reach final design grade. Mass grading engineering maps showed where soil was removed (i.e., cut areas) and where re-compacted fill was placed (i.e., fill areas). Of the 128 lots

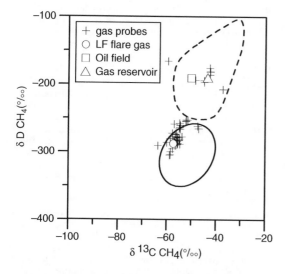

Figure 6.5.7 *Hydrogen and carbon isotopic data for CH_4 samples in case study #3. Note that several gas probe samples plot in the thermogenic part of the diagram, near samples from the oil field and the gas storage reservoir. Data from Pierce and LaFountain (2000).*

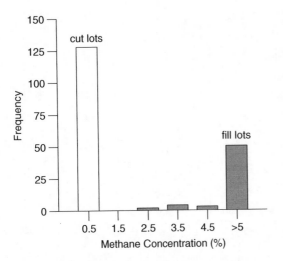

Figure 6.5.9 *Histogram of soil gas CH$_4$ results from a residential development site (case study #5). White bar shows data from cut lots. Stippled bars show data from fill lots. Data from AMEC Earth & Environmental (2001a).*

built on cut areas, none had soil gas CH$_4$ concentrations over 0.1%. In marked contrast, 85% of the 59 lots built on fill had soil gas CH$_4$ concentrations over 5%. This strong geographic relationship demonstrated that the CH$_4$ was produced from organic matter within the fill material. The fill differed from the in-place native soil only in having a higher moisture content (water was added during fill operations) and greater degree of compaction. These two factors evidently allowed anaerobic conditions to develop and methanogens to produce CH$_4$ from native organic matter. These processes were further demonstrated in laboratory column studies (AMEC Earth & Environmental, 2001b). Native soil with 4% organic matter was incubated in sealed columns for 24 weeks. Shortly after water was added in the eleventh week, the headspace CH$_4$ concentration rose sharply and reached values as high as 26% (Figure 6.5.10).

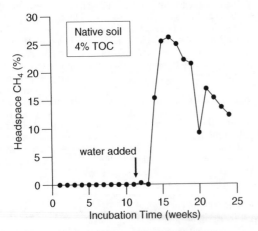

Figure 6.5.10 *Concentration of headspace CH$_4$ versus incubation time in a microcosm using native soil with 4 weight percent organic carbon from a residential development site (case study #5). Data from AMEC Earth & Environmental (2001b).*

The most important discovery of this case study was that CH$_4$ generation and accumulation could be initiated in native soils simply by changing their physical character through routine grading operations. Based on the findings of this and other cases in the area, the County planning and land use department passed an ordinance specifying soil gas assessment and mitigation requirements in housing developments where mass grading occurs. An action level was set at 12,500 ppmv CH$_4$ in soil gas. Mitigation is required on all lots with 3 m (10 feet) or more of fill, unless soil gas testing demonstrates that CH$_4$ is not present above the action level.

6.6 SUMMARY

Discovery of near-surface occurrences of CH$_4$ in soil gas can generate alarm as well as financial and legal consequences for responsible parties, property owners, developers, and other parties. Determining the origin and source of near-surface CH$_4$ is very important to the determination of environmental liability and to the selection of appropriate mitigation measures. Near-surface occurrences of CH$_4$ can have a variety of natural and anthropogenic sources. In many instances, certain sources of CH$_4$ can be ruled out by molecular compositional and/or isotopic analysis of a few samples. However, to confidently identify the source, or sources, of CH$_4$ it is generally necessary to integrate site-specific geological, land use, and forensic geochemical data on a number of samples. Spatial trends in geochemical data (vertically and laterally) are especially important. As demonstrated by the case studies presented, CH$_4$ associated with spilled petroleum is derived from the petroleum in some cases, but not in others.

6.7 ACKNOWLEDGMENTS

The generous support of the Unocal Corporation is gratefully acknowledged. I would also like to thank the following Unocal colleagues, past and present, for much of what I have learned about soil gas investigations: Todd Ririe, Bob Sweeney, Bob Haddad, and Greg Ouellette. Bill Golightly and Clay Westling helped me find information pertaining to case study #5. Chris Kitts provided references to microbiological investigations of methanogenesis.

REFERENCES

Allison, M. L., 2001, Hutchinson, Kansas: A Geologic Detective Story; Geotimes Web Feature, American Geologic Institute, http://www.agiweb.org/geotimes/Oct01/feature kansas.html

AMEC Earth & Environmental, Inc., 2001a, Methane assessment summary report, 4S Ranch, Tract 5067, San Diego County, California.

AMEC Earth & Environmental, Inc., 2001b, Methane gas column study, 4S Ranch, Ranch Bernardo, San Diego, California.

Baldassare, F. J., and Laughrey, C. D., 1997, Identifying the sources of stray methane by using geochemical and isotopic fingerprinting. *Environmental Geoscience*, Vol. 4, No. 2, pp. 85–94.

Chanton, J., and Liptay, K., 2000, Seasonal variations in methane oxidation in a landfill cover soil as determined by an in situ isotope technique. *Global Biochem. Cycles*, Vol. 14, pp. 51–60.

Chen, J., Yonchang, X., and Difan, H., 2000, Geochemical characteristics and origin of natural gas in Tarim Basin, China. *Amer. Assoc. Petrol. Geol. Bull.*, Vol. 84, pp. 591–606.

City of Huntington Beach, 1997, Methane district building permit requirements. City Specification 429.

City of Los Angeles, 1996, Building Regulations, Section 91.7104.1, High Potential Methane Zone Requirements.

Claypool, G. E., and Kaplan, I. R., 1974, The origin and distribution of methane in marine sediments. In Kaplan, I. R., ed., *Natural Gases in Marine Sediments*, Plenum Publishing Corporation, New York, pp. 99–139.

Coleman, D. D., Liu, C. L., and Riley, K. M., 1988, Microbial methane in the glacial deposits and shallow Paleozoic rocks of Illinois. *Chemical Geology*, Vol. 71, pp. 23–40.

Coleman, D. D., Risatti, J. B., and Schoell, M., 1981, Fractionation of carbon and hydrogen isotopes by methane-oxidizing bacteria. *Geochemica et Cosmochimica Acta*, Vol. 45, pp. 1033–1037.

Coleman, D. D., Liu, C. L., Hackley, K. C., and Benson, L. J., 1993, Identification of landfill methane using carbon and hydrogen isotope analysis. *Proceedings of Sixteenth International Madison Waste Conference*, University of Wisconsin, pp. 303–314.

Coleman, D. D., Liu, C. L., Hackley, K. C., and Pelphrey, S. R., 1995, Isotopic identification of landfill methane. *Environmental Geoscience*, Vol. 2, pp. 95–103.

Conrad, M. E., Templeton, A. S., Daley, P. F., and Alvarez-Cohen, L., 1999a, Isotopic evidence for biological controls on the migration of petroleum hydrocarbons. *Organic Geochem.*, Vol. 30, pp. 843–859.

Conrad, M. E., Templeton, A. S., Daley, P. F., and Alvarez-Cohen, L., 1999b, Seasonally-induced fluctuations in microbial production and consumption of methane during bioremediation of aged subsurface refinery contamination. *Environ. Sci. Technol.*, Vol. 33, pp. 4061–4058.

County of San Diego Department of Environmental Health (CSDDEH), 2004, Site assessment and mitigation manual: available at http://www.sdcounty.ca.gov /deh/lwq/sam/manual_guidelines.html#sections.

Daniels, L., Fulton, G., Spencer, R. W., and Orme-Johnson, W. H., 1980, Origin of hydrogen in methane produced by *Methanobacterium Thermoautotrophicum. Jour. Bacteriol.*, Vol. 141, pp. 694–698.

Deyo, B. G., Robbins, G. A., and Binkhorst, G. K., 1993, Use of portable oxygen and carbon dioxide detectors to screen soil gas for subsurface gasoline contamination. *Ground Water*, Vol. 31(4), pp. 598–604.

Doorn, M. R. J., and Barlaz, M. A., 1995, Estimates of global methane emissions from landfills and open dumps: Report EPA-600/R-95-019, *U. S. Environmental Protection Agency*, Washington, D.C.

Eganhouse, R. P., Baidecker, M. J., Cozzarelli, I. M., Aiken, G. R., Thorn, K. A., and Dorsey, T. F., 1993, Crude oil in a shallow sand and gravel aquifer – II. Organic geochemistry. *Applied Geochemistry*, Vol. 8, pp. 551–567.

Faure, G., 1977, *Principles of Isotope Geology*. John Wiley and Sons, New York, p. 464.

Gandoy-Bernasconi, W., Aguilar, A., and Velez, O., 2004, Detection of diesel fuel plumes using methane data in Guadalajara, Mexico. *Ground Water Monitoring and Remediation*, Vol. 24, pp. 73–81.

Groves, M., 2003, Playa Vista buyers will test capability of methane shield. *Los Angeles Times*, January 6, 2003.

Hackely, K. C., Liu, C. L., and Trainor, D., 1999, Isotopic identification of the source of methane in subsurface sediments of an area surrounded by waste disposal facilities. *Applied Geochem.*, Vol. 14, pp. 119–131.

Hoefs, J., 1980, *Stable Isotope Geochemistry*, 2nd edition: Springer-Verlag, New York, p. 208.

Inagaki, F., Tsunogai, U., Suzuki, M., Kosaka, A., Machiyama, J., Takai, K., Nunoura, T., Nealson, K. H., and Horikoshi, K., 2004, Characterization of C_1-metabolizing prokaryotic communities in methane seep habitats at the Kuroshima Knoll, southern Ryukyu Arc, by analyzing *pmoA, mmoX, mxaF, mcrA*, and 16S rRNA genes. *Applied and Envir. Microbiol.*, Vol. 70, pp. 7445–7455.

James, A. T., 1983, Correlation of natural gas by use of carbon isotopic distribution between hydrocarbon components. *Amer. Assoc. Petrol. Geol. Bull.*, Vol. 67, pp. 1176–1191.

Jenden, P. D., Drazan, D. J., and Kaplan, I. R., 1993, Mixing of thermogenic natural gases in northern Appalachian basin. *Amer. Assoc. Petrol. Geol. Bull.*, Vol. 77, pp. 980–998.

Jenden, P. D., Newell, K. D., Kaplan, I. R., and Watney, W. L., 1988, Composition and stable isotope geochemistry of natural gases from Kansas, Midcontinent, USA. *Chemical Geology*, Vol. 71, pp. 117–147.

Kaplan, I. R., 1994, Identification of formation process and source of biogenic gas seeps. *Israel. Jour. Earth Sci.*, Vol. 43, pp. 297–308.

Kerfoot, H. B., Mayer, C. L., Durgin, P. B., and D'Lugosz, J. J., 1988, Measurement of carbon dioxide in soil gases for indication of subsurface hydrocarbon contamination. *Ground Water Monitoring Review*, Spring 1988, pp. 67–71.

Khalil, M. A. K., and Rasmussen, R. A., 1983, Sources, sinks, and seasonal cycles of atmospheric methane. *Jour. Geophysical. Res.*, Vol. 88, pp. 5131–5144.

Khalil, M. A. K., and Rasmussen, R. A., 1990, Atmospheric methane: Recent global trends. *Environmental Science and Technology*, Vol. 24, pp. 549–553.

Kightley, D., Nedwell, D. B., and Cooper, M., 1995, Capacity for methane oxidation in landfill cover soils measured in laboratory-scale soil microcosms. *Applied and Environ. Microbiol.*, Vol. 61, pp. 592–601.

Kleikemper, J., Pombo, S. A., Schroth, M. H., Sigler, W. V., Pesaro, M., and Zeyer, J., 2005, Activity and diversity of methanogens in a petroleum hydrocarbon-contaminated aquifer. *Applied and Environ. Microbiol.*, Vol. 71, pp. 149–158.

Laughrey, C. D., and Baldassare, F. J., 1998, Geochemistry and origin of some natural gases in the Plateau province, central Appalachian Basin, Pennsylvania and Ohio, *Amer. Assoc. Petrol. Geol. Bull.*, Vol. 82, pp. 317–335.

Levin I., Munnich, K. O., and Weiss, W., 1980, The effect of anthropogenic CO_2 and ^{14}C sources on the distribution of ^{14}C in the atmosphere. *Radiocarbon*, Vol. 22, pp. 379–391.

Lovely, D. R., Chapelle, F. H., and Woodward, J. C., 1994, Use of dissolved H_2 concentrations to determine distribution of microbially catalyzed redox reactions in anoxic groundwater. *Environmental Science and Technology*, Vol. 28, No. 7, pp. 1205–1210.

Lundegard, P. D., Haddad, R., and Brearley, M., 1998, Methane associated with a large gasoline spill: Forensic determination of origin and source. *Environmental Geoscience*, Vol. 5, pp. 69–78.

Lundegard, P. D., Sweeney, R. E., and Ririe, G. T., 2000, Soil gas methane at petroleum contaminated sites: forensic determination of origin and source. *Environmental Forensics*, Vol. 1, pp. 3–10.

Marrin, D. L., 1987, Detection of non-volatile hydrocarbons using a modified approach to soil-gas surveying: *Proceedings of NWWA Conference on Petroleum Hydrocarbons and Organic Chemicals in Ground Water*, Houston, pp. 87–95.

Marrin, D. L., 1989, Soil gas analysis of methane and carbon dioxide: delineating and monitoring petroleum hydrocarbons. *Proceedings of NWWA Conference on Petroleum Hydrocarbons and Organic Chemicals in Ground Water*, Houston, pp. 357–367.

Marrin, D. L., 1991, Subsurface biogenic gas rations associated with hydrocarbon contamination: In Hinchee, R. E. and Olfenbuttel, R. F., eds, In situ bioreclamation: Butterworth-Heinemann, Publishers, Stoneham, MA, pp. 539–545.

Martens, C. S., Blair, N. E., Green, C. D., and Des Marais, D. J., 1986, Seasonal variations in the stable carbon isotopic signature of biogenic methane in a coastal sediment. Science, Vol. 233, pp. 1300–1303.

Pierce, J. L., and LaFountain, L. J., 2000, Application of advanced characterization techniques for the identification of thermogenic and biogenic gases. Proceedings 23rd Annual Landfill Gas Symposium, Solid Waste Association of North America, pp. 153–168.

Pine, M., and Barker, H. A., 1956, Studies on the methane fermentation. XII. The pathway of hydrogen in the acetate fermentation. Jour. Bacteriol., Vol. 71, pp. 644–648.

Quesnel-Cariboo Observer, 1997, Five killed in tragic blast. Sunday, April 20, 1997.

Revesz, K., Coplen, T. B., Baedecker, M. J., Glynn, P. D., and Hult, M., 1995, Methane production and consumption monitored by stable H and C isotope ratios at a crude oil spill site, Bemidji, Minnesota. Applied Geochemistry, Vol. 10, pp. 505–516.

Rice, D. R., and Claypool, G. E., 1981, Generation, accumulation, and resource potential of biogenic gas. American Association of Petroleum Geologists Bulletin, Vol. 65, pp. 5–25.

Rigby, D., and Smith, J. W., 1981, An isotopic study of gases and hydrocarbons in the Cooper basin. Australian Petrol. Explor. Assoc. Journal, Vol. 21, pp. 222–229.

Ririe, G. T., and Sweeney, R. E., 1995, Fate and transport of volatile hydrocarbons in the vadose zone. Conference Proceedings, Petroleum Hydrocarbons and Organic Chemicals in Ground Water, Ground Water Publishing Company, Ohio, pp. 529–542.

Robbins, G. A., Deyo, B. G., Temple, M. R., Stuart, J. D., and Lacy M. J., 1990, Soil gas surveying for subsurface gasoline contamination using total organic vapor detection instruments, Part II., Field experimentation. Ground Water Monitoring Review, Vol. 10(4), pp. 110–117.

Schoell, M., 1980, The hydrogen and carbon isotopic composition of methane from natural gases of various origins. Geochimica et Cosmochimica Acta, Vol. 44, pp. 649–661.

Schoell, M., 1988, Multiple origins of methane in the earth, in origins of methane in the earth, Schoell, M. (ed.), Chemical Geology, Vol. 71, 1–10.

Schoell, M., Jenden, P. D., Beeunas, M. A., and Coleman, D. D., 1993, Isotope analyses of gases in gas field and gas storage operations: SPE Gas Technology Conference, SPE 26171.

Schuetz, C. Bogner, J., Chanton, J., Blake, D., Morcet, M., and Kjeldsen, P., 2003, Comparative oxidation and net emissions of methane and selected non-methane organic compounds in landfill cover soils. Environ. Sci. Technol., Vol. 37, pp. 5150–5158.

Smith, J. W., Gould, K. W., Hart, G. H., and Rigby, D., 1985, Isotopic studies of Australian natural and coal seam gases. Bulletin of the Australasian Institute of Mining and Metallurgy, Vol. 290, pp. 43–51.

Snowdon, L. R., and Roy, K. J., 1975, Regional organic metamorphism in the Mesozoic strata of the Sverdrup Basin. Canadian Petrol. Geol. Bull., Vol. 23, pp. 131–148.

Stuiver, M., and Polach, H. E., 1977, Reporting of ^{14}C data. Radiocarbon, Vol. 19, pp. 355–363.

Suchomel, K. H., Kreamer, D. K., and Long, A., 1990, Production and transport of carbon dioxide in a contaminated vadose zone: A stable and radioactive carbon isotope study. Environ. Sci. Technol., Vol. 24, pp. 1824–1831.

Sugimoto, A., and Wada, E., 1995, Hydrogen isotopic composition of bacterial methane: CO_2/H_2 reduction and acetate fermentation. Geochim. Cosmochim. Acta, Vol. 59, pp. 1329–1337.

US Environmental Protection Agency (USEPA), 1996, Soil gas sampling: Standard operating procedure No. 2042, Environmental Response Team, Washington, D.C., June.

Whiticar, M. J., 1996, Isotope tracking of microbial methane formation and oxidation. Mitt. Internat. Verein. Limnol., Vol. 25, pp. 39–54.

Whiticar, M. J., Faber, E., and Schoell, M., 1986, Biogenic methane formation in marine and freshwater environments: CO_2 reduction vs. acetate fermentation—isotope evidence. Geochemica et Cosmochimica Acta, Vol. 50, pp. 693–709.

Wilson, J., and Sauerwein, K., 2003, Methane is out before school is in: Los Angeles Times, Inland Empire Edition, November 11, 2003.

Wilson, L.H., Johnson, P. C., and Rocco, J. R., 2004. Collecting and Interpreting Soil Gas Samples from the Vadose Zone: A Practical Strategy for Assessing the Subsurface Vapor-to-Indoor-Air Migration Pathway at Petroleum Hydrocarbon Sites. Final Draft. Publ. 4741. American Petroleum Institute. Washington D.C.

7 Radioactive Compounds

Arthur F. Eidson and Donna M. Beals

Contents

7.1 INTRODUCTION

Much of nuclear forensic science is concerned with identifying and quantifying radioactive materials at concentrations that might indicate human activity. This effort includes identification of man-made radionuclides or concentrations of naturally occurring radionuclides that might reflect human activity, or both. During the last six decades, the extensive literature on the measurement of radionuclides and their management and behavior in the environment have become extensive. For example, the literature related to the 1986 Chernobyl accident reflects a multinational effort to develop engineering and regulatory approaches to prevent similar accidents and to characterize the environmental and public health consequences.

The use of radiological measurements in a forensic study is generally subject to the same criteria as other scientific measurements. The quality control and quality assurance, chain of custody of samples, instrument maintenance records, analyst training records, and other criteria applied to the analysis of non-radioactive chemicals also apply to radiological measurements. Likewise, the rules for admissibility and defensibility of laboratory data in court apply to radiometric data and are described in Chapter 1. Although these rules can be expected to vary according to jurisdiction, general administrative procedures that might be applied internationally have been described (Swindle *et al.*, 2003).

This chapter presents an overview of technical and regulatory approaches used in nuclear forensic studies. The chapter cites examples of studies to monitor normal use of radioactive materials and their movement in the environment and provides references to more detailed descriptions. The chapter is limited to radioactive materials and does not address radiation generated by medical or industrial X-ray machines or particle accelerators, except as related to the radioactive waste they might produce. Radionuclides associated with commercial nuclear power generation are addressed; those associated with naval nuclear propulsion are not.

No study of radioactivity in the health and environmental sciences can be well informed without a review of *Radioactivity and Health: A History* by J. Newell Stannard (Stannard, 1988). This work begins with the isolation of ^{226}Ra in 1898 and the report of ^{226}Ra-induced skin erythema a few months later. The work includes a review of research in universities and government-sponsored laboratories, interprets literature accumulated from the early years through the Cold War, and identifies avenues of future research. The history traces the development of environmental science, analytical methods, health protection standards, and the use of radionuclides in nuclear medicine.

7.2 BASIC CONCEPTS AND TERMINOLOGY

Radioactive decay is the spontaneous emission of particles or electromagnetic radiation from an unstable atom due to a transformation in its nucleus. As a result of the transformation, alpha, beta, and neutron particles can be ejected from the nucleus, often accompanied by the emission of positron particles, gamma rays, and X-rays. The atom transforms into a new element, which can be either radioactive or stable.

Alpha particles contain two protons and two neutrons and have a +2 charge, representing the helium nucleus. Beta particles are electrons with a −1 charge that originate from a neutron in the unstable nucleus. Positrons are beta particles with a +1 charge. They have a very short half-life and are not stable in the environment. Gamma rays and X-rays are photons. According to quantum theory, these photons are emitted and absorbed as quanta of discrete light energy.

The energies of alpha particles and photons are discrete and characteristic of the unstable atom, which can allow its identification.

7.2.1 Terminology and Units

The terms and units used in this chapter are defined below. Definitions of many more terms can be found in Shleien (1992), reports published by the International Commission on Radiation Units and Measurements (ICRU, 1993, 1998), and in other references cited in context.

Activity – The quantity (A) of a radionuclide in an energy state at a given time and it is the ratio of the number of nuclei (dN) that undergo nuclear transformation (i.e., decay) in a given time interval (dt), such that $A = dN/dt$. The international unit for A is expressed as the bequerel (Bq) with units of seconds^{-1}, where 1 Bq = 1 decay per second (ICRU, 1998). Prior to the development of the International System of Units (SI), activity was expressed in units of curies (Ci) defined as 37 GBq (3.7×10^{12} Bq), approximately the rate of decay in 1 g of ^{226}Ra (Shleien, 1992). The curie is still commonly used in the United States.

Multiples of the bequerel are expressed as:

millibecquerel (mBq) = 10^{-3} Bq = 2.703×10^{-14} Ci
kilobecquerel (kBq) = 10^{3} Bq = 2.703×10^{-8} Ci
megabecquerel (MBq) = 10^{6} Bq = 2.703×10^{-5} Ci
gigabecquerel (GBq) = 10^{9} Bq = 2.703×10^{-2} Ci
terabecquerel (TBq) = 10^{12} Bq = 2.703×10^{1} Ci
petabecquerel (PBq) = 10^{15} Bq = 2.703×10^{4} Ci
exabecquerel (Ebq) = 10^{18} Bq = 2.703×10^{7} Ci

Fractions of the curie are expressed as:

millicurie (mCi) = 10^{-3} Ci = 3.7×10^{7} Bq
microcurie (μCi) = 10^{-6} Ci = 3.7×10^{4} Bq
nanocurie (nCi) = 10^{-9} Ci = 3.7×10^{1} Bq
picocurie (pCi) = 10^{-12} Ci = 3.7×10^{-2} Bq
femptocurie (fCi) = 10^{-15} Ci = 3.7×10^{-5} Bq
attocurie (aCi) = 10^{-18} Ci = 3.7×10^{-8} Bq

Activity Median Aerodynamic Diameter (AMAD) – The diameter of a unit density sphere with the same terminal settling velocity in air as that of an aerosol particle whose activity is the median for the entire aerosol (Shleien, 1992). The AMAD definition facilitates a description of the behavior of airborne particles in terms that describe their aerodynamic characteristics rather than their physical diameter or shape.

Atomic Mass Unit (u) – The unit of the atomic weight
$$= 1.660 \times 10^{-27} \text{ kg (Shleien, 1992).}$$

Dose – The term used in the literature to describe the quantity of radiation energy absorbed in a medium. If the energy is sufficient to liberate an electron from the absorbing medium, the radiation is termed ionizing radiation.

Absorbed dose (D) – The quotient of the mean energy (de) imparted by ionizing radiation to matter of mass dm, such that $D = de/dm$. Absorbed dose is expressed as the gray (Gy) with units of J kg^{-1} (ICRU, 1993). The absorbed dose rate is the absorbed dose within a time interval and is described in units of Gy per second. Prior to the development of SI units, absorbed dose was expressed as the rad, in units of 100 ergs/g, which can be converted to Gys as 1 rad = 0.01 Gy. Dose equivalent and effective dose equivalent describe absorbed dose as modified to estimate the effect of ionizing radiation on animal tissues and are expressed as the sievert (Sv). Prior to the development of SI units, the Sv was expressed as the rem, which can be converted to the Sv as 1 rem = 0.01 Sv. Rad and rem are the terms still commonly used in the United States.

Half-life – The time required for decay of one half of the quantity of a radionuclide (see Section 7.2.2).

Secular equilibrium – The equality of activities between a long-lived parent radionuclide and its short-lived daughter (see Section 7.1.2). At secular equilibrium, the daughter decays at the same rate as its formation from the parent. If the parent and daughter are separated, the in growth of the daughter resumes until equilibrium is re-established. In practice, secular equilibrium can be assumed within 7 to 10 half-lives of the daughter.

Transuranic elements – Elements with atomic numbers greater than 92 (uranium).

7.2.2 Radioactive Decay Kinetics

The quantity of a radionuclide is described by its activity according to the equation:

$$A = \lambda N \qquad (7.2.1)$$

where

A = activity (Bq)
λ = the decay constant (seconds^{-1}) which equals the probability of decay within a time interval (ICRU, 1998)
N = the number of atoms.

The activity at any time is described by the equation:

$$A = A_o e^{-\lambda t} \qquad (7.2.2)$$

where

A = activity (Bq)
A_o = initial activity at time $t = 0$
λ = the decay constant (seconds^{-1})
t = time since $t = 0$ in seconds.

When one half of the initial activity has decayed, such that $A/A_o = 1/2$, then solving Equation (7.2.2) for t results in the half-life as $T_{1/2} = 0.693/\lambda$ seconds.

The kinetics of serial decay can be described as follows:

$$A \xrightarrow{\lambda_A} B \xrightarrow{\lambda_B} C \qquad (7.2.3)$$

The number of A atoms at any time is given by Equation 7.2.2. The number of B atoms is given by the equation (Cember, 1989):

$$N_B = \frac{\lambda_A N_{Ao}}{\lambda_B - \lambda_A} \left(e^{-\lambda_A t} - e^{-\lambda_B t} \right) \qquad (7.2.4)$$

where

N_B = the number of B atoms
N_{Ao} = the number of A atoms at $t = 0$
λ_B = the decay constant of B (seconds^{-1})
λ_A = the decay constant of A (seconds^{-1})
t as defined above

When $\lambda_B \gg \lambda_A$ (i.e., the half-life of B is much shorter than the half-life of A), Equation 7.2.4 approximates $\lambda_B N_B = \lambda_A N_A$ (i.e., equal activities of A and B, or secular equilibrium).

7.3 ANALYTICAL TECHNIQUES

Because the energies of alpha and beta particles and gamma rays emitted during the decay of a radioactive atom are characteristic of each radioisotope, they allow identification of the parent atom by measuring their energies. Measuring the decay product types and energies is referred to as radiometric detection. When the half-life of the parent radioisotope is very long, there are minimal decays to measure per unit time, thus the parent may be measured directly by chemical analytical techniques. Because simpler chemical techniques cannot distinguish between isotopes of an element, mass spectrometric techniques can be used. With the advances in mass spectrometry, radioisotopes with half-lives of longer than 100 to 1000 years are often determined by mass spectrometric techniques rather than radiometric techniques (Crain and Gallimore, 1992).

Depending on the type of radioactive decay process and the half-life of the isotope of interest, many analytical techniques are available to determine the activity or concentration of a radioisotope in the environment. A useful review of radiochemistry and radioactive decay, along with methods of analysis, can be found in Radiochemistry and Nuclear Methods of Analysis by W.D. Ehmann and D.E. Vance (Ehmann and Vance, 1991). In performing environmental analyses, either the sample can be collected and brought to the laboratory or the detector can be brought to the field. Sensitivity and selectivity (precision and accuracy) are often much better for laboratory-based methods; however, the difficulty of sample collection, preservation and transportation, and cost of these analyses may not be required to meet the data quality objectives of some projects. The following discussion highlights various methods of analysis, from field screening methods to the more precise laboratory methods, which can be based on radiometric counting of the radioisotope decay or based on mass spectral analysis of the isotope itself.

7.3.1 Standard Methods

Techniques used frequently to monitor the safe use of radionuclides and to study their occurrence in the environment are described in the Multi-Agency Radiation Survey and Site Investigation Manual (MARSSIM) (USEPA, 2000). The MARSSIM was developed by four United States Government agencies (Department of Defense, Department of Energy, the Environmental Protection Agency, and the Nuclear Regulatory Commission) with contributions from individual experts and a scientific advisory board. The MARSSIM provides guidance on planning and executing an investigation and on reporting the results to meet regulatory requirements. The MARSSIM does not include requirements independent of regulatory agencies.

The Multi-Agency Radiological Laboratory Analytical Protocols (MARLAP) manual (USEPA, 2004) was developed by the Environmental Protection Agency, the Department of Energy, the Department of Homeland Security, the Nuclear Regulatory Commission, the Department of Defense, the National Institute of Standards and Technology, the United States Geological Survey, and the Food and Drug Administration. Representatives from the Commonwealth of Kentucky and the State of California also contributed to the manual. The MARLAP manual addresses radiochemical analyses, including sample preparation and analysis methods, and provides measurement performance criteria for evaluating analytical data.

7.3.2 Field Methods

Several analytical instrument types are portable and may be used for field screening methods. The most commonly known are the Gieger-Muller detectors, which respond to any radiation regardless of type or energy. These detectors are filled with a gas; incoming radiation interacts with the gas to cause an electric current, which is proportional to the amount of radiation the detector is responding to. These types of detectors have often been portrayed in television dramas when looking for radiation in the environment. These detectors respond with an audible and visual readout of radioactivity. The faster the clicking sound, the greater the radioactive field. These types of detectors are frequently used in the field and the laboratory as survey meters due to their low cost, ruggedness, and sensitivity to a variety of types of radiation. However, they offer no information about the identity of the incoming radiation. Background radiation cannot be distinguished from man-made radiation.

A slightly more sophisticated detector is based on the interaction of radiation with a solid material rather than the gas in the Gieger-Muller detector. Scintillation detectors consist of a solid material that will emit light when exposed to radiation. The light passes through the scintillating material to a photomultiplier tube, which then converts the light to an electrical signal. These detectors are most often used to detect gamma and X-ray emissions. The primary advantage of these detectors is that information about the energy of the incoming radiation is retained, thus identification of the source of radiation is possible.

The most widely used inorganic scintillation detector is the thallium "doped" sodium iodide (NaI(Tl)) detector. Several commercial sources of field portable NaI(Tl) scintillators coupled with a multi-channel analyzer are available to perform radiometric analyses in the field environment. These detectors can be used similarly to the Gieger-Muller detector to locate areas of higher radioactivity based on the measured count rate or dose rate. Identification of the major radioisotopes present can then be attempted. To quantify the amount of radioactivity present requires that the detector be carefully calibrated for efficiency of detection response with respect to radiation energy. An example of this use is demonstrated by a study completed at the Savannah River Site in South Carolina using a field portable NaI(Tl) detector to determine the extent of ^{137}Cs contamination in soil and sediment; the ^{137}Cs was released during the 50 year history of operations at the Department of Energy site (Hofstetter and Beals, 2002). The detection limit for ^{137}Cs in the near-surface soil using this detector and analysis protocol was found to be ~5–10 pCi/g soil. The human health risk benchmark limit is approximately 22 pCi/g, thus this method met the required data quality objectives of the study.

The biggest disadvantage of the NaI(Tl) crystals for gamma-ray spectrometry is their energy resolution. The best resolution that can be obtained for a NaI(Tl) crystal is about 6% for the 662 keV gamma ray of ^{137}Cs. For radioisotopes that have similar energies, such as the man-made ^{241}Am (59 keV) and the natural ^{234}Th (63 keV), quantification of one of these isotopes in the presence of the other is very difficult using NaI(Tl) detector technologies. Solid-state semiconductor detectors were developed in the 1960s and 1970s to overcome this limitation. These detectors are often made of germanium (Ge) or silicon (Si). Due to the difficulty in making pure Ge or Si, early semiconductor detectors often were doped with Li and were referred to as GeLi or SiLi detectors. Figure 7.3.1 illustrates the difference in resolution achievable with a NaI(Tl) detector versus a semiconductor (GeLi) detector. Later, manufacturing processes were improved such that high-purity germanium (HPGe) detectors with no Li drifting are commonly available.

The primary disadvantage of these semiconductor detectors is that they must operate at liquid nitrogen temperatures to minimize the thermal excitation of electrons, that is minimize the background due to electronic noise. For this reason, semiconductor detectors have, until recently, been used only in laboratory settings. The advent of thermoelectric coolers has allowed manufacture of semi-portable HPGe systems (Twomey and Keyser, 2004). These instruments are very good at providing isotopic identification in the field; however, they suffer from the same limitations as NaI(Tl) detectors when trying to quantify the amount of a particular radioisotope that is present. Again, to quantify the amount of radioactivity present requires that the detector be carefully calibrated for efficiency of detector response with respect to energy and sample geometry.

Other field portable radiation detectors exist but, like the Gieger-Muller detector, offer no information about the identity of the radiation source. Gas-filled detectors can be designed to be selective for primarily alpha radiation, with reduced beta and gamma ray sensitivity, or primarily for beta radiation with some gamma sensitivity, but minimal alpha detection. These types of instruments are often used by radiation and health protection personnel when assessing the level of fixed or transferable contamination, such as following an accidental release of radiation. They offer relative selectivity as to the type of radiation they are most sensitive to; however, they give no indication of actual quantity or identity of radioisotope present.

Some manufacturers have designed detectors for use in the field to identify an isotope by its alpha or beta emissions.

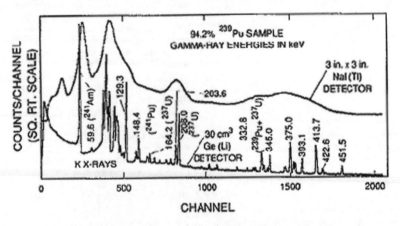

Figure 7.3.1 *NaI(Tl) Spectra* versus *GeLi Spectra. (Smith, H.A., and Lucas, M. (1991). "Gamma-ray Detectors", in Passive Nondestructive Assay of Nuclear Material, NUREG/CR-5550, eds, Reilly, Norbert Ensslin, Smith, H.A., Jr.)*

These often are rugged versions of laboratory instruments that can be placed in a mobile laboratory. Radioisotopes that emit alpha particles have similar "fingerprint" spectra to the gamma emissions. Fieldable alpha-particle detectors such as the Frisch-Grid (Ordela, Inc. Oak Ridge TN) can operate in a mobile laboratory and provide suitable spectra for isotopic identification. The resolution is nearly as good as laboratory-based instrumentation; however, the background response is higher, leading to higher detection limits than achievable using laboratory instruments. In a study completed by Sill (1995), air dusts, water, and soil were monitored directly for alpha emitters during the retrieval and disposal of buried transuranic radioactive wastes. About 100 mg of soil or dust were directly mounted and counted using a Frisch-Grid detector or 200 mL water samples were evaporated and mounted for counting. Sill (1995) was able to achieve a 28% counting efficiency. With a background count rate of 0.25 counts per minute and a 10 minute sample count, he was able to obtain a detection limit of ~40 pCi/g.

For beta-particle emitting radioisotopes, maximum beta energy is often used as a distinguishing characteristic. To determine beta energies, the system most often used is liquid scintillation counting (LSC) spectrometry. Similar to the solid scintillation detectors used for gamma-ray spectrometry, liquid scintillation detection involves the interaction of the radiation with a material that emits light. In this technique, a liquid organic scintillator is put in direct contact with the radioactive sample by dispersion in a solvent gel. The radiation emitted by the sample interacts with the solvent, which transfers the energy to a scintillator, which emits a photon of light. The photon is finally detected by a phototube. Fieldable units often have less shielding (resulting in higher detection limits) and are designed for manual sample changing. Again, because many of the naturally occurring radionuclides emit beta radiation, samples must be chemically separated to provide definitive, quantitative results.

7.3.3 Laboratory Analyses

When performing environmental forensic studies, the goal is often to determine if there is evidence of human activity. Because the crust of the earth contains naturally occurring radioactivity, primarily uranium and thorium series radionuclides and potassium (see Section 7.4.1), forensic analyses must be able to distinguish this from any anthropogenic sources. Low levels of anthropogenic radionuclides can be masked by the naturally occurring radiation when performing gamma-ray spectrometry analyses. For this reason, HPGe detectors in the laboratory are either shielded with 2–4 inches of pre-World War II lead or steel or housed in specially shielded low-background counting rooms. The shielding minimizes the contribution of naturally occurring radiation that reaches the detector, thus improving detection of the radiations originating from the sample only.

A specially designed low-level counting facility was used to monitor the contribution of anthropogenic radionuclides originating from the former Soviet Union and being transported from the coastal zones to the deep Arctic basin (Landa et al., 1998). Sea ice samples were collected from the Arctic ice pack and the debris counted by gamma-ray spectrometry. Cesium-137 (^{137}Cs) was measurable in all the samples at levels of a few to several tens of Bq/kg (0.1–2 pCi/g). Computer modeling of ocean currents suggested that the samples with the highest ^{137}Cs concentrations originated in the Kara Sea. The presence of ^{137}Cs at this level is alone not conclusive of contamination originating from the former Soviet Union because ^{137}Cs has been distributed world-wide by nuclear weapons testing. The ^{137}Cs concentrations found in the sea ice were less than those found in some soils located in otherwise pristine environments. Measurement of

^{134}Cs in addition to ^{137}Cs in one sample, and ^{60}Co and ^{152}Eu in others, indicated a recent source of contamination, probably from nuclear reactor effluents. Because of its 2-year half-life, ^{134}Cs from nuclear weapons testing in the 1960s has decayed away, as have ^{60}Co (half-life of 5 years) and ^{152}Eu (half-life of 12.7 years).

Laboratory analysis of beta-emitting radionuclides is usually accomplished by chemical separation of the radionuclide of interest and LSC spectrometry. The laboratory LSC instruments are based on the same principles as the fieldable units but typically contain additional shielding to reduce the radiation from natural background and are designed to automatically change samples for unattended analysis of large batches of similar samples. An environmental study that demonstrated the performance of a field-portable LSC versus a laboratory instrument can be found in the Proceeding of the American Nuclear Society Topical Meeting on Environmental Transport and Dosimetry. Cadieux et al. (1993) achieved a detection limit of 0.15 Bq/mL (4 pCi/mL) for tritium (^{3}H) in river-water samples following an accidental release to the Savannah River. The reported detector background was 15 to 25 counts per minute, with a detection efficiency of 50%. Samples were counted for up to 20 minutes to obtain this detection limit. Samples collected from the Savannah River were also counted in the laboratory by Beals and Dunn (1993). The laboratory detector background was ~5 counts per minute with a detection efficiency of nearly 70%. Samples were counted for 10 minutes obtaining a detection limit of less than 1 pCi/mL.

The energy range over which gamma and X-rays occur is fairly broad primarily ranging from 0 to 3000 keV. The energy range for alpha particles is generally from 3 to 8 MeV. Many of the radioisotopes in the uranium and thorium decay series decay by alpha-particle emission. A list of the common isotopes measured by alpha-particle spectrometry with their energies of decay is shown in Table 7.3.1.

Because several isotopes of different elements have similar alpha decay energies, definitive alpha-particle analysis of environmental samples requires chemical separation of the element of interest from the matrix and from other alpha-emitting elements. The individual alpha-emitting isotopes of the elements often can be distinguished if chemically separated, for example uranium isotopes (^{234}U distinguished from ^{235}U to ^{238}U, with energies ranging from 4.15 to 4.78 MeV) and thorium isotopes (^{228}Th distinguished from ^{230}Th to ^{232}Th, with energies ranging from 3.95 to 5.42 MeV).

To precisely measure very low levels of alpha-emitting radionuclides, the element of interest must be chemically purified in the laboratory and then mounted for alpha spectrometry analysis. Surface barrier semiconductor detectors are most often used to measure the alpha-particle decay. The purified sample is mounted in a standard geometry (usually by drying or electroplating a solution on a metal disk substrate) and placed in close proximity to the detector. The sample and detector are operated at very low pressures. When carefully controlled, the background of these detectors can be maintained at 0 to 2 counts per day in the energy region of interest for the radionuclide of concern (as opposed to the 0.25 counts per minute of the Frisch-Grid). Samples are often counted for 1000 to 3000 minutes (1 to 2.5 days) resulting in a detection limit of a few fCi per sample.

The energy difference between ^{239}Pu and ^{240}Pu or ^{235}U and ^{236}U (Table 7.3.1) is usually beyond the resolution capabilities of alpha spectrometry systems. However, differentiation between these isotopes is often required to access anthropogenic inputs to an environment (see Section 7.4.4). Thermal ionization mass spectrometry (TIMS) has been used to measure isotopic ratios of plutonium. The plutonium

Table 7.3.1 *Selected Alpha-Particle Decay Energies*

Isotope	Half-life	Decay energy (MeV)	Intensity (%)
Po-208	2.9 Y	5.12	100
Po-209	102 Y	4.88	99
Po-210	138 D	5.30	100
Rn-220	56 sec	6.29	100
Rn-222	3.8 D	5.49	100
Ra-223	11.4 D	5.61	24
		5.71	52
Ra-224	3.66 D	5.69	95
		5.45	5
Ra-226	1599 Y	4.78	94
		4.60	6
Th-228	1.91 Y	5.42	73
		5.34	27
Th-230	7.54 E 4 Y	4.69	76
		4.62	23
Th-232	1.4 E 10 Y	4.01	77
		3.95	23
U-232	68.9 Y	5.32	69
		5.26	31
U-233	1.59 E 5 Y	4.82	84
		4.78	13
U-234	2.45 E 5 Y	4.78	73
		4.72	28
U-235	7.04 E 8 Y	4.40	55
		4.36	11
		4.37	6
U-236	2.34 E 7 Y	4.49	74
		4.45	26
U-238	4.46 E 9 Y	4.20	77
		4.15	23
Pu-239	2.41 E 4 Y	5.16	71
		5.14	17
Pu-240	6537 Y	5.17	74
		5.12	26
Pu-242	3.76 E 5 Y	4.90	78
		4.86	22
Am-241	432 Y	5.49	85
		5.44	13
Am-243	7.37 E 3 Y	5.28	88
		5.23	11

From Lide (2001).
Similar information also available on-line at http://www.nnde. bnl.gov/nnde/nudat/
National Nuclear Data Center "Nuclear Data from NUDAT Decay Radiations"
sec   seconds
D   days
Y   years
E indicates a multiplier of 10^n, for example 7.54E 4 Y indicates 7.54×10^4 years $= 75,400$ years.

is separated from the sample similarly to the methods used for alpha spectrometry and then loaded onto specially designed filaments for measurement. The filaments are placed into the instrument and placed under vacuum. A current is passed through the filament to ionize the plutonium, which is accelerated through a magnetic field. The various isotopes are deflected by the magnetic field at different degrees according to their mass and can then be measured independently. The detection limits of these instruments are often in the order of 10^5 to 10^6 atoms. This sensitivity allows

the simultaneous determination of ^{239}Pu, ^{240}Pu, ^{241}Pu, and ^{242}Pu. Because of these very low detection limits, samples are usually prepared in clean room environments. Relative standard deviations of the major isotopic ratios are typically less than ~0.05%. Sample throughput on these instruments is usually only several samples per day.

Uranium isotopic abundances are also often measured by TIMS (Chen *et al.*, 1986). It is difficult to quantify the presence of anthropogenic ^{236}U in the presence of natural ^{235}U and, likewise, anthropogenic ^{233}U in the presence of natural ^{234}U by alpha spectrometry due to the similarities in their decay energies (Table 7.3.1). However, these isotopes are easily resolvable using TIMS. Thorium-230 (^{230}Th) and ^{232}Th also can be measured by TIMS. Thermal ionization mass spectrometry affords much lower detection limits due to the extremely long half-life of ^{232}Th (10^{10} years), which limits ^{232}Th measurement by radioactive decay counting. An example of a forensic investigation using TIMS to determine the ^{230}Th content in a uranium rod was presented by LaMont and Hall (2005).

Inductively coupled plasma-mass spectrometry (ICP-MS) provides an alternative to TIMS for the determination of isotopic abundances. In routine use, ICP-MS allows analysis of many more samples per day than TIMS, with much lower analytical costs and often simpler sample preparation. The sample must be put into solution for the ICP-MS analysis of environmental samples. Because matrix suppression of the analyte signal can occur when soil or sediment is digested and directly analyzed, chemical separation is often required when measuring radionuclides by ICP-MS. Several methods have been published for the analysis of uranium and thorium isotopes by ICP-MS (ASTM, 2005a,b). Plutonium-239 (^{239}Pu) has been measured in ocean sediments by ICP-MS (Petullo *et al.*, 1994). The routine sensitivity of quadrapole ICP-MS instruments usually is not sufficient to accurately measure ^{240}Pu or the other minor plutonium isotopes. However, a method was developed to measure ^{230}Th, ^{234}U, and ^{240}Pu in soils for an environmental restoration project in Colorado in which these isotopes were measured by ICP-MS using flow-injection preconcentration of the analytes (Hollenbach *et al.*, 1995).

Two other long-lived radionuclides that are often measured by mass spectrometry are ^{99}Tc and ^{129}I. Both isotopes are fission products. They can be found naturally associated with uranium ores (Curtis *et al.*, 1994) but more often are found in the environment as a result of processing during the nuclear fuel cycle. Technicium-99 (^{99}Tc) is a weak beta-emitting isotope which can be measured by LSC with some success. In order to obtain lower ^{99}Tc detection limits, Beals (1996) developed an analysis method using ICP-MS detection. Methods have also been published using TIMS (Anderson and Walker, 1980; Rokop *et al.*, 1990) that, like for plutonium, are able to provide a lower detection limit.

Iodine-129 (^{129}I) is a weak beta emitter and emits very low energy X-rays. Radiometric methods of analysis for ^{129}I include counting on an HPGe gamma-ray spectrometer modified to measure the very low 30–40 keV X-rays. Methods have also been developed to measure ^{129}I by neutron activation (Doshi *et al.*, 1991) and accelerator mass spectrometry (AMS) (Elmore *et al.*, 1980; Kilius *et al.*, 1992). Several studies have used AMS to measure ^{129}I in the environment to trace the movement of groundwater (Mann and Beasley, 1994) and study ocean circulation (Santschi *et al.*, 1996).

If ^{99}Tc or ^{129}I found in the environment were produced only by in situ fission, it should reach a state of secular equilibrium in which the concentration of these radionuclides is a function of the production and decay rates, both of which are well known. Due to the mobility of both technetium and iodine in the environment, it would not be unexpected to find a slight depletion in these radionuclides, as found by Curtis *et al.* (1994) in the uranium deposits at Cigar Lake and Koongarra. However, any significant excess is most likely due to anthropogenic inputs (Beals and Hayes, 1995).

7.3.4 Interferences and Detection Limits

There are many methods for the analysis of radionuclide concentrations or activities in the environment. As method detection limits are driven lower, the cost of the analysis and the required instrumentation increase. A summary of typical detection limits for radioisotopes of interest by various methods is given in Table 7.3.2. All of these detection limits for counting methods (i.e., all except mass spectrometry) vary depending on shielding and the background in which the sample is measured. If the sample is shielded, the number of counts reaching the detector is reduced, raising the detection limit for field-based methods. If the detector is shielded, as is possible in a laboratory setting, the background count rate is reduced, increasing the signal-to-noise ratio and improving detection limits. The detection limits shown in Table 7.3.2 are typical of commercial count rate meters and NaI(Tl) instruments used with no shielding in a field application. The laboratory-based HPGe is assumed to have at least 2 inches of lead shielding the detector. No additional shielding is required for alpha spectrometry analysis.

The detection limit achieved by portable count rate meters is primarily a function of the mode of decay and the associated energy. Tritium (^{3}H) is a very weak beta emitter and is not detectable by hand-held count rate meters,

as is the case for ^{129}I. Lead-210 (^{210}Pb) is also a weak beta-emitting radioisotope; however, its ^{210}Bi daughter (half-life of 5 days) is easily detectable by simple count rate meters. To measure ^{238}Pu and ^{239}Pu with a count rate meter, the meter must be designed for the measurement of alpha particles. Americium-241 (^{241}Am) also emits gamma rays and can therefore be detected with an alpha count rate meter or a beta–gamma count rate meter. Uranium-235 (^{235}U), ^{238}U, and ^{232}Th are alpha-emitting isotopes that also emit weak gamma rays, and whose daughters emit alpha, beta, and gamma radiation, thus these radioisotopes are detectable with alpha or beta–gamma count rate meters.

Increasing the counting time of the sample will provide lower detection limits. Gamma and alpha spectrometry are based on counting the number of decays of the isotope of interest per unit time. As counting time increases, the number of events reaching the detector increases, improving the signal-to-noise ratio and lowering the detection limit. Typical counting times to reach the detection limits shown in Table 7.3.2 are 30 seconds for the count rate meter, 180 seconds for the NaI(Tl) detector, and 1000 minutes for the HPGe and alpha spectrometry systems.

For laboratory-based methods, the table indicates a detection limit for the total sample mass. When reviewing laboratory results for environmental samples, sample activities will often be reported at what appears to be lower than the instrumental detection limit. For example, Landa *et al.* (1998) reported ^{137}Cs at levels of fCi/g by HPGe gamma-ray spectrometry and ^{239}Pu was reported at levels of fCi/g by alpha spectrometry. Although these concentrations appear

Table 7.3.2 *Typical Detection Limits for Isotopes of Interest by Various Methods*

Isotope	Half-life	Primary decay mode	Gas filled count rate meter	Field portable NaI(Tl)	Lab based HPGe	Lab based (1) liquid scintillation counting	Lab based (1) alpha spectrometry	Lab based (1) mass spectrometry
Naturally occurring isotopes								
U-238	4.5 E 09 Y	α	0.1 nCi	μCi	0.1 nCi (2)	NA	pCi	fCi
U-235	7.1 E 08 Y	α, γ	nCi	nCi	0.1 nCi	NA	pCi	fCi
Th-232	1.4 E 10 Y	α	0.1 nCi	μCi	0.1 nCi (2)	NA	nCi	aCi
K-40	1.3 E 09 Y	(β)	nCi	nCi	pCi	NA	NA	NA
Ra-226	1600 Y	α, γ	0.1 nCi	nCi	0.1 nCi	NA	pCi	pCi
Pb-210	22 Y	β, γ	μCi	mCi	nCi	10 pCi	NA	NA
Fission and activation products								
H-3	12.3 Y	β	NA	NA	NA	10 pCi	NA	NA
C-14	5715 Y	β	nCi	NA	NA	pCi	NA	aCi
P-32	14.3 D	β	nCi	NA	NA	pCi	NA	aCi
Co-60	5.27 Y	(β)γ	0.1 nCi	nCi	pCi	NA	NA	NA
Sr-90	30 Y	β	0.1 nCi	NA	NA	pCi	NA	NA
Tc-99	2 E 05 Y	β	0.1 nCi	NA	NA	pCi	NA	fCi
I-129	1.7 E 07 Y	β, γ	NA	mCi	nCi	nCi	NA	aCi
I-131	8 D	β, γ	0.1 nCi	nCi	pCi	pCi	NA	NA
Cs-137	30 Y	(β)	0.1 nCi	nCi	pCi	NA	NA	pCi
Transuranic elements								
Pu-238	87.7 Y	α	0.1 nCi	mCi	μCi	NA	pCi	pCi
Pu-239	2.5 E 04 Y	α	nCi	mCi	μCi	NA	pCi	fCi
Am-241	433 Y	α, γ	0.1 nCi	nCi	pCi	NA	pCi	nCi

(1) with chemical separation performed.
(2) assuming equilibrium with daughters.
NA = not applicable to this isotope.
D   days
Y   years
E indicates a multiplier of 10^n, e.g., 7.54E 4 Y indicates 7.54×10^4 years $= 75,400$ years
HPGe high purity germanium detector
NaI(Tl) thallium-doped sodium iodide detector.

to be well below the detection limits shown in Table 7.3.2, note that the sample mass is in the denominator of the units. Processing a large sample in the laboratory will, therefore, reduce the reported detection limits when expressed as activity concentration.

Very low limits of detection for the long-lived radionuclides are often achievable by mass spectrometry. Extensive chemical separation and purification of the element of interest usually precedes the analysis and then the isotope ratio is measured. In the case of ^{14}C and ^{129}I, the radioactive isotope is measured relative to the stable isotope by AMS. Isotope ratios of uranium, plutonium, and thorium may be measured by TIMS. To quantify the radionuclides requires the addition of a known isotope prior to chemical separation (isotope dilution mass spectrometry). In some cases, ICP-MS may be used with less sample preparation, but will not typically reach the detection limits shown in Table 7.3.2.

Natural background radiation (see Section 7.4.1) is one of the primary interferences and a source of bias in performing environmental forensic analyses. Natural uranium, thorium and their radioactive daughters, and ^{40}K emit alpha, beta, and gamma radiation. Many of the field portable count rate meters offer no information as to the isotope, but can provide some indication as to whether the radioactivity is due to alpha, beta, or gamma radiations. Other naturally produced radioisotopes, for example ^{14}C, ^{7}Be, or ^{3}H, are produced at such low levels, or have low energy emissions, that they typically do not interfere in the analysis of anthropogenic radionuclides.

When identifying the source of radiation, caution must be used when using low resolution gamma-ray spectrometers, such as the common NaI(Tl) spectrometers. Many elements have gamma rays of similar energy that can not be distinguished without high resolution analysis. A common example is the misidentification of the 351 keV gamma ray of ^{226}Ra as ^{131}I (365 keV) or of the 609 keV gamma ray of ^{226}Ra as ^{137}Cs (661 keV). Also, many commercial, hand-held NaI(Tl) spectrometers will identify only ^{226}Ra in uranium materials due to the much greater intensity of the ^{226}Ra daughter emissions compared to the ^{238}U emissions. Even when using high resolution HPGe spectrometers, caution must be used in evaluating the activities of uranium and thorium isotopes. For both ^{238}U and ^{232}Th, gamma rays emitted from the parent are often insufficient to accurately quantify the isotope activities. Thus, the immediate ^{238}U daughters (^{234}Th and ^{234m}Pa) and ^{232}Th daughters (^{228}Ra and ^{228}Ac) are used to calculate the activity of the parent isotope. However, this analysis assumes that the daughters are in secular equilibrium with the parent; many natural processes may disturb this equilibrium. Uranium and thorium isotope ratios and activities are best determined by chemical separation and alpha or mass spectrometric analysis. Common interferences that can bias results and alternate analytical techniques are given in Table 7.3.3.

This discussion does not include all potential interferences. Because of the complexity of gamma rays emitted by the natural uranium and thorium series, experienced analysts should carefully review any spectrum prior to reporting the presence and activity of anthropogenic radionuclides in an environmental sample. A chemical separation almost always precedes alpha spectrometry to minimize interference between isotopes of different elements; however, anthropogenic isotopes of the same element may be masked by natural isotopes, as can occur in uranium analyses. Most mass spectrometers only have unit resolution, that is they cannot distinguish anthropogenic ^{99}Tc (atomic mass = 98.906364 amu; Lide, 2001) from natural ^{99}Ru (12.7% natural abundance, atomic mass = 98.905939 amu). Therefore, any radioisotope with sufficient half-life and similar mass can bias a mass spectrometric analysis unless removed by chemical separation.

7.4 CHEMISTRY AND SOURCES

Radioactive isotopes of the elements form the same chemical compounds as do stable isotopes of the element. Therefore their transport through the environment is not affected by their radioactive properties. For example, if tritium gas (^{3}H$_2$) is released to the atmosphere, it readily oxidizes and exchanges with a stable isotope of hydrogen (^{1}H) to form tritiated water, ^{1}H^{3}HO (often written as HTO). This HTO molecule has measurable differences in vapor pressure and other physical properties related to the higher atomic mass of the tritium atom but, in practice, transport of HTO through the environment follows that of H$_2$O (NCRP, 1979). Similarly, ^{14}C is oxidized to ^{14}CO$_2$ in the atmosphere and incorporated into the carbon cycle without respect to its radioactive properties (NCRP, 1987). Potassium-40 (^{40}K) is incorporated into the body and its concentration is maintained along with the stable potassium isotopes by regulatory mechanisms that maintain a healthy sodium to potassium ratio.

Much of the effort of radionuclide forensics is directed toward distinguishing naturally occurring radionuclides from sources that could indicate nuclear power, medical, industrial, or military applications.

Table 7.3.3 *Potential Analytical Interferences and Alternative Techniques*

Analyte	Measurement Technique	Interferent	Alternative Technique
^{233}U	Alpha spectrometry	^{234}U	TIMS
^{236}U	Alpha spectrometry	^{235}U	TIMS
^{239}Pu	Alpha spectrometry	^{240}Pu	TIMS
^{241}Pu	Mass spectrometry	^{241}Am	LS Counting
^{241}Am	HPGe	^{234}Th	Alpha spectrometry
^{239}Pu	ICP-MS	UH^{a}	Alpha spectrometry or TIMS
^{99}Tc	Mass spectrometry	^{99}Ru	LS Counting

^{a} ^{238}U plus H in the plasma can form an ion at mass 239.
HGPe   High purity germanium detector
ICP-MS   Inductively coupled plasma-mass spectrometry
LS   Liquid scintillation
TIMS   Thermal ionization mass spectrometry

7.4.1 Naturally Occurring Sources

Naturally occurring radionuclides include sources with very long half-lives, often approximating the age of the earth, and are considered primordial. Many of these sources are associated with the uranium and thorium decay series. In undisturbed ore bodies, these long-lived isotopes are in secular equilibrium with several daughters that usually have shorter half-lives ranging from milliseconds to tens of thousands of years or more (Figures 7.4.1–7.4.3) (Stannard, 1988). Some primordial radionuclides with half-lives greater than 10 billion years are known but are not associated with the uranium and thorium series. These radionuclides have negligible abundance in the earth's crust or are only weakly radioactive with the exceptions of ^{40}K and ^{87}Rb isotopes, which provide significant contributions to natural background radiation (NCRP, 1987).

Nuclide	Historical name	Half-life	Major radiation energies (MeV) and intensities†					
			α		β		γ	
$^{238}_{92}$U	Uranium I	4.51 x 10⁹y	4.15	(25%)	—		—	
			4.20	(75%)				
↓								
$^{234}_{90}$Th	Uranium X₁	24.1d	—		0.103	(21%)	0.063c‡	(3.5%)
					0.193	(79%)	0.093c	(4%)
↓								
$^{234}_{91}$Pa^m	Uranium X₂	1.17m	—		2.29	(98%)	0.765	(0.30%)
99.87% ↓ 0.13%							1.001	(0.60%)
$^{234}_{91}$Pa	Uranium Z	6.75h	—		0.53	(66%)	0.100	(50%)
					1.13	(13%)	0.70	(24%)
							0.90	(70%)
↓								
$^{234}_{92}$U	Uranium II	2.47x10⁵y	4.72	(28%)	—		0.053	(0.2%)
			4.77	(72%)				
↓								
$^{230}_{90}$Th	Ionium	8.0 x 10⁴y	4.62	(24%)	—		0.068	(0.6%)
			4.68	(76%)			0.142	(0.07%)
↓								
$^{226}_{88}$Ra	Radium	1602y	4.60	(6%)	—		0.186	(4%)
			4.78	(95%)				
↓								
$^{222}_{86}$Rn	Emanation Radon (Rn)	3.823d	5.49	(100%)	—		0.510	(0.07%)
↓								
$^{218}_{84}$Po	Radium A	3.05m	6.00	(~100%)	0.33	(~0.019%)	—	
99.98% ↓ 0.02%								
$^{214}_{82}$Pb	Radium B	26.8m	—		0.65	(50%)	0.295	(19%)
					0.71	(40%)	0.352	(36%)
					0.98	(6%)		
$^{218}_{85}$At	Astatine	~2s	6.65	(6%)	?	(~0.1%)	—	
			6.70	(94%)				
↓								
$^{214}_{83}$Bi	Radium C	19.7m	5.45	(0.012%)	1.0	(23%)	0.609	(47%)
99.98% ↓ 0.02%			5.51	(0.008%)	1.51	(40%)	1.120	(17%)
					3.26	(19%)	1.764	(17%)
$^{214}_{84}$Po	Radium C'	164μs	7.69	(100%)	—		0.799	(0.014%)
$^{210}_{81}$Tl	Radium C"	1.3m	—		1.3	(25%)	0.296	(80%)
					1.9	(56%)	0.795	(100%)
					2.3	(19%)	1.31	(21%)
↓								
$^{210}_{82}$Pb	Radium D	21y	3.72	(.000002%)	0.016	(85%)	0.047	(4%)
					0.061	(15%)		
↓								
$^{210}_{83}$Bi	Radium E	5.01d	4.65	(.00007%)	1.161	(~100%)	—	
			4.69	(.00005%)				
~100% ↓ .00013%								
$^{210}_{84}$Po	Radium F	138.4d	5.305	(100%)	—		0.803	(0.0011%)
$^{206}_{81}$Tl	Radium E"	4.19m	—		1.571	(100%)	—	
↓								
$^{206}_{82}$Pb	Radium G	Stable	—		—		—	

*This expression describes the mass number of any member in this series, where n is an integer.
   Example: $^{206}_{82}$Pb (4n + 2)......4(51) + 2 = 206
†Intensities refer to percentage of disintegrations of the nuclide itself, not to original parent of series.
‡Complex energy peak which would be incompletely resolved by instruments of moderately low resolving power such as scintillators.
Data taken from: Table of Isotopes and USNRDL-TR-802.

Source: Reset from Radiological Health Handbook, PHS 1970.

Figure 7.4.1 Uranium Decay Series (Stannard, 1988). Radioactivity and Health: A History, DOE/RL/01830-T59 (DE88013791), R. W. Baalman, Jr (ed.), Pacific Northwest Laboratory, Richland, Washington. The Table of Isotopes is cited in references as Lederer (1967) and USDRL-TR-802 is cited as Hogan et al. (1964).

Nuclide	Historical name	Half-life	Major radiation energies (MeV) and intensities†					
			α		β		γ	
$^{232}_{90}$Th	Thorium	1.41×10^{10} y	3.95 4.01	(24%) (76%)	—		—	
↓ $^{228}_{88}$Ra	Mesothorium I	6.7y	—		0.055	(100%)	—	
↓ $^{228}_{89}$Ac	Mesothorium II	6.13h	—		1.18 1.75 2.09	(35%) (12%) (12%)	0.34c‡ 0.908 0.96c	(15%) (25%) (20%)
↓ $^{228}_{90}$Th	Radiothorium	1.910y	5.34 5.43	(28%) (71%)	— —		0.084 0.214	(1.6%) (0.3%)
↓ $^{224}_{88}$Ra	Thorium X	3.64d	5.45 5.68	(6%) (94%)	—		0.241	(3.7%)
↓ $^{220}_{86}$Rn	Emanation Thoron (Tn)	55s	6.29	(100%)	—		0.55	(0.07%)
↓ $^{216}_{84}$Po	Thorium A	0.15s	6.78	(100%)	—		—	
↓ $^{212}_{82}$Pb	Thorium B	10.64h	—		0.346 0.586	(81%) (14%)	0.239 0.300	(47%) (3.2%)
↓ $^{212}_{83}$Bi	Thorium C	60.6m	6.05 6.09	(25%) (10%)	1.55 2.26	(5%) (55%)	0.040 0.727 1.620	(2%) (7%) (1.8%)
$^{212}_{84}$Po	Thorium C'	304ns	8.78	(100%)	—		—	
$^{208}_{81}$Tl	Thorium C''	3.10m	—		1.28 1.52 1.80	(25%) (21%) (50%)	0.511 0.583 0.860 2.614	(23%) (86%) (12%) (100%)
$^{208}_{82}$Pb	Thorium D	Stable	—		—		—	

64.0% | 36.0%

*This expression describes the mass number of any member in this series, where n is an integer.
Example: $^{232}_{90}$Th (4n)......4(58) = 232
†Intensities refer to percentage of disintegrations of the nuclide itself, not to original parent of series.
‡Complex energy peak which would be incompletely resolved by instruments of moderately low resolving power such as scintillators.
Data taken from: Lederer, C.M., Hollander, J.M., and Perlman, I., Table of Isotopes (6th ed.; New York: John Wiley & Sons, Inc., 1967) and Hogan, O. H., Zigman, P. E., and Mackin, J. L., Beta Spectra (USNRDL-TR-802 [Washington, D.C.: U.S. Atomic Energy Commission, 1964]).

Source: Reset from Radiological Health Handbook, PHS 1970.

Figure 7.4.2 Thorium Decay Series (Stannard, 1988). Radioactivity and Health: A History, DOE/RL/01830-T59 (DE88013791), R. W. Baalman, Jr (ed.), Pacific Northwest Laboratory, Richland, Washington. The Table of Isotopes is cited in references as Lederer (1967) and USDRL-TR-802 is cited as Hogan et al. (1964).

Nuclide	Historical name	Half-life	Major radiation energies (MeV) and intensities†		
			α	β	γ
$^{235}_{92}U$	Actinouranium	7.1 x 10⁸ y	4.37 (18%) 4.40 (57%) 4.58c‡ (8%)	—	0.143 (11%) 0.185 (54%) 0.204 (5%)
$^{231}_{90}Th$	Uranium Y	25.5h	—	0.140 (45%) 0.220 (15%) 0.305 (40%)	0.026 (2%) 0.084c (10%)
$^{231}_{91}Pa$	Protoactinium	3.25x10⁴y	4.95 (22%) 5.01 (24%) 5.02 (23%)	—	0.027 (6%) 0.29c (6%)
$^{227}_{89}Ac$	Actinium	21.6y	4.86c (0.18%) 4.95c (1.2%)	0.043 (~99%)	0.070 (0.08%)
$^{227}_{90}Th$	Radioactinium	18.2d	5.76 (21%) 5.98 (24%) 6.04 (23%)	—	0.050 (8%) 0.237c (15%) 0.31c (8%)
$^{223}_{87}Fr$	Actinium K	22m	5.44 (~0.005%)	1.15 (~100%)	0.050 (40%) 0.080 (13%) 0.234 (4%)
$^{223}_{88}Ra$	Actinium X	11.43d	5.61 (26%) 5.71 (54%) 5.75 (9%)	—	0.149c (10%) 0.270 (10%) 0.33c (6%)
$^{219}_{86}Rn$	Emanation Actinon (An)	4.0s	6.42 (8%) 6.55 (11%) 6.82 (81%)	—	0.272 (9%) 0.401 (5%)
$^{215}_{84}Po$	Actinium A	1.78ms	7.38 (~100%)	0.74 (~0.00023%)	—
$^{211}_{82}Pb$	Actinium B	36.1m	—	0.29 (1.4%) 0.56 (9.4%) 1.39 (87.5%)	0.405 (3.4%) 0.427 (1.8%) 0.832 (3.4%)
$^{215}_{85}At$	Astatine	~0.1ms	8.01 (~100%)	—	—
$^{211}_{83}Bi$	Actinium C	2.15m	6.28 (16%) 6.62 (84%)	0.60 (0.28%)	0.351 (14%)
$^{211}_{84}Po$	Actinium C′	0.52s	7.45 (99%)	—	0.570 (0.5%) 0.90 (0.5%)
$^{207}_{81}Tl$	Actinium C″	4.79m	—	1.44 (99.8%)	0.897 (0.16%)
$^{207}_{82}Pb$	Actinium D	Stable	—	—	—

Branching fractions in decay chain: 98.6% / 1.4% (Ac-227); ~100% / .00023% (Po-215); 0.28% / 99.7% (Bi-211).

*This expression describes the mass number of any member in this series, where n is an integer.

Example: $^{207}_{82}Pb$ (4n + 3)......4(51) + 3 = 207

†Intensities refer to percentage of disintegrations of the nuclide itself, not to original parent of series.

‡Complex energy peak which would be incompletely resolved by instruments of moderately low resolving power such as scintillators.

Data taken from: *Table of Isotopes* and USNRDL-TR-802.

Source: Reset from *Radiological Health Handbook*, PHS 1970.

Figure 7.4.3 *Actinium Decay Series (Stannard, 1988). Radioactivity and Health: A History, DOE/RL/01830-T59 (DE88013791), R. W. Baalman, Jr (ed.), Pacific Northwest Laboratory, Richland, Washington. The Table of Isotopes is cited in references as Lederer (1967) and USDRL-TR-802 is cited as Hogan et al. (1964).*

Uranium and thorium associate with igneous rocks and weathered sedimentary deposits. Redistribution to sedimentary deposits occurs through weathering by physical processes and leaching of soluble forms such that the distribution is quite heterogeneous (NCRP, 1987). The heterogeneous nature of uranium and other naturally occurring metals in the United States has been described in detail in Hoffman and Buttleman (1994).

In the United States, uranium is associated with phosphate deposits located in parts of Idaho, Montana, Wyoming, Utah, and Nevada, a second area in central Florida, and a third area in central Tennessee and northern Alabama (NCRP, 1987). Uranium has an affinity for crude oil and its daughters are associated with brines in these deposits. Soils in the Colorado Front Range contain thorium deposits (NCRP, 1987) and uranium mining has occurred in Wyoming and in the Colorado Plateau, which includes portions of Colorado and New Mexico.

Cosmogenic radionuclides are produced in the upper atmosphere through reactions of cosmic rays or neutrons with oxygen, nitrogen, or argon atoms. The four cosmogenic sources that present appreciable radiation doses to humans are ^{3}H (tritium), ^{7}Be, ^{14}C, and ^{22}Na (NCRP, 1987). Because the concentrations of cosmogenic sources are dependent on cosmic ray flux and on atmospheric mixing and transport, their concentrations are variable with time and location on the earth's surface.

Naturally occurring radionuclide concentrations and their contributions to local background at different locations around the world have been measured (Eisenbud, 1987; UNSCEAR, 1977, 1982, 1988, 1993, 2000; NCRP, 1975, 1984, 1987).

7.4.1.1 Natural Fission Reactors

All isotopes of uranium are radioactive. At present, the natural distribution of the major isotopes by weight is 99.27% ^{238}U, 0.72% ^{235}U, and 0.0057% ^{234}U (Shleien, 1992). Of these isotopes, only ^{235}U is fissionable. A ^{235}U fission reaction occurs when the nucleus splits into two fission fragments (which are variable for each fission) and several neutrons, accompanied by the emission of gamma ray and X-ray photons. If the fission neutrons have the correct energy, they can be captured by another ^{235}U nucleus, causing a fission chain reaction. (In a forensic study, it is possible to "fingerprint" the fission products and infer the characteristics of the source using fission yield curves and other information.)

The conditions that could cause a natural uranium ore body to sustain a fission reaction in situ were described in 1956 (Kuroda, 1956a,b). In 1972, uranium ore samples from the Oklo mines in Gabon Africa were found to have anomalously low fractions of ^{235}U, which was reduced from the normal value of 0.007202 (0.7202%). Investigations of this diminished ^{235}U content led to the conclusion that natural fission reactors occurred in the Oklo vicinity (IAEA, 1975, 1978). The reactor operated for a period of approximately 1 million years (Cowan, 1976, 1978) and produced fission and activation products characteristic of a ^{235}U fission reaction (Hidaka et al., 1993, 1994).

Our experience with uranium fission reactors indicates that the uranium fission process requires fuel having a ^{235}U/^{238}U mass ratio of 3% or greater. Uranium found in geologic formations today, having a value of only ~0.7%, must be enriched to sustain a chain reaction. For this reason, a natural reactor is not possible in uranium ore deposits today. The date of the Oklo reactor was estimated from calculations using the current ^{235}U/^{238}U ratio of ~0.702%, the half-life of ^{238}U (4.51 × 10^9 years) and the shorter half-life of ^{235}U (7.1 × 10^8 years). These calculations show that a ^{235}U/^{238}U ratio of 3% occurred approximately at 2000 million years ago (Cowan, 1976). The Oklo phenomenon site is being studied to identify characteristics of an acceptable geological repository for high-level nuclear fission waste (Hagerman, 1978; Loss, 1984, 1989).

7.4.2 Uranium Mining and Milling

Mining and milling of uranium refine high concentrations of natural uranium to produce yet more concentrated sources. Because the milling process involves chemical and not nuclear reactions, this refinement of ore and other chemical processes does not affect the composition of natural uranium isotopes until the uranium enters the enrichment process.

Natural uranium isotopes include ^{238}U, ^{235}U, and ^{234}U isotopes in relatively constant ratios as given in Section 7.4.1. Because ^{238}U is the parent isotope of ^{234}U (Figure 7.4.1), the isotopes are in secular equilibrium unless separated chemically. Natural leaching of uranium from ore to groundwater has been reported to cause elevated ^{234}U concentrations in groundwater that reflect preferential leaching of ^{234}U (Cothern and Lappenbusch, 1983). The ^{234}U/^{238}U ratio in seawater has been measured at 1.15 due to this phenomenon (Ku et al., 1977). This increased solubility results from recoil of ^{234}U nuclei produced in the high energy alpha decay of ^{238}U that produces tracks of damage in the surrounding matrix, which are preferentially etched by solvent water (Fleischer, 1980).

Milling of uranium ore by chemical extraction produces "yellowcake," which is highly variable in composition. Milling processes represent the first of a series of chemical conversions of uranium ore to produce uranium compounds for various uses. Crushed ores are generally leached with H_2SO_4 to form hexavalent $[UO_2(SO_4)_3]^{4-}$ or with an aqueous Na_2CO_3/$NaHCO_3$ mixture to form $[UO_2(CO_3)_3]^{4-}$. Occasionally, an oxidizing agent is added to convert uranium from the U(IV) state to the U(VI) state (McKay, 1981). Leachates are extracted from the crushed ore slurry by liquid–liquid extraction or by ion exchange. Extracted uranium is isolated using acidic aqueous $(NH_4)_2SO_4$ and precipitated as ammonium diuranate using ammonia gas and then dried by heating. Alternatively, sodium diuranate can be precipitated using NaOH, which is usually redissolved and converted to ammonium diuranate.

Feedstocks for further uranium processing are dissolved in nitric acid to form a highly pure product that is converted to $UO_2(NO_3)_2 \cdot 6H_2O$. The $UO_2(NO_3)_2 \cdot 6H_2O$ is converted to UO_3 by heating, reduced to UO_2 with hydrogen and finally converted to UF_4 with anhydrous HF. UF_4, known as the "green salt," can be reduced to uranium metal by heating at 600 °C with calcium or magnesium metal. The UF_4 salt may be oxidized with F_2 to form UF_6, a volatile compound used in enrichment processes. Enriched UF_6 is often hydrolyzed with steam and reduced by H_2 to form enriched UO_2.

The ratios of natural uranium isotopes are maintained in the above chemical purification reactions; however, the daughter radionuclides in the ore decay series (Figures 7.4.1–7.4.3) are separated and diverted to tailings ponds. The radionuclides of concern in these waste streams are principally ^{226}Ra (from the ^{238}U series) and ^{228}Ra (from the ^{232}Th series which is often elevated in uranium ore bodies) and their daughters. The variability of mill efficiency and of the ores being processed precludes a prediction of the isotopic composition of waste. Therefore, each waste stream must be characterized individually if the composition is to be known accurately.

Although uranium may be chemically separated from its daughter products, it is important to note for the practice of forensics that secular equilibrium between ^{238}U and its ^{234}Th and ^{234m}Pa daughters is re-established relatively quickly.

Because the half-lives of ^{234}Th and ^{234m}Pa are 24.1 days and 1.17 minutes, respectively (Figure 7.4.1), the radioactivity of these daughters becomes equal to the ^{238}U activity within 6 to 8 months after chemical separation.

Two studies illustrate the characterization of environmental releases associated with uranium milling.

7.4.3 Uranium Mill Tailings Release

In July 1979, the retention dam of a tailings pond failed at the uranium mill operated by United Nuclear Corporation near Church Rock, New Mexico. The spill released approximately 1100 tons of solid material and 94 million gallons of liquid wastes (NMHED, 1983). An emergency catchments dam retained most of the solid material but liquid waste flowed down the Pipeline Arroyo to the Puerto River, then through Gallup, New Mexico, and into Arizona, where the release was contained by evaporation and infiltration into the streambed. The extent and potential health effects related to the spill were monitored and assessed by many governmental agencies including the New Mexico Environmental Improvement Division, New Mexico Scientific Laboratory Division, the Arizona Department of Health Services, the US Centers for Disease Control and Prevention, the US Nuclear Regulatory Commission, the US Department of Energy (Los Alamos National Laboratory and the Idaho Radiological and Environmental Sciences Laboratory), US Public Health Service, Navajo Area Indian Health Service, and the US Environmental Protection Agency. The Church Rock spill drew widespread attention in the news media and independent reviews of the monitoring programs were issued (Shuey and Taylor, 1982; New Mexico Conference of Churches, 1983).

Two hundred surface water samples were analyzed for uranium metal, ^{230}Th, ^{226}Ra, ^{210}Pb, ^{210}Po, and gross alpha radioactivity. One hundred and fifty groundwater samples were taken from 8 existing wells and from 18 additional groundwater wells installed in clusters of shallow and deep wells located upstream from the spill and along the Puerto River. Groundwater samples were analyzed for uranium, ^{226}Ra, and gross alpha radioactivity (NMHED, 1983).

Surface soil and sediment samples were taken every 1000 feet from the mill site to the Arizona border; 3-foot core samples were taken every 5000 feet. Additional samples were taken at areas of suspected and visible contamination to the point where flow ceased in Arizona. Surface soil and sediment samples (1440) were analyzed for uranium, ^{230}Th, ^{226}Ra, ^{210}Pb, and ^{210}Po. An additional 1550 soil samples taken from the main channel, tributary arroyos and hot spots, and 100 samples from sediment cores were analyzed for ^{238}U, ^{230}Th, ^{226}Ra, and ^{210}Pb (NMHED, 1983).

Eight air filter samples were analyzed for particles containing elemental uranium, ^{230}Th, ^{226}Ra, ^{210}Pb, and ^{210}Po, as were 22 samples of native vegetation and corn. Bone, liver, kidney, spleen, and muscle tissue from livestock collected from the Puerto River area were also analyzed. Radon gas measurements were made at existing monitoring stations operated by the New Mexico Environmental Improvement Division in the Grants Mineral Belt. An aerial survey of gamma radiation exposure rates was conducted and 252 ground gamma survey measurements were made (NMHED, 1983).

Evaluation of analytical data indicated that uranium, arsenic, ^{226}Ra, and other metals dissolved in acid waste had adsorbed to the soil as the waste was neutralized by the alkaline soil downstream. The concentrations of natural uranium and arsenic in surface water returned to pre-spill concentrations within 45 to 50 miles below the mill. Some of these deposits remained soluble and were redissolved in later thunderstorm runoff. After one month, in which two thunderstorms occurred on August 7 and 12, the river chemistry returned to approximately pre-spill conditions, although traces of the spill could be noted in 1980. By 1983, Puerto River surface water had returned to pre-spill conditions (NMHED, 1983); however, it was noted that mine dewatering effluent continued to affect surface water quality.

Contamination of Puerto River sediments was evaluated by comparison to background concentrations in samples from unaffected areas. The contamination was associated with ^{230}Th and ^{210}Pb, but not ^{238}U nor ^{226}Ra. The data showed that ^{230}Th traveled farther downstream, but ^{210}Pb was contained within a few miles below the spill. The following year, the ^{230}Th and ^{210}Pb concentrations had decreased by 70% and 50%, respectively. The decrease was attributed to mixing with uncontaminated sediments (NMHED, 1983).

The affect on groundwater was restricted to shallow groundwater in areas near the spill site. No public or private supply wells, including those drawing from deep sandstone aquifers, were affected by the spill. The increase in shallow groundwater radioactivity above background levels was attributed to uranium. The effects were variable and there was a contribution from mine dewatering effluent.

Results of airborne dust sampling during 12 weeks following the spill showed no concentrations of uranium, ^{230}Th, ^{226}Ra, or ^{210}Pb above background levels. Measurements of dust generated during cleanup activities indicated only ^{230}Th concentrations above background but not uranium, ^{226}Ra, or ^{210}Pb. Results of vegetation and produce analyses indicated slightly elevated levels of ^{230}Th and ^{226}Ra; however, they were not statistically above background concentrations. Elevated concentrations of radionuclides in tissues of livestock exposed to Puerto River water were higher than in control animals. However, the ^{226}Ra and ^{210}Pb concentrations were related to the age of the animal and the ^{210}Pb levels in bone reflected long-term accumulation in bone that probably began before the spill (NMHED, 1983).

Conclusions and recommendations based on these measurements included (NMHED, 1983 and references cited therein):

- Surface waters of the Puerto River were degraded during the spill and for a brief time afterward.
- Shallow groundwater was degraded but spill-related radionuclides did not affect deep aquifers used for domestic water.
- Mine dewatering effluent represented a more significant hazard to both surface and shallow groundwater. Therefore, use of Puerto River water or shallow groundwater for human or agricultural use was discouraged, continued monitoring of existing wells to meet requirements of the Safe Drinking Water Act was recommended, and specific engineering restrictions on installation of future wells in the Puerto River valley were recommended.
- An assessment of radiation doses to humans was made using the Uranium Dispersion and Dosimetry (UDAD) computer code (Momeni *et al.*, 1979) based on the radionuclide concentrations measured in environmental media. A conservative worst-case exposure scenario was assumed in which the person spent 100% of their time within 20 meters of the Puerto River for 1 year, that all radionuclide concentrations remained at their maximum values, that stream sediments were dry for the entire year, and that local meat and vegetables were consumed. The calculated radiation dose was far below regulatory limits.

7.4.3.1 Environmental Monitoring of Uranium Mill Tailings

Uranium mill tailings in the United States are subject to the Uranium Mill Tailings Radiation Control Act (UMTRCA) of

1978 (Public Law 95-604). Standards for cleanup under the UMTRCA are codified in Title 40 of the Code of Federal Regulations Part 192 (40CFR192). These standards specify that:

- direct gamma emissions must be below background levels.
- tailings must be covered to limit ^{222}Rn releases to less than 20 pCi/m^2-second, unless the ^{226}Ra, ^{228}Ra, ^{230}Th, and ^{232}Th concentrations averaged within 15 cm of the surface are less than 5 pCi/g above background.
- the above concentrations are less than 15 pCi/g above background when averaged over 15 cm thick layers in deeper layers.

It is implicit in these regulations that the specified concentrations of each radionuclide are protective of daughters. These standards apply only to UMTRCA sites and not to sites subject to other regulations such as the Comprehensive Environmental Response, Compensation, and Liability Act (CERCLA) (Luftig and Weinstock, 1997). Analytical methods that may be applied to measure gamma radiation, radon release rate, and ^{226}Ra, ^{228}Ra, ^{230}Th, and ^{232}Th concentrations in soil in the 5 pCi/g to 15 pCi/g range are described in Section 7.3.

The US Nuclear Regulatory Commission (USNRC) provides criteria for disposition of uranium mill tailings in Technical Criterion 6 of 10CFR40 Appendix A. These criteria include design specifications for coverings of tailings to limit ^{222}Rn releases, gamma emissions, and ^{226}Ra concentration to the above limits. The International Atomic Energy Agency (IAEA) provides methods and standards for managing uranium that are similar to the above standards (IAEA, 2002, 2003a). Cleanup of uranium mill tailings sites in the United States is nearly completed; the US Department of Energy Office of Legacy Sites, established in 2003, manages remaining cleanup and monitoring activities.

7.4.4   Military Sources

Military sources of radionuclides include production and deployment of nuclear weapons, use of depleted uranium in armor-piercing munitions, and space applications. Releases of radionuclides incidental to military uses have been studied extensively. These studies include many approaches useful to forensic science.

7.4.4.1   Testing, Production, Fabrication, and Deployment of Nuclear Weapons

7.4.4.1.1   Radionuclides Associated with Nuclear Weapons Testing

Radionuclides produced from nuclear weapons tests include fission and activation products produced in the nuclear reaction and a certain amount of unreacted fissile material, for example uranium or plutonium isotopes (NCRP, 1987). The relative abundance of these products depends on the design specifications of the device and whether it is detonated in the air, at the earth's surface, or underground. The products are primarily short-lived beta–gamma emitting radionuclides mixed with a few isotopes having half-lives of decades or more.

Radionuclide products from the early fission weapon testing were deposited in the general vicinity of the explosion. The products from atmospheric thermonuclear explosions were dispersed much more widely and are now often considered part of current background inventory (NCRP, 1987). Studies of these fission- and activation-produced radionuclides in the environment have been directed toward the long-lived radionuclides, including ^{90}Sr, ^{137}Cs, ^{3}H (tritium), ^{14}C, ^{55}Fe, ^{85}Kr, and $^{239/240}$Pu. Atmospheric thermonuclear weapons' testing prior to 1980 is estimated to have increased

the natural inventory of tritium by a factor of 200 and the inventory of ^{14}C by a factor of 2 (NCRP, 1987). Detailed studies of external radiation doses associated with fallout products have been conducted for cities in the US and Canada and for areas around the Nevada Test Site (NCRP, 1975, 1987).

The forensic literature on studies to identify radionuclides released from nuclear weapons production facilities is extensive and beyond the scope of this review. The environmental evaluations of nuclear weapons production facilities, aircraft accidents involving weapons in the late 1960s, and nuclear weapons tests are discussed in the Stannard (1988) history. The following case studies published since 1996 illustrate recent hypotheses and techniques currently used in environmental forensics.

7.4.4.1.2   Radionuclides in Soil and Surface Water (Rocky Flats Plant Site)

The Rocky Flats Plant was a plutonium fabrication facility located near Golden Colorado and built by the Atomic Energy Commission in the 1950s. Plutonium was released from leaking drums of waste stored at the 903 Pad and airborne plutonium was released by fires in 1957 and 1969. Many studies of plutonium concentrations in soil were conducted to characterize and delineate the plutonium plumes and elucidate mechanisms of transport in the local environment (Stannard, 1988). The US Department of Energy (DOE), the US Environmental Protection Agency (EPA), the Colorado Department of Public Health and Environment (CDPHE) and the State of Colorado entered into a joint agreement to implement cleanup operations at the Rocky Flats Plant site (RFCA, 1996).

The parties to the agreement formed a working group to develop action levels to protect human health and to guide the cleanup of radionuclides (ALF, 1996). The working group considered many previous studies of Rocky Flats operations and identified isotopes of plutonium (^{238}Pu, ^{239}Pu, ^{240}Pu, ^{241}Pu, and ^{242}Pu), americium (^{241}Am), and uranium (^{234}U, ^{235}U, and ^{238}U) for consideration. The working group evaluated regulatory requirements of the participating agencies including DOE orders (DOE Order 5400.5 and provisions incorporated in the proposed 10CFR834 regulations) and other applicable portions of the Code of Federal Regulations (Title 10 Parts 20, 30, 40, 50, 51, 70, 72, and 834, and 40CFR196). The decision to develop action levels that would limit the potential radiation dose to future users of the site was based on this regulatory evaluation.

The working group based action levels on 40CFR196 regulations, then still in draft form, and provided the rationale for their decision (ALF, 1996). To develop the action levels, the working group developed a conceptual model of potential uses of the land and scenarios for exposure of humans to radionuclides in surface and subsurface soil. The scenarios considered the types of work done at the site and included potential residential uses and plausible human exposure pathways. Based on applicable regulations and the conceptual model, it was decided to develop action levels for soil that would limit the radiation dose to workers at the site to 15 mrem per year above the natural background. The implementation of these and other radionuclide action levels above local background activity concentrations is specified in all of the regulations evaluated.

Action levels for individual radionuclides were calculated using the RESRAD computer model (ANL, 1993) to represent the activity concentration of each radionuclide in soil that would limit human exposure to the agreed radiation dose limits for 1000 years into the future. By using the RESRAD model, the action levels evaluated up to nine exposure pathways as appropriate for each land-use scenario including external radiation exposure, intake of radionu-clides in water, soil or airborne dust, and intake through the consumption of plant, meat, or milk products grown from the soil and irrigated by local groundwater. The model considers environmental transport from soil to air (both particulate and gas phases) and from soil to water. The model considers radioactive decay and includes radiation dose from ingrowth of daughters. The RESRAD program and others have been evaluated for similar uses (Mills *et al.*, 1997; Laniak *et al.*, 1997). Since the implementation by the ALF working group, additional versions of RESRAD have been developed and the RESRAD family of codes can be accessed online from the Argonne National Laboratory.

According to additional evaluations of the calculated action levels, the working group determined that it is only necessary to assess ^{239}Pu, ^{240}Pu, and ^{241}Am to accomplish cleanup and still be protective of the most conservative plausible exposure scenarios. The working group recognized that because the calculated action levels represent individual radionuclides, it would be possible for each radionuclide concentration in the soil to be below the action level while the radiation dose to a person from a mixture could exceed the protective dose limits. Therefore, the calculated action levels were reduced to ensure the protective dose limits would be achieved even if the soil were to contain mixtures of ^{239}Pu, ^{240}Pu, and ^{241}Am.

The development of these action levels illustrates a useful approach to remediation of a large area with a complex history. The first element of the approach was to negotiate an agreement among regulatory authorities and form a working group to address the Rocky Flats site specifically. The working group then took the following important actions:

- review of all applicable regulations, recognizing that some were still in draft form at the time, and selection of guidance applicable to the site;
- selection of radionuclides of concern from review of extensive data available;
- evaluation of plausible use scenarios and selection of assumptions designed to over estimate radiation exposure and, therefore, be conservatively protective;
- use of simplifying assumptions to address the most important radionuclides;
- documentation of assumptions and related uncertainties; and
- suggestions to modify action levels, if indicated by future research.

This approach can be used to guide other forensic studies that are intended to address health-related questions involving radioactivity at a specific site.

Since development of cleanup action levels, additional research at the Rocky Flats sites has addressed local background and the transport of plutonium in the environment. Litaor (1999) studied the spatial distribution of plutonium in soil east and southeast of the Rocky Flats Plant.

The study differentiated plutonium associated with Rocky Flats operations from plutonium deposited from global fallout that is the result of atmospheric nuclear weapons tests. Plutonium in Colorado soil from fallout has been characterized by the ^{240}Pu/^{239}Pu isotopic ratio as measured by mass spectrometry. The ratio has been measured as ≥ 0.163 (Krey, 1976), 0.169 ± 0.005 (Efurd *et al.*, 1995), 0.155 ± 0.019 (USDOE, 1995), and 0.152 ± 0.003 (Ibrahim *et al.*, 1997). The value reported by the USDOE (0.155) was chosen to represent the ^{240}Pu/^{239}Pu ratio in the Litaor (1999) study. The ^{240}Pu/^{239}Pu ratio in weapons-grade plutonium fabricated at the Rocky Flats facility was chosen as 0.051 ± 0.009 (Krey and Krajewski, 1972). Soil samples having a ^{240}Pu/^{239}Pu ratio between 0.155 and 0.051 represents a mixture of Rocky Flats Plant plutonium and global fallout.

A geostatistical spatial analysis based on TIMS analysis of 63 soil samples indicated that the ^{240}Pu/^{239}Pu isotope ratio increased in an easterly direction from the Rocky Flats Plant, from 0.06 near the plant to 0.16 away from the plant, as predicted by prevailing winds. The results showed that the open space and residential areas east of the plant were affected by Rocky Flats Plant plutonium but the area south of the plant reflected global fallout plutonium and should be considered unaffected (Litaor, 1999). The analysis showed that 96% of the plutonium activity is contained in the upper 12 cm of soil, indicating very little vertical migration during 25 years.

Santschi *et al.* (2002) investigated mechanisms of plutonium and americium resuspension from soil and dispersion by surface water. Samples of storm water runoff and water from two ponds affected by Rocky Flats Plant operations were analyzed for iron, aluminum, manganese, and uranium by ICP-MS. Dissolved organic carbon (DOC), particulate organic carbon (POC), and colloidal organic carbon (COC) concentrations were measured (Guo and Santschi, 1997). Water quality parameters such as pH, alkalinity, nutrients (phosphate, silicate, and nitrate), and suspended particulate matter (SPM) were measured by standard methods.

Water samples were filtered to separate particulate, colloidal, and dissolved forms of ^{239}Pu, ^{240}Pu, and ^{241}Am. Particulate forms $>20 \mu m$ and 0.5–$20 \mu m$ in diameter were isolated by filter cartridges. Colloidal fractions ($<0.5 \mu m$) were separated by ultracentrifugation and the fractions having molecular weights <3 kDa and <100 kDa fractions were separated using 25 cm propylene spiral-wound cartridges.

Soil samples were sieved to remove particles larger than 1 mm. Aqueous extracts of local soil containing humic acids and other organic compounds found in environmental water media were prepared. The sieved soil samples were resuspended in this solution. Additional soil resuspensions were performed in solutions of commercially available forms of environmental organic compounds such as humic or alginic acids. The aqueous phase was then fractionated to separate the $<0.5 \mu m$ and <3 kDa fractions. Each filtrate fraction was acidified and ^{239,240}Pu and ^{241}Am were separated using standard selective precipitation and ion exchange chromatography techniques (USEPA, 1980, 1982; USDOE, 1979; Yamato, 1982). The plutonium and americium activity was measured by alpha-particle counting, which does not resolve ^{239}Pu and ^{240}Pu alpha energies.

The charge and mobility of colloidal plutonium and americium species were measured using isoelectric focusing gel electrophoresis. The ultrafiltrates were mixed with ^{234}Th(IV) tracer for the actinide ions, ^{14}C-dimethyl sulfate to label hydroxyl groups of colloidal amino sugars, and ^{59}Fe(III) to label oxyhydroxide sites of colloidal clay minerals (Wolfinbarger and Crosby, 1983; Quigley *et al.*, 2002). Radioactivity isolated in the gels was measured using LSC. Most of the plutonium and americium in the water samples was found in the $>0.5 \mu m$ particulate fraction (30 to 90%). Much of the remainder was in colloidal form (>3 kDa) and 20% or less was <3 kDa, although in some samples more than 50% of the anthropogenic radioactivity was associated with colloids. The colloidal plutonium was primarily associated with organic matter and inorganic clay and iron oxide species.

Colloidal plutonium isolated from soil resuspension was more associated with the organic material than clay minerals as determined by migration of the ^{14}C-labled forms in gel electrophoresis. The organic colloids were negatively charged and were associated with plutonium in the Pu(IV) oxidation state (Santschi, 2002). The study showed that formation of organic negatively charged colloids in water is an important dispersion mechanism for plutonium in soil.

7.4.4.1.3   Radionuclides in Groundwater (Savannah River Facility)

The role of colloidal transport of plutonium from unlined waste disposal basins at the Savannah River Site in South Carolina has been investigated (Dai *et al.*, 2002). The Savannah River Site was one of the two plutonium production sites in the United States; the material produced at the Savannah River Site was sent eventually to the Rocky Flats facility for fabrication. In addition to plutonium, the waste disposal basins contained ^{243}Am, ^{244}Cm, ^{245}Cm, and ^{246}Cm, which decay to ^{239}Pu, ^{240}Pu, ^{241}Pu, and ^{242}Pu, respectively.

Groundwater samples were collected from one well located up gradient from the disposal basins (Well 1) and Wells 2–4, which define a transect at increasing distances down gradient from the basins. Samples were taken using low-flow sampling methods (approximately 15 mL/min) to minimize suspension of particles and other artifacts of the well-purging process (Kaplan *et al.*, 1993, 1994a). Samples were processed on site using cross-flow ultrafiltration methods to separate colloidal species (with nominal molecular weight >1 kDa) from permeable fractions (below 1 kDa). The colloidal plutonium was associated with fractions in the 1 kDa to 0.2 μm range. The oxidation states of the samples were maintained using a $Cr_2O_7{}^{2-}/SO_4{}^{2-}$ solution (Lovett and Nelson, 1981). Reduced Pu(III/IV) species were separated by coprecipitation with LaF_3 and ^{244}Pu(III/IV) tracer. Oxidized Pu(V/VI) species in the filtrate were reduced using Fe^{2+} ion and coprecipitated using ^{242}Pu(V/VI) tracer. Samples were analyzed using TIMS methods that had a detection limit of 10^4 atoms per kg. The precision of analysis was limited by counting statistics and the uncertainty of aliquoting tracers (±0.8%, Dai *et al.*, 2002). The measured concentrations were corrected for ingrowth of ^{239}Pu, ^{240}Pu, ^{241}Pu, and ^{242}Pu from ^{243}Am, ^{244}Cm, ^{245}Cm, and ^{246}Cm decay between the time of sample collection and separation from Am and Cm isotopes.

It was found that plutonium concentrations were lowest in water from Well 1 and that the basins did not affect this location significantly. Down gradient plutonium transport was investigated using measured ^{240}Pu/^{239}Pu ratios. The ^{240}Pu/^{239}Pu ratio increased in Wells 2–4 with increasing distance from the basins, such that the ratio in water from Well 4 was 0.13 ± 0.02, exceeding the ratio in the source basin by a factor of at least 200.

The elevated concentration of ^{240}Pu at greater distances from the source was attributed to the migration of ^{244}Cm and decay to ^{240}Pu. The decay of the more mobile ^{244}Cm isotope to ^{240}Pu can account for the reported elevated activities of $^{239+240}$Pu that were measured by alpha spectrometry (Kaplan *et al.*, 1994b), which does not resolve the separate ^{239}Pu and ^{240}Pu contributions.

Measurements of oxidized and reduced plutonium species were used to elucidate transport mechanisms. Reduced Pu(III/IV) species have distribution coefficients (Kd) for partitioning between solid and solution phases approximately 100 times greater than oxidized Pu(V/VI) species (Nelson and Lovett, 1978). Consequently, oxidized plutonium species are the more mobile forms. The plutonium isolated from the groundwater was almost exclusively in the oxidized form, as consistent with the oxidizing environment of local groundwater (Dai *et al.*, 2002).

Measurements of colloidal plutonium indicated lower concentrations than expected from earlier studies (Kaplan *et al.*, 1994b). The concentrations of ^{239}Pu or ^{240}Pu in all samples were <10^4 atoms/kg and the percentage of Pu in colloidal form was <4% in all samples; this was less than the 10 to 70% of Pu found in Savannah River Site surface water. The study concluded that transport of plutonium in colloidal forms is less than previously reported (Kaplan *et al.*, 1994b).

The authors emphasized the importance of using low-flow pumping methods with ultrafiltration, stabilizing the oxidation state during handling and use of mass spectrometry instead of alpha spectrometry to investigate plutonium transport mechanisms.

7.4.4.1.4   Radionuclides in Freshwater (FW) Environments (Ob Estuary and Kara Sea)

The role of physiochemical properties of both liquid and solid media in radionuclide transport have been investigated in studies of a river and estuarine system. The system is downstream from waste reservoirs at the Mayak Production Association facility in the Ural Mountains in the Former Soviet Union (Standring *et al.*, 2002).

The Mayak facility produced weapons-grade plutonium beginning in the 1940s and was the site of an accidental release from a waste storage tank in 1967. From 1949 to 1956, waste from the processing facility was diverted to the Techa River or to Lake Karachay. Since 1956, this waste was diverted to Reservoir 10, and then to Reservoir 11 beginning in 1963. The radionuclide inventory of Reservoir 10 has been estimated at 4 PBq ^{90}Sr, 2.5 PBq ^{137}Cs, and 82 TBq ^{99}Tc and Reservoir 11 contains approximately 0.6 PBq ^{90}Sr, 0.6 PBq ^{137}Cs, and 2.5 TBq ^{99}Tc (JNREG, 1997).

Measurements in Reservoir 10 sediments showed that 99 and 96% of ^{137}Cs and ^{60}Co, respectively, are adsorbed to sediments; ^{90}Sr appears to be more mobile with 13% in the aqueous phase. Elevated concentrations of these radionuclides have been measured in sediments downstream from both of the reservoirs. The mobilization of ^{137}Cs, ^{60}Co, ^{99}Tc, and ^{90}Sr from sediments by FW and the effect of transition from FW to Seawater (SW) on transport were investigated using native media and simulants with controlled properties.

Sediment samples from four locations near Reservoir 10 included two samples that represent old Techa River riverbed sediments and two samples from flooded areas formed by Reservoir 10 construction. The surface area of sediment particles was measured gravimetrically by the BET method using nitrogen adsorption; organic content was measured gravimetrically as mass loss on ignition to 550 °C for 12 hours. Sediments sampled from Reservoir 10 were extracted with synthetic FW and SW solutions designed to simulate the ionic composition and pH (6.5) of Techa River water and to simulate SW using a sea salt solution (~28%) at pH 8. Sediments were subjected to three consecutive 7-day FW solution extractions followed by three 7-day SW solution extractions and finally a 2-month SW solution extraction. The apparent distribution coefficient (Kd) of each radionuclide was calculated as the ratio of the activity concentration in the solid phase to the solution concentration (Standring *et al.*, 2002).

The remaining sediments were further extracted sequentially under conditions with increasing power to remove metal ions from sediment: (1) 1 M ammonium acetate at pH5 and room temperature, (2) 0.04 M hydroxylamine hydrochloride in 20% acetic acid at pH2 and 80 °C, (3) 30% hydrogen peroxide at pH2 and 80 °C, and (4) 7 M nitric acid and 80 °C, and (5) aqua regia at 80 °C.

Gamma spectrometry was used to measure ^{137}Cs and ^{60}Co concentrations directly in extracts by counting the 661 keV and 1173 keV emissions, respectively. The ^{99}Tc and ^{90}Sr were separated by column ion exchange chromatography such that ^{99}TcO$_4{}^-$ ion was retained in the column. The ^{90}Sr^{2+} eluted from the column was extracted from the solution by liquid–liquid extraction and the ^{90}Sr beta activity was measured by liquid scintillation counting of the ^{90}Y daughter. The ^{99}TcO$_4{}^-$ was eluted from the column using 7 M nitric acid and the concentration was measured by ICP-MS to a detection limit of 4 ng/L.

Results showed that ^{137}Cs remained strongly bound to sediments with ~1% mobilized when extracted with FW simulant and <10% mobilized by extraction by SW simulant. The amount of ^{137}Cs extracted in either simulant was correlated to the surface area and the organic content of the sediments, indicating that ^{137}Cs exchange probably is a surface reaction and is influenced by reaction with organic components. Cobalt-60 (^{60}Co) was more mobile than ^{137}Cs, raging from ~10 to >30% extracted in FW simulant, but ^{60}Co was more variable among sediment samples. Technicium-99 (^{99}Tc) mobilization was very low in both FW and SW simulants, including the 2 month exposure to SW simulant, indicating slow reaction kinetics. Strontium-90 (^{90}Sr) was relatively mobile with >40% extracted in the first 7 day FW simulant exposure (Standring et al., 2002).

Apparent Kd values calculated for ^{137}Cs, ^{60}Co, ^{99}Tc, and ^{90}Sr exchange between Reservoir 10 sediment and FW systems were variable for each radionuclide, reflecting the variable nature of the sediments. Technicium-99 (^{99}Tc) was the most strongly adsorbed radionuclide, which had mean Kd values decreasing as ^{99}Tc > ^{137}Cs >> ^{60}Co ~ ^{90}Sr. All exchange experiments indicated slow kinetics.

Transfer from a FW to a SW environment led to a decrease in apparent Kd, indicating increased mobility for all four radionuclides (Standring et al., 2002). Upon exposure to SW simulant, the mean Kd value for ^{137}Cs decreased by 94% and Kd values for other radionuclides also decreased significantly, ^{60}Co (77%), ^{99}Tc (27 to 70%) and ^{90}Sr (~70%). All extractions in SW, including the final 2-month extraction indicated slow exchange kinetics.

The exchange of radionuclide to sediments was described as "reversible" if mobilized by FW or SW simulants or ammonium acetate solution. Cesium-137 (^{137}Cs) and ^{99}Tc were irreversibly or slowly reversibly bound and ^{60}Co and ^{90}Sr were reversibly bound. Exchange of ^{99}Tc was strongly dependent on oxidizing conditions, such that exchange was ~80% in H_2O_2 and was <10% in all other solutions. This strong dependence on oxidizing conditions was attributed to ^{99}Tc complexation by organic compounds, which were dissolved upon oxidation.

7.4.4.1.5 Radionuclides in Marine Environments (Berents Sea)

Sources of anthropogenic radionuclides in the Berents Sea have been studied extensively (IAEA, 1999; ANWAP, 1997). The sources have been identified as fallout from atmospheric nuclear weapons testing, the Chernobyl accident, releases from nuclear facilities on the Ob and Yenisey Rivers, and European nuclear fuel reprocessing facilities.

Transport of radionuclides in a marine environment near the site of the sunken nuclear submarine Kursk has been investigated (Matishov et al., 2002). The Kursk sank on August 1, 2000 in the Berents Sea with an estimated inventory of 6 PBq each of ^{137}Cs and ^{90}Sr in the reactor core. Other fission and activation products would also be present due to reactor operation. Samples of SW, sediment, benthic organisms, and fish were collected in September 2000 to measure radionuclide levels and to estimate the potential for transport of contamination into other northern seas.

Samples were collected from near the Kursk, Kola Bay, and stations in the Berents Sea identified by longitude and latitude. Cesium-137 (^{137}Cs) was separated from SW samples using potassium ferrocyanide and measured by gamma counting using a HPGe detector with an efficiency of 25%. The uncertainty of ^{137}Cs measurements was estimated at 10 to 20%. Iodine-129 (^{129}I) was measured in SW samples using AMS (Smith et al., 1998) and an ^{127}I tracer. The standard deviation of ^{129}I measurements was estimated at 5 to 10%. Sediment and biota samples were dried and analyzed by low-level gamma spectrometry using a Ge(Li) detector.

Cesium-137 (^{137}Cs) concentrations in sediments ranged from 1–10 Bq/kg, as reported previously for Kola Bay and Berents Sea sediments (Matishov et al., 1999, 2000). These concentrations are consistent with fallout from nuclear weapons testing (Smith et al., 1995). Cesium-137 (^{137}Cs) concentrations in fish and SW indicated no significant uptake by fish within the first month of the accident. However, ^{129}I concentrations in seawater exceeded the fallout concentration range by at least a factor of 100 and exceeded levels measured in the same locations in 1992 (Raisbeck et al., 1993). These elevated ^{129}I concentrations in SW were attributed to releases from European nuclear fuel reprocessing facilities (Smith et al., 1998) rather than releases from the Kursk.

7.4.4.2 Depleted Uranium Projectiles

Naturally occurring uranium is enriched in ^{235}U for use as a fissile material. Uranium enrichment technology uses UF_6 gas to produce uranium in any degree of enrichment, although the most commonly produced materials are ~3% ^{235}U for use as nuclear power reactor fuel and 50% to 90% ^{235}U for use in other reactors. Compounds for laboratory use are available containing uranium enriched to 99% ^{235}U. Uranium-234 (^{234}U) is also enriched by most enrichment processes. The uranium feedstock for the enrichment process is either natural uranium or uranium from recycled spent nuclear reactor fuel.

The fraction of uranium that is not enriched is termed "depleted uranium" (DU) and is available in very large amounts. Because enrichment operations produce uranium with highly variable degrees of enrichment, the isotopic composition of DU is necessarily variable and does not have characteristic ^{238}U/^{235}U/^{234}U ratios analogous to natural uranium. The ^{235}U content of DU generally varies between 0.2 and 0.3% ^{235}U by weight with traces of ^{234}U (Harley et al., 1999). The typical isotopic composition is given as ^{238}U, 99.75 and ^{235}U, 0.25; and ^{234}U, 0.0005 expressed as g isotope/100 g natural uranium (Shleien, 1992).

Depleted uranium was used as ballast in aircraft until it was replaced by tungsten following an aircraft crash in the Netherlands in October 1992 (HCN, 2001). Depleted uranium is used for a variety of laboratory chemicals and is potentially highly valuable as fuel in breeder nuclear power reactors. Prior to the 1990s, DU-related forensic science was concentrated on monitoring uranium around production and handling facilities. Since 1991, the use of armor-piercing projectiles in Iraq and the Balkans has resulted in wider DU dispersal in the environment.

Depleted uranium used in projectiles is an alloy containing approximately 0.75% titanium (Royal Society, 2001) and contains variable trace amounts of ^{236}U and transuranic fission products, such as americium, plutonium, and neptunium (OSAGWI, 2000; Bhat, 2000). One DU sample obtained in Kosovo contained 0.0028% ^{236}U (HCN, 2001). Although variable amounts of transuranic residues have been identified in DU samples, their contribution to radiological dose from exposure has been estimated at ~0.1% of the dose from DU itself (Royal Society, 2001). Because of their variable concentration in DU, it is not clear whether these trace isotopes can serve as indicator compounds for DU in the environment.

The numbers of DU projectiles used during the 1991 Gulf War have been estimated (Royal Society, 2001). Tanks fired approximately 9600 large caliber (100 mm or 120 mm) rounds with masses of 4–5 kg. The total DU mass fired was approximately 44,000 kg. Approximately 78,000 30 mm 275 g rounds were fired, representing approximately 214,000 kg of DU. Approximately 10,000 30 mm rounds were fired in Bosnia in 1994–1995 and approximately 31,000 30 mm rounds were fired in Kosovo in 1999 (Royal Society, 2001; HCN, 2001). No use of large caliber DU rounds was reported by NATO forces in the Balkans conflicts (Royal Society, 2001).

A "typical" attack site in the Kosovo conflict was estimated as involving approximately 300 30 mm rounds fired from an A-10 aircraft including 60 high-explosive rounds and 240 DU rounds. The mass of DU expected at the typical site is 72 kg (WHO, 2001a). It is estimated that 10% (24) rounds hit their target (USACHPPM, 2000); 80% of the 216 rounds that missed the target (172 rounds) would land within a 100 m radius and the remaining 20% (44 rounds) would land within a radius of 1.85 km (1 nautical mile) from the target. The range of ricochet rounds has been estimated at 2–4 km from the target (AEPI, 1995).

Approximately 10–37% of the 300 g mass of a 30 mm round hitting the target would ignite in air to form aerosols containing 60–96% of the DU in respirable particles 1–10 μm in diameter (USACHPPM, 2000). Assuming that up to 50% of the DU mass is aerosolized, approximately 3.6 kg of the 24 DU rounds would be dispersed as uranium oxide dust. The remaining unoxidized DU would be dispersed as metallic fragments.

The 216 rounds missing their target would be dispersed over the ground surface or buried. There are little data on the depth of penetration in soil; however, an estimate of 30 cm in soft soils typical of the Persian Gulf and Serbia has been reported (USACHPPM, 2000; WHO, 2001a). If all of the 72 kg of DU released within one square mile of a typical site were degraded in place, the DU would add an additional amount of uranium sufficient to increase the natural uranium concentration in soil by approximately 5% (WHO, 2001a).

Chemical processes influence the fate of DU following dispersal around an attack site. Uranium oxide aerosols generated upon hitting the target also contain material from the target, such as iron, or the elements of local soil or concrete (Royal Society, 2002). The chemical composition of aerosols from spent DU rounds is, therefore, expected to be highly variable.

The uranium–oxygen system can be viewed as one of increasing oxygen incorporation into the uranium phase with certain O/U ratios representing relatively stable forms. Because the face-centered-cubic UO_2 crystal structure can absorb oxygen atoms into lattice interstices without changing the lattice symmetry, UO_2 forms hyperstoichiometric UO_{2+x} as a result of surface oxidation (Cordfunke, 1969; Manes and Benedict, 1985). Finely divided UO_2 is pyrophoric, oxidizing in air to a variety of oxide phases including U_3O_8 as the most stable phase. Above 300 °C, UO_2 absorbs oxygen to form relatively more stable UO_{2+x} phases. The U_4O_9 phase forms as UO_{2+x} where $x = 0.25$. UO_3 (UO_{2+x}, where $x = 1$) is the stable phase at one atmosphere of O_2 pressure and temperatures below 500 °C. U_3O_8 (UO_{2+x}, where $x = 0.667$) is the most stable phase between above 500 °C and 800 °C (Cordfunke, 1969) and is the form most frequently encountered under ambient conditions.

Studies of uranium dusts generated in fires (Elder and Tinkle, 1980; Hooker et al., 1983; Haggard et al., 1985; Totemeier, 1995, as cited in Royal Society, 2002) indicate that, under controlled conditions, the ignition temperature varies from 300 to 600 °C and that burning was not self-sustained unless the uranium metal was in the form of a fine dust. Approximately, 42–47% of the uranium metal was oxidized in a three-hour burn. Outdoor fires generated dusts with 62% of the mass in the respirable size range (<10 μm AMAD) compared to 14% in aerosols from indoor fires (Royal Society, 2002).

Because of the pyrophoric nature of uranium metal, use of armor-piercing penetrators generates dusts under completely uncontrolled chemical conditions. The percentage of uranium metal that is aerosolized upon impact with the target depends on the kinetic energy of the uranium penetrator and the nature of the target. Aerosolization has been estimated at 10–35% and varying up to 70% (Harley et al., 1999). The resulting dusts are primarily uranium oxides that include a variety of other elements. The respirable fraction of dust generated during depleted uranium armor penetrator tests was shown by X-ray diffraction analysis to contain predominantly U_3O_8 mixed with 18% UO_2 and an amorphous fraction estimated at 20% (Scripsick et al., 1985). More recent studies of uranium dust from battlefield sources report that target materials such as calcium, silicon, aluminum, titanium, and iron contribute to uranium-containing particles depending on the target materials (Royal Society, 2002). Properties of DU aerosols formed upon impact or in DU fires and their characteristics have been summarized; see the Royal Society document (2001, Annex G).

Dispersed, oxidized, and metallic forms of DU will be subjected to weathering according to local climatic conditions. The corrosion rate of metallic uranium is highly dependent on ambient conditions, such that highly variable rates have been reported. Under different oxidizing environmental conditions at Aberdeen Proving Ground in Maryland, Yuma Proving Ground in Arizona and at Kirkcudbright in the United Kingdom, corrosion of a mass of DU similar to a 30 mm round would release approximately 90 g of DU per year; a 120 mm round would release approximately 600 g of DU per year. These release rates indicate a lifetime of 5 to 10 years for an intact DU round in the environment (Royal Society, 2002). Under reducing (anoxic) conditions, corrosion is estimated to be very slow, such that a solid 1.4 kg (120 mm) penetrator from an anti-tank penetrator or a 300 g (30 mm) round fired from an A-10 aircraft have expected lifetimes of 2100 years and 500 years, respectively (WHO, 2001a).

As implied by the above observations, DU corrosion involves a two-step process of formation of U(IV) from metallic DU followed by oxidation of U(IV) to U(VI). The rates of these reactions are increased by the presence of water. The chemical composition and available surface area of the DU oxide and metal fragments affect corrosion rates also, as do the oxidizing potential of the environmental media and the concentration of salt and complexing agents, such as carbonates (Royal Society, 2002, Annexe G).

The stabilities of various dissolved and complexed metals (including their radioactive isotopes) under different oxidizing and pH conditions in the environment are described by the Eh-pH diagram (Brookins, 1988). The Eh-pH diagram for the uranium–carbon–oxygen–hydrogen system is shown for illustration in Figure 7.4.4 (WHO, 2001b).

The Eh axis is derived from the oxidation potential of the system and the pH axis is derived from the acid concentration. As shown in the diagram, the uranyl ion (UO_2^{2+}) is stable under oxidizing and acidic conditions (<pH5); the uranyl carbonate complexes are formed at higher pH values. Other carbonate and bicarbonate complexes that form under these conditions (Ferri et al., 1981) are not shown in Figure 7.4.4 for simplicity. Formation of solid UO_2 is favored under less oxidizing conditions (Eh ~0.2 to 0.3), indicating that insoluble, and probably less mobile, forms can be expected in the environment. Measurement of the oxidation potential and pH properties of the local environment provides useful information for simulation modeling of the fate and transport of dispersed DU and other metals in the environment.

The dissolution rate of DU oxide aerosols generated upon projectile impact, or released by DU metal corrosion, is

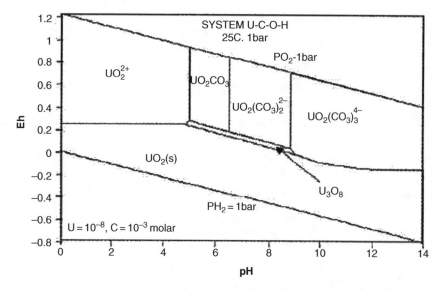

Figure 7.4.4 *The Eh-pH diagram shows the stability of uranium species under various Eh and pH conditions. Eh is an indicator of oxidation potential that is related to the presence of dissolved oxygen. pH is an indicator of acidity. The diagram represents the U-C-O-H system as published in WHO (2001b) and adapted from Brookins (1988).*

more rapid than for metallic DU, but is also highly dependent on particle size and chemical composition as influenced by the target material and local environmental conditions (Royal Society, 2002). Dissolution studies of industrial uranium oxides formed under controlled conditions have been reviewed (Eidson, 1994; Ansoborlo *et al.*, 1998, 2002; Hodgson *et al.*, 2000) and have shown multiphasic kinetics with a rapidly dissolving fraction and one or more fractions dissolving at slower rates. Although dissolution of DU oxide aerosols formed upon projectile impact or in DU fires has received less study, available information suggests that biphasic dissolution occurs with variable fractions (up to 40%) dissolving with a half-life of <1 day and the remainder dissolving with a half-life of 100–500 days (Glissmeyer and Mishima, 1979; Scripsick *et al.*, 1985). These dissolution results suggest that corrosion of DU metal fragments is the rate-limiting step for release of uranium and environmental transport from an attack site.

7.4.4.3 Space Nuclear Power Applications
Radioisotopic Thermoelectric Generators (RTGs) use heat generated by radioactive decay to generate electrical power and often use plutonium isotopes as the thermal source, although other radionuclides have been used. These devices have both civilian and military applications that require a relatively small, self-contained, low-maintenance power source that can operate in remote locations for long periods of time. As such, RTGs are ideal for use in space. The US National Aviation and Space Administration (NASA) has deployed Space Nuclear Auxiliary Power (SNAP) sources in spacecraft and satellites since the 1960s. Nuclear reactors have also been deployed in space as power generators by the US and the former Soviet Union (Johnson, 1986).

Accidental releases of radionuclides to the environment have occurred during RTG deployment in space. The single accidental release of radionuclides to the atmosphere involving a US RTG occurred in 1964 when the SNAP-9A [238]Pu source aboard a Department of Defense weather satellite burned during re-entry after failing to reach orbit

(NASA, 2004). The characterization of [238]Pu fallout has been described (Volchok, 1967).

Accidental re-entry of Soviet RTGs and reactors has released detectable radionuclides to the atmosphere (Perry, 1978; Reese and Vick, 1983; Eisenbud, 1987). Debris from the COSMOS 954 reactor fell to earth in the Northwest Territory of Canada in 1978. Removal of debris, cleanup of the site, and results of follow up monitoring of environmental media and food have been described (Gummer *et al.*, 1980; Eisenbud, 1987).

7.4.5 Civilian Sources
Civilian uses of radionuclides include nuclear power generation and waste management, medical applications, radiopharmaceuticals, mineral exploration, and industrial applications. A forensic investigation of a site must rely on all available information regarding the radionuclide inventory and their application at each site. Such information might include invoices for purchased radionuclides, standard operating procedures for their use and disposal, manifests for waste transport to disposal sites, and reports to regulatory authorities.

7.4.5.1 Commercial Nuclear Power
Among 28 Nuclear Energy Agency (NEA) member nations,[1] nuclear energy contributes an average of 23.2% of electricity production for the nation, ranging from 4.0% in the Netherlands to 77.6% in France (NEA, 2003a).

Similar percentages are reported among 441 nuclear power plants operating in 30 countries listed by the Power Reactor Information System (PRIS, 2004). Nuclear energy production in these countries contributes an average of

---

1. Australia, Austria, Belgium, Canada, Czech Republic, Denmark, Finland, France, Germany, Greece, Hungary, Iceland, Italy, Japan, Luxembourg, Mexico, Netherlands, Norway, Portugal, Republic of Korea, Slovak Republic, Spain, Sweden, Switzerland, Turkey, United Kingdom, and United States.

28.2% of electricity production, ranging from 2.2% in China to 79.9% in Lithuania.[2]

Each of these reactors contains an inventory of radionuclides determined by its design, fuel requirements, age, and time since last refueling. Methods to estimate the inventory for a nuclear power reactor are provided by the US Nuclear Regulatory Commission (USNRC, 1986). An example of the estimated nuclear reactor core inventory at the time of a hypothetical accident is shown in Table 7.4.1 (Shleien, 1992). During normal operations, certain radionuclides accumulate in the reactor coolant, which can represent the source term for the assessment of a coolant release. An example of the estimated coolant inventory of a specific reactor design and operating conditions is shown in Table 7.4.2 (Shleien, 1992).

These tables provide a list of radionuclides in the core and the coolant, their radiation activity and half-lives that can be used to identify the radionuclides of potential concern at different times following a release. However, a forensic investigation must begin with a detailed evaluation of the reactor unit in question and its operational history.

Each country has regulations for handling nuclear fuel and waste radionuclides that should be consulted during a forensic investigation in that jurisdiction. The USNRC regulates civilian nuclear reactors in the United States and requires monitoring of radionuclides as described in 10CFR20. Radiation monitoring is focused on protection of workers and the general public from radiation dose from all sources. For example, monitoring requirements are designed to ensure that "the total effective dose equivalent to individual members of the public from the licensed operation does not exceed 0.1 rem (1 mSv) in a year (10CFR20.1301), exclusive of the dose contributions from background radiation" and other sources specified elsewhere the regulations.

One way to comply with these regulations (10CFR20.1302(b)(1)) is to "demonstrate by measurement or calculation that the total effective dose equivalent to the individual likely to receive the highest dose from the licensed operation does not exceed the annual dose limit." Another way to comply is to demonstrate (10CFR20.1302(b)(2)), "The annual average concentrations of radioactive material released in gaseous and liquid effluents at the boundary of the unrestricted area do not exceed values specified in table 2 of Appendix B ... " The values shown in the Appendix B tables (not shown here) are concentrations in air and water that were calculated to limit radiation doses to the public to a total effective dose equivalent of 50 mrem (0.5 mSv), if the radionuclides were to be inhaled or ingested continuously over the course of a year. The regulations further specify that the radiation dose to an individual continuously present in an unrestricted area from external sources would not exceed 0.002 rem (0.02 mSv) in an hour and 0.50 rem (5 mSv) in a year (10CFR20.1301). Additional regulations address incineration processes (10CFR20.2004 and 10CFR50), transfer for disposal (10CFR20.2006), land disposal (10CFR61), and for geological repositories (10CFR60 and 63).

Incidents involving release of radionuclides must be reported to the USNRC according to procedures and schedules that depend on the nature of the incident (10CFR20.2202 and 20.2203). Immediate notification is required if an individual receives a total effective dose

Table 7.4.1 *Potential Reactor Core Inventory is Dependent on Reactor Operational History (Shleien, 1992)*[a]

Radionuclide	Radioactive Inventory (curies × 10^{-6})	Half-Life (days)
^{58}Co	0.0078	71.0
^{60}Co	0.0029	1920
^{85}Kr	0.0056	3950
^{85m}Kr	0.24	0.183
^{87}Kr	0.47	0.0528
^{88}Kr	0.68	0.117
^{86}kr	0.00026	18.7
^{89}Sr	0.94	52.1
^{90}Sr	0.037	11030
^{91}Sr	1.1	0.403
^{90}Y	0.039	2.67
^{91}Y	1.2	59.0
^{95}Zr	1.5	65.2
^{97}Zr	1.5	0.71
^{95}Nb	1.5	35.0
^{95}Mo	1.6	2.8
^{99m}Tc	1.4	0.25
^{103}Ru	1.1	39.5
^{105}Ru	0.72	0.185
^{106}Ru	0.25	366
^{105}Rh	0.49	1.50
^{127}Te	0.059	0.391
^{127m}Te	0.011	109
^{129}Te	0.31	0.048
^{129m}Te	0.053	0.340
^{131m}Te	0.13	1.25
^{132}Te	1.2	3.25
^{127}Sb	0.061	3.88
^{129}Sb	0.33	0.179
^{131}I	0.85	8.05
^{132}I	1.2	0.0958
^{133}I	1.7	0.875
^{134}I	1.9	0.0366
^{135}I	1.5	0.280
^{133}Xe	1.7	5.28
^{135}Xe	0.34	0.384
^{134}Cs	0.075	750
^{136}Cs	0.030	13.0
^{137}Cs	0.047	11,000
^{140}Ba	1.6	12.8
^{140}La	1.6	1.67
^{141}Ce	1.5	32.3
^{143}Ce	1.3	1.38
^{144}Ce	0.85	284
^{143}Pr	1.3	13.7
^{147}Nd	0.60	11.1
^{239}Np	16.4	2.35
^{238}Pu	0.00057	32,500
^{239}Pu	0.00021	8.9×10^6
^{240}Pu	0.00021	2.4×10^6
^{241}Pu	0.034	5,350
^{241}Am	0.000017	1.5×10^5
^{242}Cu	0.0050	163
^{244}Cu	0.00023	6,630

[a] From the Reactor Safety Study, Appendix VI WASH-1400, October 1975, Table VI. 3–1.
The Reactor Safety Study (WASH-1400) is cited in this chapter as (USNRC, 1975).

2. Armenia, Argentina, Belgium, Brazil, Bulgaria, Canada, China, Czech Republic, Finland, France, Germany, Hungary, India, Japan, Lithuania, Mexico, Netherlands, Pakistan, Republic of Korea, Romania, Russia, Slovak Republic, Slovenia, South Africa, Spain, Sweden, Switzerland, Ukraine, United Kingdom, and United States.

Table 7.4.2 *Potential Reactor Coolant Inventory is Dependent on Reactor Operational History (Shleien, 1992)[a]*

Noble Gas Fission Products		Fission Products	
Isotope	μCi/ml	Isotope	μCi/ml
^{85}Kr	1.11	^{84}Br	3.0×10^{-2}
^{85m}Kr	1.46	^{88}Rb	2.56
^{87}Kr	0.87	^{89}Rb	6.7×10^{-2}
^{88}Kr	2.58	^{89}Sr	2.52×10^{-3}
^{133}Xe	1.74×10^2	^{90}Sr	4.42×10^{-3}
^{133m}Xe	1.97	^{90}Y	5.37×10^{-5}
^{135m}Xe	0.14	^{91}Y	4.77×10^{-4}
^{138}Xe	0.36	^{92}Sr	5.63×10^{-4}
		^{92}Y	5.54×10^{-4}
Total Noble Gases	187.3[b]	^{95}Zr	5.04×10^{-4}
		^{95}Nb	4.70×10^{-4}
		^{99}Mo	2.11
Corrosion Products		^{131}I	1.55
Isotope	μCi/ml	^{132}Te	0.17
^{54}Mn	4.2×10^{-3}	^{132}I	0.62
^{56}Mn	2.2×10^{-2}	^{133}I	2.55
^{58}Co	8.1×10^{-3}	^{134}Te	2.2×10^{-2}
^{59}Fe	1.8×10^{-3}	^{134}I	0.39
^{60}Co	1.4×10^{-3}	^{134}Cs	7.0×10^{-2}
		^{135}I	1.4
Total Corrosion Products	3.7×10^{-2}	^{136}Cs	0.33
		^{137}Cs	0.43
		^{138}Cs	0.48
		^{144}Ce	2.3×10^{-4}
		^{144}Pr	2.3×10^{-4}
		Total Fission Products	12.8[b]

[a] Combination concentration corresponding to 1% filled fuel near end of fuel life [1000 MW(e) PWR at 578 deg. (from NUREG/CR-3332, 1983, Source Eichholz, 1977)].
[b] 187.3 μCi/ml = 6.9 GB q/l, μCi/ml = 0.47 GB q/l
NUREG/CR 3332 is cited in this chapter as (USNRC, 1983). The Eichholz (1977) references is cited as Eichholz, 1977.

equivalent greater than specified amounts, or if radioactive material is released in an amount that would result in an intake of radionuclides above specified levels if an individual had been present for 24 hours. Incidents that cause, or may cause, lesser specified doses must be reported within 24 hours. Incidents (including the above) that exceed occupational limits, or limits to members of the public, or in amounts exceeding yet lesser specified levels must be reported in writing within 30 days. The reports must contain the cause of the elevated exposures, dose rates, and associated concentrations.

Severe accidents have occurred at the Three Mile Island Unit 2 in Pennsylvania and at the reactor complex at Chernobyl in the Ukraine.

7.4.5.2 Three Mile Island Unit 2

The accident at Three Mile Island Unit 2 (TMI-2) near Harrisburg, Pennsylvania, occurred on March 28, 1979 and was initiated by a series of equipment failures that eventually caused loss of cooling water to the reactor core and a partial core meltdown (USNRC, 2004). The melted core and fuel materials were confined by the containment building and were not released. When coolant flow was restored, a release of radiation was made from the plant's auxiliary building to relieve pressure on the primary sys-

tem and to avoid curtailing the flow of coolant to the core (USNRC, 2004).

Studies of the TMI-2 accident have estimated that about 2.5 million Ci of noble gases (about 0.9% of the core inventory) and 15 Ci of radioiodine (about 0.00003% of the core inventory) were released to the environment. No other radioactive fission products were released (USNRC, 1996). Extensive studies of radionuclides in the environment showed that trace levels of radioiodine were detected in locally produced milk, but that no other food or water supplies were affected (USNRC, 2004). These and additional studies were made to estimate the radiation dose to the population from exposure to the accident-related releases. The maximum cumulative off-site radiation dose to an individual was less than 100 mrem (USNRC, 1980) and the average dose to about two million people in the area was estimated at approximately 1 mrem. To put this into context, the dose to the population from the natural radioactive background in the area of TMI-2 is estimated at 100–125 millirem per year (USNRC, 2004). The measurement of accident-elated radionuclides in the environment, the calculation of radiation doses to the public, and their interpretation are described in detail in USNRC documents (USNR, 1979; USNRC, 1980).

Engineering analysis of the TMI-2 accident resulted in reactor design modifications to reduce the likelihood of a similar accident occurring in the future (USNRC, 1981).

Results of environmental studies have been incorporated into the requirements for the environmental impact statement of accident scenarios, which must be included in a power plant's license renewal application (USNRC, 1996).

7.4.5.3 Chernobyl

The accident at the Chernobyl Power Complex located north of Kiev, Ukraine (about 20 km south of the Belarus border), occurred at Unit 4 on April 26, 1986 and released a large fraction of the reactor core inventory in gases and aerosols to the atmosphere over a 10-day period. It is estimated that the effluent plume represented 100% of the core inventory of xenon and krypton gases (NEA, 2003b) and 10–20% of the iodine, tellurium, and cesium. Approximately 6 tons of fragmented fuel was released, representing 3–4% of the total fuel material containing uranium, plutonium, and other transuranium isotopes. Large quantities of airborne ^{131}I, ^{134}Cs, and ^{137}Cs were eventually dispersed to the northern hemisphere, although significant contamination occurred only in the former Soviet Union and parts of Europe (NEA, 2003b).

During the first weeks following the accident, concern for the food supply in the Ukraine and Belarus, in European countries to the northwest and to the Balkans and Northern Mediterranean countries led to mapping of areas affected by ^{131}I and ^{137}Cs deposition. Among the short-lived radionuclides, ^{131}I was of concern initially because of its ability to enter the food chain and ^{137}Cs was chosen for mapping because of the ease of gamma emission measurements over wide areas. Some early maps were made using ^{137}Cs as a marker for ^{131}I based on an assumed ^{131}I/^{137}Cs ratio. Although these maps were useful in the initial assessment of deposition over very large areas, the ^{131}I/^{137}Cs ratio in the plume has been shown to vary by a factor of 5–10 (NEA, 2003b). More accurate maps of areas affected by long-lived radionuclides have been developed since, but the short half-life of ^{131}I (8 days) precluded similar mapping. Although the plume was detected in North America and Japan, very little deposition occurred outside Europe and no deposition was detected in the Southern Hemisphere (NEA, 2003b).

Radionuclide deposition on the ground was very heterogeneous as a result of enhanced deposition in some areas due to the plume passing through localized rainfall. Vertical penetration in soil was effected by rainfall at the time of deposition (Bréchignac et al., 2000). Vertical migration since initial deposition has been slow, such that ^{137}Cs and ^{90}Sr are still confined to the upper 10 cm of soil and a steady state sorption and desorption exchange with soil and sediment has apparently been established. As a result, local foodstuffs will remain contaminated much longer than initially expected (Smith et al., 2000; NEA, 2003b). Because of the strong adsorption to sediments, ^{137}Cs and ^{90}Sr remain localized in the environment rather than being flushed out with uncontaminated waters. The initial contamination levels in surface water were transferred to the sediments and remain a contamination source for fish, mollusks, and other benthic organisms. Fish from affected lakes in Sweden contain radiocesium at concentrations above Sweden's regulatory limit of 1500 Bq/kg (NEA, 2003b).

Surface waters from the Kiev, Kanev, and Kremenchung reservoirs do not pose a current health problem, but monitoring continues out of concern for future drinking water supplies (NEA, 2003b). It is estimated that ^{90}Sr measured in the soils within 30 km of the exclusion zone could contaminate drinking water above acceptable limits in 10 to 100 years (Vovk et al., 1995; NEA, 2003b). Groundwater in countries of the European Union was monitored from 1987 to 1990 and was shown to contain ^{137}Cs at or below 0.1 Bq/L, which is not a health concern (NEA, 2003b).

During the time since the accident, continued monitoring of radionuclides in the environment and uptake by people has allowed refinement of computer models for the deposition, migration, and uptake of radionuclides. The role of weather patterns, especially precipitation and other meteorological factors, has been used to develop refined models on deposition following an airborne release (NEA, 1989, 1996). Monitoring of internally deposited radionuclides in people living in affected areas has provided data to allow refinement of radiation dose models (NEA, 2003b). For example, earlier models used to predict the transport of airborne ^{131}I and ^{137}Cs to the food chain have been shown to overestimate the uptake by up to a factor of 10 (Hoffman, 1991). Continued monitoring has shown that radionuclides released to the environment have reached stable levels and the decrease in contamination appears to follow the rate of ^{137}Cs decay with a 30-year half-life (NEA, 2003b).

Because these two accidents have very little in common except that the radionuclides of interest are similar, it is apparent that any investigation of radiation releases from nuclear reactor accidents must be designed for each occurrence specifically. However, the data from monitoring programs following these accidents and improved assessment models can provide useful information to guide more specific forensic investigations.

7.4.5.4 Medical Applications

Prior to development of the cyclotron in the 1930s and the nuclear reactor in the 1940s and 1950s, medical applications were limited to natural radionuclides, such as radium or radon isotopes (NCRP, 1982; Stannard, 1988). Currently, radiopharmaceuticals are used as diagnostic imaging and therapeutic agents and as tracers of physiological processes.

The choice of each application is influenced by medical need, the associated dose to the patient and to medical personnel, and radiation measurement requirements. The isotopes in use today are primarily man-made and usually of short half-life (Table 7.4.3).

Because some isotopes have short half-lives, their use is often restricted to facilities located near the production source. Those with intermediate half-lives can be prepared and shipped to more remote hospitals for use. The ^{99}Mo-^{99m}Tc system is widely used in a variety of imaging applications at local facilities. The ^{99}Mo isotope, which has a 66 hour half-life, is prepared at a reactor site and shipped to a local facility where the ^{99m}Tc daughter, which has a 6.02 hour half-life, is separated from the parent for use. Some isotopes, such as ^{3}H, ^{14}C, and ^{32}P, are used in cell biology research at university laboratories as well as in medical applications.

Use of radionuclides in research and nuclear medicine in the United States is regulated by the USNRC (see 10CFR20, 10CFR35, 10CFR51, and 10CFR61). Disposal of waste forms containing these radionuclides usually occurs by the following methods (NCRP, 1996):

- Allowing the radionuclide to decay in storage for a minimum of 10 half-lives provided that the radioactivity of decayed waste is demonstrated to be indistinguishable from background levels. The USNRC (10CFR51) regulations limit this option to isotopes with a half-life less than 120 days. Decay in storage is a common waste disposal method used in research laboratories that use low quantities of ^{32}P. After storage for the specified time, all radioactive labels must be removed or obliterated prior to disposal as normal trash and appropriate records must be maintained.

- Specific wastes that are generated in research laboratories containing ≤0.05 μCi (1.85 kBq) of ^{3}H or ^{14}C per

CHEMISTRY AND SOURCES 133

Table 7.4.3 *Some Commonly Used Radionuclides in Nuclear Medicine and Biological Research*

Radioisotope	Half-Life	Major Emissions (MeV)
^{3}H	12.28 years	Beta-max: 0.186
^{11}C	20.48 min	Positron: 0.960
^{13}N	9.97 min	Positron: 1.199
^{14}C	5,730 years	Beta-max: 0.156
^{15}O	122.24 sec	Positron: 1.732 Gamma & X-Rays: 0.511
^{18}F	109.74 min	Positron: 0.636 Gamma & X-Rays: 0.511
^{32}P	14.29 days	Beta-max: 1.710
^{51}Cr	27.704 days	Electron: 0.00438 0.000470 Gamma & X-Rays: 0.00495
^{57}Co	270.9 days	Electron: 0.000670 0.00562 Gamma & X-Rays: 0.122
^{58}Co	70.8 days	Positron: 0.475 Electron: 0.000670 Gamma & X-Rays: 0.811
^{67}Ga	3.261 days	Electron: 0.000990 0.00753 Gamma & X-Rays: 0.0933 0.00864 0.185 0.300
^{75}Se	119.78 days	Electron: 0.00124 0.00911 Gamma & X-Rays: 0.136 0.265
^{85}Sr	64.84 days	Electron: 0.00168 0.0114 Gamma & X-Rays: 0.514 0.0134
^{99}Mo	66.02 hours	Beta-max: 1.214 0.436
^{99m}Tc	6.02 hours	Electron: 0.00162 0.00210 Gamma & X-Rays: 0.141
^{111}In	2.83 days	Electron: 0.00272 0.00210 Gamma & X-Rays: 0.245 0.171
^{123}I	13.13 hours	Electron: 0.00319 Gamma & X-Rays: 0.159
^{125}I	60.14 days	Electron: 0.00319 0.00368 Gamma & X-Rays: 0.0272 0.0275
^{129}I	1.57×10^{7} years	Beta-max: 0.152
^{131}I	8.04 days	Beta-max: 0.606 Gamma & X-Rays: 0.365
^{133}Xe	5.245 days	Beta-max: 0.346 Electron: 0.0450 0.00355 Gamma & X-Rays: 0.0810 0.0310
^{201}Tl	73.06 hours	Electron: 0.00760 0.0843 Gamma & X-Rays: 0.00999 0.0708

Sources: National Council on Radiation Protection and Measurements (NCRP) (1996); Shleien (1992, Table 8.14).

gram of used LSC medium; and similar concentrations of ^{3}H or ^{14}C per gram of animal tissue, averaged over the weight of the entire animal, may be disposed as if it were not radioactive. A USNRC licensee may not dispose of such tissue in a manner that would permit its use either as food for humans or as animal feed, and records must be maintained (10CFR20.2005).

- The radioactive material is returned to the vendor for disposal. This method is used often for disposal of ^{99}Mo/^{99m}Tc generators.
- The waste is removed to a licensed low-level waste disposal facility.
- Disposal of patient excreta and certain liquid waste forms in a sanitary sewer provided the radioactive levels are below limits specified in federal regulations.

Additional information regarding radionuclide formulations for specific medical or biological research applications may be found in specialized textbooks and reports issued by the National Council on Radiation Protection and Measurements and other organizations (NCRP, 1982, 1989, 1991; ICRP, 1988; WHO, 1976).

7.4.5.5 Industrial Applications

Radioactive sources are used in a variety of geophysical applications to characterize subsurface formations. These well logging methods include radiation measurements of geologic stratigraphy and its porosity. Passive measurements of natural gamma emissions are used to identify types of strata; clays and shales contain ^{40}K while sandstones are associated with natural uranium and thorium emissions (Driscoll, 1986).

Active gamma sources are used to develop gamma-gamma logs. A probe containing a strong gamma source, such as ^{137}Cs or ^{60}Co, and a detector located approximately 15 inches from the source and shielded from it are introduced into a borehole. Some of the source gamma rays enter the formation and are scattered back to the detector. Because the intensity of back-scatter gamma rays depends on the electron density of the formation, the formation density can be estimated and porosity can be inferred (Driscoll, 1986).

Neutron sources are used similarly in formations saturated with water or hydrocarbons such that neutrons entering the formation are scattered, lose energy, and then are either captured or returned and recorded in the detector. Because hydrogen has a high cross section (probability) for neutron capture, the amount of energy loss in the formation can be used to estimate the amount of hydrogen in the formation and infer its porosity (Driscoll, 1986 and references cited therein). Uses of radiation sources for well logging applications are regulated by the USNRC (10CFR30, 10CFR34, and 10CFR39). These regulations specify license requirements for use of sealed sources and their maintenance, storage, and disposal. Radionuclides used in a variety of industrial applications are shown in Table 7.4.4 (Shleien, 1992).

A variety of tracer compounds can be used to measure the velocity and dispersion of groundwater movement in saturated strata and the measurements can be used to calibrate computer simulation models. Regulations that apply to use of radioactive tracers and markers (10CFR39.45 and 10CFR39.47) strictly forbid injection of licensed material into FW aquifers unless specifically authorized to do so by the USNRC. The quantities of radioactive material that are exempt from USNRC licensing requirements and that may be used as markers in wells are specified in 10CFR30.71.

7.4.5.6 Naturally Occurring and Accelerator-Produced Radioactive Material

Extraction industries, such as mining or oil and gas production, necessarily handle very large quantities of natural rock and groundwater resources that contain background levels of radionuclides (Figures 7.4.1–7.4.3). Although the products of these industries are not radioactive, their processes can generate dispersed radioactivity at relatively low levels or can accumulate higher concentrations in their equipment and waste streams. Such sources are described in the literature as naturally occurring radioactive material (NORM). The radioactivity concentrations generated in NORM wastes depend on the radiological characteristics of the geologic formation and the characteristics of the production process (USEPA, 1993).

Oil production generates NORM wastes as ^{226}Ra and ^{224}Ra isotopes and their daughters, which form scale and sludge deposits in pipes and process equipment, such as oil–water separators. The radium co-precipitates with barium, strontium, and calcium as sulfates, silicates, and carbonates. These deposits reach an average concentration

Table 7.4.4 *Typical Uses of Radionuclides in Industry*

Application	Radionuclides	Typical Source Strengths
Encapsulated Sources		
Industrial radiography	^{192}Ir, ^{137}Cs, ^{170}Tm, ^{60}Co	10–100 Ci (0.4-4 TBq)
Borehole logging	^{137}Cs, ^{60}Co	10 mCi–2 Ci (~0.4–70 GBq)
	Pu-Be, Am-Be neutron sources	50 mCi–20 Ci (~1.9–700 GBq)
	^{252}Cf neutron source	100 μCi (~4 MBq)
Radiation gauges, automatic weighing equipment	^{90}Sr, ^{147}Pm, ^{144}Ce, ^{137}Cs, ^{60}Co	5–200 mCi (~0.2–7 GBq)
Smoke Detectors	^{241}Am	5 μCi (~200 kBq)
Luminous signs	^{3}H	0.5 Ci (~20 GBq)
Mössbauer analysis	^{57}Fe, ^{57}Co	2–50 μCi (~0.4 MBq)
Tracer Applications		
Hydrological tracers	^{3}H, ^{82}Br	1–100 Ci (~4 TBq)
Reservoir engineering	^{85}Kr	200 mCi (~7 GBq)

Source: Shleien (1992, Table 11.1.2).

of 1000–2000 pCi/g (USDOE, 1996). The NORM deposits encountered in natural gas production represent daughters of entrained radon gas that plate out on the interior surfaces of equipment. Because geothermal energy production involves handling large quantities of groundwater and steam, NORM wastes also involve precipitates of radium and daughters as scale and sludge in equipment (USDOE, 1996).

Mining and smelting of metal ores produces NORM waste in process tailings that contain uranium, thorium, radium, and their decay daughters (USDOE, 1996). These wastes are formed in variable amounts that reflect the radiological composition of the geologic formation and can accumulate deposits as high as 100,000 pCi/g (USEPA, 1993). Mining of phosphate deposits to produce phosphate fertilizers generates piping scale and phosphogypsum, a hydrated calcium phosphate slurry that contains ^{226}Ra and daughters (USEPA, 1993). Production of elemental phosphorous from phosphate ore using carbon and silica as catalysts in electric furnaces can generate ferrophosphorus and phosphate slag, which contain concentrated natural radium isotopes (USEPA, 1993).

Combustion of coal mined from geologic formations that contain uranium and thorium produces NORM in waste ash (USEPA, 1993) in addition to arsenic, quartz dust, and polyaromatic hydrocarbons (EPRI, 1986, 1993). The ash consists of aluminum, iron, calcium, and silicon and is collected in the bottom of power-plant combustors and in filters. Most of the waste is generated as fly ash that is entrained with the hot flue gases of the combustion process (USDOE, 1996). Radionuclides in fly ash can be measured using a method involving X-ray fluorescence, gamma spectrometry and alpha spectrometry (following chemical separation of alpha-emitting isotopes). This approach provides a sensitivity of less than 1 pCi/g with ±10% accuracy, 95% of the time (EPRI, 1986). The method includes correction for loss of ^{222}Rn during the analysis. These measurements have shown that radionuclides representing the members of uranium and thorium decay series are concentrated in airborne fly ash at approximately 50 to 100 times above background concentrations, but are well below USNRC limits.

Studies of potential plant power worker exposure to airborne fly ash showed that workers involved in maintenance of the interior of equipment located downstream from the combustion chamber can be exposed to dust having quartz and arsenic concentrations exceeding Occupational Safety and Health Administration (OSHA) Permissible Exposure Limits. Radionuclide concentrations in airborne dust in some plants exceed USNRC health standards (EPRI, 1993). Therefore, worker exposures are controlled by use of respirators and other protective equipment.

Some municipal drinking water treatment systems encounter groundwater that contains natural uranium and daughters that were leached from local geological formations. Water treatment equipment designed to remove suspended and dissolved solids from groundwater, therefore, can accumulate NORM (USEPA, 1993). Because these wastes reflect local geologic deposits, high variability can be expected.

Radionuclides that are generated during the operation of accelerators, such as in research or the production of radiopharmaceuticals, are often included in regulations under the category of Naturally Occurring and Accelerator-Produced Radioactive Material (NARM) waste. In the United States, such NARM waste is not subject to USNRC regulation and is subject to health and safety regulations of the States and other Federal agencies (Federal Register, 1995). Accelerator-produced NARM wastes contain radionuclides, most of which have short half-lives ranging from minutes to days. Because of these short half-lives, accelerator-produced material often can be stored until it has decayed to insignificant levels. Other NARM wastes are shipped for disposal to either the Envirocare Facility near Clive, Utah, or the US Ecology, Inc., site near Richland, Washington (USDOE, 1996).

7.5 HOMELAND SECURITY

Due to the perceived terrorist threat of a detonation of a nuclear weapon or radiological dispersal device (RDD) (Zimmerman and Loeb, 2004), a large number of first responders are being trained in the use of fieldable radionuclide detectors. Many police and fire departments now equip their personnel with portable radiation "pagers." These pagers are small scintillation detectors, sensitive for gamma, or gamma and neutron, radiation. Like the Geiger-Muller detectors, they offer no information as to the identity of the radioactive material but serve to warn the wearer that they are in a higher than normal radiation field.

Many of the alarms received when working with the general public are due to medical treatment cases involving either cancer treatment or diagnostic tests. Portable gamma-ray spectrometers can be used to distinguish between legitimate commerce and potential radioactive or nuclear threats. When working with normal commerce, there are a significant number of commercial items that contain NORM as well as many legitimate sources of man-made radioactive material. Some of the more common items in commerce that contain NORM are shown in Table 7.5.1; some legitimate medical isotopes that may be seen are shown in Table 7.4.3 and some commercial products are shown in Table 7.4.4.

7.6 SIGNATURE COMPOUNDS

Some radioisotopes, such as ^{99}Tc or ^{129}I, are often associated with anthropogenic activities, but can be produced naturally in unique circumstances (Curtis et al., 1994). Several other radioisotopes can indicate anthropogenic sources. Natural uranium has three naturally occurring isotopes that occur in well-characterized ratios (see Section 7.4.1). Slight differences can occur in the ^{234}U content due to its greater mobility in the aqueous environment. If ^{236}U is found to be present (when measured by a mass spectrometric technique), the result indicates an anthropogenic source of uranium. The half-life of ^{236}U is 10^7 years, thus any primordial material has decayed away since the formation of the earth. There is no production source of ^{236}U except in nuclear reactors. The well-characterized phenomenon of natural reactors (Section 7.4.1) cannot occur in geologic formations today.

Plutonium also occurs only in anthropogenic sources. Plutonium has been distributed worldwide by nuclear weapons testing, primarily in the 1960s. Due to the nuclear reactions that occur during detonation, the ^{240}Pu to ^{239}Pu ratio in worldwide fallout is ~0.16 while the weapons grade source material has a ratio of ~0.06 (see Section 7.4.4). As the burn-up of nuclear fuel increases, the ^{240}Pu to ^{239}Pu ratio also increases so that higher plutonium isotopic ratios may indicate high burn-up fuel, such as found in naval nuclear reactors. Thus, the plutonium isotopic ratio can be used to distinguish the source of plutonium found in the environment.

Many other radioisotopes have been released to the atmosphere and distributed worldwide, by nuclear weapons testing and reactor accidents such as at Chernobyl. When assessing low-level environmental forensic analyses, these sources must be taken into account as well as local potential

Table 7.5.1 Content of Naturally Occurring Radioactive Materials (NORM) in Common Commerce Products

Material	Uranium ppm	Uranium mBq/g(pCi/g)	Thorium ppm	Thorium mBq/g(pCi/g)	Potassium ppm	Potassium mBq/g(pCi/g)
Wood					11.3	3330 (90)
Marijuana						1300–1850 (35–50)
Soil						185–1850 (5–50)
Granite	4.7	63 (1.7)	2	8 (0.22)	4.0	1184 (32)
Hay						740 (20)
Clay Brick	8.2	111 (3)	10.8	44 (1.2)	2.3	666 (18)
Cocoa (Dry Powder)						481 (13)
Sandstone	0.45	6 (0.2)	1.7	7 (0.19)	1.4	4.14 (11.2)
Sandstone concrete	0.8	11 (0.3)	2.1	8.5 (0.23)	1.3	385 (10.4)
Cement	3.4	46 (1.2)	5.1	21 (0.67)	0.8	237 (6.4)
Cotton, Pure						152 (4.1)
Natural gypsium	1.1	15 (0.4)	1.8	7.4 (0.2)	0.5	148 (4)
Banana						118 (3.2)
Beef, Lean						104 (2.8)
Limestone, concrete	31 (0.8)	14 (0.4)	3	12 (0.32)	0.3	89 (2.4)
Dry wallboard						89 (2.4)
Portland Cement						74 (2)
Pyrex						63 (1.7)
Milk						48 (1.29)
Bread						33 (0.9)
Rubber						30 (0.8)
By-product gypsum	13.7	186 (5.0)	16.1	66 (1.78)	0.02	5.9 (0.2)
Steel						15 (0.4)
Aluminum						11 (0.3)
Plastic						0.4 (0.01)

www.nal.usda.gov
ppm mass concentration in parts per million

sources of radioactivity. A monitoring program performed at the US Department of Energy's Hanford Site in eastern Washington state followed the activity of many natural, cosmically produced, weapon produced, and (accidentally released) reactor produced radioisotopes over a period of three decades (Perkins *et al.*, 1990).

7.7 OTHER FORENSIC APPLICATIONS

Forensic approaches to detect and investigate illicit use of radioactive materials are similar to those described above for environmental applications, but are applied to very small samples, often to single particles (IAEA, 2003b,c). Administrative requirements and quality assurance protocols needed to support the use of modern analytical techniques for forensic analyses have been described (Swindle *et al.*, 2003).

Samples of radionuclides found on plastic and cloth stolen from a nuclear installation were analyzed to determine the ratios of uranium and plutonium isotopes to identify their source (Mayer *et al.*, 2003). Similar studies of uranium oxides by ICP-MS have measured ^{18}O to ^{16}O isotope ratios to associate the uranium oxide sample with the producing mine (Wallenius *et al.*, 2003).

Application of very sensitive techniques, such as AMS, can detect femtogram quantities of selected isotopes (Hotchkis *et al.*, 2003) and have been used to analyze wipe samples from a building suspected to contain illicit radioactivity. Such sensitive techniques can be used also to analyze air filter or water samples taken from the suspected source. Methods to apply secondary ion mass spectrometry (SIMS) to single submicron particles identified by scanning electron microscopy (SEM) have been demonstrated using round-robin interlaboratory tests to be suitable for forensics investigations (Admon *et al.*, 2003).

7.8 SUMMARY

The job of a nuclear forensic analyst is to distinguish anthropogenic activities from natural processes, which can redistribute both natural and anthropogenic radionuclides. Instruments available to measure radionuclides in the environment extend from simple count rate meters, with reasonable sensitivity but limited identification capabilities, to highly sensitive and highly specialized mass spectrometers. Natural background radioactivity is the greatest source of bias and interference in environmental radiochemistry analyses.

Many anthropogenic nuclides have been distributed worldwide due to fallout or nuclear accidents. Similar to stable isotope environmental forensic analyses, in which δ^{13}C, δ^{15}N, or δ^{18}O is used to determine the source of a material, it is often more appropriate to measure radioactive isotope ratios than absolute concentrations. Two examples include the ^{240}Pu/^{239}Pu isotope ratio, a ratio of $\sim$0.16 is indicative of fallout and a ratio of $\sim$0.06 is indicative of weapons production. Similarly, the ^{129}I/^{127}I ratio, a ratio of $\sim$10^{-12} indicates fallout while a ratio of $\sim$10^{-6} to 10^{-9} indicates releases from fuel reprocessing material. In some cases, the presence of a nuclide itself can indicate an anthropogenic source, such as ^{236}U, which is only produced in nuclear reactors.

Because radioactive isotopes of elements form the same chemical compounds as do stable isotopes, their transport through the environment is not affected by their radioactivity. This property has been used to trace planned or unplanned releases of radionuclides in the environment and elucidate geochemical, atmospheric, oceanic, and biological processes and to improve simulation models in these fields. These studies include many results that apply to forensic studies of metals. For example, plutonium, cesium, cobalt, and strontium isotopes in the environment are predominantly associated with soil and sediments and have shown irreversible or slowly reversible exchange with water resulting in slow migration over decades following a release. Partitioning of metals to water is somewhat greater in SW than FW. Ionic forms in water are often associated with organic colloids, which provide an important dispersion mechanism.

Radionuclides have precisely known decay rates, which provide a unique opportunity to study the time dimension in forensic studies. For example, the short-lived ^{134}Cs isotope is a nuclear fission product that was dispersed worldwide from atmospheric weapons testing in the 1960s. Because ^{134}Cs has since decayed away, any future measurement is of a more recent source.

One effective approach to the evaluation of radionuclide releases includes a review of all applicable regulations, selection of contaminants of interest, documentation of exposure assumptions, and identification of contaminants of greatest concern. Forensic studies of other materials can follow approaches proven effective for radioactively contaminated areas.

7.9 ACKNOWLEDGEMENTS

The authors are indebted to Dr. Philip Krey, Bruce Gallaher, and Dr. Thomas Buhl for information and helpful discussion and to Dr. Kenneth Inn for reviewing this chapter.

REFERENCES

Action Levels and Standards Framework Working Group (ALF). (1996). Action Levels for Radionuclides in Soils for the Rocky Flats Cleanup Agreement, Final, US Department of Energy, US Environmental Protection Agency, Colorado Department of Public Health and Environment, October 31.

Admon, U., Donohue, D., Aigner, H., Tamborini, G., Bildstein, O., and Betti, M. (2003). In "Proceedings of an International Conference", *Karlsruhe*, 21–23 October 2002, pp. 171–179, International Atomic Energy Agency, Vienna, Austria.

American Society for Testing and Materials (ASTM). (2005a). International, Method C1310, ASTM International, West Conshohocken PA.

American Society for Testing and Materials (ASTM). (2005b). Method C1345, ASTM International, West Conshohocken PA.

Anderson, T. J. and Walker, R. L. (1980). *Anal. Chem.*, 52: 709–713.

Ansoborlo, E., Guilmette, R. A., Hoover, M. D., Chazel, V., Houpert, P., and Henge-Napoli, M. H. (1998). *Radiat. Prot. Dosim.* 79 (1–4): 33–37.

Ansoborlo, E., Chazel, V. P., Henge-Napoli, M. H., Pihet, P., Rannou, A., Bailey, M. R., and Stradling, N. (2002). *Health Phys.*, 82 (3): 279–289.

Arctic Nuclear Waste Assessment Program (ANWAP). (1997). Radionuclides in the Arctic Seas from the Former Soviet Union: Potential Health and Ecological Risks, Layton, D., Edson, R., Varela, M., Napier, B., eds, Arctic Nuclear Waste Assessment Program, US Office of Naval Research, Washington, D. C.

Argonne National Laboratory (ANL). (1993). Manual for Implementing Residual Radioactive Material Guidelines Using RESRAD, Version 5.0, Environmental Assessment and Information Sciences Division, Argonne National Laboratory, ANL/EAD/LD-2, September.

Beals, D. M. (1996). *J. Radioanal. Nucl. Chem.*, Articles, 204: 253–263.

Beals, D. M. and Dunn, D. L. (1993). In "Proceedings of the American Nuclear Society Topical Meeting on Environmental Transport and Dosimetry," pp. 64–65 Charleston, SC, September.

Beals, D. M. and Hayes, D. W. (1995). *Science of the Total Environment*, 173/174, 01–115.

Bhat, R. K. (2000). Review of Transuranics in Depleted Uranium Armours, Tank-Automotive and Armaments Command and Army Material Command, Project Officer Bhat, R. K., Official Communication (and associated annex documentation) to Commander US Tank and Automotive Command (AMSTA-CM-PS), Warren, Michigan 48397, January 19.

Bréchignac, F., Moberg, L., and Suomela, M. (2000). Long-term Environmental Behavior of Radionuclides, CEC-IPSN Association final report.

Brookins, D. G. (1988). *Eh-pH Diagrams for Geochemistry*, Springer-Verlag, Berlin, Germany.

Cadieux, J. R., Kantelo, M. V., and Sigg, R. A. (1993). In "Proceedings of the American Nuclear Society Topical Meeting on Environmental Transport and Dosimetry," pp. 61–63, Charleston, SC, September.

Cember, H. (1989). *Introduction to Health Physics*, 2nd edition, Pergamon Press, New York.

Chen, J. H., Edwards, R. L., and Wasserburg, G. J. (1986). *Earth and Planetary Science Letters*, 80: 241–251.

Cordfunke, E. H. P. (1969). *The Chemistry of Uranium*, Elsevier, New York.

Cothern, C. R. and Lappenbusch, W. L. (1983). *Health Phys.* 45: 89–99.

Cowan, G. A. (1976). *Sci. Am.* 235 (1): 36–47.

Cowan, G. A. (1978). In "International Atomic Energy Agency, Technical Committee Meeting Proceedings," Paris, 1977, Publication 475, Vienna, p. 749.

Crain, J. S. and Gallimore, D. L. (1992). *Applied Spectroscopy*, 46, p. 547.

Curtis, D. B., Fabryka-Martin, J., Dixon, P., Aguilar, R., Rokop, D., Cramer, J. (1994). *Radiochimica Acta*, 66/67: 551–557.

Dai, M., Kelley, J. M., and Buesseler, K.O. (2002). *Environmental Science and Technology*, 36: 3690–3699.

Doshi, G. R., Joshi, S. N., and Pillai, K. C. (1991). *J. Radioanal. Nucl. Chem., Letters*, 155 (2): 115–127.

Driscoll, F. G. (1986). *Groundwater and Wells*, 2nd edition, Johnson Division, St. Paul, Minnesota.

Efurd, D. W., Poths, H. D., Rokop, D. J., Roensch, F. R., and Olsen, R. L. (1995). Isotopic Fingerprinting of Plutonium in Surface Soil Samples Collected in Colorado, Los Alamos National Laboratory, Los Alamos, New Mexico, RF91-06.

Ehmann, W. D. and Vance, D. E. (1991). *Radiochemistry and Nuclear Methods of Analysis*, John Wiley & Sons, Inc. New York.

Eichholz, G. G. (1977), *Environmental Aspects of Nuclear Power*, Ann Arbor Science Publishers, Ann Arbor, Michigan.

Eidson, A. F. (1994). *Health Phys.* 67: 1–14.

Eisenbud, M. (1987). *Environmental Radioactivity*, Academic Press.

Elder, J. C. and Tinkle, M. C. (1980). Oxidation of DU Penetrators and Aerosol Dispersal at High Temperatures, Report LA-8610-MS. Los Alamos National Laboratory, Los Alamos, New Mexico.

Electric Power Research Institute (EPRI). (1986). Techniques for Chemical Analysis of Radionuclides in Fly Ash: An Evaluation, EA-4728, Electric Power Research Institute, Palo Alto, California, May.

Electric Power Research Institute (EPRI). (1993). Fly Ash Exposure in Coal-Fired Power Plants, TR-102576,

Electric Power Research Institute, Palo Alto, California, August.

Elmore, D., *et al.* (1980). *Nature*, 286: 138–140.

Federal Register. (1995). Vol. 60, No. 140, pp. 37556–37565, July.

Ferri, D., Grenthe, I. and Salvatore, F. (1981). *Acta Chem. Scand. A*, 35: 165–168.

Fleischer, R. L. (1980). *Science*, 207: 979–981.

Glissmeyer, J. A. and Mishima, J. (1979). Characterization of Airborne Uranium from Test Firings of XM774 Ammunition, Pacific Northwest Laboratory, PNL-2944, Richland, WA.

Gummer, W. K., *et al.* (1980). COSMOS 954: The Occurrence and Nature of Recovered Debris, Minister of Supply and Services, Canada.

Guo, L. and Santschi, P. H. (1997). *Mar. Chem.*, 59: 1–15.

Hagerman, R. and Roth, E. (1978). *Radiochim. Acta*, 25, 241–247.

Haggard, D. L., Herrington, W. M., Hooker, C. D., Mishima, J., Parkhurst, M. A., Scherplez, R. I., Sigalla, L. A., and Hadlock, D. E. (1985). Hazard Classification Test of the 120 mm APFSDS-T, M829 Cartridge: Metal Shipping Container, PNL-5298, Pacific Northwest Laboratory, Richland, Washington.

Harley, N. H., Foulkes, E. C., Hilborne, L. E., Hudson, A., and Anthony, C. R. (1999). *A Review of the Scientific Literature as it Pertains to Gulf War Illnesses*, Vol. 7 Depleted Uranium, RAND Corporation.

Health Council of the Netherlands (HCN). (2001). *Health Risks of Exposure to Depleted Uranium, The Hague: Health Council of the Netherlands*, Publication No. 2001/13E, Revised Version 5, February 5.

Hidaka, H., Hollinger, P., and Matsuda, A. (1993). *Earth and Planet. Sci. Lett.*, 114: 391–396.

Hidaka, H. Sugiyana, K., Ebihra, M., and Hollinger, P. (1994). *Earth and Planet. Sci. Lett.*, 122: 173–182.

Hodgson, A., Moody, J. C., Stradling, G. N., Bailey, M. R., and Birchall, A. (2000). Application of the ICRP Human Respiratory Tract Model to Uranium Compounds Produced During the Manufacture of Nuclear Fuel, NRPB-M1156, Chilton, UK, National Radiological Protection Board.

Hoffman, F. O. (1991). The Use of Chernobyl Fallout Data to Test Model Predictions of the Transfer of ^{131}I and ^{137}Cs from the Atmosphere Through Agricultural Food Chains, Report CONF-910434-7, Oak Ridge National Lab., Tennessee.

Hoffman, J. D. and Buttleman, K. (1994). National Geochemical Data Base: Natural Uranium Resource Evaluation Data for the Conterminous United States, Digital Data Series DDS-18-A, US Geological Survey, Washington, D. C.

Hofstetter, K. J. and Beals, D. M. (2002). In "Proceedings of the Waste Management Conference 2002," Tucson AZ.

Hogan, O. H., Zigman, P. E., and Mackin, J. L. (1964). Beta Spectra, USDRL-TR-802, Atomic Energy Commission, Washington, D. C.

Hollenbach, M., Grohs, J., Kroft, M., and Mamich, S. (1995). In *Applications of Inductively Coupled Plasma-Mass Spectrometry to Radionuclide Determinations*, R. Morrow and J. Crain, eds, ASTM Publication STP 1291, p. 99.

Hooker, C. D., Hadlock, D. E., Mishima, J., and Gilchrist, R. L. (1983). Hazard Classification Test of the 120 mm APFSDS-T, XM829, PNL-4459, Pacific Northwest Laboratory, Richland, Washington.

Hotchkis, M., Child, D., and Tuniz, C. (2003). IAEA-CN-98/17. In "Proceedings of an International Conference," *Karlsruhe*, 21–23 October 2002, International Atomic Energy Agency, Vienna, Austria, pp. 109–114.

Ibrahim, S. A., Webb, S. B., and Wicker, F. W. (1997). *Health Phys.*, 72: 42–48.

International Atomic Energy Agency (IAEA). (1975). *Le Phenomenon d'Oklo, Symposium Proceedings*, Libreville, Gabon, 1975, IAEA Publication 405, Vienna, 644.

International Atomic Energy Agency (IAEA). (1978). *Technical Committee Meeting Proceedings*, Paris, 1977, Publication 475, Vienna, p. 749.

International Atomic Energy Agency (IAEA). (1999). In "Report for the International Arctic Seas Assessment Project (IASAP)," International Atomic Energy Agency, Austria, pp. 1–71.

International Atomic Energy Agency (IAEA). (2002). *Monitoring and Surveillance of Residues from the Mining and Milling of Uranium and Thorium*, Safety Reports Series No. 27, Vienna, Austria.

International Atomic Energy Agency (IAEA). (2003a) The Long-term Stabilisation of Uranium Mill Tailings, Final Report, Vienna, Austria.

International Atomic Energy Agency (IAEA). (2003b). In "Proceedings of an International Conference", Karlsruhe, 21–23 October 2002, Vienna, Austria.

International Atomic Energy Agency (IAEA). (2003c). *Safeguards Techniques and Equipment*, 2003 edition, International Nuclear Verification Series No. 1 (Revised), Vienna, Austria.

International Commission on Radiation Units and Measurements (ICRU). (1993). Quantities and Units in Protection Dosimetry, ICRU Report 51, International Commission on Radiation Units and Measurements, Bethesda, Maryland.

International Commission on Radiation Units and Measurements (ICRU). (1998). Fundamental Quantities and Units for Ionizing Radiation, ICRU Report 60, International Commission on Radiation Units and Measurements, Bethesda, Maryland.

International Commission on Radiological Protection (ICRP). (1988). *Radiation Dose to Patients from Radiopharmaceuticals*, ICRP Publication 53, Pergamon Press, Oxford.

Johnson, N. L. (1986). Nuclear Power Supplies in Orbit, Teledyne Brown Engineering, Colorado Springs, Colorado, Space Policy ISSN 0265–9646, Vol. 1, August.

Joint Norwegian Russian Expert Group (JNREG). (1997). Sources Contributing to Radioactive Contamination of the Techa River and Areas Surrounding the "Mayak" Production Association, Ural, Russia, Joint Norwegian Russian Expert Group for Investigation of Radioactive Contamination in the Northern Areas.

Kaplan, D. I., Bertsch, P. M., Adriano, D. C., and Miller, W. P. (1993). *Environ. Sci. Technol.*, 27: 1193–1200.

Kaplan, D. I., Hunter, D. B., Bertsch, P. M., Bajt, S., Adriano, D. C., and Miller, W. P, (1994a). *Environ. Sci. Technol.*, 28: 1186–1189.

Kaplan, D. I., Bertsch, P. M., Adriano, D. C., and Orlandini, K. A. (1994b). *Radiochim. Acta*, 66/67: 181–187.

Kilius, L. R., Litherland, A. E., Rucklidge, J. C., and Baba, N. (1992). *Appl. Radiat. Isot.*, 43 (1/2): 279–287.

Krey, P. W. (1976). *Health Phys.*, 30: 209–214.

Krey, P. W. and Krajewski, B. T. (1972). *Plutonium Isotopic Ratios at Rocky Flats, Health and Safety Laboratory*, US AEC, New York, HASL-249.

Ku, T. L., Knauss, K. G., and Mathieu, G. G. (1977). *Deep-Sea Res.*, 24: 1005

Kuroda, P. K. (1956a). *J. Chem. Phys.* 25 (2): 781–782.

Kuroda, P. K. (1956b). *J. Chem. Phys.* 25 (2): 1295–1296.

LaMont, S. P. and Hall, G. (2005). *J. Radioanal. Nucl. Chem.*, in press.

Landa, E. R., Reimnitz, E., Beals, D. M., Pochkowski, J. M., Winn, W. G. and Rigor, I. (1998). *Arctic*, 51: 27–39.

Laniak, G. F., Droppo, J. G., Faillace, E. R., Gnanapragasam, E. K., Mills, W., B., Strenge, D., L., Whelan, G., and Yu, C. (1997). *Risk Analysis*, 17 (2): 203–214.

Lederer, C. M., Hollander, J. M., and Perman, I. (1967). Table le of Isotopes, 6th edition, John Wiley & Sons, Inc., New York.

Lide, D. R., ed. (2001). CRC Handbook of Chemistry and Physics, 81st edition, 2002–2001, CRC Press LLC, Boca Raton, Florida.

Litaor, M. I. (1999). *Health Phys.*, 7: 171–179.

Loss, R. D., Rossman, K. J. R., and DeLaeter, J. R. (1984). *Earth Plan. Sci. Lett.*, 68: 240–248.

Loss, R. D., Rossman, K. J. R., DeLaeter, J. R., Curtis, D. B., Benjamin, T. M., Gancarz, A. L., Maeck, W. J., and Delmore, J. E. (1989). *Gabon, Chem. Geol.*, 76: 71–84.

Lovett, M. B. and Nelson, D. M. (1981). *Techniques for Identifying Transuranic Speciation in Aquatic Environments*; IAEA, Vienna, Austria, pp. 27–35.

Luftig, S. D. and Weinstock, L. (1997). Establishment of Cleanup Levels for CERCLA Sites with Radioactive Contamination, United States Environmental Protection Agency, Office Emergency and Remedial Response and Office of Radiation and Indoor Air, OSWER No. 9200. 4–18, August 22.

Manes, L. and Benedict, U. (1985). In "Actinides—Chemistry and Physical Properties, Structure and Bonding, 59/60," Springer-Verlag, New York, pp. 75–126.

Mann, L. J., Beasley, and T. M. (1994). *Journal of the Idaho Academy of Science*, 30 (2): 75–87.

Matishov, G. G., Matishov, D. G., Namjatov, A. A., Smith, J. N., and Dahle, S. (1999). *J. Environ. Rad.*, 43: 77–88.

Matishov, G. G., Matishov, D. G., Namjatov, A. A., Smith, J. N., Carrikk, J., and Dahle, S. (2000). *J. Environ. Rad.*, 48: 5–21.

Matishov, G. G, Matishov, D. G., Namjatov, A. A., Smith, J. N., Carroll, J., and Dahle, S. (2002). *Environ. Sci. Technol.*, 36: 1919–1922.

Mayer, K., Rasmussen, G., Hild, M., Zuleger, E., Ottmar, H., Abousahl. S., and Hrnecek, E. (2003). IAEA-CN-98/11. In "Proceedings of an International Conference", *Karlsruhe*, 21–23 October 2002, International Atomic Energy Agency, Vienna, Austria, pp. 63–71.

Mckay, H. A. C. (1981). In "Proceedings of the 28th Congress of the International Union of Pure and Applied Chemistry," (Vancouver, BC), Harwell, UK: Atomic Energy Research Establishment; AERE-R-10255, DE86 900833.

Mills, W. B., Cheng, J. J., Droppo, J. G., Faillace, E. R., Gnanapragasam, E. K., Johns, R. A., Laniak, G. F., Lew, C. S., Strenge, D., L., Southerland, J. F., Whelan, G., and Yu, C. (1997). Risk Analysis, 17 (2): 187–201.

Momeni, M. H., Yuan, Y., and Zielen, A. J. (1979). Argonne National Laboratory, NUREG/CR-0553.

National Aeronautics and Space Administration (NASA). (2004). Past Space Nuclear Power System Accidents, NASA Fact Sheet.

National Council on Radiation Protection and Measurements (NCRP). (1975). External Natural Background Radiation in the United States, NCRP Report No. 45, National Council on Radiation Protection and Measurements, 7910 Woodmont Ave., Bethesda, Maryland.

National Council on Radiation Protection and Measurements (NCRP). (1979). Tritium in the Environment, NCRP Report No. 62, National Council on Radiation Protection and Measurements, 7910 Woodmont Ave., Bethesda, Maryland.

National Council on Radiation Protection and Measurements (NCRP). (1982). Nuclear Medicine-Factors Influencing the Choice and Use of Radionuclides in Diagnosis and Therapy, National Council on Protection and Measurements, 7910 Woodmont Ave., Bethesda, Maryland.

National Council on Radiation Protection and Measurements (NCRP). (1984). Exposures from the Uranium Series with Emphasis on Radon and Its Daughters, NCRP Report No. 77, National Council on Radiation Protection and Measurements, 7910 Woodmont Ave., Bethesda, Maryland.

National Council on Radiation Protection and Measurements (NCRP). (1987) Exposure of the Population in the United States and Canada from Natural Background Radiation, NCRP Report No. 94, National Council on Radiation Protection and Measurements, 7910 Woodmont Ave., Bethesda, Maryland.

National Council on Radiation Protection and Measurements (NCRP). (1989). Exposure of the US Population form Diagnostic Medical Radiation, NCRP Report No. 100, National Council on Radiation Protection and Measurements, 7910 Woodmont Ave., Bethesda, Maryland.

National Council on Radiation Protection and Measurements (NCRP). (1991). Developing Radiation Emergency Plans for Academic, Medical or Industrial Facilities, NCRP Report No. 111, National Council on Radiation Protection and Measurements, 7910 Woodmont Ave., Bethesda, Maryland.

National Council on Radiation Protection and Measurements (NCRP). (1996). Sources and Magnitude of Occupational and Public Exposures from Nuclear Medicine Procedures, NCRP Report No. 124, National Council on Radiation Protection and Measurements, 7910 Woodmont Ave., Bethesda, Maryland.

Nelson, D. M. and Lovett, M. B. (1978). *Nature*, 599–601.

New Mexico Conference of Churches. (1983). Report of Rio Puerto Review Team Regarding Church Rock Uranium Mill Tailings Spill, July 1979. (The summary report issued by the New Mexico Health and Environment Department (NMHED, 1983) notes that the Puerto River that contained the spill should not be confused with the Rio Puerto, which is a tributary to the Rio Grande.)

New Mexico Health and Environment Department (NMHED). (1983). The Church Rock Uranium Mill Tailings Spill: A Health and Environmental Assessment Summary Report, Millard, J. Gallaher, B., Baggett, D., and Cary, S. New Mexico Environmental Improvement Division, Santa Fe, New Mexico, September.

Nuclear Energy Agency (NEA). (1989). The Influence of Seasonal Conditions on the Radiological Consequences of a Nuclear Accident, Proceedings of an NEA Workshop, Paris, September 1988, Organization for Economic Cooperation and Development (OECD) Publications, Paris, France.

Nuclear Energy Agency (NEA). (1996). Agricultural Issues Associated with Nuclear Emergencies, Proceedings of an NEA Workshop, June 1995, 1996. Organization for Economic Cooperation and Development (OECD) Publications, Paris, France.

Nuclear Energy Agency (NEA) (2003a). Annual Report, Organization for Economic Cooperation and Development (OECD) Publications, Paris, France.

Nuclear Energy Agency (NEA) (2003b). Chernobyl, Assessment of Radiological and Health Impacts, 2002 Update of Chernobyl: Ten Years On, Organization for Economic Cooperation and Development (OECD) Publications, Paris, France.

Office of the Special Assistant to the Deputy Secretary of Defense for Gulf War Illnesses, DoD (OSAGWI, 2000). Exposure Investigation Report, Depleted Uranium in the Gulf (II), December, available online at www.gulflink.osd.mil in the Environmental Reports Section.

Perkins, R. W., Robertson, D. E., Thomas, C. W., Young, J. A. (1990). In "Proceedings of an International Symposium on Environmental Contamination Following a Major Nuclear Accident," Vol. 1, October 1989, International Atomic Energy Agency, Vienna.

Perry, G. E. (1978). *The Royal Air Force Quarterly*, 18: (1): 60–67.

Petullo, C. F., Jeter, J., Cardenas, D., Stimmel, R., Dobb, D. E., and Hillman, D. C. (1994). In "Proceedings of the Waste Management Conference 1994," Tucson AZ, February 1994.

Power Reactor Information System (PRIS). (2004). International Atomic Energy Agency (IAEA), accessed online www.iaea.org/pris.

Quigley, M. S., Santschi, P. H., Hung, C.-C., Guo, L., and Honeyman, B. D. (2002). *Limnol. Oceanogr.* 47: 367–377.

Raisbeck, G. M., Yiou, Z. Q., Kilius, L. R., and Dahlgaard, H. (1993). In "Environmental Radioactivity in the Arctic," Strand, P., Holm, E., eds, Østerås, Norway, pp. 125–128.

Reese, R. T. and Vick, C. P. (1983). *Journal of the British Interplanetary Society*, 36: 457–462.

Rocky Flats Cleanup Agreement (RFCA). (1996). Joint Agreement Between the US Department of Energy, the US Environmental Protection Agency, and the Colorado Department of Public Health and Environment and the State of Colorado, July 19.

Rokop, D. J., Norman, J., Schroeder, C., and Wolfsberg, K. (1990). *Anal. Chem.*, 62: 1271.

The Royal Society. (2001). *The Health Hazards of Depleted Uranium Munitions*, Part I, The Royal Society, Document 6/02, London, May.

The Royal Society. (2002). *The Health Hazards of Depleted Uranium Munitions*, Part II, The Royal Society, Document 6/02, London, March.

Santschi, P. H., Schink, D. R., Corapcioglu, O., S. Oktay-Marshall, Fehn, U., and Sharma, P. (1996). *Deep-Sea Res.*, 43 (2): 259–265.

Santschi, P. H., Roberts, K. A., and Guo, L. (2002). *Environmental Science and Technology*, 36: 3711–3719.

Scripsick, R. C., Crist, K. C., Tillery, M. I., Soderholm, S. C., and Rothenberg, S. J. (1985). Preliminary Study of Uranium Oxide Dissolution in Simulated Lung Fluid, LA-10268-MS, Los Alamos National Laboratory, Los Alamos, New Mexico.

Shleien, B., ed. (1992). *The Health Physics and Radiological Health Handbook*, Revised edition, Scinta, Inc., Silver Spring, MD.

Shuey, C. and Taylor, L. S. (1982). Unresolved Issues and Recommendations for Resolution Concerning the Church Rock, New Mexico Uranium Tailings Dam Failure of July 16, 1979, Southwest Research and Information Center, Albuquerque, NM, August.

Sill, C. W. (1995). *Health Phy.*, 69 (1): 21–33.

Smith, J. N., Ellis, K. M., Kilius, L. R., Naes, K., Dahle, S., and Matishov, D. (1995). *Deep-Sea Res.*, Part II, 42(6): 1471–1493.

Smith, J. N., Ellis, K. M., and Kilius, L. R. (1998). *Deep-Sea Res.*, Part I, 45 (6): 959–984.

Smith, J. T., Comans, R. N. J., Beresford, N. A. Wright, S. M., Howard, B. J., and Camplin, W. C. (2000). *Nature*, 405: 141.

Standring, W. J. F., Oughton, D. H., and Salbu, B. (2002). *Environ. Sci. Technol*, 36: 2330–2337.

Stannard, J. N. (1988). *Radioactivity and Health, A History*, DOE/RL/01830-T59, (DE88013791), R. W. Baalman, Jr., ed., Pacific Northwest Laboratory, Richland, Washington.

Swindle, Jr., D. W., Perrin, R. E., Goldberg, S. A., and Cappis, J. (2003). IAEA-CN-98/29. In "Proceedings of an International Conference," *Karlsruhe*, 21–23 October

2002, International Atomic Energy Agency, Vienna, Austria, pp. 207–214.

Totemeier, T. C. (1995). A Review of the Corrosion and Pyrophoricity Behavior of Uranium and Plutonium, ANL/ED/95-2, Argonne National Laboratory, Illinois.

Twomey, T. R. and Keyser, R. M. (2004). In "Proceedings of the World Customs Organization Conference," Geneva Switzerland, September 2004.

United Nations Scientific Committee on the Effects of Atomic Radiation (UNSCEAR). (1977). Report of the United Nations Scientific Committee on the Effects of Atomic Radiation, Official Records of the General Assembly, 32nd Session, Supplement 40 (United Nations, New York).

United Nations Scientific Committee on the Effects of Atomic Radiation (UNSCEAR). (1982). Ionizing Radiation: Sources and Biological Effects, 1982 UNSCEAR Report to the General Assembly, with annexes (United Nations, New York).

United Nations Scientific Committee on the Effects of Atomic Radiation (UNSCEAR). (1988). Sources, Effects and Risks of Ionizing Radiation, 1988 UNSCEAR Report to the General Assembly, with annexes (United Nations, New York).

United Nations Scientific Committee on the Effects of Atomic Radiation (UNSCEAR). (1993). Sources and Effects of Ionizing Radiation, 1993 UNSCEAR Report to the General Assembly, with annexes (United Nations, New York).

United Nations Scientific Committee on the Effects of Atomic Radiation (UNSCEAR). (2000). Sources and Effects of Ionizing Radiation, 2000 UNSCEAR Report to the General Assembly, with annexes (United Nations, New York).

US Army Center for Health Promotion and Preventative Medicine (USACHPPM). (2000). Depleted Uranium, Human Exposure Assessment and Health Risk Characterization, Health Risk Assessment Consultation No. 26-MF-7555-00D, Aberdeen Maryland.

US Army Environmental Policy Institute (AEPI). (1995). Health and Environmental Consequences of DU Use in the US Army, US Army Environmental Policy Institute Technical Report.

US Department of Energy (USDOE). (1979). Procedure AS-5, RESL/ID.

US Department of Energy (USDOE). (1995). Geochemical Characterization of Background Surface Soils: Background Soils Characterization Program, Golden Colorado, EG&G, Rocky Flats Plant.

US Department of Energy (USDOE). (1996). Chapter 7. Naturally Occurring and Accelerator-Produced Radioactive Material, Integrated Data Base Report—1996: US Spent Nuclear Fuel and Radioactive Waste Inventories, Projections, and Characteristics, US Department of Energy, Office of Environmental Management, Washington, D. C.)

US Environmental Protection Agency (USEPA). (1980). Prescribed Procedures for Measurement of Radioactivity in Drinking Water, EPA Method 908.0, EPA-600/4-80-032, Washington, D. C.

US Environmental Protection Agency (USEPA). (1982). Isotopic Determination of Plutonium, Uranium, and Thorium in Water, Soil, Air and Biological Tissue, EMSL/LV, Washington, D. C. US.

US Environmental Protection Agency (USEPA). (1993). Diffuse NORM Wastes—Waste Characterization and Preliminary Risk Assessment, Office of Radiation and Indoor Air, RAE-9232/1-2, SC&A, Inc., and Rogers & Associates Engineering Corporation, Salt Lake City, Utah, May.

US Environmental Protection Agency (USEPA). (2000). *Multi-Agency Radiation Survey and Site Investigation Manual (MARSSIM)*, Revision 1 (August 2000) including the June 2001 updates, NTIS PB97-117659.

US Environmental Protection Agency (USEPA). (2004). *Multi-Agency Radiological Laboratory Analytical Protocols Manual*, EPA 402-B-04-001A, NUREG-1576, NTIS PB2004-105421.

US Nuclear Regulatory Commission (USNRC). (1975). Reactor Safety Study, WASH-1400, National Technical Information Service, Springfield, Virginia, October.

US Nuclear Regulatory Commission (USNRC). (1979). Population Dose and Health Impact of the Accident at the Three Mile Island Nuclear Station, NUREG-0558, Washington, D. C., May.

US Nuclear Regulatory Commission (USNRC). (1980). Three Mile Island—A Report to the Commissioners and the Public, Vols 1–2, NUREG/CR-1250, Washington, D. C., January.

US Nuclear Regulatory Commission (USNRC). (1981). The Development of Severe Reactor Accident Source Terms: 1957–1981, NUREG-0773, Washington, D. C., November.

US Nuclear Regulatory Commission (USNRC). (1983). Radiological Assessment, NUREG-3332, Till, J. E. and Meyer, H. R., eds, Washington, D. C.

US Nuclear Regulatory Commission (USNRC). (1986). Reassessment of the Technical Base for Estimating Source Terms, NUREG-0956, Washington, D. C., July.

US Nuclear Regulatory Commission (USNRC). (1996). Generic Environmental Impact Statement for License Renewal of Nuclear Plants, NUREG-1437, Vol. 1, Washington, D. C.

US Nuclear Regulatory Commission (USNRC). (2004). The Accident at Three Mile Island, Fact Sheet, Office of Public Affairs, Washington, D. C., March.

US Public Health Service (PHS). (1970). *Radiological Health Handbook*, Department of Health Education and Welfare, Rockville, Maryland.

Volchok, H. L. (1967). Fallout of Pu-238 from the SNAP-9A burnup, II. HASL-182, HASL Rep. 1967 July 1:I 2-, Health and Safety Laboratory.

Vovk, *et al.* (1995). In "Proceedings of an International Conference", 14–18 March, Zeleny Mys, Chernobyl, Ukraine, OECD/NEA, Paris, pp. 341–357.

Wallenius, M., Mayer, K., Tamborine, G., and Nicholl, A. (2003). IAEA-CN-98/21. In "Proceedings of an International Conference," *Karlsruhe*, 21–23 October 2002, International Atomic Energy Agency, Vienna, Austria, pp. 133–139.

Wolfinbarger, J. L. and Crosby, M. P. (1983). *J. Exp. Mar. Biol, Ecol.*, 67: 186–198.

World Health Organization (WHO). (1976) Nuclear Medicine. Report of a Joint IAEA/WHO Expert Committee on the Use of Ionizing Radiation and Radioisotopes for Medical Purposes (Nuclear Medicine). WHO Technical Report Series No. 591, World Health Organization, Geneva.

World Health Organization (WHO). (2001a). Report of the World Health Organization Depleted Uranium Mission to Kosovo, United Nations Interim Administration Mission in Kosovo (UNMIK), Secretary General, United Nations, New York, January.

World Health Organization (WHO). (2001b). Depleted Uranium: Sources, Exposure and Health Effects, WHO/SDE/PHE/01.1, United Nations World Health Organization, Department of Protection of the Human Environment, Geneva, April.

Yamato, A. J. (1982). *J. Radioanal. Nucl. Chem.*, 75 (1/2): 265–273.

Zimmerman, P. D. and Loeb, C. (2004). *Defense Horizons*, 38: 1–11.

8 Pesticides

Randy D. Horsak, Philip B. Bedient,
M. Coreen Hamilton, and F. Ben Thomas

Contents

8.1 INTRODUCTION

A pesticide is defined as a chemical agent used to destroy or control pests. The root word is the Latin word "cida" which means to kill. The generic term "pesticides" can apply to a wide spectrum of chemicals, including insecticides, rodenticides, herbicides, fungicides, biocides, and similar chemicals.

Pesticides have been extensively investigated since the 1960s, and their chemical properties, toxicological properties, and fate and transport are well known. The purpose of this chapter is to provide a brief overview of the types of pesticides, their fate in their environment, analytical considerations and forensic techniques available to age date and identify the source of the pesticide release into the environment.

8.2 TYPES OF PESTICIDES

The worldwide expenditures for pesticides is estimated to be about $31.8 billion in 2001, the latest year of full tracking (Kiely *et al.*, 2004). The United States market represents about one-third of the total world market. Chlorine and Hypochlorite count for about 52% of total United States pesticide usage, followed by conventional pesticides (18%), wood preservatives (16%), specialty biocides (7%), and other pesticides (6%). This equates to a total of about 5 billion pounds of chemical (Kiely *et al.*, 2004).

It is estimated that more than 1000 formulations of pesticides are used throughout the world (Tardiff, 1992). Pesticides may be categorized according to a number of parameters, including chemical structure, route of exposure, and method of application. Pesticides include herbicides, insecticides, fungicides, rodenticides, fumigants, and other categories (Ecobichon, 2001), as shown in Table 8.2.1. Major categories of pesticides are shown in Table 8.2.1. Typically, the most widely used pesticides in commercial applications are insecticides, herbicides, and fungicides, while more rodenticides and botanicals are used in domestic applications. Selected environmental properties of several common pesticides are listed in Table 8.2.2.

8.2.1 Organochlorines

Organochlorine pesticides are hydrocarbon compounds containing multiple chlorine substituents. One of the most widely known Organochlorine pesticides is DDT, an insecticide which was widely used during World War II. Because of its low toxicity to humans, DDT powder was widely applied by the US Army to the skin of soldiers to kill parasitic skin insects. Because of their low water solubilities and environmental persistence, DDT and other organochlorine pesticides also found widespread use as poisons for subterranean nests of termites. About 100,000 tons per year were produced in the late 1950s in the United States alone (Connell, 1997).

Organochlorine pesticides are lipophilic (i.e., they tend to accumulate in fat-rich tissues such as nerves), and they kill insects by disrupting neural function. It would be years before scientists recognized that DDT not only killed insects, but had much wider impacts in the environment. In 1962, Rachel Carson's book, *Silent Spring*, was published, which described the devastating effects of DDT on bird populations.

Carnivorous birds which consumed DDT-contaminated fish produced eggs with thinner shells than birds that had diets free of DDT-contaminated fish. Carson's book is commonly credited for raising the environmental consciousness of the American public, and eventually led to the creation of the US Environmental Projection Agency (EPA) by an Executive Order issued by President Nixon.

Not surprisingly, DDT's disturbing effects on bird populations and environmental persistence has resulted in many restrictions on its usage. Many other chlorinated pesticides are highly regulated as well.

8.2.2 Organophosphates

Organophosphate pesticides are used primarily as agricultural pesticides, and contain a chemically reactive phosphate ester side chain, consisting of a central phosphorous atom double bonded to either an oxygen or sulphur atom, and single bonded to two methoxy ($-OCH_3$) or ethoxy ($-OCH_2CH_3$) groups.

Certain nerves of insects function by releasing a chemical called acetylcholine (ACh) into the intracellular space (i.e., the synapse) where the nerve cell contacts a muscle cell (or another nerve cell). Acetylcholine stimulates the muscle cell to contract, and contraction is stopped by an enzyme called acetylcholinesterase that destroys the released ACh signal molecules. Organophosphates are acutely toxic because they chemically bind to the acetylcholinesterase enzyme in such a way that it cannot destroy ACh and the insect dies with its muscles in a prolonged state of contraction and its nervous system in a state of sustained excitation. It should be emphasized that ACh is a neurotransmitter not only in insects, but in most animal species. As a result, organophosphate pesticides are potentially toxic to a wide variety of non-target species, including humans.

Organophosphate pesticides are also associated with delayed neuropathology. Because Organophosphates are chemically reactive, they persist in the environment only for relatively short periods, especially under alkaline conditions (Sawyer *et al.*, 2003).

8.2.3 Carbamates

Carbamate pesticides generally have the formula RHNCOOR', and are relatively polar, highly soluble in water, and chemically reactive. Some Carbamates of importance are Aldicarb, Carbaryl, Carbofuran, Ferbam, and Captan (Sawyer *et al.*, 2003). Figure 8.2.1 shows annual usage of Carbofuran in the United States. Carbofuran, like many Carbamates, is primarily used in agriculture. Its usage pattern is consistent with general Carbamate pesticide application, since the Midwest and Southeast are the two regions with the greatest need for pesticides in agriculture.

Carbamates tend to hydrolyze easily, resulting in a low level of persistence in the environment, both in soil and water. Like Organophosphates, Carbamates are acetylcholinesterase inhibitors (Baird and Cann, 2005).

8.2.4 Triazines

Compounds of the s-Triazine family are among the most heavily used herbicides during the past 30 years. Although most have low acute toxicity in most animal species, chronic exposures to certain s-Triazines have been shown to be animal carcinogens. s-Triazine (for which this class is named) is a six-member ring containing alternating nitrogen and carbon atoms. Common product derivatives include Atrazine, Simazine, Terbuthylazine, Ametryne, and Terbutryne (Sawyer *et al.*, 2003).

Atrazine [2-chloro-4-(ethylamino)-6-(isopropylamino)-s-triazine] is an effective inhibitor of photosynthesis in lower plants, and accounts for 40% of applied weed killers in the United States (Baird and Cann, 2005).

8.2.5 Naturally Occurring Pesticides

Natural pesticides are pesticides that come from natural sources—generally plant or mineral derivatives. Nicotine (extracted from tobacco), Pyrethrum (extracted from chrysanthemum flowers), and Rotenone (extracted from the

Table 8.2.1 *Major Categories of Pesticides*

Category	Examples
Insecticides	
Organochlorine	• Dichlorodiphenylethanes (e.g., DDT; DDD; DMC; Methoxychlor; Dicofol; Chlorbenzylate)
	• Cyclodienes (e.g., Aldrin; Dieldrin; Endrin; Isodrin; Heptachlor; Chlordane; Endosulfan; Toxaphene)
	• Chlorinated Benzenes/Cyclohexanes (e.g., Hexachlorobenzene; Lindane)
Anticholinesterase	• Organophosphorus Esters (e.g., TEPP; Parathion; Malathion; Diazinon; Propetamphos; Chlorpyrifos, Chlorfenvinphos; also chemical warfare agents like Tabun, Sarin, Soman, VX, and CMPF)
	• Carbamic Acid Esters (e.g., Carbaryl; Aldicarb; Methomyl)
Pyrethroid Esters	• Type I Esters (e.g., Allethrin; Cismethrin; Pyrethrin I; Phenothrin; Resmethrin; Tetramethrin)
	• Type II Esters (e.g., Acrinathrin; Cycloprothrin; Cypermethrin; Deltamethrin; Fenvalerate; Fluvalinate)
Avermectins	• (e.g., Avermectin B1a; Abamectin; Ivermectin)
Other Synthetic	• Nitromethylene Heterocycles (e.g., NMI, NMTHT)
	• Chloronicotinyl (e.g., Imidacloprid)
	• Phenylpyrazoles (e.g., Fipronil)
Botanicals	• Nicotine (e.g., Nicotine Alkaloid; Nicotine Sulfate)
	• Rotenoids (e.g., Rotenone)
Metals	• (e.g., Arsenic compounds)
Herbicides	
Chlorphenoxy	• (e.g., 2,4-D; 2,4,5-T; MCPA)
Bipyridyl	• (e.g., Paraquat; Diquat)
Chloroacetanilides	• (e.g., Alachlor; Acetochlor; Amidochlor; Butachlor; Metalaxyl; Metolachlor)
Phosphonomethyl Amino Acids	• Glyphosate
	• Glufosinate
Metals & Oils	• Metals (e.g., Arsenic compounds)
	• Oils (e.g., Kerosene; Diesel fuel)
Fungicides	
Organochlorine	• (Hexachlorobenzene; Pentachlorophenol; Chlorothalonil; Dichloropropene; Quintozene)
Triazines	• (Anilizine)
Imidazole	• (Benomyl; Imazalil; Thiabendazole)
Phthalimides	• (Captan; Folpet; Captofol)
ThioCarbamates	• (EPTC; Maneb; Mancozeb; Metiram; Nabam; Thiram; Triallate; Zineb; Ziram)
Metals & Oils	• Metals (e.g., Copper-based Bordeaux mixture; Arsenic compounds; Sulfur)
	• Oils (e.g., Creosote; Used motor oil)
Fumigants	
Volatile Liquids	• (Ethylene Bromide; DBCP; Formaldehyde)
Reactive Solids	• (e.g., Zn_2P_3; AlP; NaCN; $Ca(CN)_2$)
Gases	• (e.g., Methylbromide; Hydrogen Cyanide; Ethylene Oxide)
Rodenticides	
Reactive Solids	• (e.g., Zn_2P_3)
Organofluorines	• (e.g., Fluoroacetate; Fluoroacetamide)
Phenylthioureas	• (e.g., a-Naphthylthiourea)
Anticoagulants	• (e.g., Warfarin)
Other	
Repellants	• Diazinon, Thiram, Ziram
Mating Disruptors	• Grandlure, Disparlure, Gossyplure

tuber Derris elliptica) are plant-derived, while pesticides like Boric Acid, Cryolite, and Diatomaceous Earth are mineral-derived.

The active ingredients in Pyrethrum (Pyrethrin I and II and Cinerin I and II) are nonpolar, oily, and lipophilic. These insecticidal compounds are neurotoxins. Because the Pyrethrins are unstable in sunlight, synthetic derivatives (called Pyrethroids) have been developed, which find applications in agriculture and household products (Connell, 1997; Baird and Cann, 2005).

Rotenone is also a neurotoxin, and was initially identified (along with several related compounds) in a crude plant extract used by primitive people to paralyze fish. Poisoning by Rotenone in humans is rare, although adverse effects may be seen especially following oral and respiratory ingestion. It is unstable in sunlight, giving it a short environmental half-life. Toxicity is believed to be the result of Rotenone's inhibition of substrate oxidation in the NADH2 to NAD system, which is critical for nerve function (Vogue *et al.*, 1994).

Table 8.2.2 *Classification and Properties of Common Pesticide*

Class	Category	Pesticide	PMR (Pesticide Movement Rating)	Half-life (days)	Solubility in Water (mg/l)	Molecular Weight	K_{OC} (Sorption Coeff.)
Insecticides	Organochlorines	DDT	Extremely Low	2000	0.0055	354.5	2,000,000
		Lindane	Moderate	400	7	290.8	1100
		Aldrin	Very Low	365	0.027	364.5	5000
		Pentachlorophenol	Very High	48	100,000	266.3	30
	Organophosphates	Malathion	Extremely Low	1	130	330.4	1800
		Parathion	Very Low	14	24	291.3	5000
		Diazinon	Low	40	60	304.3	1000
	Carbamates	Carbofuran	Very High	50	351	221.3	22
		Carbaryl	Low	10	120	201.2	300
		Fenoxycarb	Extremely Low	1	6	301.3	1000
	Pyrethroids	Pyrethrin	Extremely Low	12	0.001	372.5	100,000
		Permethrin	Extremely Low	30	0.006	391.3	100,000
		Cypermethrin	Extremely Low	30	0.004	416.3	100,000
Herbicides	Broad Leaf	Dinoseb	High	30	52	240.2	30
		Triclopyr ester	Low	46	23	256.5	780
	Photosynthesis Inhibitors	Atrazine	High	60	33	215.7	100
		Diuron	Moderate	90	42	233.1	480
Fungicides	DithioCarbamates	Zineb	Low	30	10	275.7	1000
		Ziram	Moderate	30	65	305.8	400
	Benzimidazoles	Thiabendazole	Low	403	50	201.2	2500

Source: Oklahoma State University Pesticide Database.

CARBOFURAN
ESTIMATED ANNUAL AGRICULTURAL USE

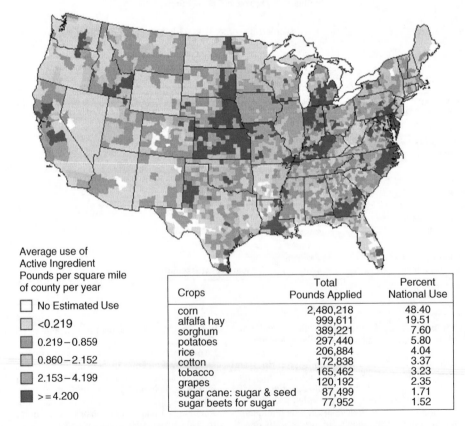

Average use of
Active Ingredient
Pounds per square mile
of county per year

☐ No Estimated Use

☐ <0.219

☐ 0.219–0.859

☐ 0.860–2.152

☐ 2.153–4.199

■ >=4.200

Crops	Total Pounds Applied	Percent National Use
corn	2,480,218	48.40
alfalfa hay	999,611	19.51
sorghum	389,221	7.60
potatoes	297,440	5.80
rice	206,884	4.04
cotton	172,838	3.37
tobacco	165,462	3.23
grapes	120,192	2.35
sugar cane: sugar & seed	87,499	1.71
sugar beets for sugar	77,952	1.52

Figure 8.2.1 *Carbofuran use. Source: USGS (1997) (see color insert).*

8.3 PHYSICAL AND CHEMICAL PROPERTIES

The behavior of pesticides in the environment is similar to that of many other chemicals, whether the pathway of concern is soils, sediments, groundwater, surface water, or the ambient air. For all environmental media, the levels of pesticides decrease naturally in the environment by two primary mechanisms: [1] natural degradation and [2] attenuation/dispersion, such that the overall net decrease of pesticide mass or concentration is expressed by the following simplified formula:

$$Decrease_{net} = Degradation_{natural}$$
$$+ Attenuation/Dispersion_{natural} \quad (8.3.1)$$

8.3.1 Natural Degradation

Natural degradation refers to the reduction in contaminant mass in a given volume of environmental medium (e.g., soil, groundwater, air, etc.) over time as a result of the contaminant being degraded by physical, chemical, biological, or other processes. The term "half-life," as subsequently discussed in more detail, refers to the time required for one-half of the contaminant mass in a given volume of environmental medium to degrade. For a given volume of medium, this also equates to the time for the concentration to be reduced by a mathematical factor of two.

8.3.2 Natural Attenuation/Dispersion

The term "natural attenuation/dispersion" refers to the reduction in contaminant concentration in a given volume of environmental medium over time as a result of the contaminant being diluted, dispersed, or irreversibly sorbed in the environmental medium. The result of natural attenuation/dispersion is a net reduction of contaminant concentration, and consequently reduced exposure and associated risk. The term "natural attenuation" generally refers to soils, sediments, and groundwater, whereas the term "dispersion" generally refers to surface water and ambient air.

8.3.3 Environmental Properties

Numerous other chemical, physical, and biological properties affect the behaviour of pesticides in the environment. Pesticides are carried by wind, move within streams or other bodies of water, partition between solid and dissolved phases, interact between the dissolved and the particulate phases within the water column of sediment, and migrate from medium to medium. These processes are governed by the individual properties of the pesticides themselves, including such conventional parameters as:

- Water solubility (mg/liter)
- Sorption coefficient (soil K_{oc})
- Organic-water partitioning coefficient (K_{ow})
- Dissolved organic carbon coefficient (K_{doc})
- Specific gravity
- Boiling point
- Vapor pressure
- Henry's Law constant
- Biodegradation rates
- Bioaccumulation factors

More importantly, pesticides can be characterized according to their:

- Fugacity and reactivity
- Pesticide movement rating (PMR) and Groundwater Ubiquity Score (GUS)
- Persistency and half-life

8.3.4 Fugacity and Reactivity

The term "fugacity" is frequently applied to the fate and transport processes associated with pesticides. In its simplest form, fugacity is the ability of a pesticide to "escape" from one phase to another, or from one environmental media/compartment to another. The compartments could include soil, water, air, and similar media, but can also be extended to include ultimate exposure points such as dust, surfaces, indoor carpeting, and food.

Fugacity is frequently used to predict the distribution of a given pesticide at equilibrium, and then to describe the fate and transport of the pesticide using kinetics. As such, it can be used to construct mass balance compartment models of any given pesticide based on its fugacity. Mathematically, fugacity is defined as:

$$F = C/Z$$

where

F = fugacity
C = concentration of the pesticide for a given phase
Z = fugacity capacity constant at a given temperature and pressure

As an example, a fugacity-based indoor residential fate model was proposed by the Lawrence Berkeley National Laboratory (Bennet *et al.*, 2002). In this model, the example is an indoor room treated with pesticides, with five surface environmental compartments being defined: the carpet (treated), vinyl flooring (treated), carpet (untreated), vinyl flooring (untreated), and walls/ceiling (untreated). Fugacity capacities, surface partition coefficients, and mass transfer rates were simulated using this mass transfer model with loss rates constants.

While the use of fugacity models presents interesting concepts, the authors acknowledged that the main challenge in using such models lies not with the models themselves, but with reliable estimates of partition coefficients, degradation rates, and transfer rates of pesticides on household surfaces (Bennet, 2002). Therefore, until more representative rates are available, the use of fugacity models may be limited.

Pesticides conform to the same general rules of reactivity as other chemicals. A pesticide's persistence is largely a function of its reactivity. More reactive pesticides are less persistent due to their greater ability to break down in the environment. Most reactions of interest take place in the presence of air, water, sunlight, and sediments, and at or near ambient temperatures, except during unique events such as forest fires. Therefore, it can be difficult to correlate reactions in the environment with such reactions found in the laboratory.

Chemical reactions occur in both air and water. Most atmospheric reactions fall within the general categories of oxidation and photolysis. Reactions in the water and sediment phases are considerably more complex, where both biological and abiotic reactions occur. Many of the biological reactions are catalyzed by enzyme processes, generally eventually oxidation and reduction. Abiotic reactions can include oxidation, reduction, hydrolysis, or proteolysis. This is illustrated for the pesticide Fipronil in Figure 8.3.1.

These reactions result in the formation of degradate compounds (sometimes referred to as "metabolites" or "degradation daughters"). Pesticide degradates are of environmental concern, as are the parent pesticide compounds, since their toxicity and persistency may be less than, equal to, or greater than that of the parent compounds.

Again, the pesticide Fipronil is a good illustration of this phenomenon. Fipronil is a phenylpyrazole-type chemical, with the trifluoromethylsulfinyl side chain being responsible for its unique insect toxicity. Fipronil acts by blocking the γ-aminobutyric acid gated chloride channel in insects

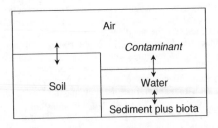

Figure 8.3.1 *Illustration of the breakdown of the insecticide Fipronil into its metabolites via oxidation, reduction, hydrolysis, and proteolysis.*

(Crosby and Ngim, 2001). Fipronil breaks down into three main degradate products. Fipronil Sulfone, Fipronil Sulfide, and Desulfinyl Fipronil. The three Fipronil metabolites have been found to be as toxic as the parent compound Fipronil and have greater environmental persistence (Schlenk *et al.*, 2001). The persistence of degradate products will be discussed later in greater detail.

In the sediments and water column there are often biologically mediated reactions that catalyze the decomposition of pesticides. First, the physical fate of pesticides might be viewed as an equal fugacity-based partitioning between the sediment, water, and the atmosphere (i.e., where the escaping tendency from each medium is equal). This is commonly referred to as the unit world concept (Baird and Cann, 2005).

Donald Mackay popularized this approach to modeling the fate of chemicals in nature (Mackay, 1991). The software "ChemCan" from Canadian Environmental Modelling Centre, Trent University, Canada, implements his concepts. Similarly, the US EPA software packages, EpiSuite and Sparc, use this fugacity concept along with various correlations to predict the physical chemical fate of chemicals.

Unfortunately, few pesticides have been studied sufficiently to model the reaction type and rate based upon fundamental chemical structure information. In lieu of specific information or reliable correlations, empirical observations are often used that have observed half-lives as single bulk numbers to use in modeling. Often first-order decay values are non-specific mechanistically, and are intended to describe the chemical fate in the specific environment.

8.3.5 Pesticide Movement Rating and Groundwater Ubiquity Score

Unlike many contaminants, pesticides have been defined by unique terminology that describes their fate and transport. Pesticide movement rating (PMR) measures a pesticide's ability to move toward groundwater. It is based on the pesticide's Groundwater Ubiquity Score (GUS), an empirically derived value which relates pesticide persistence (half-life) and sorption in soil (coefficient K_{oc}) according to the following formula:

$$GUS = Log(\text{half-life}) \times (4\text{-}LogK_{oc})$$

Movement ratings for individual pesticides range from extremely low to extremely high, as summarized in

Table 8.3.1 Pesticide Movement
Rating and Groundwater Ubiquity Score

Extremely Low	<0.1
Low	1.0–2.0
Moderate	2.0–3.0
High	3.0–4.0
Extremely High	>4.0

Source: Vogue *et al.* (1994).

Table 8.3.1 (Vogue *et al.*, 1994). While the GUS factor cannot be used indiscriminately in environmental site assessments, it does allow one to quickly assess a pesticide's mobility. As such, the GUS factor should be one of the key parameters to be reviewed in determining the potential for groundwater contamination.

8.3.6 Persistency and Half-life

Pesticide persistency in terms of half-life is another key parameter. The National Pesticide Information Center in the United States defines persistency as follows:

> The soil half-life is a measure of the persistence of a pesticide in soil. Pesticides can be categorized on the basis of their half-life as non-persistent (degrading to half the original concentration in less than 30 days), moderately persistent (degrading to half the original concentration in 30–100 days), or persistent (taking longer than 100 days to degrade to half the original concentration).

The term "persistent" is further defined by the US EPA as "the ability of a chemical substance to remain in an environment in an unchanged form. The longer a chemical persists, the higher the potential for human or environmental exposure to it . . ." The US EPA generally terms chemicals which persist in the environment for periods longer than six months as "persistent pollutants."

The term "half-life" refers to the time required for one-half of the contaminant mass in a given volume of soil, groundwater, or other environmental medium to degrade. For a given volume of medium, this also equates to the time for the concentration to be reduced by a mathematical factor of two. The mathematical formula for half-life decay is:

$$A = (I)2^{-(t/h)}$$

where

A = residual mass
I = initial mass
t = time duration
h = half-life

Some pesticides are very stable and thus have half-lives in the order of several hundred days or more. For example, DDT has a half-life of about 2000 days, Mirex about 3000 days, and Endrin about 4300 days. On the other hand, Diazinon has a half-life of about 40 days and Methyl Parathion only about 5 days, as shown in Table 8.2.2.

8.3.7 Ambient Air

Pesticides can be released to the ambient air in either vapor (gaseous) or particulate phases. In the case of particulates, the pesticide may be sprayed into the air as a liquid mist, or may be sorbed to the soil or other particles and carried with the airborne dust from the source.

Degradation of gaseous pesticides in the atmosphere is governed primarily by reaction with oxygen, free radicals, and other chemicals present in the air. Additionally, direct photolysis and hydrolysis may be significant degradation mechanisms. Degradation of particulates occurs not only via these mechanisms but also through partitioning of the sorbed pesticide into the air during transport.

Vapor phase releases follow classical gas law physics and may be transported distances of several miles or more. Large or dense particles tend to have a higher probability of particulate fallout closer to the point of release. However, if the diameters of the particles released are sufficiently small (i.e., =10 μ), they tend to behave as gases in the atmosphere and may be transported distances of several miles or more.

Airborne releases follow general wind and terrain patterns, and the resultant plumes are geometrically a function of several site-specific variables, including the wind speed and direction, release height, source and ambient temperatures, atmospheric stability and variability, and other factors. Some of these variables are combined into "Pasquill Stability Categories" which define the shape and behavior of the plume. As such, pesticide releases to the atmosphere generally follow classical principles of air dispersion modeling, such that annual average concentrations, points of maximum impact, and other factors can be simulated.

8.3.8 Groundwater

Movement into groundwater involves both hydrodynamic and abiotic processes. Hydrodynamic processes are physical mechanisms that affect the paths of groundwater flow, and include advection, diffusion, and hydrodynamic dispersion. In general, the lower molecular weight pesticides tend to have higher aqueous solubility, and only moderately sorb onto the soil particles in the groundwater. This trend of correlating adsorption coefficients and molecular weights is shown in Table 8.2.2. Thus, the lighter chemicals dissolve quickly into the groundwater and move with the natural water flow, whereas the heavier chemicals tend to be insoluble move very slowly, if at all. However, heavier parent compounds can affect groundwater quality since their degradates can be more soluble and mobile than the parent compounds themselves.

Upon entry into groundwater, lower molecular weight pesticides migrate down-gradient, at approximately the same flow velocity as the groundwater, whereas the higher molecular weight compounds are retarded and arrive either at a subsequent time frame or not at all, or degrade into compounds which move at different rates but are not monitored via routine sampling and analytical testing. This is shown in Figure 8.3.2, which displays breakthrough curves for three different pesticides with different sorption capacities.

8.3.9 Abiotic Processes

Other important methods of solute transport in groundwater are abiotic processes that include reactions between a contaminant and an aquifer medium. Sorption, including both adsorption and desorption, refers to a chemical's affinity to bind to soil particles or dissolve in water. Adsorption occurs at two media interfaces, for example, between liquid and solid phases or gas and solid phases. Adsorption coefficients, which measure a pesticide's ability to bind to soil particles, are listed in Table 8.2.2. Pesticides such as DDT have extremely high adsorption coefficients, and are therefore very persistent. Pesticides which are highly soluble in water, such as Pentachlorophenol, have corresponding low adsorption coefficients. Desorption is the mechanism by which a release of pesticides occurs from the soil surface as fresh water flows past the contaminated soil. The process can be either linear or non-linear, and is strongly related to partitioning coefficients for particular pesticides.

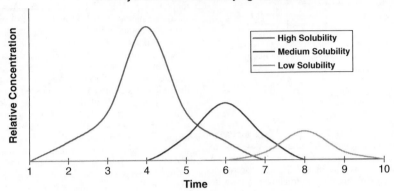

Figure 8.3.2 *Concentration over time as a function of solubility.*

Volatilization is another partitioning mechanism which occurs when aqueous phase, non-aqueous phase liquid (NAPL), or sorbed phase pesticides are directly converted to a gas phase (Schnoor, 1996). For some pesticides, up to 90% of the mass applied may be lost due to volatilization. For example, highly volatile pesticides, such as Lindane, record losses of this magnitude in about a week, while only 2% of Atrazine applications are lost to volatilization in 24 days (Bedos *et al.*, 2002).

Cosolvation is a transport mechanism in which a contaminant's mobility is enhanced by the presence of another solvent. This becomes extremely important when attempting to forensically ascertain the origin of a particular pesticide. As previously mentioned, pesticides which have high adsorption coefficients bind strongly to soil, and have shallow distributions and usually do not enter groundwater. However, the presence of a cosolvent facilitates distribution of such pesticides at depth. Chloroform and Chlorobenzene are common cosolvents which facilitate the movement of such pesticides as DDT, DDE, Toxaphene, and Dieldrin. DDT and Dieldrin will also have increased mobility in the presence of Xylene. For example, a spill of Xylene resulted in the distribution of DDT at depth. Due to Xylene's cosolvent properties, the site owner was found responsible for resulting DDT levels in the groundwater (Morrison, 1997).

8.3.10 Surface Water
Many of the same physical, biological, and chemical processes previously discussed also apply to surface water. Pesticides that are released to the land surface can invade streams, rivers, and lakes, and can be manifested in both the particulate and the dissolved phases. Most pesticides have limited solubility whereas others tend to be more soluble. In general, pesticides in surface water tend to be strongly sorbed to sediment particles.

Pesticide concentrations in surface waters may vary seasonally due to the timing of pesticide applications and runoff conditions, with concentrations being highest during spring and early summer at many sites (Ferrari *et al.*, 1998).

Pesticide degradation in water includes hydrolysis, a chemical reaction between the pesticide and water. The result of this reaction is the formation of ions and a breakdown of the pesticide into a simpler structure. Some pesticides, such as carbamates, are highly susceptible to hydrolysis. Hydrolysis is also heavily dependent on temperature and pH. Deviations from neutral pH, in either acidic

or alkaline direction, will increase rates of hydrolysis, as will temperature increases.

Pesticides are also subject to biodegradation by microorganisms present in the soil, sediment, and water. Metabolism of pesticides may result in either complete or incomplete hydrocarbon degradation depending on environmental factors such as pH, temperature, dissolved oxygen, and redox state. Lower molecular weight pesticides tend to oxidize completely to form CO_2 and H_2O. This complete breakdown of pesticides is called mineralization. Higher molecular weight pesticides tend to partially degrade, forming oxygenated metabolites such as phenolic and acid metabolites. Biodegradation of pesticides requires not only a population of microbes that are capable of metabolizing the pesticide as a carbon source, but also appropriate environmental factors to support the growth of such microbes. It is the dissolved chemicals, chemical vapors, and residual phase (adsorbed and inter-particulate) that are most amenable to natural biodegradation.

Photodegradation is the breakdown of pesticides by sunlight. Factors that influence the rate of photodegradation include the intensity and spectrum of sunlight, length of exposure, and properties of the pesticide. Although every pesticide is susceptible to photodegradation to a certain extent, some are more so than others. As previously mentioned, the instability of Pyrethrins in sunlight has spurred the development of Pyrethroids, which are not as susceptible to photodegradation. Pesticides that are applied to the soil surface or foliage are more vulnerable to photodegradation than pesticides that are incorporated into the soil.

Half-lives for pesticides can range in the order of days to years, as shown in Table 8.2.2. Data in the public domain is quite variable since the quoted decay rates are strongly a function of site-specific geochemistry and biology. Nonetheless, if the current contaminant mass, contaminant concentrations, and decay rates are known, it is possible to calculate an approximate historical contaminant mass or concentration during a time frame when the pesticide contamination first occurred.

8.4 ANALYTICAL TESTING

If there is a term that describes analytical testing of pesticides in various matrixes, the term is "complex." The analytical testing of pesticides and their degradates may require a combination of gas chromatography/mass spectrometry, high-performance liquid chromatography with diode array detection, liquid chromatography/mass spectrometry, immunoassay, and other analytical capabilities. In many instances, any given lab may not have all these capabilities, thereby requiring that multiple environmental samples be collected and shipped to several different labs for analysis.

The testing of degradates is especially important to forensic identification. For example, the US Geological Survey's analytical laboratory in Lawrence, Kansas, has as its primary goal the development of state-of-the-art analytical methods for the analysis of pesticide degradates in surface and groundwater that are vital to the study of the fate and transport of pesticides, especially as it pertains to non-point sources of contamination (Scribner *et al.*, 2000).

Analytical methods for soil, water, and biological tissue are used to determine the fate and concentration of pesticides in the environment and to provide data for studying the exposure, and ecological effects (Scribner *et al.*, 2000). Determination of pesticides in these various matrixes can be a very complex procedure. Analysis may be required for a series of related pesticides both in their original forms and as metabolites and for the "inert" ingredients also present in the commercial pesticide formulations. Inherent in this is the requirement to understand the degradation pathways of the pesticides so that appropriate compounds can be targeted in the analysis.

In addition, many pesticides are known to contain interferences that are by-products of the manufacturing process but which in themselves may be an environmental or health concern. Due to the potential range of chemical and physical properties of all of these compounds and the complex nature of the matrix of the soil, biological tissue or other samples being analyzed, analytical testing may require any combination of gas chromatography/mass spectrometry, high-performance liquid chromatography with UV or photodiode array detection, liquid chromatography/mass spectrometry, or other methods. In many cases, published analytical methods, especially for the metabolites, are not available and existing methods must be adapted to include the target compounds of interest.

Another major consideration is the need for ultra-trace laboratory capability to provide analyses at very low detection limits. Except for sites with gross contamination, most commercial laboratories do not have the necessary equipment, staff, protocols, or experience to support forensic identification of pesticides.

There are four classes of pesticides from a chemical analysis point of view, including Organochlorine (OC), Organonitrogen (ON), Organophosphorous (OP), and Carbamate (CB) pesticides. These classifications are based on the main chemical functionality of the compounds, but even within these categories, there are hundreds of individual compounds with wide-ranging chemical and physical properties.

Multi-residue methods, which screen for a large number of pesticides in a single lab test, are used on a routine basis because of their efficiency and broad applicability. Typically, a multi-residue method is focused at a particular chemical class of pesticides by employing a rigorous extraction, carrying out no cleanup to minimize the risk of losing sample components, and then using a selective detection and quantification method appropriate to the class of compounds of interest. If a sample of unknown pesticide content is being analyzed, a multi-class multi-residue analysis can be performed to screen the sample for a large range of compounds. This is achieved by analyzing the sample extract by several detection and quantification procedures.

Single residue methods are designed for the analysis of only a single compound or several related compounds.

Table 8.4.1 *Sources of Analytical Methods*

- Pesticide Analytical Manual (PAM), Vol I & II: Multi-residue Methods; 3rd edition. US Food and Drug Administration Revision October 1999. C.M. Makovi (ed.) and B.M. McMahon (Editor Emerita); US Department of Health and Human Services, Public Health Service, Food and Drug Administration (1999).
- Test Methods for Evaluating Solid Wastes, Physical/Chemical Methods (SW-846), Approved by the US Environmental Protection Agency SW-846. (1996).
- Official Methods of Analysis (AOAC), AOAC International, 16th edition, on CD-ROM 3rd Revision (1997).
- Standard Methods for the Examination of Water and Wastewater; 20th edition 1998; Edited by A.D. Eaton, L.S. Clesceri, and A.E. Greenberg; APHA, AWWA & WEF; ISBN: 8765015134 (1998).
- Method of Chemical Methods for Pesticides and Devices, 2nd edition. US Environmental Protection Agency. Office of Pesticide Programs. C.J. Stafford, E.S. Greer, A.W. Burns and D.F. Hill (eds).
- Reference for EPA Series 500 and 600 methods
- Reference for EPA Series 1600 methods

These methods are most often used when the likely residue is known or when the residue of interest cannot be determined by available multi-residue methods. This approach offers the opportunity of optimizing the method for the compounds of interest thereby resulting in better selectivity, sensitivity, and overall defensibility.

Both multi-residue and single residue methods are available from a variety of agencies and government departments. A list of the major sources of analytical methods is given in Table 8.4.1.

8.4.1 Multi-Residue Approach

The general multi-residue procedure involves extraction of the pesticides from the sample matrix into an organic solvent. To ensure that the broadest range of target compounds is retained in this extract, the extract is usually analyzed without any cleanup by one or several analysis procedures. Multi-residue capability is achieved by using gas chromatography (GC) or high-performance liquid chromatography (HPLC) to separate the individual pesticides from one another before detection and quantification. The use of a detection process of high selectivity often allows identification of a specific pesticide or class of pesticides even in the presence of interferences. Several chromatographic separation, detection, and quantification procedures can be applied to the same sample.

8.4.2 Pesticide Method Parameters

Samples are preserved or treated (e.g., pH adjustment) to avoid degradation during storage and to ensure that the target compounds are in a chemical form amenable to extraction and analysis. The details of sample preservation are analyte-specific and must be defined in consultation with the laboratory prior to sample collection. Sample-holding times before analysis may be an issue since the composition of the sample may change over time. In many cases safe-holding times have not been measured and a generic 7 days until extraction followed by 40 days until analysis is given. Since these times may be impractical due to the logistics of sample collection, immediate preservation and appropriate storage followed by extraction as soon as possible is the recommended approach.

Table 8.4.2 *Extract Cleanup Approaches*

Cleanup	Purpose
Gel permeation chromatography	Removal of lipids and other large molecules
Adsorption chromatography on Florisil, Alumina, or Silica	Removal of sample constituents of different polarity from target compounds
Adsorption chromatography on charcoal	Removal of polar constituents
Adsorption chromatography on Florisil	Fractionate sample constituents

Table 8.4.4 *Liquid Chromatographic Detection Systems*

Detection System	Applications
UV detector	Responds to all compounds that absorb UV light
Fluorescence detector	Responds to compounds that are fluorescent or have been converted to a fluorescent form such as Carbamates
Mass spectrometer	Responds to all compounds, also gives structural information

A wide range of compounds can be efficiently partitioned from a water sample into an immiscible organic solvent such as methylene chloride by liquid–liquid extraction (LLE) or liquid–solid extraction (LSE) using a cartridge or disk containing adsorbent medium. When extracting pesticides from solids, water miscible organic solvents (such as mixed solvents containing acetone) are often recommended to optimize the recovery. Alternatively, sodium sulfate can be added to the sample to absorb the moisture and then the extraction can be carried out with methylene chloride or other no-water miscible solvent. Soxhlet extraction, ultrasonic extraction, or shaker table extraction may be used, depending on the pesticide.

Once an extract of the sample has been obtained it is subjected to one or more cleanup steps to isolate the analyte compounds from other co-extracted materials that could interfere with the detection and quantification of the pesticides. There are no generally applicable cleanup steps since each pesticide will behave differently in each set of conditions. Therefore, calibration of a selected cleanup procedure must be carefully carried out to ensure that target compounds are not also lost. Typical extract cleanup procedures are listed in Table 8.4.2.

The particular combination of chromatographic separation technique and detection system gives a pesticide analysis method its specificity. Gas chromatographic methods are suitable for pesticides that are volatile and of low to medium polarity. Liquid chromatographic analysis is the method of choice for the more polar and less volatile target compounds. The detection systems most often employed for pesticide analysis are listed in Table 8.4.3 for gas chromatography and for liquid chromatography in Table 8.4.4.

The mass spectrometer is the most powerful since it can compare the compound's mass spectrum to the reference spectra in a database, which assists in identification of the components. It can also be a powerful tool in quantitative analysis since it gives separate signals for an analyte and its labeled analogue is used as an internal standard.

When an analysis is conducted using gas or liquid chromatography with a mass spectrometric detector, the observation of non-target or "tentatively identified compounds" (TICs) can provide extra information about the sample. A TIC is a compound that can be observed by the analytical method but its identity cannot be confirmed. It is usually identified by comparing the observed mass spectrum with a library of mass spectra for known compounds and finding the spectrum that most closely matches that of the unknown. The level of certainty with this approach is usually not sufficient for forensic purposes and the concentration of the TIC can only be estimated.

Table 8.4.3 *Gas Chromatographic Detection Systems*

Detection System	Applications
Electron Capture Detector (ECD)	Responds to compounds containing halogen atoms (Cl, F, Br)
Nitrogen Phosphorus Detector (NPD)	Responds to compounds containing phosphorus or nitrogen atoms such as Organophosphate and Pyrethroids
Flame Photometric Detector (FPD)	Responds to phosphorus compounds
Mass Spectrometer	Responds to all compounds, also gives structural information

Table 8.4.5 *Analytical Approaches for Major Classes of Pesticide Compounds*

Pesticide Class	Separation Technique	Detector
Organochlorine Pesticides	Gas Chromatography	Electron Capture ECD
Organophospahte Pesticides	Gas Chromatography	Flame Photometric (FPD)
Organonitrogen Pesticides	Gas Chromatography	Nitrogen Phosphorus Detector (NPD)
Organonitrogen Pesticides	Liquid Chromatography	Mass Spectrometer (MS)
N-methylCarbamates	Liquid Chromatography, derivatization	Fluorescence
Phenoxy Acid Herbicides	Gas Chromatography	Electron Capture Detector (ECD)

Analytical approaches used for each major class of pesticide compounds are shown in Table 8.4.5.

Unless a mass spectrometer is used as the detector, the possibility of false positive results exists. In pesticide analysis the use of very selective detectors keeps the analysis focused on the target class of pesticides. Some general interference issues do exist.

Sulfur in solid samples can be a serious interference in electron capture, flame photometric, mass spectrometric, and other detectors. Sulfur can be removed during the extract cleanup step, and there are a variety of procedures available such as treatment with activated copper.

The natural lipid content of biological materials can interfere in the analysis of tissue samples. The extent of this problem is variable as the total amount of lipid in a sample varies, but usually a cleanup procedure to eliminate the lipids is required.

Interference by phthalate esters can be a serious problem when using the electron capture detector since they are leached from common flexible plastic tubing and other plastic materials in the field and in the laboratory. Avoid contact with plastics or use an alternate detection system such as electrolytic conductivity or mass spectrometry.

8.5 FORENSIC TECHNIQUES

The source and age of the release of pesticides into the environment may be assessed using a variety of forensic techniques and parameters:

- Degradation compounds
- Commingled substances
- Trace substances
- Biomarkers
- Pesticide radioisotopes
- Chemical fingerprint patterns
- Spatial distribution of the contaminants at the site

8.5.1 Degradation Compounds

Parent pesticide compounds do not simply disappear in the environment, but rather naturally transform into degradate chemicals. Since pesticides naturally degrade in the environment over time, it is important to test for both the parent and the degradate compounds. The absence of the parent compound in an environmental matrix can erroneously lead the investigator to conclude that pesticides are present at de minimus levels, or not even present at all.

Degradates may be as toxic, or even more toxic, than the parent compound itself. Further, since degradate chemicals may have different characteristics than the parent compound, such as higher solubility and mobility, their fate and transport in the environment may not necessarily track that of the parent compound. In conducting forensic assessments, at least a representative number of samples should be collected and tested for degradate compounds, especially if the parent compounds normally expected to exist in a particular media are either missing or at de minimus levels.

As with the parent compound, the presence of degradate compounds at a site in sufficiently high concentrations may present both human health and ecological risks. Because of the impact degradate compounds can have on the surrounding environment, a forensic approach is being developed by agencies such as the USGS to identify and evaluate the risks of degradate compounds (Cherry, 2003).

According to studies conducted by the USGS and others, herbicide degradates are very prevalent in the groundwater and, in one study, represented 80% of the most frequently detected compounds. The detection of a given herbicide in groundwater generally increased many fold when the degradates were included. Further, a majority of the herbicides measured were in the form of degradate compounds, which ranged from 55 to over 99% of the total concentration. The data being snapshots in time, the degradate compounds were low or non-existent at the time of release (Kolpin et al., 2000).

At some sites, and especially in groundwater media, the nature and extent of contamination cannot be delineated by the presence of the parent compounds alone. In such instances, it may be critical to assess the spatial distribution of the degradate compounds. This was shown to be true in the Kolpin study referenced above.

During a 1995–1998 study conducted by the US Geological Survey (USGS), US Environmental Projection Agency (EPA), the Iowa Department of Natural Resources, and the University of Iowa Hygienic Laboratory, the distribution of herbicides and the resulting degradates present in Iowa's groundwater were analyzed and it was showed that both the parent compounds and the degradates were present. As summarized in Figure 8.5.1, eight of the ten most frequently detected compounds were degradates (USGS, 2003). Figure 8.5.2 shows that a range from 55 to 99% of the total measured concentration was in the form of degradates (USGS, 2003).

In a 1996 study conducted by the USGS National Water-Quality Assessment (NAWQA) Program, the degradation pathways of several Triazine pesticides were documented. Triazine herbicides include Atrazine, Cyanazine, Simazine, and Propazine. In 1995 alone, over 31 million kilograms of these herbicides were applied to corn and sorghum in the United States. When the parent compound of Atrazine degrades through both biotic and abiotic processes, the metabolites created include the dealkylated metabolites of de-ethylatrazine (DEA) and Deisopropylatrazine (DIA) and the hydroxylated metabolite hydroxyatrazine (HA). Before resulting the breakdown of Atrazine is complete, in the formation of carbon dioxide and nitrogen gas, at least three

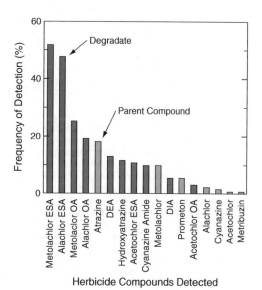

Figure 8.5.1 Graph generated from a 1995–1998 groundwater study in Iowa that shows eight of the ten most frequently detected compounds in the samples were degradate products.

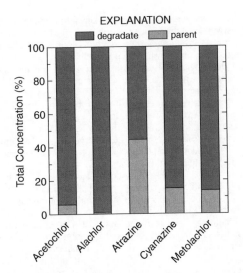

EXPLANATION

◼ degradate ▨ parent

Figure 8.5.2 Graph generated from a 1995–1998 ground-water study in Iowa that illustrates that degradate products represented from 55–99% of the total measured concentration of all contaminants measured.

degradate products will be formed in the environment. Figure 8.5.3 shows the complete degradation pathway of Atrazine (Scribner *et al.*, 2000).

Another example of science using degradation pathways was a study conducted at Greens Bayou in Houston, Texas, where the degradation rates for DDT and Lindane isomers were used to trace the timing of environmental pollutants released into the bayou sediments (Walker *et al.*, 2004). The original industrial plant mentioned in this study produced Chloral, DDT, Lindane, and other pesticides during two different periods. The first period began in 1950 and ended in 1983. When the plant was transferred to new ownership in 1983, different pesticides were produced. Studying the degradation rates of both DDT and Lindane isomers allowed scientists to evaluate the present-day contamination in the sediments and ascertain the time period at which the contaminants entered the bayou. The information generated in this study helped determine the company responsible for the contamination (Walker *et al.*, 2004).

The degradation pathways and rates of DDT and Lindane vary markedly based on environmental conditions. In anaerobic conditions, Lindane degrades rapidly into hydroquinone, dichlorophenol, and trichlorobenzene whereas DDT typically degrades into DDD and its subsequent transformation products (Walker *et al.*, 2004). Figure 8.5.4 graphically shows each step of the DDT degradation pathway. Analysis conducted during this study using EPA Method 8081A indicated rapid degradation of both DDT and Lindane in Greens Bayou, and demonstrated that nondetectable concentrations of DDT occurred in the 100–2500 ppb concentration range after 8 days and after only 15 days for Lindane. This analysis helped prove that any Lindane or DDT deposited into Greens Bayou during 1950–1983 would have degraded well before present day. This study also concluded that the current high concentrations of DDT and Lindane measured in the Bayou today (up to 70,000 ppb) most likely occurred during the 1983-present time period in concentrations high enough to be toxic to the anaerobic degradation system (Walker *et al.*, 2004).

8.5.2 Commingled Substances

Many pesticides contain commingled secondary ingredients that are advertently or inadvertently present in the mixture along with the primary pesticide chemical. These ingredients are typically termed "inert ingredients" by the US EPA, even though many are extremely toxic, and include "any ingredient in the product that is not intended to affect a target pest."

Commingled ingredients include carriers, emulsifiers, stabilizers, wetting agents, sticking agents, humectants, synergists, activators, and solvents. The US EPA Office of Pesticide Programs has prepared five lists of inert pesticide ingredients that are found in pesticide products registered by the agency:

• List 1 – Inert ingredients of toxicological concern

This list was shortened from an original list of about 50 compounds to approximately 10 compounds based on peer-reviewed studies which demonstrate carcinogenicity, adverse reproductive effects, neurotoxicity or other chronic effects, developmental toxicity, or adverse ecological effects.

• List 2 – Potentially toxic ingredients/high priority for testing

The 100 compounds in this list are structurally similar to chemicals known to be toxic, and some have data indicating such a concern.

• List 3 – Inert ingredients of unknown toxicity

This list of about 1800 compounds is under continued evaluation to determine reclassification.

• List 4A – Minimal risk inert ingredients

This list of about 250 compounds is based on minimal risk or overall safety, including the limiting of restriction of use or the amount of the compound in a pesticide product.

• List 4B – Other ingredients

This list of about 800 compounds is based on the US EPA having sufficient information to reasonably conclude that the current use pattern will not adversely affect public health or the environment.

Among the more toxic commingled ingredients are Adipic Acid, Benzene and Chlorobenzene, Bis(2-ethylhexyl) Ester, Ethylene Glycol Monoethyl Ether, Hydroquinone, Isophorone, Nonylphenol, Phenols, Phthalic Acid, Ethers, Glycols, Ketones, Petroleum Distillates, Naphtha Solvents, Met *al*s, and Chlorinated Hydrocarbons (US EPA, 1997). An example cited by the US EPA is isopropyl alcohol. Isopropyl alcohol may be an active ingredient and antimicrobial pesticide in some products; however, in other products, it functions as a solvent and may be considered an inert ingredient (US EPA, 1997).

In conducting forensic assessments, commingled ingredients may be important indicators of hazards, especially if the parent pesticide normally expected to exist in a particular media is either missing or at *de minimus* levels. The difficulty in assessment, of course, is the large number of potential compounds – testing for up to 3000 compounds can be tedious and very expensive, and may require the use of multiple laboratories.

8.5.3 Trace Contaminants

Pesticides may contain trace levels of extremely toxic substances that present unique potential health risks. These trace levels are generally not added to the pesticide product,

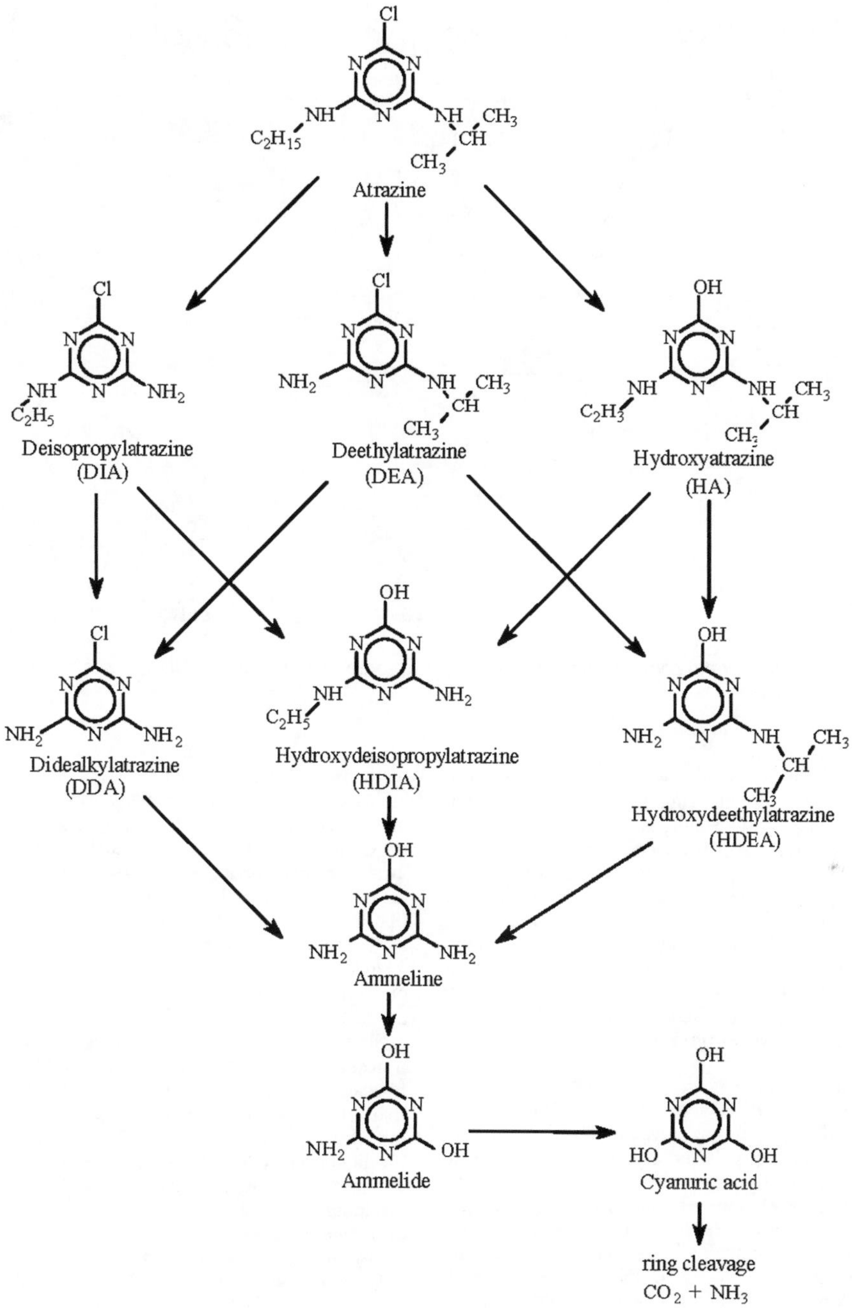

Figure 8.5.3 *Complete degradation pathway for Atrazine that can occur once the parent compound is introduced into the environment.*

as is the case with commingled substances, but rather are associated with the chemical processes used in the manufacturing of the pesticide itself. The origin of the trace contaminants may be from the various feedstocks used in the manufacturing process, or from poor quality control of chemical processing reaction temperatures or other process parameters that result in the formation of these trace substances.

Of primary concern are trace levels of Dioxins/Furnas or other chemicals with Dioxin-like properties. Initially, the US EPA reviewed approximately 161 pesticides which were suspected of having the potential to be contaminated with trace quantities of Dioxins/Furnas. As of the date of this writing, the US EPA National Center for Environmental Assessment (NCEA) has not yet finalized a database of sources of Dioxin (US EPA, 2005).

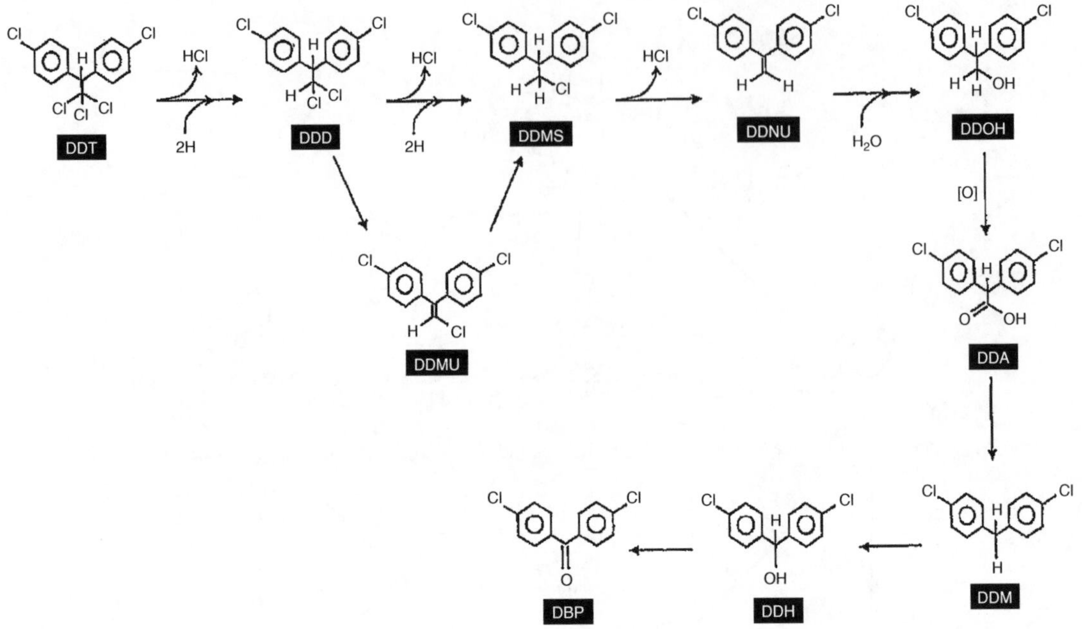

Figure 8.5.4 *A diagram of the complete bacterial anaerobic degradation pathway of DDT as seen in a study of Greens Bayou in Houston, Texas.*

As an example, the herbicides 2,4-D and 2,4,5-T were components of Agent Orange, the code name for the herbicide used by the US military in Southeast Asia (ATSDR, 2001). The combination was mixed with kerosene or diesel fuel, and then dispersed by aircraft and other means to destroy vegetation. 2,4,5-T is contaminated with 2,3,7,8-TCDD in varying amounts. In an uncontrolled manufacturing condition, TCDD contamination can be as high as 30–40 ppm, but current production standards usually limit TCDD to 0.05 ppm (FAO/UNEP, 1996). In the case of 2,4,5-T, the herbicide itself represents limited health risks (HSDB, 2003); however, the Dioxin represents a significant health concern.

Another example is Pentachlorophenol. Pentachlorophenol, although classified as a restricted use pesticide, is commonly used as a wood preservative. Exposure to Pentachlorophenol can cause liver effects, damage to the immune system, reproductive effects, and developmental effects (ATSDR, 1994). Further to this health concern is the presence of low levels (in the order of up to 100 ppm) of Dioxins/Furans in Pentachlorophenol that are associated with the manufacturing process (ATSDR, 1994). Thus, in conducting a forensic investigation of sites involving the use of either Creosote or Pentachlorophenol, various environmental media should be tested for Dioxins/Furans as well as the more conventional and obvious contaminants.

In conducting forensic assessments, trace contaminants, especially Dioxins/Furans, may be important indicators of hazards. Unlike degradates and commingled substances, however, testing for Dioxins/Furans is straightforward using high resolution GC/MS laboratory methods. Moreover, the Dioxins/Furans frequently can be chemically fingerprinted to a source, as subsequently discussed.

8.5.4 Biomarkers

Biomarkers (sometimes referred to as "biological markers") are an invaluable tool for environmental science because

of their predictive capabilities (Lam and Wu, 2003). The information that biomarker effects can yield is an extremely useful tool when managing, protecting, and conserving an environment that is dominated by industrial, agricultural, and residential pesticide use. Traditional environmental sampling and analytical methods provide direct evidence that a toxicant is present in various media, but not necessarily that a Plaintiff or target species has been exposed to a toxicologically sufficient amount to produce an adverse effect. For this reason, biomarkers are sometimes included as part of a forensic investigation (Kendall et al., 2001).

Simply stated, a biomarker is a biological chemical, fluid, process, or structure that is altered in a specific way by a toxicant. Of importance, biomarkers may provide evidence that the toxicant in the environment was actually absorbed into the body, and at a dose sufficiently high to produce the biomarker effect. In a Lam and Wu (2003) article, the definition of ecotoxicological biomarkers is simplified to "endpoints of ecotoxicological tests that register an effect on a living organism" that can indicate whether or not the environment may be polluted.

Biomarkers may be divided into several broad overlapping categories – biomarkers of exposure, biomarkers of effect, and biomarkers of susceptibility.

- *Biomarkers of exposure* include analysis of pesticide residues or reaction products in body tissues and fluids (e.g., brain, fat, red cells, plasma, breast milk, urine, etc.). In general, such biomarkers are used to estimate dose, which is then related to the continuum of changes resulting in the disease state (Nigg and Knaak, 2000; Timchalk et al., 2004). In some situations, it might be possible to extrapolate biomarker findings in samples from non-human species to their implications to human health (Bonefeld-Jorgensen, 2004).

- *Biomarkers of effect* are measurable biochemical, physiological, behavioral, or other alterations that may be recognized as an existing or potential health impairment or disease (Galloway and Depledge, 2001; Venturino *et al.*, 2003). Examples include inhibition of brain acetylcholinesterase by organophosphate or carbamate pesticides, and eggshell thinning associated with DDT. Less chemical-specific examples include various measures of DNA damage, decreased levels of circulating blood cells, and increased levels of plasma enzymes indicative of specific organ damage.

* *Biomarkers of susceptibility* are biological endpoints that are indicative of an altered physiological or biochemical state that may predispose the individual to adverse effects due to chemical, physical, or infectious agents (Costa *et al.*, 2003; Pinkney *et al.*, 2004). Examples include induction of chronic tissue injury and reparative cell proliferation, and decreases of detoxification enzymes.

The complexity of biomarker interpretation can also be illustrated by the DNA and protein adducts seen following exposures to Pentachlorophenol (PCP), a widely used biocide that has been reported to be carcinogenic in the liver of mice, but not in the rat. Several reactive metabolites of PCP have been identified and studied in these two species. Tsai *et al.* (2003) showed that both Tetrachloro-1,2-benzoquinone (1,2-TCBQ) and Tetrachloro-1,4-benzoquinone (1,4-TCBQ) formed covalent adducts with liver proteins under various oral dosing schedules. Their data suggested that 1,2-TCBQ is more important with regard to adduct formation in the mouse than in the rat.

Studies by Lin *et al.* (1999) found that the rates of adduct formation in the livers of Spraque-Dawley rats and B6C3F1 mice suggested that the production of adducts of a close metabolic relative of 1,2-TCBQ (i.e., Tetrachloro-1,2-benzosemiquinone; 1,2-TCSQ was proportionally greater at low doses of PCP (less than 4–10 mg/kg body weight) and was 40-fold greater in rats than in mice. Production of 1,4-TCBQ adducts, on the other hand, was proportionally greater at high doses of PCP (greater than 60–230 mg/kg body weight) and was 2- to 11-fold greater in mice than in rats over the entire range of dosages. In spite of much study, it remains unclear whether any type of PCP adduct is causally related to hepatocarcinogenesis in rodents, much less in man. Indeed, PCP is also known to increase the hydroxyl radical-derived DNA lesion, 8-oxodeoxyguanosine, in the liver of exposed mice (i.e., damage produced by the reaction of a reactive oxygen species with DNA, not covalent adduct formation).

Thus, while adduct formation is generally a good biomarker of exposure to a toxicant, its use as a biomarker of effect, or of susceptibility, is a matter of professional judgment and probable controversy. It is also prudent to remember that genotoxicity is only one of several recognized mechanisms of carcinogenesis.

Lam and Wu (2003) have pointed out that scientists have often confused "biomarkers" with "bioindicators." The easiest way to discern between these two terms is that "biomarkers" are responses at either the molecular, biochemical, or physiological level that can be measured in cells, body fluids, tissues, or organs whereas "bioindicators" are the accumulation of chemicals within living organisms and the responses reported within the entire organism, population, community, or ecosystem. Today, biomarkers are also being used to evaluate and record the health of humans after exposure to various pesticide-related contaminants (Depledge and Knap, 1998).

A common mechanism of toxicity for all organophosphate insecticides is the ability to inhibit the effects on cholinesterase enzymes within the nervous system. When organophosphates are released in the environment or human body, certain chemical changes occur. By understanding these chemical changes, scientists can measure and interpret biomarkers of exposure to organophosphates. One of these chemical changes in humans occurs when the liver metabolizes and transforms the double bond of the central phosphorus atom from sulfur to oxygen. This chemical reaction changes the organophosphate into a stronger inhibitor of cholinesterase enzymes (National Pesticide Information Center, 2003).

Figure 8.5.5 shows the changes that occur when chlorpyrifos (a common organophosphate) is metabolized by the human liver into chlorpyrifos-oxon. A second chemical reaction that has been documented to occur in the human body occurs when chlorpyrifos-oxon undergoes hydrolysis (or deactivation) thus producing dialkylphosphate (DEP) and 3,5,6-trichloro-2-pyridinol (TCP). Figure 8.5.6 is a chemical diagram that shows the hydrolysis of chlorpyrifos-oxon to dialkylphosphate and 3,5,6-trichloro-2-pyridinol.

Unlike the chemical reaction from chlorpyrifos to chlorpyrifos-oxon, the production of DEP and TCP will decrease the toxicity of organophosphates in the human body since both DEP and TCP do not inhibit cholinesterase enzymes (National Pesticide Information Center, 2003). The chemical structure of TCP is unique to the organophosphate chlorpyrifos thus allowing it to be utilized by scientists as a biomarker of exposure to the parent compound (Timchalk *et al.*, 2004). Therefore, when the TCP biomarker is detected in human body fluids, it could be correlated with the presence of organophosphate pesticides in various environmental media that serve as potential exposure pathways.

8.5.5 Radioisotopes

Radioisotopes can be used to trace the movement, source, and fate of pesticides throughout various media in the environment (Reddy *et al.*, 2001). The biomarkers often monitored in these studies are the stable chlorine isotope ratios

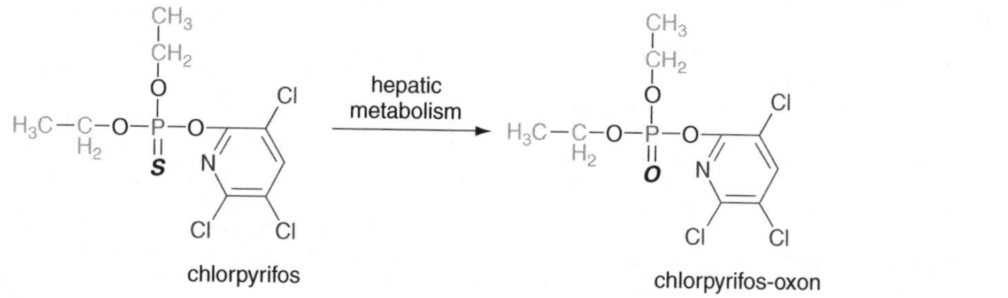

Figure 8.5.5 *Metabolism and ultimate degradation of the organophosphate chlorpyrifos into chlorpyrifos-oxon by the human body.*

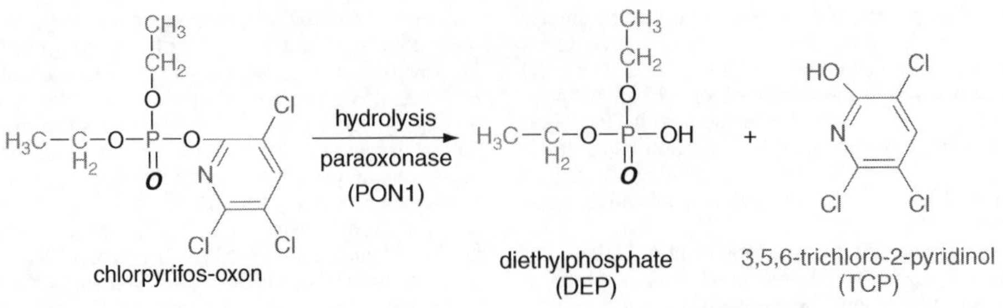

Figure 8.5.6 *Hydrolysis or deactivation of chlorpyrifos-oxon (degradate of chlorpyrifos) into the less toxic diethylphosphate and 3,5,6-trichloro-2-pyridinol degradation products.*

in chlorinated organic pesticides (Reddy *et al.*, 2001). By measuring the small variations in the ratio of two stable chlorine isotopes (^{37}Cl and ^{35}Cl) and comparing them to the ratio between ^{37}Cl and ^{35}Cl of standard mean ocean chloride (SMOC), scientists have stated it may be possible to identify the different sources of the pesticide-related contaminants (Drenzek *et al.*, 2002). For example, in a Reddy *et al.* (2000) research article, the semi-volatile pesticide DDT was studied to see how both abiotic and biotic alteration processes cause the loss of chlorine. If a distinct isotopic effect can be determined for each environmental process that affects DDT, then studying the chlorine isotope ratios may help discern and quantify the stage of DDT breakdown in the field.

In a similar study conducted by Drenzek *et al.* (2002), the isotope ^{37}Cl value in PCBs was analyzed and determined to have a vary narrow range allowing the author to determine that the ^{37}Cl isotope "may be exploited in tracer studies to invoke, and potentially quantify, specific biogeochemical processes acting upon these compounds, even from multiple sources, once introduced into the environment."

As isotope analysis for environmental studies is better understood and more widely used, scientists are applying the acquired knowledge to a wide variety of geologic, ecologic, and biologic studies (Heemskerk and Drimmie, 2000). Isotope analysis enables scientists to generate an unmistakable "fingerprint" profile of a sample. Because no two hyperfine fingerprint profiles are alike, scientists are able to track pollutants in various environmental samples and gather information about the origin and fate of specific compounds (Currie, 2000; Anon, 2002; Briggs *et al.*, 2004).

Drexler *et al.* (1998) has proposed an approach for the automated detection and identification of isotopically labeled pesticides and degradates using liquid chromatography in tandem with mass spectrometry, supplemented with pattern recognition software. The identification of isotopically labeled compounds has been used as fingerprints in complex matrices – for example, Pyrithiobac Sodium in rice leaf extract and Diuron in wheat straw extract (Drexler *et al.*, 1998).

8.5.6 Chemical Fingerprinting

Chemical fingerprinting has been used extensively in the field of environmental forensic science, and involves the comparison of unique chemicals, or patterns of chemicals, to rule in or rule out an association between two data sets, such as may represent a source and an environmental sample collected at some location.

Forensic fingerprinting of pesticide-contaminated sites can employ a wide variety of methods, ranging from simple visual comparisons to complex computer algorithms for data assessment and statistical comparison. For example, the Fingerprint Analysis of Leachaate Contaminants

(FALCON), developed by the US EPA, is an empirical data assessment and visualization tool that produces contaminant fingerprint patterns by combining data from two or more parameters to produce visually distinctive and reproducible fingerprints (Plumb, 2005).

The most common goal of fingerprinting pesticides is to correlate contamination in some environmental media to the source. Several species may be fingerprinted, depending on the pesticide and the site – parent pesticide compounds, pesticide degradates, commingled substances, trace substances, and radioisotopes.

Parent pesticide compounds and degradates are generally easy to fingerprint at pesticide-contaminated sites since they are unique chemicals. Commingled substances are more difficult to fingerprint due to their large number and relative ubiquity. Thus, except for certain circumstances, commingled substances may not offer the best method of forensic investigation, unless specific commingled substances can be identified from the manufacturer's formulation and then associated with the parent pesticide or degradates.

The fingerprints of trace contaminants associated with a pesticide may be useful in some instances. For example, the unique chemical congener patterns associated with trace Dioxin/Furan contamination in Pentachlorophenol or Agent Orange are such that their identification is generally possible in environmental samples, unless the sample contains a significant portion of confounder substances in addition to the pesticide's trace residue, which tend to blur the fingerprint pattern.

An evaluation of wetlands in Greece included the identification of current levels of organochlorine pollutants as a bioindicator. Residue levels of both Polychlorinated Biphenyls and 13 organochlorine pesticides were measured in cormorant eggs. Fingerprint and cluster analyses illustrated the overall differences in the PCB patterns, but for organochlorine pesticides the differences were smaller. Correlations of the pollutants varied considerably among areas studied and were more diverse in the organochlorine pesticides (Konstantinou *et al.*, 2000).

Fingerprinting has also been used to help evaluate whether current pesticide concentrations in soil in residential areas pose significant health risks to the residents and determine the probable source of contamination. In such cases, chemical fingerprinting techniques can show a different suite of pesticide compounds in residential areas than at the source (Sheehan and Finley, 2001). For example, Toxaphene is a complex mixture of over 200 different compounds that is difficult to sample and analyze. Since Toxaphene use was banned in 1982, the analytical fingerprint of environmental samples often differs considerably from the analytical standards due to changes over time in its chemical nature. Also, the analytical limit of detection is much higher than for other organochlorine insecticides (Majewski and Capel, 1995).

In China, large amounts of Pentachlorophenol salt were sprayed during the 1970s over vast areas to control schistosomiasis, a parasitic disease of epidemic proportions. By comparing specific congeners of the higher chlorinated Dioxins and Furans associated with the Pentachlorophenol in sediment and human tissue samples, scientists concluded that Pentachlorophenol was the source of environmental and human Dioxin exposure in the Chinese schistosomiasis area studied (Schecter et al., 1996).

As instrument detection limits improve and as newer detection techniques become more widely used, more pesticide degradates will be discovered, identified, and quantified (Koester et al., 2003).

Some pesticides, such as Arsenic compounds, have been fingerprinted using a variety of evaluation procedures for identifying the source, including assessment of spatial patterns, correlation with construction date and land use, contaminant speciation, and assessment of geochemical and depth profiles (Folkes et al., 2001).

Pesticide contamination can be manifested in a wide variety of environmental media, including surface and subsurface soils, sediments, groundwater, surface water, ambient air, and indoor dust. Even tree bark samples have been used to help study the movement of organochlorine pesticides in the atmosphere from relatively warm source regions of the world to colder, higher latitude regions where they condense onto vegetation, soil, and bodies of water. This phenomenon, known as the global distillation effect, could be the cause of such pollutants in the Arctic regions (Simonich and Hites, 1995).

8.5.7 Spatial Distribution of Contaminants

Fortunately, the fate and transport of pesticides are similar to that of other contaminants and, accordingly, the assessment of the spatial distribution of the contamination at a site may be used in many instances to identify the source and age of the release. The methodology for such assessment is well known, and will not be discussed extensively in this chapter, other than to point out that, unlike many contaminants, both the parent compounds and the degradation compounds should be assessed.

In conducting forensic investigations dealing with exposure to pesticides, the environmental expert must focus attention on the existence, nature and extent, and exposure routes of the pesticide, rather than solely on whether the observed levels resulted in, or could have resulted in, any impact to the environment or human health. Further to this point is the phenomenon of contaminant half-life in many instances; measuring current pesticide levels will yield misleading forensic information if it is not recognized that exposure levels historically may have been much higher.

At many sites, the concentration of the degradate chemicals may exceed the concentration of the parent compound. Further, seasonal variation in terms of growing season, rainfall, temperature, and other factors can occur, such that the concentrations vary significantly over time. Depending on when the environmental samples are collected, the samples may or may not be representative of usual site conditions. Since a single sampling event is simply a snapshot in time, the data generated reflect only those site conditions at the time of sampling, and may not be an accurate estimate of the exposure to organisms over a longer period of time. Snapshot sampling of pesticide-contaminated sites can, therefore, lead to improper conclusions.

Since some pesticides are extremely hydrophobic (i.e., tightly bind to organic materials and the surface of soil particles), it is possible that analytical laboratory techniques may significantly under-report the actual levels of contaminant in the matrix and the mass, of a contaminant that is bioavailable (i.e., the absorbed dose). The result is a quandary of why a certain species of organism is, or is not, impacted; why humans would experience certain illness, in lieu of low contamination levels; or why humans would not experience adverse health effects in spite of high concentrations of a toxicant.

Other pesticides, such as DDT, have an extremely low solubility in water, and the analytical testing laboratory may overreport the concentration in water, if the water sample was not filtered and the sediment analyzed separately.

8.6 CASE STUDIES

8.6.1 Case Study #1

In the Gulf Coast, rice farmers typically raise crayfish between rice growing seasons as a supplemental source of income. When a sudden decrease in crayfish population was observed in 1999, various explanations were offered for the dramatic decrease in crayfish production, including a severe seasonal drought and the application of a new insecticide, No-Weevil 2000 (fictitious name).

Rice farmers face a constant problem in controlling rice water weevils, which feed on their rice crops. In 1998, the US EPA banned the use of Furadan, which had been used to combat the rice water weevil. The replacement issued in 1999 was a pesticide sold under the brand name of No-Weevil 2000. No-Weevil 2000 is a phenylpyrazole insecticide, which has proven very successful in controlling the weevil. However, it is also highly toxic to crayfish at very low concentrations, as are its degradation metabolites.

No-Weevil 2000's toxicity to crayfish played an important role in determining the cause of the marked rise in crayfish mortality. Rice and crayfish are grown in the same fields, typically on a two-year cycle. Rice is planted and harvested the first year, and crayfish are seeded. The second year crayfish are harvested.

Rice farming practices require a constant source of water that can be drawn upon to flood fields, and subsequently receive the sediment-laden "tailwater" drained off the fields. Bayou systems provides such a source, and therefore most crayfish and rice fields use its surface water for irrigation. The continual influx and discharge of water into the same, slow-moving surface water source allows for tailwater drained off of one rice field to enter a downstream field – of either rice or crayfish—with almost no dilution, especially during dry weather. An irrigation schematic is shown in Figure 8.6.1.

The low crayfish population in 2000 appeared to correlate with the use of No-Weevil 2000 on rice crops the previous year. Public concern prompted a government-funded study in 2000 to determine No-Weevil 2000's presence throughout the region. A private study by Green University was also conducted in 2000 to analyze No-Weevil 2000 concentration in soil, water, and plant samples from specific fields.

The purpose of the study was to assess the presence and concentration of No-Weevil 2000 throughout the entire regional basin. The regional team collected soil and surface water samples throughout the regional basin, and analyzed the samples for No-Weevil 2000 and its degradation products.

Subsequent studies found that in surface water samples the highest concentrations of the parent compound and two of its degradate compounds corresponded to the release of rice field tailwater. Concentration of one metabolite was highest in June. Samples of suspended sediment revealed concentrations of No-Weevil 2000 and its degradation products to be 1–10% of the concentrations in water. The parent compound was not found to be accumulating in bed sediment, though its metabolites were accumulating.

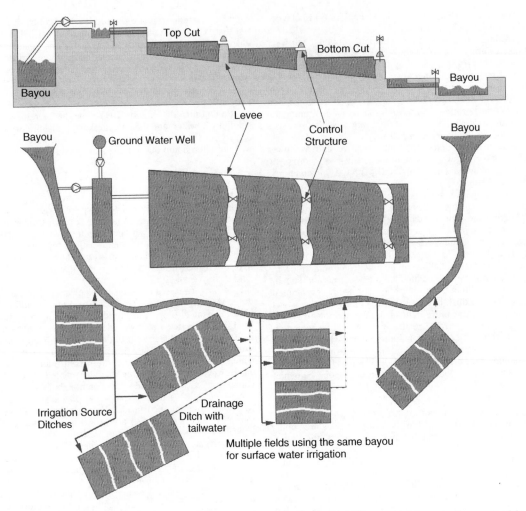

Figure 8.6.1 *Irrigation systems for rice and crayfish fields. Water source is either groundwater or surface water (shown here on the same figure for brevity). Many surface water fields draw irrigation water from the same water sources, which will also contain tailwater from other fields.*

Green University conducted additional studies to investigate the effects of No-Weevil 2000 on crayfish during April 2000. More than 100 surface soil, sediment, and surface water samples were collected and tested for both the parent compound and degradates.

The sampling strategy was to collect at least two water and two soil samples from each field to cover each of the varying field types. Farmers in the area were contacted and questioned for general information regarding rice/crayfish cycle, surface/ground water usage, and No-Weevil 2000 usage, as well as their crayfish production during the past two years.

As indicated in Table 8.6.1, the only fields which produced normal levels of crayfish in 2000 were groundwater irrigated fields which did not receive No-Weevil 2000. Fields on surface water irrigation were all impacted to some extent by the use of No-Weevil 2000 in 1999; fields that received a direct application of No-Weevil 2000 did not produce any crayfish; and the remaining fields saw a dramatic decrease in production when No-Weevil 2000-laden tailwater was introduced.

In 2003, a lawsuit was filed against the manufacturer of No-Weevil 2000. The lawsuit alleged that the application of No-Weevil 2000 to rice fields along the Gulf Coast to control the rice water weevil had also severely impacted the crayfish industry, and that the pesticide had been introduced into the Texas market without adequate testing and without considering how No-Weevil 2000 would interact with the rice/crayfish rotation system used in the Gulf Coast.

Plaintiffs alleged that the impact to the crayfish industry was a direct result of the application of No-Weevil 2000, which contributed to three primary mechanisms of contamination – the use of coated rice seed, the use of contaminated tailwater from adjacent fields, and cross-contamination of rice seed. The extent of contamination was huge – more than 300,000 acres had been impacted, and the loss of crayfish productivity over the 1999–2004 time frame totaled in the hundreds of millions of dollars.

Despite the 2000 data available from various studies, the Defendants maintained that the impact to the crayfish industry was a result of extended drought conditions, and not the use of No-Weevil 2000. The Defendants alleged that No-Weevil 2000 readily degrades such that its persistency could not be a cause for the dramatic decline in crayfish

Table 8.6.1 Green University Study. Reported No-Weevil 2000 Usage and Crayfish Production

Field #	Crop Produced in 2000	Surface/ Groundwater	Year Crayfish produced	Crayfish Production Level	Year Rice Planted	No-Weevil 2000 on Rice
1	Crayfish	G	2000	No Production	1999	Y
2	Crayfish	G	2000	No Production	1999	Y
3	Crayfish	G	2000	Normal Production	1999	N
4	Crayfish	G	2000	Normal Production	1999	N
5	Crayfish	S	2000	No Production	1999	Y
6	Crayfish	S	2000	No Production	1999	Y
7	Crayfish	S	2000	Produced some until tailwater	1999	N
8	Crayfish	S	2000	Produced little until tailwater	1999	N
9	Crayfish	S	2000	Produced some until tailwater	1999	N
10	Crayfish	S	2000	Produced some until tailwater	1999	N
11	Rice	G	1999	Normal Production	2000	N
12	Rice	G	1999	N/A	2000	N
13	Rice	G	1999	N/A	2000	Y
14	Rice	G	1999	N/A	2000	Y
15	Rice	S	1999	Produced normally until tailwater	2000	Y
16	Rice	S	1999	N/A	2000	Y
17	Rice	S	1999	Produced normally until tailwater	2000	Y
18	Rice	S	1999	Produced normally until tailwater	2000	N

production. The Defendants presented data to demonstrate that No-Weevil 2000 was not persistent, but rather dissipates in rice fields in months, and therefore could not be responsible for any long-term reduction in crayfish crops.

The technical aspects of the case focused on gathering forensic evidence to support the allegations made by the Plaintiffs. ABC Consulting, Inc., was retained by the Plaintiffs and devised multiple field programs to collect soil and sediment data from representative rice fields and test for No-Weevil 2000. It soon became apparent to the Plaintiffs that No-Weevil 2000 degrades rapidly in a rice field environment, but that the degradates were not only extremely toxic but also very persistent, with half-lives in the order of years. Due to the continued presence of No-Weevil 2000 metabolites through 2003, ABC Consulting, concluded that No-Weevil 2000 metabolites will continue to persist in the fields for up to several years into the future.

Further, samples were collected at discrete soil layers (i.e., 0–1, 1–2, 2–3, and 3–6 inches bgs) rather than from the overall 0–6 inch layer, as previously done by the Defendants. The discrete data clearly demonstrated that the highest concentrations of the metabolites were in the upper 1–3 inches of soil, and that the collection of samples from the 0–6 inch depth would bias the concentrations low. ABC Consulting, also extrapolated from the fields they tested that other fields that have similarly been treated will also have residual No-Weevil 2000 and metabolites at such depths. ABC Consulting also determined that contamination likely exists deeper than 12 inches, and that there is more contaminant bound in the soil than can be readily measured in the lab as a result of analytical laboratory limitations.

During the ensuing trial, Green University and ABC Consulting made powerful arguments concerning the persistency of the pesticide and the impact it had on the crayfish industry, and prevailed in the case.

8.6.2 Case Study #2

The second case study involved a wood preserving facility in the Southeastern United States that was located in the heart of a residential area. The facility has been used to pressure treat various lumber products since the early 1900s, including railroad ties and telephone poles, using a variety of biocides, including creosote, pentachlorophenol, and chromated copper arsenate (CCA). All three substances are classifiable as pesticides.

Historically, residents complained of noxious odors from the facility and a variety of medical ailments, including chronic respiratory problems, nervous system disorders, gastrointestinal tract illness, skin rashes, and cancers.

In 2001, the impacted citizens (Plaintiffs) pursued legal action against the ABC Wood Preserving Plant (Defendants), alleging that the facility was the primary cause of the local illnesses. Subsequent toxicological and epidemiological studies conducted by retained medical experts on behalf of the Plaintiffs confirmed a very high incidence of various illnesses in the community surrounding the ABC Plant. The plant was in the midst of a residential area; however, to prove causation, environmental forensic identification was required.

Wood preserving biocides may include creosote or metal formulations such as Chromated Copper Arsenate (CCA), Ammoniacal Copper Quat (ACQ), Ammoniacal Copper Zinc Arsenate (ACZA), Copper DimethyldithoCarbamate (CDDC), or similar compounds. Creosote is comprised of as many as 10,000 individual chemicals, including Volatile Hydrocarbons, Phenols, and Polycyclic Aromatic Hydrocarbons. Pentachlorophenol, which is normally considered a restricted use pesticide, is also frequently used. A manmade compound, Pentachlorophenol is manufactured by the chlorination of Phenol. In addition to its toxic properties, the manufacturing process also frequently results in trace quantities of other substances, including Dioxins/Furans. Wood-preserving companies generally use any or all of these biocides for special applications or to serve market niches.

The ABC Facility used all three categories of biocide throughout its history. The facility had a long history of releases of contaminants to the environment, including spills, process upsets, and general mismanagement of wastes. However, beginning in about 1990, the facility had implemented a formidable program aimed at reducing off-site releases. This program included a dramatic reduction in air emissions and releases of contaminants to the soil and groundwater. Measures were implemented to control the migration of contaminants in the groundwater, including control of a dense non-aqueous phase liquid (DNAPL) plume, from off the plant property. In brief, the plant's architecture in 2004 was substantially improved over its architecture during its 1900–1990 time frame, and the current facility featured much improved operations and was viewed as a "clean" facility by the owner.

Interestingly, over the past decade the facility's management made the business decision to receive and burn waste wood from their customers. As a result, many of their customers took advantage of the offer, and purchased new railroad ties and telephone poles and returned the waste wood to the facility as a matter of convenience. This practice proved beneficial to the facility since large volumes of waste wood were received on an annual basis and used for fuel in the plant boiler system. The waste wood was not categorized or tested, and it was highly likely that various batches contained Creosote, Pentachlorophenol, CCA, and other contaminants.

To prove their innocence in the lawsuit, the ABC Facility retained an environmental consulting firm that conducted a quick, traditional multimedia environmental site assessment of the community surrounding the facility. The assessment concluded that:

- Dense non-aqueous phase liquid was confined to the plant boundary, and dissolved plumes of Volatile Organic Compounds (VOCs) and Semi-volatile Organic Compounds (SVOCs) extended only about 500 feet beyond the plant boundary at any point.
- Wastewater discharges had been reduced by more than 90% since 1990, and were no longer a threat to the community.
- Air emissions had been reduced by more than 95% since 1990, and were no longer a threat to the community.
- The use of Pentachlorophenol was discontinued in the late 1990s, and recent sampling of soils and groundwater indicated only *de minimus* levels.

The multimedia environmental site assessment also included a formal risk assessment which evaluated various exposure pathways to the residential neighborhood, including air, soils, sediments, surface water, and groundwater. The risk assessment addressed Volatile Hydrocarbons, Phenols, and Polycyclic Aromatic Hydrocarbons and concluded that the risk to the general population was extremely low.

Despite the results of the Plaintiff's risk assessment, follow-up health studies and epidemiological studies continued to show a statistically significant and disturbing increase in medical ailments in the community near the facility, especially cancers and immune disorders. Consequently, the Plaintiffs hired a forensic consulting firm to collect additional multimedia environmental samples, including soils, sediments, surface water, groundwater, indoor household dust, and ambient air.

The results of the sampling confirmed the Defendant's conclusions that current contaminant levels in the surface soil, sediments, surface water, and groundwater were low.

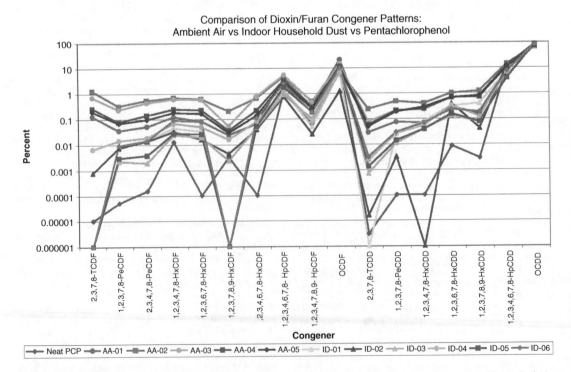

Comparison of Dioxin/Furan Congener Patterns: Ambient Air vs Indoor Household Dust vs Pentachlorophenol

Figure 8.6.2 Graph showing a comparison of Dioxin/Furan congener patterns from samples of ambient air, indoor household dust, and neat pentachlorophenol.

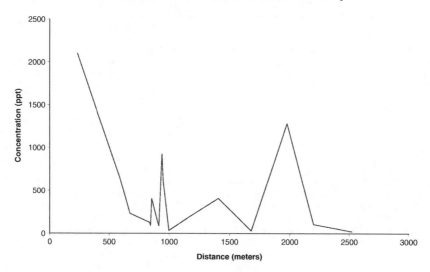

Dioxin/Furan TEQ Versus Distance From the Facility

Figure 8.6.3 *Dioxin/Furan TEQ levels in indoor dust versus the distance from the source.*

However, indoor household dust samples were collected in attics and behind appliances and indicated extremely high levels of Dioxins/Furans. Attic dust is an acceptable metric for assessing historical airborne contamination in residences (Lioy *et al.*, 2002). In some instances, the concentrations of Dioxins/Furans in the dust ranged from 50–2000 parts per trillion (ppt), significantly higher than the state regulatory risk-based Tier I remediation target level of 4 ppt for residential surface soils.

Following the finding of the high Dioxin/Furan levels in the dust samples, composite ambient air quality samples were collected using Summa canisters and high volume samplers with PUF filters, and indicated that low levels of Dioxins/Furans were still being emitted by the facility.

This data was rejected by the Plaintiffs who claimed that Dioxins/Furans are ubiquitous in the environment, and that the use of fireplaces and practice of trash burning had resulted in the Dioxin releases, not the facility. Further, the facility no longer used Pentachlorophenol, which they believed would be the only source of Dioxins/Furans, and that no chlorine was used in the plant which could result in the formation of Dioxins/Furans.

The ambient air quality samples and attic dust samples collected by the Plaintiffs' consultant were subsequently fingerprinted by comparing Dioxin/Furan congener patterns is the samples with the congener patterns from neat pentachlorophenol. The fingerprinting patterns of the indoor dust samples closely matched the patterns of neat Pentachlorophenol. Similarly, the ambient air samples, matched the patterns of the neat material as well, creating a strong case for linking the indoor dust samples to the operation of the plant. This was supported by the continued burning of waste wood from the facility's customers, some of which probably contained Pentachlorophenol. The fingerprint patterns are shown in Figure 8.6.2.

The Defendants proceeded with their flawed logic by performing classical air dispersion modeling studies that showed a decreasing concentration of Dioxins/Furans with distance from the facility. They compared the modeling results with the concentration of the Dioxins/Furans measured in the attic dust samples that were collected and tested by the Plaintiffs, which showed an inconsistent concentration gradient from the facility, as shown in Figure 8.6.3.

The Plaintiffs argued that, while it is true that an ambient air concentration gradient would be expected from the facility out into the community, the measurement of a contaminant in household attic dust does not necessarily follow such predicted gradients, since the measured concentrations can be skewed due to secondary sources of dust in residential attics. Further, the Plaintiffs conducted a confounder study to show that there was no other major sources in the immediate vicinity that could reasonably be expected to produce the concentration and spatial distribution of Dioxins/Furans observed, especially with the Pentachlorophenol congener pattern.

REFERENCES

Anon (2002) Tracing the paths of pollution. *New Agriculturist on-line.* Issue 28: 02-4. [On-Line]. Available http://www.new-agri.co.uk/02-4/focuson/focuson2html.

ATSDR. (1994) Toxicological Profile for Pentachlorophenol. *Agency for Toxic Substances and Disease Registry*, Atlanta, GA.

ATSDR. (2001) Toxicological Profile for Chlorinated Dibenzo-p-dioxins (CDDs). *Agency for Toxic Substances and Disease Registry*, Atlanta, GA.

Baird, C., and Cann, M. (2005) Environmental Chemistry. *Pesticides*, 337–343.

Bedos, C., Pierre Cellier, Raoul Clavet, Enrique Barriuso, and Benoît Gabrielle (2002) Mass Transfer of Pesticides into the Atmosphere by Volatilization from Soils and Plants: Overview. *Agronomie*, (22), 21–33.

Bennet, D.H., Furtaw, E.J. Jr and McKone, T.E. (2002) A Fugacity-Based Indoor Residential Pesticide Fate Model. *Lawrence Berkeley National Laboratory, University of California.*

Bonefeld-Jorgensen, E.C. (2004). The Human Health Effect Programme in Greenland, a review. *Sci Total Environ*, 331(1-3): 215–31.

Briggs, R., Schafer, J., Lyons, W., and Tong, W.G. (2004) Sub-Doppler high-resolution wave-mixing detection method for isotopes in environmental applications. *SPIE USE*, 4, 1–6.

Cherry, Eric M. (2003) Forensic Evaluation of Aircraft Deicing Fluids and Jet Fuel Components in Soil, Water and Air Compartments with Implications for Remediation on Surface Water and Groundwater Protection. *Presented at ISEF Virginia Beach, Virginia Workshop*, September 9–10, 2003.

Connell, D. (1997) Basic Concepts of Environmental Chemistry. *Pesticides*, 181–197.

Costa, L.G., Richter, R.J., Li, W.F., Cole, T., Guizzetti, M., and Furlong, C.E. (2003). Paraoxonase (PON 1) as a Biomarker of Susceptibility for Organophosphate Toxicity. *Biomarkers*, 8(1):1–12.

Crosby, D.G., and Ngim, K. K. (2001) Abiotic Processes Influencing Fipronil and Desthio Fipronil Dissipation in California, USA, Rice Fields. *Environmental Toxicology and Chemistry*, 20, 972–977.

Currie, Suzanne (2000) Pollution of the "Pristine" Artic. [On-Line]. Available www.sciencelives.com/artic.html.

Depledge, M., and Knap, A.H. (1998) Rapid Assessment of Marine Pollution: A Health of the Oceans Pilot Project. *Goos News*, No. 6, December 1998, 6–7.

Drenzek, Nicholas J., Tarr, C.H., Eglinton, T.I., Heraty, L.J., Sturchio, N.C., Shiner, V.J., and Reddy, C.M. (2002) Stable chlorine and carbon isotopic compositions of selected semi-volatile organochlorine compounds. *Organic Geochemistry*, 33, 437–444.

Drexler, Dieter M., Phillip R. Tiller, Sibylle M. Wilbert, Frederick Q. Branble, and Jae C. Schwartz (1998) Automated Identification of Isotopically Labeled Pesticides and Metabolites by Intelligent "Real Time" Liquid Chromatography Tandem Mass Spectrometry using a Benchtop Ion Trap Mass Spectrometer. *Rapid commun. Mass Spectrom*. 12, 1501–1507.

Ecobichon, D.J. (2001). Chapter 22: Toxic effects of pesticides. In Klaassen C.D., ed. *Casarett and Doull's Toxicology: The Basic Science of Poisons*, 6th edition (New York: McGraw-Hill, 2001), pp. 763–810.

Ferrari, M.J., Scott W. Ator, Joel D. Blomquist, and Joel E. Dysart (1998) Pesticides in Surface Water of the Mid-Atlantic Region. Water-Resources Investigations Report 97-4280. [On-Line]. Available http://md.water.usgs.gov/publications/wrir-97-4280/

Folkes, David J., Stephen O. Helgen, and Robert A. Litle (2001). Impacts of Historic Arsenical Pesticide Use on Residential Soils in Denver, Colorado. *Proceedings of the Fourth International Conference on Arsenic Exposure and Health Effects*, 18–22 June, 2000, San Diego, CA.

FAO/UNEP. (1996) Guidance for Governments: Operation of the Prior Informed Consent Procedure for Banned or Severely Restricted Chemicals in International Trade. FAO/UNEP, Rome/Geneva. [On-Line]. Available *http://www.oztoxics.org/waigani/library/dgd/2,4,5-ten.doc*.

Galloway, T.S., and Depledge, M.H. (2001). Immunotoxicity in invertebrates: measurement and ecotoxicological relevance. *Ecotoxicology*, 10(1):5–23.

Heemskerk, Richard A. and Robert J. Drimmie (2000) CF-IRMS at the Environmental Isotope Laboratory, Waterloo, Ontario, Canada. *Presented at the Geoanalysis 2000 Conference*, Special Session I: Reference Material and Analytical Development with Continuous Flow and Gas Mass Spectrometers (HCNOS).

HSDB. (2003) Hazardous Substances Data Bank [Internet]. Bethesda (MD): National Library of Medicine (US); [Last Revision Date 2003 February 14]. 2,4,5-T;

Hazardous Substances Databank Number: 1145 [about 14 p]. Available from: http://toxnet.nlm.nih.gov/cgi-bin/sis/search/f?./temp/~FYL0py:1.

Kendall, R.J., Anderson, T.A., Baker, R.J., Bens, C.M., Carr, J.A., Chiodo, L.A., Cobb (III), G.P., Dickerson, R.L., doxon K.R., Frame, L.T., Hooper, M.J., Martin, C.E., McMurry, S.T., Patino, R., Smith, E.E., and Theodorakis, C.W. (2001). Chapter 29: Ecotoxicology. In *Casarett and Doull's Toxicology: The Basic Science of Poisons*, 6th edition. New York: McGraw-Hill, 2001, 1013–1045.

Kiely, T., Donaldson, D., and Grube, A. (May 2004) Pesticide Industry Sales and Usage: 2000 and 2001 Market Estimates. *United States Environmental Protection Agency*.

Koester Carolyn, J., Staci L. Simonich, and Bradley K. Esser (2003). Environmental Analysis. *Annal. Chem.* 75, 2813–2829.

Kolpin, D., Thurman, E., and Linhart, S. (2000) Finding Minimal Herbicide Concentrations in Ground Water? Try Looking for the Degradates. *Science of the Total Environment*, 248(2–3), 115–22.

Konstantinou, I.K., Goutner, V., and Albanis, T.A. (2000). The incidence of polychlorinated biphenyl and organochlorine pesticide residues in the eggs of the cormorant (Phalacrocorax carbo sinensis): an evaluation of the situation in four Greek weblands of international importance. *Sci Total Environ*, 257(1), 61–79.

Lam, Paul, and Rudolf Wu. (2003) Use of Biomarkers in Environmental Monitoring. Presented at the Scientific and Technical Advisory Panel in December 2003.

Lin, P.H., Waidyanatha, S., Pollack, G.M., Swenberg, J.A., and Rappaport, S.M. (1999). Dose-specific production of chlorinated quinone and semiquinone adducts in rodent livers following administration of pentachlorophenol. *Toxicol Sci*, 47(1):126–133.

Lioy, P.J., Freeman, Natalie C.G., and Millette, James R. (2002) Dust: A Metric for Use in Residential and Building Exposure Assessment and Source Characterization. *Environmental Health Perspectives*, 110(10), 972.

Mackay, D. (1991) Multimedia Environmental Models: The Fugacity Approach. 1–257.

Majewski, M.S., and Capel, P.D. (1995). Pesticides in the atmosphere. Current understanding of distribution and major influences. USGS fact sheet FS-152-95. From a book by the same name (Ann Arbor Press, Inc.). Sacramento. 4pp.

Morrison, R.D. (1997) Forensic Techniques for Establishing the Origin and Timing of Contaminant Release. *Environmental Claims Journal*, 9(2), 105–122.

National Pesticide Information Center. (2003) Biomarkers of Exposure: Organophosphates. *2003 Medical Case Profile*. [On-Line]. Available http://npic.orst.edu/mcapro/archives.html.

Nigg, H.N., and Knaak, J.B. (2000). Blood cholinesterases as human biomarkers of organophosphorus pesticide exposure. *Rev Environ Contam Toxicol*, 163:29–111.

Pinkney, A.E., Harshbarger, J.C., May, E.B., and Reichert, W.L. (2004). Tumor prevalence and biomarkers of exposure and response in brown bullhead (Ameiurus nebulosus) from the Anacostia River, Washington, DC and Tuckahoe River, Maryland, USA. *Environ Toxicol Chem*, 23(3):638–647.

Plumb, R.H. (2005). Fingerprint Analysis of Contaminant Data: A Forensic Tool for Evaluating Environmental Contamination. EPA/600/S-04/054.

Reddy, Christopher M., Nicholas J. Drenzek, Timothy I. Eglinton, Linnea J. Heraty, Neil C. Sturchio, and Vernon J. Shiner (2001) Stable Chlorine Intramolecular Kinetic Isotope Effects from the Abiotic Dehydrochlorination of DDT. *Environ Sci & Pollut Res*, 8, 1–4.

Sawyer, C., McCarty, Perry L., and Parkin, Gene F. (2003) Chemistry for Environmental Engineering and Science. *Pesticides*, 279–288.

Schecter, Arnold J., Lingjun Li, Jiang Ke, Peter Frust, Christiane Frust, and Olaf Papke (1996) Pesticide Application and Increased Dioxin Body Burden in Male and Female Agricultural Workers in China. *Journal of Occupational & Environmental Medicine*, 38(9):906–911.

Schlenk, D., Huggett, D.B., Allgood, J., Bennett, E., Rimoldi, J., Beeler, A.B., Block, D., Holder, A.W., Hovinga, R., and Bedient, P. (2001) Toxicity of Fipronil and its Degradation Products to Procambarus sp.: Field and Laboratory Studies. *Archives of Environmental Contamination and Toxicology*, 41(3), 325–332.

Schnoor, J. (1996) Environmental Modeling. *Toxic Organic Chemicals*, 307–329.

Scribner, E.A., Thurman, E.M., and Zimmerman, Lisa R. (2000) Analysis of Selected Herbicide Metabolites in Surface and Ground Water of the United States. *Science of the Total Environment*, 248(2–3), 157–67.

Sheehan, P.J., and Finley, B. (2001). Assessing Residential Exposures to Pesticides and the Relation to a Former Pesticide Formulating Facility. *Presented at Society for Risk Analysis 2001 Annual Meeting*.

Simonich, Staci L., and Ronald A. Hites (1995). Organic Pollutant Accumulation in Vegetation. *Environmental Science & Technology*, 29(12), 2905.

Tardiff, R.G. ed., (1992) Scope 49: Methods to Assess Adverse Effects of Pesticides on Non-target Organisms, SGOMSEC 7-IPCS 16, Wiley, U.K. p. 304.

Timchalk, C., Poet, T.S., Kousba, A.A., Campbell, J.A., and Lin, Y. (2004). Noninvasive biomonitoring approaches to determine dosimetry and risk following acute chemical exposure: analysis of lead or organophosphate insecticide in saliva. *J Toxicol Environ Health A*, 67(8-10):635–650.

Tsai, C.H., Lin, P.H., Troester, M.A., and Rappaport, S.M. (2003). Formation and removal of pentachlorophenol-derived protein adducts in rodent liver under acute, multiple, and chronic dosing regimens. *Toxicol Sci*, 73(1):26–35; Epub 2003 Apr 15.

US EPA. (1997) Pesticide regulation (PR) notice 97-6, USEPA Office of Pesticide Programs. [On-Line]. Available http://www.epa.gov/opprd001/inerts/.

US EPA. (2005) The Inventory of Sources of Dioxin in the United States, USEPA Office of Research and Development. [On-Line]. Available (http://cfpub.epa.gov/ncea.html).

USGS (1997) Carbofuran – Insecticides: Estimated Annual Agricultural Use. *USGS Pesticide National synthesis Project*. [On-Line]. Available *http://ca.water.usgs.gov/cgi-bin/pnsp/pesticide_use_maps_1997.pl?map=W6007*

USGS (2003) Where are the Pesticides? *USGS Toxic Substances Hydrology Program*. [On-Line]. Available http://toxics.usgs.gov/highlights/herbicides_deg_gw.html.

Venturino, A., Rosenbaum, E., Caballero de Castra, A., Anguiano, O.L., Guana, L., Fonovich de Schroeder, T., and Pechen de D'Angelo, A.M. (2003). Biomarkers of effect in toads and frogs. *Biomarkers*, 8(3-4):167–186.

Vogue, P., Kerle, E.A., and Jenkins, J.J. (July 24, 1994) Oregon State University Extension Pesticide Properties Database. [On-Line]. Available (http://npic.orst.edu/ppdmove.htm).

Walker, *et al.* (2004) Use of Degradation Rates of DDT and Lindane Isomers for Determining the Timing of Release to Sediments of Greens Bayou: Houston Ship Cannel, Texas. *Environmental Forensics*, 5(1), 45–57.

Wessels, D., Barr, D.B., and Mendola, P. (2003). Use of biomarkers to indicate exposure of children to organophosphate pesticides: implications for a longitudinal study of children's environmental health. *Environ Health Perspect*, 111(16):1939–1946.

9 Perchlorate

Robert D. Morrison, Emily A. Vavricka, and P. Brent Duncan

Contents

9.1 INTRODUCTION

Potassium perchlorate was reportedly discovered in 1816 by Count Friedeick von Stadion of Vienna who created "oxygenated potassium chlorate" from a mixture of potassium chlorate and concentrated sulfuric acid. Chlorine dioxide was generated leaving a residue of potassium sulfate and potassium perchlorate. Additional research by Count von Stadion revealed that potassium perchlorate could be prepared by neutralizing perchloric acid with potassium hydroxide and electrolyzing a saturated solution of potassium chlorate between platinum electrodes. France, Germany, Switzerland, and the United States began production in the 1890s.

Perchlorate (ClO_4^-) is a contaminant whose presence in the environment and origin is of considerable environmental interest. Perchlorate is an emerging chemical of concern due to its presence in many drinking-water aquifers throughout the United States and its potentially deleterious toxicological properties (United States Environmental Protection Agency, 1998; Urbansky, 1998; Damian and Pontius, 1999; Gu *et al.*, 2000a). Perchlorate has been detected in aquifers and the environment in 34 states in the United States and Puerto Rico (Orris *et al.*, 2003; Mitguard and Mayer, 2004). Perchlorate has also been detected in surface water in the Colorado River and Lake Mead in Nevada (USEPA, 1998 Logan, 2001; Roefer *et al.*, 2004). The concentration of perchlorate in Colorado River water currently being delivered to Southern California is about 5 ug/l (California Regional Water Quality Control Board, Santa Ana Region, 2004).

While perchlorate was detected in groundwater in the United States as early as 1955 in eastern Sacramento County, California, at concentrations ranging from 1 to 18 mg/l (California Department of Water Resources, 1958; California Department of Water Resources, 1960), the ability to consistently detect perchlorate at low ug/l concentrations occurred in the United States only in 1997. Perchlorate's toxicological properties at low concentrations have also only recently been examined in detail (Lamm and Doemland, 1999; Li *et al.*, 2000a, b; Renner, 2004). The primary toxicological issue is associated with its inhibition of the thyroid gland. (Siglin *et al.*, 2000; Lawrence *et al.*, 2000, 2001; York *et al.*, 2001; Greer *et al.*, 2002; USEPA, 2004; Strawson *et al.*, 2004 National Research Council, 2005 Tellez *et al.*, 2005). The thyroid gland is deprived of iodide and becomes inactive relative to maintaining a healthy hormonal balance in the human body. This hormonal imbalance can lead to thyroid cancer.

The key physiochemical properties of perchlorate salts (ammonium, potassium, magnesium, and sodium) include their non-volatility, stability in aqueous solutions, non-reactive properties with most metal salts (Fe, Cr, Mn, Ni), the ability to form brines (i.e., $NH_4ClO_4 = 1.11\,g/cm^3$) and the high solubility of its salts, even in organic solvents (Sowinski *et al.*, 2003; Tipton *et al.*, 2003; California Environmental Protection Agency, 2004). The density, solubility in water and mass percent perchlorate as ammonium, potassium and sodium salts are summarized as follows:
These physiochemical properties have resulted in the presence of perchlorate plumes in drinking water aquifers hundreds of feet in depth and miles in length.

Perchlorate Compound	Density (g/cm^3)	Solubility ($10^3\,mg/L$)	Mass % ClO_4^-
Ammonium (NH_4ClO_4)	1.95	217–220	84.64
Potassium ($KClO_4$)	2.53	7.5–16.8	71.78
Sodium ($NaClO_4$)	2.02–2.499	2,010	81.22

This chapter presents information on the chemistry, sources of perchlorate, analytical techniques, and available forensic techniques to distinguish between naturally occurring and anthropogenic origins.

9.2 PERCHLORATE CHEMISTRY

Perchlorate (ClO_4^-) is a non-volatile anion composed of one chlorine atom surrounded by four oxygen atoms and has a molecular weight of 99.45. The central chlorine atom has an oxidation number of +7, making it the most oxidized form of chlorine and, therefore, kinetically stable. While this characteristic makes the perchlorate anion an excellent oxidizing agent, it is slow to react (perchlorate adsorbs weakly to most soil minerals) and is resistant to reduction as it travels through the subsurface. X-ray diffraction of hydronium perchlorate indicates that the perchlorate ion has a nearly perfect tetrahedral geometry with an average chlorine–oxygen bond distances of about 1.42 angstroms. Perchlorate is stable in most subsurface environments and because of its negative charge it has little to no affinity for soil minerals (low soil-water partition coefficient).

Perchlorate at ambient temperatures exhibits low to moderate decomposition at temperatures from 100 to 450 °C Perchlorate can exist in numerous forms including metal perchlorates (Ba, Cd, Cu, Mn, Ni, Ag, Pb, Zn), ammonium and alkali metal forms (NH_4, Li, Na, K, Rb, Cs), transition metal perchlorates (nitronium, nitrosyl, hydrazine, hydroxylammonium, phosphonium, selenious acid salt, perchloryl fluoride, and halogen perchlorates), and organic perchlorates (methylamine, pyridine, benzenediazonium urea and thiourea, bis(1,10-phenanthroline)copper). Other examples of organic forms of perchlorate include aromatic aldehyde, ketones, ethers, and various pyran compounds that can be combined with perchloric acid to form crystalline salts.

Perchlorate salts are the most frequently encountered form, due in part to their high solubility and stability when dissolved in water. The solubility of ammonium perchlorate is estimated to be about 200,000 mg/l; the sodium, calcium and magnesium salts are even more soluble. (Flowers and Hunts, 2000). The order of solubility of the more common perchlorate salts is sodium > lithium > ammonium > potassium (Mendiratta *et al.*, 1996). The solubility of selected alkaline earth, metal, and perchlorate salts in seven different solvents is summarized in Table 9.2.1 (Long, 2001; Schilt, 2003).

The perchlorate anion is the product of the electrochemical oxidation of chlorate (ClO_3^-) as described in the following half-reaction:

$$ClO_3^- + H_2O \rightarrow ClO_4^- + 2H^+ + 2e^-; \; E^o = -1.23\,V \quad (9.2.1)$$

The reduction pathway for perchlorate is described as follows:

$$\underset{\text{Perchlorate}}{ClO_4} \rightarrow \underset{\text{Chlorate}}{ClO_3} \rightarrow \underset{\text{Chlorite}}{ClO_2} \rightarrow \underset{\text{Hypochlorite}}{ClO^-} \rightarrow \underset{\text{Chlorine}}{Cl_2} \rightarrow \underset{\text{Chloride}}{Cl^-}$$
$$(9.2.2)$$

while a hypothesized pathway for the biological reduction of ClO_4^- is described as:

$$ClO_4^- \rightarrow ClO_3 \rightarrow ClO_2 \rightarrow Cl^- + O_2 \quad (9.2.3)$$

9.3 SOURCES OF PERCHLORATE

Sources of perchlorate include anthropogenic and naturally occurring. The bulk of anthropogenic sources are associated with military and industrial applications (see Table 9.3.1)

Table 9.2.1 *Solubility of Selected Perchlorate Salts, Metals, and Alkaline Earth Compounds in Various Solvents at 25°C (Long, 2001; Schilt, 2003)*

Solubility (grams per 100 g of solvent)	NH_4	Li	Na	K	Mg	Rb	Ca	Sr	Ba	Cs
Water	24.922	59.71	209.6	2.062	99.60	1.33	188.60	309.67	198.33	2.0
Methanol	6.862	182.25	51.36	0.105	51.83	0.06	237.38	212.01	217.06	0.093
Ethanol	1.907	151.76	14.71	0.012	23.96	0.009	166.24	180.66	124.62	0.011
n-Propanol	0.387	105.00	4.88	0.010	73.40	0.006	144.92	140.38	75.65	0.006
Acetone	2.260	136.52	51.74	0.155	42.88	0.095	61.86	150.06	124.67	0.15
Ethyl Acetate	0.032	95.12	9.64	0.001	70.91	0.016	75.62	136.93	112.95	0.0
Ethyl Ether	0.0	113.72	0.0	0.0	0.291	0.0	0.261	0.0	0.0	0.0

(Mesard and McNab, 2005). Perchlorate is also naturally occurring and is present in fertilizers and as a naturally occurring salt in geologic formations associated with arid depositional environments in the United States, Bolivia, and Chile.

9.3.1 Military and Industrial Sources

The first commercial manufacturing of ammonium perchlorate was developed in 1895 by Oscar Carlson, managing director of Stockholms Superfosfat Fabriks Aktiebolag in Sweden. This product was intended for use as an ingredient in a new type of explosive as it was found to be an excellent oxygen carrier for the explosive combustion of sulfur and various organic compounds. Ammonium perchlorate, whose products of combustion are gaseous, was found to be superior to the potassium and chloride salts that form solid potassium chloride. Patent coverage for this product was granted in 1897.

Perchlorate has been produced in the United States since the 1890s (National Organic Standards Board Technical Advisory Panel Review, 2002; National Research Council, 2005). In the 1940s and 1950s, perchlorate's strong oxidative properties were recognized and resulted in a significant global increase in production that is estimated to have increased from 1800 tons prior to the 1940s to 18,000 tons by mid-1940 (Davis, 1940; Hampel and Leppla, 1947; Medard, 1950). During this time a variety of perchlorate combinations were researched. For example, perchlorate mixtures with organic materials such as aniline, phenylenediamines, benzidine, toluidines, naphthylamines, aminoazobenzenes, antipyrine, malachite green, fuschsin, methyl violet, pyridine, quinoline, and cinchonine were found suitable for use as explosives.

Table 9.3.1 *Sources and Uses of Perchlorate Compounds (after Mesard and McNab, 2005)*

Uses and Sources		Documented Perchlorate Compounds
Formula	Name	Reported Uses
$HClO_4$	Perchloric Acid	Analytic reagent, strong acid, digesting organics, AP synthesis, electroplating, electropolishing
$C_{24}H_{48}NH_4ClO_4$	Tetra-*n*-hexylammonium Perchlorate	Electrochemical (polarographic) analysis medium
NH_4ClO_4	Ammonium Perchlorate (AP)	Propellant oxidizer, solid rocket boosters, combustible cartridge cases, flares, igniters, incendiaries, smoke-generating compositions
$Ba(ClO_4)_2$	Barium Perchlorate	Oxidizing colorant in green flares
$Cd(ClO_4)_2$	Cadmium Perchlorate	Catalyst for decomposition and explosion of AP
$Ca(ClO_4)_2$	Calcium Perchlorate	Incendiary flare compositions
$Co(ClO_4)_2$	Cobalt Perchlorate	Propellant burning rate modifier
$Cu(ClO_4)_2$	Copper Perchlorate	Propellant burning rate modifier
$N_2H_5ClO_4$	Hydrazine Perchlorate	Military explosive
$Fe(ClO_4)_2$	Iron (II) Perchlorate	Propellant burning-rate modifier
$Fe(ClO_4)_3$	Iron (III) Perchlorate	Ferric ion solution for water quality analysis
$Pb(ClO_4)_2$	Lead Perchlorate	Neutrino detection medium (astrophysics)
$LiClO_4$	Lithium Perchlorate	Oxygen candles, primary batteries
$Mg(ClO_4)_2$	Magnesium Perchlorate	Desiccant
$Hg(ClO_4)_2$	Mercury Perchlorate	Analytical reagent, catalyst for decomposition and explosion of AP
$Ni(ClO_4)_2$	Nickel Perchlorate	Detonator
$KClO_4$	Potassium Perchlorate	Pyrotechnic compositions, fusees, flares, combustible cartridge cases, igniters, photoflash compositions, smoke generators, tracers
$AgClO_4$	Silver Perchlorate	Catalyst for ignition of hydrazine monopropellant
$NaClO_4$	Sodium Perchlorate	Feedstock for manufacture of other perchlorates, flares, incendiaries, photoflash compositions, radiopharmaceuticals
$Sr(ClO_4)_2$	Strontium Perchlorate	Oxidizing colorant in red flares and pyrotechnics
$Zn(ClO_4)_2$	Zinc Perchlorate	Catalyst for decomposition, explosion of AP

Since World War II, millions of pounds of perchlorate have been synthesized for use by the military in the United States. The United States Environmental Protection Agency estimates approximately 90% of ammonium perchlorate manufactured in the United States is currently used by the Department of Defense, its contractors, and the National Aeronautics and Space Administration (General Accounting Office, 2004). Ammonium perchlorate is estimated to comprise about 70% (dry weight) of the solid propellant for the space shuttle rocket motors (aluminum dust ~30% and a small amount of iron oxide), 65–75% of the Stage I motors of the Minuteman III, and 68% of the Titan missile motors (Rogers, 1998). A single launch is of the space shuttle estimated to use up to 700,000 pounds of perchlorate propellant. The bulk of manufactured perchlorate is used as an additive to rocket and missile engine solid propellants (Mendiratta et al., 1996; Urbansky and Schock, 1999a; Logan, 2001; Vandenberg, 2004). Current perchlorate formulations for propellants generally contain ammonium perchlorate, aluminum, and a resin (i.e., carboxyl-terminated polybutadiene, glycerol sebacate) binder (Ajaz, 1995; Al-Harthi and Williams, 1998; Motzer, 2001). The thermal decomposition of ammonium perchlorate has been extensively studied given its effectiveness as a rocket propellant and explosive oxidant. Ammonium perchlorate is stable at 110°, decomposes at 130°, and explodes at 380°C. At temperatures below 300°, ammonium perchlorate decomposes according to the following reaction:

$$4NH_4ClO_4 \rightarrow 2Cl_2 + 8H_2O + 2N_2O + 3O_2 \qquad (9.3.1)$$

Above 400°C, most of the nitrogen is evolved as nitric acid.

A lithium perchlorate propellant has also been examined as it provides greater oxygen content and is thermally more stable than ammonium perchlorate. The thermal decomposition of lithium perchlorate which yields oxygen and lithium chloride occurs at 400°C.

Military sources of perchlorate other than as an additive to solid rocket fuel, include ~69 ~16, signal flares, colored and white smoke generators, numerous types of munitions, artillery tracers, incendiary delays, and railway torpedoes (Hatzinger et al., 2002). Many of these sources contain appreciable amounts of perchlorate or contain large percentages of nitrate which may have originated from Chilean deposits (Jackson et al., 2005). Historical examples of sources of potassium and ammonium perchlorate associated with World War II munitions produced by the United States and the percentage of perchlorate present by weight, for example, include emergency red signal flares (15–25%); emergency green signal flares (16%), 103-mm chaff rocket head (Bofors) (12.3%), 5-inch gun simulator (35%), booby trap illumination flare (73%), and hand-held parade flare (52%).

Non-military sources of perchlorate include:

- Match manufacturing (Ellington and Evans, 2000; Gullick et al., 2001; Mohr and Crowley, 2002; Greer et al., 2002; California EPA, 2003; Silva, 2003; United States Army Center for Health Promotion and Preventive Medicine, 2005);
- Tanneries (Motzer, 2001; Sharp and Walker, 2001; Mohr and Crowley, 2002);
- Gunpowder or flash powder (~67% potassium perchlorate with magnesium or aluminum dust ~33%) (Hussain and Rees, 1992; Rickman, 2005; Jackson et al., 2005);
- Signal and road flares, primarily ammonium and potassium perchlorate (Smith et al., 2001; General Accounting Office, 2004; Silva, 2003);
- Seismic explosives (Magnagel) (Jackson, 2004c; Jackson et al., 2005);

- Fireworks, including potassium and titanium perchlorates in firecrackers and primers, tungsten perchlorate in time delay pyrotechnics, and zirconium or boron with nitrocellulose and potassium perchlorate (Crump et al., 2000; Handy et al., 2000; Gullick et al., 2001; Smith et al., 2001; Orris et al., 2003; General Accounting Office, 2004; The Fertilizer Institute, 2004; Vandenberg, 2004);
- Fixer for fabrics and dyes (Motzer, 2001);
- Etchants made from compounds of Ce(IV) and aqueous perchloric is used to etch chromium and nichrome photoresists and masks, and for etching steels and stainless steels (Schilt, 2003);
- Paint and enamel production (California EPA, 2003; Silva, 2003);
- Desiccant (anhydrous magnesium perchlorate, barium perchlorate, and trihydrate of magnesium perchlorate) marketed under the trade names Dehydrite® and Anhydrone® $(Mg(ClO)_4)$ (Willard and Smith, 1922a; United States Environmental Protection Agency, 2002; Jackson et al., 2004a, 2005);
- Magnesium batteries; a German patent for a magnesium–magnesium dioxide battery contains 7.4–24% magnesium perchlorate, 6% carbon black, 2% barium chromate, 0.5% magnesium oxide, and 57% magnesium oxide (Augustynski et al., 1972; Logan, 2001; Motzer, 2001; Schlit, 2003);
- Lithium batteries: patents include lithium–nickel sulfide batteries that contain a lithium perchlorate solution in tetrahydrofuran as the electrolyte (Schilt, 2003);
- Air bag inflator for automobiles using potassium perchlorate along with a fuel such as azodicarbonimide and a binder; older formulations used ignition mixtures of boron, zirconium, aluminum, and/or magnesium with lithium perchlorate, sodium perchlorate, potassium perchlorate, ammonium perchlorate, and/or potassium nitrate (Smith et al., 2001; Schilt, 2003; Silva, 2003; General Accounting Office, 2004; The Fertilizer Institute, 2004; Vandenberg, 2004);
- Electropolishing of metals including mixtures of (1) acetic anhydride and perchloric acid, (2) acetic and perchloric acid, (3) acetic acid and sodium perchlorate, (4) ethanol and perchloric acid, (5) methanol and perchloric acid, (6) ethanol, ethylene glycol monobutyl ether, and perchloric acid, (7) dimethylsulfoxide and perchloric acid, and (8) glycerol, methanol and perchloric acid to polish aluminum, iron and steel, nickel and its alloys, razor blades, tin and lead alloys, zirconium and its alloys, gold, zinc, titanium, stainless steel, gadolinium, and A1-7 Si-0.3 magnesium cast alloys: (Schilt, 2003; Jackson et al., 2005);
- Chlorate defoliants (produced by electrolytic processes) (Jackson et al., 2005);
- Methamphetamine laboratories (Mohr and Crowley, 2002);[1].
- An ingredient of bleaching powder used in paper and pulp processing and calico printing (Chemical Land, 2004), and
- Cloud seeding (Vandenburg, 2004; Mohr, 2005).

The use of perchloric acid in chemical analysis is well documented and was advocated as early as 1920 for the determination of silica in silicates, although it was first synthesized

---

1. The association of perchlorate with methamphetamine laboratories is interesting. The United States Bureau of Narcotics Enforcement has revealed that large quantities of unburned highway flares and unburned matches have been found at methamphetamine facilities. The need for red phosphorous (used as a catalyst in the production of methamphetamine) is obtained by dissolving highway flare striker caps or striker pads from match books.

in 1816 via the vaccum distillation of a mixture of sulfuric acid and by the electrolysis of a saturated aqueous solution of chlorine dioxide. Perchloric acid is useful for decomposing slags, dried cement slurries, Portland cement, bauxite, and clay for identification of the analysis of silica. A summary of the uses of perchloric acid for a variety of chemical analysis applications are summarized in Table 9.3.2.

Field observations and laboratory testing suggests that perchlorate may be produced in steel tanks holding chlorinated water (~ 0.06 mg/l) that include an impressed corrosion protection system (~ 6 volts and cast iron anode). Perchlorate concentrations in water stored in a 400,000 gallon steel tank in Texas averaged 74 ug/l in a study for the Texas Commission for Environmental Quality (Tock

Table 9.3.2 *Uses of Perchloric Acid*

Application	Description
Chemical analysis of steel	Used for sample dissolution, to dehydrate the silica, and to oxidize the chromium. Determination of the presence of vanadium, chromium, chromite, and ferrochromium (Willard and Cake, 1920);
Nickel and molybdenum analysis	Used for the rapid determination of silicon, chromium, nickel, and molybdenum in steel and copper alloys (Schilt, 2003)
Manganese analysis	Used in a mixture with phosphoric and sulfuric acid to determine the concentration of manganese in tungsten and ferrotungsten, presence of chromium in chromite ores
Lead analysis	Lead analysis in chromium metal following its preliminary dissolution and oxidation.
Iron analysis	Used to determine the concentration of iron and cobalt in satellite (a non-ferrous alloy of chromium, cobalt, and carbon) based on the use of perchloric acid oxidation and the distillation of chromium as chromyl chloride to eliminate interference by chromium.
Carbonate analysis	Used as a decomposition agent in the microdiffusion method for the determination of carbonate in dolomite, siderite, calcite, and magnesite; used in the analysis of carbonate in alkali and alkaline earth carbonates;
Silica analysis	Determination of silica and R_2O_3 metals without using sodium carbonate in the mineral alunite ($K_2Al_6(OH)_{12} - (SO_4)_4$ (Smith and Taylor, 1963);
Potassium analysis	Used to precipitate and recover potassium from sea salt brine or bitten;
Trace metal analysis in organic matter	Used in the recovery and analysis of organic material containing trace metals using wet oxidation with nitric and perchloric acids (with the exception of mercury) for arsenic, cobalt, chromium, copper, manganese, lead, molybdenum, vanadium, and zinc.
Nitrogen determination (Kjeldahl method)	Perchloric and sulfuric acid used to digest samples (Bradstreet, 1954);
Cobalt	Ammonium perchlorate used to precipitate cobalt, nickel, manganese, and cadmium salts from aqueous ammonia
Phosphorous and calcium in plant and animal materials	Wet oxidation used with nitric and perchloric acid.
Sulfur content in coal, wood, paper pulp and rubber	Oxidation with nitric and perchloric acid is used to destroy organic matter with the conversion of sulfur to sulfate which is then precipitated and determined as the bariums salt (Spielholtz and Diehl, 1966);
Titrating agent	Used for titration of organic and inorganic bases; weak bases titrated with perchlorate include alkali halides, urea, and caffeine; used as a titrants for the precipitimetric or compleximetric determination of anions and organic complexing agents such as the use of mercury perchlorate for the precipitation titration of halides and pseudo-halides. Mercurous perchlorate has been recommended as a titrant for the reduction of iron (III) thiocyanate to the iron (II) state for the biamperometric determination of molybdate and gold. Thallic perchlorate has been used for the oxidimetric determination of thiourea, thiosulfate, and sulfite.
Endogeneous platelet serotonin	Used as an extraction reagent for the recovery of total endogeneous platelet serotonin and other platelet constituents such as nucleotides.
Deproteinization agent	Used as a precipitant for protein removal prior to determination of other constituents in biological fluids such as blood, milk, tissue, and urine (Seta *et al.*, 1980). Used to isolate protein-free metabolites, peptides, amines, and amino acids; used in the deproteinization of biological materials prior to the determination of formaldehyde and prior to citrate determinations.
Analysis of butterfat in milk and milk products	A combination of perchloric and glacial acetic acid is used to determine the butterfat content of milk products.
Steroid	Used as a chromogenic reagent to display different colors of steroids (Tauber, 1952).
Sodium and lithium extraction	Perchlorate is used to extract sodium lithium chloride from anhydrous mixed chlorides (Willard and Smith, 1922b).

et al., 2003). Subsequent laboratory studies indicated that high-silicon cast iron anode with DC voltage potentials greater than 6 volts were observed to generate more quantities of perchlorate than did a titanium anode operated at a 3-volt DC potential (Tock *et al.*, 2004). Materials used in anodes for the commercial production of perchlorate include platinum as the preferred anode, although manganese dioxide (MnO_2) and lead dioxide (PbO_2) are also acceptable.

The key conditions identified for promoting perchlorate production under these conditions included the following variables (Jackson, 2005):

- The presence of one or more of the following precursors: Cl^- (1–500 mg/l); ClO^- (0–5 mg/l); ClO_2^- (<0.5 mg/l); and/or ClO_3^- (<1 mg/l);
- A pH of 6–8 normal;
- An applied voltage, and
- A cast iron, titanium or niobium anode.

The overall electrolytic reaction for the electrolytic generation of perchlorate was described as:

$$Cl^- \rightarrow HOCl/ClO^- \rightarrow ClO_2^- \rightarrow ClO_3^- \rightarrow ClO_4^- \quad (9.3.2)$$

With the individual reactions described as

$$2Cl^- \leftrightarrow e^- + Cl_2 \rightarrow +2OH^- \leftrightarrow Cl^- + ClO^- + H_2O \quad (9.3.3)$$

$$6ClO^- + 3H_2O \rightarrow 2ClO_3^- + 4Cl^- + 6H^+ + 1.5O_2 + 6e^- \quad (9.3.4)$$

$$ClO_3^- + O_{ads} \rightarrow ClO_4^- \quad (9.3.5)$$

The researchers concluded that of the various nonchloride and hypochlorite ion variables, the voltage potential was a key component with the higher applied voltage rate and initial chloride concentration resulting in increased perchlorate production rate.

Perchlorate has been detected in the urine of cattle given thyreostatic drugs of the thiouracil type used in cattle fattening (Batjoens *et al.*, 1993). When given to cattle or poultry, ammonium, sodium, and potassium perchlorate extras a significant thyrostatic effect with reported weight gains of up to 20%. The perchlorate is believed to be metabolized and completely eliminated, primarily via urine, within 24–48 hours after introduction (Schilt, 2003).

9.3.2 Perchlorate in Fertilizer

The presence of perchlorate in fertilizer mined from caliche and brines with high sodium nitrate concentrations from Chile's Atacama Desert deposits have been known since 1886 (Beckurts, 1886; Dafert, 1908; Tollenaar and March, 1972; Ericksen, 1981, 1983). In 1896 potassium perchlorate was reported as a contaminant of Chilean saltpeter in amounts ranging from 0 to 6.79%; other researchers reported concentrations of perchlorate in crude and refined saltpeter of 1.5 and 1%, respectively (Schilt, 2003).

Deposits in the Atacama Desert contain the largest known natural reservoir of perchlorate in the world. These nitrogen-rich deposits have been used for fertilizer since 1830 with peak production occurring in 1930. The central Atacama Desert from which these nitrogen-rich deposits occur is the driest portion of the desert in northern Chile and consists of: the Coastal Zone, the Central Depression, and the Pre-Andean Ranges. The Central Zone receives less than 2 mm of rainfall per year and is devoid of normal hydrological processes. The perchlorate in these deposits is currently believed to have originated via atmospheric processes and subsequent deposition (Michalski *et al.*, 2004). The presence of iodate, which is believed to be of atmospheric origin, with perchlorate in groundwater samples from West Texas is also consistent with a hypothesis that perchlorate

in the Chilean deposits has an atmospheric origin (Dasgupta *et al.*, 2005). Perchlorate concentrations from five samples from the Central Depression ranged from nondetect to 250 mg/kg. No statistical association between perchlorate, nitrate, or chlorine concentrations were observed.

In the Atacama Desert, perchlorate contained in brines are pumped to the surface and concentrated in evaporation ponds. Lithium carbonate and boric acid are extracted from these ponds and the remaining salts are processed for potassium chloride, potassium sulfate, and fertilizers. The deposits mined for this fertilizer reportedly contain 6.3% nitrate, and 0.1–0.03% perchlorate (Grossling and Erickson, 1971; Urbansky *et al.*, 2001a, b; Collette *et al.*, 2003; Tian *et al.*, 2005). In the northern Chile fertilizer facilities in María Elena and Pedro of the Valdivia Sociedad Química y Minera S.A. (SQM), the company mines, refines, and markets its caliche products as Bulldog Soda through a North American subsidiary, Chilean Nitrate Corporation. Each year approximately 75,000 short tons of Chilean nitrate products are exported into the United States. Sodium nitrate (N–P–K ratio: 16–0–0) is also known as nitratine, soda niter, and nitrate of soda. The Chilean Nitrate Corporation markets the product to tobacco, citrus fruits, cotton, and some vegetable crops growers. SQM also sells this product to companies such as the Voluntary Purchasing Groups, Inc., or A.H. Hoffman, Inc., who then redistributes it to retailers such as Hi-Yield® or Hoffman® nitrate of soda. In portions of southern California, Chilean fertilizer was applied to citrus orchards at rates of about 100–500 pounds per acre. The ores and brines from these Chilean deposits have been exported throughout the world with an estimated worldwide production of about 23 million metric tons (in terms of nitrogen) (Dasgupta *et al.*, 2005).

The presence of perchlorate in groundwater originating from perchlorate-enriched deposits in the Atacama Desert is well documented (Crump *et al.*, 2000). In 1998, groundwater samples in the Atacama Desert in proximity to known nitrate deposits were tested and perchlorate concentrations were in the range of 1–10 mg/L. A well field at Agua Verde, Chile, from which groundwater samples were collected reported perchlorate concentrations in excess of 100 µg/L (Jackson *et al.*, 2004b).

In 2003, the United States Department of Agriculture analyzed 15 commercially available fertilizers from Colorado, Oklahoma, Alabama, Ohio, Texas, California, and New York for the presence of perchlorate (Hunter, 2001). The samples were extracted with sodium hydroxide rather than water which improved the extraction efficiency from 84 to 89%. The samples were analyzed using high performance liquid chromatograph (HPLC) and conductivity detector. The detection limit was about 20 ug/kg. These findings indicated that no perchlorate was present in these samples. The analyses of seven retail hydroponic nitrate fertilizer products, two liquid and five solid, were also tested for perchlorate. The test results indicated that three of the five solid products contained perchlorate ranging from about 100 to 350 mg/kg (Collette *et al.*, 2003). Other investigators have reported the presence of perchlorate in fertilizers not derived from Chilean nitrate deposits (Susarla *et al.*, 1999; Urbansky *et al.*, 2000a; Susarla *et al.*, 2000). In 2003, perchlorate was detected in a variety of commercial products at parts per billion levels (see Table 9.3.3) (Orris *et al.*, 2003). The samples listed in Table 9.3.3 were prepared by suspending 1 gram of the sample in 30 milliliters of de-ionized water in a 40-milliliter scintillation vial. The suspension was then centrifuged, filtered, and the samples analyzed by ion chromatography.

Table 9.3.3 *Perchlorate in Natural Minerals and Materials and Related Products (after Orris* et al.*, 2003)*

Samples	Concentration
Bloodmeal-Brand-1	5.4 ug/l
Bloodmeal-Brand-1-Duplicate	4.8 ug/l
Fishmeal	9.2 ug/l
Fishmeal-Duplicate	11 ug/l
Potash-Ore-1-(sylvinite)	25 mg/l
Potash-Ore-2-(sylvinite)	3741 mg/l
Potash-Ore-3-(sylvinite)	42 mg/l
Playa-Crust-1	1745 mg/l
Playa-Crust-2	560 mg/l
Playa-Crust-2-Duplicate	489 mg/l
Hanksite-1	280 mg/l
Hanksite-1-Duplicate	285 mg/l
Kelp	885 mg/l

9.3.3 Naturally Occuring Sources of Perchlorate

In addition to the presence of perchlorate in Chilean deposits used for fertilizer, other naturally occurring deposits of perchlorate have been reported. Early reports identifying the presence of perchlorate in Australian natural brine and saline deposits were later discredited (Becking *et al.*, 1958; Greenhalgh and Riley, 1960; Loach, 1962).

The presence of perchlorate in seawater was reported in 1958 by Becking *et al.*, who reported the detection of perchlorate at levels of 10 – 1000 mg/l (Becking *et al.*, 1958). Seawater samples collected 50 miles off the coast of wellington, New Zealand, in 1960 found no evidence of naturally occuring perchlorate. Subsequent studies including one in which over 30 surface- and deep-water samples were collected in the northern and southern hemispheres did not substantiate the presence of perchlorate in seawater which has been attributed to chloride interference. Perchlorate is currently not considered a significant constituent of seawater. Subsequent investigations, however, have confirmed the presence of perchlorate in the following materials:

- New Mexico and Canadian potash;
- Bolivian playa crusts;
- Searles Lake, California;
- West Texas and New Mexico, and
- Marine deposits in the Mission Valley Formation, San Diego, California.

9.3.3.1 New Mexico and Canadian Potash

Naturally occurring perchlorate has been detected in potash ore deposits in mines located near Carlsbad, New Mexico and from Saskatchewan (Canada). Potash ore samples from New Mexico contained perchlorate concentrations ranging from 25 to 3700 mg/kg. These ores are believed to be between 240 and 250 million years old (Renne *et al.*, 2001). Potash ore samples from Saskatchewan contained perchlorate at concentrations of 42 mg/kg (Orris *et al.*, 2003). Potassium chloride in the form of the mineral sylvite (potassium chloride) found in these potash deposits is also found in, ocean water, and brine from Searles Lake in California, suggesting a potential commonality with the presence of perchlorate.

9.3.3.2 Bolivian Playa Crusts

Perchlorate is present in the Bolivian playa crusts in the central Andean Altiplano (high plateau) of Bolivia. These deposits encompass an area of about 145,000 square miles. This brine evaporates in the dry season and leaves a layer of salt as thick as 20–26 feet. Two samples from these

deposits contained perchlorate in excess of 500 mg/kg (Orris *et al.*, 2003).

9.3.3.3 Searles Lake, California

Evaporative deposits from the alkali brine of Searles Lake includes the mineral hanksite ($Na_{22}K(SO_4)_9$ $(CO_3)_2Cl$), halite ($NaCl$) trona ($Na_2CO_3 \cdot NaHCO_3 \cdot 2H_2O$) borax ($Na_2B_4O_7 \cdot 10H_2O$), and glaserite ($K_3Na(SO_4)_2$). Samples from these deposits were collected and tested for perchlorate. Perchlorate was detected at concentrations of about 280 mg/kg (Orris *et al.*, 2003).

Of note is the presence of perchlorate detected in a "caliche" formation in the Death Valley region of California in 1988 (Ericksen *et al.*, 1988). The principle minerals of the caliche included many of the same minerals identified in the alkali deposits at Searles Lake including halite, trona, and borax; other distinctive minerals identified included darapskite ($Na_3(SO_4)(NO_3) \cdot H_2O$), glauberite ($Na_2Ca(SO_4)_2$), and tincalconite ($Na_2B_4O_5(OH)_4 \cdot 3H_2O$). The analysis of the nitrogen and oxygen isotopes in nitrate and the sulfur isotopes in sulfate from the Chilean and Death Valley deposits suggest the origin and accumulation of the nitrate salts from atmospheric deposition and the absence of soil leaching. The nitrogen and oxygen isotopic signature of the nitrate were reported to be significantly different from other natural or anthropogenic sources of nitrate (Bohlke *et al.*, 1997).

9.3.3.4 West Texas and New Mexico

Perchlorate has been detected in public water systems, private wells, irrigation wells, and wells monitored by the United States Geologic Survey over a 60,000 square mile area in 56 counties in West Texas (Gaines and Dawson Counties) and New Mexico (Jackson *et al.*, 2004a,b; Jackson, 2004c, 2005; Dasgupta *et al.*, 2005). Ten wells in the eastern counties of New Mexico covering an area of 6800 square miles were included in this testing.

Perchlorate concentrations (detection limit of 0.1 ug/l) in portions of the West Texas study area (Gaines and Dawson Counties) consistently measured over 20 ug/l with some samples with concentrations of nearly 60 ug/l. For 560 public water system wells sampled in 54 of the 56 Texas counties, 46% contained perchlorate >0.5 ug/l, and 18% contained perchlorate >4 ug/l. Analysis of 76 private wells resulted in 47% of the wells detecting perchlorate with 23 of the 36 wells having concentrations above 4 ug/l. All ten New Mexico wells had detectable concentrations of perchlorate (i.e., above 0.5 ug/l). Approximately 60% of the wells had concentrations above 4 ug/l.

9.3.3.5 Mission Valley Formation, San Diego, California

A source of perchlorate in a carbonate formation in southern San Diego County was identified in 2005 (Duncan *et al.*, 2005). The perchlorate was detected in the Mission Valley Formation which consists primarily of gray-colored, fine-grained marine sandstones dated at about 42 million years old. The Mission Valley Formation covers an area in excess of 23 square miles in southern San Diego County (Kennedy and Moore, 1971; Kennedy and Tan, 1997). The marine portions of this formation contain a variety of well-preserved fossils, including microfossils, bivalve and gastropod mollusks, crabs, and sea urchins. The marine depositional origin of this formation is of note as the chlorine found in Chilean nitrates, perchlorate-containing potash ores, and kelp is of marine origin (Orris *et al.*, 2003).

The composition and depositional patterns of the Mission Valley Formation exhibits similarities with other naturally occurring perchlorate sources identified in West Texas, New Mexico, and Chile. The genesis of the presence of

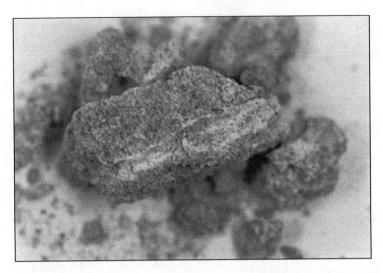

Figure 9.3.1 *Soil sample containing perchlorate from Mission Valley Formation, California (see color insert).*

perchlorate in this formation was the anomalous detection of perchlorate (EPA Method 314.0) in two shallow (<100 feet deep) groundwater monitoring wells located on a former trap and skeet shooting range. Perchlorate concentrations in groundwater samples from these wells ranged from 12 to 59 ug/l. While the source of perchlorate was originally believed to be associated with small quantities of perchlorate mixed with black powder in shotgun shells loaded by shooters, the pattern of perchlorate detected in the groundwater was inconsistent with this hypothesis. Intrusive trenching (<5 feet) in the vicinity of the two monitoring wells revealed the presence of a white, crushed fill material placed in the early 1970s as a result of the expansion of the shooting range trap houses. Further geologic investigations revealed the undisturbed presence of the same material in a nearby road cut (Figure 9.3.1).

The use of EPA Method 314.0 with sample preparation methods United States Department of Agriculture (USDA) Method 60, Chapter 6(2) & (3a), modified, resulted in a detection limit of 2.8 ug/kg. When samples were tested after using the USDA preparation method, perchlorate was detected at concentrations ranging from non-detect to 123 μg/kg. The saturated extract from the same material ranged from non-detect (detection limit of 2 ug/l) to 71 ug/l. Soil underlying the white crushed material contained perchlorate as high as 10.6 μg/kg with a saturated extract concentration from the same sample as high as 6.9 ug/l. The presence of perchlorate in the fill material and presence in the soil underlying the fill material were consistent with the transport of perchlorate through the soil column and into the shallow groundwater. The crushing of the carbonate layer into smaller particles is hypothesized to provide the greater surface area and/or disturbance to the cementation to allow the perchlorate to be leached into the groundwater. Other shallow monitoring wells in the study area located adjacent to and/or in the immediate vicinity of the carbonate layer that was consolidated, did not detect perchlorate.

An examination of two other locations in San Diego County where this carbonate-rich layer was mapped were identified in an effort to eliminate the potential impact of any blasting activities in the vicinity of the study area that may have resulted in the deposition of perchlorate.(Petrisor and Morrison, 2005). A location in which the same formation was exposed approximately 60 miles north of the study area was identified. Samples from this same marine layer were collected at this location.

Samples were extracted and analyzed in a manner identical to those used for the other soil sample. Perchlorate concentrations from the second locations were tested for wet weight, dry weight, and in the saturation extract; concentrations ranged from non-detect to 13.2 ug/kg, non-detect to 15.0 ug/kg, and non-detect to 8.5 ug/kg, respectively. The location from which these samples were collected excluded any potential impacts from blasting.

Given the similarities in depositional history (saline, marine environments) and arid settings for perchlorate detected in formations in West Texas and Chile, three samples that detected perchlorate were analyzed for total nitrate, sulfate, and chloride to examine whether a statistically significant relationship were apparent. A similar analysis of perchlorate in groundwater had been performed to examine relationships between the detection of perchlorate and field measurements (temperature, dissolved oxygen, redox potential, conductivity, total dissolved solids, and alkalinity as $CaCO_3$), anions (nitrate, nitrite, bromide, fluoride, sulfate and chloride), and cations (potassium, sodium magnesium, manganese, iron calcium and strontium) in the West Texas studies. For the West Texas studies, a statistical relationship between these anions, cations, and field measurements and perchlorate concentrations were not identified. A similar analysis of nitrate and chloride with perchlorate concentrations from soil samples collected from the Atacama Desert similarly did not identify a correlation.

An analysis of Mission Valley Formation soil samples in which perchlorate was detected was conducted for the major ions. The extracted liquid sample results from soil samples from the two locations detected these ions in the following concentrations:

Ion	Otay Mesa Road Cut	Rancho Bernardo
Calcium	77.4–556 mg/l	561–587 mg/l
Sodium	44.4–325 mg/l	110–1030 mg/l
Chloride	22–510 mg/l	23–620 mg/l
Nitrate as nitrogen	0.16–260 mg/l	0.11–0.49 mg/l
Ortho-phosphate	Non-detect	Non-detect
Sulfate	110–500 mg/l	1700–2300 mg/l
Total dissolved salts	620–3900 mg/l	3100–5300 mg/l

No statistical relationship between perchlorate concentrations and these ions was identified.

Of note was the reported correlation of perchlorate in groundwater samples in a nine-county area of Texas with iodate concentrations as analyzed by ion-pair chromatography (Gopal, et al., 2004). Iodate has been detected in a nitrate-rich vein of Chilean nitrate, as well as perchlorate and chromate (Bohlke et al., 1997). Iodate production in the trophosphere is well established (Wimschneider and Heumann, 1995; Vogt et al., 1999; Michalski et al., 2004). While the presence of chromate in the Chilean nitrate deposits is more difficult to explain, Dasgupta et al. (2005) suggested that the pathway may involve gaseous chromyl chloride (CrO_2Cl_2) that is formed under acidic and oxidizing conditions. Stable isotope evidence also exists suggesting the atmospheric origin of nitrate detected in Chile and southern California deposits (Bao and Gu, 2004). The atmospheric association of iodate, nitrate, and perchlorate therefore seems reasonable, especially given the presence of perchlorate in high altitude atmospheric sampling (Jaegle et al., 1996; Murphy and Thompson, 2000). Researchers in Texas analyzed 21 rain and 4 snow samples collected mostly in Lubbock, Texas, in 2003 and 2004 and found perchlorate in 70% of the samples at concentrations ranging from 0.02 to 1.6 ppb. Rain collected at Cocoa Beach, Fla., from Hurricane Frances contained 0.6 ppb. The presence of perchlorate with nitrate suggests deposits in an arid environment may indicate a similar atmospheric origin for the perchlorate detected in the Mission Valley formation. An atmospheric source and an arid environment may also explain the presence of perchlorate contamination in part of eastern Oregon (Lower Umatilla Basin). In 2004, sampling of 133 wells (monitoring wells, irrigation wells, domestic wells, a community well, and a livestock well) in Oregon that were known to have elevated nitrate levels were also tested for perchlorate. Perchlorate was detected in over half of the wells, with results ranging from non-detect to 25 ug/l.

The crystalline structure and mineral composition of samples from San Diego county that contained perchlorate was further examined to identify the crystalline structure of this material and possible similarities to the mineralogy of Chilean and Searles Lake deposits. Two samples were selected and examined using the following methods:

• X-ray diffraction analysis (XRD) (Berger and Cooke, 1998),
• Scanning Electron Microscope (SEM) analysis (Davis et al., 1993; Krinsley et al., 1998; Kennedy et al., 2002), and
• Polarized Light Microscopy (PLM).

The XRD results indicated that both samples were similar in composition and were composed primarily of smectite, calcite, and quartz with plagioclase, orthoclase, feldspar, and muscovite occurring to a lesser extent. A gypsum binder was also present. These findings are consistent with geologic reports describing the composition of the Mission Valley Formation (California Division of Mines and Geology, 1975).

The SEM scans of the material with atomic backscattering were similar to the findings of the XRD results with calcium and silica the major components. Figure 9.3.2 depicts a representative scan showing a mixture of calcite and quartz (jagged edge) along with plagioclase and sulfur.

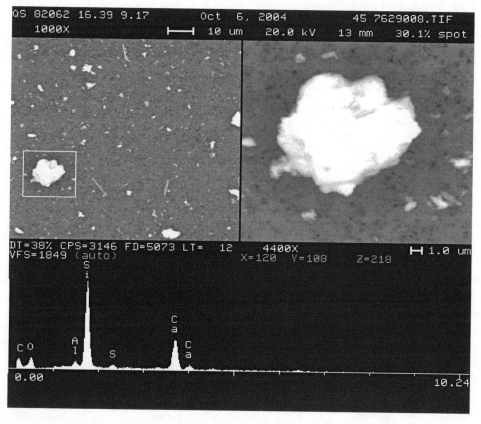

Figure 9.3.2 *Scanning Electron Microscopy (SEM) image of a sample from the Mission Valley Formation, San Diego County California containing perchlorate.*

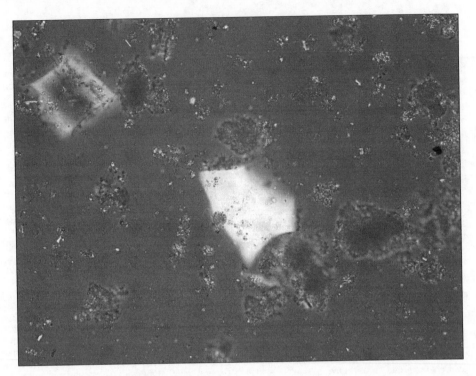

Figure 9.3.3 *Polarized light microscope image of perchlorate-enriched sample from Mission Valley Formation, San Diego County, California (see color insert).*

Polarized light microscopy was performed to examine mineral crystals and morphology on the same sample used for the SEM. A powdered sample was placed on a microscope slide with an oil of a known refractive index. The sample was then observed at 200× magnification with a polarizer, and with a second polarizer the sample was rotated slightly off 90 degrees. Figure 9.3.3 depicts this scan and includes a quartz crystal along with darker feldspar minerals. The brown material in the scan is muscovite. These results are consistent with those obtained with the XRD and SEM analysis.

No evidence of the presence of sylvinite (a sodium chloride–potassium chloride mineral with no less than 14% K_2O and small admixtures of magnesium and calcium salts) was detected in any of the XRD or SEM analysis. The presence of perchlorate in isolated samples of the mineral sylvite (potassium chloride) collected from New Mexico has been reported (Harvey, 1999). The general composition of minerals in the sample are summarized in Table 9.3.4.

Table 9.3.4 *Mineral Composition of a Perchlorate Enriched Sample from the Mission Valley Formation, San Diego, California*

Mineral	Composition	Concentration (% weight)
Smectite	$CaO_{.2}Mg_3Si_4O10(OH)_2 \cdot 4H_2O$	Major
Calcite	$CaCO_3$	Major
Quartz	SiO_2	Major
Plagioclase Feldspar	$(Na, Ca)AlSi_3O_8$	Minor
Orthoclase	$KAlSi_3O_8$	Minor
Muscovite	$KAl_2(Si_3Al)O_{10}(OH)_2$	Trace

9.4 ANALYTICAL TECHNIQUES

A variety of analytical methods are available to detect perchlorate in soil and water samples (Urbansky, 2000b). Historical techniques include the following:

- Gravimetric (Lamb and Marden, 1912; Willard and Thompson, 1930);
- Colorimetric (Junck, 1926; Feigl and Goldstein, 1958; Nabar and Ramachandran, 1959; Hayes, 1968; Shahine and Khamis, 1979);
- Atomic adsorption spectroscopy (perchlorate was complexed with copper) (Collison and Boltz, 1968; Weiss and Stanbury, 1972);
- Raman spectroscopy (Miller and Macklin, 1980; Kowalchyk et al., 1995; Gu et al., 2004b);
- Capillary electrophoresis with suppressed conductivity detectors (Avdalovic et al., 1993);
- Capillary electrophoresis with an ion-selective microelectrode (Nann, 1994);
- Ion-specific electrodes (Neuhold et al., 1996; Siswanta et al., 1997);
- Capillary electrophoresis and ion chromatography (Corr and Anacleton, 1996; Biesaga et al., 1997);
- Electrospray ionization mass spectrophometry (ESI-MS) (Urbansky et al., 2000c; Urbansky et al., 1999b);
- Ion chromatography with conductivity (Maurino and Minero, 1997; Jackson et al., 1999; Ellington and Evans, 2000).

The standard analytical technique for testing perchlorate in the United States is Environmental Protection Agency Standard Method 314.0 (Hautman and Munch, 1999). EPA Method 314.0 was published in 1999 and uses ion chromatography (IC) and conductivity. Reported detection limits for perchlorate using ion chromatography and a conductivity detector are at least 1 ug/l (Smith *et al.*, 2001). Method 314.0 requires the injection of a relatively large sample volume (1–2 ml) and uses a high anion exchange column (e.g., Dionex AS-16) to measure the retention time and conductivity. The presence of high salt contents results in significant signal suppression; as a result, low reporting limits can only be attained in samples with total dissolved solids concentrations less than 1 mg/l. Given that this method uses IC-conductivity, which is a low selectivity method, the results are influenced by sample matrix. P-chlorobenzenesulfonate, an industrial chemical, for example, can co-elute with perchlorate.

EPA Method 314.1 is an ion-chromatographic technique that uses a specialized on-line pre-concentration system that allows the use of large volumes of sample to obtain low reporting limits. After a sample (5–10 mls) is loaded into a column, it is eluted with a different mobile phase and then uses an anion exchange column to separate the perchlorate from other sample constituents and then employs a suppressor column to minimize background conductivity. The typical minimum reporting limit for this method is in the 0.2–0.3 ug/l range, even with high sample salt contents.

Recent development in liquid chromatography-mass spectrophometry (LC-MS) provide the ability to detect perchlorate at the nanogram per liter (ng/l) concentrations (Magnuson *et al.*, 2000a, b; Koester *et al.*, 2000; Roehl *et al.*, 2002; Winkler *et al.*, 2004; Zwiener and Frimmel, 2004). While LC-MS techniques for perchlorate detection is ultra sensitive, they are not widely used as ion chromatographic methods due to their cost and relative complexity, relative to IC methods. Currently, the United States EPA is assessing two proposed methods for perchlorate analysis, designated EPA Standard Methods 332.0 and 331.0. Method 332.0 is an ion chromatography-mass spectrometry (IC-MS) method. The method requires the use of a small sample (e.g., 100 ul) which is separated from other ions on a convention ion chromatographic anion exchange column. Interfering salts are routed to a matrix diversion system. The perchlorate is sent through a suppressor system to eliminate background noise and then enters the mass spectrometer through an electrospray interface. Rather than using conductivity, the intensity of mass 101 ($^{37}ClO_4^-$) and 99 ($^{35}ClO_4^-$) is using a single quadrupole mass spectrometer. This analytical technique has a minimum reporting limit of about 0.1 ug/l.

Method 331.0 is a liquid chromatography-mass spectrometry-mass spectrometry (LC-MS-MS) method. In this method, perchlorate is separated from other ions on an HPLC anion exchange column. The perchlorate sample (e.g., 100 ul) is then injected into a mass spectrometer through an electrospray unit interface. Rather than measuring the intensity of mass 101 and 99, secondary fragmentation occurs (second MS) to measure the daughter ions 85m/z ($^{37}ClO_4^-$) and 83m/z ($^{35}ClO_4^-$). The mass is then quantified against a calibration curve. An ion trap can also be used to detect these masses. While there are currently several versions of the LC-MS-MS method, the use of the isotopically labeled perchlorate as an internal standard that automatically corrects for signal suppression is of greatest interest. Reported minimum reporting limits can be as low as 0.02 ug/l. EPA Method 331.0 currently offers the greatest analytical sensitivity and specificity and absence of interferences.

9.5 FORENSIC TECHNIQUES

A number of forensic approaches have been proposed and are emerging for age dating and source identification of perchlorate. Several current approaches available for this purpose include (Vavricka and Morrison, 2005):

- surrogates analysis (nitrates, sodium, chlorides, phosphate, nitroglycerins, metals, strontium);
- stable isotopic analysis (oxygen, strontium, nitrogen, chlorine, tritium) and associated ratios ($^{37}Cl/^{35}Cl$, $^{87}Sr/^{86}Sr$, $^{16}O/^{17}O$), and
- geologic analysis (limestone, caliche, playa crusts, hanksite ($Na_{22}K(SO_4)_9(CO_3)_2Cl$), potash, phosphate ore, etc).

The presence of perchlorate with surrogate compounds associated with a known source represents a direct approach for age dating or source identification. These associations are source- and facility-specific, and ideally require an understanding of the compounds associated with the perchlorate source. For example, groundwater samples containing perchlorate from the Aerojet General Corporation Superfund site located near Sacramento, California, contains trichloroethylene, perchloroethylene, vinyl chloride, and nitrosodimethylamine as signature compounds (Motzer, 2001). Perchlorate associated with safety flares released into the environment may be distinguished from other sources of perchlorate through its association with strontium nitrate (75% by weight), sulfur (<10%), and/or oil (<10%). Strontium (brilliant red color) and barium salts, for example, provide color effects to flares while a small amount of potassium chlorate may be present with the perchlorate to provide a greater ease of ignition. Residue from the railroad safety flare materials, in association with perchlorate may provide evidence regarding the source of the perchlorate. Railroad flares historically contained vinyl chloride plastisols binders, dibutyl phthalate plasticizers, cellulose acetate butyrate, and ethyl cellulose.

Other compounds present in smaller amounts can include aromatic polycarboxylic anhydride fuel, benzene tetracarboxylic acid (dianhydride and metallic dianhydride), sodium nitrate, polyvinyl chloride, dextrin, magnesium, and red phosphorous (Silva, 2003). Potential surrogate chemicals associated with railway signal flares may include sulfur, nitrate, and carbonaceous materials that are proportioned to control the rate of burning. Another example of a surrogate approach involves the release of perchlorate from explosive fireworks, which depending on their composition may also include the following ingredients: potassium nitrate, barium chlorate, carbon and sulfur, potassium, aluminum powder, and sodium bicarbonate.

Surrogate metals associated with propellants and pyrotechnics linked with perchlorate include the following (Motzer, 2005):

> Propellants and Fuels: (Cr^{+3}, Cr^{+6}, Cu^{+2}; Zn);
> Fireworks: (Na, K, Ba, Cu^{+3}, Sb, Sr, Al, As, Cr^{+5}), and
> Flares: (Cu^{+2}, Sr, Ba, Al, Sb, Na, K).

Other metals that may be present in perchlorate containing products include zirconium, mercury, lead and iron.

Isotopic analysis can provide information regarding the source of the perchlorate as well as potentially distinguishing between anthropogenic and naturally occurring perchlorate. An isotope of an element is one that has one of two or more atoms having the same atomic number but different mass number (i.e., the same number of protons but a different number of neutrons). The advent of a new class of bifunctional anion exchange resins and resin regeneration techniques provides the ability to extract perchlorate from environmental samples (Gu, *et al.*, 2000a). The ability to concentrate small concentrations of perchlorate provides the basis to distinguish between perchlorate reagents, naturally occurring perchlorate salt deposits, salt-derived fertilizers, and perchlorate-enriched surface water and groundwater (Horita *et al.*, 2004; Erickson, 2004).

Stable isotope ratio analysis has been proposed and used as an identification technique to identify the isotopic composition of the original material (Winkler *et al.*, 2004; Coleman and Coates, 2004). Chlorine has been used for examining different sources of chlorine; because chlorine occurs primarily as a chloride ion, there are no redox reactions that induce significant changes in the bonding energy that can result in large isotopic variations in the chloride compound. Another assumption when using $\delta^{37}Cl$ is that the manufacturing process is the source of differences and not the source of the chlorine (all major sources of chlorine, whether from rock salt or marine brines, show small variations in $\delta^{37}Cl$). Therefore, if the perchlorate originates from a naturally occurring source, it would be expected to exhibit a very negative $\delta^{37}Cl$ value ($<-10‰$). Stable isotopic compositions from different perchlorate sources can exhibit a considerable range as shown in the following summary (Ader *et al.*, 2001; Motzer, 2005).

Supplier	Type	$\delta^{37}Cl(‰)$
BHD Chemical	$KClO_4$	+0.32
Fisher Scientific	$KClO_4$	+0.23
Fison Scientific	$NaClO_4$	+2.30
Chlorine Feedstock	Chlorine brine	~+0.09

Perchlorate isotope analysis for chlorine is performed by extracting and recovering perchlorate from water using bifunctional ion-exchange resin developed by Oak Ridge National Laboratory. The perchlorate is then converted to a form that can be isotopically analyzed using a gas source isotope ratio mass spectrometer as follows: CO for O_2 for oxygen isotope analysis and CH_3Cl for chlorine isotope (^{35}Cl and ^{37}Cl) analysis. The natural abundance of ^{35}Cl and ^{37}Cl is 75.77 and 24.33%, respectively (Philp, 2005). The precision of these measurements is -0.1 to $\pm0.3‰$. Figure 9.5.1 depicts the oxygen and chlorine signatures for known anthropogenic and naturally occurring sources of perchlorate (Sturchio *et al.*, 2004). The measured ä37Cl values of anthropogenic produced perchlorate range from -3.1

to $+1.6‰$, with an average of $+0.6 \pm 1.2$ which is similar to typical chlorine source values ($0\pm 2‰$) The ä37Cl values of perchlorate extracted from Atacama nitrate ore and from Chilean nitrate fertilizer products range from -14.5 to $11.8‰$ (Bohlke et al., 2005).

The stable isotopes ^{35}Cl and ^{37}Cl have been used to quantify the amount of chlorine isotope fractionation by bacterium capable of reducing perchlorate to chloride and oxygen (Romanenko *et al.*, 1976; Rikken *et al.*, 1996; Bruce *et al.*, 1999; Coates *et al.*, 1999b, 2000). The microbial reduction of perchlorate to chloride produces a dramatic shift in the isotopic composition of chloride because the kinetics of the process favors ^{35}Cl relative to ^{37}Cl. Chloride produced via microbial reduction therefore has proportionally less ^{37}Cl and a $^{37}Cl/^{35}Cl$ ratio approximately 15% less than the perchlorate from which it was produced (Coleman and Coates, 2004).

An investigation by Coleman and Coates (2004), was designed to examine the ability to isotopically distinguish between biological (anaerobic) and abiotic mechanisms (dilution, dispersion, adsorption) that resulted in perchlorate biodegradation. A perchlorate-reducing bacterium isolated from a swine waste lagoon, Dechlorosoma suillum, was selected for this purpose (Coates *et al.*, 1999a; Achenbach *et al.*, 2001; Sturchio *et al.*, 2003; Coleman *et al.*, 2003). Reported $\delta^{37}Cl$ values for perchlorate used in the biodegradation study used three samples of reagent $NaClO_4$ whose values ranged from $+0.2$ to $+2.3‰$ (parts per thousand) (Ader *et al.*, 2001).

Use of triple-oxygen isotope ratios $^{17}O/^{16}O$ and $^{18}O/^{16}O$ ratios are proposed as a means to distinguish between anthropogenic and naturally occurring sources of perchlorate (Sturchio *et al.*, 2004). As perchlorate is a non-labile oxy-anion, the oxygen atoms do not exchange with others in the ambient environment and therefore retain the oxygen signature of its sources (Hoering *et al.*, 1958). The measurement of the oxygen isotopic composition can therefore provide evidence regarding the original composition and subsequent alteration of perchlorate in the environment.

The oxygen isotope composition is defined as:

$$\delta^{18}O = (R^{18}_{sample}/R^{18}_{standard} - 1) \times 1000‰ \text{ or} \qquad (9.5.1)$$

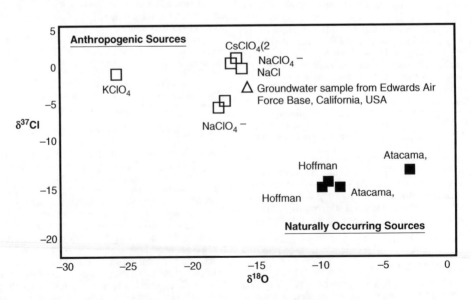

Figure 9.5.1 *Chlorine and oxygen isotopic ratios for known anthropogenic and naturally occurring sources of perchlorate (with permission, after Sturchio et al., 2004).*

$$\delta^{17}O = (R^{17}{}_{sample}/R^{17}{}_{standard} - 1) \times 1000‰ \qquad (9.5.2)$$

in which R is the abundance ratio of $^{18}O/^{16}O$ or $^{17}O/^{16}O$. The international standard for oxygen isotope composition is V-SMOW (Vienna-Standard Mean Ocean Water), which defines $\delta^{18}O = \delta^{17}O = 0‰$. For most oxygen-bearing compounds, a highly correlated relationship between their $\delta^{18}O$ and $\delta^{17}O$ values is evident due to the fact that most oxygen isotope fractionation processes are mass-dependent (Young et al., 2002) as defined by:

$$\delta^{17}O \sim 0.52 \times \delta^{18}O \qquad (9.5.3)$$

No independent information can be obtained from $\delta^{17}O$ if $\delta^{18}O$ has not been measured. Exceptions arise, however, in which the $\delta^{17}O$ and $\delta^{18}O$ relationship do not follow the previously defined mass-dependent fractionation relationship of $\delta^{17}O \sim 0.52 \times \delta^{18}O$. A parameter has therefore been developed to measure the deviation of the $\delta^{17}O$ value from the mass-dependent association as is expressed as $\Delta^{17}O$ (or ^{17}O anomaly) which is calculated by

$$\Delta^{17}O = \delta^{17}O - 1000 \times [(1 + \delta^{18}O/1000)^{0.52} - 1] \qquad (9.5.4)$$

Given that the atmospheric creation of perchlorate is known to have a high $\delta^{18}O$ as well as a high positive $\Delta^{17}O$ due to the contribution of atmospheric ozone (O_3), this knowledge provides a basis to identify naturally occurring sources of perchlorate (Johnson and Thiemens, 1997; Johnson et al., 2000). If the perchlorate originates from an atmospheric source, a positive $\Delta^{17}O$ is expected (Bao, 2005). Anthropogenic perchlorate, in contrast, produced by the electrolysis of an aqueous chlorate solution uses oxygen derived from water (i.e., $\Delta^{17}O = 0$) while perchlorate originating from a naturally occurring source will have a positive $\Delta^{17}O$ value (Grotheer and Cook, 1968).

The discovery of enrichments of ^{18}O in stratospheric ozone and the demonstration of mass-independent fractionation of oxygen isotopes (^{16}O, ^{17}O and ^{18}O) generated in laboratory and atmospheric ozone provide a means to use ozone as a signature tracer in oxygen-containing atmospheric chemistry (Thiemens et al., 2001). If perchlorate is of atmospheric origin and includes ozone (O_3), then a positive $\Delta^{17}O$ value is expected. Anthropogenic perchlorate produced by the electrolysis of an aqueous chlorate solution, however, would be expected to have a change in the $\delta^{18}O$ value along with a corresponding change in the $\delta^{17}O$ value (Grotheer and Cook, 1968; Bao and Gu, 2004). Bao and Gu (2004) used a highly selective bifunctional anion-exchange resin consisting of quaternary ammonium groups with large (C_6) and small (C_2) alkyl groups to extract and concentrate perchlorate from soils obtained from the Atacama Desert (Gu et al., 2000a). These highly selective, bi-functional anions were originally developed for removing radioactive pertechnetate (TcO_4^-) at the parts per trillion levels from groundwater at Paducah, Kentucky (Gu et al., 2000b, 2001, 2002, 2004a). Since perchlorate and pertechnetate have similar physiochemical properties (large, poorly hydrated anions), the bifunctional resin selectively sorbs perchlorate while leaving competing anions (SO_4^-, NO_3^-, Cl^-, HCO_3^-) in the water.

The results from these experiments resulted in $\delta^{18}O$ values for anthropogenic perchlorate ranging from -18.4 to $\pm 1.2‰$ as contrasted to naturally occurring perchlorate which had a variable $\delta^{18}O$ value ranging from -4.5 to $24.8‰$. Perchlorate extracted from Atacama Desert samples exhibited a large and positive ^{17}O anomaly ranging from $+4.2$ to $+9.6‰$. A comparison of total nitrate isotope compositions was also performed comparing samples from both the Atacama Desert and valleys. The $\delta^{15}N$ values for the

Table 9.5.1 Multiple Oxygen Isotope Compositions of Perchlorate from Natural and Anthropogenic Sources (after Bao and Gu, 2004)

Source of Sample	$\delta^{18}O$	$\delta^{17}O$	$\Delta^{17}O$
Anthropogenic Perchlorate			
$KClO_4$ (Allied Chemical)	−19.9	−10.5	−0.06
$KClO_4$ (Aldrich)	−17.8	−9.5	−0.12
$NaClO_4$ (EM-Science)	−17.3	−9.1	−0.06
$NaClO_4$ (Fisher Scientific)	−19.5	−10.4	−0.20
$AgClO_4$ (Aldrich)	−17.3	−9.1	−0.10
Atacama Desert Perchlorate[a]			
$CsClO_4^b$	−4.6	7.2	9.6
$CsClO_4$	−4.5	7.2	9.6
$KClO_4$	−24.8	−8.7	4.2
$CsClO_4$	−9.0	4.1	8.8

[a] First three samples obtained from the Central Depression of the Atacama Desert in Chile. The fourth sample is from Hoffman Nitrate derived from the Atacama Chile salt deposit.
[b] Major ions present in this sample included sulfate (7.4%), chloride (6.2%), and nitrate (5%) as measured by ion chromatography.

valleys and desert were found to range from $+13$ to $+25‰$, and -2.0 to $+2.0‰$, respectively (Bao, 2005). A summary of this data is found in Table 9.5.1.

All four of the Atacama perchlorate samples fall within a linear pattern when plotted along a straight line (dashed line) for the $\delta^{18}O-\delta^{17}O$ (see Figure 9.5.2 summarized in Table 9.5.1). The authors suggested that the significance of this observation is that there was probably a mixing of different natural perchlorate end-members or there was an unknown postformational alteration process.

For concentrations of perchlorate at low concentrations in groundwater, a significant amount of water is required in order to be exchanged onto the ion resin so that it can be concentrated for isotopic analysis. For perchlorate concentrations of 1 and 10 ug/l, for example, approximately 1000 and 100 gallons or more are required for isotope analysis, respectively (Personal Communication with Baohua Gu, March 1, 2005). A rule of thumb is that in order to obtain about 10 mg of perchlorate for isotopic analysis, approximately 1,000 liters of water containing 10 ug/l of perchlorate is required (Sturchio, 2005).

An isotopic method to trace the origin of perchlorate using the strontium isotope as a surrogate has been proposed, assuming that the perchlorate is associated with road flares containing strontium nitrate and/or sodium perchlorate ($NaNO_3$) associated with explosives and munitions (Hurst, 2004). The four naturally occurring Sr isotopes include ^{84}Sr (stable), ^{86}Sr, ^{87}Sr (stable radiogenic decay product of ^{87}Rb), and ^{88}Sr. Advantages associated with this approach include negligible fractionation and high analytical precision ($<0.002\% \pm 0.00002$). Differences have been observed among perchlorate sources by analyzing $^{87}Sr/^{86}Sr$ ratios. Samples from the Atacama Desert Salts in Chile, fertilizers, fireworks, road flares, and rocket propellant residue exhibit considerable differences in strontium isotope ratios (see Table 9.5.2). Strontium ratios are routinely used in archaeological forensic investigations (Graustein, 1989; Evans and Tatham, 2004).

The phytoremediation of perchlorate by poplar trees Populus deltoidex nigrawas may offer an innovative means to track the presence of perchlorate, given the appropriate environmental setting. Van Aken and Schnoor (2002) used radio-labeled $^{36}ClO_4^-$ to examine the degradation of perchlorate within the tissue of the poplar tree; results indicated

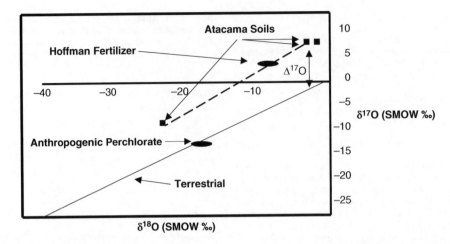

Figure 9.5.2 $\delta^{18}O - \delta^{17}O$ plot for perchlorate from natural and anthropogenic sources (with permission, after Bao and Gu, 2004).

Table 9.5.2 Isotopic Ratio of $^{87}Sr/^{86}Sr$ of Some Common Perchlorate Sources (after Hurst, 2004)

Source	Ratio $^{87}Sr/^{86}Sr$
Atacama Desert Salts (Chilean Nitrates)	~0.7070–0.7076
Fireworks (Santa Paula, California)	~0.7075–0.7078
Road Flares 30–75% $Sr(NO_3)_2$	~0.7069
Fertilizer (liquid, La Conchita, California, USA)	0.7080–0.7084
Fertilizer (Orange County, California, USA)	0.7089–0.7092
Rocket Propellant (soil extraction Vandenberg Air Force Base, California, USA	$0.70989 +/- 0.00002$

a reduction of perchlorate in an initial solution from different plant fractions of about 50% after 30 days of incubation (van Aken and Schnoor, 2002). Perchlorate reduction within the poplar tissues demonstrated that $^{36}ClO_4^-$ is partially reduced to $^{36}ClO_3^-$, $^{36}ClO_2^-$, and $^{36}Cl^-$.

A geologic analysis, especially in arid environments, can provide another means to identify potential horizons, such as caliche or other evaporates, as potential sources. In addition to direct testing of these horizons, the use of SEM and XRD may provide a means to confirm that soil enriched with perchlorate is associated with evaporate deposits and is from a marine depositional origin. Berger and Cooke (1998), for example, used XRD on Atacama Desert samples in their analysis of calcium sulfate distributions.

9.6 CONCLUSIONS

The emergence of perchlorate as a contaminant of environmental and toxicological interest has resulted in the development of techniques to distinguish between naturally occurring and anthropogenic sources. As the presence of perchlorate in additional industrial and military sources expands, greater opportunities will be available to use surrogate chemicals associated with perchlorate to distinguish co-mingled sources of anthropogenic sources of perchlorate.

The identification of naturally occurring perchlorate in a carbonate-rich marine layer within the Mission Valley Formation in addition to its presence in Texas and New Mexico suggest that large portions of the southern United States may contain natural sources of perchlorate. The need to incorporate techniques to distinguish between naturally occurring and anthropogenic sources are therefore of paramount importance in any forensic investigations dealing with the source of age and of a perchlorate release into the environment.

REFERENCES

Achenbach, L., Bruce, R., Michaelidou, U., and J. Coates, 2001. Dechloromonas agitata N.N. gen. sp. nov. two novel environmentally dominant (per)chlorate-reducing bacteria and their phylogenetic position. *International Journal of Systematic and Evolutional Microbiology.* 51, 527–533.

Ader, M., Coleman, M., Doyle, S., Stroud, M., and D. Wakelin, 2001. Methods for stable isotopic analysis of chlorine in chlorate and perchlorate compounds. *Analytical Chemistry.* 73 (20), 4946–4950.

Ajaz, A., 1995. Estimation of ammonium perchlorate in HTBP based composite solid propellants using Kjeldahl method. *Journal of Hazardous Materials.* 42 (3), 303–306.

Al-Harthi, A. and A. Williams, 1998. Effect of fuel binder and oxidizer particle diameter on the combustion of ammonium perchlorate based propellants. *Fuel.* 77 (13), 1451–1468.

Augustynski, J., Dalard, F., and J. Sohm, 1972. Etude de la pile oxide d'argent-magnesium—II Regimes lents. *Electrochimica Acta.* 17 (12), 2309–2316.

Avdalovic, N., Pohl, C., Rocklin, R., and J. Stillian, 1993. Determination of cations and anions by capillary electrophoresis combined with suppressed conductivity detection. *Analytical Chemistry.* 65, 1470–1475.

Bao, H., 2005. Atmospheric perchlorate: What we know and what we don't. Environmental Forensics: Focus on Perchlorate. *International Society of Environmental Forensics.* September 21–22, 2005. Santa Fe, New Mexico.

Bao, H. and B. Gu, 2004. Natural perchlorate has a unique oxygen isotope signature. *Environmental Science & Technology.* 38 (19), 5073–5077.

Batjoens, P., DeBrabander, H., and L. T'Kindt, 1993. Ion chromatographic determination of perchlorate in cattle urine. *Analytica Chimica Acta.* 275 (1–2), 335–340.

Becking, G., Haldane, A., and D. Izard, 1958. Perchlorate, an important constituent in sea water. *Nature*. 182 (1), 645–647.

Beckurts, H., 1886. Uber den gehalt des salpeters an chlorsaurem salz. *Archiv der Pharmazie*. 224, 333–337.

Berger, A. and R. Cooke, 1998. The origin and distribution of salts on alluvial fans in the Atacama Desert, Northern Chile. *Earth Surface Processes and Landforms*. 22 (6), 581–600.

Biesaga, M., Kwiatkowska, M., and M. Trojanowica, 1997. Separation of chlorine containing anions by ion chromatography and capillary electrophoresis. *Journal of Chromatography A*. 777, 375–381.

Bohlke, J., Erickson, G., and K. Revesz, 1997. Stable isotope evidence for an atmospheric origin of desert nitrate deposits in northern Chile and southern California, USA. *Chemical Geology*. 136 (1–2), 135–152.

Bohlke, J., Sturchio, N., Gu, B., Horita, J., Brown, G., Jackson, W., Batista, J., and P. Hatzinger. Perchlorate isotope forensics. Analytical Chemistry. (Preprint, October 12, 2005)

Bradstreet, R., 1954. Kjeldahl method for organic nitrogen. *Analytical Chemistry*. 26 (1), 185–187.

Bruce, R., Achenbach, L., and J. Coates, 1999. Reduction of (per)chlorate by a novel organism isolated from a paper mill waste. *Environmental Microbiology*. 1, 319–331.

California Department of Water Resources, 1958. Quality of ground waters in California. 1955–56. Bulletin No. 66. State of California, Department of Water Resources, Division of Resources Planning, Sacramento, CA. June 1958.

California Department of Water Resources, 1960. Quality of ground waters in California. 1957. Bulletin No. 66. State of California, Department of Water Resources, Division of Resources Planning, Sacramento, CA. April 1960.

California Division of Mines and Geology, 1975. Geology of the San Diego Metropolitan Area, California. Section A Western San Diego Metropolitan Area. Section B Eastern San Diego Metropolitan Area. Bulletin 200. Sacramento, California.

California Environmental Protection Agency, 2004. Perchlorate contamination treatment alternatives. Draft. Prepared by the Office of Pollution Prevention and Technology Development, Department of Toxic Substances Control. p. 21.

California Environmental Protection Agency, 2003. Fact Sheet: Perchlorate and Surface/Ground Water Pollution in Simi Valley. State of California, Los Angeles Regional Water Quality Control Board. September 2003. p. 2.

California Regional Water Quality Control Board, Santa Ana Region, 2004. Overview of the occurrence of perchlorate in the Santa Ana Region. Board Item 20. March 12, 2004. Santa Ana, California. pp. 373–374.

Chemical Land, 2004. Sodium Perchlorate. Chemical Land 21.com.http://www.chemicalland21.com/arokorhi/industrialchem/inorganic/sodium%20perchlorate.html.

Coates, J., Michaelidou, U., Bruce, R., O'Connor, S., Crespi, J., and L. Achenbach, 1999. The ubiquity and diversity of dissimilatory (per)chlorate-reducing bacteria. *Applied Environmental Microbiology*. 65, 5234–5241.

Coates, J., Michaelidou, U., O'Connor, S., Bruce, R., and L. Achenbach, 2000. The diverse microbiology of (per)chlorate reduction. *Environmental Science Research*. 57, 257–270. In E.D. Urbansky (ed.), Perchlorate in these Environment. Kluwer Academic/Plenum, New York, N.Y.

Coleman, M. and J. Coates, 2004. Chlorine isotopic characterization of perchlorate. Perchlorate in California's Groundwater. *The Eleventh Symposium in GRA's Series on Groundwater Contaminants*. Groundwater Resources Association. August 4, 2004. Glendale, California.

Coleman, M., Ader, M., Chaudhuri, S. and J. Coates, 2003. Microbial isotopic fractionation of perchlorate chlorine. *Applied Environmental Microbiology*. 69 (8), 4997–5000.

Collette, T., Williams, T., Urbansky, E., Magnuson, M., Hebert, G., and S. Strauss, 2003. Analysis of hydroponic fertilizer matrixes for perchlorate: Comparison of analytical techniques. *Analyst*, 128 (1), 88–97.

Collison, W. and D. Boltz, 1968. Indirect spectrophotometric and atomic adsorption spectrometric methods for determination of perchlorate. *Analytical Chemistry*. 40, 1896–1898.

Corr, J. and J. Anacleton, 1996. Analysis of inorganic species by capillary electrophoresis-mass spectrometry and ion exchange chromatography-mass spectrometry using an ion spray source. *Analytical Chemistry*. 68, 2155–2163.

Crump, C., Michaud, P., Tellez, R., Reyes, C., Gonzalez, G., Montgomery, E. Crump, K., Lobo, G., Becerra, C., and J. Gibbs, 2000. Does perchlorate in drinking water affect thyroid function in newborns or school-age children? *Journal of Occupational and Environmental Medicine*. 42 (6), 603–612.

Dafert, F., 1908. The composition of some Chilian caliche. *Monatshefte fuer Chemie*. 29, 235–244.

Damian, P. and F. Pontius, 1999. From rockets to remediation: the perchlorate problem. *Environmental Protection*. pp. 24–31.

Dasgupta, P., Martinelango, P., Jackson, A., Anderson, T., Tian, K., Tock, R., and S. Rajagopalan, 2005. The origin of naturally occurring perchlorate: The role of atmospheric processes. *Environmental Science & Technology*. Web Release Date: January 29, 2005. http://pubs.acs.org/cgi-bin/jcen?esthag/asap/html/es048612xhtml.

Davis, C., 1940. United States Number 2190703. E.I.duPont de Nemours & Company.

Davis, A., Drexler, J., Ruby, M., and A. Nicholson, 1993. Micromineralogy of mine wastes in relation to lead bioavailability, Butte, Montana. *Environmental Science & Technology*. 27, 1415–1425.

Duncan, B., Morrison, R., and E. Vavricka, 2005. Forensic identification of anthropogenic and naturally occurring sources of perchlorate. *Environmental Forensics Journal*. 6, 1–11.

Ellington, J. and J. Evans, 2000. Determination of perchlorate at parts-per-billion levels in plants by ion chromatography. *Journal of Chromatography A*. 898, 193–199.

Ericksen, G., 1981. Geology and origin of the Chilean nitrate deposits. United States Geological Survey Professional Paper 1188.

Ericksen, G., 1983. The Chilean nitrate deposits. *American Scientist*. 71, 366–374.

Ericksen, G., Hosterman, J., and P. St. Amand, 1988. Chemistry, mineralogy and origin of the clay-hill nitrate deposits, Amargosa River valley, Death Valley region, California, USA. *Chemical Geology*. 67 (1–2), 85–102.

Erickson, B., 2004. Tracing the origin of perchlorate. *Analytical Chemistry*. 76 (21), 388A–389A.

Evans, J. and S. Tatham, 2004. Defining 'local signature' in terms of Sr isotope composition using a tenth to twelfth century Anglo-Saxon population living on a Jurassic clay-carbonate terrain, Rutland, UK. Pye, KI. and D. Croft, (eds). In: *Forensic Geoscience: Principles, Techniques and Applications*. Geological Society, London. Special Publications. 232, 237–248.

Feigl, F. and D. Goldstein, 1958. Detection of perchlorate in spot test analysis. *Microchemical Journal*. 2, 105–108.

Flowers, T. and J. Hunt, 2000. Long-term release of perchlorate as a potential source of groundwater contamination. Chapter 17. In: *Perchlorate in the Environment*. (ed.) Urbansky. Kluwer Academic/Plenum Publishers, New York, NY. pp. 177–188.

Fritz, J.S., 1973. *Acid-Base Titrations in Nonaqueous Solvents.* Allyn and Bacon, Inc., Boston.

General Accounting Office, 2004. DOD Operational Ranges. More Reliable Cleanup Cost Estimates and a Proactive Approach to Identifying Contamination are Needed. Report to Congressional Requesters. GAO-04-601.

Graustein, W., 1989. $^{87}Sr/^{86}Sr$ ratios measure the sources and flow of strontium in terrestrial ecosystems. In: *Stable Isotopes in Ecological Research: Ecological Studies.* Rundel, P., Ehleringer, J., and K. Nagy, (eds) Springer Verlag, New York, NY. pp. 491–512.

Gopal, C., Jackson, A., Cobb, G., Rainwater, K., and T. Anderson, 2004. Geologic formation of perchlorate: Evidence from the presence of other anions. The Society of Environmental Toxicology and Chemistry, 25th Annual Meeting.

Greenhalgh, R. and J. Riley, 1960. Alleged occurrence of the perchlorate ion in seawater. *Nature.* 187, 1107–1108.

Greer, M., Goodman, G., Pleus, R., and S. Greer, 2002. Health effects assessment for environmental perchlorate contamination: The dose-response for inhibition of thyroidal radioiodine uptake in humans. *Environmental Health Perspective.* 10, 927–937.

Grotheer, M. and E. Cook, 1968. Mechanism of electrolytic perchlorate production. *Electrochemical Technology.* 6 (5–6), 221–224.

Grossling, B. and G. Erickson, 1971. Computer studies of the composition of Chilean nitrate ores: Data reduction, basic statistics, and correlation analysis. USGS Open File Report 1519.

Gu, B., Brown, G., Alexandratos, S., Ober, R., Dale, J., and S. Plant, 2000a. Efficient treatment of perchlorate (ClO_4^-) contaminated groundwater with bifunctional anion-exchange resins. *Environmental Science Research.* 57, 165–176.

Gu, B., Brown, P., Bonnesen, L., Liang, L., Moyer, B., Ober, R., and S. Alexandratos, 2000b. Development of a novel bifunctional anion-exchange resins with improved selectivity for pertechnetate: Column breakthrough and field studies. *Environmental Science & Technology.* 34, 1075–1080.

Gu, B., Brown, G., Maya, L., Lance, M., and B. Moyer, 2001. Regeneration of perchlorate (ClO_4^-) loaded anion exchange resins by novel tetrachloroferrate ($FeCl_4^-$) displacement technique. *Environmental Science & Technology.* 35, 3363–3368.

Gu, B., Ku, K., and G. Brown, 2002. Treatment of perchlorate-contaminated water using highly-selective, regenerable ion-exchange technology: A pilot-scale demonstration. *Remediation.* 12 (2), 51–68.

Gu, B., Ku, Y., and P. Jardine, 2004a. Sorption and binary exchange of nitrate, sulfate, and uranium on anion-exchange resin. *Environmental Science & Technology.* 38, 3184–3188.

Gu, B., Tio, J., Wang, W., Ku, Y., and S. Dai, 2004b. Raman spectroscopic detection for perchlorate at low concentration. *Applied Spectrophotometry.* 58, 741–744.

Gullick, R., LeChevallier, M., and T. Barhorst, 2001. Occurrence of perchlorate in drinking water sources. *Journal of the American Water Works Association.* 93 (1), 66–77.

Hampel, C. and P. Leppla, 1947. Production of potassium perchlorate. *Transactions of the Electrochemical Society.* 92, 10.

Handy, R., Barnett, D., Purves, R., Horlick, G., and R. Guevremont, 2000. Determination of nanomolar levels of perchlorate in water by ESI-FAIMS-MS. *Journal of Analytical Atomic Spectrometry.* 15, 907–911.

Harvey, G., Tsui, D., Eldridge, J., and G. Orris, 1999. 20th Annual Meeting Abstract Book. Society of Environmental and Toxicological Chemists. PHA015. p. 277.

Hatzinger, P., Whittier, M., Arkins, M., Bryan, C., and W. Guarini, 2002. In-situ and ex-situ bioremediation options for treating perchlorate in groundwater. *Remediation.* Spring 2002. pp. 69–86.

Hautman, D. and D. Munch, 1999. Method 314.0 Determination of perchlorate in drinking water using ion chromatography. Revision 1.0. National Exposure Research Laboratory. Office of Research and Development. United States Environmental Protection Agency. Cincinnati, Ohio.

Hayes, O., 1968. Qualitative inorganic analysis. XXXII. The systematic detection of chlorate and perchlorate. *Mikrochim. Acta.* 3, 647–648.

Hoering, T., Ishimori, F., and H. McDonald, 1958. Oxygen exchange between oxy-anions and water. II. Chlorite, chlorate and perchlorate ions. *Journal of the American Chemical Society.* 80, 3876.

Horita, J., Bohlke, J., Sturchio, N., Brown, G., and J., Batista, 2004. Environmental isotope forensics of perchlorate contamination. Geological Society of America, 2004 Denver Annual Meeting. Session 99, Environmental Geoscience II. November 7–10, 2004.

Hunter, J., 2001. Perchlorate is not a common contaminant of fertilizers. *Journal of Agronomy and Crop Science.* 187, 203–204.

Hurst, R., 2004. The role of naturally-occurring isotopes in forensic investigations of perchlorate-impacted soil and groundwater. Perchlorate in California's Groundwater. The Eleventh Symposium in GRA's Series on Groundwater Contaminants. Groundwater Resources Association. August 4, 2004. Glendale, California.

Hussain, G. and G. Rees, 1992. Combustion of ammonium perchlorate based mixture with and without black powder, studied by d.s.c and TG/DTG. *Fuel.* 71 (4), 471–473.

Jackson, A., 2004c. Occurrence and source of perchlorate in high Plains Aquifer systems of Texas and New Mexico. Perchlorate in California's Groundwater. *The Eleventh Symposium in GRA's Series on Groundwater Contaminants.* August 4, 2004. Glendale, California.

Jackson, A., 2005. Electrochemical generation of perchlorate by Cathodic protection systems. Environmental Forensics: Focus on Perchlorate. International Society of Environmental Forensics. September 21–22, 2005. Santa Fe, New Mexico.

Jackson, P., Laikhtman, M., and J. Rohrer, 1999. Determination of trace level perchlorate in drinking water and ground water by ion chromatography. *Journal of Chromatography A.* 850, 131–135.

Jackson, W., Rainwater, K., Anderson, T., Lehman, T., Tock, R., Rajagopalan, S., and M. Ridley, 2004a. Distribution and potential sources of perchlorate in the high plains region of Texas. Texas Commission on Environmental Quality, Austin, Texas. August 1, 2004.

Jackson, A., Rainwater, K., Anderson, T., Lehman, T., Ridley, M., Waldon, S., and R. Tock, 2004b. Perchlorate in groundwaters in the Southern High Plains of Texas. Presented at the 2004 AWWA Annual Conference and Exposition. Orlando, Florida.

Jackson, A., Anandam, S.K., Anderson, A., Lehman, T., Rainwater, K. Rajagopalan, S., Ridley, M., and W. Tock, 2005. Perchlorate occurrence in the Texas southern high plains aquifer system. *Groundwater Monitoring and Remediation.* 25 (1), 137–149.

Jaegle, L., Yung, Y., Toon, G., Sen, C., and J. Blavier, 1996. Balloon observations or organic and inorganic chlorine in the stratosphere: The role of $HClO_4$ production on sulfate aerosols. Geophysical Research Letters. 23, 1749–1752.

Johnson, J. and M. Thiemens, 1997. The isotopic composition of tropospheric ozone in three environments. *Journal of Geophysical Research* (Atmospheric). 102, 25395–25404.

Johnson, D., Jucks, K., Traub, W., and K. Chance, 2000. Isotopic composition of stratospheric ozone. *Journal of Geophysical Research*. 105, 9025–9031.

Junck, K., 1926. Comparison of colorimetric methods for the determination of perchlorate. Caliche. 8.

Kennedy, M. and G. Moore, 1971. Stratigraphic relations of upper Cretaceous and Eocene Formations. San Diego Coastal Area, California. *American Association of Petroleum Geologists*. Bulletin 55. pp. 709–722.

Kennedy, M. and S. Tan, 1997. Geology of National City, Imperial Beach, and Otay Mesa quadrangles, Southern San Diego Metropolitan Area, California. Map Sheet 29. California Division of Mines and Geology.

Kennedy, S., Walker, W., and B. Forslund, 2002. Speciation and characterization of heavy metal-contaminated soils using computer-controlled scanning electron microscopy. *Environmental Forensics*. 3, 131–143.

Koester, C., Beller, H., and R. Halden, 2000. Analysis of perchlorate in groundwater by electrospray ionization mass spectrometry/mass spectrometry. *Environmental Science & Technology*. 34, 1862–1864.

Kowalchyk, W., Walker, P., and M. Morris, 1995. Rapid normal Raman spectroscopy of sub-ppm oxy-anion solutions: The role of electrophoretic preconcentration. *Applied Spectroscopy*. 49, 1183–1188.

Krinsley, D., Pye, K., Bogs, S., and N. Tovey, 1998. Chapter 9. Image Analysis. In: *Backscattered Scanning Electron Microscopy and Image Analysis of Sediments and Sedimentary Rocks*. Cambridge University Press, pp. 145–172.

Lamb, A. and J. Marden, 1912. The quantitative determination of perchlorate. *Journal of the American Chemical Society*. 34, 812–817.

Lamm, S. and M. Doemland, 1999. Has perchlorate in drinking water increased the rate of congenital hypothyroidism? *Journal of Occupation and Environmental Medicine*. 41, 409–411.

Lawrence, J., Lamm, S., and L. Braverman, 2001. Low dose perchlorate (3 mg daily) and thyroid function. Thyroid. 11 (3), 295.

Lawrence, J., Lamm, S., Pino, S., Richman, K., and L. Braverman, 2000. The effect of short term low dose perchlorate on various aspects of thyroid function. *Thyroid*. 10 (8), 659–663.

Li, A., Li, F., Byrd, D., Dyhle, G., Sesser, D., Skeels, M., and S. Lamm, 2000a. Neonatal thyroxine level and perchlorate in drinking water. *Journal of Occupational and Environmental Medicine*. 42, 200–208.

Li, A., Byrd, D., Dyhle, G., Sesser, D., Skeels, M., Katkowsky, R., and S. Lamm, 2000b. Neonatal thyroid-stimulating hormone level and perchlorate in drinking water. *Teratology*. 62, 426–431.

Loach, K., 1962. Estimation of low concentrations of perchlorate in natural materials. *Nature*. 196, 754–755.

Logan, B., 2001. Assessing the outlook for perchlorate remediation. *Environmental Science & Technology*. 35 (23), 482–487.

Long, J., 2001. Perchlorate Safety: Reconciling Inorganic and Organic Guidelines. *Originally published in Chemical Health and Safety Journal*. http://www.gfschemicals.com/technicallibrary/perchloricacid.pdf).

Magnuson, M., Urbansky, E., and C. Kelty, 2000a. Determination of perchlorate at trace levels in drinking water by ion-pair extraction with electrospray ionization mass spectrometry. *Analytical Chemistry*. 72, 25–29.

Magnuson, M., Urbansky, E., and C. Kelty, 2000b. Microscale extraction of perchlorate in drinking water with low level detection by electrospray-mass spectrometry. *Talanta*. 52, 285–291.

Maurino, V. and C. Minero, 1997. Cyanuric acid-based effluent for suppressed anion chromatography. *Analytical Chemistry*. 69, 3333–3338.

Medard, L., 1950. Recent progress and tendency of the mining explosives in France. *Memorial des Poudres*. 32, 209–225.

Mendiratta, S., Dotson, R., and R. Booker, 1996. Perchloric acid and perchlorates. In: *Kirk-Othmer Encyclopedia of Chemical Technology*. 4th edition, Volume 18. pp. 157–170. J. Kroschwitz and M. Howe-Grant. (eds) John Wiley & Sons, New York, NY.

Mesard, P. and W. McNab, 2005. Anthropogenic perchlorate sources associated with groundwater chemistry. Environmental Forensics: Focus on Perchlorate. International Society of Environmental Forensics. September 21–22, 2005. Santa Fe, New Mexico.

Michalski, G., Bohlkke, J., and M. Thiemens, 2004. Long-term atmospheric deposition as the source of nitrate and other salts in the Atacama Desert, Chile: New evidence from mass-independent oxygen isotopic compositions. *Geochimica et Cosmochimica Acta*. 68, 4023–4038.

Miller, A. and J. Macklin, 1980. Matrix effects on the Raman analytical lines of oxyanions. *Analytical Chemistry*. 52, 807–812.

Mitguard, M. and K. Mayer, 2004. Perchlorate: Assessment of California Sites for an Unregulated Emerging Chemical. *2004 National Site Assessment Symposium*, United States Environmental Protection Agency. www.epa.gov/superfund/programs/siteasmt/symp04/abstracts/mitguard.htm.

Mohr, T. and J. Crowley, 2002. Perchlorate—Is it all rocket science? *Hydro-Visions Newsletter*. Groundwater Resources Association of California. Winter 2002. p. 1.

Mohr, T., 2005. Designing a forensic study of perchlorate sources in private and municipal wells in Santa Clara County, CA. Environmental Forensics: Focus on Perchlorate. International Society of Environmental Forensics. Santa Fe, New Mexico. September 21–22, 2005.

Motzer, W., 2001. Perchlorate: Problems, detection and solutions. *Environmental Forensics*. 2, 301–311.

Motzer, W., 2005. Overview of state-of these practice of perchlorate forensics. Environmental Forensics: Focus on Perchlorate. International Society of Environmental Forensics. Santa Fe, New Mexico. September 21–22, 2005.

Murphy, D. and D. Thomson, 2000. Halogen ions and NO^+ in the mass spectra of aerosols in the upper troposphere and lower stratosphere. *Geophysical Research Letters*. 27, 3217–3220.

Nabar, G. and C. Ramachandran, 1959. Quantitative determination of perchlorate ion in solution. *Analytical Chemistry*. 31, 263–265.

Nann, A., 1994. Potentiometric detection of anions separated by capillary electrophoresis using an ion-selective electrode. *Journal of Chromatography A*. 676, 437–442.

National Organic Standards Board Technical Advisory Panel Review, 2002. Chilean nitrate for general use as an adjuvant in crop production. http://www.sarep.ucdavis.edu/Organic/tap/Chileannitrate-GeneralUse.pdf).

National Research Council, 2005. Health implications of perchlorate ingestion. Committee to Assess the Health Implications of Perchlorate Ingestion. Board on Environmental Studies and Toxicology. Division on Earth and Life Sciences. National Research Council. The National Academic Press, Washington, D.C.

Neuhold, C., Kalcher, K., Cai, X., and G. Raber, 1996. Catalytic determination of perchlorate using a modified paste electrode. *Analytical Letters*. 20, 1685 and 1704.

Orris, G., Harvey, G., Tsui, D., and J. Eldridge, 2003. Preliminary analyses for perchlorate in selected natural materials and their derivative products. United States Geological Survey, United States Department of the Interior. Open File Report 03-314. p. 6.

Personal Communication with Baohua Gu, 2005. Oak Ridge National Laboratory Environmental Sciences Division, Oak Ridge Tennessee. March 1, 2005.

Petrisor, I. and R. Morrison, 2005. Identification of naturally occurring perchlorate in Southern California. Environmental Forensics: Focus on Perchlorate. International Society of Environmental Forensics. Santa Fe, New Mexico, September. 21–22, 2005.

Philp, P., 2005. Stable isotopes and biomarkers in forensic geochemistry. Fifteenth Annual AEHS Meeting and West Coast Conference on Soils, Sediments and Water. Association for the Environmental Health of Soils Workshop. March 14–17, 2005. San Diego, California.

Renne, P., Sharp, W., Montanex, I., Becker, T., and R. Zierenberg, 2001. $^{40}Ar/^{39}Ar$ dating of Lake Permian evaporites, southeastern New Mexico, USA. Earth Planet Science Letter. 193, 539–547.

Renner, R., 2004. California sets public health goal for perchlorate. Environmental Science & Technology. 38, 177A–178A.

Rickman, J., 2005. Low-level perchlorate detection methods shows promise. Los Alamos National Laboratory. News and Public Affairs. www.lanl.gov/worldview/news/releases/archive/03-095.shtml.

Rikken, G., Kroon, A., and C. van Ginkel, 1996. Transformation of (per)chlorate into chloride by a newly isolated bacterium: reduction and dismutation. Applied Microbiological Technology. 45, 420–426.

Roefer, P., Zikmund, K., and S. Snyder, 2004. Low level perchlorate sampling results in the Colorado River system and Lake Mead. Southern Nevada Water Authority Report. February 2004.

Rogers, K., 1998. Chemicals effect on crop worries Tribe. Las Vegas Review-Journal. May 20, 1998.

Roehl, R., Slingsby, R., Avdalovic, N., and P. Jackson, 2002. Applications of ion chromatography with electrospray mass spectrometric detection to the determination of environmental contaminants in water. Journal of Chromatography A. 956, 245–254.

Romanenko, V., Korenkov, V., and S. Kuznetsov, 1976. Bacterial decomposition of ammonium perchlorate. Mikrobiologiya. 45, 204–209.

Schilt, A., 2003. Perchloric acid and perchlorates. 2nd edition. G. Frederick Smith Chemical Company. Powell, Ohio. p. 300.

Seta, K., Washitake, M., and I. Tanaka, 1980. Journal of Chromatography b: Biomedical Sciences and Applications. 221 (2), 215–225.

Shahine, S. and S. Khamis, 1979. Analysis of nitrate-perchlorate mixtures by difference spectrophometry. Microchemical Journal. 24, 439–443.

Sharp, R. and B. Walker, 2001. Rocket Science. Perchlorate and the toxic legacy of the cold war. Environmental Working Group. Oakland, California. p. 44.

Siglin, J., Mattie, D., Dodd, D., Hildebrandt, P., and W. Baker, 2000. A 90-day drinking water toxicity study in rats of the environmental contaminant ammonium perchlorate. Toxicology Science. 56, 61–74.

Silva, M., 2003. Perchlorate from safety flares. A threat to water quality. Santa Clara Water District Factsheet. http://www.valleywater.org.

Siswanta, D., Takenaka, J., Suzuki, T., Saskura, H., Hisamoto, H., and K. Suzuki, 1997. Noel neutral anion ionophores based on fluoridated (poly)ether compounds as a sensory molecule for an ion-selective electrode. Chemical Letters. 195–196.

Smith, G.F., McHard, J.A., and K.L. Olson, 1936. Ind. Eng. Chem. Anal. Edl, 8, 350.

Smith, G.F. and C.A. Getz, 1937. Ind. Eng. Chem. Anal. Ed., 9, 378.

Smith, G. and W. Taylor, 1963. The dissolution of alunite employing hot concentrated perchloric acid: Determination of silica and aluminium. Talanta. 10 (10), 1107–1109.

Smith, P., Theodorakis, C., Anderson, T., and R. Kendall, 2001. Preliminary assessment of perchlorate in ecological receptors at the Longhorn Army Ammunition Plant (LHAAP), Karnack, Texas. Ecotoxicology. 10, 305–313.

Sowinski, M., Vavricka, E., and R. Morrison, 2003. An overview of perchlorate contamination in groundwater: Legal, chemical and remedial considerations. Environmental Claims Journal. 15 (3), 1–16.

Spielholtz, G. and H. Diehl, 1966. Wet ashing of coal with perchloric acid mixed with periodic acid for the determination of sulphur and certain other constituents. Talanta. 13 (7), 991–1102.

Strawson, J., Zhao, Q., and M. Dourson, 2004. Reference dose for perchlorate based on thyroid hormore change in pregnant woment as the critical effect. Regulatory and Toxicology Pharmocology. 39, 44–65.

Sturchio, N., Bohlke, J., Horita, J., Gu., B., and G. Brown, 2004. Environmental isotope forensics of perchlorate. Perchlorate in California's Groundwater. The Eleventh Symposium in GRA's Series on Groundwater Contaminants. Groundwater Resources Association. August 4, 2004. Glendale, California.

Sturchio, N., Hatzinger, P., Arkins, M., Suh, C., and L. Heraty, 2003. Chlorine isotope fractionation during microbial reduction of perchlorate. Environmental Science & Technology, 37 (17), 3859–3863.

Sturchio, N., 2005. Environmental isotope forensics of perchlorate: Source characteristics and groundwater case studies. Environmental Forensics: Focus on Perchlorate. International Society of Environmental Forensics. September. 21–22, 2005. Santa Fe, New Mexico.

Susarla, S., Collette, T., Garrison, A., Wolfe, N., and S. McCutcheon, 1999. Perchlorate identification in fertilizers. Environmental Science & Technology. 33, 3469–3472.

Susarla, S., Collette, T., Garrison, A., Wolfe, N., and S. McCutcheon, 2000. Response to Comment on "Perchlorate Identification in Fertilizers" and the Subsequent Addition/Correction. Environmental Science & Technology. 34 (20), 4454–4454.

Tauber, H., 1952. New color reaction for steroids with perchloric acid. Analytical Chemistry. 24 (9), 1494–1495.

Tellez, R., chacon, P., Abarco, C., Blunt, B., Van Landingham, C., Crump, K., and J. Gribbs, 2005. Long-term environmental exposure to perchlorate through drinking water and thyroid function durinf perchlorate through dring water and thyroid function during pergnancy and the neonatal period. Throid. 15 (9), 963–975.

The Fertilizer Institute, 2004. Perchlorate. The Fertilizer Institute, Washington D.C. http://www.tfi.org/Issues/perchloratepage.asp.

Thiemens, M., Savarino, J., Farquhar, J., and H. Bao, 2001. Mass-independent isotopic compositions in terrestrial and extraterrestrial solids and their applications. Accounts of Chemical Research. 34, 645–652.

Tian, K., Canas, J., Dasgupta, P., and T. Anderson, 2005. Preconcentration/preelution ion chromatography for the determination of perchlorate in complex samples. Talanta. 65 (13), 750–755.

Tipton, D., Rilston, D., and K. Scow, 2003. Transport and biodegradation of perchlorate in soils. Journal of Environmental Quality. 32 (1), 40–46.

Tock, R., Jackson, W., Anderson, T., and S. Arunagiri, 2004. Electrochemical generation of perchlorate ions in chlorinated drinking water. Corrosion. Corrosion Engineering Section. National Association of Corrosion Engineers. 60 (8), 757–763.

Tock, R., Jackson, W., Rainwater, R., and T. Anderson, 2003. Potential for perchlorate generation in drinking water storage tanks. Texas Water Development Board, Austin, Texas.

Tollenaar, H. and C. Martin, 1972. Perchlorate in Chilean Nitrate as the cause of leaf rugosity in soybeans plants in Chile. *Phytopathology* 62, 1164–1166.

United States Army Center for Health Promotion and Preventive Medicine, 2005. Perchlorate in drinking water. http://chppm-www.apgea.army.mil/documents/FACT/ 31-003-0502.pdf.

United States Environmental Protection Agency, 1998. Guidlines for newro-toxicity risk assessment. Federal Register. May 14, 1998. 14 (63), 26926–26954.

United States Environmental Protection Agency, 1998. Perchlorate Environmental Contamination: Toxicological Review and Risk Characterization Based on Emerging Information, External Review Draft. Washington, DC, EPA Doc. No. NCEA-1-0503.

United States Environmental Protection Agency, 2002. Perchlorate Environmental Contamination: Toxicological Review and Risk Characterization. Office of Research and Development. NCEA-1-0503. Washington, D.C.

United States Environmental Protection Agency, 2004. Perchlorate. http://www.epa.gov/swerffrr/documents/ perchlorate.htm.

Urbansky, E., 1998. Perchlorate Chemistry: Implications for Analysis and Remediation. *Bioremediation Journal.* 2 (2), 81–95.

Urbansky, E., 2000b. Quantitation of perchlorate ion: Practices and advances applied to the analysis of common matrices. *Critical Reviews of Analytical Chemistry.* 30, 311–343.

Urbansky, E., and M. Schock, 1999a. Issues in managing the risks associated with perchlorate in drinking water. *Journal of Environmental Management.* 56 (2), 79–95.

Urbansky, E., Brown, S., Magnuson, M., and C. Kelty, 2001a. Perchlorate Levels in samples of sodium nitrate fertilizer derived from Chilean caliche. *Environmental Pollution.* 112 (3), 299–302.

Urbansky, E., Magnuson, D., Freeman, D., and C. Jelks, 1999b. Quantitation of perchlorate ion by electrospray ionization mass spectrometry (ESI-MS) using stable association complexes with organic cations and bases to enhance selectivity. *Journal of Analytical Atomic Spectrometry.* 14, 1861–1866.

Urbansky, E., Magnuson, M., Kelty, C., and S. Brown, 2000c. Perchlorate uptake by salt cedar (Tamarix ramosissima) in the Las Vegas Wash riparian ecosystem. *The Science of the Total Environment.* 256 (2–3), 227–232.

Urbansky, E., Magnuson, M., Kelty, C., Gu, B., and G. Brown, 2000a. Comment on perchlorate identification in fertilizers and the subsequent addition/correction. *Environmental Science & Technology.* 34 (20), 4452–4453.

Urbansky, E., Collette, T., Robarge, W., Hall, W., Skillen, J.M., and P. Kane, 2001b. Survey of fertilizers and related materials for perchlorate (ClO_4^-). Final Report. United States Environmental Protection Agency, Office of Research and Development. EPA/600/R-01/049.

van Aken, B. and J. Schnoor, 2002. Evidence of perchlorate (ClO_4^-) reduction in plant tissues (poplar tree) using radio-labeled $^{36}ClO_4^-$. *Environmental Science & Technology.* 36 (12), 2783–2788.

Vandenburg, T., 2004. Source Identification, Causation Analysis, and the Scope of Damages in Perchlorate Groundwater Contamination Litigation. Presented at the 14th Annual West Coast Conference on Soil, Sediments and Water. *Association for Environmental Health and Sciences.* March 15–18, 2004. San Diego, CA.

Vavricka, E. and R. Morrison, 2005. Identification of naturally occuring and anthropogenic sources of perchlorate. Environmental Claims Journal. 17(2), –212

Vogt, R., Sander, R., Von Glasow, R., and P. Crutzen, 1999. Iodine chemistry and its role in halogen activation and ozone loss in the marine boundary layer: A model study. *Journal of Atmospheric Chemistry.* 32, 375–395.

Weiss, J. and J. Stanbury, 1972. Spectrophotometric determination of micro amounts of perchlorate in biological fluids. *Analytical Chemisty.* 44, 619–620.

Willard, H. and G. Smith, 1922a. The preparation and properties of magnesium perchlorate and its use as a drying agent. *Journal of the American Chemical Society.* 44 (10), 2255–2259.

Willard, H. and G. Smith, 1922b. The separation and determination of sodium and lithium by precipitation from alcoholic perchlorate solutions. *Journal of the American Chemical Society.* 44 (12), 2816–2824.

Willard, H. and J. Thompson, 1930. Determination of perchlorate. *Industrial and Engineering Chemistry.* Analytical edition. 2 (3), 272–273.

Willard, H. and W. Cake, 1920. Perchloric acid as a dehydrating agent in the determination of silica. *Journal of the American Chemical Society.* 41 (11), 2208–2212.

Wimschneider, A. and Heumann, K., 1995. Iodine speciation in size fractionated atmospheric particles by isotope dilution mass spectrometry. *Analytical and Bioanalytical Chemistry.* 353, 191–196.

Winkler, P., Minteer, M., and J. Willey, 2004. Analysis of Perchlorate and soil by electrospray LC/MS/MS. *Analytical Chemistry.* 76 (2), 469–473.

York, R., Brown, W., Girard, M., and J. Dollarhide, 2001. Oral (drinking water) developmental toxicity study of ammonium perchlorate in New Zealand white rabbits. *International Journal of Toxicology.* 20, 199–205.

Young, E., Galy, A., and H. Nagahara, 2002. Kinetic and equilibrium mass-dependent isotope fractionation laws in nature and their geochemical and cosmochemical significance. *Geochimica et Cosmochimica Acta.* 66, 1095–1104.

Zwiener, C. and F. Frimmel, 2004. LC-MS analysis in the aquatic environment and in water treatment technology—a critical review. Part 11: Applications for emerging contaminants and related pollutants, microorganisms and humic acids. *Analytical and Bioanalytical Chemistry.* 378, 862–874.

10 Polychlorinated Biphenyls

Glenn W. Johnson, John F. Quensen, III,
Jeffrey R. Chiarenzelli, and M. Coreen Hamilton

Contents

10.1 INTRODUCTION

Polychlorinated biphenyls (PCBs) are a class of chlorinated organic chemicals that were used for a variety of industrial and commercial purposes. Polychlorinated biphenyls were produced beginning in the 1930s, continuing on into the 1970s. Production ceased in the United States in 1977, in response to the 1976 passing of the Toxic Substances Control Act (Erickson, 1997). The ban on PCBs under TSCA was due primarily to mounting scientific evidence that PCBs accumulate in the environment and could adversely impact humans and other biota. Polychlorinated biphenyls do not easily degrade in the environment and because they are lipophilic, PCBs tend to bioaccumulate in adipose tissue (fat) of higher predators, including humans. Thus, even though they are no longer manufactured or used by industry, they continue to be encountered as contaminants of concern at hazardous waste sites and in the environment.

Polychlorinated biphenyl is not a single chemical, but a group of related chemicals (see Section 10.1.1). Because the relative proportions of individual PCBs varied depending on the intended use of the commercial product, analysis of PCB chemical patterns found in the environment provide clues that allow the data analyst/environmental forensics worker to infer the sources of contamination. However, this requires a broad knowledge of PCB chemistry, PCB industrial use and history, chemical compositions of various products, analytical chemistry methods, environmental alteration mechanisms, and data analysis methods. In this chapter, we present an overview of all of these considerations with the objective of providing the reader with pre-requisite background to undertake a PCB fingerprinting project.

10.1.1 Chemical Structure

Polychlorinated biphenyls were produced by direct chlorination of biphenyl, which is a molecule composed of two six-carbon phenyl rings. The biphenyl molecule may have multiple chlorine atoms attached to carbon atoms at the corners of the rings. By convention, corners of each ring are numbered (Figure 10.1.1a). There are 209 different PCB congeners; with the difference between them related to the number of chlorines attached and their location on the biphenyl ring.

There are two common naming conventions for PCB congeners. The first is based on where chlorines are attached to the numbered locations. Figure 10.1.1b shows a PCB congener with chlorines attached at the 2,4, and 5 locations on one ring and at the 2,3,4, and 5 locations on the second ring.

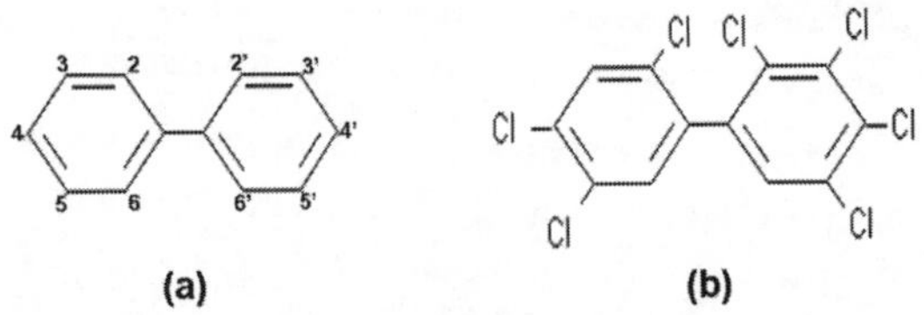

(a)   **(b)**

Figure 10.1.1 Polychlorinated biphenyl molecule. (a) shows a biphenyl molecule showing the ten possible locations where a chlorine atom may be attached to the molecule. (b) shows a PCB molecule with chlorine substitution at the 245 locations on one phenyl ring, and at the 2,3,4,5 locations on the second phenyl ring. This particular PCB congener is referred to as 245-2'3'4'5' CB or alternatively, PCB 180 (IUPAC nomenclature). Positions 2,6,2', and 6' are "ortho" positions. Positions 3,5,3', and 5' are "meta" positions. Positions 4 and 4' are "para" positions.

Under the "structural" nomenclature system, this congener is referred to as 245-2'3'4'5'-CB.

The second PCB naming convention is the "IUPAC" nomenclature (Ballschmiter and Zell, 1980) where congeners are numbered sequentially from 1 to 209. IUPAC congener numbers increase with increasing degree of chlorination. Under the IUPAC nomenclature, the congener shown in Figure 10.1.1b (245-2'3'4'5'-CB) would be referred to as PCB-180. Table 10.1.1 lists all 209 PCB congeners, in both the IUPAC and structural nomenclatures. The IUPAC nomenclature is generally accepted, but it should be noted that Frame's numbering systems (Table 10.1.1) used Guitart's (Guitart *et al.*, 1993) proposed renumbering for congeners 107, 108, and 109, which at the time of this writing had not been adopted by IUPAC.

Yet another naming convention used to identify PCBs is by homolog group. There are ten PCB homologs (Table 10.1.2). Homolog groups include all congeners that have equal number of chlorines attached to the biphenyl ring. For example, PCB-180 (Figure 10.1.1b) is a heptachlorobiphenyl homolog, because it has seven chlorines attached to the biphenyl molecule, but there are 23 other congeners that are hepta-chlorbiphenyls (Tables 10.1.1 and 10.1.2). Again, the difference between the 24 heptachlorobiphenyl congeners is the location of chlorines on the biphenyl molecule.

10.1.2 History of PCBs and Industrial Use

In contrast to polychlorinated dibenzo-*p*-dioxins (PCDDs – Chapter 3), which are primarily produced as a by-product of anthropogenic processes, PCBs were intentionally produced and commercially marketed for a variety of industrial uses. Polychlorinated biphenyls were used in transformers, capacitors, carbonless copy paper, printing inks, hydraulic fluids, and a number of other applications (de Voogt and Brinkman, 1989).

Beginning in 1929, PCBs were commercially produced by a number of companies around the world. Global production estimates range between 900,000 and 1,200,000 metric tons. In the United States, the primary producer was Monsanto Chemical Company. Monsanto's PCB products were marketed under the Aroclor® trade name. Monsanto was the clear global market leader, accounting for over half of worldwide PCB production. Estimates of Monsanto's total PCB production range from 499,000 to 635,000 metric tons (Holoubek, 2001).

Monsanto marketed a number of Aroclor products, each identified by a four-digit reference number. Aroclor reference numbers are a function of the molecule's size and the weight percent of chlorine in the formulation. For example, Aroclor 1260 is so named because the PCB molecule has **12** carbons (i.e., two six-carbon phenyl rings) and Aroclor 1260 is composed of a mixture of PCBs congeners such that the product is **60**% chlorine by weight. The one exception to this naming convention was Aroclor 1016 (which was very similar in composition to Aroclor 1242 – Section 10.1.3) with 12 carbons and a chlorine weight percentage of 41.5.

Table 10.1.3 summarizes known PCB uses for each of nine Aroclors. Their use as dielectric fluids in capacitors (~45% of sales) and transformers (~23% of sales) accounted for the majority of production in the United States (Durfee *et al.*, 1976). The term "askarel" is a generic term that refers to chlorobenzene/PCB mixtures used as dielectric fluids. Capacitor askarels included neat Aroclors 1242, 1016, and 1254 as well as mixtures of Aroclor 1254 and trichlorobenzene. Transformer askarels were most commonly mixtures of chlorobenzenes with either Aroclor 1254 or 1260 (Erickson, 1997). Table 10.1.4 shows Aroclor production estimates as determined from Monsanto US sales records.

Table 10.1.1 PCB Congeners (IUPAC and Structural Nomenclature)

Mono-Chlorobiphenyls

IUPAC	Structural (Chlorine Pos)
1	2
2	3
3	4

Di-Chlorobiphenyls

IUPAC	Structural (Chlorine Pos)
4	2-2'
5	23
6	2-3'
7	24
8	2-4'
9	25
10	26
11	3-3'
12	34
13	3-4'
14	35
15	4-4'

Tri-chlorobiphenyls

IUPAC	Structural (Chlorine Pos)
16	23-2'
17	24-2'
18	25-2'
19	26-2'
20	23-3'
21	234
22	23-4'
23	235
24	236
25	24-3'
26	25-3'
27	26-3'
28	24-4'
29	245
30	246
31	25-4'
32	26-4'
33	34-2
34	35-2'
35	34-3'
36	35-3'
37	34-4'
38	345
39	35-4'

Tetra-chlorobiphenyls

IUPAC	Structural (Chlorine Pos)
40	23-2'3'
41	234-2'
42	23-2'4'
43	235-2'
44	23-2'5'
45	236-2'
46	23-2'6'
47	24-2'4'
48	245-2'
49	24-2'5'
50	246-2'
51	24-2'6'
52	25-2'5'
53	25-2'6'
54	26-2'6
55	234-3'
56	23-3'4'
57	235-3'
58	23-3'5'
59	236-3'
60	234-4'
61	2345
62	2346
63	235-4'
64	236-4'
65	2356
66	24-3'4'
67	245-3'
68	24-3'5'
69	246-3'
70	25-34'
71	26-3'4'
72	25-3'5'
73	26-35
74	245-4'
75	246-4'
76	345-2'
77	34-3'4'
78	345-3'
79	34-3'5'
80	35-3'5'
81	345-4'

Penta-chlorobiphenyls

IUPAC	Structural (Chlorine Pos)
82	234-2'3'
83	235-2'3'
84	236-2'3'
85	234-2'4'
86	2345-2'
87	234-2'5'
88	2346-2'
89	234-2'6'
90	235-2'4'
91	236-2'4'
92	235-2'5'
93	2356-2'
94	235-2'6'
95	236-2'5'
96	236-2'6'
97	245-2'3'
98	246-2'3'
99	245-2'4'
100	246-2'4'
101	245-2'5'
102	245-2'6'
103	246-2'5'
104	246-2'6'
105	234-3'4'
106	2345-3'
107	234-3'5'
108	2346-3'
109	235-3'4'
110	236-3'4'
111	235-3'5'
112	2356-3'
113	236-3'5'
114	2345-4'
115	2346-4'
116	23456
117	2356-4'
118	245-3'4'
119	246-3'4'
120	245-3'5'
121	246-3'5'
122	345-2'3'
123	345-2'4'
124	345-2'5'
125	345-2'6'
126	345-3'4'
127	345-3'5'

Hexa-chlorobiphenyls

IUPAC	Structural (Chlorine Pos)
128	234-2'3'4'
129	2345-2'3'
130	234-2'3'5'
131	2346-2'3'
132	234-2'3'6'
133	235-2'3'5'
134	2356-2'3'
135	235-2'3'6'
136	236-2'3'6'
137	2345-2'4'
138	234-2'4'5'
139	2346-2'4'
140	234-2'4'6'
141	2345-2'5'
142	23456-2'
143	2345-2'6'
144	2346-2'5'
145	2346-2'6'
146	235-2'4'5'
147	2356-2'4'
148	235-2'4'6'
149	236-2'4'5'
150	236-2'4'6'
151	2356-2'5'
152	2356-2'6'
153	245-2'4'5'
154	245-2'4'6'
155	246-2'4'6'
156	2345-3'4'
157	234-3'4'5'
158	2346-3'4'
159	2345-3'5'
160	23456-3'
161	2346-3'5'
162	235-3'4'5'
163	2356-3'4'
164	236-3'4'5'
165	2356-3'5'
166	23456-4'
167	245-3'4'5'
168	246-3'4'5'
169	345-3'4'5'

Hepta-chlorobiphenyls

IUPAC	Structural (Chlorine Pos)
170	2345-2'3'4'
171	2346-2'3'4'
172	2345-2'3'5'
173	23456-2'3'
174	2345-2'3'6'
175	2345-2'3'5'
176	2346-2'3'6'
177	2356-2'3'4'
178	2356-2'3'5'
179	2356-2'3'6'
180	2345-2'4'5'
181	23456-2'4'
182	2345-2'4'6'
183	2346-2'4'5'
184	2346-2'4'6'
185	23456-2'5'
186	23456-2'6'
187	2356-2'4'5'
188	2356-2'4'6'
189	2345-3'4'5'
190	23456-3'4'
191	2346-3'4'5'
192	23456-3'5'
193	2356-3'4'5'

Octa-chlorobiphenyls

IUPAC	Structural (Chlorine Pos)
194	2345-2'3'4'5'
195	23456-2'3'4'
196	2345-2'3'4'6'
197	2346-2'3'4'6'
198	23456-2'3'5'
199	2345-2'3'5'6'
200	23456-2'3'6'
201	2346-2'3'5'6'
202	2356-2'3'5'6'
203	23456-2'4'5'
204	23456-2'4'6'
205	23456-3'4'5'

Nona-chlorobiphenyls

IUPAC	Structural (Chlorine Pos)
206	23456-2'3'4'5'
207	23456-2'3'4'6'
208	23456-2'3'5'6'

Deca-chlorobiphenyl

IUPAC	Structural (Chlorine Pos)
209	23456-2'3'4'5'6'

Table 10.1.2 *Polychlorinated Biphenyl Homolog and Number of Congeners Within Each Homolog Group*

Homolog	Chemical Formula	Number of Chlorines	No. of Congeners in Homolog Group	IUPAC Congener Numbering[*]
Mono-chlorobiphenyl	$C_{12}H_9Cl$	1	3	1–3
Di-chlorobiphenyl	$C_{12}H_8Cl_2$	2	12	4–15
Tri-chlorobiphenyl	$C_{12}H_7Cl_3$	3	24	16–39
Tetra-chlorobiphenyl	$C_{12}H_6Cl_4$	4	42	40–81
Penta-chlorobiphenyl	$C_{12}H_5Cl_5$	5	46	82–127
Hexa-chlorobiphenyl	$C_{12}H_4Cl_6$	6	42	128–169
Hepta-chlorobiphenyl	$C_{12}H_3Cl_7$	7	24	170–193
Octa-chlorobiphenyl	$C_{12}H_2Cl_8$	8	12	194–205
Nona-chlorobiphenyl	$C_{12}H_1Cl_9$	9	3	206–208
Deca-chlorobiphenyl	$C_{12}Cl_{10}$	10	1	209

[*] See Table 10.1 for specific congeners within each homolog group

Table 10.1.3 *Industrial Use of PCBs*

System/Category	1221	1232	1016	1242	1248	1254	1260	1262	1268
Dielectric Fluids									
Capacitors	✓		✓+	✓+		✓			
Transformers				✓		✓+	✓		
Hydraulics/Lubricants/Heat Transfer Fluids									
Heat Transfer				✓					
Hydraulic Fluids		✓		✓	✓	✓	✓		
Vacuum Pumps					✓	✓	✓		
Gas Transmission Turbines	✓			✓					
Plasticizers									
Rubber	✓	✓		✓+	✓	✓			
Synthetic Resins				✓	✓	✓	✓	✓	✓
Carbonless Copy Paper				✓+					
Miscellaneous Industrial									
Adhesives	✓	✓		✓+	✓	✓			
Wax Extenders				✓+		✓			✓
Dedusting Agents						✓	✓		
Inks						✓			
Cutting Oils						✓			
Pesticide Extenders						✓			

Table modified after Durfee *et al.*, (1976).
Information from Monsanto Industrial Chemicals Co. as reported by Durfee *et al.* (1976).
✓ denotes use of given Aroclor in a specific end-use
✓+ denotes principal use

Starting in 1971, Aroclor 1016, a new product, began to be produced and sold in the United States (Table 10.1.4) and this coincided with a dramatic drop in sales of Aroclors 1242 and 1248 (Table 10.1.4). This is not a coincidence. Aroclor 1016 was marketed as a replacement for applications that had used Aroclors 1242 and to some extent 1248. The rationale for replacement of Aroclor 1242 with Aroclor 1016 was that research in the 1970s had indicated greater persistence of tetra and higher chlorinated PCBs in biota (Burse *et al.*, 1974). While Aroclor 1016 is very similar in composition to Aroclor 1242, it has much lower percentage of congeners with more than four chlorines. The Aroclor 1016 production process reportedly involved chlorination of biphenyl to 42% chlorine (as for Aroclor 1242). This product was then fractionated to yield a distillate with approximately 41.5% chlorine (still very close in composition to Aroclor 1242 – but with fewer tetra and higher chlorinated congeners, and fewer non-ortho and mono-ortho congeners (Frame, 1999).

While Monsanto was the major producer of PCBs, and while the majority of this chapter focuses on Aroclors, it is important to note that there were other producers of commercial PCB products. Bayer AG (Germany) marketed PCBs under the trade name Clophen. Kanegafuchi Chemical Company (Japan) marketed PCBs under the trade name Kanechlor. Prodolec (France) marketed PCBs under trade names Phenoclor and Pyralene.

While chlorinated organic compounds have been observed to occur in nature, naturally occurring PCBs have not been described in the literature (Gribble, 1998, 2004). Thus, when PCBs are found in the environment they can be attributed to anthropogenic sources and ultimately to the use (or misuse) of one or a combination of the commercial PCB products discussed above.

10.1.3 Composition of Commercial PCB Products
It is important to reiterate and understand that commercial PCB products such as Aroclors are not a single chemical but, rather, complex mixtures of many congeners. Congener-specific analytical methods (see Section 10.2) can quantify most of the 209 congeners, but some congeners

Table 10.1.4 US sales (thousands of pounds) of nine Aroclors

Year	1221	1232	1016	1242	Aroclor 1248	1254	1260	1262	1268
1957	23	196	–	18,222	1,779	4,461	7,587	31	–
1958	16	113	–	10,444	2,559	6,691	5,982	184	72
1959	254	240	–	13,598	3,384	6,754	6,619	359	102
1960	103	155	–	18,196	2,827	6,088	7,330	326	189
1961	94	241	–	19,827	4,023	6,294	6,540	361	158
1962	140	224	–	20,654	3,463	6,325	6,595	432	210
1963	361	13	–	18,510	5,013	5,911	7,626	414	284
1964	596	13	–	23,571	5,238	6,280	8,535	446	190
1965	369	7	–	31,533	5,565	7,737	5,831	558	196
1966	528	16	–	39,557	5,015	7,035	5,875	768	284
1967	442	25	–	43,055	4,704	6,696	6,417	840	287
1968	136	90	–	44,853	4,894	8,891	5,252	720	280
1969	507	273	–	45,491	5,650	9,822	4,439	712	300
1970	1,476	260	–	48,588	4,073	12,421	4,890	1,023	330
1971	2,215	171	3,334	21,981	213	4,661	1,725	1	–
1972	171	–	20,902	728	807	3,495	305	–	–
1973	35	–	23,531	6,200	–	7,976	–	–	–
1974	57	–	21,955	6,207	–	6,185	–	–	–
% of Sales	0.95%	0.26%	8.8%	54.2%	7.4%	15.6%	11.5%	0.90%	0.36%

Data from Monsanto Industrial Chemicals Company, as reported by Durfee *et al.*, (1976).

"coelute" with other congeners. That is, sometimes a peak observed on a chromatogram represents the summation of more than one congener. Coeluting congeners vary between methods, in some cases even when analyses are conducted using the same nominal method. Because of this, one of the most valuable Aroclor reference data sets is that of George Frame (Frame *et al.*, 1996) who determined complete PCB congener assignments for all 209 individual congeners and for eight Aroclor formulations. This data set has become an extremely valuable reference in PCB environmental forensics, because it allows data from any lab (regardless of coelutions) to be compared to the congener patterns of suspected sources and alteration processes. Frame's 209 analytes can be summed and/or reduced to match any other analyte list. Figure 10.1.2 presents bar-graphs of Frame's data and shows congener-specific compositions of nine Aroclors (Aroclors 1221, 1232, 1016, 1242, 1248, and two variants of 1254, 1260, and 1262). While there are 209 possible PCB congeners, only about 100 to 150 are typically found at detectable levels in the ambient environment. Quantitation of, as few as, 20–30 congeners can be sufficient to infer Aroclor sources (e.g., Johnson *et al.*, 2000). Frame's data are shown in Figure 10.1.2 and here they have been reduced to the 100 most abundant congeners. Note that on these graphs, the degree of chlorination increases to the right and that the higher numbered Aroclors have higher percentages of chlorine.

Other workers have also reported congener-specific data and profiles for different Aroclor products (see Section 10.4.3). Within each Aroclor product category, there is surprising consistency between the reported patterns. Frame *et al.* (1996) reported a few notable differences between two lots of Aroclor 1248 (in particular PCB28 and PCB44) but those congener patterns were still very similar to each other. While there are often slight differences observed between different laboratories reporting the same Aroclors, these differences can often be attributed to different analytical methods and/or to different coelutions between methods. Even considering these occasional differences, there is a surprising consistency in the congener compositions of Aroclors reported in the literature.

One notable exception is Aroclor 1254. There were two distinct variants of Aroclor 1254 sold by Monsanto. George Frame was the first to report and distinguish between these two products (Frame *et al.*, 1996). A subsequent investigation by Frame (1999) indicated that the atypical variant was present in several Aroclor 1254 lots that trace back to 1974–1976 and was due to a change in Monsanto's Aroclor 1254 manufacturing process. Frame (1999) reported that the "late-production Aroclor 1254" was related to production of the new Monsanto product: Aroclor 1016 (discussed above) which began production in the early 1970s (Table 10.1.4). Aroclor 1016 was produced by a two-stage chlorination process. In the first stage, biphenyl was chlorinated to 42% (as it would be for Aroclor 1242). This product was then distilled to remove tetra- and higher chlorinated homologs, yielding the new Aroclor 1016 product, which was 41.5% chlorine. The residual by-product resulting from that process (i.e., still bottoms) were about 49% chlorine by weight (Frame, 1999), which was then further chlorinated to 54% chlorine and sold as Aroclor 1254. This product constitutes the late-production variant.

Frame *et al.*, 1996; Frame, 1999 reported that late-production 1254 was sold by Monsanto from 1974 through 1976. The reported 1974 start-date was a function of the dates attributed to the Aroclor 1254 lots analyzed in Frame's lab, as well as anecdotal evidence. Actual Monsanto production records were not available. Unpublished developments since the Frame papers push the 1974 start-date back at least two years. Larry Hansen of the University of Illinois (Hansen, personal communication) reports that a late-production Aroclor 1254 used in his laboratory (labeled Monsanto Electrical Grade, Lot KB 05-612, Aroclor 1254) was analyzed by Jack Cochran (Frame's coauthor on the 1996 paper) and Cochran confirmed this material as late-production Aroclor 1254 (Hansen, personal communication). The KB-05-612 sample has been maintained and stored in Dr Hansen's lab in the original bottle and within the original mailing container. That mailing container bears a May 30, 1972 postmark. Thus late-production Aroclor 1254 predates Frame's 1974 production start-date estimate by at least two years. Late-production Aroclor 1254 may predate 1972, but if so, it is unlikely to have been produced before

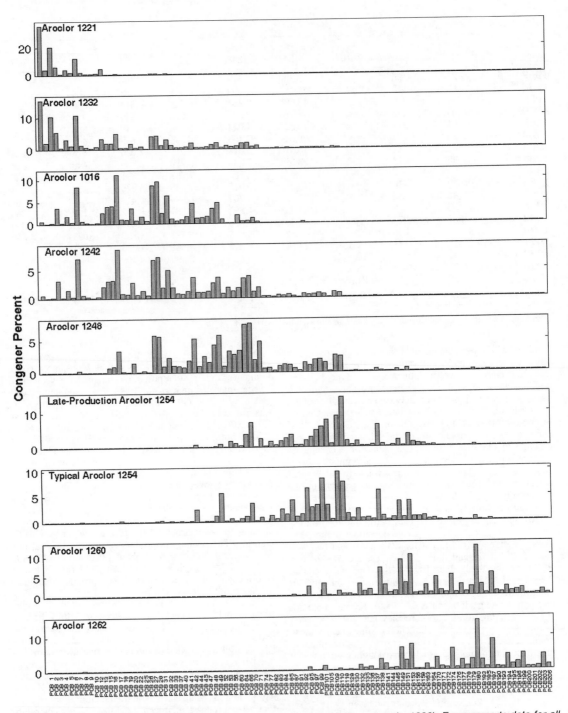

Figure 10.1.2 *Congener-specific compositions of Aroclor formulations (Frame et al., 1996). Frame reports data for all 209 congeners. Only the 100 most abundant congeners are shown in this figure (see color insert).*

1971, because its production was linked to the Aroclor 1016 production process, which began in 1971 (Table 10.1.4). Late-production Aroclor 1254 is also notable because it has higher proportions of coplanar (dioxin-like) PCBs than the typical Aroclor 1254 (Frame, 1999).

There is an admitted bias in this chapter toward Aroclors and that is primarily due to three facts: (1) Monsanto was

the market leader in production and sale of PCBs; (2) the majority of PCB research in the literature has focused on Aroclors; and (3) the authors of this chapter all reside and work primarily within North America so Aroclor case-studies represent the vast majority of the authors' collective experience. That said, a number of researchers have studied PCB compositions of commercial mixtures of PCB products

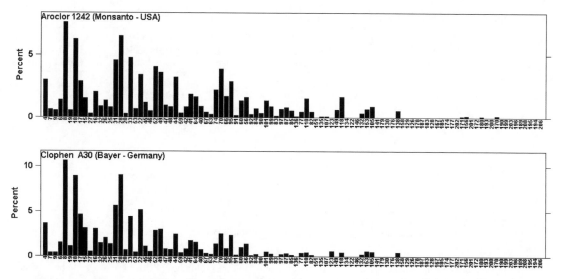

Figure 10.1.3 *Comparison of congener profiles of Aroclor 1242 (Monsanto–USA) and Clophen A30 (Bayer–Germany). The similarity between these patterns is visually evident, and the cos-θ similarity metric calculated between these two patterns is 0.96 (data from Shulz et al., 1989) (see color insert).*

manufactured outside the United States. Studies that have compared compositions of similar weight, but manufactured by different companies, have reported striking similarities between these products. Schulz *et al.* (1989) and Kannan *et al.* (1992) have demonstrated that Aroclor, Clophen, and Kanechlor products that are similar in industrial use and weight percent of chlorine exhibit similar congener patterns. This is illustrated in Figures 10.1.3 and 10.1.4 using data from Schulz *et al.* (1989). Figure 10.1.3 shows the congener profiles for Monsanto's Aroclor 1242 (USA) and Bayer's similar product Clophen A30 (Germany). Figure 10.1.4 shows the congener profiles for Monsanto's Aroclor 1260 and Bayer's similar product Clophen A30. While a review of

raw data (Schulz *et al.*, 1989) indicates that the products are not absolutely identical, the overall similarity between the profiles is obvious and striking. This similarity between the chemical compositions of these products has led workers to conclude that the production processes were similar. De Voogt and Brinkman (1989) presented a comparison of various series of commercial PCB products, and a summary of that comparison is provided as Table 10.1.5.

10.1.4 Co-contaminants in commercial PCB products
Several investigators have reported the presence of co-contaminants in commercial PCB mixtures in particular polychlorinated dibenzofurans (PCDFs – Wakimoto *et al.*,

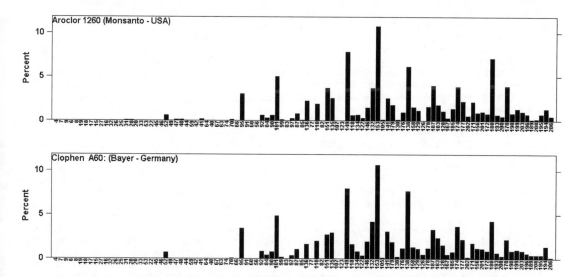

Figure 10.1.4 *Comparison of congener profiles of Aroclor 1260 (Monsanto–USA) and Clophen A60 (Bayer–Germany). The similarity between these patterns is visually evident and the cos-θ similarity metric calculated between these two patterns is 0.98 (data from Shulz et al., 1989).*

Table 10.1.5 *Polychlorinated Biphenyl Product Comparison (Analogous Products Appear on the Same Row)*

Aroclor Monsanto (USA)	Clophen Bayer (Germany)	Phenoclor Caffaro (Italy)	Pyralène Caffaro (Italy)	Kanechlor Kanegafuchi (Japan)	Fenoclor Prodelec (France)	Delor Chemko (Czechoslovakia)
1221						
1232			2000	200		
			1500			
1242 (1016)	A30	DP 3	3000	300	42	2
1248	A40	DP 4		400		3
1254	A50	DP 5		500	54	4;5
1260	A60	DP 6		600	64*	
1262						
					70*	
1268						
1270					DK	

Table modified after de Voogt and Brinkman (1989).
*Fenoclor two-digit number should indicate w% Cl. However, this does not fit within manufacturers specifications.

1988) and polychlorinated naphthalenes (PCNs—Haglund *et al.*, 1993). PCDFs have garnered the most attention (Vuceta *et al.*, 1983; Wakimoto *et al.*, 1988). Wakimoto *et al.* (1988) analyzed commercial Aroclors and Kanechlors for approximately 60 PCDF congener peaks (including the toxic 2,3,7,8 substituted congeners) and reported total PCDFs in the 0.6–26 ppm range in all mixtures. Other research has shown that PCDF concentrations can increase in PCBs as a result of heating (Hutzinger *et al.*, 1985; Erickson *et al.*, 1989). Erickson *et al.* (1989) reports that the transformation of PCB to PCDF is a single-step process and occurs optimally at temperatures of about 675 °C.

It should be noted that the studies cited above were conducted prior to Frame's identification of two different Aroclor 1254 variants (Section 10.1.3). A more recent study (Kodavanti *et al.*, 2001) focused primarily on the difference in neurochemical effects in rats exposed to the two A1254 variants. That work included congener-specific PCB analysis of both 1254 variants. They confirmed earlier reports that the late-production variant had higher proportions of dioxin-like coplanar PCBs (Frame, 1999). But in addition, Kodavanti *et al.* (2001) analyzed the two A1254 variants for co-contaminants (PCDFs and PCNs). Total PCN levels were similar between the two lots, but the late-production A1254 sample had much higher levels of PCDFs (including the 2,3,7,8-PCDFs) as compared to the typical variant.

The PCDDs are not typically encountered in neat PCB products (Wakimoto *et al.*, 1988; Kodavanti *et al.*, 2001). The reason for this is that the formation of PCDDs from PCB is much more complex and involves the severing of the carbon-carbon link between the PCB phenyl rings (Erickson, 2001). However, PCDDs have been identified as a result of combustion of PCB containing products (Schecter, 1983; O'Keefe *et al.*, 1985; Hutzinger *et al.*, 1985) but this is not attributed to combustion of PCBs themselves. Rather, it is attributed to combustion of products such as chlorobenzenes that are associated with the PCBs in the materials burned – i.e., askarels (Hutzinger *et al.*, 1985; O'Keefe and Smith, 1989; Erickson *et al.*, 1989; Erickson, 2001).

10.1.5 Chemical Properties of PCBs

Polychlorinated biphenyls have chemical and physical properties that made them very useful for industrial purposes. They have high molecular weight (a thick heavy resinous material in pure form). PCBs have a high boiling point and low flammability, thus they can withstand heat and do not easily break down. For many of the reasons that they proved useful in industrial applications (resistance to breakdown

and low water solubility), PCBs have proven to be persistent contaminants in the environment. They tend to adsorb to organic matter (particularly to organic carbon and black carbon) within soil and sediment matrices (Lohmann, 2003). They do not easily degrade. Thus they are persistent in the environment. Further, PCBs are lipophilic and tend to bioaccumulate in fatty tissues and blood lipids of humans and other higher predators.

Polychlorinated biphenyls have low water solubility and low vapor pressure (Hansen, 1999; Erickson, 1997). Table 10.1.6 shows the chemical and physical properties of several PCB congeners with differing degrees of chlorination (data from Mackay, 2001). To put these numbers in context, this table also includes physical/chemical properties of other common contaminants of concern in environmental investigations. As compared to fuel-oil related contaminants, benzene and naphthalene, PCBs exhibit lower aqueous solubility and higher octanol–water partitioning coefficients (log K_{ow}), confirming that PCBs are hydrophobic. Thus in aqueous systems, PCBs are typically found in highest concentration in sediment fraction. To the degree that PCBs are found in water, they are typically present at higher concentration adsorbed to suspended particulate matter rather than in solution.

Note also that the PCBs exhibit vapor pressures orders of magnitude lower than benzene and naphthalene (Table 10.1.6). This suggests that PCBs prefer to be bound to soils rather than in vapor phase. Erickson (2001) estimates that 99% of the global mass of environmental PCB contamination is encountered in soil and sediment. As a result, investigations of degree and extent of PCB contamination at hazardous waste sites typically focus primarily on soil and sediment.

Table 10.1.6 also includes physical/chemical properties for 1,2,3-trichlorobenzene. As discussed above, PCBs were often combined with trichlorobenzene for use as askarels in capacitors and transformers and these industrial uses account for the majority of PCBs produced. Note that trichlorobenzene is much more soluble and volatile than any of the PCBs. The vastly different chemical properties are such that one should not necessarily be surprised if PCBs from an askarel source are encountered in sediments at much higher concentrations than (or even in absence of) chlorobenzenes. The different chemical properties are such that the fate and transport of chlorobenzenes may be much different.

Table 10.1.6 also presents properties for polychlorinated dibenzo-*p*-dioxins (PCDDs) and polychlorinated dibenzofurans (PCDFs). The PCDD/F congeners shown

Table 10.1.6 *Physical and Chemical Properties of Six PCB Congeners and Seven Reference Chemicals*

Congener IUPAC	Structure	Homolog Group	Chemical Formula	Molar Mass (g/mol)	Melting Point (°C)	Vapor Pressure (Pa)	Aqueous Solubility (g/m^3)	log K$_{ow}$
PCB-7	24	Di-chlorobiphenyl	$C_{12}H_8Cl_2$	223.1	24.4	0.254	1.25	5
PCB-15	4-4′	Di-chlorobiphenyl	$C_{12}H_8Cl_2$	223.1	149	4.8×10^{-3}	6×10^{-2}	5.3
PCB-29	245	Tri-chlorobiphenyl	$C_{12}H_7Cl_3$	257.5	78	1.3×10^{-2}	1.4×10^{-1}	5.6
PCB-52	25-2′5′	Tetra-chlorobiphenyl	$C_{12}H_6Cl_4$	292.0	87.0	4.9×10^{-3}	3×10^{-2}	6.1
PCB-101	245-2′5′	Penta-chlorobiphenyl	$C_{12}H_5Cl_5$	326.4	76.5	1.1×10^{-3}	1×10^{-2}	6.4
PCB-153	245-2′4′5′	Hexa-chlorobiphenyl	$C_{12}H_4Cl_6$	360.9	103	1.2×10^{-4}	1×10^{-3}	6.9
Other common organic contaminants of concern chemical name								
Benzene				78.11	5.53	12,700	1,780	2.13
Naphthalene				128.19	80.5	10	31	3.37
1,2,3-trichlorobenzene				181.45	53	28	21	4.4
2,3,7,8-tetrachlorinated dibenzofuran (2,3,7,8-TCDD)				306	227	2×10^{-6}	4.19×10^{-4}	6.1
Octachlorinated dibenzofuran (OCDF)				443.8	258	5×10^{-10}	1.16×10^{-6}	8.0
2,3,7,8-tetrachlorinated dibenzo-*p*-dioxin (2,3,7,8-TCDD)				322	305	2×10^{-7}	1.93×10^{-5}	6.8
Octachlorinated dibenzo-*p*-dioxin (OCDD)				460	322.0	1.1×10^{-10}	7.4×10^{-8}	8.2

Data Source: Mackay (2001, Table 3.5).

here are typically reported together, as a result of the same chemical analysis (see Chapter 3). PCDD/Fs are often considered in environmental forensics investigations involving PCBs because PCDFs are known to be an impurity associated with production of PCBs (see Section 10.1.4). Note that while PCBs have low vapor pressure and aqueous solubility, they are an order of magnitude more soluble and volatile than dioxins and furans with analogous degrees of chlorination.

Within the group of PCB congeners shown in Table 10.1.6, it is evident that aqueous solubility and volatility (while low for all congeners) tends to decrease with increasing degrees of chlorination. This has important implications in later discussion of PCB volatilization. In short, empirical studies have shown that the congener pattern observed in volatile phase or aqueous phase typically shows a lower chlorinated congener pattern than is observed in the source from which it was derived (Chiarenzelli *et al.*, 1996, 2000; Jarman *et al.*, 1997).

10.1.6 Overview of PCB Fate, Transport, and Alteration

The previous discussion highlighting the low solubility and low vapor pressures of PCBs may seemingly imply that PCBs are contaminants that are limited to local (or at most regional) environmental issues, because they do not easily go into the vapor or aqueous phase. This is not the case. PCBs are global contaminants. While only a small proportion of global PCBs are present in the volatile phase in the atmosphere, that small proportion is significant. Volatile phase transport of PCBs is such that these contaminants have found their way to every corner of the globe (albeit at concentrations orders of magnitude below those encountered in urban areas and on or near sites where PCBs were spilled). The mobility of PCBs in air is such that it is gaining increased attention as an exposure pathway of concern. Specifically, recent exposure assessment studies clearly suggest that inhalation has been an under-appreciated exposure pathway in both humans and animals (DeCaprio *et al.*, 2005; Imsilp *et al.*, 2005; Casey *et al.*, 1999).

Polychlorinated biphenyls are persistent contaminants because they are recalcitrant and resistant to degradation and alteration. Few argue this point, but oversimplification has led to a common misconception that PCBs are immortal and that congener patterns do not change in the environment. Polychlorinated biphenyls (some congeners more so than others) have been shown to exhibit degradation

and weathering in the environment. This has the effect of altering PCB congener patterns observed in the ambient environment. Polychlorinated biphenyls alteration has been extensively studied. In environmental forensics investigations, it is important to be aware of these mechanisms because PCB patterns observed in ambient environmental samples may not match the composition of the original source(s) from which the PCBs were derived.

In general, the PCBs that are most susceptible to alteration are the less-chlorinated congeners, and in turn the Aroclors that are most susceptible to alteration are the "lighter" Aroclors (e.g., Aroclors 1016, 1242 and 1248). However, even the heavier Aroclor compositions such as Aroclor 1260 can be altered by microbial degradation (Bedard and Quensen, 1995) and by volatilization/vapor phase transport by air (Chiarenzelli *et al.*, 1997). These two alteration mechanisms are discussed in more detail in Section 10.3.

As indicated by octanol/water partitioning coefficients (K_{ow} – Table 10.1.6), PCBs are lipophilic. Therefore, a predator that ingests many meals with relatively low levels of PCB may ultimately exhibit PCB concentrations in blood and adipose tissue that far exceed that of individual prey (a phenomenon termed "biomagnification"). While the term "biomagnification" may imply that PCBs enter a body and remain there, this is not the case. Polychlorinated biphenyl congeners do metabolize within the body and all PCB congeners do not metabolize at the same rate. As a result, when humans or other biota are exposed to PCBs, the relative proportions of the congeners observed in blood or adipose tissue will change such that they may differ from those of the PCB source. Biological alteration and metabolism of PCBs are discussed in more detail in Section 10.3.3.

10.2 ANALYTICAL CHEMISTRY METHODS

This section provides: (1) an overview of standard regulatory methods (SW-846 and other EPA methods); (2) a discussion of their limitations in environmental forensics investigations; and (3) extensions of these methods that are commonly employed in order to obtain congener-specific PCB data sufficient for environmental forensics purposes.

Conventional analytical chemistry methods have been developed primarily within the context of environmental regulatory programs such as the Resource Conservation and Recovery Act (RCRA-1976) and the Comprehensive Environmental Response, Compensation and Liability Act

(CERCLA or "Superfund" – 1980). These methods have been developed within the context of typical objectives of site assessments conducted under RCRA (RCRA Facility Investigations – RFI) and CERCLA (Remedial Investigation/Feasibility Studies – RI/FS). Objectives of such studies are typically broad: identify contaminants of concern and degree and extent of contamination. As a result, the analytical methods used in these studies are designed to look for a wide variety of common industrial chemicals of concern and typically fall under USEPA's SW-846 Methods (USEPA, 1997a).

These methods are often specified in an RI/FS or RFI work plan and may be required for compliance with regulatory requirements at contaminated sites. Unfortunately, the broad objectives of RI/FS type investigations do not necessarily translate into data that is useful in an environmental forensics context. For detailed environmental forensics investigations, it is usually necessary to obtain congener-specific data. The standard USEPA SW-846 methods are either not congener-specific methods or they include calibration for only small subset of congeners that are insufficient to unravel complex issues such as multiple sources, dechlorination, and volatilization. This is not a situation unique to PCBs. Stout *et al.* (2002) present an analogous discussion in the context of environmental forensics investigations of petroleum hydrocarbon investigations and they advocate non-SW846 methods for environmental forensics applications.

10.2.1 EPA Methods

This section of the chapter focuses on four EPA methods commonly used for PCB analysis (Methods 8082A, 608, 680, and 1668). These methods are discussed as a general review. Each method carries with it implicit references to methods for extraction, cleanup, and data reduction. This chapter will not cover these subsidiary methods. The reader is referred to Erickson (1997) for more thorough description of SW-846 and other methods, and to US EPA (1999) for a detailed description of EPA Methods 1668 and 1668A.

There are four EPA methods available for PCB analysis: Method 608, Method 8082A, Method 680, and Method 1668. The variables among these methods include the target analytes (PCBs as Aroclor equivalents, homolog group totals, or individual PCB congeners), the procedures used for analysis (range of cleanup and separation steps), the method of detection (GC/ECD, GC/LRMS or GC/HRMS), and the method of quantification (external standard, internal standard, or isotope dilution internal standard). These various options result in differences in specificity, selectivity, detection limit, overall data quality, and in ease of analyzing small sample quantities or difficult samples.

The GC/ECD techniques have traditionally been used for chlorinated hydrocarbon (PCB and pesticide) analysis because the ECD (electron capture detector which detects electro-negative compounds, including halogenated compounds) offers some selectivity of response and good detection limits in a cost-effective manner. Surrogate standards may be added to monitor performance of the analytical procedure but they are limited in scope and restricted to similar compounds or a PCB that is not in Aroclor mixtures and therefore not expected to be found in samples. Accurate identification of PCBs is dependent on the gas chromatographic separation of PCBs from other organohalogen compounds in the sample. No other confirmation of the identity of a GC/ECD response is available other than another analysis on a different GC column. Interferences by phthalate esters introduced during sample preparation can pose a problem in PCB determinations especially when using an ECD detector.

The GC/MS techniques offer an improvement in selectivity because PCBs are identified not only by their retention time on the GC column, but also by their molecular mass. GC/MS also permits the addition of chemically labeled internal standards (usually ^{13}C). When labeled internal standards are added to the sample before extraction, cleanup, and analysis, the result is a very rigorous analytical method offering the maximum in precision and accuracy. Low resolution GC/MS, if operated in the selected ion monitoring (SIM) mode, achieves the same or better detection limits as ECD, minimizes false positives because mass spectral information is used to confirm the identity of each "hit," and produces recovery-corrected analytical results. Difficult samples are easily handled by GC/MS because co-eluting interferences producing a GC signal can be distinguished from target analytes by the MS. The state of the art in resolution and in detection limit is the high resolution MS technique. This technique provides the ultimate in analyte identification and detection limits and may be the technique of choice if sample size is limited or if one needs to accurately quantify low concentration congeners such as the non ortho- and mono ortho- substituted congeners.

Either low resolution or high resolution GC/MS method can be used to quantify individual PCB congeners. In both cases, analysis parameters must be carefully controlled to ensure that no inaccuracies are introduced by fragments of highly chlorinated PCBs being mistaken as PCB congeners of lower chlorination level and being counted twice. This is most significant when congeners of high concentration elute from the gas chromatograph at the same time as less chlorinated lower concentration congeners. If the lower concentration congener is one of the toxic congeners, a significant error could occur. This is minimized by using the appropriate high resolution gas chromatographic separation procedures. The potential for this error is more significant with low resolution MS techniques that have less mass resolution than do the high resolution MS techniques.

Some methods quantify PCBs as Aroclor equivalent concentrations. In these cases, the chromatogram is examined to determine which Aroclor pattern is present and that Aroclor is used to calibrate the instrument response, typically using a few characteristic congeners of the Aroclor selected. In some congener specific methods, total PCBs are reported as the sum of the concentration of each individual PCB congener detected in the sample. Alternately, PCBs can be reported first as homolog group totals; for example, total monochlorobiphenyls, total dichlorobiphenyls, etc. and then as a total PCB concentration by summing the mono- to decachlorobiphenyl concentrations. The homolog total approach provides some basic information about congener distribution patterns. Addition of a carbon column separation (to isolate the non-ortho and mono-ortho-substituted PCBs) with a second high resolution GC/MS run to accurately quantify PCBs 77, 126, 169 and other toxicologically important PCBs provides the most accurate quantification of all individual congeners in environmental samples. This is the recommended approach if an accurate calculation of the sample's Toxicity Equivalents (TEQ) is to be made using the World Health Organization's Toxicity Factors for PCB congeners.

10.2.1.1 EPA Method 608—Aroclors

EPA Method 608 is used for analysis of organochlorine pesticides and PCBs in wastewater by gas chromatography with electron capture detection (GC/ECD). The method does not include the use of surrogate standards for monitoring method performance but it does permit the use of an internal standard for quantification of the GC/ECD response. Polychlorinated biphenyls are identified and quantified by comparison to Aroclor standards and two analyses on different GC columns are required to confirm the identification of the PCBs. Poly-chlorinated biphenyl detection limits are in the order of 0.065 ug/L. Results are reported as Aroclor equivalent concentrations, which presents limitations for environmental forensics investigations. These limitations are discussed in more detail in Section 10.2.1.2 as part of the discussion of Method 8082 (another Aroclor reporting method).

10.2.1.2 EPA Method 8082—Aroclors

EPA Method 8082 has traditionally been used for chlorinated hydrocarbon (PCB and pesticide) analysis on solid and aqueous matrices. The scope of the most recent version of Method 8082 also includes analysis of tissue, oil, and wipe samples. Although the primary analysis technique in this method is GC with ECD or electrolytic conduction (ELCD) detection, GC/MS is allowed for confirmation analysis. The SIM mode is recommended to provide confirmatory analyses at a similar detection limit to the GC/ECD analysis. Some labs have converted to GC/MS as the primary analytical tool when applying this method. Target analytes are either Aroclors or a list of individual congeners selected as being characteristic of various Aroclors. The method allows a customized list of congeners to be analyzed, but this is typically less than 20 congeners. A surrogate standard (decachlorobiphenyl or tetrachloro–m-xylene) is added to samples to monitor the performance of the method, but it is not used in the quantification and reported results are not recovery-corrected for any inefficiencies in sample processing leading to loss of PCBs during analysis. Individual congeners are used for calibration of the PCB congener analysis and an internal standard is added immediately prior to GC/ECD or GC/MS analysis to allow correction for the effects of uncertainties in final volumes and any instrument instability. External standard quantification is normally used for Aroclor analysis. The method detection limits are not specified but are left to each lab to determine.

As with Method 608, PCB concentrations resulting from 8082 are typically reported in terms of Aroclor concentrations. These methods carry with them the implicit assumption that all samples analyzed represent unaltered Aroclors. This assumption represents these methods' greatest limitation in environmental forensics investigations. First, reported total PCB concentrations are necessarily an extrapolation given that most congeners are not actually quantified. Further, the method has no provisions for determination of non-Aroclor products (e.g., Clophens) nor can it adequately take into account alteration of Aroclors (Erickson, 1997).

A graphic example of this limitation is shown in Figure 10.2.1. These are data from a single sediment sample collected from a small stream. The sample was analyzed for PCBs by Method 8082 and yielded the following results: 30 μg/Kg Aroclor 1242; 1100 μg/Kg Aroclor 1254; and 1700 μg/Kg Aroclor 1260 (Figure 10.5a). This result was unusual because other sediment data collected from the same creek consistently indicated only Aroclor 1260 as a source. Bearing in mind that 8082 determines "Aroclor" concentrations by calibration for and quantitation of a relatively small group of congeners, Figure 10.2.1(b) shows the more complete congener profile that is implied by the 8082 results (1% Aroclor 1242; 39% Aroclor 1254; and 60% Aroclor 1260). This bar graph was constructed by creating mathematical mixtures of Aroclor data reported by Frame, in the proportions implied by the 8082 results. The same field sample was later analyzed using congener-specific Method 1668A and Figure 10.2.1(c) shows the full-congener profile of this sample. Clearly, there is a difference between the congener pattern implied by the 8082 results and the congener pattern observed in the actual sample. The most abundant peak in the sediment sample was the coeluting peak PCB 44+47+65, which is not a dominant peak in any of the three Aroclors reported by Method 8082. Why the discrepancy? It turns out that PCB-47 (24-2′4′-CB) is

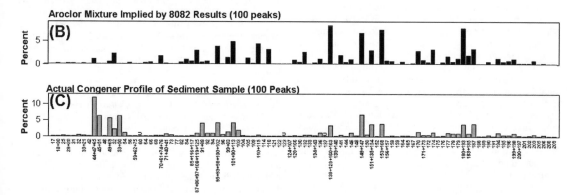

| **Method 8082 Results for Sediment Sample** | |
Aroclor	Concentration (μg/Kg)
Aroclor 1242	30
Aroclor 1248	BRL
Aroclor 1254	1,100
Aroclor 1260	1,700

Figure 10.2.1 (A) Method 8082 results for a sediment sample expressed in terms of Aroclor concentrations; (B) the congener profile implied by the 8082 results assuming the sample was indeed a mixture of unaltered Aroclors (1% A1242; 39% A1254; 60% A1260); and (C) the actual congener profile of the same sediment sample based on full congener-specific analysis.

a characteristic dechlorination product resulting from several anaerobic microbial processes reported in the literature (*Processes H and N and probably H' and M* – Bedard and Quensen, 1995), and at least the first two of these processes can operate on Aroclor 1260 impacted sediments. Thus the congener-specific data clearly suggests that this sample represents a dechlorination product rather than the mixture of unaltered Aroclors suggested by the 8082 results.

This example also illustrates a caveat in regard to total PCB concentrations, as determined by Method 8082. In the example shown in Figure 10.2.1, the 8082 results suggested a total PCB concentration of 2830 μg/Kg (2.8 ppm). In contrast, full-congener analysis (Figure 10.2.1c) of the same sample yielded a total PCB concentration of 5.65 ppm – two times higher than that indicated by Method 8082. Recall that Method 8082 determines PCB concentrations in terms of Aroclors using a small number of pre-selected peaks. In general, a total PCB concentration based on quantitation of 100 or more congeners will give a more accurate total PCB concentration. But in the case of a sample that has undergone weathering or alteration (as in the case above), the most abundant congeners may not be among the 8082 preselected peaks and the total PCB concentration reported by Method 8082 can be underestimated. In addition, any PCB method with a better (lower) detection limit is expected to give a higher total PCB concentration since more congeners will be detected and included in the total.

The above example was specifically chosen to illustrate the limitations of Aroclor methods in an environmental forensics context. However, it should be noted that as long as the forensic chemist is aware of these limitations, Methods 608 and 8082 data can hold an important place in PCB forensics investigations. Typically, when a project requires a detailed environmental forensics investigation, it is in follow-up to a series of initial investigations that have either (1) identified PCBs as a contaminant of concern and/or (2) come to preliminary conclusions on degree and extent of contamination. Usually, the PCB data upon which those preliminary conclusions are based is 608 and/or 8082 data. While such data may have to be taken at face value, it is always advisable to take these data into account before designing a sampling and analysis plan focused on more detailed forensic investigations. Any pre-existing data, however limited, can provide important clues that will impact subsequent investigations.

Aroclor methods also have a place in environmental forensics investigations because of cost. These methods provide a relatively inexpensive way to establish PCB concentrations in a large number of samples. At the time of this writing, commercial labs typically charged less than 100 dollars (US) per sample for 8082 analyses. In contrast, congener-specific analyses are much more expensive ($400–$1000 per sample). In situations where degree and extent of contamination has not yet been fully defined and large numbers of samples are to be collected and analyzed, it is often wise to establish a sampling strategy that uses the inexpensive 8082 analyses as an initial screening tool. If the 8082 results indicate the presence of PCBs above detection limits or above some threshold concentration of concern (e.g., a regulatory cleanup level), a "split" of that sample could then be analyzed by a more expensive congener-specific method. This is common and effective cost-control strategy for PCB environmental forensics investigations.

10.2.1.3 EPA Method 680—Homologs

EPA Method 680 is a GC/MS-based method used to analyze water, soil, and sediment for both organochlorine pesticides and PCBs in a single analysis. This method focuses on total PCBs and PCB homolog group totals and does not use Aroclor standards for quantification. A concentration is measured for each PCB homolog group and the total PCB concentration of the sample is determined by summing the homolog group concentrations. Two surrogate standards are added to the samples before processing (^{13}C labeled 4,4'-DDT and gamma-BHC), providing a means of monitoring method performance. Calculations of PCB concentrations by this method do not take into account the recovery of these surrogates and reported PCB concentrations are not "recovery corrected." Two internal standards are added to the samples immediately prior to GC/MS analysis to minimize the effects of uncertainties in final volumes and any instrument instability. One weakness of the method is the use of only nine PCB congeners to calibrate the instrument response, one per each homolog group (no nonachlorobiphenyl is used). This is a low resolution GC/MS analysis procedure offering some selectivity over ECD analyses with comparable detection limits if the mass spectrometer is operated in the SIM mode. The selectivity is enhanced if more than one characteristic mass spectral fragment is monitored for each homolog group and the ratio of the fragments' responses is examined and determined to be that expected for PCBs. This method is applicable to highly weathered samples or samples with more than one PCB input in which an Aroclor pattern is not evident. A concentration is measured for each PCB homolog group; total PCB concentration in each sample extract is obtained by summing isomer group concentrations. Detection limits range from 0.1 to 0.5 ug/L per congener if the MS is operated in the SIM mode and 0.5 to 2.5 ug/L if full range data acquisition is used. The lower detection limits makes this method preferable to 8082 for determination of total PCB concentrations in samples.

Concentrations are reported for each individual PCB homolog group, and the total PCB concentration in each sample extract is obtained by summing homolog group concentrations. PCB sources do exhibit characteristic proportions of homologs (Table 10.2.1). However, in terms of environmental forensics investigations, Method 680 data has a similar limitation to Method 8082. If samples have undergone alteration, an analysis of PCB homolog patterns can be misleading. An example is shown in Figure 10.2.2, which shows a composite bar graph for three patterns: (1) an unaltered Aroclor 1254 (typical A1254); (2) Aroclor 1254 which has been dechlorinated; and (3) an unaltered Aroclor 1242. These data were reported by Quensen *et al.* (1990) and are discussed in more detail in Section 10.3.2. The key point of this graphic is that when Aroclor 1254 undergoes reductive dechlorination, the resultant homolog pattern shifts toward the left and more closely resembles Aroclor 1242 than Aroclor 1254. As with Method 8082, in the presence of alteration such as dechlorination, inference of sources based on Method 680 homolog data can be misleading.

10.2.1.4 EPA Method 1668—Coplanar PCBs

Toxicologists and risk assessors concerned with health effects of PCBs have focused much of their attention on the coplanar or "dioxin-like" PCBs: congeners that do not have chlorine substitution in "ortho" positions (positions 2,6,2', and 6' on Figure 10.1.1a). As a result, these so-called "non-ortho" substituted congeners (IUPAC numbers 77, 81, 126, 169) can assume a geometric configuration where both rings of the biphenyl molecule reside in the same plane (hence the name "coplanar" PCBs), such that the coplanar PCBs have a size, shape, and polarity similar to 2,3,7,8 tetrachlorinated dibenzo-*p*-dioxin (2,3,7,8-TCDD or "dioxin"). There are also eight mono-ortho PCB congeners (105, 114, 118,

Table 10.2.1 *Homolog Distributions in Different Aroclors*

Number of Chlorines	Aroclor 1232	Aroclor 1242	Aroclor 1248	Aroclor 1254	Aroclor 1260
Mono-chlorobiphenyl	31.3	<1	<.2		
Di-chlorobiphenyl	23.7	14.7	<1	<0.1	
Tri-chlorobiphenyl	23.4	46	20.9	1.8	<0.3
Tetra-chlorobiphenyl	15.7	30.6	60.3	17.1	<0.3
Penta-chlorobiphenyl	5.8	8.7	18.1	49.3	9.2
Hexa-chlorobiphenyl		<0.3	0.8	27.8	46.9
Hepta-chlorobiphenyl			<0.3	3.9	36.9
Octa-chlorobiphenyl				<0.05	6.3
Nona-chlorobiphenyl				<0.05	0.7
Deca-chlorobiphenyl					

Table modifed after Hansen (1999), based on data from Frame *et al.* (1996).
Reproduced with permission. Copyright 1999. Kluwer Academic Publishers

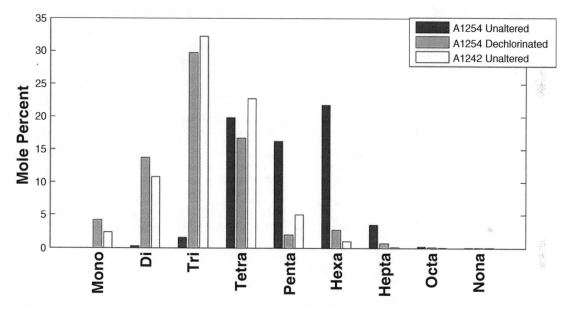

Figure 10.2.2 *Comparison of homolog data for unaltered Aroclor 1254; dechlorinated Aroclor 1254, and unaltered Aroclor 1242. The dechlorinated Aroclor 1254 is more similar to 1242 than it is to the original 1254 from which it was derived. Data from Quensen* et al.*, (1990) (see color insert).*

123, 156, 157, 167, and 189) which also assume the coplanar structure and exhibit dioxin-like behavior.

Dioxin toxicity has been linked to its ability to bind to a specific receptor (the arylhydrocarbon or "Ah" receptor). A number of studies have confirmed the "dioxin-like" effects of non-ortho and mono-ortho PCBs and established the familiar toxic equivalency factor (TEF) scheme of ranking dioxin-like chemicals with respect to 2,3,7,8 TCDD (Safe, 1990, 1994; Van den Berg *et al.*, 1998). Toxic equivalency factors are coefficients assigned to dioxin-like compounds, where 2,3,7,8-TCDD is assigned 1.0. A list of coplanar PCBs and their TEFs (as assigned by Van den Berg *et al.*, 1998) is presented in Table 10.2.2.

While there has been research to suggest that there may also be adverse health effects from non-coplanar PCBs (see Hansen, 1999), the focus on dioxin-like PCBs remains for many health effects/risk assessment studies. Because coplanar PCBs are often present at extremely low concentrations, the risk assessor's emphasis on these compounds has led to the need for analytical methods with accurate quantitation of and extremely low detection limits for the

coplanar PCBs. Accurate quantitation requires high sensitivity and specificity. The method of choice for coplanar PCB analysis has thus become high-resolution gas chromatography combined with high-resolution mass spectrometry (HRGC/HRMS Method 1668).

Prior to about 1997, analysis for coplanar PCBs was done primarily by modification of existing SW-846 methods. There was no specific EPA method for analysis of coplanar PCBs. In the 1990s, EPA developed Method 1668 for determination of coplanar PCBs in water, soil, sediment, sludge, tissue, and other sample matrices by HRGC/HRMS (USEPA, 1997b). The method was for use in monitoring programs associated with RCRA and CERCLA and was based on a compilation of methods from the literature (Ahlborg *et al.*, 1994; McFarland, 1994) and on EPA Method 1613 (USEPA, 1997b). In 1999, USEPA came out with a revision of the method (1668A) that expanded 1668 to include congener-specific determination of more than 150 PCB congeners, retaining the coplanar congeners (USEPA, 1999). The remaining congeners are quantified as co-eluting

Table 10.2.2 *Toxic Equivalency Factors for Coplanar PCBs in Humans/Mammals*

Congener	TEF
77	0.0001
81	0.0001
105	0.0001
114	0.0005
118	0.0001
123	0.0001
126	0.1
156	0.0005
157	0.0005
167	0.00001
169	0.01
189	0.0001

From Van den Berg *et al.* (1998)

groups of congeners. Method 1668A is discussed in more detail in Section 10.2.2.

It should be noted that Methods 1668 and 1668A are USEPA methods, but they are not technically part of SW-846. Further, these methods are "performance-based." That is, the laboratory is permitted to omit any step or modify any procedure provided that all performance requirements in the method are met (USEPA, 1999).

HRGC/HRMS methods (such as 1668) have become the method of choice for determining the concentrations of individual dioxin-like congeners (Rushneck *et al.*, 2004). However, if method 1668 is used (or any method that produces data for only the coplanar PCBs), the results may be insufficient for environmental forensics purposes. This is illustrated in Figure 10.2.3. As in Figure 10.1.2 earlier, this figure shows the congener profiles for PCBs in Aroclor prod-

ucts (data from Rushneck *et al.*, 2004), but in this case, we have limited ourselves to coplanar PCBs only. While there is a clear difference between the coplanar PCB compositions of Aroclor 1260 as compared to the others, the coplanar congener patterns for Aroclors 1016 through 1254 are very similar. The distinctive differences between congener patterns for these Aroclors (as is evident in Figure 10.1.2) are much more muted when one looks only at the coplanar PCBs. Coplanar PCBs, while important in toxicological studies and risk assessment, are of limited utility when the focus of an investigation turns to source identification. Method 1668A, which can provide quantification of all 209 congeners including the coplanars, is the preferred approach in such situations.

10.2.2 Congener-Specific Methods

The EPA methods described above have limited utility in environmental forensics investigations. For detailed environmental forensics investigations focused on inference of PCB sources in complex environmental systems, it is usually necessary to obtain congener-specific data with quantitation of 60 to 160 peaks. In general, there are three choices: GC/ECD methods, HRGC/Low Resolution MS, and HRGC/High Resolution MS. All of these can provide excellent congener data for environmental forensics purposes.

HRGC/HRMS (which includes Method 1668A – an extension of Method 1668) is the most expensive of these options (~$1000 per sample), but because its development was based on the need for good quantitation and detection of coplanar PCBs, this method is recommended for projects where toxicological and risk assessment aspects are an important consideration (in addition to forensics). Method 1668A is a high resolution MS method applicable to any type of aqueous, solid, or biological sample and provides state of the art in both specificity and detection limit. This method

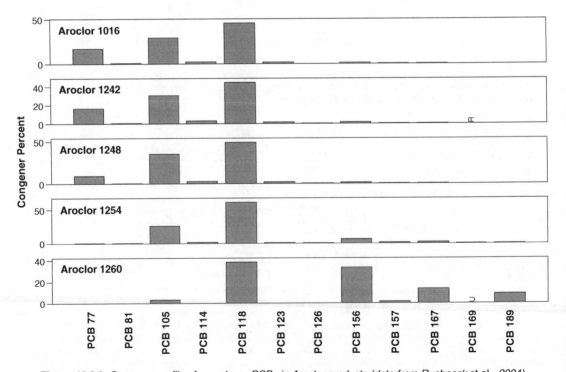

Figure 10.2.3 *Congener profiles for coplanar PCBs in Aroclor products (data from Rushneck* et al.*, 2004).*

employs multiple ^{13}C-labelled surrogate standards, added before the samples are extracted, so that data are recovery corrected for losses in extraction and cleanup. Key analytes are quantified against their labeled analogues and quantification by isotope dilution provides data of high accuracy and precision. It is the technique of choice if sample size is limited, high caliber defensible data are required to quantify low concentrations of non-ortho and mono-ortho substituted PCB congeners in the presence of high levels of other (interfering) congeners. Although low resolution MS can be used for PCB congener determinations, high resolution MS minimizes interferences and provides higher quality individual congener data. Options to report total PCB, homolog groups, 209 individual congeners, The World Health Organization (WHO) toxic congener list, WHO/NOAA congener list, and any custom list are possible with this method.

Extensions and variations of methods such as 8082 (GC/ECD) as well as GC/MS methods can also be used to analyze PCBs as individual congeners. High quality congener-specific PCB data can be obtained using a variety of GC columns (e.g., DB5, DB1 or SPB Octyl), detectors (ECD, MS), and calibration standards (Aroclors, individual congeners, etc.) Frame (1997a,b) presents a summary and comparison of 27 different GC-ECD and HRGC-MS systems (i.e., systems with different columns and different detectors) comprising 20 different stationary phases. Some congener-specific methods have gained wide-spread use and these typically imply specific laboratory setup. The "Green Bay Method," for example (Mullin, 1985), has been used in many Great Lakes PCB studies and reference to this method implies GC-ECD using a DB5 column and calibration using a mixture of Aroclors 1232, 1248, 1262 in the 28:18:18 proportions.

No single column resolves all 209 congeners and therefore the column of choice for a particular application depends on which congeners are of most interest. Some coelutions are inevitable, even using all columns at one's disposal. When a congener of interest co-elutes with another congener, the concentration is reported as a total for the two congeners. To do this, an assumption is made that the two congeners have the same response factor in the analytical system. This is a reasonable assumption in GC/MS analysis where all co-elutions have the same number of chlorines since any congeners not separated by the gas chromatography will be given separate signals in the mass spectrometer if they have differing numbers of chlorine atoms. If properly handled, co-elutions will not significantly affect the accuracy of the measured total PCB concentration.

However, it is often important to separately resolve all of the toxic congeners as each has its own toxicity. The quantification of the toxic PCB congeners (non-ortho and mono-ortho substituted congeners) requires special consideration. Because they are present in much lower concentration than many of the abundant congeners, they can be difficult to quantify even if they do not exactly co-elute with other congeners. They can appear as a small shoulder on a much larger peak or they can be interfered with by a mass spectral fragment of a more highly chlorinated and more concentrated congener that is not separated by the gas chromatograph. One approach is to use a GC column (SPB octyl) that optimizes the separation of these low concentration toxic congeners. However, even this approach does not always adequately resolve CBs 126 and 169 from their closely eluting interferences. Further processing of the sample extract, specifically isolation of the toxic congeners from the bulk of the PCBs by chromatographic separation on carbon, followed by a separate GC run is required to produce accurate quantification of all the toxic PCBs. This is the most reliable approach to measuring toxic PCB concentrations and obtaining PCB TEQ concentrations. In such a

situation, total PCB concentrations would be obtained from an initial GC run (either GC/LRMS or GC/HRMS) and then a carbon column cleanup performed on the extract followed by a second GC run (GC/HRMS to obtain the required specificity and detection limits).

10.2.3 Stable Isotope Methods

Over the past 10 years, improvements in analytical instrumentation have made possible measurement of the relative proportions of stable isotopes of carbon (^{12}C and ^{13}C) and chlorine (^{35}Cl and ^{37}Cl). With regard to carbon, this has been accomplished on a congener-specific basis. By convention, stable carbon isotope values are expressed not as a simple ratio of these two isotopes, but rather using the delta (δ) notation in parts per thousand (‰) as compared to a standard (std), as per the following equation:

$$\delta^{13}C = \left[\frac{\left(^{13}C / _{12}C \right)_{sample}}{\left(^{13}C / _{12}C \right)_{std}} - 1 \right] \times 1000 \qquad (10.2.1)$$

By convention, the standard for stable carbon isotope work is Pee Dee Belemnite (PDB), which is a calcium carbonate fossil cephalopod from the Cretaceous Pee Dee Formation of South Carolina, USA.

Jarman et al. (1998) first reported compound-specific isotope analysis (CSIA) values in commercial PCB products (Aroclors, Clophens, Kanechlors, and Phenoclors). Jarman reported that within each PCB product, there was a wide range of congener-specific δ^{13}C values, but that variation was systematic in that the δ^{13}C values generally showed ^{13}C depletion as a function of increased degree of chlorination of the congeners. When comparing δ^{13}C values for a single congener, across the various commercial products, there were distinct differences. This suggested that δ^{13}C data could prove useful in environmental forensics investigations.

A subsequent study by Yanik et al. (2003) reported results of CSIA in four unaltered Aroclor products and in biota samples from the Housatonic River near Pittsfield, Massachusetts (an area known to have been impacted primarily by Aroclor 1260 (Bedard et al., 1997). The δ^{13}C results for unaltered Aroclors were consistent with Jarman's results. Congener-specific δ^{13}C values were generally consistent between Aroclor 1260 and the tissue samples (≤ 1‰ deviation). This suggests that isotopic ratios may be maintained through the food chain. There were, however, some notable exceptions. For example, PCB-180 (the most abundant congener in Aroclor 1260) showed a 8.0‰ difference in δ^{13}C values when Aroclor 1260 data were compared to duck muscle.

Taking a different approach to PCB stable isotope studies, Reddy et al. (2000) looked at stable chlorine isotopes. There are two naturally occurring stable isotopes of chlorine (^{35}Cl and ^{37}Cl) and the ratio between them can be expressed in a δ notation similar to that for carbon:

$$\delta^{37}Cl = \left[\frac{\left(^{37}Cl / _{35}Cl \right)_{sample}}{\left(^{37}Cl / _{35}Cl \right)_{std}} - 1 \right] \times 1000 \qquad (10.2.2)$$

Reddy's group reported δ^{37}Cl values for unaltered PCB products (five Aroclors, three Clophens, and one Phenoclor) as well as sediments from the Hudson River (New York), New Bedford Harbor (Massachusetts), and Turtle River Estuary (Georgia). The isotopic variability among the source materials was small (e.g., δ^{37}Cl for all Aroclors was -2.78 ± 0.39‰). Further, in contrast to stable carbon isotope results discussed above, there was no correlation between degree of chlorination and δ^{37}Cl. While most sediment samples were within two standard deviations of mean Aroclor value, there was much more variability in the δ^{37}Cl values in sediments, suggesting that such data might ultimately be of value in the study of PCB fate and transport studies.

As compared to more traditional PCB chemical fingerprinting approaches (e.g., congener pattern analysis), the use of stable isotope in PCB environmental forensic applications is in its early days. Studies such as those cited above clearly indicate that stable isotope analysis offers the promise of a powerful tool in the PCB environmental forensics toolbox. However, additional research (both field and lab studies) are needed before these techniques become a commodity service provided by commercial labs in support of environmental forensics investigations. At present, stable isotope analysis can be an important aspect of an environmental forensics investigation, but in most cases it should be done in conjunction with the more traditional forensic approach (GC-ECD and GC-MS congener data in conjunction with congener pattern comparisons – Philp, 2002).

10.3 PCB ALTERATION MECHANISMS

While PCBs were manufactured as Aroclors, Clophens and other specific commercial products with known congener distributions, residual PCB patterns in the environment do not necessarily resemble the original mixtures from which they were derived. Although PCBs are more resistant to weathering than other common contaminants of concern (e.g., PAHs), congener patterns can be altered in the environment. This weathering may take the form of dechlorination, volatilization/dissolution, or metabolism. This section discusses these three alteration mechanisms.

10.3.1 Volatilization
10.3.1.1 Overview
The study of Haque et al. (1974) was one of the first to clearly demonstrate that PCBs could volatilize. Haque and coworkers verified volatile loss of Aroclor 1254 from thin-films and sand. Over the 30 years that followed, the combination of improved analytical chemistry methods and development of high volume air sampling techniques has been crucial in improving the ability to determine the concentration and composition of PCB mixtures in air. In short, these studies have confirmed that PCB volatilization and transport by air are issues of concern. Key findings are summarized as follows:

- Polychlorinated biphenyls can be readily measured in air samples and when measured, they are generally found predominantly (>90%) in the vapor phase (Hillery et al., 1997);
- PCBs and related semi-volatile compounds can be readily transported and deposited via air, often many hundreds or thousands of kilometers from their point of origin or use (Wania and MacKay, 1996);
- The congener-specific pattern of PCBs in air varies considerably and often includes a wide range of congeners, including those that are highly chlorinated (Chiarenzelli et al., 2001);
- Mass balance studies, particularly those done in the Great Lakes region, suggest that atmospheric processes including the wet and dry deposition of PCBs and their volatilization from the lake surface are the major fluxes controlling PCB concentrations in various environmental media associated with the lakes (Eisenreich et al., 1981);
- PCB concentrations in ambient air are generally high in areas adjacent to sites of their manufacture, use, or disposal. Ambient air in urban areas also tends to have

higher PCB concentrations than rural and remote areas; however, even in remote areas concentrations are still measurable. Indoor air concentrations generally exceed outdoor air concentrations of PCBs, even those measured near highly contaminated areas (Vorhees et al., 1997).
- PCBs are a global contaminant. While only a small proportion of global PCBs are present in the volatile phase in the atmosphere, transport of PCBs in air is such that these contaminants have found their way to every corner of the globe (albeit at concentrations orders of magnitude below those encountered in urban areas and on or near sites where PCBs were spilled).

It is now well accepted that semi-volatile compounds such as PCBs readily undergo redistribution in the environment by a variety of processes, including volatilization, atmospheric transport, and wet and dry deposition. It is further recognized that volatilization of PCBs in the environment can be enhanced by a variety of physical, chemical, and biological processes.

10.3.1.2 Multi-Phase Partitioning of PCBs.
The individual physicochemical characteristics of PCB congeners vary over several orders of magnitude (Table 10.1.6) and thus provide ample opportunity for partitioning between phases. Lightly chlorinated congeners tend to move more easily into the aqueous or vapor phase and, indeed, atmospheric PCB mixtures tend to have higher proportions of lower chlorinated congeners (Bidleman and Olney, 1974; Hansen, 1999). It has been further shown that ortho-rich PCB congeners tend to be more easily volatilized that non-ortho congeners (Falconer and Bidleman, 1994; Monosmith and Hermanson, 1996; Chiarenzelli et al., 1996, 1997).

A considerable amount of research has been done to quantify and establish predictive models for movement of chemicals from one phase to another (e.g., soil bound to vapor phase). A general term for this approach is "fugacity modeling." Fugacity is quantitative measure of the "escaping" tendency of a chemical from one phase to another (Mackay, 2001). The objective is to predict the concentrations of a chemical in multiple phases (e.g., air, soil, and water) based on equilibrium partitioning equations (Mackay, 2001). The underlying scientific disciplines involved are thermodynamics and phase equilibrium physical chemistry. Such methods are extremely useful and widely used as a predictive tool to determine the concentrations of multiple PCB congeners in various media (e.g., Harner et al., 1995; Hermanson et al., 2003). For example, Harner et al. (1995) suggest that the decreasing fugacity of PCBs in the atmosphere has led to widespread "degassing" from soil repositories as the concentrations of PCBs in the atmosphere have decreased.

10.3.1.3 Links between Volatilization and Evaporation
One of the most important factors in volatilization appears to be the presence of water. Two studies suggesting this were by Chiarenzelli et al. (1996, 1997). These studies are discussed here in context of general issues of PCB volatilization, but the impetus for these studies was an investigation of PCB sources, fate, and transport at Superfund sites near the Akwesasne Mohawk Nation near Massena, New York. The full context of that case study is discussed further in Section 10.6.2.

It has been shown empirically that soils and sediments that are intermittently wetted and then dried release significantly more PCBs into the vapor phase than purely dry media (Chiarenzelli et al., 1996, 1997). As expected based on their physical properties (Section 6.1.5), lower chlorinated congeners were observed in higher proportions in vapor-phase samples. The rate of PCB volatile loss was rapid until dryness (ambient room humidity – 25%) was reached and then drastically less PCBs were volatilized. When additional water was reapplied to the dried sediments, PCB volatile losses resumed at a rapid rate, although not as rapid as was observed during initial saturation (Chiarenzelli *et al.*, 1996). These results suggest that areas that experience periodic submersion (flood plains, tidal flats, seasonal rivers, etc.) are ideal for enhanced volatilization of semi-volatile compounds.

10.3.1.4   Amount of PCBs Lost Through Volatilization

In volatilization experiments involving contaminated natural sediments, as much as 75% volatile loss of the original PCB mass occurred for small samples (0.25–1.0 g) over a 7-day drying period (Chiarenzelli *et al.*, 1996). In a series of Aroclor volatilization experiments (Chiarenzelli *et al.*, 1997), up to 62% of the original PCB mass was lost to the volatile phase (Aroclor 1242 experiment). It should be noted that volatile loss rates of this magnitude are not likely in natural settings where mass transfer considerations in thicker sediment and/or water layers would ultimately control the availability of PCBs. However, it is possible that the seeds of a remedial technology exist in these observations. For instance, the procedures employed (wetting, composting, heating, tilling, etc.) at aerobic bioremediation sites may inadvertently promote rapid volatile loss of semi-volatile compounds. Hence, designing facilities for the enhanced volatile loss and capture of semi-volatile compounds may provide a low-cost, low-energy option to entombment or alternative remedial technologies.

It is also important to note that the volatile loss of PCBs during these experiments would require solubilization of the PCBs from the sediment into the water, transport to the air–water interface, and co-evaporative loss. In these experiments, both Henry's Law and the solubility of individual congeners within the mixtures would influence volatile loss. It is theorized that a dynamic equilibrium occurs in which

the congener-specific composition of PCBs in all media (sand/sediment, water, air) change continually with time in response to the loss of PCB molecules from the system at the air–water interface (Chiarenzelli *et al.*, 1997).

10.3.1.5   Congener Profiles

An issue relevant to environmental forensics concerns how the congener patterns in the volatile phase and sediment changed relative to the unaltered Aroclor patterns. The data generated by Chiarenzelli's 24-hour volatilization experiments (Chiarenzelli *et al.*, 1997) have proven extremely useful in interpreting ambient data collected in the field. While one may intuitively understand the implications of volatilization research discussed above, when faced with the task of interpreting PCB data collected in the field, that understanding only goes so far. The use of reference data set to provide direct congener pattern comparisons is extremely important in environmental forensics investigations.

An example is provided as Figure 10.3.1. This series of bar graphs shows the congener-specific data from Chiarenzelli's Aroclor 1242 volatilization experiment. The top graph is unaltered Aroclor 1242. The second graph is the "Evaporation fraction" or the vapor-phase fraction recovered after evaporation of the solvent that had been used to originally dissolve the PCBs and sorb them to the sand medium. The next two graphs show vapor-phase fraction volatized from subaqueous sand at the 0–1 hour interval and the 8–24 hour time intervals. One surprise observed in the Aroclor 1242 data is that the vapor-phase congener profiles were, on average, more highly chlorinated than the unaltered Aroclor 1242 technical mixture (3.41 vs 3.28 chlorines per biphenyl molecule). This is evident in the visual comparison between unaltered Aroclor 1242 and the 8–24 hour interval.

The "residual" 1242 pattern represents the PCB pattern remaining in the sand, after conclusion of the volatilization experiment. Finally, the bottom pattern shown in this figure shows the congener pattern of unaltered Aroclor 1248. This congener pattern is provided primarily for comparison to the

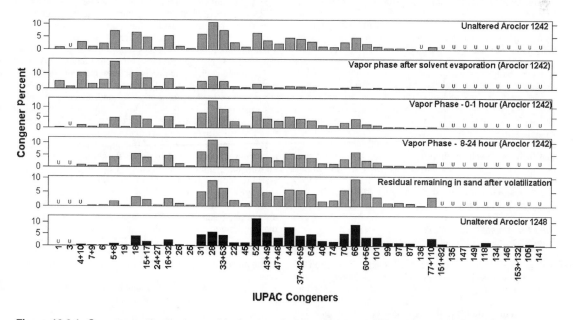

Figure 10.3.1 *Congener patterns observed in Aroclor 1242 volatilization experiments (Chiarenzelli et al., 1997). An unaltered Aroclor 1248 pattern is shown in the bottom graph for comparison to the Aroclor 1242 residual pattern. These data are included in Appendix, along with data from Chiarenzelli's other Aroclor volatilization experiments (see color insert).*

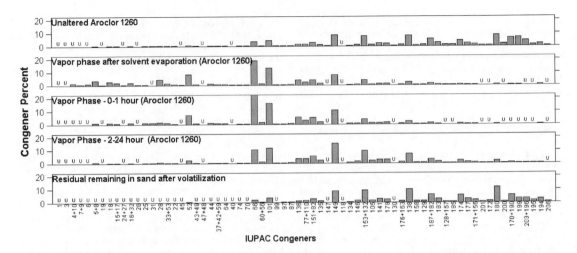

Figure 10.3.2 *Congener patterns observed in Aroclor 1260 volatilization experiments (Chiarenzelli et al., 1997). These data are included in Appendix, along with data from Chiarenzelli's other Aroclor volatilization experiments.*

residual 1242 pattern. Clearly, the residual 1242 congener pattern is more similar to unaltered Aroclor 1248 than it is to unaltered Aroclor 1242. We can see this visually, but we can also quantify the observed similarity as will be discussed in Section 10.4.4. In the field, if one were to base an interpretation on pattern similarities alone, without knowledge of the effects of volatilization one could falsely conclude that the medium in question was contaminated by Aroclor 1248. It is important to evaluate congener pattern comparisons in full context of all available information.

A subset of Chiarenzelli's Aroclor 1260 volatilization series is shown in Figure 10.3.2. Once again, we observe the presence of lighter congeners in the volatile phase samples. In contrast to the Aroclor 1242 experiment, however, the residual 1260 sample still looks similar to the unaltered Aroclor 1260 from which it was derived. The reason for this is simple. Because Aroclor 1260 is composed of more highly chlorinated congeners that have lower aqueous solubility and lower vapor pressures (Table 10.1.5), the residual pattern observed in sediments after volatilization retains a congener pattern similar to the original Aroclor 1260. Also, the percentage of the total PCB mass lost through volatilization is far less than measured in experiments with Aroclors 1242 and 1248, resulting in less change. Thus, in soil and sediment studies with Aroclor 1260 as a major source, volatilization will not as dramatically alter the congener patterns observed in sediment. That does not mean, however, that there is no volatilization from Aroclor 1260. The vapor phase samples clearly show that there is a lighter component, including substantial amounts of congeners with five, six, and seven chlorines that can volatilize from heavier Aroclors such as 1260.

10.3.1.6 Volatilization Summary

Knowledge of these phenomena are important in environmental forensics investigations. Drying of wet sediment and soil will enhance the rate of volatile loss of PCBs and other semi-volatile compounds. This is likely to occur during both remedial actions (e.g., dredging, excavation, outdoor storage, etc.) and as a result of natural processes (intermittent wetting and drying of flood plains, irrigated or tilled fields, tidal flats, etc.). Initially high volatilization rates are likely to slow because of mass transfer considerations and depletion of the most volatile congeners; however, substantial losses may occur where conditions enhance

evaporative losses (i.e., at the surface of contaminated water bodies, Bush *et al.*, 1987) and PCBs are continually replenished. In particular, sediment that has undergone anaerobic microbial dechlorination contains lower ortho-chlorinated congeners that are substantially more mobile in environmental media including water and air. Experimental work reveals that the volatile losses are greatest from the least chlorinated Aroclors and mixtures. However, measurable volatile loss from heavier sources such as Aroclor 1260 can be observed in the field and highly chlorinated congeners can be measured in air samples.

Congener-specific analyses allow the degree of partitioning of PCBs to be investigated and inferred based on empirical data. Depending on the original congener mixture and the extent of volatile loss, it may be possible to infer the original PCB source through analysis of congener patterns in comparison to experimental volatilization reference data sets. However, there are clearly instances where volatilization can result in ambiguous inference of source (e.g., the example of the Aroclor 1242 residual being more similar to unaltered Aroclor 1248). When such a situation is encountered, confident attribution of sources may have to come from lines of evidence other than congener profile analysis (e.g., a facility's production history) combined with a knowledge of the effects of physical, chemical, and biological processes on the original material introduced into the environment.

Because of the realization of the importance of air as PCB migration pathway, PCB air sampling studies have increasingly been conducted and reported in the literature (e.g., Hermanson and Hites, 1989; Cotham and Bidleman, 1995; Chiarenzelli *et al.*, 2000, 2001; Kohler *et al.*, 2002). Air sampling can be valuable in PCB studies. However, the utility of air sampling data in environmental forensics investigations may be limited. A typical objective of environmental forensics investigations is to decipher and reconstruct historical conditions. Air data provides information only on PCB levels and patterns during the limited time that air samples were collected.

In situations where present day levels and patterns in air are thought to be different from historical trends, there is an alternative approach. The lipid tissue in tree bark is able to absorb gas phase PCBs from the atmosphere and retain them. Thus tree bark serves as a long-term integrator of atmospheric PCB concentrations over the life-span of the tree (Meredith and Hites, 1987), and is an excellent passive sampler for historical PCB levels and patterns in air. Even if gas phase PCBs decline over time, the bark will retain the signal from the time of high concentrations (Hermanson and Hites, 1990). On a local scale, the bark from trees from different locations in a study area, and of different ages, can be analyzed for congener-specific PCB analyses, thus allowing inference of historical PCB levels and patterns in air near PCB contaminated sites (Hermanson and Johnson, 2005). Such data can also be useful on a global scale. Trees of various ages and types have been used to identify the global transport and partial fate of organochlorine contaminants (Simonich and Hites, 1995).

10.3.2 Dechlorination
10.3.2.1 Overview
Brown and colleagues first proposed that PCBs could be reductively dechlorinated by microorganisms in anaerobic sediments after examining chromatograms of samples collected from various locations, initially in the upper Hudson River (Brown et al., 1984; 1987a,b). They noted that, compared to the Aroclors originally input to these sediments, the chromatograms showed a depletion of later eluting, more heavily chlorinated congeners, and an enrichment of early eluting, lesser chlorinated congeners. They gave numerous letter designations to the observed dechlorination patterns, based in part on the original Aroclor(s) and in part on apparent congener selectivity inferred from the putative dechlorination products. Presumably, a different microbial species or enzyme was responsible for each dechlorination process.

Quensen et al. (1988) first demonstrated microbially mediated reductive dechlorination of an Aroclor in laboratory experiments. Microorganisms from the Hudson River dechlorinated Aroclor 1242 to give a PCB congener profile similar to that observed in the environment. During the next several years, they and other researchers expanded this line of research to show the dechlorination of other Aroclors by organisms from the Hudson River and other locations, and to determine factors influencing the dechlorination process. In reviewing this body of research, Bedard and Quensen (1995) tried to bring some order to the results by inferring a minimal set of dechlorination processes and reinterpreting results as to how experimental conditions affected each individual dechlorination process. They proposed six primary processes, designated H, H′, M, N, P, and Q, and explained other observed dechlorination patterns as combinations of two or more of these six. Since that time, two other basic processes have been recognized from laboratory experiments: LP (Bedard et al., 1997) and T (Wu et al., 1997a). In all cases, dechlorination occurs from the meta and/or para positions. None of these processes are capable of removing ortho chlorines.

Brown et al. (1987a,b) originally described patterns F and G in Silver Lake (MA) samples as the result of dechlorination of Aroclor 1260 from ortho, meta, and para positions, but the ortho dechlorination was questioned by Bedard and Quensen after they learned that Aroclor 1254 was a major contaminant of these sediments. Nevertheless, ortho dechlorination is possible because it has been observed in laboratory experiments with single congeners (Williams, 1994; Wu et al., 1996, 1997b; Cutter et al., 1998) and in one case Aroclor 1260 was dechlorinated from ortho positions by a culture obtained by enrichment on a single congener (Wu et al., 1998).

10.3.2.2 Recognizing Dechlorination Patterns
The various dechlorination patterns can be best recognized by the products that accumulate as a result of their respective dechlorination processes. The dechlorination process

descriptions discussed above have proven sufficient to explain dechlorination patterns observed in ambient environmental samples (e.g., Imamoglu et al., 2002; Magar et al., 2005a,b).

Dechlorination products are dependent on the initial PCBs (Aroclors) present. That is, patterns depend not just on the congener selectivity of the responsible microorganisms/enzymes, but also on the starting material (which Aroclor or mixture of Aroclors). Product formation has not been determined for all combinations of Aroclors and dechlorination processes, partly from a lack of research, and also because the dechlorination processes vary in their ability to dechlorinate highly and lesser chlorinated congeners. For example, Process M dechlorination of Aroclor 1260 has not been observed, while Processes N and P do dechlorinate Aroclor 1260.

The microbially mediated dechlorination of PCBs is a conservative process in the sense that the products are lesser chlorinated congeners; the biphenyl moiety itself is not degraded, and in fact there is little evidence for the production of biphenyl from Aroclor dechlorination. For this reason, it is useful to present concentrations on a molar basis (averaged in the case of co-eluting congeners) rather than by weight. This allows the percent of chlorine removed to be easily calculated and difference plots as in Figure 10.3.3 to be used as an aid in recognizing the various dechlorination processes (see Quensen and Tiedje, 1997) for further description of the methods involved.

Process M dechlorination removes meta chlorines from both flanked and unflanked positions (i.e., whether or not there is an adjacent chlorine). It results in high accumulations of 2-CB, 2,2′-CB and congeners substituted in the ortho and para positions. It is more active on lightly chlorinated Aroclors, although it can apparently, in combination

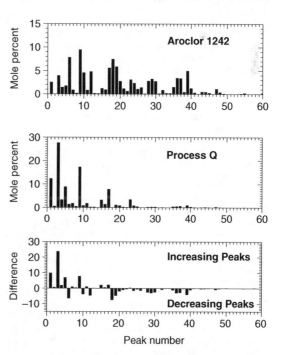

Figure 10.3.3 Congener profile and difference plot for the dechlorination of Aroclor 1242 by Process Q. The Aroclor profile was subtracted from the Process Q profile to obtain the difference plot in the bottom panel. Increasing peaks represent dechlorination products. Data taken from Table A.2.

with other processes, dechlorinate Aroclor 1254 (Quensen *et al.*, 1990). Only Processes M and Q result in high accumulations of 2-CB (Bedard and Quensen, 1995).

Process Q removes para chlorines from both flanked and unflanked positions and the meta chlorine from rings substituted at 2 and 3 positions (and possibly from rings substituted at the 2,3, and 4 positions). It results in high accumulations of 2-CB, 2,2'-CB, 2,3'-CB, and other congeners substituted in the ortho and meta positions. Like Process M, it is more active on lesser chlorinated Aroclors, although in combination with other processes, it can apparently dechlorinate some congeners present in Aroclor 1254 (Bedard and Quensen, 1995; Quensen and Tiedje, 1997).

Process H removes chlorines from flanked and doubly flanked positions and the meta chlorine from rings substituted in the 2,3, and 4 positions. As a result, the accumulation of congeners substituted at the ortho and meta positions, and those with unflanked para chlorines is characteristic. Process H' is different in that it can also remove meta chlorines from rings substituted at the 2 and 3 positions, and possibly from those substituted at the 2, 3, and 6 positions. These processes are active on all Aroclors and the most abundant products depend on the initial PCB mixture (see Table 10.11 – Bedard and Quensen, 1995).

Process N dechlorination removes meta chlorines from both flanked and doubly flanked positions. It is active on all Aroclors and high proportions of 2,2',4,4'-CB are characteristic from the dechlorination of Aroclors 1254 and 1260 (Bedard and Quensen, 1995).

Process P dechlorination has been described for Aroclor 1260 only. It removes para chlorines from flanked and doubly flanked positions. As a result, the predominant products from the dechlorination of Aroclor 1260 are 2,2',4,5'-CB and congeners substituted at the 2,2', and meta positions (Bedard *et al.*, 1996).

Process LP is similar to Process Q in that it removes flanked and unflanked para chlorines and the meta chlorine from rings substituted at the 2 and 3 positions (Bedard, *et al.*, 2005). It has not been observed in isolation, but only in combination with Process N. It further dechlorinated Aroclor 1260/Process N products mainly from para positions so that the predominant products were 2,2',4-CB, 2,2',5-CB, and 2,2',6-CB (Bedard *et al.*, 1997).

Process T dechlorination is very restricted in its dechlorination activity. It removes the chlorine at position 3 from the rings of hepta- and octa-chlorobiphenyls substituted at the 2, 3, 4, and 5 positions. It has been observed only in laboratory experiments conducted at temperatures of 50–60 °C, possibly because the high temperatures suppress other dechlorination processes, and is unlikely to contribute significantly to PCB dechlorination observed in environmental samples (Wu *et al.*, 1997a).

In the environment, PCBs are often dechlorinated by some combination of the above processes. For example, Patten C is a combination of Processes M and Q. In this case, congeners substituted only at the ortho positions are dominant. Processes M, Q, H, H', and N have all occurred in experiments with upper Hudson River sediments (Bedard and Quensen, 1995) and combinations of Processes P and N have been observed in experiments with Housatonic River and Woods Pond (MA) sediments (Van Dort *et al.*, 1997; Wu *et al.*, 1997a; Deweerd and Bedard, 1999). When the extent to which the contributions of different dechlorination processes vary among samples, multivariate methods such as polytopic vector analysis (PVA-Section 10.4.2) may be useful in identifying the individual processes.

10.3.3 Biological Alteration

While PCBs bioaccumulate in lipids, and biomagnify in the food chain, all PCB congeners are not metabolized equally. Some PCB congeners are more readily metabolized and passed out of the body as hydroxy or methyl sulfonyl metabolites (Letcher *et al.*, 2000). As a result, given time, the congener pattern observed in biota samples will not necessarily resemble that of the source to which the organism was exposed. PCBs 153 and 138 are typically the most abundant congeners found in biota samples, across a wide range of environments (Morrison *et al.*, 1996; Fisk *et al.*, 1998; Ikonomou *et al.*, 2002). In particular, Fisk *et al.* (1998) went beyond the observation of PCBs 153 and 138 as the most abundant congener in biota and empirically determined biomagnification factors (BMF) for 25 organochlorine compounds, including 16 PCB congeners. PCBs 153 and 138 exhibited the highest BMFs. Part of the reason for the high BMFs of PCB-153 (245-2'4'5'-CB) and PCB138 (234-2'4'5'-CB) is the 2,4,5 chlorine substitution pattern. Congeners with this substitution pattern have been shown to be more recalcitrant to metabolism (Porte and Albaigés, 1993).

These trends are also observed in humans. Hansen (2001) reported a list of the more metabolism resistant congeners, which include those typically observed in high proportions in biological media (in particular, PCB-153 and 138 are the most recalcitrant; PCB-180 and 118 are not far behind). Hansen (2001) also reported a list of "transient" congeners: those congeners that are metabolized much more rapidly, and thus are not typically encountered in biological samples, unless sampled shortly after exposure.

In a recent publication by the US Center for Disease Control (CDC, 2003), CDC presented exposure data for 116 environmental chemicals in the civilian population of the United States. This study included reporting of concentrations of 25 PCB congeners in human blood. Data were reported as percentile concentrations for a number of demographic groups. In general, PCB concentrations increased with increasing age, but the overall congener pattern (i.e., relative proportions of the congeners) remained fairly consistent (again, with PCB-153, 138, and 180 the most abundant congeners). A graphic summarizing the NHANES data is provided as Figure 10.3.4. In this figure, the thick shaded bar represents 50th percentile value. The thin line extending above the shaded bar represents the 95th percentile concentration. Where a "U" is plotted at the 95th percentile level, this indicates that the 95th percentile concentration was below the limit of detection (LOD), and the value indicated is the LOD. Note that metabolism resistant congeners PCB-153, 138, and 180 are typically the most abundant congeners in human blood with PCB-153 the most abundant.

10.4 DATA ANALYSIS

10.4.1 Data Preparation

PCB environmental forensics investigations often involve collection of hundreds, if not thousands of samples. If such data are analyzed by congener-specific methods, each sample will have associated with it 50 to 209 chemical measurements. Such large chemical data sets translate to major data management and data analysis challenges. Data preparation is unglamorous and time-consuming but crucial. It is not unusual for data preparation to account for 60–80% of the so-called "analysis time." It often requires relatively mundane tasks to be performed repeatedly. Such an effort, while crucial, may not at all be evident to the reader of the resultant paper or report. Nonetheless, it is an essential step. Some typical data preparation tasks necessary in PCB forensics projects are discussed below.

Commonly, PCB data from a site are available only as hardcopy in old reports. In such cases, a time-consuming

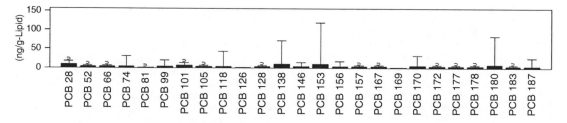

Figure 10.3.4 *Concentration percentiles for 25 PCB congeners in blood of US population – 20 years and older. Units are ng/g of lipid (ppb). The dark shaded bar shows the 50th percentile for the lipid adjusted concentration of that congener. The thin line extends to the level of the 95th percentile. For congeners that were reported as "non-detect" at the 95th percentile, a "U" is plotted. Data from CDC, 2003.*

first step is hand-entry of data into a spreadsheet and digitization of sample locations off of hard-copy maps such that the data can be analyzed in geographic context.

Analysis of PCB congener data from different labs is another common challenge. Often, PCB data are available at a site, but samples may have been analyzed by different labs, perhaps by different methods (typically with different coelutions and different congener lists). Alternatively, all data from the site may be from a single lab, but a key reference data set which the analyst wishes to compare were run by a different method. The ability to do congener pattern comparisons between samples analyzed by different labs is dependent on (1) determining which reported analytes are most comparable between data sets and (2) reducing both data sets to a congener list that is most comparable. The solution to such a challenge is usually project-specific, taking into account the analytical methods being compared, the most likely suspected sources, age of the data, etc. That said, it is almost always helpful to have knowledge of the general abundance of PCBs in sources (e.g., Aroclors) and in environmental media in general. For example, PCB-153 coelutes with different congeners on different laboratory systems. On a GC-ECD system using a DB1 column, PCB-153 often coelutes with PCB-184. In contrast, on a GC/MS system using a DB5 column, PCB-153 often coelutes with PCBs 168 and 132 (Frame, 1997a). Fortunately, when comparing data from two labs running these two different systems, it is not unreasonable to compare these two peaks as "PCB 153," because PCB 153 is more abundant in Aroclors and more persistent in the environment and biota. The key to compiling comparable analytes lists is a peak-by-peak comparison between the two data sets, done in context of (1) known or suspected sources and alteration processes in the field area and (2) the media being sampled.

Another common data-preprocessing challenge is how to deal with low concentration samples and congeners with numerous non-detects. A data analyst would like to have a robust data set with few non-detects. But clearly, when 99% of reported congeners in a sample are non-detect, that sample will be of limited utility when it comes to congener pattern comparisons. In the experience of these authors, there are no hard and fast rules on when to remove samples from a data set and when to remove congeners. These decisions are usually project-specific, with logic and practical considerations being the key considerations.

10.4.2 Analysis

Once the data preprocessing step is completed and data are in a format suitable for analysis, there is still a major data analysis challenge to consider. The data analyst is often working with large multivariate data sets. It is not at all uncommon to be working with hundreds of samples, 50–100 congeners, and data from multiple environmental

media (soil, sediment, air, water, biota, etc.). Polychlorinated biphenyl environmental forensics is inherently a multivariate problem.

If a PCB source and alteration processes exhibit characteristic congener patterns, then given a suite of samples with congener-specific analyses, one can make inferences of source and alteration through use of multivariate statistical methods. In an environmental forensics context, the objective of a multivariate approach to chemical fingerprinting is to determine (1) the number of congener patterns present in the system; (2) the multivariate chemical composition (congener pattern) of each fingerprint; and (3) the relative contribution of each fingerprint in each sample. Numerical methods have been developed to handle such data analysis problems. A detailed review of such methods (applied to environmental forensics and using PCB congener data as an example data set) is presented by Johnson *et al.* (2002) and Johnson and Ehrlich (2002). Methods applied to PCBs in the literature include principal components analysis (Ikonomou *et al.*, 2002), principal components with non-negative constraints (Rachdawong and Christensen, 1997; Imamoglu *et al.*, 2002, 2004), SIMCA (Schwartz *et al.*, 1987), PVA (Johnson *et al.*, 2000; Magar *et al.*, 2005a), and alternating least squares analysis (Salau *et al.*, 1997).

All of these methods can provide useful results and the results from one multivariate method should be consistent with those from another method. Johnson *et al.* (2002) described a number of these methods and applied principal components analysis and four receptor modeling/multivariate curve resolution (MCR) methods to a three-component congener-specific PCB data set. All four methods resolved reasonable estimates of contributing PCB source patterns. All of these methods are effective, assuming a clean data set. Regardless of methods chosen, a crucial step in this process is vigilant outlier detection and data cleaning.

While such multivariate methods are well established in the literature and widely used in scientific studies, they can present a unique challenge in context of an environmental forensics investigation. These methods are sophisticated and require considerable background in both numerical analysis methods and PCB chemistry. Unfortunately, in many environmental forensics investigations, the audience is not made up of scientists, let alone scientists with a background in multivariate statistics. Rather, the audience is a cross section of laypeople: lawyers, judges, and jurors. Fortunately, the results of a good multivariate analysis can and should be confirmed through direct inspection of individual samples and raw data. Multivariate methods may expedite identification of different patterns in a data set, but these patterns should also be evident through direct inspection of raw data: individual sample bar graphs and/or chromatograms. Therefore, even if sophisticated numerical methods are used to identify congener patterns in a suite of samples, it may ultimately be more effective to frame conclusions about a data set in context of direct inspection of individual samples. While the chemistry involved may still present a communication challenge, it is clearly easier for the layman to grasp than when presented in context of things like eigenvectors, principal components, least squares, etc.

An example of this is provided in Figure 10.4.1. These data are from the study of Magar *et al.* (2005a) which used the multivariate curve resolution method, PVA, to identify PCB congener patterns in sediment samples from Lake Hartwell/12 Mile Creek, South Carolina. Four "end-member" congener patterns were resolved and the relative proportions of those four fingerprints were determined in 211 sediment samples. Polytopic vector analysis resolved four primary congener patterns in the data set: two source patterns (Aroclor 1254 and an Aroclor 1242 which had weathered such that it looked more similar to Aroclor 1248) and two microbial dechlorination patterns (Processes C and H'). While the data analysis method expedited this interpretation, in retrospect, the inferences of source and alteration is evident by direct inspection of individual sample congener patterns. As an example, Panel A in Figure 10.12 shows the congener pattern of one of the Lake Hartwell sediment samples. Polytopic vector analysis resolved this sample as mixture of Aroclor 1254, devolatilized Aroclor 1242, and slightly dechlorinated by Process C. Panel B shows the congener pattern for mixture of Aroclors 1248 (very similar to devolatilized Aroclor 1242 – Section 10.2.1) and Aroclor 1254 (data from Frame *et al.*, 1996). Panel C shows the congener pattern for Process C dechlorination (Data from Quensen *et al.*, 1990). The Lake Hartwell sediment sample clearly shows a mixed contribution from both of these reference data set patterns. While multivariate statistical methods were useful in expediting interpretation of this large data set, the results and inferences were ultimately evident by direct inspection of individual sample patterns in comparison to reference data sets. This individual sample comparison approach is often more effective when communicating

with an audience that is skeptical of and/or unfamiliar with multivariate statistics.

10.4.3 Reference Data Sets

Regardless of data analysis methods used (either multivariate statistical analysis or direct inspection of individual samples), PCB fingerprinting ultimately reduces to a comparison of congener patterns observed in the field to congener patterns that represent prospective sources or alteration mechanisms. Therefore, the data analyst must ultimately answer these questions:

- Are the congener patterns observed in the field similar to one or more PCB source patterns (e.g., Aroclors)?
- If not, are they similar to common alteration patterns (e.g., dechlorination, volatilization)?

In order to address these questions, the data analyst needs to compare their field data to one or a series of reference data sets. The availability of these data sets varies. In some cases, full congener data are reported by the authors within the manuscript (e.g., Frame *et al.*, 1996; Schulz *et al.*, 1989). In other cases, the data are not included in the original manuscripts, but are available on the World Wide Web (e.g., Chiarenzelli *et al.*, 1997; CDC, 2003). In some cases, the raw data may be included neither in the published manuscript nor on the web (e.g., many of the dechlorination studies summarized by Bedard and Quensen, 1995). Such data can sometimes (but not always) be obtained via direct request of the paper's corresponding author. In the case of studies more than a decade old, such data may no longer be available or accessible. Reasons for this vary, but it is often due to changes in data storage formats and/or outdated technology. In these instances, it is sometimes still possible to obtain a reference congener pattern from such a study by digitizing congener pattern bar graphs or chromatograms directly from a manuscript.

Table 10.4.1 presents information on ten commonly cited reference data sets in the literature. Because there is sparse

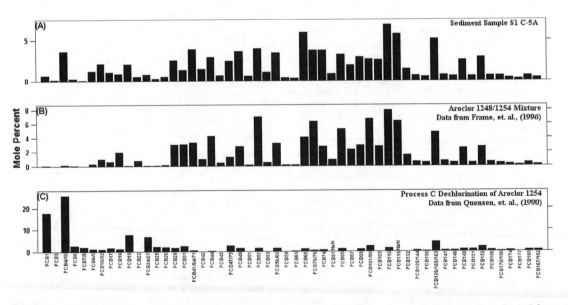

Figure 10.4.1 *Individual sample bar graph (A) showing the congener pattern of a sediment sample collected from 12 Mile Creek/Lake Hartwell South Carolina (Magar et al., 2005). The multivariate curve resolution method polytopic vector analysis classified this (and 211 other samples) in terms of proportions of four end-members (two sources and two dechlorination processes). Note, however, that the congener pattern of this sample is also interpretable by direct comparison of congener pattern (A) to source and alteration patterns reported in the literature (B and C).*

Table 10.4.1 *PCB Reference Data Sets*

Literature Citation	Data Reported	Analytical Method(s)	Comments	Data Availability
Data Sets Reporting Congener Patterns of Unaltered Sources				
Frame et al. (1996)	Congener-specific data for 8 Aroclors	3 HRGC Systems: (1) 30m DB-XLB capillary column w/ MS-SIM (2) 60m DB-XLB capillary w/full scan ion trap MS; and (3) Parallel dual column DB-17 & series coupled HP5/HT5 ECD	All 209 congeners reported. Allows data to be reduced, analytes summed to match any laboratory's coelutions/analyte list.	Hardcopy tables included in paper
Frame, et al., 1996	Congener-specific data for 15 Aroclor 1254 PCB lots.	Single DB-XLB column – GC-MS-SIM detection	144 congeners. Reports congener data for multiple lots of both typical and late-production Aroclor 1254.	Hardcopy tables included in paper
Rushneck et al., 2004	Congener-specific data for 9 Aroclors	EPA Method 1668A (HRGC/MS)	209 congeners – 159 peaks. Excellent detection limits. Excellent quantation of coplanar PCBs	Hardcopy tables not include in paper. Available at http://www.epa.gov/superfund/resources/pcb/
Schulz et al., 1989	Congener-specific data for 4 Aroclors and 4 analagous Clophens	GC-ECD	129 congeners reported. Historically significant. One of the first congener-specific data sets	Chromatograms and hardcopy tables included in paper.
Erickson, 1997	Chromatograms for 9 Aroclors	GC-ECD (DB-5 column)	Hardcopy chromatograms. DeVoogt data tables (see below) also included as an Appendix.	Hardcopy chromatograms in Appendix D DeVoogt et al. (1990) data in Appendix B.
DeVoogt, et al., 1990	Congener Compositions of 4 Clophens, 4 Aroclors, Delor D103, & a Kanechlor Mixture.	GC-ECD	62 congeners – 58 peaks Data compiled from seven different sources.	Hardcopy tables in paper and reproduced by Erickson, 1997
Data Sets Reporting Volatilization, Dechlorination and Metabolic Alteration Patterns				
Bedard and Quensen, 1995	Characteristic dechlorination products for 12 microbial reductive dechlorination processes.	Review paper – methods vary	Summarizes large volume of literature on microbial dechlorination of PCBs. Describes twelve dechlorination processes observed in field and lab.	Characteristic dechlorination products (congeners) are reported in tabular form for each process, but no congener-specific data reported. Congener data may be available by direct request of cited authors.
Johnson et al., (This chapter)	Congener data for Processes C, H, M, N, Q, H', and P	GC-ECD	Data from published study (Quensen et al., 1990) and previously unpublished data	Appendix A of this chapter.
Chiarenzelli et al., 1997	Congener-specific data for 4 Aroclor volatilization experiments.	GC-ECD (DB-5 column)	89 congeners—68 peaks.	Data not in paper, but available as ES&T supporting information on ACS web site. http://pubs.acs.org/journals/esthag/index.html
CDC (NHANES), 2003	Congener-specific data for PCBs in blood of a random cross-section of the US population.	GC-MS (DB5 column – Mass spec detector)	25 congeners (3 coplanar PCBs) Coelutions not reported. Detection limits generally high. New NHANES study in progress will report more congeners with lower detection limits.	Data available at CDC web site in SAS format. http://www.cdc.gov/nchs/about/major/nhanes/NHANES99_00.htm

congener-specific data available for various dechlorination processes, data for seven common dechlorination patterns are attached to this chapter as Appendix. There may well be other good reference data sets in the literature, but these nine provide a useful cross-section of congener patterns discussed in this chapter, ranging from unaltered source material (e.g., Aroclor and Clophens) to congener patterns representative of alteration processes (e.g., dechlorination, volatilization, and metabolism).

The reference data sets discussed above are limited to PCB congener data representative of sources and alteration processes. Where available, it is often also useful to compare site-specific PCB data to some sort of regional reference data set. Examples of such regional data sets include the San Francisco Estuary Institute's (SFEI) Regional Monitoring Program for Trace Substances (RMP), and the Integrated Atmospheric Deposition Network (IADN) database. The RMP was initiated in 1992 and monitors contaminant concentrations in water, sediments, and biota in the San Francisco Estuary in California. The program is funded by the major dischargers to the estuary with the objectives: (1) obtain baseline data on organic and inorganic contaminants (including PCBs); (2) determine seasonal, annual, and long-term trends in chemistry and water quality; (3) determine whether water quality and sediment quality in the estuary meets pre-defined objectives; and (4) provide a database that is compatible with other programs in the region. The RMP and analysis of congener-specific PCB data resulting from that program are summarized by Jarman et al. (1997) and Johnson et al. (2000). The RMP data can be obtained via the World Wide Web at http://www.sfei.org/rmp/data.htm.

The IADN is a joint effort between Environment Canada and the United States EPA to monitor atmospheric deposition of persistent organic pollutants (including PCBs) in the Great Lakes Region. The IADN operates five air-sampling stations in rural areas near each of the five Great Lakes. The stations collect vapor and particle-phase air samples every 12 days for 24 hours. Samples are analyzed for PAHs, organochlorine pesticides, and 56 PCB peaks. These data have proven valuable in regional studies where the IADN data are the main focus (e.g., Buehler et al., 2001). It has also proven to be a valuable reference data set for other PCB air studies conducted in and around the Great Lakes region (e.g., Chiarenzelli et al., 2001). The IADN data may be obtained by filling out a data request on the World Wide Web at http://www.msc-smc.ec.gc.ca/iadn/index_e.html.

10.4.4 Pattern Comparison and Interpretation

An important question at the end of any congener profile analysis is determining which sample patterns match which patterns in one's reference data set(s). Such a comparison is best done using a combination of visual (qualitative) analysis and numerical methods based on similarity indices. For quantitative comparisons of congener patterns, we prefer the cosine theta (cos-θ) metric (Jöreskog et al., 1976; Davis, 1986). In PCB environmental forensics applications, this metric would be used to compare two PCB patterns, by treating each as a multi-dimensional vector, and to calculate the cosine of the angle-θ between the two vectors. For each multivariate sample, where the number of variables (congeners) is equal to n, the sample composition may be thought of as an n-dimensional vector. The angle between two sample vectors is a function of the similarity in the two chemical compositions. If the two samples are identical, the vectors will be coincident, the angle between the vectors will be 0°, and the cosine of that angle is 1.0. Similarly, if two samples share no common congeners, the angle defined

between them is 90° and the cosine of that angle is zero. Thus cosine theta (cos-θ) is bounded between zero and one.

Other similarity indices are available, including Pearson and Spearman correlations. In some respects, cos-θ is conceptually similar to a correlation coefficient, r (i.e., perfect correlation is assigned an $r = 1.0$, no correlation yields $r = 0$). However, the correlation coefficient can be less than zero, if there is an inverse relationship (negative correlation). The concept of an inverse relationship is useful in the context of statistical methods such as regression, for general comparison of variables. It is not at all intuitive, however, when comparing the similarity of two samples. For example, if one compares a sample from the field to Aroclor 1242 using a correlation coefficient and gets an $r = -1.0$, is it safe to conclude that sample is the opposite or inverse of Aroclor 1242? It is not clear what this means. If another sample yields $r = 0.0$ in comparison to Aroclor 1242, what is conceptual distinction between it and the sample that yielded $r = -1.0$? Cosine theta is a more intuitive metric because it is always a positive number and the range of possible interpretations vary from completely dissimilar (Cosine θ = 0) to identical (cos-θ = 1.0).

Cosine theta is also preferable because it is simply a geometrical calculation between two vectors. Thus, it does not carry inherent implications of a linear regression diagnostic (as do correlation coefficients). Calculation of the metric is simple and its conceptual meaning of cos-θ is intuitive. Cosine theta has proven very useful for comparison of sample patterns in a number of earth and environmental science disciplines (Jöreskog et al., 1976; Davis, 1986; Gary et al., 2005; Magar et al., 2005a).

When comparing two samples (samples i and j), with reported concentrations of n chemicals measured in each sample, cos-θ is calculated as follows (Jöreskog et al., 1976, Davis, 1986):

$$\cos\theta = \frac{\sum_{k=1}^{n} x_{ik}x_{jk}}{\sqrt{\sum_{k=1}^{n} x_{ik}^2 \sum_{k=1}^{n} x_{jk}^2}} \qquad (10.4.1)$$

As an example of the use of this metric, we refer the user back to Figure 10.3.1. In the discussion of the Aroclor 1242 volatilization experimental data of Chiarenzelli et al. (1997), it was noted that the residual Aroclor 1242 sample looked more similar to unaltered Aroclor 1248 than it did to Aroclor 1242. While obvious by direct inspection of the two bar graphs, cos-θ can be used to quantify that statement. The cos-θ between unaltered A1242 and the Residual-1242 is 0.87. The cos-θ between unaltered 1248 and Residual-1242 is 0.96.

While there may be some comfort in reliance on an objective, quantitative similarity metric, it does not replace qualitative visual pattern analysis. The human eye remains one of the most effective pattern recognition tools available and often picks up similarities and differences in congener patterns that cannot be communicated by a scalar similarity metric. A similarity metric may assign a number on the degree of similarity between patterns, but no scalar measure will provide information on why or how two patterns differ. A visual pattern analysis allows the user to take qualitative information into account, which ultimately is just as relevant to an interpretation. Thus, regardless of the numerical methods used for pattern recognition or pattern comparison, the data analyst is well advised to spend time doing direct, visual inspection of congener patterns and simple graphical analysis.

10.5 CASE-STUDY: AKWESASNE, NEW YORK

The Mohawk Nation at Akwesasne is a Native American community in upstate New York located just down river of Massena (Figure 10.5.1). The Mohawks have lived on the banks of the St Lawrence River and its tributaries for generations. In 1957, the Robert Moses Power Dam was built on the St Lawrence River, approximately 6 miles upstream from Akwesasne. The availability of cheap electricity resulted in the construction of three aluminum foundries (Reynolds Aluminum, General Motors, and ALCOA) upstream from Akwesasne. All three foundries used PCBs as hydraulic fluids, primarily Aroclor 1248 (Fitzgerald *et al.*, 1995).

In the early 1980s, elevated concentrations of PCBs were measured in breast milk of Mohawk women (Fitzgerald *et al.*, 1995; Lean, 2000). This was the impetus for subsequent studies involving researchers from several universities (Carpenter *et al.*, 2002). That work has resulted in a series of papers focused on fate and transport of PCBs, as well as routes of human exposure and health effects (Sloan and Jock, 1990; Skinner, 1992; Sokol *et al.*, 1994; Fitzgerald *et al.*, 1995; Fitzgerald *et al.*, 1998; Cho *et al.*, 2000; Chiarenzelli *et al.*, 2000; Carpenter *et al.*, 2002; Schell *et al.*, 2003; DeCaprio *et al.*, 2005). A subset of that research is summarized below with a focus on aspects of environmental forensics that have been highlighted in this chapter, specifically inference of PCB sources, alteration of congener patterns in various environmental media (sediment, air, and biota).

10.5.1 PCB Source and Dechlorination Patterns in Sediments

The study of Sokol *et al.* (1994) is particularly relevant to the topics discussed in this chapter. These researchers collected sediment cores in St Lawrence River in close proximity to the three foundries. Comparison of congener-specific data from sediments to Aroclor reference standards confirmed that Aroclor 1248 was the primary PCB source in sediments (consistent with PCB purchase records from the three foundries that indicated that Aroclor 1248 was the primary PCB product used). However, a subset of sediment samples collected near the Reynolds facility showed a very clear pattern match to unaltered Aroclor 1260, suggesting a secondary Aroclor 1260 source.

There was also a strong dechlorination signal in sediments near the three sites. The extent of dechlorination varied widely from site to site and even from core to core. Sokol *et al.* (1994) described the observed dechlorination patterns in detail, but did not use the lettered nomenclature discussed here (e.g., Processes C, M, Q, etc. Section 10.3.2) to describe the observed dechlorination patterns. Sokol reported that the characteristic dechlorination products in this pattern were 2-CB, 2,2'-CB + 2,6-CB, 2,4'-CB + 2,3-CB, 2,2', 4-CB + 4,4'-CB, and 2,4', 5-CB + 2,4,4'-CB.

For instructive purposes and to illustrate how to interpret PCB data using reference data sets and methods described in this chapter, we have compared the Sokol sediment patterns that showed the highest degree of dechlorination (a core taken near the GM facility) to the various dechlorination processes described in Section 10.3.2 (and presented in Appendix). First, the various dechlorination processes described by Bedard and Quensen (1995) were reviewed in comparison to the congener patter reported from the GM core. Based on that comparison, the GM001 dechlorination pattern described by Sokol *et al.* (1994) had many of the congener-pattern characteristics of Process M. Process M has largely been observed associated with Aroclor 1242 sources (Quensen *et al.*, 1990; Ye *et al.*, 1992; Bedard and Quensen, 1995), but Bedard and Quensen indicated that this process should also act on Aroclor 1248.

To make a more quantitative comparison, we compared the GM core pattern to the Process M reference data from Quensen *et al.* (1990-see Appendix A). This pattern was originally identified as "N" by Quensen *et al.* (1990) but the interpretation was later revised as Process M. Quensen *et al.* (1990) observed the pattern in a laboratory dechlorination

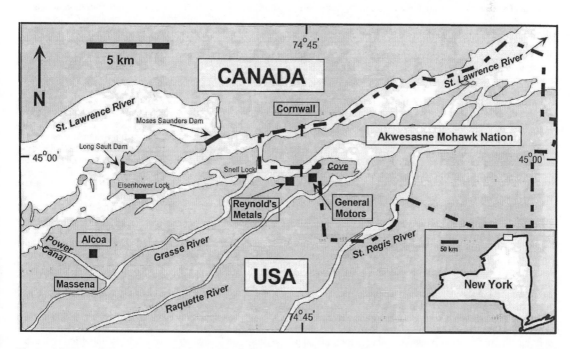

Figure 10.5.1 Location of Akwesasne Mohawk Nation. Contaminant Cove location is shown, as well as the locations of three nearby aluminum foundries (General Motors, Reynolds, and Alcoa).

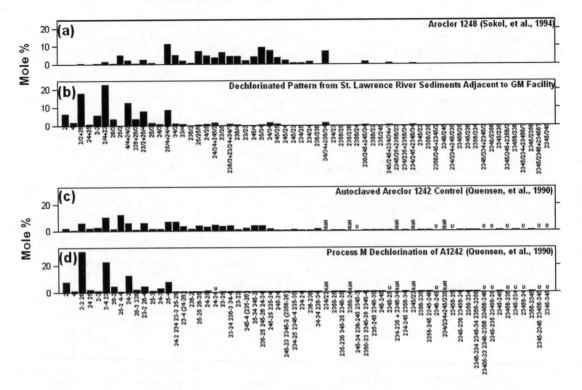

Figure 10.5.2 *Comparison graphs for (a) Unaltered Aroclor 1248 – the primary PCB source at foundries near Akwesasne (data from Sokol et al., 1994); (b) Dechlorinated congener pattern observed in sediments (data from Sokol et al., 1994); (c) Unaltered Aroclor 1242 (data from Quensen et al., 1990); and (d) Process M dechlorination pattern observed after dechlorination of Aroclor 1242 with Silver Lake, Mass organisms (data from Quensen et al., 1990). The cos-θ between pattern (b) and pattern (d) is 0.94.*

experiment with sediment inoculated with microorganisms from Silver Lake sediments (Pittsfield, MA) and incubated with Aroclor 1242. The analyte lists between the two were not the same (a common issue, discussed in Section 10.4.1). Thus, it was necessary to reduce Quensen's analyte list (88 peaks) to match that of Sokol (62 peaks). Figure 10.5.2 shows a comparison between (a) Aroclor 1248 reported by Sokol *et al.* (1994); (b) Sokol's GM001 dechlorination pattern (c) Aroclor 1248 as reported by Quensen *et al.* (1990); and (d) Quensen's Process M dechlorination pattern. As is evident by simple visual comparison and by a cos-θ of 0.94, the GM001 pattern of Sokol *et al.* (1994) is quite consistent with Process M. The one notable difference (2,2'-CB + 2,6-CB is the highest peak in the Quensen *et al.*, reference pattern; 2,4'-CB + 2,3-CB is the highest peak in Sokol's GM001 pattern) may be due to the different starting Aroclor compositions.

10.5.2 PCBs in Air

In the early studies at Akwesasne, ingestion was assumed to be the primary exposure route concern. Such a default assumption was logical because the impetus for the Akwesasne research was the finding of PCBs in breast milk of Mohawk women (which clearly implies ingestion as a route of exposure in infants). Polychlorinated biphenyls were also shown to be present in fish and wildlife near Akwesasne (Sloan and Jock, 1990; Skinner, 1992; Fitzgerald *et al.*, 1995), which the Mohawks had traditionally depended upon for food. Finally, of the three major possible routes of exposure (ingestion, inhalation, and dermal contact), conventional wisdom has held that ingestion is the most likely route

of PCB exposure in humans. Polychlorinated biphenyls are known to biomagnify in the food chain and that implies ingestion as the primary route of exposure. The rationale for assuming minimal exposure by dermal contact is that except for occupational exposures, most humans do not come in direct contact with PCB product. Finally, inhalation has not traditionally been considered an important route of exposure because PCBs have low vapor pressure (James, 2001).

The essence of environmental forensics is scientific detective work. The most interesting and valuable parts of forensics studies are not those instances where we confirm what we had already assumed, but rather when the data leads us to an unexpected conclusion. In the late 1990s, studies by Chiarenzelli and colleagues began to raise questions regarding air as a PCB migration pathway and as a potential route of exposure. Chiarenzelli *et al.* (1996) conducted a series of volatilization experiments using sediment from "contaminant cove," a small embayment of the St Lawrence River, adjacent to both the Akwesasne Nation and the General Motors industrial landfill. Contaminant cove sediment samples were placed in a sealed glass chamber and purified air was pulled through the chamber at a consistent rate. The PCBs lost by volatilization were recovered on Florisil columns. Key findings of this research included:

- As long as the sediment was saturated with water, PCB volatile loss was rapid.
- When the sediments became dry (with respect to ambient humidity – 25%) PCB volatilization was much less.
- When additional water was reapplied to dried sediments, PCB volatile losses resumed at a rapid rate, although less were lost overall during the same time period.

- An exponential correlation was found between the initial moisture content and the amount of PCBs volatilized and recovered on the columns. Dry sediment lost only 0.31% of its PCBs to volatilization, whereas submerged sediment lost 27.23% in 24 hours. These percentages correspond to volatile loss rates of 2.1–184.2 ng/hr from 0.25 grams of sediment.

These data strongly suggest that PCB-contaminated sediments that experience periodic submersion and drying (flood plains, tidal flats, seasonal rivers, etc.) provide conditions that enhance volatilization of PCBs (Bush *et al.*, 1987). This work led directly to the laboratory PCB volatilization experiments discussed in Section 10.3.1 and as reported by Chiarenzelli *et al.* (1997). That study provided a valuable reference data set of congener-specific data for four Aroclors, the vapor-phase congener patterns that result from the volatilization and the residual congener patterns that remain in sediments after volatilization. Vapor phase and residual PCBs remaining in sediment can be substantially different from the original Aroclor mixture. In addition, the composition of the vapor phase changes with time, reflecting the relative abundance of congeners within the mixture. Thus, the Aroclor volatilization data of Chiarenzelli is a valuable resource for those involved in environmental forensics investigation focused on inference of sources of PCBs and their fate and transport.

These data clearly suggested the possibility of an alternative route of exposure at Akwesasne, and that prompted further investigations of air at Akwesasne. An ambient air sampling program had been conducted at Akwesasne in 1993. Chiarenzelli *et al.* (2000) reported those data which showed that concentrations were comparable to PCBs in air in highly contaminated areas such as the Fox River/Green Bay (Hornbuckle *et al.*, 1993). Concentrations were highest in the summer months at the contaminant cove sampling station, falling to background levels during the winter months when the temperature was low and the ground and rivers were frozen. The congener patterns of summer contaminant-cove air samples were compared to the volatile fractions of Aroclor 1248 (Chiarenzelli *et al.*, 1997) and to the volatile fraction of dechlorinated sediments from contaminant cove. The patterns in air were more consistent with the volatile fraction of unaltered Aroclor 1248 (Chiarenzelli *et al.*, 2000). As Aroclor 1248 was the primary PCB mixture used at the nearby aluminum foundries, these data clearly suggested air as a potential migration pathway. The Chiarenzelli *et al.* (2000) study clearly illustrated that congener patterns observed in laboratory volatilization experiments were indeed reflected in the ambient air data.

10.5.3 Routes of Human Exposure

As these air and volatilization studies were being completed, a group of researchers at the University at Albany were undertaking a study involving analysis of PCBs in blood of over 700 adult Mohawks. The samples were collected between 1998 and 2000, and analyzed for congener-specific PCBs using GC-ECD methods (DeCaprio *et al.*, 2000). The resulting Mohawk blood data are presented by DeCaprio *et al.* (2005). As a group, the PCB levels in blood of this cohort were elevated with respect to the general US population (CDC, 2003). The congeners present in the highest proportions in Mohawks were PCBs 153, 138, 180, and 118, in that order. This dominance of PCBs 153, 138, 180, and 118 is typical (CDC, 2003). The reason these congeners appear so consistently in blood is that they are more metabolism-resistant in both humans and other biota (Hansen, 2001). The Akwesasne population shows a general dominance of the typically persistent PCBs, but DeCaprio *et al.* (2005)

also demonstrated that there were subsets of Mohawks that exhibited blood congener patterns that differed dramatically from the typical pattern. The multivariate data analysis method PVA was used to identify these congener patterns. The reader is referred to DeCaprio *et al.* (2005) for a complete discussion. In the context of this chapter, two atypical and unexpected congener patterns were observed in blood from this cohort. One of these was much lighter than the typical 153/138 congener profile (Figure 10.3.4). This pattern was observed only in a subset of the cohort and many of the lighter, lower chlorinated congeners present in this pattern are known to be metabolized much more rapidly than more typical congeners, PCBs 153, 138, 180, and 118 (Hansen, 2001). This suggested a more recent exposure.

In addition, the presence of lighter congeners in this subset of subjects suggests that (at least in this group of people) inhalation might indeed be a significant route of exposure. Figure 10.5.3 shows a comparison of (a) the blood-congener pattern described above; (b) unaltered Aroclor 1248 (the primary PCB source at Akwesasne); (c) ambient air from the contaminant cove sampling station (Chiarenzelli *et al.*, 2000); and (d) the volatile fraction of Aroclor 1248 (from Chiarenzelli *et al.*, 1997). Clearly the pattern observed in blood is a reasonable match to the ambient air and volatile fraction A1248 samples, providing further evidence that inhalation is a significant route of exposure in at least a subset of Mohawks.

These results should not be taken to imply that inhalation is the only route of exposure at Akwesasne. This was only one of the five congener patterns observed in blood of the Akwesasne adult cohort. However, the Akwesasne data clearly suggests that inhalation is a more important exposure pathway than conventional wisdom has indicated. It certainly suggests that common practice of a default dismissal of inhalation as a route of exposure is unjustified.

The second unexpected congener pattern observed in blood was very similar to typical Aroclor 1254. As with the pattern discussed above, this congener pattern was unexpected because of the presence of less metabolism resistant congeners, CBs 70, 95, and 110 (Hansen, 2001), which are prominent in A1254. Thus, this pattern was not believed to be the result of differential metabolism, but rather a recent exposure. This pattern was also unexpected because, to the extent that it suggests an Aroclor 1254 as source at Akwesasne, this was a surprise. All previous literature had indicated Aroclor 1248 as the primary PCB source in the study area, with Aroclor 1260 as a possible minor source.

As a PCB environmental forensics case study, Akwesasne story has many of the elements discussed in this chapter. There is evidence of multiple Aroclor sources (multiple facilities that used Aroclor 1248 and evidence suggesting Aroclor 1260 as well as Aroclor 1254 sources). There has been alteration of the original Aroclor source patterns by dechlorination, volatilization, and metabolism. From the perspective of data analysis, these data required considerable preprocessing, as congener-specific data in sediment, air, and human blood were generated by different investigators at different labs, using different methods. Finally, the use of congener-specific data in conjunction with multivariate analyses, simple inspection of the raw data, and comparison to source and alteration reference data sets provided new insight into unexpected migration and exposure pathways.

10.6 ACKNOWLEDGEMENTS

We acknowledge the Superfund Basic Research Program and the Agency for Toxic Substances and Disease Registry for funding of the Akwesasne case-study.

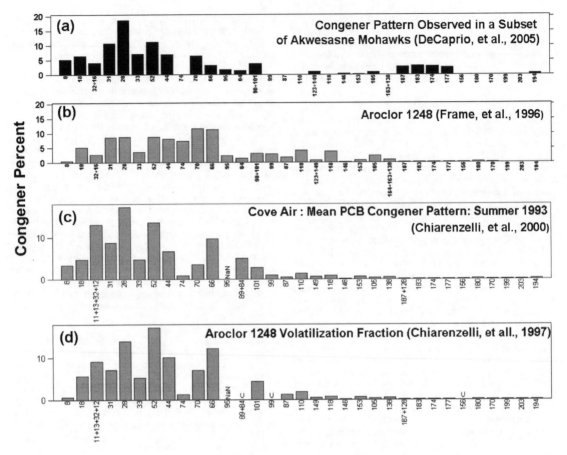

Figure 10.5.3 *Congener patterns observed in (a) a subset of Akwesasne Mohawk blood samples; (b) unaltered Aroclor 1248—the primary PCB source at aluminum foundries near Akwesasne; (c) a composite ambient air sample collected from contaminant cove–adjacent to Akwesasne and the GM foundry landfill; and (d) the volatile fraction of Aroclor 1248 as observed in Aroclor volatilization experiments.*

Thank you to the many colleagues who provided input to the chapter and/or collaboration on the studies referenced. These include Donna Bedard, David Carpenter, Anthony DeCaprio, Greg Durrell, Larry Hansen, Mark Hermanson, Wally Jarman, Randy Horsak, Victor Magar, Larry Schell, Paul Philp, and Bob Wagner.

REFERENCES

Ahlborg, U.G., Becking, G.C., Birnbaum, L.S., Brouwer, A., Derks, H.J.G.M., Feeley, M., Golor, G., Hanberg, A., Larsen, J.C., Liem, A.K.D., Safe, S.H., Schlatter, C., Waern, F., Younes, M., and Yrjänheikki, E. (1994). Toxic Equivalency Factors for Dioxin-like PCBs. *Chemosphere* **28**: 1049–1067.

Ballschmiter, K.R. and Zell, M. (1980). Analysis of polychlorinated biphenyls (PCB) by glass capillary gas chromatography. Composition of technical Aroclor and Clophen PCB mixtures. *Fresenius. J. Anal. Chem.* **302**: 20–31.

Bedard, D.L. and Quensen, J.F. (1995). Microbial reductive dechlorination of polychlorinated biphenyls. In *Microbial Transformation and Degradation of Toxic Organic Chemicals*. Young, L.Y. and Cerniglia, C.E. (eds). Wiley-Liss, New York, pp. 127–216.

Bedard, D.L., Bunnell, S.C., and Smullen, K.A. (1996). Stimulation of microbial para dechlorination of polychlorinated biphenyls that have persisted in Housatonic River sediment for decades. *Environ. Sci. Technol.* **30**: 687–694.

Bedard, D.L., Van Dort, H.M., May, R.J., and Smullen, L.A. (1997). Enrichment of microorganisms that sequentially meta, para-dechlorinate the residue of Aroclor 1260 in Housatonic River sediment. *Environ. Sci. Technol.* **31**: 3308–3313.

Bedard D.L., Pohl, E.A., Bailey, J.J., and Murphy, A. (2005). Characterization of the PCB substrate range of microbial dechlorination process LP. *Environ. Sci. Technol.* **39**: 6831–6838.

Bidleman, T.F. and Olney, C.E. (1974). Chlorinated hydrocarbons in the in the Sargasso Sea atmosphere and surface water. *Science.* **183**: 516–518.

Brown, J.F., Wagner, R.E., Bedard, D.L., Brennan, M.J., Carnahan, J.C., May, R.J., and Tofflemire, T.J. (1984). PCB transformations in upper Hudson sediments. *Northeastern Environ. Sci.* **3**: 167–179.

Brown, J.F., Bedard, D.L., Brennan, M.J., Carnahan, J. C., Feng, H., and Wagner, R.E. (1987a). Polychlorinated biphenyl dechlorination in aquatic sediments. *Science.* **236**: 709–712.

Brown, J.F., Jr, Wagner, R.E., Feng, H., Bedard, D.L., Brennan, M.J., Carnahan, J.C., and May, R.J. (1987b).

Environmental dechlorination of PCBs. *Environ. Toxicol. Chem.* **6**: 579–593.

Buehler, S.S., Basu, I., Hites, R.A. (2001). A comparison of PAH, PCB, and pesticide concentrations in air at two rural sites on Lake Superior. *Environ. Sci. Technol.* **35**: 2417–2422.

Burse, V.W., Kimbrough, R.D., Villanueva, E.C., Jennings, R.W., Lnder, R.E., and Sovocool, G.W. (1974). Polychlorinated biphenyls; storage, distribution, excretion, and recovery: liver morphology after prolonged dietary ingestion. *Arch. Environ. Health.* **29**: 301–307.

Bush, B., Shane, L., Whalen, M., and Brown, M. (1987). Sedimentation of 74 PCB congeners in the Upper Hudson River: Vol. 16, pp. 537–545.

Carpenter, D.O., Tarbell, A., Fitzgerald, E., Kadlec, M.J., O'Hehir, D., and Bush, B. (2002). University-community partnership for the study of environmental contamination at Akwesasne. In *Biomarkers of Environmentally Associated Disease: Technologies, Concepts and Perspectives.* S.H. Wilson and W.A. Suk (eds). Lewis Publishers (CRC Press), Boca Raton, pp. 507–523.

Casey, A.C., Berger, D.F., Lombardo, J.P., Hunt, A., and Quimby, F. (1999) Aroclor 1242 inhalation and ingestion by Sprague-Dawley rats. *J. Toxicol. Environ. Health* **A56**: 311–342.

Center for Disease Control (CDC). (2003). Second National Report on Human Exposure to Environmental Chemicals. Center for Disease Control and Prevention. Dept. Health & Human Services. NCEH Pub. 02-0716. January, 2003.

Chiarenzelli, J., Pagano, J., Scrudato, R., Falanga, L., Migdal, K., Hartwell, A., Milligan, M., Battalagia, T., Holsen, T.M., Hopke, P., and Hsu, Y. (2001). Enhanced airborne polychlorinated biphenyl (PCB) concentrations and chlorination downwind of Lake Ontario. *Environ. Sci. Technol.* **31**: 597–602; **35**: 280–3286.

Chiarenzelli, J., Scrudato, R., Arnold, G., Wunderlich, M., and Rafferty, D. (1996). Volatilization of polychlorinated biphenyls from sediment during drying at ambient conditions: *Chemosphere.* **33**: 899–911.

Chiarenzelli, J., Scrudato, R., and Wunderlich, M. (1997). Volatile loss of PCB Aroclors from subaqueous sand: *Environ. Sci. Technol.* **31**: 597–602.

Chiarenzelli, J., Bush, B., Casey, A., Barnard, E., Smith, R., O'Keefe, P., Gilligan, E., and Johnson, G. (2000). Defining the sources of airborne polychlorinated biphenyls: evidence for the influence of microbially dechlorinated congeners from river sediment? *Canadian J. Fish. Aquati. Sci.* **57**: 86–94.

Cho, Y.-C., Kim, J., Sokol, R.C., and Rhee, G.-Y. (2000). Biotransformation of polychlorinated biphenyls in St. Lawrence River sediments: Reductive dechlorination and dechlorinating microbial populations. *Can. J. Fish. Aquat. Sci.* 57(Suppl. 1): 95–100.

Cotham, W.E. and Bidleman, T.F. (1995). Polycyclic aromatic hydrocarbons and polychlorinated biphenyls in air at an urban and a rural site near Lake Michigan. *Environ. Sci. Technol.* **29**: 282–2789.

Cutter, L., Sowers, K.R., and May, H. (1998). Microbial dechlorination of 2,3,5,6-tetrachlorobiphenyl under anaerobic conditions in the absence of soil or sediment. *Appl. Environ. Microbiol.* **64**: 2966–2969.

DeCaprio, A.P., Tarbell, A.M., Bott, A., Wagemaker, D.L., Williams, R.L., and O'Hehir, C.M. (2000). Routine analysis of 101 polychlorinated biphenyl congeners in human serum by parallel dual-column gas chromatography with electron capture detection. *J. Anal. Toxicol.* **24**: 403–420.

DeCaprio, A.P., Johnson, G.W., Tarbell, A.M., Carpenter, D. O., Chiarenzelli, J.R., Morse, G.S., Santiago-Rivera, A.L., and Schymura, M.J. (2005). Serum PCB Congener Profiles as Exposure Biomarkers in an Adult Native American Population. Accepted. *Environ. Research.* September 2005. Accepted. In Press.

Davis, J.C. (1986). *Statistics and Data Analysis in Geology.* John Wiley and Sons. New York. p. 646.

De Voogt, P., and Brinkman, U.A.T. (1989). Production properties and usage of polychlorinated biphenyls. In *Halogenated Biphenyls, Terphenyls, Naphthalenes, Dibenzodioxins and Related Products.* R.D. Kimbrough and A.A. Jensen (eds). Elsevier, New York. pp. 3–45.

De Voogt, P., Wells, D.E., Reutergàrdh, L., and Brinkman, U.A.Th.(1990). Biological activity determination and occurrence of planar mono and di-ortho PCBs. *Int. J. Environ. Anal. Chem.* **40**: 1–46.

Deweerd, K.A. and Bedard, D.L. (1999). Use of halogenated benzoates and other halogenated aromatic compounds to stimulate the microbial dechlorination of PCBs. *Environ. Sci. Technol.* **33**: 2057–2063.

Durfee R.L., Contos, G., Whitmore, F.C., Barden, J.D., Hackman, E.E. III, Westin, R.A. (1976). PCBs in the United States—industrial use and environmental distribution. US Environmental Protection Agency, EPA publication no. 560/6-76-005:88. Washington, DC.

Eisenreich, S.J., Lonney, B.B., and Thornton, J.D. (1981). Airborne organic contaminants in the Great Lakes ecosystem. *Environ. Sci. Technol.* **15**: 30–38.

Erickson, M.D., Swanson, S.E., Flora, J.D., and Hinshaw, G.D. (1989). Polychlorinated dibenzofurans and other thermal combustion products and dielectric fluids containing polychlorinated biphenyls. *Environ. Sci. Technol.* **23**: 46–469.

Erickson, M.D. (1997). *Analytical Chemistry of PCBs.* CRC Press. Boca Raton, FL.

Erickson, M.D. (2001). Introduction: PCB properties, uses, occurrences and regulatory history. In *PCBs: Recent Advances in Environmental Toxicology and Health Effects.* L.W. Robertson and L.G. Hansen (eds). University of Kentucky Press, Boca Raton, FL, pp. xi–xxx.

Falconer, R.L. and Bidleman, T.F. (1994). Vapor pressures and predicted particle/gas distributions of polychlorinated biphenyl congeners as functions of temperature and ortho-chlorine substitution. *Atmos. Environ.* **28**: 547–554.

Fisk, A.T., Norstrom, R.J., Cymbalisty, C.D., and Muir, D.C.G. (1998). Dietary accumulation and depuration of hydrophovic organochlorines: Bioaccumulation parameters and their relationship with the octanol/water partition coefficient. *Environ.Toxicol. Chem.* **17**: 951–961.

Fitzgerald, E.F., Hwang, S.-A., Brix, K.A., Bush, B., Cook, K., and Worswick, P. (1995). Fish PCB concentrations and consumption patterns among Mohawk women at Akwesasne. *J. Expo. Anal, Environ. Epidemiol.* **5**: 1–19.

Fitzgerald, E.F., Hwang, S.-A., Bush, B., Cook, K., and Worswick, P. (1998). Fish consumption and breast milk PCB concentrations among Mohawk women at Akwesasne. *Am. J. Epidemiol.* **148**: 164–172.

Frame, G.M., Cochran, J.W., Bowadt, S.S. (1996). Complete PCB congener distributions for 17 Aroclor mixtures determined by 3 HRGC systems optimized for comprehensive, quantitative congener specific analysis. *J. High Resol. Chromatogr.* **19**: 657–668.

Frame, G.M. (1997a). A collaborative study of 209 PCB congeners and 6 Aroclors on 20 different HRGC columns: 1. Retention and coelution database. *Fresenius J. Anal. Chem.* **357**: 701–713.

Frame, G.M. (1997b). A collaborative study of 209 PCB congeners and 6 Aroclors on 20 different HRGC columns: 2. Semi-quantitative Aroclor congener distributions. *Fresenius J. Anal. Chem.* **357**: 701–713.

Frame, G.M. (1999). Improved procedure for single DB-XLB column GC-MS-SIM quantitation of PCB congener distributions and characterization of two different preparations sold as "Aroclor 1254." *J. High Resol. Chromatogr.* **22**: 533–540.

Gary, A.C., Johnson, G.W., Ekart, D.D., Platon, E., and Wakefield, M.I. (2005). A Method for Two-Well Correlation using Multivariate Biostratigraphical Data. In *Exploration Biostratigraphy*. A.J. Powell and J.B. Riding (eds). British Geological Society Special Publication. In Press.

Gribble, G.W. (1998). Naturally occurring organohalogen compounds. *Acc. Chem. Res.* **31**: 141–152.

Gribble, G.W. (2004). Amazing organohalogens. *American Scientist.* **92**: 342–348.

Guitart, R., Puig, P., Gómez-Catalán, J. (1993). Requirement for a standardized nomenclature criterium for PCBs: Computer assisted assignment of correct congener denomination and number. *Chemosphere.* **27**: 1451–1459.

Haglund, P., Jakobsson, E., Asplund, M., Athanasiadou, and Bergman, Å. (1993). Determination of polychlorinated naphthalenes in polychlorinated biphenyl products via capillary gas chromatography – mass spectrometry after separation by gel permeation chromatography. *J. Chromatogr.* **634**: 79–86.

Hansen, L.G. (1999). *The Ortho Side of PCBs: Occurrence and Disposition*. Kluwer Academic Publishers, Norwell, Massachusetts.

Hansen, L.G. (2001). Identification of steady state and episodic PCB congeners from multiple pathway exposures. In *PCBs: Recent Advances in Environmental Toxicology and Health Effects*. L.W. Robertson and L.G. Hansen (eds). University of Kentucky Press, pp. 17–26.

Harner, T., Mackay, D., and Jones, K. (1995). Model of the long-term exchange of PCBs between soil and the atmosphere in Southern U.K. *Environ. Sci. Technol.* **29**: 1200–1209.

Haque, R., Schmedding, D., and Freed, V. (1974). *Environ. Sci. Technol.* **8**: 139–142.

Hermanson, M.H. and Hites, R.A. (1989). Long-term measurements of atmospheric polychlorinated biphenyls in the vicinity of Superfund dumps. *Environ. Sci. Technol.* **23**: 1253–1258.

Hermanson M.H. and Hites, R.A. (1990) Polychlorinated biphenyls in tree bark. *Environ. Sci. Technol.* **24**: 666–671.

Hermanson, M.H. and Johnson, G.W. (2005) Polychlorinated biphenyls in tree bark from Anniston, Alabama. In *Recent Advances in Environmental Toxicology and Health Effects of PCBs*. L.E. Hanson (ed.). Urbana: University of Illinois Press (in Press).

Hermanson, M.H., Scholten, C.A., and Compher, K. (2003). Variable air temperature response of gas-phase atmospheric polychlorinated biphenyls near a former manufacturing facility. *Environ. Sci. Technol.* **37**: 4038–4042.

Hillery, B.R., Basu, I., Sweet, C.W., and Hites, R.A. (1997). Temporal and spatial trends in a long-term study of gas-phase PCB concentrations near the Great Lakes. *Environ. Sci. Technol.* **31**: 1811–1816.

Hornbuckle, K.C., Achman, D.R., and Eisenreich, S.J. (1993). Over-water and over-land polychlorinated biphenyls in Green Bay, Lake Michigan. *Environ. Sci. Technol.* **27**: 87–98.

Holoubek, I. (2001). Polychlorinated biphenyl (PCB) contaminated sites worldwide. In *PCBs: Recent Advances in Environmental Toxicology and Health Effects*. L.W. Robertson and L.G. Hansen (eds). University of Kentucky Press, pp. 17–26.

Hutzinger, O., Choudry, G.G., Chittim, B.G., and Johnston, L.E. (1985). Formation of polychlorinated dibenzofurans and dioxins during combustion, electrical equipment fires and PCB incineration. *Environ. Health Perspect.* **60**: 3–9.

Ikonomou, M.G., Fernandez, M.P., Knapp, W., and Sather, P. (2002). PCBs in Dungeness crab reflect distinct source fingerprints among harbor/industrial sites in British Columbia. *Environ. Sci. Technol.* **36**: 2545–2551.

Imamoglu, I., Li, K., and Christensen, E.R. (2002). PCB Congener Patterns in dated sediments from Ashtabula River, Ohio, USA. *Environ. Toxicol. Chem.* **21**: 2283–2291.

Imamoglu, I., Li, K., Christensen, E.R. and McMullin, J.K. (2004) Sources and dechlorination of polychlorinated biphenyl congeners in the sediments of Fox River, Wisconsin. *Environ. Sci. Technol.* **38**: 2574–2583.

Imsilp, K., Schaeffer D.J., Hansen, L.G. (2005) PCB disposition and different biological effects in rats following direct soil exposures vs. PCBs off-gassed from the soil. *Toxicol. Environ. Chem.* (in press).

James, M.O. (2001). Polychlorinated biphenyls: Metabolism and metabolites. In *PCBs: Recent Advances in Environmental Toxicology and Health Effects*. L.W. Robertson and L.G. Hansen (eds). University of Kentucky Press, pp. 35–46.

Jarman, W.M., Johnson, G.W., Bacon, C.E., Davis, J.A., Risebrough, R.W., and Ramer, R. (1997). Levels and patterns of polychlorinated biphenyls in water collected from the San Francisco Bay and Estuary, 1993–1995. *Fresnius J. Anal. Chem.* **359**: 254–260.

Jarman, W.M., Hilkert, A., Bacon, C.E., Collister, J.W., Ballschmiter, K., and Risebrough, R.W. (1998). Compound-specific carbon isotopic analysis of Aroclors, Clophens, Kaneclors, and Phenoclors. *Environ. Sci. Technol.* **32**: 833–836.

Johnson, G.W., and Ehrlich, R. (2002). State of the Art Review: Multivariate Statistical Methods in Environmental Forensics. *Environmental Forensics.* **3**: 59–79.

Johnson, G.W., Jarman, W.M., Bacon, C.E., Davis, J.A., Ehrlich, R., and Risebrough, R. (2000). Resolving polychlorinated biphenyl source fingerprints in suspended particulate matter of San Francisco Bay. *Environ. Sci. Technol.* **34**: 552–559.

Johnson, G.W., Ehrlich, R., and Full, W. (2002). Principal components analysis and receptor models in environmental forensics. In *An Introduction to Environmental Forensics*. R. Morrison and B. Murphy (eds). Academic Press. San Diego. pp. 461–515.

Jöreskog, K.G., Klovan, J.E., and Reyment, R.A. (1976). *Geological Factor Analysis*. Elsevier Scientific Publishing Company, Amsterdam. pp. 178.

Kannan, N., Schulz, D.E., Petrick, G., and Duinker, J.C. (1992). High resolution PCB analysis of Kanechlor, Phenoclor and sovol mixtures using multidimensional gas chromatography. *Int. J. Environ. Anal. Chem.* **47**: 201–215.

Kodavanti, P.R., Kannan, N., Yamashita, N., Derr-Yellin, E.C., Ward, T.R., Burgin, D.E., Tilson, H.A., Birnbaum, L.S. (2001). Differential effects of two lots of Aroclor 1254: Congener-specific analysis and neurochemical end points. *Environ. Health Perspect.* **109**: 1153–1161.

Kohler, M., Zennegg, M., Waeber, R. (2002). Coplanar Polychlorinated Biphenyls (PCB) in Indoor Air. *Environ. Sci. Technol.* **36**: 4735–4740.

Lean, D.R.S. (2000). Some secrets of a great river: An overview of the St. Lawrence River supplement. *Can. J. Fish. Aquat. Sci.* 57(Suppl. 1): 1–6.

Letcher R.J., Klasson-Wehler, E., Bergman, Å. 2000. Methyl sulfone and hydroxylated metabolites of polychlorinated biphenyls. In *The Handbook of Environmental Chemistry 3K*. J. Paasivirta (ed.). Springer-Verlag, Berlin. pp. 315–359.

Lohmann, R. (2003). The emergence of black carbon as a super-sorbent in environmental chemistry: The end of octanol? *Environ. Forensics.* **4**: 161–165.

Magar, V.S., Johnson, G.W., Brenner, R., Durell, G., Quensen, J.F., III, Foote, E., Ickes, J.A., Peven-McCarthy, C. (2005a). Long-term recovery of PCB-contaminated sediments at the Lake Hartwell Superfund Site: PCB Dechlorination I–End-Member Characterization. *Environ. Sci. Technol.* **39**: 3538–3547.

Magar, V.S., Brenner, R., Johnson, G.W., Quensen, J.F. III (2005b). Long-term recovery of PCB-contaminated sediments at the Lake Hartwell Superfund Site: PCB Dechlorination II – Rates and Extent. *Environ. Sci. Technol.* **39**: 3548–3554.

Meredith M.L. and Hites R.A. (1987) Polychlorinated biphenyl accumulation in tree bark and wood growth rings. *Environ. Sci. Technol.* **21**: 709–12.

McFarland, M. (1994). Analysis of Coplanar PCBs. Axys Environmental Systems Ltd., Fax from Mary McFarland to Dale Rushneck dated November 25, 1994, available from the EPA Sample Control Center, operated by Dyn-Corp I&ET, 300 N. Lee St, Alexandria, VA 22314.

Mackay, D. (2001). *Multimedia Environmental Models: The Fugacity Approach.* Second Edition. Lewis Publishers. Boca Raton.

Monosmith, C.L. and Hermanson, M.H. (1996). Spatial and temporal trends of atmospheric organochlorine vapors in the central and upper Great Lakes. *Environ. Sci. Technol.* **30**: 3464–3472.

Morrison, H.A., Gobas, F.A.P.C., Lazar, R., and Haffner, G.D. (1996). Development and verification of a bioaccumulation model for organic contaminants in benthic invertebrates. *Environ. Sci. Technol.* **30**: 3377–3384.

Mullin, M.D. (1985). PCB Workshop; USEPA Large Lakes Research Station, Goose Isle, MI. June 1985.

O'Keefe, P.W., Silkworth, J.B., Gierthy, J.F., Smith, R.M., DeCaprio, A.P., Turner, J.N., Eadon, G., Hilker, D.R., Aldous, K.M., Kaminsky, and Collins, D.N. (1985). Chemical and biological investigations of a transformer accident at Binghamton, N.Y. *Environ. Health Perspect.* **60**: 201–209.

O'Keefe, P.W., and Smith, R.M. (1989). PCB capacitor/transformer accidents. In *Halogenated Biphenyls, Terphenyls, Naphthalenes, Dibenzodioxins and Related Products.* R.D. Kimbrough and A.A. Jensen (eds). Elsevier, New York. pp. 417–444.

Philp, R.P. (2002). Application of stable isotopes and radioisotopes in environmental forensics. In *An Introduction to Environmental Forensics.* R. Morrison and B. Murphy (eds). Academic Press. San Diego. pp. 99–136.

Porte, C. and Albaigés, J. (1993). Bioaccumulation patterns of hydrocarbons and polychlorinated biphenyls in bivalves, crustaceans and fishes. *Arch. Environ. Contam. Toxicol.* **26**: 273–281.

Quensen, J.F. III, Tiedje, J.M., and Boyd, S.A. (1988). Reductive dechlorination of polychlorinated biphenyls by anaerobic microorganisms from sediments. *Science.* **242**: 752–754.

Quensen, J.F. III, Boyd, S.A., and Tiedje, J.M. (1990). Dechlorination of four commercial polychlorinated biphenyl mixtures (Aroclors) by anaerobic microorganisms from sediments. *Appl. Environ. Microbiol.* **56**: 2360–2369.

Quensen, J.F. III, and J.M. Tiedje. (1997). Methods for evaluation of PCB dechlorination in sediments. In *Protocols for Bioremediation.* D. Sheehan (ed.). vol. 2. Humana Press, Towana, NJ. pp. 241–253

Rachdawong, P. and E.R. Christensen. (1997). Determination of PCB Sources by a Principal Component Method with Nonnegative Constraints. *Environ. Sci. Technol.* **31**: 2686–2691.

Reddy, C.M., Heraty, L.J., Holt, B.D., Sturchio, N.C., Eglinton, T.I., Drenzek, N., Lake, J., and Maruya, K. (2000). Stable chlorine isotopic compositions of Aroclors and Aroclor-contaminated sediments. *Environ. Sci. Technol.* **34**: 2866–2870.

Rushneck, D.R., Beliveau, A., Fowler, B., Hamilton, C., Hoover, D., Kaye, K., Berg, M., Smith, T., Telliard, W.A., Roman, H., Ruder, E., and Ryan, L. (2004). Concentrations of dioxin-like PCB congeners in unweathered Aroclors by HRGC/HRMS using EPA Method 1668A. *Chemosphere.* **54**: 79–87.

Safe, S. (1990). Polychlorinated biphenyls (PCBs), dibenzo-p-dioxins (PCDDs) and dibenzofurans (PCDFs) and related compounds: Environmental and mechanistic considerations which support the development of toxic equivalency factors (TEFs). *C.R.C. Crit. Rev. Toxicol.* **21**: 51–88.

Safe, S. (1994). Polychlorinated biphenyls (PCBs): Environmental impact, biochemical and toxic responses, and implications for risk assessment. *C.R.C. Crit. Rev. Toxicol.* **24**: 87–149.

Salau, J.S., Tauler, R., Bayona, J.M., and Tolosa, I. (1997). Input characterization of sedimentary organic contaminants and molecular markers in the Northwestern Mediterranean Sea by exploratory data analysis. *Environ. Sci. Technol.* **31**: 3482–3490.

Schecter, A., (1983). Contamination of an office building in Binghamton, New York by PCBs, dioxins, dibenzofurans and biphenylenes after an electrical panel transformer incident. *Chemosphere.* **12**: 669–680.

Schell, L.M., Hubicki, L.A., DeCaprio, A.P., Gallo, M.V., Ravenscroft, J., Tarbell, A., Jacobs, A., David, D., Worswick, and the Akwesasne Task Force on the Environment. (2003). Organochlorines, lead and mercury in Akwsesasne Mohawk youth. *Environ. Health Perspect.* **111**: 954–961.

Schulz, D.E., Petrick, G., and Duinker, J.C. (1989). Complete characterization of polychlorinated biphenyl congeners in commercial Aroclor and Clophen mixtures by multidimensional gas chromatography – electron capture detection. *Environ. Sci. Technol.* **23**: 852–859.

Schwartz, D.L. Stalling, D.L., and Rice, C.L. (1987). Are polychlorinated biphenyl residues adequately described by Aroclor mixture equivalents? Isomer-specific principal components analysis of such residues in fish and turtles. *Environ. Sci. Technol.* **21**: 72–76.

Simonich, S.L. and Hites, R.A. (1995). Global distribution of persistent organochlorine compounds. *Science.* **269**: 1851–1854.

Skinner, L. (1992). Chemical contaminants in the wildlife from the Mohawk Nation at Akwesasne and the vicinity of the General Motors Corporation/Central Foundry Division, Massena, NY plant. New York State Department of Environmental Conservation. Albany, NY.

Sloan, R. and Jock, K. (1990). Chemical contaminants in fish from the St. Lawrence River drainage of lands on the Mohawk Nation at Akwesasne near the General Motors Corporation/Central Foundry Division, Massena, NY plant. New York State Department of Environmental Conservation. Albany, NY.

Sokol, R., Kwon, O.-S., Bethoney, C., and Rhee, G.-y. (1994). Reductive dechlorination of polychlorinated biphenyls in St. Lawrence River sediments and variations in dechlorination characteristics. *Environ. Sci. Technol.* **28**: 2054–2064.

Stout, S.A., Uhler, A.D., McCarthy, K.J., and Emsbo-Mattingly, S. (2002). Chemical Fingerprinting of Hydrocarbons. In *An Introduction to Environmental Forensics.* R. Morrison and B. Murphy (eds). Academic Press. San Diego, pp. 139–260.

US Environmental Protection Agency (USEPA). (1997a). Test methods for evaluating solid waste (SW-846). Update III. United States Environmental Protection Agency, Office of Solid Waste and Emergency Response. Washington, D.C.

US Environmental Protection Agency (USEPA). (1997b). Method 1668: Toxic Polychlorinated Biphenyls by Isotope Dilution High Resolution Gas Chromatography/High Resolution Mass Spectrometry. US Environmental Protection Agency. Office of Water. Office of Science and Technology. Engineering and Analysis Division. Draft 1997.

US Environmental Protection Agency (USEPA). (1999). Method 1668, Revision A: Chlorinated Biphenyl Congeners in Water, Soil, Sediment, and Tissue by HRGC/HRMS, EPA-821-R-00-002. Available from the EPA Sample Control Center operated by DynCorp I&ET, 6101 Stevenson Av, Alexandria, VA 22304, <scc@dyncorp.com>;

Van den Berg, M., Birnbaum, L., Bosveld, A.T.C., Brunstrom, B., Cook, P., Feeley, M., Giesy, J.P., Hanberg, A., Hasegawa, R., Kennedy, S.W., Kubiak, T., Larsen, J.C., van Leeuwen, F.X.R., Djien Liem, A.K., Nolt, C., Peterson, R.E., Poellinger, L., Safe, S., Schrenk, D., Tillit, D., Tysklind, M., Younes, M., Waern, F., Zacharewski, T. (1998). Toxic equivalency factors (TEFs) for PCBs, PCDDs, PCDFs for humans and wildlife. *Environ. Health Perspect.* **106**: 775–792.

Van Dort, H.M., Smullen, L.A., May, R.J., and Bedard, D. L. (1997). Priming microbial meta-dechlorination of polychlorinated biphenyls that have persisted in Housatonic River sediments for decades. *Environ. Sci. Technol.* **31**: 3300–3307.

Vorhees, D.J., Cullen, A.C., and Altshul, L.M., 1997. Exposure to polychlorinated biphenyls in residential indoor and outdoor air near a superfund site. *Environ. Sci. Technol.* **31**: 3612–3618.

Vuceta, J., Marsh, J.R., Kennedy, S., Hildemann, W.S. (1983). *State-of-the-art Review: PCDDs and PCDFs in Utility PCB Fluid.* Electric Power Research Institute, Palo Alto, CA, USA.

Wakimoto, T., Kannan, N., Ono, M., Tatsukawa, R., and Masuda, Y. (1988). Isomer specific determination of

Table A.1 Dechlorination Data from Quensen et al. *(1990)*

Peak #	IUPAC #	Structure	Ortho	Total	Molecular Weight	Hudson River 1242 Pattern C Autoclaved	Live	Difference	Hudson River 1248 Pattern C Autoclaved	Live	Difference
1	1	2	1	1	188.7	2.52	35.41	32.89	0.00	28.69	28.69
2	3	4	0	1	188.7	0.00	3.79	3.79	0.00	0.00	0.00
3	4 10	2-2 26	2	2	223.1	3.94	46.18	42.24	0.63	34.61	33.99
4	7 9	24 25	1	2	223.1	1.41	0.20	−1.21	0.22	1.84	1.62
5	6	2-3	1	2	223.1	1.75	0.74	−1.01	0.28	3.21	2.93
6	8 5	2-4 23	1	2	223.1	7.84	2.88	−4.96	1.47	3.41	1.94
7	19	26-2	3	3	257.5	0.86	3.02	2.16	0.28	6.20	5.92
8	12 13	34 3-4	0	2	223.1	0.16	0.10	−0.06	0.00	0.32	0.32
9	18 15	25-2 4-4	1.51	2.75	249	9.46	0.35	−9.11	4.79	0.90	−3.89
10	17	24-2	2	3	257.5	4.55	0.65	−3.90	1.48	0.43	−1.05
11	27 24	26-3 236	2	3	257.5	0.81	1.40	0.59	0.26	4.30	4.04
12	16 32	23-2 26-4	2	3	257.5	4.80	0.34	−4.46	2.08	0.84	−1.24
13	34 (54)	35-2 (26-26)	1	3	257.5	0.13	0.19	0.06	0.07	0.00	−0.07
14	29	245	1	3	257.5	0.11	0.02	−0.09	0.05	0.00	−0.05
15	26	25-3	1	3	257.5	1.17	0.33	−0.85	0.46	2.00	1.54
16	25	24-3	1	3	257.5	0.92	0.45	−0.46	0.26	1.11	0.85
17	31	25-4	1	3	257.5	5.57	0.75	−4.83	3.98	5.29	1.31
18	28 (50)	24-4 (246-2)	1	3	257.5	7.48	0.35	−7.13	3.92	0.00	−3.92
19	33 21 20 53	34-2 234 23-3 25-26	1.12	3.06	259.5	5.84	0.26	−5.58	3.17	0.00	−3.17
20	22 (51)	23-4 (24-26)	1	3	257.5	2.70	0.06	−2.64	1.39	0.21	−1.18
21	45	236-2	3	4	292	1.16	0.04	−1.11	1.36	0.04	−1.32
22	46	23-26	3	4	292	0.60	0.00	−0.60	0.78	0.00	−0.78
23	52 73	25-25 26-35	2	4	292	3.07	0.24	−2.83	5.76	0.43	−5.33
24	49	24-25	2	4	292	2.31	0.19	−2.12	3.53	0.46	−3.07
25	47	24-24	2	4	292	1.05	0.27	−0.78	3.59	0.56	−3.03
26	48 75	245-2 246-4	2	4	292	1.40	0.00	−1.40	0.00	0.00	0.00
27	35	34-3	0	3	257.5	0.08	0.00	−0.08	0.06	0.00	−0.06
28	44	23-25	2	4	292	2.77	0.05	−2.72	4.85	0.07	−4.78
29	42 59 37	23-24 236-3 34-4	0.84	3.42	272.4	3.21	0.09	−3.12	3.39	0.09	−3.30
30	71 41 64 72	26-34 234-2 236-4 25-35	1.75	4	292	2.71	0.06	−2.65	4.33	0.08	−4.25
31	96	236-26	4	5	326.4	0.08	0.00	−0.08	0.10	0.00	−0.10
32	40	23-23	2	4	292	0.84	0.05	−0.79	1.34	0.18	−1.16
33	57 70 100 103	235-3 245-3 246-24 246-25	1.4	4.2	298.9	0.19	0.01	−0.18	0.21	0.09	−0.12
34	58 63	23-35 235-4	1	4	292	0.17	0.08	−0.09	0.09	0.09	−0.01
35	74 (94)	245-4 (235-26)	1	4	292	1.49	0.09	−1.40	2.70	0.32	−2.38
36	70 76	25-34 345-2	1	4	292	3.40	0.06	−3.34	6.87	0.23	−6.64
37	95 102 66	236-25 245-26 24-34	1.08	4.04	293.5	3.33	0.17	−3.16	6.89	0.35	−6.53
38	55 91	234-3 236-24	2.9	4.95	324.7	0.40	0.04	−0.36	1.59	0.00	−1.59

polychlorinated dibenzofurans in Japanese and American polychlorinated biphenyls. *Chemosphere.* **4**: 743–750.

Wania, F. and MacKay, D. 1996. Tracking the distribution of persistent organic pollutants – control strategies for these contaminants will require a better understanding of how they move around the globe. *Environ. Sci. Technol.* **30**: 390A–396A.

Williams, W.A. (1994). Microbial reductive dechlorination of trichlorobiphenyls in anaerobic sediment slurries. *Environ. Sci. Technol.* **28**: 630–635.

Wu, Q.Z., Bedard, D.L., and Wiegel, J. (1996). Influence of incubation temperatures on the microbial reductive dechlorination of 2,3,4,6-tetrachlorobiphenyl in two freshwater sediments. *Appl. Environ. Microbiol.* **62**: 4174–4179.

Wu, Q., Bedard, D.L., and Wiegel, J. (1997a). Temperature determines the pattern of anaerobic microbial dechlorination of Aroclor 1260 primed by 2,3,4,6-tetrachlorobiphenyl in Woods Pond sediment. *Appl. Environ. Microbiol.* **63**: 4818–4825.

Wu, Q.Z., Bedard, D.L., and Wiegel, J. (1997b). Effect of incubation temperature on the route of microbial reductive dechlorination of 2,3,4,6-tetrachlorobiphenyl in polychlorinated biphenyl (PCB)-contaminated and PCB-free freshwater sediments. *Appl. Environ. Microbiol.* **63**: 2836–2843.

Wu, Q., Sowers, K.R., and May, H.D. (1998). Microbial reductive dechlorination of Aroclor 1260 in anaerobic slurries of estuarine sediments. *Appl. Environ. Microbiol.* **64**: 1052–1058.

Yanik, P.J., O'Donnell, T.H., Macko, S.A., Qian, Y., Kennicutt, M.C. (2003). Source apportionment of polychlorinated biphenyls using compound specific isotope analysis. *Organic Geochem.* **34**: 239–251.

Ye, D., Quensen, J.F. III, Tiedje, J.M., and Boyd, S.A. (1992). Anaerobic dechlorination of polychlorobiphenyls (Aroclor 1242) by pasteurized and ethanol-treated microorganisms from sediments. *Appl. Environ. Microbiol.* **58**: 1110–1114.

| Hudson River 1254 Pattern C | | | Hudson River 1260 Process H | | | Silver Lake 1242 Process M | | | Silver Lake 1260 Process N | | |
Autoclaved	Live	Difference	Autoclaved	Live	Difference	Autoclaved	Live	Difference	Autoclaved	Live	Difference
0.00	16.66	16.66	0.00	0.00	0.00	1.38	6.69	5.31	0.01	0.00	−0.01
0.00	0.00	0.00	0.00	0.00	0.00	0.32	0.54	0.22	0.00	0.00	0.00
0.00	23.69	23.69	0.00	0.00	0.00	4.69	26.47	21.78	0.00	0.47	0.47
0.00	1.85	1.85	0.00	0.00	0.00	1.45	1.67	0.23	0.02	0.00	−0.02
0.00	2.32	2.32	0.00	0.00	0.00	1.85	0.48	−1.37	0.02	0.00	−0.02
0.20	1.14	0.94	0.00	0.00	0.00	8.33	19.79	11.46	0.05	0.06	0.02
0.00	6.87	6.87	0.00	0.00	0.00	1.02	3.48	2.46	0.00	0.19	0.19
0.00	0.00	0.00	0.00	0.00	0.00	0.18	0.68	0.51	0.00	0.00	0.00
0.40	0.80	0.39	0.00	0.00	0.00	9.84	0.21	−9.62	0.02	0.12	0.10
0.16	1.25	1.09	0.00	0.00	0.00	4.79	10.59	5.80	0.01	0.31	0.30
0.00	6.17	6.17	0.00	0.00	0.00	0.78	1.49	0.71	0.01	0.89	0.88
0.19	0.62	0.43	0.00	0.00	0.00	4.87	3.45	−1.43	0.02	0.70	0.68
0.00	0.00	0.00	0.00	0.00	0.00	0.11	0.17	0.06	0.00	0.00	0.00
0.00	0.00	0.00	0.00	0.00	0.00	0.09	0.00	−0.09	0.00	0.36	0.36
0.05	1.64	1.60	0.00	0.42	0.42	1.10	0.82	−0.28	0.02	0.03	0.01
0.00	1.94	1.94	0.00	0.10	0.10	0.84	2.30	1.46	0.03	0.10	0.07
0.47	2.03	1.56	0.05	0.04	−0.01	5.36	6.58	1.23	0.01	0.00	−0.01
0.35	1.58	1.23	0.05	0.00	−0.05	7.29	10.14	2.85	0.01	0.83	0.83
0.32	0.00	−0.32	0.03	0.00	−0.03	5.72	0.30	−5.42	0.00	3.63	3.63
0.10	0.03	−0.08	0.00	0.00	0.00	2.56	0.21	−2.35	0.00	0.00	0.00
0.06	0.10	0.04	0.00	0.00	0.00	1.13	0.25	−0.88	0.00	0.00	0.00
0.00	0.11	0.11	0.00	0.00	0.00	0.60	0.05	−0.55	0.00	0.19	0.19
2.62	1.14	−1.48	0.28	7.57	7.29	3.10	0.37	−2.72	0.00	1.43	1.43
1.10	1.25	0.14	0.02	3.32	3.30	2.37	0.52	−1.85	0.05	2.13	2.09
0.56	2.41	1.85	0.00	0.48	0.48	3.40	0.00	−3.40	0.00	18.11	18.11
0.00	0.00	0.00	0.00	0.00	0.00	0.00	0.00	0.00	0.00	0.00	0.00
0.00	0.00	0.00	0.00	0.00	0.00	0.07	0.01	−0.06	0.00	0.10	0.10
1.18	0.09	−1.09	0.05	0.31	0.25	2.71	0.04	−2.67	0.03	0.00	−0.03
0.43	0.16	−0.27	0.00	0.14	0.14	3.12	0.28	−2.83	0.01	0.04	0.03
1.22	0.42	−0.80	0.03	0.05	0.02	2.57	0.64	−1.93	0.02	0.17	0.16
0.02	0.03	0.01	0.00	0.02	0.02	0.04	0.01	−0.04	0.00	0.55	0.55
0.18	0.70	0.52	0.00	1.36	1.36	0.81	0.08	−0.73	0.01	1.30	1.29
0.04	0.23	0.20	0.00	0.60	0.60	0.18	0.03	−0.15	0.01	2.55	2.55
0.00	0.21	0.21	0.00	0.00	0.00	0.07	0.07	0.00	0.00	0.08	0.08
1.39	0.43	−0.96	0.03	0.00	−0.03	1.35	0.08	−1.27	0.01	0.00	−0.01
4.56	0.35	−4.20	0.10	0.00	−0.10	3.20	0.02	−3.17	0.04	0.00	−0.04
4.96	0.83	−4.13	1.77	2.07	0.31	3.20	0.17	−3.03	2.05	0.65	−1.40

(Continued)

Table A.1 *(Continued)*

Peak #	IUPAC #	Structure	Molecular			Hudson River 1242 Pattern C			Hudson River 1248 Pattern C		
			Ortho	Total	Weight	Autoclaved	Live	Difference	Autoclaved	Live	Difference
39	56 60	23-34 234-4	1	4	292	4.95	0.16	−4.79	8.63	0.41	−8.22
40	101 90	245-25 235-24	2	5	326.4	1.21	0.14	−1.07	4.33	0.29	−4.05
41	99	245-24	2	5	326.4	0.30	0.04	−0.25	0.96	0.16	−0.80
42	150 112 119	236-246 2356-3 246-34	2.67	5	326.4	0.01	0.00	0.00	0.02	0.00	−0.01
43	97 86 (152)	245-23 2345-2 (2356-26)	2	5	326.4	0.46	0.04	−0.43	1.67	0.04	−1.63
44	87 115 111	234-25 2346-4 235-35	1.67	5	326.4	0.44	0.06	−0.38	1.68	0.17	−1.51
45	85	234-24	2	5	326.4	0.22	0.03	−0.19	0.80	0.06	−0.74
46	136	236-236	4	6	360.9	0.02	0.00	−0.02	0.11	0.00	−0.11
47	77 110	34-34 236-34	1.38	4.69	315.8	1.09	0.15	−0.94	4.29	0.23	−4.06
48	151	2356-25	3	6	360.9	0.37	0.05	−0.32	1.20	0.19	−1.02
49	135 124 144	235-236 345-25 2346-25	2.74	5.87	356.3	0.06	0.00	−0.06	0.32	0.08	−0.24
50	118 149 106	245-34 236-245 2345-3	1.64	5.32	337.5	0.00	0.03	0.03	0.00	0.00	0.00
51	134 143 114	2356-23 2345-26 2345-4	2.24	5.62	347.8	0.00	0.00	0.00	0.06	0.02	−0.04
52	122 131 133	345-23 2346-23 235-235+	1.45	5.3	336.8	0.02	0.00	−0.02	0.14	0.00	−0.14
53	146 161	235-245 2346-35	2	6	360.9	0.00	0.00	0.00	0.16	0.05	−0.11
54	153	245-245	2	6	360.9	0.07	0.04	−0.03	0.53	0.22	−0.31
55	132 105	234-236 234-34	2.24	5.62	347.8	0.21	0.04	−0.17	0.00	0.00	0.00
56	141	2345-25	2	6	360.9	0.01	0.00	−0.01	0.13	0.05	−0.08
57	179	2356-236	4	7	395.3	0.00	0.00	0.00	0.08	0.03	−0.05
58	137	2345-24	2	6	360.9	0.00	0.00	0.00	0.43	0.21	−0.22
59	176	2346-236	4	7	395.3	0.00	0.00	0.00	0.03	0.01	−0.02
60	138 163	234-245 2356-34	2	6	360.9	0.23	0.14	−0.09	1.15	0.58	−0.57
61	158	2346-34	2	6	360.9	0.02	0.03	0.01	0.15	0.07	−0.08
62	178	2356-235	3	7	395.3	0.00	0.03	0.03	0.09	0.03	−0.05
63	175	2346-235	3	7	395.3	0.00	0.02	0.02	0.00	0.00	0.00
64	187 182	2356-245 2345-246	3	7	395.3	0.00	0.00	0.00	0.12	0.10	−0.02
65	183	2346-245	3	7	395.3	0.00	0.00	0.00	0.05	0.03	−0.02
66	167	245-345	1	6	360.9	0.01	0.00	−0.01	0.08	0.03	−0.05
67	185	23456-25	3	7	395.3	0.00	0.00	0.00	0.01	0.00	−0.01
68	174 181	2345-236 23456-24	3	7	395.3	0.00	0.00	0.00	0.09	0.05	−0.03
69	177	2356-234	3	7	395.3	0.00	0.00	0.00	0.04	0.02	−0.01
70	171 156 202	2346-234 2345-34 2356-2356	2.41	6.71	386.4	0.00	0.00	0.00	0.09	0.05	−0.04
71	173 200 204	23456-23 2346-2356 23456-246	3.87	7.87	425.4	0.00	0.00	0.00	0.01	0.00	−0.01
72	172 192	2345-235 23456-35	2	7	395.3	0.00	0.00	0.00	0.00	0.00	
73	180	2345-245	2	7	395.3	0.00	0.00	0.00	0.17	0.16	−0.01
74	193	2356-345	2	7	395.3	0.00	0.00	0.00	0.01	0.00	−0.01
75	191	2346-345	2	7	395.3	0.00	0.00	0.00	0.00	0.01	0.00
76	199	23456-236	4	8	429.8	0.00	0.00	0.00	0.00	0.00	0.00
77	170	2345-234	2	7	395.3	0.00	0.00	0.00	0.05	0.03	−0.02
78	190	23456-34	2	7	395.3	0.00	0.06	0.06	0.00	0.00	0.00
79	201	2356-2345	3	8	429.8	0.02	0.03	0.01	0.00	0.13	0.13
80	196 203	2345-2346 23456-245	3	8	429.8	0.00	0.00	0.00	0.00	0.00	0.00
81	189	2345-345	1	7	395.3	0.00	0.00	0.00	0.00	0.00	0.00
82	195	23456-234	3	8	429.8	0.00	0.00	0.00	0.02	0.00	−0.02
83	208	23456-2356	4	9	464.2	0.00	0.00	0.00	0.00	0.00	0.00
84	194	2345-2345	2	8	429.8	0.00	0.00	0.00	0.05	0.05	0.01
85	205	23456-345	2	8	429.8	0.00	0.00	0.00	0.00	0.00	0.00
86	206	23456-2345	3	9	464.2	0.00	0.00	0.00	0.03	0.05	0.01
87	OCN	Internal Standard				0.00	0.00	0.00	0.00	0.00	0.00
88	209	23456-23456	4	10	498.6	0.00	0.00	0.00	0.00	0.00	0.00

Hudson River 1254 Pattern C			Hudson River 1260 Process H			Silver Lake 1242 Process M			Silver Lake 1260 Process N		
Autoclaved	Live	Difference	Autoclaved	Live	Difference	Autoclaved	Live	Difference	Autoclaved	Live	Difference
1.71	0.54	−1.17	0.05	2.01	1.96	0.39	0.13	−0.26	0.02	3.30	3.28
3.32	1.30	−2.02	0.74	10.10	9.36	4.61	0.28	−4.33	0.84	3.36	2.52
9.46	1.96	−7.50	3.37	4.17	0.80	1.17	0.11	−1.05	3.54	7.79	4.25
2.44	0.75	−1.69	0.06	0.26	0.19	0.29	0.03	−0.26	0.05	0.54	0.50
0.02	0.01	−0.01	0.00	0.02	0.02	0.01	0.00	−0.01	0.01	0.00	0.00
3.05	0.32	−2.73	0.17	0.03	−0.14	0.37	0.02	−0.35	0.26	0.47	0.21
4.08	0.54	−3.54	0.44	0.05	−0.39	0.43	0.05	−0.38	0.54	0.41	−0.13
1.72	0.27	−1.45	0.02	0.00	−0.02	0.20	0.01	−0.19	0.02	0.00	−0.02
0.43	0.19	−0.24	1.41	1.52	0.11	0.02	0.01	−0.01	1.50	1.20	−0.29
12.41	1.31	−11.10	1.87	4.27	2.40	1.17	0.18	−0.99	2.09	1.13	−0.96
2.06	0.86	−1.21	4.39	5.17	0.79	0.30	0.04	−0.26	4.56	4.34	−0.22
1.28	0.42	−0.86	2.94	5.44	2.51	0.07	0.03	−0.04	2.85	0.98	−1.87
2.04	0.83	−1.21	6.61	5.96	−0.64	0.00	0.06	0.06	7.28	2.63	−4.65
0.43	0.24	−0.19	0.47	1.97	1.49	0.01	0.01	0.00	0.46	0.74	0.27
0.45	0.10	−0.34	0.15	0.47	0.32	0.04	0.00	−0.04	0.14	0.26	0.11
0.99	0.44	−0.56	1.73	3.12	1.39	0.03	0.02	−0.01	1.62	1.02	−0.60
5.04	1.84	−3.19	9.44	2.96	−6.48	0.08	0.03	−0.05	9.79	3.14	−6.66
0.00	0.00	0.00	1.15	0.74	−0.40	0.00	0.01	0.01	1.27	0.27	−1.00
0.90	0.31	−0.59	2.30	0.36	−1.94	0.01	0.00	−0.01	2.28	0.52	−1.76
0.19	0.13	−0.06	2.30	2.76	0.46	0.01	0.01	0.00	2.37	2.10	−0.27
4.27	2.01	−2.27	0.41	0.15	−0.26	0.05	0.00	−0.05	0.39	1.55	1.16
0.17	0.06	−0.11	0.29	0.19	−0.10	0.00	0.00	0.00	0.33	0.11	−0.21
13.85	4.23	−9.62	14.35	8.13	−6.22	0.15	0.08	−0.07	12.84	5.50	−7.34
1.69	0.42	−1.27	1.28	0.25	−1.03	0.01	0.00	−0.01	1.38	0.36	−1.03
0.57	0.21	−0.36	1.05	2.08	1.03	0.01	0.01	0.00	1.02	1.08	0.07
0.00	0.00	0.00	0.37	0.15	−0.22	0.00	0.00	0.00	0.35	0.13	−0.21
0.48	0.40	−0.08	3.20	4.04	0.83	0.01	0.02	0.00	3.57	1.72	−1.85
0.36	0.21	−0.16	2.74	0.80	−1.95	0.00	0.00	0.00	2.83	1.16	−1.67
1.49	0.35	−1.14	0.39	0.07	−0.32	0.01	0.00	−0.01	0.41	0.09	−0.32
0.08	0.03	−0.05	0.97	0.24	−0.73	0.00	0.03	0.03	0.89	0.36	−0.53
0.46	0.27	−0.18	4.05	1.17	−2.89	0.01	0.01	0.00	4.10	2.03	−2.07
0.23	0.14	−0.09	2.11	2.53	0.42	0.01	0.01	0.00	2.17	0.89	−1.29
1.26	0.53	−0.73	2.01	0.52	−1.48	0.01	0.01	0.00	2.00	0.95	−1.05
0.20	0.08	−0.12	0.32	0.23	−0.09	0.00	0.00	0.00	0.30	0.18	−0.12
0.16	0.13	−0.03	1.12	0.55	−0.56	0.00	0.00	0.00	1.04	0.61	−0.43
1.15	0.81	−0.34	9.18	3.27	−5.91	0.02	0.01	0.00	9.05	4.39	−4.66
0.02	0.02	−0.01	0.31	0.18	−0.13	0.00	0.00	0.00	0.26	0.15	−0.10
0.02	0.02	0.00	0.07	0.02	−0.05	0.00	0.00	0.00	0.05	0.03	−0.02
0.00	0.00	0.00	0.16	0.08	−0.08	0.00	0.00	0.00	0.12	0.09	−0.02
0.56	0.40	−0.15	2.94	0.97	−1.97	0.01	0.00	0.00	2.92	1.47	−1.45
0.00	0.00	0.00	1.23	0.72	−0.50	0.00	0.00	0.00	1.06	0.89	−0.17
0.14	0.00	−0.14	2.42	1.46	−0.96	0.03	0.07	0.04	2.31	1.98	−0.32
0.09	0.09	0.01	2.71	1.49	−1.22	0.00	0.00	0.00	2.55	2.03	−0.52
0.01	0.02	0.00	0.07	0.03	−0.04	0.00	0.00	0.00	0.06	0.03	−0.03
0.02	0.04	0.02	0.97	0.58	−0.39	0.00	0.00	0.00	0.97	0.63	−0.34
0.00	0.00	0.00	0.06	0.05	0.00	0.00	0.00	0.00	0.05	0.04	−0.01
0.09	0.11	0.03	1.91	1.13	−0.78	0.00	0.00	0.00	1.83	1.38	−0.45
0.00	0.00	0.00	0.07	0.05	−0.02	0.00	0.00	0.00	0.07	0.03	−0.04
0.04	0.00	−0.04	1.21	0.98	−0.23	0.00	0.00	0.00	1.14	0.91	−0.23
0.00	0.00	0.00	0.00	0.00	0.00	0.00	0.00	0.00	0.00	0.00	0.00
0.00	0.00	0.00	0.00	0.00	0.00	0.00	0.00	0.00	0.00	0.00	0.00

Table A.2 Dechlorination Data (Previously Unpublished)

Column groupings: Columns "Ortho–Total–Weight" are grouped under **Molecular**. The three "1242 Process" blocks (M, Q, H) are **Data from Quensen (previously unpublished)**; each has Autoclaved / Live / Difference. The "1260 Process P" block (0 Time / 28 Days / Difference) and the Peak–Congener columns are **Data from Bedard (previously unpublished)**.

Peak #	IUPAC #	Structure	Ortho	Total	Mol. Weight	M Autoclaved	M Live	M Difference	Q Autoclaved	Q Live	Q Difference	H Autoclaved	H Live	H Difference	Peak	Congener	P 0 Time	P 28 Days	P Difference
1	1	2	1	1	188.7	2.52	7.90	5.37	2.52	12.30	9.77	2.03	3.47	1.43	1		0.0000	0.0000	0.0000
2	3	4	0	1	188.7	0.00	1.21	1.21	0.00	0.52	0.52	0.57	1.24	0.67	2	2	0.0000	0.0000	0.0000
3	4 10	2-2 26	2	2	223.1	3.94	23.18	19.24	3.94	27.64	23.69	3.69	5.99	2.30	3		0.0000	0.0000	0.0000
4	7 9	24 25	1	2	223.1	1.41	1.48	0.07	1.41	3.26	1.85	1.42	2.56	1.14	4	4	0.0000	0.0000	0.0000
5	6	2-3	1	2	223.1	1.75	3.15	1.41	1.75	8.76	7.01	1.79	11.99	10.20	5		0.0000	0.0000	0.0000
6	8 5	2-4 23	1	2	223.1	7.84	19.75	11.91	7.84	1.37	-6.47	7.98	14.17	6.19	6		0.0000	0.0000	0.0000
7	19	26-2	3	3	257.5	0.86	3.13	2.27	0.86	1.74	0.88	0.85	1.51	0.65	7	7	0.0335	0.0000	-0.0335
8	12 13	34 3-4	0	2	223.1	0.16	0.68	0.52	0.16	0.42	0.26	0.19	0.94	0.75	8	8	0.0995	0.0000	-0.0995
9	18 15	25-2 4-4	1.51	2.75	249	9.46	0.43	-9.03	9.46	17.20	7.74	9.22	19.11	9.89	9		0.0000	0.0000	0.0000
10	17	24-2	2	3	257.5	4.55	9.71	5.15	4.55	0.77	-3.78	4.56	8.65	4.09	10	26-2	0.0084	0.0000	-0.0084
11	27 24	26-3 236	2	3	257.5	0.81	1.59	0.78	0.81	1.79	0.98	0.84	1.65	0.81	11		0.0000	0.0000	0.0000
12	16 32	23-2 26-4	2	3	257.5	4.80	4.33	-0.47	4.80	0.19	-4.61	4.80	8.56	3.76	12		0.0000	0.0000	0.0000
13	34 (54)	35-2 (26-26)	1	3	257.5	0.13	0.16	0.03	0.13	0.15	0.02	0.14	0.48	0.33	13		0.0000	0.0000	0.0000
14	29	245	1	3	257.5	0.11	0.04	-0.07	0.11	0.01	-0.10	0.13	0.05	-0.08	14	25-2/4-4	0.2199	1.1292	0.9093
15	26	25-3	1	3	257.5	1.17	1.82	0.64	1.17	3.22	2.04	1.20	6.19	4.98	15	24-2	0.0418	0.0491	0.0073
16	25	24-3	1	3	257.5	0.92	2.72	1.80	0.92	1.37	0.46	0.96	6.64	5.68	16	26-3	0.0090	0.1214	0.1125
17	31	25-4	1	3	257.5	5.57	4.26	-1.31	5.57	7.76	2.19	5.51	11.54	6.03	17	26-4	0.0693	0.1597	0.0904
18	28 (50)	24-4 (246-2)	1	3	257.5	7.48	8.19	0.72	7.48	0.00	-7.48	7.33	15.35	8.01	18		0.0000	0.0000	0.0000
19	33 21 20 53	34-2 234 23-3 25-26	1.12	3.06	259.5	5.84	0.44	-5.40	5.84	1.09	-4.75	5.74	2.49	-3.25	19	26-26	0.0056	0.2004	0.1949
20	22 (51)	23-4 (24-26)	1	3	257.5	2.70	0.47	-2.23	2.70	0.82	-1.88	2.77	3.74	0.97	20		0.0000	0.0000	0.0000
21	45	236-2	3	4	292	1.16	0.07	-1.09	1.16	0.22	-0.94	1.16	1.85	0.69	21	25-3	0.1186	0.7728	0.6542
22	46	23-26	3	4	292	0.60	0.07	-0.53	0.60	0.09	-0.51	0.61	1.06	0.45	22	24-3	0.0421	0.1430	0.1009
23	52 73	25-25 26-35	2	4	292	3.07	0.32	-2.75	3.07	3.38	0.31	3.01	5.70	2.69	23	25-4	0.1003	0.1497	0.0494
24	49	24-25	2	4	292	2.31	0.46	-1.85	2.31	0.77	-1.54	2.32	4.29	1.97	24	24-4	0.1131	0.1309	0.0177
25	47	24-24	2	4	292	1.05	2.06	1.01	1.05	0.41	-0.63	1.51	3.90	2.39	25	25-26	0.1291	1.4524	1.3233
26	48 75	245-2 246-4	2	4	292	1.40	0.00	-1.40	1.40	0.00	-1.40	1.00	0.00	-1.00	26	24-26	0.0964	0.9985	0.9021
27	35	34-3	0	3	257.5	0.08	0.01	-0.07	0.08	0.01	-0.06	0.09	0.06	-0.03	27		0.0000	0.0000	0.0000
28	44	23-25	2	4	292	2.77	0.17	-2.60	2.77	0.22	-2.55	2.70	3.57	0.87	28		0.0000	0.0000	0.0000
29	42 59 37	23-24 236-3 344	0.84	3.42	272.4	3.21	0.26	-2.95	3.21	0.28	-2.94	3.22	3.97	0.75	29	23-26	0.0086	0.0810	0.0725
30	71 41 64 72	26-34 234-2 236-4 25-35	1.75	4	292	2.71	0.00	-2.71	2.71	0.27	-2.45	2.91	2.49	-0.42	30		0.0000	0.0000	0.0000
31	96	236-26	4	5	326.4	0.08	0.00	-0.08	0.08	0.01	-0.07	0.05	0.07	0.03	31A	25-25	1.0951	6.6804	5.5853
32	40	23-23	2	4	292	0.84	0.13	-0.71	0.84	0.07	-0.77	0.87	1.47	0.60	31B	26-35	0.0134	0.1572	0.1438
33	57 70 100 103	235-3 245-3 246-24 246-25 (235-26)	1.4	4.2	298.9	0.19	0.02	-0.17	0.19	0.03	-0.16	0.21	0.10	-0.10	32	24-25	0.2054	1.3848	1.1794
34	58 63	23-35 235-4	1	4	292	0.17	0.05	-0.12	0.17	0.01	-0.15	0.06	0.10	0.04	33	24-24	0.1724	0.8607	0.6884
35	74 (94)	245-4 (235-26)	1	4	292	1.49	0.11	-1.38	1.49	0.40	-1.09	1.52	0.57	-0.96	34		0.0000	0.0000	0.0000

36	70 76	25-34 345-2	1	4	292	3.40	0.06	-3.34	3.40	0.46	-2.94	3.31	0.71	-2.60	35		0.0000	0.0000	0.0000
37	95 102 66	236-25	1.08	4.04	293.5	3.33	0.20	-3.13	3.33	0.61	-2.72	3.38	1.71	-1.66	36		0.0000	0.0000	0.0000
38	55 91	245-26 24-34	2.9	4.95	324.7	0.40	0.12	-0.28	0.40	0.08	-0.32	0.43	0.83	0.39	37A		0.0000	0.0081	0.0081
39	56 60	234-3 236-24	1	4	292	4.95	0.21	-4.74	4.95	0.96	-3.99	4.94	1.63	-3.31	37B	246-26	0.3477	2.2387	1.8910
40	101 90	23-34 234-4 / 245-25 / 235-24	2	5	326.4	1.21	0.12	-1.09	1.21	0.27	-0.94	0.82	0.46	-0.35	38	23-25 / 23-24	0.3899	0.2023	-0.1877
41	99	245-24	2	5	326.4	0.30	0.04	-0.26	0.30	0.09	-0.20	0.37	0.21	-0.16	39		0.0917	0.5520	0.4603
42	150 112 / 119	236-246 / 2356-3 / 246-34	2.67	5	326.4	0.01	0.00	0.00	0.01	0.00	0.00	0.01	0.01	0.01	40/41	26-34/25-35 / 24-35/236-26	0.0074	0.1548	0.1474
43	97 86 / (152)	245-23 / 2345-2 / (2356-26)	2		326.4	0.46	0.02	+0.44	0.46	0.08	-0.38	0.44	0.22	-0.22	42	23-23	0.0367	0.1633	0.1266
44	87 115 / 111	234-25 / 2346-4 / 235-35	1.67	5	326.4	0.44	0.09	-0.35	0.44	0.14	-0.30	0.53	0.47	-0.06	43	246-25	0.0586	0.5465	0.4879
45	85	234-24	2	5	326.4	0.22	0.04	-0.18	0.22	0.06	-0.15	0.25	0.25	0.00	44	246-24	0.0014	0.0174	0.0161
46	136	236-236	4	6	360.9	0.02	0.01	-0.01	0.02	0.01	-0.01	0.02	0.04	0.01	45	235-4	0.0000	0.0000	0.0000
47	77 110 / 151	34-34 236-34 / 2356-25	1.38	4.69	315.8	1.09	0.27	-0.82	1.09	0.25	-0.84	1.19	1.64	0.45	46A	235-26	0.0346	0.3864	0.3518
48	135 124 / 144	235-236 / 345-25 / 2346-25	3	6	360.9	0.37	0.07	-0.30	0.37	0.07	-0.30	0.39	0.46	0.07	47	25-34	0.0749	0.0166	-0.0582
49			2.74	5.87	356.3	0.06	0.02	-0.04	0.06	0.03	-0.04	0.09	0.08	0.00	48A	24-34	0.0000	0.0000	0.0000
50	118 149 / 106	245-34 / 236-245 / 2345-3	1.64	5.32	337.5	0.00	0.05	0.05	0.00	0.03	0.03	0.00	0.12	0.12	48B	236-25/245-26	2.7905	2.8707	0.0802
51	134 143 / 114	2356-23 / 2345-26 / 2345-4	2.24	5.62	347.8	0.00	0.01	0.01	0.00	0.00	0.00	0.01	0.02	0.02	49	236-24	0.0518	0.5508	0.4990
52	122 131 / 133	345-23 2346-23 235-235+	1.45	5.3	336.8	0.02	0.00	-0.02	0.02	0.01	-0.01	0.04	0.02	-0.01	50	23-34/234-4	0.0431	0.0011	-0.0421
53	146 161	235-245 2346-35	2	6	360.9	0.00	0.01	0.01	0.00	0.01	0.01	0.02	0.03	0.01	51	235-25/236-23	0.7884	3.5480	2.7596
54	153	245-245	2	6	360.9	0.07	0.03	-0.04	0.07	0.03	-0.04	0.10	0.08	-0.02	52	245-25/235-24	0.0000	0.0000	0.0000
55	132 105	234-236 234-34	2.24	5.62	347.8	0.21	0.03	-0.18	0.21	0.07	-0.13	0.24	0.14	-0.10	53		3.8751	1.5873	-2.2878
56	141	2345-25	2	6	360.9	0.01	0.00	-0.01	0.01	0.01	-0.01	0.02	0.01	-0.01	54	245-24/236-35	0.1419	0.8057	0.6638
57	179	2356-236	4	7	395.3	0.00	0.00	0.00	0.00	0.00	0.00	0.01	0.01	0.00	55	246-34/236-246	0.1291	0.8846	0.7555
58	137	2345-24	2	6	360.9	0.00	0.03	0.03	0.00	0.03	0.03	0.05	0.05	0.00	56A	235-23	0.0000	0.0041	0.0041
59	176	2346-236	4	7	395.3	0.00	0.00	0.00	0.00	0.00	0.00	0.00	0.01	0.00	57A	245-23/2356-26	0.1339	0.1979	0.0640

(Continued)

Table A.2 (Continued)

| | | | Molecular | | | Data from Quensen (previously unpublished) | | | | | | | | | Data from Bedard (previously unpublished) | | | | |
| | | | | | | 1242 Process M | | | 1242 Process Q | | | 1242 Process H | | | | | 1260 Process P | | |
Peak #	IUPAC #	Structure	Ortho	Total	Weight	Autoclaved	Live	Difference	Autoclaved	Live	Difference	Autoclaved	Live	Difference	Peak	Congener	0 Time	28 Days	Difference
60	138 163	234-245 / 2356-34	2	6	360.9	0.23	0.09	-0.14	0.23	0.06	-0.17	0.19	0.20	0.01	58A	234-25/ 235-35	1.0940	0.9758	-0.1182
61	158	2346-34	2	6	360.9	0.02	0.01	-0.01	0.02	0.01	-0.01	0.02	0.03	0.00	59		0.0000	0.0000	0.0000
62	178	2356-235	3	7	395.3	0.00	0.00	0.00	0.00	0.00	0.00	0.01	0.01	0.00	60	236-236	1.6670	1.5242	-0.1428
63	175	2346-235	3	7	395.3	0.00	0.00	0.00	0.00	0.00	0.00	0.00	0.00	0.00	61	236-34	2.5697	2.1528	-0.4169
64	187 182	2356-245 / 2345-246	3	7	395.3	0.00	0.01	0.01	0.00	0.00	0.00	0.00	0.01	0.01	62	245-246	0.0063	0.1233	0.1170
65	183	2346-245	3	7	395.3	0.00	0.00	0.00	0.00	0.00	0.00	0.00	0.00	0.00	63	2356-25	0.0000	0.0000	0.0000
66	167	245-345	1	6	360.9	0.01	0.00	0.00	0.01	0.00	0.00	0.01	0.02	0.01	64	235-236	2.8072	3.0172	0.2100
67	185	23456-25	3	7	395.3	0.00	0.00	0.00	0.00	0.00	0.00	0.00	0.00	0.00	65	345-25	1.4314	2.7495	1.3181
68	174 181	2345-236	3	7	395.3	0.00	0.00	0.00	0.00	0.00	0.00	0.00	0.01	0.01	66	2346-25	1.3219	1.0271	-0.2948
69	177	23456-24 / 2356-234	3	7	395.3	0.00	0.00	0.00	0.00	0.00	0.00	0.00	0.00	0.00	67	2356-24/ 235-34/ 234-35	0.0461	0.3820	0.3359
70	171 156 202	2346-234 / 2345-34 / 2356-2356	2.41	6.71	386.4	0.00	0.00	0.00	0.00	0.00	0.00	0.01	0.01	0.00	68		0.0000	0.0000	0.0000
71	173 200 204	23456-23 / 2346-2356 / 23456-246	3.87	7.87	425.4	0.00	0.00	0.00	0.00	0.00	0.00	0.00	0.00	0.00	69	236-245/ 245-34	7.4898	4.6905	-2.7993
72	172 192	2345-235 / 23456-35	2	7	395.3	0.00	0.00	0.00	0.00	0.00	0.00	0.00	0.00	0.00	70		0.0000	0.0000	0.0000
73	180	2345-245	2	7	395.3	0.00	0.01	0.01	0.00	0.01	0.01	0.01	0.02	0.00	71	2356-23/ 2345-26	0.6664	0.6831	0.0167
74	193	2356-345	2	7	395.3	0.00	0.00	0.00	0.00	0.00	0.00	0.00	0.00	0.00	72	2346-23/ 235-235	0.0906	0.2988	0.2082
75	191	2346-345	2	7	395.3	0.00	0.00	0.00	0.00	0.00	0.00	0.00	0.00	0.00	73	235-245/ 234-34/ 234-236	1.3809	1.6151	0.2342
76	199	23456-236	4	8	429.8	0.00	0.00	0.00	0.00	0.00	0.00	0.00	0.00	0.00	74		3.4996	3.2294	-0.2702
77	170	2345-234	2	7	395.3	0.00	0.00	0.00	0.00	0.00	0.00	0.00	0.00	0.00	75	245-245	10.3144	5.6051	-4.7093
78	190	23456-34	2	7	395.3	0.00	0.00	0.00	0.00	0.00	0.00	0.05	0.00	-0.05	76		0.0000	0.0000	0.0000
79	201	2356-2345	3	8	429.8	0.02	0.04	0.02	0.02	0.01	-0.01	0.06	0.07	0.01	77	2345-25	2.8813	0.9923	-1.8890
80	196 203	2345-2346 / 23456-245	3	8	429.8	0.00	0.01	0.01	0.00	0.00	0.00	0.01	0.01	0.00	78	2356-236	1.3482	1.5010	0.1528
81	189	2345-345	1	7	395.3	0.00	0.00	0.00	0.00	0.00	0.00	0.00	0.00	0.00	79/80/81	2345-24/ 234-235/ 2346-236	0.9052	1.3102	0.4050

Table (left section, compounds 82–88):

No.	Ref	Structure	Cl₁	Cl₂	MW								
82	195	23456-234	3	8	429.8	0.00	0.00	0.00	0.00	0.00	0.00	0.00	0.00
83	208	23456-2356	4	9	464.2	0.00	0.00	0.00	0.00	0.00	0.00	0.00	0.00
84	194	2345-2345	2	8	429.8	0.00	0.00	0.00	0.00	0.00	0.00	0.00	0.00
85	205	23456-345	2	8	429.8	0.00	0.00	0.00	0.00	0.00	0.00	0.00	0.01
86	206	23456-2345	3	9	464.2	0.00	0.00	0.01	0.01	0.01	0.01	0.00	0.00
87	OCN	Internal Standard				-0.01	3.96	3.96	3.96	0.00	0.00	0.00	0.00
88	209	23456-23456	4	10	498.6	0.00	0.00	0.00	0.00	0.00	0.00	0.00	0.00

Table (right section, compounds 82–117):

No.	Structure			
82	234-245/ 2356-34/ 236-345	13.4171	8.1080	-5.3092
83	2346-34	0.4071	0.3453	-0.0618
84		0.0000	0.0000	0.0000
85	2356-235	0.7385	1.0219	0.2834
86		0.0000	0.0000	0.0000
87		0.0000	0.0000	0.0000
88	2356-245	4.2383	4.3262	0.0879
89	234-234	0.0222	0.0464	0.0242
90	2346-245	1.9811	1.9019	-0.0792
91	245-345	0.1840	0.1019	-0.0821
92	23456-25	0.4692	0.3312	-0.1380
93	2345-236	3.4564	2.3211	-1.1353
94	2356-234	1.8238	1.9136	0.0898
95	2345-34/ 2346-234	1.4201	1.1506	-0.2695
96	2356-2356/ 234-345	0.2478	0.2476	-0.0002
97		0.0000	0.0000	0.0000
98		0.0000	0.0000	0.0000
99	2346-2356	0.1659	0.1610	-0.0049
100	2345-235	0.7797	0.6246	-0.1551
"101/102"	2345-245/ 2346-2346	0.0000	0.0000	0.0000
		8.8472	7.1189	-1.7284
103	2356-345	0.1824	0.1404	-0.0421
104	2346-345	0.0530	0.0334	-0.0196
105	23456-236	0.1931	0.1480	-0.0451
106	2345-234	3.9469	2.8833	-1.0636
107	23456-34	0.5106	0.4185	-0.0921
108		0.0000	0.0000	0.0000
109	2345-2356	1.1828	1.1289	-0.0540
110	2345-2346/ 23456-245	1.1898	1.1462	-0.0436
111		0.0000	0.0000	0.0000
112	23456-234	0.6635	0.6176	-0.0460
113		0.0000	0.0000	0.0000
114		0.0000	0.0000	0.0000
115	2345-2345	1.5347	1.4173	-0.1174
116		0.0000	0.0000	0.0000
117	23456-2345	1.2341	1.1984	-0.0357

11

Microbial Forensics

Ioana G. Petrisor, Rachel A. Parkinson, Jacqui Horswell,
James M. Waters, Leigh A. Burgoyne, David E.A. Catcheside,
Winnie Dejonghe, Natalie Leys, Karolien Vanbroekhoven,
Priyabrata Pattnaik, and Duane Graves

Contents

11.1   INTRODUCTION

Microbial forensics is the focusing of microbiology, virology, biochemistry, and molecular biology for use in environmental forensic investigations. Microbial forensics provides a means by which a microbial signature is used to trace a contaminant source, similar to the use of DNA in criminal forensics (Budowle *et al.*, 2003).

The purpose of this chapter is to:

- provide an overview of the relationship between soils and microbial forensics, which represent the most common media from which microbial samples are obtained;
- present traditional microbial techniques available to trace specific microbial source and to use the changes recorded in microbial community to possibly track contamination, and
- present an overview of emerging microbial techniques (DNA fingerprinting) available to identify the source and potentially the age of a contaminant release.

Examples of the application of these techniques in forensic investigations are subsumed in these chapter sections, including their successful use in criminal forensic and in environmental studies related to the effect of different contaminants (metals) on the structure and function of microbial communities.

11.2   SOIL AND MICROBIAL FORENSICS

Soil is the primary media via which microbial analysis is available in forensic investigations (Stotzky, 1997). The specific composition of soil varies widely due to the presence of these components in different proportions at different geographical locations (Liesack *et al.*, 1997; Murray and Tedrow, 1992). For many soils, the organic component does not exceed 5% by volume. Water makes up about 20–30% of the average soil volume and is essential for the survival of the soil microbial community and plant life (Wood, 1995).

Analysis of soil microorganisms has historically been ignored by the environmental forensic scientific community. This is primarily due to the limitations of traditional culturing techniques, which allow only a small subset of organisms to be isolated and characterized (Amann *et al.*, 1995). A limited number of studies have been performed that use soil microbiology for environmental forensic purposes, using culturing (van Dijck and van de Voorde, 1984), enzymatic analysis (Thornton and McLaren, 1975), and functional diversity analysis (Omelyanyuk *et al.*, 1999). However, the rapid growth of molecular biology has resulted in techniques that can circumvent the requirement to isolate and culture microorganisms as a prerequisite to identification. Soil microbial diversity is now routinely characterized using simple molecular techniques based on variations in microbial DNA (Tiedje *et al.*, 1999). To date, molecular analysis of microbial diversity and community composition has been used to analyze microbial populations in many diverse environments including marine and fresh waters (Donachie *et al.*, 2002), soils (Buckley and Schmidt, 2001), composts (Nogales *et al.*, 2001; LaMontagne *et al.*, 2002), and landfills.

Bacteria are the most numerous component of the soil microbial population. It has been estimated that there may be as many as 10^9 bacterial cells per gram of soil (Harris, 1994), but it is widely accepted that the majority of soil bacteria, possibly as many as 99%, cannot be cultured using traditional laboratory media–based techniques (Amann *et al.*, 1995). This is due to the complex nature of the environments that soil provides, which is impossible to simulate in the laboratory.

The number of bacteria found in soil has been shown to decrease with depth (Zhou *et al.*, 2002). The vast majority are found in the top 10 cm or root-zone, with a 1000-fold decrease in numbers at a depth of 135 cm (Wood, 1995) where there is a significant reduction in nutrients, particularly in carbon availability and quality.

11.3   TRADITIONAL AND EMERGING MICROBIAL ANALYSIS

Numerous microbiological techniques, both traditional and emerging, are available to track a specific microbial source or to use changes in microbial communities to identify a particular contaminant. Microorganisms are ideal indicators of contamination as they have a simple cellular structure that enables them to adapt easily and survive in any environment (such as high temperature, salinity, extreme pHs, high concentrations of contaminants, lack of most nutrients, etc.). Microbial communities are also able to assimilate a large variety of chemicals (such as PCBs or MTBE) into their structure (Truper, 1992).

In addition to studying effects of inorganic pollution on specific soil microbial processes by classic tools, more recently, a broad range of cultivation-independent techniques has found its way in microbial community studies (Griffiths *et al.*, 1997; Sandaa *et al.*, 1999; Müller *et al.*, 2001; Feris *et al.*, 2003; Turpeinen *et al.*, 2004). These culture-independent techniques use DNA structure and/or signal molecules to describe microbial communities and their dynamics. Signal molecules reflecting microbial community structure are lipids and nucleic acids.

11.3.1   Traditional Microbial Source Tracking Methods

Traditional microbiological techniques applicable in environmental forensic investigations include:

- Indicator bacteria and virus identification;
- Microbial ratios;
- Phenotypic methods (MAR and immunological approaches);
- Soil microbial biomass;
- Soil microbial biomass growth indicators;
- Community level physiological profiling (CLPP); and
- Phospholipid fatty acid analysis (PLFA).

These classic methods all based on culturing different microbial groups on artificial media. As a result, the use of these methods for forensic purposes can be limited to those microbial communities that grow on artificial media (i.e., cultures common to humans and warm-blooded animals).

11.3.2   Indicator Bacteria and Virus Identification

A simple way to track pollution is to identify indicator microorganisms. Ideally, an indicator microorganism should be nonpathogenic, simple and rapidly detectable, and with similar survival characteristics as the pathogens of concern, being strongly associated with the presence of pathogenic microorganisms (Scott *et al.*, 2002). The detection of an indicator microorganism can predict the presence of a pathogenic organism and evaluate the potential risk associated with pathogenic microbes. The method consists in the isolation and cultivation of the selected indicator microorganism from the environmental sample.

Colony morphology, phage sensitivity, growth on selective medium, fermentation of or growth on various sugars, fats, or proteins are classical microbiology techniques for distinguishing bacterial strains. For some potential agents, these phenotypic characteristics are well known for several distinct strains. For others, the suite of known distinguishing phenotypic characteristics may be very limited and of little use. For example, *B. anthracis* strains have almost no distinguishing phenotypic characteristics. Phenotypic characterization of viruses is more difficult because of the complexity of growing viruses; however, serological characteristics are often useful for screening virus strains.

Protein composition investigated by one- and two-dimensional separation techniques may reveal proteins that are unique to specific strains. Allelic variability among the same enzyme expressed in different strains has been detected using electrophoretic techniques. Similarly, immunoblotting, the reaction of electrophoretically separated proteins to specific antisera or monoclonal antibodies, to detect individual target proteins may separate strains of some bacteria (Grunow *et al.*, 2000; Steichen *et al.*, 2003). However, these techniques must be used with caution since growth conditions can dramatically affect protein expression. Further minor changes in amino acid sequence in a protein may not affect immunological reactivity or electrophoretic mobility. Native clinical and environmental samples may not contain enough of the agent for testing without growth and isolation of the suspect agent. With the widespread availability and greater reliability of genetic subtyping techniques, protein analysis finds increasingly limited use.

The most common examples of using these traditional microbiological techniques come from tracking fecal pollution of water, in order to distinguish between the animal and the human sources. For such purposes, microorganisms such as *Escherichia coli*, *Enterocossus sp.*, and *Clostridium perfrigens* are used to distinguish between different sources of feces in water. The use of each of these microorganisms has advantages and limitations. Thus, *E. coli* which is a very well-known species of non-human pathogenesis may not be a reliable indicator of fecal pollution in tropical and subtropical climates due to its ability to replicate in contaminated soils (Desmarais *et al.*, 2002). The use of different *Enterococcus* species, usually including *Enterococcus faecalis*, *Enterococcus faecium*, *Enterococcus durans*, *Enterococcus galinarum*, and *Enterococcus avium*, implies the possibility to perform the cultivation in saline (6.5% NaCl) and alkaline (pH 9.6) conditions. The *Enterococcus* species are also resistant to high temperatures (45 °C). Yet, environmental reservoirs of these microbial species exist and their re-growth may be possible once introduced in the environment (Desmarais *et al.*, 2002). *Clostridium perfrigens* is used as an indicator particularly in situations where the prediction of the presence of viruses or fecal pollution is required (Fujioka and Shizumura, 1985). Another indicator bacterial species is *Bifidobacterium spp.*, an obligate anaerobic, non-spore-forming bacterium. The bacteria from this group are used as indicator of human sources of fecal pollution, as they represent a major component of human intestine and are rarely found in animals. Even when they are found in animals, they tend to be isolated at different frequencies than from humans (Gavini *et al.*, 1991). Apart from their abundance in human feces, the ability to ferment sorbitol of only the human isolates is also of advantage for their forensic use. Additionally *Bifidobacteria* are strictly anaerobic, which means that once released in the environment they are not able to reproduce, so their presence may be used as evidence of recent pollution events. Their poor survival ability is also a limitation for their forensic use. New techniques with increasing sensitivity and specificity of detection should be developed.

Human enteric viruses can be used to track fecal pollution of water. Many such viruses are specifically associated with the human gastrointestinal tract. Valuable information for water quality could be obtained by monitoring directly for human pathogens. However, the limitation is related to our gaps in knowledge about cultivation conditions for many such species. The need for cultivation is related to the estimation of viable species of threat to human health. Thus, the inability to detect an enteric virus or any other selected microorganism cannot be considered evidence of its absence.

Non-human viruses such as the bacteriophage of *Bacteroides fragilis* (an obligatory anaerobic Gram-negative bacterium) can also be used for tracking fecal pollution in water. The mentioned bacterium is commonly found in both humans and animals (intestines). However, one *B. fragilis* strain HSP40 is found in 10% of human samples and not detected in animal samples. Thus, the bacteriophage that specifically infected this strain could be used as indicator for human feces pollution. Jofre *et al.* (1989) found a significant correlation between the numbers of *B. fragilis* phages and human enteric viruses. The presence of such phages in an environmental sample is a clear indication of human pollution sources as the phages do not replicate in the environment. The method is highly specific for sources associated with human pollution. The main limitations of this method are related to the absence of *B. fragilis* phage in highly polluted waters of some areas and the inherent difficulty in performing the assay (Havelaar, 1993; Puig *et al.*, 1999).

F-specific RNA coliphages represent another example of viruses that are used in bacterial source tracking. The apparent differences in host tropism for the various groups of F+ RNA coliphage have been utilized to predict the presence of fecal contamination based on the presence or absence of a particular group of coliphage. *F-specific RNA coliphage* are viruses that infect *E. coli* (attack only bacteria that possess an F plasmid). Coliphages are viral indicators of enteric pathogens in environmental samples. The F + RNA *(Leviviridae)* family of coliphages can distinguish human and animal waste contamination by typing isolates into one of four subgroups using genetic hybridization or serological methods. The coliphage use in differentiating sources of fecal pollution is based on the observation that animal and human feces contain different serotypes of RNA coliphages (Furuse *et al.*, 1981; Hsu *et al.*, 1995). Thus, there are four main subgroups of coliphages (group I through IV), members of the groups II and III are highly associated with human fecal contamination and domestic sewage, while group IV is usually associated with animals and livestock. Group I coliphages are common to both humans and animals. The detected phages can be further characterized as being human or animal derived by immunological or genetic methods. In general, F + RNA bacteriophage predominate in domestic sewage (Scott *et al.*, 2002). While it may be possible to distinguish between human and most animal wastes by serotyping F+ coliphage isolates, there is a problem with separation between human serotypes and serotypes associated with pigs, which can contain some groups II and III. In addition, some animal groups may not be detected by this method alone because some do not have F+ coliphage associated with their *E. coli*. The main limitation of using coliphage for bacterial source tracking in water relates to the fact that the coliphage persists in the environment for less than a week and survival is a function of sunlight and water temperature. Ultraviolet light denatures the virus and below 25 °C F-pilus synthesis ceases. The coliphage does not replicate in the environment, only in the presence of F-pilus *E. coli*, and is not found in sediments, just in the water column. DNA fingerprinting of F + RNA coliphages may be able to resolve some of the problems with serological typing. Another limitation relates to the presence of these phages in only a limited percentage of human fecal samples. Due to these limitations, this technology has only been applied in a few locations worldwide.

11.3.3 Microbial Ratios

The ratio of different microorganisms are used in forensic investigations to track the source and potentially the age of a release. Environmental conditions will determine the proportion of different microbial groups present. Thus, the most adapted groups will prevail in a certain environment. The introduction of a certain contaminant into an environment will trigger a change in the proportion of different microbial groups, in the favor of those capable of metabolizing the contaminant. A main advantage of using this method is the ability to provide rapid results with minimal expertise. While the traditional fecal coliform–fecal streptococcus ratio is no longer considered reliable for accurate source identification, their ratios may be useful as a general indicator of human vs non-human fecal bacterial contamination, and the ratio concept could become more reliable if other microbes were used in developing the ratios (e.g., *Bacteroides*, *Prevotella*, and *Clostridium*). The most common application is related to tracking fecal pollution in water. Thus the ratio of fecal coliforms to fecal streptococci has been used to differentiate faecal contamination from human and animal sources (Feachem, 1975). If the ratio is higher than 4.0, this is usually indicative of human pollution (Geldreich and Kenner, 1969).

The main limitation in this technique is that only very few microbial species can be analyzed to provide useful information from a forensic standpoint. The approach may also be unreliable in different environmental conditions due to variable survival rates of streptococci species, variation in detection methods, and sensitivity to water treatment (Clesceri et al., 1998). Samples should be collected soon after the contamination has occurred and more than 100 fecal streptococci/100 mL should be present in the sample. Moreover, due to the potential growth of coliforms in tropical environments, the method application in such areas is questionable.

The application of the ratios between different microbial groups for tracking contaminant passage through an environment holds great potential. In order to detect changes in the microbial structure caused by a contaminant, we need first to isolate, identify, and characterize the specific microbial groups that are most impacted by the presence of contamination. The most useful of these will be the species with ability to grow and consume as direct substrate or co-metabolite the contaminant.

11.3.4 Phenotypic Methods

The use of phenotypic methods for microbial source tracking is based on the existence of phenotypic differences within different lineages of microorganisms. Such differences are usually connected with traits acquired from exposure to different environments, such as multiple antibiotic resistance patterns, cell surface or flagellar antigens, as well as biochemical variations in the utilization of different substrates from a particular environment (Scott et al., 2002). Several commonly used phenotypic methods to discriminate between different microbial groups include multiple antibiotic resistance, immunological methods, biochemical tests, outer membrane protein profiles, phage susceptibility, fimbriation, bacteriocin production and susceptibility (Barenkamp et al., 1981). The main limitation of phenotypic discriminatory methods is that phenotypes can be unstable,

and that the technique can have low sensitivity at intra-species level (Scott et al., 2002).

11.3.4.1 Multiple Antibiotic Resistance (MAR)

One of the simplest phenotypic characterization tools for bacteria involves challenging the bacterium to a suite of antibiotics. Multiple Antibiotic Resistance testing is a simple and powerful tool (Wiggins, 1996). Different strains of the same species may display different antibiotic sensitivities. While differential antibiotic sensitivity is not always present among strains of the same bacterial species, the occurrence of this characteristic, when it does occur, provides a simple method to compare samples from multiple locations for the presence of the same strain. This method should not be used as an absolute test to discriminate among strains; however, for some potential agents, it represents a useful screening tool (Cavallo et al., 2002; Coker et al., 2002).

By subjecting the microbial species from an environment to different types, concentrations, and frequencies of antibiotics (commonly associated with human and animal therapy), over time, selective pressure discerns (put in evidence) for the microbiota that possess specific "fingerprints" of antibiotic resistance. An example is the application of pig slurry to soil where pigs have been fed with antibiotics.

The method is based on the isolation and culturing of a target organism, followed by the replica plating of isolates on media with various antibiotics at different concentrations. Usually, each microbial isolate is tested on 30 to 70+ antibiotic concentrations. Fecal bacteria are grown in wells in microtiter trays and then replica-plated onto a series of agar plates, each containing one specific antibiotic concentration. After incubation, each isolate is scored for growth or no growth on each plate and the resultant resistance pattern can be used in source differentiation. Thus, the antibiotic resistance profile of the cultured microorganisms is obtained and specific fingerprints are generated. The results are analyzed by discriminate (cluster) analysis and compared to a reference database for establishing the source.

MAR often uses *E. coli* from known fecal samples for comparison with unknown *E. coli* bacteria in water samples. To determine the resistance profile, the *E. coli* bacteria are exposed to numerous antibiotics at different concentrations and susceptibility (growth or lack of growth) to the antibiotic is noted, a resistance pattern emerges that can be used to identify the source. Bacteria from human sources are more resistant to antibiotics than bacteria from wild animal sources because humans more commonly undergo antibiotic therapy. Bacteria from domestic animals have intermediate MAR indices and different MAR patterns than those from humans.

Successful applications of this technique in discriminating *E. coli* or fecal streptococci isolated from specific animal species have been reported (Hagedorn et al., 1999; Wiggins et al., 1999; Harwood et al., 2000). While domestic and wildlife *E. coli* bacteria have significantly less resistance to antibiotics, the type of animal can often be determined by analyzing the type of antibiotic resistance and the concentration of antibiotic necessary to cause resistance. Moreover, Kaspar and Burgess (1990) reported a larger MAR of *E. coli* isolated from urban area, than from rural ones, confirming the presence of human isolates.

A limitation of MAR is associated with cultivating the microbial species and therefore being restricted to species able to grow on artificial media, which represent a tiny proportion of microbial variability in nature. The method is time-intensive and uses a complicated and costly procedure. Parveen et al. (1999) stated that antibiotic resistance patterns of bacteria are influenced by selective pressure and therefore may be different temporally and geographically. Certain antibiotic sensitivity may not be useful when the studied isolates show no significant resistance patterns, although they originate from different animal species. Another limitation is that antibiotic resistance is often carried on plasmids that can be lost from the cells through cultivation/storage. Variations in specific sensitivities of strains from different locations may also appear due to variable antibiotics used for human and livestock species (Scott et al., 2002). However, large databases compiled from multiple organisms from a large geographic area may overcome these limitations.

11.3.4.2 Immunological Methods

The basis of immunological approaches is the observation that different serotypes of bacterium (e.g., of E. coli) are associated with different animal sources, even when many serotypes are shared among humans and animals (Hartly et al., 1975; Bettelheim et al., 1976). A serotype is an antigenic property of a cell (e.g., bacteria) or virus identified by serological methods. Microorganism serogrouping is based on the presence of various somatic (O) antigenic determinants.

Microbial serogrouping has been successfully used to differentiate E. coli from different sources (Crichton and Old, 1979; Gonzalez and Blanco, 1989), especially human vs animal. The use of F-specific RNA coliphages to differentiate between human and animal waste contamination (described before) is an example of the practical application of this method to track sources of fecal pollution in water. An important limitation of this method is the necessity for a large bank of antisera. It is suggested to use this method in conjunction with other methods (e.g., ribotyping), allowing the testing of a limited number of serotypes.

Immunological methods are useful for detecting agents but often lack the discriminating power to identify specific strains (Fatah et al., 2001; De et al., 2002). The general operating principle of immunological techniques is that biomolecules (proteins, lipoproteins, glycoproteins, polysaccharides, etc.) will elicit an immunological response when presented to a host animal. The antibodies produced in response to exposure to these biomolecules (antigens) can be isolated and used to detect the agent. Antibodies are often linked to a chromofluor or enzyme directly or indirectly through an anti-antibody. In any of several permutations, the agent-specific antibody will bind with and permit detection of the agent; however, the amplifying power, strain specificity, and detection limits do not match those of even the least sophisticated PCR-based DNA fingerprinting methods.

11.3.5 Soil Microbial Biomass and Basal Respiration

Soil microbial biomass, which plays an important role in nutrient cycling and ecosystem sustainability, has been found to be sensitive to the presence of particular contaminants, including heavy metals (Giller et al., 1998; Huang and Khan, 1998). Carbon dioxide evolution, the major product of aerobic catabolic processes in the carbon cycle, is also commonly measured and indicates the total carbon turnover. The metabolic quotient qCO_2 (i.e., the ratio of basal respiration to microbial biomass) is inversely related to the efficiency with which the microbial biomass uses the indigenous substrates (Anderson and Domsch, 1990) and can be a sensitive indicator to reveal heavy metal toxicity under natural conditions (Wardle and Ghani, 1995). Yao et al. (2003) studied the microbial biomass and activity of seven Chinese paddy soils with different heavy metal concentrations in the vicinity of a Cu-Zn smelter. The plots situated 1 (available

[Cu]: $45\,\mu g/g$; available [Zn]: $64\,\mu g/g$), 1.5 (available [Cu]: $40\,\mu g/g$; available [Zn]: $56\,\mu g/g$), and 5 km (available [Cu]: $23\,\mu g/g$; available [Zn]: $32\,\mu g/g$) from the smelter contained nearly 10 times more microbial biomass-C (C_{mic}) than the most polluted soil. The C_{mic}/organic C (C_{org}) ratio varied widely and showed an increased trend with the decrease of heavy metal content. Basal respiration was lowest in the most polluted plot and highest in the plots situated 0.4 (available [Cu]: $166\,\mu g/g$; available [Zn]: $140\,\mu g/g$) and 0.6 km (available [Cu]: $116\,\mu g/g$; available [Zn]: $72\,\mu g/g$) from the smelter, but was not correlated to organic carbon or microbial biomass C. The metabolic quotient was six times higher in the most polluted soil than in the remotest soil (5 km), and was significantly correlated with both Cu and Zn concentrations in all soils (Yao et al., 2003). Brookes and McGrath (1984) also found a higher qCO_2 on polluted, as compared to uncontaminated, soils. Chander and Brookes (1991) and Bardgett and Saggar (1994) reported a doubling of qCO_2 upon heavy metal contamination through sewage sludge amendment, and Ortiz and Alcaniz (1993) found an elevated qCO_2 after sewage sludge amendment which partially may be attributed to heavy metals. However, Bååth et al. (1991) found a slight decrease of qCO_2 with increased heavy metals. In the study of Bååth et al. (1991) the mean qCO_2 was 3.6 for slightly polluted soils ($<400\,\mu g\,Cu/g$), and only 3.2 for heavily polluted soils ($\sim 10000\,\mu g\,Cu/g$). It was suggested that it is possible that at low rates of contamination and short-term exposure the microflora may compensate by higher turnover (respiration), while at high rates of contamination and long-term pollution action the nature of the response is different, e.g., through a community shift. Therefore the qCO_2 can only be used for assessing heavy metals effect only with similar soils (Anderson and Domsch, 1990; Bååth et al., 1991).

The growth characteristics of soil microbial biomass after substrate addition have been suggested as a potentially useful forensic indicator of metal toxicity (Haanstra and Doelman, 1984). Nordgren et al. (1988) investigated soils polluted by smelter emissions and found that the lag time before the onset of exponential microbial growth upon glutamic acid addition increased with increasing heavy metal concentrations, whereas the effect on the specific microbial growth rate was less clear-cut. Dahlin et al. (1997) did not only notice an effect of the heavy metals on the lag time before the exponential growth began but also a higher specific growth rate compared with microbial growth in the control soil that was not polluted with Cd, Cr, Cu, Pb, and Zn. An increase in the lag time before the onset of exponential microbial growth, upon addition of an easily degradable substrate, can so be seen as a sensitive microbial indicator of metal pollutions of soil (Haanstra and Doelman, 1984; Nordgren, 1988; Palmborg and Nordgren, 1996).

The effects of metal contamination on the structure of soil microbial communities have been shown by using bioindicators such as the isolation and identification of bacterial strains (Barkay et al., 1985), groups of organisms and microbial mediated processes (Dahlin et al., 1997). The population size of rhizobia has been found to decrease as a result of metal contamination. For example, in a long-term sewage sludge experiment on a near-neutral soil, there were less than 10^2 cells/g soil of Rhizobium leguminosarum bv. trifolii in a soil with $226\,\mu g/g$ Zn but 10^4 cells/g soil in the uncontaminated control plot (Chaudri et al., 1993). In an experiment at Woburn (UK), Giller et al. (1989) noticed that as a result of past sludge application the total nitrification system of the soil was lost due to the loss of the effective strains of Rhizobium leguminosarum bv. trifolii. The acetylene reduction potential of heterotrophic N_2-fixing soil bacteria like rhizobia and blue-green algae was shown to be sensitive to heavy metals (Brookes *et al.*, 1986; Lorenz *et al.*, 1992; Dahlin *et al.*, 1997). Several studies using plate count techniques have demonstrated an increase in Gram-negative bacteria (Doelman and Haanstra, 1979; Barkay *et al.*, 1985) and a shift in the composition of fungal species toward a more metal-tolerant community in metal-contaminated soils (Jordan and Lechevalier, 1975). However, investigations which involve the cultivation of microorganisms on agar plates are limited as only a minor part of the community is studied and as the majority of soil microorganisms are nonculturable with current techniques.

The urgent problem of the continuous enrichment of soil with heavy metals has also led to exploit different types of enzyme activities as biochemical indicators of soil pollution and soil quality (Nannipieri, 1995). Tyler (1981) reported a negative correlation between enzyme activities in soils and their heavy metal contents, especially Cu, Pb, and Zn. In contrast, in a study of Brunner and Schinner (1984), Pb and Cd caused increasing and decreasing effects on xylanase, urease, and phosphatase activities depending on pollution level and duration of the laboratory experiment. Also in a study of Kandeler *et al.* (1996), enzyme activities and microbial biomass decreased with increasing heavy metal pollution, but the amount of decrease in enzymatic activity differed among the enzymes. More particularly, when three soil types were experimentally contaminated with heavy metals at four different levels (light pollution: 300 mg Zn/kg soil, 100 mg Cu/kg soil, 50 mg Ni/kg soil, 50 mg V/kg soil, and 3 mg Cd/kg soil; medium pollution: twofold concentrations; heavy pollution: threefold concentrations; uncontaminated control), univariate statistical analysis revealed that soil contamination with 300 ppm Zn (and other metals) significantly reduced microbial biomass and soil enzyme activities involved in the N-, P-, and S-cycling. In particular, arylsulfatase and phosphatase activities are dramatically affected and decrease to a level of a few percent of their activities in the corresponding unpolluted controls. In contrast, respiration and enzymes involved in C-cycling (cellulose, xylanase, β-glucosidase) did not change significantly at the lowest contamination level. Cellulase and xylanase are produced mainly by saprophytic fungi in aerobic environments (Alexander, 1977). Since microbial biomass was reduced and since cellulase and xylanase were not significantly influenced, light heavy metal contamination probably shifted the balance of the soil microbial community from bacteria to fungi. This hypothesis was confirmed in another study conducted by Kandeler *et al.* (2000), on the same soils but using PLFA to study the impact of different metal loadings on the soil microbial community. The higher ratio of fungal to bacterial PLFAs of the medium polluted soils clearly demonstrated that heavy metal–resistant fungi can survive in the medium polluted soil (Kandeler *et al.*, 2000). The data suggest that aside from the loss of rare biochemical capabilities heavy metal-contaminated soil lose common potential biochemical properties necessary for the functioning of the ecosystem. Although Kandeler *et al.* (2000, 1996) obtained similar results in the three soil types tested, the microbial response to predominantly Cd was different in sandy than in the finer textured soils in a study conducted by Renella *et al.* (2003). Both phosphatase activities and ATP content were more sensitive in sandy soils. In studies by Renella *et al.* (2005, 2004) it was concluded that exposure to high Cd concentrations led to a less efficient metabolism, which was responsible for lower enzyme activity in Cd-contaminated soils. While it can be concluded that heavy metals have a negative effect on soil enzyme activity (Deng and Tabatai, 1995; Renella *et al.*, 2003), depending on the class of enzymes that is monitored this statement can be negated. In general, the degree of inhibition varies with the concentration and form of added heavy metal, the soil investigated, and the enzyme assayed (Ladd 1985; Nannipieri, 1995) so that no single microbial parameter can be used universally and only a combination of microbial parameters may provide a more sensitive indication of heavy metal pollution (Kandeler *et al.*, 1996).

11.3.6 Community Level Physiological Profiling (CLPP)

Microbial community structure has been recommended as a biological indicator of heavy metal stress. Sole carbon–source utilization, also designated community-level physiological profiles (CLPPs), is based on the utilization patterns of individual carbon substrates generated with commercially available 96-well Biolog microtiter plates. The assay is based on measuring oxidative catabolism of the substrates as indicated by a color formation after the transformation of a redox-dye to generate patterns of potential sole carbon–source utilization (Garland and Mills, 1991). The Biolog procedure offers the opportunity to study microbial community structure because it is based on community function, and it does not require isolation of individual bacterial strains (Kelly and Tate, 1998). Several authors have used the Biolog system to study the effect of heavy metals on the activity of the soil microbial community (Kelly and Tate, 1998; Pennanen, 2001; Yao *et al.*, 2003). In a study that Kelly and Tate (1998) performed on soils obtained along a gradient influenced by a Zn smelter, Biolog metabolic profile analysis did differentiate between metal-contaminated (field-parts B ([Zn]: 2600 mg/kg), C ([Zn]: 4000 mg/kg) and D ([Zn]: 13600 mg/kg)) and uncontaminated (field parts A ([Zn]: 550 mg/kg)) field soils as well as between contaminated soils with varying degrees of metal loading. Two aspects of the Biolog procedure were impacted by metal loading; the rate of color development and the metabolic profiles. The most contaminated site D samples tended to have higher scores for succinic acid, which means that site D samples were better able to oxidize succinic acid on the Biolog plates. Site D samples also showed a greater ability to oxidize a number of carboxylic acids and amino acids, and a lesser ability to oxidize several carbohydrates, as compared with the relatively uncontaminated samples from site A. In addition to the separation of sites, site D had the highest degree of variability in its Biolog profiles. This suggests the possibility that the metal contamination may be resulting in a community that is more variable and less stable (Kelly and Tate, 1998) and which can be detected by CLPP analysis.

11.3.7 Phospholipid Fatty Acid Analysis (PLFA)

An often used approach to detect possible changes in the soil microbial community, in a nonselective way, is analysis of the PLFA composition in the soil (Frostegard *et al.*, 1993; Pennanen *et al.*, 1996; Griffiths *et al.*, 1997; Pennanen *et al.*, 1998; Kelly *et al.*, 1999; Pennanen, 2001; Shi *et al.*, 2002; Perkiömäki *et al.*, 2003; Turpeinen *et al.*, 2004). Changes in PLFA profiles are indicative of changes in the overall structure of microbial communities (Frostegard *et al.*, 1996) and signature PLFAs can provide information on specific groups of microorganisms present in a community (Frostegard *et al.*, 1993). PLFAs are located in membranes of the cells and are only detected in living microorganisms (Tunlid and White, 1992). PLFAs can be isolated as such from the environment so it is possible to characterize the features of the microbial community directly in a natural habitat, without an initial isolation (Pennanen *et al.*, 1996). Different subsets of microorganisms have different PLFA patterns: Gram-positive bacteria have usually relatively more iso-, anteiso-, or otherwise branched fatty acids (O'Leary and Wilkinson, 1988); Gram-negative bacteria have more mono-unsaturated or cyclic fatty acids (Wilkinson, 1988); actinomycetes often have a methyl group in the tenth carbon atom from the carboxyl end of the chain (Kroppenstedt, 1985); and fungi have more long-chain polyunsaturated fatty acids than bacteria (Federle, 1986; Lösel, 1988). However, the PLFA profile does not give an actual species composition but instead gives an overall picture of the community structure.

In some cases, changes in the concentrations of certain PLFAs can be correlated to changes in more specific groups of organisms (Pennanen et al., 1996). Analysis of the microbial community structure based on PLFA and terminal restriction fragment length polymorphism (TRFLP) profiles by Turpeinen et al. (2004) revealed considerable differences between field samples taken from three former wood impregnated plants located in southern Finland. The three sites were contaminated with different concentrations of a mixture consisting of arsenic (As), copper (Cu), and chromium (Cr). PLFA analyses revealed a decrease in several iso- and anteiso-branched PLFAs, all commonly found in Gram-positive bacteria, in the metal-contaminated soils. This observation indicated the predominance of Gram-negative over Gram-positive bacteria in these contaminated soils, which was also evident from plate counts and identification of the As-resistant isolates (Turpeinen et al., 2004). This finding was consistent with earlier studies (Jordan and Lechevalier, 1975; Doelman and Haanstra, 1979; Hiroki, 1992). A further indication that such a shift had occurred was shown by the increase in cy17:0, which is considered to be typical for Gram-negative bacteria (Lechevalier, 1977). In addition, a relative increase in 18:2ω6,9, an indicator fatty acid for fungi (Guckert et al., 1985), was observed at the Finnish As, Cu, and Cr contaminated sampling sites. This proportional increase in fungi at contaminated sites is likely due to the fact that fungi tend to be more resistant to heavy metals than bacteria as is also noticed by other authors (Jordan and Lechevalier, 1975; Hiroki, 1992; Khan and Scullion, 2000). A last group of PLFAs, namely the methyl-branched 10Me16:0, 10Me17:0, and 10Me18:0 which are found almost exclusively in actinomycetes (Lechevalier, 1977), varied in response to the observed As, Cu, and Cr metal contamination. At contaminated sites, an increase was found in 10Me18:0 while a decrease was found for 10Me16:0 and 10Me17:0. Other results in the literature also indicate that different actinomycetes can respond differently to elevated heavy metal concentrations (Williams et al., 1977; Frostegard et al., 1993). Therefore, the mixed results observed in the study of Turpeinen et al. (2004) may be the result of different responses to elevated metal concentrations by different members of the actinomycete population. Although the results in the study of Turpeinen et al. (2004) corresponded to several other studies in literature (Jordan and Lechevalier, 1975; Williams et al., 1977; Doelman and Haanstra, 1979; Hiroki, 1992; Frostegard et al., 1993; Khan and Scullion, 2000), the study of Pennanen et al. (1996) indicates that care is recommended when interpreting PLFAs on the effect of heavy metals on the microbial community. Pennanen et al. (1996) noticed that the increase in mycorrhizal fungal 18:2ω6c, which has been observed in field samples, is the opposite of what they observed in the laboratory. In their study the amount of 18:2ω6c decreased in copper-contaminated soil as compared to non-contaminated soil. In addition, the Gram-positive bacteria dominated over the Gram-negative bacteria in the heavy metal polluted soil which is unusual compared to other studies in literature (Timoney et al., 1978). The authors explained these two

phenomena as due to the effects of metal contaminants on tree roots. Pennanen et al. (1996) suggested that damage to fine roots may have resulted in loss of rhizosphere habitats for mycorrhizal fungi and Gram-negative bacteria in the soil used in the laboratory tests.

Phospholipid fatty acid analysis has been used to evaluate the microbial population composition of environmental samples (Crocker et al., 2000). The same techniques can be applied to pure cultures or bulk agent samples. Distinct PLFA profiles may be observed between species (Leung et al., 1999). However, similar to protein analysis, culture conditions and sample preparation can influence the results and yield data that may be misinterpreted. This technique has not been used for strain identification of bioterrorism agents, apparently due to lack of reliable discriminating power.

11.4 EMERGING MICROBIAL FORENSIC TECHNIQUES

Of the emerging microbial forensic techniques, DNA analysis represents the dominant technology. Examples of the use of DNA analysis in environmental forensic investigations comprise the identification of the origin of a microbial sample and include:

- identify the presence or absence of *Enterococcus faecium* and *Bacteroides* indicator of human sources in a sample;
- the presence or absence of human fecal viruses (i.e., enteroviruses); as indicator of human sources
- the presence or absence of *E. coli* and *Bacteroides* from cattle sources; indicating animal source
- the presence or absence of cattle fecal viruses (i.e., bovine enteroviruses); indicator for animal sources
- comparative analysis of the *E. coli* isolates from fecal and water or sediment sources; and
- DNA fingerprinting of *Giardia* species and strains, genetic enteric virus detection, and others.

11.4.1 Overview of DNA (Deoxyribonucleic Acid) and the Genetic Code

DNA is a polymer consisting of a large repetition of monomer sequences. The monomer units of DNA are nucleotides. Each nucleotide consists of a deoxyribose (a 5-carbon sugar), a nitrogen containing base attached to the sugar and a phosphate group.

Deoxyribose and phosphate components are common for all nucleotides while the nitrogen containing bases may vary, being of four types. These bases belong to two main classes: purinic (Adenine and Guanine) and pyrimidinic (Cytosine and Thymine). Their number is equal. It is this distinct arrangement of adenine, guanine, thymine and cytosine that regulates the production of specific proteins and enzymes in a cell. The nucleotides are given one letter abbreviations (shorthand for the four bases): A for adenine; G is for guanine, C for cytosine; and T for thymine. DNA is the basic genotype (genetic identity) of an organism, which in turn determines the phenotype (physical features). Subsequently, particular DNA profiles can be ascribed to particular organisms. The DNA profile thus constitutes a unique fingerprint that is specific to each individual. Spatial conformation – a DNA molecule consists of two strands which are coiled around each other in a double heilx. The two DNA strands run in opposite direction and form a helical spiral, winding around a heilx axis in a right-handed spiral. The bases of the individual nucleotides are on the inside of the helix, stacked on top of each other like the steps of a spiral staircase, while the phosphates are on the outside. The bases are joined together in pairs, a single base from one chain being hydrogen-bounded to a single base from the other chain, so that the two lie side by side. One of the pair must be purine and the other pyrimidine for the bonding to occur. Only specific pairs of bases can bond together. These pairs are: adenine (purine) with thymine (pyrimidine), and guanine (purine) with cytosine (pyrimidine). So, the information responsible for the proper function of a cell and organism is stored into DNA, as a recipe in a recipe book. Basically, DNA encodes the information determining the protein synthesis in an organism.

The sequence of each 3 nucleotides in the polymeric chain of DNA codes for a particular amino acid from the polymeric chain of proteins. The protein synthesis process involves the reading of the genetic code stored in DNA which is accomplished by transcribing DNA into another molecule called RNA or ribonucleic acid. The RNA is similar to the DNA molecule except it contains U (uracyl) as a base in the nucleotide structure instead of T (thymine). The RNA inserts itself inside the DNA molecule that has partially unzipped to reveal the protein code (nucleotide sequence). It reads the code by matching its molecules with the corresponding DNA molecules in a process called transcription.

With the exact code for a particular protein transcribed in its structure, the RNA proceeds to a cellular structure called the ribosome, which is the place where proteins are synthesized by putting together the amino acids according to the code within RNA. More amino acids will continue to accumulate until the end of the RNA code is reached. There are distinct molecules of RNA that contain the transcribed information and those that bring and put together the amino acids into the protein chain. With the amino acids linked together, the end result is a protein.

11.4.2 PCR (Polymerase Chain Reaction)

Polymerase Chain Reaction is a powerful technique which effectively amplifies DNA. The technique was relatively recently developed (in 1987). PCR requires tiny amounts of sample DNA, which makes it of great potential for forensic investigations usually limited by the amount of sample that can be recovered. PCR can produce multiple copies of DNA segments from an initial very limited amount of DNA (as little

as 50 molecules), enabling a DNA fingerprint to be made from a single hair for example. PCR allows the duplication of minute quantities of DNA into billion and even trillions of copies.

To initiate this analysis, primers are required. Primers are small snippets of DNA homologous to different regions in a target gene. For any targeted DNA gene at least two primers are necessary to delimitate a DNA variable segment targeted to be amplified. This segment comprised at each end by a primer and amplified during the PCR reaction is called "amplicon." The selection of appropriate primers represents the key for a successful PCR amplification of a target gene. Complete failure of amplification reactions or amplification of nonspecific targets can frequently be attributed to poor primers. In order to design such primers, a collection of sequences for the target gene from many different genetic backgrounds is necessary. At least two conserved sequence regions must exist in the gene of interest in order to provide priming sites in a gene from a broad range of organisms (Kitts, 2001). Primers should be spaced far enough apart to obtain a multiplied DNA fragment with sufficient sequence divergence. The primer regions should not be too far apart, because the long resulted amplicons may create (after digestion with restriction enzymes within DNA fingerprinting applications) some fragments too large to be analyzed. An optimal amplicon length is between 400 and 700 base pairs (bp) (Kitts, 2001).

The PCR process (Figure 11.4.1) requires three main steps:

1. Denaturation of the DNA sample at 94 °C (the DNA sample is first heated to unravel and split or unzip the double DNA helix);
2. Annealing at 54 °C—along with the selected primers, primase and polymerase enzymes are used to identify DNA sequences and produce a copy of them; ionic bonds are constantly formed and broken between the single-stranded primer and the single-stranded template. On the piece of double-stranded DNA (template and primer), the polymerase attaches itself and starts copying the template. Once a few bases are built in, the ionic bond is so strong between the templates and the primer does not break.

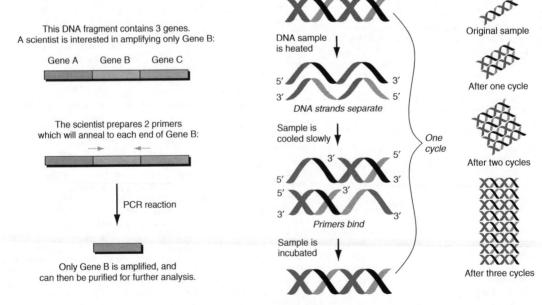

Figure 11.4.1 PCR Method.

3. Extension at 72 °C—the bases (complementary to the template) are coupled to the primer on the 3′ side (the polymerase adds dNTPs (nucleotides) from 5′ to 3′, reading the template from 3′ to 5′ side) complementary to the template.

The cycle is repeated a number of times, each time doubling the sequence samples.

Primer/probe kits are commercially available for *B. anthracis*, *Brucella* species, *Francisella tularensis*, *Yersinia pestis*, and *Clostridium botulinum*. These reagent sets allow samples to be analyzed in the field or laboratory in about 2 hours (Fatah *et al.*, 2001). A positive detection is influenced by the probability of carrying at least one copy of the target gene sequence all the way through the extraction and sample preparation procedure so it is available for amplification during the PCR step. The PCR method was publicly field proven during the clean-up of *B. anthracis* spore–contaminated postal facility in Washington, DC. The method was used to test thousands of samples for the presence of *B. anthracis* spores. The dispersion of spores was tracked using the method, suspect samples were tested for the presence of *B. anthracis*, and preliminary evidence of successful decontamination was also generated using the PCR method (D. Graves, unpublished data).

PCR does not discriminate between DNA from living or dead organisms. Any organism in the sample is analyzed, as long as it contains the target gene (primer sites). The inability of PCR to discriminate between living and dead organism's DNA could represent a limitation. A scenario in which living organisms are exposed to contamination may therefore be required. This potential limitation may be overcome by cultivating the respective organisms, assuming that the type of organism that is tracked is known and well-characterized or one can work on the RNA instead of the DNA level. RNA is only present in living organisms, thus enabling discrimination between living and dead organisms. For the potential applicability of PCR-based fingerprinting techniques in tracking a contaminant's passage through different environments, the nondiscrimination between living and dead organism's DNA is a significant advantage. This is because the changes in microbial communities induced while contaminants are present could still be visible in DNA fingerprints a long time after the contamination has gone through a certain environment.

Most common PCR-based DNA fingerprinting techniques

The main of DNA analysis techniques with potential applicability to environmental forensic studies include the following:

- Terminal restriction fragment length polymorphism (TRFLP);
- amplified fragment length polymorphism (AFLP);
- Single stranded conformation polymorphism (SSCP);
- Thermal and denaturing gradient gel electrophoresis (TGGE and DGGE);
- Amplified ribosomal DNA restriction analysis (ARDRA); and
- Randomly Amplified Polymorphic DNA (RAPD).

Other DNA fingerprinting techniques

DNA fingerprinting offers a powerful tool to examine a wide diversity of microbial communities rather than targeting a single species, and the subsequent changes that may occur due to contamination and any other environmental stress. The following text describes the primary concepts and principles associated with DNA that are applicable to its use as an emerging technique in environmental forensic investigations.

Because of the 2001 anthrax attack in the United States, the relevance and intensity of research on anthrax forensic techniques has outstripped work with other potential agents. *B. anthracis* has proven to be a challenging agent with regard to the application of molecular forensic techniques for strain identification. *B. anthracis* strains are remarkably more monomorphic in their nucleotide sequence than most other bacteria (Read *et al.*, 2002). The nucleotide sequence polymorphism among strains of *B. anthracis* is less than 1 % of the total genetic sequence. (*Francisella tularensis* seems similarly monomorphic [Grunow *et al.*, 2000].) The lack of polymorphisms reduces the usefulness of most of the more common strain discrimination techniques such as specific and nonspecific amplified restriction fragment length polymorphisms (i.e., TRFLP and AFLP), DNA typing methods (pulsed-field gel electrophoresis [PFGE] typing), and gene or whole genome sequence comparisons. The poor performance of many of the common DNA evaluation techniques has led to the development and use of less common and new methods for strain discrimination. The techniques described below are currently considered to be the most appropriate for *B. anthracis* strain discrimination.

- *Multiple-locus variable number tandem repeat sequence typing*. The premise of this method is that certain nucleotide sequences repeat themselves in series in the bacterial genome (Keim *et al.*, 2000). The purpose of these repeated sequences and the significance of the variable number of repeats is unclear. The number of times a sequence is repeated varies among strains, providing a target for flanking primer PCR amplification. PCR produces DNA fragments of varying lengths depending on the strain-specific number of repeats. To date 15 loci have been identified and determined to be variable enough to serve as strain discriminators. Depending on the locus, two to nine distinct alleles have been observed. Using a suite of VNTRs, strain-specific profiles have been generated for *B. anthracis* strains. Extension of the method to many geographically diverse strains produces a bank of VNTR-discriminated strains that can be compared with an unknown strain to determine its relationship to known strains. This method identified all *B. anthracis* samples collected from the 2001 anthrax attack as the Ames strain.
- *Single nucleotide polymorphisms (SNPs) and indels*. SNPs are nucleotide sequences that have single base pair variations. A total of 38 SNPs potentially suitable for *B. anthracis* strain subtyping have been identified (Read *et al.*, 2002). SNPs are identified by direct comparison of the entire genome sequence of multiple strains (Pearson *et al.*, 2004). In order to establish the reliability of the occurrence of an SNP, each potential SNP is sequenced three or more times to confirm that the nucleotide variation is real and not an error of the sequencing process. The initial identification of SNPs is time-consuming and expensive. However, once SNPs are identified along with their bracketing sequences, PCR primers are synthesized and used to efficiently amplify SNP-bearing sequences for comparison against strains with known polymorphisms in the sequence.
- Whole genome sequence analysis also revealed sequences that have been altered by nucleotide insertions and deletions (indels) (Read *et al.*, 2002). Once indels are identified, PCR primers are synthesized to efficiently amplify indel regions. Indel regions are genetic sequences that are characterized by the presence of strain-specific DNA insertion or deletion mutations. Sequencing of amplified indels allows strain-to-strain comparisons. Read *et al.* (2002) detected three large indels that varied among strains.
- These indels were used in conjunction with 38 SNPs and 8 VNTRs to subtype the *B. anthracis* strain used in the 2001 bioterrorism attack (Read *et al.*, 2002). This method

indicated the 2001 bioterrorism strain, known as Florida, was derived from the Ames strain. The Ames strain was originally isolated from a dead cow in Texas and used by the US Army for bioweapons research and development.

- *Non-coding sequence interval polymorphisms.* Areas of an agent's genome that do not encode proteins or ribosomal RNA are more prone to accumulate mutations because no selective pressure will result from the mutation. Intergenic spacer DNA (the DNA that separates individual genes) has been used to subtype *Yersinia pestis* (plague agent) (Drancourt *et al.*, 2004). Once variable intergenic spacer DNA sequences are identified, they can be amplified in PCR reactions with sequence comparison among the amplified DNA to discriminate between strains.

11.4.3   Terminal Restriction Fragment Length Polymorphism (TRFLP)

The most commonly used DNA-fingerprinting technique in both criminal and environmental applications is the TRFLP technique. TRFLP is a microbial community profiling method typically targeting the 16S rRNA gene, which can be used with universal or species-specific primers, depending on the resolution required.

Terminal Restriction Fragment Length Polymorphism is based on the PCR amplification of the target gene from the extracted DNA using a fluorescently end-labeled primer, followed by the digestion of resulted fragments with restriction enzymes (that only cuts at a particular sequence of nucleotides) and by the electrophoresis separation of resulted cut fragments with the visualization of only the terminal (labeled) segments. Because these amplified fragments of DNA originate from different organisms, the genes have different sequences. Due to sequence variations, the terminal restriction site for each species in the community should be different thereby resulting in a unique DNA profile for each microbial community. TRFLP uses the length difference from different resulting DNA terminal fragments to differentiate and discern specific microbial community patterns.

The 16S rRNA gene is the most commonly used gene in TRFLP analysis, due to the existence of well-established primers of conserved regions of this gene and also the existence of highly variable regions generating polymorphism. For environmental forensic purposes, genes other than 16s may be targeted, once primers for such genes are identified and designed. There is no limit to the types and variety of genes targeted by the TRFLP method if changes in a microbial community due to the presence of a particular contaminant is known. Thus, the gene responsible for the synthesis of a certain enzyme that degrades (metabolizes) the contaminant of interest can be targeted to examine changes in the community patterns due to the presence and passage of a certain contaminant.

The main steps of TRFLP analysis (Figure 11.4.2) are:

- DNA extraction;
- PCR amplification of the target gene with fluorescently labeled primers (only used if automated);
- Digestion of the amplified fragments with one or more restriction enzymes;
- Separation and visualization of terminal-labeled fragments (using automated capillary-based electrophoresis); and
- Data analysis for peak identification (species identification in some cases).

Since there is no microbial cultivation, the DNA extracted from the sample is the sum of DNA from living organisms and possibly dead ones in the sample. The extracted DNA is then subjected to PCR in order to amplify a certain target gene. The aim is to target a gene for which there are at least two regions of conserved sequence for

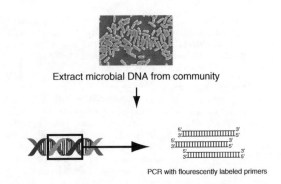

Extract microbial DNA from community

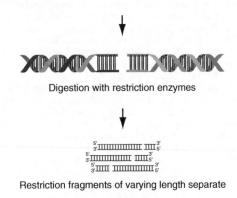

PCR with flourescently labeled primers

Conserved portion of the 16S rRNA gene is targeted for PCR

Digestion with restriction enzymes

Restriction fragments of varying length separate

ABI Prism 310 Genetic Analyzer

Fragments are detected on an automatic capillary-based electrophoresis system (ABI 310)

Figure 11.4.2 *TRFLP Procedure.*

a broad range of organisms. Such regions provide priming sites. In addition, these conserved regions must be at certain distances between each other in the poly-nucleotidic DNA chain, avoiding being too close (for sufficient divergent DNA segment in the flanked amplicon and reflect the true diversity of a sample) or too distant (to avoid creating fragments that are too large for electrophoresis analysis). According to Kitts (2001), an optimal amplicon length is between 400 and 700 base pairs (bp), which may allow for best estimates of microbial diversity in an environment. One key for the success of this technique is the selection of right primers, able to show the targeted and expected differences in microbial communities, induced by the presence of contamination. For the primer design, a collection of sequences of the target gene from many different organisms is needed. Primer selection can dramatically alter the picture that is presented in a terminal restriction fragment (TRF) pattern, because only a fraction of microbial 16S rRNA sequences in a sample will be amplified. One primer is labeled with a fluorescent molecule at the 5′ end (if automated).

The amplified (by PCR) DNA fragments are then digested with a restriction enzyme. In general, enzymes used have a tetra-nucleotide recognition sequence (Kitts, 2001). The digestion enzyme used should be chosen to best reproduce the diversity expected in a sample, taking into account the predicted fragments from 16S rRNA genes in a sequence database. The most used enzymes are *Hha*I and *Msp*I or their isoschizomers. For a targeted gene with templates without an extensive database of sequences available, a process of trial and error must be used to choose the appropriate restriction enzyme. The use of multiple restriction enzymes, rather than of just one, may generate more accurate characterization patterns and is usually preferred (Clement *et al.*, 1998).

The obtained digested amplicons are subjected to electrophoresis in either polyacrilamide gel or a capillary gel apparatus. The apparatus contains a DNA sequencer with a fluorescence detector, which insures that only the fluorescently labeled TRF are visualized. Usually, an automated fragment analysis program is used. This program calculates the TRF length in base pairs after comparing TRF peak retention time to a DNA size standard. The output is digital and provides information on the size of the product in base pairs (i.e., species) and the intensity of fluorescence or relative abundance of the various community members. Finally, the resulted TRF peaks are numerically compared between samples using multivariate statistical methods. At the same time, individual peaks of a pattern can be compared to a clone library and identified. Fragment analysis software exists for most DNA sequencing machinery which will automatically digitize the electrophoresis output and export a tab delimited text file that can be loaded into standard statistical or spreadsheet packages.

Many laboratories use TRFL patterns to characterize the microbial communities from different environments, thereby allowing the determination of spatial and temporal shifts in community structures. Such shifts may be used in forensic studies as evidence for the presence of certain contaminants through an environment, or to track their past passage. Also, the toxicity of contaminants to bacterial communities coupled with their use as nutrient sources (such as oil that may be used by bacteria as carbon source) are likely selective pressures on bacterial community composition and diversity. Thus, the characterization of such microbial diversity and its response to contamination through simple automated techniques such as TRFLP could help predict the fate of contaminants in the environment.

Terminal restriction fragment patterns are used to characterize functional diversity in bacterial communities, including reports on the functional diversity of important processes such as nitrification, nitrogen fixation, de-nitrification, and mercury resistance investigated in a large variety of environments from marine sediments to termite intestines. For example, TRFLP analyses revealed that *Nitrosospira*-like organisms were one of the major contributors to ammonia oxidation in a full-scale aerated-anoxic Orbal reactor (Park *et al.*, 2002).

Terminal restriction fragment length polymorphism has the ability to rapidly analyze large amounts of information due to available automated systems that make it easy to obtain robust TRF data, generating hundreds of reproducible TRF patterns. Microbial communities can be monitored on a scale and with a resolution never achievable before. There is a potential to use TRF data to search existing databases for matching sequences and identification of individual organisms in a community profile. TRFLP, and particularly TRFLP using a capillary electrophoresis sequencer, seems to have the advantages of higher throughput and reproducibility in monitoring the bacterial community (Osborn *et al.*, 2000).

Clement *et al.* (1998) identified limitations with this method, which included the following:

- the tendency of PCR to differentially amplify templates, thereby preventing quantification of relative abundance in the community;
- individual TRF detection is limited by the electrophoresis technology;
- patterns resulted from a single enzyme digest may not be accurate for community characterizations. It is possible that phylogenetically distant bacteria, for example, could produce TRF of different lengths when digested with one restriction enzyme, but of the same length when digested with a different enzyme; and
- challenges with incomplete digestion of amplicons creating artifactual peaks in TRF patterns (Kitts, 2001).

Another disadvantage is the complexity of identification of the organisms responsible for a particular element in a profile. This is due to the fact that TRFLP are destructively sampled and the DNA cannot be reclaimed. DGGE however allows for southern blotting or direct cloning of profile elements. In order to overcome or minimize different biases of the technique, the following procedures can be implemented:

- multiple DNA extractions from each well-homogenized sample;
- combine amplicons from multiple PCRs;
- load only as much polyacrylamide gel from each DNA digestion as is practical to allow detection of less abundant fragments; and
- combine TRFLP data after applying digestion with more than three different restriction enzymes, thereby potentially increasing the amount of information available for each community.

Other biases relate to the limited coverage insured by the PCR primers used (only a fraction of microbial 16S rRNA sequences in a sample will be amplified) and to the possibility of TRF length to overlap.

The potential of molecular techniques for measuring microbial community diversity in soils as a forensic soil comparison method was first realized by Horswell *et al.* (2002), who used TRFLP technology to profile soil bacterial communities for comparative forensic purposes. TRFLP is a simple and robust technique that applies well to soil forensic analysis because of its visual output (the electropherogram). Using TRFLP to generate profiles from forensic soil samples has the additional advantage of using equipment and technology already in use in most forensic DNA laboratories worldwide. Only basic molecular biological expertise is required to perform the profiling and analyses, making the technique ideal as a routine soil analysis tool for forensic laboratories to perform.

Horswell *et al.* (2002) established that bacterial community DNA profiling can be used to compare soil samples of the size likely to be encountered in real forensic situations. A simple comparison index, the Sorenson's Index, was used to compare profiles in which a value of 1 indicates identical profiles and 0 indicates the profiles that share no common fragment sizes (peaks).

In one scenario, a shoe print was made at hypothetical scene A and adhering soil was recovered from the tread of the shoe outsole (sample A1, Figure 11.4.3). Soil was also collected from the shoe print (sample A2, Figure 11.4.3). Eight months later the location of the shoe print was re-visited and a further soil sample was taken (sample A3, Table 11.4.1). Reference soils (approximately 50 g wet weight) were collected from four other locations for analysis (samples B, C, D, E, Figure 11.4.4).

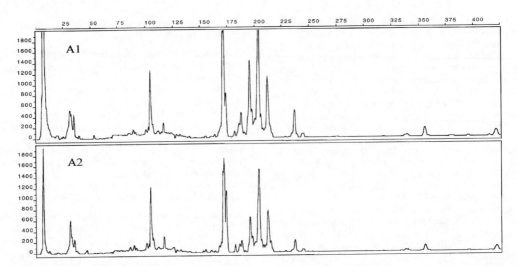

Figure 11.4.3 *Profiles of the soil microbial communities from Scenario 1 – footwear impression; soil collected from shoe (A1) and soil collected from the shoe print in soil (A2) (from Horswell et al. 2002 Reprinted, with permission, from the* Journal of Forensic Sciences, *Vol. 47, No. 2, copyright ASTM International, 100 Barr Harbor Drive, West Conshohocken, PA 19428).*

Table 11.4.1 *Similarity Index for Forensic and Reference Soil Samples in Scenario 1 – footwear impression (from Horswell et al. 2002. Reprinted, with permission, from the* Journal of Forensic Sciences, *Vol. 47, No. 2, copyright ASTM International, 100 Barr Harbor Drive, West Conshohocken, PA 19428)*

Site	A1	A2	A3	B	C	D	E
A1							
A2	0.91						
A3	0.62	0.70					
B	0.54	0.53	0.56				
C	0.59	0.67	0.61	0.63			
D	0.56	0.64	0.57	0.59	0.73		
E	0.48	0.64	0.64	0.57	0.62	0.50	

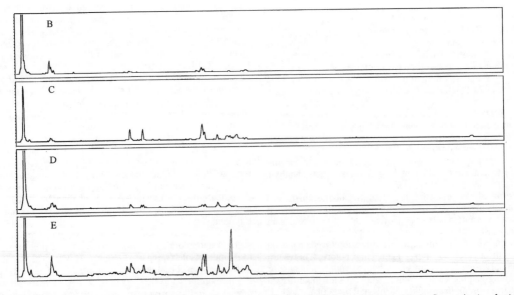

Figure 11.4.4 *Profiles of the soil microbial community profiles from reference soil samples in Scenario 1 – footwear impression (from Horswell et al. 2002. Reprinted, with permission, from the* Journal of Forensic Sciences, *Vol. 47, No. 2, copyright ASTM International, 100 Barr Harbor Drive, West Conshohocken, PA 19428).*

Table 11.4.2 *Summary of Some Commonly Used 16S rRNA Gene Bacterial Community Analysis Techniques*

	Differentiation based on	Advantages	Disadvantages	Selected References
Denaturing gradient gel electrophoresis (DGGE)	Melting temperature/base composition in denaturant gradient	Bands can be excised for sequencing Can hybridize profile with probes	PCR biases PCR products larger than 500 bp not separated well Different sequences can have similar melting properties	(Muyzer *et al.*, 1993; Kozdroj and van Elsas, 2000)
Temperature gradient gel electrophoresis (TGGE)	Melting temperature/base composition in temperature gradient	Bands can be excised for sequencing Can hybridize profile with probes	PCR biases PCR products larger than 500 bp not separated well Different sequences can have similar melting properties	(Heuer *et al.*, 1997)
Single-strand conformation polymorphism (SSCP)	Single-stranded secondary structure mobility	Bands can be excised for sequencing	PCR biases Only fragments between 150–400 nucleotides can be used Electrophoretic variables have strong influence	—
Amplified ribosomal DNA restriction analysis (ARDRA) or restriction fragment length polymorphism (RFLP) analysis	Restriction fragment length	Straightforward No expensive equipment	PCR biases Diversity difficult to estimate due to multiple fragments per gene copy	—
Length heterogeneity-PCR (LH-PCR)	Gene fragment length	Simple and rapid Products can be excised and sequenced	PCR biases	—
Terminal Restriction Fragment Length Polymorphism (TRFLP)	Restriction fragment length	Can be automated (high resolution) Potential to be quantitative Standards allow controlled comparison Reproducible	PCR biases Cannot be sequenced without use of cloning	(Liu *et al.*, 1997; Osborn *et al.*, 2000)
Ribosomal RNA-targeted nucleic acid probes	Sequence	Not subject to PCR biases	rRNA can be naturally low in abundance Time consuming/ complex Some sequence knowledge required for probe design	—

The microbial community profiles for soil from the shoe print and soil collected from the shoe itself produced very similar electropherograms and a very high (>0.9) similarity index (Figure 11.4.3 and Table 11.4.2). In contrast, there were major differences between the profiles of the reference soils and those of the crime scene, and the suspect's shoe, with a similarity index of <0.6 (Figure 11.4.4 and Table 11.4.3). When the soil sample collected 8 months later (A3) was compared with the microbial community DNA profile of soil collected at the time of original sampling (A2), differences were observed, but overall, there was a considerable similarity between the profiles and a similarity index of 0.70 was obtained. Differences in profiles with time are not an unexpected result as seasonal fluctuations, e.g., rainfall and temperature, are likely to impact on the microbial community, causing population shifts.

For the second scenario, an imprint was made in the soil at a second site by kneeling, wearing a clean pair of jeans. Soil was sampled from the impressions made by each knee in the soil (left knee and right knee). The jeans were taken back to the laboratory and areas of soil staining removed for analysis.

The microbial community profiles of the soil from the left knee impression in the soil and the soil extracted from the left knee of the jeans were very similar with a Sorenson's Index of 0.82.

In 2002, Horswell *et al.* demonstrated that a soil microbial DNA profile can be obtained from a small sample of soil recovered from the sole of a shoe, i.e., with sample sizes likely to be encountered in forensic casework. Profiles were also obtained from stains on clothing, e.g., denim jeans. Using the simple Sorenson's Similarity Index, profile comparisons indicated that soil samples from the same location had a greater degree of similarity than soils from a different location. The graphical output of the TRFLP method means that visual comparison can give a reasonable idea of the similarity between two profiles.

TRFLP has been used to track a contaminant source. The majority of recent studies use *Eubacteriaceae* (eubacterial

Table 11.4.3 *Summary of some Microbial Community Diversity Analysis Techniques that do not use 16S rRNA Gene*

	Gene used	Differentiation based on	Advantages	Disadvantages
Randomly Amplified Polymorphic DNA (RAPD)	Random DNA	Annealing of primers in different places in different genomes	No sequence information required at all	Some questions raised over reproducibility. PCR biases. No phylogenetic information can be inferred
Ribosomal intergenic spacer analysis (RISA)	Intergenic spacer (IGS)	Length polymorphism	Can be automated (ARISA) Good phylogenetic resolution	Limited IGS sequence data available PCR biases/operon heterogeneity
Total genomic cross-DNA hybridization	Total DNA	Reannealing between two different samples whole DNA	Not subject to PCR biases	Does not give information on relative abundance of individual genomes Requires large amounts of high quality DNA
Whole DNA fractionation	Total DNA	Density gradient/base composition	Not subject to PCR biases	Low taxon resolution. Requires large amounts of high quality DNA
Low-molecular-weight (LMW) RNA analysis	5S rRNA and tRNA	Length polymorphism	Not subject to PCR biases	Rapid degradation of small RNA molecules leading to gel artifacts Limited information in small RNA

communities ranging from those present in pig intestines to those from marine bacterio-plankton) because this microbial community has been well characterized (Brunk *et al.*, 1996; Liu *et al.*, 1998; Flynn *et al.*, 2000; Franklin *et al.*, 2001).

TRFLP was also used to characterize and compare subsurface and surface soil bacterial communities in California grassland. Results showed the low diversity of bacterial communities at depths, consistent with species-energy theory (LaMontagne *et al.*, 2003). Such ecological studies are important in nutrient and carbon cycling. The TRFLP technique was efficiently used to discern bacterial population changes in feces during feeding of rats with *Lactobacillus acidophilus* NCFM (Kaplan *et al.*, 2001).

Clement *et al.* (1998) used TRFLP method and principal components analysis (PCA) to characterize and compare bacterial communities in deer fecal pellets, petroleum hydrocarbon-contaminated sands, and pristine sand. They obtained accurate characterizations reflecting the expected bacterial community biology only by using more than one enzyme for obtaining digestion patterns.

Other studies observe taxonomic diversity in other microbial groups, such as archaebacterial communities in soil and fish intestines (van der Maarel *et al.*, 1998; Fey and Conrad, 2000; Leuders and Friedrich, 2000). Another interesting application was marked by Derakshani *et al.* (2001) for planctomycetes.

Marsh *et al.* (1998) used 18S rRNA gene primers to describe fungal communities in sewage sludge. Brodie *et al.* (2003) studied alterations in soil micro-fungal community structure along a transect between a semi-natural upland grassland and an agriculturally improved enclosure using an

indirect measurement of active fungal biomass (ergosterol), together with TRFLP and DGGE (another fingerprinting technique). In a recent study related to ectomycorrhizal fungal diversity, TRFLP was used as a tool for species identification to investigate the vertical niche differentiation of ectomycorrhizal hyphae in soil (Dickie *et al.*, 2002).

The TRFLP technique was also used to investigate bacterial community structure and dynamics during a bioremediation experiment in a land treatment facility (land farming) contaminated with highly weathered petroleum hydrocarbons (C_{10}–C_{32} range) at the Guadelupe Oil Field located in Central California, USA (Kaplan and Kitts, 2004). Shifts in the bacterial community structure were examined and linked to TPH (total petroleum hydrocarbon) degradation rates in the soil. TRF patterns revealed a series of sample clusters describing bacterial succession along with the identification of specific polytypes of bacteria that reflected different phases of petroleum degradation in the land treatment unit. Such findings support the potential use of TRFLP to monitor and improve the procedures used for degradation of petroleum hydrocarbons at a land treatment facility. This same process may have similar applications for tracking the rate of biodegradation of petroleum hydrocarbons in the soil profile and/or groundwater as an indicator of both source or relative age of the release.

The comparative analysis of the 16S rRNA gene pool provides a sophisticated tool for molecular-based analysis of bacterial diversity in soil. This approach, however, does have some methodological uncertainties. DNA from soil can be particularly difficult to work with because of the presence of PCR inhibitors such as humic acids. These can prevent amplification or could potentially bias PCR results away from the original ratios of DNA fragments, if not removed completely from the DNA preparation before PCR. Bias may also be introduced during the PCR step. Any molecular method that uses PCR is subject to its inherent biases, so an awareness of this is required when interpreting results. Another factor that can complicate interpretation of environmental genetic data is that the number of copies of the 16S rRNA gene can range between 1 and 14 in different bacterial species (Cole and Girons, 1994). These copies can have some variation between them, and this may artificially increase the diversity seen in a profile (Clayton *et al.*, 1995).

A significant challenge associated with TRFLP is the presence of pseudo terminal restriction fragments (pseudo T-RFs) or false peaks in the community profiles. Pseudo T-RFs are single-stranded DNA fragments formed during the PCR. The height and area of the pseudo T-RFs increase with the number of PCR cycles indicating that they are a PCR artifact. Restricting the number of PCR cycles to the minimum required will reduce the effect of the pseudo T-RFs, but is unlikely to eliminate them completely. Hence, the presence of these peaks should be considered when evaluating species diversity in a mixed community using TRFLP. Incomplete restriction enzyme digestion of PCR products can also lead to false peaks appearing in profiles but careful optimization of digestion protocols can avoid this (Kitts, 2001).

11.4.4 Amplified Fragment Length Polymorphism (AFLP)

Amplified fragment length polymorphism (AFLP) is a DNA fingerprinting method that uses the difference in resulting DNA fragment lengths to discriminate and produce typical DNA patterns. No prior sequence information is required. AFLP or its fluorescent version (fAFLP) is a PCR-based derivative method of the Restriction Fragment Length Polymorphism (RFLP) method.

AFLP is based on the selective amplification of a subset of genomic restriction fragments using PCR. AFLP consists of the restriction of extracted DNA that generates specific sequences, followed by the ligation of adaptors complementary to the restriction sites and selective PCR amplification of a subset of the adapted restriction fragments. Thus, the technique selectively amplifies sequences (AFLPs) using primers from restriction digested genomic DNA (Karp *et al.*, 1997; Matthes *et al.*, 1998).

Different events can change AFLP profiles, including a loss of an amplified fragment, an increase in fragment size, a gain of a fragment, or a decrease in fragment size. Vos *et al.* (1995) described the AFLP technique as being based on the detection of restriction fragments by PCR amplification and argued that the reliability of the RFLP technique is combined with the power of the PCR technique.

The primary steps associated with performing an AFLP analysis (Figure 11.4.5) are:

- DNA extraction;
- DNA digestion with (usually two) restriction enzymes;
- Ligation of adapters to the restriction fragments;
- PCR amplification of the restricted fragments using primers complementary to each of the adaptor sequences; selective nucleotides are added in the extension of restriction fragment at 3′ ends determining the selectivity and complexity of the amplification; and
- Separation and visualization of the resulted AFLP fragments.

In the first step, extracted DNA is cut with restriction enzymes (endonucleases) to generate specific sequences, which are then amplified. The DNA is usually digested with two restriction enzymes, preferably a hexa-cutter and a tetra-cutter, although a wide scope of restriction enzymes can be used. For example, a restriction enzyme that cuts frequently (*Mse*I, 4 bp recognition sequence) and one that cuts less frequently (*Eco*RI, 6 bp recognition sequence) may be used for this purpose. The resulting restriction fragments are ligated to end-specific adaptor molecules. The sequence of the adapters and the adjacent restriction site serve as primer binding sites for subsequent amplification of the restriction fragments by PCR.

A preselective PCR amplification is then performed using primers complementary to each of the two adaptor sequences, except for the presence of one additional base at the 3′ end, which is chosen by the user. Thus, the PCR consists of the amplification of a subset of the restriction fragments using two primers complementary to the adapter and restriction site sequences, and extended at their 3′ ends by "selective" nucleotides. Therefore, only restriction fragments in which the nucleotides flanking the restriction site match the selective nucleotides will be amplified. The selective nucleotides are added in the extension of restriction fragments at 3′ ends determining the selectivity and complexity of the amplification. All feasible combinations of selective nucleotides could be used in the process. If sequence information is available, an in-silico analysis can be performed to select the most informative enzyme and primer combinations.

The final step is the separation and visualization of the resulted AFLP fragments. The resulted amplified DNA fragments are separated and visualized on denaturing polyacrylamide gels through either autoradiographic or fluorescence methodologies. When the DNA fingerprints of related samples are compared, common bands as well as differing bands will be observed. These differences are referred to as DNA polymorphisms.

AFLP is used for "basic" diversity and genetic variation studies. The availability of many different restriction enzymes and corresponding primer combinations enables the direct manipulation of AFLP fragment generation for defined applications such as polymorphism screening, QTL analysis, or genetic mapping. Russell *et al.* (1999) investigated the genetic variation of *Calycophyllum spruceanum* (Rubiaceae), a fast-growing pioneer tree of the Amazon Basin. Other studies have more specific goals such as investigations into introgression and hybridization, e.g., Rieseberg *et al.* (1999) examined introgression between cultivated sunflowers and a sympatric wild sunflower *Helianthus petiolaris* (Asteraceae), while Beismann *et al.* (1997) studied the distribution of two *Salix* species and their hybrid.

The method is reliable and efficient for detecting molecular markers. AFLP markers are fast becoming a molecular standard for investigations ranging from systematic to population genetics. The DNA polymorphisms resulted by AFLP are typically inherited in Mendelian fashion and may be used for typing, ID of molecular markers, and mapping of genetic loci. Xu *et al.* (1999) suggested that using AFLP is the most efficient way to generate a large number of markers that are linked to target genes. AFLP markers have been used at the level of the individual for application in paternity analyses and gene-flow investigations, e.g., Krauss and Peakall (1998) analyzed paternity in natural populations of *Persoonia mollis* (Proteaceae), a long-lived fire-sensitive shrub from southern Australia. Another application in paternity analysis was described by Krauss (1999).

Several studies have used AFLP markers in phylogenetic analyses. Heun *et al.* (1997) used a phylogenetic analysis of the allele frequency at different AFLP loci to suggest that *Triticum monococcum* subsp. *boeticum* (Poaceae) was

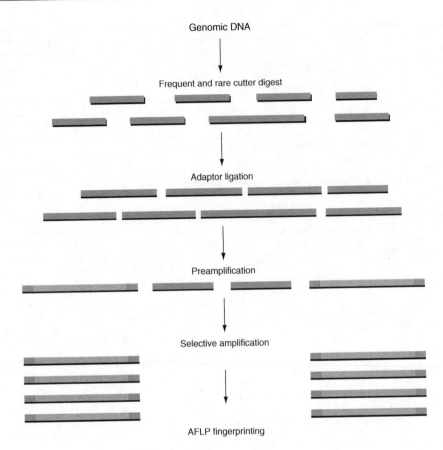

Genomic DNA

Frequent and rare cutter digest

Adaptor ligation

Preamplification

Selective amplification

AFLP fingerprinting

Figure 11.4.5   AFLP Method.

the likely progenitor of cultivated einkorn wheat varieties. Aggarwal *et al.* (1999), who investigated the phylogenetic relationships among *Oryza* species using AFLP markers, and Kardolus *et al.* (1998), who applied AFLP to *Solanum* taxonomy, concluded that AFLPs were "an efficient and reliable technique for evolutionary studies." However, other studies are using AFLP to create dendrograms, but then suggest evolutionary hypotheses and correlate AFLP pattern similarity with phylogenetic closeness (e.g., Aggarwal *et al.*, 1999; Mace *et al.*, 1999). Applications in gene-flow experiments and for plant variety registration were also marked in the literature (Law *et al.*, 1998). This technique was frequently applied to mapping studies, e.g., of species such as *Oryza* (Zhu *et al.*, 1998), *Zea* (Xu *et al.*, 1999), and *Solanum* (Bradshaw *et al.*, 1998).

An advantage of AFLP is the large number of polymorphisms that the method generates, being highly sensitive to polymorphism detection at the total genome-level. Other advantages are: (1) no sequence information is required; (2) the PCR technique is fast; and (3) a high multiplex ratio is possible (Rafalski *et al.*, 1996). The lack of sequence information needed by the AFLP method is similar to that of the randomly amplified polymorphic DNA (RAPD) technique. This is contrary to RFLP or TRFLP that need a high degree of characterization of the target gene.

Compared to other marker technologies, e.g., randomly amplified polymorphic DNA (RAPD), restriction fragment-length polymorphism (RFLP), or microsatellites (not discussed here), AFLP provides in general greatly enhanced performance in terms of reproducibility, resolution, and time

efficiency. By using AFLP, it is possible to evaluate more loci than with RFLP or RAPD. AFLP-based assays are cost-effective and can be automated.

Homology is perhaps the greatest challenge in AFLP analysis. While it is usually assumed that co-migrating bands are homologous, there is no *a priori* reason to accept this assumption. The problems of non-homology and non-independence of the AFLP data can result in erroneous estimates regarding similarity and distance. Additional problems include scoring, bias introduced by dominance, reproducibility problems, and the effect of polyploids.

Additionally, being a PCR-based technique, AFLP is limited by all PCR biases. The choice between restriction enzyme and primer can affect the number of AFLP polymorphisms detected. For example, more polymorphisms are detected in barley with the combination of restriction enzymes *Pst*I/*Mse*I than with *Eco*RI/*Mse*I (Ridout and Donini, 1999). The level of polymorphism that different markers reveal is important. If the marker reveals little variation, then it may not be possible to discriminate taxa.

11.4.5   Single Stranded Conformation Polymorphism Analysis (SSCP)

The single-strand conformation polymorphism (SSCP) technique distinguishes DNA molecules of the same size but with different nucleotide sequences. The principles of the method are based on the unique three-dimensional (3D) features of single-stranded DNA, which makes it possible to observe small changes in the nucleotidic sequence. Thus, single DNA strands have specific 3D structures based on their specific nucleotidic sequence. In the absence of a complementary strand, the single DNA strand may experience intra-strand base pairing, resulting in loops and folds that give the single strand a unique three-dimensional structure that will have specific and different motilities through an electrophoresis gel (polyacrylamide). In this context, a single nucleotide change can affect the strand's mobility through a gel by altering the intra-strand base pairing and its resulting three-dimensional conformation (Melcher, 2003).

For double-stranded DNA, mobility is affected by the length because all double strands have similar helical three-dimensional arrangement. On the contrary, the single DNA strands display different three-dimensional arrangements based on the succession of nucleotides in the single strain (so sequence) in order to allow different bindings between such nucleotides. The mobility of a single strand is thus noticeably affected by even small changes in sequence.

SSCP analysis requires that DNA is first extracted from the sample. Then PCR is conducted with a phosphorylated and a nonphosphorylated primer, both binding to conserved regions in usually the rRNA target genes. The double-stranded PCR products are converted to single strands by lambda exonuclease digestion of the phosphorylated strand. With SSCP, using this single-strand approach, community patterns are obtained from a diverse range of environmental samples, such as rhizosphere, compost, or soil. The patterns in the polyacrylamide gel can be visualized by silver-staining. Single bands of silver-stained profiles can be cut out of the gels, re-amplified by PCR, and sequenced directly or after cloning in *E. coli*.

The primary steps associated with SSCP are:

- DNA extraction;
- Amplification of DNA fragments by PCR using a phosphorylated and a nonphosphorylated primer for the target gene (usually rRNA);
- Denaturation of the amplified fragments to a single-strand form by exonuclease digestion;
- Separation of the denaturated amplified fragments, based on their polymorphisms by polyacrylamide gel-electrophoresis; and
- Visualization of the separated fragments by silver-staining or autoradiography and interpretation of the resulted pattern by hybridization with either DNA fragments or RNA copies synthesized on each strand as probes.

SSCP analysis is an inexpensive, convenient, and sensitive method for determining genetic variation (Sunnucks *et al.*, 2000). The method can be applied for the cultivation/independent analysis of microbial community diversity in environmental samples, based on PCR-amplified small sub-unit (SSU) rRNA gene sequences from directly extracted DNA. Under optimal conditions, approximately 80 to 90% of the potential base exchanges are detectable by SSCP (Wagner, 2002).

SSCP analysis can detect DNA polymorphisms and mutations at multiple places in DNA fragments. Many candidate genes are the same length but contain DNA sequence differences known as single nucleotide polymorphisms (SNP). SSCP analysis can be applied to search effectively and rapidly for point mutations or polymorphisms in mitochondrial DNA (Suomalainen *et al.*, 1992). Sarkar *et al.* (1992) have evaluated SSCP for the detection of single-base mutations in the factor IX gene. SSCP is more often used to analyze the polymorphisms at single loci, especially when used for medical diagnoses (Sunnucks *et al.*, 2000). Kuhn *et al.* (2004) investigated the use of high throughput SSCP (using a 16 capillary genetic analyzer) to separate alleles in three size classes and at four different temperatures as an alternative to the more expensive SNP techniques. The SSCP technique can be used in cloning studies. Thus, the PCR-based

SSCP procedure with degenerate oligonucleotide primers was used to identify and clone nine aquaporin genes in cucumber (Xie *et al.*, 2002).

SSCP analysis is more sensitive and efficient for discriminating different clones than restriction enzyme analysis (Xie *et al.*, 2002). Other advantages of using SSCP are that the PCR amplicons are smaller (200–300 bp) and easier to amplify when polymorphisms are detected and that the amplicons are of optimal size to sequence using an ABI automated sequencer.

Limitations to SSCP include the absence of mobility differences that are not correlated to the amount of sequence differences. Under some SSCP conditions, cytochrome b PCR products from one species are connected better to that from a different species with little or no correlation within subspecies of the same main species. Thus, the only information that can be gained from SSCP is whether PCR amplicons are "identical" or not. Second, the optimal amplicon size for detection of most point mutations is rather small, involving complicated strategies to deal with this limitation (e.g., dideoxy fingerprinting or cutting amplicons with restriction enzymes).

Single-stranded DNA mobility is also dependent on temperature and pH. For best results, gel electrophoresis must be run at a constant temperature and under lower pH. The fragment length may also affect SSCP analysis. For optimal results, DNA fragment size should fall within the range of 150 to 300 bp, although SSCP analysis of RNA allows for a larger fragment size (Wagner, 2002). Such limitations may be overcome. Thus, adding glycerol to the polyacrylamide gel lowers the pH of the electrophoresis buffer, more specifically the Tris-borate buffer, resulting in increased SSCP sensitivity and clearer data, as well as allowing for larger fragment size analysis. If the specific nucleotide responsible for the mobility difference is known, a similar technique called SNP may be applied.

11.4.6 Thermal and Denaturating Gradient Gel Electrophoresis (TGGE, DGGE)

Molecular fingerprinting techniques based on nucleic acids like PCR-DGGE have recently gained increased attention for the study of the impact of environmental stresses like presence of metals on the microbial community. The technique is based on the PCR amplification of short 16S ribosomal RNA (16S rRNA) or other gene sequences and involves the separation of amplicons with the same length but a different basepair sequence on polyacrylamide gels containing a linear gradient of a DNA denaturing agent. The genetic fingerprints obtained provide mostly a complex band profile which yields a picture of the genetic structure of the community as a whole (Muyzer *et al.*, 1993).

Denaturing gradient gel electrophoresis (DGGE) and temperature gradient gel electrophoresis (TGGE) can discriminate between DNA molecules generating specific patterns on the basis of differences in thermal stability caused by differences in base sequence. DGGE and TGGE techniques allow PCR products of the same length but of different sequence composition to be separated in gradient gels according to the melting behavior of DNA. The incomplete separation (unzipping) of the DNA molecule is achieved during these techniques as compared with the complete separation into single strands encountered for SSCP technique. When a portion of the DNA molecule melts, a drop in electrophoresis mobility occurs, resulting in the formation of a structure that is both helical and random, with a specific mobility in the electrophoreses gel.

While DGGE and TGGE are based on the same principle consisting in differentiating DNA based on thermal behavior of different sequences, the two techniques differ in the procedure by which the incomplete separation (denaturation) of the two-stranded DNA molecule is induced. Thus, DGGE is using a linearly increasing gradient of formamide and urea for inducing separation of the double-stranded amplification DNA products, while TGGE is using a linearly increasing temperature gradient for the same purpose. The targeted amplified gene may be 16S rRNA, 23S rRNA or species/genera specific. These analyses generate a community fingerprint, but sequence information can also be obtained since individual bands can be recovered and analyzed following DGGE/TGGE analysis.

The following steps are required in TGGE/DGGE testing:

- DNA extraction;
- Amplification of DNA fragments by PCR using primers for the target gene (usually rRNA);
- Incomplete denaturation of the amplified fragments (using two main modalities as described before corresponding to DGGE and TGGE respectively) and separation (based on fragment polymorphism) in gels containing a linear gradient of DNA denaturing agent or temperature; and
- Visualization of the separated fragments.

After the DNA is extracted from the sample, PCR amplification of the extracted DNA by PCR reaction is performed targeting usually the 16S rRNA genes for which primers (conserved regions) are well established. The primers used to amplify the 16S rRNA genes are designed to contain a GC clamp and to prevent the complete denaturation of DNA fragments (that may result in single strands such as in SSCP technique described above). The final step is the denaturation (separation of the two DNA strands), separation, and visualization of denaturated DNA fragments. Gel electrophoresis (in gels containing a linear gradient of DNA denaturing agent) is used as with most DNA fingerprinting techniques. As the DNA fragments move down the gradient they will denature except for the terminal GC clamp. The denaturation or melting dramatically reduces the mobility of the DNA fragments in the gradient gel. Amplification of the 16S rRNA genes within a community and subsequent analysis by DGGE give rise to a banding pattern in which each band may correspond to a single species. Such bands may be recovered from the gel for further analysis such as sequencing.

Analysis of PCR-amplified 16S rDNA gene fragments from environmental samples by denaturing gradients of chemicals or heat within polyacrylamide gels is a popular tool in microbial ecology (Bruggemann et al., 2000). These techniques are ideal for analyzing complex bacterial communities. Some studies have determined the reproducibility of the profiles generated by DGGE when using soil. Most find similarity index (SI) values of replicate analyses to be 0.90 or above (Duineveld et al., 1998; Kozdroj and van Elsas, 2000; Girvan et al., 2003). However, one study found within-group similarity to be predominantly comparable to between-group similarity for three distinct grassland soils (McCaig et al., 2001). The SI values obtained for replicate fingerprints of one of these grassland soils was 0.58, only slightly higher than comparisons of the most distant profiles (0.53; $p = 0.12$ as determined by Student's t-test) and the three soils could not be distinguished by principal component analysis (PCA) (McCaig et al., 2001). However, canonical variate analysis (CVA), a subjective technique that looks at the overall pattern of variation across the entire data set and maximizes between-group differences, did distinguish the three grassland soils. DGGE/TGGE techniques are similar to SSCP and are used to detect mutations and in general genetic variations. DGGE is efficient as a method for detection of

DNA sequence differences. Due to strand dissociation phenomena, however, its use has been limited to the analysis of sequences with a relatively low content of GC pairs. DGGE is a convenient tool for analyzing community shifts that involves only a small region (~200–300 bp) of the 16S rRNA gene.

Víglasky et al. (2000) used TGGE technique as a tool for detecting various conformational modifications of plasmid DNAs. The early melting temperature of a structural transition for any topoisomer is dependent on the value of superhelicity. Supercoiled topoisomers represent a system of molecules that is sensitive to changes in temperature. Basically, the authors showed that the topoisomers with the highest absolute value of superhelicity melt earlier than the topoisomers with lower values.

Heuer et al. (1997) developed a group-specific primer for comparison of sequences of genes encoding 16S rRNA (16S rDNA) for the detection of actinomycetes in the environment with PCR and temperature or denaturing gradient gel electrophoresis (TGGE or DGGE, respectively). These DNA fingerprinting methods were applied to monitor actinomycete community changes in potato rhizosphere and to investigate actinomycete diversity in different soils.

Müller et al. (2001) studied the impact of long-term mercury pollution on the soil microbial community in three soils originating from different locations and containing up to 511 mg Hg/kg. The highest polluted soil markedly affected the soil microbial community as resulted from DGGE analyses of 16S rRNA gene fragments and substrate utilization patterns. More particularly, an inverse relation between number of bands and mercury concentration was demonstrated, whereas the number of substrates utilized did not differ significantly between the three different soils. In addition, fungal biomass (measured as chitinase activity) did not significantly differ between different soils, whereas the size of bacterial and protozoal populations was reduced in the most contaminated soil (Müller et al., 2001).

Renella et al. (2004) studied the long-term effect of high cadmium concentration (ranging from 0 to 40 mg/kg) in sandy soils on the bacterial community structure and on biochemical functions including enzyme activities, respiration, and microbial biomass. These features were studied under two regimes, one cultured with maize and one as background. Since DGGE analyses showed slight changes in the maize cropped soils containing high Cd contents in comparison with the control soil, it was concluded that high Cd concentrations induced mainly physiological adaptations (as shown by higher respiration rates, lower biomass, and either reduced or unaffected enzyme activities) rather than selection for metal-resistant culturable soil microflora (Renella et al., 2004).

MacNaugthon et al. (1999) used a PCR-DGGE to target eubacterial 16S rRNA genes to investigate the effect of the addition of a mixture of toxic metals (500 mg/kg Cd, Cs, Sr and 1800 mg/kg Cs) on the prokaryotic community in sandy soil samples. The DGGE profile of the indigenous eubacterial community changed within 1 week compared to that of the pristine controls. The major changes in community structure in the metal-polluted microcosms consisted of the appearance of four novel bands not detected in the controls. Sequence analysis of these bands suggested that they related to species belonging to the genus Acinetobacter (2 bands) and Burkholderia (2 bands), both belonging to the Gram-negative division of the Prokaryotic organisms (MacNaugthon et al., 1999). In another study, Gremion et al. (2004) compared the microbial community present in uncontaminated soil samples with the one found in samples artificially contaminated mainly by Zn (5 mg/kg soil). Both 16S rRNA gene fragments of bacteria and β-proteobacteria and *amoA* (encoding the α-subunit of ammonia monooxygenase) gene fragments obtained from all different soil samples were targeted for analyses. Principal component analyses of DGGE data obtained for all amplified fragments revealed a significant difference between uncontaminated and contaminated samples. Since PCR-DGGE reveals differences in the total eubacterial soil community (both uncultured and cultured), but is limited in detecting bacteria constituting less than 0.01% of the bulk population (Heuer and Smalla, 1997), Gremion *et al.* (2004) emphasized the necessity of studying the microbial ecology by a combined approach of culture-independent and conventional methods. Therefore, the substrate utilization pattern of the soil microbial communities was studied using Biolog plates and it was indeed noticed that the bacterial functional abilities were, as is the case for the community structure, heavily affected by heavy metal contamination. However, Ellis *et al.* (2003) also used the combined approach of 16S rRNA gene-DGGE and plate counting of the culturable community and found results opposite to those mentioned above. The proportion of culturable plate-counted bacteria from the five studied soil samples (principal metal concentrations ranging from 671–16300 mg Zn/kg, 925–18800 mg Pb/kg, and 516–10300 mg Cu/kg) varied between 0.08 and 2.2% and these values correlated inversely with metal concentrations. The DGGE data obtained directly from different soil samples were highly similar. In addition, sequencing of DNA-bands extracted from the DGGE-gel revealed that bacteria belonged to 16 different genera including *Pseudomonas, Flavobacterium,* and *Bacillus* and this as well in the contaminated as noncontaminated soil samples. However, when the culture-based and culture-independent populations obtained from the different soils were compared, data showed that metal contamination did not have a significant effect on the total genetic diversity present but rather affected its physiological status. Therefore, Ellis *et al.* (2003) suggested that plate counts may be a more appropriate method for determining the effect of metals on soil bacteria rather than culture-independent approaches. Although most of the studies were performed on the 16S rRNA gene level, Sandaa *et al.* (1999) investigated the effect of the presence of heavy metals on the archaean population. More particularly, the impact of four different sludge amendments containing different metals (but mainly Zn at 102 to 359 mg/kg dm) at different concentrations on the microbial community present in soils was studied by DGGE of 16S rRNA gene fragments amplified from the total archaean DNA. Different banding patterns were found for the soils containing different metal loads indicating a difference in the archaean community structure in the soils, but no relationship between number of bands or archaeal diversity and metal concentration was found. In addition, analysis of cloned 16S rRNA gene fragments showed close similarities to a unique and globally distributed lineage of the kingdom *Crenarchaeota,* i.e., the group of terrestrial nonthermophilic strains that are phylogenetically distinct from currently characterized crenarchaeotal species. Furthermore, fluorescent *in situ* hybridization (FISH) was applied on the soil samples and showed a decrease in the percentage of stained *Archaea* cells from 1 to 3% in untreated soil to below detection limit in heavy metal amended soils. As for the eubacterial community, a negative effect of the heavy metals on the archaean community was observed by Sandaa *et al.* (1999).

The diversity of a microbial community in North Sea sediments was studied by three different PCR-DGGE protocols using identical general eubacterial primers, traditional plate counts, and chitinase activity after exposing the sediment samples to Cu (50 μg/l) (Gillan, 2004). Plate counts and chitinase activity measurements have suggested limited effects of Cu on growth rate and cell metabolism. The three DGGE protocols indicated that Cu had no immediate effect on the genetic diversity of the community. However, Cu-sensitive bacterial populations were detected by one of the DGGE protocols. Gillan cautioned that care must be taken for the interpretation of results obtained by an established DGGE protocol (Gillan, 2004).

Feris *et al.* (2003) studied the microbial community structure in sediments containing metal concentrations of Zn (6.77–433 mg/kg), Pb (2.59–69.8 mg/kg), Cu (1.14–332 mg/kg), and As (2.01–68.9 mg/kg). The sediment samples were examined microbiologically by using three different molecular techniques: (i) PLFA, (ii) PCR-DGGE, targeting part of the 16S rRNA gene in combination with cloning and sequencing of DNA-bands obtained from the DGGE-gel, and (iii) quantitative PCR, using four different group-specific primers targeting groups most closely related to α-, β-, and γ-Proteobacteria and cyanobacteria. The PCR-DGGE data revealed a significant linear relationship between the microbial community structure and sediment metal load. As observed by other researchers (MacNaugthon *et al.*, 1999; Sandaa *et al.*, 1999; Wenderoth and Reber, 1999), Feris *et al.* (2003) showed the predominance of Gram-negative bacteria in the microbial community after phylogenetic analysis of partial 16S rRNA gene sequences recovered from the different heavy metal–loaded sediments originating from different sediment samples. Sediment metal content gradients were positively correlated with group III abundance (γ-Proteobacterial types) and negatively correlated with group II (β-Proteobacterial types) abundance (Feris *et al.*, 2003). The seasonal dynamics of the hyporheic-zone microbial community structure in these sediment samples was evaluated for more than one year. DGGE pattern analysis showed a linear relation between differences in microbial community composition and differences in heavy metal content of the sediments throughout the year, whereas correlations between metal content and abundance of the four groups α-, β-, and γ-Proteobacteria and cyanobacteria only appeared during fall and winter (Feris *et al.*, 2004b). Feris *et al.* (2004a) determined the rates of change in hyporheic microbial communities in response to metal contamination and the differences in resiliency in response to metals for these four different groups. For all four groups, heavy metal treatment negatively affected their abundance, although all groups were able to recover from the metal treatment, each to a different extent and at a unique rate. From these studies, it can be concluded that hyporheic microbial communities are a sensitive indicator of heavy-metal contamination in streams and might be used as a tool for environmental quality analyses (Feris *et al.*, 2004a).

One limitation of DGGE and TGGE is that the limited resolution of the acrylamide gels used to separate bands may lead to under-representation of the number of genotypes. McCaig *et al.* (2001) found only 77 bands in DGGE fingerprints when clone libraries of the same soil DNA contained ~130 distinct 16S rRNA gene sequences. This underrepresentation may result from the inability to detect minor components of the profile on gels, or from the co-migration of distinct gene sequences resulting in single bands containing multiple sequences. This may underestimate community diversity.

There are limitations associated with any soil DNA analysis technique, regardless of the technology used to generate and analyze profiles. However, the solution is to choose the technology best suited for its intended application. It has been suggested that due to the increased sensitivity, ease of operation, and more objective comparisons associated with the automation of its analysis, TRFLP is better suited to highly diverse communities. Examples of these communities include low or moderate complex systems, such as found in marine or freshwater samples (Lukow *et al.*, 2000). This claim is supported by difficulties (smears and low resolution of bands) encountered by researchers when applying DGGE to bulk soil communities (Ovreas and Torsvik, 1998; Alvey *et al.*, 2003). DGGE and TGGE analysis have the advantage over TRFLP of offering the ability to determine sequence information from profiles. This can be achieved by hybridizing profiles with labeled probes or by recovering DNA from the gels for sequencing, by excising bands. Such procedures are not so readily applicable in post-T-RFLP analysis.

11.4.7 Amplified Ribosomal DNA Restriction Analysis (ARDRA)

Amplified ribosomal DNA restriction analysis (ARDRA) is based on restriction endonuclease digestion of amplified ribosomal DNA (coding usually for 16S rRNA gene) (Massol-Deya *et al.*, 1995). The technique consists of PCR amplification of the ribosomal DNA-intervening internal transcribed spacer regions with universal primers followed by the analysis of the PCR products by the principle of single-strand conformation polymorphism (SSCP) method.

The following are the main testing steps for ARDRA analysis:

* Cultivation of the targeted microorganisms and microbial isolates;
* DNA extraction from microbial isolates;
* DNA amplification by PCR, using 16S rRNA gene universal primers;
* Digestion of amplified DNA fragments by endonuclease restriction with one or more restriction enzymes (selection of restriction enzymes should be on the basis of simulated digests of the complete 16S rRNA gene sequences of chosen reference strains);
* Separation and visualization of the digested DNA fragments based on their polymorphisms, by agarose gel electrophoresis (gels stained with ethidium bromide); and
* DNA visualization by transillumination with ultra-violet light after the gels are stained with ethidium bromide.

Amplified ribosomal DNA restriction analysis has been used for rapid and unambiguous species differentiation within the genus *Acinetobacter* (Seifert *et al.*, 1997). The PCR-ARDRA procedure is used as a reliable and rapid method for identifying *Lactobacillus* species from intestinal and vaginal microflora at species and subspecies level (Ventura *et al.*, 2000). ARDRA is also used to classify *Aspergillus Section Flavi*. ARDRA is capable of classifying 67 of the 68 *Aspergillus flavi* strains tested into the following four groups, regardless of origin: *A. flavus*/*A. oryzae, A. parasiticus*/*A. sojae, A. tamarii,* and *A. nomius.*

De Baere *et al.* (2002) combined automated culture with rapid genotypic identification ARDRA for the detection and identification of all mycobacterial species, instead of attempting direct PCR-based detection from clinical samples of *M. tuberculosis* only. The assessment of microbial community structure changes by amplified ribosomal DNA restriction analysis can also be performed (Gich *et al.*, 2000).

Rahman *et al.* (2003) used molecular and culture-based methods to characterize thermophilic bacteria associated with the subsurface soil environment in Northern Ireland. They were screened by amplified ribosomal DNA restriction analysis prior to 16S rRNA gene sequencing. Gigh *et al.* evaluated the suitability of this method for detecting and monitoring changes in activated sludge systems. The method was efficiently applied to detect differences in activated sludge bacterial communities fed on domestic or industrial wastewater, and subjected to different operational conditions.

A disadvantage of ARDRA is that it is culture-dependent. Thus more time is required for this analysis. ARDRA is also limited to a small number of microbial species that can be isolated and grown in the lab. However, culture is needed after all to assess the antibiotic susceptibility of the strains, too (De Baere *et al.*, 2002). In addition, ARDRA procedure is not suited for typing analysis, as it overlooks intra-specific differences.

11.4.8 Randomly Amplified Polymorphic DNA (RAPD)

Randomly amplified polymorphic DNA (RAPD) is based on the differences between DNA sequences. RAPD reactions are PCR reactions, but they amplify segments of DNA which are essentially not a *priori* known (randomly chosen). RAPD consists of fishing for the sequence using random amplification. As compared with standard PCR where a primer sequence (designed based on preexisting knowledge of the target sequence) is used specifically to amplify a known sequence of an organism's genome, in RAPD random primer sequences may be used in organisms where a specific genome sequence is not known. Thus, RAPD can be applied to create a biochemical fingerprint of an organism.

The following are the primary testing steps for RAPD analysis:

* DNA extraction;
* Amplification of DNA fragments by PCR using randomly chosen primers (about 10 bp) at low annealing temperatures at multiple loci;
* Separation of the amplified fragments based on their polymorphisms by agarose gel electrophoresis (two-dimensional gel);
* Visualization of the separated fragments;
* Possible cloning and sequencing of the DNA band of interest to develop longer specific PCR primer pairs targeting reproducible amplification and detection of the differences.

The genomic DNA is cut and amplified using PCR, with short single randomly chosen primers at low annealing temperatures, resulting in amplification at multiple loci. The primers have arbitrary sequences, typically 10 base pairs that under non-stringent PCR conditions have a moderate-to-high probability of annealing to the target DNA in forward and reverse orientations within the "amplification range" (approximately 3 Kilobases of each other). A single random primer can produce one or more products by PCR, depending on the number of target sites present in DNA from a particular species or individual. Then, by running an agarose gel electrophoresis, it is possible to examine if DNA segments were amplified when the arbitrary primer was used. To compare the results a two-dimensional electrophoresis gel is used to determine the change in sequence pattern by superimposing the two gels. One extension of the RAPD method is to clone and sequence RAPD fragments that differ between species or within species. The identification of the band of interest may follow; the gel is cut and the DNA of interest from that particular band is isolated and sequenced. Such resultant sequences can be used to develop longer specific PCR primer pairs targeting reproducible amplification and detection of the differences. In addition, these sequences make it possible to determine which known genes are most similar to the RAPD fragment.

The RAPD method is typically used for genetic mapping and studies of population genetic structure. RAPD provides an approach to find polymorphisms within species, or genetic differences between species. RAPD is a method of producing a biochemical fingerprint of a particular species. Relationships between species are determined by comparing their unique fingerprint information.

The RAPD technique has infrequently been used to infer phylogeny. The main reason is the difficulty to know if bands of the same size (amplified using a particular primer) are homologous across species. Shianna *et al.* (1998) used the technique to determine whether variations exist among bovine strains isolated from a localized geographic area (watershed of the Red River) to compare 16 isolated strains to each other along with two human and two calf strains from Australia. A statistical analysis of the data indicated that the isolates belonged to four different groups of strains.

In order to test for genetic differentiation in fish populations within the Rio Dolce Basin of southeastern Brazil (in the freshwater fish *Hoplias malabaricus* (trahira), a widespread predatory characin and one of the few resilient native fishes) for bio-geographic relationships among populations of this species in other basins, a study (Dergam *et al.*, 2002) was conducted using RAPD analysis. In another application, Krane *et al.* (1999) used the RAPD profile-based measures of genetic diversity in crayfish for correlation with known environmental impacts. Benecke (1998) used RAPD to identify arthropods and support classical morphological and medico-legal analysis of maggots on a human corpse. Maggots on the inside of a body bag were identical with maggots found on the outside of the bag and the pupae found on the floor under the corpse.

The greatest advantage of RAPD is the relative speed (a complete analysis in one day) and ease with which results can be obtained, without the need for blotting or radioactive hybridization. RAPD analysis is inexpensive relative to RFLPs. Romalde *et al.* (2002) found the RAPD procedure to be rapid and easier to perform as compared to ribotyping and pulsed-field gel electrophoresis techniques (for molecular typing purposes).

The RAPD protocol is more efficient than other techniques for generating random DNA markers. Cloning of RAPD fragments is rapidly accomplished after the simple recovery of ethidium bromide-detected bands, and cloned RAPD loci will not contain repetitive sequences. RAPD also lacks specificity, due to low annealing temperatures and easier reaction conditions. RAPD can, however, provide a starting point for further DNA-based studies.

By RAPD, a polymorphism detected between one pair of strains may not translate into use for another pair of strains. Moreover, it is not possible to distinguish animals that are heterozygous at any locus from those that are homozygous for the '+' allele. Thus, on average, only half of the RAPD polymorphisms detected between two strains would be mappable. A difficulty of RAPD markers is that most of them are dominant, and this makes it difficult to use the data in studies of population genetics. RAPD is sometimes not reproducible. An explanation for this difficulty is that it can be sensitive to the amount of DNA template and free Mg present in the PCR reaction. However, with careful standardization, it is possible to identify RAPD markers that are reproducible.

11.4.9 Non-PCR DNA Fingerprinting Techniques with Applicability in Forensic Studies

Although non-PCR DNA fingerprinting methods do not benefit from all the advantages of PCR, they have potential for forensic studies, with proven applicability for tracking sources of water pollution. The two non-PCR DNA methods described are restriction fragment length polymorphism (RFLP) and ribotyping. While their applicability to environmental forensic investigations is emerging, an understanding of these methods is important for identifying future environmental forensic approaches.

11.4.9.1 Restriction Fragment Length Polymorphisms (RFLP)

Prior to the advent of PCR, RFLP was a popular tool for inferring evolutionary trees or assessing the phylo-geographic structure of populations. The RFLP method is a non-PCR alternative to the AFLP method.

The primary process steps associated with RFLP are:

- DNA extraction;
- Digestion of DNA by endonuclease restriction;
- Separation of the resulted fragments by agarose gel-electrophoresis;
- Transference of the fragments to a nylon filter by Southern blotting;
- Hybridization of fragments to a locus-specific radiolabeled DNA probe; and
- Visualization of the polymorphisms by autoradiography.

The sample DNA is extracted and cut into segments with special restriction enzymes (endonucleases). The resulted segments are radioactively tagged to produce a visual pattern known as "DNA fingerprint" on X-ray film and are separated by electrophoresis, which sorts the fragments by length. There is no need for PCR amplification of DNA in this method.

RFLPs are used to identify the origins of a particular species for the purpose of mapping its evolution. An important use of RFLPs is to characterize a large number of individuals, given the preexisting knowledge of the sequence variation. For example, when there are three different species of a particular nematode genus in a region, the sequence data of these three species can likely be used to determine the species based on restriction digestion.

RFLP mapping offers several advantages over conventional genetic analysis, including a higher degree of polymorphism and the possibility of obtaining an almost infinite number of probes which are evenly distributed over the whole genome.

The main advantages of this technique are:

- High stability and reproducibility, giving constant results over time, and location;
- High genomic abundance;
- Random distribution throughout the genome;
- Band profiles interpretation in terms of loci and alleles;
- Co-dominance of alleles; and
- Producing of semi-dominant markers, allowing determination of homozygosity or heterozygosity.

Limitations with the RFLP method of DNA fingerprinting are that the results do not specifically indicate the chance of a match between two organisms and the process is very expensive. Restriction fragment patterns cannot be considered valid markers for inferring phylogeny. This is because fragment patterns for a variable segment of DNA are often not composed of independent characters. For example, for the same evolutionary event, the possible gain of a restriction site within a region of DNA will cause two bands to be produced for only one DNA fragment. Such two bands may falsely appear as two characters supporting shared ancestry for the individuals or species having them, which clearly violates the assumption of character independence. This limitation is overcome by mapping restriction sites rather than fragments, which is not always easy but allows the investigator to use the presence/absence of restriction sites as characters in phylogenetic analysis.

Disadvantages include:

- Large quantities of purified, high molecular weight DNA required ($5–10\,\mu g$);
- Laborious and technically demanding;
- High costs;

- Additional development costs in case suitable probes are unavailable;
- Not amenable to automation;
- Very long methodology before results are gained;
- Must frequently work with radio isotopes;
- Many probes are not available depending on species;
- Too many polymorphisms may be present for a short probe; and
- Low frequency of desired polymorphisms in polyploid plants (e.g., wheat).

11.4.9.2 Ribotyping (with or without PCR)

Ribotyping is a DNA fingerprinting technique that analyzes the DNA coding for the makeup of the cell structures called ribosomes, where proteins are synthesized. These structures are composed of ribosomal DNA and various proteins. During protein synthesis, the messenger RNA and the amino acids are transferred to the ribosome, which is gradually moved along the messenger RNA to place the amino acids in the coded order and form a specific protein. The ribosomes are conserved within species. Subsequently, ribosomal genes are highly conserved, with little variation within the same species. However, small variations in the ribosomal DNA structure (variable fragments) exist within the same species of bacteria such as *E. coli*. This characteristic allows for a greater ability to distinguish between different bacterial strains (based on their ribosomal DNA polymorphism).

Traditional ribotyping is modeled as RFLP analysis and is typically used to analyze bacteria and other microorganisms. Ribotyping consists of the digestion of chromosomal DNA with restriction enzymes followed by the analysis of the digested fragments obtained. This analysis could be performed by two main methods: Southern blotting or PCR.

Southern blotting with rRNA probes in order to generate DNA banding patterns allows subtype differentiation of bacterial isolates beyond the species and subspecies levels. It uses the so-called universal probes targeted at specific conserved domains of ribosomal RNA coding sequences.

Depending on the protocols used, the resultant band patterns are differentiated by length and compared with known species and strains of organisms to determine genetic and evolutionary relationships (Ge and Taylor, 1998).

The following are the main ribotyping steps:

- Cultivation of targeted organism (optional step);
- DNA extraction;
- Digestion of extracted DNA with restriction enzymes;
- Analysis of the digested DNA fragments by Southern blotting or PRC followed by separation in an electrophoresis gel; and
- Discriminant analysis of the resulted banding patterns for strain identification.

DNA extraction is the first step and consists of the extraction and purification of DNA from the sample. In the case of tracking sources of fecal pollution in water, the cultivation of the targeted organisms is firstly required (such as *E. coli*) to insure that the resulted DNA pattern is an expression of whatever is alive, thus responsible for pollution at that moment, in that water.

The second step is the digestion of DNA with restriction enzymes followed by cutting the extracted DNA into many fragments with restriction enzymes. Restriction enzymes always make their cuts in the same place (for example between the C and G), not discriminating among different organisms as long as the DNA is in an accessible an form. The next step is the analysis of the digested DNA fragments. The two main procedures to analyze the resulted DNA fragments are PCR and Southern Blotting.

For an analysis by PCR, the resulted digested DNA is amplified with primers specific for ribosomes (to amplify conserved rRNA coding regions) and then transferred and separated onto an electrophoresis polyacrylamide gel.

If sufficient quantities of DNA are present, the Southern blotting technique can be used. The DNA is probed with ribosomal DNA probes and the banding patterns will appear on the gel (with PCR) or blotter (with Southern Blotting). Discriminant analysis is then used to compare these resulted banding patterns with the banding patterns of other known strains of the same species that were initially cultivated (e.g., *E. coli*). A statistically acceptable database (sufficiently large and varied) for the studied strains is required for this comparison analysis. Ribotyping represents a molecular subtyping method with widespread applications, including in subtyping of foodborne pathogens and of microorganisms important for fermentation and food spoilage. PCR-based ribotyping is also a useful technique for genotyping methicillin-sensitive *S. aureus* strains (Oliveira and Ramos, 2002).

Major limitations of most molecular typing methods include:

- a lack of standardization;
- high cost per isolate;
- discrimination is dependent upon the restriction enzyme and the probe used;
- time-consuming procedure;
- a need for highly skilled technical staff; and
- access to specialized equipment.

In environmental forensics, ribotyping has been shown to be an effective tool for bacterial source tracking (BST). Ribotyping is an excellent tool for indicating major sources of biological pollutants since most BST projects only need to determine broad groups of contamination. In the context of BST investigations, ribotyping is an effective tool for discriminating between human and non-human forms of *E. coli*. It has also been shown to be effective in distinguishing *E. coli* from major animal groups. The statistical probabilities for distinguishing the major animal groups are greatly enhanced when comparison samples are submitted.

A broad range of applications are related to tracking the source of fecal pollution in water. Fecal pollution affects the quality and safety of water systems used for drinking, recreation, and in the harvesting of seafood. It can originate from numerous sources, including sewage treatment plant discharges, failing septic systems, agricultural and urban runoff, improper disposal of wastes from boats, and wildlife. Ribotyping can now be used to track *E. coli* that are re-growing and proliferating in a watershed. The study of Scott *et al.* (2004) summarizes one of the few case studies in which ribotyping was successfully utilized to resolve a "true-life" water quality and management problem consisting in tracking sources of fecal pollution in a watershed in South Carolina. The ability of ribotyping to differentiate between human and nonhuman sources of fecal pollution (generally *E. coli*) in water was demonstrated in the study of Parveen *et al.* (1999) who applied ribotyping on a total of 238 *E. coli* isolates from human sources (HS) and non-human sources (NHS) collected from Apalachicola National Estuarine Research Reserve, from associated sewage plants, and directly from animals. The discriminant analysis of ribotyping profiles showed that 97% of the HS isolates and 100% of the animal fecal isolates were correctly classified.

Carson *et al.* (2001) also used ribotyping to successfully identify individual host sources of fecal *E. coli*, and distinguish between fecal *E. coli* of human and nonhuman origin, such as fecal *E. coli* ribotype patterns from human and seven nonhuman hosts. Such application could assist in formulation of pollution reduction plans.

In another study also related to sources of fecal pollution of water, Scott *et al.* (2003) were applying ribotyping analysis to *E. coli* isolates obtained from humans, beef cattle, dairy cattle, swine, and poultry from different locations in Florida. The study indicated that using a single restriction enzyme (*Hind*III) ribotyping cannot differentiate *E. coli* isolates from different animal species (at least those sampled in the mentioned study). A possible conclusion of the study was that a combination of geographic and environmental variation may play a significant role in affecting the ability of ribotyping for identification of sources of *E. coli* in the environment. The same study indicated and confirmed, however, that ribotyping procedure can be used effectively to differentiate *E. coli* between human and non-human sources when applied to organisms isolated from a large geographic region.

Fontana *et al.* (2003) have used Automated Ribotyping and Pulsed-Field Gel Electrophoresis for the rapid identification of multi-drug-resistant *Salmonella* serotype Newport (Fontana *et al.*, 2003). A project conducted at University of Georgia, Athens, GA, described at the website http://alpha.marsci.uga.edu/coastalcouncil/hartel_ribotyping.htm, was aimed at applying ribotyping to determine the host origin of fecal contamination in Georgia's coastal waters. As a result of the project, 400 *Enterococcus faecalis* ribotypes (200 each from birds and humans) were also added to this host origin database.

11.4.10 Forensic Interpretation of DNA Data

The various DNA techniques that generate DNA community fingerprints should be amenable to the same range of similarity analyses. To date, a range of similarity indices have been used (Liu *et al.*, 1997; Dunbar *et al.*, 2001; Horswell *et al.*, 2002; Blackwood *et al.*, 2003), including the simple Sorenson's Similarity Index (also known as the Jaccard Coefficient) and more quantitative methods that make use of more of the profile information. The best choice of index depends largely on the molecular profiling method used, e.g., a gel-based method as opposed to a capillary sequencer method (as is increasingly used in TRFLP).

Sorenson's Index is a simple calculation based on the presence or absence of peaks in a profile (Figure 11.4.6) and ranges from 1 for identical profiles to 0 for profiles that share no common bands. It utilizes no other information, unlike other quantitative methods that take the band intensities or peak heights into account.

Horswell *et al.* (2002) used the Sorenson's Similarity Index and cautioned that it may be too simplistic to use as evidence in a court of law. It was soon realized, after the introduction of human DNA evidence to the courtroom, that statistical interpretation of comparisons is crucial and that it must be backed up with statistically sound validation.

A disadvantage of the Sorenson's Index is that a very high similarity index value can be determined if the same

$$SI = \frac{2n(AB)}{n(A) + n(B)}$$

SI = Sorenson's similarity Index

$n(AB)$ = Number of matching peaks

$n(A)$ = Number of peaks in profile A

$n(B)$ = Number of peaks in profile B

Figure 11.4.6 *Equation for Sorenson's Index calculation between two data sets.*

peaks are present in two samples, even if they have different heights. This situation would arise where communities from similar soils or from nearby sites have very similar populations, but in widely differing proportions. The problem created by differing proportions of bacterial species is particularly obvious when profiles have only a few peaks. Here, two profiles may appear different, but the Sorenson's Index indicates that they are very similar. Another disadvantage to Sorenson's Index is that all peaks are weighted equally, regardless of whether they are major peaks, accounting for a large percentage of the total fluorescence, or minor ones and only small contributors. Unavoidable experimental errors, such as small variations in the amount of template DNA put into a PCR, can affect whether small peaks are above, or fall below, detection thresholds. Running duplicate profiles from a single sample might result in some minor peaks being detected in one profile but not the other, while the majority of peaks remain relatively constant. The absence of even these minor peaks will result in a disproportionately low similarity index value using the Sorenson's Index.

There is debate over whether simplistic presence/absence comparison methods or more quantitative techniques, which take into account peak height or area, are more appropriate for forensic comparison. A drawback to the use of quantitative methods is that variation in peak heights of replicate TRFLP profiles electrophoresed on the same gel can average ~10% (range of 4.2–18.3%) of the peak height (Osborn *et al.*, 2000) or 7% after rigorous standardization of the data (Dunbar *et al.*, 2001). When compounded over a large number of peaks within each profile, such variation can lead to an appreciably reduced similarity between replicate profiles. The first paper describing the TRFLP technique (Liu *et al.*, 1997) utilized both peak-area sensitive analysis and the Sorenson's Similarity Index to analyse TRFLP profiles generated from four microbial environments. The Sorenson's Index analysis discriminated the four microbial communities with similarity indices ranging from 0.38 to 0.48 and grouped two replicate control samples (size references) with a similarity of 0.98. The area sensitive analysis was also capable of discriminating the microbial communities, but the similarity indices were higher (0.5–0.6 similarity) and the control samples showed reduced similarity (0.95).

In forensic applications, high similarity indices for replicates and low similarity indices for distinct microbial communities are desired. Because evaluation of different comparison indices for forensic analysis is still in its infancy, a definitive method is yet to be established as the most likely to be used in evidential analysis.

One factor for consideration when using molecular techniques is the geographical soil microbial variation as the physical and chemical characteristics of soil vary greatly over short distances (Wood, 1995; Junger, 1996; Prosser, 1997; Stotzky, 1997). Soil bacterial communities also vary between different sites (Horswell *et al.*, 2002; Zhou *et al.*, 2002). The degree of variation in these communities between sample sites in close proximity, however, is not well established. It has been shown that microbial biomass and activity are spatially dependant at scales less than 1 m in response to nutrient availability (Robertson *et al.*, 1988; Smith *et al.*, 1994), but variation in the actual community structure is less well known. A study performed by Cavigelli *et al.* (1995) suggested that spatial variation in a cultivated field was at the scale of individual soil aggregates or of the rhizosphere of individual plants. Using temperature gradient gel electrophoresis (TGGE), Felske and Akkermans (1998) determined that soil samples collected 1 m apart had the same prominent bacteria present, but did not investigate the more minor variation. Saetre and Baath (2000) showed that microbial variation in a forest environment is influenced by the location of trees and hence there is less spatial variation over small distances. Grundmann and Debouzie (2000) examined variation in NH_4^+ and NO_2^- oxidizing bacteria at millimeter intervals along 10 cm transects, and discovered that heterogeneity occurs even at this level. Spatial variation is an important factor to consider when comparing soil samples for forensic purposes, because if variation is high between sites in close proximity, sampling the exact spot where the sample in question originated from would be necessary to get a conclusive result. If variation in soil bacterial communities is low over short distances, but higher in more geographically distinct areas, then linking a sample with a large area is possible.

11.5 ISOTOPIC TESTING AND CORRELATION TO CONTAMINANT SOURCE

Stable light isotopes in culture water and medium components are accumulated in growing bacterial cells. Radioactive isotope accumulation by growing cells or viral particles can provide an internal clock for estimating when a microorganism was produced. Tritium naturally occurs at low levels and decays with a half life of 12.41 years. Similar to carbon dating of antiquities, tritium dating can be used to date recent materials. This approach has been used to date groundwater from the time of entry in the earth (isolation from the atmosphere) to the time of analysis (Poreda et al., 1988).

Stable isotopes of carbon and nitrogen and their ratio in microbial cells carry a strong geographic signature. Stable isotope ratio analysis of pathogens can provide information about the condition in which the agent was produced, complementing the information obtained from genotyping and phylogenetic trace analysis isotope ratio as a tool in microbial forensics that has been explored recently (Kreuzer-Martin et al., 2003). Kreuzer-Martin and the co-workers (2004a,b) demonstrated that a source of a micro-organism (spores of Bacillus subtilis used in the study) carries a specific signature of the environment where in they were grown. The authors demonstrated that microbial isotopic composition is a function of growth medium (Kreuzer-Martin et al., 2004a) and isotopic variation among different growth media can be used as a tool for sourcing origin of bacterial cells or spores (Kreuzer-Martin et al., 2004b). Also, Kreuzer-Martin et al. (2003) correlated the stable oxygen and hydrogen isotope composition of B. subtilis spores with the isotope composition of spatially divergent water sources. Given a uniform medium and variable water sources, the deuterium/hydrogen ($^2H/^1H$) and $^{18}O/^{16}O$ ratios of spores varied with the composition of the water source. The water-influenced stable isotope composition of spores has the potential to define the point of production of a microorganism. However, the medium also influences the hydrogen and oxygen isotope composition of the agent. The hydrogen, carbon, and nitrogen stable isotope composition of culture medium was shown to affect the isotope composition of spores (Kreuzer-Martin et al., 2004a,b). The physiological differences between C_3 and C_4 (main photosynthetic paths) plants are the basic cause of isotope variability in biological material. This difference results in a distinct $\delta^{13}C$ value in their organic molecules which is incorporated to animal tissues when animal feed on these plant products. The isotope ratio values of bacterial culture media show the variations based on biological source of the media components (animal or plant origin). The range of variation of ^{13}C, ^{15}N, and ^2H content of bacteriological media can yield differences in microbe isotope ratios which are readily measurable (Kreuzer-Martin et al., 2004b).

11.6 CONCLUSIONS

Soil is frequently encountered as trace evidence in environmental forensic investigations, but because of the limitations of current analytical techniques, this evidence is rarely utilized. Recent applications of molecular biology have provided tools (different methods known under the generic name of DNA fingerprinting) to describe and characterize soils, based on cultivation-independent fingerprinting of their bacterial community DNA. Moreover, the DNA fingerprinting techniques are successfully applied to distinguish sources of water contamination. These molecular microbiological techniques have the potential to enhance the range of tools available to forensic scientists. This chapter presents the main classic and emerging (DNA fingerprinting) microbiological tools available, emphasizing on the DNA fingerprinting of soils which represent the basis for developing future forensic methods.

Soil DNA profiling or fingerprinting is based on the analysis of DNA extracted directly from environmental samples. They are generally PCR-based techniques (such as TRFLP, DGGE/TGGE, AFLP, RAPD, ribotyping, etc.), analysing PCR-amplified genes from microbial community DNA, targeting usually the 16S rRNA gene, and resulting in fingerprinting of community DNA. A summary of some commonly used bacterial community analysis techniques with or without targeting 16S rRNA gene are synthesized in Tables 11.4.2 and 11.4.3, respectively. Like all amplification-based methods, these techniques are subject to the limitations and shortcomings of PCR. In addition, soil is a highly complex and dynamic medium with considerable variability, making DNA extraction and analyses problematic. However, after solving the technical problems associated with working with soils and recognising the limitations of molecular biology techniques (such as PCR bias), DNA fingerprinting of soils has the potential to become a valuable tool in forensic soil profiling.

Direct research into the applicability of community DNA profiling for forensic investigations has historically been limited. However, the basic requirements for a forensic technique, such as robustness and reproducibility, have been demonstrated in numerous environmental-based studies. With further validation, soil DNA fingerprinting techniques will emerge as an invaluable asset to environmental forensic scientists by making forensic soil analysis simple, reliable, routine, and able to withstand detailed scrutiny when used as evidence in a court of law. Molecular tools for soil analysis offer the significant advantage of using equipment and concepts that are already available and in use in most forensic DNA laboratories worldwide. Such techniques can provide the basis for developing innovative forensic methods tracking the migration of a contaminant through the soil or any other environment and for identifying the source and possible the age of contamination.

11.7 ACKNOWLEDGEMENTS

The authors wish to thank Dr Arpad Vass, Dr Tom Speir, and Dr Geoff Chambers for their contributions, and also the University of Tennessee's Law Enforcement Innovation Center.

The authors are also addressing their thanks to Krissy Lovering from DPRA Inc. for her valuable help with illustrations.

REFERENCES

Aggarwal, R.K., Brar, D.S., Nandi, S., Huang, N., and Khush, G.S. 1999. Phylogenetic relationships among *Oryza* species revealed by AFLP markers. *Theoretical and Applied Genetics.* **98**: 1320–1328.

Alexander, M. 1977. *Introduction to soil microbiology.* John Wiley & Sons, New York.

Alvey, S., Yang, C.H., Buerkert, A., and Crowley, D.E. 2003. "Cereal/legume rotation effects on rhizoshpere bacterial community structure in West African soils." *Bio. Fertil. Soils.* **37**: 73–82.

Amann, R.I., Ludwig, W., and Schleifer, K.H. 1995. "Phylogenetic identification and in situ detection of individual microbial cells without cultivation." *Microbiological Reviews.* **59**: 143–169.

Anderson, T.H., and Domsch, K.H. 1990. Application of eco-physiological quotients (qCO_2 and qD) on microbial biomass from soils of different cropping histories. *Soil Biol. Biochem.* **22**: 251–255.

Bååth, E., Arnebrandt, K., and Nordgren, A. 1991. Microbial biomass and ATP in smelter-polluted forest humus. *Bull. Environ. Contam. Toxicol.* **47**: 278–282.

Bardgett, R.D., and Saggar, S. 1994. Effects of heavy metal contamination on the short-term decomposition of labelled [^{14}C] glucose in pasture soil. *Soil Biol. Biochem.* **26**: 727–733.

Barenkamp, S.J., Munson, R.S. Jr., and Granoff, D.M. 1981. Subtyping isolates of *Haemophilus influenzae* type b by outer-membrane protein profiles. *J. Infect. Dis.* **143**: 668–676.

Barkay, T., Tripp, S.C., and Olson, B.H. 1985. Effect of metal-rich sewage sludge application on the bacterial communities of grass-lands. *Appl. Environ. Microbiol.* **49**: 333–337.

Beismann, H., Barker, J.H.A., Karp, A., and Speck, T. 1997. AFLP analysis sheds light on distribution of two *Salix* species and their hybrid along a natural gradient. *Molecular Ecology.* **6**: 989–993.

Benecke, M. 1998. Random Amplified Polymorphic DNA (RAPD) typing of necrophageous insects (diptera, coleoptera) in criminal forensic studies: Validation and use in practice. *Forensic Science International.* **98**: 157–168.

Bettelheim, K.A., Ismail, N., Sinebaum, R., Shooter, R.A., Moorhouse, E., and O'Farrel, S. 1976. The distribution of serotypes of *Escherichia coli* in cow pats and other animal materials compared with serotypes of *E. coli* isolated from human sources. *J. Hyg.* **76**: 403–406.

Blackwood, C.B., Marsh, T., Kim, S.H., and Paul, E.U. 2003. "Terminal restriction fragment length polymorphism data analysis for quantitative comparison of microbial communities." *Appl. Environ. Microbiol.* **69**: 926–932.

Bradshaw, J.E., Hackett, C.A., Meyer, R.C., Milbourne, D., McNichol, J.W., Philips, M.S., and Waugh, R. 1998. Identification of AFLP and SSR marker associated with quantitative resistance to *Globodera pallida* (Stone) in tetraploid potato (*Solanum tuberosum* subsp. *tuberosum*) with a view to marker-assisted selection. *Theoretical and Applied Genetics.* **97**: 202–210.

Brodie, E., Edwards, S., and Clipson, N. 2003. Soil fungal community structure in a temperate upland grassland soil. *FEMS Microbiology Ecology.* **45** (2): 105–114.

Brookes, P.C., and McGrath, S.P. 1984. Effects of metal toxicity on the size of the microbial biomass. *J. Soil. Sci.* **35**: 341–346.

Brookes, P.C., McGrath, S.P., and Heiijnen, C.E. 1986. Metal residues in soils previously treated with sewage sludge and their effects on growth and nitrogen fixation by blue-green algae. *Soil Biol. Biochem.* **18**: 345–353.

Bruggemann, J., Stephen, J.R., Chang, Y.J., Macnaughton, S.J., Kowalchuk, G.A., Kline, E., and White, D.C. 2000. Competitive PCR-DGGE analysis of bacterial mixtures: An internal standard and an appraisal of template enumeration accuracy. *J. Microbiol Methods.* **40** (2): 111–123.

Brunk, C.F., Avaniss-Aghajani, E., and Brunk, C.A. 1996. A computer analysis of primer and probe hybridization potential with bacterial small subunit rRNA sequences. *Appl. Environ. Microbiol.* **62**: 872–879.

Brunner, I., and Schinner, F., 1984. Einfluss von blei und cadmium auf die mikrobielle aktivität eines bodens. *Die Bodenkultur.* **35**: 1–12.

Buckley, D.H., and Schmidt, T.M., 2001. "The Structure of Microbial Communities in Soil and the Lasting Impact of Cultivation." *Microbial Ecology.* **42**: 11–21.

Budowle, B., Schutzer, S.E., Einseln, A., Kelley, L.C., Walsh, A.C., Smith, J.A., Marrone, B.L., Robertson, J., and Campos, J. 2003. Building microbial forensics as a response to bioterrorism. *Science.* **301**: 1852–1853.

Carson, C.A., Shear, B.L., Ellersieck, M.R., and Asfaw, A. 2001. Identification of Fecal *Escherichia coli* from Humans and Animals by Ribotyping. *Appl. Environ. Microbiol.*, April 2001: 1503–1507.

Cavallo, J.-D., Ramisse, F., Girardet, M., Vaissaire, J., Mock, M., and Hernandez, E. 2002. Antibiotic susceptibilities of 96 isolates of bacillus anthracis isolated in France between 1994 and 2000. *Antimicrobial Agents and Chemotherapy.* **46** (7): 2307–2309.

Cavigelli, M.A., Robertson, G.P., *et al.* 1995. "Fatty acid methyl ester (FAME) profiles as measures of soil microbial community structure." *Plant and Soil.* **170**: 99–113.

Chander, K., and Brookes, P.C. 1991. Microbial biomass dynamics during decomposition of glucose and maize in metal-contaminated soils. *Soil Biol. Biochem.* **23**: 917–925.

Chaudri, A.M., McGrath, S.P., Giller, K.E., and Sauerbeck, D. 1993. Enumeration of indigenous *Rhizobium leguminosarum* biovar *trifolii* in soils previously treated with metal-contaminated sewage sludge. *Soil Biol. Biochem.* **25**: 301–309.

Clayton, R.A., Sutton, G., Hinkle, P.S., Bult, C., and Fields, C. 1995. "Intraspecific variation in small-subunit rRNA sequences in GenBank: Why single sequences may not adequately represent prokaryotic taxa." *International Journal of Systematic Bacteriology.* **45**: 595–599.

Clement, B.G., Kehl, L.E., DeBord, K.L., and Kitts, C.L. 1998. Terminal restriction fragment patterns (TRFPs), a rapid, PCR-based method for the comparison of complex bacterial communities. *Journal of Microbiological Methods.* **31**: 135–142.

Clesceri, L.S., Greenberg, A.E., and Eaton, A.D. (ed.). 1998. Standard methods for the examination of water and wastewater, 20th ed. American Public Health Association, Washington, D.C.

Coker, P.R., Smith, K.L., and Hugh-Jones, M.E. 2002. Antimicrobial Susceptibilities of Diverse *Bacillus anthracis* Isolates. *Antimicrobial Agents and Chemotherapy.* **46** (12): 3843–3845.

Cole, S.T., and Girons, I.S. 1994. "Bacterial genomics." *FEMS Microbiology Reviews.* **14**: 139–160.

Crichton, P.B., and Old, D.C. 1979. Biotyping of *Escherichia coli. J. Med. Microbiol.* **12**: 473–486.

Crocker, F.H., Fredrickson, J.K., White, D.C., Ringelberg, D.B., and Balkwill., D.L. 2000. Phylogenetic and physiological diversity of *Arthrobacter* strains isolated from unconsolidated subsurface sediments. *Microbiology.* **146**: 1296–1310.

Dahlin, S., Witter, E., Martensson, A., Turner, A.R., and Baath, E. 1997. Where's the limit? Changes in the microbiological properties of agricultural soils at low levels of metal contamination. *Soil Biol. Biochem.* **29**: 1405–1415.

De, B.K., Bragg, S.L., Sanden, G.N., Wilson, K.E., Diem, L.A., Marston, C.K., Hoffmaster, A.R., Barnett, G.A., Weyant, R.S., Abshire, T.G., Ezzell, J.W., and Popovic, T. 2002. Two-Component Direct Fluorescent-Antibody Assay for Rapid Identification of Bacillus anthracis. *Emerging Infectious Diseases*. **8** (10): 1060–1065.

De Baere, T., de Mendonça, R., Claeys, G., Verschraegen, G., Mijs, W., Verhelst, R., Rottiers, S., Van Simaey, L., De Ganckand, C., and Vaneechoutte, M. 2002. Evaluation of amplified rDNA restriction analysis (ARDRA) for the identification of cultured mycobacteria in a diagnostic laboratory. *BMC Microbiology*. **2**: 4.

Deng, S., and Tabatai, M.A. 1995. Cellulase activity of soils: effects of trace elements. *Soil Biol. Biochem*. **27**: 977–979.

Derakshani, M., Lukow, T., and Liesack, W. 2001. Novel bacterial lineages at the (sub)division level as detected by signature nucleotide-targeted recovery of 16S rRNA genes from bulk soil and rice roots of flooded rice microcosms. *Appl. Environ. Microbiol*. **67**: 623–631.

Dergam, J.A., Paiva, S.R., Schaeffer, C.R., Godinho, A.L., and Vieira, F. 2002. Phylogeography and RAPD-PCR variation in *Hoplias malabaricus* (Bloch, 1794) (Pisces, Teleostei) in southeastern Brazil. *Genet. Mol. Biol*. **25** (4): 1415–4757.

Desmarais, T.R., Solo-Gabriele, H.M., and Palmer, C.J. 2002. Influence of soil on fecal indicator organisms in a tidally influenced subtropical environment. *Appl. Environ. Microbiol*. **68**: 1165–1172.

Dickie, I.A., Xu, B., and Koide, R.T. 2002. Vertical niche differentiation of ectomycorrhizal hyphae in soil as shown by T-RFLP analysis. *New Phytologist*. **156**: 527–535.

Doelman, P., and Haanstra, L. 1979. Effects of lead on the soil bacterial microflora. *Soil Biol. Biochem*. **11**: 487–491.

Donachie, S.P., Christenson, B.W., *et al*. 2002. "Microbial community in acidic hydrothermal waters of volcanically active White Island, New Zealand." *Extremophiles*. **6**: 419–425.

Drancourt, M., Roux, V., Dang, L.V., Tran-Hung, L., Castex, D., Chenal-Francisque, V., Ogata, H., Fournier, P.-E., Crubezy, E., and Raoult, D. 2004. Genotyping, Orientalislike Yersinia pestis, and Plague Pandemics. *Emerging Infectious Diseases*. **10** (9): 1585–1592.

Duineveld, B.M., Rosado, A.S., *et al*. 1998. "Analysis of the dynamics of bacterial communities in the rhizosphere of the chrysanthemum via denaturing gradient gel electrophoresis and substrate utilization patterns." *Appl. Environ. Microbiol*. **64**: 4950–4957.

Dunbar, J.M., Ticknor, L.O., and Kuske, C.S. 2001. "Phylogenetic specificity and reproducibility and new method for analysis of terminal restriction fragment profiles of 16S rRNA genes from bacterial communities." *Appl. and Environ. Microbiol*. **67**: 190–197.

Ellis, R.J., Morgan, P., Weightman, A.J., and Fry., J.C. 2003. Cultivation-Dependent and -Independent Approaches for Determining Bacterial Diversity in Heavy-Metal-Contaminated Soil. *Appl. Environ. Microbiol*. **69**: 3223–3230.

Fatah, A.A., Barrett, J.A., Arcilesi, Jr., R.D., Ewing, K.J., Lattin, C.H., and Moshier, T.F., 2001. An Introduction to Biological Agent Detection Equipment for Emergency First Responders. US Department of Justice, Office of Justice Programs, National Institute of Justice. NCJ 190747.

Feachem, R.G. 1975. An improved role for faecal coliform to faecal streptococci ratios in the differentiation between human and non-human pollution sources (Note). *Water Research*. **9**: 689–690.

Federle, T.W. 1986. Microbial distribution in soil-new techniques, pp. 493–498, *In* F. Megusar and M. Gantar, eds. *Perspectives in microbial ecology*, Ljubljana, Jugoslavia.

Felske, A., and Akkermans, A.D. 1998. "Spatial Homogeneity of Abundant Bacterial 16S rRNA Molecules in Grassland Soils." *Microbial Ecology*. **36**: 31–36.

Feris, K., Ramsey, P., Frazar, C., Moore, J.N., Gannon, J.E., and Holben. W.E. 2003. Differences in hyporheiczone microbial community structure along a heavy-metal contamination gradient. *Appl. Environ. Microbiol*. **69**: 5563–5573.

Feris, K.P., Ramsey, P.W., Rillig, M., Moore, J.N., Gannon, J.E., and Holben, W.E. 2004a. Determining rates of change and evaluating group-level resiliency differences in hyporheic microbial communities in response to fluvial heavy-metal deposition. *Appl. Environ. Microbiol*. **70**: 4756–4765.

Feris, K.P., Ramsey, P.W., Frazar, C., Rillig, M., Moore, J.N., Gannon, J.E., and Holben. W.E. 2004b. Seasonal dynamics of shallow-hyporheic-zone microbial community structure along a heavy-metal contamination gradient. *Appl. Environ. Microbiol*. **70**: 2323–2331.

Fey, A., and Conrad, R. 2000. Effect of temperature on carbon and electron flow and on the archaeal community in methanogenic rice field soil. *Appl. Environ. Microbiol*. **66**: 4790–4797.

Flynn, S.J., Loffler, F.E., and Tiedje, J.M. 2000. Microbial community changes associated with a shift from reductive dechlorination of PCE to reductive dechlorination of cis-DCE and VC. *Environ. Sci. Technol*. **34**: 1056–1061.

Fontana, J., Stout, A., Bolstorff, B., and Timperi, R. 2003. Automated ribotyping and pulsed-field gel electrophoresis for rapid identification of multidrug-resistant *salmonella* serotype newport. *Emerging Infections Diseases*. **9** (4). http://www.cdc.gov/ncidod/EID/vol9no4/02-0423.htm.

Franklin, R.B., Garland, J.L., Bolster, C.H., and Mills, A.L. 2001. Impact of dilution on microbial community structure and functional potential: Comparison of numerical simulations and batch culture experiments. *Appl. Environ. Microbiol*. **67**: 702–712.

Frostegard, A., Tunlid, A., and Baath, E., 1993. Phospholipid fatty acid composition, biomass, and activity of microbial communities from two soil types experimentally exposed to different heavy metals. *Appl. Environ. Microbiol*. **59**: 3605–3617.

Frostegard, A., Tunlid, A., and Baath, E. 1996. Changes in microbial community structure during long-term incubation in two soils experimentally contaminated with minerals. *Soil. Biol. Biochem*. **28**: 55–63.

Fujioka, R.S. and Shizumura, L.K. 1985. *Clostridium perfrigens*: a reliable indicator of stream water quality. *J. Pollut. Control Fed*. **57**: 986–992.

Furuse, K., Ando, A., Osawa, S., and Watanabe, I. 1981. Distribution of ribonucleic acid coliphage in raw sewage from treatment plants in Japan. *Appl. Environ. Microbiol*. **41**: 1139–1143.

Garland, J.L., and Mills., A.L. 1991. Classification and characterization of heterotrophic microbial communities on the basis of patterns of community-level-sole-carbon-source-utilization. *Appl. Environ. Microbiol*. **57**: 2351–2359.

Gavini, F., Pourcher, A.M., and Neut, C. 1991. Phenotypic differentiation of bifidobacteria of human and animal origins. *Int. J. Syst. Bacteriol*. **41**: 548–557.

Ge, Z., and Taylor, D.E. 1998. *Helicobacter pylori*: Molecular genetics and diagnostic typing. *Br Med Bull*. **54** (1): 31–38.

Geldreich, E.E., and Kenner, B.A. 1969. Concepts of fecal streptococci in stream pollution. *J. Water Pollut. Control Fed*. **41**: R336–R352.

Gich, F.B., Amer, E., Figueras, J.B., Abella, C.A., Balaguer, M.D., and Poch, M. 2000. Assessment of microbial community structure changes by amplified ribosomal DNA restriction analysis (ARDRA). *Internatl. Microbiol.* **3**: 103–106.

Gillan, D.C. 2004. The effect of an acute copper exposure on the diversity of a microbial community in North Sea sediments as revealed by DGGE analysis—the importance of the protocol. *Marine Pollution Bulletin.* **49**: 504–513.

Giller, K.E., McGrath, S.P., and Hirsch., P.R. 1989. Absence of nitrogen fixation in clover grown in soil subject to long-term contamination with heavy metals is due to survival of only ineffective *Rhizobium. Soil Biol. Biochem.* **21**: 841–848.

Giller, K.E., Witter, E., and McGrath., S.P. 1998. Toxicity of heavy metals to microorganisms and microbial processes in agricultural soils: A review. *Soil Biol. Biochem.* **30**: 1389–1414.

Girvan, M.S., Bullimore, J., Pretty, J.N., Osborn, A.M., and Ball, A.S. 2003. "Soil type is the primary determinant of the composition of the total and active bacterial communities in arable soils." *Appl. and Environ. Microbiol.* **69**: 1800–1809.

Gonzalez, E.A., and Blanco, J. 1989. Serotypes and antibiotic resistance of verotoxigenic (VTEC) and necrotizing (NTEC) *Escherichia coli* strains isolated from calves with diarrhea. *FEMS Microbiol. Lett.* **60**: 31–36.

Gremion, F., Chatzinotas, A., Kaufmann, K., Von Sigler, W., and Harms., H. 2004. Impacts of heavy metal contamination and phytoremediation on a microbial community during a twelve-month microcosm experiment. *FEMS Microbiology Ecology.* **48**: 273–283.

Griffiths, B.S., Riaz-Ravina, M., Ritz, K., McNicl, J.W., Ebblewhite, N., and Baath, E. 1997. Community DNA hybridisation and %G+C profiles of microbial communities from heavy metal polluted soils. *FEMS Microbiology Ecology.* **24**: 103–112.

Grundmann, G.L., and Debouzie, D. 2000. "Geostatistical analysis of the distribution of NH_4^+ and NO_2^- oxidising bacteria and serotypes at the millimeter scale along a soil transect." *FEMS Microbiological Ecology.* **34**: 57–62.

Grunow, R., Splettstoesser, W., McDonald, S., Otterbein, C., O'Brien, T., Morgan, C., Aldrich, J., Hofer, E., Finke, E.-J., and Meyer, H. 2000. Detection of Francisella tularensis in Biological Specimens Using a Capture Enzyme-Linked Immunosorbent Assay, and Immunochromatographic Handheld Assay, and a PCR. *Clinical and Diagnostic Laboratory Immunology.* **7** (1): 86–90.

Guckert, J., Antworth, C., Nichols, P., and White, D.C. 1985. Phospholipid ester-linked fatty acid profiles as reproducible assays for changes in prokaryote community structure of estuarine sediments. *FEMS Microbiol. Ecol.* **31**: 147–158.

Haanstra, L., and Doelman, P. 1984. Glutamic acid decomposition as a sensitive measure of heavy-metal pollution in soil. *Soil Biol. Biochem.* **16**: 595–600.

Hagedorn, C.S., Robinson, S.L., Filtz, J.R., Grubbs, S.M., Angier, T.A., and Reneau, R.B. Jr. 1999. Using antibiotic resistance patterns in the fecal streptococci to determine sources of fecal pollution in a rural Virginia watershed. *Appl. Environ. Microbiol.* **65**: 5522–5531.

Harris, D. 1994. Analyses of DNA extracted from microbial communities. *Beyond the Biomass.* Ritz, K., Dighton, J., and Giller, K.E., Wiley-Sayce.

Hartly, C.L., Howne, K., Linton, A.H., Linton, K.B., and Richmond, M.H. 1975. Distribution of R plasmids among O-antigen types of *Escherichia coli* isolated from human and animal sources. *Antimicrob. Agents Chemother.* **8**: 122–131.

Harwood, V.J., Whitlock, J., and Withington, V.H. 2000. Classification of the antibiotic resistance patterns of indicator bacteria by discriminant analysis: Use in predicting

the source of fecal contamination in subtropical Florida waters. *Appl. Environ. Microbiol.* **66**: 3698–3704.

Havelaar, A.H. 1993. Bacteriophages as models of human enteric viruses in the environment. *ASM News.* **59**: 614–619.

Heuer, H., and Smalla, K. 1997. Application of denaturing gradient gel electrophoresis and temperature gradient gel electrophoresis for studying soil microbial communities, pp. 353–373, *In* J.D. Van Elsas, *et al.*, eds. *Modern Soil Microbiology.* Marcel Dekker, Inc., New York.

Heuer, H., Krsek, M., Baker, P., Smalla, K., and Wellington, E.M. 1997. Analysis of actinomycete communities by specific amplification of genes encoding 16S rRNA and gel-electrophoretic separation in denaturing gradients. *Appl. Environ. Microbiol.* **63** (8): 3233–3241.

Heun, M., Schäfer-Pregl, R., Klawan, D., Castagna, R., Accerbi, M., Borghi, B., and Salamini, F. 1997. Site of einkorn wheat domestication identified by DNA fingerprinting. *Science.* **278**: 1312–1314.

Hiroki, M. 1992. Effects of heavy metal contamination on soil microbial population. *Soil Sci. Plant Nutr.* **38**: 141–147.

Horswell, J., Cordiner, S.J., Mass, E.W., Martin, T.M., Sutherland, B.W., Speier, T.W., Nogales, B., and Osborn A.M. 2002. "Forensic comparison of soils by bacterial community DNA profiling." *Journal of Forensic Sciences.* **47**: 350–353.

Hsu, F.-C., Shieh, Y.-S., van Duin, J., Beekwilder, M.J., and Sobsey, M.D. 1995. Genotyping male-specific RNA coliphages by hybridization with oligonucleotide probes. *Appl. Environ. Microbiol.* **61**: 3960–3966.

Huang, C.Y., and K.S. Khan. 1998. Effects of cadmium, lead and their interaction on the size of microbial biomass in a red soil. *Soil Environ.* **1**: 227–236.

Huysman, F., Verstraete, W., and P.C. Brookes. 1994. Effect of manuring practices and increased copper concentrations on soil microbial populations. *Soil Biol. Biochem.* **26**: 103–110.

Jofre, J., Blasi, M., Bosch, A., and Lucena, F. 1989. Occurrence of bacteriophages infecting *Bacteriodes fragilis* and other virus in polluted marine sediments. *Water Sci. and Technol.* **21**: 15–19.

Jordan, M., and Lechevalier, M.P. 1975. Effects of zinc-smelter emissions on forest soil microflora. *Can. J. Microbiol.* **21**: 1855–1865.

Junger, E.P. 1996. "Assessing the unique characteristics of close-proximity soil samples: Just how useful is soil evidence." *Journal of Forensic Sciences.* **41**: 27–34.

Kandeler, E., Kampichler, C., and Horak, O. 1996. Influence of heavy metals on the functional diversity of soil microbial communities. *Biol. Fertil. Soils.* **23**: 299–306.

Kandeler, E., Tscherko, D., Bruce, K.D., Stemmer, M., Hobbs, P.J., Bardgett, R.D., and Amelung, W. 2000. Structure and function of the soil microbial community in microhabitats of a heavy metal polluted soil. *Biol. Fertil. Soils.* **32**: 390–400.

Kaplan, C.W., Astaire, J.C., Sanders, M.E., Reddy, B.S., and Kitts, C.L. 2001. 16S Ribosomal DNA terminal restriction fragment pattern analysis of bacterial communities in feces of rats fed *lactobacillus acidophilus* NCFM. *Appl. Environ. Microbiol.* April 2001: 1935–1939.

Kaplan, C.W., and Kitts, C.L. 2004. Bacterial succession in a petroleum land treatment unit. *Appl. and Environ. Microbiol.* March 2004: 1777–1786.

Kardolus, J.P., van Eck, H.J., and van den Berg, R.G. 1998. The potential of AFLPs in biosystematics: A first application in Solanum taxonomy (Solanaceae). *Plant Systematics and Evolution.* **210**: 87–103.

Karp, A., Kresovich, S., Bhat, K.V., Ayand, W.G., and Hodgkin, T. 1997. Molecular tools in plant genetic resources conservation: A guide to the technologies. *IPGRI Technical Bulletin* No. 2, International Plant Genetic Resources Institute, Rome, Italy. Available at http://198.93.227.125/publicat/techbull/TB2.pdf.

Kaspar, C.W., and Burgess, J.L. 1990. Antibiotic resistance indexing of *Escherichia coli* to identify sources of fecal contamination in water. *Canadian Journal of Microbiology.* **36**: 891–894.

Keim, P. 2003. *Microbial Forensics: A Scientific Assessment.* American Academy of Microbiology. Washington D.C.

Keim, P., Price, L.B., Klevytska, A.M., Smith, K.L., Schupp, J.M., Okinaka, R., Jackson, P.J., and Hugh-Jones, M.E. 2000. Multiple-locus variable-number tandem repeat analysis reveals genetic relationships within *Bacillus anthracis. Journal of Bacteriology.* **182**: 2928–2936.

Kelly, J.J., and Tate, R.L. 1998. Effects of heavy metal contamination and remediation on soil microbial communities in the vicinity of a zinc smelter. *J. Environ. Qual.* **27**: 609–617.

Kelly, J.J., Hägglblom, M., and Tate, R.L. 1999. Changes in soil microbial communities over time resulting from one time application of zinc: A laboratory microcosm study. *Soil Biol. Biochem.* **31**: 1455–1465.

Khan, M., and Scullion, J. 2000. Effect of soil on microbial responses to metal contamination. *Environ. Pollut.* **110**: 115–125.

Kitts, C.L. 2001. Terminal restriction fragment patterns: A tool for comparing microbial communities and assessing community dynamics. *Curr. Issues Intest. Microbiol.* **2** (1): 17–25.

Kozdroj, J. and van Elsas, J.D. 2000. "Application of Polymerase chain reaction-denaturing gradient gel electrophoresis for comparison of direct and indirect extraction methods of soil DNA used for microbial community fingerprinting." *Biol. Fertil. Soils.* **31**: 372–378.

Krane, D.E., Sternberg, D.C., and Burton, G.A. 1999. Randomly amplified polymorphic DNA profile-based measures of genetic diversity in crayfish correlated with environmental impacts. *Environ. Toxicol. Chem.* **18**: 504–508.

Krauss, S.L. 1999. Complete exclusion of nonsires in an analysis of paternity in a natural plant population using amplified fragment length polymorphism (AFLP). *Molecular Ecology.* **8**: 217–226.

Krauss, S.L., and Peakall, R. 1998. An evaluation of the AFLP fingerprinting technique for the analysis of paternity in natural populations of *Persoonia mollis* (Proteaceae). *Australian Journal of Botany.* **46**: 533–546.

Kreuzer-Martin, H.W., Chesson, L.A., Lott, M.J., Dorigan, J., and Ehleringer, J.R. 2004a. Stable Isotope Ratios as a Tool in Microbial Forensics—Part 1. Microbial Isotopic Composition as a Function of Growth Medium. *Journal of Forensics Science.* **49** (5): 1–7.

Kreuzer-Martin, H.W., Chesson, L.A., Lott, M.J., Dorigan, J., and Ehleringer, J.R. 2004b. Stable Isotope Ratios as a Tool in Microbial Forensics—Part 2. Isotopic Variation Among Different Growth Media as a Tool for Sourcing Origins of Bacterial Cells or Spores. *Journal of Forensics Science.* **49** (5): 8–14.

Kreuzer-Martin, H.W., Lott, M.J., Dorigan, J., and Ehleringer, J.R. 2003. Microbe forensics: Oxygen and hydrogen stable isotope ratios in bacillus subtilis cells and spores. *Proceeding of the National Academy of Science.* **100** (3): 815–819.

Kroppenstedt, R.M. 1985. Fatty acid and menaquinone analysis of actinomycetes and related organisms, pp. 173–199, *In* M. Goodfellow and D.E. Minnikin, eds. *Chemical methods in bacterial systematics.* Academic Press, London.

Kuhn, D.N., Borrone, J., Meerow, A.W., Motamayor, J., Brown, J.S., and Schnell Ii, R.J. 2004. Single Strand Conformation Polymorphism Analysis Of Candidate Genes For Reliable Identification Of Alleles By Capillary Array Electrophoresis. Submitted to *Electrophoresis.* http://www.ars.usda.gov/research/publications/Publications.htm?seq_no_115=164237.

LaMontagne, M.G., Michel, Jr., F.C., *et al.* 2002. "Evaluation of extraction and purification methods for obtaining PCR-amplifiable DNA from compost for microbial community analysis." *Journal of Microbiological Methods.* **49**: 255–264.

Ladd, J.N. 1985. Soil enzymes. *In* Vaughan, D. and Malcolm, E. (eds). Soil Organic Matter and Biological activity. Martinus Nijhoff Dr H Junk Publishers, Dordrecht, Netherlands, pp. 175–221.

LaMontagne, M.G., Schimel, J.P., and Holden, P.A. 2003. Comparison of subsurface and surface soil bacterial communities in California grassland as assessed by terminal restriction fragment length polymorphisms of PCR-Amplified 16S rRNA Genes. *Microb. Ecol.* **46**: 216–227.

Law, J.R., Donini, P., Koebner, R.M.D., Jones, C.R., and Cooke, R.J. 1998. DNA profiling and plant variety registration III: The statistical assessment of distinctness in wheat using amplified fragment length polymorphisms. *Euphytica.* **102**: 335–342.

Lechevalier, M.P. 1977. Lipids in bacterial taxonomy-a taxonomist's view. *Critic. Rev. Microbiol.* **5**: 109–210.

Leuders, T., and Friedrich, M. 2000. Archaeal population dynamics during sequential reduction process in rice field soil. *Appl. Environ. Microbiol.* **66**: 2732–2742.

Leung, K.T., Chang, Y.J., Gan, Y.-D. Peacock, A.D., Macnaughton, S.J., Stephen, J.R., Burkhalter, R.S., Flemming, C.A., and White, D.C. 1999. Detection of Sphingomonas spp. in soils by PCR and sphingolipid biomarker analysis. *J. Industrial Microbiology* **23**: 252–260.

Liesack, W., Janssen, P.H., *et al.* 1997. Microbial diversity in soil: The need for a combined approach using molecular and cultivation techniques. *Modern Soil Microbiology.* van Elsas, J.D., Trevors J.T., and Wellington. E.M., New York, Marcel Dekker.

Liu, W., Marsh, T.L., and Forney, L.J. 1998. Determination of the microbial diversity of anaerobic-aerobic activated sludge by a novel molecular biological technique. *Water Sci. and Technol.* **37**: 417–422.

Liu, W.T., Marsh, T.L., Cheng, H., and Forney, L.J. 1997. "Characterization of microbial diversity by determining terminal restriction fragment length polymorphisms of genes encoding 16S rRNA." *Appl. Environ. Microbiol.* **63**: 4516–4522.

Lorenz, S.E., McGrath, S.P., and Giller, K.E. 1992. Assessment of free-living nitrogen fixation activity as a biological indicator of heavy metal toxicity in *soil. Soil Biol. Biochem.* **24**: 601–606.

Lösel, D.M. 1988. Fungal lipids, pp. 699–806, *In* C.C. Rattledge and Wilkinson, S.G., eds. *Microbial lipids,* Vol. 1. Academic Press, London.

Lukow, T., Dunfield, P.F., and Liesack, W. 2000. "Use of the T-RFLP technique to assess spatial and temporal changes in the bacterial community structure within an agricultural soil planted with transgenic and nontransgenic potato plants." *FEMS Microbiological Ecology.* **32**: 241–247.

Mace, E.S., Gebhardt, C.G., and Lester, R.N. 1999. AFLP analysis of genetic relationships in the tribe Datureae (Solanaceae). *Theoretical and Applied Genetics.* **99**: 634–641.

MacNaugthon, S., Stephen, J.R., Chang, Y.-J., Peacock, A., Flemming, C.A., Leung, K.T., and White, D.C. 1999. Characterization of metal-resistant soil eubacteria by polymerase chain reaction-denaturing gradient gel electrophoresis with isolation of resistant strains. *Can. J. Microbiol.* **45**: 116–124.

Marsh, T.L., Liu, W., Forney, L.J., and Cheng, H. 1998. Beginning a molecular analysis of the eukaryal community in activated sludge. *Water Sci. and Technol.* **37**: 455–460.

Massol-Deya, A.A., Odelson, D.A., Hickey, R.F., and Tiedje, J.M., 1995. Bacterial community fingerprinting of amplified 16S and 16-23S ribosomal DNA gene sequences and restriction endonuclease analysis (ARDRA). *In Molecular Microbial Ecology Manual.* Akkermans, A.D.L., van Elsas, J.D., and de Brujin, F.J. (eds). Dordrecht: Kluwer Academic Publications, 1–8.

Matthes, M.C., Daly, A., and Edwards, K.J. 1998. Amplified fragment length polymorphism (AFLP). *In*: Karp A, Isaac, P.G., Ingram, D.S. (eds). *Molecular Tools for Screening Biodiversity.* Chapman and Hall, London, pp. 183–190.

McCaig, A.E., Glover, L.A., and Prosser, J.I. 2001. "Numerical analysis of grassland bacterial community structure under different land management regimens by using 16S ribosomal DNA sequence data and denaturing gradient gel electrophoresis banding patterns." *Appl. and Environ. Microbiol.* **67**: 4554–4559.

Melcher, U. 2003. SSCPs.<http://opbs.okstate.edu/~melcher/MG/MGW1/MG11129.html>.

Müller, A.K., Westergaard, K., Christensen, S. and Sorensen, S.J. 2001. The effect of long-term mercury pollution on the soil microbial community. *FEMS Microbiol. Ecol.* **36**: 11–19.

Murray, R.C., and Tedrow, J.C.F., 1992. *Forensic geology.* New Jersey, Prentice Hall.

Muyzer, G., Dewaal, E.C., and Uitterlinden, A.G. 1993. Profiling of complex microbial populations by denaturing gradient gel electrophoresis analysis of polymerase chain reaction amplified genes coding for 16S ribosomal RNA. *Appl. Environ. Microbiol.* **59**: 695–700.

Nannipieri, P. 1995. The potential use of soil enzymes as indicators of productivity, sustainability and pollution, pp. 238–244, *In* C.E. Pankhurst, *et al.*, eds. Soil Biota: Management in Sustainable Farming systems. CSIRO, East Melbourne, Victoria, Australia.

Nogales, B., Moore, E.R., Llobet-Brossa, E., Rossello-Mora, R., Amann, R., and Timmis, K.N. 2001. "Combined use of 16S ribosomal DNA and 16S rRNA to study the bacterial community of polychlorinated biphenyl-polluted soil." *Appl. Environ. Microbiol.* **67**: 1874–84.

Nordgren, A. 1988. Apparatus for the continuous long-term monitoring of soil respiration rates in a large number of samples. *Soil Biol. Biochem.* **20**: 955–957.

Nordgren, A., Baath, E., and Söderström, B., 1988. Evaluation of soil respiration characteristics to assess heavy metal effects on soil microorganisms using glutamic acid as substrate. *Soil Biol. Biochem.* **20**: 949–954.

O'Leary, W.M., and Wilkinson. S.G., 1988. Gram-positive bacteria, pp. 117–202, *In* Ratledge, C., and Wilkinson, S.G., eds. *Microbial lipids,* Vol. 1. Academic Press, London, UK.

Oliveira, A.M., and Ramos, M.C. 2002. PCR-based ribotyping of *Staphylococcus aureus. Braz J Med Biol Res,* February 2002, **35** (2): 175–180.

Omelyanyuk, G.G., Alekseev, A.A., *et al.* 1999. *The Possibility of Application of the Multisubstrate Testing Method in Soil Criminalistic Investigation. Proceedings of the 15th International Association of Forensic Science.*

Ortiz, O. and Alcaniz, J.P., 1993. Respiration potential of microbial biomass in a calcareous soil treated with sewage sludge. *Geomicrobiol. J.* **11**: 333–340.

Osborn, A.M., Moore, E.R., and Timmis, K.N. 2000. An evaluation of terminal-restriction fragment length polymorphism (T-RFLP) analysis for the study of microbial community structure and dynamics. *Environ. Microbiol.* **2**: 39–50.

Ovreas, L., and Torsvik, V. 1998. "Microbial diversity and community structure in two different agricultural soil communities." *Microbial Ecology,* **36**: 303–315.

Palmborg, C., and Nordgren, A. 1996. Partitioning the variation of microbial measurements in forest soils into heavy metal and substrate quality dependent parts by use of near infrared spectroscopy and multivariate statistics. *Soil Biol. Biochem.* **28**.

Park, H.-D., Regan, J.M., and Noguera, D.R. 2002. Molecular analysis of ammonia-oxidizing bacterial populations in aerated-anoxic Orbal processes. *Water Sci. and Technol.* **46** (1–2): 273–280.

Parveen, S., Portier, K.M., Robinson, K., Edmiston, L., and Tamplin, M.L. 1999. Discriminant analysis of ribotype profiles of *Escherichia coli* for differentiating Human and Nonhuman sources of fecal pollution. *Appl. and Environ. Microbiol.* **65** (7): 3142–3147.

Pearson, T., Busch, J.D., Ravel, J., Read, T.D., Rhoton, S.D., U'Ren, J.M., Simonson, T.S., Kachur, S.M., Leadem, R.R., Cardon, M.L., Van Ert, M.N., Huynh, L.Y., Fraser, C.M., and Keim, P. 2004. Phylogenetic Discovery Bias in Bacillus anthracis using single-nucleotide polymorphisms from Whole-Genome Sequencing. *Proceeding of the National Academy of Science.* **101** (37): 13536–13541.

Pennanen, T. 2001. Microbial communities in boreal coniferous forest humus exposed to heavy metals and changes in soil pH-a summary of the use of phospholipid fatty acids, Biolog[R] and ^{3}H-thymidine incorporation methods in field studies. *Geoderma.* **100**: 91–126.

Pennanen, T., Frostgard, A., Fritze, H., and Baath, E. 1996. Phospholipid fatty acid composition and heavy metal tolerance of soil microbial communities along two heavy metal polluted gradients in coniferous forests. *Appl. Environ. Microbiol.* **62**: 420–428.

Pennanen, T., Perkiömäki, J., Kiikkilä, O., Vanhala, P., Neuvonen, S., and froitze, H. 1998. Prolonged, simulated acid rain and heavy metal deposition: Separated and combined effects on forest soil microbial community structure. *FEMS Microbiol. Ecol.* **27**: 291–300.

Perkiömäki, J., Tom-Petersen, A., Nybroe, O., and Fritze, H. 2003. Boreal forest microbial community after long-term field exposure to acid and metal pollution and its potential remediation by using wood ash. *Soil Biol. Biochem.* **35**: 1517–1526.

Poreda, R.J., Cerling, T.E., and Solomon, D.K. 1988. Tritium and helium isotopes as hydrologic tracers in a shallow unconfined aquifer. *Journal of Hydrology.* **103**: 1–9.

Prosser, J.I. 1997. Microbial Processes within the Soil. *Modern Soil Microbiology.* van Elsas, J.D., Trevors, J.T., and Wellington, E.M. New York, Marcel Dekker.

Puig, A., Queralt, N., Jofre, J., and Araujo, R. 1999. Diversity of Bacteriodes fragilis strains in their capacity to recover phages from human and animal wastes and from fecally polluted wastewater. *Appl. Environ. Microbiol.* **65**(4): 1772–1776.

Rafalski, J.A., Vogel, J.M., Morgante, M., Powell, W., Andre, C., and Tingey, S.V. 1996. Generating and using DNA markers in plants. *In* Birren, B., and Lai, E. (eds). *Non-Mammalian Genomic Analysis: A Practical Guide.* Academic Press, London. pp. 75–134.

Rahman, T.J., Marchant, R., and Banat, I.M. 2003. Distribution and molecular investigation of highly thermophilic bacteria associated with cool soil environments. *Biochem. Soc. Trans.* (2004). **32**: 209–213 (Printed in Great Britain).

Read, T.D., Salzberg, S.L., Pop, M., Shumwat, M., Umayam, L., Jiang, L., Holtzapple, E., Busch, J.D., Smith, K.L., Schupp, J.M., Solomon, D., Keim, P., and Fraser, C.M.

2002. Comparative genome sequencing for discovery of novel polymorphisms in *Bacillus anthracis*. *Science*. **296**: 2028–2033.

Renella, G., Ortigoza, A.L.R., Landi, L., and Nannipieri, P. 2003. Additive effects of copper and zinc on cadmium toxicity on phosphatase activities and ATP content of soil as estimated by the ecological dose (ED50). *Soil Bio. Biochem*. **35**: 1203–1210.

Renella, G., Mench, M., van der Lelie, D., Pietramellara, G., Ascher, J., Ceccherini, M.T., Landi, L., and Nannipieri, P. 2004. Hydrolase activity, microbial biomass and community structure in long-term Cd-contaminated soils. *Soil Biol. Biochem*. **36**: 443–451.

Renella, G., Mench, M., Landi, L., and Nannipieri, P. 2005. Microbial activity and hydrolase synthesis in long-term Cd-contaminated soils. *Soil Biol. and Biochem*. **37**: 133–139.

Ridout, C.J. and, Donini, P. 1999. Use of AFLP in cereals research. *Trends in Plant Science*. **4**: 76–79.

Rieseberg, L.H., Kim, M.J., and Seiler, G.J. 1999. Introgression between the cultivated sunflower and a sympatric wild relative, *Helianthus petiolaris* (Asteraceae). *International Journal of Plant Science*. **160**: 102–108.

Robertson, G.P., Huston, M.A., Evans, F.C., and Tiedje, J.M. 1988. "Spatial variability in a successional plant community: Patterns of nitrogen availability." *Ecology*. **69**: 1517–1524.

Romalde, J.L., Castro, D., Magarinos, B., Lopez-Cortes, L., and Borrego, J.J. 2002. Comparison of ribotyping, randomly amplified polymorphic DNA, and pulsed-field gel electrophoresis for molecular typing of *Vibrio tapetis*. *Systematic and Applied Microbiology*. **25** (4): 544–550.

Russell, J.R., Weber, J.C., Booth, A., Powell, W., Sotelo-Montes, C., and Dawson, I.K. 1999. Genetic variation of *Calycophyllum spruceanum* in the Peruvian Amazon Basin, revealed by amplified fragment length polymorphism (AFLP) analysis. *Molecular Ecology*. **8**: 199–204.

Saetre, P., and Baath, E. 2000. "Spatial variation and patterns of soil microbial community structure in a mixed spruce-birch stand." *Soil Bio. Biochem*. **32**: 909–917.

Sandaa, R.-A., Torsvik, V., Enger, O., Daae, F.L., Castberg, T., and Hahn, D. 1999. Analysis of bacterial communities in heavy metal-contaminated soils at different levels of resolution. *FEMS Microbiol. Ecol*. **30**: 237–251.

Sarkar, G., Yoon, H.S., and Sommer, S.S. 1992. Screening for mutations by RNA single-strand conformation polymorphism (rSSCP): Comparison with DNA-SSCP. *Nucleic Acids Research*. **20** (4): 4871–4878.

Scott, T.M., Rose, J.B., Tracie, M.J., Farrah, S.R., and Lukasik, J. 2002. Microbial source tracking: Current methodology and future directions. *Appl. Environ. Microbiol*. December 2002: 5796–5803.

Scott, T.M., Parveen, S., Portier, K.M., Rose, J.B., Tamplin, M.L., Farrah, S.R., Koo, A., and Lukasik, J. 2003. Geographical Variation in Ribotype Profiles of *Escherichia coli* Isolates from Humans, swine, Poultry, Beef, and Dairy Cattle in Florida. *Appl. Environ. Microbiol*. February 2003: 1089–1092.

Scott, T.M., Caren, J., Nelson, G.R., Jenkins, T.M., and Lukasik, J. 2004. Tracking Sources of Fecal Pollution in a South Carolina Watershed by Ribotyping *Escherichia coli*: A Case Study. *Environmental Forensics*. **5**: 15–19.

Seifert, H., Dijkshoorn, L., Gerner-Smidt, P., Pelzer, N., Tjernberg, I., and Vaneechoutte, M. 1997. Distribution of *Acinetobacter* species on human skin: Comparison of phenotypic and genotypic identification methods. *J. Clin. Microbiol*. **35**: 2819–2825.

Shi, W., Becker, J., Bischoff, M., Turco, R.F., and Konopka, A.E. 2002. Association of microbial community composition and activity with lead, chromium and hydrocarbon contamination. *Appl. Environ. Microbiol*. **68**: 3859–3866.

Shianna, K.V., Rytter, R., and Spanier, J.G. 1998. Randomly amplified polymorphic DNA PCR analysis of bovine cryptosporidium parvum strains isolated from the watershed of the Red River of the North. *Appl. Environ. Microbiol*. **64** (6): 2262–2265.

Smith, J.L., Halvorson, J.J., and Bolton, H. 1994. "Spatial relationships of soil microbial biomass and C and N mineralisation in a semi-arid shrub-steppe ecosystem." *Soil Biol. Biochem*. **26**: 1151–1159.

Steichen, C., Chen, P., Kearney, J.F., and Turnbough, Jr., C.L. 2003. Identification of the Immunodominant Protein and Other Proteins of the *Bacillus anthracis* Exosporium. *Journal of Bacteriology*. **185** (6): 1903–1910.

Stotzky, G. 1997. Soil as an environment for environmental life. *Modern Soil Microbiology*. van Elsas, J.D., Trevors, J.T., and Wellington, E.M. New York, Marcel Dekker.

Sunnucks, P., Wilson, A.C.C., Beheregaray, L.B., Zenger, K., French, J., and Taylor, A.C. 2000. SSCP is not so difficult: The application and utility of single-stranded conformation polymorphism in evolutionary biology and molecular ecology. *Molecular Ecology*. 2000 (9): 1699–710.

Suomalainen, A., Ciafaloni, E., Koga, Y., Peltonen, L., DiMauro, S., and Schon, E.A. 1992. Use of single strand conformation polymorphism analysis to detect point mutations in human mitochondrial DNA. *J Neurol Sci*. **111** (2): 222–226.

Thornton, J.I., and McLaren, A.D., 1975. "Enzymatic Characterisation of Soil Evidence." *Journal of Forensic Sciences*. **20**: 674–692.

Tiedje, J.M., Asuming-Brempong, S., Nusslein, K., Marsh, T.L., and Flynn, S.J. 1999. "Opening the black box of soil microbial diversity." *Applied Soil Ecology*. **13**: 109–122.

Timoney, J.F., Port, J., Goiles, J., and Spanier, J. 1978. Heavy-metal and antibiotic resistance in bacterial flora of sediments of New York Bight. *Appl. Environ. Microbiol*. **36**: 465–472.

Truper, H.G. 1992. Prokaryotes—an overview with respect to biodiversity and environmental importance. *Biodiv. Cons*. **1**: 227–236.

Tunlid, A., and White, D.C. 1992. Biochemical analysis of biomass, community structure, nutritional status, and metabolic activity of microbial communities in soil. *Soil Biochem*. **7**: 229–262.

Turpeinen, R., Kairesalo, T., and Häggblom, M.M. 2004. Microbial community structure and activity in arsenic-, chromium- and copper-contaminated soils. *FEMS Microbiol. Ecol*. **47**: 39–50.

Tyler, G. 1981. Heavy metals in soil biology and biochemistry, pp. 371–413, *In* E.A. Paul and J.N. Ladd, eds. *Soil biochemistry*, Vol. 5.

Van der Maarel, M.J.E.C., Artz, R.R.E., Haanstra, R., and Forney, L.J. 1998. Association of marine archaea with the digestive tracts of two marine fish species. *Appl. Environ. Microbiol*. **64**: 2894–2898.

van Dijck, P.J., and van de Voorde, H. 1984. "Evaluation of microbial soil identity in forensic science." *Zeitschrift Rechtsmedizin*. **93**: 71–77.

Ventura, M., Casas, I.A., Morelli, L., and Callegari, M.L. 2000. Rapid amplified ribosomal DNA restriction analysis (ARDRA) identification of *Lactobacillus* spp. isolated from fecal and vaginal samples. *Syst Appl Microbiol*. **23** (4): 504–509.

Víglasky, V., Antalík, M., Adamcík, J., and Podhradský D. 2000. Early melting of supercoiled DNA topoisomers observed by TGGE. *Nucleic Acids Research*. **28** (11): E51–E51.

Vos, P., Hogers, R., Bleeker, M., Reijans, M., van de Lee, T., Hornes, M., Frijters, A., Pot, J., Peleman, J., Kuiper, M., and Zabeau, M. 1995. AFLP: A new technique for DNA fingerprinting. *Nucleic Acids Research*. **23**: 4407–4414.

Wagner, J. 2002. Screening methods for detection of unknown point mutations. <http://www-users.med.cornell.edu/~jawagne/screening_for_mutations.html#Single-Strand.Conformational.Polymorphism>.

Wardle, D.A., and Ghani, A. 1995. A critique of the microbial metabolic quotient (qCO_2) as a bioindicator of disturbance and ecosystem development. *Soil Biol. Biochem.* **27**: 1601–1610.

Wenderoth, D.F., and Reber, H.H. 1999. Correlation between structural diversity and catabolic versatility of metal-affected prototrophic bacteria in soil. *Soil Biol. Biochem.* **31**: 345–352.

Wiggins, B.A., 1996. Discriminant analysis of antibiotic resistance patterns in faecal streptococci, a method to differentiate human and animal sources of faecal pollution in natural waters. *Appl. Environ. Microbiol.* **62** (11): 3997–4002.

Wiggins, B.A., Andrews, R.W., Conway, R.A., Corr, C.L., Dobratz, E.J., Dougherty, D.P., Eppard, J.R., Knupp, S.R., Limjoco, M.C., Mettenburg, J.M., Rinehardt, J.M., Sonsino, J., Torrijos, R.L., and Zimmerman, M.E. 1999. Use of antibiotic resistance analysis to identify nonpoint sources of fecal pollution. *Appl. Environ. Microbiol.* **65** (8): 3483–3486.

Wilkinson, S.G. 1988. Gram-negative bacteria, pp. 299–488, *In* C. Ratledge and S.G. Wilkinson, eds. *Microbial lipids*, Vol. 1. Academic Press, London, UK.

Williams, S.T., McNeilly, T., and Wellington, E.M. 1977. The decomposition of vegetation growing on metal mine waste. *Soil Biol. Biochem.* **9**: 271–275.

Wood, M. 1995. *Environmental Soil Biology*. Glasgow, Chapman & Hall.

Xie, J., Wehner, T.C., and Conkling, M.A. 2002. PCR-based Single-strand Conformation Polymorphism (SSCP) analysis to clone nine aquaporin genes in cucumber. *J. Amer. Soc. Hort. Sci.* **127** (6): 925–930.

Xu, M.L., Melchinger, A.E., Xia, X.C., and Lübberstedt, T. 1999. High-resolution mapping of loci conferring resistance to sugarcane mosaic virus in maize using RFLP, SSR and AFLP markers. *Molecular and General Genetics.* **261**: 574–581.

Yao, H., Xu, J., and Huang, C. 2003. Substrate utilization pattern, biomass and activity of microbial communities in a sequence of heavy metal-polluted paddy soils. *Geoderma.* **115**: 139–148.

Zhou, J., Xia, B., Treves, D.S., Wu, L.-Y., Marsh, T.L., O'Neill, R.V., Palumbo, A.V., and Tiedje, J.M. 2002. "Spatial and resource factors influencing high microbial diversity in soil." *Appl. Environ. Microbiol.* **68**: 326–334.

Zhu, J., Gale, M.D., Quarrie, S., Jackson, M.T., and Bryan, G.J. 1998. AFLP markers for the study of rice biodiversity. *Theoretical and Applied Genetics.* **96**: 602–611.

12

Chlorinated Solvents

Robert D. Morrison, Brian L. Murphy, and Richard E. Doherty

Contents

12.1 INTRODUCTION

Chlorinated solvents are one of the contaminants most frequently encountered in environmental forensic investigations. This chapter provides a presentation of the chemistry of the most commonly used chlorinated solvents, degradation pathways for these compounds, a historical perspective regarding their use and production, and forensic techniques available for source identification and age dating. While there are many chlorinated solvents of interest encountered in forensic investigations, the focus of this chapter is on the following five:

- Trichloroethylene (TCE);
- Perchloroethylene (PCE or tetrachloroethylene);
- 1,1,1-Trichloroethane (TCA);
- Carbon tetrachloride (CT); and
- Methylene chloride (MC).

The detection of additives that are associated with a particular compound or time period, isotopic analysis and molar ratio analysis of the degradation products of these chlorinated solvents are the primary forensic techniques presented in this chapter.

12.2 CHLORINATED SOLVENT CHEMISTRY

The chemistry and physical characteristics of chlorinated solvents are well understood and documented in the literature (Bouwer and McCarty, 1983; Wilson and Wilson, 1984; Kleopfer et al., 1985; Barrio-Lage et al., 1986; Fogel et al., 1986; Egli et al., 1987; Vogel et al., 1987; Fathepure et al., 1987; Vogel and McCarty, 1987; Janssen et al.,

1988; McCarty, 1994; Benker et al., 1994; Morrison et al., 1998; Morrison, 1999). The intent of this section is not to present a detailed summary of this information but rather to share physical and chemical properties of chlorinated solvents of most direct application in environmental forensic investigations.

The most commonly encountered chlorinated solvents in environmental forensic investigations are TCE, TCA, PCE, CT, and MC. Physical and chemical properties of these solvents are summarized in Table 12.2.1. The high solubility of most chlorinated solvents is of special interest given their preference to dissolve into soil pore water and groundwater. Similarly, the vapor pressure of these solvents is important relative to their susceptibility to be used as a tracer in soil gas investigations. The last column in Table 12.2.1 summarizes information on the boiling points of the solvents. This is particularly important relative to the suitability of a particular solvent for vapor degreasing. Methylene chloride and Chlorofluorocarbon-113 (CFC-113) have low boiling points while PCE's boiling point is the highest. While other chlorinated solvents are volatilized in a degreaser using hot water pipes, PCE requires steam pipes or an electric heater. The maximum recommended boiling temperatures for PCE, TCE, TCA, and MC in a distillation still, for example, are 270, 210, 200, and 125°F, respectively (Ethyl Corporation, no date). The third column shows the vapor pressure for each of the solvents. This property is also important in vapor degreasing where the solvents are heated to their boiling points. The higher the boiling point, the higher the energy needed to boil the solvents. The higher the latent heat of vaporization, the higher the energy needed to keep the solvent at its boiling point.

Table 12.2.1 *Summary of Physical and Chemical Properties of Selected Chlorinated Solvents at 25°C*

Compound	Molecular Weight	Vapor Pressure (p°, torr)	Solubility (S, mg/L)	Henry's constant (H, atm-m^3/mol)	Relative Vapor Density	Boiling Point (°C)
dichloromethane	84.9	415	20000	0.00212	2.05	41
chloroform	119.4	194	8000	0.00358	1.80	62
bromodichloromethane	163.8	64.2	4500	0.00206	1.39	90
dibromochloromethane	208.3	17	4000	0.00115	1.14	119
trichlorofluoromethane	137.4	796	1100	0.0888	4.91	23.8
carbon tetrachloride	153.8	109	825	0.0298	1.62	76.7
1,1-dichloroethane	99	221	5100	0.00543	1.70	57.3
1,2-dichloroethane	99	82.1	8500	0.0015	1.26	83.5
1,1,1-trichloroethane	133.4	124.6	1300	0.0167	1.59	
1,1,2-trichloroethane	133.4	24.4	4400	0.00108	1.12	113.7
1,1,2,2-tetrachloroethane	167.9	6.36	2900	0.000459	1.04	146.4
1,1-dichloroethylene	97	603	3350	0.0255	2.86	31.9
cis-1,2-dichloroethylene	97	205	3500	0.00374	1.63	60
trans-1,2-dichloroethylene	97	315	6300	0.00916	1.97	48
trichloroethylene (TCE)	131.5	75	1100	0.00937	1.35	86.7
tetrachloroethylene (PCE)	165.8	18.9	200	0.0174	1.12	121.4
1,2-dichloropropane	113	52.3	2800	0.00262	1.20	96.8
trans-1, 3-dichloropropylene	110	34	2800	0.0013	1.10	112
bis(chloro)methylether	115	30	22000	0.00021	1.09	104
bis(2-chloroethyl)ether	143	1.11	10200	0.00013	1.004	178
bis(2-chloroisopropyl)ether	171	0.73	1700	0.00011	1.003	189
2-chloroethylvinylether	106.6	34.3	15000	0.00025	1.010	108
chlorobenzene	112.6	11.7	500	0.00390	1.04	132
o-dichlorobenzene	147	1.39	140	0.00198	1.01	179
m-dichlorobenzene	147	2.25	119	0.00325	1.01	172

12.3 DEGRADATION REACTIONS AND PATHWAYS

Degradation pathways of chlorinated solvents are important in understanding the fate and transport of these chemicals in the subsurface and form the basis for the use of these relationships in forensic investigations. The degradation of chlorinated solvents in soil and groundwater occurs by chemical (abiotic) and microbial (biotic) processes and is well understood (see Figures 12.3.1 and 12.3.2).

Numerous biochemical and abiotic reactions are potentially involved in chlorinated solvent degradation. Details regarding dechlorination mechanisms leading, for example, to the sequential degradation of TCE to *cis*-1,2-dichloroethylene (*cis*-1,2-DCE) are often treated in a nonspecific manner. Four types of reactions are commonly recognized as responsible for organic compound degradation; substitution, dehydrohalogenation, oxidation, and reduction reactions (Schwarzenbach *et al.*, 1985).

Dehydrohalogenation reactions involve the elimination of HCl and the creation of a C=C double bond in place of a single bond (C–C). Oxidation processes may involve a number of mechanisms that add oxygen to the structure of the organic molecule. Epoxidation, where the C=C bond is replaced by a single bond and a mutually bonded oxygen, is possibly the most common. The potential for oxidation decreases as the degree of chlorination (number of Cl atoms on the molecule) increases. Thus only the simpler degradation products (single and double chlorine molecules

such as 1,2-DCE and vinyl chloride) can effectively be oxidized.

Reduction reactions typically involve hydrogenolysis (where a C–Cl bond is broken and Cl⁻ is replaced by H⁺) and dihalo-elimination (where two chlorines are removed, and a C=C bond is produced from a C–C bond). Abiotic and biotic reductive dechlorination is reported to occur for many chlorinated solvents. Biotically mediated processes comprise the focus of degradation under reduced conditions. Reduction reactions are more favored to occur for the more highly chlorinated compounds such as TCE and PCE that contain three and four chlorine atoms, respectively (McCarty and Semprini, 1994).

The most commonly recognized degradation pathways for chlorinated solvents are those that occur biotically under anaerobic conditions and for the less chlorinated compounds (one or two chlorine atoms) under aerobic conditions. The exception is TCA, which degrades abiotically by hydrolysis with time scales of interest under typical aquifer conditions (Vogel *et al.*, 1987; McCarty, 1993). The compounds are presented vertically in Figures 12.3.1 for PCE and TCE and 12.3.2 for TCA as a function of the degree of chlorination for each compound.

The degradation reactions and ultimately the efficiency of degradation processes vary as a function of the environmental conditions and reaction types. *First-order* rate processes are used to describe rates of transformation of inorganic chemical species in aqueous solution. This approach is often

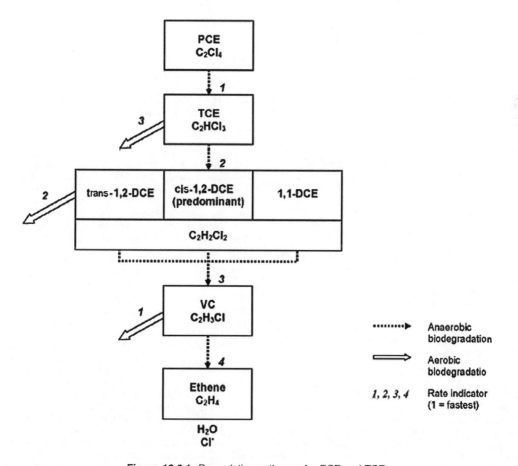

Figure 12.3.1 *Degradation pathways for PCE and TCE.*

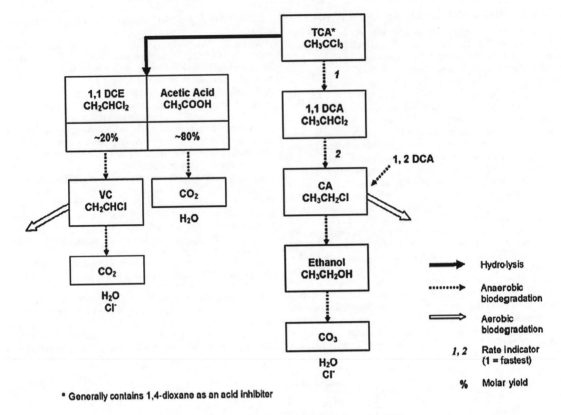

* Generally contains 1,4-dioxane as an acid inhibitor

Figure 12.3.2 *Degradation pathways for 1,1,1-TCA.*

used as a first approximation to evaluate biotic and abiotic degradation rates. The rate of change of a chemical in solution is given in terms of half-lives. The concentration varies as a (negative exponential) function of e^{-kt}, where t is the elapsed time and k is the rate constant. This degradation rate process is easily calculated and is commonly used in contaminant fate and transport models.

Temperature effects are significant in biotic and abiotic processes. Metabolic and enzymatic reaction rates that drive biotic degradation processes generally increase with temperature. Using TCA as a case example for abiotic degradation, the abiotic transformation rate of TCA to 1,1-DCE ($CH_3CCl_3 \Rightarrow CH_2=CCl_2 + H^+ + Cl^-$; forming approximately 20% of the degradation product) can be estimated as a function of temperature (McCarty, 1993). At 10°C, the degradation half-life of TCA is about 12 years. The observed half-life decreases to 4.9 years at 15°, and to 0.95 years at 20° (Note that acetic acid, where $CH_3CCl_3 + 2H_2O \Rightarrow CH_3COOH + 3H^+ + 3Cl^-$, forms approximately 80% of the degradation product.). These data suggest that extrapolation of data between sites or experiments must carefully consider temperature effects.

12.4 ANALYTICAL METHODS

Analytical methodology for measuring the concentration of chlorinated solvents is mature and well developed. In the United States, volatile organic compounds are analyzed via EPA Standard Method 8260B/624. For drinking water samples, EPA Standard Method 524.4 is employed. Both methods rely upon gas chromatography.

12.5 HISTORICAL SOURCES AND COMPOSITION OF CHLORINATED SOLVENTS

Chlorinated solvents were used in a wide variety of 20th-century industries. Most noteworthy in terms of volume were the aerospace, military, metal-working, and dry-cleaning industries. The popularity of these solvents in these industries derived from their low flammability and reactivity, ease of evaporation, and strong dissolving power.

Potential sources of releases to the environment varied over the course of time for each of the solvents. For example, in dry cleaning, CT was the first solvent to be used, followed by TCE and finally PCE (Morrison, 2003). In metal cleaning ("cold cleaning") and vapor degreasing, TCE was initially popular, but was eclipsed in the 1970s by TCA, which became the dominant solvent for this purpose (Wolf, 1997). The trend reversed when TCA was phased out in the 1990s, resulting in a resurgence in TCE use. The evolution of uses of CT, TCE, PCE, and TCA in the 20th-century United States was summarized in 2000 in two review articles by Richard Doherty (Doherty, 2000a, b).

During the first 60 years of the 20th century, the use of these solvents was influenced primarily by economic conditions and wartime demand. Increased demand during wartime years was generally followed by a period of post-war oversupply. During the latter 40 years, environmental regulation played an increasing role. Federal regulations and standards arising from the 1977 Clean Water Act, the 1980 Resource Conservation and Recovery Act, and the 1990 Clean Air Act Amendments directly affected the use and handling of chlorinated solvents. The Clean Air Act facilitated the United States' implementation of the 1990 amendments to the Montreal Protocol, which set a schedule for an international ban of most uses of TCA and CT.

In pure form and in appropriate containers, chlorinated solvents can be stored for extended periods without degradation. However, in commercial use, the solvents are exposed to a number of environmental factors that can cause degradation and/or diminish solvent effectiveness. These factors include exposure to oxygen, light, metals/metal salts, water, high temperatures, strong bases, and oxidizing agents (Solvay Chemicals, 2002). The degradation of chlorinated solvents produces hydrogen chloride (HCl), which degrades metal surfaces, sometimes yielding products that initiate further degradation. Additives were utilized with each of the solvents to varying degrees to address these issues.

Historical information regarding the five most commonly encountered chlorinated solvents in forensic investigations (CT, PCE, TCE, TCA, and MC) is presented below.

12.5.1 Carbon Tetrachloride (CT)

Carbon tetrachloride was the first of the five chlorinated solvents to come into general use. Production of commercial quantities in Europe began in approximately 1900 or earlier, and in the United States between 1905 and 1908. Although CT was used for a wide variety of commercial and industrial applications, its major uses in terms of quantity were as a cleaner, and as a raw material in the manufacture of other chemicals, particularly CFCs. The latter constituted the largest single use of CT since World War II. Other CT applications of note included it's use in fire extinguishers, dry cleaning (before the use of PCE became prevalent), grain fumigation, military smokescreens, and as a solvent for lacquers (Doherty, 2000a).

Although CT production in the United States peaked in the 1970s, roughly 80–95% of the CT produced during that time was consumed in the manufacture of CFC-11 and CFC-12. Increasing awareness of toxicity and bans on specific uses were factors in the production decline that began in the mid-1970s and continued in the following decades. Carbon tetrachloride was banned in consumer products in 1970, in aerosol products in 1978, and in grain fumigation in 1985 (Holbrook, 1991; NIH, 1999).

The earliest commercial-scale process for manufacturing CT employed chlorination of carbon disulfide. As a result, small quantities of carbon disulfide (about 1 part per million (ppm) in technical grades and 100 ppm in commercial grades) can be found in CT produced by this method. Trace concentrations of bromine, chloroform, and hydrochloric acid may also be present in CT (Brallier, 1949; Holbrook, 1991).

Beginning in the 1950s, production of CT using the pyrolytic chlorination of methane or propane (also known as chlorinolysis) began. This method soon became predominant; however, the carbon disulfide process continued to be used commercially into the 1990s (Doherty, 2000a).

In metal cleaning applications, CT was replaced by TCE and other compounds due to its high toxicity, and because it tended to leave metal surfaces susceptible to corrosion (Brallier, 1949). Carbon tetrachloride was usually shipped in galvanized tin or lead-lined containers for this reason. Low concentrations of corrosion inhibitors were used with some CT formulations in an effort to address this problem. Chemicals used included diphenylamine (0.34–1%), ethyl acetate (to protect copper), alkyl cyanamides, ethyl cyanide (up to 1%), and thiocarbamide (DeShon, 1979; Holbrook, 1991). However, other literature state that commercial grades of CT rarely contained inhibitors (DeForest, 1979). Despite these inconsistencies in the literature, most researchers agree that concentrations of additives in CT formulations were generally low relative to other chlorinated solvents, such as TCE and TCA.

12.5.2 Tetrachloroethylene (PCE)

Most commonly known for its widespread use in dry-cleaning, PCE has several other important uses. PCE was used for metal cleaning and degreasing, particularly for cleaning aluminum prior to the development of stabilized TCA formulations, and for the removal of wax and resin residues. Tetrachloroethylene is also used in cleaning small, low-mass parts because the condensed solvent contact time, before the part reaches the vapor temperature, is longer than with other solvents. Other uses included automotive brake cleaning, rubber dissolution, paint removal, sulfur recovery, printing ink bleeding, soot removal, and catalyst regeneration (Lowenheim and Moran, 1975). PCE was used in various textile operations as a scouring solvent, a carrier medium, and for spot removal. The primary use of PCE after 1996 was in the production of fluorinated compounds such as CFC-113 and HFC-134a.

By the late 1940s, PCE had surpassed CT as the predominant non-petroleum dry-cleaning solvent. Peak years for PCE production in the United States ranged from the late 1960s to the early 1980s, when approximately 600–700 million pounds per year were produced. The effects of environmental regulation and significant improvements in the dry-cleaning process resulted in an overall decreased demand for PCE by the 1980s (Doherty, 2000a). In many countries, this decrease in demand resulted in the cessation of PCE manufacturing. In Australia, for example, PCE manufacturing ceased in 1991 (NICNAS, 2001). In Western Europe, the production and sales of PCE more than halved between 1986 and 1994 (Linn, 2002). In the United States, the use of PCE for dry cleaning in 2000 has decreased to about one-sixth of the levels of the 1970s.

Tetrachloroethylene production processes are capable of producing product of 99.9% purity (Lowenheim and Moran, 1975; Gerhartz, 1986). The primary method of PCE production prior to the 1970s involved the chlorination of acetylene to produce both PCE and TCE. Subsequent production methods could produce TCE as a co-product (e.g., high-temperature chlorination, oxychlorination) or CT as a co-product (chlorinolysis) (Hickman, 1991). Ethanol is a reported impurity in PCE.

Because PCE is a relatively stable molecule, small concentrations of stabilizer additives are needed relative to other chlorinated solvents. Reported antioxidant (i.e., amine or phenolic compound) concentrations for metal-cleaning grades range from 50 to 200 ppm. Concentrations of acid acceptors, such as an epoxide, for PCE range from 0.2 to 0.7% (Archer and Stevens, 1977).

Tetrachloroethylene used for dry cleaning is usually of high purity. In the 1960s, Dow Chemical offered a dry-cleaning grade of PCE (DOWPER-C-S), which reportedly contained six additives. These included a redeposition agent, a water-soluble detergent, a corrosion inhibitor, an anti-static compound, "hand agent" additives, and a scavenger for fatty acid control. Dow tested the product in 1963, and brought it to market in approximately 1967 (Chemical and Engineering News, 1963; Dow Chemical, 1970, 1971a, b, 1973). The intent of the product was to save the dry cleaner from the effort of pre-mixing PCE with detergents, particularly for use in coin-operated dry-cleaning machines, which were becoming popular at that time.

Tetrachloroethylene used for vapor degreasing typically had a higher concentration of additives than most dry-cleaning grades (Von Grote, 2003; Dow Chemical, 2005a). Classes of chemicals used include acid acceptors, antioxidants, and ultraviolet (UV) light stabilizers. Acid acceptors were only required for PCE when used in high-temperature or other "stressful" applications (Archer, 1996). Alkylamines and other hydrocarbons were added to early

PCE formulations; later stabilizers included morpholine derivatives (Gerhartz, 1986). Epoxides, esters, and phenols have also been used as PCE additives (Mohr, 2001; Morrison, 2003) (see Table 12.5.1).

12.5.3 Trichloroethylene (TCE)

Trichloroethylene's solvent properties resulted in its widespread use in commerce and industry. TCE has been used in the electronics, defense, chemical, rail, adhesive, automotive, boat, textile, food processing, and dry-cleaning industries (ASTR, 1997). It is an excellent solvent because of its aggressive action on oils, greases, waxes, tars, gums, and certain polymers, and has historically been the major solvent used in industrial vapor degreasing and cleaning applications (Wood, 1982; ASM, 1996). Trichloroethylene has also been used in inks, paints, paint removers, adhesives, fire extinguishers, lubricants, pesticides, polishes, pipe and drain cleaners, medical and dental anesthetics, and many other industrial and consumer products (Doherty, 2000b).

Commercial-scale TCE production in the United States began in approximately 1921, and increased as it became widely used in dry cleaning and vapor degreasing. Usage of TCE in dry cleaning decreased in the early 1950s after it was found to degrade cellulose acetate dyes. By 1952, it was estimated that 92% of TCE was used in vapor degreasing (Chemical Week, 1953). Production grew steadily until 1970, when annual production peaked at approximately 600 million pounds (Doherty, 2000b). The decline in production that began in 1970 was the result of increasing evidence of toxicity, economic factors, and increased environmental regulation. Its use as a solvent experienced a rebound in the 1990s when it was listed as a recommended substitute for other solvents (such as TCA) banned under the Montreal Protocol and the Clean Air Act Amendments (Kirschner, 1994). In 1991, Sweden issued an ordinance that banned the sale, transfer or use of chemical products containing TCE. This ordinance became effective for consumers and industry in 1993 and 1996, respectively (NICNAS, 2000). Austria and Switzerland also have regulations banning certain chlorinated solvent applications (KEMI, 1995).

Prior to 1950, TCE was produced almost exclusively from acetylene. By 1978, the acetylene production method was no longer in commercial-scale use in the United States. During the 1960s and 1970s, TCE was increasingly produced from ethylene or 1,2-dichloroethane (1,2-DCA) using chlorination processes. Different countries have unique production chronological histories; in Australia, for example, TCE was manufactured by ICI at their Botany chemicals facility in Sydney between 1948 and 1977 (ICI Botany Operations, 1996).

Unstabilized TCE is typically greater than 99% pure (Gerhartz, 1986); however, it is particularly vulnerable to oxidation when exposed to air, light, or heat. Stabilizing compounds were used in TCE to prevent chemical breakdown. Degradation can occur especially rapidly in the presence of aluminum, producing significant quantities of HCl gas. Typical stabilizer formulations included an acid acceptor, a metal stabilizer, and/or an antioxidant. These additives typically comprised from 0.1 to 0.5% of the solvent, but concentrations reportedly ranged as high as 2%. In vapor degreasing grades, concentrations at the higher end of the range were typical, and additional compounds might be added to enhance thermal stability.

A variety of chemicals have been used as TCE additives, including alcohols, amines, ethers, esters, epoxides, substituted phenols, and heterocyclic nitrogen compounds. The earliest stabilizers for PCE and TCE were gasoline and other unsaturated hydrocarbons (Shepherd, 1962). Until 1954, the most commonly used acid acceptors in TCE were amines, including trimethylamine, triethylamine, triethanolamine, aniline, and diisopropylamine (Chemical Engineering, 1961). One source cites the typical trimethylamine concentrations as 20 ppm by weight (Lowenheim and Moran, 1975). Kircher reported that typical concentrations ranged from 10 to 100 ppm (Kircher, 1957). In the mid-1950s, amines began to be replaced by non-alkaline formulations, particularly a pyrrole-based, six-to-seven component mixture developed by DuPont. Metal stabilizers used with TCE included epoxides such as 1,2-butylene oxide and epichlorohydrin. The use of the latter was discontinued in the 1980s due to its toxicity (Mertens, 1991). Additives for thermal stability primarily included cyclohexene, diisobutylene, and amylene, although many others were used (Shepherd, 1962). Table 12.5.2 lists acid inhibitors, metal inhibitors, antioxidants, and light inhibitors associated with TCE (Hardie, 1964; Mohr, 2001; Morrison, 2003).

Solvent stabilizers and inhibitors for TCE can be found at significantly different concentrations in still bottom residue "muck" than in virgin TCE. Stabilizers were found to be

Table 12.5.1 *Tetrachloroethylene Additives (after Morrison, 2000c; Mohr, 2001)*

Acid Inhibitors	Metal Inhibitors	Light Inhibitors	Antioxidants
Acetylenic alcohols	Alcohols	Amines	Acetylene ethers
Acetylenic carbinols	Aromatic hydrocarbons	Cyanide	Phenols
Acetylenic esters	Cyclic trimers	Hydroxyl aromatic	Pyrrole
Alcohols	Esters	compounds	Thiocyanates
Aliphatic amines	Lactone	Nitriles	
Aliphatic monohydric	Oxazoles	Organo-metallic	
alcohols	Oximes	compounds	
Amides	Sulfones		
Amines	Sulfoxide		
Azo aromatic			
compounds			
Epoxides			
Hydroxyl aromatic			
compounds			
Ketones			
Nitroso compounds			
Pyridines			

Table 12.5.2 *Trichloroethylene Additives (after Morrison, 2000c)*

Acid Inhibitors	Metal Inhibitors	Antioxidant	Light Inhibitor
Acetylenic alcohols	Alcohols	Amides	Aromatic benzene nuclei
Alcohols	Amides	Amines	Boranes
Aliphatic amines	Amines	Aromatic carboxylic acids	Ethers
Aliphatic monohydric	Aromatic hydrocarbons	Alkyl pyrroles	Guanidine
alcohols	Complex ethers and	Aryl stibine	Hydroxyl-aromatic compounds
Alkaloids	oxides	Boranes	Organo-metallic compounds
Alky Halides	Cyanide	Butylhydroxyanisole	
Amines	Cyclic Ethanes	Phenols	
Azines	Cyclic trimers	Pyridines	
Azirdines	Epoxides	Pyrrole	
Azo-aromatic	Esters	Thiocyanates	
compounds	Ethers		
Epoxides	Ketones		
Essential oils	Olefins		
Hydroxyl-aromatic	Peroxides		
compounds	Pyridines		
Nitroso compounds	Oxazoles		
Olefins	Oxazolines		
Organic Substitute	Oximes		
NH_4 hydroxides	Sulfones		
Oxirane	Sulfoxide		
Phenols	Thiophene		
Pyridines			
Pyrrole			
Quatenary			
Ammonium			

retained in still bottoms in excess of 35% of their concentration in original TCE (Joshi *et al.*, 1989). For vapor degreasers, as much as 50% of the still bottom residue is solvent (United States Environmental Protection Agency, 1979). The significance of this information is that while the original concentrations of these stabilizers in the virgin TCE may be less than 1%, when accumulated in still bottoms and then released into the environment, they may be present at significantly elevated concentrations. Another example of different concentrations of TCE stabilizer concentrations as a function of use is summarized in Table 12.5.3 for butylene oxide, epichlorohydrin, ethyl acetate, and methyl pyrrole (Hardie, 1964; Mohr, 2001).

12.5.4 1,1,1-Trichloroethane (TCA)

The primary applications of TCA (methyl chloroform) were in cleaning and degreasing, where it served as a less toxic replacement for TCE, PCE, and other solvents. Similar to TCE, TCA was used in a variety of industries and purposes. The aircraft, automotive, electronic, and missile industries were significant users of TCA. Among the many products

that included TCA were pesticides, drain cleaners, aerosol propellants, and carpet glue.

Although occurred TCA production in the United States in the mid-1930s, the chemical did not see significant commercial use as an end product until the mid-1950s (Halogenated Solvents Industry Alliance, 1994). Its acceptance was tied to the development of suitable stabilizer formulations. Its production increased steadily throughout the 1960s and 1970s, and first surpassed the production of TCE in 1973 (Doherty, 2000b). Production peaked in the mid-1980s and then began to decrease due to increased environmental regulation and a heightened awareness of environmental impacts. Because TCA has a lower boiling point than either TCE or PCE, it had special applications for cleaning items such as computer boards and electric components, all of which can be damaged by high temperatures (Soble, 1979; Warner and Mertens, 1991). In 1989, the industrial use of TCA was summarized as: metal degreasing (32%); cold cleaning (19%); aerosols (11%); adhesives (9%); chemical intermediate (9%); electronics (7%); coatings and inks (6%); textiles (3%); and miscellaneous (4%). The 1990s marked the beginning of the end of TCA's use as a solvent, as its ozone-depleting

Table 12.5.3 *Stabilizer Concentrations in New and Spent TCE (after Mohr, 2001)*

Sample Description	Stabilizer Concentration (Weight Fraction)			
Sample	Butylene Oxide ($\times 10^3$)	Epichlorohydrin ($\times 10^3$)	Ethyl Acetate ($\times 10^3$)	Methyl Pyrrole ($\times 10^4$)
New TCE	1.64	1.66	3.46	1.59
Spent TCE	0.685	1.69	2.85	2.18
TCE Distillate	0.718	1.61	2.58	1.66
Carbon Adsorbed TCE	0.44	1.31	2.65	0.90

potential caused it to be phased out under the 1990-amended Montreal Protocol and the 1990 Clean Air Act Amendments. Most emissive uses in the United States were phased out by the end of 1995.

The commercialization of TCA as a metal-cleaning and metal-degreasing solvent was hindered by the lack of effective stabilizers. Prior to the mid-1960s, TCA was found to be unacceptable for use in heated degreasing due to the absence of these inhibitors (solvents progressively deteriorate due to exposure to ultra-violet light and heat). Uninhibited TCA in contact with aluminum forms aluminum chloride, 2,2,3,3-tetrachlorobutane, 1,1-DCE, HCl, and magnesium. Improperly stabilized TCA can also decompose in the presence of magnesium. The use of TCA in vapor degreasing involves condensing the vapors with cold water pipes or a refrigeration unit near the top of the tank so that the liquid TCA is returned to the solvent bath. In the process water is condensed from the air. Uninhibited TCA hydrolyzes when boiled with water to produce hydrochloric and acetic acid (Manufacturing Chemists' Association, 1965; United States Environmental Protection Agency, 1979).

As a result, unstabilized TCA could not be used for many metal-cleaning applications, particularly in the high-temperature environment of a vapor degreaser. When heated to a range of 360–440 °C, TCA decomposes to 1,1-DCE and HCl. Stabilizers used with TCE were found to be only partially effective with TCA; therefore, significant efforts were expended by chemical companies to find new stabilizers for use with TCA. Until suitable stabilizing formulations were developed, TCA use with metals was largely limited to cold cleaning (Bachtel, 1958; United States Army, 1978; Jordan, 1979). In the 1960s Dow Chemical Company marketed a solvent based on TCA that was specifically designed for spray-cleaning railroad equipment (Chemical and Engineering News, 1962).

Similar to TCE, stabilizer formulations used for TCA typically included an acid acceptor and a metal stabilizer. However, unlike TCE and PCE, an antioxidant was not typically needed due to TCA's stability to oxidation. However, antioxidants were sometimes added to TCA (particularly vapor-degreasing grades) to prevent degradation of other additives (Jordan, 1979).

Due to TCA's greater reactivity with metals, the overall concentration of additives was typically higher in TCA than in TCE. Concentrations of stabilizing chemicals ranged between 3 and 8% (Lowenheim and Moran, 1975), but were frequently in the range of 4–6%. Typical concentrations in United States vapor-degreasing grades were reported as 2–3.5% 1,4-dioxane, 1–2% sec-butanol, 1% 1,3-dioxalane, 0.4–0.7% nitromethane, and 0.5–0.8% 1,2-butylene oxide (Archer, 1984; United Nations Industrial Development Organization, 1994).

Additive formulations varied between manufacturers and between grades produced by the same manufacturer. Typically, epoxides, ethers, amines, and alcohols were used as acid acceptors, and nitro- and cyano-organo compounds were used as metal stabilizers. A wide variety of chemicals has been reportedly used in TCA stabilizer formulations, including 1,4-dioxane; 1,3-dioxolane (synonyms include 1,3-dioxolane, glycol formal, 1,3-dioxole, dioxolane, Glycol methylene ether, dihydroethylene, glycol formal, and formal glycol); 1,2-butylene oxide (synonyms include 1,2-dpoxybutane, EBU, propyl oxirane, epoxybutane, and 2-ethyloxirane); epichlorohydrin (synonyms include chloromethyloxirane, glycidyl chloride; chloropropylene oxide, glycerol, epichlorohydrin, 1,2-epoxy-3-chloropropane, 3-chloro-1,2-epoxypropane, gamma-chloropropylene oxide, 1-chloro-2,3-epoxypropane, and 2,3-dpoxypropyl chloride); nitromethane (synonyms include

NMT and nitrocarbol); 1,2-epoxybutane; methyl ethyl ketone; n-methyl pyrrole; ethyl acetate; tetraethyl lead; n-methyl pyrrole; acrylonitrile; isopropyl alcohol; monohydric acetylenic alcohols; glycol diesters; nitriles; butyl alcohols; tetrahydrofuran (synonyms include THF, 1,4-epoxybutane, cyclotetra-methylene oxide, oxacyclopentane, and oxolane); morpholine; toluene; and dialkyl sulfoxides, sulfides, and sulfites (Irish, 1963; Lowenheim and Moran, 1975; Jordan, 1979; Archer, 1982). Several hundred TCA stabilizer formulations have been patented worldwide (Snedecor, 1991).

Dow Chemical, the first and only major TCA manufacturer in the United Sates until 1962, introduced the Chlorothene brand of TCA in 1954 (USPTO). Chlorothene contained inhibitors that allowed its use as an aerosol propellant (Chemical Week, 1956), but it was not recommended for use with aluminum (Barber, 1957). In patents issued in 1954 and 1955, Dow registered the use of 1,4-dioxane (synonyms included DX, 1,4-diethylene-dioxide, diethylene oxide, p-dioxane, tetrahydro-1,4-dioxan, dioxyethylene-ether, and glycolethylene ether) and a non-primary alkonol, the first effective TCA stabilizer system for use with aluminum (Bachtel, 1957). Chloroethene NU, introduced by Dow in May 1960, utilized the 1,4-dioxane-based stabilizer system (Chemical and Engineering News, 1962). However, as of 1962, TCA was not recommended for use in vapor degreasers (ASTM, 1962). Further patents incorporated modified formulations to address this need by the later 1960s. Reported impurities found in unstabilized TCA include 1,2-DCA, 1,1-DCA, chloroform, CT, TCE, 1,1,2-TCA, and 1,1-DCE.

An important TCA stabilizer is 1,4 dioxane. 1,4-dioxane was produced in the United States by Ferro Corporation, Dow Chemical (also imported 1,4-dioxane), and Stephan company. Dow Chemical applied for a US patent to stabilize TCA in 1954 with 1,4 dioxane at a concentration of 3.5%. In 1962, Dow Chemical applied for a patent that described stabilizing TCA with a combination of 1–10% 1,4 dioxane and 0.001–1% n-methyl pyrrole. The presence of dioxane in TCA is designed to prevent corrosion of aluminum, zinc, and iron surfaces by neutralizing hydrochloric acid. Approximately 90% of the 1,4-dioxane produced in 1985 was used as a stabilizer for chlorinated solvents, especially for TCA (United States Environmental Protection Agency, 1995). While TCA is associated with other industrial and commercial uses (fumigants, an additive in antifreeze, cosmetics, a wetting and dispersion agent in textile processes, a solvent in paper manufacturing, in liquid scintillation counters, and an "inert" ingredient in herbicides [Roundup®, Pondmaster®, Rattler®, Rodeo®]), its primary association is as a stabilizer in chlorinated solvents (Italia and Nunes, 1991; Scalia et al., 1992).

12.5.5 Methylene Chloride (MC)

Methylene chloride (dichloromethane) did not become an important industrial chemical until the years immediately after World War II, when production increased fivefold. The peak years of production were the late 1970s and early 1980s. A 1985 National Toxicology Program study indicating that MC caused cancer in mice was a factor in the decreased demand (NIOSH, 1986).

One of MC's first uses was in paint strippers, and it remained the predominant use for many decades. Methylene chloride paint strippers remove many types of finishes from a variety based of surfaces. Concentrations of methylene chloride is paint strippers range from 10 to 90%. Methylene chloride was also used in metal cleaning, polyurethane foam production, in aerosol products formulation, adhesives (where it served as a replacement for TCA), and as an extraction solvent (e.g., in laboratories and in the pharmaceutical industry) (Archer, 1996; Dow Chemical, 2005b). Methylene chloride was used in oven cleaners, tar removers, refrigerants, and in the production of spices, beer hops, and decaffeinated coffee. Other uses included metal degreasing and chemical processing. Due to its relatively low boiling point, MC could be used for vapor degreasing of temperature-sensitive materials.

Unstabilized MC reacts with lighter metals, particularly aluminum. Commercial grades of MC reportedly almost always contained stabilizers (DeForest, 1979). Reported concentrations of stabilizers in technical grades of methylene chloride range from 0.0001 to 1% (IPCS, 1987). Similar to other chlorinated solvents, grades of MC intended for use in vapor degreasers typically have higher concentrations of stabilizing chemicals. Stabilizers included not only cyclohexane, but propylene oxide (for aerosol and vapor degreasing formulations), methanol, ethanol, thymol, hydroquinone, phenols and amines (e.g., tertiary butylamine) (DeForest, 1979; IPCS, 1987, 1996). Amylenes are reported to be the most widely used stabilizers in laboratory-grade MC (Hsu *et al.*, 2005). Reported impurities in commercial grades of MC include methyl chloride, chloroform, 1,1-DCA, and *trans*-1,2-DCA (IPCS, 1987).

Methylene chloride paint strippers can contain varying concentrations of other solvents such as methanol. Other chemicals incorporated into MC paint strippers include amines, acids, ammonium hydroxide, detergents, paraffin wax, and other alcohols.

12.6 FORENSIC TECHNIQUES

Forensic techniques commonly employed to discriminate and potentially age-date chlorinated solvent releases include (Morrison, 2000a,b,c; 2001; 2003; Morrison and Heneroulle, 2000; Morrison, 2005):

- The presence of additives and/or chemical surrogates associated with chlorinated solvent releases;
- Isotopic analysis;
- Ratio analysis using measured parent compound–degradation product concentrations; and
- Age-dating based on groundwater plume length.

While each environmental forensic investigation is unique, the use of these methods, or combinations thereof, often provide the answers regarding the source and age of a chlorinated solvent release.

12.6.1 Solvent Stabilizers

As discussed earlier in this chapter, chlorinated solvents marketed for use in metal cleaning, degreasing, electronics, and textile cleaning require stabilizers (acid receptors, metal inhibitors, and antioxidants), so that the solvents can function as intended. Acid receptors are neutral (epoxides) or slightly basic (amines) compounds that react with hydrochloric acid, which is commonly produced when solvents and oil decompose. Alcohol is normally formed in this process (Archer, 1984). Metal inhibitors deactivate metal surfaces and complex metal salts that might form. Antioxidants are added to solvents to reduce their potential to form oxidation products (Joshi *et al.*, 1989).

Both TCE and TCA require metal inhibitors and acid acceptors whereas TCE requires only an oxidant. The TCE vapor-degreasing solvent marketed as NEU-TRI®, for example, is highly stabilized to prevent the accumulation of acid (Mohr, 2001). Tetrachloroethylene is considered to be relatively stable and has only minor amounts of acid inhibitors when used for degreasing and no metal inhibitors (Keil, 1978). Methylene chloride is considered to be stable and requires less than 0.1% of acid inhibitors.

The use and identification of solvent stabilizers to identify the origin and age-date a contaminant plume requires detailed information regarding the original composition of the solvent released into the environment. In addition, unless concentrated in still bottom residue or other circumstances, many of these additives are present in the original additive package at low concentrations, and are therefore difficult to detect in an environmental sample. Since formulations of solvent packages vary depending on the country where the solvent was synthesized, the origin of the original solvent is important (Morrison, 2000c).

An exception to these challenges is when a solvent stabilizer is recalcitrant to degradation and is present or is accumulated at a detectable concentration. 1,4-dioxane, is an example of such a stabilizer that is available to trace the release of 1,1,1-TCA (Nyer *et al.*, 1991; Barone *et al.*, 1992; Duncan *et al.*, 2004). 1,4-dioxane can become significantly concentrated in recycled or used TCA (Mohr, 2001). Because 1,4-dioxane can accumulate during use and unlike TCA, it does not readily degrade, it can act as an indicator of a TCA spill even when TCA is not present, for example when TCA has degraded to a concentration below its detection limit. In addition, 1,4-dioxane is hardly retarded in groundwater, unlike TCA and its principal chlorinated degradation products, hence the plume front from a TCA spill may only contain 1,4-dioxane. When using additives as tracers, it is important to note that depending on the purity and whether the solvent was recycled, contaminants in the solvent that are not intentionally part of the additive package may be present. For example, the analysis of contaminants and additives in a TCA sample tested in 1989 contained the following chemicals:

- butylene oxide (inhibitor);
- 1,2-DCE (contaminant);
- TCE (contaminant);
- ethylene dichloride (contaminant);
- 1,4-dioxane (inhibitor);
- nitromethane (contaminant);
- nitroethane (contaminant); and
- 1,1,2-TCA (contaminant).

Surrogate chemicals used to associate the release of a chlorinated solvent also include chemicals other than additives. For example, in the United States PCE releases into the environment are often associated with dry-cleaning operations. The release of other compounds with chlorinated solvents (leather dyes, soaps, etc.) may provide evidence regarding the age of a release if the surrogate compound was used within a discrete time.

12.6.1.1 Accumulation of Less Volatile Compounds During Degreasing Operations

When less volatile trace contaminants are present in a degreasing solvent they can accumulate during degreasing operations and become enriched in the spent solvent. The phenomenon is illustrated by calculation in this section for PCE as a minor contaminant in TCE. Even at sites where PCE was ostensibly never used it may be found in the environment when spent TCE is discharged.

Perchloroethylene can be present in TCE, particularly in degreasing grades, because the two solvents are produced by the same process and are separated by fractional distillation. Specifications at www.rbchemtrade.com indicate that other chlorinated compounds comprise 3–4% of degreasing-grade TCE.

The boiling points of PCE and TCE are 121.2 and 87.2 °C, respectively. Trichloroethylene degreasers usually use hot water below the boiling point of 100 °C. Because of this

Table 12.6.1 *Example TCE Degreaser Parameter Values*

Parameter	Value
PCE Impurity Level	1%
TCE Volume During Initial Operation	350 gallons
TCE Addition	55 gallons/week
Cleaning Frequency	4 times/year
Spent Solvent Discharge Each Cleaning	100 gallons

difference in boiling points, TCE volatilizes into the air and is lost from the degreaser at a more rapid rate than PCE.

There are two steps to estimate the accumulation of PCE in TCE. First, calculate how much PCE enters the degreaser along with the TCE and, second, estimate how much of the PCE volatilizes. For example, one can calculate the accumulation of PCE in a vapor degreaser using the properties shown in Table 12.6.1. The volume of spent solvent in Table 12.6.1 that is discharged during cleaning is intended to represent the volume below the hot water pipes used to boil the TCE, less any solids that have collected. Trichloroethylene above the hot water pipes would typically be conserved by being boiled back to a separate tank. From Table 12.6.1 the amount of TCE added between cleanings is found to be $350 + 55 \times 12 = 1015$ gallons (solvent is not added just before cleaning). The amount of PCE introduced to the degreaser is thus 10.15 gallons.

In estimating how much of the PCE would evaporate, it is necessary to estimate the vapor pressure of PCE in the degreaser. Assuming that the solvent temperature during operation is approximately the TCE boiling point temperature of 87.2 °C, the vapor pressure of PCE at this temperature is about 0.34 atmospheres. This estimate is based on the Clausius-Clapeyron equation:

$$\ln P_v = \text{constant} \left(\frac{1}{T_B} - \frac{1}{T} \right), \qquad (12.6.1)$$

where P_v is the vapor pressure in atmospheres and T_B is the boiling point temperature in degrees Kelvin, which is 394.35 K (121.2 °C). Perry's *Chemical Engineer's Handbook* (1999) gives the vapor pressure as 200 mg mercury or 0.263 atmospheres at $T = 352.95$ K (79.8 °C). From this we find the constant in Equation 12.6.1 and find that the vapor pressure for PCE at the TCE boiling point temperature of 360.35 K (87.2 °C) is about 0.34 atmospheres.

The loss of PCE is then estimated as a ratio to the TCE loss. Prior to cleaning the degreaser, 100 gallons of TCE are assumed to be present in the vapor degreaser and 200 gallons are present in a boil back tank.[1] Thus the total TCE volatilization loss is $1015 - 300 = 715$ gallons.

The amount of PCE in the evaporated TCE would be 7.15 gallons and the portion evaporated is estimated as the ratio of the PCE to the TCE vapor pressure, the latter being one atmosphere at the boiling point. Thus the estimated loss is $0.34 \times 7.15 = 2.43$ gallons. This estimate assumes that the only difference between PCE and TCE evaporation

is the relative vapor pressures.[2] Thus the amount of PCE present along with the 100 gallons of TCE in the main tank is the amount introduced minus the evaporative loss or $10.15 - 2.43 = 7.72$ gallons.

The percentage of PCE in the remaining 100 gallons of TCE is about 7.7% indicating an enhancement from the original level of about 7.7 times. Note that in making this estimate TCE and PCE that was evaporated and recondensed into the boil back tank has been treated like solvent that simply evaporated into the air. If the 100 gallons of TCE/PCE is discharged to the environment, the resulting PCE levels in the environment may be high enough that it will appear that PCE was also used as a degreasing fluid on-site, rather than a minor contaminant in the TCE.

This enhancement phenomenon also occurs with other relatively nonvolatile solvent additives, such as 1,4-dioxane. However, some additives are consumed during degreasing operations and need to be considered.

12.6.2 Isotopic Analyses

Compound-specific isotope analysis (CSIA) represents a mature methodology used in environmental forensic investigations to (1) distinguish between different contaminant sources and/or (2) to demonstrate that biodegradation is occurring (Hunkeler *et al.*, 1997, 1999; Dayan *et al.*, 1999; Sherwood Lollar *et al.*, 1999; Sturchio *et al.*, 1999; Drenzek *et al.*, 2002; Barth *et al.*, 2004). These techniques have been successfully used to examine the biodegradation of chlorinated solvents under either anaerobic (Slater *et al.*, 2001) or aerobic (Barth *et al.*, 2002) environments as evidence of natural attenuation and for source differentiation at field sites throughout the world (Hunkeler *et al.*, 1999, 2003; Dayan *et al.*, 1999; Sturchio *et al.*, 1998; Bloom *et al.*, 2000; Slater *et al.*, 2001; Bill *et al.*, 2002; Mancini *et al.*, 2002; Song *et al.*, 2002; Kirkland *et al.*, 2003; Hunkeler *et al.*, 2004; Kloppmann, *et al.*, 2005).

Heraty *et al.* in 1999 investigated the isotopic fractionation of carbon and chlorine during the aerobic degradation of dichloromethane by MC8N, a gram-negative methylotropic organism related to the genera *methylobacterium* or *ochrobactrum*, and found this technique a useful indicator of microbial degradation. The current use of isotopic analyses for chlorinated solvents has focused upon the ability to use isotopic information as an indicator of biodegradation and to distinguish between different manufacturers and/or sources. The use of isotopic analyses as evidence of degradation is especially intriguing not only for chlorinated solvents but for other compounds such as methyl tertiary butyl ether (MTBE) and n-alkanes (Kelly *et al.*, 1997; Stehmeier *et al.*, 1999; Pearson and Eglinton, 2000; Gray *et al.*, 2002; Pond *et al.*, 2002; Reddy *et al.*, 2002).

The use of isotopic analyses to distinguish between sources of chlorinated solvent releases is premised on the assumption that a wide range of isotopic signatures exist for different manufacturers (Tanaka and Rye, 1991; Poulson and Drever, 1999). This assumption implies that carbon isotopic fractionation is not expected to occur during synthesis unless there are incomplete reactions and/or recycling of by-products during the manufacturing process. The primary $\delta^{13}C$ variability is therefore likely associated with differences in the isotopic signature of the original carbon materials (Ertl *et al.*, 1998). Halogenated compounds are

---

1. This feature may not always be present, for example in degreasers equipped with solvent distillation. Its purpose is to preserve the bulk of the degreasing fluid for reuse while what remains in the main tank contains oil, grease, and dirt, and is disposed.

2. The relative rate of solvent loss is controlled by a "Henry's Law" type constant, which for dilute solutions is equal to the ratio of vapor pressure to solubility. However, assuming that PCE is infinitely miscible in TCE, this reduces the relative rate of solvent loss being approximately proportional to the ratio of vapor pressures.

expected to exhibit a wide range of manufacturer-dependent isotopic signatures due to various chemical reactions, which may include dehydrochlorination or dehydrogenation reactions and production conditions (e.g., temperature differences, catalysts used, engineering design, etc.) as well as use of different feedstocks. The difference in bond strength results in chlorine isotope fractionation due to temperature and pressure differences during synthesis (Tanaka and Rye, 1991). The ^{37}Cl isotope fraction in organic solvents is bound more tightly to carbon than are ^{35}Cl atoms (Bartholomew et al., 1954).

Isotopes used for distinguishing between different chlorinated solvent sources include ^{13}C, ^{35}Cl, and ^{37}Cl (Clark and Fritz, 1997). Van Warmerdam et al. (1995) examined the isotopic ratios for ^{13}C/^{12}C and ^{37}Cl/^{35}Cl for chlorinated solvents from four manufacturers. Beneteau et al. (1996, 1999) continued this work using ^{13}C and ^{37}Cl for comparison between two batches from the Van Warmerdam et al. study and five pure-phase chlorinated solvents manufactured by Dow Chemical and PPG. Of note is that significant differences in the δ^{13}C of chlorinated solvents obtained from Dow Chemical and PPG and analyzed by van Warmerdam et al. in 1995 and by Beneteau et al. in 1999 exist. Of note is that the δ^{13}C signatures of PCE, TCE, and TCA between batches analyzed in 1995 and 1999 from the same manufacturers (Dow Chemical and PPG) were found to be variable. For TCA, δ^{37}Cl values between batches supplied by PPG in 1995 and 1999 ranged from -2.90 to -0.36‰, respectively, suggesting that the use δ^{37}Cl values for TCA may not be appropriate. δ^{13}C results for TCA, however, were similar between 1995 and 1999 PPG batches (-25.86 and -25.78‰).

In 2003 Shouakar-Stash examined δ^{13}C values for TCE from five different manufacturers used in the van Warmerdam et al. (1995) investigations. One conclusion of the subsequent study was that the samples had not degraded from the 1995 study. Contrary to the results of Beneteau et al., (1999), Shouakar-Stash found that δ^{13}C values for TCE and TCA did have characteristic isotopic signatures associated with the respective manufacturer. The difference in results is probably due to the use of the Parr Oxygen Bomb methodology in the earlier research, which resulted in low yields and loss of sample (Jendzejewski et al., 1997). A similar pattern was found for δ^{37}Cl values for different manufacturers regardless of whether the compound was TCE or TCA. The consistent δ^{37}Cl signature for batches from different years is consistent with the conclusion that each manufacturer has a characteristic δ^{37}Cl value. An explanation for the variation in δ^{37}Cl values between manufacturers is due to isotopic fractionation occurring during the processing of the source brines used to produce chlorine gas. Other researchers indicated that a more negative δ^{37}Cl value implied a biodegraded chlorinated solvent (Sturchio et al., 1998, 2002; Reddy et al., 2000; Numata et al., 2002). These results suggest that δ^{13}C values may be used to distinguish between TCA manufactures.

Shouakar-Stash identified differences in δ^2H attributable to different manufacturers although large errors are associated with this analysis due to impurities in the sample. One aspect of the δ^2H analysis was its use to distinguish between manufactured TCE and TCE resulting from the dechlorination from PCE. The authors found that the hydrogen atom on TCE does not undergo an exchange reaction with the surrounding water, thus any change in the hydrogen signature should be associated with the degradation of TCE, via carbon isotope fractionation. An analysis of TCE from different manufacturers exhibited a completely different δ^2H signature than from TCE produced from PCE dechlorination. Manufactured and dechlorinated TCE (degraded from PCE) δ^2H values ranged from $+466.9$ to $+681.9$‰ and -351.9 to -320.0‰, respectively. An extension of this relationship is if a linear relationship between δ^2H and δ^{13}C exists for different manufactures, then a means exists to identify the source of TCE in a series of samples collected along the axis of a contaminant plume. Table 12.6.2 summarizes the results of the δ^{13}C values for different manufactured chlorinated solvents.

From a forensic perspective, releases of chlorinated solvents from different manufacturers may be difficult to establish due to the precision of the analyses, effects of dissolution, sorption, and volatilization, whose extent are difficult to quantify, as well as significant differences in the δ^{13}C from the same supplier. Researchers have postulated that a TCE sample obtained in the field that may be affected by any of these processes is impossible to isotopically differentiate between sources that differ by less than 1.0 or ±0.5‰ (Slater, 2003). It is also difficult to reliably interpret small variations in isotopic composition to different sources as viewed from a confidence interval perspective. The standard deviation of multiple, independent analyses of a sample can range from 0.1 to 0.3‰ (represents a confidence interval of 68% or one standard deviation). In order to delineate a 95% confidence level, two standard deviations are required, which correspond to a range of 0.2–0.6‰.

Table 12.6.2 Comparison Between Recent δ^{13}C Value for TCE and TCA from Different Sources and Years (after Shouakar-Stash et al., 2003)

Compound	Shouakar-Stash et al., 2003			van Warmerdam et al., 1995			Beneteau et al., 1999		
	n	Mean $\delta^{13}C_{VPDB}$ (‰)	STDEV 1σ	n	Mean $\delta^{13}C_{VPDB}$ (‰)	STDEV 1σ	n	Mean $\delta^{13}C_{VPDB}$ (‰)	STDEV 1σ
TCE DOW 92	7	-31.57	0.01	2	-31.90	0.05	—	—	—
TCE DOW 95	4	-29.33	0.10		—	—	3	-29.84	0.07
TCE PPG 93	4	-27.37	0.09	2	-27.80	0.01	—	—	—
TCE PPG 95	4	-31.12	0.06		—	—	3	-31.68	0.01
TCE ICI 93	4	-31.01	0.09	3	-31.32	0.03	—	—	—
TCE StanChem 93	3	-29.19	0.14		—			—	—
TCA ICI 93	5	-26.48	0.18	4	-26.64	0.09	—	—	—
TCA PPG 93	3	-26.17	0.12	3	-25.80	0.46	—	—	—
TCA PPG 95	5	-25.84	0.14		—	—	4	-25.78	0.13
TCA Vulcan 593	3	-28.54	0.17	3	-28.42	0.07	—	—	—
TCA StanChem 93	6	-27.39	0.10		—	—	—	—	—

An issue in interpreting isotopic data for chlorinated solvent source differentiation is temporal source variation as well as the assumption that the isotopic characteristics of the released chlorinated solvent(s) are known (Slater, 2003). In practice, it is rare that a historical sample of a chlorinated solvent is available to provide a baseline isotopic signature. If a release of TCE into the environment is from a single release, then the source $\delta^{13}C$ is the $\delta^{13}C$ of the TCE that entered the subsurface. If the TCE released into the subsurface is the result of multiple releases over time, then the $\delta^{13}C$ value of the TCE is an isotopic mass balance of the total amount of TCE released. In a field investigation of TCE released into an anaerobic, unconfined aquifer located in a glacial till, additional challenges in data interpretation included the inability to identify discrete sources unless neither degradation nor isotopic fractionation was occurring, temporal variability, data density issues, and difficulties in quantifying the relationship between isotopic fractionation and degradation (Slater et al., 2000, 2001).

Kloppmann et al. (2005) used carbon specific isotopic analysis analyses for PCE, TCE, and cis-1,2-DCE in addition to tritium analyses and major ion chemistry to characterize a chlorinated solvent plume at the Plaine des Bouchers site, which is an old industrial area of Strasburg in use since 1910 (Zwank et al., 2003; Kloppmann et al., 2005). The site is situated on highly permeable alluvial deposits of the Rhine Valley. $\delta^{13}C$ Values of PCE in monitoring wells were found to be in a narrow range of $-24.5 \pm 0.48‰$; this information along with concentration data allowed the identification of two PCE source areas. The TCE and cis-1,2-DCE analysis exhibited enriched $\delta^{13}C$ with respect to PCE and was comparable to measured ranges for unaltered industrial products (Beneteau et al., 1999; Jendrzewski et al., 2001). The observed TCE and cis-1,2-DCE $\delta^{13}C$ values were found to range from -27.6 to $8.54‰$ and -14.7 to $13.5‰$, respectively, which are not representative of unaltered industrial values and therefore are evidence of biodegradation of TCE.

Issues of temporal variability are frequently encountered in environmental forensic investigation and represent tremendous challenges, especially where historical information regarding the release history and chemical usage is unknown. The ability to discriminate between $\delta^{13}C$ values resulting from multiple releases of product from different manufacturers with different $\delta^{13}C$ values that are co-mingled introduces significant difficulty interpreting isotopic data for source discrimination purposes.

Sherwood Lollar et al. (1999) proposed criteria when evaluating isotopic data to distinguish between differences due to degradation versus source differentiation. These criteria are that the isotopic values must be sufficiently different so as to distinguish between different sources, that the isotopic differences are greater than the precision at which the isotopic compositions of the compounds of interest are known, and that the isotopic behavior of the compound must be predictable (i.e., the effect of environmental processes on the isotopic values of the chlorinated solvents must be known). A suggested approach to resolving some of these issues is to perform laboratory studies in controlled conditions to simulate the expected field environment as a baseline for providing discrimination for many of the interpretative challenges when evaluating isotopic data.

12.6.3 Ratio Analysis For Age-Dating a TCA Spill

Gauthier and Murphy (2003) describe a method of age-dating a TCA spill to groundwater based on the ratio of daughter degradation products to the parent compound. The method is based on the fact that the rate of hydrolysis of TCA to 1,1-DCE appears to depend only on groundwater

temperature and not on other factors including sorption. The variation of the hydrolysis molar rate constant k with groundwater temperature is significant, occurring through the Arrhenius equation:

$$k = Ae^{\frac{E}{RT}}, \qquad (12.6.2)$$

where

A is the Arrhenius constant,
E is the activation energy,
T is the groundwater temperature in degrees Kelvin (K), and
R is the gas constant ($8.3145 \times 10^{-3}\,kJ/mol - K$).

Based on a review of uncertainties and biases in the laboratory data, Gauthier and Murphy recommend values $A = 8.7 \times 10^{13}\,sec^{-1}$ and $E = 122.8\,kJ/mol - K$. Groundwater temperature can either be measured or, for shallow groundwater, can be estimated as equal to the annual average air temperature. For deeper groundwater it may be necessary to consider that except in geothermal regions there is about a $1\,°C$ increase in temperature for every 40 meters of depth because of the earth's thermal gradient.

As shown in Figure 12.3.2, TCA also biodegrades anaerobically to 1,1-DCA and then to chloroethane. Because anaerobic dechlorination is faster for the first step than the second, 1,1-DCA accumulates. Similarly, if conditions are not too reducing, the conversion of TCA to 1,1-DCE is faster than the loss of 1,1-DCE to vinyl chloride. In that case the effective[3] molar rate constant for biodegradation, k_b, can be estimated from the hydrolysis rate constant, k, and the molar concentrations of 1,1-DCE and 1,1-DCA:

$$k_b = k\frac{[1,1\,DCA]}{[1,1\,DCE]}, \qquad (12.6.3)$$

where square brackets indicate molar values.

The kinetic equation is given by Gauthier and Murphy (2003) as:

$$t = \frac{1}{(k + k_b)} \ln\left(1 + \frac{1}{\alpha}\left(\frac{[1,1\,DCE]}{[TCA]}\right) + \left(\frac{[1,1\,DCA]}{[TCA]}\right)\right) sec, \qquad (12.6.4)$$

where t is the length of time since hydrolysis began and α is the molar fraction of 1,1-DCE produced by TCA hydrolysis estimated by Gauthier and Murphy (2003) to be about 0.21. Equations 12.6.2 through 12.6.4 and parameter values are sufficient to calculate the time t. When time series concentrations are available, plotting the logarithm in equation (12.6.4) vs t gives a direct estimate of $k + vk_b$, so use of Equation (12.6.2) is unnecessary. An example, provided by Wing (1997) is shown as Figure 12.6.1. For this example, the spill to a shallow aquifer occurred on August 15, 1984.

Gauthier and Murphy (2003) estimated the uncertainty in t as about $\pm 25\%$ when laboratory data are used to estimate rate constants. For the data used by Wing (1997), Gauthier and Murphy estimate the uncertainty as $\pm 25\%$ for one groundwater monitoring well and $\pm 36\%$ for another well.

An important observation is the interpretation of t. Only near the contaminant plume front is the rate of hydrolysis the same as the time since TCA entered groundwater. In particular, when pure solvent is introduced to groundwater as a dense nonaqueous phase liquid (DNAPL) one expects hydrolysis to begin with dissolution of the outer surface of

---

3. The actual rate constant may vary in space and time. For example, biodegradation may only occur in an anaerobic region near the source.

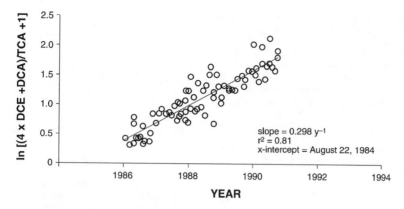

Figure 12.6.1 *Wing (1997) Analysis for t (x-intercept) and k (slope).*

the DNAPL, which is also when transport and biodegradation begins. As noted by Gauthier and Murphy (2003), laboratory experiments demonstrate that only the dissolved TCA hydrolyzes. Thus, as successive DNAPL layers are dissolved, hydrolysis begins at a later time. The net effect is that the data curve in Figure 12.6.1 should decrease in slope and become a horizontal line. This "bending over" in fact is what is observed at later times in the data used by Wing. When the logarithm is constant, the time calculated should correspond to the transport time from the DNAPL region to the well in which the data were recorded.

12.6.3.1 Ratio Analysis for Age-Dating Other Chlorinated Solvents

The question naturally arises whether ratios, for example of 1,2-DCE to TCE can be reliably used to age-date releases to groundwater. Because the degradation product in this case is a result of biological activity, the situation is clearly more complex than the use of a simple exponential decay model to develop a half-life degradation to estimate the date of the chlorinated solvent release. In addition to temperature, additional variables, such as the type and population of microbes present, the presence of nutrients or electron acceptors, the time required for microbial acclimatization to a spill, percent organic carbon, mass and chemical composition of the degradation products in the original mass released and soil texture. The concentration of the chlorinated solvents (dissolved or free phase) released into the subsurface is also important when considering this approach as the ability of micro-organisms to biodegrade chlorinated hydrocarbons at

high concentrations is considerably reduced. It may be possible in some unique circumstances to bracket a release date from ratios but, in general, consistency with age estimates developed by other means is probably more accurate.

To demonstrate consistency, biodegradation rates can be determined by comparing results of a transport and biodegradation model, such as Biochlor, with field data. Through trial and error, effective biodegradation rates can be estimated. The rates are "effective" because they represent an integration over space and time. These biodegradation rates can then be compared with literature values to demonstrate consistency. Of course, there is still an issue as to which literature values, field or laboratory, best represent conditions at the site of interest.

12.6.4 Ratio Analysis for Source Identification

Although using chlorinated ethane ratios for age-dating is usually problematic, these ratios can be used to identify additional sources. Figure 12.6.2 shows an example from the Tutu well field in St. Thomas in the Virgin Islands. The plot shows the molar ratio of 1,2-DCE to the sum of TCE and PCE. In this case, PCE is believed to be the parent compound. Molar ratios are used so that numerator and denominator both correspond to number of molecules that are or were formerly PCE, but that are different levels in the anaerobic decay chain depicted in Figure 12.3.1.

In this example there is a sudden ratio reduction at about 1500 feet downgradient where there is a dry-cleaning establishment. In general, in order to reliably determine a source location the change in ratio should be statistically significant

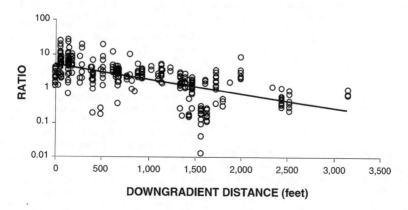

Figure 12.6.2 *[DCE]/{[TCE[+[PCE]} vs. Downgradient Distance at the Tutu Wellfield, St. Thomas, Virgin Islands.*

compared with upgradient fluctuations. In Figure 12.6.2 the downward trend in the ratio may be due to aerobic degradation or preferential loss of some compounds in the bedrock matrix. Because this method is phenomenological, it is not necessary to know the underlying processes.

In the example shown in Figure 12.6.2 the change in ratios occurs at only one groundwater monitoring well. There is a groundwater divide near monitoring the dry-cleaner location so that wells shown in Figure 12.6.2 as downgradient may in fact not be relative to the drycleaner discharge location. In general, other reasons that only one or a few wells may be affected include the release being very small or very recent. In other cases one would expect a number of downgradient wells to be affected.

12.6.5 Age-Dating a Chlorinated Solvent Release from the Position of the Plume Front

The age dating of a chlorinated solvent release via the location of the leading edge of the contaminant plume in groundwater is frequently used for age-dating chlorinated solvent plumes. Chlorinated solvent plumes are less subject to biodegradation than petroleum product constituents such as benzene, toluene, ethylbenzene, and xylene (BTEX): often there is a source region where anaerobic degradation occurs but then the plume enters an aerobic region where the more heavily chlorinated compounds such as PCE or TCE biodegrade little if at all. Thus the position of the plume front for the parent compound is more likely to be determined by travel-time considerations than biodegradation processes.

This forensic technique is based on an analytical approach suitable for a homogeneous and isotropic medium. When the medium is inhomogeneous or anisotropic, numerical modeling should lead to more accurate results than a simple analytical formulation. However, numerical modeling will not reduce the uncertainty in plume age resulting from uncertainties in the basic parameter values.

In this example, assume that the length of the chlorinated solvent plume is L. Some portion of L, ΔL, is due to longitudinal dispersion. If the groundwater velocity is u and the retardation factor for the solvent in question is R, so that the solvent transport velocity is u/R, then the release date, t, can be estimated as:

$$t = \frac{(L - \Delta L)R}{u} \qquad (12.6.5)$$

The longitudinal dispersion can be estimated as:

$$\Delta L = 2\sqrt{\frac{Dt}{R}} = 2\sqrt{\frac{\alpha_L u t}{R}} = 2\sqrt{\alpha_L(L - \Delta L)}, \qquad (12.6.6)$$

where D is the coefficient of hydrodynamic dispersion, α_L is the longitudinal dispersivity, and molecular dispersion has been neglected. The factor of two is included because this appears as part of the characteristic length scale in two- and three-dimensional analytical solutions (Domenico and Schwartz, 1997). The approximate solution for ΔL is:

$$\Delta L \cong 2\sqrt{\alpha_L L} - 2\alpha \cong 2\sqrt{\alpha_L L} \qquad (12.6.7)$$

Thus combining Equations (12.6.5) and (12.6.6) gives:

$$t = \left(\frac{R}{u}\right)\left(L - 2\sqrt{\alpha L}\right), \qquad (12.6.8)$$

which shows that the importance of correcting for longitudinal dispersion decreases with the length of the plume.

The groundwater velocity and retardation factor can be written in terms of measurable quantities as:

$$u = \frac{KI}{\theta_e} \qquad (12.6.9)$$

and

$$R = 1 + \frac{\rho_b K_d}{\theta_e}, \qquad (12.6.10)$$

where K is the hydraulic conductivity, I is the hydraulic gradient over the plume extent, θ_e is the effective porosity or interconnected pore space of the aquifer, ρ_b is the aquifer material dry bulk density, and K_d is the adsorption distribution coefficient.

The adsorption distribution coefficient K_d is often written as:

$$K_d = K_{oc} f_{oc}, \qquad (12.6.11)$$

where K_{oc} is the distribution coefficient for partitioning of the solvent between water and organic carbon and f_{oc} is the fraction organic carbon in the soil.

With substitutions from Equations (12.6.9–12.6.11) the expression for the elapsed time since a release to groundwater becomes:

$$t = \left(\frac{\theta_e}{K\,i}\right)\left(1 + \frac{\rho_b K_{oc} f_{oc}}{\theta_e}\right)\left(L - \sqrt{2\alpha L}\right) \qquad (12.6.12)$$

For determining the chlorinated solvent plume length, there are generally two cases. First, the chlorinated solvent may not be detected at a distant groundwater monitoring well but may be detected at the next closest well to the source. Second, the chlorinated solvent may be detected at all distant wells but with a declining concentration at the most distant well. In both cases it is important to determine that the observed decline in concentration represents a contaminant front and not just a decline due to lateral dispersion or biodegradation. It may be helpful to plot centerline concentration versus distance downgradient to determine if the decline in concentration is more rapid than would be caused by dispersion or degradation alone.

If degradation is occurring along the entire extent of the groundwater contaminant plume (rather than just in a source region) it may be helpful to use molar concentrations and to add the concentrations of daughter products to the concentration of the original solvent. An example of degradation along the entire plume length is hydrolysis of 1,1,1-TCA to 1,1-DCE.

Another parameter used in this methodology is hydraulic conductivity. Hydraulic conductivity (units of length over time, LT^{-1}) is related to permeability $k(L^2)$ as follows:

$$K = \frac{\rho g}{\mu} k, \qquad (12.6.13)$$

where ρ is the fluid density (ML^{-3}), g is the acceleration of gravity (LT^{-2}), and μ is the kinematic viscosity ($ML^{-1}T^{-1}$).

Most groundwater texts provide ranges and typical values for K and k for various rock and soil types (Freeze and Cherry, 1979).

Often a few hydraulic conductivity measurements are available. According to Zheng and Bennett (1995), "ample evidence has suggested that within a given hydrogeological unit, hydraulic conductivity often follows a logarithmic normal (or lognormal) distribution. They cite Law (1944) and Bennion and Griffiths (1966) as a basis. This suggests that the geometric mean, rather than the arithmetic mean, is a better representation. Beacause the geometric mean is always less than or equal to the arithmetic mean, using the geometric mean rather than the arithmetic mean has the effect of weighting low conductivity regions more heavily. The effect is similar to an electrical circuit with resistances in parallel where the more resistive elements control the current flow.

Porosity can be measured in the laboratory, as the ratio of pore volume to bulk volume (Zheng and Bennett, 1995). However, θ, appearing in the above equations, is the effective or connected porosity, which will be smaller. Effective porosity will be much smaller if a high percentage of the flow occurs through a small fraction of the pore space.

Bulk density is the mass of dry soil divided by bulk volume. In homogeneous media it is related to porosity and particle density by $\rho_b = \rho_p(1 - \theta)$, where ρ_p is the particle density. A typical particle density is that of silicon dioxide, $2.65\,g/cm^3$.

Gelhar et al. (1992) estimated dispersivities from field data at a number of sites. Their data are classified by the degree of reliability assigned by the authors. Without regard to reliability, the data indicate α increasing with scale (distance). However, when only data of high reliability are considered, dispersivity appears to change very little with scale. In contrast, the high reliability data are clustered at intermediate scales making a trend difficult to discern. Gelhar et al. also noted no significant difference between porous media and fractured rock. Gelhar et al. also analyzed a smaller number of observations in which the ratio of longitudinal to transverse or vertical dispersivity could be determined.

Changing groundwater flow direction has some effect on apparent longitudinal dispersivity but it is much less than the effect on apparent transverse dispersivity (Goode and Konikow, 1990).

The ith parameter value is denoted as x_i. When parameter values are uncorrelated, the uncertainties are normally distributed,[4] and the uncertainty in each, δx_i, is small so that $\delta x_i/x_i \ll 1$, then the uncertainty in elapsed time is (Mandel, 1964)

$$\delta t = \left(\sum_{i=1}^{n} \left(\frac{\partial t}{\partial x_i} \right)^2 V(x_i) \right)^{\frac{1}{2}}, \qquad (12.6.14)$$

where $V(x_i) = \delta(x_i)^2$ is the variance of the ith variable. Equation (12.6.14) can also be written as:

$$\frac{\delta t}{t} = \sqrt{ \sum_{i=1}^{n} \left(x_i \frac{\partial \ln t}{\partial x_i} \right)^2 \frac{V(x_i)}{x_i^2} } \qquad (12.6.15)$$

For normally distributed parameters $\frac{\sqrt{V(x_i)}}{x_i} = \frac{\sigma_i}{x_i}$, where σ_I is the standard deviation of the ith variable. Even when the inequality $\delta x_i/x_i \ll 1$ is not satisfied, Equation 12.6.15 provides a useful way of assessing which parameters are most responsible for the fractional uncertainty in elapsed time $\delta t/t$.

The values of $\frac{V(x_i)}{x_i}$ depend on site-specific circumstances. However, it is clear that in general, significant uncertainty exists. Suppose, for example, that each of the three factors in Equation 12.6.15 has a normally distributed uncertainty with a value of $\frac{\sigma(x_i)}{x_i} = 0.3$. Then the uncertainty in $\frac{t}{t}$ is about 0.52. This would correspond, for example, to a range of 5–15 years with a central estimate of 10 years. In many cases this degree of uncertainty will prevent detailed forensic conclusions, such as "during whose 'tenure' did the release occur" from being concluded. The most precise conclusions that can then be drawn are confirmatory, namely that with reasonable parameter values the plume extent either is, or is not, consistent with other information. Thus, it may be possible to disprove other estimates that are independently arrived at.

In general, collecting additional hydraulic conductivity information, seasonal hydraulic gradient information, etc. may reduce uncertainties. Additional observations may also reduce uncertainty in specific circumstances as the following examples illustrate.

When retardation is small, the value of R is only slightly larger than one. An example is the stabilizer 1,4-dioxane found in 1,1,1-TCA. According to Mohr (2001) the estimated K_{oc} for 1,4-dioxane is 1.23. This means 1,4-dioxane is nearly unretarded in groundwater transport. Basing the elapsed time estimate on 1,4-dioxane rather than 1,1,1-TCA therefore eliminates the uncertainties associated with the retardation factor. Of course, this reduction in uncertainty could be partially or totally cancelled if the position of the 1,4-dioxane plume front is less precisely determined than the 1,1,1-TCA plume front. In contrast, because it is the fractional uncertainty $\frac{\sqrt{V(L-\sqrt{\alpha L})}}{L-\sqrt{\alpha L}}$ that is of interest, the larger value of L for 1,4-dioxane relative to 1,1,1-TCA may actually reduce this quantity.

(Similarly, MTBE may be used to date gasoline plumes largely avoiding the uncertainties associated with retardation and biodegradation affecting the position of the plume front. In the special case where an MTBE plume lags behind a BTEX plume this indicates that there was an earlier spill of non-MTBE petroleum.)

If the date one chemical entered groundwater t^* can be reliably estimated, the position of the plume front L^* for that chemical may be used to estimate the retarded groundwater velocity, u^*/R^*, for that chemical. The date that other chemicals entered groundwater is then given by:

$$t = t^* \frac{u^* R}{u \, R^*} \frac{\left(L - \sqrt{\alpha L} \right)}{\left(L^* - \sqrt{\alpha L^*} \right)}, \qquad (12.6.16)$$

In writing Equation 12.6.16, we have not formally recognized the possibility that α is a function of downgradient distance L.

When a multi-component contaminant has been spilled there may be multiple plume fronts. An example would be the various polycyclic aromatic hydrocarbon (PAH) plume fronts from a creotote release to groundwater. Because the higher ring-number PAHs are more heavily retarded than the lower ring-number ones, chromatographic separation occurs. In this case the estimated plume age can be calculated. The estimated value of $L - \sqrt{\alpha L}$ is plotted versus the estimated values of $1/R$ for each component. The resulting best-fit straight line is an estimate of ut and the scatter about the line provides an estimate of the uncertainty in ut.

Finally, we mention a special case observed by one of the authors where a pumping well installed as part of a remedy left a "signature" in the plume at the capture radius. Observing the subsequent progress of this plume "back end" provided a direct estimate of the retarded transport velocity, albeit neglecting the effects of longitudinal dispersion.

---

4. Note that this is not the same thing as a normally distributed parameter. According to the Central Limit Theorem the uncertainty in an unbiased estimate of the mean for most distributions is approximately normal.

REFERENCES

Abrajano, T., and B. Sherwood Lollar, 1999. Introduction Note: Compound-specific isotope analysis: Tracing organic contaminant sources and processes in geochemical systems. Organic Chemistry 30(8a), v–vii.

Archer, W., 1966. Industrial Solvents Handbook. Marcel Dekker, Inc., New York, NY.

Archer, 1982. Aluminum-1,1,1-Trichloroethane. Reactions and inhibition. Industrial Engineering Chemistry Product Research Development. 21(4):670–672.

Archer, W., 1984. A laboratory evaluation of 1,1,1-trichloroethane-metal-inhibitor systems. *Werkstoffe and Korrosion*. 35:50–69.

Archer and Stevens, 1977. Comparison of Chlorinated, Aliphatic, Aromatic, and Oxygenated Hydrocarbons as Solvents. *Industrial & Engineering Chemistry*. 16(4), 319.

Aravena, R., Beneteau, K., Frape, S., Butler, B., Abrajano, R., Major, D., and E. Cox, 1998. Application of isotopic fingerprinting for biodegradation studies of chlorinated solvents in groundwater. In *Risk, Resource, and Regulatory Issues: Remediation of Chlorinated and Recalcitrant Compounds* (Wickramanayake, G. and Hinchee, R., Eds), pp. 66–71. Battelle Press, Columbus, OH.

ASM, 1996. *Guide to vapor degreasing and solvent cold cleaning*. Materials Park, OH.

ASTM, 1962. Handbook of Vapor Degreasing, ASTM Committee D-26 on Halogenated Organic Solvents, ASTM Special Technical Publication No. 310.

ASTR, 1997. Toxicological profile for trichloroethylene. United States Department of Health and Human Services. Public Health Service, Agency for Toxic Substances and Disease Registry. Atlanta, GA.

Bachtel, H., 1957. Methyl Chloroform Inhibited with Dioxane, United States Patent 2,811,252, assigned to Dow Chemical Company. October 29, 1957.

Bachtel, H., 1958. Inhibited Methyl Chloroform, 1958. United States Patent 2,838,458, assigned to Dow Chemical Company. June 10, 1958.

Barone, F., Rowe, R., and R. Quigley, 1992. A laboratory estimation of diffusion and absorption coefficients for several volatile organics in a natural clayey soil. *Journal of Contaminant Hydrology*. 10:225–250.

Barber, J., 1957. Chloroethene in Aerosols. Soap Chemical Spec. February, 1957, 33, 99.

Barrio-Lage, G., Parsons, F., Nassar, R., and P. Lorenzo, 1986. Sequential dehalogenation of chlorinated ethenes. *Environmental Science & Technology*. 20(1), 96–99.

Barth, J., Slater, C., Schuth, G., Bill, M., Downey, A., Larkin, M., and R. Kalin, 2004. Carbon isotope fractionation during aerobic biodegradation of trichloroethylene by *Burkholderia cepacia* G4: A tool to map degradation mechanisms. *Applied Environmental Microbiology*. 68, 1728–1734.

Bartholomew, R., Brown, F., and M. Lounsbury, 1954. Chlorine isotope effect in reactions of tert-butyl chloride. *Canadian Journal of Chemistry* 32, 979–983.

Beneteau, K., Aravena, R., Frape, S., Abragano, T., and R. Drimmie, 1996. Chlorinated solvent fingerprinting using ^{13}C and ^{37}Cl stable isotopes. American Geophysical Union, Section 121, H31B-12. Spring AGU Meeting, Baltimore, MD.

Beneteau, K., Aravena, R., and S. Frape, 1999. Isotopic characterization of chlorinated solvents – laboratory and field results. *Organic Geochemistry* 30, 739–753.

Benker, E., Davis, G., Appleyard, S., Berry, D., and T. Power, 1994. Groundwater contamination by trichloroethene (TCE) in a residential area of Perth: Distribution, mobility, and implications for management. In: Proceedings of the Water Down Under 94, 25th Congress of IAH, Adelaide, South Australia, November 21–25, 1994.

Bill, M., Schuth, C., Barth, J., and R. Kalin, 2001. Carbon isotope fractionation during abiotic reductive dehalogenation of trichloroethene (TCE). *Chemosphere*. 44, 1281–1286.

Bloom, Y., Aravena, R., Hunkeler, D., Edwards, E., and S. Frape, 2000. Carbon isotope fractionation during microbial dechlorination of trichloroethene, cis-1,2-dichloroethene and vinyl chloride: Implications for assessment of natural attenuation. *Environmental Science & Technology*. 34, 2768–2772.

Bouwer, E., and P. McCarty, 1983. Transformations of 1 and 2 carbon halogenated aliphatic organic compounds under methanogenic conditions. *Applied & Environmental Microbiology*. 45(4), 1286–1294.

Brallier, P., 1949. Carbon tetrachloride, *in Encyclopedia of Chemical Technology*, Vol. 3, Kirk, R. and Othmer, D. eds, New York, NY. Interscience Encyclopedia.

Chemical and Engineering News, 1962, Solvent Introduced for Railway Equipment Cleaning, Oct. 1, 40:52.

Chemical and Engineering News, 1963. New Dry-Cleaning System Under Field Test, Nov. 25, 41:57.

Chemical Engineering, 1961. Competition Sharpens in Chlorinated Solvents, Oct. 30, 68, 22, 62–66.

Chemical Week, 1953. Tri, Per, and Carbon Tet. May 2, 1953.72, 56.

Chemical Week, 1956. Accent's on New Products, Chem. Week, Dec. 15, 79, 87.

Chur, K., Mahendra, S., Song, D., Conrad, M., and L. Alvarez-Cohen, 2004. Stable carbon isotope fractionation during aerobic biodegradation of chlorinated ethenes. *Environmental Science & Technology*. 38, 3126–3130.

Clarke, I., and P. Fritz, 1997. *Environmental Isotopes in Hydrogeology*. CRC Press, Boca Raton, FL.

Dayan, H., Abrajano, T., Sturchio, N., and L. Winsor, 1999. Carbon isotope fractionation during reductive dehalogenation of chlorinated ethenes by metallic iron. *Organic Geochemistry*. 30(8a), 755–763.

DeShon, H., 1979. Carbon Tetrachloride, in Kirk-Othmer Encyclopedia of Chemical Technology, 3rd edition, Vol. 5, Grayson, M. and Eckroth, D. eds, New York, NY. Wiley & Sons.

DeForest, E., 1979. Chloromethanes, in *Encyclopedia of Chemical Processing and Design*, McKetta, J. and Cunningham, W. eds, New York, NY. Marcel Dekker.

Doherty, R., 2000a A History of the Production and Use of Carbon Tetrachloride, Tetrachloroethylene, Trichloroethylene and 1,1,1-Trichloroethane in the United States: Part 1: Historical Background, Carbon Tetrachloride and Tetrachloroethylene. *Journal of Environmental Forensics.*, 1, 69–81.

Doherty, R., 2000b. A History of the Production and Use of Carbon Tetrachloride, Tetrachloroethylene, Trichloroethylene and 1,1,1-Trichloroethane in the United States: Part 2: Trichloroethylene and 1,1,1-Trichloroethane. *Journal of Environmental Forensics*. 1, 83–93.

Domenico, P., and F. Schwartz, 1997. *Physical and Chemical Hydrogeology*, Wiley, New York.

Dow Chemical Company, 1970. Success Story with lots of DOW-PER C-S, *Spot News*, Summer 1970, p. 2.

Dow Chemical Company, 1971a. *Advertisement in Spot News*, Fall 1971, p. 18.

Dow Chemical Company, 1971b. Beverly-Wilshire stresses service, *Spot News*. Summer 1971, p. 10.

Dow Chemical Company, 1973. Huck Finn plants get clothes 'Whiter than white' with NEW complete service solvent, *Spot News*. Summer 1973, p. 5.

Dow Chemical Company, 2005a, Global Chlorinated Organics Business, Perchloroethylene, accessed April 2005 at http://www.dow.com/gco/eu/prod/perchlor/

Dow Chemical Company, 2005b, Global Chlorinated Organics Business, Methylene chloride, accessed April 2005 at http://www.dow.com/gco/na/prod/meth_ch/grades/vapor_de.htm

Drenzek, N., Tarr, C., Eglinton, T., Heraty, L., Sturchio, N., Shiner, V., and C. Reddy, 2002. Stable chlorine and carbon isotopic compositions of selected semi-volatile organochlorine compounds. *Organic Geochemistry*. 33, 437–444.

Duncan, B., Vavricka, E., and R. Morrison, 2004. A forensic overview of 1,4-Dioxane. *Environmental Claims Journal.* 16(1), 69–79.

Egli, C., Scholtz, R., Cook, A., and T. Leisinger, 1987. Anaerobic dechlorination of tetrachloromethane and 1,2 dichloroethane to degradable products by pure cultures of

Ertl, S., Selbel, F., Eichinger, L., Frimmel, F., and A. Kettrup, 1998. The $^{13}C/^{12}C$ and $^{2}H/^{1}H$ ratios of trichloroethene, tetrachloroethene and their metabolites. *Isotopic Environmental Health Studies.* 34, 245–253.

Ethyl Corporation (no date). Vapor degreasing with chlorinated solvents. Trichloroethylene. Perchloroethylene. 1,1,1-Trichloroethane. Methylene Chloride. The Ethyl Corporation Way. Ethyl Corporation, Industrial Chemicals Division. Baton Rouge, LA.

Fathepure, B., Nengu, J., and S. Boyd, 1987. Anaerobic bacteria that dechlorinate perchloroethene. *Applied & Environmental Microbiology.* 53(11), 2671–2674.

Fogel, M., Taddero, A., and S., Fogel, 1986. Biodegradation of chlorinated ethenes by a methane utilizing mixed culture. *Applied & Environmental Microbiology.* 51(4), 720–724.

Freeze, A., and J. Cherry, 1979. *Groundwater*, Prentice Hall, New York.

Gauthier, T., and B. Murphy (2003) Age dating groundwater plumes based on the ratio of 1,1-dichloroethylene to 1,1,1-trichloroethane: An uncertainty analysis. *Environmental Forensics.* 4(3), 205–213.

Gelhar. L.W., Welty, C., and K.W. Rehfeldt, 1992. A critical review of data on field scale dispersion in aquifers, *Water Resources Research.* 28(7), 1955–1974.

Gerhartz, W., ed., 1986. *Ullman's Encyclopedia of Industrial Chemistry*, 5th edition. New York, Weinheim.

Goode, D.J., and L.F. Konikow, 1990 Apparent dispersion in transient groundwater flow. *Water Resources Research.* 26(10), 2339–2351.

Halogenated Solvents Industry Alliance, 1994. White paper on methyl chloroform (1,1,1-Trichloroethane). February 1994. Washington, D.C.

Hardie, D., 1964. Trichloroethylene, in *Kirk-Othmer Encyclopedia of Chemical Technology.* 2nd edition. Vol. 5, pp. 183–195. John Wiley & Sons. New York, NY.

Heraty, L., Fuller, M., Huang, L., Abrajano, T., and N. Sturchio, 1999. Isotope fractionation of carbon and chlorine by microbial degradation of dichloromethane. *Organic Geochemistry.* 30(8). 793–799.

Hickman, J., 1991. Tetrachloroethylene, in *Kirk-Othmer Encyclopedia of Chemical Technology*, 4th edition, Kroschwitz, J. and Howe-Grant, M. eds, New York, NY. Wiley & Sons.

Holbrook, M., 1991. Carbon Tetrachloride, in *Kirk-Othmer Encyclopedia of Chemical Technology*, 4th edition, Kroschwitz, J. and Howe-Grant, M. eds, New York, NY. Wiley & Sons.

Hsu, C., Sherman, P., and P. Bouis, 2005. "The Quest for Purity I: High Purity Solvents," JTBaker.com Technical Library, accessed April 2005 at http://www.jtbaker.com/techlib/documents/quest1.html.

Huang, L., Sturchio, N., Abrajano, T., and B. Holt, 1999. Carbon and chlorine isotope fractionation of chlorinated aliphatic hydrocarbons by evaporation. *Organic Geochemistry.* 30(8a), 777–785.

Hunkeler, D., Hoehn, E., Hohener, P., and J. Zeyer, 1997. ^{222}Rn as a partitioning tracer to detect diesel fuel contamination in aquifer: Laboratory study and field observations. *Environmental Science and Technology.* 31(11), 3180–3187.

Hunkeler, D., Aravena, R., and B. Butler, 1999. Monitoring microbial dechlorination of tetrachloroethene (PCE) in groundwater using compound-specific stable carbon isotope ratios: Microcosm and field studies. *Environmental Science and Technology.* 33(16), 2733–2738.

Hunkeler, D., Aravena, R., Parker, B., Cherry, J., and X. Diao, 2003. Monitoring oxidation of chlorinated ethenes by permanganate in groundwater using stable isotopes: Laboratory and field studies. *Environmental Science & Technology.* 37, 798–804.

Hunkeler, D., Chollet, N., Pittet, X., Aravena, R., Cherry, J., and B. Parker, 2004. Effect of source variability and transport processes on carbon isotope ratios of TCE and PCE in two sandy aquifers. *Journal of Contaminant Hydrology.* 74, 265–282.

ICI Botany Operations, July 1996. Manufacturing Histroy 1942 to 1996. ICI Botany pamphlet. www.oztoxics.org/research3000_hcbweb/hcb2/site5_1.html

Italia, M., and M. Nunes, 1991. Gas chromatography determination of 1,4-dioxane at the parts-per-million level in consumer shampoo products. *Journal of the Society of Cosmetics Chemistry.* 43:97–104.

IPCS, 1987. International Programme on Chemical Safety, "Methylene Chloride Health and Safety Guide," United Nations Environment Programme, World Health Organization, Geneva.

IPCS, 1996. International Programme on Chemical Safety, Methylene Chloride, 2nd edition, Environmental Health Criteria 164, World health Organization.

Irish, D., 1963. Halogenated Hydrocarbons, Aliphatic, in *Industrial Hygiene and Toxicology*, Vol. 2, Patty, F.A., ed., pp. 1287–90.

Janssen, D., Grobben, G., Hoekstra, R., Oldenhuis, R., and B. Witholt, 1988. Degradation of trans-1,2-dichloroethene by mixed and pure cultures of methanotrophic bacteria. *Applied Microbiology and Biotechnology.* 29:392–299.

Jendrzejewski, N., Eggenkamp, H., and M. Coleman, 2001. Characterization of chlorinated hydrocarbons from chlorine and carbon isotopic compositions: scope of application to environmental problems. *Applied Geochemistry.* 16:1021–1031.

Jordan, J., 1979. Chloroethanes, in *Encyclopedia of Chemical Processing and Design*, McKetta, J. and Cunningham, W. eds, New York, NY. Marcel Dekker.

Joshi, S., Donahue, B., Tarrer, A., Guin, J., Rahman, M., and B. Brady, 1989. Methods for monitoring solvent condition and maximizing its utilization. STP 1043. *American Society of Testing Materials.* Philadelphia, PA. pp. 80–103.

Keil, S., 1978. Tetrachloroethylene. In: *Encyclopedia of Chemical Technology.* 3rd edition. Volume 5, Kirk, R., Othmer, D., Grayson, M. and Eckroth. E. eds, John Wiley & Sons, New York, NY. pp. 754–762.

KEMI, 1995. Cholrine and cholrine compounds. Use, occurrence and risks-the need for action. Solna, The Swedish National Chemicals Inspectorate.

Kircher, C., 1957. Solvent Degreasing – What Every User Should Know, ASTM Bulletin. January 1957, 44.

Kirschner, E., 1994. Environment, Health Concerns Force Shift, in *Use of Organic Solvents*, Chemical Engineering News, June 20, 72:13.

Kirtland, B., Aelion, C., Stone, P., and D. Hunkeler, 2003. Isotopic and geochemical assessment of in situ biodegradation of chlorinated hydrocarbons. *Environmental Science & Technology.* 37, 4205–4212.

Kleopfer, R., Easley, D., Haas, B., Deihi, T., Jackson, D., and C. Wurray, 1985. Anaerobic degradation of trichloroethylene in soil. *Environmental Science & Technology.* 19; 277–280.

Kloppmann, W., Hunkeler, D., Aravena, R., Elsass, P., and D. Widory, 2005. Organic solvent contamination of a complex industrial site: Insights from compound specific isotope analysis combined with classical isotopic tools. *Geophysical Research Abstracts*. 7, 08004. European Geosciences Union.

Kueper, B., Abbot, W., and G. Farquhar, 1989. Experimental observations of multiphase flow in heterogeneous porous media. *Journal of Contaminant Hydrology*. 5, 83–95.

Linn, B., 2002. Reported leaks, spills and discharges at Florida dry cleaning sites. Florida department of Environmental Protection, Florida, USA.

Lowenheim, F. and M. Moran, 1975. Faith, Keyes and Clark's Industrial Chemicals, 4th edition. New York, Wiley & Sons.

Mancini, S., Lacrampe-Couloumme, A., Jonker, H., Van Breukelen, B., Groen J., Volkering, F., and B. Sherwood Lollar, 2002. Hydrogen isotope enrichment: An indicator of biodegradation at a petroleum hydrocarbon contaminated field site. *Environmental Science & Technology*. 36: 2464–2470.

Manufacturing Chemists' Association, 1965. Chemical Safety Data Sheet SD-90. Properties and Essential Information for Safe Handling and Use of 1,1,1 Trichloroethane. Adopted 1965. Washington, D.C.

McCarty, P., 1994. An overview of anaerobic transformation of chlorinated solvents. In Symposium on Natural Attenuation of Ground Water, August 30–September 1, 1994. EPA/600/R-94/162. United States Environmental Protection Agency. Office of Research and Development.

McCarty, P., and L. Semprini, 1994. Ground-water treatment for chlorinated solvents. Section 5. Handbook of Bioremediation. Eds. Robert Norris. Lewis Publishers, Boca Raton, FL. pp. 87–116.

McCarty, P., 1994. In situ bioremediation of chlorinated solvents, In *Handbook of Bioremediation*, Norris *et al.*, eds, pp. 323–330. Lewis Publishers, Baton Raton, FL.

Mertens, J., 1991. Trichloroethylene, in *Kirk-Othmer Encyclopedia of Chemical Technology*, 4th edition, Kroschwitz, J. and Howe-Grant, M. eds, New York, NY. Wiley & Sons.

Mohr, T., 2001. Solvent stabilizers. White Paper. Prepublication Paper. June 14, 2001. Santa Clara Valley Water District. Underground Storage Tank Program, Water Supply Division. p. 52.

Morrison, R., 1999. *Environmental Forensics: Principles & Applications*. CRC Press, Boca Raton, FL. p. 333.

Morrison, R., 2000a. Critical review of environmental forensic techniques. Part I. *Environmental Forensics*. 1(4): 157–173.

Morrison, R., 2000b. Applications of forensic techniques for age dating and source identification in environmental litigation. *Journal of Environmental Forensics*. 1(3): 131–153.

Morrison, R., 2000c. Environmental Forensics. Chapter 10, In *Environmental Science* Deskbook. eds. James Conrad. West Group. St. Paul, MN. pp. 10:1–104.

Morrison, R., 2001. Chlorinated solvents and source identification. *Environmental Claims Journal*. 13(3), 95–104.

Morrison, R., 2003. PCE contamination and the dry cleaning industry. *Environmental Claims Journal*. 15(1), 93–106.

Morrison, R., 2005. *Environmental Forensics*. Chapter 10. In: Environmental Science Deskbook. James Conrad (ed.) pgs. 10–2 to 10–115. Thompson West Publishing.

Morrison, R., Hartman, B., Erickson, R., Jones, J., and R. Beers, 1998. Chlorinated Solvents: Legal and Technical Considerations. Argent Publishing Company, Foresthill, CA. p. 332.

Morrison, R., and K. Heneroulle, 2000. Chlorinated Solvents: Use of synonyms and additives for age dating and source identification. *Environmental Claims Journal*. 13(1), 93–105.

NICNAS, March 2000. Trichloroethylene, Priority Existing Chemical Assessment Report No.8. Commonwealth of Australia.

NICNAS, June 2001. Tetrachloroethylene, Priority Existing Chemical Assessment Report No.15. Commonwealth of Australia.

NIOSH, National Institute of Safety and Health, 1986. "Current Intelligence Bulletin 46: Methylene Chloride," April 18, 1986.

Numata, M., Nakamura, N., Koshikawa, H., and Y. Terashima, 2002. Chlorine isotope fractionation during reductive dechlorination of chlorinated ethenes by anaerobic bacteria. *Environmental Science & Technology*. 36, 4389–4398.

Nyer, E., Kramer, V., and N. Valkenburg, 1991. Biodegradation effects on contaminant fate and transport. *Ground Water Monitoring Review*. 11:80–82.

Pearson, A., and T. Eglinton, 2000. The origin of n-alkanes in Santa Monica Basin surface sediments: A model based on compound specific C-14 and delta C-13 data. *Organic Geochemistry*. 31, 1103–1116.

Perry R., and D. Green, 1999 Perry's *Chemical Engineers' Handbook*, McGraw-Hill, New York.

Philp, P., 2005. Stable isotopes and biomarkers in forensic geochemistry. The 15th Annual AEHS Meeting & West Coast Conference on Soils, Sediments and Water. AEHS Workshop. March 15, 2005. San Diego, California.

Poulson, M., and B. Kueper, 1992. A field experiment to study the behavior of tetrachloroethylene in unsaturated porous media. *Environmental Science and Technology*. 26, 889–895.

Poulson, S., and Drever, J., 1999. Stable isotope (C, Cl and H) fractionation during vaporization of trichloroethylene. *Environmental Science and Technology*. 33(20), 3689–3694.

Scalia, S., Testoni, F., Frisina, G., and M. Guarnerij, 1992. Assay of 1,4-dioxane in cosmetic products by solid-phase extraction and GC-MS. *Journal of the Society of Cosmetic Chemists*. 43:207–213.

Schwarzenbach, R., Gschwend, P., and D. Imboden., 1993. *Environmental Organic Chemistry*. John Wiley and Sons, Inc., New York, NY. 681 pages.

Shepherd, C., 1962. Trichloroethylene and Perchloroethylene, in *Chlorine: Its Manufacture, Properties, and Use*, Sconce, J.S., ed. American Chemical Society, Reinhold Publishing Corp., New York.

Slater, G., 2003. Stable isotope forensics – when isotopes work. *Environmental Forensics*. 4(1), 13–23.

Sherwood Lollar, B., Slater, G., Ahad, J., Sleep, B., Spivack, J., Brennan, M., and P. MacKenzie, 1999. Contrasting carbon isotope fractionation during biodegradation of trichloroethylene and toluene: Implications for intrinsic bioremediation. *Organic Geochemistry*. 30(8a), 813–820.

Shouakar-Stash, O., Frape, S., and R. Drimmie, 2003. Stable hydrogen, carbon and chlorine isotope measurements of selected organic solvents. *Journal of Contaminant Hydrology*. 30(3–4), 211–228.

Slater, G., Dempster, H., Sherwood-Lollar, J., Brennan, M., and P. MacKensie, 1998. Isotopic tracers of degradation of dissolved chlorinated solvents, in *Natural Attenuation: Chlorinated and Recalcitrant Compounds*, Wickramanayake, G. and Hinchee, R. eds, pp. 133–138. Battelle Press, Columbus, OH.

Slater, G., Ahad, J., Sherwood-Lollar, B., Allen-King, R., and B. Sleep, 2000. Carbon isotope effects resulting from equilibrium sorption of dissolved VOCs. *Analytical Chemistry*. 72:5669–5672.

Slater, G., Sherwood-Lollar, B., Edwards, E., and B. Sheep 2001. Variability in carbon isotopic fractionation during biodegradation of chlorinated ethenes: Implications for field applications. *Environmental Science & Technology.* 35, 901–907.

Snedecor, G., 1991. 1,1,1-Trichloroethane, in *Kirk-Othmer Encyclopedia of Chemical Technology*, 4th edition, Kroschwitz, J. and Howe-Grant, M. eds, Wiley & Sons. New York, NY.

Soble, R., 1979. Solvent5 cleaning of printing wiring assemblies. Printed circuit techniques. Insulation/Circuits. October 1979. pp. 25–29.

Solvay Chemicals, 2002, "Chlorinated Solvents Stabilization," accessed April 2005 at www.solvaychemicals.com.

Song., D., Conrad, M., Sorenson, K., and L. Alvarez-Cohen, 2002. Stable carbon isotope fractionation during enhanced in situ bioremediation of trichloroethene. *Environmental Science & Technology.* 36, 2262–2268.

Sturchio, N., Clausen, J., Heraty, L., Huang, L., Holt, B., and T. Abrajano, 1998. Chlorine isotope investigation of natural attenuation of trichloroethene in an aerobic aquifer. *Environmental Science & Technology.* 32, 3037–3042.

Sturchio, N., Heraty, L., Holt, B., Huant, L., and T. Abrajano, 1999. Stable isotope investigations of the chlorinated aliphatic hydrocarbons. *9th Annual V. M. Goldschmidt Conference*, August 22–27, 1999. Cambridge, MA.

Sturchio, N., Hatzinger, P., Arkins, M., Suh, C., and L. Hearty, 2003. Chlorine isotope fractionation during microbial reduction of perchlorate. *Environmental Science & Technology.* 37, 3859–3863.

Tanaka, N., and D., Rye, 1991. Chlorine in the stratosphere. *Nature* 353, 707.

Reddy, C., Hearty, L., Holt, B., Sturchio, N., Eglinton, T., Drenzek, N., Xu, L., Lake, J., and K. Maruya, 2000. Stable chlorine isotopic compositions of Aroclors and Aroclor-contaminated sediments. *Environmental Science & Technology.* 34, 2866–2870.

Reddy, C., Pearson, A., Xu., L., McNichol, A., Benner, B., Wise, S., Klouda, G., Currie, L., and T. Eglinton, 2002. Radiocarbon as a tool to apportion the sources of polycyclic aromatic hydrocarbons and black carbon in environmental samples. *Environmental Science & Technology.* 36, 1774–1782.

United Nations Industrial Development Organization, 1994. Environmental management in the electronics industry. Semiconductor manufacture and assembly. UN E94-III-D2. Vienna International Centre. Vienna, Austria.

United States Army, 1978. Evaluation and control of vapor degreasing operations. United States Army Environmental Hygiene Agency. Aberdeen Proving Grounds. Aberdeen, MD.

United States Environmental Protection Agency, 1979. Source Assessment: Solvent Evaporation-Degreasing Operations. EPA-600/2-79-019f.

United States Environmental Protection Agency, 1995. Office of Pollution Prevention and Toxics Fact Sheet for 1,4-dioxane. EPA 749-F-95-010.

USPTO, 1954. US Patent and Trademark Office, Chloroethene, Trademark Text and Image Database, Washington, D.C.

Vaillancourt, J., 1998. Chlorine and carbon isotopic trends during the degradation of trichloroethylene with zero valent metal. MSc Thesis. Department of Earth Sciences, University of Waterloo.

Van Warmerdam, Frape, E., Aravena, S., Drimmie, R., Flatt, R., and J. Cherry, 1995. Stable chlorine and carbon isotope measurements of selected chlorinated organic solvents. *Applied Geochemistry.* 10(5), 547–552.

Vogel, T., and P. McCarty, 1987. Abiotic and biotic transformations of 1,1,1 trichloroethane under methanogenic conditions. *Environmental Science & Technology.* 21:1208–1213.

Vogel, T., Criddle, C., and P. McCarty, 1987. Transformation of halogenated aliphatic compounds. *Environmental Science & Technology.* 21, 722–736.

Von Grote, J., 2003. Occupational exposure assessment in metal degreasing and dry cleaning-influences of technology innovation of legislation. Submitted to the Swiss Federal Institute of Technology (dissertation). Zurich, Switzerland.

Ward, J., Ahad, J., LaCrampe-Coulome, G., Slater, G., Edwards, E., and B. Sherwood Lollar, 2000. Hydrogen isotope fractionation during methanogenic degradation of toluene: Potential for direct verification of bioremediation. *Environmental Science and Technology.* 34(21), 4577–4581.

Warner, D., and J. Mertens, 1991. Replacing 1,1,1-trichloroethane: consider other solvents. *Plating & Surface Finishing.* November 1991. pp. 60–62.

Wilson, J., and B. Wilson, 1984. Biotransformation of trichloroethylene in soil. *Applied & Environmental Microbiology.* 49, 242–243.

Wilson, J., Weaver, J., and D. Kampbell., 1994. Intrinsic bioremediation of TCE in ground water at an NPL site in St. Joseph, Michigan. In: US EPA. Symposium on Natural Attenuation of Ground Water, Denver, CO, August 30-September 1. EPA/600/R-94/162. pp. 116–119., 1,1-trichloroethane in groundwater following a transient release. *Chemosphere.* 34, 771–781,

Wing, M.R., 1997. Apparent first-order kinetics in the transformation of 1.

Wolf, K., 1997. Chlorinated Solvents: The decline and fall. White Paper. Institute for Research and Technical Assistance. January 17, 1997. Santa Monica, California. USA. p. 38.

Wood, W., 1982. *Metals Handbook Ninth Edition*, Vol. 5, *Surface Cleaning, Finishing, and Coating.* American Society of Metals. Metals Park, Ohio.

13

Arsenic

Jennifer K. Saxe, Teresa S. Bowers, and Kim Reynolds Reid

Contents

13.1 INTRODUCTION

Arsenic occurs naturally in the environment, and arsenic in the environment is ultimately of natural origin. Localized increases in arsenic concentrations in soil due to human activities are the result of humans moving and redistributing naturally occurring arsenic. Observed arsenic levels in any particular area can result from a combination of natural processes, historical, widely dispersed human sources, and current human activities. This combination of processes clearly has the potential to complicate forensic analysis of arsenic sources.

Arsenic has held our fascination for generations, in part because it has long been known for its poisonous properties. It was often considered a "practical" way to murder an individual, because it was so difficult to detect arsenic after death. Today murder by arsenic is easier to recognize, but identifying sources of arsenic in the environment has become the new forensic challenge.

13.2 CHEMISTRY AND SOURCES

13.2.1 Chemistry

Arsenic, with an atomic number of 33 in the periodic table of the elements, is the 50th most abundant element in the earth's crust (Mason, 1966) and the 22nd most abundant element in seawater (Bearman, 1989). Arsenic has only one stable isotope (atomic mass of 75), which accounts for 100% of its natural abundance. Numerous short-lived radioactive isotopes of arsenic have been observed, but are not useful for the discussion here.

Arsenic is found naturally in the minerals arsenopyrite (FeSAs), orpiment (As_2S_3), and realgar (AsS), also as various oxide minerals, but most commonly as a trace element in other minerals. Weathering of arsenic-bearing minerals leads to trace levels of arsenic in soil, and leaching of arsenic from minerals and soil leads to trace levels of arsenic in groundwaters and surface waters.

Arsenic typically occurs in either a +3 or +5 oxidation state; inorganic forms of arsenic are referred to as arsenite and arsenate, respectively. Arsenite is the more mobile and toxic form of arsenic. Arsenic, as arsenite or arsenate, is commonly found adsorbed to the surfaces of soil particles, in particular iron, or to natural organic matter in soil, such as humus. The form of arsenic in soil often has more to do with

the physical and chemical characteristics of the soil than with the original source of the arsenic. These controls are discussed further, in Section 13.2.3. Arsenic from various sources may be indistinguishable in soil, complicating any forensic analysis.

The structure and nomenclature of various arsenic compounds are shown in Figure 13.2.1.

13.2.2 Sources

Arsenic contributions to soil from historical human activities have occurred for more than 5000 years due to the use of arsenic in the production of tools, ornaments, pigments, and cosmetics, the use of coal-fired furnaces, the use of arsenic as backing for mirrors, the use of arsenic in glass manufacturing, in tanneries, its medicinal uses and, in the 20th century, its use in pesticides (Carey et al., 1976; Azcue and Nriagu, 1994). As a result of this multitude of uses, arsenic is the second most common inorganic constituent found at the original 1000 National Priority List (NPL) sites in the United States (Davis et al., 2001).

13.2.2.1 Pristine Background

No discussion of arsenic can be complete without first considering natural or "pristine" background sources of arsenic. The term "background" is often used to include anthropogenic as well as natural sources of a contaminant, and can be a useful concept when trying to distinguish a local point source of contamination from natural and anthropogenic background sources.

Arsenic occurs naturally in the environment, and there is a large variability in the background levels of arsenic in soil and other environmental media due to natural variations in soil types and source rocks. Unfortunately, it is often difficult to distinguish between natural and anthropogenic influences on observed background levels as there can be significant overlap in their concentration ranges.

Shacklette and Boerngen (1984) and Boerngen and Shacklette (1981) analyzed soils in the United States that they judged not to have been influenced by human activities. They found that natural background concentrations of arsenic in soil ranged from less than 1 to 97 mg/kg. Natural background concentrations are greater in soils overlying mineral formations that are naturally high in arsenic.

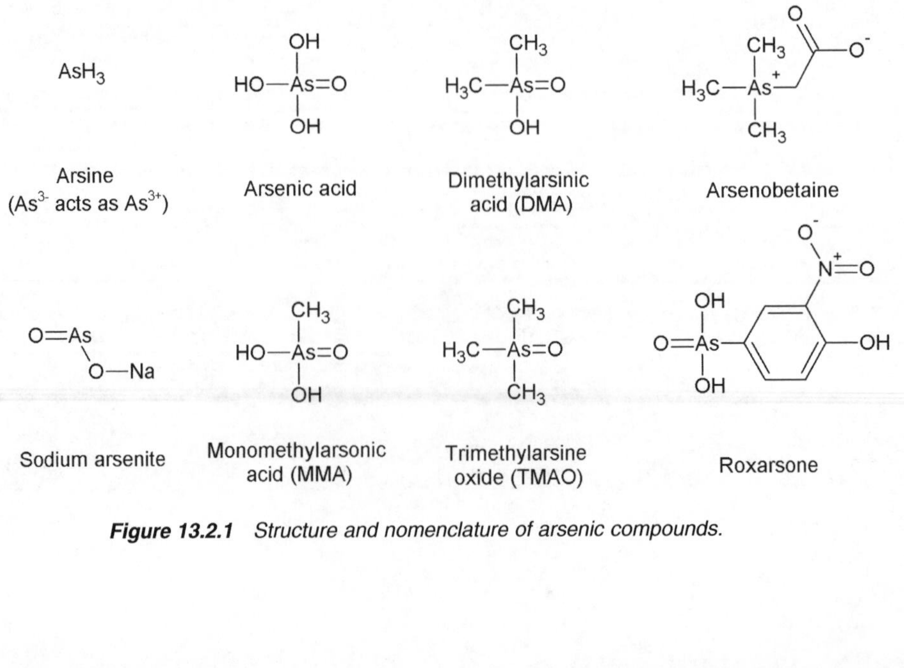

Figure 13.2.1 Structure and nomenclature of arsenic compounds.

Soil-forming rocks with higher concentrations of arsenic are shales, clays, and specifically clays that were originally formed in the sea but are now on land (deep-sea clays), although igneous rocks contain a wide range of arsenic concentrations as well (Dragun and Chiasson, 1991). Sands typically have the lowest levels of arsenic.

Drever (1982) reports natural background levels of arsenic of $2\,\mu g/L$ in streams and $3\,\mu g/L$ in ocean waters.

13.2.2.2 Pesticides and Fertilizers

Probably the most widespread source of anthropogenic arsenic in our environment stems from the use of pesticides and fertilizers. The use of arsenic-containing fertilizers and pesticides represents a historic and a continuing addition to background concentrations of arsenic in soils. Early in the 20th century, pesticides including lead arsenate and calcium arsenate were commonly applied to turf grass (e.g., golf courses, sod farms) and agricultural crops (e.g., apple orchards, vegetable fields) (Alden, 1983; Welch et al., 2000). Lead arsenate was used to control a grasshopper infestation in the Midwest during the 1930s and 1940s (USEPA, 1985). Due to the general immobility of arsenic in soil, greater background concentrations of arsenic are still measured in many historic orchards and fields today.

More recently, organic compounds containing arsenic have replaced the use of pesticides such as lead arsenate (Murphy and Aucott, 1998). The current use of pesticides that contain arsenic is regionalized, with most US use in the Southern United States. The organic arsenic-containing pesticides currently registered for use in the United States include monosodium methylarsonate (MSMA), disodium methylarsonate (DSMA), calcium acid methanearsonate (CAMA), and cacodylic acid (DMA), each of which contains almost 50% arsenic by weight. These herbicides are primarily applied to cotton and turf grass.

In southern states, in the United States, higher arsenic concentrations in soil occur near former cattle dip vats where the use of arsenic-containing pesticides was mandated by law in the early 1900s to eradicate ticks in livestock (FDEP, 2002). Cattle dip vats each contained approximately 1500 gallons of arsenic solution through which cattle were driven for a brief bath, and it has been estimated that there are 3400 abandoned vats in Florida alone (Thomas, 1998).

Fertilizers that are currently used by consumers contain arsenic at significant concentrations (e.g., 75 mg/kg) (WSDE, 1999). Natural fertilizer materials such as manure from livestock and seaweed also contribute substantial arsenic to the soil environment. Poultry and swine feed commonly contain arsenic compounds such as roxarsone and arsanilic acid as additives to prevent illness. The manure of these animals contains substantial arsenic concentrations (e.g., Jackson and Bertsch, 2001). Seaweed used as fertilizer contains arsenosugars (usually arsenoribofuranosides), which have been shown to degrade in soil to form inorganic arsenic (Castlehouse et al., 2003).

13.2.2.3 Mining and Smelting

Arsenic is mined and occurs with many other metals which are mined. Arsenopyrite is the chief ore of arsenic, and the primary ores for both copper and silver include arsenic-bearing minerals. Arsenic is also typically high in ores for gold, tin, tungsten, and lead. Most rock removed during mining operations is waste rock, left behind in tailings and waste piles. In some areas of the western United States, residential and other construction has occurred in close proximity to mining wastes, and some communities have elevated levels of arsenic throughout their soils.

Smelting, the process of separating metals from their ores, has contributed arsenic to the environment. Emissions from copper, silver, and gold smelters have resulted in the deposition of arsenic to soils. Areas such as Anaconda, Montana, have soil arsenic levels as high as several hundred parts per million in the proximity of the smelters.

13.2.2.4 Electronics

The electronics industry uses arsenic as a component of solid-state devices such as microchips. Pure silicon crystals used as microchips are poor conductors of electricity, and, as a result, controlled amounts of impurities, called dopants, must be added to obtain the desired electrical properties. Arsenic, like other semi-metals, is used as a dopant in silicon chips (USEPA, 1995). Additionally, silicon-free gallium arsenide semiconductor crystals are manufactured for use in diodes, field-effect transistors (FETs), and integrated circuits due to their unique physical properties: they are less noisy than silicon chips, operate at higher frequencies, and readily produce light in LEDs or lasers. (e.g., NIST, 1993). Because of these unique properties, gallium arsenide chips are used in mobile phones and for military equipment. Crystal-growth and cleaning solutions and dusts from grinding and sawing contribute to arsenic-containing wastes from the electronics industry (Sheehy and Jones, 1993). The amount of arsenic used and disposed of from electronic chip-making industries is small, because the industry generally reclaims its high purity arsenic wastes (Loebenstein, 1994).

13.2.2.5 Wood Treating

Lumber products have been treated with chromated copper arsenate (CCA) to enhance their durability by preventing damage due to fungi, termites, and marine boring organisms. CCA-treated wood was developed in 1933, and has been used in industrial applications beginning before 1940 and in residential applications beginning around 1974 (DeVenzio, 1998). The CCA that is used to pressure-treat wood is a water-based mixture containing 0.6–6.0% (by weight) of chromic acid, copper oxide, and arsenic acid (USDA, 1980). During the treatment process, CCA, at pH 1.6–2.5, is infused into wood at elevated pressure (AWPA, 2002). The resulting treated wood contains arsenic(V), postulated to be in the form of chromium(III) arsenate (Bull, 2001) or chromium dimer–arsenic clusters that are stable over long periods of time (Nico et al., 2004).

There has been considerable attention to the issue of arsenic leaching to soils from decks and other wooden structures made from wood treated with CCA. For example, Stilwell and Gorny (1997) report large variations in the soil arsenic levels below seven different treated decks, ranging from concentrations within local background levels to as high as 350 mg/kg beneath one of the investigated decks. However, the Stilwell and Gorny (1997) results may be reflective of construction debris such as sawdust or wood shavings, because the results are in conflict with other results (reviewed in Lebow et al., 2004) showing lower concentrations in soil when a history of the sampling location is known, so that the inclusion of wood debris in soil can be ruled out.

13.2.2.6 Miscellaneous Additional Sources

Arsenic is used in the manufacture of glass. The peak use of arsenic in glass manufacturing in the 20th century in the United States occurred in the 1970s, during which about 1900 tons of arsenic were used per year (Loebenstein, 1994). Arsenic trioxide was formerly used in glass manufacturing, but now substitute ingredients and changes in the manufacturing process have resulted in a drop in manufacturing losses between 1968 and the late 1980s from almost

600 metric tons per year to about 20 metric tons per year (Loebenstein, 1994).

Paris Green (copper acetoarsenite, $C_4H_6As_6Cu_4O_{16}$) is an emerald green powder that was formerly used as a pigment, insecticide and fungicide. Its use as a pigment in wallpaper was implicated in the mysterious illness of Clare Booth Luce when she served as US Ambassador to Italy in the 1950s. Arsine gas, generated by the interaction of fungal organisms present in mildew with Paris Green in the wallpaper of the bedroom of her Italian villa was blamed. Some debate centers on whether wallpaper, rather than deliberate poisoning, was the source of arsenic found in Napoleon's hair.

Older cemeteries represent an additional source of arsenic in the environment. In the late 19th century, typical embalming techniques used between 4 ounces and 12 pounds of arsenic to preserve human remains (Konefes and McGee, 2001). This practice was prevalent from about 1880 to 1900, and was phased out completely by 1905 (Spongberg and Becks, 2000). Groundwater underlying older cemeteries in Iowa and New York contained elevated arsenic concentrations that researchers attributed to the cemeteries (Konefes and McGee, 1996 cited in USEPA, 2000a).

Arsenic metal is also used as an alloying element in ammunition and solders, as an anti-friction additive to metals used for bearings, and to strengthen lead-acid storage battery grids (USGS, 2005).

13.2.3 Transformation and Transport in the Environment

If arsenic were immobile in soil, its forensic analysis would be less complicated. There are several factors that govern the transport and fate of arsenic in soil, including the chemical form of arsenic, the chemical and physical characteristics of the soil such as soil pH and redox potential, and the presence of other soil constituents which may limit arsenic mobility by adsorption or precipitation. These factors should be considered or investigated for their effect on arsenic transport in order to reduce uncertainty associated with any forensic analysis.

Arsenate is typically less mobile in soil, and arsenite is more mobile; organic compounds that contain arsenic are generally not mobile in soil (ATSDR, 2000). The water solubility of arsenic is a key factor in determining its mobility. The solubility of the various chemical forms of arsenic depend in part on soil pH, which naturally varies between soils and within a soil over time. Soil pH can be altered by outside influences, such as acid rain or fertilizers. However, the presence of clay minerals and natural organic matter in soil can minimize or prevent such pH changes (i.e., buffer the pH) (Rieuwerts et al., 1999).

The amount of oxygen in soil influences the water solubility of arsenic. Water-saturated, muddy soils often have very little free oxygen, whereas aerated soil like sand contain oxygen, leading to changes in the identity of minerals that are present. The oxygen level in a soil also changes in deeper soils because the degree of water saturation generally increases with depth, leaving less space in soil pores for air.

An example of how pH and redox potential can affect arsenic mobility is seen in the transformation of arsenite in water into arsenate. If very little free oxygen is present in water, arsenite will transform to arsenate only at high pH values (alkaline conditions), but when more oxygen is present, the arsenite-to-arsenate transformation will take place at low pH values (acidic conditions).

In general, arsenic is least mobile in soil when the soil's pH is near neutral or slightly acidic (approximately pH 4–8),

but arsenic mobility is also dependent upon soil composition. Arsenate is the most common form of inorganic arsenic in soil. Arsenite, or arsenic(III), tends to exist under reducing redox conditions. Because arsenite is more mobile in soil than arsenate, it may leach through surface soils with rainwater percolation (Matera and Le Hecho, 2001). Manganese oxides (e.g., the mineral birnessite) in soil can oxidize arsenite, transforming it into arsenate, and the arsenate then binds strongly to soil particles (Manning et al., 2002), thus limiting arsenic mobility.

Arsenic can also form an insoluble precipitate in oxygen-containing, alkaline soils by combining with common elements such as calcium and manganese. In soils containing very little available oxygen, arsenic can form insoluble precipitates with sulfur. Some of these precipitates may re-dissolve with shifts in soil pH resulting from seasonal or other climatic changes. This also affects arsenic mobility and solubility in soil (Matera and Le Hecho, 2001).

Arsenic may adsorb, or bind to the surface of soil particles. Adsorption is the principal mechanism through which arsenic is retained in soil, thus limiting its mobility. Arsenic is bound via adsorption to soil minerals, particularly iron oxides and aluminum oxides, often found in clay-rich soils. Arsenic can also be bound to organic matter in soil. The quantity of arsenic bound to soil through adsorption depends upon the types of adsorbing minerals present in the soil (e.g., clays), the soil pH, and other components that compete for adsorption sites (Matera and Le Hecho, 2001). Chen et al. (2002) found a significant correlation between the amount of iron or aluminum in soil and the amount of arsenic in soil. Presumably, the arsenic is bound to the aluminum- or iron-oxides that exist as clay particles in the soil. Correlations exist between clay mineral content and arsenic content in soil, but the relationships are not strong (correlation coefficients (r^2) ranged from 0.18 to 0.45) because other factors in soil also contribute substantially to arsenic fate.

Competition among chemicals exists for soil adsorption sites. Arsenate can be displaced from binding sites on clay by phosphate or sulfate, thus increasing arsenic's mobility. The effects of competition depend on the relative concentrations of these constituents in soil as well as soil pH. For example, phosphate is found at naturally high concentrations in soils from some areas of the country (e.g., Florida, North Carolina, Idaho, and Utah where it is mined) (FIPR, 2001). Phosphate is also a component of most fertilizers. Phosphate is similar to arsenate and has been shown to compete with arsenate for the same binding sites in soil. However, phosphate binds more strongly with most soils than arsenate, so the addition and depletion of phosphate over time, as with fertilizer applications, can serve to cyclically mobilize and immobilize arsenate (Matera and Le Hecho, 2001). An example of the possible result of this scenario of cyclical "release and reattachment" in soil is that arsenic may percolate with rainwater into deeper soils after phosphate fertilizer has been applied, but then become bound to those deeper soils and remain immobile.

In addition to the abiotic transport and transformation phenomena described above, arsenic can undergo biologically mediated transformation. Arsenic ingested by mammals undergoes methylation, resulting in the formation of MMA, DMA, and TMAO (Figure 13.2.1) which are excreted in urine (ATSDR, 2000). In the environment, microorganisms in soil and natural waters are capable of both methylation and demethylation of arsenic compounds (Woolson et al., 1982). A large body of research reviewed by the National Research Council maintains that such biological transformations in soil can only occur to compounds that are soluble in soil pore water (NRC, 2003). Thus, when arsenic compounds are irreversibly sorbed to soil particles or present in a precipitate, transformation is not expected to occur. Studies of the demethylation of organic arsenic compounds show biphasic behavior, with transformation occurring initially after the compound is introduced to soil, slowing dramatically after a few days to a few weeks (Woolson *et al.*, 1982; Shariatpanahi *et al.*, 1981; Abdelghani *et al.*, 1977; Woolson and Kearney, 1973; Von Endt *et al.*, 1968). This provides further evidence that arsenic-containing compounds sorbed to soil particles are not available for biological transformations (e.g., Abdelghani *et al.*, 1977).

13.3 ANALYTICAL TECHNIQUES

Numerous standard analytical methods are published by the United States Environmental Protection Agency for arsenic. The majority of these methods analyze arsenic in its total or "elemental" form in routine matrices such as soils and groundwater. However, in the past decade or so, laboratory instrumentation has been significantly improved so that the various species of arsenic may also be analyzed (e.g., arsenite, arsenate, MMA, and DMA) in a variety of matrices. In addition, powerful characterization techniques such as Scanning Electron Microscopy (SEM) are being used to "fingerprint" arsenic sources in more unique media.

Choosing the most appropriate analysis method greatly depends on sample matrix, whether physical and/or chemical characterization is necessary to achieve the objective of the investigation, the form of arsenic to be studied (total *vs.* speciated), the presence of other chemical constituents (see interferences), and detection limit needs. Table 13.3.1 summarizes the various US EPA digestion and analysis methods currently published for arsenic.

13.3.1 Standard Analytical Techniques

Many standard analytical techniques for quantifying arsenic concentrations in environmental media require that the sample be extracted or digested in acid prior to instrumental analysis. Determining the appropriate digestion method used is generally dependent on the analytical method chosen; US EPA's Test Methods for Evaluating Solid Waste, Physical/Chemical Methods, SW-846 (USEPA, 1986), includes numerous digestion methods. Arsenic analyses in aqueous and soil matrices are most commonly performed using either Graphite Furnace Atomic Absorption (GFAA), or Inductively Coupled Plasma Atomic Emission Spectrometry (ICP-AES) methodologies. The US EPA has published standard GFAA and ICP methods in SW-846. In GFAA, aqueous samples or digestates are placed into a graphite tube in the furnace, evaporated to dryness, charred, and atomized (USEPA, 1998a). The use of smaller sample volumes or detection of lower concentrations of elements is possible in GFAA compared to atomic absorption using a flame, since a greater percentage of available analyte atoms is vaporized and dissociated for absorption in the tube rather than the flame (USEPA, 1998a). In ICP, arsenic's characteristic emission spectrum is measured *via* optical spectrometry. Aqueous samples or digestates are nebulized and the resulting aerosols are transported to a plasma torch. Element-specific emission spectra are produced by a radio-frequency ICP, dispersed by a grating spectrometer, and the intensities of the emission lines are monitored by photosensitive devices (USEPA, 2000b). Background correction is required for trace determination (USEPA, 2000b). Arsenic detection limits for these methods generally range from 1–5 μg/L for aqueous samples to 1–5 mg/kg for solid samples.

In recent years, ICP has been further optimized by combining the instrumentation with more sensitive detectors

or chromatographic separation technologies. For example, ICP-Mass Spectroscopy (ICP-MS) can achieve much lower detection limits than the conventional ICP methods (as low as 0.1 μg/L for aqueous samples, 1 mg/kg for soil samples); the ICP-MS method also incorporates internal standards which can eliminate troublesome interferences. When hydride-generation (HG), ion chromatography (IC), or high-performance liquid chromatography (HPLC) is coupled with ICP-MS, it is possible to analyze very low concentrations of total inorganic arsenic as well as the individual species of arsenic (e.g., arsenite, arsenate, MMA, and DMA) in aqueous, soil, and biological matrices, which can be critical in certain forensic applications (see Section 13.4.3). Arsenic species may also be determined in tissues and water using HG coupled with furnace AA spectrometry as in US EPA Method 1632A (USEPA, 2001). The detection range in water is 0.01–50 μg/L; in tissue, 0.10–500 mg/kg.

13.3.2 Scanning Electron Microscopy

Several arsenic-rich minerals can be identified under a petrographic microscope. These include arsenopyrite and additional copper, silver, and other metal ores which contain arsenic. Observation of these minerals generally indicates a mining source, or natural background in mining districts. However, powerful microscopes provide for identification of much more.

A scanning electron microscope is capable of producing high-resolution images of a sample surface (e.g., Morin *et al.*, 2003). Due to the manner in which the image is created, SEM images have a characteristic three-dimensional quality which allows for high-definition visual characterization of particle morphology and, when coupled with detectors (EDS or WDS, explained below), semi-quantitative chemical analysis is also possible. The method is useful in determining arsenic sources on an atomic level.

When SEM is coupled with chemical detectors, the result is a powerful tool for determining the nature and potential origin of environmental samples. Two types of detection methods available that are coupled with SEM are x-ray wavelength dispersion spectrometry (WDS) and energy dispersive spectroscopy (EDS), which allow quantitative (WDS) and semiquantitative (EDS) elemental analysis of individual particles in an SEM image. Figure 13.3.1 shows an SEM image of several particles collected from outdoor surfaces where an arsenic source is present. EDS analysis of the individual particles allows differentiation between particles from the arsenic source and those from other origins, such as soil. For example, clues to the identity of the numbered particles in Figure 13.3.1 are evident in examining the abundance of arsenic relative to iron, silicon, and aluminum (elements that are common in soil but not the arsenic source material), shown in Table 13.3.2.

The data in Table 13.3.2 suggest that particle 1 is an iron-rich clay mineral. The particle was collected in an area of the United States where iron and kaolinite are common in soil. Kaolinite's ideal structure is $Al_2Si_2O_5(OH)_2$, but ferric iron may be readily substituted for aluminum in this structure. The increased abundance of arsenic relative to iron, silicon, and aluminum suggests that particle 2 originated from the arsenic source material. Particle 3 is likely a silt grain containing quartz, according to the data, because of the relatively high abundance of silicon when compared with the remaining elements included in Table 13.3.2.

13.3.3 Laser Ablation ICP-MS

Laser ablation ICP-MS is a relatively new but powerful analytical technique used for the *in situ* analysis of ultra trace elements in solid samples. Combining the fine resolution of

Table 13.3.1 *Examples of US EPA Arsenic Analysis Methods*

Analytical Method	Method Number	Comments	Source/Reference
EPA 200 Series Methods			
ICP-AES	200.7	Multi-analyte method. Determines total or dissolved arsenic in water, wastewater, and solid wastes. Includes sample digestion procedures.	600/R-94-111
ICP-MS	200.8	Multi-analyte method. Determines total or dissolved arsenic in water, wastewater, and solid wastes. Includes sample digestion procedures.	600/R-94-111
Arsenic, AA Furnace	206.2	Similar to SW-846 Method 7060A.	600/4-79-020, NEMI
Arsenic, AA Gaseous Hydride	206.3	Determines arsenic in drinking water and fresh and saline waters in the absence of high concentrations of chromium, cobalt, copper, mercury, molybdenum, nickel, and silver. Sample digestion by EPA Method 206.5 (Sample Digestion Prior to Total Arsenic Analysis by Silver Diethylthiocarbamate or Hydride Procedures).	600/4-79-020, NEMI
Arsenic, Spectrophotometric-Silver Diethyldithio-carbamate (SDDC)	206.4	Determines arsenic in drinking water and most freshwaters and saline waters in the absence of high concentrations of Cr, Co, Cu, Hg, Mo, Ni, and Ag. Industrial wastes may be analyzed after digestion by EPA Method 206.5.	600/4-79-020, NEMI
SW-846 Methods			
ICP-AES	6010C	Multi-analyte method. Determines trace arsenic in solution/digestates. Requires digestion methods such as 3005A, 3010A, 3050B, or 3051.	SW-846
ICP-MS	6020A	Multi-analyte method. Determines sub μg/L arsenic in solution/digestates. Requires digestion methods such as 3005A, 3010A, 3050B, or 3051.	SW-846
Graphite Furnace Atomic Absorption	7010	Multi-analyte method. Digestion procedures described in Chapter 3 of SW-846.	SW-846
Arsenic, AA, Furnace	7060A	Includes digestion procedure for aqueous samples; soil samples require digestion by SW-846 Method 3050B. Method 7010 may also be used.	SW-846
Arsenic–Flame AA, Gaseous Hydride	7061A	Approved only for matrices that do not contain high Cr, Cu, Hg, Ni, Ag, Co, and Mo.	SW-846
Antimony and Arsenic, AA-Borohydride Reduction	7062	Determines arsenic in wastes, mobility procedure extracts, soils, and groundwater. Can tolerate high concentrations (up to 4000 mg/L) of Co, Cu, Fe, Hg, or Ni. Digestion performed using methods 3010 (aqueous) or 3050 (solids).	SW-846
Arsenic in Aqueous Samples & Extracts by Anodic Stripping Voltammetry (ASV)	7063	Determines free dissolved arsenic or As(III) and As(V) in drinking water, surface water, seawater, wastewater, soil extracts.	SW-846
Other EPA Methods			
Arsenic in Water & Tissue by HGC/FAA	1632A	"Clean" method that determines inorganic arsenic, As(III), As(V), MMA, and DMA in filtered and unfiltered water and tissue.	821/R-01-006, SEL2-TRA
ICP-AES and ICP-MS	ILM05.2	Multi-analyte methods; include digestion methods and analysis procedures for ICP and ICP-MS.	http://www.epa.gov/superfund/programs/clp/index.htm

a laser probe with the sensitivity, speed, and accuracy of ICP-MS detection, the method complements electron microprobe analysis, typically measuring arsenic at concentrations less than 1 ppt. A distinct advantage of this procedure is that sample digestion is not necessary, and the method can be considered nearly nondestructive since such a small amount of material is ablated. Solid particles of nearly any size or shape are physically ablated due to the interaction

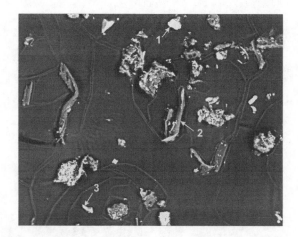

Figure 13.3.1 *Particles identified on the basis of semiquantitative EDS analysis of arsenic and other elements within a 240 × SEM image. Particle 1 – clay; Particle 2 – particle from arsenic source material; Particle 3 – quartz. Figure reproduced courtesy of Battelle (see color insert).*

Table 13.3.2 *Elemental Ratios in Particles Shown in Figure 13.3.1, Detected Using EDS*

Particle No. (Figure 13.3.1)	As:Fe	As:Si	As:Al	Likely particle type
1	0.20	0.12	0.16	Iron-rich clay mineral
2	1.3	0.70	1.1	Arsenic source material
3	0.73	0.10	2.9	Quartz-containing particle

of a high power laser beam with the surface of the sample. The particles are carried in a stream of inert gas (helium or argon) into an argon plasma where they are ionized prior to measurement in a quadrupole mass spectrometer. Arsenic's isotope (As^{75}) is then measured to determine the elemental concentration.

This method is extremely sensitive since the ability to focus the laser beam on an exceptionally small area allows for direct sampling of individual microscopic features on a sample surface. In addition to soil analysis and bulk analysis of metals, it is possible to analyze fluid inclusions contained within the microscopic voids of solid samples (Ghazi, 2005).

13.3.4 Time of Flight Secondary Ion Mass Spectrometry

Time-of-flight secondary ion mass spectrometry (TOF SIMS), like laser ablation ICP-MS, is a surface analytical technique that uses an ion beam to ablate small numbers of atoms from the outermost atomic layer of a surface. In TOF SIMS, energetic primary ions, typically ions such as Ga+, Cs+, or O−, bombard the sample resulting in the emission of secondary elemental or cluster ions. Smaller secondary particles travel faster than larger ones, and thus have a shorter time-of-flight to the mass spectrometer, allowing elements to be distinguished (e.g., Zhu *et al.*, 2001). Arsenic can be analyzed on solid samples using this technique. Figure 13.3.2A shows a total ion map of dust samples collected from outdoor surfaces where an arsenic source is present, and Figure 13.3.2B shows an arsenic map of the same sample. Figure 13.3.2 provides evidence that arsenic is relatively evenly dispersed throughout the sample, suggesting that the particles are homogeneous on the scale shown. Furthermore, the weak signal response suggests that arsenic is not deposited preferentially on the surface of

the particles, but rather is contained within the bulk of each particle.

13.3.5 X-Ray Fluorescence

X-Ray Fluorescence (XRF) is generally used for analysis of arsenic in bulk solids, with a detection limit as low as 5 mg/kg in soil, but typically ranging between 10 and 50 mg/kg. The basis of XRF spectrometry is the detection and measurement of x-rays emitted from the atoms of an irradiated sample. A beam of x-rays is directed into a sample, exciting some of the atoms in the sample to energy levels above their ground state. The intensity of the fluorescent radiation depends on several factors, but is related to the concentration of the element in the sample. It can also be used for micro samples, thin samples, aerosols, and liquids, with detection limits of 2–20 ng/cm^2 for most elements. The XRF analysis has the additional advantage that it does not require dissolution of a sample, thereby eliminating concern about insoluble residues. The XRF instruments can be field portable, which is useful for rapid, real-time measurements.

13.3.6 X-Ray Absorption Fine Structure Spectroscopy

X-Ray absorption fine structure spectroscopy (XAFS) is a powerful analytical technique in that it is nondestructive, requires very small sample sizes (e.g., less than one gram of soil), requires no special sample preparation, and allows the identification of distinct arsenic species and compounds by providing information on the coordination chemistry of arsenic in a sample. In XAFS, a light source is focused on the sample, and the pattern of x-ray energy absorption or of fluorescent emissions at specific characteristic energy is used to identify arsenic species. Identification is facilitated by comparison to spectra for known species and compounds.

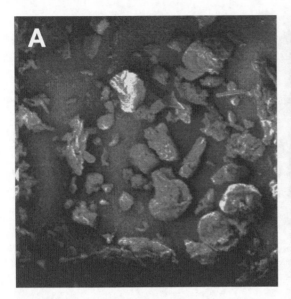
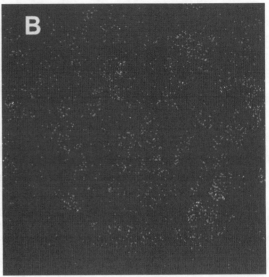

Figure 13.3.2 *TOF SIMS image of a 300 μm × 300 μm area containing mixed particles collected from outdoor surfaces where an arsenic source is present. (A) shows a total ion map and (B) shows an arsenic map. Figure reproduced courtesy of Battelle (see color insert).*

In XAFS, an incident beam of continually increasing x-ray energy is absorbed uniformly by a sample until reaching a characteristic energy level where there is a sudden increase in absorption. The energy level where this increase occurs is the absorption edge, where an electron from the arsenic atom is ejected. The absorption edge of arsenic is between 11,860 and 11,890 eV (Foster, 2001). As the incident x-ray beam's energy increases beyond the edge, absorption decreases and then fluctuates, as a cascade of electrons move to fill the void after the initial ejection. The energy and pattern of the spectrum beginning just before the edge to 50 eV past the edge is described as the x-ray absorption near edge structure (XANES) region. The information in this region is used primarily to determine the valence of arsenic, although some qualitative information regarding the coordination chemistry can also be gleaned from XANES data (e.g., Smith *et al.*, 2005). The spectrum at energy levels beyond the XANES region is called the extended x-ray absorption fine structure (EXAFS). The information in the EXAFS region of the spectrum is primarily used to determine the coordination chemistry of arsenic. That is, the identity and quantity of the nearest neighbor and next nearest neighbor atoms can be determined from the EXAFS, thus identifying the molecular structure surrounding arsenic atoms present in samples.

The XAFS analysis requires a high powered light source, thus XAFS is not performed in typical laboratories, but through the use of large synchrotron light source facilities. Obtaining reliable speciation information in the XANES region typically requires an arsenic concentration between 10 and 100 mg/kg (Foster, 2001). Obtaining reliable coordination chemistry information from the EXAFS spectrum can require higher concentrations, depending on the homogeneity of the sample and the compound present.

13.3.7   Interferences and Biases

Most published analytical methods describe the interferences that may occur during arsenic measurement. Conventional GFAA and ICP analyses may be affected by spectral, physical, chemical and/or memory interferences (USEPA, 1998a, 2000b). Chemical interferences are fairly uncommon and are usually characterized by molecular compound formation, ionization effects, or solute vaporization effects. Spectral interferences in GFAA occur when an absorbing wavelength of an element present in the sample but not being determined falls within the width of the absorption line of the element of interest. In ICP, spectral interferences are caused by background emission from continuous or recombination phenomena, stray light from the line emission of high concentration elements, overlap of a spectral line from another element, or unresolved overlap of molecular band spectra (USEPA, 1998a, 2000b). The results of the determination affected by spectral interference may be erroneously high. Physical interferences are generally caused by high dissolved solids or acids content, which can affect sample viscosity and thus absorption, nebulization, and/or sample transport. Finally, memory interferences can result when analytes in a previous sample contribute to the signals measured in a new sample. Background correction techniques (e.g., Zeeman background correction in GFAA; interelement corrections in ICP) can usually compensate for the various types of interferences encountered.

In GFAA, arsenic can also be affected by nonspecific absorption and light scattering caused by matrix components during atomization (USEPA, 1998a). Arsenic analysis is particularly susceptible to this interference because of its fairly low analytical wavelength (193.7 nm). Simultaneous background correction must be employed to avoid false positives or erroneously high results. In addition, aluminum can act as a significant positive interferent in arsenic analysis; the use of Zeeman or another appropriate method of background correction will help to prevent this.

In ICP-MS, arsenic is prone to isobaric polyatomic ion interferences, which occur when ions consisting of more than one atom have the same nominal mass-to-charge ratio as the isotope of interest, and which cannot be resolved by the mass spectrometer (USEPA, 1998b). Appropriate corrections must be made to the data when these interferences occur. In addition, memory interferences can result when isotopes of elements in a previous sample contribute to the signals measured in a new sample (USEPA, 1998b). Most ICP-MS physical interference effects may be compensated by implementing internal standardization techniques.

High concentrations (>4000 mg/L) of chromium, cobalt, copper, mercury, molybdenum, nickel, or silver may cause analytical interferences in AA-gaseous hydride and silver diethyldithiocarbamate (SDDC) spectroscopy methodologies (see Table 13.3.1) due to generation of arsine.

13.4 FORENSICS TECHNIQUES
13.4.1 Elemental Ratios
Often, another metal found associated with arsenic can give a clue to the ultimate origin of the arsenic. Arsenic in soil that resulted from smelter emissions is generally accompanied by the primary smelted metals, such as lead, copper, or zinc. For example, Kimbrough and Suffet (1995) describe the use of metal ratios, including arsenic, to identify soils affected by lead smelter emissions. Arsenic resulting from a pesticide or herbicide source may also be accompanied by lead, but the concentrations of arsenic relative to lead differ when the ultimate source is a pesticide compared to when the ultimate source is smelter emissions. Arsenic that results from wood-treatment may be accompanied by chromium and copper. Uncertainty exists in any forensic analysis involving metals ratios because the picture can be clouded by other sources of the accompanying metals, such as lead-based paints, and by the passage of time, allowing for differential transport and fate of metals. Despite the uncertainty, we describe three examples here where metal ratios proved useful in establishing an ultimate source of arsenic.

Folkes et al. (2001) described an investigation of elevated off-site soil arsenic levels surrounding the Globe Smelter in Denver, Colorado. Arsenic was measured in soil at residential sites near the smelter in order to delineate areas impacted by historic smelter emissions. However, as an unexpected result, investigators found that, in many yards, elevated arsenic concentrations in soil were due to historical pesticide application rather than, or in addition to, the smelter. Folkes et al. identified the herbicide PAX as

the likely contributor. PAX was a crabgrass killer used extensively in the Denver area in the 1950s and 1960s. The authors showed a favorable comparison of the relationship between lead and arsenic in several soils distant from the smelter to the ratio of lead and arsenic that would be expected based on the PAX formulation. This allowed them to conclude that PAX was the likely source of elevated arsenic distant from the smelter, in some yards contributing to arsenic concentrations in excess of 1000 mg/kg.

Soils surrounding the site of another historic smelter operation in the mid-continent of the United States area showed significant correlation between lead and arsenic (both believed to have been emitted in smelter emissions). Figure 13.4.1 shows the relationship of lead and arsenic in 1140 soil samples taken in the residential neighborhoods geographically contiguous to the facility. Figure 13.4.1 shows a strong linear correlation between lead and arsenic for the majority of the samples, with a slope that mirrors the slope seen in samples taken from the facility soils (not shown). However, there are also a number of elevated arsenic concentrations that are unaccompanied by elevated lead. This observation suggests the potential for an additional source of arsenic in some soils, possibly due to residential use of pesticides. The pesticide PAX, identified by Folkes et al. (2001) in the Denver area, has a Pb/As ratio of 0.289. Using this ratio as an example, a hypothetical yard with a background soil Pb concentration of 250 mg/kg and a variable rate of PAX application will show Pb/As concentrations that trend along the leftmost line shown in Figure 13.4.1 (labeled "PAX"). This hypothetical line representing Pb/As concentrations consistent with PAX applications appears to roughly describe some of the elevated arsenic levels (note the proximity of the symbols to the lines). Figure 13.4.2 shows the relationship of lead to arsenic on a residential yard basis for the ten yards with the highest arsenic outliers (arsenic concentrations above 100 mg/kg). This figure shows a within-yard trend that is roughly consistent with the PAX lead–arsenic ratio, suggesting that in-yard variation may stem from variable application rates over time. The yards which show curves in Figure 13.4.2 that are displaced to the right, that is to higher Pb concentrations, are those presumably with a greater influence of the historic site operations. Thus this graphical analysis is helpful in

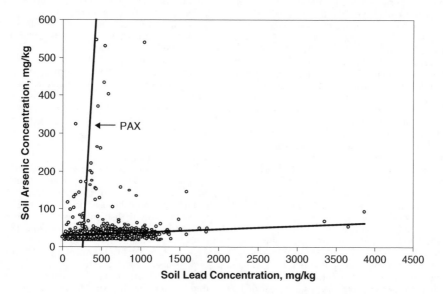

Figure 13.4.1 *Residential soil lead and arsenic in community surrounding a historic smelter site.*

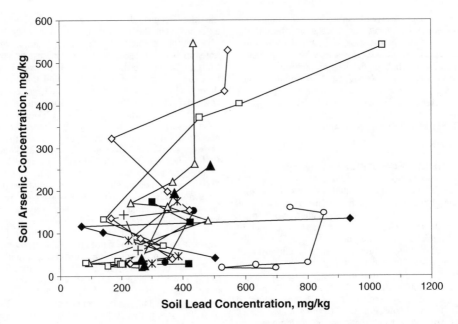

Figure 13.4.2 *Soil lead and arsenic by residential property in community surrounding a historic smelter site. Each symbol corresponds to a residential property, while lines connect multiple samples per property.*

distinguishing the relative contributions of two sources of arsenic: smelter emissions and pesticide applications.

Spills of the CCA solution used to treat wood have occurred at wood treating sites. Additionally, lumber treated with the wood preservative CCA has been widely used to construct outdoor structures such as decks. Researchers have attempted to link arsenic concentrations in soil with these activities mostly by historical knowledge of the location of CCA-treated wood or treatment facilities, followed by arsenic analysis in nearby soils (e.g., Lebow *et al.*, 2004; Stilwell and Gorny, 1997). Attempts to identify CCA as a source of arsenic in soil based solely on the co-location or the ratios of arsenic, copper, and chromium in soil are likely to be hampered by the differing environmental fate of these elements in soil. The use of elemental ratios to identify CCA is challenging due to differences in the environmental mobility of arsenic, copper, and chromium. For instance, arsenic(V) and chromium(III) adhere to clay minerals in soil, while copper has a greater affinity for natural organic matter, and if soil oxidation/reduction and pH conditions cause arsenic and chromium to change valence states to arsenic(III) and chromium(VI), these forms are more highly mobile in soil (reviewed by McLean and Bledsoe, 1992). There are even differences between the types of soils in which arsenic(V) and chromium(III) will be least mobile; arsenic(V) generally is immobilized to the greatest extent in iron-containing soils, while chromium(III) tends to be immobile in most soils regardless of iron content (McLean and Bledsoe, 1992). These differences in the mobility of arsenic, copper, and chromium make it impossible to assign a specific ratio of the three elements that should be present in soil as a marker that CCA-treated wood is the arsenic source in soil. However, if bits of wood (e.g., sawdust, shavings) are present in soil as the chief arsenic source, the relative abundance of arsenic, copper, and chromium would be similar to that in the parent CCA-treated wood, assuming that the elements remain predominantly fixed within the wood.

An example of the behavior of CCA in soil is provided by Lund and Fobian (1991), who performed a highly detailed examination of cores over 2.5 m deep of two Danish soils at wood treating sites where CCA solutions had been historically spilled. Elevated chromium concentrations were entirely retained in the top 35 cm of soil, while elevated arsenic and copper concentrations existed at the surface and at some additional distinct depths, extending as deep as about 2.2 m. The soils' permeability as well as composition (i.e., organic carbon, and iron-, aluminum-, and manganese-oxide content) correlated to the locations where elevated arsenic and copper concentrations were found. These same soil characteristics have been implicated in laboratory studies as the soil characteristics most important for retaining the components of CCA (e.g., Carey *et al.*, 1996; Balasoiu *et al.*, 2001). In another example, Lebow *et al.* (2004) investigated soil alongside and below CCA-treated wood posts in test plots in Wisconsin and Mississippi, which had been maintained outdoors under controlled conditions (i.e., no construction debris was present). Lebow's work showed that arsenic, copper, and chromium concentrations due to leaching near the CCA-treated wood were not correlated with the amount of CCA impregnated in the wood, and that the location of arsenic, copper, and chromium relative to the posts was not predictable, apparently changing with soil properties and local weather (e.g., precipitation amount and pH, freeze–thaw cycles).

13.4.2 Statistical Techniques

Statistical techniques are sometimes effective in identifying whether an environmental data set includes samples affected by one or more sources of a contaminant. Such techniques cannot be used to identify the source of contamination, only to distinguish between sources. However, these techniques are often used to distinguish background levels from contamination.

Log-probability plots of environmental data are commonly used to distinguish sources. The plots are described by Gilbert (1987) and Ott (1995), among others, and are often used to determine the extent to which an environmental data set is lognormally distributed. Use of log-probability plots in separating contamination from background is described by Singh *et al.* (1994). A data set that is perfectly lognormally distributed will appear as a straight line on a log-probability plot. Data sets may depart from perfect lognormal distributions as a result of measurement error (typically at low concentrations approaching the detection limit). Ott (1990) notes that data sets may depart from perfect lognormality at high concentrations as well because there is an upper limit on concentration, either the concentration of the source itself or, theoretically, at a concentration of 100%. As a result, even a data set that is well described by the lognormal distribution may show some curvature at the low and high concentrations when viewed on a log-probability diagram.

Another reason that data sets may show departure from a perfect lognormal distribution is that the data may represent more than one source. Data sampled from two populations (i.e., two sources, such as background and contamination) with concentration ranges that do not overlap, if plotted together on a single log-probability diagram, will appear approximately as two straight line segments with a point of intersection sometimes termed "a point of inflection". More commonly, two populations of data will have concentration ranges that overlap, and the log-probability plot will show two line segments that join in a curve. The curved portion of the plot corresponds to the range of overlapping concentrations between the two populations. When visual inspection of a log-probability plot is insufficient to make judgments about populations of data, Singh *et al.* (1994) have proposed various mathematical techniques to classify samples into two or more populations.

An example is shown in Figure 13.4.3, which shows arsenic concentration in 635 soil samples from a residential area adjacent to a site involved in a historic manufacturing process that used arsenic. Some residential areas received arsenic-contaminated fill from the site several decades ago. Figure 13.4.3 shows that, at concentrations above approximately 20 mg/kg, the data conform to an approximately straight line segment, although with downward curvature at the very highest concentrations. This downward curvature likely represents the upper limit on the source concentration, as described by Ott (1990). Concentrations below approximately 20 mg/kg can also be approximated by a straight line, with a slope that clearly differs from the line segment representing concentrations above 20 mg/kg. Some departure from a straight line is seen at concentrations below about 4 mg/kg, which may be a result of measurement error. The figure suggests that there may be two populations of data present in the data set (and therefore possibly two sources of arsenic), corresponding to concentrations above and below approximately 20 mg/kg. Given what we know about typical natural background levels of arsenic, this data set can be used to distinguish between those residential yards (or samples) affected only by background levels of arsenic *vs.* those affected by contamination.

13.4.3 Abundance of Arsenic Compounds

The relative abundance of the arsenic compounds listed in Figure 13.2.1 can be used, in part, to distinguish the probable source of arsenic in environmental samples, under certain circumstances. Arsenic compounds are interconverted in the environment through the metabolism of microorganisms, which are able to methylate and demethylate arsenic compounds (e.g., Castlehouse *et al.*, 2003; Shariatpanahi *et al.*, 1981). Methylation and demethylation of arsenic compounds do not occur through abiotic means (e.g., hydrolysis, photolysis). Thus, in order for interconversion to occur, the arsenic compound must be bioavailable. In natural surface waters, arsenic compounds are bioavailable and the action of microorganisms results in seasonal arsenic cycling,

with the occurrence of organic forms often associated with algal blooms (ATSDR, 2000). In groundwater, these types of microorganisms are generally absent, so interconversion does not occur. In soil, microorganisms are present, and interconversion can occur to the portion of arsenic that is free, and thus bioavailable, in pore water. However, the portion bound to soil (e.g., *via* irreversible sorption) is not bioavailable and thus is not interconverted. In most soils, especially those containing iron and aluminum oxide clays (McLean and Bledsoe, 1992), arsenic compounds are readily sorbed and thus made unavailable for metabolism (e.g., NRC, 2003). Thus in groundwater, and to some extent in soil, one can identify the predominant arsenic compounds and use the information as evidence of the original source material.

This technique would be useful for distinguishing an anthropogenic organic arsenic source from background arsenic concentrations, because natural background arsenic in soil and groundwater is typically inorganic. Some potential organic arsenic sources for which this method would be useful are arsenic in seaweed applied to land as fertilizer (i.e., arsenosugars), arsenic in animal manures applied to land as fertilizer (i.e., roxarsone or arsanilic acid), or arsenic in herbicides such as MSMA, CAMA, or cacodylic acid.

Identifying a likely arsenic source by assessing the relative abundance of the arsenic compounds present in a sample requires the ability to separately distinguish and quantify arsenic compounds in environmental samples. In soil, the use of gentle extraction followed by separation of the various compounds and conventional analysis (e.g., water extraction, ion chromatographic separation, and detection by ICP-AES) may be used. Numerous specific methods of this type have been developed and used by researchers to distinguish arsenic forms in environmental samples (e.g., Bissen and Frimmel, 2000; Naidu *et al.*, 2000; Gallardo *et al.*, 2001; Vassileva *et al.*, 2001; Mazan *et al.*, 2002; Milstein *et al.*, 2002; Pizzaro *et al.*, 2003; Kahakachchi *et al.*, 2004). If the total arsenic concentration is relatively high – greater than 100 mg/kg in soil – then the use of XAFS methods described in Section 13.3.6 may be possible.

An example of the use of this technique is provided by Bednar *et al.* (2002) who investigated soils in Mississippi and Arkansas where the herbicide MSMA had been used or where spills were suspected. Organic arsenical herbicides of low toxicity such as MSMA have remained registered for use in the Unites States even after the more toxic inorganic arsenical pesticides such as lead arsenate were banned. When MSMA is released into the environment and is in the presence of water, it dissociates to form MMA^- and Na^+. Bednar *et al.* (2002) measured significant MMA concentrations in puddles on soil at an airstrip where MSMA had been historically spilled during loading into crop dusters, indicating that transformation of parent MMA was slow or nonexistent and thus forensic identification of the arsenic source was possible. In these same samples, inorganic arsenic concentrations measured were within background levels, consistent with the conclusion that elevated arsenic levels in the form of MMA resulted from MSMA spills and minimal transformation of MMA occurred. Bednar *et al.* investigated one soil where significant MMA was found in puddles on the soil, but only inorganic arsenic was found in significant quantities when the soil was extracted using several different liquids in the laboratory (69 µg/L of arsenic was the maximum concentration, when the soil was extracted using a phosphate solution). The authors attributed this result to the transformation of MMA to inorganic arsenic. However, that 69 µg/L in the soil extract translates to 0.69 mg/kg of arsenic removed from the soil. That concentration is well within the range

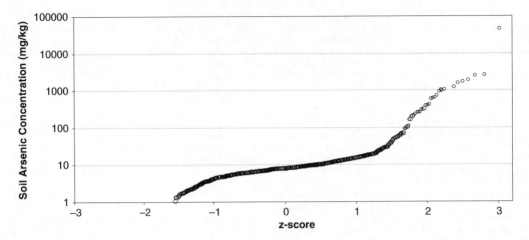

Figure 13.4.3 *Log-probability plot of soil arsenic concentrations in residential soils surrounding a historic manufacturing site.*

of natural background arsenic concentrations in soil and is consistent with the conclusion that these arsenic compounds are sufficiently stable in soil to be useful in forensic analysis.

The persistence of MMA in the environment was also demonstrated at the Crystal Chemical site, located in West Houston, texas, one of the first National Priority List sites in the United States. Crystal Chemical, a former arsenical herbicide manufacturer, was closed in 1980 after several documented spills of MSMA, DSMA, DMA, and TMAO, which were all compounds that were handled and stored at the site. In 1997, 17 years after Crystal Chemical's closure, Kuhlmeier (1997) analyzed soil samples obtained from the site and found that soil contained a mixture of arsenic compounds, including 34.5% MMA and 1.7% DMA. The amount and types of transformation of arsenic compounds that may have taken place at the site are unclear, however, because the relative proportion of MMA, DMA, and inorganic arsenic forms that were originally spilled in soil are unknown. This finding does indicate that organic arsenic compounds can persist in soil for long periods of time, allowing environmental researchers to use the identity of arsenic compounds found in soil as one line of evidence for the initial arsenic source.

There are specific examples of organic arsenic compounds that are added to soil through human activities for which this method could be useful for forensic determinations. Roxarsone, an additive to chicken feed, which is excreted largely unchanged in chicken litter, may also be detectable in soil extracts, allowing identification of the arsenic source when total arsenic concentrations in soil are elevated. Roxarsone is stable in dry or sterilized compost of chicken waste, but can be demethylated in moist conditions favoring microbial activity (Garbarino *et al.*, 2003). Evidence from the laboratory suggests that roxarsone is susceptible to photolysis (Bednar *et al.*, 2003), but photolysis is only relevant in the topmost surface of soil, because light does not effectively penetrate into soil. Seaweed used as fertilizer contains arsenosugars (usually arsenoribofuranosides). Analytical methods for the detection of organic arsenic compounds may be less useful in identifying arsenosugars, however, because they have been shown to degrade in soil to form inorganic arsenic (Castlehouse *et al.*, 2003).

REFERENCES

Abdelghani, A., Anderson, A., Englande, A.J., Mason, J.W., and Dekernion, P. (1977). Demethylation of MSMA by soil microorganisms. In *Trace Substances in Environmental Health-Part XI* (D.D. Hemphill, ed.). Columbia: University of Missouri, pp. 419–426.

Agency for Toxic Substances and Disease Registry (ATSDR). (2000). *Toxicological Profile for Arsenic.*

Alden, J.C. (1983). The continuing need for inorganic arsenical pesticides. In *Arsenic: Industrial, Biomedical, Environmental Perspectives* (W.H. Lederer and R.J. Fensterheim, eds), New York: Van Nostrand Reinhold Company, pp. 63–71.

American Wood Preservers' Association (AWPA). (2002). *Standards 2002.*

Azcue, J.M. and Nriagu, J.O. (1994). Arsenic: Historical perspectives. In *Advances in Environmental Science and Technology, Arsenic in the Environment Part 1: Cycling and Characterization* (J.O. Nriagu, ed.). New York: John Wiley & Sons, Inc., pp. 1–16.

Balasoiu, C.F., Zagury, G.J., and Deschenes, L. (2001). Partitioning and speciation of chromium, copper, and arsenic in CCA-contaminated soils: Influence of soil composition. *Sci. Total Environ.*, 280: 239–255.

Bearman, G. (ed.). (1989). *Seawater: Its Composition, Properties and Behaviour.* The Open University, Pergamon Press.

Bednar, A.J., Garbarino, J.R., Ranville, J.F., and Wildeman, T.R. (2002). Presence of organoarsenicals used in cotton production in agricultural water and soil of the southern United States. *J. Agric. Food Chem.*, 50: 7340–7344.

Bednar, A.J., Garbarino, J.R., Ferrer, I., Rutherford, D.W., Wershaw, R.L., Ranville, J.F., and Wildeman, T.R. (2003). Photodegradation of roxarsone in poultry litter leachates. *Sci. Total Environ.*, 302: 237–245.

Bissen, M. and Frimmel, F.H. (2000). Speciation of As(III), As(V), MMA and DMA in contaminated soil extracts by HPLC-ICP/MS. *Fresenius J. Anal. Chem.*, 367: 51–55.

Boerngen, J.G. and Shacklette, H.T. (1981). *Chemical Analysis of Soils and Other Surficial Materials of the Conterminous United States (Report and diskette data).* US Geological Survey, USGS Open-File Report 81–197.

Bull, D.C. (2001). The chemistry of chromated copper arsenate II. Preservative-wood interactions. *Wood Science and Technology*, 34: 459–466.

Carey, A.E., Wiersma, G.B., and Tai, H. (1976). Pesticide residues in urban soils from 14 United States cities, 1970. *Pest. Monit.*, 10 (2): 54–60.

Carey, P.L., McLaren, R.G., Cameron, K.C., and Sedcole, J.R. (1996). Leaching of copper, chromium, and arsenic through some free-draining New Zealand soils. *Aust. J. Soil Res.*, 34: 583–597.

Castlehouse, H., Smith, C., Raab, A., Deacon, C., Meharg, A.A., and Feldmann, J. (2003). Biotransformation and accumulation of arsenic in soil amended with seaweed. *Environ. Sci. Technol.*, 37 (5): 951–957.

Chen, M., Ma, L.Q., and Harris, W.G. (2002). Arsenic concentrations in Florida surface soils: Influence of soil type and properties. *Soil Sci. Soc. Am. J.*, 66: 32–640.

Davis, A., Sherwin, D., Ditmars, R., and Hoenke, K.A. (2001). An analysis of soil arsenic Records of Decision. *Environ. Sci. Technol.*, 35 (12): 2401–2406.

DeVenzio, H. (1998). Happy birthday, Sonti: A marketing history of CCA-treated wood. In *Proceedings: Ninety-fourth Annual Meeting of the American Wood-Preservers' Association, Marriott's Camelback Inn*, Scottsdale, Arizona, May 17–19, 1998, pp. 243–248.

Dragun, J. and Chiasson, A. (1991). *Elements in North American Soils*. Hazardous Materials Control Resources Institute, Greenbelt, Maryland.

Drever, J.I. (1982). *The Geochemistry of Natural Waters*. Englewood Cliffs: Prentice-Hall, Inc.

Florida Department of Environmental Protection (FDEP). (2002). *Stormwater Non-Point Source Management. Cattle Dipping Vats*. Bureau of Waste Cleanup, Tallahassee, Florida.

Florida Institute of Phosphate Research (FIPR). (2001). *Is Florida the only place to get phosphate?* Available at http://www.fipr.state.fl.us/southb_is_florida_the_only_place_to_get_phosphate.htm.

Folkes, D.J., Kuehster, T.E., and Litle, R.A. (2001). Contributions of pesticide use to urban background concentrations of arsenic in Denver, Colorado. *Env. Forensics*, 2 (2): 127–140.

Foster, A.L. (2001). *Synchrotron—based spectroscopic studies of metal species in solid phases: The case of arsenic*. US Geological Survey Workshop: Arsenic in the environment. February 21–22, 2001. Denver, Colorado.

Gallardo, M.V., Bohari, Y., Astruc, A., Potin-Gautier, M., and Astruc, M. (2001). Speciation analysis of arsenic in environmental solids Reference Materials by high-performance liquid chromatography-hydride generation-atomic fluorescence spectrometry following orthophosphoric acid extraction. *Anal. Chim. Acta*, 441: 257–268.

Garbarino, J.R., Bednar, A.J., Rutherford, D.W., Beyer, R.S., and Wershaw, R.L. (2003). Environmental fate of roxarsone in poultry litter. I. Degradation of roxarsone during composting. *Environ. Sci. Technol.*, 37: 1509–1514.

Ghazi, A.M. (2005). Laser Ablation ICP-MS: A New Elemental and Isotopic Ratio Technique in Environmental Forensic Investigation. *Env. Forensics*, 6: 7–8.

Gilbert, R.O. (1987). *Statistical Methods for Environmental Pollution Monitoring*. New York: John Wiley & Sons, Inc.

Jackson, B.P. and Bertsch, P.M. (2001). Determination of arsenic speciation in poultry wastes by IC-ICP-MS. *Environ. Sci. Technol.*, 35, 4868–4873.

Kahakachchi, C., Uden, P.C., and Tyson, J.F. (2004). Extraction of arsenic species from spiked soils and standard reference materials. *Analyst*, 129: 714–718.

Kimbrough, D.E. and Suffet, I.H. (1995). Off-site forensic determination of airborne elemental emissions by multi-media analysis: A case study at two secondary lead smelters. *Environ. Sci. Technol.*, 29 (9): 2217–2221.

Konefes, J.L. and McGee, M.K. (1996). Old cemeteries, arsenic, and health safety. *Cultural Resource Management*, 19 (10): 15–18.

Konefes, J.L. and McGee, M.K. (2001). Old cemeteries, arsenic, and health safety. In *Dangerous Places, Health, Safety, and Archaeology* (D.A. Poirer and K.L. Feder, eds). Westport: Bergin and Garvey.

Kuhlmeier, P.D. (1997). Partitioning of arsenic species in fine-grained soils. *J. Air & Waste Management Association*, 47: 481–490.

Lebow, S., Foster, S., and Evans, J. (2004). Long-term soil accumulation of chromium, copper, and arsenic adjacent to preservative-treated wood. *Bull. of Contam. and Toxicol.*, 72: 225–232.

Loebenstein, J.R. (1994). *The Materials Flow of Arsenic in the United States*. US Department of the Interior. Bureau of Mines Information Circular 9382.

Lund, U. and Fobian, A. (1991). Pollution of two soils by arsenic, chromium and copper, Denmark. *Geoderma*, 49: 83–103.

Manning, B.A., Fendorf, S.E., Bostick, B., and Suarez, D.L. (2002). Arsenic(III) oxidation and arsenic(V) adsorption reactions on synthetic Birnessite. *Environ. Sci. Technol.*, 36: 976–981.

Mason, B. (1966). *Principles of Geochemistry*. New York: John Wiley & Sons, Inc.

Matera, V. and Le Hecho, I. (2001). Arsenic behavior in contaminated soils: Mobility and speciation. In *Heavy Metals Release in Soils* (H.M. Selim and D.M. Sparks, eds). Boca Raton: CRC Press, pp. 207–235.

Mazan, S., Cretier, G., Gilon, N., Mermet, J.-M., and Rocca, J.-L. (2002). Porous graphitic carbon as stationary phase for LC-ICPMS separation of arsenic compounds in water. *Anal. Chem. (Wash.)*, 74: 1281–1287.

McLean, J.E. and Bledsoe, B.E. (1992). Behaviour of Metals in Soils. United States Environmental Protection Agency. Groundwater Issue. EPA/540/S-92/018.

Milstein, L.S., Essader, A., Pellizzari, E.D., Fernando, R.A., and Akinbo, O. (2002). Selection of a suitable mobile phase for the speciation of four arsenic compounds in drinking water samples using ion-exchange chromatography coupled to inductively coupled plasma mass spectrometry. *Environ. Int.*, 28: 277–283.

Morin, G., Juillot, F., Casiot, C., Bruneel, O., Personne, J.-C., Elbaz-Poulichet, F., Leblanc, M., Ildefonse, P., and Calas, G. (2003). Bacterial formation of tooeleite and mixed arsenic(III) or arsenic(V)-iron(III) gels in the carnoulès acid mine drainage, France. A XANES, XRD, and SEM study. *Environ. Sci. Technol.*, 37: 1705–1712.

Murphy, E.A. and Aucott, M. (1998). An assessment of the amounts of arsenical pesticides used historically in a geographic area. *Sci. Total Environ.*, 218: 89–101.

Naidu, R., Smith, J., McLaren, R.G., Stevens, D.P., Sumner, M.E., and Jackson, P.E. (2000). Application of capillary electrophoresis to anion speciation in soil water extracts: II. Arsenic. *Soil Sci. Soc. Am. J.*, 64: 122–128.

National Institute of Standards and Technology (NIST). (1993). Advanced Technology Program Status Report: *Gallium Arsenide: A Faster Alternative to Silicon for Microprocessors and Telecommunications Applications*. Available at http://statusreports-atp.nist.gov/basic_form.asp.

National Research Council (NRC). (2003). *Bioavailability of Contaminants in Soils and Sediments*. Washington, DC: The National Academies Press.

Nico, P.S., Fendorf, S.E., Lowney, Y.W., Holm, S.E., and Ruby, M.V. (2004). Chemical structure of arsenic and chromium in CCA-treated wood: Implications of environmental weathering. *Environ. Sci. Technol.*, 38: 5253–5260.

Ott, W.R. (1990). A physical explanation of the lognormality of pollutant concentrations. *J. Air Waste Manage. Assoc.*, 40: 1378–1383.

Ott, W.R. (1995). *Environmental Statistics and Data Analysis.* Boca Raton: Lewis Publishers.

Pizzaro, I., Gomez, M., Camara, C., and Palacios, M.A. (2003). Arsenic speciation in biological and environmental samples: Extraction and stability studies. *Anal. Chim. Acta.*, 495: 85–98.

Rieuwerts, J.S., Thornton, I., Farago, M.E., Ashmore, M.R., Fowler, D., Hall, J., Kodz, D., Nemitz, E., Lawlor, A., and Tipping, E. (1999). Critical loads of metals in UK soils: An overview of current research. In *Geochemistry of the Earth's Surface* (H. Armannsson, edition). Amsterdam: A.A. Balkema.

Shacklette, H.T. and Boerngen, J.G. (1984). *Element concentrations in soils and other surficial materials of the conterminous United States.* US Geological Survey. USGS Professional Paper 1270.

Shariatpanahi, M., Anderson, A.C., and Abdelghani, A. (1981). Microbial demethylation of monosodium methanearsonate. In *Trace Substances in Environmental Health-Part X* (D.D. Hemphill, edition). Columbia: University of Missouri, pp. 383–387.

Sheehy, J.W. and Jones, J.H. (1993). Related Articles, Links Assessment of arsenic exposures and controls in gallium arsenide production. *Am. Ind. Hyg. Assoc. J.*, 54 (2): 61–69.

Singh, A., Singh, A.K., and Flatman, G. (1994). Estimation of background levels of contaminants. *Mathematical Geology*, 26: 361–388.

Smith, P.G., Koch, I., Gordon, R.A., Mandoli, D.F., Chapman, B.D., and Reimer, K.J. (2005). X-ray absorption near-edge structure analysis of arsenic species for application to biological environmental samples. *Environ. Sci. Technol.*, 39: 248–254.

Spongberg, A.L. and Becks, P.M. (2000). Inorganic soil contamination from cemetery leachate. *Water Air Soil Pollut.*, 117 (1–4): 313–327.

Stilwell, D.E. and Gorny, K.D. (1997). Contamination of soil with copper, chromium and arsenic under decks built form pressure treated wood. *Bull. Environ. Contam. Toxicol.*, 58: 22–29.

Thomas, J.E. (1998). *Distribution, movement, and extraction of arsenic in selected Florida soils.* Ph.D. Dissertation, University of Florida.

United States Department of Agriculture (USDA). (1980). The Biologic and Economic Assessment of Pentachlorophenol, Inorganic Arsenicals, Creosote. Volume I: Wood Preservatives. Cooperative Impact Assessment Report. Technical Bulletin Number 1658-1.

United States Environmental Protection Agency (USEPA). (1985). EPA Superfund Record of Decision: Morris Arsenic Dump. EPA/ROD/R05-85/015.

United States Environmental Protection Agency (USEPA). (1986). Test Methods for Evaluating Solid Waste, Physical/Chemical Methods, Updates I, II, IIA, IIB, III, and IIIA. SW-846. NTIS Publication No. PB97-156111 or GPO Publication No. 955-001-00000-1. Office of Solid Waste. Washington, DC.

United States Environmental Protection Agency (USEPA). (1995). EPA Office of Compliance Sector Notebook Project Profile of the Electronics and Computer Industry. EPA/310-R-95-002. Office of Enforcement and Compliance Assurance. Washington, DC.

United States Environmental Protection Agency (USEPA). (1998a). Method 7010. Graphite Furnace Atomic Absorption Spectrophotemetry. In SW-846, January.

United States Environmental Protection Agency (USEPA). (1998b). Method 6020A. Inductively Coupled Plasma—Mass Spectrometry. In SW-846, January.

United States Environmental Protection Agency (USEPA). (2000a). Arsenic Occurrence in Public Drinking Water Supplies. EPA-815-R-00-023. Office of Water. Washington, DC.

United States Environmental Protection Agency (USEPA). (2000b). Method 6010C. Inductively Coupled Plasma Atomic Emission Spectrometry. In SW-846, November.

United States Environmental Protection Agency (USEPA). (2001). Method 1632A. Chemical Speciation of Arsenic in Water and Tissue by Hydride Generation Quartz Furnace Atomic Absorption Spectrometry. EPA-821-R-01-006. January.

United States Geological Survey (USGS). (2005). Mineral Commodity Summary: Arsenic. Available at http://minerals.usgs.gov/minerals/pubs/commodity/arsenic/.

Vassileva, E., Becker, A., and Broekaert, J.A.C. (2001). Determination of arsenic and selenium species in groundwater and soil extracts by ion chromatography coupled to inductively coupled plasma mass spectrometry. *Anal. Chim. Acta.*, 441, 135–146.

Von Endt, D.W., Kearney, P.C., and Kaufman, D.D. (1968). Degradation of MSMA by soil microorganisms. *J. Agr. Food Chem.*, 16 (1): 17–20.

Washington State Department of Ecology (WSDE). (1999). Final Report Screening Survey for Metals and Dioxins in Fertilizer Products and Soils in Washington State. Ecology Publication No. 99–309.

Welch, A.H., Westjohn, D.B., Helsel, D.R., and Wanty, R.B. (2000). Arsenic in ground water of the United States: Occurrence and geochemistry. *Ground Water*, 38: 589–604.

Woolson, E.A. and Kearney, P.C. (1973). Persistence and reactions of 14C-DMA in soils. *Environ. Sci. Technol.*, 7: 47–50.

Woolson, E.A., Aharonson, N., and Iadevaia, R. (1982). Application of the high-performance liquid chromatography-flameless atomic absorption method to the study of alkyl arsenical herbicide metabolism in soil. *J. Agric. Food Chem.*, 30, 580–584.

Zhu, Y.-J., Olson, N., and Beebe, Jr., T.P. (2001). Surface chemical characterization of 2.5 μm particulates (PM2.5) from air pollution in Salt Lake City using TOF-SIMS, XPS, and FTIR. *Environ. Sci. Technol.*, 35: 3113–3121.

14

Dioxins and Furans

Walter J. Shields, Yves Tondeur, Laurie Benton,
and Melanie R. Edwards

Contents

The term "dioxins" refers to chlorinated dibenzo-*p*-dioxins (CDDs) and "furans" to chlorinated dibenzofurans (CDFs). Sometimes "dioxin" also refers to the most studied and toxic of the dioxins, 2,3,7,8-tetrachlorodibenzo-*p*-dioxin (2,3,7,8-TCDD). CDD/CDFs are not created intentionally, but are produced inadvertently by a number of human activities including chemical manufacturing and incomplete combustion, as well as by natural processes such as forest fires and volcanoes.

CDD/CDFs are among the most studied chemicals in terms of their formation processes, environmental occurrence, and toxicity. Chemical fingerprinting studies have been published for the purposes of identifying potential sources of CDD/CDFs in air, soil, sediments, water, and tissue samples.

In this chapter, the physical and chemical properties of CDD/CDFs and the toxicity equivalent (TEQ) methods are reviewed. We describe the historical development of analytical methods and present some of the pitfalls of using historical data in chemical fingerprinting. We provide an overview of the natural and anthropogenic sources of CDD/CDFs then outline the general approaches used in forensic studies illustrated with three case studies.

14.1 PHYSICAL AND CHEMICAL PROPERTIES

CDDs and CDFs have a triple-ring structure that consists of two benzene rings connected by either one or two oxygens (CDFs and CDDs respectively). (Figure 14.1.1). These molecules have eight possible positions where substitution by a halogen such as chlorine can occur. Of environmental interest are the CDD/CDFs with four or more chlorines and specifically, those molecules with chlorine atoms on the 2,3,7, and 8 ring positions because of their toxicity. CDD/CDF homologues refer to compounds with the same number of chlorine atoms, regardless of position. For example, as shown in Table 14.1.1, there are 22 possible tetrachlorodibenzo-*p*-dioxin (TCDD) isomers within the TCDD homologue class. Only one of these, the 2,3,7,8-TCDD isomer, is considered toxic. The term "congener" refers to any individual CDD/CDF compound, regardless of homologue class. There are 75 possible congeners of CDDs and 135 possible CDF congeners.

CDD/CDFs are strongly lipophilic with very low water solubility. They have similar physical and chemical properties and are typically found as complex mixtures. They are primarily associated with particles and organic matter in the water column and with organic matter in soils and sediments. Once sorbed to soil, CDD/CDFs exhibit little potential for significant leaching or volatilization. In the atmosphere, CDD/CDFs partition between the particles and the gas phase, with higher vapor pressure congeners (i.e., the less chlorinated congeners) found to a greater extent in the gas phase. CDD/CDFs are very stable compounds under most environmental conditions, with the exception of atmospheric photooxidation and photolysis of nonsorbed species in the gaseous phase or at the soil or water–air interface.

CDD/CDF concentrations are often expressed as 2,3,7,8-TCDD TEQ concentrations by multiplying the concentration of each of the 17 2,3,7,8-substituted congeners by its respective toxicity equivalent factor (TEF) as shown in Table 14.1.2. Studies conducted in the 1990s, and even some current laboratory studies, use the so-called "international toxicity equivalent factors" (I-TEFs) adopted by the US Environmental Protection Agency (EPA) in 1989 (US EPA, 1989) to calculate TEQs (designated as I-TEQs). In 1998, the TEFs were revised by a consensus scientific committee sponsored by the World Health Organization (WHO). These TEFs (Van den Berg *et al.*, 1998) are recommended by US EPA (2003) and the resultant TEQs are termed "WHO-TEQs" in this chapter. For a comprehensive review of the literature on the environmental chemistry of CDD/CDFs, the reader is referred to a 2003 draft publication by EPA titled *Exposure and Human Health Reassessment of 2,3,7,8-Tetrachlorodibenzo-*p*-dioxin (TCDD) and Related Compounds*.

14.2 ANALYTICAL METHODS

When measuring CDD/CDFs, the motion of the field of analytical determinations has not been a linear succession of simple steps so much as it has been, and continues to be, a stimulating evolution (Buser, 1991). Initially, exposure-focused interests centered on a single molecule, 2,3,7,8-TCDD, as did the measurement techniques. Over the years, as an appreciation for the significant role played

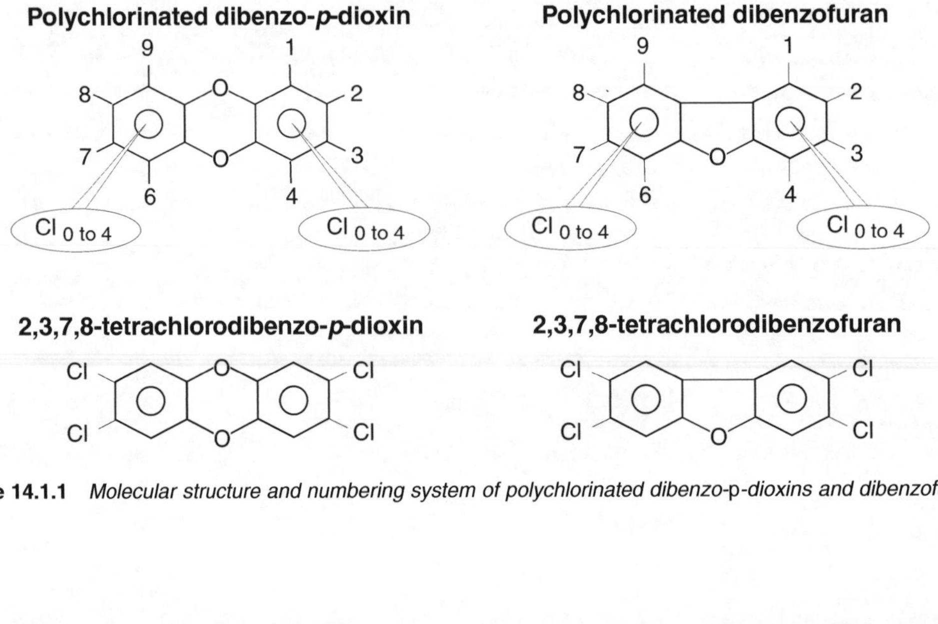

Figure 14.1.1 *Molecular structure and numbering system of polychlorinated dibenzo-p-dioxins and dibenzofurans.*

Table 14.1.1 *Homologue Classes of CDD/CDFs*

Homologue Class	Abbreviation	Chlorines	Number of Isomers
Polychlorinated Dibenzo-*p*-dioxins (CDDs)			
Monochlorodibenzo-*p*-dioxins	MCDD	1	2
Dichlorodibenzo-*p*-dioxins	DCDD	2	10
Trichlorodibenzo-*p*-dioxins	TrCDD	3	14
Tetrachlorodibenzo-*p*-dioxins	TCDD	4	22
Pentachlorodibenzo-*p*-dioxins	PeCDD	5	14
Hexachlorodibenzo-*p*-dioxins	HxCDD	6	10
Heptachlorodibenzo-*p*-dioxins	HpCDD	7	2
Octachlorodibenzo-*p*-dioxin	OCDD	8	1
Total			75
Polychlorinated Dibenzofurans (CDFs)			
Monochlorodibenzofurans	MCDF	1	4
Dichlorodibenzofurans	DCDF	2	16
Trichlorodibenzofurans	TrCDF	3	28
Tetrachlorodibenzofurans	TCDF	4	38
Pentachlorodibenzofurans	PeCDF	5	28
Hexachlorodibenzofurans	HxCDF	6	16
Heptachlorodibenzofurans	HpCDF	7	4
Octachlorodibenzofuran	OCDF	8	1
Total			135

Table 14.1.2 *2,3,7,8-Substituted CDD/CDF Congeners, TEFs, and Homologue Classes*

Congener	NATO-89 I-TEFs	WHO-98 TEFs	Figure Order[a]	Homologue Class	Figure Order[a]
2,3,7,8-TCDD	1	1	1	Total TCDDs	1
1,2,3,7,8-PeCDD	0.5	1	2	Total PeCDDs	2
1,2,3,4,7,8-HxCDD	0.1	0.1	3	Total HxCDDs	3
1,2,3,6,7,8-HxCDD	0.1	0.1	4	Total HpCDDs	4
1,2,3,7,8,9-HxCDD	0.1	0.1	5	OCDD	5
1,2,3,4,6,7,8-HpCDD	0.1	0.01	6	Total TCDFs	6
OCDD	0.001	0.0001	7	Total PeCDFs	7
2,3,7,8-TCDF	0.1	0.1	8	Total HxCDFs	8
1,2,3,7,8-PeCDF	0.05	0.05	9	Total HpCDFs	9
2,3,4,7,8-PeCDF	0.5	0.5	10	OCDF	10
1,2,3,4,7,8-HxCDF	0.1	0.1	11		
1,2,3,6,7,8-HxCDF	0.1	0.1	12		
1,2,3,7,8,9-HxCDF	0.1	0.1	13		
2,3,4,6,7,8-HxCDF	0.1	0.1	14		
1,2,3,4,6,7,8-HpCDF	0.01	0.01	15		
1,2,3,4,7,8,9-HpCDF	0.01	0.01	16		
OCDF	0.001	0.0001	17		

Notes: TEF—toxicity equivalence factor
NATO-89—refers to the so-called "international TEFs" (I-TEFs) reported in NATO/CCMS (1988)
WHO-98—refers to the World Health Organization TEFs (WHO-TEFs) reported in Van den Berg *et al.* (1998)
[a]Figure order refers to the order of these congeners and homologue classes shown on the horizontal axes of Figure 14.4.1.

by other compounds developed, a shift in the scope of the analytical methodologies occurred to include dioxin-like compounds (DLC), which included the co-planar polychlorinated biphenyls (PCBs). When a subset of the DLCs attracted special attention from an exposure standpoint, the TEQ concept emerged. By the time this concept was introduced, most of the toxic CDD/CDF congeners had been synthesized in both native and ^{13}C fully labeled forms.

From a regulatory point of view, the early rules were primarily based on 2,3,7,8-TCDD. Regulations evolved in parallel with regulatory and scientific interests and the simultaneous evolution of methodologies. Today, most international standards are TEQ based. To support the

various international rules and regulations, a number of analytical protocols describing sampling, sample preparation, analysis, and validation procedures were developed throughout the 1980s and 1990s (US EPA, 1986, 1994, 1999a,b; Ballmacher, 2001; Environment Canada, 1997; JIS, 1999). Their current form benefited from years of research and development, as well as refinements with contributions from scientists, who over the past two decades, assisted with important developments in sample preparation and analytical techniques. Thus far, the best analytical results are achieved when pre-analytical sample cleanup, high-resolution gas chromatography (HRGC), and high-resolution mass spectrometry (HRMS) are combined. These

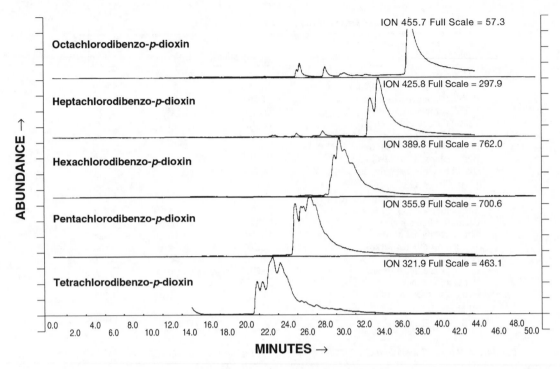

Figure 14.2.1 *Example of low-resolution gas chromatography column technology.* Source: *Eiceman* et al. *(1980). Reprinted with permission. Copyright (1980) American Chemical Society.*

developments were necessary for meeting new demands for improved sensitivity and specificity, which were mostly driven by risk assessment requirements.

Sophisticated and elaborate cleanups involving chemical treatments, gravity flow, and low-pressure chromatography columns became integral parts of the analytical methods. Their function is to isolate the target analytes from co-extractants while removing potential interferences to ensure that extremely small quantities of "dioxins" can be reliably analyzed. When not removed during sample fractionation, for instance, polychlorinated diphenylethers coeluting with CDFs are a source of positive bias even under high-resolution mass spectrometric conditions.

Early studies regarding sample introduction to the gas chromatograph were conducted using packed gas chromatographic (GC) column technology. Low-resolution GC (Figure 14.2.1) is unable to achieve isomer specificity and is therefore unsuited for TEQ determinations. As the risk assessment-driven need for isomer specificity increased, GC capillary columns (Figure 14.2.2) became widely accepted. Currently, no single GC column is capable of achieving isomer-specificity for each of the relevant 2,3,7,8-substituted congeners. Thus, at least two separate GC/MS analyses are required for accurate congener determinations.

Current improvements in chromatographic separations and time-saving procedures will become the norm in the future. For example, the application of molecular recognition technologies (e.g., molecular cavities, baskets, and other nests or hosts in which molecules or "guests" fit and bind selectively and reversibly) and molecularly imprinted polymers as chromatographic phases might become routine analytical tools in standard protocols. Other approaches, based on two-dimensional (2D) GC × GC and comprehensive 2D GC (Ligon and May, 1984; Marriott *et al.*, 2003), provide advanced separation between congeners

while considerably reducing the run time to achieve powerful separations as illustrated in Figure 14.2.3. These new technologies will also require powerful computation, more effective data acquisition, and faster GC/MS systems.

In addition to sample preparation and introduction techniques, detection methods have also evolved from the electron capture detector (ECD), which was hampered by interferences, to the current use of the mass spectrometer (Richardson, 2001). At first, MS was used for confirmatory purposes. Eventually, the success and appeal of combining GC with MS elevated the technology to the method of choice for the analysis of organic pollutants by EPA; a consequence of the 1976 consent decree between EPA and environmental activists, and the need for more scientifically and legally defensible data.

The practice of analytical chemistry evolved from the development of instrumentation into more of a multidisciplinary and interactive, problem-solving discipline. In particular, by means of its versatility for interfacing with other analytical methods, MS turned into a universal method of analysis. It plays a central role in the work surrounding "dioxins," helping the scientific community develop a better understanding of pollution and its consequences, and identifying prevention and treatment options.

The foundations of MS were laid more than a century ago by the pioneering research of J.J. Thomson, who envisioned it as a tool for investigating the structures of molecules. Early GC/MS methods for "dioxins" were based on low-resolution MS (LRMS) using full-scan or selected ion recording techniques. Although early methodologies represented the state-of-the-art technology, their limitations were quickly recognized. For instance, risk assessors were not satisfied with the high detection limits obtained with GC-LRMS (e.g., part per trillion in water; part per billion in soil). LRMS is also susceptible to false positives and

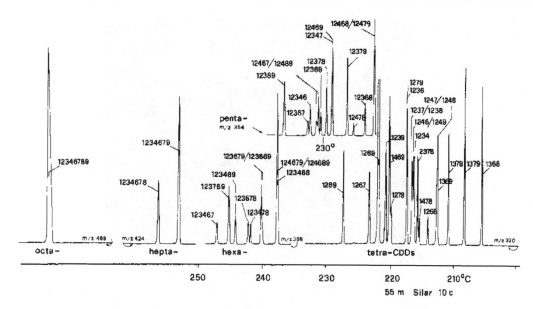

Figure 14.2.2 *Example of high-resolution gas chromatography (GC) based on capillary GC column technology.* Source: *Buser and Rappe (1984). Reprinted with permission. Copyright (1984) American Chemical Society.*

false negatives (Figure 14.2.4). The lack of mass separation further hinders the use of critical standards to monitor the extraction-fractionation efficiencies of specific toxic congeners. For instance, the use of ^{13}C-labeled HxCDs is not likely with LRMS because of interferences caused by native HxCDDFs, which can behave differently from their HxCDF counterparts during sample fractionation. Finally, LRMS does not offer the means to monitor the GC/MS system performance each and every second of the analysis, which can be done in high-resolution GC (HRGC) through the use of the quality control check ions and selected ion current traces (Tondeur *et al.*, 1984).

Baughman and Meselson (1973) demonstrated the value of HRMS for the detection of 2,3,7,8-TCDD in fish tissue extracts. The sample extract was introduced via a solid probe while performing a voltage sweep with the mass spectrometer over the molecular ion region of TCDD. A separation between TCDD and DDT/PCB (commonly found interferences) was possible with a resolving power of 10,000.

The optimization of GC/MS protocols is currently based on the integration of sample extraction and fractionation procedures, sample introduction techniques, and the operation of the mass spectrometer at high resolutions. When considered in the context of stable isotope dilution, these key elements are the basis for the high level of accuracy and reliability as well as the low detection limits (e.g., 1 to 5 pg/L or parts per quadrillion in 1 L water; sub-pg/g in 10 g soil; 0.5 to 5 pg/g lipid-based in 20–30 mL human serum; and 0.01 pg/g in 25 g fish tissue) characteristic of today's methodologies.

Risk assessment drives the need for improved sensitivity, specificity, and selectivity, as well as accuracy and reliability of CDD/CDF analyses. Despite its high cost, isotope-dilution HRMS remains the method of choice. Other techniques based on tandem MS/MS have been shown valuable in some specific applications where the accuracy of HRMS was challenged (Charles *et al.*, 1989). Other MS technologies have been recently described for CDD/CDFs (e.g., quadruple ion storage and time-of-flight) (Focant *et al.*, 2004).

The past several years have seen a renewed interest in rapid and low-cost screening bioassays (e.g., Denison *et al.*, 2002). Even though the significance of bioassays as an analytical tool was first recognized in the 1960s when antibodies were used to measure insulin in plasma by radioimmunoassay, they never achieved the status of recognition and adoption that has occurred following the 1999 Belgian dioxin scare over the contamination of livestock from tainted animal feed. Food safety has special requirements that are not comparable to conventional environmental testing of soil/sediment and wastewater matrices. Time, frequency, and cost are understandably of the essence; thus, the observed push for rapid screening tools. A cost-effective and reliable food safety monitoring program requires a balanced combination of screening and confirmatory assays because bioassays cannot provide molecular-level information. Furthermore, bioassays are incapable of accounting for actual losses of target compounds occurring during the extraction and fractionation procedures of a particular sample. Other limitations are associated with the ongoing assessment of co-extractants that bind the aryl hydrocarbon receptor as well as dioxin antagonists.

14.3 NATURAL AND ANTHROPOGENIC SOURCES

The presence of CDD/CDFs has been documented in "... practically all media including air, soil, meat, milk, fish, vegetation, and human biological samples" (Travis and Hattemer-Frey, 1991). The widespread occurrence of these persistent compounds is likely the result of atmospheric dispersion and deposition of particles resulting from combustion processes, from forest fires to waste incineration to auto exhaust. CDD/CDFs are also unwanted by-products in the manufacture of chlorinated organic compounds such as herbicides and wood preservatives.

The environmental forensic scientist investigating the potential sources of CDD/CDFs must understand the wide variety of sources as well as the range of "normal background" concentrations that would be expected in environments similar to those being evaluated. The purpose of this

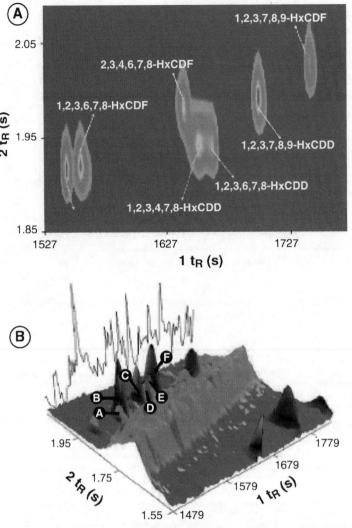

Note:
(A) Expanded section of a GC x GC contour plot of a standard solution containing 1 ng of HxCDD/Fs. The deconvoluted ion current (DIC is based on the sum of the molecular ions corresponding to HxCDD/Fs (m/z 390 + 374). (B) Expanded section of the HxCDD/F region of a GC x GC shade surface plot after injection of the clean-up fraction containing PCDD/Fs isolated from a real fish sample. DIC based on m/z 390 and 374. Concentrations are in the range of 2–3 pg µL^{-1} (A: 1,2,3,4,7,8-HxCDF; B: 1,2,3,7,8,9-HxCDF; C: 2,3,4,6,7,8-HxCDF; D: 1,2,3,4,7,8-HxCDD; E: 1,2,3,6,7,8-HxCDD; F: 1,2,3,7,8,9-HxCDD).

Figure 14.2.3 *Example of two-dimensional gas chromatography contour plots.* Source: *Focant* et al., *(2004). Reprinted with permission. Copyright (2004) Elsevier.*

section is to provide a summary of sources and to present a range of background concentrations reported in environmental media.

There have been many excellent literature reviews and compilations of CDD/CDF sources (e.g., Fiedler *et al.*, 1996; Rappe, 1994; US EPA, 1997, 2000, 2003). US EPA (2005) at the time of writing had updated their "Inventory of sources and environmental releases of dioxin-like compounds in the United States" with data through the year 2000. This is the most comprehensive resource of CDD/CDF sources currently available, with more than 800 references cited, although at the time of this writing the document was available only as an "external review draft" and

the final report may include additional data provided during the peer review and public comment process. The final report and future updates will be posted on EPA's National Center for Environmental Assessment (NCEA) Web site (http://cfpub.epa.gov/ncea/).

The ranking of sources from the 1987, 1995, and 2000 inventories is shown in Figure 14.3.1. The number one category in 2000 was emissions from backyard burn barrels (US EPA, 2005), accounting for approximately 32% of estimated emissions, followed by medical waste incinerators (24%) and municipal waste combustors (5%).

The EPA also maintains a downloadable database of CDD/CDF sources (Table 14.3.1) on the NCEA website.

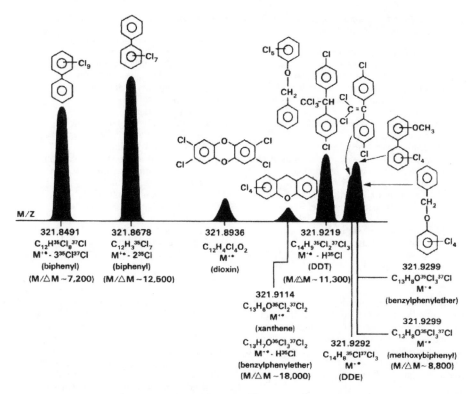

Figure 14.2.4 *Commonly encountered interferences for TCDD when poor separation techniques are employed (e.g., insufficient and ineffective cleanups combined with GC/LRMS). The structures, elemental compositions, and resolving powers necessary to resolve TCDD from the interferences are given.* Source: *Tondeur et al., (1987). Reprinted with permission. Copyright (1987) John Wiley and Sons, Ltd.*

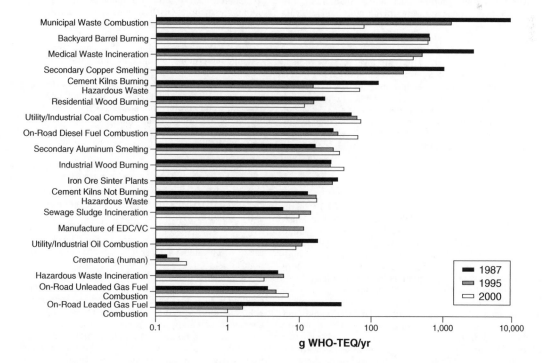

Figure 14.3.1 *Estimates of annual air emissions of CDD/CDF (WHO-TEQ) for 1987, 1995, and 2000.* Source: *Figure 1.9 from US EPA (2005).*

Table 14.3.1 *Sources of CDD/CDFs Emissions and Emission Included within EPA's Database as of March 2001*

Bleached chemical pulp and paper mills	Power generating facilities
Cement kilns burning hazardous waste	Coal-fired electric generating plants
Inlet temperature to APCD >4, 500 °F	Oil-fired electric generating plants
Inlet temperature to APCD <4, 500 °F	Primary ferrous metal smelting
Cement kilns not burning hazardous waste	Sinter production
Crematoria	Coke production
Drum and barrel reclamation facilities	Primary non-ferrous metal smelting
Ferrous foundries	Petroleum refining catalyst regeneration
Hazardous waste incinerators	Residential oil combustion
Industrial boilers burning hazardous waste	Secondary non-ferrous metal smelting
Kraft black liquor recovery boilers	Secondary aluminum smelting
Motor vehicles	Secondary copper smelting
Powered with unleaded gasoline	Secondary lead smelting
Powered with leaded gasoline	Sewage sludge incineration
Diesel powered heavy duty trucks	Scrap electric wire recovery
Municipal solid waste incinerators	Tire combustion
Medical waste incinerators	Industrial wood combustion

Source: US EPA, (2001).

The availability of this electronic database is a valuable resource for source identification studies, but the primary articles should be obtained and reviewed so that the original data can be evaluated for variability of profiles, handling of detection limits, and potential problems (e.g., coelution of isomers in older data sets). Several of the key source categories are discussed later with an emphasis on combustion.

14.3.1 Metals Smelting, Refining, and Process Sources

CDD/CDFs can be formed during various types of ferrous and non-ferrous smelting (both primary and secondary). US EPA (2005) provides congener profiles for emissions from secondary aluminum, copper, and lead smelters, iron ore sinter plants, a scrap wire incinerator, and a drum incinerator.

14.3.2 Manufacturing Sources

Trace amounts of CDD/CDFs can form as by-products from the manufacture of chlorine-bleached wood pulp, chlorinated phenols (e.g., pentachlorophenol [PCP]), PCBs, phenoxy herbicides (e.g., 2,4-dichlorophenoxyacetic acid [2,4-D] and 2,4,5-trichlorophenoxyacetic acid [2,4,5-T]), and chlorinated aliphatic compounds (e.g., ethylene dichloride). Congener profiles are provided in US EPA (2005) for pulp, sludge, and effluent from mills using chlorine bleach process, for technical grade PCP, for 2,4-D salts and esters, and for sewage sludge.

14.3.3 Combustion

Combustion is the primary source of CDD/CDFs to the global environment. CDD/CDFs are generated in waste incineration (e.g., municipal solid waste, sewage sludge, medical waste, and hazardous wastes), fuel combustion (e.g., oil, gasoline, diesel, coal, and wood), other high-temperature sources (such as cement kilns), and poorly controlled or uncontrolled combustion sources (e.g., forest fires, building fires, and open burning of wastes).

Cleverly et al. (1997) reported that combustion sources typically emit all 2,3,7,8-substituted CDD/CDFs, although the relative congener concentrations vary. These authors found that 2,3,7,8-TCDD is usually 0.1 to 1% of total CDD/CDFs in combustion source emissions, with the exception of stack emissions from industrial oil-fired boilers, where the available data indicate that 2,3,7,8-TCDD constitutes an average of 7% of total CDD/CDF emissions.

In evaluating congener profiles for the EPA inventory of sources, Cleverly et al. (1997) noted that OCDD is the dominant congener in some, but not all, combustion emissions. OCDD dominates emissions from mass-burn municipal waste combustors (MWCs) that have dioxin emission controls. It also dominates emissions from industrial oil-fired boilers, industrial wood-fired boilers, unleaded gasoline combustion, diesel fuel combustion in trucks, and sewage sludge incinerators. These authors reported that the dominant congeners for other combustion sources are 1,2,3,4,6,7,8-HpCDF in emissions from mass-burn MWCs equipped with hot-sided electrostatic precipitators, hazardous waste incineration, and secondary aluminum smelters; OCDF in emissions from medical waste incineration and industrial/utility coal-fired boilers; 2,3,4,7,8-PeCDF in cement kilns burning hazardous waste; and 2,3,7,8-TCDF in cement kilns not burning hazardous waste.

Congener profiles for the following combustion sources are provided in US EPA (2005): MWC, medical waste incinerators, furnaces burning hazardous waste, crematoria, sewage sludge incinerators, a tire combustor, vehicle exhaust, wood combustion, power boilers (coal, wood, and oil), cement kilns (burning hazardous and non-hazardous waste), petroleum catalytic reformer units, black liquor recovery boilers, cigarette smoke, landfill flares, and forest fires.

Open burning of yard waste and household trash is not only the number one source of CDD/CDFs in the national inventory (US EPA, 2005), but can also be a source of locally elevated CDD/CDFs immediately downwind from the burn barrel (MOEP 1997; Wevers et al., 2003). CDD/CDF source identification studies in rural communities should include emissions from backyard burning as a potentially significant cause of elevated CDD/CDF concentrations in soils and house dust.

Unlike combustion of diesel or municipal solid waste in a controlled incinerator, the types of waste and combustion conditions in a backyard burn barrel are extremely variable. Therefore, it is difficult to predict what the CDD/CDF concentrations and profile would be in the surface soils or indoor dust of a house next to a burn barrel or a burn pit.

Available studies of CDD/CDF congeners in emissions from backyard burning (Ikeguchi and Tanaka, 2000; Gullet et al., 2001; Lemieux et al., 2003; Wevers et al., 2003; Gönczi et al., 2005; US EPA, 2005) indicate a wide range of congener patterns and more than four orders of magnitude range of TEQ concentrations. For example, in Wevers *et al.* (2003), the congener patterns for burning garden waste and household waste were both dominated by OCDD and HpCDD, but the household waste emissions were also characterized by high relative concentrations of the low to mid-weight CDFs and one HpCDF. In contrast, the profiles from Lemieux *et al.*, (2003) were very low in OCDD and HpCDD.

According to Lemieux *et al.* (2003), "[m]any possible parameters could have a significant influence on CDD/CDF emissions from burn barrels. Many of these parameters could be caused by variations in practice-related variables that would vary from homeowner to homeowner. Some of these parameters include physical condition of waste in the barrel (e.g., fullness of the barrel, degree of compression of the waste, distribution of waste components within the barrel), chemical composition of the waste (e.g., wetness, trace metal content, Cl content, organic vs. inorganic Cl), and combustion conditions resulting from variations in the previously mentioned physical and chemical characteristics."

In the US EPA (2005) review of burn barrel data, the authors state that " . . . the wide variability in test results (from less than 10 to more than 6000 ng I-TEQ$_{DF}$/kg) also indicates that a high degree of CDD/CDF emission variation can be expected due to factors that are not wholly related to waste composition or burning practice, such as waste orientation."

14.3.4 Background Levels

One of the first steps in CDD/CDFs source evaluation studies is to determine if the concentrations in the media of concern are actually above background levels. Site-specific background samples are always preferred, but often unavailable. If site-specific background data are not available, concentrations need to be compared to published values. US EPA (2003) provides a comprehensive review of background

data for soil, sediment, ambient air, water, fish tissue, and a variety of food items as summarized in Table 14.3.2. In addition, background data for indoor house dust from urban neighborhoods (Berry *et al.*, 1993; Wittsiepe *et al.*, 1996; Saito *et al.*, 2003) indicate a WHO-TEQ range of 2.1–270 ng/kg for 27 samples, with an arithmetic mean of 45 ng/kg (undetected results set to one half the detection limit). It is important to note that domestic laundry activities (e.g., dryerlist) can be an important source of CDD/Fs in indoor house dust.

14.4 FORENSIC TECHNIQUES

Environmental forensic investigations are most effective when they rely upon multiple tools. Chemical fingerprinting is one tool and is often useful in identifying (or eliminating) potential sources. However, there must be a plausible transport pathway from the source to a receptor in order to conclude that a fingerprint is a result of that source.

Chemical fingerprinting is a well-established technique for distinguishing different sources of contamination in the environment. It is particularly well suited for work with families of organic compounds that occur together, such as PCBs, polycyclic aromatic hydrocarbons, and CDD/CDFs. This is because (1) these classes contain many individual compounds, which together comprise a compositional pattern, also referred to as a "profile," a "signature," or a "fingerprint" and (2) the relative concentrations and/or ratios of an individual profile can be used as a marker of the original source material. However, because organic compounds can be transformed in the environment through chemical weathering and biological degradation, segments of the original patterns can be altered. Another complicating factor is that multiple sources often mix together and mask the individual signatures.

Environmental transformations must be considered. The environmental forensic scientist needs to take these "real

Table 14.3.2 *Summary of North American CDD/CDF WHO-TEQ Levels in Environmental Media and Food*

Media	Units	Sample Size (n)	Mean[a]	Minimum	Maximum
Urban soil	ppt	270	9.3	2	21
Rural soil	ppt	354	2.7	0.1	6
Sediment	ppt	11	5.3	<1	20
Urban air	pg/m^3	106	0.12	0.03	0.2
Rural air	pg/m^3	60	0.013	0.004	0.02
Freshwater fish and shellfish	ppt	289	1.0	NA	NA
Marine fish and shellfish	ppt	158	0.26	NA	NA
Milk[b]	ppt	8 composites	0.018	NA	NA
Dairy[c]	ppt	8 composites	0.12	NA	NA
Eggs[d]	ppt	15 composites	0.081	NA	NA
Beef	ppt	63	0.18	0.11	0.95
Pork	ppt	78	0.28	0.15	1.8
Poultry	ppt	78	0.068	0.03	0.43
Vegetable fats	ppt	30	0.056	NA	NA

Notes:
CDD—chlorinated dibenzo-*p*-dioxin;
CDF—chlorinated dibenzofuran;
NA—not available;
TEQ—toxicity equivalent;
WHO—World Health Organization.
Source: From Table 1.4 in US EPA (2003).
[a] Values are the arithmetic mean TEQs. Non-detects were set to one-half the limit of detection, except for soil and CDD/CDFs in vegetable fats for which non-detects were set to zero. CDD/CDFs in water were generally undetected except for a small proportion of samples with detected values of OCDD and OCDF.
[b] Each composite for CDD/CDF was composed of 40+ US regional samples.
[c] Dairy concentrations calculated from milk lipid concentrations and then assuming a fat fraction for dairy.
[d] Each composite for CDD/CDF data was composed of 24 eggs.

world" factors into account when interpreting chemical source fingerprints. For example, the chemical fingerprint of a sample from an environmental medium such as soil, sediment, or indoor house dust typically represents many decades of input of CDD/CDFs that may have been chemically transformed during transport from their original source and after deposition and/or mixed with other sources.

Comparison of weathered environmental samples to fresh industrial sources can sometimes be made. A comparison of these weathered mixed profiles (e.g., soil, sediment, indoor dust) to published profiles of an industrial material or unweathered emission samples taken from another location can be made in some situations. If a known industrial source is the dominant source and the profile of emissions from that source is well established, stable over time, distinct from ambient background, and not subject to significant transformations en route from source to receptor, then it can be quite useful in source identification. For example, Peek *et al.* (2002) found that the CDD/CDF profile of stack emissions from a pulp mill distinctly matched the soils in an adjacent forested hillside. However, that profile did not match soils from nearby residential areas, which were subject to other more immediate sources such as burn barrels, as well as background soils in the area. The authors concluded that the forest was affected by only one source and the source profile was so distinct from other sources (i.e., very low OCDD) that a plausible linkage could be made.

Analytical data must be reliable. Chemical fingerprinting methods, whether based on sample-specific comparisons of profiles, analyses of "diagnostic ratios," or multivariate statistical analyses, are dependent on good quality analytical chemistry data and are thus vulnerable to data quality problems. Analytical data for the same compounds, but from different laboratories or derived by different methods, can introduce uncertainty in the comparisons. The frequency of "non-detects" in the data set and how these results below the method reporting limit are handled can sometimes be critical to the data analyses and can potentially bias chemical fingerprinting results, particularly when compounds, such as CDD/CDFs, occur at extremely low concentrations. Another analytical problem, particularly when older laboratory data are used, is "coelution" (i.e., when non-target analytes elute in a GC column at the same time as a target compound and are thus analyzed en masse), which can lead to the misidentification of analytes or inaccurate quantification. Coelution of certain CDD/CDF congeners was common in the analyses done in the 1980s before high-resolution techniques became more common. The problem of coelution led to increasing dependence on more definitive analytical methods such as HRGC.

There are many chemical fingerprinting methods, which range from simple profile comparisons of individual samples to sophisticated multivariate analyses. Individual profile comparisons can be useful when the profiles are clearly different. However, the human eye has difficulty detecting subtle patterns between histogram plots of 17 CDD/CDF congeners or, as an extreme example, the peaks of 80 semivolatile compounds on a gas chromatogram. Multivariate statistical analyses such as hierarchical cluster analysis and principal components analysis (PCA) are often used to condense and simplify a complex set of variables. These widely used and accepted techniques are scientifically defensible, although the underlying mathematics are complex.

14.4.1 Individual Profile Evaluation

The first fingerprinting step, and sometimes the only step, is the evaluation of individual samples by comparison of profiles of the relative concentrations of either the commonly reported 17 2,3,7,8-substituted congeners or the 10 homologue classes. Concentrations of CDD/CDFs found in environmental media can vary by orders of magnitude; therefore, standardization of the results is necessary so that the congener or homologue profiles from different locations and different media can be compared.

14.4.1.1 Standardization Methods

Four types of standardization methods are commonly used. We define them as the "2,3,7,8-sum," "relative homologue," "relative TEQ," and "total homologue" methods. In Figure 14.4.1, profiles using these four standardization methods are shown for individual environmental samples (e.g., soil, dust, sediment, and air) and source samples (e.g., various combustion sources, Kraft mill sludge, soils impacted by PCP, and sediments impacted by 2,4,5-T). As illustrated in Figure 14.4.1, samples of similar media or sources may have common 2,3,7,8-sum profiles, but very different profiles using the other standardization methods. Each of these methods provides a different, yet equally valid, view of the relative concentrations of the congeners and homologue classes. Consideration of multiple standardization methods for both visual comparisons and exploratory data analyses provides a more rigorous analysis than using just one standardization method. Each of these standardization methods is discussed in the following subsections.

14.4.1.2 "2,3,7,8-Sum" Standardization Method

Each reported 2,3,7,8-substituted congener is divided by the sum of the 2,3,7,8-substituted congeners reported. This is a common standardization method and is similar to dividing each congener by the total CDD/CDFs (US EPA, 2005) and takes advantage of the detail provided in the congener-specific results. However, there are two problems with this method: CDD/CDF profiles can be altered by weathering and bioaccumulation (this is partially addressed by "relative homologue standardization" method) and the profile comparisons can be limited by lack of detection of specific congeners (this is partially addressed by the "total homologue standardization" method). Another challenge is the problem of viewing the low concentrations that are masked when the relative concentrations of OCDD or others are extremely high. Presenting the relative concentrations on a truncated linear scale or a logarithmic scale allows examination of the pattern of the low concentration congeners. However, if the logarithmic scale is expanded too much to show the differences in the low-percentage congeners, the differences between the major contributing congeners are difficult to see. Moreover, the differences between the low concentration congeners can be exaggerated to an extent not justified by the analytical uncertainty.

14.4.1.3 "Relative Homologue" Standardization Method

Each 2,3,7,8-substituted congener is divided by its respective homologue class (e.g., 2,3,7,8-TCDD is divided by the total TCDDs, 1,2,3,4,7,8-HxCDF is divided by the total HxCDFs). OCDD and OCDF are divided by the total 2,3,7,8-substituted dioxins and furans, respectively. This method, first proposed by Hagenmaier *et al.* (1994), somewhat neutralizes the effects of differential weathering and bioaccumulation resulting from the degree of chlorination (i.e., homologue class) among environmental samples (e.g., soil, sediments, and dust). Differences in weathering and bioaccumulation between 2,3,7,8-substituted congeners and non-2,3,7,8-substituted congeners with the same degree of chlorination probably also occur; yet fewer changes are likely within the homologue class than between them. This

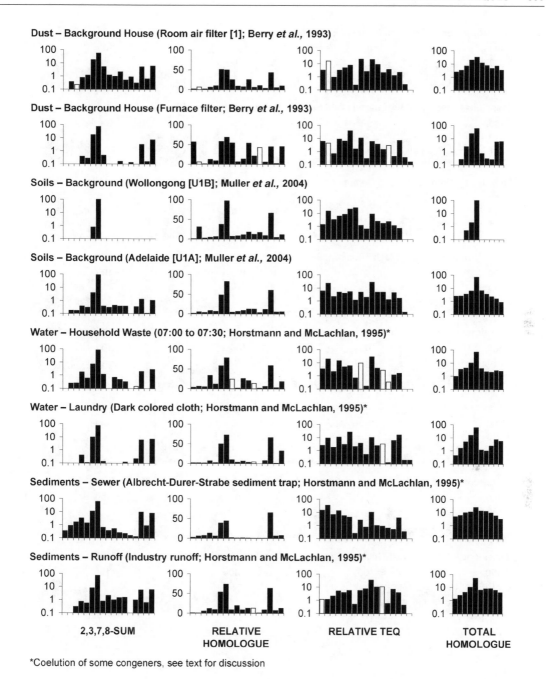

Figure 14.4.1 *Congener and homologue profiles of individual environmental and source samples using four different standardization methods. All values shown on the vertical axes are percents. Refer to Table 14.1.2 for the order of congeners and homologues shown on the horizontal axes. Open bars represent calculations performed on non-detects using 1/2 detection limit.*

method, however, does not address the radical shift in congener patterns resulting from ultraviolet radiation and subsequent photodegradation that occurs in sunlight within minutes and hours after combustion or aerial spraying of herbicides containing trace amounts of CDD/CDFs (Karch *et al.*, 2004; Podoll *et al.*, 1986).

14.4.1.4 "Relative TEQ" Standardization Method
The TEQ for each reported 2,3,7,8-substituted congener is divided by the TEQ for the sample (Fiedler *et al.*, 1996; MDEQ, 2003). This standardization method takes advantage of the detail provided in the congener-specific results, and it also provides information on which congeners contribute

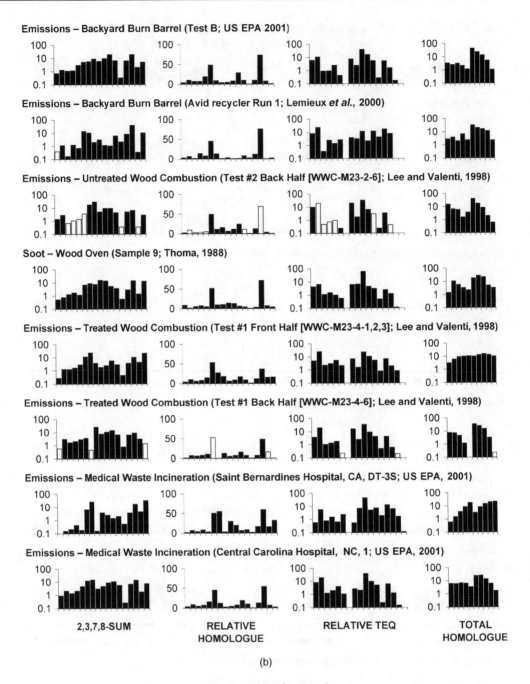

(b)

Figure 14.4.1 *(Continued)*

to toxicity. Another advantage of this method is that the dominance of OCDD (often orders of magnitude greater than some of the lower chlorinated congeners) does not mask the patterns of the low-concentration congeners. However, this standardization method has the same two problems as the 2,3,7,8-sum method: CDD/CDF profiles of similar environmental media can be altered by weathering and bioaccumulation (this is partially addressed by the "relative homologue" standardization method), and the profile comparisons can be limited by lack of detection of specific congeners (this is partially addressed by the "total homologue" standardization method).

14.4.1.5 "Total Homologue" Standardization Method
Each homologue class is divided by the total CDD/CDFs (e.g., total PeCDFs divided by the total CDD/CDFs). This is a common standardization method and a convenient way of showing gross differences in profiles. However, unlike the congener-specific methods, subtle but potentially important differences among the low-concentration congeners are not

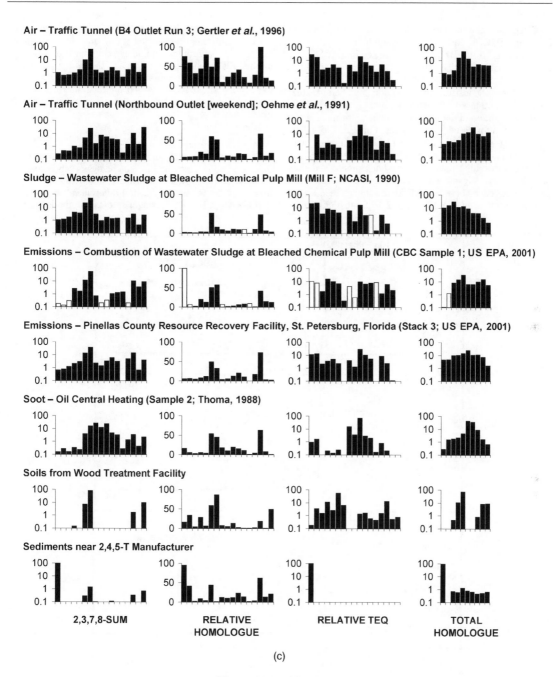

Figure 14.4.1 *(Continued)*

presented. Also, unlike the "relative homologue" method, it does not account for any significant dechlorination of environmental samples resulting from weathering or differential uptake by organisms.

14.4.1.6 Selected Ion Current Profiles (SICPs)

For the analysis of CDD/CDFs using HRGC/HRMS, SICPs are generated by the laboratory for each homologue class. These are analogous to the "mass fragmentograms" that Rappe (1994) presented for a variety of CDD/CDF homologue classes from different types of environmental samples

and the HRGC patterns presented for different environmental and source samples by Swerev and Ballschmiter (1989). The SICP is a visual representation of the measurements of a particular family of organic compounds and provides information on the presence or absence of CDD/CDF congeners and other organic compounds within a given homologue class, not just the 2,3,7,8-substituted congeners. For example, only the three 2,3,7,8-substituted HxCDD congeners (i.e., 1,2,3,4,7,8-HxCDD, 1,2,3,6,7,8-HxCDD, and 1,2,3,7,8,9-HxCDD) are usually reported, yet there are a total of 10 HxCDD congeners (Table 14.1.1). The responses of all

HxCDDs that may be present in a sample are measured and a total HxCDD concentration would be reported by the laboratory as "Total HxCDD." The SICP for HxCDD, for example, could potentially show a response (a peak on the graph) for any one or up to all 10 congeners if they were present above the instrument's detection limit. Samples that have been influenced by different sources of CDD/CDFs would show different dominant peaks and patterns of peaks on the SICPs for certain homologues. For example, SICPs for HxCDDs are shown in Figure 14.4.2 for chimney soot, fly ash, PCP, and hexachlorocyclohexane. These SICPs provide the distribution pattern for all ten isomers and thus provide the investigator with a more detailed evaluation of pattern differences between individual samples than do the standard profiles of the 17 target 2,3,7,8-substituted con-

geners that are usually reported. The SICPs are most useful in evaluating samples collected at the same time, submitted to the same laboratory, and analyzed under the same conditions. SICPs should be used qualitatively (i.e., for visual pattern comparisons) rather than quantitatively because stable isotope standards (required for quantitative analyses) are typically only available for the 2,3,7,8-substituted congeners and several others.

14.4.2 Exploratory Data Analysis
To complement visual comparisons and/or ratio analyses described above, exploratory data analyses can be used to evaluate sources. Mathematical methods can be used to identify patterns (similarities and differences) in groups of multivariate CDD/CDF congener data. Methods used to

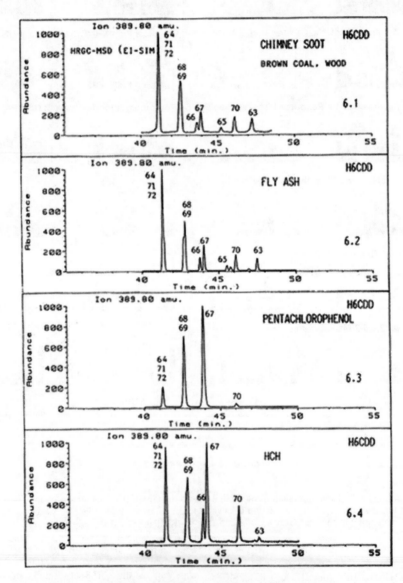

Figure 14.4.2 *Selected ion current profiles of HxCDD isomers for chimney soot, fly ash, pentachlorophenol, and hexachlorocyclohexane (HCH). The 2,3,7,8-substituted isomers are shown as peak numbers 66 (1,2,3,4,7,8-HxCDD), 67 (1,2,3,6,7,8-HxCDD), and 70 (1,2,3,7,8,9-HxCDD) from the SIL 88 HRGC column. Source: Swerev and Ballschmiter (1989). Reprinted with permission. Copyright (1989) Elsevier.*

assist in source identification of CDD/CDFs include, but are not limited to, double ratio plots (Horstmann *et al.*, 1993), hierarchical cluster analysis (e.g., Hagenmaier *et al.*, 1994; Fiedler *et al.*, 1996; Gotz and Lauer, 2003), discriminant analysis (Peek *et al.*, 2002; Gotz and Lauer, 2003), PCA (Creaser *et al.*, 1990; Tysklind *et al.*, 1993; Wenning *et al.*, 1993a; Rappe, 1994; Fiedler *et al.*, 1996; Gotz *et al.*, 1998; Masunaga *et al.*, 2001; Fattore *et al.*, 2002; Abad *et al.*, 2003; Bakoglu *et al.*, 2005; Kim *et al.*, 2005; Lee *et al.*, 2005; Sun *et al.*, 2005; Watanabe *et al.*, 2005), neural networks (Gotz and Lauer, 2003), and polytopic vector analysis (PVA) (Wenning *et al.*, 1993b; Ehrlich *et al.*, 1994; Huntley *et al.*, 1998; Barabas *et al.*, 2004).

The statistical theory for these techniques is discussed elsewhere (e.g., Johnson *et al.* [2002] for PCA and PVA; Morrison [1976] for discriminant analysis; SYSTAT [2004], S-Plus [2005], and StatMost [2002] for hierarchical and other cluster analysis; SPSS [1995] and StatSoft [2005] for neural networks). Some important practical considerations when applying these methods to a variety of CDD/CDF data sets are discussed below.

Data sets often have a large proportion of undetected congeners. The analyst needs to describe not only the screening criteria used to include or exclude these data but also, if included, what value was used to represent the estimated concentration. Typically, one-half the detection limit is used, but sometimes, zero, the full detection limit, or some other surrogate value is used. The analyst should conduct and describe sensitivity analyses with regard to differences and similarities in the conclusions depending on the data censoring method applied.

As shown in Figure 14.4.1, congener profiles can vary significantly depending on the standardization method. The multivariate analyses should be applied to all four common standardization methods if the data are available (sometimes only the 2,3,7,8 congeners are available and no homologue class data is reported; sometimes the converse is true). The results of the exploratory data analysis are more robust if they are consistent across different standardization methods.

Historical data must be carefully screened for high detection limits and coelution of key congeners. For example, there are very few reports of the specific congeners found in historic PCP. The most commonly used data to represent PCP source profiles are congener concentrations for two PCP samples and two sodium pentachlorophenate samples reported by Hagenmaier and Brunner (1987). However, the original report clearly shows that two of the 2,3,7,8-substituted congeners coelute with another congener. Specifically, 1,2,3,7,8-PeCDF coeluted with 1,2,3,4,8-PeCDF and 1,2,3,4,7,8-HxCDF coeluted with 1,2,3,4,7,9-HxCDF. Isomer-specific analyses of CDD/CDFs from the 1980s commonly had coelution problems for these congeners (e.g., Christmann *et al.*, 1989a,b; Buser, 1991). For historic data with such coelution problems, congener-specific analyses would not be reliable and the analyst should use the total homologue standardization method if these profiles need to be included for source determination.

To illustrate the use of several of these exploratory data analysis methods, we present three case studies. Case study 1 addresses the use of discriminant analysis to determine if aerial emissions from a power boiler at a pulp mill caused elevated CDD/CDFs in downwind drinking water supplies. Case study 2 presents the use of a variety of data analysis methods to determine the predominant sources of CDD/CDFs in the sediments of Hamburg Harbor and the Elbe River in Germany. Case study 3 shows the use of PCA in evaluating potential sources of CDD/CDFs in human milk in northern China and in Tokyo, Japan.

14.4.3 Case Study 1

Homes and small businesses downwind from a pulp mill used roof catchment systems to collect rainwater into cisterns for domestic uses including drinking water. Analyses of the sediments in some of these cisterns showed elevated CDD/CDF concentrations. Peek *et al.*, (2002) conducted air dispersion modeling and chemical fingerprinting to determine if the CDD/CDFs in the cistern sediments resulted from historical deposition of stack emissions from the mill's power boiler. Dewatered chlorinated sludge and salt-water soaked wood and bark were combusted with fuel oil and thus the fly ash had elevated CDD/CDF concentrations. Data evaluated included fly ash from the mill boilers when the mill was in operation, cistern sediment samples, soil samples from areas outside the impact of aerial deposition from the mill, and soil samples near the mill potentially impacted by aerial deposition. Also included were literature samples for several other potential sources, including auto exhaust, burn barrels, wood burning, oil heat, and fertilizers.

Data used in analysis were the homologue class concentrations divided by the total CDD/CDF concentrations (i.e., the "total homologue" standardization method).

Analyses included discriminant analysis and plots comparing the mean and standard deviation of each total homologue proportion between the different sources. A single plot for each homologue class allowed the different sources to be compared relative to the variability within each source.

In this application, discriminant analysis was used to generate linear functions of the relative homologue proportions that best separate the different groups or sources included in a "training data set" of source terms. These discriminant functions were then used to classify samples of unknown origin (e.g., offsite soils and cistern sediment samples) into one of these original groups. Additionally, the new variables generated by the discriminant functions, known as canonical variables, were plotted to show the relative positions of samples, similar to PCA. The advantage of this method over standard PCA is that the initial discriminating functions can be evaluated regarding their quality using the percent of samples correctly classified. Standard statistical output for discriminant analysis included F-statistic values for testing whether the initial groups could be distinguished from one another. These values were then compared to a critical F-value for a specified alpha level and converted to p-values using the appropriate degrees of freedom based on the number of groups and samples included.

This method requires that there be sufficient samples to characterize each initial group or source available for the "training data set." Inclusion of too few or too many groups in the initial training data set can affect the final results. Too few initial groups or incomplete characterization of the included groups could result in the samples of unknown origin being incorrectly classified, because the method requires them to be classified as one of the initial groups. Too many groups included initially could result in misclassifications because the variability within groups overrides the differences between them. This ought to be detected during initial evaluations of the discriminant functions as a high rate of misclassifications between the overlapping groups. If the overlapping groups do not include a critical source and a background group, this may not affect the overall final conclusions.

The discriminant functions were developed based on the training data set of samples of the mill fly ash, background soil, and literature source samples including emissions from trash burning, oil furnaces, wood burning, automobile exhaust, and fertilizers. All of these samples represented potential sources of CDD/CDFs to the cistern sediment. Plotting the resulting canonical variables showed the cistern sediment and soil from a nearby developed area

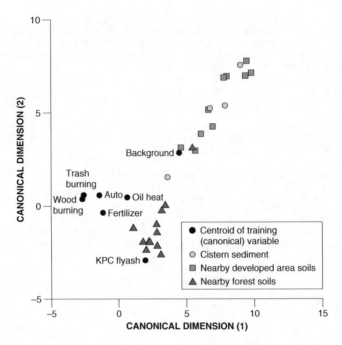

Figure 14.4.3 *Canonical variable scores of the site data set. The first two canonical variables are shown. The centroid of the training data sets are shown as filled circles and labeled. The site test data are categorized by symbol. Note: Symbols for pulp mill area soils removed from figure for clarity.* Source: *Peek* et al. *(2002). Reprinted with permission. Copyright (2002) American Chemical Society.*

as indistinguishable from the local background soil (see Figure 14.4.3). Onsite soils showed a mixture of fly ash and background soils indicating multiple sources. (Note that the symbols for the many onsite soils samples were removed from the original figure in Peek *et al.* [2002] so that the remaining symbols could be clearly seen in Figure 14.4.3.) The forest soils were strongly associated with fly ash. This was not surprising because these samples were collected in an undisturbed forest within the area of maximum deposition of the stack emissions (as shown by the dispersion modeling). The measured CDD/CDF concentration in the forest soils was consistent with the estimated deposition (Shields *et al.*, 1999). Additionally, a high percentage (10 of 12) of the forest soil samples from near the mill but within the aerial deposition plume were classified by the discriminant functions as fly ash.

14.4.4 Case Study 2
The Elbe River flows through Germany from the Czech border north into the North Sea. CDD/CDF concentrations are elevated in the sediments of the Elbe River, its tributaries, and the harbor of Hamburg. The objective of the investigation by Gotz *et al.* (1998) was to test whether the industrial center of the former German Democratic Republic, the Bitterfeld region, contributed to the elevated concentrations. Analysis included multiple possible sources as well as other sampled media from the general area. The categories of data included represent possible external sources (PCP, pesticides, and chloralkali process); metallurgical processes; municipal waste incinerators and air samples; sewage sludge; sediment and suspended particulate from the Elbe River and Hamburg harbor, and soil samples from the flooding areas of the Elbe and Dove Elbe rivers; and sediment, suspended particulate, surface water,

and soil from the Bitterfeld region. A total of 143 samples were included.

Data used in these analyses were the 2,3,7,8-Substituted congener concentrations divided by their homologue class concentration (i.e., standardized by the "relative homologue" method). OCDD and OCDF were each divided by the total CDD and CDF, respectively. Also included as a variable was the relative proportion of CDDs divided by the total CDD/CDFs.

In this case study, data were analyzed using several multivariate statistical methods, specifically hierarchical and *K*-means cluster analysis, factor analysis (i.e., PCA), and combinations of factor analysis with each of these clustering methods. Later analysis (Gotz and Lauer, 2003) also included a Kohonen neural network analysis. The final cluster groups were interpreted with regard to the relationships between sources and sample types in addition to the similarities and differences between the multiple analytical methods.

Overall, the various methods used in this case study provide a similar story. Soil from the Bitterfeld region was the most significant source of CDD/CDFs to the sediments of the Elbe River and Hamburg harbor as well as to the other floodplain soils of the Elbe. Elevated CDD/CDFs in the Bitterfeld soils were attributed to both chemical production and metallurgical processes.

Results were presented for two of the linkage methods used with hierarchical clustering but the results are similar, as are the results using the *K*-means cluster analysis. PCP, PCB, organochlorine pesticide production, and chloralkali process samples each cluster separately. (Note that the PCP profiles are represented by the Hagenmaier and Brunner [1987] data with the coelution problem.) The metallurgical, Elbe River, Hamburg harbor, and Bitterfeld region samples cluster together, as do the samples from the flooding areas of the Elbe and Dove Elbe rivers. The *K*-means and the hierarchical clustering based on Ward's linkage method show a separation for the magnesium production samples and the flooding area of the Dove Elbe River. The remaining source samples and the upstream Elbe River samples group together to some degree.

The groupings identified in the plot of the first two factors from the factor analysis show the Elbe River, Hamburg harbor, Bitterfeld region, and metallurgical source samples closely grouped and the other external sources predominantly clustered away from these, the more specific sources more independently (PCP and the chloralkali plant, for example). Additionally, the upstream Elbe River samples were separate from the rest of the river samples. Hierarchical cluster analysis of the first three factors resulted in much the same groupings, though the copper slag samples separated from the rest of the Bitterfeld region cluster, and the upstream Elbe River samples were considered a separate cluster from the other external sources. The groupings based on factor analysis are remarkably similar to the cluster analysis methods, especially considering that the first three factors account for only 48.3% of the original variability in the relative concentrations.

In a later analysis of data from the same area, Gotz and Lauer (2003) used neural networks, specifically a Kohonen network or self-organizing feature map. This method is classified as an unsupervised learning technique because it requires only input data, as opposed to "learning" relationships from a data set that includes both the input variables (i.e., relative homologue concentrations) and the correct classifications (sample groups). This method uses an iterative process to adjust a starting set of clusters to better reflect the structure of the data (StatSoft, 2005). Each iteration of the fitting process re-centers the group or cluster nearest a new sample to reflect inclusion of the new sample in that group. The final results are similar to a cluster analysis approach. An advantage of this method over cluster analysis is that the final structure can be used to classify future samples similar to discriminant analysis. The advantage of the neural network over discriminant analysis is that future samples can be classified as not belonging to any of the original source groups. Discriminant analysis requires a new sample to be classified as one of the original groups analyzed, whereas Kohonen networks allow an acceptance threshold to be set such that samples not meeting the criteria are classified as undecided with regard to their group membership.

Application of the Kohonen network to the larger Elbe River data set, augmented with data collected between the earlier publication and 2003, resulted in much the same results as the earlier reported cluster and factor analysis. This analysis resulted in a Bitterfeld–Elbe cluster containing the Bitterfeld soil region, Elbe River, Hamburg harbor, and flooding area samples. All clustering was identical to the hierarchical cluster analysis conducted on the larger data set.

14.4.5 Case Study 3

Human milk samples were collected from 41 breastfeeding mothers in the Hebei Province in northern China, 30 from Shijiazhuang city (industrialized), and 11 from the Tanshan countryside (agricultural). For comparison, 20 samples were obtained from mothers in Tokyo, Japan. All samples were analyzed for CDD/CDFs and for dioxin-like polychlorinated biphenyls (dl-PCBs) by HRGC/HRMS. The objectives of the investigation by Sun *et al.* (2005) were to evaluate potential dietary risk factors and to determine possible sources.

The mean CDD/CDFs and dl-PCBs in Hebei were 3.6 and 1.9 pg TEQ/g fat, respectively, which were about one-fourth of the levels in Japan. There was no significant difference in mean CDD/CDFs between the urban and rural areas in Hebei. Based on dietary interviews of the Chinese mothers, freshwater fish consumption was found to correlate with the body burden of CDD/CDFs. There were no dietary interviews reported for the Japanese mothers.

The average congener patterns (2,3,7,8-substituted congeners divided by the total CDD/CDFs) were roughly similar for the Hebei and Tokyo data sets, with two exceptions: the relative percentage of OCDD was about 65 percent for the Hebei samples and about 45 percent for the Tokyo samples. Also, the relative percentage of 1,2,3,6,7,8-HxCDD was about 3 percent for the Hebei samples and about 20 percent for the Tokyo samples. The five most abundant isomers for the Hebei samples were, in order, OCDD, 1,2,3,4,6,7,8-HpCDD, 2,3,4,7,8-PeCDF, 1,2,3,4,7,8-HxCDF, and 1,2,3,6,7,8-HxCDD. For the Tokyo samples, the five most abundant isomers were, in order, OCDD, 1,2,3,4,6,7,8-HpCDD, 1,2,3,6,7,8-HxCDD, 1,2,3,7,8-PeCDD, and 2,3,4,7,8-PeCDF.

In this case study, the investigators used PCA to evaluate the similarities or differences in congener patterns of the 61 human milk samples. Congener data from pooled human milk samples from an area in China that had been sprayed with sodium pentachlorophenate (Na-PCP) and congener data from an area in China with no recorded Na-PCP use were also included in the PCA.

The PCA results showed that the first three components accounted for about 77 percent of the total variance (the first, second and third components accounted for 36.6, 31.7 and 8.3 percent, respectively). The PCA score plot for the first and second factors showed that the CDD/CDF congeners for the urban and rural regions of Hebei have quite similar patterns. However, they were clearly different from those for the Japanese samples. The pooled sample from the unsprayed Na-PCP area occurred in the cluster of Hebei samples. However, the sample from Na-PCP sprayed area was closer to the cluster of Tokyo samples. The authors concluded that past use of Na-PCP in Japan may have contributed to the distinct separation of the groups in the PCA score plots.

There are several sources of uncertainties associated with this case study. First, because the sample fractionation method described in the paper may not have removed polychlorinated diphenyl ethers (PCDPEs), some of the CDF results may reflect coelution with PCDPEs, which are common in tissue matrices. The extent of coelution could be evaluated by examining the SICPs showing the CDFs and PCDPEs simultaneously. Second, another potential coelution problem with the method described is between 1,2,3,7,8-PeCDD and PCB-169. Third, given the sample volume available, the reported concentrations for some of the congeners are very near the instrument detection limit and thus care must be taken in making fingerprinting determinations at these low concentrations where analytical uncertainty is high. Fourth, it is difficult to draw conclusions regarding the contribution of Na-PCP with only two data points. It would be useful to consider other potential exposure differences between the Hebei and Tokyo populations as well. For example, the Tokyo group may have had a higher component of marine fish in their diet than the inland Chinese groups.

14.5 SUMMARY

Identification of CDD/CDF sources in environmental and biological media is challenging because of low concentrations and associated analytical uncertainties; the potential need to compare data from different laboratories and different methodologies; the presence of multiple sources that are difficult to distinguish, particularly at low concentrations; and the confounding effects of chemical and biological transformations, particularly for airborne sources.

Multiple lines of evidence should be used to minimize, or at least to understand, these uncertainties. Appropriate background data is necessary to determine if the subject media is, in fact, "elevated" and if the chemical fingerprint in the subject media is substantially distinct from typical background in that area. Chemical fingerprinting should be conducted using a variety of methods such as discussed the Section 14.4. The fingerprinting results should be consistent with a conceptual model that provides plausible and verified pathway(s) from the alleged source(s) to the receptors. The spatial distribution of CDD/CDF concentrations needs to be consistent with the conceptual site model. For example, soil concentrations resulting from particle deposition from a single stack should decrease with distance from the stack in the predominant wind direction and the concentration pattern should be gradual rather than groups of "hot spots".

REFERENCES

Abad, E., F. Perez, J.J. Llerena, J. Caixach, and J. Rivera. 2003. Evidence for a specific pattern of polychlorinated dibenzo-*p*-dioxins and dibenzofurans in bivalves. *Environ. Sci. Technol.* 37 (22): 5090–5096.

Bakoglu, M., A. Karademir, and E. Durmusoglu. 2005. Evaluation of PCDD/F levels in ambient air and soils and estimation of deposition rates in Kocaeli, Turkey. *Chemosphere.* 59: 1373–1385.

Ballmacher, H. 2001. EN 1948: Reference for monitoring legal dioxins limit values and reference for long-term measurements. *Anal. Sci.* 17: 1551.

Barabas, N., P. Goovaerts, and P. Adriaens. 2004. Modified polytopic vector analysis to identify and quantify a dioxin dechlorination signature in sediments. 2. Application to the Passaic River. *Environ. Sci. Technol.* 38 (6): 1821–1827.

Baughman, R. and M. Meselson. 1973. An analytical method for detecting TCDD (dioxin): Levels of TCDD in samples from Vietnam. *Environ. Health Perspect.* 5: 27–35.

Berry, R.M., C.E. Luthe, and R.H. Voss. 1993. Ubiquitous nature of dioxins: A comparison of the dioxins content of common everyday materials with that of pulps and papers. *Environ. Sci. Technol.* 27: 1164–1168.

Buser, H.R. 1991. Volume 11—Review of methods of analysis for polychlorinated dibenzodioxins and dibenzofurans. pp. 105–146. In: *Environmental Carcinogens—Methods of Analysis and Exposure Assessment.* First Edition. C. Rappe, H.R. Buser, B. Dodet, and I.K. O'Neill (eds). International Agency for Research on Cancer, Lyon, France.

Buser, H.R. and C. Rappe. 1984. Isomer-specific separation of 2378-substituted polychlorinated dibenzo-*p*-dioxins by high-resolution gas chromatography/mass spectrometry. *Anal. Chem.* 56: 442–448.

Charles, M.J., B. Green, Y. Tondeur, and J.R. Hass. 1989. Optimization of a hybrid-mass spectrometer method for the analysis of polychlorinated dibenzo-*p*-dioxins and polychlorinated dibenzofurans. *Chemosphere.* 19: 51–57.

Christmann, W., K.D. Kloppel, H. Partscht, and W. Rotard. 1989a. Determination of PCDD/PCDF in ambient air. *Chemosphere.* 19: 521–526.

Christmann, W., K.D. Kloppel, H. Partscht, and W. Rotard. 1989b. PCDD/PCDF and chlorinated phenols in wood preserving formulations for household use. *Chemosphere.* 18: 861–865.

Cleverly, D.H., D.J. Schaum, G. Schweer, J. Becker, and D. Winters. 1997. The congener profiles of anthropogenic sources of chlorinated dibenzo-*p*-dioxins and chlorinated dibenzofurans in the United States. *Organohal. Comp.* 32: 430–435.

Creaser, C.S., A.R. Fernandes, S.J. Harrad, and E.A. Cox. 1990. Levels and sources of PCDDs and PCDFs in urban British soils. *Chemosphere.* 21: 931–938.

Denison, M.S., S.R. Nagy, M. Ziccardi, G.C. Clark, M. Chu, D.J. Brown, G. Shan, Y. Sugawara, S.J. Gee, J. Sanborn, and B.D. Hammock. 2002. Bioanalytical approaches for the detection of dioxin and related halogenated aromatic hydrocarbons. In: *Biomarkers of Environmentally Associated Disease, Technologies, Concepts, and Perspectives.* S.H. Wilson and W.A. Suk (eds). Lewis Publishers.

Ehrlich, R., R.J. Wenning, G.W. Johnson, S.H. Su, and D.J. Paustenbach. 1994. A mixing model for polychlorinated dibenzo-*p*-dioxins and dibenzofurans in surface sediments from Newark Bay, New Jersey using polytopic vector analysis. *Arch. Environ. Contam.* 27 (4): 486–500.

Eiceman, G.A., A.C. Viau, and F.W. Karasek. 1980. Ultrasonic extraction of polychlorinated dibenzo-*p*-dioxins and other organic compounds in fly ash from municipal incinerators. *Anal. Chem.* 52 (9): 1492–1496.

Environment Canada. 1997. Reference Method for the analysis of polychlorinated biphenyls (PCBs). Report EPS 1/RM/31. Environment Canada, Environmental Technology Advancement Directorate, Analysis and Methods Division.

Fattore, E., L. Viganò, G. Mariani, A. Guzzi, E. Benfenati, and R. Fanelli. 2002. Polychlorinated dibenzo-*p*-dioxins and dibenzofurans in River Po sediments. *Chemosphere.* 49 (7): 749–754.

Fiedler, H., C. Lau, L.-O. Kjeller, and C. Rappe. 1996. Patterns and sources of polychlorinated dibenzo-*p*-dioxins and dibenzofurans found in soil and sediment samples in southern Mississippi. *Chemosphere.* 33: 421–432.

Focant, J.-F., E.J. Reiner, K. MacPherson, T. Kolic, A. Sjödin, D.G. Patterson, Jr., S.L. Reese, F.L. Dorman, and J. Cochran. 2004. Measurement of PCDDs, PCDFs, and non-*ortho*-PCBs by comprehensive two-dimensional gas chromatography-isotope dilution time-of-flight mass spectrometry (GC × GC – IDTOFMS). *Talanta.* 63 (2004): 1231–1240.

Gertler, A.W., J.C. Sagebiel, W.A. Dippel, and L.H. Sheetz. 1996. A study to quantify on-road emissions of dioxins and furans from mobile sources: Phase 2. Final Report. Prepared for American Petroleum Institute, Washington, DC. Energy and Environmental Engineering Center, Desert Research Institute, Reno, NV. November 1996. p. 89.

Gönczi, M., M. Guannarsson, B.Hedman, N. Johansson, M.Näslund, and S. Marklund. 2005. Emission of PCDD/F and PCB from uncontrolled combustion of domestic waste in Sweden. *Organohal. Comp.* :2033–2036.

Gotz, R. and R. Lauer. 2003. Analysis of sources of dioxin contamination in sediments and soils using multivariate statistical methods and neural networks. *Environ. Sci. Technol.* 37: 5559–5565.

Gotz, R., B. Steiner, P. Friesel, K. Roch, F. Walkow, V. Maass, H. Reincke, and B. Stachel. 1998. Dioxin (PCDD/F) in the River Elbe—Investigations of their origin by multivariate statistical methods. *Chemosphere.* 37: 1987–2002.

Gullett, B.K., P.M. Lemieux, C.C. Lutes, C.K. Winterrowd, and D.L. Winters. 2001. Emissions of PCDD/F from uncontrolled, domestic waste burning. *Chemosphere.* 43: 721–725.

Hagenmaier, H. and H. Brunner. 1987. Isomer specific analysis of pentachlorophenol and sodium pentachlorophenate for 2,3,7,8-substituted PCDD and PCDF at sub-PPB levels. *Chemosphere.* 16: 1759–1764.

Hagenmaier, H., C. Lindig, and J. She. 1994. Correlation of environmental occurrence of polychlorinated dibenzo-*p*-dioxins and dibenzofurans with possible sources. *Chemosphere* 29: 2163–2174.

Horstmann, M. and M.S. McLachlan. 1995. Concentrations of polychlorinated dibenzo-*p*-dioxins (PCDD) and dibenzofurans (PCDF) in urban runoff and household wastewaters. *Chemosphere.* 31 (3): 2887–2896.

Horstmann, M., M.S. McLachlan, and M. Reissinger. 1993. Investigations of the origin of PCDD/F in municipal sewage sludge. *Chemosphere.* 27: 113–120.

Huntley, S.L., H. Carlson-Lynch, G.W. Johnson, D.J. Paustenbach, and B.L. Finley. 1998. Identification of historical PCDD/F sources in Newark Bay Estuary subsurface sediments using polytopic vector analysis and radioisotope dating techniques. *Chemosphere.* 36 (6): 1167–1185.

Ikeguchi, T. and M. Tanaka. 2000. Dioxin emission from an openburning-like waste incinerator: Small incinerators for household use. *Organohal. Comp.* 46: 298–301.

JIS. 1999. K 0311: Method for determination of tetra- through octa-chlorodibenzo-*p*-dioxins, tetra- through octa-chlorodibenzofurans and coplanar polychlorobiphenyls in stationary source emissions, and K 0312: Method for determination of tetra- through octa-chlorodibenzo-*p*-dioxins, tetra- through octa-chlorodibenzofurans and coplanar polychlorobiphenyls in industrial water and wastewater. Japan Industrial Standards Committee.

Johnson, G.W., R. Ehrlich, and W. Full. 2002. Principal components analysis and receptor models in environmental forensics. Chapter 12. pp. 462–515. In: *Introduction to Environmental Forensics.* B.L. Murphy and R.D. Morrison (eds). Academic Press, New York, NY.

Karch, N.J., D.K. Watkins, A.L. Young, and M.E. Ginevan. 2004. Environmental fate of TCDD and Agent Orange and bioavailability to troops in Vietnam. *Organohal. Comp.* 66: 3689–3694.

Kim, B.-H., S.-J. Lee, S.-J. Mun, and Y.-S. Chang. 2005. A case study of dioxin monitoring in and around an industrial waste incinerator in Korea. *Chemosphere.* 58: 1589–1599.

Lee, S.Y. and J.C. Valenti. 1998. Products of incomplete combustion from direct burning of pentachlorophenol-treated wood wastes. Prepared for US Environmental Protection Agency, Research Triangle Park, NC. *Acurex Environmental Corp.*, Durham, NC.

Lee, C.C., L.H.L. Chen, H.J. Su, Y.L. Guo, and P.C. Liao. 2005. Evaluation of PCDD/Fs patterns emitted from incinerator via direct ambient sampling and indirect serum levels assessment of Taiwanese. *Chemosphere.* 59: 1465–1474.

Lemieux, P.M., C.C. Lutes, J.A. Abbott, and K.M. Aldous. 2000. Emissions of polychlorinated dibenzo-*p*-dioxins and polychlorinated dibenzofurans from the open burning of household waste in barrels. *Environ. Sci. Technol.* 34 (3): 377–384.

Lemieux, P.M., B.K. Gullett, C.C. Lutes, C.K. Winterrowd, and D.L. Winters. 2003. Variables affecting emissions of PCDD/Fs from uncontrolled combustion of household waste in barrels. *J. Air Waste Manage. Assoc.* 53: 523–531.

Ligon, M.V. and R.J. May. 1984. Isomer specific analysis of selected chlorodibenzofurans. *J. Chromatogr.* 294: 87–98.

Marriott, P.J., P. Haglund, and R.C.Y. Ong. 2003. A review of environmental toxicant analysis by using multidimensional gas chromatography and comprehensive GC. *Clinica Chimica Acta* 328 (1–2):1–19.

Masunaga, S., Y. Yao, I. Ogura, T. Sakurai, and J. Nakanishi. 2001. Quantitative estimation of dioxin sources on the basis of congener-specific information. *Organohal. Comp.* 51: 22–25.

MDEP. 1997. State of Maine 1997 backyard trash burning (BYB) study. Main Department of Environmental Protection, Augusta, Maine. 48pp.

MDEQ. 2003. Appendix J: Dioxin congener profile charts. Missouri Department of Environmental Quality. p. 47.

Morrison, D.F. 1976. *Multivariate statistical methods.* Second Edition. A.A. Arthur, L.A. Young (eds). McGraw-Hill, New York, NY.

Muller, J., R. Muller, K. Goudkamp, M. Shaw, M. Mortimer, and D. Haynes. 2004. Dioxins in soil in Australia: Technical Report No. 5. Australian Department of the Environment and Heritage, Australia.110 pp..

NATO/CCMS. 1988. Scientific basis for the development of the international toxicity equivalency factor (I-TEF) method of risk assessment for complex mixtures of dioxins and related compounds. Report No. 178, December 1988. North Atlantic Treaty Organization, Committee on Challenges of Modern Society.

NCASI. 1990. USEPA/paper industry cooperative dioxin study: The 104 mill study. Technical Bulletin No. 590. National Council of the Paper Industry for Air and Stream Improvement, Inc., New York, NY.

Oehme, M., S. Larssen, and E.M. Brevik. 1991. Emission factors of PCDD and PCDF for road vehicles obtained by tunnel experiment. *Chemosphere.* 23: 1699–1708.

Peek, D.C., M.K. Butcher, W.J. Shields, L.Y. Yost, and J.A. Maloy. 2002. Discrimination of aerial deposition sources of polychlorinated dibenzo-*p*-dioxin and polychlorinated dibenzofuran downwind from a pulp mill near Ketchikan, Alaska. *Environ. Sci. Technol.* 36 (8): 1671–1675.

Podoll, R.T., H.M. Jaber, and T. Mill. 1986. Tetrachlorodibenzodioxin: Rates of volatilization and photolysis in the environment. *Environ. Sci. Technol.* 20 (5): 490–492.

Rappe, C. 1994. Dioxin, patterns and source identification. *Fresenius' J. Anal. Chem.* 348: 63–75.

Richardson, S. 2001. Mass spectrometry in environmental sciences. *Chem. Rev.* 101: 211–254.

Saito, K., M. Takekuma, M. Ogawa, S. Kobayashi, Y. Sugawara, M. Ishizuka, H. Nakazawa, and Y. Matsuki. 2003. Extraction and cleanup methods of dioxins in house dust from two cities in Japan using accelerated solvent extraction and a disposable multi-layer silica-gel cartridge. *Chemosphere.* 53: 137–142.

Shields, W., J.A. Maloy, L. Yost, and D. Peek. 1999. Comparison of soil concentrations of dioxins and furans with predictions based on aerial deposition modeling. *Organohal. Comp.* 41: 455–458.

S-Plus. 2005. *S-Plus 7 user's guide.* Insightful Corp., Seattle, WA.

SPSS. 1995. *Neural connection user's guide.* SPPS Inc. and Recognition Systems Inc., Chicago, IL.

StatMost. 2002. *StatMost user's guide.* DataXiom Software Inc., Los Angeles, CA.

StatSoft. 2005. *Neural Networks, Electronic Statistics Textbooks*, StatSoft, Tulsa, OK. Available at: www.statsoftinc.com/textbook/stathome.html.

Sun, S-J, J-H. Zhao, H-J Liu, D-W. Liu, Y-X. Ma, L. Li, H. Horiguchi, H. Uno, T. Iida, M. Koga, Y. Kiyonari, M. Nakamura, S. Sasaki, H. Fukatu, G. C. Clark and F. Kayama. 2005. Dioxin concentration in human milk in Hebei province in China and Tokyo, Japan: Potential dietary risk factors and determination of possible sources. Chemosphere *In Press, Available online September 6, 2005. www.elsevier.com/locate/chemosphere. 9 pp.*

Swerev, M. and K. Ballschmiter. 1989. Pattern analysis of PCDDs and PCDFs in environmental samples as an approach to an occurrence/source correlation. *Chemosphere*. 18: 609–616.

SYSTAT. 2004. *SYSTAT 11 Statistics I*. SYSTAT Software, Inc., Richmond, CA.

Thoma, H. 1988. PCDD/F concentrations in chimney soot from house heating systems. *Chemosphere*. 17 (7): 1369–1379.

Tondeur, Y., P.W. Albro, J.R. Hass, D.J. Harvan, and J.L. Schroeder. 1984. Matrix effect in determination of 2,3,7,8-tetrachlorodibenzo-*p*-dioxin by mass spectrometry. *Anal. Chem.* 56: 1344–1347.

Tondeur, Y., W.J. Niederhut, and S.R. Missler. 1987. A hybrid HRGC/MS/MS method for the characterization of tetrachlorinated-*p*-dioxins in environmental samples. *Bio. Med. Environ. Mass Spectr.* 14: 449–456.

Travis, C.C. and H.A. Hattemer-Frey. 1991. Human exposure to dioxin. *Sci. Tot. Environ.* 104: 97–127.

Tysklind, M., I. Fangmark, S. Marklund, A. Lindskog, L. Thaning, and C. Rappe. 1993. Atmospheric transport and transformation of polychlorinated dibenzo-*p*-dioxins and dibenzofurans. *Environ. Sci. Technol.* 27: 2190–2197.

US EPA. 1986. Method 8290: Analytical procedures and quality assurance for multimedia analysis of polychlorinated dibenzo-p-dioxins and dibenzofurans by high-resolution gas chromatography/high-resolution mass spectrometry. US Environmental Protection Agency, Environmental Monitoring Systems Laboratory, Las Vegas, NV.

US EPA. 1989. Interim procedures for estimating risks associated with exposures to mixtures of chlorinated dibenzo-*p*-dioxins and -dibenzofurans (CDDs and CDFs) and 1989 update. EPA/625/3-89/016. US Environmental Protection Agency, Risk Assessment Forum, Washington, DC.

US EPA. 1994. Method 1613: Tetra- through octachlorinated dioxins and furans by isotope dilution HRGC/HRMS. Revision B. US. Environmental Protection Agency, Office of Water, EAD, Washington, DC.

US EPA. 1997. Locating and estimating air emissions from sources of dioxins and furans. Dcn No. 95-298-130-54-01. US Environmental Protection Agency, Office of Air Quality Planning and Standards, Research Triangle Park, NC.

US EPA. 1999a. Compendium Method TO-9A: Determination of polychlorinated, polybrominated and brominated/chlorinated dibenzo-*p*-dioxins and dibenzofurans in ambient air. In: Compendium of Methods for the Determination of Toxic Organic Compounds in Ambient Air. EPA/625/R-96/010b. US Environmental Protection Agency, Office of Research and Development, Center for Environmental Research Information, Cincinnati, OH.

US EPA. 1999b. Standards of performance for new stationary sources. Test Method 23. 40 CFR Part 60. US Environmental Protection Agency.

US EPA. 2000. Exposure and human health reassessment of 2,3,7,8-tetrachlorodibenzo-*p*-dioxin (TCDD) and related compounds. Draft. Part 1: Estimating exposure to dioxin-like compounds. EPA/600/P-00/001B(b-d). US Environmental Protection Agency, National Center for Environmental Assessment, Office of Research and Development, Washington, DC.

US EPA. 2001. Database of sources of environmental releases of dioxin like compounds in the United States (Version 3.0): Reference years 1987 and 1995. EPA/600/C-01/012. Available at http://cfpub.epa.gov/ncea/cfm/recordisplay.cfm?deid=20797. US Environmental Protection Agency, National Center for Environmental Assessment, Washington, DC.

US EPA. 2003. Exposure and human health reassessment of 2,3,7,8-tetrachlorodibenzo-*p*-dioxin (TCDD) and related compounds. Part 1: Estimating exposure to dioxin-like compounds. Volume 2: Properties, environmental levels, and background exposure. NAS review draft. US Environmental Protection Agency, Office of Research and Development, Washington, DC. Available at: www.epa.gov/ncea/dioxin.

US EPA. 2005. The inventory of sources and environmental releases of dioxin-like compounds in the United States: The Year 2000 update. EPA/600/p-03/002A. External review draft. Available from: National Technical Information Service, Springfield, VA, and online at http://epa.gov.ncea. US Environmental Protection Agency, Washington DC.

Van den Berg, M., L. Birnbaum, A.T.C. Bosveld, B. Brunstrom, P. Cook, M. Feeley, J.P. Giesy, A. Hanberg, R. Hasegawa, S.W. Kennedy, T. Kubiak, J.C. Larsen, F.X. van Leeuwen, A.K. Liem, C. Nolt, R.E. Peterson, L. Poellinger, S. Safe, D. Schrenk, D. Tillit, M. Tysklind, M. Younes, F. Waern, and T. Zacharewski. 1998. Toxic equivalency factors (TEFs) for PCBs, PCDDs, PCDFs for humans and wildlife. *Environ. Health Perspect.* 106 (12): 775–792.

Watanabe, M., H. Iwata, M. Watanabe, S. Tanabe, A. Subramanian, K. Yoneda, and T. Hashimoto. 2005. Bioaccumulation of organochlorines in crows from an Indian open waste dumping site: Evidence for direct transfer of dioxin-like congeners from the contaminated soil. *Environ. Sci. Technol.* 39: 4421–4430.

Wenning, R.J., D.J. Paustenbach, M.A. Harris, and H. Bedbury. 1993a. Principal components analysis of potential sources of polychlorinated dibenzo-*p*-dioxin and dibenzofuran residues in surficial sediments from Newark Bay, New Jersey. *Arch. Environ. Contam.* 24: 271–289.

Wenning, R., D. Paustenbach, G. Johnson, R. Ehrlich, M. Harris, and H. Bedbury. 1993b. Chemometric analysis of potential sources of polychlorinated dibenzo-*p*-dioxins and dibenzofurans in surficial sediments from Newark Bay, New Jersey. *Chemosphere*. 27: 55–64.

Wevers, M., R. De Fre, and M. Desmedt. 2003. Effect of backyard burning on dioxin deposition and air concentrations. *Chemosphere*. 54: 1351–1356.

Wittsiepe, J., U. Ewers, P. Schrey, and F. Selenka. 1996. PCDD/F in house dust. *Organohal. Comp.* 30: 80–84.

15

Polycyclic Aromatic Hydrocarbons (PAHs)

Paul D. Boehm

Contents

15.1 INTRODUCTION

Polycyclic aromatic hydrocarbons (PAHs)—sometimes referred to as polynuclear aromatic hydrocarbons (PNAs), condensed ring aromatics, or fused ring aromatics—are a class of organic compounds consisting of two or more fused aromatic rings (Figures 15.1.1 and 15.1.2). Naphthalene, consisting of two fused benzene rings, is the simplest PAH. Commonly, PAHs are depicted without labeling

the carbon and hydrogen atoms. Polycyclic aromatic hydrocarbons most commonly encountered in the environment contain two (naphthalene) to seven (coronene) fused benzene rings, though PAHs with greater number of rings are also found (Sander and Wise, 1997). The "ultimate" PAH is graphite, an inert material comprised of planes of fused benzene rings. Like all hydrocarbons, PAHs contain only hydrogen and carbon. However, closely related

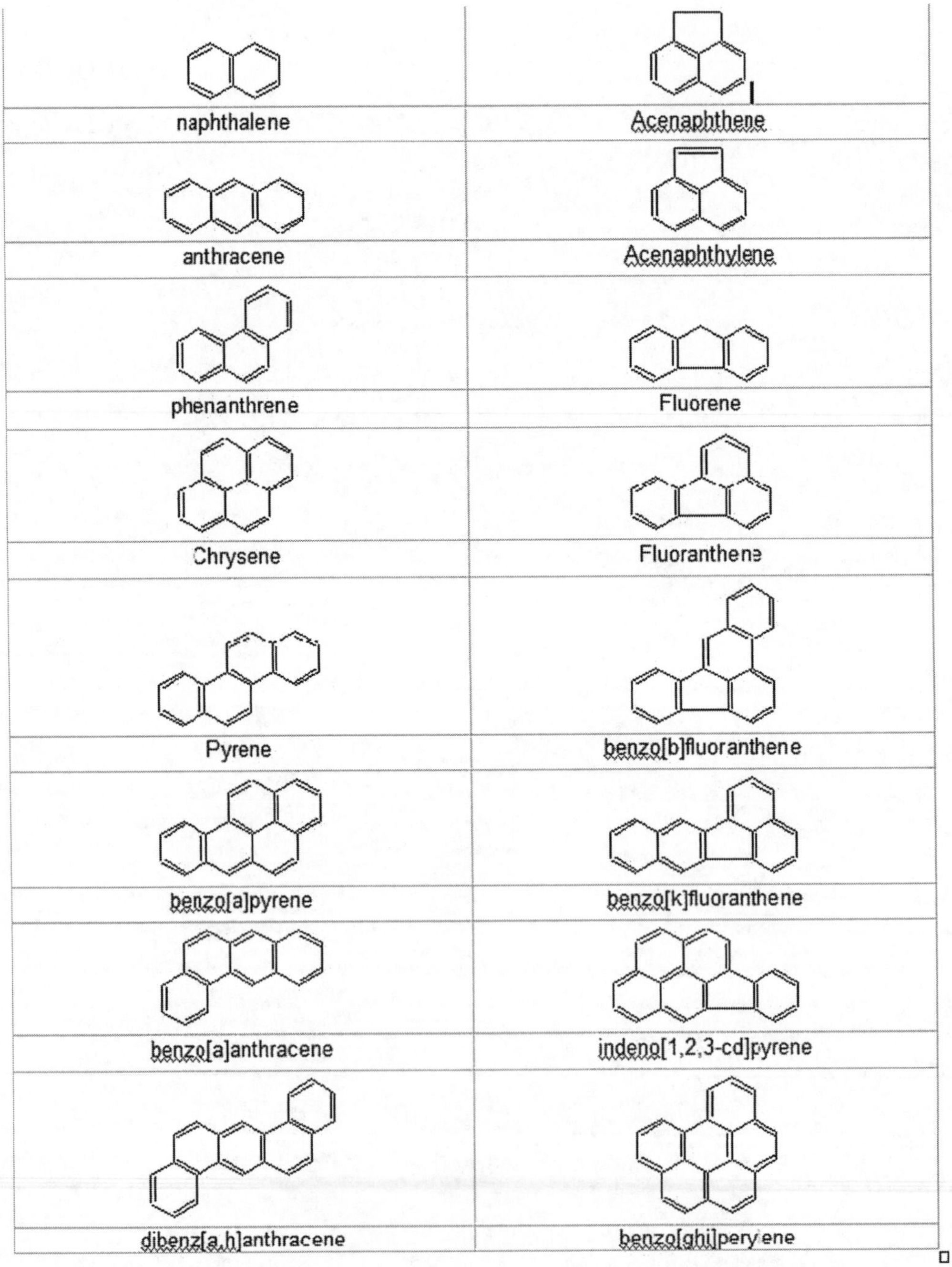

naphthalene	Acenaphthene
anthracene	Acenaphthylene
phenanthrene	Fluorene
Chrysene	Fluoranthene
Pyrene	benzo[b]fluoranthene
benzo[a]pyrene	benzo[k]fluoranthene
benzo[a]anthracene	indeno[1,2,3-cd]pyrene
dibenz[a,h]anthracene	benzo[ghi]perylene

Figure 15.1.1 *Environmental protection agency "priority pollutant" PAH compounds (non-alkylated).*

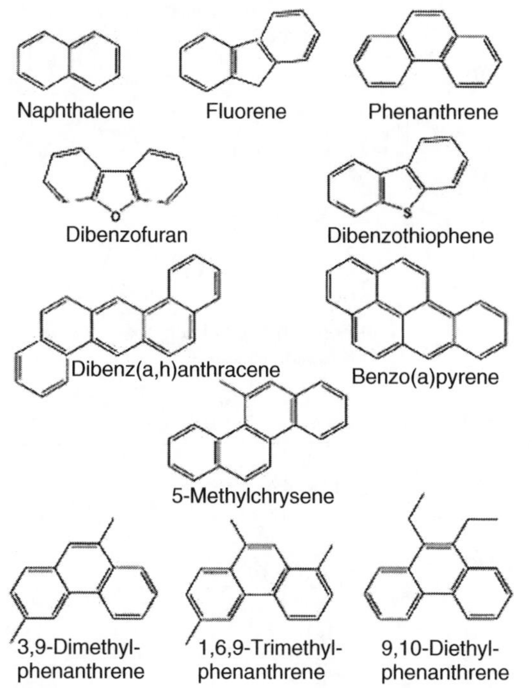

Figure 15.1.2 Some common PAH structures—unsubstituted (parent), heterocycle, and alkylated homologs.

compounds called heterocyclic aromatics, or polycyclic aromatic compounds (PACs), in which an atom of nitrogen, oxygen, or sulfur replaces one of the carbon atoms in a ring, are commonly found with PAHs from most sources. Dibenzothiophene, for example, is a sulfur-containing heterocycle. Polycyclic aromatic hydrocarbons often occur with aliphatic (straight chain) hydrocarbons attached to the rings at one or more points. These compounds are referred to as "branched" or "alkylated" PAHs. The aliphatic chains are depicted as lines attached to the PAH with the end of the line representing a methyl group ($-CH_3$) and an angle at an intermediate carbon ($-CH_2-$). Thus, as particular examples, methylnaphthalene and ethylpyrene are depicted as

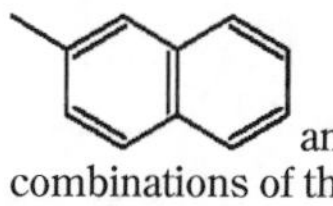

and 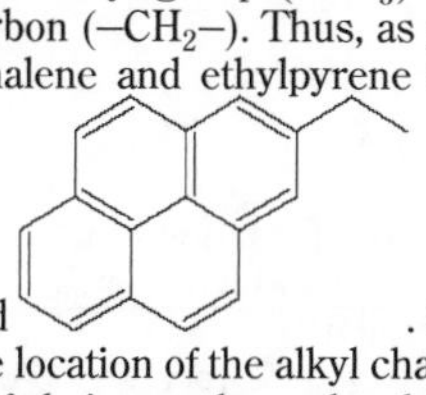 . Since numerous combinations of the location of the alkyl chain on the parent PAH, the number of chains on the molecule, and the length of the chains are possible, alkylated PAHs are often classified by the number of alkyl carbons they contain. Thus methylnaphthalene as depicted above is a C_1-naphthalene, while ethylpyrene is a C_2-pyrene.

15.2 POLYCYCLIC AROMATIC HYDROCARBON SOURCES

Hundreds of PAH compounds present in nature have been identified and named (Bjorseth, 1983, 1985; Sander and Wise, 1997). Polycyclic aromatic hydrocarbons are produced by natural processes and by anthropogenic processes. Similar compounds are introduced by both sets of processes, and these similarities, along with key diagnostic tools to differentiate sources, must be carefully

understood and considered in any PAH environmental forensics investigation.

Polycyclic aromatic hydrocarbons are produced through four generalized pathways:

1 Relatively rapid (days to years), low temperature ($<70\,°C$) transformation or *diagenesis* of organic matter as part of the changes undergone by biomolecules and related organic matter after initial deposition in sediments.
2 Slow, long-term, moderate temperature ($100-300\,°C$) formation of fossil fuels—petroleum and coal (i.e., *petrogenic*).
3 Rapid, high temperature ($>500\,°C$), incomplete or inefficient (i.e., oxygen-starved) combustion of organic biomass (pyrolysis) such as that which occurs naturally in forest and grass fires and anthropogenically in, for example, fossil fuel combustion (i.e., *pyrogenic*).
4 The biosynthesis by plants and animals of individual PAH compounds or relatively simple mixtures.

Natural sources of *petrogenic* PAHs arise from oil seepages and erosion of petroliferous shales (NRC, 1985; Bence and Kvenvolden, 1996) while natural sources of PAHs from combustion or pyrolysis (i.e., *pyrogenic* sources) include PAHs from incomplete (i.e., insufficient oxygen availability) combustion of wood and biomass via forest and grass fires (Hites et al., 1977).

While diagenesis and biosynthesis are solely natural processes, anthropogenic sources of PAHs arise through a multitude of pathways. Anthropogenic (pollution) related PAHs inputs can result in similar, but not identical, PAH compounds and assemblages of PAHs to those of natural origin. Anthropogenic inputs of PAH arise from the release into the environment of petrogenic PAHs through accidental acute petroleum spillages and through chronic non-point source and point-source inputs such as urban (stormwater) runoff and municipal waste treatment plane discharges (NRC, 1985). The most common and ubiquitous sources of anthropogenic PAHs, however, are those associated with pyrogenic inputs.

One of the most widespread categories of pyrogenic PAH inputs relates to the high temperature combustion of motor (automobile), bunker (shipping), and power plant (coals and petroleum) fuels (Bjorseth, 1985). The combustion processes introduce large amounts of PAHs globally, but in more concentrated amounts in urban areas. Residential burning of wood is one of the largest sources of atmospheric pyrogenic PAHs. Other important sources in indoor air include environmental tobacco smoke, unvented radiant and convective kerosene space heaters, and gas cooking and heating appliances. Stationary sources account for about 80% of total annual PAH emissions; although mobile sources (vehicular exhaust) are often the major atmospheric sources in urban or suburban areas (ATSDR, 1995).

Pollutant pyrogenic inputs also arise from the high temperature processing of fossil fuels, such as coal tar and tar products (e.g., creosote) formed from coals (Emsbo-Mattingly and Boehm, 2003) and from fugitive emissions released from aluminum smelters with most of these emissions being released from smelters that use the Horizontal Stud Söderberg process (Naes and Oug, 1998). Other uses of these products provide other potential entry routes. For example, the use of creosote in the treatment of wood has resulted in the association of wood treatment facilities with potential release of creosote-related (pyrogenic) PAHs (Walker et al., 2004).

What primarily differentiates the origins of petrogenic and pyrogenic PAHs, whether *natural sources* of PAHs or *anthropogenic sources* is the temperature of formation. During the generation of PAHs, the degree of alkylation in a given

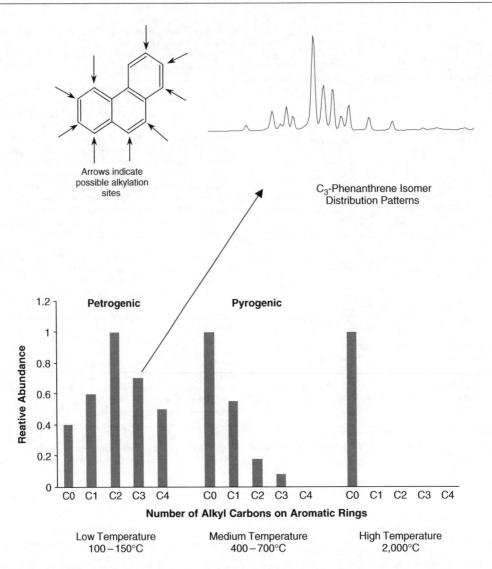

Figure 15.2.1 *Representative distribution of alkylated PAHs formed at different temperatures within the phenanthrene homologous series (C₀ =phenanthrene).*

PAH assemblage is inversely proportional to the temperature of formation (Blumer, 1976; Figure 15.2.1). Because the degree of alkylation and the resulting distribution of PAHs depend on the temperature of formation, the characteristic compositional profiles of these different sources can be used to help distinguish among different sources of PAHs in the environment (see Section 15.4). Once produced, PAHs are introduced into the environment through a number of pathways, which are described in Section 15.1.6.

15.2.1 Diagenic PAHs
Diagenic PAHs are those produced by natural processes that occur when organic matter is deposited in nature—in soils or sediments. These processes, collectively called diagenesis, begin shortly after deposition of the organic matter (Wakeham *et al.*, 1980). These are low temperature processes that occur after oxygen is depleted, and are believed to involve microorganisms, such as bacteria, though nonbiological processes may occur in tandem. These processes are generally termed "aromatization" reactions and produce

a range of aromatic biomarkers that can be found in some recent sediments, but which increase in both amount and complexity in fossil fuels.

These early diagenic processes produce relatively few individual PAHs. One of the most notable PAHs produced in this manner is the 5-ringed PAH, perylene. Perylene is commonly found in sediments of rivers, lakes, and oceans at a depth in the sediment where oxygen is reduced. Diagenesis of organic matter derived from diatoms and other plant material is thought to be a major source of perylene in anoxic marine sediments (Gschwend *et al.*, 1983; Louda and Baker, 1984; Venketesan, 1998).

Retene (1-methyl-7-isopropylphenanthrene) and smaller amounts of other C_2 through C_4 alkyl phenanthrenes occur frequently at trace concentrations in near shore marine sediments, particularly adjacent to forested shorelines. These phenanthrenes and some chrysenes are derived from the dehydrogenation of abietic acids, pimaric acids, and diterpenoid and triterpenoid precursors that are abundant in pine rosin, terrestrial plants, and wood ash (Tan and

Heit, 1981). The PAH assemblages produced by early diagenesis of these biological precursors have a simple composition involving only a few PAH species and, for this reason, can be distinguished from the complex multi-species petrogenic and pyrogenic PAH assemblages, described below. Retene has also been shown to be of biogenic origin (Wen *et al.*, 2000)

15.2.2 Fossil Fuel (Petroleum and Coal) PAHs

Over geological time and within petroleum reservoirs and coal beds in geological structures, another type of PAH is produced—*petrogenic* PAHs. Coal is derived from the remains of land plants that accumulated at first as peat. Upon burial, peat is converted to coal by millions of years of exposure to elevated temperatures and pressures during which diagenesis converts plant debris into a highly aromatic, three-dimensional structure (Teichmuller, 1987). The number of fused aromatic rings per structural unit within the coal structure varies; most coals contain three to five rings per structural unit, with some individual units containing up to 10 aromatic rings (Davidson, 1982; Berkowitz, 1988).

While PAHs represent a small percentage of the overall hydrocarbon content of petroleum (e.g., Figure 15.2.2), petrogenic PAHs are nevertheless a significant contributor to the PAH assemblages of most environmental samples. A typical crude petroleum may contain from 0.2% to more than 7% PAHs (Table 15.2.1). Petrogenic PAHs are formed at elevated pressures (and at higher temperatures than diagenic PAHs) within deeply buried layers of sediments. Petrogenic

PAHs are formed, for example, when biological organic matter from plankton is converted to petroleum. Petroleum or petrogenic PAHs and coal-derived PAHs are "fossil fuel" PAHs. The nature of the processes and their dependence on initial organic matter, burial conditions of temperature and pressure, subsurface migration, and biodegradation, which convert organic matter to fossil fuels, involves semi-random chemical processes. These facets result in the great molecular complexity of petroleum itself, the many chemical and physical properties and compositions of crude oil and coals, and the complexity of the many PAH structures that are found in fossil fuels. Hundreds to thousands of individual PAHs may be produced by nature during the processes that form fossil fuels. Nevertheless, while their compositions vary greatly, fossil fuels (crude oils and coal) have in common the existence of 2–6+ ringed PAHs. The abundance of aromatic hydrocarbons in petroleum usually decreases markedly with increasing molecular weight with a preponderance of alkylated structures associated with the 2–4 ringed compounds (see Figure 15.2.2).

The types of PAHs formed as fossil fuels are produced include a complex variety of parent (i.e., unsubstituted) and alkylated PAHs. Series of PAHs comprised of parent and substituted PAHs form many families or *homologous series* of PAHs. The phenanthrene homologous series of PAHs includes, for example, phenanthrene itself, plus a series of alkylated homologues of phenanthrene with many alkyl substitutions. The relative abundance of the alkylated PAHs within the petrogenic PAHs far

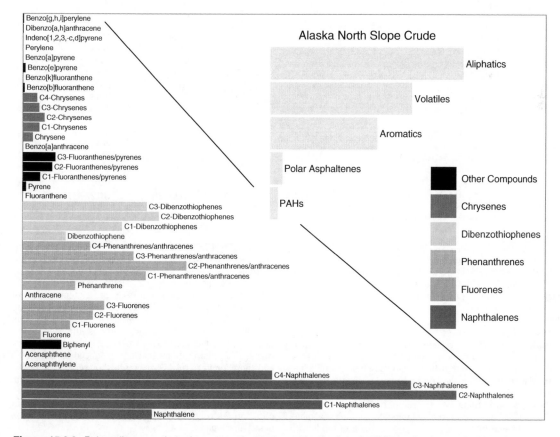

Figure 15.2.2 *Polycyclic aromatic hydrocarbons in Alaska north slope crude oil (from Boehm et al., 2001a) (see color insert).*

Table 15.2.1 *Concentrations of Individual and Total PAHs in a Typical Crude OIL, Coal, Coal Tar and Creosote. Concentrations are mg/kg. From Neff et al. (1998).*

PAH	Crude Oil	Coal	Coal Tar	Creosote
Naphthalene	1268	21	4044	60,274
C_1-Naphthalenes	3886	46	1814	17,987
C_2-Naphthalenes	4511	74	1193	8229
C_3-Naphthalenes	2988	54	619	2987
C_4-Naphthalenes	1000	31	213	775
Biphenyl	233	2.0	411	NA
Acenaphthylene	ND	ND	45	5248
Acenaphthene	47	ND	3817	22,699
Dibenzofuran	54	NA	3053	NA
Fluorene	267	1.8	4761	18,774
C_1-Fluorenes	521	4.3	896	2735
C_2-Fluorenes	682	5.6	485	1295
C_3-Fluorenes	420	7.3	5.7	ND
Anthracene	ND	3.7	5217	7073
Phenanthrene	370	18	16,231	44,572
C_1-Phenanthrenes/Anthracenes	718	32	1733	12,605
C_2-Phenanthrenes/Anthracenes	716	36	1856	4372
C_3-Phenanthrenes/Anthracenes	460	23	526	1425
C_4-Phenanthrenes/Anthracenes	154	16	1383	ND
Dibenzothiophene	29	3.1	926	6103
C_1-Dibenzothiophenes	63	5.1	278	1671
C_2-Dibenzothiophenes	83	10	172	876
C_3-Dibenzothiophenes	49	6.5	79	428
Fluoranthene	14	3.2	10,988	29,232
Pyrene	18	3.5	8517	21,131
C_1-Fluoranthenes/Pyrenes	101	16	5399	12,106
C_2-Fluoranthenes/Pyrenes	137	17	NA	NA
C_3-Fluoranthenes/Pyrenes	99	16	NA	NA
Benz(a)anthracene	2	4.4	4218	5149
Chrysene	32	3.2	4032	4108
C_1-Chrysenes	51	13	2151	2228
C_2-Chrysenes	67	17	687	605
C_3-Chrysenes	38	15	260	ND
C_4-Chrysenes	25	7.6	ND	ND
Benzo(b)fluoranthene	9	3.0	3848	2414
Benzo(k)fluoranthene	ND	ND	1524	2159
Benzo(e)pyrene	12	1.3	1894	NA
Benzo(a)pyrene	ND	3.1	2932	2222
Perylene	ND	ND	658	NA
Indeno(1,2,3-c,d)pyrene	ND	0.53	1597	718
Dibenz(a,h)anthracene	ND	ND	469	208
Benzo(g,h,i)perylene	4	1.8	1355	574
Total PAHs	**19,156**	**526**	**103,287**	**302,981**

ND Not detected. NA Not analyzed.

exceeds the abundance of the parent (i.e., unsubstituted) compound or C_0-phenanthrene. The fact that alkylated PAHs ≫ parent PAHs is a main feature of petrogenic PAHs. This is illustrated in the first chart of Figure 15.2.1 for a typical alkyl homologue distribution.

15.2.3 Pyrogenic PAHs

High temperature natural and anthropogenic processes also form PAHs. During these higher temperature processes, those organic compounds that escape complete combustion (oxidation to carbon dioxide and water) include the *pyrogenic* PAHs. Grossly similar *pyrogenic* PAHs are produced during wood burning (stoves, campfires) and the combustion of fossil fuels in internal combustion engines (gasoline, diesel, heated motor oils, etc.). The smoking of foods with certain woods produces pyrogenic PAHs, which become incorporated in the smoked food. Included in this pyrogenic

category are the products of high temperature processing of coals in coal gasification processes. The residuals of the coal gas processes are termed *coal tars*, and they are rich in the pyrogenic PAHs (Emsbo-Mattingly *et al.*, 2001).

Because the high temperature processes tend to destroy the more reactive alkylated PAHs, the production of unsubstituted PAHs are favored in pyrogenic processes. Thus, a major feature of pyrogenic PAHs is that the parent PAHs ≫ alkylated PAHs (Figures 15.2.1 and 15.2.3). As a result, PAH distributions that are produced by the hotter and more rapid pyrolysis or combustion-related processes are markedly different from those produced by petrogenic processes (see Figures 15.2.1 and 15.2.2). Also, pyrogenic PAHs are characterized by the higher abundance of the 4, 5, and 6+ ringed PAHs relative to PAHs found in most petroleum types.

During the combustion of organic matter, the organic compounds in the biomass/fuel are fragmented forming

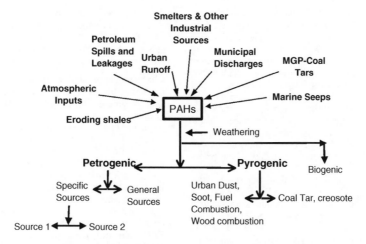

Figure 15.2.3 *Generalized PAH forensics flow chart.*

smaller compounds that react and build aromatic ring systems based on a series of benzene rings. The formation of PAHs and soot particles during the combustion processes are interrelated (Richter and Howard, 2000) with high molecular weight PAHs acting as molecular precursors to soot. The generation of PAHs during combustion appears to be independent of the actual fuel source (Jenkins *et al.*, 1996), the composition being more dependent on temperature and oxygen. Polycyclic aromatic hydrocarbons are produced in wood burning, both in natural forest fires and in emissions from wood stoves in residential heating. The amounts of PAH emissions can vary by 2 orders of magnitude depending on the type of vegetation burnt (Jenkins *et al.*, 1996). The PAH compositional differences attributable to fuel *types* vary only slightly although fuel type can affect the *amount* of PAHs produced (Ramdahl *et al.*, 1982; Jenkins *et al.*, 1996; Lima *et al.*, 2005). The composition of PAHs produced from biomass and/or fossil fuel burning depends on burning conditions (e.g., oxygen content, moisture content, etc.) with oxygen content being the most important factor determining the amount of PAHs produced. However, it appears that temperature is the most important factor in determining the composition (fingerprint) of pyrogenic PAHs produced in combustion processes. Lower burning temperatures, such as those which characterize some forest fires and cigarette smoking, generate alkyl-enriched pyrogenic PAH assemblages, while higher temperature favor the production of unsubstituted or parent compounds (Blumer, 1976). In diesel fuel combustion emissions, Jenkins and Hites (1983) showed that alkylated PAHs increased with decreasing combustion temperature.

Automobile emissions result in pyrogenic PAH emissions as a result of various factors: the type of fuel, the type of engine, age of the vehicle, and other factors. Typically uncombusted fuel in emissions of the more inefficient diesel engines contribute petrogenic PAHs to the pyrogenic emissions (Miguel *et al.*, 1998; Williams *et al.*, 1986, 1989), and pyrogenic PAHs also accumulate in waste crankcase oil, resulting in a mixture of petrogenic and pyrogenic PAHs (Pruell and Quinn, 1988).

15.2.4 Biogenic PAHs

Certain PAH precursor compounds are biosynthesized in nature, although the importance of direct biosynthesis of PAHs *per se*, remains uncertain. Though the contribution of microorganisms to the production of PAHs in nature has been reported, their contribution may be more to the oxygen-containing aromatic compounds rather than to PAHs themselves. It is well known that other PAH precursor compounds (e.g., abietic acid) exist in abundance in certain tree resins (e.g., conifer resins) and that specific PAHs are formed from the diagenesis or the combustion of these resins. For example, *retene*, a specific C_4-phenanthrene isomer (1-methyl-7-(1-methylethyl)-phenanthrene), is ubiquitous in residues from these plants and can be found in high relative abundance in sediments of pristine northern environments. Retene has also been shown to have an algal and/or bacterial origin (Wen *et al.*, 2000). *Simoneltite*, a substituted PAH compound, is also found in abundance where organic conifer residues exist. These specific, singular PAHs are found in coastal sediments around the world.

15.3 POLYCYCLIC AROMATIC HYDROCARBON SOURCE ASSEMBLAGES

Once they are produced by petrogenic and pyrogenic processes, PAHs may be introduced into the environment through a number of pathways (Figure 15.2.3). These include:

- Natural oil seeps (NRC, 1985);
- Erosion of source rocks (shales) (Boehm *et al.*, 2001a);
- Petroleum spills and releases (NRC, 1985);
- Urban and municipal runoff (NRC, 1985) inclusive of atmospheric deposition of combustion products;
- Industrial emissions—e.g., smelters (Naes and Oug, 1997, 1998);
- Historical-coal gasification (i.e., manufactured gas plant residues) seepages and runoff (EPRI, 2001);
- Creosote preservation-based leaching and runoff (Emsbo-Mattingly *et al.*, 2001);
- Biomass burning (wood waste, controlled burns; trash and waste burning) (ASTDR, 1995).

Polycyclic aromatic hydrocarbons can enter the environment on local, regional, and global scales (Neff, 1979). Point sources, such as municipal or industrial outfalls, are on the local scale, and are generally made up of mixtures of PAHs that are combustion- or oil-related. Non-point sources (e.g., rainfall runoff or atmospheric deposition) are found on regional scales, and are also made up of PAHs from multiple primary sources. Wide-field atmospheric deposition is a global source that distributes primarily pyrogenic PAHs (Ohkouchi, *et al.*, 1999) to remote regions of the earth. Airborne transport of PAHs on soot particles from forest fires and the combustion of coal and oil have been established as a major mechanism for the distribution and delivery of PAHs to soils and sediments on the regional and global scales.

The source(s) of PAH in coastal and urban sediments was first investigated in the early to mid-1970s. The origins of PAH in sediments were shown to be both non-alkylated and alkylated PAH distributions (Giger and Blumer, 1974; Youngblood and Blumer, 1975), rather than individual, non-alkylated PAH. These petrogenic and pyrogenic source categories could be readily distinguished on the basis of their alkyl PAH distributions (Youngblood and Blumer, 1975; Lee *et al.*, 1977; Laflamme and Hites, 1978). The petrogenic and pyrogenic PAH can be derived from both "point" and "non-point" sources.

In the latter half of the 18th and first half of the 20th centuries, manufactured gas plants (MGP) produced gas from oil and from coal, resulting in residues collectively referred to as "coal tar". These coals tars contain large quantities (up to 70–80% by weight) of pyrogenic PAHs as a result of the high temperature processing of coal in the plants, largely dominated by the 2- and 3-ringed compounds (e.g., naphthalene, anthracene) in fresh coal tars and 3- to 6-ringed PAHs in more weathered material. These PAHs have entered the environment via groundwater flow and runoff to coastal rivers and sediments.

Other point sources might include direct or indirect discharges from former or existing industrial facilities (e.g., petroleum handling, aluminum smelting, manufactured gas production, tar distillation, rail yards, etc.) as well as loading/unloading facilities (e.g., creosote pilings), marinas, discharge canals, and storm water outfalls. Storm water runoff from paved urban and suburban roadways and developments, as they are discharged into surface waters, serve as local, point sources for PAH that are derived from 'non-point' sources, for example, atmospheric (combusted) particulates and fugitive (dripped and leaked) petroleum washed from the surrounding urban roadways, parking lots, vegetation, and structures during rainfall events. Other non-point sources of PAH can include recreational boat traffic, commercial ship traffic, general runoff (i.e., not entering from a specific outfall location), and direct atmospheric particulate deposition to the waterway.

Understanding how and where PAHs enter the environment is important in conducting environmental forensic studies and especially in determining the PAH background into which emissions from a specific PAH source are added. On a global basis and in areas remote from urban influence, PAHs from pyrogenic processes transported over large distances are the principal source of background concentrations—more important than petrogenic PAH inputs—though the levels tend to be very low. On more localized scales, background PAH concentrations are generally much higher, and PAHs from urban runoff together with combustion-related PAH inputs are very important contributors to most receiving environments. In selected geologically active environments, oil seeps and erosion from oil source rocks and coal result in elevated concentrations from natural sources of PAHs. These background concentrations of PAH are of particular significance when the potential effects (i.e., incremental addition) of PAHs from new oil and gas exploration projects or effects of oil spills are being evaluated as part of new project plans or environmental impact and damage assessments.

15.4 ANALYTICAL STRATEGIES

The choice of analytical methods for PAHs starts with a consideration of the end use of the analytical results, which, in turn, drives the analytical strategy and the analytical methodology. For many environmental applications the end uses depend on determining the specific target PAH analytes of focus or concern and then determining the concentrations of these PAHs compounds. In general, the basic needs for PAH data are focused on the following:

- Characterization and delineation—which uses data to determine the nature and extent of PAH contamination—soils, sediments. Typically "total PAHs" are used for these evaluations.
- Exposure assessments—in an ecological risk assessment context (US EPA, 1992) or in a human health risk assessment context. In the former case "total PAHs" and the specific concentrations of key PAH compounds for which benchmarks may exist (e.g., water quality criteria, sediment quality guidelines, wildlife toxicology, etc.). For human health risk assessments, PAH concentrations in fish, soils, household dust, etc. usually focus on the concentrations of individual compounds such as the 5-ringed PAH, benzo(a)pyrene, and "toxicity equivalent factors" applied to concentrations of other PAHs. Such TEQs report "concentrations" in terms of the relative toxicity of particular compounds (in the case of PAHs usually compared to benzo(a)pyrene) rather than absolute concentrations.
- Source identifications and forensics—where the source(s) of a single compound are of interest or the apportionment or allocation to various source(s) of the full assemblage of PAHs is the goal of the analyses.

For characterization/delineation and risk assessments, a full suite of PAHs may not be absolutely necessary for data end uses. However, PAH environmental forensics relies on a more comprehensive set of analyte targets, which define the "chemical fingerprint." Therefore, as a general, integrated and cost-effective analytical strategy, where questions involve characterization as well as source determinations, data acquisition plans should start with the more demanding needs of PAH forensics.

The number and list of possible PAHs in environmental samples is on the order of hundreds of compounds (Neff, 1979; Bjorseth, 1983). However, historically, the PAHs of principal environmental concern, as designated by environmental regulators, have been those 16 listed on the United States Environmental Protection Agency's (US EPA) Priority Pollutant List (Figure 15.1.1; Table 15.4.1). However, the assemblages of PAHs which are essential to forensic investigations and which address PAH source determinations include a wider list (Table 15.4.1; Boehm *et al.*, 1983, 1997, 2001b; Boehm and Farrington, 1984; Sauer and Boehm, 1991, 1995; Stout *et al.*, 2002; Federal Register 40CFR300). When executed properly, these methods can be used collectively to define source signatures for PAH allocation purposes.

All of the priority pollutant PAH are "parent" or unsubstituted PAHs and are "non-alkylated." They contain no carbon side chains. However, many "alkylated" PAH exist that contain carbon side chains of varying number, length, and location. These C_1 (one carbon side chain) to C_4 (four carbons on various side chains) PAHs and other parent (C_0) PAHs (e.g., biphenyl) are not on the EPA's Priority Pollutant List, and therefore are not included in standard EPA Method 8270 (US EPA, 1986). Furthermore, the incorporation of key heterocyclics (i.e., polycyclic aromatic compounds, or PACs), such as dibenzofuran and dibenzothiophene, are important compounds in an overall target analyte list developed to discern the sources of PAH assemblages. It is both the parent and alkylated PAH (and PAC) (C_0- to C_4-PAH) that when combined with other compounds are most useful in environmental forensic investigations (Table 15.4.1).

Table 15.4.1 *Comparison of PAH Analytes Commonly Used in Environmental Forensic Investigations (ΣPAH_{50}) to the US EPA "Priority Pollutant" List (ΣPAH_{16})*

Analyte/Analyte Group	Abbr.	Analyte/Analyte Group	Abbr.
Naphthalene	N0	**Pyrene**	PY
C1-naphthalenes	N1	C1-fluoranthenes/pyrenes	FP1
C2-naphthalenes	N2	C2-fluoranthenes/pyrenes	FP2
C3-naphthalenes	N3	C3-fluoranthenes/pyrenes	FP3
C4-naphthalenes	N4	**Benz(a)anthracene**	BaA
Biphenyl	Bph	**Chrysene**	C0
Acenaphthylene	Acl	C1-chrysenes	C1
Acenaphthene	Ace	C2-chrysenes	C2
Dibenzofuran	DbF	C3-chrysenes	C3
Fluorene	F0	C4-chrysenes	C4
C1-fluorenes	F1	Benzo(a)fluoranthene	BaF
C2-fluorenes	F2	**Benzo(b)fluoranthene**	BbF
C3-fluorenes	F3	**Benzo(j,k)fluoranthene**	BkF
Anthracene	AN	Benzo(e)pyrene	BeP
Phenanthrene	P0	**Benzo(a)pyrene**	BaP
C1-phenanthrenes/anthracenes	P1	Perylene	Per
C2-phenanthrenes/anthracenes	P2	**Indeno(1,2,3-c,d)pyrene**	ID
C3-phenanthrenes/anthracenes	P3	**Dibenzo(a,h)anthracene**	DA
C4-phenanthrenes/anthracenes	P4	**Benzo(g,h,i)perylene**	BgP
Dibenzothiophene	D0	Dibenzo(a,e)pyrene	DeP
C1-dibenzothiophenes	D1	Dibenzo(a,h)pyrene	DhP
C2-dibenzothiophenes	D2	Dibenzo(a,l)pyrene	DlP
C3-dibenzothiophenes	D3	Dibenzo(a,i)pyrene	DiP
C4-dibenzothiophenes	D4	Dibenzo(a,e)fluoranthene	DeF
Fluoranthene	FL	Anthanthrene	A

bold - 16 Priority Pollutant PAH

In addressing the "concentration of PAHs in the environment" the very first decision to be made is "which PAHs"? There are a number of PAH target analyte lists that are used—the US EPA Priority Pollutant List (US EPA, 1986), the NOAA Status and Trends PAH list (NOAA, 1993), the "oil spill" alkylated PAH target lists (Sauer and Boehm, 1991, 1995), and others. Different lists produce different results. There are two general PAH analytical approaches that have been used.

The first approach is the analyses of the US EPA's 16-PP PAHs, which consist of 2 to 6 rings (see Table 15.4.1). As part of the regulatory environment, the US EPA has specified methods to identify and measure industrial chemicals of concern using gas chromatography-mass spectrometry (GC/MS) methods. Those most frequently cited are the US EPA's SW-846 Methods (US EPA, 1986). This list, comprising "total PAHs" as a sum of the 16 Priority Pollutants (i.e, ΣPAH_{16}), has been used extensively around the world. It has also been expanded slightly for certain large-scale environmental studies such as the US and International Mussel Watch and Status and Trends monitoring programs (NRC, 1980; NOAA, 1993). Essentially, this approach represents only a screening for important PAHs, because many PAHs that can be found in environmental samples are not included.

The second analytical approach is to target an "extended list" of PAHs, again from 2 to 6 rings, but also inclusive of alkylated PAHs and selected heterocyclic compounds (also referred to as "polycyclic aromatic compounds" or PACs). This extended list is shown in Table 15.4.1 (i.e., ΣPAH_{50}). The target list includes not only all of the priority pollutant PAHs, but also other important compounds, such as alkylated PAH and dibenzothiophenes, found in petroleum, coal tars, and other sources (Federal Register 40CFR300, Sauer and Boehm, 1991, 1995; Boehm *et al.*, 2001b; Stout *et al.*, 2001, 2002; Wang and Fingas, 2003). This approach has

been used for petroleum and other fossil fuel related PAH studies for over 20 years and is now used fairly routinely for modern PAH monitoring and forensics investigations studies. The use of such an approach is essential to many PAH source determinations especially those focused on urban harbor and river sediments. The "extended PAH method" (i.e., GC/MS method [Method 8270, Semivolatile Organic Compounds by Gas Chromatography/Mass Spectrometry (GC/MS); US EPA. (1986)], as modified to include alkylated PAHs per Federal Register 40CFR300, is now considered standard practice (Boehm *et al.*, 1995; Boehm *et al.*, 1997; Stout *et al.*, 2002; Neff *et al.*, 2005). Intermediate extended PAH lists (e.g., ΣPAH_{33}) have been recently adapted for ecological studies (Ozretich *et al.*, 2000). Beyond the PAHs themselves, the target list for PAH forensic investigations focuses on other potential indicator analytes (other heterocyclics; cyclic alkanes, for example the triterpanes), which may be helpful in discerning PAH sources (Boehm *et al.*, 2001b; Emsbo-Mattingly *et al.*, 2001; Stout *et al.*, 2002). The importance of using the alkylated PAH list (Table 15.4.1) is apparent from an evaluation of the data results from the PAH analysis of a petrogenic PAH source sample analyzed by the two different lists. As shown in Figure 15.4.1 a focus on only the ΣPAH_{16} yields a smaller group of analytes and does not yield an optimal set of PAH data for PAH forensics studies.

The use of these extended PAH methods in an environmental forensic laboratory requires strict adherence to requirements that are both included and excluded from the written method. For example, the method requires a five-level calibration curve to which the environmental forensic laboratory will add mass discrimination controls, numerous indicator analytes, and reference materials. The recommended specific use of these methods is presented in the Federal Register (40CFR Subchapter J, Part 300, Subpart L, Appendix C, par. 4.6.3 to 4.6.5).

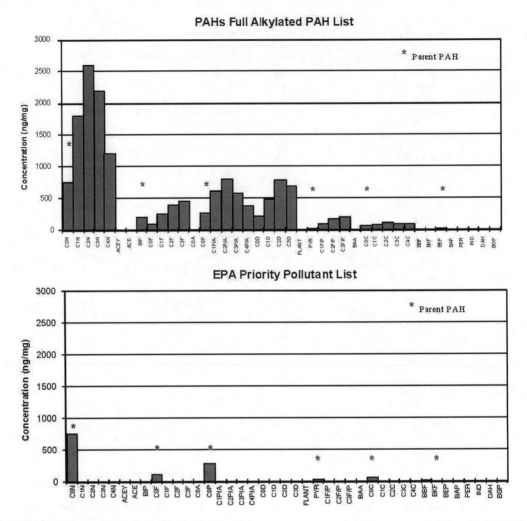

Figure 15.4.1 *Polycyclic aromatic hydrocarbon analyses of crude oil sample: A) Full parent and alkylated list (ΣPAH_{50}); B) Priority pollutant target PAH list (ΣPAH_{16}).*

Over the past two decades the use of the term "total PAH" has been inconsistent in the literature. While some refer to "total PAH" as the sum of only the 16 Priority Pollutants, or ΣPAH_{16}, others, especially in a petroleum and/or PAH environmental forensics context, use "total PAH" to represent a wider suite of target analytes (e.g., ΣPAH_{50}) (Boehm *et al.*, 2005). It is clear from the environmental forensics literature (e.g., Boehm *et al.*, 1997; 2005; Stout *et al.*, 2002) that "total PAHs" or ΣPAH_{50} are now considered to represent those analytical targets shown in Table 15.2.1. Other lists (e.g., Ozertich *et al.*, 2000; ΣPAH_{33}) focus on other groups of PAHs. All too often, the results from "total PAH" determinations from one list are inappropriately compared to "total PAH" values from other lists. Depending upon the assemblage of PAHs in environmental samples, the comparison of ΣPAH_{16} to ΣPAH_{50} may range from about $\Sigma PAH_{16} \leq 0.2[\Sigma PAH_{50}]$, where petrogenic PAH dominate in the samples, to $\Sigma PAH_{16} \geq 0.8[\Sigma PAH_{50}]$, where pyrogenic PAH dominate (Figure 15.4.2).

15.4.1 Quantitative Analyses

Quantitative analysis of PAHs is a key to accurate and precise environmental forensics studies. While ratios of key

compounds and patterns of PAHs are at the heart of source determinations and apportionment, as will be seen below, accurate concentrations of individual constituents must be determined before such ratios are constructed (Sauer and Boehm, 1995; Douglas *et al.*, 2004). Quantitative analysis requires the calculation of absolute concentrations of individual PAHs through a series of steps as described in EPA Method 8270, including the use of surrogate standards, added at the start of analysis; quantitation standards, added prior to instrumental analysis; application of GC/MS response factors to each analyte; and correction of results for recovered constituents (i.e., "surrogate correction").

The PAHs (and other organic compounds) in an environmental sample (water, sediment, tissue, etc.) are initially transferred to an organic solvent. An important component in this first set of steps is the addition of "surrogate PAHs" (i.e., known quantities of known compounds) to the sample. These surrogates are then used to insure that the entire analytical method and the results are valid and to assist in quantification of the amounts of PAHs, which are extracted. Data for PAH forensics is typically "surrogate-corrected" wherein the PAH analyte results are corrected to 100% recovery using the percent recovery determinations from the

- **Petrogenic**
  - ➤ Alkyl > Parent
  - ➤ Little 4 to 6 Ring

- ## Pyrogenic – Type 1
  - ➤ Parent > Alkyl
  - ➤ High 2 and 3 Ring

- ## Pyrogenic – Type 2
  - ➤ Parent > Alkyl
  - ➤ High 4 to 6 Ring

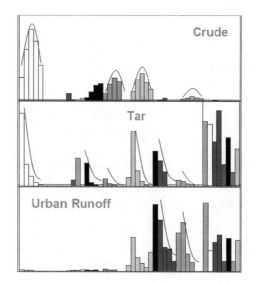

Figure 15.4.2 *Typical characteristics of PAH assemblages for petrogenic and pyrogenic sources (see color insert).*

PAH surrogates (Page *et al.*, 1995a, b; Douglas *et al.*, 2004; Federal Register 40CFR300).

The sample extract in an organic solvent is then reduced in volume and subjected to one of several separation or fractionation procedures, which both eliminates many interfering organic chemicals, and concentrates the PAHs in one fraction. Detection limits may be decreased by increasing sample size and by concentrating the extracts to smaller volumes (Douglas *et al.*, 2004).

The concentrated sample fraction is injected into a GC/MS instrument and the individual parent and alkylated PAHs are detected, identified, and quantified by the computer-assisted mass spectrometer. Different MS configurations (i.e., full scan or selected ion monitoring (SIM)) can be applied to the analyses. Response factors used in the GC/MS analyses are those determined for the PAH parent compounds in the homologous series. So, for example, the instrumental response factor determined for phenanthrene in the instrumental standards is applied to phenanthrene and all of the alkylated phenanthrenes in the sample (40CFR Subchapter J, Part 300, Subpart L, Appendix C, par. 4.6.3 to 4.6.5).

The US EPA Standard Methods specify detection limits that are appropriate for screening of samples for the presence of large amounts of PAHs, but are generally too high for ecological risk or related impact studies. Many regulatory limits for PAHs are such that it is important to be able to detect and reliably quantify PAHs at low levels. For example, the current limit for total PAHs in ambient water in the State of Alaska is 10 parts per billion. Such a benchmark requires that individual PAHs be quantified reliably at levels 1/10th to 1/50th of the total PAH limit (i.e., 0.2–1 ppb) (Douglas *et al.*, 2004) (see Table 15.5.2). Of great significance from the environmental forensics aspects of analyses is the need to be able to optimize detection levels (i.e., decrease them) so as to facilitate the detection and accurate quantification of key PAH compounds. The application of factors to "censured data" (i.e., data below the regulatory minimum detection limit (MDL)) such as ½ of the detection limit as is often applied to analyte concentrations below the MDL are not useful for forensics studies which require that accurate concentrations be obtained for ratio and or statistical data analysis methods.

It is important to have explicit QA/QC plans for all PAH environmental forensics analyses. The components of a QA/QC plan should include those elements recommended by the US EPA. These elements include data quality objectives such as:

- Instrument calibration criteria
- Practical quantitation limits
- Method detection limits
- Use of procedural blanks and
- Use of laboratory matrix spikes and matrix spike duplicates.

QA/QC plans should explicitly contain: the use of Standard Reference Materials (SRMs) from NIST or other appropriate certifying agencies; the use of "check samples" such as a standard oil sample containing all of the alkylated PAH analytes (see Sauer and Boehm, 1991 for additional details).

15.5 POLYCYCLIC AROMATIC HYDROCARBON ENVIRONMENTAL FORENSICS

The questions which drive the need for the use of PAH environmental forensics include several that are central to the environmental forensics practice overall. Since PAHs are multisourced and are ubiquitous in the environment, one of the most important central issues in PAH forensics is the identification of the background PAH fingerprint and source regime. Release of coal tar into a river, the spill of petroleum into a coastal area, the burning of wood and the use of creosote at a wood treatment plant with possible deposition onto adjacent soils, and the emissions from an aluminum smelting operation all introduce different assemblages of PAHs into the environmental. However, all environmental forensic investigations of these and similar releases must consider the pre-existing baseline or background due to both global scale inputs of PAHs (LaFlamme and Hites, 1978) and local inputs into coastal regions, industrialized areas, and all areas of human influence (Page *et al.*, 1999).

Beyond an elucidation of the pre-release background or existing baseline, the second major driver is the differentiation of multiple inputs and specific PAHs sources from known inputs or discharges. Since PAHs are ubiquitous, analysis of PAH assemblages must have the ability to determine both the major source types (i.e., petrogenic and pyrogenic) comprising the assemblage, and also the contribution of specific sources. The analysis must lead to the determination not only if a specific PAH source is present in a sample, but also what percentage of that source comprises the PAH assemblage.

This is the ultimate challenge for the PAH forensic environmental chemist—to identify and quantify a specific PAH source in a "study area" (as represented by a set of samples), sometimes long after it has been introduced to the environment (and therefore altered by weathering), and after it becomes mixed with similar PAH assemblages. In urban areas the PAH forensic challenge may be represented, for example, by the need to identify and apportion pyrogenic PAHs in sediments from a historical manufactured gas plant, from a combined PAH assemblage that consists of pyrogenic PAHs, from a city storm water discharge, and weathered petrogenic PAH from shipping and petroleum terminal operations. In rural areas the analogous challenge may be to determine the source of a single carcinogenic PAH (e.g., benzo(a)pyrene or B(a)P) in household dust through the matching of the composition of the dust to multi-compound PAH sources assemblages from: 1) a wood treatment (creosote) plant operation, 2) local diesel and auto particulate emissions, and 3) backyard burn barrels, all of which contain B(a)P. The use of robust analytical chemical and data analysis strategies and methods applied to a well-designed collection of an unbiased set of environmental samples and complete set of candidates source samples are the keys to meeting the challenges.

The goals of comprehensive PAH forensics investigations are to: a) chemically characterize an environmental assemblage of PAHs; b) determine potential source types comprising the assemblage; c) determine specific source(s) of the PAHs potentially contributing to the observed assemblage; d) "apportion" or "allocate" fractions of the PAH assemblage to those specific sources. Although questions related to the age of PAHs in contaminated sites or sediments arise and "age dating" has become one facet of PAH forensics, such age-dating of PAHs is almost always approached from either parallel lines of investigations related to historical practices (i.e., "industrial archeology") and/or PAH forensic chemistry supported by sediment coring and sediment age-dating. Direct measures of contemporary PAH age from chemical measurements alone are infeasible, although, as discussed below, radiocarbon (^{14}C) dating can be used to discern "old" fossil carbon (e.g., oil) from "newer" contemporary carbon (e.g., wood).

The ability of PAH forensic chemists and statisticians to address these goals depends on several factors. First is a consideration of the type of questions being posed and addressed and the design of the study to address those questions. Second, and very importantly, is the quality of the samples—both source samples and environmental media samples. Has careful attention been paid to identifying the possible "suspect" sources and to the collection of a set of these sources? Has a representative set of environmental samples been obtained? Third is the set of analytical specifications applied to the problems. Are the data generated focused on the problems posed? Are the data of high enough quality (i.e., meeting pre-specified data quality objectives—e.g., detection limits, reproducibility, low blanks levels, etc.). The fourth consideration is the set of data analysis techniques applied to the PAH data. Are the techniques aimed at differentiating and quantitatively allocating an environmental assemblage of PAHs to its component sources within a known and acceptable degree of uncertainty?

15.5.1 Source, Environmental, and Background Sampling

Forensic environmental chemists are usually quick to discuss environmental analytical chemistry and chemical techniques and the power of those techniques to answer questions. However, a high quality investigation begins with a well-developed sampling strategy, which includes source samples, environmental samples that are believed to have been impacted, and background samples taken at locations believed not to have been impacted by a specific point source or set of sources.

Source sampling may be a relatively straightforward process or it may be quite difficult and reliant on historical or surrogate samples. In the case of an oil spill, sampling of the cargo oil and other possible suspect cargoes may be sufficient for determining the sources of petrogenic PAHs in sediment or biota samples, for example. In areas of multiple historical petrogenic PAH inputs (e.g., Prince William Sound Alaska) multiple candidate source samples have been required to sort out the influence of spilled *Exxon Valdez* oil versus others sources (Kvenvolden *et al.*, 1995; Boehm *et al.*, 2001b). In the case of an historical and/or an ongoing release of coal tar, for example, from an urban harbor site, source samples may only be available from ongoing seepages from that site or from a set of representative (not a single sample) subsurface contaminated soil samples taken from the onshore site itself.

In some studies, where a temporal understanding of the history of PAH source inputs is of relevance, it is necessary to collect sediment core samples and to conduct radiogenic isotopic age-dating (e.g., ^{137}Cs or ^{210}Pb) along with PAH "fingerprinting." Historical source samples in aquatic environments are often available in sediment cores if the cores are undisturbed over time, for example, removed from areas of historical sediment dredging. Such cores contain materials, which were deposited over time—the deeper the older—and may contain a fairly well-preserved record of sources (Page *et al.*, 1995a) (see Figure 15.5.1) that can be dated. An added feature of using cores to reconstruct a historical source record is that sediment core sections containing a source fingerprint or a changing set of fingerprints over time can be dated. This permits an evaluation of the PAH signatures present in sediments deposited at different time intervals that, along with good historical information, can be used to identify candidate historical sources that may have existed along a waterway.

Multiple source samples are often called for in complex site investigations. Multiple source inputs to an urban waterway in one coal tar forensic investigation (Boehm, unpublished) have included sediment residues taken from a manhole close to the terminus of a storm sewer; coal tar seepage from shoreline banks; tar residues taken from various manholes on an historic manufactured gas plant site; and offshore seepages of oily material. Another study area in an adjacent waterway consisted of site petroleum source samples from machinery used at the site; sediments from manholes contributing to a drainage ditch into the waterway; directional particulate air samples indicative of industrial plant emissions; and creosote pilings on a loading pier adjacent to the site.

15.5.2 Source Type Differentiation (Petrogenic vs. Pyrogenic)

The basic PAH source challenge, and the one longest studied, has been the separation of a PAH assemblage into petrogenic and pyrogenic source types (Figure 15.4.2). These two source categories can be readily distinguished on the basis of their alkyl PAH distributions (Youngblood and Blumer, 1975; Hites *et al.*, 1977; Lee *et al.*, 1977; Laflamme and

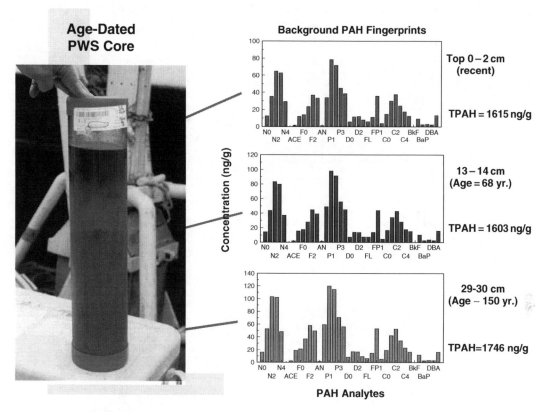

Figure 15.5.1 *Historical reconstruction of PAHs in sediments of Prince William Sound (from page et al., 1995a) (see color insert).*

Hites, 1978). This differentiation is usually performed as a first step in a forensic PAH allocation study. The use of source ratios for this and other facets of the PAH forensics science is presented in Table 15.5.1. Blumer (1976) was the first to describe the basic source type differentiation. Boehm and Farrington (1984) used PAH differences to quantitatively ascribe a PAH assemblage into its petrogenic and pyrogenic component parts through an index, the "Fossil Fuel Pollution Index" (FFPI). In this approach, each PAH analyte (Table 15.4.1) is assigned to either a petrogenic, pyrogenic, or mixed category. This classification is based on the general features of petrogenic and pyrogenic source materials as described earlier in this chapter, and the expected weathering they endure upon release into the environment. As previously mentioned, the petrogenic source materials (in spite of weathering) are generally enriched in the alkylated PAH, rather than the non-alkylated parent PAH (Figure 15.4.2). The opposite is true for the pyrogenic source materials (Figure 15.4.2). In addition, petrogenic sources are generally enriched in 2- and 3-ring PAH whereas pyrogenic sources are generally enriched in 4- to 6-ring PAH. (Clearly, some coal-derived liquids, such as coal tar and creosote, are exceptions to this generalization) (Figure 15.5.2) This approach provides a "gross" level of separation—between the petrogenic and pyrogenic PAH groups. Several PAH (e.g., phenanthrene) originate in both petrogenic and pyrogenic sources and therefore are considered to have 'mixed' sources. Thus in this simple allocation method, half of each of these 'mixed' PAH is apportioned to each PAH category. This basic approach is still used today (Figure 15.5.3) along with a variation developed for marine

monitoring studies (NOAA, 1993), which focuses on a less refined version—high molecular weight PAHs (4-5 ringed) versus low molecular weight PAHs (2- to 3-ringed) (HMWPAH/LMWPAH). More recently, more refined ratios consisting of fewer compounds have been used successfully to describe the pyrogenic versus petrogenic components of a PAH assemblage (Table 15.5.2).

15.5.3 Petrogenic PAH Differentiation

Petroleum fingerprinting is a well-established methodology dating back to the 1970s (Bentz, 1976) that relies on a combination of chemical characteristics revealed by a group of analytical methods that examine bulk chemical and molecular similarities and differences to match petroleum in the environment with its source(s). The data from these techniques have been used to allocate petroleum hydrocarbons in environmental samples to their component sources. Contemporary petroleum fingerprinting (Overton *et al.*, 1981; Boehm *et al.*, 1983; Sauer and Boehm, 1995; Boehm *et al.*, 1997; Stout *et al.*, 2002; Wang and Fingas, 2003. Relies on molecular characteristics of the hydrocarbons and on specific classes of hydrocarbons in particular, depending on the specific petroleum product of concern (i.e., refined products versus crude oil, for example).

Many of the molecular techniques have been based on GC/MS identifications of specific compounds. In particular, differentiation of petrogenic PAHs has been a well-studied facet of PAH forensics for many years (Sporstol *et al.*, 1983; Brown and Boehm, 1993; Stout *et al.*, 2001; Wang and Fingas, 2003). The PAH forensics focusing on differentiation of petrogenic sources have largely focused on key source

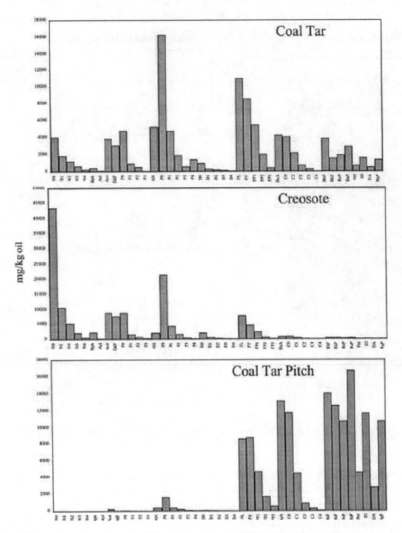

Figure 15.5.2 *Polycyclic aromatic hydrocarbon histograms representing various coal tar derived pyrogenic sources of PAH.*

ratios within the PAH assemblage, which are both characteristic of the sources and are different among the various source (Overton *et al.*, 1981; Boehm *et al.*, 1983). Specific ratios of key alkylated PAH components have been used in "double ratio plots" to differentiate sources of petrogenic PAH (Boehm *et al.*, 1983; Brown and Boehm, 1993) and create allocation models for source apportionment (Boehm *et al.*, 1997) (see Table 15.5.2 and Figure 15.5.4). While in this approach the sum of all of the isomers in any one alkyl-homolog group are summed (e.g., C_3 dibenzothiophenes), each alkyl homolog group actually consists of a number of individual compounds, thus providing additional detail for PAH forensic identifications (e.g., Figure 15.5.5) (Boehm *et al.*, 2001b).

Very often petrogenic PAH forensics methods rely on other diagnostic compounds such as the molecular "biomarkers" (e.g., steranes, triterpanes, triaromatic steranes, etc.) that have been developed for petroleum geochemical uses. In combination with each other and with PAH ratios, biomarkers have been applied to PAH environmental forensics problems (Bence *et al.*, 1996; Boehm *et al.*, 1997; Page *et al.*, 2002; Wang and Fingas, 2003).

15.5.4 Pyrogenic PAH Source Differentiation

Significant sources of pyrogenic PAH to urban waterways include direct atmospheric deposition of combustion particles, which are also found in stormwater runoff. These urban particulates contain gasoline and diesel combustion particulates as well as, in the case of stormwater, oily (petrogenic) roadway runoff (crankcase oil dripping). Also of prominence in some urban areas are discharges from aluminum smelting operations (particularly those employing Soderburg processing, for example, Naes and Oug, 1998) and the products and byproducts of manufactured gas production. The aluminum smelting industry produces pyrogenic PAH in the course of heating mixtures of petroleum coke and coal tar pitch (i.e., potliner) along with the aluminum ore, which yields abundant PAH-laden particulates, scrubber sludge, and "spent" potliner. Manufactured gas production (MGP) yielded coal- and petroleum-derived liquid tar residues (coal tar and petroleum tar) that were produced in the course of heating coal or oil during gas production (Gas Research Institute, 1987). These tarry byproducts of MGP were often further processed (distilled) into additional liquid materials enriched in pyrogenic PAH (e.g., creosote, which is/was

Table 15.5.1 *Diagnostic PAH Source Ratios*

Diagnostic Ratio	Application	Reference
FFPI	Petrogenic/pyrogenic PAH allocation	Boehm and Farrington (1984)
(Fluoranthene+pyrene) ($C_2 + C_3 + C_4$ phenanthrenes)	Fraction of pyrogenic PAHs	Page *et al.* (2004)
[Acenaphthene+acenaphthylene+anthracene+ uoranthene,+pyrene,+benzo(a)anthracene+ chrysene+benzo(b)fluoranthene+ benzo(k)fluoranthene+benzo(e)pyrene+ benzo(a)pyrene+indeno(1,2,3-cd)pyrene+ dibenzo(a,h)anthracene+benzo(g,h,i)perylene] ÷ (total PAH)	Pyrogenic PAH fraction	Boehm *et al.* (2004)
Σ4-6 ringed PAH/Total PAH	Pyrogenic PAH fraction	Kennicutt (1998)
ΣMP/P	Differentiation of fossil fuel sources (oil vs, coal); Petrogenic vs. pyrogenic PAHs	Gschwend and Hites (1981); Garrigues *et al.* (1995)
1,7-DMP/2.6-DMP	Differentiate PAHs from burning of biomass vs. fossil fuel combustion	Benner *et al.* (1995)
Double Ratio Plots (C_2Dibenzothiphene/C_2 Phenanthrene) vs. C_3Dibenzothiphene/C_3 Phenanthrene	Petrogenic source differentiations	Boehm *et al.* (1983; 1997); Page *et al.* (1995b); Brown and Boehm (1993)
Double Ratio Plots (C_3Dibenzothiphene/C_3 Phenanthrene) vs. (C_3Dibenzothiphene/C_3 Chrysene)	Weathered oil differentiation	Douglas *et al.* (1996); Sauer and Boehm (1995)
P/A	Differentiation of fossil fuel sources (oil vs. coal): Petrogenic: P/A >10 Pyrogenic: P/A <10	Gschwend and Hites (1981); Budzinski *et al.* (1997)
4.5DMP/ΣMP	Oil vs. coal	Gschwend and Hites (1981); Garrigues *et al.* (1995)
FL/PY	Pyrogenic source differentiation	Gschwend and Hites (1981); Emsbo-Mattingly and Boehm (2003)
Double ratio plot: P/A vs. FL/PY	Pyrogenic PAH source differentiation	Budzinski *et al.* (1997)
BaP/BeP; BbF/BkF; BaA/C	Pyrogenic PAH source types (wood, auto emissions, coal)	Dickhut *et al.* (2000) Costa and Sauer (2005)

P = phenanthrene; DBT = dibenzothiophene; C = chrysene; MP = methylphenanthrene, DMP = dimethylphenathrene; A = anthracene; FL = fluoranthene; PY = pyrene; BaP and BeP = benzo(a) and benzo(e)-pyrene; BbF and BkF = benzo(b) and benzo(k) fluoranthene; BaA = benzo(a) anthracene, C_1, C_2, C_3, C_4 = alkyl homologues, 1 through 4 carbons

used for wood preservation) and the residues from distillation (e.g., pitch). In fact, creosote-soaked piling for docks and other shoreline structures (railroad ties) are common and can, in some instances, become localized sources of pyrogenic PAH to urban sediments.

The general characteristics of the PAH in these pyrogenic materials is demonstrated in Figure 15.5.2, which shows the PAH distributions for a typical unweathered coal tar, creosote, and coal tar pitch. These materials are enriched in higher molecular weight PAH, include several 5- and 6-ring PAH. Within any given homolog series (C_0- to C_4-) of PAH there is a dominance of the unalkylated (parent) PAH and decreasing abundance of PAH with increasing degree of alkylation. This invokes a characteristic "sloped" profile in pyrogenic source materials (as compared to the "bell-shaped" profile of the petrogenic materials in Figure 15.4.2). Of note is the high concentration of PAH in the pyrogenic materials as compared to petroleum products. The coal tar, creosote, and coal tar pitch shown in Figure 15.5.2

contain 103,000, 142,000, and 141,000 mg/kg of total PAH (i.e., 10.3–14.2% by weight). These concentrations are much higher than that which occur in most petrogenic source materials, with total PAH concentrations typically in the 1–5% (by weight) range. A chronic composite source of pyrogenic PAH to urban sediments includes urban runoff (O'Connor and Beliaeff, 1995). The sources of PAH in urban runoff vary, but the most common sources are (1) urban dust containing combustion-related PAH (principally arising from internal combustion engines, especially diesel-based [e.g., Harrison *et al.*, 1996]), (2) street runoff containing traces of lubricating oils (principally arising from releases from automobiles), and (3) illegal or unintentional discharging of waste oil and petroleum products into storm drain systems. Although urban runoff has a petroleum component, its PAH sources are typically dominated by pyrogenic PAH (Eganhouse *et al.*, 1982).

In rural areas, wood burning and burning of other biomass materials, whether accidental (e.g., forest and grass

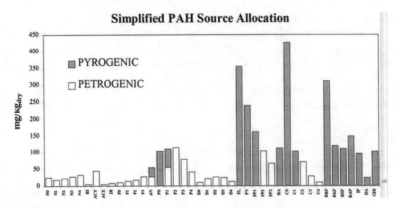

Figure 15.5.3 *Histogram showing a simplified PAH source allocation obtained by assigning individual PAHs to a source category (from Boehm and Farrington, 1984).*

Table 15.5.2 *Recommended Detection Limits for PAHs in Environmental Samples*

	Method 8270 (EPA SW-846)	Modified Method 8270
Medium	Estimated Quantitization Limits	Lower Detection Limits Individual PAHs or Isomer Groups
Water	10 µg/L	0.010 µg/L
Sediment/Soils	660 µg/kg(wet weight)	1 µg/kg(dry weight)

fires) or planned/deliberate (e.g., wood burning stoves, burn barrels) provide common sources of pyrogenic PAHs to soils and dusts in areas remote from urban centers. Emissions from wood treatment plants can also provide localized additional industrial inputs of PAHs.

Differentiation between compositionally similar, high temperature pyrogenic sources is the most challenging problem in the PAH forensics field and requires a more focused analysis of PAH compounds. Ratios such as the phenanthrene/anthracene (P/A) and fluoranthene/pyrene (Fl/Py) along with ratios of isomeric pairs of PAHs (e.g., B(b)F/B(k)F) have been used to differentiate wood burning-sourced PAH emissions, for example, from automotive (fossil fuel burning) emissions. The basis for the ratio differences lies in the relative stability of the isomer pairs. For example, Budzinski et al., (1997) found that the P/A ratio varied from 5.6 at 1000 K to 49 at 300 K. An illustration of the use of the Fl/Py ratio to differentiate very similar pyrogenic assemblages found in roadway runoff via storm-sewers and those in coal tar in the Thea Foss (Tacoma, WA) estuary is shown on Figure 15.5.6.

Additional enhancements of the application of parent (non-alkylated) PAH ratios and the use of sets of double ratios have been researched and applied by Costa et al., (2004) and Stout et al., (2004), and are summarized by Costa and Sauer (2005). This approach has to be carefully applied and the ratios used must be valid throughout a range of weathering of the source material (see Section 15.5.5). Such weathering tests include the evaluation of the stability of specific 4- or 5-ringed parent PAH ratios across a set of samples with varying degrees of weathering. One way of evaluating the use of PAH ratios is to look at the constancy of the ratio of interest among a group of similar samples (e.g., samples taken from the same area) across a range of weathering states as measured by the low molecular weight PAH (2- and 3-ringed PAHs) to high molecular weight (4- to 6-ringed PAHs) (e.g., $\sum$LMW/$\sum$HMW) ratio.

The use of radiocarbon (^{14}C) dating is another effective means for differentiating PAHs assemblage from wood burning from those associated with fossil carbon (oil- and coal-based burning). Radiocarbon (^{14}C) is produced from ^{14}N in the atmosphere and ^{14}CO$_2$ is assimilated by plants during photosynthesis. Radioactive decay of ^{14}C results in the loss of ^{14}C with a half-life of 5730 years. Thus radiocarbon dating can be a tool for differentiation of PAH particulate source generic source types (i.e., biomass or fossil carbon sourced). Particulate emissions are sampled and analyzed with petroleum-based combustion sources being identified by the existence of "dead carbon" or ^{14}C-free (Reddy et al., 2002), while combustion particulates associated with biomass burning containing roughly atmospheric ratios of ^{14}C and ^{12}C carbon (Eglinton et al., 1996). Compound-specific radiocarbon analysis (CSRA) can be performed by concentrating quantities of specific PAH compounds preparative gas chromatography prior to radiocarbon analysis by accelerator mass spectrometry. Compound-specific radiocarbon analysis has been used to evaluate the origin of PAHs (Eglinton et al., 1997; Lichtfouse et al., 1997; Reddy et al., 2002; Mandalakis et al., 2004). However, application of these CRSA techniques can be limited by the amount of material available. Typically 20–50 ug of carbon are required for such analyses (Reddy, personal communication).

15.5.5 Effects of Weathering on Source Indicators

The overall PAH distributions, concentrations, and individual homologous series patterns described in the previous section provide useful guidelines in distinguishing between pyrogenic and petrogenic sources of PAH in urban sediments. The diagnostic ratios shown in Table 15.5.2 are very effective in source determinations and, in general, the use of diagnostic source ratios can be very powerful, but should be used with care and applied to specific situation after careful evaluation. However, the use of ratios must be undertaken with an understanding that PAHs and PAH compositions may vary with environmental conditions and that the stability of these ratios may change over time and with differential weathering/degradation of specific PAH components of the ratio.

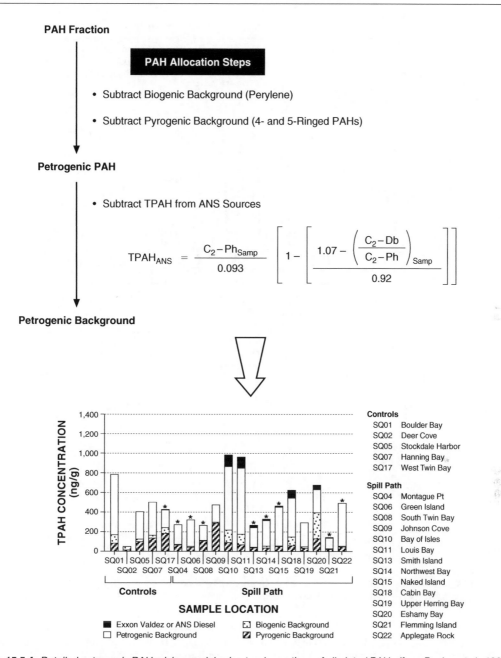

Figure 15.5.4 *Detailed petrogenic PAH mixing model using two key rations of alkylated PAHs (from Boehm et al., 1997).*

"Weathering" is a group of processes that refers to changes that can occur after a PAH is released into the environment. These changes are brought about by a combination of factors including: evaporation/volatilization; biodegradation by microorganisms; and solubilization. Evaporation is a significant process with regard to the 2-ringed PAHs, but the majority of the PAH are essentially non-volatile. Though solubilization may be important with regard to the 2- and 3-ringed PAHs, weathering of PAH in sediments and soils and groundwater is primarily limited to biodegradation. The effects of biodegradation on PAH compounds have been studied in both field and laboratory studies. The rates of biodegradation vary widely among

PAHs. Biodegradation relies on a number of factors including: oxygen content; microbial populations in receiving waters, sediments, and soils; hydrocarbon source "loading"; and, the chemical structure of the PAHs themselves. The greater the surface area available to mediate the biodegradation process, the more rapid the biodegradation. Thus PAHs sorbed to particles degrade more rapidly than do those in intact non-aqueous phase liquids (NAPL) plumes. It is generally accepted that biodegradation rates of PAH decrease with an increase in the number of aromatic rings. Thus, 2-ring PAH are more easily biodegraded than 4-ring PAH. Further, biodegradation rates are typically higher for parent PAH than for their alkylated equivalents (Douglas *et al.*,

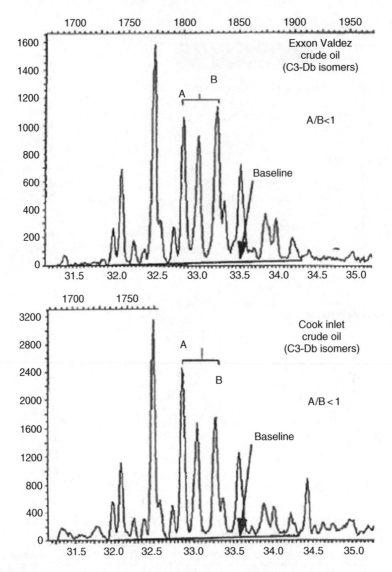

Figure 15.5.5 *Difference in alkyl homologue PAH patterns within a homologous series (e.g., C_3-Dibenzothiophenes) present opportunity for detailed chemical fingerprinting of petrogenic PAHs. (Note: Difference in ration of two isomers A and B.)*

1992). Overall the C_3-C_4 alkyl PAH of the 4-ring PAH should be considered the least biodegradable PAH. And these PAH components they should retain their original distributions throughout biodegradation.

Water-washing is predominantly controlled by the aqueous solubility of the individual PAH. Solubility generally decreases with increasing ring number, for example, naphthalenes have solubilities that are an order of magnitude higher than other PAH. Also, the solubilities of PAH decrease with increasing alkylation (Neff, 1979). Some exceptions are notable. Chrysene, for example, is less soluble than its alkylated C_1- or C_2-chrysene homologs. Linear PAH (for example, anthracene) are less soluble than angular equivalents (e.g., phenanthrene). Douglas *et al.*, (1992) report that because of their relatively high water solubility, the 2- and 3-ring PAH concentrations in the water soluble fraction of coal tar exceed the concentration of the 4-

and 5-ring PAHs by an order of magnitude, implying that, regardless of their original distributions in the source contaminant, contaminated water will almost always contain predominantly 2- and 3-ring PAH.

In turn, it would be expected that the contaminant remaining in sediment should be proportionally depleted in the smaller PAH relative to the source material. In summary, PAH weathering considerations lead to the following generalizations:

1) Parent (non-alkylated) PAHs are more susceptible to biodegradation than are the alkylated PAHs, and the lesser alkylated PAHs degrade faster than the higher alkylated PAHs.

2) Weathering preferentially reduces the proportion of low molecular weight (2- and 3-ring) PAHs, thereby increasing the proportion of 4- to 6-ring PAHs; and

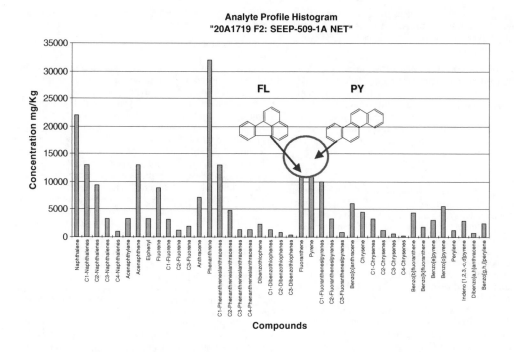

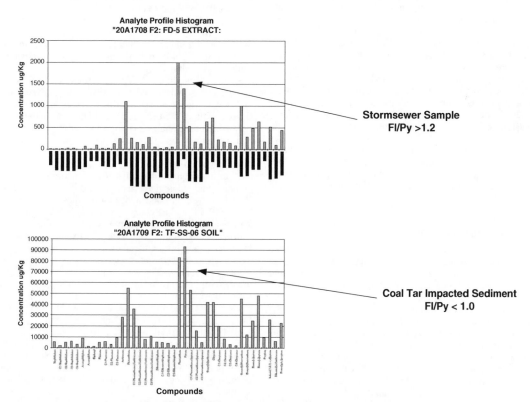

Diagnostic Ratio of Fluoranthene to Pyrene

Figure 15.5.6 Use of Fluoranthene to Pyrene ratios to distinguish differences in similar PAH assemblages (see color insert).

3) Weathering preferentially reduces the proportion of non-alkylated PAHs relative to alkylated PAHs, thereby increasing the proportion of more alkylated PAHs (Neff, 1979; Douglas *et al.*, 1992).

The implication of these changes for PAH forensics is that the PAH signature of a contaminant dominated by 2- and 3-ring PAHs (e.g., fuels oil, crude oil, creosote, etc.) ultimately may be transformed to a PAH signature dominated by 4- to 6-ring PAHs due to severe weathering. Therefore, the higher molecular weight PAHs are more robust keys to source identification over longer periods of time and/or greater weathering.

15.5.6 Isotopic Methods

Bulk and compound-specific carbon isotope techniques can help to identify the specific sources of PAHs compounds. Numerous applications of stable isotopic analyses have been applied to biogeochemical (Hayes *et al.*, 1990) and environmental forensics investigations (O'Malley *et al.*, 1994, 1996; Okuda *et al.*, 2002a, b; Fabbri *et al.*, 2003). The development of compound-specific GCIRMS allows for the analysis of the relative abundance of the carbon isotopes of individual PAH compounds (Freeman *et al.*, 1990; O'Malley *et al.*, 1994; Philp, 2002).

Carbon exists as two stable isotopes, ^{12}C and ^{13}C, with the natural abundance of the two being about 99:1. Stable isotope ratios are computed as a ratio of ^{13}C to ^{12}C ("R") or more specifically as $\delta^{13}C = (R_{sample}/R_{standard} - 1) \times 1000$ in units designated as "per mil." The nature of the original organic matter, or biomass, and the specifics of its formation (climate, temperature, photosynthesis) set the ratio of the two isotopes. The PAHs that are formed from this biomass during fossil fuel (petroleum or coal formation) generation acquire an isotopic composition or signature as the biomass is converted to fossil carbon over time, as mediated by temperature and other parameters. Isotopic compositions of bulk fossil fuel and other materials can be discerned as can the composition of individual PAH compounds, by gas chromatography-isotope ratio mass spectrometry (GCIRMS). These isotopic techniques are becoming additional and complementary lines of analytical support to environmental forensics investigations, but, to date, molecular investigations have been the main analytical approach with isotopic analyses playing supportive and confirmatory roles.

15.6 ALLOCATING SOURCES OF PAH

The PAH environmental forensics questions range from determination which of two possible sources is the source of contamination (e.g., oil spills)—a two end-member problem—to assignment and apportionment (i.e., allocation) of a PAH assemblage to percentages of multiple source materials. Allocation of PAHs has been approached using chemical-based mixing models (double ratio plots and mixing models), exploratory statistical techniques, such as principal component analysis (Wold *et al.*, 1987; Johnson *et al.*, 2002), and quantitative allocation techniques, such as partial least squares (PLS) or constrained least squares (CLS) (Burns *et al.*, 2005). The techniques used to approach these problems range from use of mixing models of sources with differing PAH ratios (e.g., Boehm *et al.*, 1997) to assignment of percentages of sources from numerous source possibilities (Boehm *et al.*, 2002; Page *et al.*, 2002) using least squares analysis (Burns *et al.*, 1997; Burns *et al.*, 2005).

Most often in urban environments, no single source of anthropogenic PAH contributes to an observed PAH assemblage. The relative contributions and quantitative allocation to different sources of PAH can rigorously be determined through:

- The collection of a robust set of source and environmental samples.
- The collection of detailed chromatographic fingerprints showing the boiling range and distribution of all PAHs present (including non-PAH hydrocarbons).
- Determination of the non-alkylated and alkylated PAHs concentrations (Table 15.4.1) in the samples.
- Evaluating concentration gradients (spatial relationships) of PAHs relative to known or suspected point sources identified through historical research.
- Evaluation of specific source types over time.
- Statistical analyses of PAH data to assess compositional similarities of environmental distributions to specific, well-characterized, and representative source materials.
- Use of ratio mixing models and/or statistical models to allocate the observed PAH assemblage to its component source types and/or specific sources.

Apportioning the contribution from multiple sources has historically been a qualitative exercise. In 1984, the first attempt to quantify the proportions of 'pyrogenic' *versus* 'petrogenic' PAH in sediments relied upon a simple expression using PAH homolog concentrations (Boehm and Farrington, 1984). These authors developed a "Fossil Fuel Pollution Index" (FFPI), which has the same basis as the apportionment method that is described below. This simple PAH classification and allocation technique has been used to show that a sediment sample contains PAH derived from a specific percentage of petrogenic source(s) and a specific percentage from pyrogenic source(s). A large number of sediment samples analyzed in the same manner can be used to "map" the distribution of petrogenic and pyrogenic PAH within a study area's sediments.

Numerical methods of PAH apportionment can be facilitated using sophisticated numerical analyses techniques. The most widely used is principle component analysis (PCA) (Wold *et al.*, 1976; Naes and Oug, 1997, 1998). Principle component analysis is one of several types of ordination techniques also known as "factor analyses" by which multivariate data sets are explored, reduced, interpreted and/or studied further. Principle component analysis is used in many types of studies (e.g., environmental, demographic, epidemiological, genetic) and has been applied to PAH fingerprinting and allocation studies (e.g., Naes and Oug, 1997). Principle component analysis is an exploratory statistical technique that permits a comparison of samples with suspected source materials. An example is shown in Figure 15.6.1. This figure shows a theoretical output of a PCA analysis for sediments from an urban waterway (Stout *et al.*, 2001) in which three sources of PAH (creosote, urban (combustion) background, stormwater (petrogenic) sources) were recognized. The source samples tend to plot as clusters at or near the apices of the trends revealed by the PCA factor score plot (Figure 15.6.1). However, many other samples tended to plot in locations intermediate between the three end-members indicating that they contain a mixture of sources. Spatial relationships among samples on a PCA score plot (such as Figure 15.6.1) can be used to estimate rough proportions of each end-member in each sediment sample.

More detailed quantitative assessments of contributions use other statistical techniques (e.g., least-squares, Polytopic vector analysis) to develop quantitative mixing models of possible multiple end-member sources. The least squares method of Burns *et al.*, (1997, 2005) was applied to an 18 source (5 source types) model for PAHs in Alaska and the result was a quantitative apportionment (Figure 15.6.2)

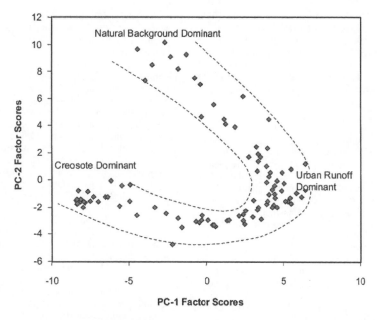

Figure 15.6.1 *Principal component analysis (PCA) factor score plot for sediment PAH data. The PCA revealed three dominant PAH sources (natural background, urban runoff, and creosote). Samples plotting between the apices indicate a mixture of these dominant sources (from Stout* et al., *2002).*

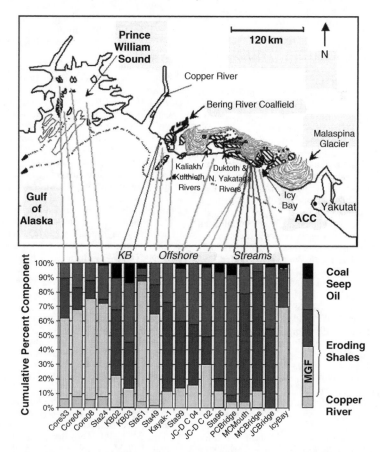

Figure 15.6.2 *Example of result of least squares apportionment method (Burns* et al., *1997) for PAHs in the vicinity of prince william sound (from Boehm* et al., *2001) (see color insert).*

(Boehm *et al.*, 2001a, 2002). Polytopic vector analysis (PVA) is an analogous technique, but one which computes end-members (sources) from environmental sample data rather than using known sources as a starting point for apportioning PAHs. Regardless of what quantitative apportionment method is used, an understanding of the uncertainty of the results is a key to defensibility of the approach.

REFERENCES

Agency for Toxic Substances and Disease Registry (ATSDR). 1995. Toxicological profile for polycyclic aromatic hydrocarbons (PAHs): US Department of Health and Human Services, Public Health Service. Atlanta, GA.

Bence, A.E., K.A. Kvenvolden, and M.C. Kennicutt, II, 1996. "Organic geochemistry applied to environmental assessments of Prince William Sound, Alaska, after the Exxon Valdez oil spill—A review." *Organic Geochemistry*. 24: 7–42.

Benner, B.A., S.A. Wise, L.A. Curie, G.A. Klouda, D.B Klinedinst, R.B. Zweidinger, R.K. Stevens and C.W. Lewis. 1995. Distinguishing the contribution of residential wood combustion and mobile source emissions using relative concentrations of dimethylphenanthrene isomers, *Environmental Science and Technology*. 29: 2382–2389.

Bentz, A.P. 1976. Oil spill identification. *Analytical Chemistry*. 48: 454A–472A.

Berkowitz, N. 1988. *Coal Aromaticity and Average Molecular Structure*. In: L.B. Eberg, ed., Polynuclear Aromatic Compounds. Advances in Chemistry Series 217. American Chemical Society, Washington, DC. pp. 217–233.

Bjorseth A, ed. 1983. *Handbook of PAH*. New York, NY: M. Dekker, Inc., 507.

Bjorseth, A. 1985. Sources of emissions of PAH. In. Bjorseth, A. and T. Ramdahl, eds, *Handbook of Polycyclic Aromatic Hydrocarbons*. Marcel Dekker, New York. pp. 1–20.

Blumer, M. 1976. Polycyclic aromatic hydrocarbons in nature. *Scientific American*. 234: 35–45.

Boehm, P.D. and Farrington, J.W. 1984. Aspects of the polycyclic aromatic hydrocarbon geochemistry of recent sediments in the Georges Bank region. *Environmental Science and Technology*. 18: 840–845.

Boehm, P.D., D.L. Fiest, I. Kaplan, P. Mankiewicz, and G.S. Lewbel. 1983. A natural resources damage assessment study: The Ixtoc-1 blowout, In Proceedings, 1983 Oil Spill Conference, American Petroleum Institute, Washington, D.C. pp. 507–515.

Boehm, P.D., D.S. Page, E.S. Gilfillan, W.A. Stubblefield, and E.J. Harner. 1995. Shoreline ecology program for Prince William Sound, Alaska, following the *Exxon Valdez* oil spill: Part 2—Chemistry. In Wells P.G., Butler J.N., Hughes J.S., eds, *Exxon Valdez Oil Spill: Fate and Effects in Alaskan Waters, ASTM Special Technical Publication # 1219*. American Society for Testing and Materials, Philadelphia, P.A., USA. pp. 347–397.

Boehm, P.D., G.S. Douglas, W.A. Burns, P.J. Mankiewicz, D.S. Page, and A.E. Burns. 1997. Application of petroleum hydrocarbon chemical fingerprinting and allocation techniques after the *Exxon Valdez* oil spill. *Marine Pollution Bulletin*. 34: 599–613.

Boehm, P.D., D.S. Page, W.A. Burns, A.E. Bence, P.J. Mankiewicz, and J.S. Brown. 2001a. Resolving the origin of the petrogenic hydrocarbon background in Prince William Sound, Alaska. *Environmental Science and Technology*. 35: 471–479.

Boehm, P.D., C.P.Loreti, A.B. Rosenstein, and P. M. Rury. 2001b. A Guide to Polycyclic Aromatic Hydrocarbons for the Non-Specialist, Publication Number 4714, American Petroleum Institute, Washington, D.C., p. 32.

Boehm, P.D., W.A. Burns, D.S. Page, A.E. Bence, P.J. Mankiewicz, J.S. Brown, and G.S Douglas. 2002. Total organic carbon, an important tool in a holistic approach to hydrocarbon fingerprinting, *Journal of Environmental Forensics*, 3: 243–250.

Boehm, Paul D., Cynda L. Maxon, Frederick C. Newton, John S. Brown, and Yakov Galperin. 2005. "Aspects of polycyclic aromatic hydrocarbons in offshore sediments in the Azeri sector of the Caspian Sea." In Armsworthy, S.L., P.J. Cranford and K. Lee eds. Offshore Oil and Gas Environmental Effects Monitoring: Approaches and Technologies. Battelle Press, Columbus, Ohio. p. 631.

Brown, J. and P.D. Boehm. 1993. The use of double-ratio plots of polynuclear aromatic hydrocarbon (PAH) alkyl homologues for petroleum source identification, In *Proceedings, 1993 Oil Spill Conference*, American Petroleum Institute, Washington, D.C.

Budzinski, H., I. Jones, J. Bellocq, C. Pierard, and P. Garrigues. 1997. Evaluation of sediment contamination by polycyclic aromatic hydrocarbons in the Gironde Estuary. *Marine Chemistry*. 58, 85–97.

Burns, W.A., P.J. Mankiewicz, A.E. Bence, D.S. Page, and K.R. Parker. 1997. A principal-component and least-squares method for allocating polycyclic aromatic hydrocarbons in sediment to multiple sources. *Environmental Toxicology and Chemistry*. 16: 1119–1131.

Burns, W.A., S.M. Mudge, A.E. Bence, P.D. Boehm, J.S. Brown, D.S. Page, K.R. Parker. 2005. Source Allocation by Least-Squares Hydrocarbon Fingerprint Matching, *Environmental Science and Technology*, in press.

Costa, H.J., K.A. White, and J.J. Ruspantini. 2004. Distinguishing PAH background and MGP residues in sediments of a freshwater creek. *Environmental Forensics*. 5: 171–182.

Costa, H.J. and T.C. Sauer. 2005. Forensic approaches and considerations in identifying PAH background, *Environmental Forensics*. 6: 9–16.

Davidson, R.M. 1982. Molecular structure of coal. In: M.L. Gorbaty, J.W. Larsen, and I. Wender, eds., *Coal Science*, Vol. 1, Academic Press, New York. pp. 84–160

Dickhut, R., E. Canuel, K. Gustafson, K. Liu, K. Arzayus, S. Walker, G. Edgecombe, M. Gaylor, and E. Macdonlad. 2000. Automotive sources of carcinogenic polycyclic aromatic hydrocarbons associated with particulate matter in the Chesapeake Bay Region. *Environmental Science and Technology*. 34: 4635–4640.

Douglas, G.S., A.E. Bence, R.C. Prince, S.J. McMillen, and E.L. Butler. 1996. Environmental stability of selected petroleum hydrocarbon source and weathering ratio. *Environmental Science and Technology*. 30: 2332–2389.

Douglas, G.S., W.A. Burns, A.E. Bence, D.S. Page, and P.D. Boehm. 2004. Optimizing detection limits for the analysis of petroleum hydrocarbons in complex environmental samples, *Environmental Science and Technology*. 38: 3958–3964.

Douglas, G.S., K.J. McCarthy, D.T. Dahlen, J.A. Seavey, W.G. Steinhauer, R.C. Prince, and D.L. Elmendorf. 1992. The use of hydrocarbon analyses for environmental assessment and remediation. In "Contaminated Soils: Diesel Fuel Contamination" (eds P.T. Kostecki and E.J. Calabrese), 1–21. Lewis Publishers, Boca Raton.

Eganhouse, R.P., D.L. Blumfield, and I.R. Kaplan. 1982. Petroleum hydrocarbons in stormwater runoff and municipal wastes: Input to coastal waters and fate in marine sediments. *Thalassia Jugoslavica*. 18: 411–431.

Eglinton, T.I., L.I. Aluwihare, J.E. Bauer, E.R.M. Druffel, and A.P. McNichol. 1996. Gas chromatographic isolation of individual compounds from complex matrices for radiocarbon dating. *Analytical Chemistry*. 68: 904–912.

Eglinton, T.I., B.C. Benitez-Nelson, A. Pearson, A.P. McNichol, J.E. Bauer, and E.R.M. Druffel. 1997. Variability in radiocarbon ages of individual organic compounds from marine sediments. *Science.* 277: 796–799.

Electric Power Research Institute (EPRI). 2001. Sediments Guidance Compendium, Report No. 1005216, Electric Power Research Institute, Palo Alto, California.

Emsbo-Mattingly, S. and P.D. Boehm. 2003. Identifying PAHs from Manufactured Gas Plant Sites, Technical Report No. 1005289, Electric Power Research Institute, Palo Alto, California.

Emsbo-Mattingly, S., A. Uhler, S.A. Stout, K.S. McCarthy, G.S. Douglas, J.S. Brown, and P.D. Boehm. 2001. Polycyclic aromatic hydrocarbon (PAH) chemistry of MGP tar and source identification in sediment, pp. 1–1 to 1–41, In *Sediments Guidance Compendium*, Report No. 1005216, Electric Power Research Institute, Palo Alto, California.

Fabbri, D., I. Vassura, C.G. Sun, C.E. Snape, C. McRae, and A.E. Fallick. 2003. Source apportionment of polycyclic aromatic hydrocarbons in a coastal lagoon by molecular and isotopic characterization. *Marine Chemistry.* 84: 123–135.

Federal Register 40CFR Subchapter J, Part 300, Subpart L, Appendix C, paragraphs 4.6.3 to 4.6.5. Freeman, K., J. Hayes, J.-M. Trendel, and P. Albrect. 1990. Evidence from carbon isotope measurements for diverse origins of sedimentary hydrocarbons. *Nature.* 343: 254–256.

Garrigues, P., H. Budzinski, M.P. Manitz, and S.A. Wise. 1995. Pyrolytic and petrogenic inputs in recent sediments: A definitive signature through phenanthrene and chrysene compound distribution, Polycyclic Aromatic Compounds. 7: 275–284.

Gas Research Institute. 1987. Management of manufactured gas plant sites. GRI-87/0260.1, Gas Research Institute. p. 82.

Giger, W. and M. Blumer, 1974. Polycyclic aromatic hydrocarbons in the environment: Isolation and characterization by chromatography, visible, ultraviolet, and mass spectrometry. *Analytical Chemistry.* 46: 1663–1671.

Gschwend, P.M. and R.A. Hites. 1981. Fluxes of polycyclic aromatic hydrocarbons to marine and lacustrine sediments in the northeastern United States. *Geochimica et Cosmochimica Acta.* 45: 2359–2367.

Gschwend, P.M., P.H. Chen, and R.A. Hites. 1983. On the formation of perylene in recent sediments: Kinetic models. *Geochimica et Cosmochimica Acta.* 47: 2115–2119.

Harrison, R.M., D.J.T. Smith, and L. Luhana, 1996. Source apportionment of atmospheric polycyclic aromatic hydrocarbons collected from an urban location in Birmingham, U.K. *Environmental Science and Technology.* 30: 825–832.

Hayes, J.M., Freeman, K.H., Popp, B.N. and C.H. Hoham. 1990. Compound-specific isotope analysis: A novel tool for reconstruction of ancient biogeochemical processes. *Organic Geochemistry.* 16: 1115–1128.

Hites, R.A., R.E. Laflamme, and J.W. Farrington, 1977. Sedimentary polycyclic aromatic hydrocarbons: A historical record. *Science.* 198: 829–831.

Jenkins, B.M., A.D. Jones, S.Q. Turn, and R.B. Williams. 1996. Particle concentrations, gas partitioning, and species intercorrelations for polycyclic aromatic hydrocarbons (PAH) emitted during biomass burning. *Atmospheric Environment.* 30: 3825–3835.

Jenkins, T.E. and R.A. Hites. 1983. Aromatic diesel emissions as a function of engine conditions. *Analytical Chemistry.* 55: 594–599.

Johnson, G.W., R. Ehrlich, and W. Full. 2002. Principal components analysis and receptor models in environmental forensics, In: Murphy, B.L. and R.D. Morrison, eds. *Introduction to Environmental Forensics.* Academic Press, New York. pp. 461–516.

Kennicutt II, M.C. 1998. The effect of biodegradation on crude oil bulk and molecular composition, Oil and Chemical Pollution. 4: 89–112.

Kvenvolden, K.A., F.D. Hostettler, P.R. Carlson, J.B. Rapp, C.N. Threlkeld, and A. Warden, 1995. Ubiquitous tar balls with a California-source signature on the shorelines of Prince William Sound, Alaska. *Environ. Sci. Technol.* 29: 2684–2694.

Laflamme, R.E. and R.A. Hites, 1978. The global distribution of polycyclic aromatic hydrocarbons in recent sediments. *Geochim. Cosmochim. Acta.* 42: 289–303.

Lee, M.L., G.P. Prado, J.B. Howard, and R.A. Hites. 1977. Source identification of urban airborne PAHs by GC/MS and high resolution MS. *Biomedical Mass Spectrometry.* 4: 182–186.

Lichtfouse, E., H. Budzinski, P. Garrigues, and T.I. Eglinton. 1997. Ancient polycyclic aromatic hydrocarbons in modern soils: C-13, C-14 and biomarker evidence. *Organic Geochemistry.* 26: 353–369.

Lima, A.C., J.W. Farrington, and C.M. Reddy. 2005. Combustion-derived polycyclic aromatic hydrocarbons in the environment–a review. *Environmental Forensics.* 6: 109–131.

Louda, W.J. and E.W. Baker. 1984. Perylene occurrence, alkylation and possible sources in deep ocean sediments. *Geochimica et Cosmochimica. Acta.* 48: 1043–1058.

Mandalakis, M., O. Gustafsson, C.M. Reddy, and L. Xu. 2004 Radiocarbon apportionment of fossil versus biofuel combustion resource of polycyclic aromatic hydrocarbons in the Stockholm metropolitan area. *Environmental Science and Technology.* 38: 5344–5349.

Miguel, A., T. Kirchstetter, and R. Harley. 1998. On-road emissions of particulate polycyclic aromatic hydrocarbons and black carbon from gasoline and diesel vehicles. *Environmental Science and Technology.* 32: 450–455.

Naes, K. and E. Oug. 1997. Multivariate approach to distribution patterns and fate of polycyclic aromatic hydrocarbons in sediments from smelter-affected Norwegian fjords and coastal waters. *Environmental Science and Technology.* 31: 1253–1258.

Naes, K. and E. Oug. 1998. The distribution and environmental relationships of polycyclic aromatic hydrocarbons (PAHs) in sediments from Norwegian smelter-affected fjords. *Chemosphere.* 36: 561–576.

National Oceanic and Atmospheric Administration (NOAA). 1993. Sampling and Analytical Methods of the National Status and Trends Program National Benthic Surveillance and Mussel Watch Project. Volume III. NOAA Technical Memorandum NOS ORCA 71 National Oceanic and Atmospheric Administration, Silver Spring, MD.

National Research Council (NRC). 1980. The International Mussel Watch. National Academy of Sciences, Washington D.C., p. 248.

National Research Council (NRC). 1985. *Oil in the Sea: Inputs, Fates, and Effects*, National Academy Press, Washington, D.C., p. 601.

Neff, J.M. 1979. "Polycyclic Aromatic Hydrocarbons in the Aquatic Environment", Applied Science Publishers Ltd., London, p. 262.

Neff, J.M., S.A. Ostazeski, S.C. Macomber, L.G. Roberts, W. Gardiner, and J.Q. Word. 1998. Weathering, Chemical Composition and Toxicity of Four Western Australian Crude Oils. Report to Apache Energy Ltd., Perth, Western Australia, Australia.

Neff, J.M., S.A. Stout, and D.G. Gunster. 2005. Ecological risk assessment of polycyclic aromatic hydrocarbons in sediments: Identifying sources and ecological hazard. *Integrated Environmental Assessment.* 1: 22–33.

O'Connor, T.P. and F.J. Beliaeff. 1995. *Recent trends in environmental quality: Results from the Mussel Watch*

Project (and data therein) US Department of Commerce, National Oceanic and Atmospheric Administration, Silver Spring, MD.

Ohkouchi, N., K. Kawamura, and H. Kawahata. 1999. Distributions of three-to seven ring polynuclear aromatic hydrocarbons on the deep seafloor of the Central Pacific, *Environmental Science and Technology*. 33: 3086–3090.

Okuda, T., H. Kumata, M.P. Zakaria, H. Naraoka, R. Ishiwatari, and H. Takada. 2002a. Source identification of Malaysian atmospheric polycyclic aromatic hydrocarbons nearby forest fires using molecular and isotopic compositions. *Atmospheric Environment*. 36: 611–618.

Okuda, T., H. Kumata, M.P. Naraoka, and H. Takada. 2002b. Origin of atmospheric polycyclic aromatic hydrocarbons (PAHs) in Chinese cities solved by compound specific stable carbon isotope analyses. *Organic Geochemistry*. 33: 1737–1745.

O'Malley, V., T. Abrajano Jr., and J. Hellou 1994. Determination of 13C/12C ratios of individual PAH from environmental samples: Can PAH sources be apportioned? *Organic Geochemistry*. 21: 809–822.

O'Malley, V.P., T.A. Abrajano, and J. Hellou. 1996. Stable carbon isotope apportionment of individual polycyclic aromatic hydrocarbons in St. Johns Harbour, Newfoundland. *Environmental Science and Technology*. 30: 634–639.

Overton, E.B., J.A. McFall, S.W. Mascarella, C.F. Steele, S.A. Antoine, I.R. Politzer, and J.L. Laseter. 1981. Identification of petroleum residue sources after a fire and oil spills. In, *Proceedings 1981 Oil Spill Conference*, American Petroleum Institute, Washington, D.C. pp. 541–546.

Ozretich, R.J., S.P. Ferraro, J.O. Lamberson, and F.A. Cole. 2000. Test of Σ polycyclic aromatic hydrocarbons model at a creosote contaminated site, Elliot Bay, Washington, USA, *Environmental Toxicology and Chemistry*. 19: 2378–2389.

Page, D.S., P.D. Boehm, G.S. Douglas, and A.E. Bence. 1995a. Identification of Hydrocarbon Sources in the Benthic Sediments of Prince William Sound and the Gulf of Alaska Following the *Exxon Valdez* Oil Spill, *Exxon Valdez Oil Spill: Fate and Effects in Alaskan Waters*, ASTM STP #1219, Peter G. Wells, James N. Butler, and Jane S, Hughes, eds., American Society for Testing and Materials, Philadelphia, PA., pp. 41–83.

Page, D.S., E.S. Gilfillan, P.D. Boehm, and E.J. Harner. 1995b. Shoreline Ecology Program for Prince William Sound, Alaska, Following the *Exxon Valdez* Oil Spill: Part 1—Study Design and Methods, *Exxon Valdez Oil Spill: Fate and Effects in Alaskan Waters*, ASTM STP #1219, Peter G. Wells, James N. Butler, and Jane S, Hughes, eds., American Society for Testing and Materials, Philadelphia, PA. pp. 263–295.

Page, D.S., P.D. Boehm, G.S. Douglas, A.E. Bence, W.A. Burns, and P.J. Mankiewicz. 1999. Pyrogenic polycyclic aromatic hydrocarbons in sediments record past human activity: A case study in Prince William Sound Alaska, *Marine Pollution Bulletin*. 38: 247–260.

Page, D.S., A.E. Bence, W.A. Burns, P.D. Boehm, J.S. Brown, and G.S. Douglas. 2002 Holistic approach to hydrocarbon source allocation in the subtidal sediments of Prince William Sound Embayments, *Journal of Environmental Forensics*. 3: 331–340.

Philp, P. 2002. Application of stable isotopes and radioisotopes in environmental forensics, In, B. Murphy and Morrison, R., eds., *Introduction to Environmental Forensics*. Academic Press, New York, pp. 99–139.

Pruell., R. and J.G. Quinn. 1988. Accumulation of polycyclic aromatic hydrocarbons in crankcase oil. *Environmental Pollution*. 49: 89–97.

Ramdahl, T., I. Alfheim, S. Rustad, and T. Olsen. 1982. Chemical and biological characterization of emissions

from small residential stoves burning wood and charcoal. *Chemosphere*. 11: 601–611.

Reddy, C.M., A. Pearson, L. Xu, A.P. McNichol, B.A. Benner, S.A. Wise, G.A. Klouda, L.A. Curry, and T.I. Eglinton. 2002. Radiocarbon as a tool to apportion the sources of polycyclic aromatic hydrocarbons and black carbon in environmental samples. *Environmental Science and Technology*. 36: 1774–1782.

Richter, H. and J. Howard. 2000. Formation of polycyclic aromatic hydrocarbons and their growth to soot—A review of chemical reaction pathways. *Progress in Energy and Combustion Science*. 26: 565–608.

Sander, L.C. and S. Wise. 1997. Polycyclic Aromatic Hydrocarbons Structure Index, National Institute of Standards and Technology (NIST) Special Publication 922, Chemical Science and Technology Laboratory, National Institute of Standards and Technology, Gaithersburg, MD.

Sauer, T.C. and P.D. Boehm. 1991. The use of defensible analytical chemical measurements for oil spill natural resource damage assessments, In *Proceedings, 1991 International Oil Spill Conference*, American Petroleum Institute, Washington, D.C., pp. 363–369.

Sauer, T.C. and P.D. Boehm. 1995. Hydrocarbon Chemistry Analytical Methods for Oil Spill Assessments. Marine Spill Response Corporation, Washington, D.C. MSRC Technical Report Series 95-032, p. 114.

Sporstol, S.N., N. Gjos, R.G. Lichtenthaler, K.O. Gustaveson, K. Urdall, and F. Oreld. 1983. Source identifcation of aromatic hydrocarbons in sediments using GC/MS. *Environmental Science and Technology*. 17: 282–286.

Stout, S.A., A.D. Uhler, and P.D. Boehm, 2001. Recognition of and allocation among sources of PAH in Urban Sediments. *Environmental Claims Journal*. 13: 141–158.

Stout, S.A., A.D. Uhler, K.J. McCarthy, and S. Emsbo-Mattingly. 2002. Chemical Fingerprinting of Hydrocarbons. In: B. Murphy and Morrison, R., eds., *Introduction to Environmental Forensics*. Academic Press, New York, pp. 137–260.

Stout, S.A., A.D. Uhler, and S.D. Emsbo-Mattingly. 2004. Comparative evaluation of background anthropogenic hydrocarbons in surficial sediment from nine urban waterways. *Environmental Science and Technology*. 38: 2987–2994.

Tan, Y.L. and M. Heit. 1981. Biogenic and abiogenic polycyclic aromatic hydrocarbons in sediments from two remote Adirondack lakes. *Geochimica et Cosmochimica Acta*. 45: 2267–2279.

Teichmuller, M. 1987. Recent advances in coalification studies and their application to geology. In: A.C. Scott, ed., Coal and Coal-Bearing Sequences: Recent Advances. *Geol. Soc.* London, Special Publ. 32. London, pp. 127–169.

US EPA (US Environmental Protection Agency). 1986. *SW-846 Manual for Waste Testing*. EPA: Washington, D.C., 1986; Vols. 1B and 1C.

US EPA (US Environmental Protection Agency). 1992. *Framework for Ecological Risk Assessment*. USEPA Risk Assessment Forum, Washington, DC. EPA/630/R-92/001.

Venkatesan, M. I. 1988. Occurrence and possible sources of perylene in marine sediments—A review. *Marine Chemistry*. 25: 1–27.

Wakeham, S., C. Schaffner, and W. Giger 1980. Polycyclic aromatic hydrocarbons in recent lake sediments—II. Compounds derived from biogenic precursors during early diagenesis. *Geochimica et Cosmochimica Acta*. 44: 415–4129.

Walker, S.E., R. M. Dickhut, and C. Chisholm-Brause‡. 2004. Polycyclic aromatic hydrocarbons in a highly industrialized urban estuary: Inventories and trends, *Environmental Toxicology and Chemistry*. 23: 2655–2664.

Wang, Z. and Fingas, M.F. 2003. Development of oil hydrocarbon fingerprinting and identification techniques. *Marine Pollution Bulletin*. 47: 423–452.

Wen, Z., W. Ruiyong, M. Radke, W. Qingyu, S. Guoying, and L. Zhili. 2000. Retene in pyrolysates of algal and bacterial organic matter, *Organic Geochemistry*. 31: 757–762.

Williams, P.T., K.D. Bartle, and G.E. Andrews. 1986. The relation between polycyclic aromatic hydrocarbons in diesel fuels and exhaust particulates. *Fuel*. 65: 1150–1158.

Williams, P.T., M.K. Abbass, G.E. Andrews, and K.D. Bartle. 1989. Diesel particulate emissions: The role of uncombusted fuels. *Combustion and Flame*. 75: 1–24.

Wold, S. 1976. Pattern recognition by means of disjoint principal component models. *Pattern recognition*. 8: 127–139.

Wold, S., K. Esbensen, and P. Geladi. 1987. Principal Component Analysis. *Chemometrics and Intelligent Laboratory Systems*. 2: 37–52.

Youngblood, W.W. and M. Blumer. 1975. Polycyclic aromatic hydrocarbons in the environment: homologous series in soils and recent marine sediments. *Geochimica et Cosmochimica Acta*. 39: 1303–1314. Table 11-1. Concentrations of individual and total PAHs in a typical crude oil, coal, coal tar, and creosote. Concentrations are mg/kg. From Neff *et al.*, (1998).

16

Crude Oil and Refined Product Fingerprinting: Principles

Zhendi Wang, Merv Fingas, Chun Yang, and Jan H. Christensen

Contents

16.1 INTRODUCTION

The word *petroleum* means "rock oil" from the Greek words *petros* (rock) and *elaion* (oil). Petroleum is a complex mixture of hydrocarbons that exist naturally in gaseous (natural gas), liquid (crude oil), and solid (asphalt) states. It is derived from a variety of organic materials that are chemically converted over long periods of time (hundreds of million of years) under different geological and thermal conditions. Crude oil is composed mainly of carbon and hydrocarbon, but also minor amounts of sulfur, oxygen, and nitrogen as well as trace amounts of metals (e.g., nickel and vanadium) are present. Refined petroleum products are fractions derived by distillation from crude oil. Thus, due to variations in crude oil feedstocks and in the refining process individual oil samples have unique *chemical fingerprints*, which provide a basis for distinguishing oils and identifying the source(s) of spilled oil(s).

Liquid petroleum (crude oil and the products refined from it) plays a pervasive role in our modern society. For example, about 286,000 tonnes of oil and petroleum products are used in Canada every day. The United States uses about 10 times this amount and, worldwide, about 11 million tonnes are used per day. Exploration, transportation, and widespread use of petroleum inevitably result in intentional and accidental releases to the environment. In addition, natural seepage of crude oil from geologic formations below the seafloor to the sea surface also makes a contribution to pollution of the marine environment.

The average yearly petroleum spills into North American waters is about 260,000 tonnes, and annual worldwide estimates of petroleum input to the sea exceed 1,300,000 tonnes (National Research Council, 2002). In Canada, about 12 spills of more than 4000 L are reported each day, of which about one spill is into navigable waters. Most spills take place on land, including oil spills from pipelines, underground storage tanks, and aboveground storage containers. In the US, about 25 such spills occur each day into navigable waters and about 75 occur on land (Fingas, 2001). Although large oil spills from tankers (such as 1989 *Exxon Valdez* spill, 1999 *Erika* spill, and 2002 *Prestige* spill) occur in the marine environment, these spills only make up about 5% of oil entering the sea. Most of oil pollution in the oceans comes from the run-off of oil and fuel from land-based sources rather than from accidental spills.

Oil spills of unknown origins occur frequently in open water and navigable waterways. These often occur due to accidental or intentional discharges from vessels of oiled ballast water, bilge waters, or tank wash slops. The significant liability is often associated with small volume oil spills. In addition, because oils and refined products are often transported large distances from their source along heavily traveled shipping lanes, the potential for significant number of candidate "source" vessels is common.

Oil spills can cause extensive damage to our environment and ecosystem, and have resulted in legal battles in suits amounting annually to billions of dollars in casualty and punitive payments. Agencies and potentially responsible parties often request characterization of spilled oil. Hence, characterization of spilled oils and defensibly linking to potential sources are important for environmental damage assessment; understanding the fate and behavior and predicting the potential long-term effects of spilled oils on the impacted ecosystems; selecting appropriate spill responses; and taking effective cleanup measures. In addition, successful forensic investigation and analysis of oil hydrocarbons at contaminated sites and receptors yield a wealth of chemical fingerprinting data. These data, in combination with historic, geological, environmental, and any related information on the contaminated site can, in many cases, help to settle legal liability and to support litigation against the responsible party.

Biological markers or biomarkers are one of the most important hydrocarbon groups in petroleum for chemical fingerprinting. They are complex molecules derived from formerly living organisms. Biomarkers found in crude oils, rocks, and sediments show little or no changes in structures from their parent organic molecules or so-called biogenic precursors (e.g., hopanoids, sterols, and steroids) in living organisms. In comparison with the concentrations of the biogenic precursors, biomarker concentrations in oil are low, often in the range of several to less than a hundred parts per million (ppm).

Biomarkers are useful for chemical fingerprinting because they retain all or most of the original carbon skeleton of the original natural product, and this structural similarity reveals more information about oil source than do other compounds in oil. Excellent reviews on fundamentals of biomarker characterization, their application in geochemistry, and interpretation of biomarker data for oil exploration have been published (Peters and Moldowan, 1993). Biomarker fingerprinting has historically been used by petroleum geochemists in characterization of oils in terms of:

- oil-to-oil and oil-to-source rock correlation;
- the type(s) of precursor organic matter present in the source rock (e.g., bacteria, algae, marine algae, microorganisms, and high plants, because each type of organism may contain different biomarkers);
- effective ranking of the relative thermal maturity of petroleum (i.e., immature, mature, postmature) and assessment of thermal history of oil throughout the entire oil-generation window (i.e., early, peak, or late generation);
- evaluation of migration and the degree of in-reservoir biodegradation based on the loss of *n*-alkanes, isoprenoids, aromatics, terpanes, and steranes during biodegradation;
- determination of depositional environmental conditions (e.g., marine, terrestrial, deltaic or hypersaline environments, and anoxic or suboxic depositional environments, because different depositional environments can also result in characteristic and even subtle difference of the biomarker composition); and
- providing information on the age of the source rock for petroleum. For example, oleanane ($C_{30}H_{52}$) is a biomarker characteristic of angiosperms (flowering plants) found only in *Tertiary* and *Cretaceous* (<130 million years) rocks and oils, while dinosterane is a marker for marine dinoflagellates, possibly distinguishing Mesozoic and Tertiary from Paleozoic (~660 million years) source input.

Biomarkers can be detected in low quantities (ppm and sub-ppm level) in the presence of a wide variety of other types of petroleum hydrocarbons by the use of the gas chromatography-mass spectrometry (GC-MS). Relative to other hydrocarbon groups such as alkanes and most aromatic compounds, biomarkers are highly degradation-resistant in the environment. Furthermore, due to the wide variety of geological conditions and ages under which oil has formed, every crude oil may exhibit an essentially unique biomarker fingerprint. Therefore, chemical analysis of biomarkers generates information of great importance to environmental forensic investigations in terms of determining the source of spilled oil, differentiating and correlating oils, and monitoring the degradation process and weathering state of oils under a wide variety of conditions. They have also proven useful in identification of petroleum-derived contaminants in the marine and aquatic environments (Wang *et al.*, 1994a,b, 1999; Kvenvolden *et al.*, 1995, 2002; Bence *et al.*, 1996; Boehm *et al.*, 1997; Volkman *et al.*, 1997; Hostettler *et al.*, 1999a; Zakaria *et al.*, 2000; Stout *et al.*, 2002) and in indicating chronic industrial and urban releases. (Volkman *et al.*, 1992; Kaplan *et al.*, 1997; Stout *et al.*, 1998.

In this chapter we will focus our discussion on biomarker chemistry (including the "isoprene rule" and biomarker families and labeling system and nomenclature of biomarkers), biomarker genesis, overview of analytical methodologies for biomarker separation and analysis, identification of biomarkers, biomarker distributions in crude oils and various petroleum products, and sesqueterpane and diamondoid biomarkers in oils and lighter petroleum products.

16.2 OIL CHEMISTRY

16.2.1 Chemical Composition of Oil

Crude oils consist of complex mixtures of hydrocarbons. By strict definition, a hydrocarbon contains carbon and hydrogen only. Petroleum hydrocarbons range from small, volatile compounds to large, non-volatile ones. More than 300 compounds have been identified in the Alberta Sweet Mixed Blend using GC-MS and by comparison of GC retention data with authentic standards and calculation of retention index values (Wang *et al.*, 1994a). In general, petroleum hydrocarbons are characterized and classified chemically by their structures, including saturates, olefins, aromatics, polar compounds (wide variety of compounds containing sulfur, oxygen, and nitrogen), and asphaltenes.

16.2.1.1 Saturates or Alkanes

Saturates are a group of hydrocarbons composed of only carbon and hydrogen with no carbon–carbon double bonds. In other words, carbon atoms in saturated hydrocarbons are bonded exclusively by covalent *sigma* (σ) *bonds*. Methane (CH_4) is the simplest saturated hydrocarbon. In methane, carbon is at the center of the regular tetrahedron and the four hydrogens are at the corners. Each of four σ bonds is formed by overlap of one of four equivalent hybrid *sp3* orbitals of carbon with the *1s* orbital of hydrogen. The angle between any two bonds is the tetrahedral angle 109.5° and the bond length is 1.09 Å.

Saturates are the predominant class of hydrocarbons in crude oil. Saturates include straight chain and branched chain (saturated hydrocarbons with straight and branched chains are also called *paraffins*), and cyclo alkanes (also called *naththenes*). The general formula for alkanes is C_nH_{2n+2}. As the number of carbon atoms in hydrocarbons increases, the carbon atoms may be connected in continuous chains or connected as branches with more than two carbon atoms linked together (branched hydrocarbons). The hydrocarbons with the same molecular formula but different connections among their atoms are known as *structural* or *constitutional isomers*, or simply as *isomers*. For example, there are two constitutional isomers of butane (C_4H_{10}: *n*-butane and isobutane) and three isomers of pentane (C_5H_{12}: *n*-pentane, isopentane, and neopentane).

The cycloalkanes or naphthenes are a special class of alkanes with ring structure. Naphthenic hydrocarbons have the formula C_nH_{2n} and all C–H bonds are saturated. As such, naphthenic hydrocarbons in petroleum are also relatively stable compounds. The nomenclature of cycloalkanes is similar as that for alkanes with the addition of a *cyclo* prefix. For example, a five-carbon and six-carbon alkane ring is cyclopentane and cyclohexane respectively. Cycloalkane compounds can exhibit configurational isomerism (i.e., the attached groups can differ in their position relative to the ring). The cycloalkane prefix of *cis* denotes that two groups (other than hydrogen atoms) attached to the ring both lie

either above or below the plane of the ring. *Trans* denotes that one of the two groups lies above and the other lies below the plane of the ring. The *cis-trans* isomers (for example, *cis-* and *trans*-1,2-dimethylcyclopentane) are also called *geometric isomers*. Geometric isomers are one special type of stereoisomers. The *cis-trans* isomers are not constitutional isomers because their atoms are bonded in the same sequence.

- Straight chain saturates (normal alkanes or normal paraffins) ranging from *n*-C_5 to *n*-C_{40} are often the most abundant constituents in crude oils. Large *n*-alkanes (>C_{18}) are often referred to as waxes.
- Branched chain saturates (isoalkanes or iso-paraffins) are hydrocarbons containing branched carbon chains, which are also major constituents of oil. The five most abundant isoprenoid compounds in oil are farnesane (*i*-C_{15}: 2,6,10-trimethyl-dodecane), trimethyl-tridecane (*i*-C_{16}), norpristane (*i*-C_{18}: 2,6,10-trimethyl-pentadecane), pristane (*i*-C_{19}: 2,6,10,14-tetramethyl-pentadecane) and phytane (*i*-C_{20}: 2,6,10,14-tetramethyl-hexadecane).
- Cycloalkanes (cyclo-paraffins or naphthenes) consist of carbon atoms joined by single bonds in a ring structure. The most abundant cycloalkanes in oil are the single-ring cyclopentane (C_5H_{10}) and cyclohexane (C_6H_{12}), and their alkylated (from C_1 to C_{14}) homologues (alkyl-cyclopentanes and alkyl-cyclohexanes).
- Terpanes and steranes are branched cyloalkanes consisting of multiple condensed five- or six-carbon rings. Sesquiterpanes and diamondoids are bicyclic biomarkers which can be particularly valuable for identification and differentiation of lighter petroleum products.

16.2.1.2 Alkenes

Alkenes, commonly referred to as olefins, are partially unsaturated straight-chain hydrocarbons characterized by one or two double carbon–carbon bonds. The most simple alkene hydrocarbon is ethylene (CH_2CH_2). The carbon atoms in ethylene are linked by two bonds: a strong σ bond formed from the hybrid *sp2* orbitals and a weaker π bond formed from overlap of *p* orbitals. This type of bond is called a "double bond." Each of the carbon–hydrogen bonds (σ bonds) is formed by overlap of a *sp2* hybrid orbital of carbon with the *1s* orbital of a hydrogen atom. The C–H bond length in ethylene is slightly shorter (1.08 C) than the C–H bond in ethane (1.09 C), and the angle between C–H bonds is about 120°.

The general formula for the alkene family is C_nH_{2n}. Simple alkenes are named much like alkanes, using the root name of the longest chain containing the double bond. The ending is changed from *-ane* to *-ene*. For example, "ethane" becomes "ethene," "propane" becomes "propene," and "cyclopentane" becomes "cyclopentene." Olefins are found only in some refined products, produced primarily from larger molecules in cracking processes.

16.2.1.3 Aromatic Hydrocarbons

Aromatic hydrocarbons are cyclic, planar compounds that resemble benzene in electronic configuration and chemical behavior. Benzene has the molecular formula C_6H_6 and is the simplest aromatic hydrocarbon. The carbon atoms in benzene are linked by six equivalent Φ bonds and six B bonds. The six *p* electrons are shared equally or delocalized among the donating carbon atoms, forming a continuous ring of orbitals above and below the plane of the carbon atoms. These delocalized B bonds are more stable than isolated double bonds. All the carbon–carbon bonds in benzene are the same length (1.397 Å), and all of the bond angles are 120°. A wide range of aromatic compounds has benzene rings located in ortho positions. These are called condensed or fused rings. Aromatic compounds with two or more fused aromatic benzene rings are called polycyclic aromatic hydrocarbons (PAH) and they have the general formula $C_{4r+2}H_{2r+4}$ for rings without substituents, where r=number of rings.

Aromatics in petroleum include the mono-aromatic hydrocarbons such as BTEX (the collective name for benzene, toluene, ethylbenzene, and o-, m-, and p-xylene isomers) and other alkyl-substituted benzene compounds (C_n-benzenes), and PAHs, including oil-characteristic alkylated C_0 to C_4-PAH (naphthalene, phenanthrene, dibenzothiophene, fluorene, and chrysene) homologous series and US EPA priority PAHs. Benzene is the simplest one-ring aromatic compound. The commonly analyzed PAH compounds range from 2-ring PAHs (i.e., naphthalene) up through 6-ring PAHs (e.g., benzo[g,h,i]perylene). BTEX and PAHs are of concern because of their acute toxic and carcinogenic potential.

16.2.1.4 Polar Compounds

Polar compounds are those with distinct regions of positive and negative charge, as a result of bonding with atoms such as nitrogen, oxygen, or sulfur. Heavy oils generally contain greater proportions of higher-boiling, more aromatic, and heteroatom-containing (–, O–, S–, and metal-containing) constituents. The *polarity* or charge that the molecules carry result in behavior that, under some circumstances, is different from that of non-polar compounds. In the petroleum industry, the smaller polar compounds are called "resins," which are largely responsible for oil adhesion. The large polar compounds are called "asphaltenes" because they often make up the largest percentage of the asphalt commonly used for road construction.

The polar heteroatom cyclic compounds are present in oils at low concentrations. However, they become concentrated with weathering because they are generally biorefractory and persistent in the environment. A large number of individual alkyl homologues of N-, O-, and S-containing compounds, including alkylated carbazole, quinoline, pyridine, thiophene, and dibenzothiophene, have been identified in many oils. These compounds may provide important clues for potential sources of hydrocarbons in the environment and for tracing petroleum molecules back to their biological precursors.

Resin compounds include heterocyclic hydrocarbons (such as nitrogen, oxygen, and sulfur containing PAHs), phenols, acids, alcohols, and monoaromatic steroids. Because of their polarity, these compounds are more soluble in polar solvents. Sulfur compounds are among the most important heteroatomic constituents of petroleum and may be present in several forms, including elemental sulfur, hydrogen sulfide, mercaptans, thiophenes (thiophene and its alkylated homologues), benzothiophenes and dibenzothiophenes (benzothiophene, dibenzothiophene and their alkylated homologues), and naphthobenzothiophenes. The sulfur content in most crude oils varies from about 0.1–3% to 5–6% for some heavy oils and bitumen. Most organic nitrogen in crude oils are present as alkylated aromatic heterocycles with a predominance of neutral pyrrole and carbazole structures over basic pyridine and quinoline forms. They are chiefly associated with high boiling fractions, and much of the nitrogen in petroleum is in the asphaltenes. The nitrogen-containing compounds are important in their roles as natural surfactants. The concentrations of these compounds in crude oil have a great influence on the chemical and physical activities of crude oil, on metal/oil interface and ground/oil interface. Oxygen reacts with hydrocarbons to form various oxygen-containing compounds, such as phenols, cresols, and benzofurans. Compared to PAHs, the concentrations of these nitrogen and oxygen containing compounds in crude oil are generally very low, varying from 0.1–3%. Generally, the oxygen content in crude oil fractions increases with the boiling point interval of the fraction.

Asphaltenes are a class of very large compounds (Berkowitz, 1997; Speight, 1999). They are not dissolved in petroleum but are dispersed as colloids. Asphaltene is derived from the word "asphaltu" (meaning "to split"), and adopted by the Greeks, signifying "firm," "stable," or "secure." Asphaltenes are generally defined, based on the solution properties of petroleum residues in various solvents, as the oil constituents precipitated from oils and bitumen by natural processes or in laboratory by addition of excess n-pentane or n-hexane. If abundant in crude oil, they have a significant effect on oil behavior.

Despite a considerable volume of relevant analytical data, very little is known about molecular configurations of asphaltenes. From X-ray diffraction patterns of solid asphaltenes, it has been inferred that crystallographic organization can be represented by an asphaltene "macromolecule," in which clusters of partly ordered aromatic matter and carrying aliphatic chains of varying length are associated in micelles or particles. A more recent study of dispersions of asphaltenes in liquids by small-angle X-ray scattering, a classic method for characterizing colloid system, confirms the earlier conclusions, but also provides some evidences for colloidal polydispersion in a marked solvent-dependence of average cluster, micelle, and particle sizes (Ravey *et al.*, 1988). Strausz *et al.* characterized the Alberta's Athabasca asphaltenes by various analytical techniques including solvent fractionation, chromatographic separation, chemical and spectroscopic analyses of products from pyrolysis and degradation by liquid oxidants (Strausz *et al.*, 1992). The Alberta's Athabasca asphaltenes contained alkyl carbazoles, variously substituted thiophenes, benzothiophenes, dibenzothiophenes, benzocarbazoles, dibenzocarbazoles, and many other substituted alkyl aromatics. Proceeding from the database developed, the average empirical molecular formula of the Athabasca asphaltenes was modeled as $C_{420}H_{496}N_6S_{14}O_4V$ (Strausz *et al.*, 1992).

Porphyrins are complex derivatives of the basic material porphine. Porphine consists of four pyrrole units joined by methine, $-C=$, bridges; the methine bridges establish conjugated linkages between the component pyrrole nuclei, forming a more extended resonance system. Although the resulting structure retains much of the inherent character of the pyrrole components, the larger conjugated system gives increased aromatic character to the porphine molecule. The *porphyrin* compounds are degradation products of the chlorophyll (photosynthetic pigments of plants and some bacteria). Most of the porphyrin material in crude oils is chelated with metal, of which vanadium is the most important, followed by nickel. Iron and copper-porphyrin chelates may also be present in oil. Porphyrins are not usually considered among the usual nitrogen-containing constituents of petroleum, nor are they considered a mellallo-containing organic material. Conversely, they are often classified as a unique class of biomarker compounds because they may establish a link between compounds found in the geosphere and their corresponding biological precursors (Figure 16.2.1).

Crude oils and bitumens contain small amounts of vanadyl and nickel porphyrins. In general, mature, lighter oils contain less of these compounds, whereas heavy oils may contain larger amounts of vanadyl and nickel porphyrins.

A simplified classification system based on the bulk chemical composition to classify petroleum has been proposed

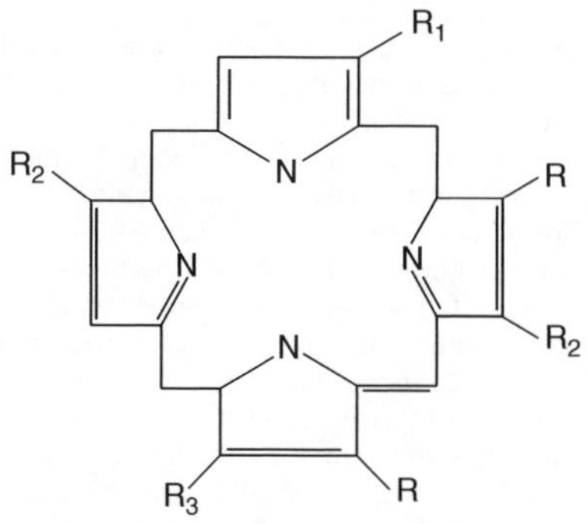

Figure 16.2.1 *The basic structural unit of porphyrins.*

(Speight, 1999) with paraffinic, naphthenic, aromatic, and asphaltic petroleum as extremes (Simanzhenkov and Idem,

2003). By this simplified classification system, petroleum can be characterized semi-quantitatively. Based on density, oils can be qualitatively classified into light, medium, heavy, and very heavy oil. Table 16.2.1 summarizes the typical bulk composition of some oils and petroleum products.

16.2.2 Physical Properties of Oil

Physical properties of crude oils are generally correlated with their chemical composition. The properties briefly discussed here are density, specific gravity, solubility, viscosity, flash point, carbon distribution, distillation, and interfacial tension. Table 16.2.2 lists the typical oil physical properties.

16.2.2.1 Density

Density is the mass of a given volume of oil and is typically expressed in grams per cubic centimeter (g/cm^3). It is the property used by the petroleum industry to define light or heavy crude oils. Density is also important because it indicates whether an oil will float or sink in water. As the density of water is $1.0\,g/cm^3$ at $15\,°C$ and the density of crude oils and refined products generally ranges from

Table 16.2.1 *Typical Composition of Common Oils and Petroleum Products (%)*

Group	Compound Class	Gasoline	Diesel	Light Crude	Heavy Crude	IFO	Bunker C
Saturates		50 to 60	65 to 95	55 to 90	25 to 80	25 to 45	20 to 40
	alkanes	45 to 55	35 to 45				
	cyclo-alkanes	~5	30 to 50				
	waxes		0 to 1	0 to 20	0 to 10	2 to 10	5 to 15
Olefins		5 to 10	0 to 10				
Aromatics		25 to 40	5 to 25	10 to 35	15 to 40	40 to 60	30 to 50
	BTEX	15 to 35	0.5 to 2	0.1 to 2.5	0.01 to 2	0.05 to 1	0 to 1
	PAHs		0.5 to 5	0.5 to 3	1 to 4	1 to 5	1 to 5
Polar Compounds			0 to 2	1 to 15	5 to 40	15 to 25	10 to 30
	resins		0 to 2	0 to 10	2 to 25	10 to 15	10 to 20
	asphaltenes			0 to 10	0 to 20	5 to 10	5 to 20
Sulphur		<0.05	0.05 to 0.5	0 to 2	0 to 5	0.5 to 2	2 to 4
Metals (ppm)				30 to 250	100 to 500	100 to 1000	100 to 2000

Table 16.2.2 *Typical Oil Properties*

Property	Units	Gasoline	Diesel	Light Crude	Heavy Crude	Intermediate Fuel Oil	Bunker C	Crude Oil Emulsion
Viscosity	m.Pa.s at $15\,°C$	0.5	2	5 to 50	50 to 50,000	1,000 to 15,000	10,000 to 50,000	20,000 to 100,000
Density	g/mL at $15\,°C$	0.72	0.84	0.78 to 0.88	0.88 to 1.00	0.94 to 0.99	0.96 to 1.04	0.95 to 1.0
Flash Point	°C	−35	45	−30 to 30	−30 to 60	80 to 100	>100	>80
Solubility in Water	ppm	200	40	10 to 50	5 to 30	10 to 30	1 to 5	—
Pour Point	°C	not relevant	−35 to −1	−40 to 30	−40 to 30	−10 to 10	5 to 20	>50
API Gravity		65	35	30 to 50	10 to 30	10 to 20	5 to 15	10 to 15
Interfacial Tension	mN/m at $15\,°C$	27	27	10 to 30	15 to 30	25 to 30	25 to 35	not relevant
Distillation Fraction	% distilled at							
	100 °C	70	1	2 to 15	1 to 10	—	—	not relevant
	200 °C	100	30	15 to 40	2 to 25	2 to 5	2 to 5	
	300 °C		85	30 to 60	15 to 45	15 to 25	5 to 15	
	400 °C		100	45 to 85	25 to 75	30 to 40	15 to 25	
	Residual			15 to 55	25 to 75	60 to 70	75 to 85	

0.7 to $0.99\,g/cm^3$, spilled oil will float on water. As the density of seawater is $1.03\,g/cm^3$, even heavier oils will usually float on it. Only certain bitumen and very heavy residual oils such as Bunker C have densities greater than water and may submerge in water. The density of oil increases with time, as the light fractions evaporate.

Another measure of density is specific gravity, which is an oil's relative density compared with that of water at $15\,°C$. The American Petroleum Institute (API) uses the API gravity as a measure of density for petroleum:

$$\text{API gravity} = [141.5/(\text{density at } 15.6\,°C)] - 131.5$$

Water has an API gravity of $10°(10\,°C)$. Oils with progressively lower specific gravity have higher API gravity. The scale is commercially important for ranking oil quality. Heavy inexpensive oils are $<25°$ API; medium oils are 25 to 35° API; and light commercially valuable oils are 35 to 45° API. API gravities vary inversely with viscosity, asphaltic matter content (which increase from 4–8% at 40° to ~50% at 10–15° API), and N-content (which rises from 0.08–0.20% to ~1% over the same interval).

Oil density can be measured using an acoustic cell density meter following the American Society for Testing and Materials (ASTM) method D5002. Conventional crude oil and heavy oil have been defined generally in terms of API gravity. Using this definition, heavy oils are those petroleum-type materials that have API gravity less than 20°, with the heavy oils falling into the API gravity range of 10–15° and bitumen falling into the 5–10° API range.

16.2.2.2 Solubility

Solubility in water is the measure of how much of an oil will dissolve in the water column at a known temperature and pressure. The more polar the compound, the more soluble it is in water. BTEX compounds are so frequently encountered in groundwater in part due to their high water solubility. BTEX solubility in water is dependent on the nature of the multi-component mixture, such as gasoline, diesel, or crude oil. The solubility of a constituent within a multi-component mixture may be orders of magnitude lower than the aqueous solubility of the pure constituent in water. Oil is a complex mixture of compounds, each of which partitions uniquely between oil and water; therefore, the water solubility varies between oils. The solubility of oil in water is low, generally less than 100 ppm. However, solubility is an important issue after oil spill accidents because dissolved oil components are often acutely toxic to aquatic life.

16.2.2.3 Carbon Distribution

Carbon distribution according to volatility is one of the main properties of petroleum. Theoretically, any fractionating column with sufficient number of theoretical plates may be used for recording a curve in which the boiling points of each fraction is plotted against the percentage by weight. A method has been developed by the Environment Canada Oil Spill Research Program to determine the carbon distribution of petroleum by measurement of a simulating boiling point distribution by high temperature GC-FID analyzer. By correlation with known standard, the retention times of a high-temperature gas chromatograph are matched with the temperature cuts of a distillation curve. This results in a mass distribution of boiling points from 40 to $750\,°C$, corresponding roughly to a range of hydrocarbons from C_6 to C_{120}.

16.2.2.4 Viscosity

Viscosity is the resistance to flow in a liquid; thus the lower the viscosity, the more readily the liquid flows. The viscosity of oil is a function of its composition; therefore crude oil has a wide range of viscosities. For example, the viscosity of Federated oil from Alberta is 5 mPa, while a Sockeye oil from California is 45 mPa at $15\,°C$. In general, the greater the fraction of saturates and aromatics and the lower the amount of asphaltenes and resins, the lower the viscosity. Evaporation of lighter oil components during evaporative weathering leads to increased viscosity.

As with other physical properties, viscosity is affected by temperature, with a lower temperature giving a higher viscosity. For most oils, the viscosity varies significantly as the logarithm of the temperature. Oils that flow readily at high temperature can become a slow-moving, viscous mass at low temperature. In terms of oil spill cleanup, viscous oils do not spread rapidly, do not penetrate soils rapidly, and affect the ability of pumps and skimmers to handle oil. The dynamic viscosity of oil (in mPa) is conveniently measured by a viscometer using a variety of cup-and-spindle sensors at strictly controlled temperatures.

16.2.2.5 Flash Point

The flash point of oil is the temperature at which the vapor over the liquid will ignite upon exposure to an ignition source. A liquid is considered to be flammable if its flash point is less than $60\,°C$. Flash point is an important factor in relation to the safety of spill cleanup operations. Gasoline and other light fuels can ignite under most ambient conditions and therefore are a serious hazard when spilled. Many freshly spilled crude oils also have low *flash points* until the lighter components have evaporated or dispersed. On the other hand, Bunker C and heavy crude oils generally are not flammable when spilled. The flash point of low viscosity oil is measured by ASTM method D1310, while that of heavier oils is measured by ASTM method D93.

16.2.2.6 Pour Point

The pour point of oil is the temperature at which no flow of oil is visible over a period of five seconds from a standard measuring vessel. The pour point of crude oils generally varies from -60 to $30\,°C$. Lighter oils with low viscosities generally have lower pour points. As oils are made up of hundreds of compounds, some of which may still be liquid at the pour point, the pour point is not the temperature at which oil will no longer pour. The pour point represents a consistent temperature at which oil will pour very slowly and therefore has limited use as an indicator of the state of the oil. For example, waxy oils can have very low pour point, but may continue to spread slowly at that temperature and can evaporate to a significant degree. The pour point of oil is measured by ASTM method D97.

16.2.2.7 Distillation Fractions

Distillation fractions of an oil represent the fraction (generally measured by volume) of an oil that is boiled off at a given temperature. This data is obtained on most crude oils so that oil companies can adjust parameters in their refineries to handle the oil. This data also provides environmentalists with useful insights into the chemical composition of oils. For example, while 70% of gasoline will boil off at $100\,°C$, only 5% of a crude oil, and an even smaller amount of Bunker C, will boil off at that temperature. Distillation fractions correlate strongly to the composition as well as to other physical properties of oil.

16.2.2.8 Oil/Water Interfacial Tension

The oil/water interfacial tension, sometimes called surface tension, is the force of attraction or repulsion between the surface molecules of oil and water. The SI units for interfacial tension are milliNewtons per meter (mN/m). Together with viscosity, surface tension is an indication of how rapidly, and to what extent, oil will spread on water. The lower interfacial tensions with water, the greater the extent of oil spreading. In actual practice, the interfacial tension must be considered along with the viscosity because it has been found that interfacial tension alone does not account for spreading behavior. Surface tensions vary only to a small degree between oils, but larger changes can accompany changes in temperature. Interfacial tensions can be measured by ASTM method D971, using a Krüss K-10 Tensionmeter by the de Noüy ring method.

16.2.2.9 Odor

Odor of oil is a quality parameter, not a quantitative parameter. Oils that contain significant amount of unsaturates, certain types of nitrogenous compounds, and/or sulfur-containing compounds such as mercaptans tend to possess a pervasive H_2S-like odor. In contrast, oils mainly composed of light hydrocarbons, containing high proportions of aromatics, or mix of paraffins and naphthenes possess a sweet gasoline-like odor.

16.3 CHEMISTRY OF BIOMARKER COMPOUNDS

To better understand the structure and bonding of complex biomarker compounds, we start from reviewing some basic principles and concepts of molecular structures of simpler isoprenoid compounds, and then the labeling system and nomenclature of biomarker compounds.

16.3.1 The "Isoprene Rule" and Biomarker Families

16.3.1.1 The Isoprene Rule and Terpenoids

In 1887, German chemist Otto Wallach determined the structures of several terpenes and discovered that all of them were composed of two or more five-carbon units of isoprene [2-methyl-1,3-butadiene, $CH_2=C(CH_3)-CH=CH_2$]. The isoprene unit maintains its isopentyl structure in a terpene, usually with modification of the isoprene double bonds (Figure 16.3.1).

Organic chemists and geochemists have long realized that *isoprene* is the basic structural unit of many natural products and all oil biomarker compounds (Peters and Moldowan, 1993). Compounds composed of isoprene subunits (i.e., to obey the "isoprene rule") are called terpenoids or isoprenoids. The triterpenoids constitute a large diverse group of natural products (Connolly, 1972). Terpenoids are the most widespread and ubiquitous in higher and lower plants, and now characterized to an increasing extent from the animal kingdom. A few have been known for centuries but in recent decades the level of research and activity in isolating and studying new substances has shown no sign

of abating; the discovery of completely new carbon skeleton among the naturally occurring plant and animal terpenoids is a frequent occurrence.

The immense variety of structural types found in the terpenoids was rationalized by the isoprene rule of Ruzicka (Connolly and Hill, 1991). However, the number of exceptions to the regular arrangement of isoprene units led to the biogenetic isoprene rule, which encompassed the possibility of rearrangements during biosynthesis. Terpenoids are thus seen as being formed from linear arrangements of isoprene units followed by various cyclizations and rearrangements of the carbon skeleton. They can also be biosynthetically modified by the loss or addition of carbon atoms. Cholesterol is an example of terpenoid that has lost some of the isoprenoid carbon atoms.

Terpenoids can be classified according to the number of isoprene units from which they are biogenetically derived, even though some carbons may have been added or lost (Connolly and Hill, 1991). The isoprene rule states that biosynthesis of these compounds occurs by polymerization of appropriately functionized C_5-isoprene subunits. Unlike other biopolymers such as proteins, terpenoids are not readily depolymerized because they are joined together by covalent carbon–carbon bonds. Terpenoids are grouped according to the numbers of the isoprene units from which they are biogenetically derived.

Terpenoid Groups	Number of carbons
Hemiterpenoids	5
Monoterpenoids	10
Sesquiterpenoids	15
Diterpenoids	20
Sesterterpenoids	25
Triterpenoids	30
Tetraterpenoids	40
Polyterpenoids	$5n (n > 8)$

As for the oil saturated terpenoids, they are generally categorized into families based on the approximate number of isoprene subunits they contain. The various oil terpane families are composed of a wide variety of acyclic and cyclic structures (Peters and Moldowan, 1993):

Hemiterpane (C_5)	containing one isoprene subunit
Monoterpanes (C_{10})	containing two isoprene subunits
Sesquiterpanes (C_{15})	containing three isoprene subunits
Diterpanes (C_{20})	containing four isoprene subunits
Sesterterpanes (C_{25})	containing five isoprene subunits
Triterpanes and steranes (C_{30})	containing six isoprene subunits
Tetraterpanes (C_{40})	containing eight isoprene subunits
Polyterpanes ($C_{5n(n>8)}$)	containing nine or more isoprene subunits

16.3.1.2 Acyclic Terpenoids or Isoprenoids

One of the most important discoveries in petroleum chemistry and organic geochemistry was the detection of a large number of aliphatic isoprenoid hydrocarbons in oils, coals, shales, and dispersed organic materials. The variety of isoprenoid compounds is incomparably large.

The linkage between isoprene subunits can vary from regular (head-to-tail) to irregular (differing in the order of attachment of the isoprene subunits, such as head-to-head or tail-to-tail) linkages. Phytane ($C_{20}H_{42}$), which is one of

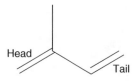

Figure 16.3.1 Molecular structure of isoprene (C_5).

the most abundant isoprenoids in oil and has been widely used for estimation of the degree of oil biodegradation in the environment, is a typical example of a regular, acyclic isoprenoid consisting of four head-to-tail linked isoprene units. Squalane ($C_{30}H_{62}$) and Botryococcane ($C_{34}H_{70}$) are examples of irregular isoprenoids. Squalane contains six isoprene subunits with one tail-to-tail linkage, while irregular Botryococcane is a highly specific biomarker for lacustrine sedimentation.

Degraded, rearranged, or homologous structures can be categorized into their corresponding parent terpenoid family. The precise number of carbon atoms in a given terpenoid family varies due to differences in source materials, diagenesis, thermal maturity, and in-reservoir biodegradation. For example, pristane ($C_{19}H_{40}$), another isoprenoid compound widely used for environmental oil biodegradation studies, contains one less methylene group ($-CH2-$) than phytane ($C_{20}H_{42}$), but it is still classified as an acyclic diterpane. Other examples include pseudohomologous series of regular isoprenoids from farnesane (C_{15}) through C_{16} (trimethyl-C_{13}) and C_{18} (norpristane), which are also quite abundant in oil.

Figure 16.3.2 show molecular structures of some representative acyclic terpenoid compounds in oil including 2,6-dimethyloctane (monoterpane, C_{10}); farnesane (2,6,10-trimethyldodecane, sesquiterpane, C_{15}); pristane and phytane (diterpane, C_{19} and C_{20}); 2,6,10,15,19-pentamethyleicosane (sesterterpane, C_{25}); botryococcane and squalane (triterpane, C_{30}); and bisphytane (tetraterpane, C_{40}).

16.3.1.3 Cyclic Terpenoids

The most common cyclic terpenoids in oil are terpanes, steranes (irregular cyclic terpenoid compounds), and aromatic steranes. Although cyclic terpenoids containing almost any number of carbons can occur in theory, only those containing combination of five or six carbons (cyclopentyl or cyclohexyl) occur commonly in petroleum.

The terpanes include sesqui- (C_{15}, bicyclic), di- (C_{20}, largely tricyclic), and triterpanes (C_{30}, mainly pentacyclic, and some tricyclic and tetracyclic), which are found in most crude oils. The terpanes comprise several homologous

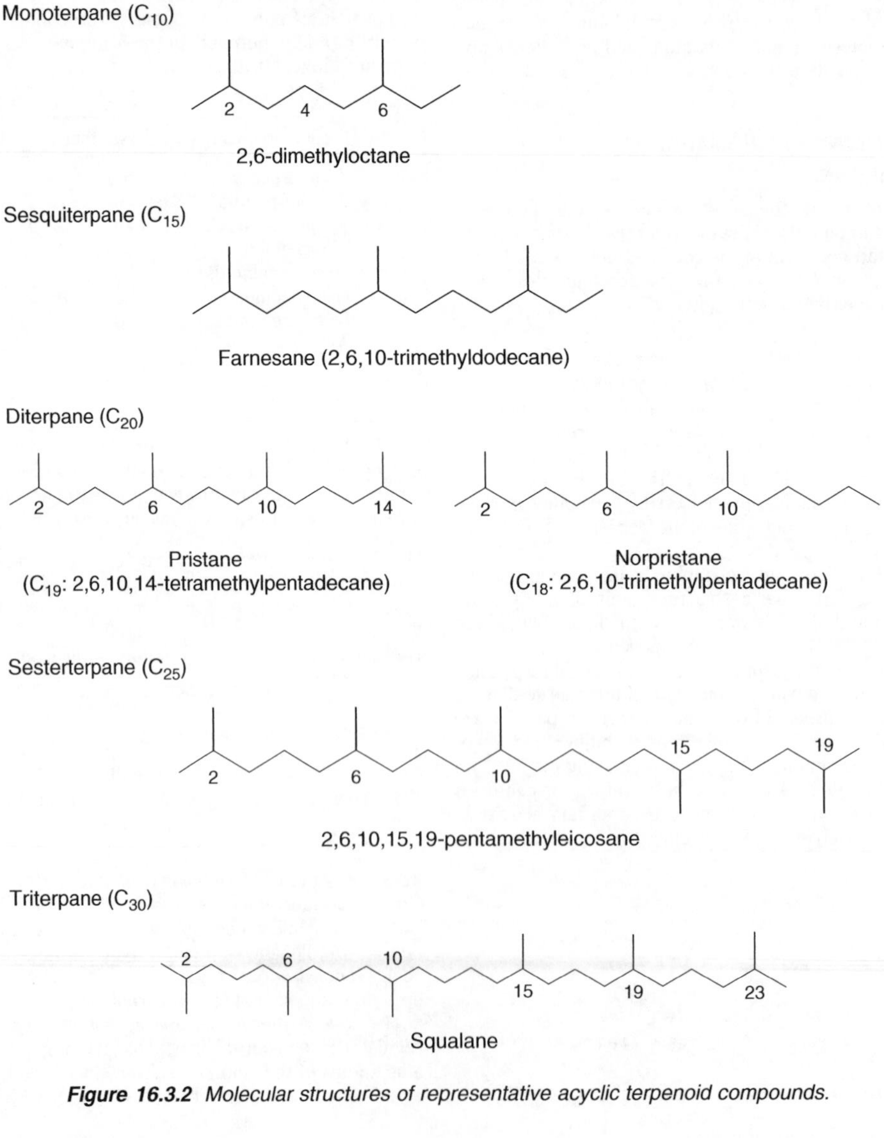

Monoterpane (C_{10})

2,6-dimethyloctane

Sesquiterpane (C_{15})

Farnesane (2,6,10-trimethyldodecane)

Diterpane (C_{20})

Pristane
(C_{19}: 2,6,10,14-tetramethylpentadecane)

Norpristane
(C_{18}: 2,6,10-trimethylpentadecane)

Sesterterpane (C_{25})

2,6,10,15,19-pentamethyleicosane

Triterpane (C_{30})

Squalane

Figure 16.3.2 Molecular structures of representative acyclic terpenoid compounds.

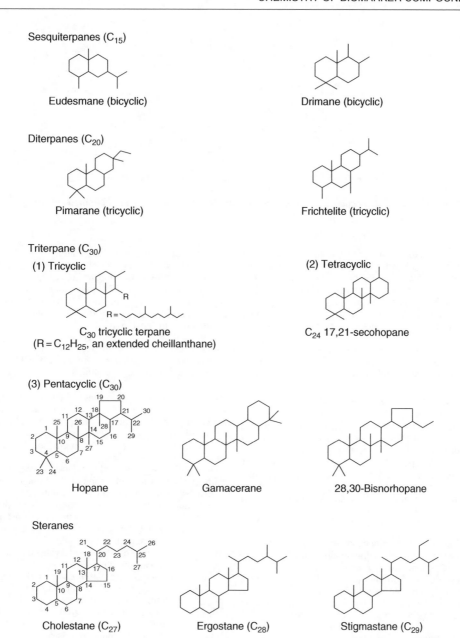

Figure 16.3.3 *Molecular structures of representative cyclic terpenoid compounds.*

series, including bicyclic, tricyclic, tetracyclic, and penta-cyclic compounds. Hopanes are pentacyclic triterpanes commonly containing 27 to 35 carbon atoms in a naphthenic structure composed of four six-membered rings and one five-membered ring (Figure 16.3.3). Hopanes with the 17 αβ-configuration in the range of C_{27} to C_{35} are characteristic of petroleum because of their large abundance and thermodynamic stability compared to other epimeric (ββ and βα) series.

The four-ringed steranes are a class of biomarkers containing 21 to 30 carbons, including regular steranes, rearranged diasteranes, and mono- and tri-aromatic steranes. Among them, the regular C_{27}–C_{28}–C_{29} homologous sterane series (cholestane, ergostane, and stigmastane) are the

most common steranes and are useful for chemical fingerprinting because of their high source specificity. These sterane homologue series do not contain an integral number of isoprene subunits, and thus only approximate the isoprene rule. However, they still show some terpenoid character and can be categorized into the corresponding cyclic terpenoid families.

Aromatic steranes are another group of biomarker compounds found in the oil aromatic hydrocarbon fraction. These compounds can also provide valuable information on organic matter input for oil-to-oil and oil-to-source rock correlation and as supporting evidence for assessment of thermal maturity. The C-ring monoaromatic (MA) steranes are characterized by a series of 20R and 20S

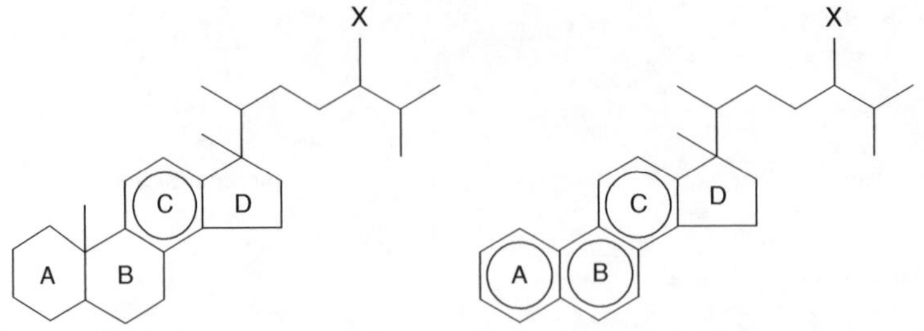

Figure 16.3.4 Molecular structures of representative monoaromatic and triaromatic steranes.

C_{27}–C_{28}–C_{29} 5α- and 5β-cholestanes, ergostanes, and stigmastanes. The ABC-ring triaromatic steranes are formed from aromatization of C-ring monoaromatic steranes involving the loss of a methyl group at the A/B ring junction. This fraction is composed mainly of C_{20} and C_{21}, and C_{26}–C_{27}–C_{28} homologous triaromatic steranes. Examples of monoaromatic and triaromatic steranes are shown in Figure 16.3.4.

As a summary, Table 16.3.1 lists important biomarker terpane, sterane, and aromatic sterane compounds, used frequently for oil spill identification.

16.3.2 Labeling System and Nomenclature of Biomarkers

The chemical structures of terpenoids are more complicated than that of normal alkanes and isoalkanes, in particular the three-dimensional structure. The system used for the nomenclature of terpenoids has evolved over many years. For many terpenoid classes, several names have been proposed for the carbon skeleton, but the basic rules of the IUPAC (International Union of Pure and Applied Chemistry) system are used for nomenclature of biomarkers, and the essential rules are (Morrison and Boyd, 1977, Wade, 2003):

1) Select the longest continuous chain as the parent structure and use the name of this chain as the base name of the compound, and then consider the compound to have been derived from this structure by replacement of hydrogen atom by various alkyl group.
2) In numbering the parent carbon chain, start at whichever end that results in the use of the lowest numbers.
3) Indicate by a number of the carbon to which an alkyl group is attached.
4) If the same alkyl group occurs more than once as a side chain, indicate this by the prefix *di-, tri-, tetra-*, etc., and indicate the position of each group.
5) If there are several different alkyl groups attached to the parent chain, name them in order of increasing size or in alphabetical order.

Figure 16.3.5 shows the IUPAC labeling of carbon atoms in phytane (2,6,10,14-tetramethylhexadecane) and 2,6,10-trimethyl-7-(3-methylbutyl)-dodecane (i.e., acyclic isoprenoids).

The labeling system for cyclic triterpanes and steranes is shown in Figure 16.3.6 and described as follows:

1) Each carbon atom and the rings in biomarker molecules are labeled systematically. Rings are specified in succession from left to right as the A-ring, B-ring, C-ring, D-ring, and so on.

2) A capital "C" followed immediately by a subscript number refers to the number of carbon atoms in a particular compound (e.g., C_{30} hopane and C_{27} sterane mean that they contain 30 and 27 carbon atoms, respectively).
3) A capital "C" followed by a dash and numbers refers to a particular position within the compound (e.g., C-17 and C-21 in the 17α(H), 21β(H)-hopane are the carbon atoms at positions 17 and 21).
4) Prefixes are used to indicate the changes to the normal biomarker carbon skeleton, which include the prefix *nor-, seco-,* and *neo-,* and others.

The prefix *nor-* is used to indicate loss of carbons from a carbon skeleton. For example, 25-norhopane is identical to hopane except that a methyl group at the C-25 position has been lost from its point of attachment at the C-10 position. If two or three carbons are lost, the prefix "*bisnor-*" or "*trisnor-*" is used respectively.

The prefix "*seco-*" is used to indicate cleavage of a bond, with the locants for both ends of the broken bond given, e.g., 3,4-secoeudesmane and 17,21-secohopane.

For more information, see *Appendix IV* in the current *Chemical Abstract Index Guide. Chemical Abstract* uses these prefixes extensively for some classes of terpenoids.

Table 16.3.2 summarizes the nomenclature used to modify the structural specification of cyclic biomarkers. For example, if a carbon atom is absent or present in a biomarker named after a parent compound, the prefix *nor* or *homo* is used, preceded by the number of the absent or present atom. Thus, 25-norhopanes are identical to C_{30} hopane except that a methyl group (at C-25) has been removed from its point of attachment at the C-10 position. If two or three carbon atoms are missing, the prefix *bisnor-* or *trisnor-* is used respectively. Thus, 28, 30-bisnorhopanes have two methyl groups (at C-28 and C-30) to be removed from their parent C_{30} hopane. The 30-homohopanes are identical to C_{30} hopane except that a methyl group has been added at C-30 position. The 17,21-secohopane indicates that the bond between carbon number 17 and 21 in the E-ring of C_{30} hopane has been broken, resulting in the formation of a new tetracyclic terpane.

16.3.2.1 α and β Stereoisomers

Isomers, as mentioned in Section 16.2.1, are different compounds that have the same molecular formula but the atoms are attached in different ways. There are two classes of isomers (Figure 16.3.7): *constitutional isomers* and *stereoisomers*. *Constitutional isomers* (or structural isomers) differ in their bonding sequence, and their atoms are connected differently. The number of *constitutional* isomers increases

Table 16.3.1 *Source-Specific Target Biomarker Terpane and Sterane Compounds for Oil Spill Studies*

Peak	Compound	Code	Empirical Formula	Target ions
	Sesquiterpanes (BicyclicTerpanes)			
	Drimanes and drimane-structure terpanes		$C_{14}H_{26}$, $C_{15}H_{28}$, $C_{16}H_{30}$	123
	Adamantanes and Diamantanes			
	Adamantanes		$C_{10}H_{16}$, alkyl-$C_{10}H_{15}$	136, 135, 149, 163, 177
	Diamantanes		$C_{14}H_{20}$, alkyl-$C_{14}H_{19}$	188, 187, 201, 215, 229
	Terpanes			
1	C_{19} tricyclic terpane	TR19	$C_{19}H_{34}$	191
2	C_{20} tricyclic terpane	TR20	$C_{20}H_{36}$	191
3	C_{21} tricyclic terpane	TR21	$C_{21}H_{38}$	191
4	C_{22} tricyclic terpane	TR22	$C_{22}H_{40}$	191
5	C_{23} tricyclic terpane	TR23	$C_{23}H_{42}$	191
6	C_{24} tricyclic terpane	TR24	$C_{24}H_{44}$	191
7	C_{25} tricyclic terpane (a)	TR25A	$C_{25}H_{46}$	191
8	C_{25} tricyclic terpane (b)	TR25B	$C_{25}H_{46}$	191
9	triplet: C_{24} tetracyclic terpane + C_{26} (S + R) tricyclic terpanes	TET24 + TR26A + TR26B	$C_{24}H_{42}$ + $C_{26}H_{48}$	191
10	C_{28} tricyclic terpane (a)	TR28A	$C_{28}H_{52}$	191
11	C_{28} tricyclic terpane (b)	TR28B	$C_{28}H_{52}$	191
12	C_{29} tricyclic terpane (a)	TR29A	$C_{29}H_{54}$	191
13	C_{29} tricyclic terpane (b)	TR29B	$C_{29}H_{54}$	191
14	Ts: $18\alpha(H),21\beta(H)$-22,29,30-trisnorhopane	Ts	$C_{27}H_{46}$	191
15	$17\alpha(H),18\alpha(H),21\beta(H)$-25,28,30-trisnorhopane	TH27	$C_{27}H_{46}$	191, 177
16	Tm: $17\alpha(H),21\beta(H)$-22,29,30-trisnorhopane	Tm	$C_{27}H_{46}$	191
17	C_{30} tricyclic terpane 1	TR30A	$C_{30}H_{52}$	191
18	C_{30} tricyclic terpane 2	TR30B	$C_{30}H_{52}$	191
19	$17\alpha(H),18\alpha(H),21\beta(H)$-28,30-bisnorhopane	H28	$C_{28}H_{48}$	191
20	$17\alpha(H),21\beta(H)$-25-norhopane	NOR25H	$C_{29}H_{50}$	191, 177
21	$17\alpha(H),21\beta(H)$-30-norhopane	H29	$C_{29}H_{50}$	191
22	$18\alpha(H),21\beta(H)$-30-norneohopane (C_{29}Ts)	C29Ts	$C_{29}H_{50}$	191
23	$17\alpha(H)$-diahopane	DH30	$C_{30}H_{52}$	191
24	$17\alpha(H),21\beta(H)$-30-norhopane (normoretane)	M29	$C_{29}H_{50}$	191
25	$18\alpha(H)$ and $18\beta(H)$-oleanane	OL	$C_{30}H_{52}$	191, 412
26	$17\alpha(H),21\beta(H)$-hopane	H30	$C_{30}H_{52}$	191
27	$17\alpha(H)$-30-nor-29-homohopane	NOR30H	$C_{30}H_{52}$	191
28	$17\beta(H),21\alpha(H)$-hopane (moretane)	M30	$C_{30}H_{52}$	191
29	$22S$–$17\alpha(H),21\beta(H)$-30-homohopane	H31S	$C_{31}H_{54}$	191
30	$22R$–$17\alpha(H),21\beta(H)$-30-homohopane	H31R	$C_{31}H_{54}$	191
31	Gammacerane	GAM	$C_{30}H_{52}$	191, 412
32	$17\beta(H),21\beta(H)$-hopane	(IS)	(Internal standard)	191
33	$22S$–$17\alpha(H),21\beta(H)$-30,31-bishomohopane	H32S	$C_{32}H_{56}$	191
34	$22R$–$17\alpha(H),21\beta(H)$-30,31-bishomohopane	H32R	$C_{32}H_{56}$	191
35	$22S$–$17\alpha(H),21\beta(H)$-30,31,32-trishomohopane	H33S	$C_{33}H_{58}$	191
36	$22R$–$17\alpha(H),21\beta(H)$-30,31,32-trishomohopane	H33R	$C_{33}H_{58}$	191
37	$22S$–$17\alpha(H),21\beta(H)$-30,31,32,33-tetrakishomohopane	H314S	$C_{34}H_{60}$	191

Table 16.3.1 *(Continued)*

Peak	Compound	Code	Empirical Formula	Target ions
38	22R–17α(H),21β(H)-30,31,32,33-tetrakishomohopane	H34R	$C_{34}H_{60}$	191
39	22S-17α(H),21β(H)-30,31,32,33,34-pentakishomohopane	H35S	$C_{35}H_{62}$	191
40	22R–17α(H),21β(H)-30,31,32,33,34-pentakishomohopane	H35R	$C_{35}H_{62}$	191
	Steranes			
41	C_{20} 5α(H),14α(H),17α(H)-sterane	S20	$C_{20}H_{34}$	217 & 218
42	C_{21} 5α(H),14β(H),17β(H)-sterane	S21	$C_{21}H_{36}$	217 & 218
43	C_{22} 5α(H),14β(H),17β(H)-sterane	S22	$C_{22}H_{38}$	217 & 218
44	C_{27} 20S–13β(H),17α(H)-diasterane	DIA27S	$C_{27}H_{48}$	217 & 218, 259
45	C_{27} 20R–13β(H),17α(H)-diasterane	DIA27R	$C_{27}H_{48}$	217 & 218, 259
46	C_{27} 20S–13α(H),17β(H)-diasterane	DIA27S2	$C_{27}H_{48}$	217 & 218, 259
47	C_{27} 20R–13α(H),17β(H)-diasterane	DIA27R2	$C_{27}H_{48}$	217 & 218, 259
48	C_{28} 20S–13β(H),17α(H)-diasterane	DIA28S	$C_{28}H_{50}$	217 & 218, 259
49	C_{28} 20R–13β(H),17α(H)-diasterane	DIA28R	$C_{28}H_{50}$	217 & 218, 259
50	C_{29} 20S–13β(H),17α(H)-diasterane	DIA29S	$C_{29}H_{52}$	217 & 218, 259
51	C_{29} 20R–13α(H),17β(H)-diasterane	DIA29R	$C_{29}H_{52}$	217 & 218, 259
52	C_{27} 20S–5α(H),14α(H),17α(H)-cholestane	C27S	$C_{27}H_{48}$	217 & 218
53	C_{27} 20R–5α(H),14β(H),17β(H)-cholestane	C27bbR	$C_{27}H_{48}$	217 & 218
54	C_{27} 20S–5α(H),14β(H),17β(H)-cholestane	C27bbS	$C_{27}H_{48}$	217 & 218
55	C_{27} 20R–5α(H),14α(H),17α(H)-cholestane	C27R	$C_{27}H_{48}$	217 & 218
56	C_{28} 20S–5α(H),14α(H),17α(H)-ergostane	C28S	$C_{28}H_{50}$	217 & 218
57	C_{28} 20R–5α(H),14β(H),17β(H)-ergostane	C28bbR	$C_{28}H_{50}$	217 & 218
58	C_{28} 20S–5α(H),14β(H),17β(H)-ergostane	C28bbS	$C_{28}H_{50}$	217 & 218
59	C_{28} 20R–5α(H),14α(H),17α(H)-ergostane	C28R	$C_{28}H_{50}$	217 & 218
60	C_{29} 20S–5α(H),14α(H),17α(H)-stigmastane	C29S	$C_{29}H_{52}$	217 & 218
61	C_{29} 20R–5α(H),14β(H),17β(H)-stigmastane	C29bbR	$C_{29}H_{52}$	217 & 218
62	C_{29} 20S–5α(H),14β(H),17β(H)-stigmastane	C29bbS	$C_{29}H_{52}$	217 & 218
63	C_{29} 20R–5α(H),14α(H),17α(H)-stigmastane	C29R	$C_{29}H_{52}$	217 & 218
64	C30 steranes	C30S	$C_{30}H_{54}$	217 & 218
	Monoaromatic Steranes			253
	Triaromatic Steranes			231

exponentially with the increase of carbon atoms in each compound. For example, Butane (C_4H_{10}) has two isomers, *n*-butane and isobutane (methyl-propane), while dodecane ($C_{10}H_{22}$) and Eicosane ($C_{20}H_{42}$) can have 75 and 355 possible isomers respectively. *Stereoisomers* are discussed in detail in the following sections.

Asymmetric or Chiral Carbons

A carbon atom that is bonded to four different groups is called *asymmetric* carbon or a *chiral* carbon atom, and is often designed by a*. For example, the possible chiral centers for steranes are at C-5, C-14, C-17, C-20, and C-24 (for C_{28} and C_{29} steranes).

(1)

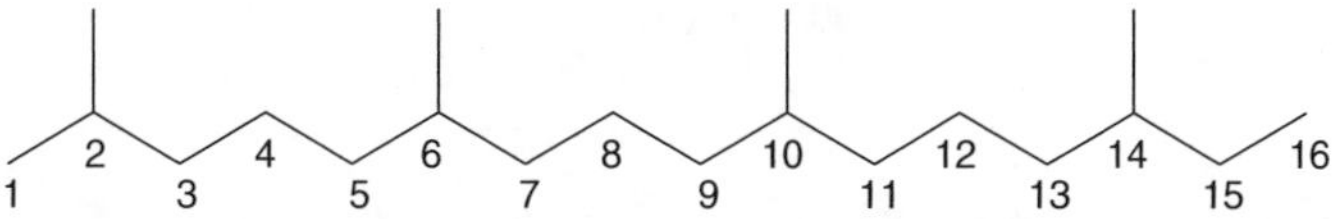

Phytane (2,6,10,14-tetramethylhexadecane)

(2)

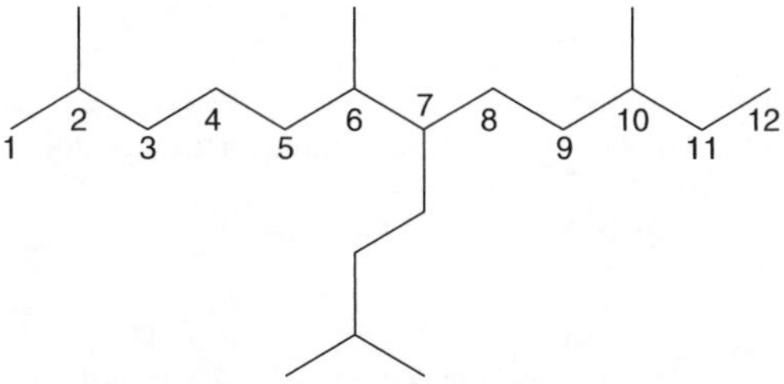

2,6,10-trimethyl-7-(3-methylbutyl)-dodecane

Figure 16.3.5 *The chemical structures of phytane and 2,6,10-trimethyl-7-(3-methylbutyl)-dodecane and labeling for carbon atoms.*

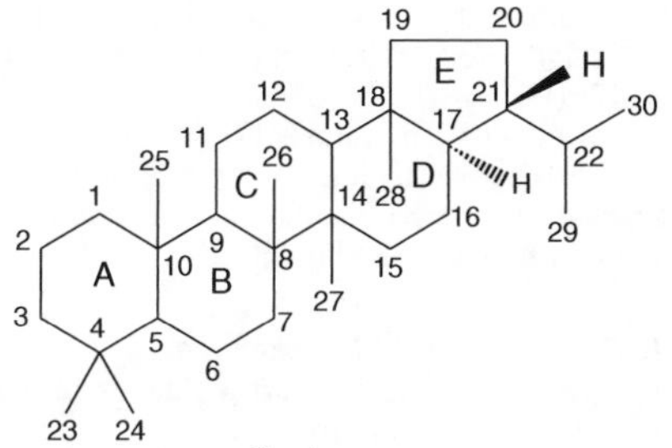

C_{30} hopane

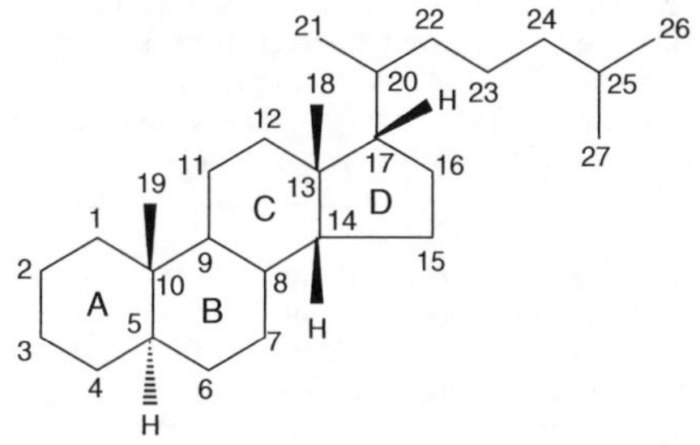

C_{27} sterane

Figure 16.3.6 *Labeling system for carbon atoms in C_{30} hopane (left) and C_{27} sterane (right).*

Table 16.3.2 *Common Modifiers and other Nomenclatures Related to Cyclic Biomarkers*

Modifier	Description	Example Biomarker
homo-	one additional carbon on the parent molecular structure	C_{31} 17α(H),21β(H)-30-homohopane
bis-, tris-, tetrakis-, pentakis- (also di-, tri-, tetra-, and penta-)	two to five additional carbons on the parent molecular structure	C_{32} 17α,21β-30,31-homohopane C_{33} 17α,21β-30,31,32-homohopane C_{34} 17α,21β-30,31,32,33-homohopane
seco-	cleaved C–C bond	C_{24} 17,21-secohopane (tetracyclic)
nor-	one less carbon on the parent molecular structure	25-norhopane
bisnor-	two less carbons on the parent molecular structure	28,30-bisnorhopane
trisnor-	three less carbons on the parent molecular structure	25,28,30-trisnorhopane
neo-	methyl group shifted from C-18 to C-17 position on hopanes	C_{29}Ts: 30-norneohopane
α	asymmetric carbon in ring with "H" down	17α(H),21β(H)-hopane
β	asymmetric carbon in ring with "H" up	17β(H),21β(H)-hopane
R	asymmetric carbon in acyclic moiety of biomarkers obeying convention in a clockwise direction	C_{27} 20R cholestane
S	asymmetric carbon in acyclic moiety of biomarkers obeying convention in a clockwise direction	C_{27} 20S cholestane

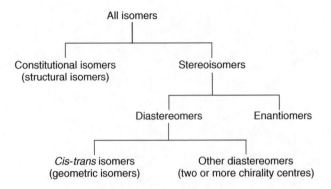

Figure 16.3.7 *Types of isomers.*

Stereochemistry is the part of organic chemistry dealing with molecular structures in three dimensions. One important aspect of stereochemistry is *stereoisomerism*. *Stereoisomers* are isomers whose atoms are bonded together in the same sequence but differ from each other in the orientation of the atoms in space. The *cis-trans* geometric isomers (such as *cis-* and *trans*-1,2-dimethylcyclopentane) are special types of *stereoisomers*. Differences in special orientation might seem unimportant, but stereoisomers often have remarkably different physical, chemical, and biological properties.

As described in Table 16.3.2, hydrogen atoms which are attached to an *asymmetric* or *chiral* carbon in a ring structure and are below the plane of the molecule are called α hydrogens, and the bond is drawn with a dashed line and designated as having the α-configuration. Conversely, hydrogen atoms located above the plane of the molecule are called β hydrogens, and the bond is drawn with a wedge bond and designated as having β-configuration. In many common ring systems the α hydrogen atoms found at ring junctions are generally omitted for clarity. For example, in 17α(H), 21β(H)-hopane ($C_{30}H_{52}$, Figure 16.3.5) the hydrogens at carbon number 17 and 21 are down and up; while in 5α(H), 14β(H), 17β(H)-cholestane ($C_{27}H_{48}$, Figure 16.3.5) the hydrogen attached to carbon number 5, 14, and 17 are down, up and up.

Hopanes exist as three stereoisomers: 17α(H), 21β(H)-hopane, 17β(H), 21β(H)-hopane, and 17β(H), 21α(H)-hopane (Waples and Machihara, 1991; Peters and Moldowan, 1993). Hopanes in the βα series are also called moretanes. Hopanes with the 17αβ-configuration in the range of C_{27} to C_{35} are characteristic of petroleum because of their greater thermodynamic stability compared to other epimeric series (ββ and αα).

Hopanoids produced by living organisms have generally a ββ-configuration. With increasing maturity the thermodynamically less stable ββ-hopanes are lost or converted to αβ- and βα-hopanes. The ββ series is, generally, not found in petroleum because it is thermally unstable. The αα series were not considered as natural products and since their stability is low compared to the αβ and βα series it is unlikely that they occur in more than trace levels in petroleum (Waples and Machihara, 1991; Peters and Moldowan, 1993). However, mechanics calculations have shown that the α-hopanes should be less stable than αβ- and βα-hopanes, but more stable than αα-hopanes. Recently, Nytoft and Bojesen-Koefoed (2001) showed that moderate quantities of 17α(H), 21α(H)-hopanes are present in several sediments and oils. The ratios of C_{30} 17α(H), 21α(H)-hopane to C_{30} 17α(H), 21β(H)-hopane are typically 0.02–0.04 in crude oils and mature sediments, but ratios up to 0.10 have been found in immature sediments.

16.3.2.2 R and S Stereoisomers

Stereoisomers that are mirror images of each other (i.e., differing in the same manner as right and left hands) are called *enantiomers*; while all other stereoisomers, which are not mirror images, are *diastereomers*. Most diastereomers are either geometric isomers or compounds containing two or more chiral centers (usually asymmetric carbons).

Enantiomer molecules are not superimposable. Many pairs of biomarkers with the same molecular formula (such as 22R and 22S homohopane homologous series in C_{31} to C_{35} range) are *enantiomers*. Since, chiral molecules are not superimposable on their *mirror images*, chirality is a necessary and sufficient condition for existence of enantiomers. Thus, a compound with at least one chiral carbon atom can exist as enantiomers, whereas a compound without chirality cannot exist as enantiomers.

Specification of Configuration: R and S

The arrangement of atoms that characterizes a particular stereoisomer is called its *configuration*. How can we

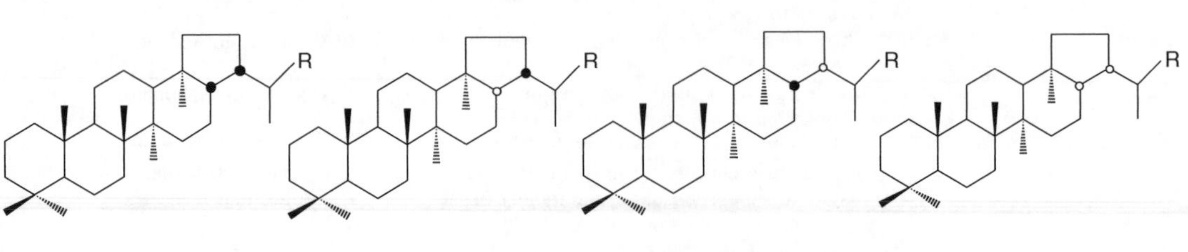

βΒ   αΒ   βα   αα

Figure 16.3.8 *Molecular structures of* ββ, αβ, βα, *and* αα-*hopanes.*

specify a particular configuration in a simple and convenient way to distinguish between enantiomers and give each of them a unique name? The Cahn-Ingold-Prelog convention procedure proposed by R. S. Cahn (the Chemical Society, London), C. Ingold (University College, London), and V. Prelog (Eidgenössiche Technische Hochschule, Zurich) (Cahn *et al.*, 1966) is the most widely accepted system for naming the configurations of chiral centers. Each asymmetric carbon atom is assigned a letter (R) or (S) based on its three-dimensional configuration.

The *sequence rule* described below is used to determine the priority sequence of atoms or groups of atoms bonded to the asymmetric carbon atom:

1) If the four atoms or groups of atoms attached to the chiral center are all different, *priority* depends on atomic number, with the atom of higher atomic number getting higher priority. If two atoms are isotopes of the same element, the atom of higher mass has the higher priority. Examples of priority for atoms bonded to an asymmetric carbon are the following:

$$I > Br > Cl > S > F > O > N > {}^{13}C > {}^{12}C > Li$$
$$> {}^{3}H > {}^{2}H > {}^{1}H$$

2) If the relative priority of two groups cannot be decided by (1), it shall be determined by comparison of the next atoms in the group and so forth. Thus, if two atoms attached to the chiral center are identical, we compare the atoms attached to each of these first atoms.

To determine the stereoisomeric (R or S) configuration, two steps are involved:

1) Following the *sequence rules* the *sequence of priority* is assigned to the four atoms or groups of atoms bonded to the asymmetric carbon atom. In the case of bromochloroiodomethane (CHClBrI), the four atoms attached to the chiral center are all different and priority depends on the atomic number, the atom of higher number having higher priority, thus, the sequence of priority is I, Br, Cl, H.

2) The molecule is oriented so that the group of lowest priority is directed away from the viewer. Subsequently, the remaining groups are arranged. If proceeding the remaining groups in a clockwise direction, that is, from the group of the highest priority to the group of second priority and then to the third, the configuration is specified **R** (Latin: *rectus*, meaning right); if counterclockwise, the configuration is specified **S** (Latin: *sinister*, meaning left). Thus, the configurations of CHClBrI are viewed like this (Figure 16.3.9) and are specified R and S enantiomers, respectively.

For the isoprenoid compound 3-isopropyl-heptane ($C_{10}H_{22}$), three of the atoms attached to the chiral center (carbon atom

at position 3 of the parent chain) are carbon. In i-C_3H_7 (isopropyl), the second atoms are C, C, H; in n-C_4H_9 (n-butyl), the second atoms are C, H, H. Since carbon has higher atomic number than hydrogen, isopropyl has higher priority than n-butyl. For the same reason, n-butyl has higher priority than C_2H_5 (ethyl). Thus, a complete sequence of priority for 3-isopropyl-heptane is isopropyl, n-butyl, ethyl, and hydrogen. 3-isopropyl-heptane has R-configuration and its three-dimensional structure is shown in Figure 16.3.10.

R and S Nomenclature of Biomarkers

For biomarkers, the use of R and S nomenclature is generally restricted to carbon atoms which are not part of a ring, while the use of α versus β nomenclature is used to describe asymmertric configurations at ring carbons.

The steranes including the ones most abundant in oils, cholestanes ($C_{27}H_{48}$), ergostanes ($C_{28}H_{50}$), and stigmastanes ($C_{29}H_{52}$), can have R- and S-configurations at the acyclic (chain position) carbon atom C-20, resulting in two homologue series with 20R (20R $\alpha\alpha\alpha$ and 20R $\alpha\beta\beta$) and 20S (20S $\alpha\alpha\alpha$ and 20S $\alpha\beta\beta$) configurations.

Hopanes with 30 carbons or less show asymmetric centers at C-21 and all ring-juncture carbons including C-5, C-8, C-9, C-10, C-13, C-14, C-17, and C-18. Common homohopanes (C_{31} to C_{35}) have an extended side chain with an additional asymmetric center at C-22, resulting in two homologues with 22R and 22S configurations. These two homologous homohopanes (22R and 22S) can be well separated by GC-MS as well-resolved doublet peaks, prominent in gas chromatograms of oils.

The R and S and α versus β designations are a useful means of describing the relative configuration of biomarker compounds. It should, however, be noted that these designations are determined strictly on the basis of the convention as described by Cahn *et al.* (Cahn *et al.*, 1966) without reference to optical rotation.

16.3.3 Biomarker Genesis

The origin and formation of petroleum is complicated, and the details of petroleum genesis have long been a topic of interest. It is now generally accepted that petroleum is formed from biological substances of various classes, mostly algae, bacteria (especially lipid of cell membranes), phyto- and zooplankton as well as higher plants. The conversion of precursor biochemical compounds from living organisms into petroleum hydrocarbons creates a large suite of compounds in crude oils with distinct structures. The precursor biochemical compounds from which petroleum is generated include (Speight, 1999):

- *Hydrocarbons* (occurring in living plants).
- *Lipids* (can be hydrolyzed to long-chain carboxy acids and subsequently decarboxylated to form alkanes).
- *Fats* (mixtures of glycerides).

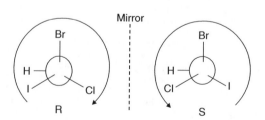

Figure 16.3.9 *Enantiomers of an asymmetric carbon atom. Four different atoms (I, Br, Cl, and H) bonded to carbon atom. These two mirror images are nonsuperimposable.*

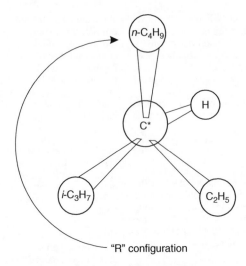

Figure 16.3.10 *Stereochemistry of the asymmetric center in 3-isopropyl-heptane ($C_{10}H_{22}$, a acyclic monoterpane). The three dimensional drawing shows an R configuration at the asymmetric carbon atom.*

- *Fatty acids* (a ready source of petroleum constituents because they can be converted to hydrocarbons by elimination of carbon dioxide from the molecule).
- *Carbohydrates* (ranging from simple sugars to polysaccharides, such as glycogen, starch, and the cellulose).
- *Proteins* (complex polymeric substances composed of one or several of some 25 amino acids linked through a carboxyl carbon and nitrogen).
- *Lignin* (a mixture of complex, high-molecular-weight, amorphous substances which forms the cell wall structure of plants; these substances are characterized by an abundance of aromatic and phenolic units).
- *Sterols* and *terpenoids*.
- *Porphyrins*.

The conversion and transformation of biochemical compounds into petroleum is completed through the *maturation processes* generally referred to as *diagenesis, catagenesis*, and *metagenesis* (Petrov, 1987; Speight, 1999; Peters and Moldowan, 1993). These processes are combinations of a series of bacteriological actions and low-temperature reactions. Furthermore, geological migration of the liquid product from the source sediment to the reservoir rock may also occur. More specifically, biomarker hopanes are converted from the biogenic precursors bacteriohopanetetrol (a hopanoid), steranes from cholesterol (a steroid), and porphyrins, pristane, and phytane from chlorophyll (a tetrapyrrole pigment), respectively.

- *Diagenesis* or *sedimentogenesis*—the process referring to the physical, chemical, and biological (e.g., aerobic and anaerobic degradation) alteration of organic matter in sediments prior to significant changes caused by heat. Diagenetic processes proceed at relatively moderate temperatures, and are not accompanied by extensive destruction of organic matter. The source material, animal, and vegetable deposits accumulate on the bottom of inland seas and are then decomposed by bacteria under oxic or anoxic conditions. The carbohydrates and the bulk of protein are converted into water-soluble material or gases and thus removed from the site. The fats, waxes, and other fat-soluble and stable materials remain.
- *Catagenesis*—the most important process in petroleum formation by which the organic matter in rocks (i.e., kerogen) from the diagenesis stage is thermally altered at temperatures in the range of 60 to 150°C under typical burial conditions during millions of years. In the process of catagenesis the maturation and rearrangement of organic molecules, their stereochemical modification and, most importantly, a separation between the mineral part and the organic molecules occur. High temperatures and pressures cause carbon dioxide to evolve from compounds containing a carboxyl group, and water is produced from the hydroxy-acids and alcohols to leave a bituminous residue. Continued application of the heat and pressure causes light cracking, producing a liquid product with high olefin content (*protopetroleum*). Moreover, a statistical rupture of C–C bonds in long aliphatic chains, resulting in the petroleum hydrocarbon homologues occur. During catagenesis, biomarkers undergo structural changes, and these changes can be used to estimate the extent of heating of their host sediments and oils migrated from these sediments.
- *Metagenesis*—the process by which the organic molecules are further altered at temperatures in the range of about 150 to 200°C. In this stage, the unsaturated components of the *protopetroleum* are polymerized under the influence of contact catalysts and thus the polyolefins are converted into paraffins and/or cycloparaffins. Aromatics are presumed to be formed either directly during cracking, by cyclization through condensation reactions, or even during the decomposition of proteins. During this process, biomarkers are severely reduced in concentration or even absent because of their instability under these conditions.

16.3.3.1 Genesis of Hopanes

Hopanes present in sediments range from C_{27} to C_{35}, usually with the C_{30} isomers as the predominant. The origins of most hopanes are the C_{35} bacteriohopanetetrol (tetrahydroxybacteriohopane) and related bacteriohopanes (Figure 16.3.11) found in the lipid membranes of prokaryotic organisms (Peters and Moldowan, 1991). Bacteriohopanetetrol is amphipathic in the molecular structure because one end of the molecule is polar, while the other is nonpolar.

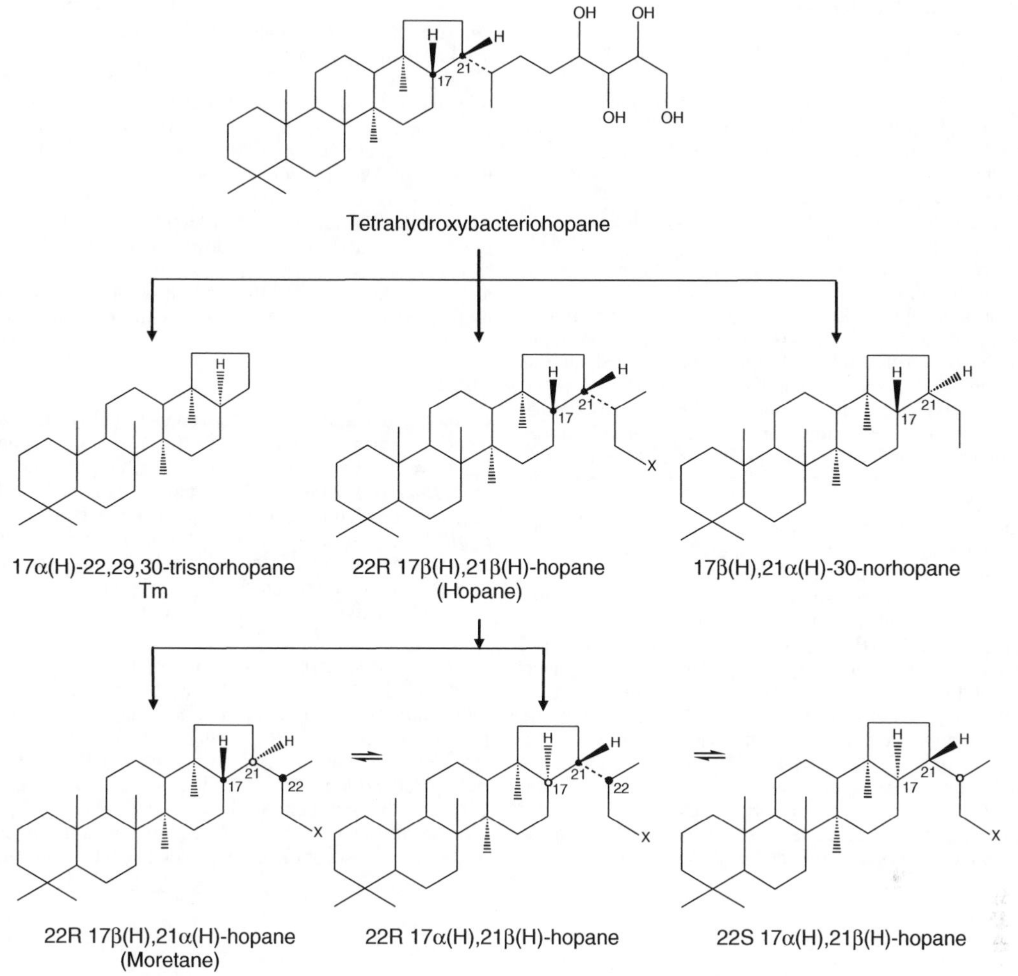

Figure 16.3.11 *Transformation of hopanes in sediments. The biological configuration [17β(H), 21β(H), 22R] imposed on bacteriohopanetetrol and its immediate saturated product, the 17β(H), 21β(H), 22R configuration, is unstable during catagenesis process and undergoes isomerization to geological configuration βα (22R), αβ (22R), and αβ (22S) hopanes.*

Because the stereochemical arrangement is thermodynamically unstable, diagenesis and catagenesis of bacteriohopanetetrol result in a transformation of the 17β(H), 21β(H), 22R precursors to the 17α(H), 21β(H)-hopanes (22R) and 17β(H), 21α(H)-hopanes (22R, moretanes). Thus, C_{29} and higher hopane homologues are isomerized from the natural 17β(H), 21β(H) stereochemistry to 17α(H), 21β(H) and 17β(H), 21α(H) isomers at an early stage of diagenesis, and the ratio of αβ/(αβ + ββ) reaches 1 with the transformation of the ββ configuration due to increasing maturation. Similarly, the biological 22R configuration found in bacteriohopanetetrol transforms into an equilibrium mixture of 22S and 22R αβ-homohopanes, because the stability of 22S epimers is slightly higher. Thermal equilibrium between the 22S and the 22R configuration of hopane homologues is reached at about 0.57 to 0.62 for most crude oils (Seifert and Moldowan 1980). Thus, predominance of 22S over 22R indicates high thermal maturity, while predominance of 22R over 22S indicates low thermal maturity.

For example, Simons *et al.* have demonstrated that the hopane fingerprints of Pasquia Hill samples (from the Western Interior Seaway, Canada) are dominated by 17β, 21β-hopanes with the predominance of 22R over 22S configuration, clearly indicating these samples are thermally immature. In contrast, the hopane fingerprints of Mill Creek samples are dominated by 17α, 21β-hopanes with predominance of 22S over 22R configuration, indicating that these samples are thermally more mature than the Pasquia Hill samples (Simons *et al.*, 2003).

C_{28} hopane compounds formed by processes involving loss of a methyl group from the hopanoid ring system would represent another compound class. For example, loss of the C-25 methyl group from the hopane results in a 25-norhopane. C_{27} hopanes have no side chains.

Oleananes occur widely in the plants and are formed in sediments through diagenetic and catagenetic alteration of oleananes and 3β-functionalized angiosperm triterpenoids, which are found in higher terrestrial plants (Alberdi and Lopez, 2000). Reduction of the double bounds of oleananes gives a mixture of 18α(H)- and 18β(H)-oleananes. Increased maturation favors the 18α(H)-configuration, because it is more stable than the 18β(H)-configuration.

16.3.3.2 Genesis of Steranes

The tetracyclic steranes originate from naturally occurring four-ringed alcohols called sterols. Sterols are widespread in animals and plants including phytoplankton, zooplankton, and vascular plants, where they are important in membranes of eukaryotic cells (Mackenzie *et al.*, 1982; de Leeuw *et al.*, 1989; Peters and Moldowan, 1993). Because of the large number of asymmetric centers in sterols, crude oil comprises complex mixtures of stereoisomers. The most abundant isomers are C_{27} cholesterol (R = H) and its homologues (campesterol with R = CH_3 and β-sitosterol with R = C_2H_5), C_{28} ergosterol, and C_{29} stigmasterol. As bacteriohopanetetrol, sterols are amphipathic.

Campesterol, β-sitosterol, and stigmasterol are commonly named phytosterols. Phytosterols have been detected in all higher plants. Ergosterol is commonly encountered in yeast and fungi, and is the most important of the provitamins D. Cholesterol is a characteristic sterol of higher animals, but also found in planktonic algae. Diagenesis of sterols in sediments ultimately leads to steranes (Figure 16.3.13), which is one of the largest classes of biomarkers in petroleum.

The steroid skeleton contains several asymmetric centers; in particular those at C-14, C-17, and C-20 are of great importance to the geochemists because of their usefulness in oil maturity studies. Due to their origin from enzyme controlled biosynthesis, they have a fixed stereochemistry: 5*, 14α(H), 17β(H), 20R. The 5* double bond is reduced at the early stage of diagenesis, giving mixtures of the 5α(H) and 5β(H) isomers. The biological sterol configuration (14α(H), 17α(H), 20R) imposed on the sterol precursor and its immediate saturated product (the stereoisomer 5α(H), 14α(H), 17α(H), 20R configuration) is unstable during catagenesis and undergoes isomerization to more stable geological configurations. With increasing maturation, the stereoisomers reach an equilibrium ratio of ααα 20R, ααα 20S, αββ 20R, and αββ 20S of approximately

1:1:3:3 (Peters and Moldowan, 1993). Thus, the ratios of (5α, 14β, 21β/total steranes) and 20S/(20S + 20R) can be used as maturity parameters.

Figure 16.3.12 shows the C-20 in the sterol side chain, relatively free from steric effects imposed by the cyclic system, thus the biologically derived 20R isomer is converted to a near-equal mixture of 20R and 20S at equilibrium (i.e., the ratio of 20R/20S isomers is about 50:50).

The C-ring monoaromatic (MA) steranes may be derived exclusively from sterols with the side chain double bond during early diagenesis (Moldowan and Fago, 1986; Riolo *et al.*, 1986), while the ABC-ring triaromatic (TA) steranes can be generated by aromatization and loss of a methyl group from the A/B ring-juncture of monoaromatic steranes.

16.3.3.3 Genesis of Adamantanes

Diamondoid compounds (adamantanes and diamantanes) in petroleum are believed to be the result of carbonium ion rearrangements of suitable cyclic precursors (such as multi-ringed terpene hydrocarbons) on clay superacids in the source rock during oil generation (Wingert, 1992; Chen *et al.*, 1996; Dahl *et al.*, 1999). The higher homologues of diamondoids are considered to be formed from lower homologues under extreme temperature and pressure conditions (Grice *et al.*, 2000).

16.3.3.4 Genesis of Pristane and Phytane

Phytol (E-3, 7(R), 11(R), 15-tetramethylhexadec-2-enol) is the major precursor of isoprenoid compounds including pristane and phytane. Phytol (Figure 16.3.14) is biosynthesized by enzymes in living organism and it possesses a rigidly defined R-configuration at the chiral centers C-7 and C-11 (which correspond to the centers C-10 and C-6 in isoprenoid alkanes). With respect to the spatial arrangement of the methyl substituents at C-7 and C-11, phytol in the

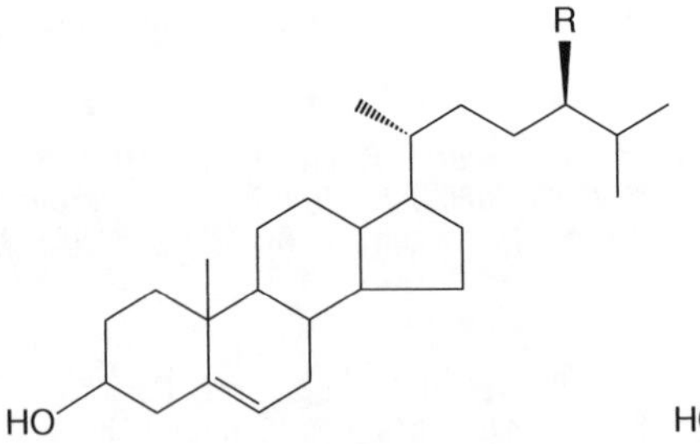

1. Cholesterol (R = H)
   Campesterol (R = CH_3)
   β-Sitosterol (R = C_2H_5)

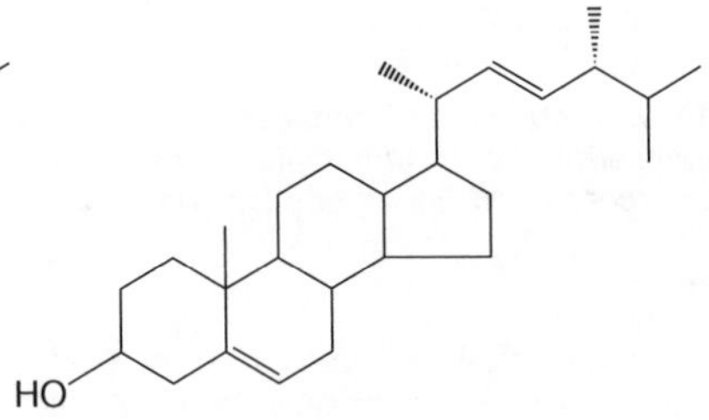

2. Ergosterol

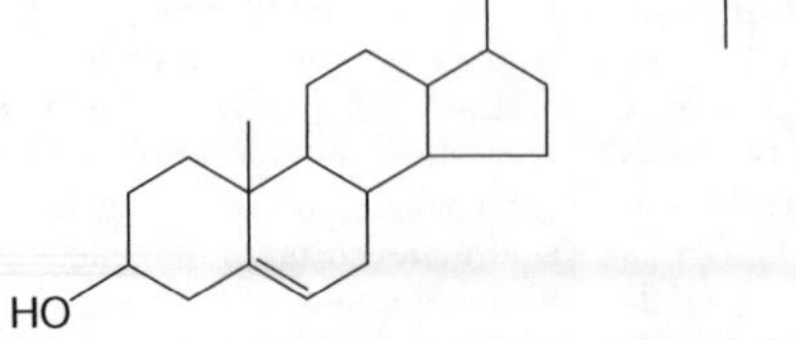

3. Stigmasterol

Figure 16.3.12 *Molecular structures of cholesterol (1, R = H), campesterol (1, R = CH_3), β-sitosterol (1, R = C_2H_5), ergosterol (2), and stigmasterol (3).*

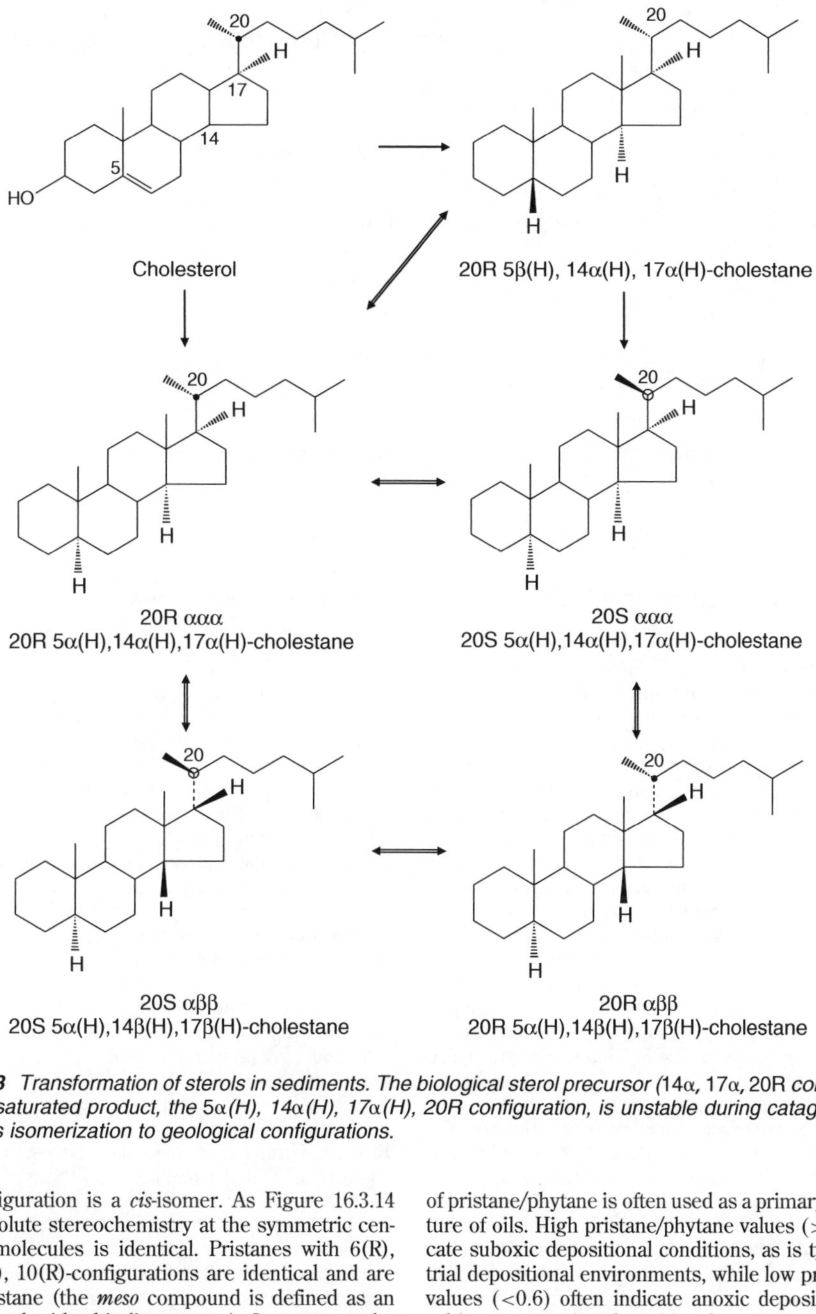

Figure 16.3.13 *Transformation of sterols in sediments. The biological sterol precursor (14α, 17α, 20R configuration) and its immediate saturated product, the 5α(H), 14α(H), 17α(H), 20R configuration, is unstable during catagenesis process and undergoes isomerization to geological configurations.*

biological configuration is a *cis*-isomer. As Figure 16.3.14 shows, the absolute stereochemistry at the symmetric centers in these molecules is identical. Pristanes with 6(R), 10(S)- and 6(S), 10(R)-configurations are identical and are called *meso*-pristane (the *meso* compound is defined as an achiral compound with chirality centers). In contrast, the 6(R), 10(R)- and 6(S), 10(S)-configuration pristanes are mirror images (i.e., they are chiral enantiomers but not asymmertric).

Most regular isoprenoid compounds are considered to originate primarily from the phytyl side chain of chlorophylls during diagenesis. Under anoxic conditions in sediments, the phytyl side chain of chlorophylls is cleaved to yield phytol. Subsequently, phytane is derived by dehydration and reduction of phytol, while pristane is derived by oxidation and decarboxylation of phytol. Since chlorophyll is pervasive in all natural environments, pristane and phytane are initially present in all crude oils, though they can be depleted or removed from oils by biodegradation. The ratio

of pristane/phytane is often used as a primary molecular feature of oils. High pristane/phytane values (>3.0) often indicate suboxic depositional conditions, as is typical in terrestrial depositional environments, while low pristane/phytane values (<0.6) often indicate anoxic deposition, as is typical in some marine depositional environments (Peters and Moldowan, 1993).

16.4 BIOMARKER ANALYSIS

16.4.1 Oil Analysis Methods

In the last two decades, a wide variety of instrumental techniques have been developed and used for the analysis of petroleum hydrocarbons. The methods used for the study of oil spills can, in general, be divided into two categories depending on the level of analytical detail: non-specific methods and specific methods for detailed chemical component analysis.

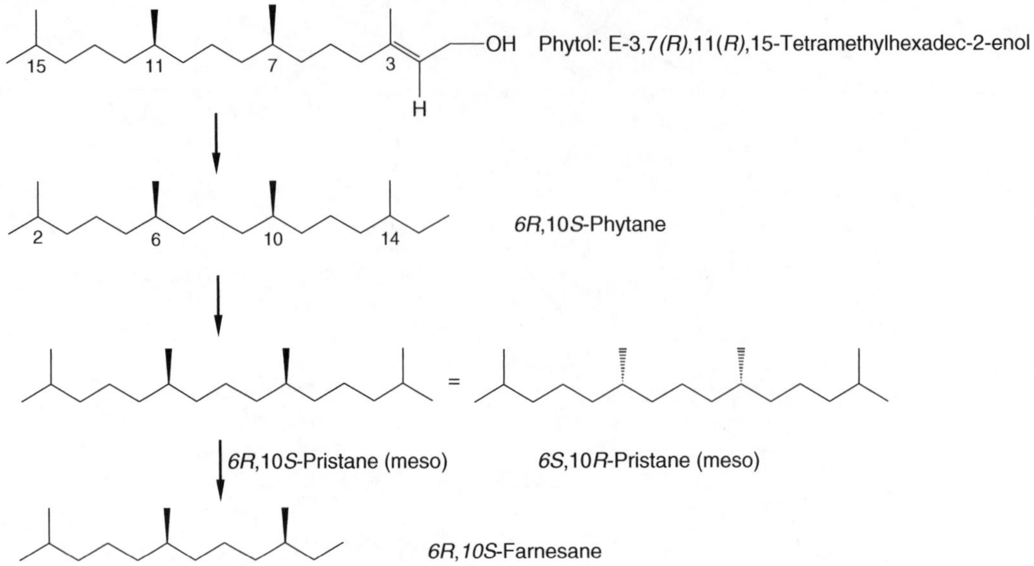

Figure 16.3.14 *Molecular structure of phytol, the precursor of pristane and phytane.*

The conventional non-specific methods include field-screening gas chromatography equipped with flame ionization detection (GC-FID) and photo ionization detection (GC-PID); gravimetric determinations and infrared spectroscopy (IR) (e.g., US Environmental Protection Agency (EPA) Method 418.1 and Method 9071, and ASTM Method 3414 and 3921); fluorescence spectroscopy and thin layer chromatography (TLC) used for oil component class (saturated, aromatic, resin, and asphaltene fraction) characterization; high-performance liquid chromatography (HPLC); size-exclusion chromatography with fluorescence detection; supercritical fluid chromatography (SFC); and ultraviolet spectroscopy (UV). These non-specific methods require shorter preparation and analytical time and are less expensive compared to the specific methods. The techniques have been used to: screen sediments for petroleum saturate and aromatic compounds, measure total petroleum hydrocarbons (TPH), assess site contamination and remediation, determine the presence and type of petroleum products, and to qualitatively examine and compare oil weathering. The major shortcoming associated with the non-specific methods is that data lack detailed individual component and petroleum source-specific information.

Specific methods for oil analysis include GC-MS, gas capillary column GC-FID, and gas chromatography-isotope ratio mass spectrometry (GC-IRMS). These methods can be used to quantify a broad range of individual petroleum hydrocarbons. Compared to non-specific methods, these specific methods can provide far more information directly useful for characterization and quantification of oil hydrocarbons and for oil spill identification. A variety of diagnostic ratios, especially ratios of PAH and biomarker compounds, for interpreting chemical data from oil spills have been proposed.

Comprehensive two-dimensional gas chromatography (GC × GC) is another hyphenated technique where two different chromatographic separation mechanisms act in concert to further improve component separation and identification. In recent years, the GC × GC method has been used to analyze light, middle distillate petroleum products (Gains *et al.*, 1999; Frysinger *et al.*, 2003) and unresolved oil complex mixtures (Reddy *et al.*, 2002).

16.4.1.1 Selected US EPA Methods and Their Limitations for Oil Analysis

Table 16.4.1 summarizes selected EPA methods, major applications, and their limitations for oil analysis. The EPA methods have been used as routine procedures for determination of volatile and semivolatile aromatic hydrocarbons in spilled oil and petroleum product samples. The methods were originally designed to measure a wide variety of industrial chemicals in wastewater and industrial waste. The fundamental shortcoming is that none of these standard EPA methods can provide detailed information on the chemical composition of crude oil and petroleum products. Hence, data generated from these methods are insufficient to answer questions important in oil spill liability investigation such as oil type, oil source, weathering status of spilled oil, candidate sources, and so forth. Only 20 of the more than 160 EPA priority pollutant organic compounds determined by these methods are petroleum-related hydrocarbons. Furthermore, only half of these compounds are found in significant quantities in oils and petroleum products. PAHs in petroleum are dominated almost exclusively by the C_1 to C_4 alkylated homologues of the parent PAH, in particular, the alkylated homologues of naphthalene, phenanthrene, dibenzothiophene, fluorene, and chrysene, none of which are measured by the standard EPA methods. Other important classes of petroleum hydrocarbons (e.g., aliphatics and biomarkers) are not measured by these methods at all.

Another example is the use of the EPA 418.1 method to determine TPH content. The EPA 418.1 method, based on measuring the absorption of C–H bond in the 3200 to 2700 wavenumber range, was originally intended for use only with liquid waste. However, prior to its demise because of the use of chloro-fluoro carbon extractant, it has been widely used for the determination of TPH in soils. For some site assessments, Method 418.1 was the sole criterion for verification of site cleanup. However, there were some problems associated with this method such as inherent inaccuracy in the method (i.e., positive or negative biases caused by various factors) and the lack of effective reference standards when working with an unknown.

Table 16.4.1 *Major Applications and Limitations of Standard EPA Methods for Oil Analysis*

EPA Standard Method	Target Compounds and Application	Limitation for Oil Work
EPA 418.1	TPH by IR spectroscopy	—Inherent inaccuracy of the method (positive or negative biases) —Subject to various interferences —Lack of effective reference standards
EPA 1664	*n*-Hexane extractable materials and silica gel treated *n*-hexane extractable material by extraction and gravimetry	—Only measures total extractable materials in aqueous matrices —Heavy interference —Low molecular weight hydrocarbons could be lost during distillation
EPA 600 series (method standards for waste water)		
602	Purgeable aromatics, by GC/FID	—600 and 8000 series were originally designed for waste water and industrial waste
610	16 polycyclic aromatics, by HPLC/GC	—Cannot provide detailed composition information of spilled oil
624	Purgeable volatiles, by GC/MS	—Only BTEX measured, do not measure over 100 important oil hydrocarbons
625	Semi-volatiles and pesticides, by GC/MS	—Do not measure dominated alkylated PAH homologues, aliphatics, and biomarkers in oils —Provide little diagnostic source information
EPA 8000 series (method standards for solid waste, SW 846)		
8015	Non-halogenated volatiles by GC/FID	
8020	Aromatic volatiles by GC	
8100	Volatiles by Capillary GC/MS	
8260	24 PAHs by GC/FID	
8270	Semi-volatiles by capillary GC/MS	

TPH: Total petroleum hydrocarbons; PAH: polycyclic aromatic hydrocarbons.

In recent years, many EPA and ASTM methods have been modified (such as the modified EPA method 8015, 8260, and 8270; and the modified ASTM methods 3328-90, 5037-90, and 5739-95) to improve specificity and sensitivity for measuring spilled oil and petroleum products in soil and water. For example, the EPA Method 8270 has been modified to increase the analytical sensitivity and to expand the list of analytes to include petroleum-specific compounds such as the alkylated PAHs, and sulfur and nitrogen-containing PAHs. The main modification to EPA Method 8270 is the use of high resolution GC-MS with selected ion monitoring (SIM). The modified EPA Method 8270 (i.e., EPA 8270C and 8270D), combined with column-cleanup and rigorous quality assurance (QA) measures, has been used to identify and quantify low levels of hydrocarbons by many environmental laboratories.

16.4.1.2 Selection of Source-Specific Target Analytes

The selection of appropriate target oil analytes for oil spill identification and specific site investigation is dependent mainly on the type of oil spilled, the particular environmental compartments being assessed, and expected needs for current and future data comparison. In general, the major petroleum-specific target analytes that may be needed in oil spill identification and site assessment studies include:

- Individual saturated hydrocarbons including *n*-alkanes (*n*-C_8 through *n*-C_{40}), selected isoprenoids pristane and phytane and, in some cases, farnesane (2,6,10-trimethyl-C_{12}), 2,6,10-trimethyl-C_{13}, and norpristane (2,6,10-trimethyl-C_{15}).

- Alkyl (C_1–C_{14}) cyclo-hexane homologous compound series. This homologue series exhibits a characteristic distribution pattern in *m/z* 83 mass chromatograms for different types of fuels.
- The volatile hydrocarbons including BTEX and alkylated benzenes (C_3- to C_6-benzenes), and volatile paraffins and isoparaffins.
- EPA priority parent PAHs and, in particular, the petroleum-specific alkylated (C_1 to C_4) homologues of selected PAHs (i.e., alkylated naphthalene, phenanthrene, dibenzothiophene, fluorene, and chrysene homologous series).
- Biomarker terpane and sterane compounds (see details in Table 16.3.1). Analysis of selected ion chromatograms generates information of great importance in determining source(s) and weathering state.
- Bulk hydrocarbon group information including TPH, unresolved complex mixtures (UCM), total saturates, total aromatics, total GC-resolved peaks, asphaltenes, and resin contents.
- Petroleum product additives such as alkyl lead additives; oxygenates including ethanol, methanol, methyl tertiary butyl ether (MTBE), ethyl tertiary butyl ether (ETBE), and tertiary amyl methyl ether (TAME); fuel dyes used for differentiation among fuel grades; and anti-oxidant compounds added to fuels to retard autooxidation.
- Stable carbon isotope ratios (*^{13}C) for hydrocarbon groups, individual *n*-alkanes, and isoprenoids.
- Nitrogen and oxygen containing heterocyclic hydrocarbons. These heterocyclic hydrocarbons are generally only present in oils at quite relatively low concentrations compared to PAHs. However, their relative concentrations

increase as weathering proceeds. The application of nitrogen and oxygen-containing heterocyclic hydrocarbons in source identification is still in its infancy, and more research is clearly needed.

16.4.2 Capillary Gas Chromatography-Mass Spectrometry (GC-MS)

A mass spectrometer produces ions from gaseous neutral molecules or species, the ions are accelerated by an electric field, and then separated according to their mass-to-charge ratio (m/z, where z is the number of elemental charges) and the number of ions. The mass spectrum is a display of detector response versus m/z.

The mass spectrometer has long been recognized as the most powerful detector for a gas chromatograph due to its high sensitivity and specificity, and its capability to elucidate compound structures. Mass spectral data provide both qualitative and quantitative information for identification and characterization of sample components that is lacking in other GC detectors. Today, GC-MS (e.g., benchtop quadrupole GC-MS, high resolution GC-MS, GC-ion trap MS, and GC-MS-MS) have become the principal and routine techniques used in most oil and environmental forensics laboratories to analyze a wide range of petroleum hydrocarbons. Computerized capillary column GC-MS, which combines chemical separation by GC, spectral resolution by MS, and computerized data preprocessing, enables separation, identification, and quantification of trace amounts of biomarker compounds in oil.

Early use of mass chromatograms in organic geochemistry was pioneered at Chevron and led to a stereochemical understanding of steroids and the first practical method of oil fingerprinting based on terpanes and steranes (Seifert, 1977).

16.4.2.1 Benchtop Quadrupole GC-MS

The quadrupole is the most common mass separator in use today. The benchtop quadrupole GC-MS systems, although lacking the high-resolution capabilities of larger and more expensive magnetic-sector instruments, have sufficient sensitivity and selectivity for most purposes of biomarker analysis.

Most benchtop GC-MS systems use a quadrupole mass filter to separate ions. In the high vacuum, ions pass down the lengths of four parallel metal rods to which are applied both a constant voltage and a radio-frequency oscillating voltage. The electric field deflects ions in complex trajectories as they migrate from the ionization chamber toward the detector, allowing only ions with one particular m/z ratio to reach the detector at any instant. Other nonresonant ions collide with the rods and are lost before they reach the detector. By rapidly varying the applied voltages, ions of different masses are selected to reach the detector. A wide range of masses can be recorded in less than 1 second. In this way many mass spectra are taken and stored on a computer as the components of the sample pass from the chromatographic column into the mass spectrometer. Benchtop quadrupole GC-MS can be operated in various modes, including scan and selected ion monitoring (SIM).

Scan Mode

In scan mode, the mass spectrometer is used to scan the entire spectrum of ions generated in the ion source. Full scan records hundreds of ions per scan but with lower sensitivity in comparison with the SIM mode. Among ions generated from the scan, there are always several ions being the most characteristic and diagnostic of the molecule or the compound type, and the most abundant ion in the mass spectrum is called the *base peak*.

A mass spectrum is generated as the MS detector scans through a predefined mass range (e.g., 50–700 amu; the larger the mass range, the fewer scans per second). Identification and characterization of petroleum hydrocarbons are largely based on the full mass spectral data for structural elucidation, comparison of GC retention data with that of reference standards, recognition of distribution pattern, calculation of retention indexes (RI), and comparison with literature RI values.

Selected Ion Monitoring

In the SIM mode, only limited number of characteristic ions (e.g., the *base peaks* 191, 217, and 218, and those m/z values diagnostic for molecule structural elucidation for target terpanes and steranes) are monitored. The *selected ion mass chromatogram* is the plot of peak intensity for a specific m/z versus the GC retention time. For quantification of individual target compounds, the SIM mode is used most frequently, since it shows several advantages compared to the scan mode: (1) SIM only records a few selected m/z per scan, resulting in a much longer dwell time for each monitored ion (usually between 25 and 100 milliseconds, depending on the number of m/z selected) than in the scan mode; (2) method detection limits for target analytes are generally lower by almost an order of magnitude than those produced by the Full Scan GC-MS; (3) the use of the SIM mode is often less noisy and the linear quantification range is increased for trace analytes.

The following examples briefly describe the analytical GC-MS conditions used by the Environment Canada Oil Spill Research Lab and Petrobras Geochemistry Laboratory (Barbanti, 2004).

Example Benchtop GC-MS Conditions (EC Oil Spill Research Laboratory)

Analyses of biomarkers are performed on an Agilent 6890 GC coupled with an Agilent 5973 mass selective detector (MSD). System control and data acquisitions are achieved with the Agilent G1701 BA MSD ChemStation. A 30 m × 0.25 mm i.d. (0.25 μm film thickness) HP-5MS fused-silica capillary column is used. The chromatographic conditions are as follows: Carrier gas, helium (1.0 mL/min); injection mode, splitless; injector and detector temperature, 280 and 300 °C respectively. The temperature program employed for biomarkers and alkylated PAHs is: 50 °C hold for 2 min, then ramp at 6 °C/min to 300 °C and hold for 20 min. Prior to sample analysis, the GC-MS is tuned with perfluorotributylamine (PFTBA). The total run time is 60 min.

Example Benchtop GC-MS Conditions (Petrobras Geochemistry Laboratory)

A 60 m × 0.25 mm i.d. (0.25 :m film thickness) HP-5MS or equivalent 60 m capillary column is used to achieve improved resolution for biomarkers (Barbanti, 2004). The temperature program is as follows: 55 °C hold for 2 min, ramp at 20 °C/min to 150 °C and then 1.5 °C to 310 °C and hold for 15 min. The total run time is 128 min.

It has been demonstrated that the 30-m capillary column provides sufficient peak resolutions for most oil spill works. However, the 60-m capillary column with a slow temperature rate and longer running time can offer further improved resolution for some paired biomarker isomers which may not be well-resolved by the use of a 30-m column.

16.4.2.2 GC-Ion Trap MS

The quadrupole ion-trap mass spectrometer is a compact device that is well suited as a chromatography detector. It consists of two end-caps normally held at ground potential with a central ring electrode between them. High radiofrequency voltages are applied to electrodes to create a cage (cavity) for the ions in the closed quadrupole field.

In the internal ionization quadrupole ion trap MS, substances emerging from the GC column enter the cavity of the mass analyzer through a heated transfer line. The gate electrode periodically admits electrons from the filament. Molecules undergo electron ionization in the cavity. A constant radio frequency voltage applied to the central ring electrode causes ions to circulate in stable trajectories around the cavity. By increasing the amplitude of the radio-frequency voltage, ions of a particular m/z value are expelled. Ions expelled through the lower end-cap are captured by the electron multiplier and detected with high sensitivity. Scans from m/z 10 to 650 can be conducted eight times per second. In other mass analyzers, only a small fraction of ions reaches the detectors. With an ion trap, however, about half of the ions of all m/z value reach the detector, giving the ion trap mass spectrometer considerably higher sensitivity than the transmission quadrupole.

16.4.2.3 GC-MS-MS

The combination of two or more MS analyzers, commonly known as MS-MS or tandem mass spectrometry, is a highly specific means of separating mixtures and studying molecule fragments. In the first MS, one ion is isolated and subsequently in the second MS, reactions of that ion are studied further. GC-MS-MS includes linked and de-linked double focusing and triple quadrupole mass spectrometry. Triple quadrupole mass spectrometers are the most common type of tandem mass spectrometers. The first (or parent) and the third (or daughter) quadrupole are MS-1 and MS-2, whereas the second quadrupole in the middle is acting as collision cell. In the collision cell, the transmitted ions formed in the ion source and selected by or passed through MS-1 undergo low energy collision with an inert gas such as argon. The fragment ions formed in the collision cell are selectively monitored by the daughter quadrupole and recorded using an electron multiplier. Because of the use of three linked quadrupoles, triple quadrupole mass spectrometry offers improved selectivity for oil analysis. Triple quadrupole mass spectrometers can be operated in three GC-MS-MS modes:

- Precursor (parent) ion scan mode
- Product (daughter) ion scan mode
- Neutral loss scan mode

For the product (daughter) ion scan, only one precursor (parent) ion is selected and the second analyzer scans for the product (daughter) ions. This type of scan is often used to analyze the fragmentation pattern of a component with specific molecular weight. For the precursor (parent) ion scan, the first quadrupole is scanned and only the selected product ion (such as the major daughter ions 191 for terpanes and 217 for steranes) is recorded. This approach has been successfully used for product screening and distribution pattern recognition of biomarkers. In the neutral loss scan, both analyzers are scanned with the selected mass difference, and for reaction monitoring only one precursor and one product ion species is permitted to travel through MS-1 and MS-2, respectively. This approach can be particularly useful in the search for specific compounds derived from a certain precursor compound.

16.4.3 Sample Collection, Extraction, Cleanup, and Fractionation

16.4.3.1 Sampling

Oil sample collection begins with a well-designed sampling plan. The sampling objective should be clearly stated in the sampling plan. Proper collection and preservation of oil and oil-contaminated samples is one important step for accurate analytical results and should follow the guidelines described in authorized standard methods (e.g., EPA 624, 8100, and 8270).

Oils and various oil-contaminated samples are best collected in pre-cleaned glass containers with Teflon-lined caps. Usually few milliliters of oil are sufficient for oil analysis including characterization of biomarkers. Each sample container must be properly sealed and labeled to ensure that the oil is not tainted from contact with other materials.

For most purposes a comparison is made between the spilled oil and suspected source candidates. In the case of a ship being the suspected source, it is essential that reference samples are taken of all the oils carried on board the vessel, which might include cargo oils, fuels, lubricating oils, and waste oils. Careful and detailed examination of the contaminated site such as beaches should be made to determine the uniformity of the spilled oil deposits. Any apparent variation in the type of oil should be sampled, the extent noted and supported by photographs.

To collect oil-contaminated soil samples, common tools such as shovels, trowels, scoops, and hand-operated auger coring devices are suitable for the top 30 cm. From 30 to 100 cm, one can manually remove the top layer of soil and then use the common tools as described above. For oil deposited on solid surfaces such as wood, rock, and concrete, it can be scrapped off the solid surfaces and placed directly into a sample container. On prolonged weathering at sea, oil tends to form blackish, semi-solid tar balls. These tar balls can be collected by hand and placed into sample containers. If freshly spilled oils or refined products have been absorbed and/or penetrated into sand or soil, representative oil-contaminated sand or sediment samples from various sites and varying depths should be collected.

When oil has spread to a thin film on the surface of the water, it is often difficult to obtain a representative sample. In the absence of specialized equipment, oil-absorbing materials and wide-necked glass jars can be used to skim samples from the water surface. Highly viscous oils and emulsions tend to be concentrated on the sea surface and can usually be scooped up fairly easy.

Oil samples should be stored refrigerated in clean glass bottles or jars with Teflon lined caps at $c.5\,°C$ until analysis. Bottled oils do not appear to be affected by biodegradation over periods of years under typical storage conditions.

In addition, adequate records for any procedure must be complete and should include information such as:

- The precise sampling location from which the sample was collected.
- The character of the bulk material at the time of sampling.
- Sampling protocols.
- The date and the amount of the sample that was originally placed into storage.
- The methods used for measurements and analyses of the sample and the name of the analysts.
- The log sheet showing the names of the persons who removed the sample from storage and the amount of each sample that was removed for testing.

16.4.3.2 Sample Preparation

There are a variety of common sample extraction techniques, including sonication, Soxhlet, supercritical fluid, and accelerated solvent extraction, which have been promulgated by the US EPA as soil extraction methods. Solvent extractions and separations of oil hydrocarbons are based on the "like-dissolve-like" principle. Surrogates, compounds are not found in samples, are often added to monitor extraction efficiency. Glass columns are used for sample cleanup and fractionation prior to analysis by GC-FID and GC-MS. The oil-related sample preparation procedures used by the Environment Canada Oil Spill Research Laboratory are described below.

Oil Samples

Approximately 0.8 g of crude oil or petroleum products are weighed, dissolved in hexane and made up to a final volume of 10.00 mL. A 200 :L of the oil solutions containing ~16 mg of oil is spiked with appropriate amounts of surrogates (100 :L 200 ppm of o-terphenyl and a 100 :L mixture of deuterated acenaphthene, phenanthrene, benz[a]anthracene, and perylene, 10 ppm each), and then transferred into a 3-g silica gel chromatographic column for sample cleanup and fractionation.

Oil-Contaminated Sediment Samples

Oil-contaminated sediment samples are weighed, dried with anhydrous sodium sulphate, and then spiked with appropriate amounts of surrogate standards (o-terphenyl, and a mixture of d_{10}-acenaphthene, d_{10}-phenanthrene, d_{12}-benz[a]anthracene, and d_{12}-perylene). The samples are extracted ultrasonically with 1:1 (v/v) hexane/dichloromethane (DCM), and then twice with DCM alone (if there is visible color in the third extraction, additional extraction should be performed); or Soxhlet-extracted with dichloromethane for 24 hours/overnight. Extracts are combined and concentrated by rotary evaporation, solvent-exchanged to hexane, and evaporated under a gentle stream of dry nitrogen to a final volume of 2 to 10 mL (the final volume depends on the concentration of hydrocarbons in the final extract). An aliquot of the concentrated extract (0.5–1.0 mL) is taken and blown down with N_2 to dryness and the residue is weighed on a microbalance to obtain a total solvent extractable material weight (TSEM, expressed as mg/g of sample). TSEM provides an equal basis for quantitative comparison of petroleum hydrocarbon groups between samples. Based on the TSEM value, appropriate aliquot of the concentrated extract (containing approximate 20 mg oil) is transferred to a 3-g silica gel chromatographic column for sample cleanup and fractionation.

Oil-Contaminated Water Samples

Water samples are commonly extracted using liquid–liquid extraction (EPA method 3510). Oil-contaminated water samples are weighed and spiked with appropriate amounts of surrogate standards (o-terphenyl, and a mixture of d_{10}-acenaphthene, d_{10}-phenanthrene, d_{12}-benz[a]anthracene, and d_{12}-perylene). The sample is then serially extracted three times with DCM using a separatory funnel. Extracts are combined, filtered, and dried by passage through anhydrous sodium sulfate, then concentrated by rotary evaporation, solvent-exchanged to hexane, and evaporated under a gentle stream of dry nitrogen to a final volume of 1 to 5 mL (the final volume depends on the concentration of hydrocarbons in the final concentrated extract). An aliquot of the concentrated extract (containing about 20 mg oil) is transferred to a 3-g silica gel chromatographic column for sample cleanup and fractionation.

16.4.3.3 Silica Gel Column Cleanup, and Fractionation

Silica gel is used frequently for cleanup and fractionation of oil extracts. The procedure used by the Environment Canada Oil Research Laboratory to fractionate samples into saturated and aromatic fractions and to remove polar components and other interferences is described in the following text.

A chromatographic column with a Teflon stopcock (200 × 10.5 mm i.d.) is plugged with Pyrex glass wool at the bottom, serially rinsed with methanol, hexane, and dichloromethane, and allowed to dry. The column is dry-packed with 3 g of activated silica gel and topped with about 1-cm anhydrous granular sodium sulfate. Silica gel:

100–200 mesh, pore size 150 Å, pore 1.2 cm^3/g, active surface 320 m^2/g. Before use, the silica gel is serially rinsed with acetone, hexane, and dichloromethane, completely dried in a fumehood, and activated for 20 hours at 160–180 °C. Columns are then preconditioned using 20 mL of hexane. Just prior to exposure of the sodium sulfate layer to air, appropriate volumes of the oil solutions or concentrated oil extracts are transferred quantitatively to the column (Wang et al., 1994a; ETC Method, 2002).

Saturated hydrocarbons are eluted with 12 mL of hexane (Fraction 1, labeled F1). Aromatic hydrocarbons are eluted with 15 mL of hexane:dichloromethane (v/v, 1:1, Fraction 2, labeled F2). Saturated biomarkers are eluted with other saturates in F1. Aromatic steranes are eluted in the aromatic fraction, F2.

16.4.4 Internal Standards and Quantification of Biomarkers

For each sample, half of F1 is used for analysis of the total GC-detectable saturates, n-alkanes and isoprenoids, and biomarker compounds; and half of F2 is used for analysis of alkylated PAH homologous and other EPA priority parent PAHs, and aromatic steranes. The remaining halves of the F1 and F2 are combined into one fraction (Fraction 3, labeled F3) and used for the determination of TPH and UCM. The three fractions are concentrated under a stream of nitrogen to appropriate volumes, spiked with appropriate internal standards, and then adjusted to an accurate pre-injection volume (0.50–1.00 mL) for GC-FID and GC-MS analyses.

Quantitative analysis requires standards, either external or internal. In many laboratories, the concentrations of individual biomarkers and other petroleum hydrocarbons are determined using the internal standard (IS) methods. In the Environment Canada Oil Research Laboratory the GC-FID is calibrated using a standard solution composed of n-C_8 through n-C_{36}, pristane and phytane. 5-α-androstane ($C_{19}H_{32}$, a short-chained sterane) is used as the internal standard. A five-point calibration curve that demonstrates the linear range of the analysis is established. The relative response factor (RRF) for each hydrocarbon component in the standard solution is calculated relative to the internal standard using the following equation:

$$\mathrm{RRF} = (A_S\, C_{IS})/(A_{IS}\, C_S) \qquad (16.4.1)$$

A_S Response area for the target analyte in the standard solution to be measured.

A_{IS} Response area for the internal standard in the standard solution.

C_S Concentration of the target analyte in the standard solution.

C_{IS} Concentration of the internal standard in the standard solution.

The GC-MS is calibrated using the NIST (the National Institute of Standards and Technology) SRM 1491 PAH standard mixture. d_{14}-Terphenyl is used as the internal standard to quantify the PAH compounds in oil. The terpane standards include C_{27} 17α(H)-22,29,30-trisnorhopane, C_{29} 17β(H), 21α(H)-30-norhopane, and C_{30} 17β(H), 21α(H)-hopane. The sterane standards include C_{21} 5β(H)-pregnane, C_{22} 20-methyl-5α(H)-pregnane, and the series of C_{27}, C_{28}, and C_{29} steranes. The C_{30} 17β(H), 21β(H)-hopane is used as internal standard. The average RRF are determined relative to the internal standard C_{30} 17β(H), 21β(H)-hopane. In most cases, the average RRF for C_{30} 17β(H), 21α(H)-hopane at m/z 191 is used for quantification of C_{30} 17α(H), 21β(H)-hopane and other terpanes (in the range of C_{19} to C_{35}). For quantification of steranes, the average RRF of C_{29} 20R-αααα-ethylcholestane at m/z 217 relative to the internal standard is used to estimate the concentrations of sterane compounds. The deuterated d_3 monoaromatic steranes [$5\alpha(H)/5\beta(H)$, $C_{21}H_{27}D_3$] are used as internal standards for quantification of monoaromatic and triaromatic steranes. The deuterated d_{18}-decahydronaphthalene (*cis*) and d_{16}-adamantane are used as internal standard for quantification of sesquiterpanes and diamondoid compounds respectively.

16.4.5 Quality Assurance and Quality Control

The reliability of analytical methods is dependent on the quality control (QC) procedures followed. In order to support oil spill forensic investigations, quality management (including quality assurance and quality control system, updated standard operational procedures, personnel training program and record, up-to-date methodology, equipment management, sample management, and data management) is a fundamental element of the analytical quality assurance (QA) program. The chemical measurements must be conducted within the framework of highly stringent, defensible, and reliable QC and QA programs.

The QA/QC programs used by different laboratories may differ more or less, but the principle and purposes are similar. The QA/QC program for oil analysis used by the Environment Canada (EC) Oil Spill Research Laboratory is described briefly below.

Prior to TPH analysis or quantification of individual components, a five-point response calibration curve is established to demonstrate the linear-range of the analysis. Control standards (at the mid-point of the established calibration curves) are analyzed before and after each batch of samples (7–10 samples) to validate the integrity of the initial calibration. If the response of the control standards varies from the historical average response by more than 25%, the test should be repeated using a fresh calibration standard. The RRF stability is a key factor in maintaining a high analytical quality. A control chart for RRF values should be monitored. The RRFs for n-C_8 to n-C_{34}, and pristane and phytane should be 0.95 ± 0.1 relative to the internal standard 5-α-androstane. Mass discrimination for high molecular weight n-alkanes in the injection port must be carefully monitored. If there is a problem with mass discrimination, it can be minimized by trimming the column and by replacing the liner. All samples and quality control samples (procedural blank, matrix spike samples, duplicate, and reference oil sample) are spiked with appropriate surrogates to measure individual sample matrix effects associated with sample preparation and analysis. Surrogate and matrix spike recoveries should be within 60 to 120%, and duplicate relative reference values should be less than 25%. Method detection limits (MDLs) are determined according to the procedure described in the EPA protocol entitled *Definition and Procedure for the Determination of the Method Detection Limit* (Code of Federal Regulations 40CFR Part 136).

Besides these routine QC measures required by standard EPA and ASTM methods, some refinements may be further implemented for the purposes of unambiguous spill source identification and for environmental samples with low hydrocarbon concentrations. The key refinements may include the following:

- Applying more rigorous calibration control standards of $\pm 15\%$;
- Quantification of alkylated PAH homologues using RRFs obtained directly from authentic alkylated PAH standards, rather than the standard parent PAHs;
- Quantification of alkylated isomer PAH series at various alkylation levels by manually setting the integration baselines;

- Increasing sample size to reduce the sample extract pre-injection volume for those sediment samples with low biomarker or PAH concentrations.

16.4.6 Analysis of Oil Compositional Data

16.4.6.1 Oil Spill Identification Protocol

The oil spill identification system currently used is largely based on GC-FID and GC-MS techniques. Data produced from these two methods are used to compare spill samples with samples taken from suspected sources. Recently, SINTEF Applied Chemistry of Norway and Battelle of the USA published the "Improved and standardized methodology for oil spill fingerprinting" (Stout *et al.*, 2001; Daling *et al.*, 2002; Faksness *et al.*, 2002), which includes four "levels" of analyses and data treatment. The recommended methodology is a result of documented and analytical improvements, a more quantitative treatment of analytical data from GC-FID and GC-MS, and the operational experiences over past few years among the participating forensic laboratories.

Figure 16.4.1 presents the modified "Protocol/decision chart for the oil spill identification methodology." The oil spill identification is based on GC-FID screening of all involved samples (*Level 1*), and then GC-MS fingerprinting of spill and candidate source samples (*Level 2*), from which a number of diagnostic ratios (>20) of selected PAHs and biomarkers are extracted and evaluated (*Level 3*). Only those diagnostic indices that can be measured precisely and are resistant to weathering effect should be evaluated for comparing and correlating candidate sources to the spill oil. The evaluation of the diagnostic indices is based on relative standard deviation of triplicate measurements to identify those ratios variability. The final assessment can be *positive match, probable match, inconclusive* or *non-match*. These categories represent degrees of differences between the analyses of two oils according to the present criteria (e.g., ASTM Method D3328):

1) *Positive match*: The chromatographic patterns of the samples submitted for comparison are virtual identical, and the observed differences between the spill sample and suspected source are caused, and can be explained, by the acceptable analytical variance and/or weathering effects.
2) *Probable match*: The chromatographic patterns of the spill sample is similar to that of the samples submitted for comparison, except: (a) for obvious changes which could be attributed to weathering, or (b) differences attributable to specific contamination.
3) *Inconclusive*: The chromatographic patterns of the spill sample are somewhat similar to that of the sample submitted for comparison, except for certain differences that are of such magnitude that it is impossible to ascertain whether the unknown is the same oil heavily weathered or a totally different oil.
4) *Non-match*: Unlike the samples submitted for comparison.

16.4.6.2 Tiered Analytical Approach

Tiered analytical approaches (Wang *et al.*, 1999; Stout *et al.*, 2002) have been increasingly applied for oil spill identification and hydrocarbon forensic analysis in recent years. Depending on the needs of spilled oil characterization, support for biological studies, monitoring weathering effects on chemical composition changes, or source differentiation, the tiered analytical approaches may vary. The tiered approach used by the Environment Canada Oil Spill Research Program (Wang and Fingas, 2003) includes the following:

- Tier 1, determination of hydrocarbon groups in oil residues;
- Tier 2, product screening and determination of n-alkanes and TPH;

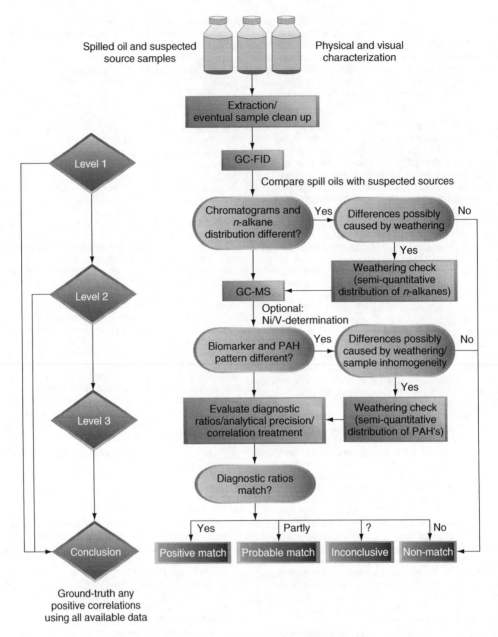

Figure 16.4.1 *Protocol/decision chart for the oil spill identification methodology.*

- Tier 3, fingerprinting and distribution pattern recognition of volatile hydrocarbon (including BTEX and alkylbenzenes), target PAHs, and biomarker components (sometimes the volatile hydrocarbons are monitored);
- Tier 4, determination and comparison of diagnostic ratios of the "source-specific marker" compounds with the potential source oil and with the corresponding data from database;
- Tier 5, identification and characterization of major unknown peaks (*e.g.*, additives); and
- Tier 6, determination of weathered percentages of the residual oil.

In this tiered analytical approach, capillary GC-FID analysis is applied to evaluate hydrocarbon groups (including TPH, the total saturates, the total aromatics, the total resolved peaks, and UCM), to determine concentrations of *n*-alkanes and major isoprenoid compounds, and to characterize the product types in fresh to highly weathered oil samples. The GC-MS analysis provides data on the "source-specific" marker compounds including the target alkylated PAH homologues and other EPA priority PAHs, and biomarker terpane and sterane compounds to support the GC-FID results and to provide additional information for low-concentration hydrocarbon contaminated samples.

16.5 IDENTIFICATION OF BIOMARKERS

Mass spectra produced by GC-MS are one of the most valuable tools for identification of unknown compounds. In addition to molecular formula, the mass spectrum provides structural information of a given molecule. An electron with typical energy of 70 eV (1610 kcal/mol or 6740 kJ/mol) has far more energy than needed to ionize a molecule. The ion impact forms the radical cation, and it often breaks a bond to give a cation and a radical. The resulting cation is observed by the mass spectrometer, but the uncharged radical is not accelerated or detected. This bond breaking does not occur randomly, it tends to form the most stable (or the most characteristic) fragments, resulting in a mass spectrum showing the mass of a given molecule and the masses of fragments from that molecule. By knowing what stable fragments result from different kinds of compounds, we can recognize structural features and use the mass spectra to confirm a proposed structure.

$$\textit{Ionization} \qquad R_1:R_2 + e^- \rightarrow \underset{\substack{\text{radical cation} \\ \text{(molecular ion)}}}{[R_1 \cdot R_2]^{+\bullet}} + 2e^-$$

$$\textit{Fragmentation} \qquad \underset{}{[R_1 \cdot R_2]^{+\bullet}} \rightarrow \underset{\substack{\text{cation fragment} \\ \text{(observed)}}}{R_1^+} + \underset{\substack{\text{radical fragment} \\ \text{(not observed)}}}{\cdot R_2}$$

For example, the mass spectrum of n-hexane $(CH_3–CH_2–CH_2–CH_2–CH_2–CH_3, MW = 86)$ shows several characteristics typical of n-alkanes. The *base peak* at m/z 57 corresponds to loss of an ethyl group, giving an ethyl radical and a butyl ion with m/z 57. The neutral ethyl radical is not detected by the mass detector because it is uncharged, and hence neither accelerated nor deflected.

$$\underset{\text{hexane radical cation, M}^+}{[CH_3–CH_2–CH_2–CH_2–CH_2–CH_3]^{+\bullet}}$$

$$= \underset{\substack{\text{butyl cation} \\ \text{detected at } m/z \text{ 57}}}{CH_3–CH_2–CH_2^+} + \underset{\substack{\text{ethyl radical (29)} \\ \text{not detected by MS}}}{{}^\bullet CH_2–CH_3}$$

Similarly, other fragmentations give a characteristic propyl cation at m/z 43, ethyl cation at m/z 29, and much weaker pentyl cation at m/z 71 (this is because the methyl radical is less stable than a substituted radical).

For the similar mechanism, the mass spectrum of n-octane (MW = 114) shows the base peak at m/z 43, and other major characteristic cation peaks at 29, 57, 71, and 85.

Much work on the isolation and identification of individual biomarker components in oils and sediment extracts has been done by geochemists. The methods used for identification of unknown biomarker compounds comprise comparison of the mass spectrum of the unknown compound to those of reference standard molecules of known structure, as well as comparison to published mass spectra.

The Chevron Biomarker Laboratory developed a so-called *coinjection and mass spectra matching* technique, in combination with other analytical techniques, for provisional identification of unknown biomarker compounds (Peters and Moldowan, 1993). Using this technique, a synthesized or commercial standard compound is mixed with the sample (this process is called "spiking") containing the compound to be identified. If the standard and the unknown compound coelute, the relative peak intensity of the unknown compound on GC chromatograms of the mixture will be higher than the unspiked sample. Note, however, that coelution of a standard with the unknown peak only support, but does not prove, that the unknown compound is identical to the standard. After coelution is established, matching of mass spectra of the unknown to the standard can then be used to

infer that the unknown is identical to the standard. Application of a chromatographic column with a different stationary phase and the use of additional instrumental techniques (e.g., Nuclear Magnetic Resonance (NMR) or x-ray) can be performed to verify the proposed structure.

Mass spectra for common petroleum biomarkers used in environmental forensic studies are shown in Figure 16.5.1. These figures clearly show that common features of the mass spectra of terpanes, steranes, monoaromatic steranes, and triaromatic steranes are a large parent ion (M^+), an important parent minus a methyl ion $(M^+ -15)$, and a base peak at m/z 191, 217 and 218, 253, and 231, respectively. For example, $C_{30}17\beta(H), 21\beta(H)$-hopane has characteristic parent ion, parent minus methyl ion, and base peak at 412, 397, and 191, respectively. C_{27} 20R $\alpha\alpha\alpha$-cholastane has characteristic parent ion, parent minus methyl ion, and base peak at 372, 357, and 217, respectively; while C_{27} 20R $\alpha\beta\beta$-cholastane has characteristic parent ion, parent minus methyl ion, sterane-characteristic ion, and base peak at 372, 357, 217, and 218, respectively. As for admantane, it has the base peak (it is parent ion too) at m/z 136.

The m/z 191 fragment is often the base peak of mass spectra of cyclic terpanes (Figure 16.5.2). It is derived from rings (A+B) of the molecule, but rings (D+E) may also be the source.

The m/z 177 fragment is most likely derived from rings (A+B) of triterpane molecules which have lost a methyl group from position 10, that is, 25-norhopanes (Volkman *et al.*, 1983a,b; Grahl-Nielsen and Lygre, 1990). The notable feature of mass spectra for 25-demethylated hopanes is that the m/z 177 fragment has higher intensity than the m/z 191 fragment. Demethylated triterpanes contain different information than the triterpanes and have been suggested as markers for biodegradation (Volkman *et al.*, 1983a). The other triterpanes do also give the m/z 177 fragment upon electron impact in the mass spectrometer, but in lower abundance than the m/z 191 fragment. The fragment is formed by the loss of CH_2 from the m/z 191 fragment, and can be seen in all mass spectra of triterpanes.

The biomarker C_{29} 18α, 21β-30-norneuhopane (or C_{29}Ts), which elutes immediately after C_{29} $\alpha\beta$-hopane, has a greater abundance of the m/z 177 ion than the m/z 191 ion. The proposed mechanism is that m/z 177 ion is derived from the (D+E) ring fragment and stabilized by methyl group at position 14 (Gallegos, 1971; Killops and Howell, 1991). Increased stability of the (D+E) ring fragment producing the m/z 177 ion relative to that generating the m/z 191 ion may relate to ring junction configuration or to methyl substitution position in the (D+E) ring. For the similar mechanism, C_{29} 17β, 21α-30-norhopane (normoretane) also shows significantly greater abundance of the m/z 177 ion than the m/z 191 ion.

The m/z 217 and 218 fragment ions (Figure 16.5.2) are derived from rings (A+B+C) of most $14\alpha(H)$- and $14\beta(H)$-steranes. The $\beta\alpha\alpha$ and $\alpha\alpha\alpha$ steranes have a base peak at m/z 217, while the base peak of $\alpha\beta\beta$ steranes is m/z 218. The relative intensities of m/z 149 to m/z 151 fragment ions in the mass spectra of steranes have been used to distinguish between 5α- and 5β-stereo isomers (Gallegos, 1971). Note that the only significant difference between the mass spectra of 5α- and 5β-epimers is that the m/z 149 fragment is more abundant than the m/z 151 moiety for the 5α-epimer (e.g., 5α-cholestane vs 5β-cholestan, $C_{27}H_{48}$). Furthermore, the GC retention time of the 5β epimer is shorter than that of the 5α isomer. Hence, the stereo configuration of 5α- and 5β-steranes having the same parent ion, a parent ion minus a methyl ion, and a base peak at m/z 217 can be determined from the peak ratio at m/z 149 to m/z 151. If the ratio of

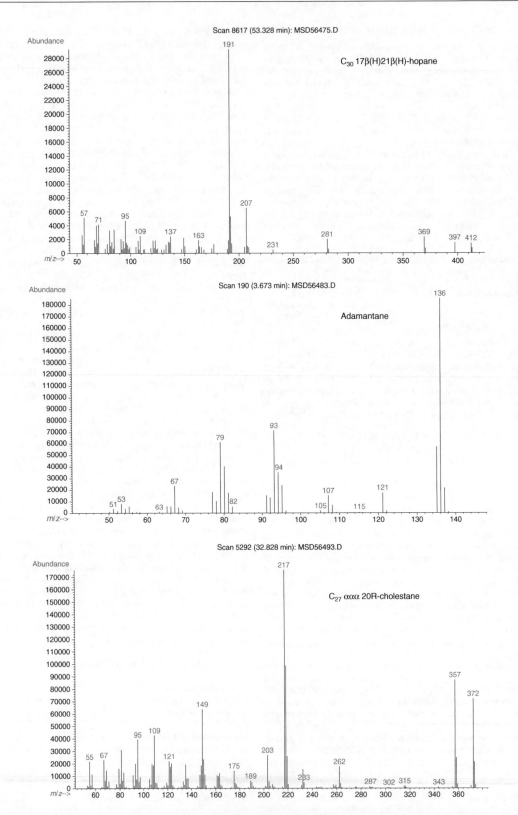

Figure 16.5.1 *Mass spectra of common biomarker compounds: (a) C_{30} $\beta\beta$ hopane, (b) adamantane, (c) C_{27} $\alpha\alpha\alpha$ 20R-cholestane, (d) C_{27} $\beta\alpha\alpha$ 20R-cholestane, (e) C_{27} $\alpha\beta\beta$ 20R-cholestane, (f) C_{29} monoaromatic sterane, (g) C_{26} triaromatic sterane, (h) C_{27} 17α-22,29,30-trisnorhopane, (i) C_{29} 17α, 21β-30-norhopane, and (j) C_{29} 17β, 21α-30-norhopane.*

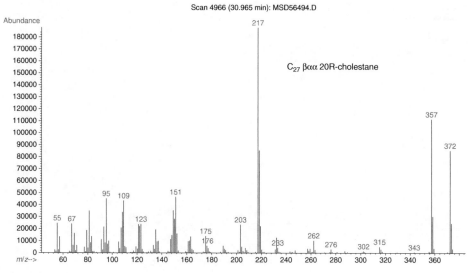

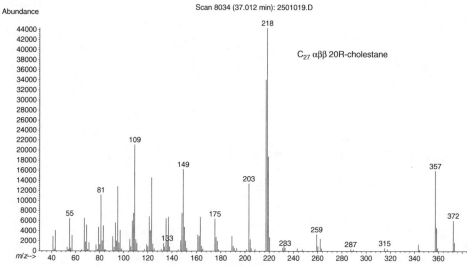

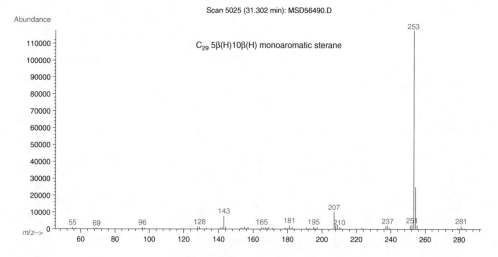

Figure 16.5.1 (Continued)

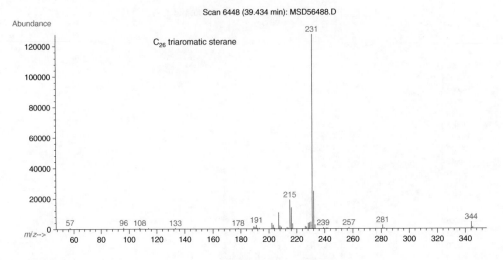

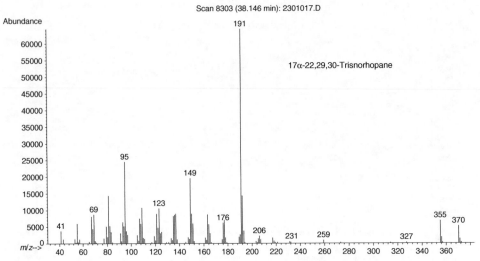

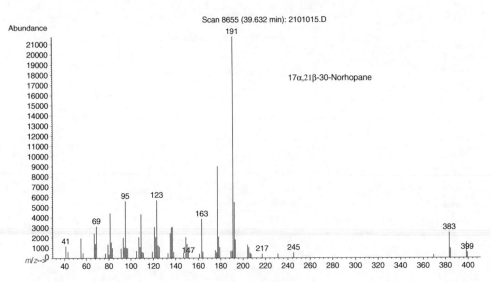

Figure 16.5.1 (Continued)

Scan 8782 (40.168 min): 2201016.D
Abundance
177
191
17β,21α-30-Norhopane
95
69
123
149
193
41
219
245
287
313
342
383
399
m/z-->

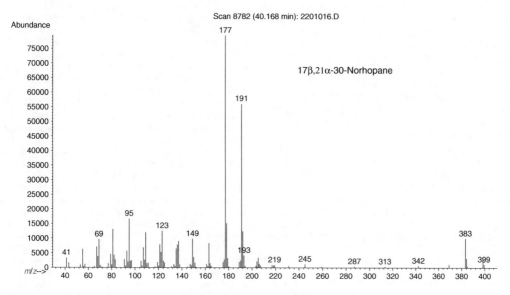

Figure 16.5.1 *(Continued)*

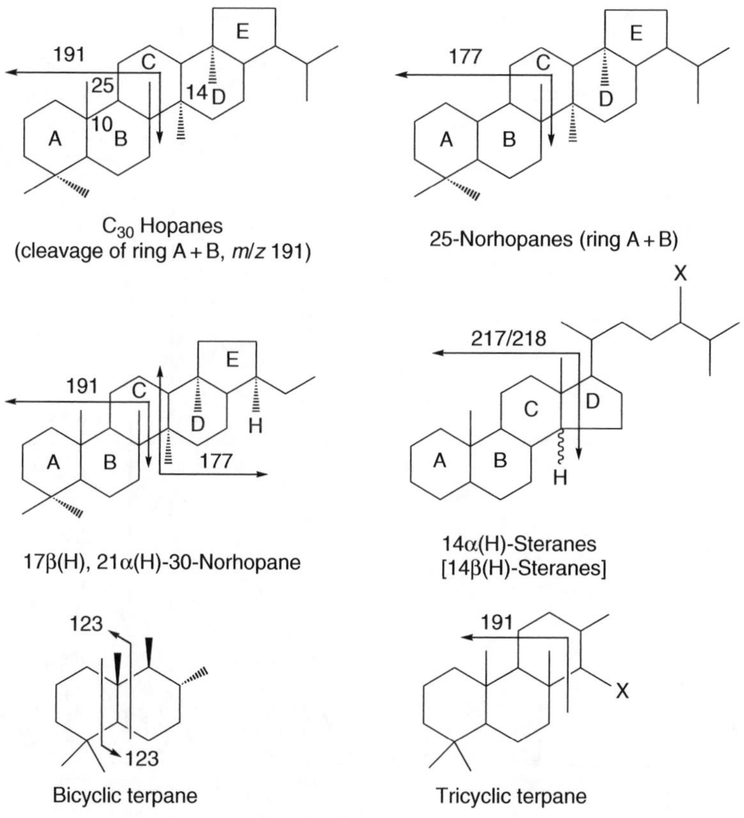

191
C
25
10
B
A
14
D
E
177
C
D
E
A
B
C30 Hopanes
(cleavage of ring A + B, m/z 191)
25-Norhopanes (ring A + B)
X
217/218
191
C
E
D
H
A
B
177
C
D
A
B
H
17β(H), 21α(H)-30-Norhopane
14α(H)-Steranes
[14β(H)-Steranes]
123
123
191
X
Bicyclic terpane
Tricyclic terpane

Figure 16.5.2 *Mass spectrometric fragmentation of common biomarkers.*

the m/z 149 ion to the m/z 151 ion is greater than 1, it is 5α-sterane; otherwise, it is 5-β-sterane.

SIM chromatograms (i.e., mass fragmentograms) are very useful not only for quantification but also for biomarker identification. The SIM chromatogram of one ion of given m/z with the GC retention time is often diagnostic of a class of homologous compounds with similar structures but different carbon numbers and isomerism, and can be used for identification. As an example, Figure 16.5.3 shows the SIM chromatograms for common biomarker classes (terpanes at 191 and steranes at 217 and 218) in a Kuwait oil obtained by using a 60 m column (Barbanti, 2004). Thirty-eight terpanes from C_{19} tricyclic terpane to C_{35} homohopanes (m/z 191) and nineteen steranes from C_{21} to C_{29} steranes (m/z 217 and 218) in total have been unambiguously identified and characterized in this Kuwait oil (refer Table 16.3.1 for peak identity). Less abundant C_{30} steranes can be clearly recognized as well. Paired biomarker isomers (H29 and C29Ts, H30 and NOR30H) and triplet (TET24 + TR26A + TR26B) are well-resolved.

Identification of Vegetation Biomarkers

In addition to petroleum biomarkers, oil-contaminated sediment samples may contain biogenic biomarker compounds. Identification of these biogenic biomarkers (e.g., major unknown peaks in SIM chromatograms) of sediment extracts can often provide valuable information about the nature and source of samples. For example, during the years 1970 to 1972 the Nipisi, Rainbow, and Old Peace River pipeline spills occurred in the Lesser Slave Lake area of northern Alberta. The Nipisi spill was by far the largest of the three spills and is also one of the largest land-spills in Canadian history. The last field survey was conducted in 1995 in order to determine which cleanup method has been the most successful, and to provide up-to-date information about any changes in residual oil and vegetative recovery 25 years after the spills. A total of 34 samples were collected for characterization of petroleum hydrocarbons (Wang *et al.*, 1998a). The background samples showed typical biogenic *n*-alkane distribution in the range of C_{21} to C_{33} with abundances of odd-carbon-number *n*-alkanes being much higher than that of even-carbon-number *n*-alkanes. The biogenic cluster was also obvious and no UCM was observed. Three vegetation biomarker compounds with significant abundances were detected and positively identified as 12-oleanene ($C_{30}H_{50}$, MW = 410.7, RT = 42.27 min), 12-ursene ($C_{30}H_{50}$, MW = 410.7, RT = 42.74 min), and 3-friedelene ($C_{30}H_{50}$, MW = 410.7, RT = 44.26 min). Formation of a six-membered ring E from the baccharane precursor leads to the oleanane group. Oleananes and their derivatives form the largest group of triterpenoids and occur widely in the plant kingdom (Connolly and Hill, 1991). The friedelene-type triterpenoids arise by increasing degrees of backbone rearrangement of the oleanene skeleton. Methyl migration in ring E of the oleanene precursor leads to the ursene skeleton (Connolly and Hill, 1991).

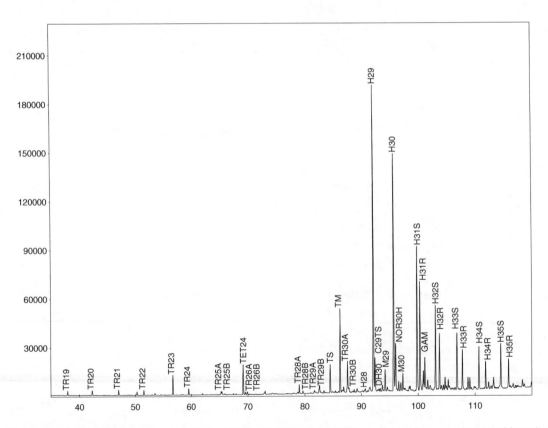

Figure 16.5.3 *The SIM chromatograms at* m/z *(a) 191, (b) 217, and (c) 218 for common terpane and sterane biomarker classes in a Kuwait oil (by courtesy of Barbanti, 2004).*

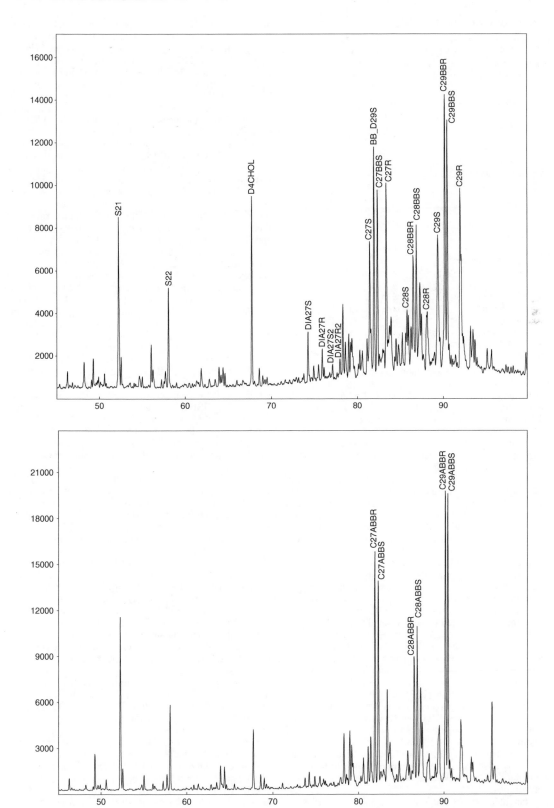

Figure 16.5.3 *(Continued)*

16.6 BIOMARKER DISTRIBUTIONS

Characterization of n-alkanes are achieved using GC-FID and GC-MS at m/z 85, 71, and 57, while characterization of major biomarker groups is achieved using GC-MS at the following MS fragment ions:

- alkyl-cyclohexanes m/z 83
- methyl-alkyl- m/z 97
  cyclohexanes
- isoalkanes and m/z 113, 183
  isoprenoids
- sesquiterpanes m/z 123
- adamantanes m/z 135, 136, 149, 163,
                             177, and 191
- diamantanes m/z 187, 188, 201, 215,
                             and 229
- tri-, tetra-, penta-cyclic m/z 191
  terpanes
- 25-norhopanes m/z 177
- 28,30-bisnorhopanes m/z 163, 191
- steranes m/z 217, 218
- 5α(H)-steranes m/z 149, 217, 218
- 5β(H)-steranes m/z 151, 217, 218
- diasteranes m/z 217, 218, 259
- methyl-steranes m/z 217, 218, 231, 232
- monoaromatic steranes m/z 253
- triaromatic steranes m/z 231

16.6.1 Distribution of Biomarkers in Crude Oils

Crude oil compositions vary widely (Figure 16.6.1), depending on the oil sources and the geological migration conditions. Crude oils can have: (1) large differences in distribution patterns of the n-alkane and cyclic-alkanes as well as UCM profiles; (2) significantly different relative ratios of isoprenoids to normal alkanes; and (3) large differences in distribution patterns and concentrations of alkylated PAH homologues and biomarkers.

The distribution patterns of biomarkers are, in general, different from oil to oil and from oil to refined products. Various biomarkers can occur in different carbon ranges of crude oils (Figure 16.6.2). In general, GC-MS chromatograms of terpanes (m/z 191) are characterized by the terpane distribution in a wide range from C_{19} to C_{35} often with $C_{29}αβ$- and $C_{30}αβ$-pentacyclic hopanes and C_{23} and C_{24} tricyclic terpanes being the most abundant. As for steranes (at m/z 217 and 218), the dominance of C_{27}, C_{28}, and C_{29} 20S/20R homologues, particularly the epimers of $αββ$-steranes, among the C_{20} to C_{30} steranes is often apparent.

GC-MS chromatograms at m/z 191, 217, and 218 for representative light (API $>$35) to medium (API: 25–35) crude oils including Alaska North Slope (ANS), Arabian Light, Federated (Alberta), Scotia Light oil (Nova Scotia), and Cook Inlet are shown in Figure 16.6.3. GC-MS chromatograms at m/z 191, 217, and 218 for representative heavy oils (API $<$ 25) Orinoco Bitumen (Venezuela), Cold Lake Bitumen (Canada), California API-11, Sockeye (California), and Boscan crude (Venezuela) are shown in Figure 16.6.4. For comparison, Table 16.6.1 summarizes the quantification results of target biomarkers in these light, medium, and heavy oils.

Figures 16.6.3 and 16.6.4 and Table 16.3.4 demonstrate that the relative biomarker distributions and absolute concentrations of terpanes and steranes are very different between oils.

Different from most crude oils, the Scotia Light (API = 59) only contain trace amount of biomarkers (the total concentration of target biomarkers (i.e., terpanes and steranes) is only 29:μg/g oil), far lower than the corresponding values

for other crude oils. The ANS oil contains a wide range of terpanes from C_{20} tricyclic terpane to C_{35} pentacyclic terpanes with the $C_{30}αβ$ hopane as the most abundant, followed by $C_{29}αβ$ hopane. The triplet C_{24} tetracyclic $+C_{26}$ (S + R) tricyclic terpanes is highly abundant as well. In contrast, the terpanes in Arabian Light and Cook Inlet are largely located in the C_{27} to C_{35} pentacyclic hopane range containing only small amounts of tricyclic terpanes. In addition, the abundance of $C_{29}αβ$ hopane is higher than that of $C_{30}αβ$ hopane in Arabian Light crude oil.

The steranes are present in all five light to medium crude oils but with different distribution patterns. The characteristic V-shaped C_{27}–C_{28}–C_{29} regular $αββ$ sterane (m/z 218) distribution is clearly demonstrated, which indicates high thermal maturity. The relative abundances of C_{27}–C_{28}–C_{29} steranes in oils reflect the carbon number distribution of the sterols in the organic matter in the source rocks for these oils. In general, a dominance of C_{27} over C_{29} steranes specifies marine algae organic matter input, while a predominance of C_{29} steranes over C_{27} steranes may indicate a preferential higher plant input (Peters and Moldowan, 1993; Gürgey, 2002). Compared with other oils, the ASN oil contains higher amounts of diasteranes as well as C_{21} and C_{22} regular steranes. The Arabian Light has much lower concentrations of steranes in total (the total of C_{27}–C_{28}–$C_{29}αββ$ steranes is only 110:μg/g oil), but displays significantly higher concentration of $C_{29}αββ$ steranes than $C_{27}αββ$ cholestane series and C_{28} ergostane series (Table 16.6.1).

The dominance of C_{28} 17β(H), 18α(H), 21β(H)-28,30-bisnorhopane (BHN28) is particularly prominent in Sockeye and California API-11 (both oils are from California), and its abundance is even higher than C_{30} and C_{29} 17α(H), 21β(H)-hopane (Figure 16.6.4). A high concentration of C_{28} 17α(H), 18α(H), 21β(H)-28,30-bisnorhopane is typical of petroleum from highly reducing to anoxic depositional environments (Mello et al., 1990). The California API-11 oil demonstrates higher concentration of C_{31} to C_{35} homohopanes than the Sockeye oil. Also, it has significantly higher concentration of C_{35} homohopanes (22S+22R) than C_{34} homohopanes (22S+22R), further indicating a highly reducing marine environment of deposition with no available free oxygen (Peters and Moldowan, 1993). For Orinoco Bitumen, C_{23} terpane is the most abundant, followed by the C_{30} and C_{29} hopane. The presence of triplets with different relative distributions is apparent for most heavy oils. Different from three other heavy oils which have somewhat the V-shaped C_{27}–C_{28}–C_{29} regular $αββ$ sterane (m/z 218) distribution, California API-11 and Sockeye oil contain far higher levels of steranes (Table 16.6.1), with a more abundant C_{28} ergostane than C_{27} and C_{29} sterane series. This is also the case for several other heavy California oils including California API-15 and Platform Irene (data not shown here). The high relative levels of C_{28} ergostane may be related to increased diversification of phytoplankton assemblages in the Jurassic and Cretaceous.

The distribution of petroleum biomarkers has been applied to investigations of oil spill accidents (Page et al., 1988; Kvenvolden et al., 1993; Wang et al., 1994b, 1995, 1998b; Bence et al., 1996; Barakat et al., 1999; Zakaria et al., 2000, 2001) and to trace the record of hydrocarbon input to San Francisco Bay (Hostettler et al., 1999a). Recently (from 1997 to 2003), biomarkers, together with PAHs and other hydrocarbon characterization results, have been extensively applied to the controversial issue over the origin of petrogenic hydrocarbon background in Prince William Sound (PWS) of Alaska, site of the 1989 Exxon Valdez oil spill (i.e., is the petrogenic hydrocarbon background mainly from eroding tertiary shales and residues of natural oil seepage, or mainly from Berling River coals and oil from the Katalla

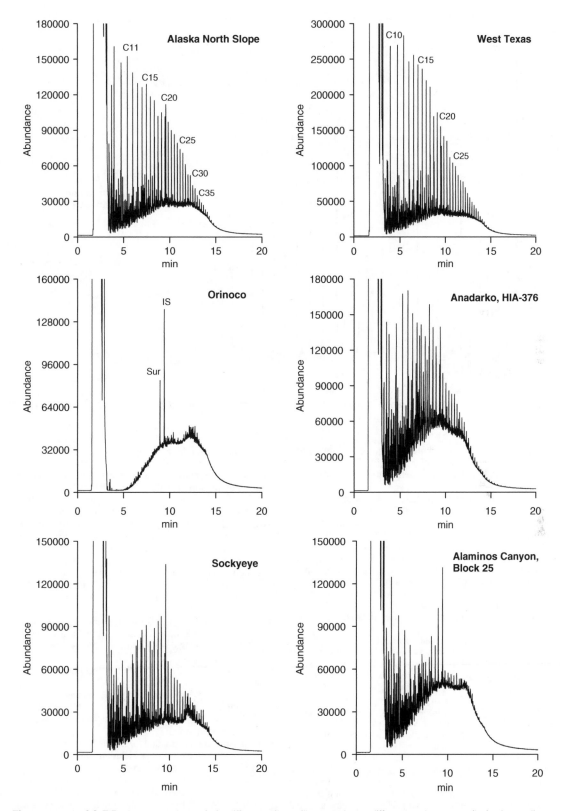

Figure 16.6.1 *GC-FID chromatograms of six different oils to illustrate large differences between oils in the n-alkane distributions and UCMs.*

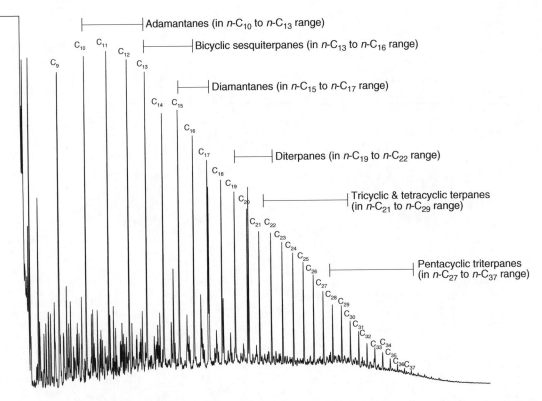

Figure 16.6.2 *Carbon number range distribution of common cyclic biomarker classes in crude oil and petroleum products.*

area?) by two groups of scientists (Page *et al.*, 1995, 1996, 2002; Bence *et al.*, 1996; Short and Heintz, 1997; Short *et al.*, 1999; Hostettler *et al.*, 1999b; Boehm *et al.*, 2001, 2002).

16.6.2 Distribution of Biomarkers in Petroleum Products

Refined petroleum products are obtained from crude oil through a variety of refining processes including distillation, cracking, catalytic reforming, isomerization, alkylation, and blending (Olah and Molnar, 1995; Speight, 2002; Simanzhenkov and Idem, 2003). Depending on the chemical composition of their "parent" crude oil feedstocks, varying refining approach and conditions, wide range of applications, regulatory requirements, and economic requirements, refined products can have wide variety in chemical compositions.

Light distillates are typically products in the C_4 to C_{13} carbon range. They include aviation gas (gasoline-type jet fuel which has a wider boiling range than kerosene-type jet fuel and include some gasoline fractions), naphtha (a liquid petroleum product that boils from about 30°C to approximately 200°C, the term "petroleum solvent" is often used synonymously with naphtha), and automotive gasoline. Light-end, resolved hydrocarbons and a minimal UCM dominate the GC trace of fresh light distillates. The composition of gasoline is well characterized. There are more than 200 individual components in the C_4 to C_{13} boiling range. These hydrocarbons include 50–60% straight, branched, and cyclic alkanes, 5–10% of alkenes (olefins), and 25–40% aromatics (Table 16.2.1). The major components of gasoline that are of environmental concern include BTEX, C_3-benzenes, and naphthalene. Gasoline and other light distillates do not contain any terpane and sterane biomarker compounds.

Mid-range distillates are typically products in a relatively broader carbon range (C_6 to C_{26}) and include kerosene (a flammable pale yellow or colorless oily liquid with a characteristic odor and intermediate in volatility between gasoline and diesel oil that distills between 125 and 260°C), aviation jet fuels, and lighter diesel products. Jet fuel is kerosene-based aviation fuel used for aviation turbine power units and has usually the same distillation characteristics and flash point as kerosene. Jet fuels are similar in gross composition, and compositional differences are attributable to additives designed to control some fuel parameters such as freeze and pour point characteristics. The main types of Jet fuels in North America are:

- Jet-A: a narrow cut kerosene-type product, which is the standard commercial and general Jet fuel available in the US. It usually contains no additives but anti-icing chemicals are sometimes added,
- Jet-A1: identical with Jet-A with the exception of its freezing point,
- Jet-B: a wide cut kerosene-type product with lighter gasoline-type naphtha components, which is widely used in Canada. It contains a static dissipator and has a very low flash point,
- JP-4: a military designation for a fuel like Jet-B, which contains a full additive package including corrosion inhibitor, anti-icing, and static dissipator,
- JP-5: a military fuel, with higher flash point than JP-4, which was designed for use by the US Navy on board aircraft carriers. It contains anti-icing and collosion inhibitors, and
- JP-8: similar to Jet-A1 with a full additive package.

(a)

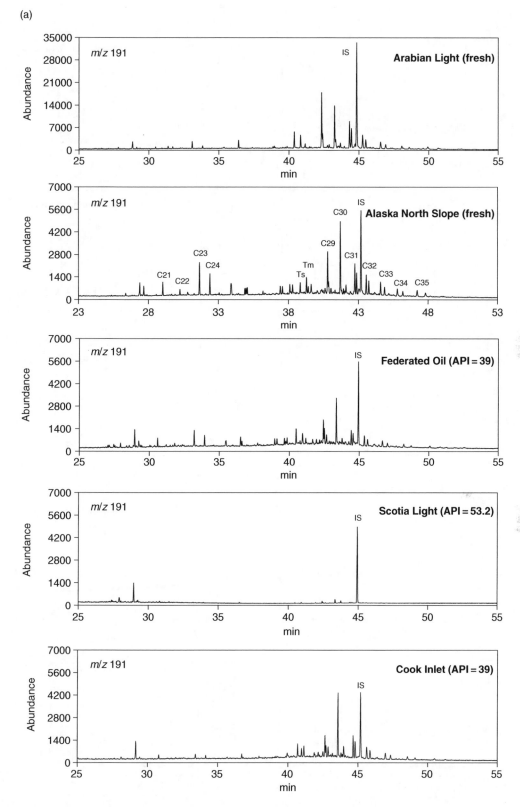

Figure 16.6.3 *GC-MS chromatograms at (a)* m/z *191 and (b)* m/z *217 and 218 for representative light (API > 35) to medium (API: 25-35) crude oils including Alaska North Slope (ANS), Arabian Light, Federated (Alberta), Scotia Light oil (Nova Scotia), and Cook Inlet.*

(b)

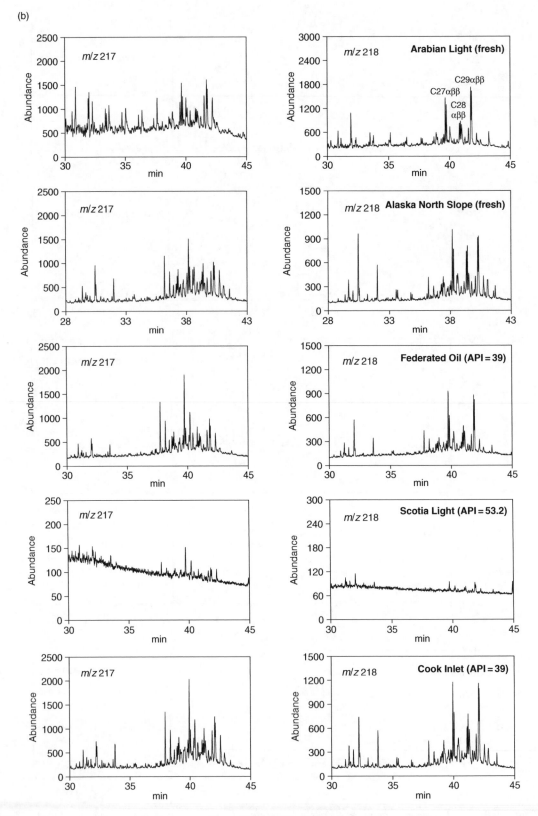

Figure 16.6.3 (Continued)

(a)

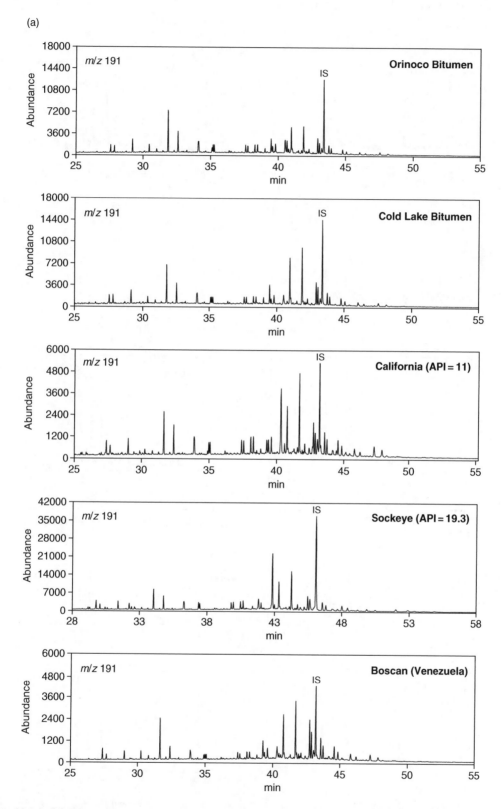

Figure 16.6.4 *GC-MS chromatograms at (a) m/z 191 and (b) m/z 217 and 218 for representative heavy oils (API < 25) Orinoco Bitumen (Venezuela), Cold Lake Bitumen (Canada), California API-11, Sockeye (California), and Boscan crude (Venezuela).*

(b)

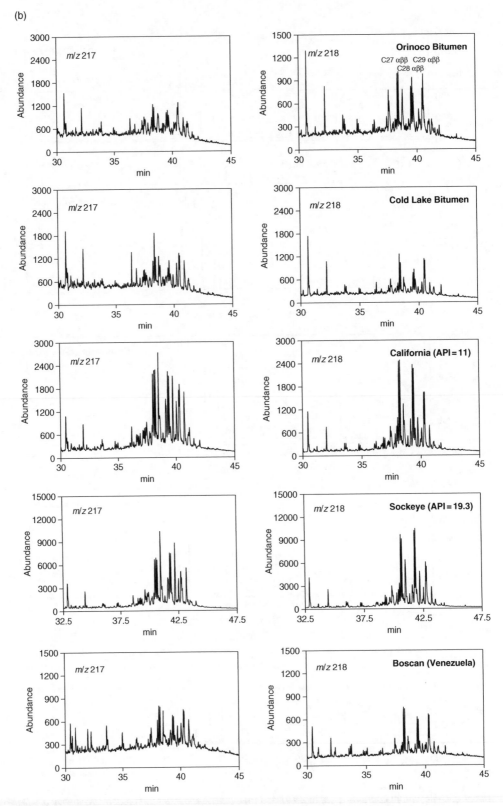

Figure 16.6.4 (Continued)

Table 16.6.1 *Quantitation Results of Target Biomarkers in Ten Light, Medium and Heavy Oils*

Oil samples	Alaska North Slope (μg/g)	Arabian Light (μg/g)	Federated oil (μg/g)	Scotia Light (μg/g)	Cook Inlet (μg/g)	Orinoco bitumen (μg/g)	Cold Lake bitumen (μg/g)	California (API = 11) (μg/g)	Sockeye (API = 19.3) (μg/g)	Boscan (Venezuela) (μg/g)
Compounds										
C_{21}	18.7	4.47	11.5	0.00	7.12	35.9	33.3	22.5	19.1	15.5
C_{22}	8.65	4.73	4.48	0.00	2.77	21.1	16.1	8.86	17.7	11.5
C_{23}	49.6	17.7	26.1	0.87	9.88	121	97.1	56.5	46.2	61.2
C_{24}	31.6	6.60	15.2	0.61	6.16	56.8	50.4	39.3	31.3	20.4
C_{29} αβ	69.3	152	36.7	3.32	45.0	79.2	115	69.3	61.2	71.4
C_{30} αβ	112	125	71.0	5.79	125	83.6	144	109	99.5	89.9
C_{31} (S)	48.9	79.9	24.6	1.74	45.0	42.7	58.2	46.1	38.7	65.5
C_{31} (R)	35.8	65.7	18.0	1.24	35.6	30.4	41.4	32.7	40.6	43.5
C_{32} (S)	37.4	48.1	18.1	0.95	29.0	27.2	32.0	32.5	27.5	33.4
C_{32} (R)	24.6	29.8	13.4	0.79	20.1	16.2	20.7	22.0	18.9	20.7
C_{33} (S)	24.2	27.0	12.9	0.00	16.6	17.4	20.9	25.1	18.8	23.5
C_{33} (R)	16.1	17.8	9.88	0.00	12.3	10.5	14.7	17.6	12.8	16.6
C_{34} (S)	19.1	14.4	8.14	0.00	10.5	13.0	14.7	17.9	8.40	17.6
C_{34} (R)	11.2	8.80	4.29	0.00	7.48	5.96	7.97	11.6	5.70	9.0
C_{35} (S)	17.7	14.7	5.73	0.00	5.98	13.0	11.9	23.0	12.1	16.4
C_{35} (R)	15.0	7.80	3.40	0.00	4.29	9.71	10.2	20.8	9.15	13.4
Ts	16.2	42.6	22.8	1.40	22.7	15.6	17.1	9.08	6.90	7.74
Tm	25.2	36.5	21.3	1.66	23.4	42.2	46.4	20.7	35.4	32.2
C_{27} αββ-steranes	124	35.1	113	2.84	184	52.4	55.4	438	208	122
C_{28} αββ-steranes	121	20.1	48.6	2.77	113	66.1	37.7	427	260	115
C_{29} αββ-steranes	152	55.1	113	5.20	232	67.2	55.4	289	152	112
Total	979	814	603	29.2	958	827	901	1738	1129	918
Diagnostic ratios										
C_{23}/C_{24}	1.58	2.68	1.71	1.42	1.60	2.13	1.93	1.44	1.48	3.00
C_{23}/C_{30} αβ	0.45	0.14	0.37	0.15	0.08	1.45	0.67	0.52	0.46	0.68
C_{24}/C_{30} αβ	0.28	0.05	0.21	0.11	0.05	0.68	0.35	0.36	0.31	0.23
C_{29} αβ/C_{30} αβ	0.62	1.22	0.52	0.57	0.36	0.95	0.80	0.64	0.62	0.79
$C_{31}(S)/C_{31}$ (S+R)	1.36	1.22	1.37	1.40	1.26	1.40	1.41	1.41	0.95	1.50
$C_{32}(S)/C_{32}$ (S+R)	1.52	1.61	1.35	1.20	1.44	1.67	1.54	1.48	1.46	1.62
Ts/Tm	0.64	1.17	1.07	0.84	0.97	0.37	0.37	0.44	0.19	0.24
C_{27} αββ-steranes/ C_{29} αββ-steranes	0.82	0.64	1.00	0.55	0.79	0.78	1.00	1.52	1.37	1.09
$C_{30}/(C_{31}+C_{32}+C_{33}+C_{34}+C_{35})$	0.45	0.40	0.60	1.23	0.67	0.45	0.62	0.44	0.52	0.35

For example, the chromatogram (Figure 16.6.5) of a commercial Jet fuel (Jet A) is dominated by GC-resolved n-alkanes in a narrow range of n-C_7 to n-C_{18} with maximum around n-C_{11} and a well-defined UCM.

Diesel fuels are originally straight-run products obtained from the distillation of crude oil. Currently, diesel fuel may also contain varying amounts of selected cracked distillates to increase the available volume. The boiling range of diesel fuel is approximately 125–380 °C. Diesel fuel depends on the nature of the original crude oil, the refining processes, and the additive (if any) used. Furthermore, the specification for diesel fuel can exist in various combinations of characteristics, such as volatility, ignition quality, viscosity, gravity, and stability. One of the most widely used specifications (ASTM D-975) covers three grades of diesel fuel oils: diesel fuel No. 1, diesel fuel No. 2, and diesel fuel No. 4. Grades No. 1 and No. 2 are distillate fuels, they are most

commonly used in high-speed engines of the mobile type, in medium speed stationary engines, and in railroad engines. Grade No. 4 diesel covers the class of more viscous distillates and, at times, blends of these distillates with residual fuel oils. The marine fuel specifications have four categories of distillate fuels and fifteen categories of fuels containing residual components (ASTM D-2069).

Diesels consist of hydrocarbons in a carbon range of C_8 to C_{28} and contain high levels of n-alkanes, alkyl-cyclohexane, and PAHs. The properties of a given diesel are largely a function of the crude oil feedstock. The GC chromatogram of diesel fuel No. 2 (Figure 16.6.5) is dominated by a nearly normal distribution of n-alkanes with maxima around n-C_{11} to n-C_{14}. Also, a central UCM hump is obvious.

Classic heavy fuel types include fuel No. 5 and No. 6 (also known as Bunker C) fuel. The heavy residual fuels are largely used in marine diesel and industrial power generation.

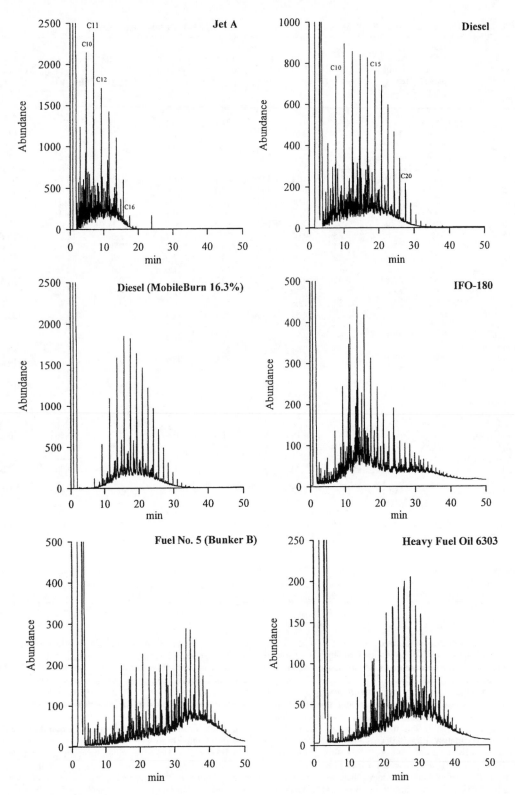

Figure 16.6.5 *The GC-FID chromatograms of common petroleum products. The differences in the chromatographic profiles, carbon ranges, the shapes of UCM, distributions of n-alkanes and major isoprenoids, and diagnostic ratios of target alkanes among these products are considerable.*

For years the term "Bunker C fuel oil" has been widely used to designate the most viscous residual fuels for general land and marine use. The chemical composition of Bunker C (or IFO 380) can vary widely, depending on production oilfields, production years, and processes it has undergone. Currently, many Bunker-type fuels are produced by blending residual oils with diesel fuels or other lighter fuels in various ratios to produce residual fuel oil of acceptable viscosity for marine or power plant use. For comparison, the chromatograms of IFO 180, a lighter residual fuel No. 5 (also called Bunker B) and a Heavy Fuel Oil 6303 (also called Bunker C or Land Bunker, from Imperial Oil Ltd., Nova Scotia, Canada) are also shown in Figure 16.6.5. The differences in the chromatographic profiles, carbon range, shapes of UCM, n-alkane and isoprenoid distributions, and diagnostic ratios of target alkanes (such as n-C17/pristane and n-C18/phytane) among these products are obvious.

GC-MS chromatograms of m/z 191, 217, and 218 for common light petroleum products Jet fuel and diesels from various sources are shown in Figure 16.6.6; while chromatograms of m/z 191, 217, and 218 for heavier fuels including IFO-180 (Fuel No. 4), Fuel No. 5 (or called Bunker B sometimes), Fuel No. 6 (Bunker C), HFO 6303 (also called Bunker C), and a spilled Bunker C fuel from Quebec (2002) are shown in Figure 16.6.7. Table 16.6.2 summarizes the quantification results of target biomarkers in common light to heavy petroleum products.

The differences in the concentrations and relative distributions of terpanes and steranes between light refined and heavy refined products and between different heavy refined products are apparent. No target terpane and sterane compounds are detected in Jet A fuel. Only traces of smaller terpanes are detected in Diesel No. 2 (2002, Ottawa) and Diesel No. 2 (1994 Mobile Burn, 16.3% weathered). For the 2002 Quebec spill diesel sample, both terpanes and steranes with a wider carbon range (C_{20} to C_{30}) were detected. Relative to other light petroleum products, biomarkers are highly abundant in Korea Diesel No. 2, which also contains a much wider carbon number range of biomarkers. Clearly, diesels from different manufacturers (i.e., using different crude oil feedstocks and refining process) have correspondingly varying biomarker fingerprints.

The GC-MS chromatogram of terpanes for the Venezuela Fuel No. 6 is characterized by a distribution in a wide range from C_{19} to C_{35} with C_{23} tricyclic terpanes being the most prominent. As for steranes, the dominance of C_{27}, C_{28}, and C_{29} 20S/20R homologues is apparent. The relative proportion of $C_{27}-C_{28}-C_{29}$ $\alpha\beta\beta$ steranes shows a consistent decrease with increasing carbon number ($C_{27} > C_{28} > C_{29}$). The suspected spill source candidate from Quebec has different biomarker distribution patterns with the C_{29} and C_{30} hopane and $C_{27}-C_{28}-C_{29}$ sterane homologous series being prominent. C_{23} and C_{24} tricyclic terpanes are the most abundant terpanes in HFO 6303, while the level of homohopanes (i.e., from C_{31} to C_{35}) is close to the detection limit. Furthermore, as opposed to most Bunker C type oils, the concentrations of terpanes and steranes are quite low in HFO 6303 (total target biomarker concentration is 255:μg/g oil). For IFO-180, the presence of diasteranes is significantly higher than in other heavy fuel oils. As for Fuel No. 5, the dominance of C_{29} sterane peaks, in particular the 20S C_{29} $\alpha\alpha\alpha$ sterane, in SIM chromatogram (m/z 217) is pronounced.

16.6.3 Distribution of Biomarkers in Lubricating Oils

Small-scale spills of lubricating oil are common due to their wide application. Lubricating oil is a class of refined products used to reduce friction and wear between bearing metallic surfaces. It is distinguished from other fractions of crude oil by its high ($>340\,^\circ$C) boiling point region. Lubricating

oils can be divided into two categories: *petroleum derived* or *synthetic* based on the main ingredient (i.e., lubricating base).

Petroleum derived lubricating oil is a mixture produced by atmospheric and vacuum distillation of selected paraffinic and naphthenic crude oils, after which chemical changes may be required to produce the desired properties of the refined product. The production of lubricating oils is well established and consists of four basic procedures (Speight, 2002):

- Distillation and deasphalting to remove lighter constituents of the feedstock;
- Solvent refining and/or hydrogen treatment to remove non-hydrocarbon constituents and to improve the feedstock quality;
- Dewaxing to remove the wax constituents and to improve the low-temperature properties;
- Clay treatment or hydrogen treatment to prevent instability of the product.

Typical carbon-range of lubricating oils is C_{20} to C_{45+} and n-alkanes are usually removed by solvent extraction. Petroleum-derived lubricating oils are the most commonly used for both automotive and industrial applications.

Synthetic lubricating oils are created by the chemical reaction of several ingredients. Two main classes are used for lubricants: esters and synthetic hydrocarbons, in particular polyalphaolefins manufactured from ethylene. These products have excellent physical properties and thermal stability. Semi-synthetic lubricating oils are obtained from mixing petroleum-derived and synthetic base lubricating oils.

Lubricating oils may also be divided into categories according to the type of services and applications, such as motor oil, transmission oil, hydraulic fluid, crankcase oil, cutting oil, turbine oil, heat-transfer oil, electrical oil, and many others. However, there are two main groups of lubricating oils classified according to the area of their application: (1) oils used in intermittent service, such as motor and aviation oils, and (2) oils designed for continuous service, such as turbine oils. The most developed standards system of classification of industrial lubricating oils is the series MS ISO 6794 "Lubricant materials, industrial oils and related products (Class I): Classification of groups." The classification includes 18 groups of products according to the area of application.

Lubricating oil additives are synthetic active substances and are often added to base oil, e.g., to improve the oil oxidation and corrosion resistance. The main lubricating oil additives include:

- Viscosity index improvers: class of polymer additives introduced into the lubricating base to produce a relative greater increase in viscosity;
- Antiwear additives: include alkyl-zinc dithiophosphates and phosphorus derivatives;
- Antioxidants: eliminate or slow down oxidation of lubricants and include dithiophosphates, substituted phenols, and aromatic amines;
- Corrosion inhibitors: these prevent the corrosion of ferrous metal under the combined effects of water, atmospheric oxygen, and a number of oxides formed during combustion processes. Mainly alkaline or alkaline-earth sulfonates, neutral or alkaline salts, fatty acids or amines, alkenylsuccinic acids and their derivatives are used as corrosion inhibitors;
- Antifreeze agents: the main function of these additives is to enable lubricating oils to retain fluidity at low temperatures (-15 to $-45\,^\circ$C). The main classes of antifreeze agents are polymethacrylates, maleate-styrene copolymers, naphthalene waxes, and vinylacetate-fumarate polyesters.

(a)

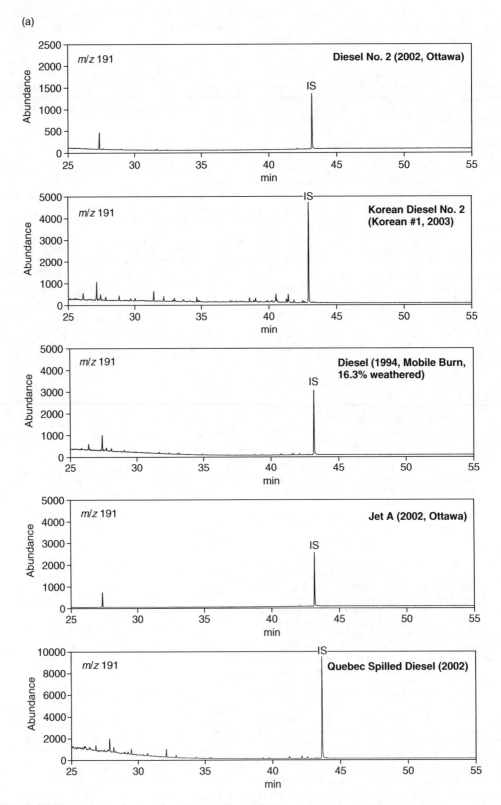

Figure 16.6.6 GC-MS chromatograms at m/z (a) 191 and (b) 217 and 218 for common light petroleum products Jet fuel and diesels from various sources.

(b)

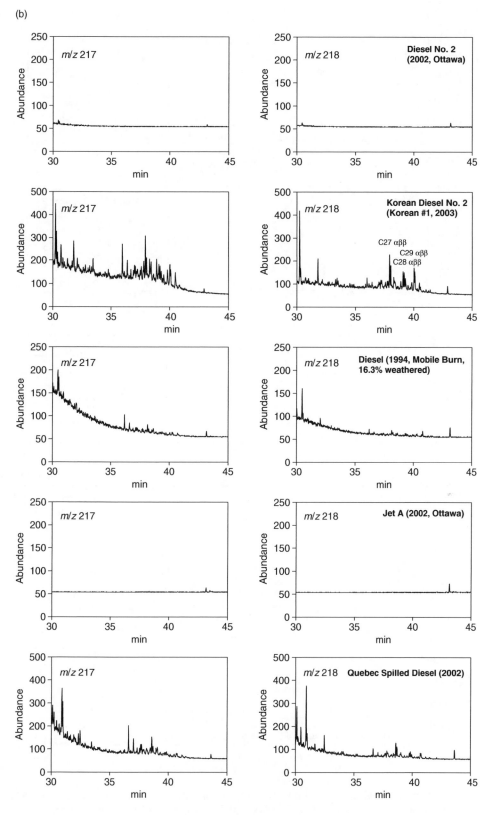

Figure 16.6.6 (Continued)

(a)

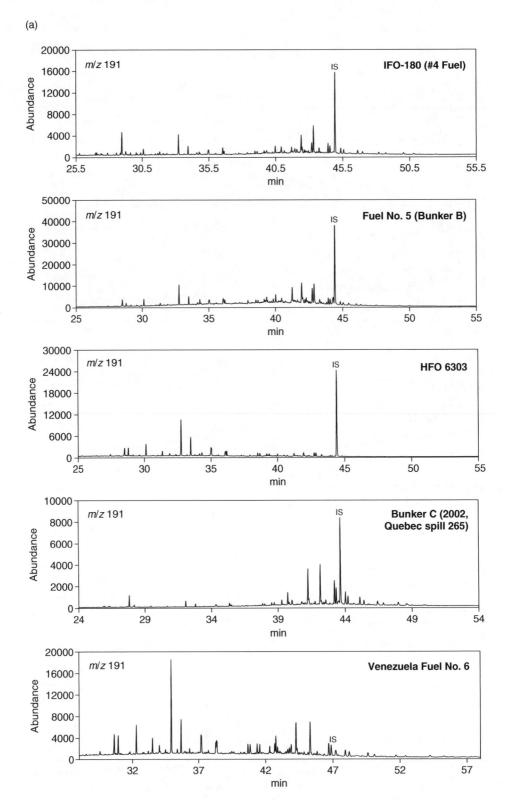

Figure 16.6.7 *GC-MS chromatograms at* m/z *(a) 191 and (b) 217 and 218 for heavy fuels including IFO-180 (Fuel No. 4), Fuel No. 5 (or called Bunker B sometimes), Fuel No. 6 (Bunker C), HFO 6303 (or called Bunker C), and a spilled Bunker C fuel from Quebec, 2002.*

(b)

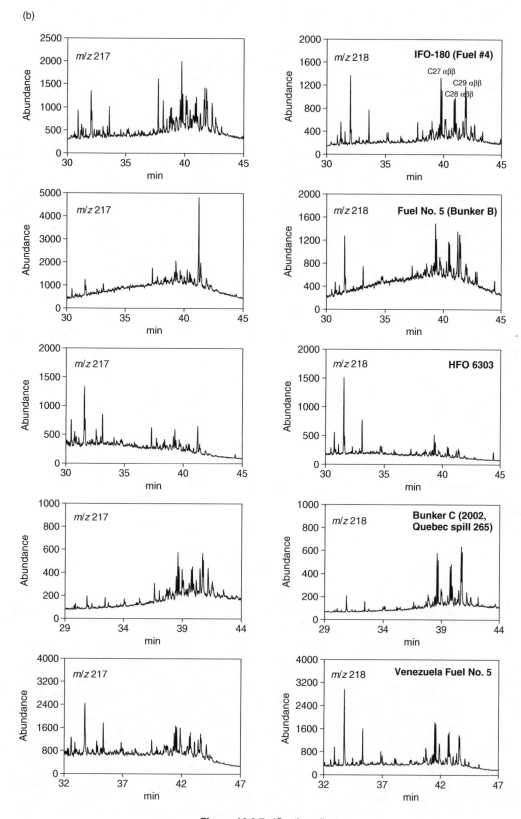

Figure 16.6.7 (Continued)

Table 16.6.2 *Quantitation Results of Target Biomarkers in Ten Common Petroleum Products*

Oil samples	Diesel-02 (µg/g)	Korean diesel #1 (2002) (µg/g)	Diesel (Mobile Burn, 16.3%) (µg/g)	Jet A (2002, Ottawa) (µg/g)	Quebec spill diesel (2002) (µg/g)	IFO-180 (Fuel #4) (µg/g)	Fuel No. 5 (Bunker B) (µg/g)	HFO 6303 (µg/g)	Bunker C (2002, Quebec spill 265) (µg/g)	Venezuela Fuel No. 6 (µg/g)
Compounds										
C_{21}	3.11	12.2	4.76	0.00	4.17	9.07	18.4	30.2	1.91	59.0
C_{22}	1.42	6.68	1.14	0.00	2.35	4.39	6.67	12.0	1.18	30.3
C_{23}	3.85	22.0	3.40	0.00	6.33	30.5	49.7	92.3	7.55	192
C_{24}	1.39	9.74	3.18	0.00	2.17	11.5	20.1	45.9	3.66	65.0
C_{29} $\alpha\beta$	0.00	22.5	4.45	0.00	2.01	29.7	71.0	14.4	51.6	72.6
C_{30} $\alpha\beta$	0.00	24.4	4.63	0.00	2.82	40.2	67.6	11.5	56.5	87.7
C_{31} (S)	0.00	5.52	0.70	0.00	0.68	14.8	17.6	3.70	31.6	37.1
C_{31} (R)	0.00	3.77	0.84	0.00	0.54	11.7	16.9	3.30	23.5	36.9
C_{32} (S)	0.00	3.85	1.57	0.00	0.59	9.93	11.5	3.20	20.9	23.7
C_{32} (R)	0.00	2.12	0.35	0.00	0.31	7.55	10.2	2.00	12.7	17.8
C_{33} (S)	0.00	1.37	0.00	0.00	0.00	8.26	7.00	1.50	13.5	16.8
C_{33} (R)	0.00	1.00	0.00	0.00	0.00	5.77	4.40	1.20	9.13	11.3
C_{34} (S)	0.00	0.00	0.00	0.00	0.00	4.87	4.00	1.10	12.0	10.1
C_{34} (R)	0.00	0.00	0.00	0.00	0.00	3.33	2.30	0.90	6.08	6.70
C_{35} (S)	0.00	0.00	0.00	0.00	0.00	4.49	3.15	0.78	7.96	11.8
C_{35} (R)	0.00	0.00	0.00	0.00	0.00	3.15	2.60	0.44	7.65	6.38
Ts	0.00	10.4	0.00	0.00	0.91	10.6	21.7	5.70	7.34	101
Tm	0.00	9.40	0.00	0.00	0.84	12.0	16.2	1.60	15.8	197
C_{27} $\alpha\beta\beta$-steranes	0.00	48.9	0.00	0.00	1.44	58.8	17.2	10.7	169	30.6
C_{28} $\alpha\beta\beta$-steranes	0.00	46.0	0.00	0.00	0.67	46.9	13.7	6.96	15.1	28.5
C_{29} $\alpha\beta\beta$-steranes	0.00	39.4	0.00	0.00	0.71	55.5	18.0	5.50	21.4	24.7
Total	9.77	269	25.0	0.00	26.5	383	400	255	344	1068
Diagnostic ratios										
C_{23}/C_{24}	2.78	2.25	1.07	*NA*	2.92	2.66	2.47	2.01	2.06	2.96
C_{23}/C_{30} $\alpha\beta$	*NA*	0.90	0.73	*NA*	2.25	0.76	0.74	8.03	0.13	2.19
C_{24}/C_{30} $\alpha\beta$	*NA*	0.40	0.69	*NA*	0.77	0.29	0.30	3.99	0.06	0.74
C_{29} $\alpha\beta/C_{30}$ $\alpha\beta$	*NA*	0.92	0.96	*NA*	0.71	0.74	1.05	1.25	0.91	0.83
$C_{31}(S)/C_{31}$ (S+R)	*NA*	1.47	0.84	*NA*	1.26	1.27	1.04	1.12	1.35	1.01
$C_{32}(S)/C_{32}$ (S+R)	*NA*	1.81	4.48	*NA*	1.90	1.31	1.13	1.60	1.64	1.33
Ts/Tm	*NA*	1.10	*NA*	*NA*	1.08	0.89	1.34	3.56	0.47	0.51
C_{27} $\alpha\beta\beta$-steranes/ C_{29} $\alpha\beta\beta$-steranes	*NA*	1.24	*NA*	*NA*	2.03	1.06	0.96	1.95	0.79	1.24
$C_{30}/(C_{31}+ C_{32}+C_{33}+ C_{34}+C_{35})$	*NA*	1.39	1.34	*NA*	1.33	0.54	0.85	0.63	0.39	0.49

It is anticipated that in order to meet the demand for the development of new lubricants and for protection of the environment, additive development in this new century will change from the synthesis of complex mixtures based on petroleum-derived components to the use of naturally occurring or bio-derived components that have antioxidant and corrosion-resistant properties.

The GC-FID chromatograms of eight lubricating oils, demonstrating the variability among this group of petroleum products, are shown in Figure 16.6.8. Marinus Turbine Oil and Marinus Valve Oil produced by a Canadian company are synthetic-based refined products. Marinus Turbine Oil is a synthetic lubricant specifically designed for the lubrication of hydro-electric turbine generator bearings, while the Marinus Valve Oil is a custom-formulated synthetic fluid designed to meet the demands of the gas pipeline industry. Therefore, the GC chromatograms of these two lubricant oils are quite unique and largely comprised of a few selected hydrocarbons.

Figure 16.6.9 shows SIM chromatograms of biomarker distributions in lubricating oils. Table 16.6.3 summarizes the quantification results of target biomarkers in these five representative lubricating oils.

In general, lubricating oils have broad GC profiles in the carbon range of C_{20} to C_{40} with boiling points greater than 340 °C, whereas residual stocks may contain hydrocarbons with 50–60 carbon atoms. Lubricating oil does not, generally, contain the low boiling fraction of petroleum hydrocarbons. Lubricating oil is largely composed of saturated hydrocarbons, and its GC trace is often dominated by a large UCM with few resolved peaks. In lubricating oil such as hydraulic fluid, for example, the PAH concentrations can be low, while the biomarker concentrations are, generally, high.

Significant features of the biomarker distribution in petroleum-derived lubricating oils include:

• Biomarkers are predominantly located in the high carbon number end. This is because the refining processes have

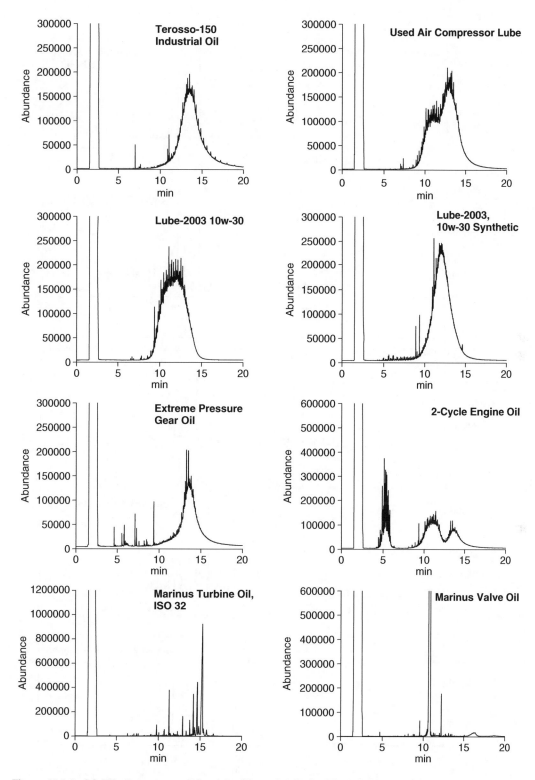

Figure 16.6.8 GC-FID chromatograms for eight different lubricating oils by using the high temperature program.

(a)

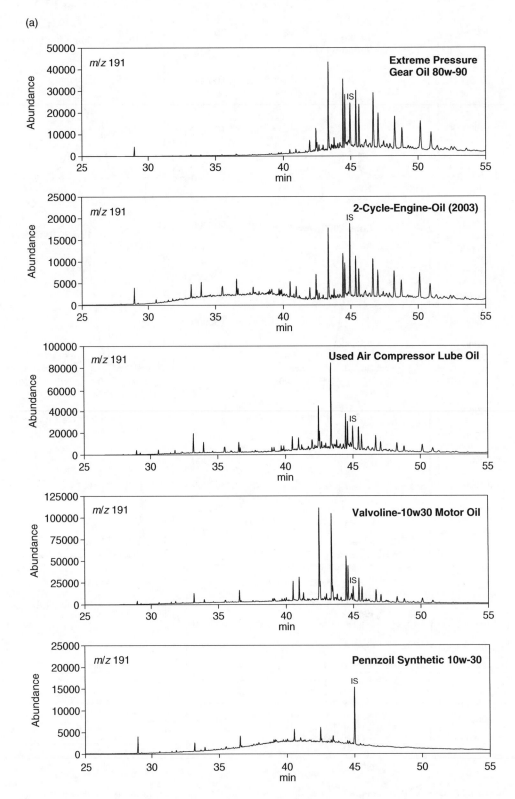

Figure 16.6.9 *GC-MS chromatograms at* m/z *(a) 191 and (b) 217 and 218 for biomarker distributions in representative lubricating oils.*

(b)

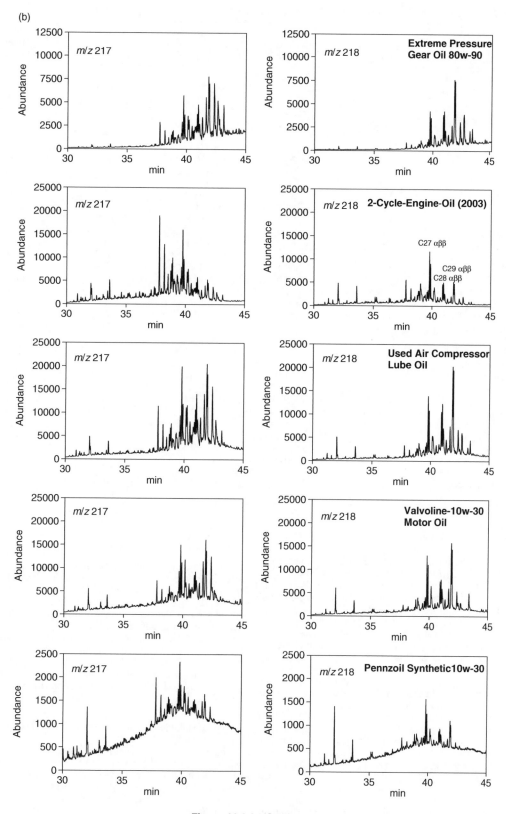

Figure 16.6.9 (Continued)

Table 16.6.3 *Quantitation Results of Target Biomarkers in Five Representative Lubricating Oils*

Oil samples	Extreme pressure gear oil (μg/g)	2-Cycle engine oil (μg/g)	Used air compressor oil (μg/g)	Valvoline 10w-30 motor oil (μg/g)	Pennzoil synthetic 10w-30 (μg/g)
Compounds					
C_{21}	0.57	5.57	17.1	11.6	3.38
C_{22}	0.45	3.72	14.3	15.2	3.39
C_{23}	2.84	22.3	86.7	68.2	15.2
C_{24}	2.42	21.5	45.0	25.5	6.78
C_{29} $\alpha\beta$	55.9	39.6	190	864	28.3
C_{30} $\alpha\beta$	233	111	414	718	15.7
C_{31} (S)	185	76.7	180	385	5.98
C_{31} (R)	142	60.1	148	305	5.78
C_{32} (S)	181	77.4	142	238	3.74
C_{32} (R)	134	57.9	96.1	164	2.48
C_{33} (S)	206	92.1	104	140	2.30
C_{33} (R)	134	62.0	69.5	91.7	1.27
C_{34} (S)	152	72.0	78.3	77.6	0.00
C_{34} (R)	93.1	47.0	43.1	51.6	0.00
C_{35} (S)	155	83.3	72.5	85.7	0.00
C_{35} (R)	113	56.1	46.5	47.6	0.00
Ts	9.72	23.0	61.9	148	23.2
Tm	13.4	21.2	74.8	215	11.8
C_{27} $\alpha\beta\beta$-steranes	142	460	437	525	57.9
C_{28} $\alpha\beta\beta$-steranes	145	230	384	363	23.9
C_{29} $\alpha\beta\beta$-steranes	313	245	761	778	36.6
Total	2413	1867	3466	5318	248
Diagnostic ratios					
C_{23}/C_{24}	1.17	1.03	1.93	2.68	2.24
C_{23}/C_{30} $\alpha\beta$	0.01	0.20	0.21	0.09	0.97
C_{24}/C_{30} $\alpha\beta$	0.01	0.19	0.11	0.04	0.43
C_{29} $\alpha\beta/C_{30}$ $\alpha\beta$	0.24	0.36	0.46	1.20	1.80
$C_{31}(S)/C_{31}(S+R)$	1.31	1.28	1.22	1.26	1.03
$C_{32}(S)/C_{32}(S+R)$	1.35	1.34	1.48	1.45	1.51
Ts/Tm	0.73	1.08	0.83	0.69	1.97
C_{27} $\alpha\beta\beta$-steranes/ C_{29} $\alpha\beta\beta$-steranes	0.45	1.88	0.57	0.67	1.58
$C_{30}/(C_{31}+C_{32}+ C_{33}+C_{34}+C_{35})$	0.16	0.16	0.42	0.45	0.73

removed low MW biomarkers and concentrated high MW biomarkers from the corresponding crude oil feedstocks. For example, almost no biomarkers smaller than C_{27} were detected in the *Extreme Pressure Gear Oil 80W-90*;

- Lubricating oils, in general, contain high levels of target terpane and sterane compounds in comparison with most crude oils and petroleum products;
- The dominance of characteristic pentacyclic C_{31} to C_{35} homohopanes is particularly prominent. For example, the concentrations of 22S/22R C_{31} to C_{33} homohopanes in the *Extreme Pressure Gear Oil 80W-90* are only slightly lower than the most abundant C_{30} $\alpha\beta$ hopane;
- The dominance of C_{27}, C_{28}, and C_{29} 20S/20R homologues is apparent too. In addition, the *Extreme Pressure Gear Oil 80W-90* contains high concentrations of C_{30} regular steranes, while the *2-Cycle-Engine-Oil* contains high amounts of diasteranes. The relative sterane distribution of $C_{27} < C_{28} < C_{29}$ $\alpha\beta\beta$ epimers is clear for the *Extreme Pressure Gear Oil 80W-90*. The dominance of C_{27} $\alpha\beta\beta$ over C_{29} $\alpha\beta\beta$ steranes is apparent for *2-Cycle-Engine-Oil* and *Pennzoil Synthetic 10W-30* oil. Conversely, the used Air Compressor lubricating oil shows clear relative dominance of C_{29} $\alpha\beta\beta$ over C_{27} $\alpha\beta\beta$ steranes;
- The synthetic lubricating oil has a different biomarker distribution pattern from petroleum-derived lubricating

oils. The concentration of target biomarkers is low, and the unresolved humps are particularly pronounced.

Lubricating oil contamination through engine exhaust and through leakage and spillage occurs everywhere (Kaplan *et al.*, 2001; Stout *et al.*, 2002). Bieger *et al.*, 1996 has reported the use of terpane biomarker fingerprints of refined oils and motor exhausts to indicate the presence of, and trace the origin of diffuse, lubricating oil contamination in plankton and sediments around St John's of Newfoundland, Eastern Canada. Different types of lubricating oils and motor exhausts were found to consistently feature distinct terpane distributions. Variable inputs of automotive and outboard motor oils were clearly recognized through characterization and recognition of the terpane distributions.

16.6.4 Distribution of Aromatic Steranes in Oils and Petroleum Products

Aromatic steranes are another group of biomarker compounds that are highly resistant to biodegradation and can be used for oil-to-oil correlation and oil source tracking. Figures 16.6.10 and 16.6.11 show the mass chromatograms of the monoaromatic (MA, *m/z* 253) and triaromatic steranes (TA, *m/z* 231) in the aromatic hydrocarbon fractions of representative crude oils (California Platform Irene and

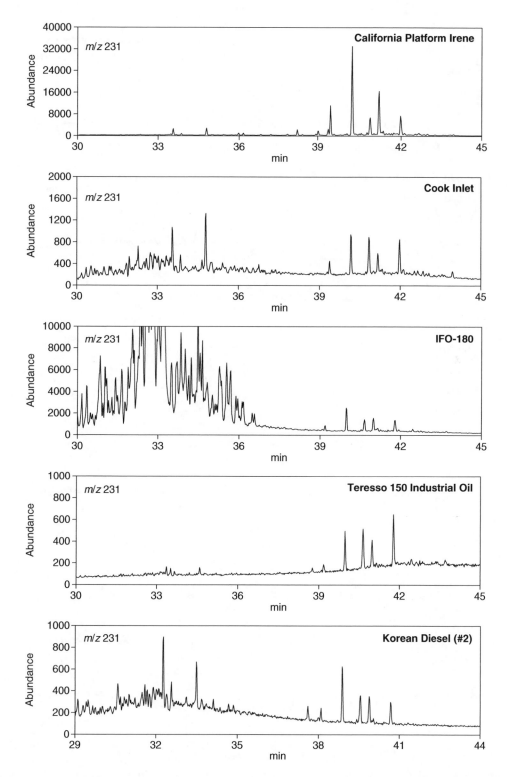

Figure 16.6.10 *GC-MS chromatograms of the triaromatic steranes (TA, m/z 231) in the aromatic hydrocarbon fractions of representative crude oils (California Platform Irene and Cook Inlet) and refined products (IFO-180, a lube oil, and a Diesel No. 2 from Korea).*

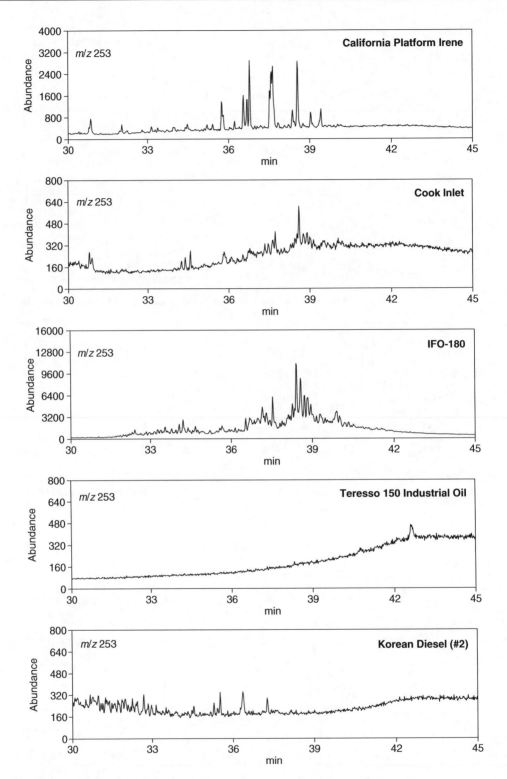

Figure 16.6.11 GC-MS chromatograms of the monoromatic steranes (MA, m/z 253) in the aromatic hydrocarbon fractions of representative crude oils (California Platform Irene and Cook Inlet) and refined products (IFO-180, a lube oil, and a Diesel No. 2 from Korea).

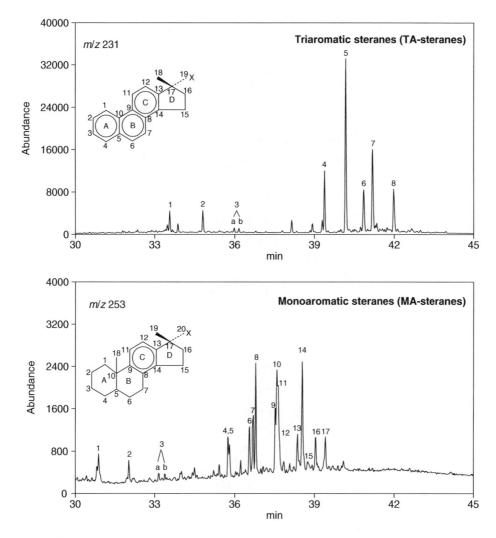

Figure 16.6.12 *Peak identification of the triaromatic (m/z 231) and monoaromatic (m/z 253) steranes in the NIST SRM 1582 oil.*

Cook Inlet) and refined products (IFO-180, a lubricating oil, and a Diesel No. 2 from Korea), respectively.

The *m/z* 231 mass chromatograms are characterized by series of 20R and 20S C_{26}–C_{27}–C_{28} triaromatic steranes (TA-cholestanes, TA-ergostanes, and TA-stigmastanes) plus C_{20} to C_{22} TA-steranes. Peak identification is summarized in Figure 16.6.12 and Table 16.6.4. Figure 16.6.12 shows that all target TA-steranes are well separated under the present GC conditions except that the C_{26} 20R isomer coelutes with the C_{27} 20S isomer.

The *m/z* 253 mass chromatograms are featured by series of 20R and 20S C_{27}–C_{28}–C_{29} 5β(H) and 5α(H) MA steranes as well as rearranged ring-C 20S and 20R MA-diasteranes. Peak identification is summarized in Figure 16.6.12 and Table 16.6.4 as well. The structure of rearranged MA steranes has been established as 10-desmethyl 5α- and 5β-methyl (20S and 20R) MA-diasterane isomers (Riolo *et al.*, 1985; Moldowan and Fago, 1986).

The differences in the relative distributions and absolute concentrations of TA- and MA-steranes between oils and refined products are apparent. Lubricating oils demon-

strate significantly lower concentrations of TA-steranes than crude oils. The Korea Diesel No. 2 also contains a relatively large quantity of high carbon number TA-steranes. Furthermore, triaromatic steranes are much more abundant than monoaromatic steranes for all oils studied. In many lighter oils such as Cook Inlet, Federated, West Texas, and Scotia Light, only trace (or no) MA-steranes are detected. This implies that TA-steranes are more valuable marker compounds than MA-steranes.

Similarly, lubricating oils only contain trace (or no) levels of MA-sterane compounds. Synthetic lubricants (such as Marinus Turbine oil and Valve oil), as expected, do not contain any TA- or MA-sterane compound. However, GC-MS analyses show that apparent TA-steranes are present in the *Synthetic 10W-30* lubricating oil, indicating that this lubricating oil may not be 100% synthesized and may be composed of portion of petroleum-derived hydrocarbons.

Barakat *et al.* have recently reported a case study in which oil residues were correlated to a fresh crude oil sample of the Egyptian Western Desert-sourced oil by fingerprinting monoaromatic and triaromatic steranes

Table 16.6.4 *Peak Identification of Triaromatic and Monoaromatic Steranes in the NIST SRM 1582 oil*

Peak No.	Compounds	Molecular formula
\multicolumn{3}{l}{Triaromatic steranes (TA-steranes, m/z 231)}		
1	C_{20} TA-sterane (X = ethyl)	$C_{20}H_{20}$
2	C_{21} TA-sterane (X = 2-propyl)	$C_{21}H_{22}$
3	C_{22} TA-sterane (X = 2-butyl)	$C_{22}H_{24}$
	(a and b are epimers at C-19)	
4	C_{26} TA-chloestane (20S)	$C_{26}H_{32}$
5	C_{26} TA-chloestane (20R)	$C_{26}H_{32}$
	+ C_{27} TA-ergostane (20S)	$C_{27}H_{34}$
6	C_{28} TA-stigmastane (20S)	$C_{28}H_{36}$
7	C_{27} TA-ergostane (20R)	$C_{27}H_{34}$
8	C_{28} TA-stigmastane (20R)	$C_{28}H_{36}$
\multicolumn{3}{l}{Monoaromatic steranes (MA-steranes, m/z 253)}		
1	C_{21} MA-sterane (X = ethyl)	$C_{21}H_{30}$
2	C_{22} MA-sterane (X = 2-propyl)	$C_{22}H_{32}$
3	C_{23} MA-sterane (X = 2-butyl)	$C_{23}H_{34}$
	(a and b are epimers at C-20)	
4	C_{27} 5β(H) MA-cholestane (20S)	$C_{27}H_{42}$
5	C_{27} MA-diacholestane (20S)	$C_{27}H_{42}$
6	C_{27} 5β(H) MA-cholestane (20R)	$C_{27}H_{42}$
	+ C_{27} MA-diacholestane (20R)	
7	C_{27} 5α(H) MA-cholestane (20S)	$C_{27}H_{42}$
8	C_{28} 5β(H) MA-ergostane (20S)	$C_{28}H_{44}$
	+ C_{28} MA-diaergostane (20S)	
9	C_{27} 5α(H) MA-cholestane (20R)	$C_{27}H_{42}$
10	C_{28} 5α(H) MA-ergostane (20S)	$C_{28}H_{44}$
11	C_{28} 5β(H) MA-ergostane (20R)	$C_{28}H_{44}$
	+ C_{28} MA-diaergostane (20R)	
12	C_{29} 5β(H) MA-stigmastane (20S)	$C_{29}H_{46}$
	+ C_{29} MA-diastigmastane (20S)	
13	C_{29} 5α(H) MA-stigmastane (20S)	$C_{29}H_{46}$
14	C_{28} 5α(H) MA-ergostane (20R)	$C_{28}H_{44}$
15	C_{29} 5β(H) MA-stigmastane (20R)	$C_{29}H_{46}$
16	C_{29} 5α(H) MA-stigmastane (20R)	$C_{29}H_{46}$
17	C_{30} 5β(H) MA-sterane (20S)	$C_{30}H_{48}$

and by determination and comparison of molecular ratios of the target MA- and TA-sterane compounds (Barakat *et al.*, 2002).

16.6.5 Sesquiterpanes in Oils and Lighter Petroleum Products

Polymethyl-substituted decalins (i.e., C_{14}–C_{16} sesquiterpanes) were first reported in 1974 (Bendoraitis, 1974) and discovered in crudes of the Loma Novia and Anastasievsko-Troyitskoe deposits (Petrov, 1987). Alexander *et al.* later identified and confirmed the existence of 8β(H)-drimane and 4β(H)-eudesmane in most Australian oils (Alexander *et al.*, 1983). Noble (1986) identified a series of C_{14} to C_{16} sesquiterpane isomers using synthesized standards and mass spectral studies. Various sesquiterpanes, with the greatest enrichment in condensate, were also identified by Simoneit *et al.* from fossil resins, sediments, and crude oils (Simoneit *et al.*, 1986), and by Chen and He from a great offshore condensate field of Liaodong Bay, Northern China (Chen and He, 1990).

Bicyclic biomarker sesquiterpanes with the drimane skeleton are ubiquitous components of crude oils and ancient sediments. Most sesquiterpanes probably originate not only from higher plants, but also from algae and bacteria (Alexander *et al.*, 1984; Philip, 1985; Fan *et al.*, 1991).

During the thermal evolution, the relative concentration of C_{14} sesquiterpanes decreases with increasing maturation of organic matters. The concentration of C_{14} bicyclic sesquiterpanes is higher at the immature stage, while those of C_{15} drimane and C_{16} homodrimane are relatively lower. As a result of the dehydroxylation and chemo-dynamics of their higher molecular weight precursors, the concentrations of drimane (C_{15}) and homodrimanes (C_{16}) gradually increase, but the concentration of C_{14} sesquiterpanes becomes lower (Cheng *et al.*, 1991).

Though biomarker sesquiterpanes and diamondoids in recent years have been used frequently for petroleum exploration, there are only few examples on their use in environmental forensics. As opposed to terpanes and steranes that are generally absent in lighter petroleum products (e.g., jet fuels and diesels), the bicyclic sesquiterpanes are abundant. Thus, comparison of SIM chromatograms of these bicyclic biomarkers at m/z 123, 179 (the ion after sesquiterpane $C_{14}H_{26}$ loses CH_3), 193 (the ion after $C_{15}H_{28}$ loses CH_3), and 207 (the ion after $C_{16}H_{30}$ loses CH_3) can provide a comparable and highly diagnostic means for chemical fingerprinting of lighter petroleum products.

The sesquiterpanes ranging from C_{14} to C_{16} elute between n-C_{13} and n-C_{16} in the SIM chromatogram of saturated hydrocarbon fraction (Figure 16.6.13). Peaks 1 and 2, peaks

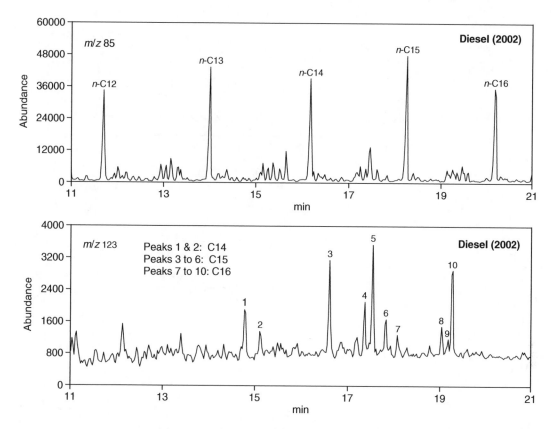

Figure 16.6.13 Total ion GC-MS (SIM) chromatograms of sesquiterpanes that elute out between the n-C_{13} and n-C_{16} range. Peaks 1 and 2, peaks 3 to 6, and peaks 7–10 are identified as C_{14}, C_{15} and C_{16} sesquiterpanes, respectively.

3 to 6, and peaks 7 to 10 are identified as C_{14}, C_{15}, and C_{16} sesquiterpanes, respectively. Among these ten compounds, peaks 5 and 10 are identified to be 8β(H)-drimane and 8β(H)-homodrimane, respectively.

Figure 16.6.14 shows SIM chromatograms of sesquiterpanes at m/z 123 for representative light (API >35), medium (API: 25–35), and heavy (API < 25) crude oils including ANS, Arabian Light, Scotia Light oil (Nova Scotia), West Texas, and California API 11. Figure 16.6.15 compares GC-MS chromatograms of sesquiterpanes at m/z 123 for representative petroleum products from light kerosene to heavy fuel oil.

The presence of common sesquiterpanes with the drimane skeleton in all studied oils is apparent. However, the sesquiterpane distribution and concentrations vary between oils from different sources. The ANS, Arabian Light, and Scotia Light contain high levels of sesquiterpanes with peak 10 (C_{16} homodrimane) as the most abundant for the ANS and Arabian Light, and peak 3 (C_{15} sesquiterpane) the most abundant for Scotia Light. The Arabian Light has the lowest concentration of C_{14} sesquiterpanes (peaks 1 and 2), indicating that this oil is highly mature. On the contrary, the heavy California API 11 oil contains the highest level of C_{14} sesquiterpane (peak 1) among all sesquiterpane peaks, indicating that this oil may be relatively immature.

Light kerosene and heavy lubricating oils do not contain sesquiterpanes (Figure 16.6.15), whereas the refined products IFO-180 and HFO 6303 (Bunker C type) contain high levels of sesquiterpanes. Furthermore, it is noticed that an unknown bicyclic biomarker compound (between peak 2 and peak 3) is abundant in these two products.

The GC-MS chromatograms at m/z 123 for a jet fuel, Diesel No. 2 (2002, from an Ottawa Gas Station), Diesel No. 2 (for 1994 Mobile Burn, 16.3% weathered), Diesel No. 2 (2003, from Korea), 1998-spilled-diesel (from Quebec), and 1998-suspected-source-diesel (from Quebec) are shown in Figure 16.6.16. Peaks 1, 3, and 5 are the most abundant in Jet A, while peak 10 is the most abundant in the diesel (2002, Ottawa) followed by peaks 5 and 3. The different concentrations and distributions of sesquiterpanes between three diesels (the Ottawa diesel, the weathered diesel used for the 1994 Mobile Burn study, and the Korean diesel) and between diesels and the jet fuel are apparent.

As for 1998 Quebec spill samples (collected at a sewer outlet flowing into the Lachine Canal of Quebec in a spill case study, 1998), only small amounts (<10:μg/g oil) of low molecular weight C_{19}–C_{24} tricyclic terpanes and regular C_{20}–C_{22} steranes and trace of diasteranes were detected (Wang et al., 2000). However, these spill samples contain considerable amounts of sesquiterpanes (Figure 16.6.16). Table 16.6.5 summarizes diagnostic ratios of selected sesquiterpanes in two representative 1998 Quebec spill diesel samples and one suspected source diesel oil. Figure 16.6.16 and Table 16.6.5 illustrate that the spill samples and the suspected source have similar SIM chromatograms and diagnostic sesquiterpane ratios. The only difference is that the spilled samples showed slightly higher abundances than the suspected source diesel sample, caused by weathering. These similarities, in combination with other hydrocarbon quantification results (e.g., hydrocarbon groups, n-alkane distributions, diagnostic ratios of

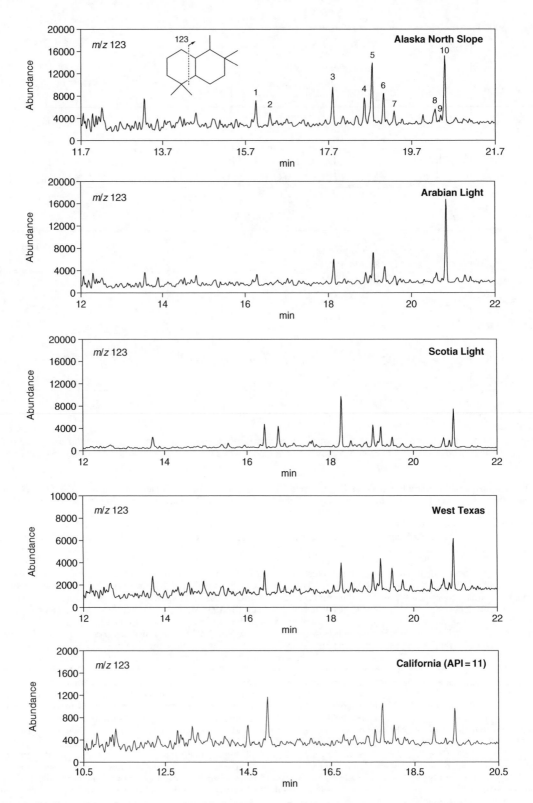

Figure 16.6.14 *GC-MS chromatograms of sesquiterpanes at* m/z *123 for representative light (API > 35), medium (API: 25–35) and heavy (API < 25) crude oils including Alaska North Slope (ANS), Arabian Light, Scotia Light oil (Nova Scotia), West Texas, and California API 11.*

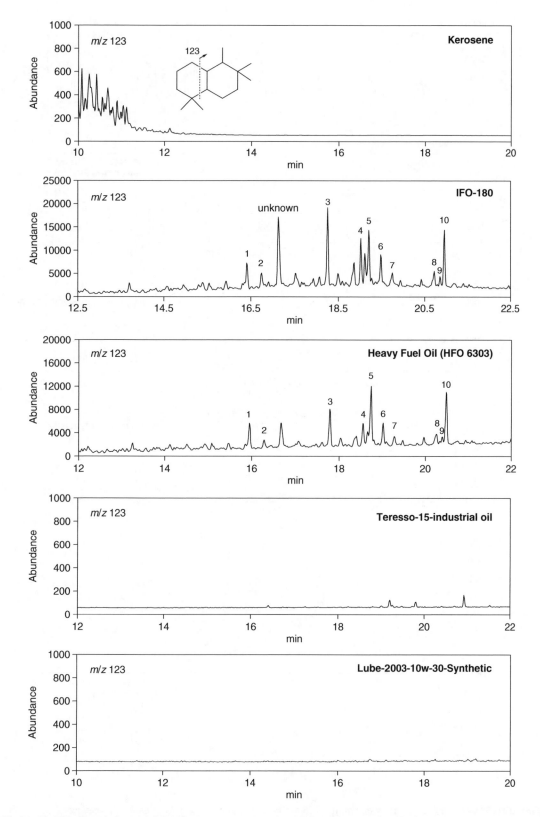

Figure 16.6.15 *Comparison of GC-MS chromatograms of sesquiterpanes at* m/z *123 for representative petroleum products from light kerosene to heavy fuel oil.*

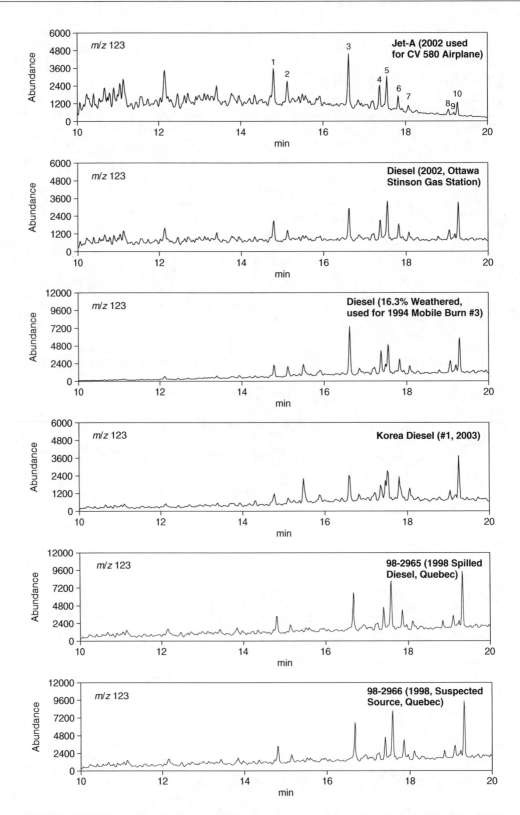

Figure 16.6.16 *GC-MS chromatograms at* m/z *123 for sesquiterpane analysis for a jet fuel, Diesel No. 2 (2002, from an Ottawa Gas Station), Diesel No. 2 (for 1994 Mobile Burn, 16.3% weathered), Diesel No. 2 (2003, from Korea), 1998-spilled-diesel (from Quebec) and 1998-suspected-source-diesel (from Quebec).*

Table 16.6.5 *Diagnostic Ratio Values of Selected Sesquiterpanes Determined from Two Representative 1998-Spill-Diesel and One Suspected-Source-Diesel Samples*

Diesel samples	2994 (spill sample)	2995 (spill sample)	2996 (suspected source)
P5:P3	1.30	1.33	1.28
P10:P3	1.30	1.35	1.29
P8:P10	0.28	0.28	0.29
P2:P1	0.48	0.47	0.50
P1:P3:P5:P10	0.54:1.00:1.30:1.30	0.57:1.00:1.34:1.36	0.58:1.00:1.28:1.29

source-specific PAHs), argued strongly that the spilled oil was a diesel fuel, and the suspected diesel collected from the pumping station (close to the spill site) was the source of the spilled diesel.

16.6.6 Diamondoid Compounds in Oils and Lighter Petroleum Products

The group of diamondoids (collective term for adamantane (C_{10}), diamantane (C_{14}), and their alkyl homologous series) is another group of low boiling cyclic biomarkers of interest to environmental forensic. Admantane was first discovered in crude oils in the early 1930s. Diamondoids are rigid, three-dimensionally fused cyclohexane-ring alkanes that have a diamond-like (cage-like) structure (Chen *et al.*, 1996; Dahl *et al.*, 1999; Grice *et al.*, 2000) and have been identified in crude oils (Williams *et al.*, 1986; Petrov, 1987; Wingert, 1992). Poly-amantanes, including triamantane, tetramantane, pentamantane, and hexamantane, have also been reported in petroleum (Lin and Wilkes, 1995). The diamondoids found in petroleum are thought to be formed from rearrangements of suitable organic precursors (such as multi-ring terpene hydrocarbons) under thermal stress with strong Lewis acids (in natural environments the only real catalysts of this kind are aluminosilicates, that is, clays) acting as catalysts during oil generation (Chen *et al.*, 1996; Dahl *et al.*, 1999).

The diamond structure endows these molecules with a high thermal stability and resistance to biodegradation. The laboratory thermal cracking experiments (Dahl *et al.*, 1999) showed that diamondoids have a higher thermal stability than most other hydrocarbons during thermal cracking of oil; therefore, diamondoids become increasingly enriched in the residual oil or condensate. The increase in methyl-diamantane (C_{15}) concentration is directly proportional to the extent of cracking, indicating that, under the conditions of the experiments, diamondoids are neither destroyed nor created. Instead, they are conserved and concentrated, and hence can be considered a naturally occurring *internal standard* by which the extent of oil loss can be determined. The extent of cracking is equal to:

$$\left(1 - \frac{C_0}{C_C}\right) \times 100\% \qquad (16.6.1)$$

where C_0 is the concentration of methyldiamantanes in the uncracked samples and C_C the methyldiamantane concentration of any cracked samples derived from the same starting oil. This principle can also be applied to determine the weathering percentages of light refined products (e.g., diesels). In this case C_0 and C_C represent the concentrations of selected methyldiamantane in the fresh and weathered oil, respectively.

Adamantanes and diamantanes elute in the ranges of n-C_{10} and n-C_{13} and n-C_{15} and n-C_{17}, respectively, in the GC-MS chromatogram of saturated hydrocarbon fraction. As an example, Figure 16.6.17 shows the total ion chromatograms of the diamondoid compounds in Prudhoe Bay

oil. At least 26 diamondoid compounds can be identified and among these 17 are adamantanes and 9 diamantanes. Peak assignments are presented in Table 16.6.6. Identification of diamondoid hydrocarbons are based on comparison with standards, assignments in the literature (Wingert, 1992; Chen *et al.*, 1996), and evaluation of mass spectra.

Figure 16.6.17 reveals the following:

- The differences in concentrations and relative distributions of adamantanes are apparent. 1,3,5,7-tetramethyl-adamantane (peak 5) has the lowest concentration among the adamantane series. This is most likely due to the fact that 1,3,5,7-tetramethyl-adamantane has four methyl groups which could affect each other and cause the molecule structure to be thermally unstable.

- The group of methyladamantanes contains only two isomers (peak 2: 1-methyl-adamantane and peak 6: 2-methyl-adamantane) due to their structural symmetry. 1-methyl-adamantane that have only one methyl group attached to the bridgehead position (i.e., at carbon position 1) is the most abundant. Similarly, 4-methyl-diamantane (peak 19), also a bridgehead methylated compound, is the most abundant compound in the diamantane series. The reason is that the methyl substitution in adamantane or diamantane at a bridgehead position (i.e., position of a tertiary carbon in the ring structure) creates a more stable molecule than substitution at a secondary carbon atom (carbon position 2) as the later produces additional skew-butane repulsions that are not imposed by the bridgehead attachment (Wingert, 1992). Therefore, 1-methyl-adamantane has a higher thermal stability than 2-methyl-adamantane. Likewise, 4-methyl-diamantane has a higher thermal stability than 1-methyl-diamantane (peak 21) and 3-methyl-diamantane (peak 25). Stable hydrocarbons will gradually increase in relative abundance over the less stable isomeric ones with increasing thermal stress. Hence, the relative distribution of alkyl-substituted diamondoid hydrocarbons may be used for assessing the maturity, especially for highly mature petroleum. Two diamondoid hydrocarbon ratios [methyl adamantane index (MAI) and methyl diamantane index (MDI), defined as 1-MA/(1 + 2-MA) and 4-MD/(1 + 3 + 4-MD), respectively] have been developed and used as novel high maturity indices to evaluate the maturation and evolution of crude oils, and to determine the thermal maturity of thermogenic gas and condensate in several Chinese basins, the maturity of which may be difficult to assess using routine geochemical techniques (Chen *et al.*, 1996).

- The elution of alkyladamantanes (i.e., the sequence of their boiling points) is quite peculiar. All methyl adamantanes substituted at the bridgehead (that is, at position 1) have much lower boiling points than adamantanes with at least one of the methyl groups not situated at the bridgehead (such as 2-methyl-adamantane, 1,2-dimethyl-adamantane, 1,4-dimethyl-adamantane, and 1,3,4-trimethyl-adamantane). The difference in the boiling

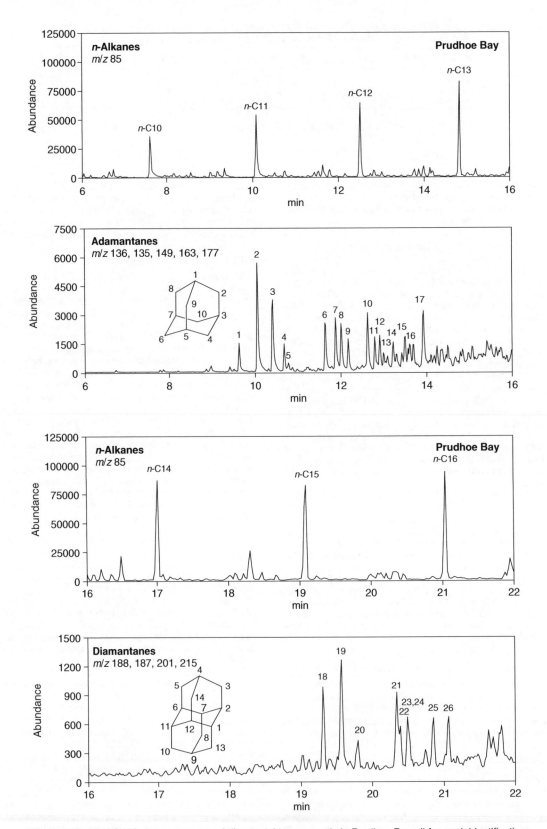

Figure 16.6.17 *GC-MS chromatograms of diamondoid compounds in Prudhoe Bay oil for peak identification.*

Table 16.6.6 *Peak Identification of Diamondoid Compounds in Prudhoe Bay Oil*

Peak No.	Compounds	Abbreviations	Base peak	M⁺ (m/z)	Formula
Adamantanes					
1	Adamantane	A	136	136	$C_{10}H_{16}$
2	1-Methyladamantane	1-MA	135	150	$C_{11}H_{18}$
3	1,3-Dimethyladamantane	1,3-DMA	149	164	$C_{12}H_{20}$
4	1,3,5-Trimethyladamantane	1,3,5-TMA	163	178	$C_{13}H_{22}$
5	1,3,5,7-Tetramethyladamantane	1,3,5,7-TeMA	177	192	$C_{14}H_{24}$
6	2-Methyladamantane	2-MA	135	150	$C_{11}H_{18}$
7	1,4-Dimethyladamantane, *cis*	1,4-DMA, *cis*	149	164	$C_{12}H_{20}$
8	1,4-Dimethyladamantane, *trans*	1,4-DMA, *trans*	149	164	$C_{12}H_{20}$
9	1,3,6-Trimethyladamantane	1,3,6-TMA	163	178	$C_{13}H_{22}$
10	1,2-Dimethyladamantane	1,2-DMA	149	164	$C_{12}H_{20}$
11	1,3,4-Trimethyladamantane, *cis*	1,3,4-TMA, *cis*	163	178	$C_{13}H_{22}$
12	1,3,4-Trimethyladamantane, *trans*	1,3,4-TMA, *trans*	163	178	$C_{13}H_{22}$
13	1,2,5,7-Tetramethyladamantane	1,2,5,7-TeMA	177	192	$C_{14}H_{24}$
14	1-Ethyladamantane	1-EA	135	164	$C_{12}H_{20}$
15	1-Ethyl-3-methyladamantane	1-E-3-MA	149	178	$C_{13}H_{22}$
16	1-Ethyl-3,5-Dimethyladamantane	1-E-3,5-DMA	163	192	$C_{14}H_{24}$
17	2-Ethyladamantane	2-EA	135	164	$C_{12}H_{20}$
Diamantanes					
18	Diamantane	D	188	188	$C_{14}H_{20}$
19	4-Methyldiamantane	4-MD	187	202	$C_{15}H_{22}$
20	4,9-Dimethyldiamantane	4,9-DMD	201	216	$C_{16}H_{24}$
21	1-Methyldiamantane	1-MD	187	202	$C_{15}H_{22}$
22	1,4 & 2,4-Dimethyldiamantane	1,4 & 2,4-DMD	201	216	$C_{16}H_{24}$
23	4,8-Dimethyldiamantane	4,8-DMD	201	216	$C_{16}H_{24}$
24	Trimethyldiamantane	TMD	215	230	$C_{17}H_{26}$
25	3-Methyldiamantane	3-MD	187	202	$C_{15}H_{22}$
26	3,4-Dimethyldiamantane	3,4-DMD	201	216	$C_{16}H_{24}$

points of these adamantanes is so large that 2-methyl-admantane (C_{11}) elutes later than 1,3,5,7-tetramethyl-adamantane (C_{14}).

Adamantane and its alkyl-substituted homologues are analyzed at *m/z* 136 (for adamantane), 135 (for methyl-adamantanes), 149 (for dimethyl-adamantanes), 163 (for trimethyl-adamantanes), and 177 (for tetramethyl-adamantanes); while diamantane and its alkyl substituted homologues are measured at *m/z* 188 (for diamantane), 187 (for methyl-diamantanes), 201 (for dimethyl-diamantanes), and 215 (for trimethyl-diamantanes). Chromatograms of adamantanes (*m/z* 136, 135, 149, and 163) in a Korean diesel (#4, 2003) are shown in Figure 16.6.18. Figure 16.6.19 compares the distributions of diamantanes (*m/z* 187 and 201) in two Korean diesels (#4 and #1). The differences in the distribution of diamantanes between two diesels from different sources are apparent.

The sesquiterpane and diamondoid biomarkers are useful for source correlation and differentiation of lighter refined products, in particular for the weathered refined products, because of their stability and resistance to biodegradation. In general, light to medium weathering has little effect on distribution pattern and diagnostic ratios of sesquiterpanes and diamondoids. Figure 16.6.20 presents the normalized abundance (*m/z* 123) of ten sesquiterpanes (relative to peak 3; i.e., the response of peak 3 = 1.00) in the Diesel No. 2 (2002, Ottawa Stinson gas station) at four weathering percentages: 0, 7.18, 14.20, and 21.95%. Table 16.6.7 summarizes index values of target sesquiterpanes and diamondoids in the weathered Diesel No. 2 samples. Clearly, no apparent depletion was observed for sesquiterpanes and diamondoids. Almost all target sesquiterpanes were concentrated in proportion with the increase of the weathered percentages, resulting in the sesquiterpane indexes remain constant for

diesels from 0 to 22% weathered. However, the preferential losses of the lower molecular weight and early-eluted adamantane and methylated adamantanes are observed as samples are further weathered.

16.7 CONCLUSIONS

Biomarkers retain all or most of the original carbon skeleton of the original natural product, and this structural similarity reveals more information about oil source than do other compounds in oil. Therefore, chemical analysis of source-characteristic and environmentally persistent biomarkers generates information of great importance to environmental forensic investigations in terms of determining the source of spilled oil, differentiating and correlating oils, and monitoring the degradation process and weathering state of oils under a wide variety of conditions.

In this chapter we briefly overviewed and discussed biomarker chemistry, biomarker genesis, analytical methodologies for biomarker separation and analysis, identification of biomarkers, biomarker distributions in crude oils and various petroleum products, and distribution of sesquiterpane and diamondoid biomarkers in oils and lighter petroleum products. In the next chapter we will discuss diagnostic ratios and cross-plots of biomarkers, unique biomarkers, weathering effects on biomarker distribution, and application of biomarker fingerprinting techniques for spill source identification, oil correlation and differentiation using univariate and multivariate methods. We believe that advancements in spilled oil fingerprinting techniques will continue and these developments will further enhance the utility and defensibility of oil hydrocarbon fingerprinting.

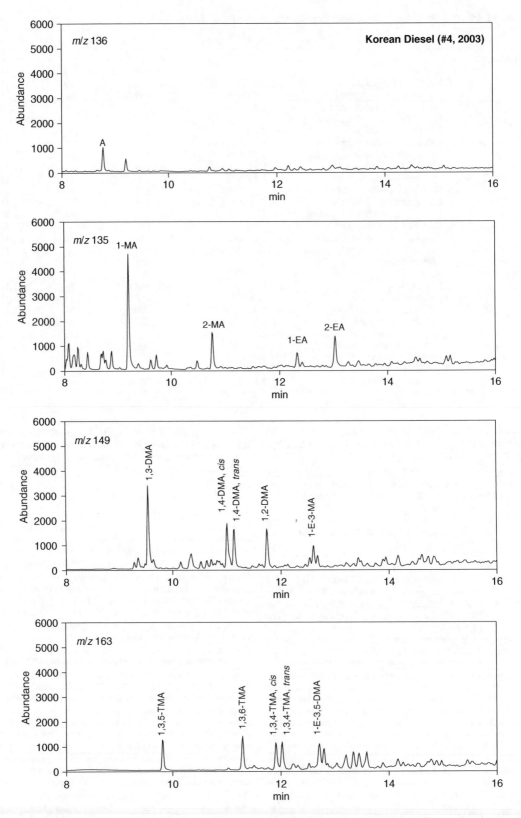

Figure 16.6.18 *Distribution of adamantanes (at m/z 136, 135, 149, and 163) in a Korean diesel.*

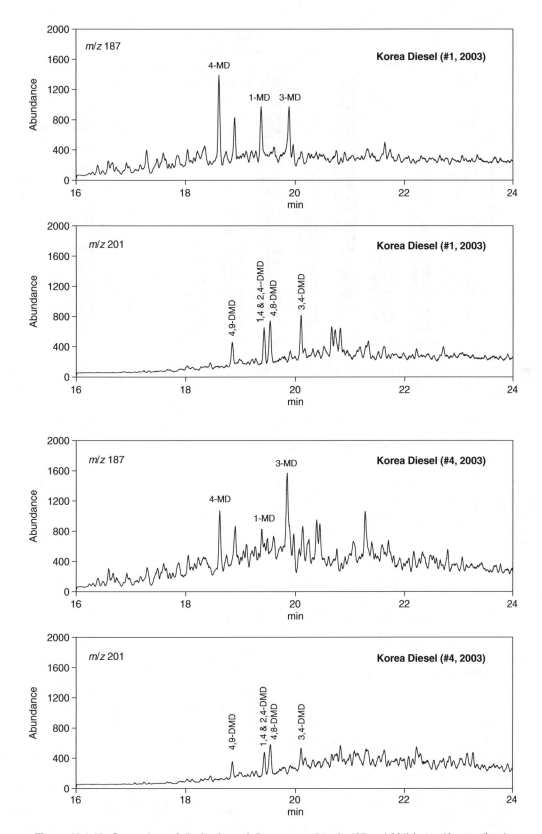

Figure 16.6.19 Comparison of distributions of diamantanes (at m/z 187 and 201) in two Korean diesels.

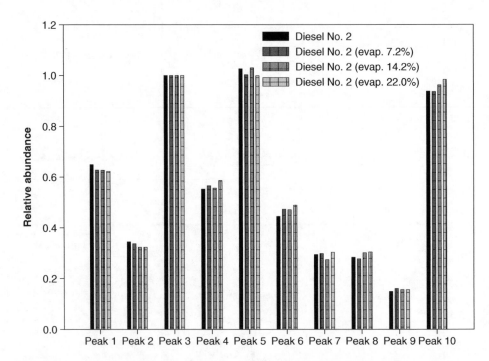

Figure 16.6.20 *The relative GC-MS response (at* m/z *123) of ten sesquiterpanes (relative to peak 3; that is, the response of peak 3 = 1.00) in the Diesel No. 2 (2002, Ottawa Stinson gas station) at four weathering percentages: 0, 7.18, 14.20, and 21.95%.*

Table 16.6.7 *Index Values of Target Sesquiterpanes and Diamondoids in the Weathered Diesel No. 2 Samples*

Diesel series	Diesel-02 (fresh)	Diesel-02 (Evap. 7.2%)	Diesel-02 (Evap. 14.2%)	Diesel-02 (Evap. 22.0%)
Sesquiterpanes				
STI-1 (P5:P3)	1.03	1.00	1.03	1.00
STI-2 (P10:P3)	0.94	0.94	0.96	0.98
STI-3 (P8:P10)	0.30	0.30	0.31	0.31
STI-4 (P2:P1)	0.53	0.54	0.52	0.52
STI-5 (P1:P3:P5:P10)	0.65:1.00:1.03:0.94	0.63:1.00:1.03:0.94	0.63:1.00:1.03:0.96	0.62:1.00:1.00:0.98
Diamondoids*				
MAI	0.73	0.73	0.72	0.68
DMAI	0.57	0.57	0.56	0.56
TMAI	0.47	0.47	0.47	0.47
EAI	0.68	0.68	0.68	0.69
MDI	0.37	0.38	0.37	0.37

*MAI = 1-MA/(1- + 2-MA); DMAI = 1, 4-DMA, *cis*/(1, 4-DMA, *cis* + 1, 4-DMA, *trans*); TMAI = 1, 3, 4-TMA, *cis*/(1, 3, 4-TMA, *cis* + 1, 3, 4-TMA, *trans*); EAI = 1-EA/(1- + 2-EA); and MDI = 4-MD/(1- + 3- + 4-MD).

16.8 ACKNOWLEDGEMENT

We thank Chiron, Norway for providing some biomarker standards.

REFERENCES

Alberdi, A. and L. Lopez, Biomarker 18∀(H)-oleanane, a geochemical tool to assess Venezuelan petroleum systems, *J. South American Earth Sciences*, 2000, 13: 751–759.

Alexander, R., R. Kagi, and R. A. Noble, Identification of bicyclic sesquiterpanes drimane and eudesmane in petroleum, *J. C. S. Chem. Comm.*, 1983, 226–228.

Alexander, R., R. Kagi, R. A. Noble, and J. K. Volkman, Identification of some bicyclic alkanes in petroleum, In: *Advances in Organic Geochemistry 1983* (edited by P. A. Schenck, J. W. De Leeuw, and G. W. M. Lijmbach), Pergamon Press, Oxford, 1984.

Barakat, A. O., A. R. Mostafa, J. Rullkotter, and A. R. Hegazi, Application of a multimolecular marker approach to fingerprint petroleum pollution in the marine environment, *Mar. Pollut. Bull.*, 1999, 38: 535–544.

Barakat, A. O., Y. Qian, M. Kim, and M. C. Kennicutt II, Compositional changes of aromatic steroid hydrocarbons in naturally weathered oil residues in Egyptian Western Desert, *Environmental Forensics*, 2002, 3: 219–226.

Barbanti, S. M., *Personal communication*, 2004.

Bence, A. E., K. A. Kvenvolden, and M. C. Kennicutt II, Organic geochemistry applied to environmental assessments of Prince William Sound, Alaska, after the Exxon Valdez oil spill—a review, *Organic Geochemistry*, 1996, 24: 7–42.

Bendoraitis, J. G., Hydrocarbons of biogenic origin in petroleum: aromatic triterpanes and bicyclic sesquiterpanes, In: *Advance in Organic Geochemistry 1973* (edited by B. Tissot, and F. Bienner), Editions Technip, Paris, 1974, 209–224.

Berkowitz, N., *Fossil hydrocarbons: chemistry and technology*, Academic Press, San Diego, CA, 1997.

Bieger, T., J. Helou, and T. A. Abrajano Jr, Petroleum Biomarkers as tracers of lubricating oil contamination, *Mar. Pollut. Bull.*, 1996, 32: 270–274.

Boehm, P. D., G. S. Douglas, W. A. Burns, P. J. Mankiewicz, D. S. Page, A. E. Bence, Application of petroleum hydrocarbon chemical fingerprinting and allocation techniques after the Exxon Valdez oil spill, *Mar. Pollut. Bull.*, 1997, 34: 599–613.

Boehm, P. D., D. S. Page, W. A. Burns, A. E. Bence, P. J. Mankiewicz, and J. S. Brown, Resolving the origin of the petrogenic hydrocarbon background in Prince William Sound, Alaska. *Environ. Sci. Technol.*, 2001, 35: 471–479.

Boehm, P. D., W. A. Burns, D. S. Page, A. E. Bence, P. J. Mankiewicz, J. S. Brown, and G. S. Douglas, Total organic carbon, an important tool in holistic approach to hydrocarbon source fingerprinting, *Environmental Forensics*, 2002, 3: 243–250.

Cahn, R. S., C. Ingold, Sir, and V. Prelog, Specification of molecular chirality, *Angewandte Chemie International Ed.*, 1966, 5: 385–415.

Chen, J. and B. He, The character and genesis of condensate in the north of Liaodong Bay, China, *Organic Geochemistry*, 1990, 6: 561–567.

Chen, J., J. Fu, G. Sheng, D. Liu, and J. Zhang, Diamondoid hydrocarbon ratios: Novel maturity indices for highly mature crude oils, *Organic Geochemistry*, 1996, 25: 179–190.

Cheng, K., W. Jin, Z. He, and J. Chen, Application of sesquiterpanes to the study of oil-gas source: the gas-rock correlation in the Qiongdongnan Basin, *J. Southeast Asian Earth Science*, 1991, 5: 189–195.

Connolly, J. D., in Chemistry of Terpenes and Terpenoids (Newman, A. A., ed.), Academic Press, London, 1972, 207.

Connolly, J. D. and R. A. Hill, Dictionary of Terpenoids, Chapman and Hall, London, 1991.

Dahl, J. E., J. M. Moldowan, K. E. Peters, G. E. Claypool, M. A. Rooney, G. E. Michael, M. R. Mello, and M. L. Kohnen, Diamondoid hydrocarbons as indicators of natural oil cracking, *Nature*, 1999, 399: 54–57.

Daling, P. S., L. G. Faksness, A. B. Hansen, and S. A. Stout, Improved and standardized methodology for oil spill fingerprinting. *Environmental Forensics*, 2002, 3: 263–278.

de Leeuw, J. W., H. C. Cox, G. Van Graas, F. W. Van de Meer, T. M. Peakman, J. M. A. Baas, and V. Van de Graaf, Limited double bond isomerization and selective hydrogenation of steranes during early diagenesis, *Geochimica et Cosmochimica Acta*, 1989, 53: 903–909.

ETC Method (updated version), *Analytical Methods for Determination of Oil Components* (by Wang, Z. D.), ETC Method No.: 5.3/1.3/M, 2002, Environmental Technology Centre, Environment Canada, Ottawa, Ontario, 2002.

Faksness, L. G., P. S. Daling, A. B. Hansen, Round Robin Study—Oil Spill Identification, *Environmental Forensics*, 2002, 3: 279–292.

Fan, P., Y. Qian, and B. Zhang, Characteristics of biomarkers in the recent sediments from Qinghai Lake, Northwest China, *J. Southeast Asian Earth and Science*, 1991, 5: 113–128.

Fingas, M. *The Basics of Oil Spill Cleanup* (2nd edition), Lewis Publishers, New York, 2001.

Frysinger, G. S., R. B. Gains, L. Xu, and C. M. Reddy, *Environ. Sci. Technol.*, 2003 37: 1653–1662.

Gains, R. B., G. S. Frysinger, M. S. Hendrick-Smith, and J. D. Stuart, *Environ. Sci. Technol.*, 1999, 33: 2106–2112.

Gallegos, E. J. Identification of new steranes, terpanes, and branched paraffins in Great River Shall by combined capillary GC and MS, *Anal. Chem.*, 1971, 43: 1151–1160.

Grahl-Nielsen, O. and T. Lygre, Identification of samples of oil related to two spills, *Mar. Pollut. Bull.*, 1990, 21: 176–183.

Grice, K., R. Alexander, and R. I. Kagi, Diamondoid hydrocarbon ratios as indicators of biodegradation in Australian crude oils, *Organic Geochemistry*, 2000, 31: 67–73.

Gürgey, K., An attempt to recognize oil populations and potential source rock types in Paleozoic sub- and Mesozoic-Cenozoic supra-salt strata in southern margin of the Pre-Caspian basin, Kazakhstan republic, *Organic Geochemistry*, 2002, 33: 723–741.

Hostettler, F. D., W. E. Pereira, K. A. Kvenvolden, A. Green, S. N. Luoma, C. C. Fuller, and R. Anima, A record of hydrocarbon input to San Francisco Bay as traced by biomarker profiles in surface sediment and sediment cores, *Marine Chemistry*, 1999a, 64: 115–127.

Hostettler, F. D., R. J. Rosenbauer, and K. A. Kvenvolden, PAH refractory index as a source discriminant of hydrocarbon input from crude oil and coal in Prince William Sound, Alaska, *Organic Geochemistry*, 1999b, 30: 873–879.

Kaplan, I. R., S. T. Lu, H. M. Alomi, and J. MacMurphey, Fingerprinting of high boiling hydrocarbon fuels, asphalts and lubricants, *Environmental Forensics*, 2001, 2: 231–248.

Kaplan, I. R., Y. Galperin, S. Lu, and R. P. Lee, Forensic environmental geochemistry differentiation of fuel-types, their sources, and release time, *Organic Geochemistry*, 1997, 27: 289–317.

Killops, S. D. and V. J. Howell, Complex series of pentacyclic triterpanes in a lacustrine sourced oil from Korea Bay Basin, *Chem. Geol.*, 1991, 91: 65–79.

Kvenvolden, K. A., F. D. Hostettler, J. B. Rapp, and P. R. Carlson, Hydrocarbon in oil residues on beaches of islands of Prince William Sound, Alaska, *Mar. Pollut. Bull.*, 1993, 26: 24–29.

Kvenvolden, K. A., F. D. Hostettler, P. R. Carleson, J. B. Rapp, C. N. Threlkeld, and A. Warden, Ubiquitous tar balls with a California-source signature on the shorelines of Prince William Sound, Alaska, *Environ. Sci. Technol.*, 1995, 29: 2684–2694.

Kvenvolden, K. A., F. D. Hostettler, R. W. Rosenbauer, T. D. Lorenson, W. T. Castle, and S. Sugarman, Hydrocarbons in recent sediment of the Monterey Bay, National Marine Sanctuary, *Marine Geology*, 2002, 181: 101–113.

Lin, R. and Z. A. Wilkes, Natural occurrence of tetramantane, pentamantane and hexamantane in a deep petroleum reservoir, *Fuel*, 1995, 74: 1512–1521.

Mackenzie, A. Z., S. C. Brassell, G. Eglinton, and J. R. Maxwell, Chemical fossil: the geological fate of steroids, *Science*, 1982, 217: 491–504.

Mello, M. R., E. A. M. Koutsoukos, M. B. Hart, S. C. Brassell, and J. R. Maxwell, Late cretaceous anoxic events in the Brazilian continental margin, *Organic Geochemistry*, 1990, 14: 529–542.

Moldowan, J. M. and F. J. Fago, Structure and significance of a novel rearranged monoaromatic steroid hydrocarbon in petroleum, *Geochimica et Cosmochimica Acta*, 1986, 50: 343–351.

Morrison, R. T. and R. N. Boyd, *Organic Chemistry* (3rd edition), Allyn and Bacon Inc., Boston, 1977.

National Research Council, *Oil in the Sea III: Inputs, Fates, and Effects*, The National Academies Press, Washington, DC, 2002.

Noble, R. A., A geochemical study of bicyclic alkanes and diterpenoid hydrocarbons in crude oils, sediments, and coals, *Ph. D. Thesis* (365 pages), Department of Organic Chemistry, University of Western Australia, 1986.

Nytoft, H. P. and J. A. Bojesen-Koefoed, *17∀(H), 21∀(H)-hopane: Natural and Synthetic*, Organic Geochemistry, 2001, 32: 841–856.

Olah, G. A. and Á. Molnar, *Hydrocarbon Chemistry*, Wiley-Interscience, New York, 1995.

Page, D. S., J. D. Foster, P. M. Fickett, and E. S. Gilfillan, Identification of petroleum sources in an area impacted by the Amoco Cadiz oil spill, *Mar. Pollut. Bull.*, 1988, 3: 107–115.

Page, D. S., P. D. Boehm, G. S. Douglas, and A. E. Bence, Identification of hydrocarbon sources in the benthic sediments of Prince William Sound and the Gulf of Alaska following the Exxon Valdez spill, In: P. G. Wells, J. N. Butler, J. S. Hughes (eds), *Exxon Valdez Oil Spill: Fate and Effects in Alaska Waters*. ASTM, Philadelphia, PA, 1995, 41–83.

Page, D. S., P. D. Boehm, G. S. Douglas, A. E. Bence, W. A. Burns, and P. J. Mankiewicz, The natural petroleum hydrocarbon background in subtidal sediments of Prince William Sound, Alaska, USA, *Environ. Toxicol. Chem.*, 1996, 15: 1266–1281.

Page, D. S., A. E. Bence, W. A. Burns, P. D. Boehm, and G. S. Douglas, A holistic approach to hydrocarbon source allocation in the subtidal sediments of Prince, William Sound, Alaska, Embayments, *Environmental Forensics*, 2002, 3: 331–340.

Peters, K. E. and J. M. Moldowan, Effects of source, thermal maturity, and biodegradation on the distribution and isomerization of homohopanes in petroleum, *Organic Geochemistry*, 1991, 17: 47–61.

Peters, K. E. and J. W. Moldowan, *The Biomarker Guide: Interpreting Molecular Fossils in Petroleum and Ancient Sediments*, Prentice Hall, New Jersey, 1993.

Petrov, A. A., *Petroleum Hydrocarbons*, Springer-Verlag, Berlin, Germany, 1987.

Philip, R. P., *Fossil Fuel Biomarkers, Application and Spectra*, Elsevier, New York, 1985.

Ravey, J. C., G. Goucouret, and D. Espinat, Asphaltene macrostructure by small angle neutron scattering, *Fuel*, 1988, 67: 1560–1567.

Reddy, C. M., T. I. Eglinton, A. Hounshell, H. K. White, L. Xu, R. B. Gains, and G. S. Frysinger, The West Falmouth oil spill after thirty years: The persistence of petroleum hydrocarbons in marsh sediments, *Environ. Sci. Technol.*, 2002, 36: 4754–4760.

Riolo, J. and P. Albrecht, Novel arrangement ring C monoaromatic steroid hydrocarbons in sediments and petroleums, *Tetrahedron Letters*, 1985, 26: 2701–2704.

Riolo, J., G. Hussler, P. Albrecht, and J. Connan, Distribution of aromatic steroids in geological samples: Their evaluation as geochemical parameters, *Organic Geochemistry*, 1986, 10: 981–990.

Seifert, W. K., Source rock/oil correlations by C27-C30 biological marker hydrocarbons, In: *Advances in Organic Geochemistry 1974* (R. Campos and J. Goni, eds), Enadimsa, Madrid, 1977, 21–44.

Seifert, W. K. and J. M. Moldowan, The effect of thermal stress on source-rock quality as measured by hopane stereochemistry, *Physics and Chemistry of the Earth*, 1980, 12: 229–237.

Short, J. W. and R. A. Heintz, Identification of Exxon Valdez oil in sediments and tissues from Prince William Sound and the Northwestern Gulf of Alaska based on a PAH weathering model, *Environ. Sci. Technol.*, 1997, 31: 2375–2384.

Short, J. W., K. A. Kvenvolden, P. R. Carlson, F. D. Hostettler, R. J. Rosenbauer, and B. A. Wright, Natural hydrocarbon background in benthic sediments of Prince William Sound, Alaska: oil vs. coal, *Environ. Sci. Technol.*, 1999, 33: 34–42.

Simanzhenkov, V. and R. Idem, *Crude Oil Chemistry*, Marcel Dekker, Inc., New York, 2003.

Simoneit, B. R. T., J. O. Grimalt, and T. G. Wang, A review on cyclic terpanoids in modern resinous plant debris amber, coal, and oils, *Annual Research Report of Geochemistry Laboratory*, Institute of Geochemistry, Academia Sinica, People's Publishing House, Guizhou, China, 1986.

Simons, D. H., F. Keniq, and C. J. Schroder-Adams, An organic geochemical study of Cenomanian-Turonian sediments from the Western Interior Seaway, Canada, *Organic Geochemistry*, 2003, 34: 1177–1198.

Speight, J. G., *The Chemistry and Technology of Petroleum*, Marcel Dekker, Inc., New York, 1999.

Speight, J. G., *Handbook of Petroleum Product Analysis*, Wiley-Interscience, Hoboken, NJ, 2002.

Stout, S. A., A. D. Uhler, T. G. Naymik, and K. J. McCarthy, Environmental forensics: Unravelling site liability, *Environ. Sci. Technol.*, 1998, 32: 260A–264A.

Stout, S. A., A. D. Uhler, and K. J. McCarthy, A strategy and methodology for defensibly correlating spilled oil to source candidates, *Environmental Forensics*, 2001, 2: 87–98.

Stout, S. A., A. D. Uhler, K. J. McCarthy, and S. Emsbo-Mattingly, Chapter 6: Chemical Fingerprinting of Hydrocarbons, In: *Introduction to Environmental Forensics* (B. L. Murphy and R. D. Morrison, eds), Academic Press, London, 2002, 139–260.

Strausz, O. P., T. W. Mojelsky, and E. M. Lown, The molecular structure of asphaltene: An unfolding story, *Fuel*, 1992, 71: 1355–1363.

Volkman, J. K., R. Alexander, R. I. Kagi, and G. W. Woodhouse, Demethylated hopanes in crude oils and their applications in petroleum geochemistry, *Geocim. Cosmochim. Acta*, 1983a, 47: 785–794.

Volkman, J. K., R. Alexander, R. I. Kagi, and J. Rüllkötter, GC-MS characterization of C27 and C28 triterpanes in sediments and petroleum, *Geocim. Cosmochim. Acta*, 1983b, 47: 1033–1040.

Volkman, J. K., D. G. Holdsworth, G. P. Neill, and Jr H. J. Bavor, Identification of natural, anthropogenic and petroleum hydrocarbons in aquatic environments, *Sci. to. Environ.*, 1992, 112: 203–219.

Volkman, J. K., A. T. Revil, and A. P. Murray, Application of biomarkers for identifying sources of natural and pollutant hydrocarbons in aquatic environments, In: *Molecular Markers in Environmental Geochemistry* (R. P. Eganhouse, ed.), American Chemical Society, Washington DC, 1997, 83–99.

Wade, L. G. Jr, *Organic Chemistry* (5th edition), Prentice Hall, Upper Saddle River, NJ, 2003.

Wang, Z. D., M. Fingas, and K. Li, Fractionation of ASMB Oil, Identification and Quantitation of Aliphatic, Aromatic and Biomarker Compounds by GC/FID and GC/MSD, *J. Chromatogr. Sci.*, 1994a, 32: 361–366 (Part I) and 367–382 (Part II).

Wang, Z. D., M. Fingas, and G. Sergy, Study of 22-year-old Arrow Oil Samples Using Biomarker Compounds by GC/MS, *Environ. Sci. Technol.*, 1994b, 28: 1733–1746.

Wang, Z. D., M. Fingas, and G. Sergy, Chemical Characterization of Crude Oil Residues from an Arctic Beach by GC/MS and GC/FID, *Environ. Sci. Technol.*, 1995, 29: 2622–2631.

Wang, Z. D., M. Fingas, S. Blenkinsopp, G. Sergy, M. Landriault, and L. Sigouin, Study of the 25-year-old Nipisi Oil Spill: Persistence of Oil Residues and Comparisons between Surface and Subsurface sediments, *Environ. Sci. Technol.*, 1998a, 32: 2222–2232.

Wang, Z. D., M. Fingas, M. Landriault, L. Sigouin, B. Castel, D. Hostetter, D. Zhang, and B. Spencer, Identification and Linkage of Tarballs from the Coasts of Vancouver Island and Northern California Using GC/MS and Isotopic Techniques, *J. High Resolut. Chromatogr.*, 1998b, 21: 383–395.

Wang, Z. D., M. Fingas, and D. Page, Oil Spill Identification, *J. Chromatogr.*, 1999, 843: 369–411.

Wang, Z. D. and M. Fingas, and L. Sigouin, Characterization and Source Identification of an Unknown Spilled Oil Using Fingerprinting Techniques by GC-MS and GC-FID, *LC-GC*, 2000, 10: 1058–1068.

Wang, Z. D. and M. Fingas, Development of oil hydrocarbon fingerprinting and identification techniques, *Mar. Pollut. Bull.*, 2003, 47: 423–452.

Waples, D. W. and T. Machihara, *"Biomarkers for Geologists,"* American Association of Petroleum Geologists, AAPG Methods in Exploration Series No. 9 (91 pages), 1991.

Williams, J. A., M. Bjoroy, D. L. Dolcater, and J. C. Winters, Biodegradation in South Texas Eocene oil-effects on aromatics and biomarkers, *Organic Geochemistry*, 1986, 10: 451–462.

Wingert, W. S., GC-MS analysis of diamondoid hydrocarbons in Smackover petroleums, *Fuel*, 1992, 71: 37–43.

Zakaria, M. P., A. Horinouchi, S. Tsutsumi, H. Takada, S. Tanabe, and A. Ismail, Oil pollution in the Straits of Malacca, Malaysia: Application of molecular markers for source identification, *Environ. Sci. Technol.*, 2000, 34: 1189–1196.

Zakaria, M. P., T. Okuda, and H. Takada, PAHs and hopanes in stranded tar-balls on the coast of Peninsular Malaysia: Applications of biomarkers for identifying source of oil pollution, *Mar. Pollut. Bull.*, 2001, 12: 1357–1366.

17

Crude Oil and Refined Product Fingerprinting: Applications

Zhendi Wang and Jan H. Christensen

Contents

17.1 INTRODUCTION

In the past decade, use of biomarker fingerprinting techniques to study spilled oils has rapidly increased and biomarker parameters have been playing a prominent role in almost all oil spill-related environmental forensic studies and investigations. In Chapter 16, we briefly discussed the basics of biomarker chemistry. In this Chapter we will discuss diagnostic ratios of biomarkers and cross-plots of biomarkers, unique biomarkers, weathering effects on biomarker distribution, and application of biomarker fingerprinting techniques for spill source identification, oil correlation and differentiation using univariate and multivariate methods.

Chemical fingerprinting techniques to identify the sources of oil spilled in the environment have been used since the 1980s and major advances have been made in recent years to address difficult oil spill issues (such as heavily weathered samples, multi-sourced samples, and large fingerprinting data set from large number of samples) on source identification and, in particular, quantitative allocation of complex hydrocarbons to multiple sources. Among target compounds selected, petroleum biomarkers have been used frequently.

Data analysis is an important part of chemical fingerprinting and a broad collection of statistical techniques has been used for evaluation of data in real oil spill cases. Data evaluation techniques within environmental forensics and specifically for spill/source identification, oil correlation, and differentiation have traditionally centered on univariate methods and comparison of diagnostic ratios and complex profiles for subjective pattern matching such as those applied in the tiered approaches by Wang et al. (1999a) and the modified Nordtest method (Daling et al., 2002). However, in spill cases with a large number of candidate sources (such as spills in the harbor area), subjective pattern matching and univariate analysis becomes increasingly difficult to use for defensibly linking spilled oil with its generic source. In particular, when the spilled oil(s) and the suspected spill source candidates are closely related, the use of multivariate statistical methods would be beneficial.

Multivariate methods have been used for data analysis in organic geochemistry since the 1980s (Øygard et al., 1984; Telnaes and Dahl, 1986) and has recently been adopted for oil spill identification within environmental forensic science (Lavine et al., 1995; Aboul-Kassim and Simoneit, 1995a, b; Burns et al., 1997; Stout et al., 2001; Mudge, 2002; Stella et al., 2002; De Luca et al., 2004; Christensen et al., 2004, 2005).

Lavine et al. (1995) employed pattern recognition and principal components analysis (PCA) to study spilled jet fuels which had undergone weathering in subsurface environment, and then classified these fuels into types: JP-4, Jet-4, JP-5, JP-8, and JPTS. Aboul-Kassim and Simoneit have used a variety of statistical techniques for source oil identification (Aboul-Kassim and Simoneit, 1995a, b). In their analysis of the aliphatic and aromatic compositions in particulate fallout samples (PFS) in Alexandria (Aboul-Kassim and Simoneit, 1995a), multivariate statistical analyses, including extended Q-mode factor analysis and linear programming, were performed in order to reduce the hydrocarbon data set into a meaningful number of end members (sources). Their analyses indicated that there were two significant end members explaining 90% of the total variation among samples and confirming petrochemical (79.6%) and thermogenic/pyrolytic (10.4%) sources in the PFS model. In a study of sediment samples in the Eastern Harbour (EH) of Alexandria (Aboul-Kassim and Simoneit, 1995b), a similar multivariate statistical approach, including factor analysis and linear programming techniques, was used to determine the

end member compositions and evaluate sediment partitioning and transport in the EH area. They found that untreated sewage was the main source of petroleum hydrocarbons in the EH area rather than direct inputs from boating activities or urban run-off.

Burns et al. (1997) used PCA and a least-squares iterative matching procedure to allocate PAHs in intertidal and subtidal sediment samples from the Prince William Sound (PWS) of Alaska to 30 potential sources. They used PCA to identify 18 possible sources, including diesel oil, diesel soot, spilled crude oil in various weathering states, natural background, creosote, and combustion products from human activities and forest fires. Subsequently, the least-squares model was used to estimate the source mix, with the best least-squares fit of 36 PAH analytes including the parent and alkylated homologues of naphthalene, phenanthrene, fluorene, dibenzothiophene, and chrysene. Isomers were grouped by the number of carbons in side groups and PAH family and treated as individual analytes. Stout et al. (2001) analyzed a suite of diagnostic PAH and biomarker ratios with PCA. The ratios were selected on the basis of high analytical precision and low susceptibility to weathering. The analysis helped to identify the prime suspects for a heavy fuel oil spill of unknown origin from 66 candidate sources.

In a recent attempt to resolve the origin of background hydrocarbons in the sediments of PWS and the Gulf of Alaska, Mudge (2002) used partial least squares regression (PLS). The percentage distribution of five possible sources – coal, seep oil, shales and input from two rivers – to the hydrocarbon loading in the Gulf of Alaska was estimated and the analysis suggests mixed sources whose contributions varied significantly across the sampling area. Christensen et al. (2004) presented an integrated methodology for forensic oil spill identification based on GC-MS analysis, chromatographic data processing, variable-outlier detection, multivariate data analysis, estimation of uncertainties, and statistical evaluation. The combination of PCA and statistical evaluation of sample similarities ensured not only simultaneous analysis of a large number of diagnostic ratios but also an objective matching of oil spill samples with suspected source oils as well as classification into positive match, probable match, and non-match.

Recently, Christensen et al. (2005) presented a methodology for objective analysis of sections of GC-MS chromatograms of biomarker compounds (i.e., steranes m/z 217). The method consisted of GC-MS analysis, preprocessing of chromatograms (including derivatization, normalization, and chromatographic alignment), and PCA of selected chromatographic regions. The resolution power of the PCA model (i.e., the ability of the model to distinguish samples from different sources) was enhanced by deselecting the most uncertain variables or scaling them according to their uncertainty, using a weighted least squares criterion (WLS). Furthermore, the four principal components, found to be the optimal number of components, was interpreted as: boiling point range (PC1), clay content (PC2), carbon number distribution of sterols in the source rock (PC3), and thermal maturity of the oil (PC4) from a priori knowledge on biomarkers.

17.2 BIOMARKER DISTRIBUTION, DIAGNOSTIC RATIOS, AND CROSS-PLOTS OF BIOMARKERS

Biomarker diagnostic parameters have been long established and are widely used by geochemists for oil correlation (oil-source rock correlation and oil-oil correlation), determination of organic input and precursors, depositional environment, assessment of thermal maturity, and evaluation of oil in-reservoir biodegradation. Many diagnostic biomarker ratios currently used in oil spill studies and environmental forensics originate from geochemistry.

17.2.1 Comparison of Biomarker Distribution for Oil Correlation and Differentiation

Comparison of biomarker distribution patterns and profiles is widely used for oil correlation and differentiation in environmental forensic studies. As described in Chapter 16, the distribution pattern and profiles of biomarkers are, in general, different from oil to oil and from oil to refined products. Various biomarkers can occur in different carbon ranges of crude oils and petroleum products. Also, the abundances or concentrations of individual biomarkers could be markedly different. Therefore, qualitative and quantitative comparisons of biomarker distribution are important for spill/source identification. In general, comparison of biomarker distribution for oil correlation and differentiation involves the following steps:

- Whether target biomarkers detected in spill samples can be found in the same defined carbon range of suspected source candidates.
- Whether the distribution patterns and profiles of biomarkers are matching.
- Whether the abundances of target biomarkers are matching.
- Whether there are any unique or unknown biomarker compounds.
- Whether the diagnostic ratios of the major biomarkers are matching.

In most cases, no matching of biomarker distribution is strong evidence for lack of correlation between spill sample(s) and suspected source(s). Matching may be an indication of correlation of spill sample(s) and suspected source(s), but is not necessarily a "proof" that samples are from the same source.

Based on analysis of triterpane distribution patterns and determination of two pentacyclic C_{27} triterpanes, Shen (1984) distinguished four Arabian crudes, which in their weathered forms were extremely similar. Volkman *et al.* (1992a) determined the distribution of various biomarker compounds in a range of aquatic sediment samples to confirm the presence of oil contamination and identify possible oil sources. Among a number of pollution sources, lubricating oils were identified as a major source of hydrocarbon pollution in many estuaries and coastal areas around Australia. Currie *et al.* (1992) proposed utilization of triterpanes to distinguish tarballs originated from Southeast Asia from those of Australian petroleum sources.

Barakat *et al.* (1997) studied the biomarker distribution within five crude oils from the Gulf of Suez, Egypt. The results revealed significant difference in biomarker distribution within the oils and the oils can be categorized into three groups. Type 1 oils show a high relative abundance of gammacerane indicating a marine saline-source depositional environment. Furthermore, these oils have a predominance of C_{35} over C_{34} 17α(H)-homohopanes. Type 2 oils have an oleanane content of more than 20% of the concentration of C_{30} $\alpha\beta$ hopane, indicating they originated from an angiosperm-rich, Tertiary source rock. Type 3 oil has geochemical characteristics intermediate between type 1 and 2 oils.

Wang *et al.* (1999b) studied oil spilled after a fire broke out at a carpet factory in Acton Vale, Quebec, on June 29, 1998. An explosion caused by a natural gas leak was identified as the probable cause of the accident. On August 26 two months after the incident, there were reports of large amounts of oil on the surface of the river near the factory and samples of the oil sheen were collected. Additional samples collected inside the factory, included water at a retention basin, water from the tank room, and oil from the heat exchange equipment near the boilers. GC characterization showed that the GC-FID chromatograms of the spill samples were markedly different from the suspected-source sample. Spill samples were highly weathered (e.g., the *n*-alkanes were nearly completely lost with the abundances of pristane and phytane greatly reduced and only a hump of unresolved complex mixture (UCM) was seen in the chromatograms). However, both spill samples and suspected sources were recognized as Bunker C type oil (Wang *et al.*, 1999b), and GC-MS biomarker analysis demonstrated that the distribution pattern of biomarker terpanes and steranes were nearly identical for the highly weathered spill and suspected-source samples (Figure 17.2.1). In addition to the presence of the regular biomarkers from C_{21} to C_{35}, high abundances of C_{30}-$\beta\alpha$ hopane were also observed. This feature is quite unique and is rarely observed in crude oils and petroleum products. The conclusion based on the biomarker distribution, diagnostic ratios of biomarkers as well as the PAH distribution was:

1. The residual oil in the spill samples was a Bunker C type fuel.
2. The oil in water samples collected from the river and from a retention basin in the factory all matched with the oil in the heat exchange equipment near the boiler, suggesting the oil in the river came from the burned factory.
3. The residual oil in the river oil–water samples QR-1, QR-2, and QR-3 had been heavily weathered.
4. Biodegradation had begun in the river oil–water samples during the two-month period between the date of fire incident and the date of sampling, as evidenced by the preferential loss of *n*-alkanes over isoprenoid compounds and specific PAH compounds within certain isomeric PAH families.

17.2.2 Diagnostic Ratios of Biomarkers

Genetic oil-oil correlations are based on the concept that the composition of biomarkers in spill samples does not significantly differ from those of the candidate source oils. Most biomarkers in spill samples and source oils, in particular those homologous isomers of biomarkers with similar structure, show little or no changes in their diagnostic ratios. An important benefit of comparing diagnostic ratios of spilled oil and suspected source oils is that concentration effects are minimized. In addition, the use of ratios tends to induce a self-normalizing effect on the data since variations due to the fluctuation of instrument operating conditions day-to-day, operator, and matrix effects are minimized. Therefore, comparison of diagnostic ratios reflects more directly differences of the target biomarker distribution between samples.

Diagnostic ratios can be calculated from either quantitative (i.e., compound concentrations) or semi-quantitative data (i.e., peak areas or heights). Ratios of compound concentrations or double normalized diagnostic ratios as those proposed by Christensen *et al.* (2004) describe the relative chemical composition. Three types of diagnostic ratios are listed in Equations (17.2.1)–(17.2.3).

$$DR = a_n^S / a_{n*}^S \tag{17.2.1}$$

$$DR = a_n^S / \left(a_n^S + a_{n*}^S \right) \tag{17.2.2}$$

$$DR = \frac{DR^S}{\left(DR^S + DR^R \right)}$$

$$DR^S = a_n^S / a_{n*}^S \quad \text{or} \quad DR^S = a_n^S / \left(a_n^S + a_{n*}^S \right)$$

$$DR^R = a_n^R / a_{n*}^R \quad \text{or} \quad DR^R = a_n^R / \left(a_n^R + a_{n*}^R \right) \tag{17.2.3}$$

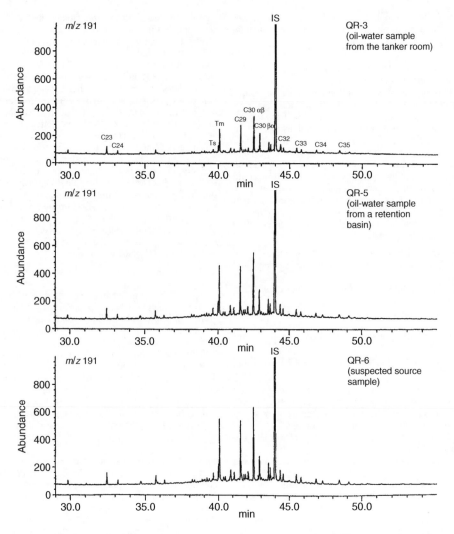

Figure 17.2.1 *Distribution pattern and profile of biomarker terpanes for the highly weathered spill and suspected-source samples (Quebec, 1998).*

where a_n^S is the peak area, peak height, or concentration of compound n or the sum of several compounds in an oil sample and a_n^R the peak area, peak height, or concentration of the corresponding compound(s) in the reference oil. Ratios of the type defined in Equation (17.2.1) range between 0 and infinity whereas ratios defined in Equation (17.2.2) range between 0 (when compound(s) n is absent) and 1 (when compound(s) n^* is absent) and circumvent problems with values rapidly approaching infinity. For example, diagnostic ratios of the former type cannot be calculated when compound(s) n^* is absent whereas this is only the case for diagnostic ratios of the latter type when both a_n^S and $a_{n^*}^S$ equals zero.

The use of semi-quantitative analyses as opposed to quantitative ones has been a matter of discussion in the scientific community. In this chapter, diagnostic ratios based on compound concentrations are used for oil spill identification in case studies 1 and 2 (Section 17.5), whereas a semi-quantitative approach is used in case studies 3 and 4 (Section 17.7). The ability to defensibly correlate spilled oil to candidate sources could be undermined if the discriminative power (its ability to distinguish oils from dif-

ferent sources) of the diagnostic ratios are not considered in some way. Proper selection of diagnostic ratios for chemical fingerprinting is discussed in Section 17.6.4.

Diagnostic biomarker ratios frequently used as defensible indices by the environmental chemists for spilled oil identification, correlation, and differentiation are summarized in Table 17.2.1. These ratios consist of alkanes, terpanes, steranes, sesquiterpanes, and diamondoids. Ratios are defined from Equation 17.2.1 for simplicity, but can readily be redefined using Equations 17.2.2 and 17.2.3.

Selection of diagnostic biomarker ratios employed in oil spill studies is mainly based on source-specific variables. It is important to realize that the suit of diagnostic ratios as listed in Table 17.2.1 is neither inclusive nor appropriate for all oil spill identification cases. In some spill cases, it may be prudent to include some particularly characteristic ratios. In other situations, the abundance of some biomarkers may be too low to obtain reliable diagnostic ratios. Thus, maintaining flexibility in the selection of diagnostic ratios to be used in specific cases is important.

The triplet ratio (see Table 17.2.1) generally varies in oils from different sources and is dependent upon sources,

Table 17.2.1 *Diagnostic Ratios of Biomarkers Frequently used for Environmental Forensic Studies*

Biomarker Classes	Diagnostic Ratios	Code
Acyclic Isoprenoids	pristane/phytane	pri/phy
	pristane/n-C_{17}	pri/C_{17}
	phytane/n-C_{18}	phy/C_{18}
Terpanes	C_{21}/C_{23} tricyclic terpane	TR21/TR23
	C_{23}/C_{24} tricyclic terpane	TR23/TR24
	C_{23} tricyclic terpane/C_{30} $\alpha\beta$ hopane	TR23/H30
	C_{24} tricyclic terpane/C_{30} $\alpha\beta$ hopane	TR24/H30
	C_{24} tertracyclic/C_{26} tricyclic (S)/C_{26} tricyclic (R) terpane	triplet ratio
	C_{27} 18α,21β-trisnorhopane/C_{27} 17α,21β-trisnorhopane	Ts/Tm
	C_{28} bisnorhopane/C_{30} $\alpha\beta$ hopane	H28/H30
	C_{29} $\alpha\beta$-25-norhopane/C_{30} $\alpha\beta$ hopane	NOR25H/H30
	C_{29} $\alpha\beta$-30-norhopane/C_{30} $\alpha\beta$ hopane	H29/H30
	oleanane/C_{30} $\alpha\beta$ hopane	OL/H30
	moretane(C_{30} $\beta\alpha$ hopane)/C_{30} $\alpha\beta$ hopane	M30/H30
	Gammacerane/C_{30} $\alpha\beta$ hopane	GAM/H30
	tricyclic terpanes (C_{19}–C_{26})/C_{30} $\alpha\beta$ hopane	Σ(TR19-TR26)/H30
	C_{31} homohopane (22S)/C_{31} homohopane (22R)	H31S/H31R
	C_{32} bishomohopane (22S)/C_{32} bishomohopane (22R)	H32S/H32R
	C_{33} trishomohopane (22S)/C_{33} trishomohopane (22R)	H33S/H33R
	Relative homohopane distribution	H31:H32:H33:H34:H35
	Σ(C_{31}–C_{35})/C_{30} $\alpha\beta$ hopane	Σ(H31–H35)/H30
	homohopane index	H31/Σ(H31–H35) to H35/Σ(H3l–H35)
Steranes	Relative distribution of regular C_{27}–C_{28}–C_{29} steranes	C27:C28:C29 steranes
	C_{27} $\alpha\beta\beta$/C_{29} $\alpha\beta\beta$ steranes (at m/z 218)	C27$\beta\beta$(S+R)/C29$\beta\beta$(S+R)
	C_{28} $\alpha\beta\beta$/C_{29} $\alpha\beta\beta$ steranes (at m/z 218)	C28$\beta\beta$(S+R)/C29$\beta\beta$(S+R)
	C_{27} $\alpha\beta\beta$/(C_{27} $\alpha\beta\beta$ + C_{28} $\alpha\beta\beta$ + C_{29} $\alpha\beta\beta$) (at m/z 218,	C27$\beta\beta$/(C27+C28+C29)$\beta\beta$
	C_{28} $\alpha\beta\beta$/(C_{27} $\alpha\beta\beta$ + C_{28} $\alpha\beta\beta$ + C_{29} $\alpha\beta\beta$) (at m/z 218,	C28$\beta\beta$/(C27+C28+C29)$\beta\beta$
	C_{29} $\alpha\beta\beta$/(C_{27} $\alpha\beta\beta$ + C_{28} $\alpha\beta\beta$ + C_{29} $\alpha\beta\beta$) (at m/z 218)	C29$\beta\beta$/(C27+C28+C29)$\beta\beta$
	C_{27}, C_{28}, and C_{29} $\alpha\alpha\alpha$/$\alpha\beta\beta$ epimers (at m/z 217)	C27$\alpha\alpha$/C27$\beta\beta$
		C28$\alpha\alpha$/C28$\beta\beta$
		C29$\alpha\alpha$/C29$\beta\beta$
	C_{27}, C_{28}, and C_{29} 20S/(20S+20R) steranes	C27 (20S)/C27 (20R)
		C28 (20S)/C28 (20R)
		C29 (20S)/C29 (20R)
	C_{30} sterane index: C_{30}/(C_{27} to C_{30}) steranes selected diasteranes/regular steranes	C30/(C27 to C30) steranes
	Regular C_{27}–C_{28}–C_{29} steranes/C_{30} $\alpha\beta$-hopanes	C27–C28–C29 steranes/H30
Sesquiterpanes	Relative distribution of sesquiterpanes	
	Peak 5/Peak 3	
	Peak 10/Peak 3	
	Peak 1/Peak 3	
Diamondoids	Methyl adamantane index: 1-MA/(1+2-MA) 1,4-DMA, *cis*/1,4-DMA, *trans*	MAI
	Dimethyl admantane index: 1,3-DMA/(1,3+1,4+1,2-DMA) 1,3,4-DMA, *cis*/1,3,4-DMA, *trans*	DMAI
	Trimethyl admantane index: 1,3,4-DMA, *cis*/(1,3,4-DMA, *cis* + 1,3,4-DMA, *trans*)	TMAI
	Ethyl admantane index: 1-EA/(1-+2-EA)	EAI
	Methyl-diamantane index: 4-MD/(1-+3-+4-MD)	MDI
	Relative distribution of diamantanes: C_0-D:C_1-D:C_2-D:C_3-D	
Triaromatic Steranes	C_{20} TA/(C_{20} TA + C_{21} TA)	
	C_{26} TA (20S)/sum of C_{26} TA (20S) through C_{28} TA (20R)	
	C_{27} TA (20R)/C_{28} TA (20R)	
	C_{28} TA (20R)/C_{28} TA (20S)	
	C_{26} TA (20S)/[C_{26} TA(20S) + C_{28} TA (20S)]	
	C_{28} TA (20S)/[C_{26} TA (20S) + C_{28} TA(20S)]	
Monoaromatic Steranes	C_{27}–C_{28}–C_{29} monoaromatic steranes (MA) distribution.	

*The diagnostic ratios in Table 17.2.1 consist of alkanes, terpanes, steranes, sesquiterpanes, and diamondoids. Ratios are defined from Equation (17.2.1) for simplicity, but can be readily redefined using Equations (17.2.2) and (17.2.3). For example, in accordance with Equation (17.2.2), the ratio of C_{29} $\alpha\beta$-30-norhopane/C_{30} $\alpha\beta$ hopane (H29/H30) can be readily redefined as H29/(H29+H30).

depositional environment and maturity. The ratio was first used by Kvenvolden *et al.* (1985) to study a North Slope crude, in which the ratio was ~2. The spilled Exxon Valdez oil (an Alaska North Slope crude) and its residues also have triplet ratios of ~2. Conversely, many tarballs and residues collected from the shorelines of the PWS Sound were similar to each other but chemically distinct from the spilled Exxon-Valdez oil with triplet ratios of ~5. The triplet ratio, combined with other diagnostic biomarker ratios and isotopic compositions, revealed that these non-Valdez tarballs originated from California with a likely source being the Monterey Formation (Kvenvolden *et al.*, 1995).

During the Arrow oil spill work, the ratio of the most abundant C_{29} and C_{30} hopane was defined and used as a reliable source indicator (Wang *et al.*, 1994). Similar approaches, combined with determinations of a number of other "source-specific marker" ratios, were applied to characterize oil samples from the Arctic Baffin Island spill (Wang *et al.*, 1995a), oil on birds (Wang *et al.*, 1997), the 25-year-old wetland Nipisi spills (Wang *et al.*, 1998a), a mystery spill in Quebec (Wang *et al.*, 2001a), and the Detroit River oil spill (Wang *et al.*, 2004).

A spill case concerning identification and linkage of tarballs from the coasts of Vancouver Island (British Columbia) and Northern California is described in more detail in the following to illustrate the applications, usefulness, and limitation of diagnostic biomarker parameters.

During January and February 1996, a significant number of tarball/patty incidents occurred along the coasts of Vancouver Island, Washington, Oregon, and California. Samples of the tarballs were collected from the affected beaches and characterized by GC-FID and GC-MS using a multi-criterion analytical approach (Wang *et al.*, 1998b). Selected samples were further analyzed using a carbon isotopic technique.

The distribution pattern and profile of biomarker terpanes and steranes are identical for BC-1 and BC-2. Also, the concentrations of the most abundant biomarker compounds are very close to each other for these two samples. The CA-1 sample has a very similar distribution pattern of biomarkers to the BC samples; however, the concentrations of all target biomarkers are slightly higher than those of the BC samples. However, the biomarker distribution of

the tarball samples and the suspected source Alaska North Slope (ANS) crude are significantly different.

Diagnostic ratios of biomarker terpanes C_{23}/C_{24}, Ts/Tm, C_{29}/C_{30}, $C_{32}(22S)/C_{32}(22R)$, and $C_{33}(22S)/C_{33}(22R)$ for the BC and CA samples are similar, but the CA-1 sample has noticeably lower ratios of C_{23}/C_{30} and C_{24}/C_{30} than the corresponding BC samples (Table 17.2.2). The difference in the ratios of target biomarkers implies that the CA samples were more heavily weathered (which leads to some depletion of smaller biomarkers C_{23} and C_{24}). These indications do no defensively settle that the CA samples are from another source than the BC samples.

However, diagnostic ratios of PAHs, in particular the diagnostic ratios of selected paired PAH isomers within the same alkylation groups, are nearly identical for BC-1 and BC-2, but strikingly different from the CA-1 sample (Table 17.2.2). Together with evidence from GC-FID analysis and carbon-isotope compositions, it could be concluded that:

1. California/Oregon samples were chemically similar and consistent with the same source. They were identified to be bunker type fuel.
2. Tarball samples collected from British Columbia and Ocean Shores, Washington, were chemically similar and consistent with the same source (also bunker type fuel). The latter samples were similar to the California/Oregon samples but may have a source different than the California/Oregon samples.
3. The source of the tarball/patty samples was neither ANS oil nor California Monterrey Miocene oil.
4. The spilled oil samples have been highly weathered since release and the California tarball samples were more heavily weathered than the British Columbia tarball samples.

During the application of diagnostic ratios of biomarkers for spill studies, it is important to acknowledge that regardless of diagnostic parameters used, a basic rule applied to all correlations and differentiations should be:

• No matching of biomarker distribution and/or diagnostic ratios is strong evidence for lack of a correlation between spill sample(s) and suspected source(s).

Table 17.2.2 *Comparison of Diagnostic Ratios of Target Biomarker and PAH Compounds for BC and CA Tarball Samples*

Samples	BC-1	BC-2	CA-1	ANS
Aliphatics				
C17/pristane	3.70	3.82	/*	2.39
C18/phytane	1.87	1.93	/	1.82
Pristane/phytane	0.51	0.84	/	0.98
Biomarkers				
C_{23}/C_{24}	2.15	2.19	2.07	1.69
$C_{29}\alpha\beta/C_{30}\alpha\beta$	0.84	0.84	0.84	0.61
Ts/Tm	0.31	0.33	0.29	0.61
$C_{32}(S)/C_{32}(R)$	1.46	1.49	1.52	1.46
$C_{33}(S)/C_{33}(R)$	1.56	1.56	1.52	1.44
$C_{27}\,\alpha\beta\beta/C_{29}\,\alpha\beta\beta$	1.11	1.09	1.14	0.84
Target PAHs				
3 C_3N isomers	4.37:3.42:1.00	4.46:3.50:1.00	3.17:2.35:1.00	4.58:3.16:1.00
(3-+2-m-P)/(4-/9-+1-m-P)	1.37	1.42	0.89	0.74
2 C_2P isomers	3.11	3.00	4.85	4.10
2 C_4P isomers	1.43	1.48	0.95	0.61
3 C_1F isomers	1.00:1.02:0.43	1.00:1.00:0.42	1.00:1.42:0.53	1.00:1.94:0.42
4-:2/3-:1-m-DBT	1.00:0.92:0.55	1.00:0.93:0.53	1.00:0.92:0.60	1.00:0.64:0.32

*The *n*-alkanes and isoprenoids were completely lost for CA-1.

- Matching may be an indication of correlation of spill sample(s) and suspected source(s), but is not necessarily a "proof" for identity.

Hence, in order to make more reliable and defensible correlations, the use of a "multi-criteria approach" is often a prerequisite. In a multi-criteria approach, the final conclusion is based on analysis and evaluation of the distribution of more than one suite of petroleum compounds (e.g., terpanes, steranes, PAHs, and alkanes) (Peters and Moldowan, 1993; Daling *et al.*, 2002; Stout *et al.*, 2002; Christensen *et al.*, 2004; Wang *et al.*, 1999a, 2004).

17.2.3 Cross-Plots of Biomarkers

Cross-plots (i.e., plot of one diagnostic biomarker ratio vs another ratio) are frequently used in oil geochemistry for oil-oil correlation and determination of oil source and depositional environment.

Cross-plots have, for example, been used extensively by Li *et al.* in the project of Bakken/Madison petroleum systems in the Canadian Williston Basin (Jiang *et al.*, 2001; Li and Jiang, 2001; Jiang and Li, 2002) for oil-source rock correlations and for distinguishing two source rocks with distinct distribution patterns of biomarkers.

Cross-plots are more sensitive to lithological changes than single diagnostic ratios. Gürgey analyzed 56 rock and 28 crude oil samples from the sub-salt and supra-salt section of the southern Pre-Caspian Basin. Source-specific biomarker ratios reveal two populations: Population 1 and Population 2. Population 1 oils are subgrouped into two populations: Population 1A and 1B (Gürgey, 2002). Based on plots of C_{24} tetracyclic/C_{26} tricyclic terpanes vs C_{29}/C_{30} hopane, the author illustrates a clear lithological separations between clay-rich source for oil Population 1A and 2, and a carbonate-rich source for oil Population 1B (Figure 17.2.2).

Cross-plots of C_{29} $\alpha\beta\beta/(\alpha\beta\beta + \alpha\alpha\alpha)$ sterane vs C_{29} $20S/(20S + 20R)$ steranes are particularly effective in describing the thermal maturity of source rocks or oil (Seifeit and Moldowan, 1986). The plots can be used to cross-check one maturity parameter vs another. Isomerization at the C-14 and C-17 positions in the 20S and 20R C_{29} steranes causes an increase in the ratio of $\alpha\beta\beta/(\alpha\beta\beta + \alpha\alpha)$ from small nonzero values to the equilibrium values (about 0.66–0.71). This ratio appears to be independent of source organic input, thus making it effective to describe the thermal maturity of oil. This ratio, if approaching to the equilibrium values, suggest that the oils had reached peak oil window maturity.

Cross-plots of $C_{30}/(C_{27}$ to $C_{30})$ steranes vs oleanane/C_{30} $\alpha\beta$ hopane give better assessment of marine vs terrestrial input to petroleum than either biomarker ratio parameter alone. Based on the relative abundances of marine (C_{30} 24-propylcholestanes) and terrestrial (oleanane) markers (i.e., based on plots of C_{30} sterane/(C_{27} to C_{30}) steranes (%) vs oleanane/C_{30} $\alpha\beta$ hopane (%)), oils and seep oils from Columbia are categorized into five groups. Oil groups IIA, IIB, and IIC have relatively high oleanane and low C_{30}-sterane levels suggesting a deltaic source with strong terrestrial input; while oil group IA has lower oleanane and higher C_{30} sterane levels, like oils from Venezuela, suggesting a marine source which is more remote from terrestrial influences (Peters and Moldowan, 1993). Based on a plot of C_{30} sterane/(C_{27} to C_{30}) steranes vs C_{34} or $C_{35}/(C_{31}$ to $C_{35})$ homohopanes, Moldowan *et al.* (1992a) proposed that oils derived from source rocks deposited under restricted saline to hypersaline lagoonal conditions have lower $C_{30}/(C_{27}$ to $C_{30})$ sterane ratios than those from open marine systems.

Based on cross-plots of relative abundance (at *m/z* 123) of various isomeric sesquiterpanes (C_{15} and C_{16} drimanes) vs relative abundances of bicadinanes to C_{30} hopane on the *m/z* 412 mass chromatogram (bicadinane-T/C_{30}-hopane), crude oils from the eastern Pearl River Mouth Basin were classified into two groups (Zhang *et al.*, 2003).

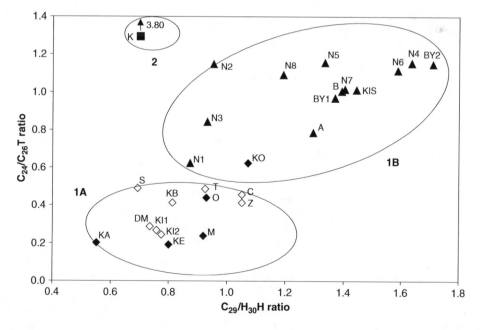

Figure 17.2.2 *Cross-plot of the $C_{29}H/C_{30}H$ vs $C_{24}/C_{26}T$ hopane ratios for oils from the southern margin of the Pre-Caspian Basin. Oil populations: ♦, Population 1A; ◊, Biodegraded Population 1A; ▲, Population 1B; ■, Population (from Gürgey, 2002) 2.*

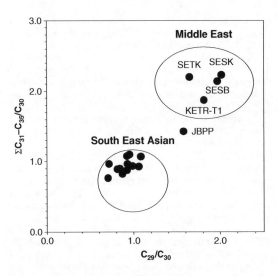

Figure 17.2.3 *Application of cross-plots of C_{29} $\alpha\beta/C_{30}$ $\alpha\beta$ hopane ratio vs the homohopane index $\Sigma(C_{31}-C_{35})/C_{30}$ hopane as key biomarker indicators to distinguish large number of tarball samples originated from southeast Asian crude oil sources from those of Middle East sources (from Zakaria et al., 2001).*

Malaysian coasts are subjected to various threats of petroleum pollution including deliberate and accidental oil spills from various sources. The identification of detailed sources of the oil pollution, therefore, is essential to reduce the oil pollution through effective regulation. Based on chemical evidences that Middle East crude oils were characterized by C_{29} 17α, 21β-norhopane and $C_{31}-C_{35}$ homohopanes, whereas these compounds were deplete in southeast Asian crude oils, Zakaria et al. proposed utility of the cross-plots of C_{29} $\alpha\beta/C_{30}$ $\alpha\beta$ hopane ratio vs the homohopane index $\Sigma(C_{31}-C_{35})/C_{30}$ hopane (Figure 17.2.3) as key biomarker indicators and successfully distinguished large number of tarball samples originated from southeast Asian crude oil sources from those of Middle East sources (Zakaria et al., 2000, 2001). Application of the source-specific biomarkers to sediment and mussel samples showed the Middle East oil signature.

17.3   UNIQUE BIOMARKER COMPOUNDS

Biomarker terpanes and steranes are common constituents of crude oils. However, some biomarker compounds including several geologically rare acyclic alkanes are found to exist only in certain oils and can, therefore, be used as unique markers for oil spill identification. These ratios can furthermore provide additional diagnostic information on the types of organic matter that give rise to the crude oil. Some unique biomarkers, which have been used to identify new class of oils by geochemists and environmental chemists, are summarized below:

Botryococcane $(C_{34}H_{70})$

Botryococcane is an irregular C_{34}-isoprenoid compound $(C_{34}H_{70})$, which is derived from botryococcene known in only one living organism, *Botryococcus braunii*, fresh to brackish water lacustrine green alga. Botryococcane eluates immediately before n-C_{29} in the oil saturate fraction and is measured using m/z 183 chromatogram. The occurrence of botryococcane in petroleum is limited to a few geographic regions, including Australia (McKirdy et al.,

1986) and Sumatra (Seifert and Moldowan, 1981). Therefore, characterization of geologically rare botryococcane can be highly specific. McKirdy et al. (1986) identified botryococcane in a new class of Australian non-marine crude oils and detected exceptionally high concentrations of botryococcane (C_{34} botr/n-C_{29} = 0.3-0.7) in coastal bitumens from southern Australia (McKirdy et al., 1994). The presence of this biomarker in the Australian coastal bitumen suggested that the parent oils were generated from lacustrine source beds under anoxic to suboxic conditions. Hwang et al. (2002) used botryococcane as an effective key source indicator to genetically separate the highly similar Sumatra oils into five groups. On the contrary, commonly used source indicators including homologous distribution of steranes, diasteranes, and hopanes are unable to genetically group most of the studied Sumatra oils.

Methyl-hopane $(CH_3-C_{30}H_{51})$

Rings A and B methyl-hopane were first detected in a series of Jurassic oils from the Middle East (Seifert and Moldowan, 1978). The ring A/B fragment of the hopane series, normally m/z 191, is increased to m/z 205 by the additional methyl group. The most prominent series of these compounds identified in Precambrian oils has been identified as 2α-methyl-17α(H), 21β(H)-hopane (Summons and Jahnke, 1992). Based on the fact that methyl-hopanes are major polycyclic biomarkers in some Middle East oils from carbonate source but are uncommon in Australian oils, Volkman et al. (1992) identified that lubricating oils, which contaminated many estuaries and coastal areas around Australia, originated from Middle East crudes.

β-Carotane $(C_{40}H_{78})$

This compound is a fully saturated C_{40}-dicyclane. It elutes after C_{35} homohopane in the oil saturate fraction. It is highly specific for anoxic, saline, lacustrine deposition of algal organic matter. It is measured at fragment m/z 125 and/or at molecular ion m/z 558. β-carotane has been detected in several Chinese oils and the Mississippian Alberta shale. The presence of a significant amount of β-carotane and gammacerane relative to the hopanes has recently been detected, suggesting that the source rocks of the oil from the Liaohe Basin of China were probably deposited in a highly stratified, strongly reducing environment (Wang et al., 1996).

Extended Hopanes beyond C_{40}

Hopanes with 30 or fewer carbon atoms are often interpreted as diagenetic products of C_{30} hopanoids, while the extended hopanes ($C_{31}-C_{35}$) have been related to C_{35} precursors, such as bacteriohopane polyols, aminopolyols, and a number of composite hopanoids (Ourisson and Albrecht, 1992; Rohmer et al., 1992). A series of side-chain extended 17α(H), 21β(H)-hopanes and 17β(H), 21α(H)-hopanes (or moretanes) up to C_{44} has been identified in crude oils and source rock extracts in the Liaohe Basin, Northeast China (Wang, et al., 1996). These compounds may be viewed as the representatives of a new class of molecules whose natural distribution and biological significance remains to be determined.

Irregular Isoprenoids

Long-chain isoprenoids are common constituents in many oils, for example isoprenoids extended to at least C_{42} and possibly higher have been identified in most Taiwan oils (Oung and Philp, 1994). However, irregular isoprenoids only exist in certain oils. The irregular isoprenoid compound crocetane (2,6,11,15-tetramethyl-hexadecane) is a C_{20} tail-to-tail isoprenoid identified as a major component in the hydrocarbon fractions of sediments derived from recent and ancient methane vent ecosystem and methane-rich mud volcanic sediments. It has been identified (based on the GC-MS chromatograms at m/z 183) in a series of

West Australian crude oils from Canning and Perth Basins (Barber *et al.*, 2001). The structurally related C_{25} irregular isoprenoid 2,6,10,15,19-pentamethylicosane (PMI) was found to co-occur with crocetane in the same samples. In addition, a newly observed series of C_{22}–C_{24} irregular isoprenoids possessing the same carbon skeleton have also been tentatively identified in these crude oils. This may indicate a source for crocetane from the diagenesis of 2,6,10,15,19-pentamethylicosane. This is the first reported occurrence of isoprenoid biomarkers attributable to methanogenic archaea in crude oils.

Bicadinanes ($C_{30}H_{52}$)

Bicadinanes are C_{30}-pentacyclic biomarker compounds and have three configurations labeled as W (cis-cis-trans-bicadinane), T (trans-trans-trans-bicadinane), and R (cadinane). The mass spectra of bicadinane contain prominent m/z 191 and 217 fragments, while peaks can appear in corresponding chromatograms of both hopanes and steranes. All three bicadinanes (W, T, and R form) elute prior to C_{29} hopane in the m/z 191 chromatogram. But it can be conveniently monitored with little interference using the m/z 412 mass chromatogram. They are believed to originate from the angiosperm Dammar resin and are thus highly specific for resinous input from higher plants (van Aarssen *et al.*, 1990). Crude oils from the North, Central, and South Sumatra basins, Indonesia, have been characterized and the biomarker compositions of the crude oils have been used to classify samples into three types (Sosrowidjojo *et al.*, 1994). In the study by Sosrowidjojo *et al.*, 1994, Group II oils were distinguished from Group 1 oils by their high abundance of bicadinanes relative to C_{30} hopane on the m/z 412 mass chromatogram.

C_{30} 17α(H)-Diahopane ($C_{30}H_{52}$)

Moldowan *et al.* (1991) identified two rearranged hopanes from Prudhoe Bay oil, C_{30} 17α(H)-diahopane (C_{30}^*) and 18α(H), 21β(H)-30-norneohopane (C_{29}Ts), by x-ray crystallography and advanced NMR methods. C_{30}^* elutes right after C_{29} αβ norhopane and C_{29} 18α-30-norneohopane in the m/z 191 mass chromatogram. The occurrence of C_{30}^* may be related to bacterial hopanoid precursors that have undergone oxidation and rearrangement by clay-mediated acidic catalysis under oxic to suboxic conditions. C_{30} 17α(H)-diahopane has been regarded as a possible terrestrial marker. Recently, El-Gayar *et al.* (2002) characterized seven oils representing the different petroleum-bearing basins in the Western Desert, Egypt. The characterization indicated that type 2 and type 3 oils are similar and show relative high pristane/phytane ratios, paucity of C_{30} steranes, and high relative abundance of C_{30} 17α(H)-diahopane, all suggesting that they probably originated from source rock containing significant proportions of higher plant material.

18α(H)-Oleanane

Oleanane has two isomers: 18α(H)-oleanane and 18β(H)-oleanane. The α type configuration has the highest thermodynamic stability, thus it is the predominant configuration in mature crude oils and rocks (Riva *et al.*, 1988). 18α(H)-Oleanane is a biomarker characteristic of angiosperms (flowering plants) found only in Tertiary and Cretaceous (<130 million years) rocks and oils. No examples of oleananes have been found in oils known to be older than Cretaceous. 18α(H)-Oleanane is highly specific for higher plant input. The presence of 18α(H)-oleanane in benthic sediments in PWS, combined with its absence in Alaska North Slope crude and specifically in Exxon Valdez oil and its residue, confirmed another petrogenic source (Kvenvolden *et al.*, 1993; Bence *et al.*, 1996; Page *et al.*, 1996). Characterization of 18α(H)-oleanane in oils from the Anaco area and Maturin sub-basin, Venezuela, has been used for organic type and age indicator for assessment of Venezuelan petroleum system (Alberdi and Lopez, 2000).

4-Methyl Steranes

The 4-methyl steranes can be divided into two major classes: (1) C_{28}–C_{30} analogues of the steranes at positions 4 and 24 (e.g., the C_{30} sterane is 4α-methyl-24-ethyl-cholestane) and (2) C_{30} dinosteranes (e.g., 4α,23, 24-trimethyl-cholestanes). 4α-Methyl sterane can be formed from dehydration and hydrogenation of 4α-methyl sterol in dinoflagellates, prymnesiophyte microalgae, or even in bacteria (Wolff *et al.*, 1986; Volkman *et al.*, 1990). 4α-Methyl-24-ethyl-cholestanes often occur in relatively high abundance in Tertiary source rocks and related oils from China (Fu *et al.*, 1992). For example, almost all of the oils from the eastern Pearl River Mouth Basin contain significant amounts of 4-methyl-steranes (Zhang *et al.*, 2003). Similar to C_{29} steranes, C_{30} 4-methyl-sterane has four types of epimers detectable including ααα (20S), αββ (20R), αββ (20S), and ααα (20R). In general, the αββ type has a much greater abundance than the ααα type, the former has the base peak at m/z 232 and the latter at m/z 231.

Hu (1991) found that 4-methyl steranes (in the range of C_{28}–C_{30}) are unusually rich (which comes up to 20–40% of the total steranes) in certain oils from terrestrial facies within the South China Sea. C_{30} 4-methyl-sterane ($M^+ = 414$) is particularly abundant among the 4-methyl-steranes. The extremely high peak of 4-methyl-sterane is an important characteristic of both terrestrial crude and Eocene source rock in the South China Sea, suggesting that the former originated from the later.

Rogers *et al.* (1999) has investigated oil seeps and stains from the East Coast Basin, New Zealand. They found that oils sampled from Hawkes Bay and Wairarapa have higher abundances of C_{30} regular steranes and 28,30-bisnorhopane than oil seeps from the Raukumara Peninsula and Marlborough. These biomarker analysis results have been used to distinguish at least two distinct sources of hydrocarbons in the basin.

Dinosterane has only been reported in petroleum younger than Triassic age (Summons *et al.*, 1992). The presence of dinosterane in relatively high concentrations in asphaltic bitumens from southern Australia (McKirdy *et al.*, 1994) suggests that their source is no older than mid-Triassic.

In a study to re-evaluate the petroleum prospective potential in southeast Australia, Volkman *et al.* (1992b) examined ten bitumen samples collected between 1880 and 1915. The high proportions of C_{27} steranes and presence of C_{30} steranes including dinosteranes suggested that the bitumens were derived from a marine source rock containing mainly marine organic matter. The unusual absence of 2α-methylhopanes and low abundance of tricyclic terpanes ruled out carbonates or the tasmanite oil shales respectively as sources.

Macrocyclic Alkanes

Most of the biomarkers in oil are polycyclic and based commonly on five- or six-carbon membered rings. However, macrocyclic hydrocarbon ring systems not based on isoprene (i.e., non-isoprenoidal) are much less commonly reported. In 1994 Muurisepp *et al.* first reported the presence of non-isoprenoidal macrocyclic-alkanes in sedimentary material and tentatively identified these cyclic hydrocarbons of the cyclododecane and cyclohexadecane series in the non-aromatic hydrocarbon fractions of the semi-coking oil from an Estonian oil shale (Muurisepp *et al.*, 1994).

In 2001, Audino *et al.* have unambiguously identified for the first time a new class of cyclic hydrocarbon biomarker, macrocyclic-alkanes and their methylated analogues in a *Botryococcus braunii* rich sediment (torbanite) of Late Carboniferous age and in two Indonesian crude oils (Minas and

Duri, Tertiary). The compounds consist of a homologous series of macrocyclic-alkanes in a wide range from C_{15} to C_{34} and their methylated derivatives (ranging from C_{17} to C_{26}). The distribution of macrocyclic-alkanes was measured at the characteristic ion m/z 111. The macrocyclic alkanes appear to be novel markers of *B. braunii* and add to the catalogue of the characteristic hydrocarbons derived from this alga. More importantly, these compounds could be original marker specific to highly resistant algaenan of *B. braunii* in sediments and crude oils, unlike *n*-alkanes, which can have multiple origins (Figure 17.3.1).

The search for unique geochemical biomarkers continues to be a fertile area of research for fingerprinting similar sources of petroleum. Note, however, that reliable biomarker interpretation is usually based on a whole biomarker distribution chromatogram and a series of biomarker parameters. There is no single parameter which can be exclusively used for unambiguous source identification of unknown spills. Individual unique biomarker parameters become valuable and meaningful only when they are used together and agree with other biomarker parameters.

17.4 WEATHERING EFFECTS ON OIL AND BIOMARKER DISTRIBUTION

When oil is spilled, whether on water or land, a number of processes occurs and they can be referred to as the *behaviour* of oil. In general, these processes can be categorized into two groups: the first group of processes is related to the "movement" of oil in the environment (such as spreading, movement of oil slicks, sinking, and over-washing) and the second group of processes changes the oil composition, which is termed as "weathering processes." Weathering and movement processes can overlap, with weathering strongly influencing how oil is moved in the environment and vice versa. These processes highly depend on the type of oil spilled and the weather conditions during and after the spill (e.g., temperature, wave movements, and sun-incidence).

Weathering is defined as the combination of a wide variety of physical, chemical, and biological processes which affect the composition of spilled oil in the environment (Jordan and Payne, 1980; Wang and Fingas, 1995b). Weathering processes include (1) evaporation, (2) emulsification, (3) natural dispersion, (4) dissolution, (5) microbial degradation, (6) photooxidation, and (7) other processes such as sedimentation, adhesion onto surface of suspended particulate materials, and oil-fine interaction. Weathering can be further divided into three types:

1. Physical processes (evaporation, emulsification, natural dispersion, dissolution, and other processes such as sedimentation, adhesion onto surface of suspended particulate materials, and oil-fine interaction)
2. Chemical processes (e.g., photooxidation)
3. Biological processes (microbial degradation)

While the physical processes result in redistribution of oil components in different compartments of the environment, the photooxidation and microbial processes lead to chemical transformation and degradation of petroleum hydrocarbons, respectively.

In the following sections, the weathering processes most affecting oil composition after an oil spill are briefly described.

Evaporation

In the short term after an oil spill, evaporation is usually the single most important and dominant weathering process, in particular for the light petroleum products such as gasoline. Evaporation has the greatest effect on the amount of oil remaining on water or land after a spill. In the first few days following a spill, the loss can be up to 70 and 40% of the volume of light crudes and petroleum products, and gasoline can evaporate completely above 0 °C. For heavy or residual oils such as Bunker C oil, the losses due to evaporation comprise only a few percentages of the total volume. The rate at which oil evaporates depends primarily on the oil composition. The more volatile components an oil or fuel contains, the greater the extent and rate of its evaporation.

The rate of evaporation is rapid immediately after an oil spill and then slows considerably. The properties of oil can change significantly with the extent of evaporation. If about 40% (by weight) of an oil evaporates, its viscosity could increase by as much as thousand folds, its density could rise by as much as 10%, and its flash point by as much as 400%. The extent of evaporation is often the most important factor for determining oil properties at a given time after the spill and for changing the behavior of the oil.

Emulsification

Emulsification is the process by which water is dispersed into oil in the form of small droplets. Water droplets can remain in an oil layer in a stable form and the properties of the emulsified oil is very different from the starting oil. The mechanism of water-in-oil emulsion formation is not yet fully understood, but most likely it starts with sea energy forcing the entry of small water droplets, about 10 to 25 μm in size, into the oil. Emulsions contain about 70% water and thus when emulsions are formed, the volume of spilled oil triples. In general, water can be present in oil in four ways (Fingas and Fieldhouse, 2003):

1. *Soluble*: Some oils contain about 1% or less water as soluble water and this water does not significantly change the physical properties and chemical composition of the oil.
2. *Unstable emulsion*: Water droplets are "entrained" in the oil by its viscosity to form an unstable emulsion. In the marine environment, unstable emulsions break down into water and oil within minutes or a few hours once the sea energy diminishes.
3. *Semi- or meso-stable emulsion*: These are formed when the small droplets of water are stabilized to a certain extent by a combination of the viscosity of the oil and the interfacial action of asphaltenes and resins. For this to happen, the asphaltene or resin content of the oil must be at least 3% by weight. The viscosity of meso-stable emulsions is 20 to 80 times higher than that of the starting oil. These emulsions generally break down into oil and water or sometimes into water, oil, and stable emulsion within a few days.
4. *Stable emulsions*: They form in a way similar to meso-stable emulsion except that the oil must contain at least 8% of asphaltenes. The viscosity of stable emulsion is 500 to 1200 times higher than the viscosity of the starting oil and the emulsion will stay stable for weeks and even months after formation. Stable emulsions are reddish-brown in color and appear to be nearly solid. These emulsions do not spread and tend to remain in lumps or mats on the sea or shore.

The formation of emulsions is an important event following an oil spill. It substantially increases the actual volume and viscosity of the spilled oil, making cleanup operations more difficult. Emulsion formation also changes the fate of the oil. It has been noted that when oil forms stable or meso-stable emulsions the rate of evaporation slows down considerably. Microbial degradation also appears to slow down. The dissolution of soluble components from oil may also cease once emulsification has occurred.

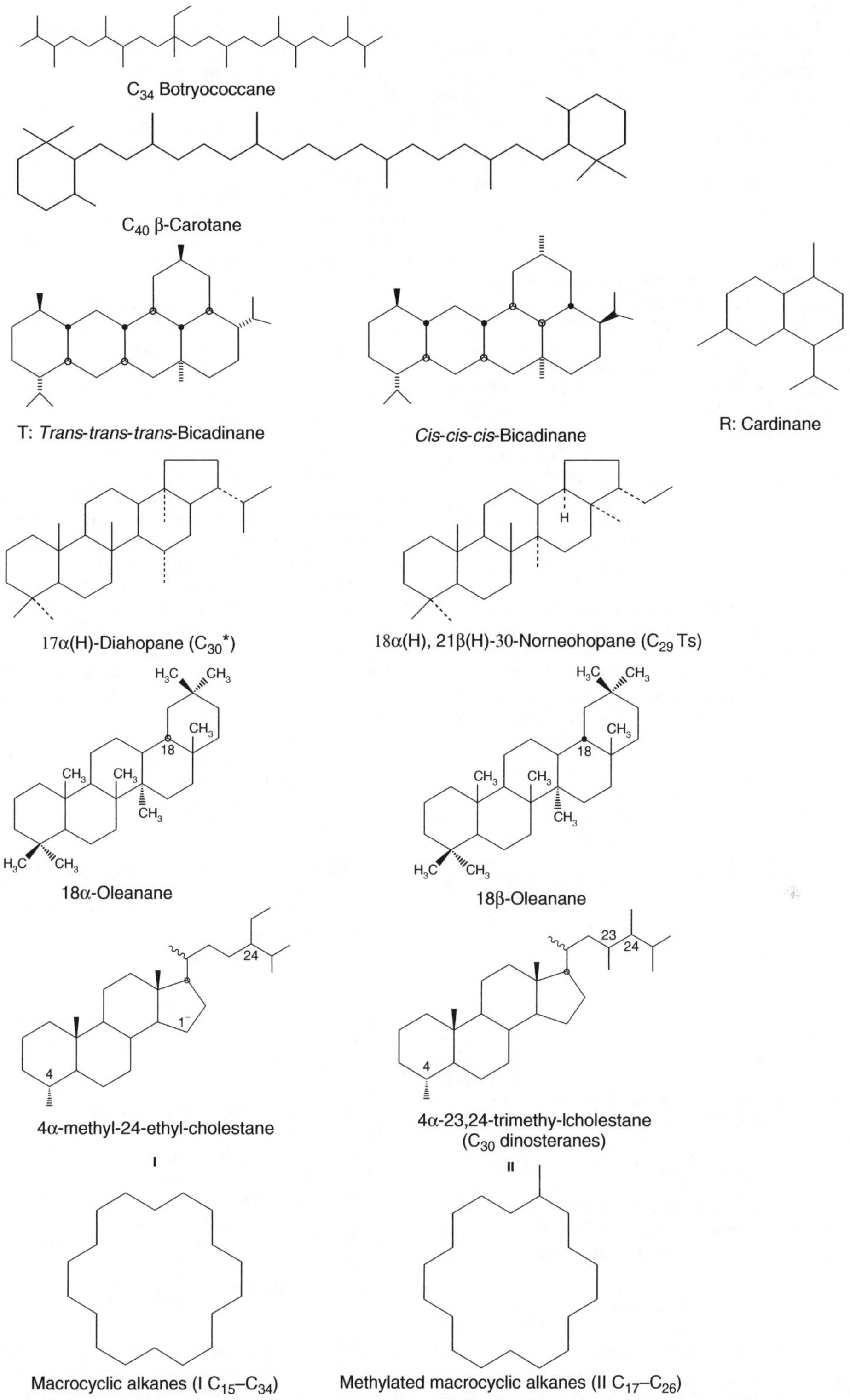

C$_{34}$ Botryococcane

C$_{40}$ β-Carotane

T: *Trans-trans-trans*-Bicadinane

Cis-cis-cis-Bicadinane

R: Cardinane

17α(H)-Diahopane (C$_{30}$*)

18α(H), 21β(H)-30-Norneohopane (C$_{29}$ Ts)

18α-Oleanane

18β-Oleanane

4α-methyl-24-ethyl-cholestane

I

4α-23,24-trimethy-lcholestane
(C$_{30}$ dinosteranes)

II

Macrocyclic alkanes (I C$_{15}$–C$_{34}$)

Methylated macrocyclic alkanes (II C$_{17}$–C$_{26}$)

Figure 17.3.1 *Molecular structures of a selection of unique biomarkers: Botryococcane (C$_{34}$H$_{70}$), β-Carotane (C$_{40}$H$_{78}$), and bicadinanes, C$_{30}$ 17α(H)-Diahopane, 18α(H),21β(H)-30-norneohopane (C$_{29}$Ts), 18α(H)-oleanane, 18β(H)-oleanane, 4α-methyl-24-ethyl-cholestane, C$_{30}$ dinosteranes (4α,23,24-trimethyl-cholestanes), macrocyclic alkanes, and methylated macrocyclic alkane.*

Natural Dispersion

Natural dispersion occurs when fine droplets of oil are transferred into the water column by wave action or turbulence. Small droplets ($<20\,\mu m$) are relatively stable in water and will remain so for long periods of time. Large droplets tend to rise and droplets larger than $100\,\mu m$ will not stay in the water column for more than a few seconds. The effect of natural dispersion varies depending on oil type and weather conditions and can thus have an insignificant effect or remove the bulk of the oil.

Dissolution

Solubility is defined as the amount of a substance (solute) that dissolves in a given amount of another substance (solvent). Through the process of dissolution, petroleum compounds are transferred to the water column depending, for example, on their solubilities. The amount of an individual compound dissolving in the water phase from oil slicks in a given time largely depend on kinetic and equilibrium conditions affected by, for example, molecular structure and polarity. In general (1) the aromatic hydrocarbons are more soluble than aliphatic hydrocarbons, (2) the solubility increases as the degree of alkylation of benzenes and PAHs decreases, (3) the lower molecular weight hydrocarbons are more soluble than the high molecular weight hydrocarbons in each class of petroleum compounds, and (4) the more polar S, N, and O-containing compounds are more soluble than hydrocarbons. Hence BTEX, lighter alkyl-benzene compounds, and PAHs with few rings such as naphthalene are particularly susceptible to dissolution or *waterwashing*. Dissolution is an important weathering process, which affects the bioavailability of individual compounds. Consequently, the soluble fraction is more assessable to microbes for degradation and has an acute toxicity to fish and other aquatic life.

Biodegradation

Biodegradation of hydrocarbons by natural populations of microorganisms represents the primary mechanisms by which petroleum and other hydrocarbon pollutants are eliminated from the environment (Leahy and Colwell, 1990; Prince, 1993). Biodegradation of petroleum in the environment is generally a long-term and complex process, whose quantitative and qualitative aspects depend on the composition of the microbial community (e.g., indigenous bacteria and other microorganisms are often the best adapted and more effective at degrading oil as they are acclimatized to the temperature and other conditions of the area): the type, nature, and amount of oil, and the ambient and seasonal environmental conditions (such as temperature, oxygen, nutrients, salinity, and pH). Petroleum hydrocarbons differ in their susceptibility to microbial attack. Transformations of petroleum hydrocarbons by biodegradation occur stepwise, producing alcohols, phenols, aldehydes, and carboxylic acids in sequence by phase 1 and phase 2 metabolic pathways. The compounds can eventually be completely metabolized to carbon dioxide and water or the polar metabolites may be spread to the surrounding water or accumulated in the residual oil.

Photooxidation

Photooxidation is a potentially significant process in degradation of crude oil spilled at sea, but the effects of photooxidation on the oil composition following oil spills are not yet well understood. In general, photooxidation is considered to be a factor involved in the transformation of crude oil or its products released into the marine environment (Garrett *et al.*, 1998). The photooxidation is dependent on the thickness of the oil slick as well as sun-incidence. The photochemical degradation yields a variety of oxidized compounds including alcohols, aldehydes, ketones, and acids,

which are more soluble in water than the starting compounds. Photodegradation affects the oil composition, different than is the case for microbial degradation and can hence complicate the observed weathering patterns for spills in areas with large sun incidence.

Sedimentation and Oil-mineral Aggregation

Sedimentation is the process by which oil is deposited on the bottom of the sea or other water body. Once oil is on the bottom, it is usually covered by additional sedimentation and degraded very slowly. Oil-mineral aggregates (OMA) result from interactions among the oil residues, fine mineral particle, and seawater. The OMA formation has been identified as an important process that facilitates the natural removal of oil stranded in coastal sediments (Bragg and Owens, 1994; Owens and Lee, 2003). The OMA formation is enhanced by physical processes such as wave, energy, tides, or currents. It has recently been noted that oil biodegradation may be enhanced by OMA formation.

17.4.1 Chemical Composition Changes of Oils due to Weathering

Weathering causes considerable changes in the physical properties and the chemical composition of spilled oil. The "degree" (lightly, moderately, or severely weathered) and *rate* of weathering is different for each spill and is controlled by a number of spill conditions and natural processes such as: (1) type of the spilled oil and the chemical composition and concentrations of oil components (heavy oils are generally weathered at a much slower rate than lighter oils); (2) spill site and environmental conditions (such as temperature, pH, water level, salinity, soil type, air, and nutrients); and (3) natural population of indigenous bacteria. Oil samples affected by a combination of physical, chemical, and biological processes show complex weathering patterns. Weathered oil samples can generally be divided into three groups based on degree of changes in their chemical composition:

1. Lightly weathered oils ($<15\%$ naturally weathered): the abundances of low-end *n*-alkanes are significantly reduced. Furthermore, BTEX and C_3-benzene compounds are lost.

2. Moderately weathered oils (~15–30% weathered): a significant loss of *n*-alkanes and low-molecular-weight isoprenoid compounds has occurred. BTEX and C_3-benzenes could be completely lost (Wang *et al.*, 1995c) and the loss of lighter C_0 and C_1- naphthalenes can be significant. The ratio of GC-resolved peaks to UCM is decreased considerable due to the preferential loss of resolved hydrocarbons over the unresolved complex hydrocarbons.

3. Severely weathered oils: not only *n*-alkanes but also branched and cyclo-alkanes are heavily or completely lost and the UCM is pronounced. The BTEX and alkyl benzene compounds are completely lost. PAHs and their alkylated homologous series could also be highly degraded, resulting in development of a profile in each alkylated PAH family with the distribution of C_0– $<C_1$– $< C_2$– $<C_3$–.

Hence, it is difficult and often an impossible task to identify severely weathered oil samples through recognition of *n*-alkane and PAH distribution patterns. However, the biomarker distribution patterns are often unaltered even in severely weathered oil samples. Thus, fingerprinting terpane and sterane biomarkers provide a powerful tool for tracking the source and correlation and differentiation of weathered oils.

17.4.2 Effects of Physical Weathering on Oil Chemical Composition

The chemical composition changes of numerous oils and petroleum products with various weathering percentages by evaporation have been thoroughly studied as part of the Environment Canada Oil Spill Research Program. The studied oils include the Alberta Sweet Mix Blend (ASMB) oil (0 to 45% weathered) (Wang *et al.*, 1995b); 19 EPA oils (Wang *et al.*, 2002; EPA, 2003), and six US Mineral Management Service (MMS) oils.

Figure 17.4.1 compares *n*-alkane and PAH distribution for the ASMB oil at weathered percentage of 0 and 44.5% to illustrate the effects of evaporative weathering on oil chemical composition changes. The loss of lower molecular weight *n*-alkanes and PAHs and alterations in the relative distributions of target *n*-alkanes and PAHs are apparent. Conversely, no evaporative losses were observed for biomarkers. As an

example, Figures 17.4.2a and 17.4.2b show the SIM chromatograms of the terpane (at *m/z* 191) and sterane (at *m/z* 217 and 218) distribution for the 0 and 31% weathered ANS oil samples. Figure 17.4.3 compares the chromatograms at *m/z* 231 for triaromatic (TA) steranes for the heavy Platform Elly oil samples with evaporative weathering percentages at 0 and 13.3%, respectively.

The main chemical composition changes caused by evaporative weathering include (Wang *et al.*, 1995a):

- Loss of low-molecular-weight (MW) *n*-alkanes and relative increase in the concentration of less volatile high MW *n*-alkanes.
- n-C_{17}/pristane, n-C_{18}/phytane, and pristane/phytane are virtually unaltered because these compounds have about the same volatility.

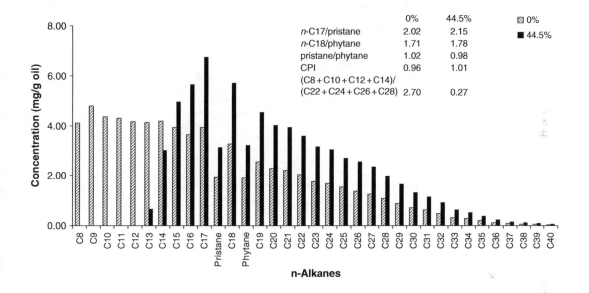

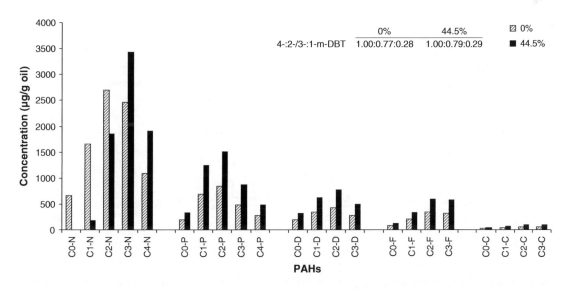

Figure 17.4.1 *Comparison of* n-*alkane and PAH distribution for the ASMB oils at 0 and 44.5% weathered percentages.*

- The values of Carbon Preference Index (CPI, the total amount of n-alkanes with odd carbon number divided by the total amount of n-alkanes with even carbon number in the carbon range of C_8–C_{40}), are virtually unchanged.
- The abundances of BTEX, C_3-benzenes, and two-ring naphthalene are reduced until completely lost. Conversely, the high-ring number alkylated PAHs are gradually concentrated.
- Physical weathering processes do not cause preferential loss of one isomer over another within alkylated PAH homologous series. Thus, excellent consistency of the relative ratios within the same PAH isomeric series can be demonstrated.
- Biomarker terpanes and steranes are not depleted during evaporative weathering. All target biomarker compounds in the range C_{19}–C_{35} are concentrated in proportion with the increase of the weathered percentages. Both terpanes and steranes show a great consistency in the relative ratios of paired biomarker compounds and biomarker compound classes.
- A great consistency in the distribution patterns of TA-steranes and MA-steranes is observed and diagnostic ratios of target TA-steranes and MA-steranes are unaltered (e.g., for heavy Platform Elly oil as the evaporative weathering percentages increased from 0 to 13.3%).

17.4.3 Effects of Biological Weathering on Oil Chemical Composition

Microbial degradation of petroleum hydrocarbons represents one of the main mechanisms by which oil and oil-related hydrocarbons are removed from contaminated sites. The degree and rate of biodegradation is different for each spill and is controlled by a greater number of variables than in physical weathering. These variables could include a number of spill conditions and natural processes such as type of the oil spilled, spill site, the ambient and seasonal environmental conditions (such as temperature, oxygen, nutrients, water activity, salinity, and pH), and composition and activities of the autochthonous microbial community. Several biodegradation studies reveal that the effects of microbial degradation on the oil composition are different from those of physical weathering processes. In general, biodegradation affects the oil composition in the following patterns (Leahy and Colwell, 1990; Prince, 1993; Wang *et al.*, 1998c):

- Smaller hydrocarbons are degraded faster than larger hydrocarbons
- Straight-chain n-alkanes degrades faster than branched alkanes
- GC-resolved compounds are degraded more than GC-unresolved complex hydrocarbons
- Small aromatics are degraded faster than high molecular weight aromatics
- Increase in alkylation level within their alkylated homologous families significantly decreases susceptibility to microbial attack
- Microbial degradation is often isomer specific. For example, 2-/3-methyl dibenzothiophene biodegrades at the fastest rate within its isomeric series. This preferential degradation effect is mirrored in the significant decrease

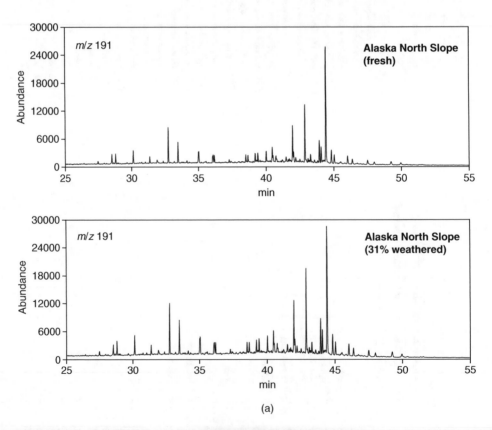

(a)

Figure 17.4.2 GC-MS chromatograms of (a) the terpane (at m/z 191) and (b) sterane (at m/z 217 and 218) distribution for the ANS oil samples with evaporative weathering percentages at 0 and 31%.

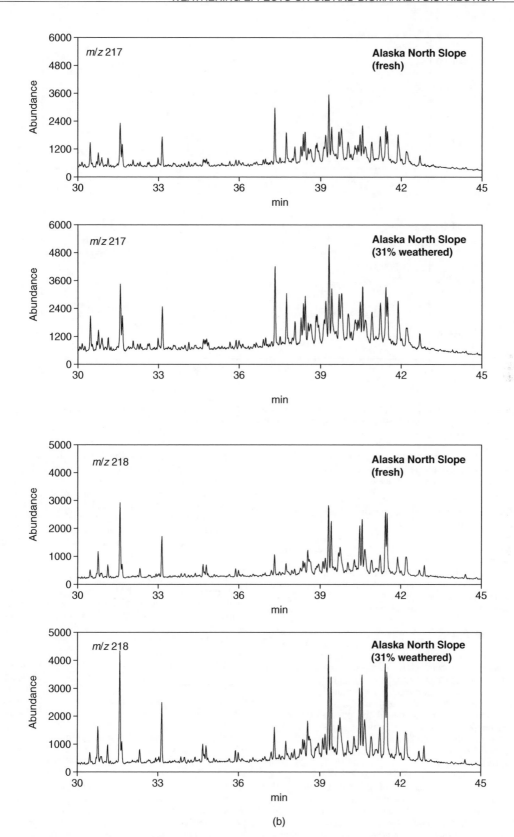

(b)

Figure 17.4.2 (Continued)

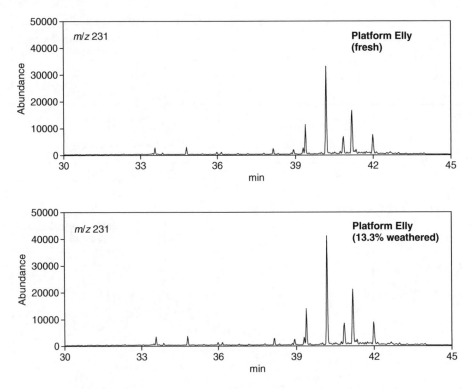

Figure 17.4.3 *Comparison of GC-MS chromatograms at m/z 231 for TA-sterane distribution for the heavy Platform Elly oil samples with evaporative weathering percentages at 0 and 13.3%.*

in the relative isomeric distribution of compounds such as $(3 + 2$-methyl phenanthrenes) and $(4$-$/9$- $+ 1$-methyl phenanthrenes) and other alkylated PAH homologous series.

A general sequence of biodegradation of oil hydrocarbon classes is summarized below:

n-alkanes > BTEX and other monoaromatic compounds > branched and cyclo-alkanes > PAHs (lighter PAHs are more susceptible than larger PAHs and increase in alkylation level in the same PAH series decreases susceptibility to microbial attack) > biomarker terpanes and steranes

This sequence only represents a general biodegradation trend and does not mean that the more resistant class of hydrocarbons starts to be biodegraded only after the less resistant class be completely degraded. Hence, some components of more easily degradable compound classes that may remain after biodegradation has already begun on the next more resistant class of compounds.

In order to better understand the effects of biodegradation on petroleum hydrocarbons, a number of laboratory biodegradation studies have been conducted under controlled laboratory conditions (Atlas *et al.*, 1992; Prince, 1993; Blenkinsopp *et al.*, 1996; Swannel *et al.*, 1996; Foght *et al.*, 1998; Wang *et al.*, 1998c). Figures 17.4.4 and 17.4.5 show GC-MS chromatograms of *m/z* 85 for Prudhoe Bay (PB) biodegradation series incubated at $10\,°C$ in fresh water using a laboratory standard inoculum consisting of 6 individual strains (three aliphatic and three aromatic degraders) (Wang *et al.*, 1998c) to examine the composition changes with time.

Figure 17.4.4 illustrates the changes in the relative *n*-alkane contents in the absence (negative control) and presence (positive control, PC) of nutrients. Figure 17.4.5 shows representative GC-MS chromatograms of C_2-naphthalenes $(m/z = 156$, after 14 days of incubation), C_3-naphthalenes $(m/z = 170$, after 28 days), C_1-phenanthrene $(m/z = 192$, after 14 or 28 days), C_1-dibenzothiophenes $(m/z = 198$, after 14 or 28 days), and C_1-fluorenes $(m/z = 180$, after 28 days) in the sterile control PB oil and the corresponding positive controls.

The encircled peaks in Figure 17.4.5 indicate preferential biodegradation of these PAH isomers over isomers that have not been encircled within the same isomeric group. Among the identified alkylated PAH compounds, bacteria preferentially degraded 1,3- and 1,6-dimethylnaphthalene; C_3-, $\beta\beta$-ethylmethyl-, 1,3,7-, 1,3,6-, 1,3,5-, and 1,2,5-trimethylnaphthalene; methyl-fluorene (completely gone after 28 days); 2-methyl-phenanthrene (after 14 days) and then 3- and 1-methyl-phenanthrene (after 28 days); and 2-/3-methyl-dibenzothiophene (DBT) (significantly decreased in abundance after 14 days and completely removed after 28 days), resulting in great changes in the relative ratios of (1,3-methyl- naphthalene + 1, 6-dimethyl- naphthalene)/$\sum C_2$-naphthalenes, methyl-fluorene/ $\sum C_1$-fluorenes, (3-methyl-phenenanthrene + 2-methyl-phenanthrene)/ (9-methyl-phenanthrene+ 1-methyl-phenanthrene), and 2-/3-methyl-dibenzothiophene/4-methyl-dibenzothiophene.

Preferential degradation of certain isomers within isomeric groups were observed not only for PB oil incubated for different period of times (7, 14, and 28 days) at different temperatures (4, 10, 15, and $22\,°C$), but also for other tested oils under varying incubation conditions and for nine

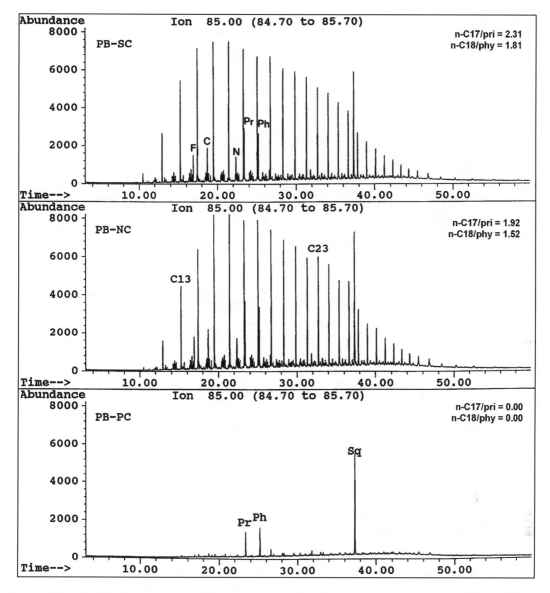

Figure 17.4.4 *GC-MS chromatograms (m/z 85) for Prudhoe Bay (PB) oil biodegradation series incubated at 10°C in fresh water using a laboratory standard inoculum.*

Alaskan oils (Blenkinsopp *et al.*, 1996). Conversely, no preferential degradation can be observed for physical weathering processes such as evaporation, and oil composition changes due to biodegradation can be readily distinguished from the changes due to physical weathering processes (e.g., evaporation).

No observable sign of alteration in the composition of biomarkers was observed after 28 days incubation under the current experimental conditions, regardless of the oil type (light or heavy), incubation conditions, with and without the presence of nutrients. The concentrations of terpanes and steranes in the tested oils were consistent and the diagnostic ratios of paired terpanes and steranes remained constant. As an example, Table 17.4.1 summarizes biomarker ratios for four biodegraded oils. The average of the sum of eight diagnostic biomarker ratios from 70 biodegradation samples of

the ASMB oil inoculated under various inoculum conditions during 1994 was 8.2 ± 0.2 with relative standard deviation less than 4%.

As opposed to the biomarker compounds, *n*-alkanes, pristane, and phytane were greatly reduced in the positive controls (Figure 17.4.4) and $n\text{-}C_{17}$/pristane, $n\text{-}C_{18}$/phytane, and pristane/phytane were significantly altered, indicating degradation of pristane and phytane had also occurred.

Compared to the laboratory-controlled biodegradation, biodegradation under field conditions is generally a long-term weathering process. Alterations in chemical composition of naturally weathered oil spill samples are caused by the combination of various weathering processes.

Figures 17.4.6 and 17.4.7 show the GC chromatograms for TPH and *n*-alkane analysis, alkylated benzenes, alkylated PAHs, and biomarker terpane analysis, respectively,

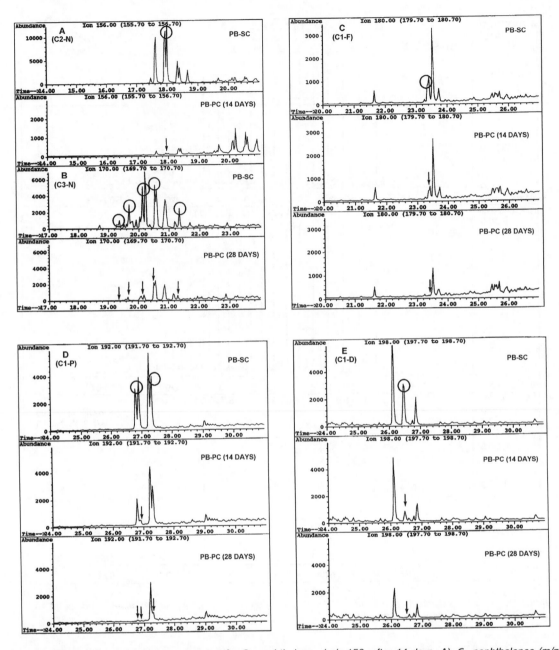

Figure 17.4.5 *Extracted ion chromatograms for C_2-naphthalenes (m/z 156, after 14 days, A), C_3-naphthalenes (m/z 170, after 28 days, B), C_1-fluorenes (m/z 180, after 14 and 28 days, C), C_1-phenanthrenes (m/z 192, after 14 and 28 days, 10D), and C_1-dibenzothiophenes (m/z 198, after 14 and 28 days, 6E) in the source sterile control PB oil and the corresponding positive controls illustrating distinct and consistent changes in the PAH composition due to microbial degradation. Encircled regions indicate isomer-specific degradation (i.e., preferentially degradation of some isomers).*

for 25-year-old Nipisi oil spill samples collected from the same location and site but with different sampling depths. Figures 17.4.6 and 17.4.7 demonstrate the effects of weathering under natural conditions and sample depths on the chemical composition of the spilled oil during the time period after the spill. That is, after 25 years, the surface oil (0–2 cm) has been heavily weathered (featured by complete depletion of *n*-alkanes, isoprenoids including pristane and phytane, and light BTEX and C_3-benzene compounds and by significant reduction in abundance of alkylated C_0 to C_4-

naphthalenes, phenanthrenes, dibenzothiophenes, and fluorenes). Conversely, the subsurface residual oil (>30–40 cm) is still almost unaffected by weathering, with GC chromatographic profiles similar to the lightly weathered reference oil. The microbial degradation of subsurface oil has been slow since the peat in this wetland habitat is acidic and water saturated, that is, largely anaerobic (Wang *et al.*, 1998a) and because evaporation is less important in subsurface samples. Also the microbial activities are slow most of the year because of the northern location of this site with a

Table 17.4.1 *Comparison of Diagnostic Ratios of Target Biomarkers for Four Biodegraded Oils (Under Freshwater Inoculum Conditions, 28 Days at 10°C)*

Oil		C_{32}/C_{24}	Tm/Ts	C_{29}/C_{30}	$C_{32}(S)/$ $C_{32}(R)$	$C_{33}(S)/$ $C_{33}(R)$	C_{23}/C_{30}	C_{24}/C_{30}	C_{27} ααββ/C_{29} αββ steranes	Total ratios
ASMB oil	W	1.73	0.98	0.53	1.58	1.52	0.44	0.25	0.97	8.0
	SC	1.67	1.23	0.51	1.60	1.47	0.47	0.28	0.93	8.2
	PC1	1.71	1.21	0.53	1.61	1.48	0.47	0.28	1.01	8.3
	PC2	1.75	1.22	0.50	1.59	1.37	0.48	0.27	1.02	8.2
	PC3	1.70	1.06	0.52	1.52	1.43	0.47	0.28	1.02	8.0
	NC1	1.71	0.93	0.53	1.59	1.53	0.45	0.26	1.01	8.0
	NC2	1.77	0.99	0.52	1.56	1.53	0.46	0.26	0.92	8.0
CLB oil	W	1.85	3.70	0.75	1.44	1.49	0.53	0.29	0.94	11.0
	SC	1.82	3.72	0.77	1.50	1.43	0.46	0.25	0.76	10.7
	PC1	1.87	4.19	0.76	1.59	1.43	0.45	0.24	0.74	11.3
	PC2	1.84	4.20	0.77	1.47	1.48	0.46	0.25	0.83	11.3
	PC3	1.80	3.82	0.74	1.51	1.48	0.46	0.25	0.92	11.0
	NC1	1.83	4.16	0.77	1.44	1.49	0.46	0.25	0.85	11.3
	NC2	1.89	3.84	0.76	1.38	1.52	0.47	0.25	0.88	11.0
NW oil	W	1.27	2.68	0.65	1.49	1.58	0.63	0.47	0.68	9.5
	SC	1.29	2.96	0.65	1.48	1.46	0.59	0.46	0.71	9.6
	PC1	1.31	2.65	0.62	1.52	1.42	0.55	0.42	0.70	9.2
	PC2	1.33	2.77	0.62	1.59	1.57	0.56	0.42	0.78	9.6
	PC3	1.32	2.73	0.61	1.63	1.58	0.65	0.49	0.71	9.7
	NC1	1.34	2.56	0.64	1.58	1.49	0.51	0.38	0.70	9.2
	NC2	1.29	2.58	0.65	1.48	1.46	0.65	0.50	0.72	9.3
TN oil	W	0.86	1.41	0.32	1.44	1.46	0.05	0.06	1.54	7.1
	SC	0.94	1.17	0.34	1.48	1.48	0.05	0.06	1.64	7.2
	PC1	0.91	1.19	0.34	1.42	1.51	0.05	0.06	1.51	7.0
	PC2	0.84	1.18	0.33	1.53	1.55	0.05	0.06	1.54	7.1
	PC3	0.81	1.20	0.35	1.49	1.53	0.05	0.06	1.65	7.1
	NC1	0.82	1.30	0.33	1.45	1.47	0.05	0.06	1.48	7.0
	NC2	0.84	1.34	0.35	1.43	1.57	0.05	0.06	1.54	7.2

Notes
ASMB: Alberta Sweet Mixed Blend; CLB: Cold Lake Bitumen; NW: Norman Wells; TN: Terra Nova; W: weathered oil; SC: sterile control; PC: positive control; NC: negative control.

mean annual temperature of 1.7°C, based on 22 years of weather records from 1970 to 1992.

Conversely, the biomarker composition is nearly unaffected (Figure 17.4.7). The accumulation of terpanes relative to the reference oil during the 25-year period of weathering is apparent, especially for the severely weathered surface sample N2-1A (e.g., the concentration of C_{30}-αβ hopane in sample N2-1A was approximately 1.8 times more than found in the reference oil. Also, five diagnostic biomarker ratios (C_{23}/C_{24}, Ts/Tm, C_{29}/C_{30}, C_{32} 22S/22R, and C_{33} 22S/22R) are consistent between samples.

17.4.4 Biodegradation of Biomarkers
Although terpanes and steranes are highly resistant to biodegradation, several studies have shown that they can be degraded to certain degree under severe weathering conditions (i.e., extensive microbial degradation) (Seifert *et al.*, 1984; Chosson *et al.*, 1991; Munoz *et al.*, 1997; Prince *et al.*, 2002).

The Arrow oil spill (Wang *et al.*, 1994) and BIOS oil spill (Wang *et al.*, 1995a) studies have demonstrated degradation of C_{23} and C_{24} tricyclic terpanes. In addition, Tm is degraded faster relative to Ts even though Ts chromatographically elutes out earlier than Tm, resulting in an increase of the Ts/Tm ratio for heavily degraded oil samples.

Based on several geochemical studies, Peters and Moldowan (Peters and Moldowan, 1993) have created a "quasi-stepwise" sequence for assessing the extent to which biomarkers are degraded in the reservoir.

n-paraffins (most susceptible to biodegradation) > acyclic isoprenoids > hopane (25-norhopanes present) ≥ steranes > hopanes (no 25-norhopanes) ~ diasteranes > aromatic steroids > porphyrins (least susceptible)

Steranes 20R ααα and 20R αββ are preferentially degraded compared to their 20S epimers (McKirdy *et al.*, 1983). Similar to the 20R epimers in the steranes, the 22R epimers of the 17α, 21β(H)-homohopanes are more susceptible to biodegradation than their 22S configuration homohopanes (Requejo and Halpern, 1989). In March 1986, sections of peaty mangrove in a tropical ecosystem were polluted by Arabian Light crude oil. Eight years later, Munoz *et al.* (1997) found that isoprenoids were severely degraded and the biomarker distribution altered (Munoz *et al.*, 1997). Norhopanes were the most biodegradation-resistant among the studied terpane and sterane groups and the C30-αβ hopane appeared more sensitive to weathering than its higher homologues.

Three coastal sites, heavily oiled from the 1974 Metula oil spill in the Strait of Magellan, Chile, were examined during May 1998 to determine the long-term fate and persistence of Metula oil in a marine marsh environment (Wang *et al.*, 2001b). Among the characterized samples, the asphalt pavement samples were the most heavily weathered (Figure 17.4.8), evidenced by a complete loss of n-alkanes from n-C_8 to n-C_{41} and by depletion of more than 98% of the alkylated PAHs. Even the most refractory

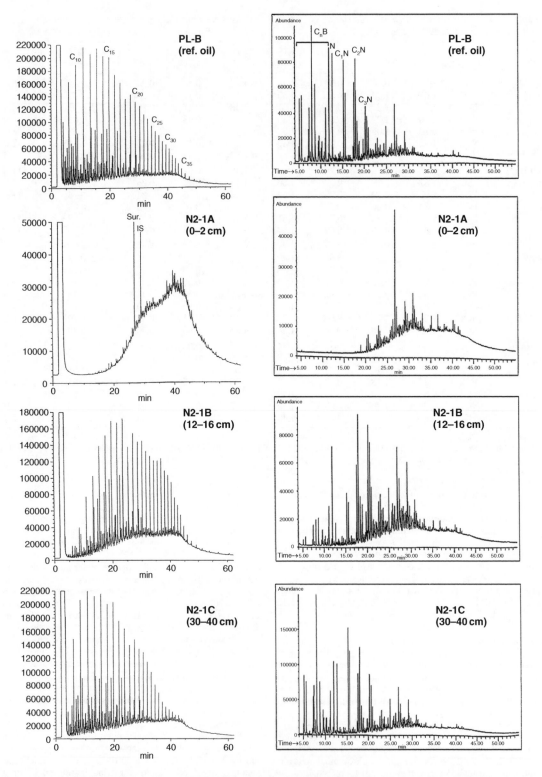

Figure 17.4.6 *GC-FID (left panel) and GC-MS-SIM (right panel) chromatograms of saturate and aromatic fractions for TPH and n-alkane analysis, alkylated benzenes and alkylated PAH distribution analysis for 25-year-old Nipisi spill oil samples collected from the same location and site but with different sampling depths. Figure 17.4.6 illustrates the effects of field weathering conditions and sample depths on chemical composition changes of the spilled oil. Sur and IS represent surrogate and internal standard compound.*

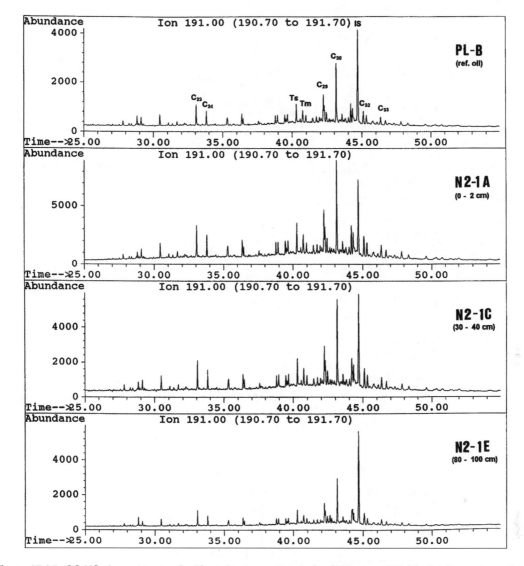

Figure 17.4.7 *GC-MS chromatograms for biomarker terpane analysis of 25-year-old Nipisi oil spill samples collected from the same location and site but at varying sampling depths.*

biomarker compounds were affected to varying degrees (Figures 17.4.9, 17.4.10 and 17.4.11). Table 17.4.2 tabulates diagnostic ratios of major target biomarker compounds for the most and the least degraded samples. For comparison, the ratios for the heavily degraded sample EE-T1 from the East Marsh plot are also listed in Table 17.4.2.

The chemical fingerprinting data of the Metula spill samples revealed that the degree of biomarker biodegradation was strongly correlated with molecular structures. The degradation trends of biomarkers were summarized as follows:

- Biomarkers were generally altered in the declining order of importance as: diasteranes > C_{27} steranes > tricyclic terpanes > pentacyclic terpanes > norhopanes(C_{29}Ts) ~ C_{29} αββ steranes
- The degradation of steranes was in the order of $C_{27} > C_{28} > C_{29}$ with the stereochemical degradation sequence $20R\alpha\alpha\alpha$ steranes > $20(R+S)\alpha\beta\beta$ steranes > $20S\alpha\alpha\alpha$ steranes

- For the pentacyclic terpanes, degradation of $C_{35} > C_{34} > C_{33} > C_{32} > C_{31}$ was apparent with a significantly preferential degradation of the 22R epimers over the 22S epimers
- C_{30}-αβ-hopane appeared more degradable than the 22S epimers of C_{31} and C_{32} homohopanes, but had roughly the same biodegradation rate as the 22R epimers of C_{31} and C_{32} homohopanes and was significantly more resistant to degradation than the 22S and 22R epimers of C_{34} and C_{35} homohopanes
- C_{29}-18α(H), 21β(H)-30-norneohopane and C_{29}-αββ 20R and 20S stigmastanes were the most resistant terpane and sterane compounds, respectively, among the studied target biomarkers.

17.4.5 Determination of Weathered Percentages Using Biomarkers

As discussed above, oil weathering is a complex process and the weathering degree and the rate are dependent on a large number of factors. Traditionally, the ratios

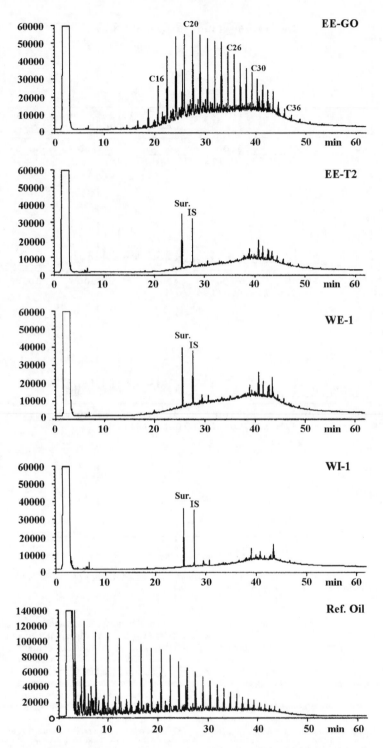

Figure 17.4.8 *Comparison of GC-FID chromatograms of the reference Arabian Light oil and the degraded Metula spilled oil samples EE-GO, EE-T2, WE-1, and WI-1.*

n-C_{17}/pristane and n-C_{18}/phytane have been widely used for estimating the degree of biodegradation and for comparing the weathering state of spilled oil. For lightly and some moderately weathered oils, these ratios may provide a useful tool for estimation of the weathering degree and for oil-source identification and differentiation. In severely weathered oils, however, the n-alkanes and even the isoprenoids (including pristane and phytane) may be partially

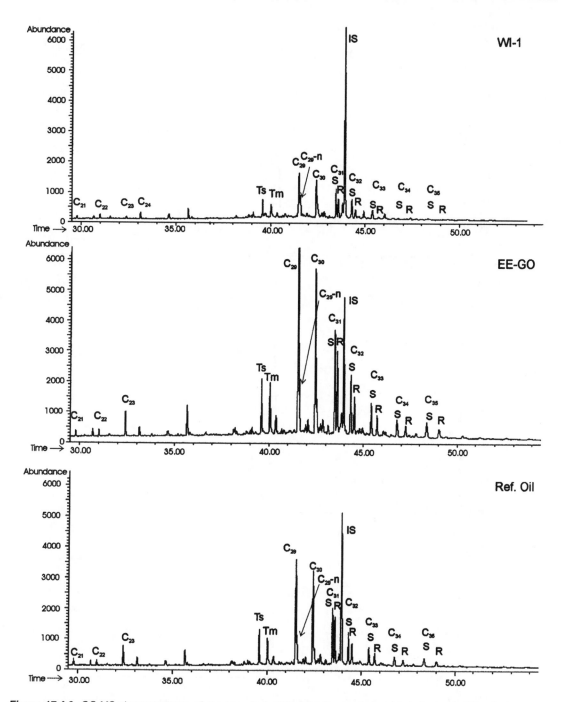

Figure 17.4.9 *GC-MS chromatograms of terpanes (m/z 191) from the most degraded oil sample WI-1 (top), the less degraded oil sample EE-GO (middle) after 24 years of weathering, and the reference oil (bottom). IS: internal standard C_{30}-$\beta\beta$-hopane.*

or completely lost as shown in Figures 17.4.6 and 17.4.8. Under such circumstances, the use of the traditional measure of n-C_{17}/pristane and n-C_{18}/phytane might substantially underestimate the extent of biodegradation. Recently, highly degradation-resistant oil components such as C_{30} $\alpha\beta$ hopane or $C2_{29}$ $\alpha\beta$ norhopane have been applied as conserved "internal standards" for estimation of the weathering degree and extent of the spilled residual oil (Butler

et al., 1991; Douglas *et al.*, 1994; Prince *et al.*, 1994; Wang *et al.*, 1995a):

$$P(\%) = 1 - \frac{C_S}{C_W} \times 100\% \qquad (17.4.1)$$

where P is the weathered percentages of the weathered samples, and C_s and C_w are the concentrations of

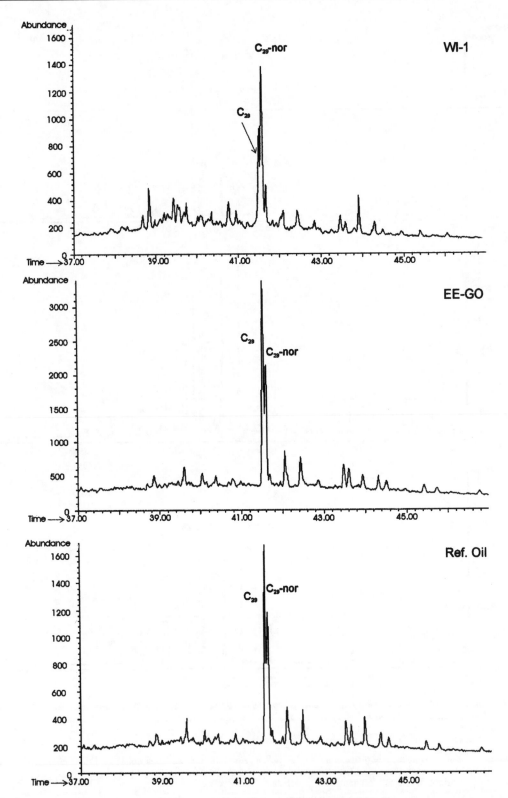

Figure 17.4.10 *Comparison of GC-MS chromatogram (m/z 177) of 25-demethylated pentacyclic terpanes from the most degraded oil sample WI-1 (top), the less degraded oil sample EE-GO (middle) after 24 years of weathering, and the reference oil (bottom) illustrating differences in degradation and persistence of biomarkers C_{29}-17α(H),21β(H)-30-norhopane and C_{29}-18α(H),21β(H)-30-norneohopane.*

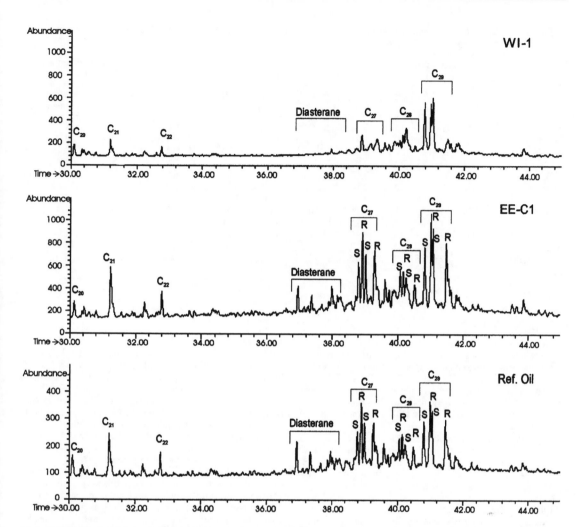

Figure 17.4.11 *Comparison of GC/MS chromatogram (m/z 217) of steranes including C_{27} and C_{28} diasteranes from the most degraded oil sample WI-1 (top), the less degraded oil sample EE-C1 (middle), and the reference oil (bottom) illustrating the degradation of steranes in the order of diasteranes $>C_{27}$ steranes $>C_{28}$ steranes $>C_{29}$ steranes. The preferential depletion of C_{27}–C_{29} steranes in the order of 20R $\alpha\alpha\alpha$ $>20(S+R)\alpha\beta\beta$ $>20S$ $\alpha\alpha\alpha$ steranes, as a function of their stereochemistry, is also illustrated.*

Table 17.4.2 *Comparison of Diagnostic Ratios of Major Target Biomarkers for the Less Degraded Samples EE-CI, EE-C2, and EE-GO and the Most Degraded Samples WI-1 and WI-2*[a]

| | Less degraded | | | | The most degraded | | |
Samples	EE-C1	EE-C2	EE-GO	Heavily degraded, EE-T1	WI-1	WI-2	Ref. Oil, AL
C_{23}/C_{29}-nor	0.51	0.48	0.48	0.43	0.08	0.24	0.67
C_{24}/C_{29}-nor	0.18	0.15	0.18	0.15	0.25	0.17	0.28
C_{29}/C_{29}-nor	4.96	4.69	4.54	4.24	2.30	2.90	4.14
C_{30}/C_{29}-nor	3.96	3.81	3.72	3.51	1.92	2.51	3.71
$C_{31}(S)/C_{29}$-nor	2.44	241	2.31	2.20	1.41	1.64	2.32
$C_{31}(R)/C_{29}$-nor	1.94	1.95	1.88	1.72	0.92	1.23	1.88
$C_{32}(S)/C_{29}$-nor	1.45	1.43	1.35	1.33	1.03	1.12	1.48
$C_{32}(R)/C_{29}$-nor	1.03	0.96	0.89	0.88	0.52	0.68	0.99
$C_{33}(S)/C_{29}$-nor	0.94	0.91	0.83	0.88	0.49	0.64	0.88
$C_{33}(R)/C_{29}$-nor	0.65	0.83	0.57	0.53	0.11	0.24	0.60
Ts/C_{29}-nor	1.24	1.26	1.20	1.10	0.84	0.99	1.21
Tm/C_{29}-nor	1.33	1.28	1.23	1.09	0.70	0.83	1.20
C_{27} $\alpha\beta\beta$/C_{29}-nor	0.51	0.46	0.43	0.33	0.07	0.1	0.55

[a] C_{29}-nor: C_{29} 18α(H), 21β(H)-30 norhopane; C_{27} $\alpha\beta\beta$: C_{27} 20(S + R) $\alpha\beta\beta$-steranes.

Table 17.4.3 *Quantitation Results of Selected Biomarkers and Estimation of Weathered Percentages of Residual Oil in Long-Term Weathered BIOS Samples*

Sample	Concentration (µg/g of TSEM)				Diagnostic ratios							Weathered percentages (%)		
	C_{23}	C_{24}	C_{29} $\alpha\beta$	C_{30} $\alpha\beta$	$C_{23}/$ C_{24}	$Ts/$ Tm	$C_{29}/$ C_{30}	$C_{32}(S)/$ $C_{32}(R)$	$C_{33}(S)/$ $C_{33}(R)$	$C_{23}/$ C_{29}	$C_{24}/$ C_{29}	Based on C_{29}-hopane	Based on C_{30}-hopane	*av.*
BIOS-3	109	72	131	139	1.53	0.22	0.94	1.57	a	0.83	0.55	11.5	12.2	12
BIO-5	267	138	287	303	1.94	0.28	0.95	1.57	1.59	0.93	0.48	60.0	60.0	60
BIOS-6	185	116	289	308	1.61	0.33	0.94	1.59	1.60	0.64	0.40	60.0	60.4	60
BIOS-7	222	109	221	231	2.02	0.27	0.95	1.61	1.57	1.00	0.49	47.5	47.2	47
BIOS-2	174	81	145	157	2.15	0.23	0.93	1.58	1.56	1.20	0.56	20.0	22.3	21
BIOS-11	180	83	150	160	2.16	0.23	0.94	1.58	1.62	1.20	0.55	22.7	23.8	23
BIOS-1	163	75	129	136	2.18	0.24	0.95	1.56	1.54	1.26	0.58	10.1	103	10
BIOS-4	179	84	145	153	2.14	0.24	0.95	1.59	1.57	1.23	0.58	20.0	20.3	20
BIOS-9	158	74	126	132	2.14	0.24	0.95	1.60	1.61	1.25	0.58	8.0	7.6	8
BIOS-10	166	77	132	138	2.16	0.24	0.96	1.57	1.56	1.26	0.58	12.1	11.6	12
BIOS-8	166	78	136	143	2.12	0.23	0.95	1.56	1.60	1.22	0.57	14.7	14.7	15
Weathered Lago	156	72	125	131	2.17	0.22	0.95	1.57	1.53	1.25	0.58	7.2	6.9	7
Fresh Lago	142	65	116	122	2.17	0.22	0.95	1.60	1.52	1.23	0.56			

a The peaks are too small to be accurately intergrated.

$C_{30}\alpha\beta$-hopane in the source oil and weathered samples, respectively. Hopanes $C_{29}\alpha\beta$ and $C_{30}\alpha\beta$ are abundant in most oils, thus the weathering percentage can be readily estimated by the use of Equation 17.4.1. Table 17.4.3 represents quantification results of selected biomarkers and estimation of weathered percentages of residual oil in the long-term weathered Baffin Island Oil Spill samples (BIOS) (Wang *et al.*, 1995a).

For lighter refined products, such as diesel and jet fuel samples, which may not contain significant quantities of biomarker terpane and sterane compounds, the sesquiterpanes as well as a selection of the more conservative PAHs with a high degree of alkylation such as C_3 or C_4-phenanthrenes can be used as an alternative internal standard for estimating the degree of weathering.

Note, however, that the weathered percentages can be underestimated by using C_{30}-$\alpha\beta$-hopane as an internal oil reference for extremely degraded oil samples because C_{30}-$\alpha\beta$-hopane under such circumstances is itself partially depleted, such as in the case of the Metula oil spill (Wang *et al.*, 2001b). For example, the weathered percentages of three East Marsh untreated plot samples were 30-38% if C_{29}-18α(H), 21β(H)-30-norneohopane was used as the internal reference. These percentages are higher than the values of 25–32% obtaining when using C_{30}-$\alpha\beta$-hopane as internal reference. However, in most cases, C_{30}-$\alpha\beta$-hopane is the preferred choice for estimating weathered percentages, because C_{30}-$\alpha\beta$-hopane is often the most abundant among C_{19} to C_{35} biomarkers and can thus be quantified more accurately.

17.5 APPLICATION OF BIOMARKERS FOR SPILL SOURCE IDENTIFICATION, OIL CORRELATION, AND DIFFERENTIATION

17.5.1 Case Study 1: Multiple Criteria Approach for Fingerprinting Unknown Oil Samples Having Very Similar Chemical Composition

On March 28, 2001, three unknown oil samples were received from Montreal for product characterization, correlation, and differentiation. Figure 17.5.1 shows the GC-FID chromatograms of these three unknown samples. Samples #1 and #2 show nearly identical GC profiles and *n*-alkane

and isoprenoid distribution patterns. Sample #3 has similar GC profile as samples #1 and #2. The GC-FID chromatograms including the UCM of three samples are significantly different from those of crude oils and most refined products. Petroleum hydrocarbons are mainly distributed in a carbon range from C_{20} to C_{37} with the retention time ranging between 30 and 50 minutes. No compounds were detected prior to the retention time of 16 minutes. Also, the amount of *n*-alkanes is very low and the relative ratios of low abundant hydrocarbons *n*-C_{17}/*n*-C_{18}, *n*-C_{17}/pristane, and *n*-C_{18}/phytane are similar (Wang *et al.*, 2002).

The GC trace features suggest that the oils are *hydraulic fluid type products*. The questions, at this stage, are whether (1) these three oil samples come from the same source; (2) sample #3 is from the same source as #1 and #2; and (3) if the observed differences in the chemical composition of sample #3, #1, and #2 respectively can be caused by contamination or weathering? In order to unambiguously differentiate and identify these samples, the multi-criterion approach was adopted; that is, analysis and evaluation of more than one suite of petroleum compounds.

Distribution and diagnostic biomarker ratios

Figure 17.5.2 compares biomarker terpane distributions (*m/z* 191) for samples #1, #2, and #3. Table 17.5.1 summarizes quantification results and diagnostic ratios for the target terpanes and C_{27} and $C_{29}\beta\beta$-steranes. From Figure 17.5.1 and Table 17.5.1, it can be seen that the chromatographic pattern of biomarker terpanes of samples #1 and #2 is almost identical. The total target biomarker concentration was 4701, 4759, and 5464 µg/g oil for samples #1, #2, and #3, respectively, which is high compared to most crude oils.

Sample #3 has a different biomarker distribution than samples #1 and #2. Concentrations of C_{29} and C_{30} hopanes in sample #3 match those in samples #1 and #2 (Table 17.5.1). Conversely, concentrations of C_{23} and C_{24}, and the sum of C_{31} to C_{35} homohopanes are significantly lower (39 vs 145-152 µg/g oil for C_{23} and 36 vs 83-87 µg/g oil for C_{24}) and higher (2072 vs 1358-1376 µg/g oil for C_{31-35}) than those of the corresponding compounds in samples #1 and #2, respectively.

The diagnostic ratios of target biomarker terpanes are similar for samples #1 and #2 as well as the triplet ratios.

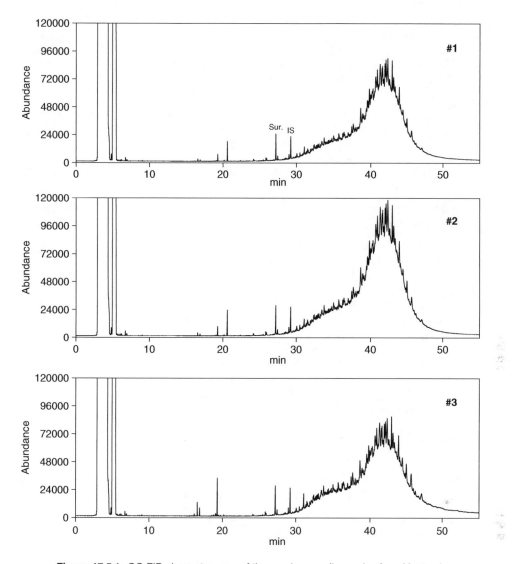

Figure 17.5.1 *GC-FID chromatograms of three unknown oil samples from Montreal.*

Sample #3 is indeed different and is from another source than #1 and #2, evidenced by a difference not only in the biomarker concentrations, but also in diagnostic ratios of target biomarkers. All these observations point toward the conclusion that samples #1 and #2 are identical, while sample #3 comes from another source.

Distribution of PAHs and Identification of Major Unknown Additives

In order to further confirm conclusion obtained from the characterization of biomarkers, the PAH distribution was analyzed.

Figure 17.5.3 shows the total ion GC-MS chromatograms of these three samples for analyses of BTEX compounds and target PAHs. The UCM in 35 to 45 minutes are pronounced for samples #1 and #2, but much smaller for sample #3. Table 17.5.2 summarizes quantification results of the alkylated PAH homologue series and other EPA priority PAHs. Clearly, samples #1 and #2 show not only nearly identical PAH distribution, but also similar diagnostic ratios of source-specific PAHs. Conversely, sample #3 has a different PAH distribution profile and diagnostic ratios than samples #1 and #2.

Three unknown compounds with high abundances in the aromatic fractions (F2) were positively (>96%) identified. They are 2, 6-bis (1,1-dimethylethyl)-phenol at 15.49 min, butylated hydroxytoluene (BHT) or 2,6-di-*tert*-butyl-4-methylphenol at 17.01 min, and N-phenyl-1-naphthalenamine at 27.72 min.

Samples #1 and #2 contained these three compounds at comparable concentration. Conversely, BHT was not detected in sample #3 which contained only 2, 6-bis (1,1-dimethylethyl)-phenol and N-phenyl-1-naphthalenamine with a higher abundance than the former.

The three identified compounds are antioxidants. Hydraulic oils are often stabilized with 0.5–1.0% of zinc dialkyldithiocarbamates in combination with 0.5–1.0% of a

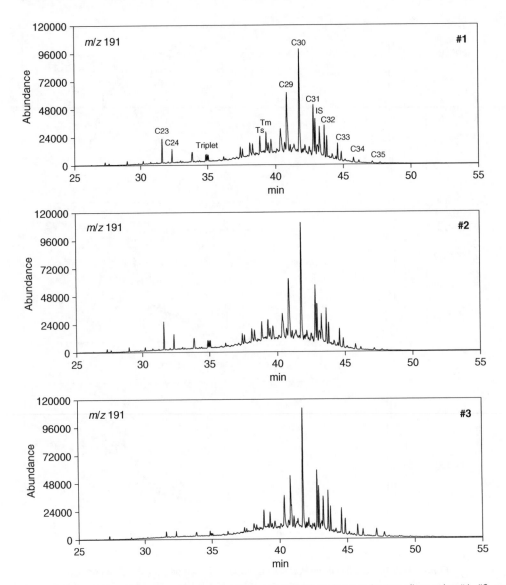

Figure 17.5.2 *Comparison of biomarker terpane distribution (m/z 191) for three unknown oil samples #1, #2, and #3.*

phenolic antioxidant (such as BHT) or with an aminic antioxidant (alkylated diphenylamine, or phenyl-naphthalamine or its derivatives). This supports the general conclusion described above.

In summary, the chemical fingerprinting evidences and data interpretation results for this case study reveal the following:

- Oils are most likely hydraulic fluid type oil.
- The three oils are pure and largely composed of saturated hydrocarbons. The aromatic fractions in the TPH were only 4% for samples #1 and #2.
- There is no clear sign indicating that oils have been weathered.
- PAH concentrations are extremely low ($<10\,\mu g/g$ oil) and the biomarker concentrations are unusually high (4700–$5500\,\mu g/g$ oil).

- Three major unknown compounds in the oil samples were positively identified as 2, 6-bis (1,1-dimethylethyl)-phenol at 15.49 min, butylated hydroxytoluene at 17.01 min, and N-phenyl-1-naphthalenamine at 27.72 min. These compounds are antioxidants added to oils.
- Samples #1 and #2 are identical and come from the same source.
- Sample #3 is different from samples #1 and #2 and does not come from the same source as samples #1 and #2, but its bulk hydrocarbon compositions, such as TPH and total saturates, are similar to samples #1 and #2.

This case study illustrates that in cases where two oils may exhibit similar or even nearly identical *n*-alkane and isoprenoid distributions, their biomarker distribution may be markedly different. Thus, the successful forensic investigation will require detailed analysis and comparison of not

Table 17.5.1 *Quantitation Results and Diagnostic Ratios of Major Biomarkers in Samples #1, #2, and #3*

Samples	#1	#2	#3
Biomarkers (μg/g oil)			
C_{23}	145	152	38.6
C_{24}	82.5	87.0	35.9
C_{29}	550	554	524
C_{30}	1050	1055	1165
C_{31} (S)	416	427	504
C_{31} (R)	259	263	334
C_{32} (S)	231	225	331
C_{32} (R)	149	148	196
C_{33} (S)	126	129	220
C_{33} (R)	73.7	75.0	137
C_{34} (S)	45.1	48.3	131
C_{34} (R)	22.0	22.1	63.8
C_{35} (S)	19.7	22.3	87.2
C_{35} (R)	15.2	16.8	68.4
Ts	134	137	137
Tm	148	148	129
C_{27} $\alpha\beta\beta$-steranes	529	534	597
C_{29} $\alpha\beta\beta$-steranes	705	716	767
Sum of C_{31} to C_{35} homohopanes	1358	1376	2072
Total	4701	4759	5464
Diagnostic Ratios			
C_{23}/C_{24}	1.76	1.74	1.08
C_{23}/C_{30}	0.14	0.14	0.03
C_{24}/C_{30}	0.08	0.08	0.03
C_{29}/C_{30}	0.52	0.53	0.45
Triplet (RT = ~35 min)	1.14:1.08:1.00	1.14:1.12:1.00	2.22:1.09:1.00
C_{31}(S)/C_{31}(R)	1.60	1.63	1.51
C_{32}(S)/C_{32}(R)	1.55	1.52	1.68
C_{33}(S)/C_{33}(R)	1.71	1.72	1.61
C_{34}(S)/C_{34}(R)	2.05	2.19	2.05
C_{35}(S)/C_{35}(R)	1.30	1.33	1.28
$C_{30}/(C_{31}+C_{32}+C_{33}+C_{34}+C_{35})$	0.77	0.77	0.56
Ts/Tm	0.90	0.92	1.06
C_{27} $\alpha\beta\beta$-steranes/C_{29} $\alpha\beta\beta$-steranes	0.75	0.75	0.78

only the concentrations but also the diagnostic ratios of biomarkers among similar product types.

17.5.2 Case Study 2: Characterization and Identification of the Detroit River Mystery Oil Spill (2002)

An oil spill to the Rouge River and Detroit River was discovered and reported in the second week of April 2002. The spill impacted approximately 43 k of US and Canadian shorelines. The presence of sheen over the majority of the impacted river area was observed. Environment Canada Ontario Region collected a number of spill samples from various spots and sent 11 samples to the Oil Research Laboratory of the Emergencies Science and Technology Division (ESTD), Environment Canada, for characterization of the chemical composition and determination of the type, nature, and origin of the Detroit River mystery oil spill samples (Wang *et al.*, 2004).

Product Type Screen and Determination of Hydrocarbon Groups
In general, the product type and chemical composition features can be illustrated by qualitative and quantitative examination of GC-FID chromatograms. Figure 17.5.4 shows the GC-FID chromatograms for TPH and *n*-alkane analysis for

three representative Detroit River spill samples #1, #2, and #3. Table 17.5.3 summarizes the hydrocarbon group analysis results (gravimetrically determined TSEM, GC-TPH values) of the spill samples.

The major chemical composition features of TPH and saturate hydrocarbons in the samples are summarized as follows:

- The GC-FID chromatogram of Fractions 1 and 3 (F1 and F3) of the spill samples are dominated by large UCM (located in the n-C_{18}-n-C_{36} range) with almost no *n*-alkanes detected above n-C_{20} (retention time ~28min). The ratios of the GC-resolved peaks to the total GC area were only 0.06 for the three samples. Furthermore, ratios of the total saturate to the GC-TPH were approximately 90%, significantly higher than most crude oils. These chromatographic features suggest that the major portion of the spilled oil might be a lubricating oil.
- The resolved *n*-alkanes are mainly distributed in the diesel carbon range (C_8–C_{27}). No *n*-alkanes less than C_{10} and greater than C_{24} were detected. The total *n*-alkanes including pristane and phytane concentration were only 9.3, 10.4, and 8.6 mg/g of TSEM for samples #1, #2, and #3 respectively.
- It is suspected that the spill samples contain a diesel fraction. Using the estimation value of 120 mg *n*-alkanes

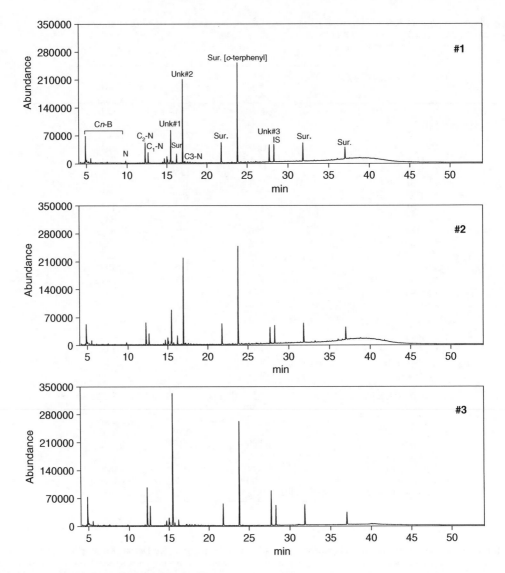

Figure 17.5.3 *The GC-MS total ion chromatograms of the unknown samples for analyses of BTEX compounds and target PAHs.*

per gram diesel (the concentrations of *n*-alkanes in most diesels are in the range of ~120 mg/g diesel), the percentage of diesel in the spill samples was estimated not exceeding 20% of the total hydrocarbons detected.

- The three samples had almost identical chromatographic profiles, *n*-alkane distribution, and diagnostic ratios of n-C_{17}/pristane, n-C_{18}/phytane, and pristane/phytane. This implies that these samples are most likely from the same source and the small differences are most likely caused by analytical variability and weathering.
- The *n*-alkanes with the lowest carbon number detected in samples #1 and #2 are n-C_{11} (0.07 mg/g TSEM) and n-C_{10} (0.10 mg/g TSEM), respectively. Sample #3 collected from N. Boblo Island showed complete loss of n-C_8 to n-C_{12}, indicating that the diesel portion in this sample has been more weathered (most probably by more evaporation and water-washing in its longer journey from spill source to the destination) than samples #1 and #2.

Determination of Alkylated PAH Homologues and their Diagnostic Ratios
Shows the distribution of alkylated PAH homologues and other EPA priority PAHs. The comparison results of the PAH distributions between samples indicate the following:

- The relative distribution patterns and profiles of alkylated PAHs are almost identical for the spilled samples, in particular for samples #1 and #2.
- The concentration of five target alkylated PAH homologue series and other EPA priority PAHs were 1404, 1479, and 1028 µg/g TSEM, and 250, 257, and 167 µg/g TSEM for samples #1, #2, and #3, respectively. Compared to crude oils and most refined products such as jet fuel and diesel (>10,000 µg/g for most oils), the PAH concentrations in these spill samples are quite low. The dominance of alkylated naphthalenes and phenanthrenes among the five target alkylated PAH homologous series is pronounced for all three samples.

Table 17.5.2 *Quantitation Results of the Alkylated PAH Homologues Series and other EPA Priority PAHs*

Samples	#1	#2	#3
Alkylated PAHs (µg/g oil)			
Sum of naphthalenes (C_0–C_4)	4.53	4.57	5.91
Sum of phenanthrenes (C_0–C_4)	0.94	0.97	0.31
Sum of dibenzothiophenes (C_0–C_3)	0.51	0.53	0.10
Sum of fluorenes (C_0–C_3)	0.27	0.28	0.14
Sum of chrysenes (C_0–C_3)	0.17	0.18	0.02
Total alkylated PAHs	6.42	6.51	6.48
Other EPA priority PAHs (µg/g oil)			
Biphenyl	0.04	0.04	0.10
Acenaphthylene	0.01	0.01	0.02
Acenaphthene	0.02	0.02	0.03
Anthracene	0.00	0.00	0.00
Fluoranthracene	0.00	0.00	0.00
Pyrene	0.01	0.01	0.00
Benz[a]anthracene	0.00	0.00	0.00
Benzo[b]fluoranthene	0.00	0.00	0.00
Benzo[k]fluoranthene	0.00	0.00	0.00
Benzo[e]pyrene	0.01	0.00	0.00
Benzo[a]pyrene	0.01	0.00	0.00
Perylene	0.00	0.00	0.00
Indeno[1,2,3-cd]pyrene	0.00	0.00	0.00
Dibenz[ah]anthracene	0.00	0.00	0.00
Dibenzo[ghi]perylene	0.00	0.00	0.00
Total EPA priority PAHs	0.09	0.08	0.16
Diagnostic ratios			
2-m-N:1-m-N	1.83	1.88	1.90
(3+2-m-phen)/(4-/9-+1-m-phen)	0.86	0.85	0.80
Phens/Dibens	1.84	1.83	3.10
(C_2D/C_2P):(C_3D/C_3P)	0.69:0.15	0.68:0.16	0.25:0.01
C_0N:C_1N:C_2N:C_3N:C_4N	0.34:8.9:6.2:3.3:1.0	0.33:8.7:6.0:3.0:1.0	0.20:23.0:11.1:4.1:1.0
Naphs:Phens:DBTs:Fluors:Chrys	4.8:1.0:0.54:0.29:0.18	4.7:1.0:0.55:0.29:0.19	19.1:1.0:0.32:0.45:0.06

- Sample #2 contained small amounts of BTEX and C_3-benzene compounds. In comparison, almost no BTEX and other alkyl benzene compounds were detected in samples #1 and #3, which further demonstrates that sample #2 is the least weathered one. The loss of lighter molecular weight PAHs (i.e., naphthalene and C_1- and C_2-naphthalenes) was pronounced for all three samples, resulting in development of the relative distribution of C_0-N < C_1-N < C_2-N < C_3-N. This distribution pattern is particularly pronounced for the more weathered sample #3. The concentrations of biphenyl, acenaphthylene, and acenaphthene are also lower in the more weathered sample #3. Lubricating oils contain low levels of PAHs and the concentrations are high in diesel. Obviously, the small portion of diesel in spill samples largely contributed to PAHs in these spill samples

- A number of diagnostic PAH ratios revealed the following: (a) the relative distribution of PAH isomers 4-, 2-/3-, and 1-methyl dibenzothiophene at *m/z* 198 (1.00:0.65:0.27, 1.00:0.65:0.28, and 1.00:0.65:0.27 for #1, #2, and #3 respectively) and (3- + 2-methyl-phenanthrene) to (4-/9- + 1-methyl-phenanthrene) at *m/z* 192 (1.57, 1.59, 1.66 for #1, #2, and #3 respectively) matched closely and (b) the double ratios (C2D/C2P : C3D/C3P) were also nearly identical (0.22:0.31, 0.22:0.30, and 0.23:0.30 for samples #1, 2, and 3 respectively)

Input of Pyrogenic PAHs to the Spill Samples

Another pronounced feature of the PAH distribution is that the parent PAH phenanthrene, fluorene, and chrysene are the most abundant among the homologue series. The concentrations of parent PAHs are higher than the corresponding alkylated homologous (Figure 17.5.5). In particular, the highest abundance of parent chrysene over its alkyl-substituted homologues and the decrease in relative abundances with increasing level of alkylation (C_0-C > C_1-C > C_2-C > C_3-C) were pronounced. This type of PAH distribution profile are termed "skewed" or "sloped" distribution. Furthermore, the "Pyrogenic Index," Σ(other 3–6 ring EPA priority PAHs)/Σ(the 5 target alkylated PAHs), was high (0.16) and the relative ratios of chrysene to benz[a]anthracene were close to 1.0 for the three spill samples, far higher than the same ratios for crude oils and refined products. All these features indicate input of pyrogenic PAHs (Wang *et al.*, 1999c).

The most likely source of pyrogenic PAHs in used motor oils is combustion blow-by past the piston rings of exhaust gasses directly into the crankshaft cavity. Excessive heat in the motor lubrication process can also increase the concentration of PAHs, in particular the high molecular weight PAHs, in used lubricating oil. Therefore, it is reasonable to conclude that the pyrogenic PAHs found in the spilled oil were most probably produced from combustion and the

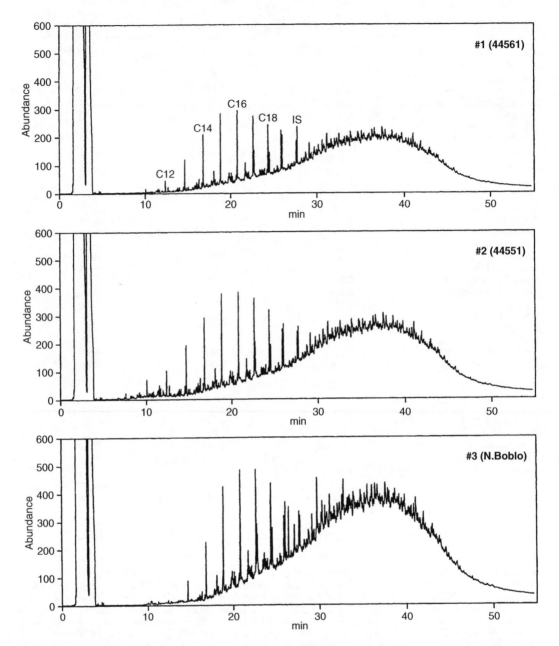

Figure 17.5.4 *GC-FID chromatograms of three representative Detroit River spill samples #1, #2, and #3. The GC traces are featured by the dominance of large UCM with small amounts of resolved peaks being detected in the lubricating oil carbon range (retention time: 24–50 min).*

motor lubrication process, and the lubricating oil in these spill samples was waste lubricating oil.

Characterization of the Biomarker Distribution
Figures 17.5.6 and 17.5.7 show the GC-MS distribution profiles of biomarker terpane and sterane compounds in the spill samples at m/z 191 and m/z 218, respectively. Table 17.5.4 summarizes the quantification results of target biomarker compounds.

- The samples show nearly identical distribution patterns of biomarker terpane and sterane compounds and these

biomarkers were mostly from the lubricating oil portion of the spill samples. Diesels do not contain high molecular weight biomarkers and only traces of low molecular weight biomarker compounds.
- The total concentration of target biomarker compounds were 1103, 941, and 941 µg/g TSEM for samples #1, #2, and #3. The biomarker concentrations in these three samples were high and matched closely.
- The diagnostic biomarker ratios C_{23}/C_{24} tricyclic terpane, $C_{29}\alpha\beta$-hopane/$C_{30}\alpha\beta$-hopane, Ts/Tm, $C_{31}(22S)/C_{31}(22S+22R)$, $C_{32}(22S)/C_{32}(22S+22R)$, $C_{33}(22S)/C_{33}(22S+$

Table 17.5.3 Hydrocarbon Group and Diagnostic Ratio Analysis Results for the three Representative Spill Samples #1, #2, and #3

Samples	#1 (44561)	#2 (44551)	#3 (N.Boblo)
GC-TPH (mg/g oil)	997	990	959
GC-saturates (mg/g oil)	956	955	861
GC-aromatics (mg/g oil)	41	36	98
$\dfrac{\text{GC-saturates}}{\text{GC-TPH}}$ (%)	96	96	90
$\dfrac{\text{GC-aromatics}}{\text{GC-TPH}}$ (%)	4	4	10
$\dfrac{\text{Resolved peaks}}{\text{GC-TPH}}$	0.04	0.04	0.05
$\dfrac{\text{UCM}}{\text{GC-TPH}}$	0.96	0.96	0.95

22R), $C_{34}(22S)/C_{34}(22S + 22R)$, $C_{35}(22S)/C_{35}(22S + 22R)$, and $C_{31}/(C_{31}$ to $C_{35})$ were also similar. This point toward the conclusion that the three spill samples came from the same source.

In summary, the chemical fingerprinting evidences and data interpretation results reveal the following:

- The Detroit spill samples were largely composed of used lubricating oil mixed with a smaller amount of diesel fuel.
- The diesel has been weathered.
- The diesel in sample #3 collected from N. Boblo Island was more weathered than samples #1 and #2.
- The chemical compositions of oil hydrocarbons in the three samples were the same and the spill oil samples came from the same source.
- Most PAH compounds were from the diesel portion in the spill samples, while the biomarker compounds were largely from the lubricating oil portion.
- Input of pyrogenic PAHs to the spill samples was demonstrated. The pyrogenic PAHs were most probably produced from combustion and motor lubrication processes.

17.6 A GUIDE TO MULTIVARIATE STATISTICAL ANALYSIS FOR BIOMARKER FINGERPRINTING

The literature on multivariate statistical methods is comprehensive and several books and articles give a thorough description of the theory of the most commonly applied methods such as PCA and PLS (Wold *et al.* 1984, 1987; Jolliffe 1986; Martens and Martens, 2001; Brereton 2003). However, the literature is rarely accessible for the non-specialist reader with limited knowledge on linear algebra and mathematics in general. Furthermore, publications often focus on theoretical aspects and when practical aspects are treated these applies quite different data than those used for biomarker fingerprinting for spill/source identification. Some of the most common practical aspects considered in the literature include pattern recognition of spectroscopic data (UV-VIS, fluorescence, IR, and NIR) and curve resolution of chromatographic peaks from, for example, LC-DAD and GC-MS.

Numerous commercial softwares for unsupervised multivariate statistical analysis are available on the market, which enables the forensic chemist to apply a range of multivariate statistical methods on a complex data set. These methods can be used by the chemist as a "black-box" without prior knowledge on data preprocessing, interpretation of results, and how the latter depends on initial preprocessing and variable selection. This may lead to far from optimal models with low resolution powers and more severely to misinterpretation of the outputs (e.g., loading and score plots). Hence, multivariate methods have to be used and interpreted with care, but if some general considerations are made during the modeling, these methods can considerably improve the analysis of complex fingerprinting data of petroleum biomarkers.

In the following sections, we discuss the advantages and limitations of multivariate statistical methods for environmental forensic compared to traditional methods for spill/source identification, correlation, and differentiation, based on comparison of biomarker distributions (subjective pattern matching) and cross-plots of diagnostic biomarker ratios. Furthermore, subsequent sections present a dedicated investigation and practical description of how to apply multivariate statistical methods, specifically PCA, for chemical fingerprinting of GC-MS biomarker data, based on either diagnostic ratios or sections of GC-MS chromatograms. The steps involved in proper multivariate statistical analysis of complex fingerprinting data are described and can be used as a tutorial for chemical fingerprinting of complex data by the use of multivariate statistical methods. The steps include (1) building an oil database, (2) preprocessing of chromatographic data, (3) variable selection or scaling, (4) choosing software, (5) centering of the data matrix, (6) scaling of the data matrix, (7) detection of outliers, (8) choosing the correct number of components, (9) interpretation of score plots, (10) interpretation of loading plots, (11) analysis of spill samples not included in the model, and (12) matching of suspected source and spill sample. In the subsequent sections, two recent case studies applying PCA to diagnostic biomarker ratios and digitized sections of GC-MS chromatograms are presented.

17.6.1 Limitations and Advantages of Multivariate Approaches

Before going into detail with the practical aspects of multivariate statistical methods for biomarker fingerprinting, it is important to acknowledge that these methods all work by approximating the "best" model of data (often in a least-squares sense). Hence, this model can be a more or less appropriate model of data and the parts not described by the model are left in the residuals. The size of these residuals can vary from sample to sample and between variables (e.g., diagnostic ratios) leading to a more or less adequate

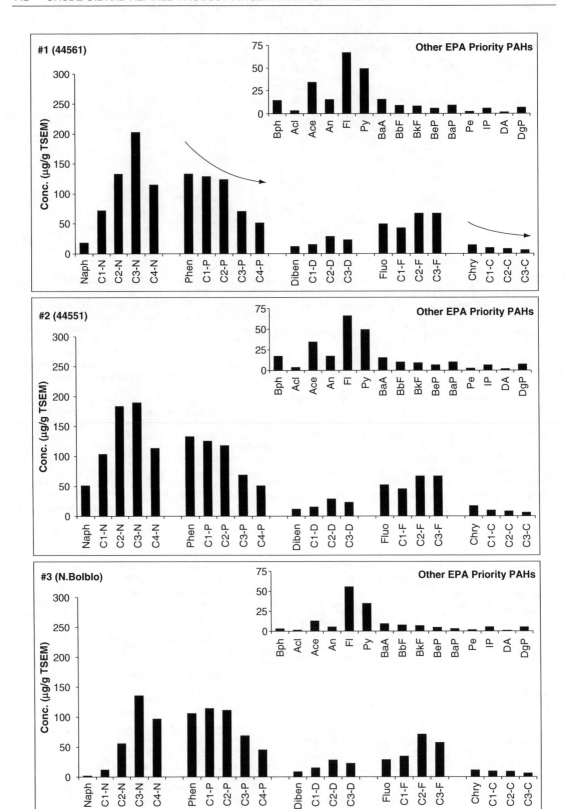

Figure 17.5.5 *Distribution of alkylated PAHs in the Detroit River spill samples. The distribution of other EPA priority PAHs are shown in the left insets. The input of pyrogenic PAHs is clearly demonstrated.*

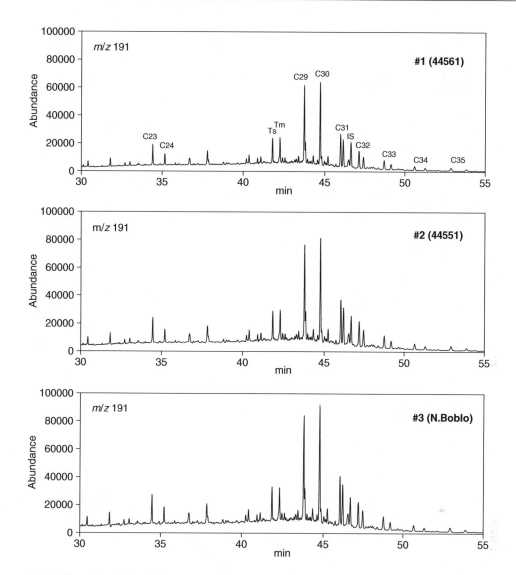

Figure 17.5.6 *Comparison of distribution of biomarker terpane compounds (m/z 191) in the Detroit River spill samples.*

description of individual samples and variables (e.g., samples and variables with high residuals are poorly described by the model). It is straightforward that if a multivariate model does not describe data sufficiently, the interpretation can lead to inadequate and even misleading conclusions on the similarity and dissimilarity of oil samples and fingerprinting variables.

On the other hand, if the applied multivariate statistical model (e.g., the PCA model) offers an appropriate description of data, and the data preprocessing and variable selection is adequate, the multivariate approach greatly improves the objectivity and resolution power compared to that of standard methods. One advantage of multivariate compared to univariate statistical methods is the ease by which relations between multiple samples and variables (e.g., compound concentrations or diagnostic ratios) can be resolved and visualized by score and loading plots. In standard evaluation methods, only a few oil samples (e.g., visual comparison of GC-MS fingerprints) or variables (e.g., double plots of two biomarker ratios) can be investigated simultaneously.

Hence, in cases where there are a large number of suspected spill/source candidates, univariate analyses and visual comparison of biomarker distributions are time-consuming and subjective.

Conversely, multivariate statistical methods enable simultaneous analysis of many samples and variables and can as such be used to explore correlations in data (between samples and variables) and to evaluate similarities between oil samples. In case study 3, for instance, a data set comprised of 101 oil samples and 1231 variables (chromatographic abundancies) were used for spill/source identification. Specifically, PCA was applied to sections of preprocessed chromatograms of m/z 217 (tricyclic and tetracyclic steranes) to correlate the chemical composition of biomarkers in oil spill samples and potential source oils.

Another advantage of multivariate methods is noise reduction. Noise reduction is obtained when more than one variable describe the same phenomenon (e.g., correlated variables). Examples of such diagnostic ratios are n-C_{17}/pristane and n-C_{18}/*phytane*, which describe the same

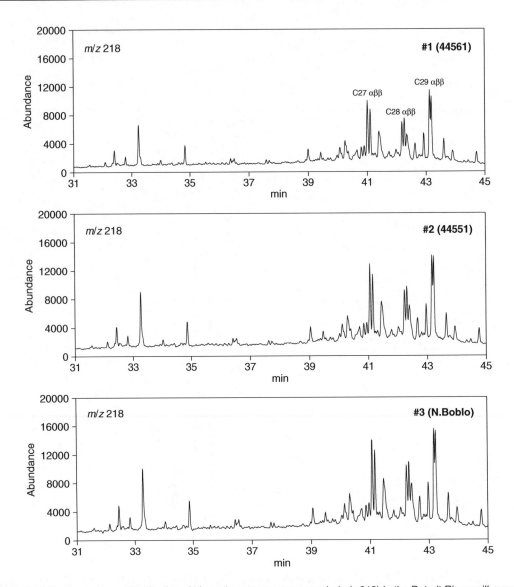

Figure 17.5.7 *Comparison of distribution of biomarker sterane compounds (m/z 218) in the Detroit River spill samples.*

phenomenon, namely biodegradation. These ratios describe the degree of microbial degradation an oil sample has been exposed to. The ratios are highly selective to biodegradation because its physicochemical properties (e.g., boiling point and water solubility) are almost identical, but the branched alkanes are less susceptible to microbial degradation than the *n*-alkanes. Likewise, if several diagnostic ratios separate oil samples due to the same underlying phenomenon (e.g., depositional environment, thermal maturity, in-reservoir degradation) the distinction becomes less and less affected by noise. Reduction of the uncertainties due to correlated variables and modeling of these variations correspond to taking the mean of replicate samples. The more replicates that is used to calculate the mean the more certain it becomes ($\bar{x}$ approaches μ, which is the true mean). In PCA the information in many correlated variables are summarized into few principal components (PCs) that are weighted sums of the original ones, where the weights are the loadings. The approach of including several

variables in the data analysis that in essence describe the same phenomenon opposes univariate data analysis where a few descriptive ratios are selected from a suite of ratios potentially describing the same phenomenon. In fact, the benefit of PCA compared to univariate analysis increases as the amount of colinearity in the data set increases.

Several methods for model validation, quality assurance, outlier detection, and chemical interpretation can be part of a multivariate statistical analysis. Especially, outlier detection and chemical interpretation of the results are fairly easy in multivariate compared to univariate analyses. In case study 4, for instance, the covariance of the complex mixture of compounds present in the chromatograms helped in interpreting the four significant principal components and hence the chemical explanation for the observed separation of oil samples in score plots. Also the observed correlation of variables can be used to describe whether the model indeed describes significant systematic variation or merely noise.

Table 17.5.4 *Quantitation Results of Target Biomarker Compounds for the Three Representative Detroit River Spill Samples*

Samples	#1 (44561)	#2 (44551)	#3 (N.Boblo)
Biomarkers (μg/g TSEM)			
C_{23}	45.5	40.6	42.3
C_{24}	25.1	22.7	24.0
C_{29}	184	165	161
C_{30}	208	190	192
C_{31} (S)	89.2	87.3	84.7
C_{31} (R)	79.0	77.2	75.6
C_{32} (S)	56.8	53.4	51.8
C_{32} (R)	39.6	37.0	36.2
C_{33} (S)	32.6	31.5	30.2
C_{33} (R)	20.9	20.1	18.8
C_{34} (S)	16.4	15.5	14.7
C_{34} (R)	11.2	9.75	9.52
C_{35} (S)	14.5	14.5	13.2
C_{35} (R)	8.10	8.22	7.24
Ts	54.4	45.5	45.4
Tm	51.8	43.0	43.7
C_{27} αββ-steranes	41.6	36.8	37.9
C_{29} αββ-steranes	58.1	50.6	53.2
Total	1035	949	941
Diagnostic Ratios			
C_{23}/C_{24}	1.82	1.79	1.76
C_{23}/C_{30}	0.22	0.21	0.22
C_{24}/C_{30}	0.12	0.12	0.13
C_{29}/C_{30}	0.89	0.87	0.84
Ts/Tm	1.05	1.06	1.04
$C_{31}(S)/C_{31}(S+R)$	0.53	0.53	0.53
$C_{32}(S)/C_{32}(S+R)$	0.59	0.59	0.59
$C_{33}(S)/C_{33}(S+R)$	0.61	0.61	0.62
$C_{34}(S)/C_{34}(S+R)$	0.61	0.61	0.61
$C_{35}(S)/C_{35}(S+R)$	0.64	0.64	0.65
$C_{30}/(C_{31}+C_{32}+C_{33}+C_{34}+C_{35})$	0.56	0.54	0.56
C_{27} αββ-steranes C_{29} αββ-steranes	0.72	0.73	0.71

17.6.2 The Oil Database

We recommend that data be divided into at least two sets. One set containing all source oils in the oil spill database including the suspected source oils ("the calibration set") and another containing all collected spill samples of the present investigation ("the test set"). An oil database should include a wide selection of crude oils, petroleum products (i.e., heavy fuel oils, light fuel oils, and lubricating oils), and mixture samples (including sludge and bilge samples) most often treated in real spill cases. Oils in the database is used in the subsequent matching of spill samples with suspected sources and, generally, the more samples included in the database the more knowledge is obtained about the variability of oils from different sources, the analytical uncertainty, and effects of weathering processes.

17.6.3 Preprocessing of Chromatographic Data

The defensibility of spill/source identification, correlation, and differentiation is highly dependent on the specific identity of variables used in the analysis as well as the type of preprocessing applied. Hence, it is important to remove variations unrelated to the chemical composition prior to spill/source matching. Chemical fingerprinting of GC-MS data will, for instance, be affected by the analytical variability associated with the extraction, cleanup, and detection steps.

Furthermore, as described in previous sections, weathering processes such as evaporation, dissolution, photooxidation, and biodegradation affect the compound distribution in spill samples compared to source oils, which can lead to models with low resolution power or worse to incorrect matching or classification of samples. In addition, concentrations are irrelevant to the spill/source matching process based on diagnostic ratios and relative biomarker distribution, thus variable normalization is a prerequisite. Otherwise, the important compositional information may be hidden by concentration dependent variations.

Hence, a method or a collection of methods for preprocessing data is a necessity to reduce data variability unrelated to the chemical composition. In the following sections, two types of fingerprinting variables are treated: diagnostic ratios and normalized GC-MS chromatograms.

17.6.3.1 Diagnostic Ratios

The calculation of diagnostic ratios has been described thoroughly in Section 17.2.2 and will not be further discussed here. The data set in Christensen *et al.* (2004) comprised 137 double normalized diagnostic ratios calculated by Equation 17.3. The ratios were based on peak areas and normalized to the laboratory reference oil consisting of 1:1

crude oil and heavy fuel oil. The double normalized diagnostic ratios had low analytical standard deviations (between 0.05 and 3.2%), comparable or smaller than those obtained by a quantitative method based on concentrations (Stout *et al.*, 2001).

17.6.3.2 Digitized Chromatograms

Computerized data analysis of digitized chromatograms is an objective alternative to the standard method of visual comparison of complex chromatographic profiles for subjective pattern matching. Here the compositional information in selected chromatographic regions is analyzed without prior compound selection and quantification. An initial high data quality as well as appropriate data preprocessing is a prerequisite since variations unrelated to the chemical composition need to be removed. A method for preprocessing chromatographic data was proposed by Christensen *et al.* (2005) and this same method is used in case study 4. Time shifts, baselines, concentration effects, and sensitivity changes are the three main issues that need to be taken care of before digitized chromatographic data can be analyzed properly by PCA. In the procedure outlined here, three preprocessing methods are applied: derivatization, normalization, and retention time alignment.

Baseline removal
Baselines can negatively affect the data analysis and should thus be removed (Christensen *et al.*, 2005). Derivatization and polynomial or piecewise-linear baseline fits are two methods that can be employed for removing baselines. While the latter are subjective and uncertain when peaks are not baseline separated, derivatization is highly objective and data varies around zero (i.e., the first derivative is zero at peak maximum). A disadvantage of derivatization is, however, that it amplifies noise (reduces the signal-to-noise ratio). Although combining derivatization with smoothing (e.g., Satitsky-Golay) reduces noise, we do not recommend this – to avoid unnecessary alterations of the original chromatographic data.

There exist numerous ways of calculating the first derivative. In case study 4, it was calculated numerically, as the difference of consecutive points, which makes integration easy by cumulative summation.

Normalization
As emphasized previously, concentrations are of less importance compared to the relative compound distribution. Normalization to constant area is a common procedure used to compensate for concentration effects and sensitivity changes. In case study 4, Equation 17.6.1 was used for normalization.

$$x_{nj}^N = \frac{x_{nj}}{\sqrt{\sum_{j=1}^{J} x_{nj}^2}}$$ (17.6.1)

where x_{nj} is the first derivative of the nth chromatogram at the jth retention time and J is the total number of retention times. This method implies the so-called "closure" of the data set, that is if one peak increases, the size of the other peaks decrease (Johansson *et al.*, 1984). In situations where the amount of information is limited (i.e., few peaks), this may lead to artificial correlations in data, which makes chemical interpretation of the results more difficult. In such cases, more complex normalization schemes by using only a limited set of retention times referring to specific peaks could be adopted. To reduce the uncertainty and closure effects, these peaks should preferably be large and relatively constant in oil samples of different origin. Under such

circumstances the denominator in Equation 17.6.1 is modified to the sum of the squared first derivatives of selected retention times.

Chromatographic alignment
The most severe impediment to multivariate statistical analysis of digitized chromatograms is, however, the inevitable retention time shift caused largely by deterioration of the capillary column (Fraga *et al.*, 2000). Hence, PCA will model even small time shifts as a source of variation, which will be confounded not only with the first, but also with subsequent components, which prevents any sensible interpretation of the PCA without alignment. A number of alignment methods have been proposed to correct for retention time shifts in chromatographic data (Wang and Isenhour, 1987; Nielsen *et al.*, 1998; Fraga *et al.*, 2000; Johnson *et al.*, 2003). Dynamic time warping (DTW) and correlation optimized warping (COW) are methods that seem to work for a broad range of chromatograms (Nielsen *et al.*, 1998; Pravdova *et al.*, 2002; Eilers, 2004; Tomasi *et al.*, 2004; Christensen *et al.*, 2005).

In case study 4, COW was used to align chromatograms of steranes (m/z 217) but DTW showed comparable results (Christensen *et al.*, 2005). With the COW procedure, a target chromatogram is selected and divided into segments; then the optimal boundary positions for corresponding segments are determined separately for each of the remaining chromatograms ("sample chromatograms"). All combinations of segment boundaries are evaluated and the one that maximizes the sum of the correlations between corresponding segments in a sample and the target chromatogram gives the optimal chromatographic alignment (Nielsen *et al.*, 1998). When the segment length in a sample and the target chromatogram is different, the former is linearly interpolated to the same number of points as the latter. Consequently, the chromatograms are aligned along the time axis by local compression or expansion. Together with the segment length the so-called "slack parameter," which determines the number of data points each boundary is allowed to move, is the parameter that is used as input to the COW algorithm and the value of these parameters affects the quality of the alignment. Correlation optimized wraping is robust with respect to the choice of segment length and slack (Tomasi *et al.*, 2004), but some optimization is necessary. This is not further elaborated here but the use of COW for aligning chromatograms is described in several publications (Nielsen *et al.*, 1998; Tomasi *et al.*, 2004). The COW algorithm is not yet included in any of the standard software packages available for multivariate statistical analysis. The preprocessing including time-alignment using COW can be performed in MATLAB ® 6.5 (The MathWorks) and algorithms for COW can be downloaded from the Internet, for example from the website www.models.kvl.dk.

17.6.4 Variable Selection or Scaling

Although data have been preprocessed to reduce information unrelated to the chemical composition, some diagnostic ratios or retention times are more informative than others and some may even lead to incorrect conclusions when they are kept in the data set. Hence, a proper variable selection or scaling is important in order to keep the uncertainties to a minimum and yield reliable results. This is not only the case when multivariate statistical data analysis is used to compare the chemical composition of oil samples, but should always be an important part of a chemical fingerprinting approach.

However, two of the main purposes of using multivariate statistical methods for chemical fingerprinting are to reduce the noise and increase the "resolution power" of the analysis by using a larger part of the compositional information.

Furthermore, the method should be more or less unsupervised, which increases the objectivity (decrease the bias and uncertainty) of the chemical fingerprinting. The resolution power of a PCA model is defined in Christensen *et al.* (2004), as its ability to distinguish dissimilar samples (Christensen *et al.*, 2004). Generally, the resolution power increases when more variables are included in the analysis due to noise reduction and because more independent variations may be included in the data set (e.g., marine vs terrestrial source, thermal maturity, and boiling point range). However, when more uncertain variables are included in the data set as well as variables affected by weathering, the resolution power decreases and the latter can lead to misclassification of oil samples.

In general, three factors affect the diagnostic power (DP) of a specific variable and its contribution to the resolution power of the PCA model: (1) analytical precision, (2) variability between oils from different sources, and (3) susceptibility to weathering.

The analytical precision is an important factor for selecting the most diagnostic variables, whether it is a diagnostic ratio or a relative abundance in chromatograms. The precision of a diagnostic ratio depends on the robustness of the GC-MS analysis, and low robustness, which show up as poor reproducibility may lead to diagnostic ratios with low precision. If the reproducibility varies along retention time and for different m/z chromatograms (e.g., changes in sensitivity), this leads to systematic changes in the ratios. Furthermore, the precision of a diagnostic ratio depends on signal-to-noise ratio, peak resolution, and baseline distortion. Whereas the precision increases when the signal-to-noise ratio and peak resolution increases, it decreases when the baseline becomes more distorted. When the diagnostic ratios are based on quantitative data, the precision also depends on the variability in the spiking of surrogate standards, the uncertainties of the calibration curve, and how well the chemical properties of the surrogate standards correspond to that of the measured compounds. For the digitized chromatograms, the robustness are of equal or even larger importance. In particular, variations in peak shape (e.g., from symmetrical to tailing) during column deterioration will affect the multivariate data analysis negatively due to changes in intensity distribution of adjacent retention times within a peak region. Peak quantification and hence diagnostic ratios is less affected by these factors since peak areas and heights are less dependent on peak shape.

The selection process is further complicated by the fact that some variables may be heavily affected by weathering. As opposed to analytical precision, weathering may systematically affect the distribution of biomarkers in oil spill samples. As described in previous sections of this chapter, weathering rarely affects the distribution of biomarkers in the initial phase following an oil spill. However, for heavily affected spill samples collected weeks or months after oil spills, the biomarker composition may have been affected by weathering (see Section 17.4.4).

The third factor, which affects the DP of a variable, is its variability in oils from different sources. As described in previous sections of this chapter, some diagnostic biomarker ratios describe the source rock whereas others describe the thermal maturity. In that regard, the DP is high for variables with large variability between oils from different sources. Conversely, the DP is low for variables with low variability between oils from different sources. Hence, a variable with low analytical precision may have the same DP as a variable with higher analytical precision as long as its variability between oils from different sources increases accordingly.

Two criteria for variable selection are presented in the following sections. The first criteria have been suggested by Stout *et al.* (2001) and includes identification of diagnostic ratios based upon PAHs and biomarkers that were independent of weathering; and precisely measured, both of which was determined by statistical analysis of the data. In the criteria suggested by Stout *et al.* (2001), diagnostic ratios with analytical relative standard deviation (RSD_A) larger than 5% was excluded from the PCA (Equation (17.6.2)). In this way, 19 of the original 45 ratios were selected for PCA. Christensen *et al.* (2004) suggested another approach, applying Equation (17.6.3) to calculate the DP of each ratio and subsequently excluded or downscaling the ones with lowest DP.

$$RSD_A > 5\% \qquad (17.6.2)$$

$$DP = \frac{RSD_V}{RSD_A} \quad \text{or} \quad DP = \frac{RSD_V}{RSD_S} \qquad (17.6.3)$$

where RSD_A is the random error arising from the chemical analysis, while the relative sampling standard deviation (RSD_S) is the combined random errors from the chemical analysis as well as the sample variation (relative sampling standard deviation). The latter includes the sample heterogeneity and the effect of weathering. DP is defined as the relative standard deviation of a diagnostic ratio in oils with different origin (RSD_V) divided by RSD_A or RSD_S. The distribution within each diagnostic ratio calculated from replicate analysis of a laboratory reference oil (RSD_A) and replicate spill samples (RSD_S) can be evaluated based on their skewness and kurtosis, since normal distributed data facilitates the use of parametric statistical methods.

In case study 3, variable-outlier detection is introduced as an integrated part of a forensic oil spill identification methodology. The DP is used for variable selection optimizing the resolution power of the PCA model. Specifically, the resolution power was assessed by visual inspection of the scores of PCA models leaving out an increasing number of ratios starting with those with the lowest DP. In case study 4, a related selection method is applied to select retention times based on their analytical precision (RSD_A). Retention times are sorted with regard to their variation in replicates and subsequently an increasing number of retention times are excluded from the PCA, starting from the one with the highest relative standard deviation and progressively excluding more uncertain ones. The optimization are in this study done by minimizing the variance of replicates for each oil sample in the multivariate analysis compared to the total variance explained by the statistical model describing the variations of the entire data set.

Instead of deselecting uncertain variables prior to PCA, variables can be scaled in regard to their uncertainty. Hence, variables with high uncertainty are downscaled compared to variables with low uncertainty. This approach is more objective than deselecting variables, because by using the latter method the model is dependent on the number of variables included in the model and hence the optimization method. Weighted least squared fitting of the principal components are used in both case studies to scale variables in accordance to their precision.

17.6.5 A Hint of Theory

Preprocessed data exposed to variable selection can be collected in a two-way data matrix ($\mathbf{X}$) with samples as rows and variables as columns (Figure 17.6.1). Multivariate statistical methods, specifically two-way decomposition methods such as PCA works on two-way data ordered in a matrix.

Multivariate statistical methods based on bilinear modeling consist of a broad range of decomposition and calibration methods, for example singular value decomposition (SVD),

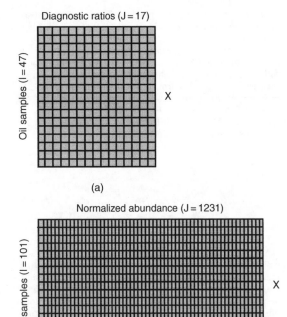

Figure 17.6.1 *Two-way data matrix (**X**) with oil samples as rows and variables (diagnostic ratios) as columns. The size of **X** is* I *(total number of oil samples)* × J *(total number of diagnostic ratios). The data matrix illustrated in a) is the one used in case study 3 with* I= 47 *and* J= 17, *whereas b) is used in case study 4 with* I= 101 *and* J= 1231.

principal component analysis (PCA), principal component regression (PCR), and partial least squares regression (PLS). These methods are well established for analyzing multivariate two-way data matrixes comprised of samples × variables (e.g., oil samples × diagnostic ratios) (Stout *et al.*, 2001; Christensen *et al.*, 2004). As mentioned previously, this tutorial focuses on PCA, which works by bilinearly decomposing the data matrix (X) of size I (oil samples) × J (variables) into products of scores, t ($I \times 1$), and loading

vectors, p^T ($1 \times J$) (i.e., transposed loading vectors), plus residuals, E ($I \times J$). Principal Component analysis can be described in the terminology of linear algebra by vectors and matrices, and readers not familiar with the basic theory of linear algebra are referred to introductory textbooks on linear algebra.

The decomposition of X into products of scores and loadings are shown graphically in Figure 17.6.2 for a two-component PCA model.

It is important to acknowledge that multivariate statistical decomposition methods such as PCA summarizes the information in many correlated variables into few so-called "PCs" that are weighted sums of the original ones. A one-component bilinear model ($t_1 \times p_1^T$) is fitted by a least-squares criterion so that it describes the most prominent trend in data (i.e., direction of largest variation). The bilinear one-component model can be illustrated graphically in the simple case of three correlated diagnostic ratios essentially showing the same generic phenomenon (Figure 17.6.3). The example describes the analysis of three diagnostic ratios of terpanes that distinguish between marine and terrestrial sources $C_{23}/C_{30}\alpha\beta$-hopane, $C_{24}/C_{30}\alpha\beta$ hopane, and (C_{23} + C_{24})/($C_{29}\alpha\beta$ hopane + $C_{30}\alpha\beta$ hopane), for 100 oil samples.

The straight line in Figure 17.6.3a is the first principal component (PC1) found by PCA and corresponds to the least-squares solution (i.e., minimization of the squared residuals) of a bilinear model using SVD to estimate the model. Scores ($t_{i,1}$) can then be obtained by projecting each sample onto this new component, illustrated by dotted lines in Figure 17.6.3a. Loadings can graphically be interpreted as the cosine of the angle between the original variables and the PC. Hence, the loading for a diagnostic ratio increases the smaller the angle between the linear fit (PC1) and the original axis in the coordinate system ($x, y,$ or z). The loading values are between 0 and 1 (the loading vectors are orthonormal), where the latter is the case when the original variable (axes) equals the PC. The loading value is equal to zero when the original axes are orthogonal to the PC and hence show different phenomena (e.g., depositional environment, thermal maturity, in-reservoir degradation). In the example shown in Figure 17.6.3, the corresponding loadings for the three diagnostic ratios are 0.63, 0.48, and 0.61, respectively.

In Figure 17.6.3b, oil samples have been projected onto PC1 and samples located close to one another on the line have similar chemical composition based on the three diagnostic ratios modeled by the straight line. Principal component analysis seeks to extract all systematic patterns of variation in data, leaving the unsystematic in the residuals. In the above example, a one-component model was sufficient to describe the systematic variations. However, when

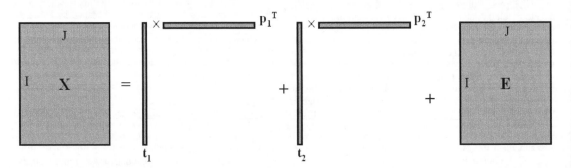

Figure 17.6.2 *Decomposition of the two-way data matrix (**X**) into two principal components. The size of **X** is* I *(total number of samples)* × J *(total number of variables) and the size of* t_k *and* p_k^T *is (*1 × *1) and (*1 × J*), respectively.*

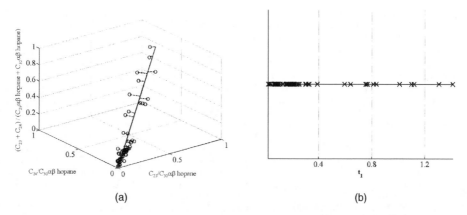

(a)   (b)

Figure 17.6.3 *Graphical illustration of the bilinear one-component model of three diagnostic ratios: C_{23}/C_{30} $\alpha\beta$ hopane, C_{24}/C_{30} $\alpha\beta$ hopane, and $(C_{23} + C_{24})/(C_{29}$ $\alpha\beta$ hopane $+ C_{30}$ $\alpha\beta$ hopane) in 100 oil samples (**O**). (a) The regression line is the first principal component (PC1), which in this example is a weighted linear combination of the three diagnostic ratios ($\mathbf{p}_1 =[0.63, 0.48, 0.61]$). It explains most of the variation in the data set **X** of size 100 $\times$ 3. Dotted lines illustrate the projection of samples onto PC1. (b) The projected values for each oil sample (i.e., the scores) are plotted on PC1 using ($\times$). Oil samples which are located close along PC1 have similar composition of the three original variables.*

analyzing more complex data sets with more variables and samples, additional components describing systematic variation can be extracted from the residuals. The residuals are calculated by subtracting the first component $\mathbf{t}_1 \times \mathbf{p}_1^T$ from **X** (see Equation 17.6.4).

$$\mathbf{X} - \mathbf{t}_1 \times \mathbf{p}_1^T = \mathbf{E}_{-1} \qquad (17.6.4)$$

where $\mathbf{E}_{-1}$ is the residuals after subtracting the first component. A second component $\mathbf{t}_2 \times \mathbf{p}_2^T$ can subsequently be estimated from the residuals, $\mathbf{E}_{-1}$, and so forth for additional components.

The successive components (PC2, PC3, etc.) all describe additional orthogonal trends in decreasing order of importance. Bilinear models have an intrinsic rotational freedom and uniqueness is attained via ad hoc mathematical constraints, such as orthogonality. Owing to the orthogonality constraint, it is unlikely that PCA components describe underlying chemical spectra (for spectral data) or sources of information. With no constraints, an infinite number of straight lines can be drawn in the multidimensional space, which give identical least-squares solution (rotational freedom). The complete decomposition of a data matrix **X** to **T**, **P**, and **E** is illustrated in Figure 17.6.4 and defined in matrix formulation in Equation 17.6.5.

$$\mathbf{X} = \mathbf{T} \times \mathbf{P}^T \qquad (17.6.5)$$

The scores (**T**), loadings (**P**), and residuals (**E**) contain all the information in **X**.

As mentioned earlier, scores are weighted sums of the original variables (e.g., the three diagnostic ratios in the example in Figure 17.6.3) with loadings as weights. Thus, the scores can also be expressed as a linear combination of the original variables. The scores of 100 samples based on the three diagnostic ratios of biomarkers, $C_{23}/C_{30}\alpha\beta$ hopane, $C_{24}/C_{30}\alpha\beta$ hopane, and $(C_{23} + C_{24})/(C_{29}\alpha\beta$ hopane $+ C_{30}\alpha\beta$ hopane) are given in Equation 17.6.6.

$$t_{1,k} = p_{1,k} \cdot x_{1,1} + p_{2,k} \cdot x_{1,2} + p_{3,k} \cdot x_{1,3}$$
$$t_{2,k} = p_{1,k} \cdot x_{2,1} + p_{2,k} \cdot x_{2,2} + p_{3,k} \cdot x_{2,3}$$
$$\cdots\cdots\cdots\cdots\cdots\cdots\cdots\cdots\cdots\cdots$$
$$t_{100,k} = p_{1,k} \cdot x_{100,1} + p_{2,k} \cdot x_{100,2} + p_{3,k} \cdot x_{100,3} \quad (17.6.6)$$

where k is the considered PC ($k = 1$ for the first principal component, $k = 2$ for the second, and so forth). Hence, the scores of sample 2 ($t_{2,k}$) is a weighted sum of the three diagnostic ratios for sample 2 ($x_{2,1}$, $x_{2,2}$, and $x_{2,3}$), where the weights are the corresponding loadings ($p_{1,k}$, $p_{2,k}$, and $p_{3,k}$).

The systematic information in the data is usually retained within the first few principal components whereas the unsystematic (e.g., noise) are left in the residuals, (**E**). Thus, score

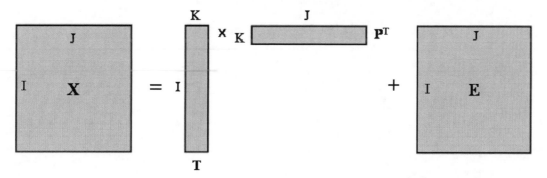

Figure 17.6.4 *Decomposition of the two-way data matrix (**X**) by PCA. A data matrix (**X**) is decomposed into a product of a loading matrix (**P**) and a score matrix (**T**), plus a matrix of residuals (**E**). Dimensions I, J, and K are the number of samples, variables, and principal components, respectively.*

and loading plots retrieve and summarize the systematic information and mutual variation of samples and variables, respectively, which is a great advantage, especially for complex data sets, as the one in case study 4 (101×1231).

17.6.6   Chemometric Software
A large number of commercial softwares which require limited experience with computer programming and multivariate statistical analysis is available. SIMCA from Umetrics (www.umetrics.com) and The Unscrambler from Camo Technologies Inc (www.camo.com) are both all-in-one multivariate data analysis software packages that contain all standard chemometric methods and free trial versions can be obtained from their Internet sites. Furthermore, PCA is included in most statistical software packages such as SAS/STAT® (www.sas.com).

However, for more advanced multivariate statistical analysis and if you are familiar with computer programming, the Matlab® programming environment from Mathworks (www.mathworks.com) is highly recommended. Matlab is a computing language and interactive environment for algorithm development, data visualization, data analysis, and numerical computation, and a large variety of algorithms for multivariate data analysis can be purchased from different Internet sites, for example (www.eigenvector.com) and (www.chemometrics.com). Furthermore, numerous noncommercial sources (i.e., individual scientists and research groups) provide matlab toolboxes, which include wide variety of multivariate statistical tools; some useful sites are www.models.kvl.dk, www.acc.umu.se, www.shef.ac.uk, and www.its.chem.uva.nl/research/pac. The use of chemometric toolboxes for matlab is more flexible than commercial software and enables the forensic chemist to collect a suite of matlab files (i.e., m-files) for specific problems, such as those met in chemical fingerprinting. Hence, the chromatographic preprocessing and statistical tests applied in case study 3 and 4 are, at the time of writing, not yet an integrated part of commercial softwares.

17.6.7   Centering of the Data Matrix
In multivariate data analysis in general and specifically in chemical fingerprinting, the variations around the mean is of prime importance (i.e., patterns of variation). Thus, it can, generally, be recommended to mean-center variables such as diagnostic ratios prior to bilinear modeling, which is a built-in function in chemometric software's. Mean-centering is done algebraically by subtracting the mean from $\mathbf{X}$ (Equation 17.6.7).

$$\mathbf{X}_{mc} = \mathbf{X} - \mathbf{1}\bar{\mathbf{x}} \qquad (17.6.7)$$

where $\mathbf{X}_{mc}$ is the mean-centered data and $\bar{\mathbf{x}}$, a row vector consisting of the means of each variable. For consistency, $\bar{\mathbf{x}}$ is multiplied by a column vector of ones ($\mathbf{1}$) with length equal to the number of samples. Without mean centering, the loadings of the first principal component is often equal to the means of variables, and variations more relevant to chemical fingerprinting are described by higher components (PC2, PC3, and so forth). Mean centering is illustrated graphically in Figure 17.6.5 using the data set of 100 samples and three diagnostic biomarker ratios.

The straight line in Figure 17.6.5a is PC1 (or $\mathbf{t}_1$) for the mean-centered data and in Figure 17.6.5 oil samples have been projected onto PC1. Note that some samples have negative scores while others have positive ones. In fact, due to the mean centering the sum of scores for the 100 oil samples is zero.

17.6.8   Scaling of the Data Matrix
Large noise levels in some variables may hide small but important variations in other if the variables have not been properly weighted *a priori*. Such effects are especially pronounced if the input data cover widely different ranges and may be handled by re-scaling variables in regard to *a priori* knowledge. The standard scaling method consists of scaling each individual variable with regard to the inverse of the total standard deviation of that variable. Such standardization by scaling to unit variance is termed "autoscaling" when it is performed together with mean centering of the data matrix and is the standard method for standardizing the data prior to the analysis.

However, autoscaling is not recommended as a standardization method for chemical fingerprinting data. Diagnostic ratios and digitized chromatograms do not cover different ranges and autoscaling amplifies noisy variables (e.g., less diagnostic ratios and noisy regions of chromatograms). In addition, autoscaling may reduce the influence of highly diagnostic variables in the modeling (e.g., diagnostic ratios with high DP as well as peak regions), because these often have large standard deviations between samples of different origin (RSD_V).

An alternative method, which is a more reasonable way of scaling chemical fingerprinting data, is to scale with respect

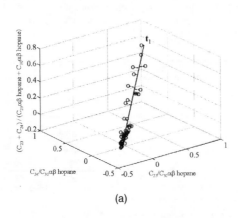

(a)

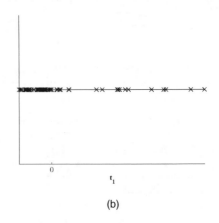

(b)

Figure 17.6.5 *Graphical illustration of the bilinear one-component model in Figure 17.6.3 with mean centering. (a) The regression line is the first principal component (PC1) of the mean-centered data set. Dotted lines illustrate the projection of samples onto PC1. (b) Score values are plotted on PC1 using (×).*

to the analytical or sampling uncertainties. By doing this, variables with large analytical or sampling uncertainties are downscaled compared to variables with low uncertainties. Weighted least squares estimation of the principal components based on the sampling and analytical uncertainties is used in both case studies to improve the resolution power of the PCA model. For chemical fingerprinting, we recommend to use either standard PCA and scaling by the error standard deviations (RSD_A or RSD_S) (which can be done in commercial softwares) or a WLS algorithm such as MILES-PCA (maximum likelihood via iterative least-squares PCA) (www.models.kvl.dk) or weighted PCA (www-its.chem.uva.nl/research/pac). The analytical or sampling uncertainty of oil samples varies depending on, for example, the signal-to-noise ratios of individual peaks, and the chromatographic and mass spectrometric conditions. In the two case studies the error standard deviations are determined from replicate reference samples and replicate oil spill samples exposed to varying degrees of weathering.

17.6.9 Detection of Outliers

In forensic oil spill identification, three types of possible sample outliers can, generally, be present in data:

- Oil samples for which the measured data (diagnostic ratios or normalized chromatograms) contain artifacts/abnormal values, due to instrumental malfunction, lack of proper preprocessing, or effects of heavy weathering.
- Oil samples that have been contaminated from external sources after its release to the environment, thereby changing its composition compared to its source (e.g., pyrogenic input).
- Oil samples with unique biomarker composition.

Also, some variables may be less diagnostic and their presence may introduce noise in the PCA model. However, with proper variable selection (or scaling) such outlying variables are removed prior to the PCA modeling or its importance is made negligible by scaling. This emphasizes the importance of objective and systematic treatment of variables. Furthermore, since the presence of unique biomarkers can be essential for matching oil spill samples with their source(s), as has been thoroughly described previously, oil samples containing unique biomarkers are not true outliers and should hence not be removed.

A large number of methods exist for detecting outliers in data that may have a detrimental effect on the PCA model, and hence affect the fingerprinting, negatively. Thus, it is important that the PCA model describes the variation between oil samples in the data set sufficiently and not solely the abnormal composition of an outlying sample. In the latter case, the resolution power of the PCA model is low. Hence, optimizing the PCA model (e.g., the variables and samples selected for the analysis) based on the resolution power can be used as a tool to detect outliers. Furthermore, sensible data handling and quality control measures will reduce the risk of measurement artifacts and are an important part of tiered oil spill identification methods.

Secondly, a wide range of measures related to the multivariate data analysis can be used to detect outlying samples. Some of the most frequently used methods, that are part of commercial softwares, include plots of the residual **X**-variance vs sample leverage (Figure 17.6.6) and inspection of score and loading plots. The residual **X**-variance of an oil sample is a measure of how well the model describes the sample, whereas the leverage shows how extreme the position of a sample is inside the model (what is its importance). Hence, an oil sample with abnormal high residual variance (e.g., the Lago Medio crude oil in Figure 17.6.6) is poorly described by the model and may thus be an outlier. If this is the case, additional steps have to be taken (e.g., repeated quality control of data) to decide whether or not the sample is an outlier. On the other hand, oil samples with

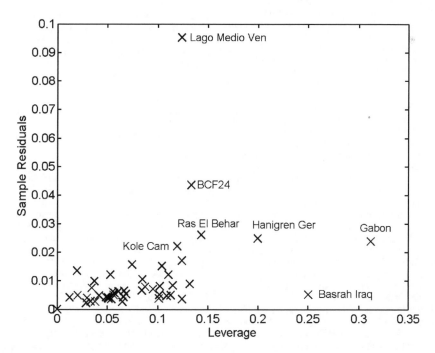

Figure 17.6.6 *Residual x-variance vs sample leverage. The plot is based on data from 47 source oils and 17 diagnostic biomarker ratios listed in Table 17.6.1.*

high leverage (such as crude oils from Gabon and Basrah in Figure 17.6.6) describe no new phenomenon, but abnormal levels of a known phenomenon. Thus, a sample with high leverage may be important for the model, and it should only be removed if it has a detrimental effect on the resolution power and if the sample is unimportant for the fingerprinting (if its composition is very different from that of the spill samples). In the example in Figure 17.6.6, which is part of the initial outlier detection in case study 3, all samples were kept in the analysis.

Visual inspection of score plots for outlier detection is comparable to the leverage plot since abnormal samples have abnormally large score values. However, as opposed to the leverage measure, which offers a general evaluation based on all significant PCs, oil samples can be evaluated along each individual component by visual inspection of score plots. For example, an oil sample may not look like having one or several measurement artifacts on the leverage plot. However, such artifacts often stand out in the higher PCs still containing systematic information (e.g., PC3 or PC4), and their identity (e.g., specific diagnostic ratio affected) can be determined in the corresponding loading plots.

17.6.10 Choosing the Correct Number of Components

The optimal number of PCs to keep in a multivariate model depends on the amount of independent systematic information present in the data set. It is not straightforward to determine the optimal number and numerous methods have been proposed in the literature. For chemical fingerprinting, it is important not to include too many components in the subsequent statistical evaluation because it may decrease the resolution power of the PCA model. On the other hand, it is also important not to leave out information that is relevant for distinguishing closely related samples. Generally, the more samples and types of variations included in the analysis the more difficult it is to determine the correct number of components (i.e., independent chemical variations). Full cross-validation (leave-one-out cross validation) are the most frequently applied method for finding the optimal number of components. In full cross-validation, one sample is left out in each segment and the PCA model is calculated. Other segment sizes (dividing the data into segments and analyzing these segments separately) can be used for cross validation. However, test set validation (i.e., validating the model using an independent data set) is, however, recommended since it is less biased.

A model containing the optimal number of components results in the highest explained variance for the left out samples. Conversely, for models with too few or too many components, the explained variances are lower because of under- or overfitting of the data matrix ($\mathbf{X}$). Hence, the optimal number of components corresponds to the number of PCs where a sharp bend is observed in the explained variance plot when too many components are included.

In case study 3, the optimal number of components was established from cross validation in combination with visual inspection of score and loading plots. In addition to cross validation, a number of formulas has been proposed to help in determining the optimal number of components and these can be used especially under circumstances where it is difficult to determine the correct number based on cross validation and scores and loading plots.

Additional tools not often used in the literature but which in indirect ways make it easier to determine the optimal number of components concern refinements of the data analysis. These refinements consist of applying local PCA models, *a prior* scaling, or weighted least squares PCA. The risk of excluding variables explaining small but important

variations decreases. This is especially true for local PCA models where a subset of samples with similar composition is analyzed separately and hence the total amount of variation in $\mathbf{X}$ decreases (Christensen *et al.*, 2004).

17.6.11 Interpretation of Score Plots

As previously stated, one of the advantages of multivariate statistical methods is the ease by which relations between multiple samples and variables can be resolved and visualized by score and loading plots. Score plots map the main relationships between samples based on the original variables. In Figure 17.6.7, the first principal component PC1 is plotted vs the second principal component PC2 for 47 source oils and 4 samples from the Round Robin exercise based on 17 diagnostic biomarker ratios listed in Table 17.6.1.

Since score plots map the relationships between samples, their interpretation are straightforward. Hence, oil samples located close in score plots have similar chemical composition based on the original variables (17 diagnostic ratios in case study 3). Conversely, oil samples located far apart in score plots have dissimilar chemical composition and this dissimilarity increases as the distance increases. Note that PCs are ordered according to their explained variance, with the first component describing most of the variation and subsequent components (e.g., PC2, PC3, etc.) explaining a decreasing amount of variation. Hence, differences along the first components are more significant than along higher components.

Although the higher components describe a lesser percentage of the total variance, they are not necessarily of lesser importance for the separation of dissimilar oil samples. Especially for large data sets (i.e., many samples and variables), an important separation between similar samples are often found in higher components. Christensen *et al.* (2004) used local PCA models (model of a subset of closely located samples) to focus the data analysis on separating closely related samples with similar biomarker composition. Hence, the first few components described variations relevant for the specific oil spill case by separating related samples in the first few PCs instead of in higher PCs. The latter are more affected by noise since they describe only a small percentage of the total variation in $\mathbf{X}$. In the score plot in Figure 17.6.7, oil samples with similar chemical biomarker composition is grouped together. Closely related oils from the Middle East (Iran Heavy, Iran Light, Dubai Iraq, Basrah Iraq), Venezuela (Lago Medio and Tia Juana), and Russia (Ural and Romanskino) form four separate clusters in the scores plot. Furthermore, the round-robin spill samples are clustered with three of the suspected source oils (A, B, and C).

17.6.12 Interpretation of Loading Plots

Loading plots map the main relationships between original variables and are thus of lesser importance for chemical fingerprinting than scores plots. Loading plots are mainly used for chemical interpretation of the sample correlations observed in score plots and for validating the PCA model. Figure 17.6.8 shows the loading plot (PC1 vs PC2) corresponding to the scores plot in Figure 17.6.7.

Diagnostic ratios located close to one another along a PC co-vary along this component. Thus, since C24TT/(29ab+30ab) and C24TT/30ab are located close in the loading plot (at high positive PC1 and negative PC2) they co-vary. For mean-centered data, ratios located at similar value along a component but with opposite signs countervary (high positive and negative values, respectively).

A chemical explanation of similarities and dissimilarities between samples can be obtained from loading plots. Thus,

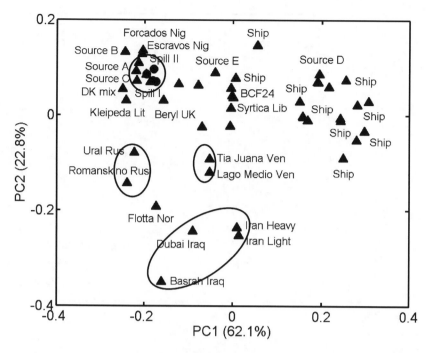

Figure 17.6.7 *Score plot (PC1 vs PC2) of 47 source oils, 4 spill samples from the Round Robin exercise based on 17 diagnostic biomarker ratios. The four significant components (PC1–PC4) describe 93.7% of the variation in **X**.*

Table 17.6.1 *Application of the Variable-Outlier Detection Method to 17 Biomarker Ratios. RSD_S are the Relative Sampling Standard Deviations in Percentages, Estimated from 24 Baltic Carrier Oil Spill Samples, Whereas RSD_V are the Variations Between Source Oils. DP is the Diagnostic Power Based on RSD_S. The Normalization Factor is Omitted in the Description for Brevity.*

Diagnostic Ratios	Code	RSD_S(%)	RSD_V(%)	DP
C_{24} tricyclic terpane/(C_{24} tricyclic terpane + C_{30} αβ hopane)	C24TT/30ab	1.5	37.7	25.7
C_{29} αβ-25-norhopane/(C_{29} αβ-25-norhopane + C_{30} αβ + C_{30} αβ hopane)	%25nor	2.2	43.3	20.0
C_{29} αβ-25-norhopane/(C_{29} αβ-25-norhopane + Ts + Tm)	%25norTsTm	1.9	33.5	17.8
C_{24} tricyclic terpane/(C_{24} tricyclic terpane + C_{29} αβ + C_{30} αβ hopane)	C24TT/(29ab + 30ab)	2.2	34.6	15.9
(C_{29} αβ + C_{30} αβ hopane)/[C_{29} αβ + C_{30} αβ hopane + C_{27} αββ(S + R) + C_{28}αβ (S + R) steranes]	Hop/S	0.4	4.8	11.8
Gammacerane/(Gammacerane + C_{30} αβ hopane)	%30G	1.7	19.3	11.6
Ts/(Ts + Tm)	Ts/Tm	1.2	12.7	10.5
C_{27} (R)/[C_{27} (R) + C_{28} (R) triaromatic steranes]	%TA27	1.0	8.3	8.5
C_{32} αβ hopane (S)/[C_{32} αβ (S) + C_{32} αβ hopane (R)]	%32abS	0.3	2.7	8.0
C_{29} αβ/(C_{29} αβ + C_{30} αβ hopane)	29ab/30ab	1.5	10.0	6.9
C_{29} αββ(S + R)/[C_{29} αββ(S + R) + C_{29} ααα (S + R) steranes]	%29bb	1.0	6.9	6.9
C_{27} diasteranes (R + S)/[C_{27} diasteranes (R + S) + C_{27} αββ(R + S)]	%27dia	2.7	15.3	5.7
C_{29} ααα(S)/(C_{29} ααα(S) + C_{29} ααα(R)	%29aaS	1.7	8.4	5.0
C_{28} αβ hopane/(C_{28} αβ + C_{30} αβ hopane)	%28ab	6.7	29.0	4.3
C_{29} αββ(S + R)/[C_{27} αββ(S + R) + C_{28} αββ(S + R) + C_{29} αββ(S + R) steranes]	%29bbSt	1.6	6.3	3.9
C_{27} αββ(S + R)/[C_{27} αββ(S + R) + C_{28} αββ(S + R) + C_{29} αββ(S + R) steranes]	%27bbSt	1.5	3.5	2.4
C_{28} αββ(S + R)/[C_{27} αββ(S + R) + C_{28} αββ(S + R) + C_{29} αββ(S + R) steranes]	%28bbSt	1.8	4.0	2.2

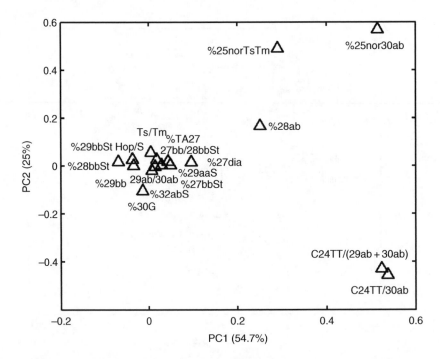

Figure 17.6.8 *Loading plot (PC1 vs PC2) of 47 source oils, 4 spill samples from the Round Robin exercise based on 17 diagnostic biomarker ratios. The four significant components (PC1-PC4) describe 93.7% of the variation in* **X**.

oil samples located at high PC1 in the score plot have high values of C24/(29ab+30ab), C24TT/30ab, and %25nor30ab, whereas the values of these diagnostic ratios are low in samples located at high negative scores (e.g., Basrah Iraq). This is used in case study 4 to determine the chemical meaning of PC1–PC4, which describe independent phenomenon related to the refining process, oil-source rock, and thermal maturity.

Furthermore, loadings can be used to validate the PCA model by comparing the covariance of variables with *a priori* knowledge. Diagnostic ratios, which describe the same phenomenon from *a priori* knowledge, should co-vary in loading plots. If this is not the case, the respective PC either describes noise or the *a priori* knowledge is incorrect. This can also be used for outlier detection of variables if these co-vary with variables by which they are not expected to do.

17.6.13 Analysis of Spill Samples
To minimize bias of the PCA model, we suggest that the PCA model be calculated using only the calibration set consisting of source oils with different origin. The scores of oil spill samples included in the so-called "test set" ($\mathbf{T}_{new}$) should be calculated by projecting the data ($\mathbf{X}_{new}$) on the loadings ($\mathbf{P}$) using Equation 17.6.8.

$$\mathbf{T}_{new} = \mathbf{X}_{new}\mathbf{P}(\mathbf{P}\mathbf{P}^T)^{-1} = \mathbf{X}_{new}\mathbf{P} \qquad (17.6.8)$$

Projection of the test set on the loadings from the calibration set increases the objectivity of the analysis since the model is less biased. Furthermore, it enables evaluation of the residuals as a means to determine whether the model describes the test set samples satisfactorily.

17.6.14 Matching of Suspected Source Oils and Spill Samples
A range of methods can be used to classify and match suspected source oils and spill samples based on PCA. The

three methods described in the following section are based on the scores of the significant components.

The first and most straightforward method for evaluating the similarity of oil samples is by visual comparison of spill samples and suspected sources from score plots. Although the data analysis can comprise a large number of oil samples and variables, this procedure is somewhat subjective and depends on the distribution and type of oils in the oil database. Consider a database comprised of 50 heavy fuel oils of similar chemical composition and one light fuel oil with very different chemical composition. Under such circumstances, spill/source identification of a light fuel oil will most likely result in positive match of the light fuel oil in the database. Although this oil may have a rather different biomarker composition, it is the oil in the database with most similar composition.

The objectivity of the matching process can be improved by comparing samples using similarity indices such as the correlation coefficient. Thus, the similarity of oil samples can be calculated based on the scores (Equation 17.6.9).

$$r = \frac{\sum\limits_{f=1}^{F}(t_{1f} - \bar{t}_1)(t_{2f} - \bar{t}_2)}{\sqrt{\sum\limits_{f=1}^{F}(t_{1f} - \bar{t}_1)^2 \sum\limits_{f=1}^{F}(t_{2f} - \bar{t}_2)^2}} \qquad (17.6.9)$$

Where r is the correlation coefficient, t_{1f} and t_{2f} the scores for the first and second sample of the fth component, and $\bar{t}_1$ and $\bar{t}_2$ the mean scores for the first and second sample. A range of similarity measures can be applied to compared scores as well as the preprocessed variables.

The third evaluation method is described thoroughly in Christensen *et al.* (2004) and is used to compare two spill samples (spill I and spill II) to 47 source oils in case study 4 based on four groups of diagnostic ratios (biomarker, PAH, C_2-phenanthrene, and C_3- and C_4-naphthalene ratios).

One of these groups was comprised of the 17 diagnostic biomarker ratios listed in Table 17.6.1. The method consists of statistical evaluation based on the overall null-hypothesis (H_0) that the spilled oil and the tested source oil are identical. The optimal number of PCs in a PCA model (i.e., the retained PCs) is tested independently using the inequality in Equation (17.6.10) accepting a certain error level (often a 5% error level is used, $\alpha = 0.05$). If the inequality is false in at least one of these tests, the overall H_0 is rejected and the tested source oil is "beyond reasonable doubt", not the source of the spill.

$$\frac{\left| \bar{t}_k^{(spill)} - \bar{t}_k^{(source)} \right|}{s_k^{(pooled)} \sqrt{\frac{1}{n_{spill}} + \frac{1}{n_{source}}}} \leq q_{\alpha, d.f.} \qquad (17.6.10)$$

Where $s_k^{(pooled)}$ is the pooled standard deviation which can be calculated from either the analytical or the sampling standard deviation. In addition, n_{spill} and n_{source} are the number of replicates used to calculate the mean scores along the kth PC of the spilled oil ($\bar{t}_k^{(spill)}$) and a source oil ($\bar{t}_k^{(source)}$). $q_{\alpha, d.f.}$ is the α-quantile from t-student's distribution with $d.f.$ degrees of freedom and $s_k^{(pooled)}$ the pooled standard deviation for the kth PC.

Since the method includes multiple comparisons, the risk for an overall type one error increases (i.e., that H_0 is rejected when it is true). One way to compensate for this is to apply the Bonferroni correction to the alpha value (Massart et al., 1997).

$$\alpha = \frac{\alpha_{overall}}{K_{all}}; \qquad K_{all} = \sum_{g=1}^{G} K_g \qquad (17.6.11)$$

where K_g is the optimal number of PCs for the gth group of diagnostic ratios, G the total number of groups, α the quantile for the individual comparisons, and $\alpha_{overall}$ the one for the entire set of K_{all} comparisons. There are three possible outcomes of the classification of a source oil with respect to the spilled oil in the multiple tests. Below is shown the criteria used in Christensen et al. (2004); however, other criteria can be used.

Positive match: H_0 is acceptable (5% error level) for the tested source oil and the spill sample and H_0 is rejected for all other source oils in the data set.
Probable match: H_0 are acceptable for the tested source oil and spill sample, but the same holds for other source oils.
Non-match: H_0 is rejected.

17.7 APPLICATION OF MULTIVARIATE STATISTICAL METHODS FOR BIOMARKER FINGERPRINTING

The following sections describe two case studies where multivariate statistical analysis has been used for chemical fingerprinting of petroleum biomarkers in oil. The methods applied in the two case studies follow the same concept based on:

1. Chemical analysis and development of an oil database
2. Preprocessing of chromatographic data
3. Variable selection or scaling
4. Multivariate statistical analysis (PCA or WLS-PCA)
5. Matching of oil samples.

However, the methods applied in the two case studies are very different and uses different types of variables (i.e., diagnostic ratios and digitized chromatograms) and preprocessing methods.

17.7.1 Oil Samples, Analysis, and Chromatographic Data Preprocessing

Oil samples from the oil database at the forensic oil spill laboratory, National Environmental Research Institute, Denmark, was used in both studies. Oils in the database are stored at $-20\,^{\circ}C$ in airtight vials at a total oil concentration of 2000 mg/l in dichloromethane (Rathburn, HPLC grade). The oil database consisted mainly of a large selection of crude oils and refined petroleum product's ship samples (i.e., heavy fuel oils, light fuel oils, and lubricating oils) send to the forensic laboratory. A subgroup of oils from the oil database consisted of extracts of environmental oil spill samples collected after the Baltic Carrier oil spill, Denmark 2001, and simulated spill samples and "suspected source oils" from the Nordtest round-robin oil spill exercise (Faksness et al., 2002).

Baltic Carrier oil spill samples were randomly sampled 0–14 days after the accident from the area affected by the spill. The oil spill occurred on March 29, 2001 when the oil tanker, MT Baltic Carrier collided with the bulk carrier Tern in the Baltic Sea, southeast of Falster (Kadetrenden) between Denmark and Germany. The collision resulted in the release of about 2700 tons of heavy fuel oil (bunker oil). Due to currents and wind conditions open sea response to the spill was impossible and the oil slick drifted toward the southern part of Denmark, polluting the coastal and near-coastal areas of Falster and the isles of Møn, Bogø, and Farø.

The round-robin exercise was an artificial oil spill scenario consisting of two artificially evaporated spill samples and five suspected sources. The two spill samples were oil/water emulsions with approximately 75% water (spill I and spill II). The five source oils were crude oils from Oseberg East (Source A), Oseberg southeast (Source B), Oseberg Field Centre (Source C), and two heavy fuel oil IF 180-Shell refinery (Source D) and IF 180-Esso refinery (Source E).

Oil samples were analyzed on a HP-6890 gas chromatograph (Agilent Technologies) equipped with a 60 m HP-5MS capillary column (0.25 mm ID × 0.25 μm film) and interfaced to a HP-5973 quadrupole mass spectrometer (Agilent Technologies) operating in EI mode. Forty-eight mass fragments were analyzed in six groups of 14–15 ions using SIM. The focus of case study 3 is on the use of biomarker data from m/z 191 (triterpanes), 217 and 218 (steranes), and 231 (triaromatic steroids) for oil spill identification.

In the analytical sequence, a blank and laboratory reference oil was analyzed between every five oil samples. The replicate references are used for external normalization of diagnostic ratios, estimation of analytical uncertainties, optimization of warping parameters, and for quality control. The reference sample was a mixture of oil types (i.e., 1:1 mixture of Brent crude oil (North Sea crude) and the heavy bunker oil from the Baltic Carrier oil spill). Reference samples was analyzed in-between every five oil samples in the analytical sequence.

The case studies present two fingerprinting methods that consist of a collection of procedures. Case study 3 is based on multivariate analysis of diagnostic ratios, where peak quantification is a prerequisite. Conversely, in case study 4 sections of digitized chromatograms of biomarkers (m/z 217) are analyzed without any peak quantification. Instead chromatograms are preprocessed by derivatisation (to remove baseline effects), normalization (to remove sensitivity and concentration effects), and subsequent retention time alignment using a time warping procedure. The two methods for forensic oil spill identification are extensions of the standard methods based on double plots of diagnostic ratios and subjective pattern matching of complex chromatographic profiles.

17.7.2 Case Study 3 - Chemical Fingerprinting of Biomarker Ratios

This case study concerns chemical fingerprinting of source oils and spill samples using diagnostic biomarker ratios and multivariate statistical analysis. The procedure used in this case study is elaborated in Christensen *et al.* (2004). The rationale of the suggested fingerprinting method is that identity prevails if no significant chemical differences within selected diagnostic biomarker ratios can be demonstrated. The selected ratios are highly robust and discriminating and by comparing these ratios in multiple oil samples (inclusive suspected source oils and spill samples) instead of proving an all-encompassing similarity, evaluation of identity becomes viable and can be tested statistically.

Specifically, the method consists of chromatographic data preprocessing, variable-outlier detection, multivariate statistical analysis, statistical evaluation of dissimilarities, and refinements of the analysis (i.e., local PCA and weighted least squares PCA). The spilled oil is compared statistically with source oils and identity prevails if no statistical significant differences can be observed and this does not hold for other source oils. A flowchart of the method is shown in Figure 17.7.1.

17.7.2.1 Data

Data from 47 source oils, 4 round-robin spill samples, 21 replicate laboratory reference samples, and 24 Baltic Carrier spill samples were used in the case study. The 47 source oils were used to calculate the PCA models, whereas round-robin spill samples were projected onto the model afterward and subsequently matched to the source oils. Diagnostic ratios were calculated by Equation 17.3 and the replicate analyses of the reference oil were used for the external normalization (DR^R). Finally, the Baltic Carrier oil spill samples were applied to calculate the sampling uncertainties for each diagnostic ratio and subsequently to estimate the standard deviation of the scores of the individual oil samples. The calibration data matrix ($\mathbf{X}$) was of size 47×17 (samples $\times$ diagnostic ratios) and the test set ($\mathbf{X}_{new}$) of size 4×17. The 17 diagnostic biomarker ratios used in this study have been used frequently for chemical fingerprinting (Wang *et al.*, 1999a; Stout *et al.*, 2001; Daling *et al.*, 2002; Faksness *et al.*, 2002). In Christensen *et al.* (2004), three additional groups were used for the oil spill identification.

17.7.2.2 Variable-Outlier Detection

A variable-outlier detection method was used to deselect diagnostic ratios with low sampling uncertainty. Since the compositions of biomarkers in oils are affected only to a minor degree of weathering, the sampling standard deviation (RSD_S) are similar to the analytical standard deviation (RSD_A) calculated from replicate references. Here, RSD_S were calculated for each diagnostic ratio from the 24 replicate Baltic Carrier spill samples and the DP was calculated from Equation 17.6.3. RSD_S, RSD_V, and DP for the 17 diagnostic fingerprinting ratios are listed in Table 17.7.1. Since either of the ratios was influenced by large measurement errors or weathering, they were all selected for analysis.

17.7.2.3 Principal Component Analysis

Principal component analysis with mean centering was applied to four different groups of diagnostic ratios using the 47 source oils. The optimal number of components was for each group of diagnostic ratios established via cross validation (two oils per validation segment) to 4 (biomarker ratios), 4 (PAH ratios), 3 (C_2-phenanthrene ratios), and 3 (C_3-C_4-naphthalene ratios), respectively. The root mean squared error in cross validation showed a sharp bend at the optimal number in all four cases. These choices were further verified by evaluating the loading and scores.

17.7.2.4 Statistical Evaluation

The $s_k^{(pooled)}$ is calculated from the standard deviation of replicate spill samples from the Baltic Carrier oil spill. The statistical tests comprised 14 individual comparisons per H_0 $(4 + 4 + 3 + 3 = 14)$ using 23 *d.f.* (since, 24 spill samples were used to calculate the uncertainty) for both $s_k^{(spill)}$ and $s_k^{(source)}$ ($\alpha_{overall} = 0.05$, $\alpha = 3.57 \times 10^{-3}$, $q_{a,46} = 2.816$). The statistical analysis showed that H_0 was acceptable for Source A and Spill I as well as for Source C and Spill II if the C_2-phenanthrene ratios were left out of the analysis (i.e., their uncertainty have been clearly underestimated). Consequently, Sources A and C are classified as positive matches and all other potential sources as nonmatches.

17.7.2.5 Refinements of the Data Analysis

A prerequisite for defensibly linking the spilled oil to a ship is that the data set contains all possible source oils from the suspected ship(s), that the spill is composed of a single source and that the RSD_S have been adequately estimated. When this is fulfilled the data analysis is completed when source oils are classified as positive match or nonmatch. However, when source oils are classified as probable match the data analysis needs to be refined to determine whether the spilled oil originate from one of them.

The probability of yielding an inconclusive answer (probable match) increases with the sampling uncertainty, the size of the data set, and its heterogeneity. Using WLS fitting of the PCs thereby downscaling the influence of the most uncertain DRs may reduce the effect of high sampling uncertainties. When closely related source oils are present in a large data set, it is likely that major trends, which are represented by the first PCs, mask the differences between these oils. The components that describe these minor differences may not be included in the optimal PCA model as they represent a minimal variation compared to the total (Jolliffe, 1986).

If the differences between closely related source oils are described by few components compared to K_{all}, the Bonferroni adjustment may become too large leading to an increase in the number of probable matches and in the risk of a type 2 error (i.e., H_0 is accepted when it is false). To ensure that small but important variation is not disregarded as well as decrease the risk of a type 2 error, PCA can be applied to a subset of source oils that lie close to the spilled oil (e.g., the probable matches) along the components retained in the original PCA model ("local" PCA).

Although the present data set is small and positive matches were found using standard PCA, the effects of WLS-PCA and local PCA were tested. Figure 17.7.2a, b shows the score plots (PC1 vs PC2) for local PCA using a set of 21 PAH ratios (their specific identity are not shown) with least-squares fitting and WLS fitting of the PC model.

The optimal number of PCs was two describing 56.6% and 19.2%, respectively, of the variation in the data set. The RSDs of the 21 diagnostic ratios varied between 0.08 and 3.3% and the example illustrates nicely how the resolution power of the PCA can be improved when using the additional information gained from the uncertainty of the diagnostic ratios in the fitting of the PCA model. The ellipses shown in the figures are the sampling uncertainty of individual samples estimated from RSD_S of replicate Baltic Carrier spill samples. The two fitting methods performed almost identically for the 17 diagnostic biomarker ratios since their uncertainty (RSD_S) are of the same magnitude (Christensen *et al.*, 2004).

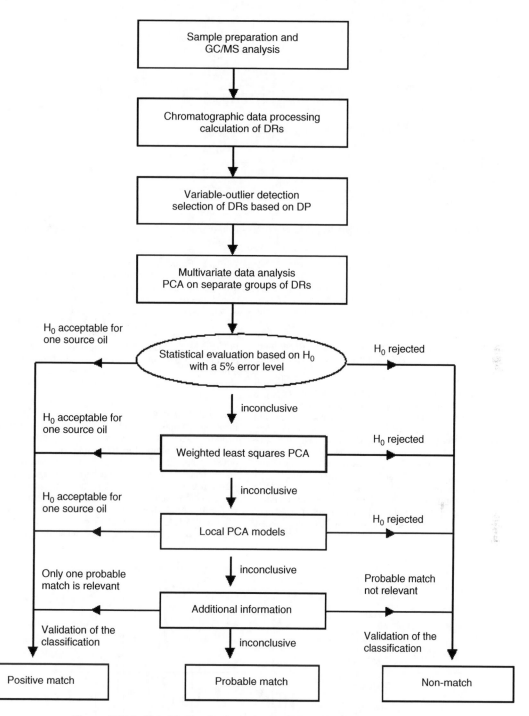

Figure 17.7.1 *Flowchart for the forensic oil spill identification methodology.*

Weathering is considered implicitly in the statistical evaluation. Diagnostic ratios affected by weathering are either left out in the initial outlier detection or it is partially accounted for by the WLS approach. The main assumptions in the approach for multivariate statistical analysis of diagnostic fingerprinting ratios concern the estimation of the analytical and sampling uncertainties. Hence, it is important to obtain good estimates of the uncertainties and one should always try to obtain replicate analyses of source oils and spill samples to estimate the true analytical uncertainties. However, it is often impossible to obtain the true sampling uncertainties for individual oil samples, since they can be obtained only from spill samples exposed to different degrees of weathering, which are rarely available in spill cases.

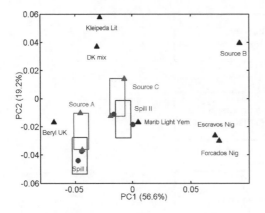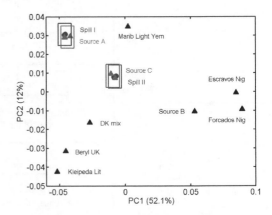

Figure 17.7.2 *Local PCA illustrating the statistical testing by squares. (a) Ordinary PCA; (b) WLS-PCA.*

17.7.3   Case Study 4–Chemical Fingerprinting of Digitized GC-MS Chromatograms

In this case study, 101 digitized chromatograms of m/z 217 were used for forensic oil spill identification (Figure 17.7.3). This m/z chromatogram comprises tricyclic and tetracyclic steranes (Moldowan *et al.*, 1992a) and other compounds yet unidentified. Many peaks coelute and hence only a fraction of these is commonly employed for forensic oil spill identification (Wang *et al.*, 1994; Barakat *et al.*, 1999; Daling *et al.*, 2002).

The data is composed of 51 source oils, 16 weathered Baltic Carrier oil spill samples, 18 replicate references, 6 spill samples from the round-robin exercise (spill I and spill II analyzed in triplicate) and 10 replicate analyses of selected source oils (101 oil samples in total). The chromatograms comprised 2510 data points which after preprocessing were reduced to 1231 (i.e., retention time window from 26 to 42 minutes) by omitting the parts without chemical information. The data were divided into a *calibration set* of 61 chromatograms used to calculate the PCA (61×1231), a *reference set* containing the 18 references (18×1231), and a *test set* comprised of the replicate Baltic Carrier spill samples and the two round-robin spill samples analyzed in triplicate (22×1231).

17.7.3.1   Preprocessing of Data

Combinations of segment lengths from 25 to 225 data points with increments of 25 and slacks between 1 and 4 were tested to find the optimal warping parameters. A reference oil with average time shifts was selected as a target chromatogram to reduce the need for correction. The best alignment based on the explained variance of a one-component PCA model is obtained with a segment length of 175 data points and a slack of 3. Figure 17.7.4 illustrates the effect of time warping on a section of the derivatized and normalized chromatograms (35.5–38.5 minutes).

17.7.3.2   Principal Component Analysis

Principle component analysis was applied to the mean-centered calibration set with and without warping. Figure 17.7.5a shows the scores plot of PC1 vs PC2 without warping. The 16 Baltic Carrier oil spill samples and the 18 replicate references form a pattern typical of situations where the retention time shift is the main cause of systematic variation. Hence, the PCA describes the misalignment rather than the chemical composition of the oil samples. Figure 17.7.5b shows the score plot of PC1 vs PC2 after warping. The improvement in the resolution power of the

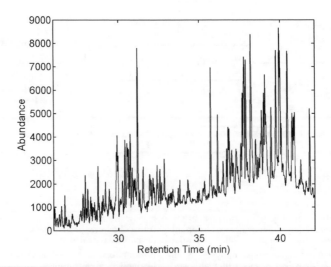

Figure 17.7.3 *Chromatographic profile of m/z 217. Among others, this profile contains tricyclic steranes (eluting between 26 and 34 minutes) and tetracyclic steranes (eluting between 35 and 42 minutes).*

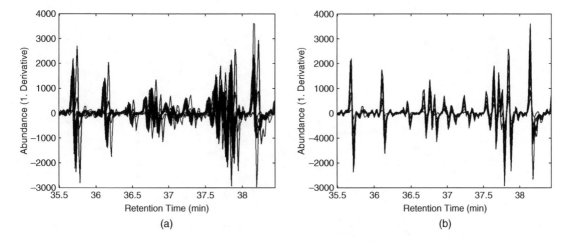

Figure 17.7.4 *First derivative of a section of m/z 217 for five references and five oil samples: (a) before warping and (b) after warping using COW with segment length of 175 data points and a slack of 3 points.*

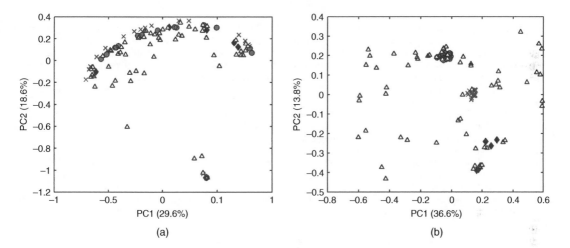

Figure 17.7.5 *PCA score plots of PC1 vs PC2: (a) without alignment, (b) with alignment. Baltic Carrier oil spill samples (•), replicate references (×), triplicate round-robin oil spill samples (♦), and oil samples in the calibration set (△).*

PCA is evident: both replicate references and Baltic Carrier oil spill samples are clustered and the Baltic Carrier source oil can be assigned to the cluster of the spill samples. The clustering of replicate reference samples of the Baltic Carrier spill samples holds for the first four components, whereas subsequent PCs to some extent describe noise and residual misalignment. Thus, the chemical information in additional components is confounded with the residual misalignment.

17.7.3.3   Variable Selection or Weighted Least Squares PCA

Similar to case study 3 the multivariate analysis can be refined by deselecting the most uncertain variables or scale variables with regard to their analytical uncertainties. These two approaches have been successfully used to identify spilled oils (Christensen *et al.*, 2004) and it was found that both approaches improved the resolution power of the PCA by increasing the variance described by the model compared with the variability of replicate samples. It was concluded that although variable selection improved the model, the selection process is not trivial and it contradicts the original aim of minimizing the subjectivity in the data analysis. Furthermore, it was concluded that fitting the PCA model according to a WLS criterion represents a more objective alternative and such an approach also improves the resolution power of the PCA. Both approaches have advantages and selection of the best approach depends generally on the specific case.

However, the results from the WLS-PCA with mean-centering shows that, in spite of weathering processes (mainly evaporation and water washing) for up to 14 days, the Baltic Carrier oil spill samples and the corresponding source oil are clustered in PC1 through PC4. Likewise, the round-robin spill samples, spill I and spill II, are grouped in the plot with the corresponding sources, Oseberg East (E) and Oseberg Field Centre (FC) (Faksness *et al.*, 2002). Oseberg southeast (SE) lies close to Oseberg E along PC1, PC2 and PC3, but are well separated along PC4 which is a minor component describing only 6.3% of the total variation in the calibration set (Figure 17.7.6).

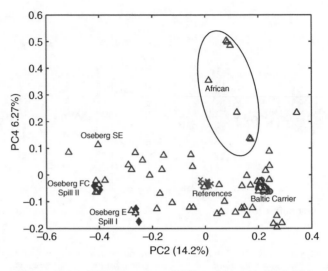

Figure 17.7.6 *PCA score plots of PC2 vs PC4 using WLS-PCA.*

17.7.3.4 Chemical Interpretation of Loading Plots

Chemical interpretation of the fingerprinting results facilitates the correlation between source oils and spill samples, and contributes to the advantages of multivariate statistical analysis of digitized chromatograms.

Retention times that contribute the most to a PC are associated with large negative or positive coefficients in the corresponding loadings. Panels a and b in Figure 17.7.7 show part of the cumulative sum of the PC2 and PC4 loadings, respectively, for the chemical fingerprinting study in Christensen *et al.* (2005) based on preprocessed chromatograms of petroleum biomarkers (i.e., steranes).

Retention times that contribute the most to a PC are associated with large negative or positive coefficients in the corresponding loading. Figure 17.7.7 shows part of the cumulative sum of the PC2 and PC4 loadings.

The PC1 (not shown) describes the boiling point range. Tetracyclic steranes have positive loading coefficients, whereas most of the lower boiling-range compounds in the first peak-cluster (26 to 34 minutes) have negative ones. Correspondingly, lubricants have positive scores for these components, whereas the scores for light fuel oils are negative (see Figure 17.7.6).

The loading coefficients of PC2 are negative for diasteranes (DAS), whereas they are positive for the rearranged steranes (RS). Ratios of the type DAS/RS are commonly used to distinguish oil originating from source rocks with different clay content (Peters and Moldowan, 1993). Low DAS/RS indicate anoxic clay-poor, carbonate source rock, whereas high DAS/RS indicate source rocks containing abundant clays. Hence, PC2 can be interpreted as a source parameter where oil samples with high PC2 (e.g., the Baltic

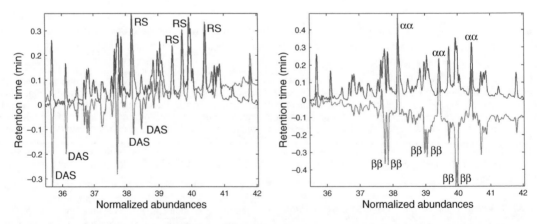

Figure 17.7.7 *Integrated mean-chromatogram (blue) and integrated loadings of PC2 (red) for MILES-PCA. 17-36a) (1) 13β,17α,20S-cholestane (DAS), (2)13β,17α,20R-cholestane (DAS), (3) 5α,14α,17α,20R-cholestane (RS) (4) 24-ethyl-13β,17α,20R-cholestane (DAS), (5) 24-ethyl-13α,17β,20S-cholestane (DAS), (6) 24-methyl-5α,14α,17α,20R-cholestane (RS), (7) 24-methyl-5α,14α,17α,20S-cholestane (RS), (8) 24-ethyl-5α,14α,17α,20R-cholestane (RS). 17-36b) (9) 5α,14β,17β,20R-cholestane (ββ), (10) 5α,14β,17β,20S-cholestane (ββ), (11) 5α,14α,17α,20R-cholestane (αα), (12) 24-methyl-5α,14β,17β,20R-cholestane (ββ), (13) 24-methyl-5α,14β,17β,20S-cholestane (ββ), (14) 24-methyl-5α,14α,17α,20R-cholestane (αα), (15) 24-ethyl-5α,14β,17β,20R-cholestane (ββ), (16) 24-ethyl-5α,14β,17β,20S-cholestane (ββ), (17) 24-ethyl-5α,14α,17α,20R-cholestane (αα).*

Carrier oil) are derived from a source rock containing less clay than oil samples with low PC2 (e.g., North Sea crude oils). Consequently, it appears that the Baltic Carrier oil and North Sea crudes are distinguished by the clay content of their source rocks.

In PC3 (not shown) large positive coefficients of 5α, 14α,17α,20R-Cholestane (27ααR) and 24-methyl-5α,14α,17α,20R-Cholestane (28ααR) compared to 24-ethyl-5α,14α,17α,20R-Cholestane (29ααR) suggest that PC3 is a source parameter reflecting the carbon number distribution of sterols in the organic matter of the source rock (Peters and Moldowan, 1993).

The PC4 describes the thermal maturity of oils. The ββ-isomers of C_{27} to C_{29}-regular steranes have negative coefficients whereas the αα-isomers have positive ones. The ratio of C_{29}-regular steranes (ββ/(ββ + αα)) is a highly specific parameter for maturity and appears to be independent of source organic matter input (Peters and Moldowan, 1993). The ββ have a higher thermal stability compared to the αα isomers, thus the above ratio increases with thermal maturity. African crude oils (Gabon, Kole-Cameroon, Escravos-Nigeria) have positive scores for this component. Consequently, the former appears to have a lower thermal maturity. Analogous conclusions can be drawn from additional maturity parameters, for example 20S/(20S + 20R) C29-regular steranes.

17.7.3.5 Statistical Evaluation
The scores of source oils and spill samples were not compared statistically. However, several methods can be used to test the significant components (four in this case study). In this case the variable uncertainties can be calculated from either the replicate reference samples or the Baltic Carrier spill samples in the same way as in case study 3. Consequently, instead of calculating the uncertainties on the scores from uncertainties on of all diagnostic ratios and their corresponding loading values, the uncertainties on the scores can be calculated from the uncertainties of each individual retention time variable.

The described method allows for analyses of chromatograms using a fast and highly objective procedure. Once the PCA model is constructed, the complete data analysis of a new oil sample (derivatization, normalization, alignment, and PCA) requires few seconds. Furthermore, as long as the variation between oils in the calibration set is sufficient, the PCA can distinguish coeluting peaks. The same would be far more difficult in standard quantification procedures. For example, peaks 3 and 4 in Figure 17.7.7a are highly overlapping in the mean chromatogram, but their loading coefficients have different signs. This is consistent with the interpretation of PC2 that regular steranes have positive coefficients and diasteranes have negative ones.

17.8 CONCLUSIONS

The petroleum hydrocarbon fingerprinting and data interpretation methods and approaches have made great advances in the last two decades, now allowing for detailed qualitative and quantitative characterization of spilled oils. Chemical fingerprinting is a powerful tool for hydrocarbon source identification and differentiation, when it is applied properly.

Among many fingerprinting techniques, the characterization of biomarkers by GC-MS and data evaluation using various statistical approaches remains the cornerstone of environmental forensic investigations. It should be noted, however, that in any case, particularly for complex hydrocarbon mixtures or extensively weathered and degraded oil residues, there is no single fingerprinting technique which

can fully and readily meet the objectives of forensic investigation and quantitatively allocate hydrocarbons to their respective sources. Combined and integrated multiple tools are often necessary under such situations.

Advancements in spilled oil fingerprinting techniques including use of emerging instrumental techniques such as isotope ratio MS, two-dimensional GC, and characterization of UCM of hydrocarbons and further development of multivariate statistical tools for data evaluation will continue. It can be anticipated that these developments will further enhance the utility and defensibility of oil hydrocarbon fingerprinting.

REFERENCES

Aboul-Kassim, T. A. T. and B. R. T. Simoneit, Aliphatic and aromatic hydrocarbons in particulate fallout of Alexandria, Egypt: Sources and implications, *Environ. Sci. Technol.*, 1995a, 29, 2473–2483.

Aboul-Kassim, T. A. T. and B. R. T. Simoneit, Petroleum hydrocarbon fingerprinting and sediment transport assessed by molecular biomarker and multivariate statistical analyses in the Eastern Harbour of Alexandria, Egypt, *Mar. Pollut. Bull.*, 1995b, 30: 63–73.

Alberdi, A. and L. Lopez, Biomarker 18α(H)-oleanane, a geochemical tool to assess Venezuelan petroleum systems, *J. South American Earth Sciences*, 2000, 13: 751–759.

Atlas, R. M. and R. Bartha, Hydrocarbon biodegradation and oil spill bioremediation, In: *Advances in Microbial Ecology*, K. C. Marshall, (ed.), Plenum Press, New York, 1992, 12: 287–3382.

Audino, M., K. Grice, R. Alexander, and R. Kagi, Macrocyclic-alkanes: A new class of biomarker, *Org. Geochem.*, 2001, 32: 759–763.

Audino, M., K. Grice, R. Alexander, and R. Kagi, Macrocyclic-alkanes in crude oils from algaenan of Botryococcus braunii, *Org. Geochem.*, 2002, 33: 978–984.

Barakat, A. O., A. Mostafa, M. S. El-Gayar, and J. Rullkotter, Source-dependent biomarker properties of five crude oils from the Gulf of Suez, Egypt, *Org. Geochem.*, 1997, 26: 441–450.

Barakat, A.O., A.R. Mostafa, J. Rullkotter, and A.R. Hegazi, Application of a multimolecular marker approach to fingerprint petroleum pollution in the marine environment, *Mar.Pollut.Bull.*, 1999, 38: 535–544.

Barber, C. J., K. Grice, T. P. Bastow, R. Alexander, and R. I. Kagi, The identification of crocetane in Australian crude oils, *Org. Geochem.*, 2001, 32: 943–947.

Bence, A. E., K. A. Kvenvolden, and M. C. Kennicutt II, Organic geochemistry applied to environmental assessments of Prince William Sound, Alaska, after the Exxon Valdez oil spill—a review, *Organic Geochemistry*, 1996, 24, 7–42.

Blenkinsopp, S., Z. Wang, J. Foght, to D. W. S. Westlake, G. Sergy, M. Fingas, L. Sigouin, and K. Semple Assessment of the freshwater biodegradation potential of oils commonly transported in Alaska, *Final Repor to Alaska Government, ASPS 95-0065*, Environment Canada, Ottawa, 1996.

Bragg J. R. and E. H. Owens, Clay-oil flocculation as a natural cleansing process following oil spill: Part 1—studies of shoreline sediments and residues from past spills, In: *Proceedings of the 17th Arctic and Marine Oil Spill Program (AMOP) Technical Seminar*, Environment Canada, Ottawa, Ontario, 1994, 1–25.

Brereton, R.G., *Chemometrics—Data Analysis for the Laboratory and Chemical Plant*, John Wiley & Sons Ltd, 2003.

Burns W. A., P. J. Mankiewicz, A. E. Bence, D. S. Page, and K. R. Parker, A principal-component and least-squares method for allocating polycyclic aromatic hydrocarbons in sediment to multiple sources, *Environ.Toxicol.Chem.*, 1997, 16: 1119–1131.

Butler, E. L., G. S. Douglas, W. S. Steinhauter, R. C. Prince, T. Axcel, C. S. Tsu, M. T. Bronson, J. R. Clark, and J. E. Lindstrom, Hopane, a new chemical tool for measuring oil biodegradation, In: R. E. Hinchee, R. F. Olfenbuttel (eds), *On-site Reclamation*, Butterworth-Heinemann, Boston, MA, 1991, 515–521.

Chosson, P., C. Lanau, J. Connan, and D. Dessort, Biodegradation of refractory hydrocarbon biomarkers from petroleum under laboratory conditions, *Nature*, 1991, 351: 640–642.

Christensen, J. H., A. B. Hansen, J. Mortensen, G. Tomasi, and O. Andersen, Integrated methodology for forensic oil spill identification, *Environ. Sci. Technol.*, 2004, 38: 2912–2918.

Christensen, J. H., G. Tomasi, and A. B. Hansen, Chemical fingerprinting using time warping and PCA, *Environ. Sci. Technol.*, 2005, 39: 255–260.

Currie, T. J., R. Alexander, and R. I. Kagi, Coastal bitumens from Western Australia-long distance transport by ocean currents, *Org. Geochem.*, 1992, 18: 595–601.

Daling, P. S., L. G. Faksness, A. B. Hansen, and S. A. Stout, Improved and standardized methodology for oil spill fingerprinting. *Environmental Forensics*, 2002, 3: 263–278.

De Luca G., A. Furesi, R. Leardi, G. Micera, A. Panzanelli, P.C. Piu, and G. Sanna, Polycyclic aromatic hydrocarbons assessment in the sediments of the Porto Torres Harbor (Northern Sardinia, Italy), *Mar. Chem.*, 2004, 86: 15–32.

Douglas, G. S., R. C. Prince, E. L. Butler, and W. G. Steinhauer, The use of internal chemical indicator in petroleum and refined products to evaluate the extent of biodegradation, In: *Hydrocarbon Bioremediation* (R. E. Hinchee, B. C. Hoeppel, and R. N. Miller, eds), Lewis Publishers, Boca Raton, FL, 1994, 219–236.

Eilers, P. H. C., Parametric time warping, *Anal. Chem.*, 2004, 76: 404–411.

El-Gayar, M. S., A. R. Mostafa, A. E. Abdelfattah, and A. O. Barakat, Application of geochemical parameters for classification of crude oils from Egypt into source-related types, *Fuel Processing Technology*, 2002, 79: 13–28.

EPA, US EPA Report: Characteristics of spilled oils, fuels, and petroleum products: 1. Composition and properties of selected oils, EPA/600/R-03/072, National Exposure Research Laboratory (NERL), EPA, 2003. Also, the EPA website: www.epa.gov/epahome/recentadditions.htm.

Faksness, L. G., P. S. Daling, and A. B. Hansen, Round robin study—oil spill identification, *Environ. Forensics*, 2002, 3: 279–291.

Fingas, M. F. and B. Fieldhouse, Studies of the formation of water-in-oil emulsions. *Mar. Pollut. Bull.*, 2003, 47: 369–396.

Foght, J., K. Semple, C. Gauthier, D. W. S. Westlake, S. Blenkinsopp, G. Sergy, Z. D. Wang, and M. Fingas, Development of a standard bacterial consortium for laboratory efficacy testing of commercial freshwater oil spill bioremediation agents, *Environmental Technology*, 1998, 20: 839–849.

Fraga C. G., B. J. Prazen, and R. E. Synovec, Comprehensive two-dimensional gas chromatography and chemometrics for the high-speed quantitative analysis of aromatic isomers in a jet fuel using the standard addition method and an objective retention time alignment algorithm. *Anal. Chem.*, 2000, 72: 4154–4162.

Fu, J., C. Pei, G. Sheng, and D. Liu, A geochemical investigation of crude oils from eastern Pearl River mouth basin, South China, *J. Southeast Asian Earth Science*, 1992, 7: 271–272.

Garrett, P. M., I. J. Pickerring, C. E. Haith, and R. C. Prince, Photooxidation of crude oils, *Environ. Sci. Technol.*, 1998, 32: 3719–3723.

Gürgey, K., An attempt to recognize oil populations and potential source rock types in Paleozoic sub- and Mesozoic-Cenozoic supra-salt strata in southern margin of the Pre-Caspian basin, Kazakhstan republic, *Org. Geochem.*, 2002, 33: 723–741.

Hu, G., Geochemical characterization of steranes and terpanes in certain oils from terrestrial facies within South China Sea, *J. Southeast Asian Earth Science*, 1991, 5: 241–247.

Hwang, R. J., T. Heidrick, B. Mertani, Qivayanti, and M. Li, Correlation and migration studies of North Central Sumatra oils, *Org. Geochem.*, 2002, 33: 1361–13792.

Jiang, C. and M. Li, Bakken/Madison petroleum systems in the Canadian Williston Basin, Part 3: Geological evidence for significant Bakken-derived oils in Madison Group reservoirs, *Org. Geochem.*, 2002, 33: 761–787.

Jiang, C., M. Li, K. G. Osadetz, L. R. Snowdon, M. Obermajer, and M. G. Fowler, Bakken/Madison petroleum systems in the Canadian Williston Basin. Part 2: Molecular markers diagnostic of Bakken and Lodgepole source rocks, *Org. Geochem.*, 2001, 32: 1037–1054.

Johansson, E., S. Wold, and K. Sjodin, Minimizing effects of closure on analytical data, *Anal. Chem.* 1984, 56: 1685–1688.

Johnson K. J., B. W. Wright, K. H. Jarman, and R. E. Synovec, High-speed peak matching algorithm for retention time alignment of gas chromatographic data for chemometric analysis, *J. Chromatogr. A*, 2003, 996: 141–155.

Jolliffe I.T., *Principal Component Analysis*, Springer Verlag, 1986.

Jordan, R. E. and J. R. Payne, Fate and weathering of petroleum spills in the marine environment: A literature review and synopsis, Ann Arbor Science Publishers, Ann Arbor, Michigan, 1980.

Kvenvolden, K. A., J. B. Rapp, and J. H. Bourell, In: L. B. Magoon, G. E. Claypool (eds), *Alaska North Slope Oil/Rock Correlation Study*, American Association of Petroleum Geologists Studies in Geology, No. 20, 1985, 593–617.

Kvenvolden, K. A., F. D. Hostettler, J. B. Rapp, and P. R. Carlson, Hydrocarbon in oil residues on beaches of islands of Prince William Sound, Alaska, *Mar. Pollut. Bull.*, 1993, 26: 24–29.

Kvenvolden, K. A., F. D. Hostettler, P. R. Carleson, J. B. Rapp, C. N. Threlkeld, and A. Warden, Ubiquitous tar balls with a Carlifonia-source signature on the shorelines of Prince William Sound, Alaska, *Environ. Sci. & Tech.*, 1995, 29: 2684–2694.

Lavine, B. K., H. Mayfield, P. R. Kromann, and A. Faruque, Source identification of underground fuel spills by pattern recognition analysis of high-speed gas chromatograms. *Anal. Chem.*, 1995, 67: 3846–3852.

Leahy, J. G. and R. R. Colwell, Microbial degradation of hydrocarbons in the environment, *Microbial Rev.*, 1990, 54: 305–315.

Li, M. and C. Jiang, Bakken/Madison petroleum systems in the Canadian Williston Basin, Part 1: C_{21}-C_{26} 20-n-alkylpregnanes and their triaromatic analogs as indicators for Upper Devonian-Mississippian epicontinental black shale derived oils, *Org. Geochem.*, 32: 667–675, 2001.

Martens, H. and M. Martens, *Multivariate Analysis of Quality: An Introduction*, John Wiley & Sons Ltd, 2001.

Massart D., B. G. M. Vandeginste, L. M. C. Buydens, S. De Jong, P. J. Lewi, and J. Smeyers-Verbeke, *Handbook of Chemometrics and Quilimetrics: Part A*, Elsevier, Amsterdam, 1997.

Mckirdy, D. M., A. K. Aldridge, and P. J. M. Ypma, A geological comparison of some crude oils from Pre-Ordovician carbonate rocks, In: *Advances in Organic Geochemistry 1981* (M. BjorRy et al., eds), J. Wiley & Sons, New York, 1983, 99–107.

Mckirdy, D. M., R. E. Cox, J. K. Volkman, and V. J. Howell, Botryococcane in a new class of Australian non-marine crude oils, *Nature*, 1986, 320: 57–59.

Mckirdy, D. M., R. E. Summons, D. Padley, K. M. Serafini, C. J. Boreham, and H. I. M. Struckmeyer, Molecular fossils in coastal bitumens from southern Australia: Signature of precursor biota and source rock environments, *Org. Geochem.*, 1994, 21: 265–286.

Moldowan, J. M., F. J. Fago, R. M. K. Carlson, D. C. Young, G. V. Duyne, J. Clardy, M. Schoell, C. T. Phillinger, and D. S. Watt, Rearranged hopanes in sediments and petroleum, *Geochimica et Cosmochimica Acta*, 1991, 55: 3333–3353.

Moldowan, J. M., P. Sundararaman, T. Salvatori, A. Alajbeg, B. Gjukic, C. Y. Lee, and G. I. Demaison, Source correlation and maturity assessment of selected oils and rocks from the Central Adriatic Basin, In: *Biological Markers in Sediments and Petroleum* (J. M. Moldowan, P. Albrecht, and R. P. Philp, eds), Prentice Hall, Englewood Cliffs, NJ, 1992a, 370–4012.

Moldowan, J. M., P. Albrecht, and R. P. Philp, *Biological Markers in Sediments and Petroleum*, Prentice Hall, Englewood Cliffs, New Jersey, 1992b.

Mudge, S. M., Reassessment of the hydrocarbons in Prince William Sound and the Gulf of Alaska: Identifying the source using partial least-squares, *Environ. Sci. Technol.*, 2002, 36: 2354–2360.

Munoz, D., M. Guiliano, P. Doumenq, F. Jacquot, P. Scherrer, and G. Mille, Long term evolution of petroleum biomarkers in mangrove soil, *Marine Pollution Bulletin*, 1997, 34: 868–874.

Murrisepp, A. M., K. Urof, M. Liiv, and A. Sumberg, A comparative study of non-aromatic hydrocarbons from kukersite and dictyonema shale semicoking oils, *Oil Shale*, 1994, 11: 211–216.

Nielsen, N. P. V., J. M. Carstensen, and J. Smedsgaard, Aligning of single and multiple wavelength chromatographic profiles for chemometric data analysis using correlation optimised warping. *J. Chromatogr. A*, 1998, 805: 17–35.

Oung, J. N. and R. P. Philp, Geochemical characteristics of oils from Taiwan, *J. Southeast Asian Earth Science*, 1994, 9: 193–206.

Ourisson, G. and P. Albrecht, Hopanoids, 1. Geohopanoids: The most abundant natural products on Earth? *Acc. Chem. Rev.*, 1992, 25: 398–402.

Ownes, E. H. and K. Lee, Interaction of oil and mineral fines on shorelines: Review and assessment, *Marine Pollution Bulletin*, 2003, 47: 397–405.

Øygard K., O. Grahl-Nielsen, and S. Ulvøen, Oil/oil correlation by aid of chemometrics. *Organic Geochemistry*, 1984, 6: 561–567.

Page, D. S., P. D. Boehm, G. S. Douglas, A. E. Bence, W. A. Burns, and P. J. Mankiewicz, The natural petroleum hydrocarbon background in subtidal sediments of Prince William Sound, Alaska, USA, *Environ. Toxicol. Chem.*, 1996, 15: 1266–1281.

Peters, K.E. and J.M. Moldowan, *The Biomarker Guide: Interpreting Molecular Fossils in Petroleum and Ancient Sediments*, Prentice Hall, Englewood Cliffs, New Jersey, 1993.

Pravdova V., B. Walczak, and D. L. Massart, A comparison of two algorithms for warping of analytical signals, *Anal. Chim. Acta*, 2002, 456: 77–92.

Prince, R. C., Petroleum spill bioremediation in marine environment, *Crit. Rev. Microbial*, 1993, 36: 724–728.

Prince, R. C., D. L. Elmendorf, J. R. Lute, C. S. Hsu, C. E. Haith, J. D. Senius, G. J. Dechert, G. S. Douglas, and E. L. Butler, $17\alpha(H)$, $21\beta(H)$-Hopane as a conserved internal marker for estimating the biodegradation of crude oil, *Environ. Sci. Technol.*, 1994, 28: 142–145.

Prince, R. C., E. H. Owens, and G. A. Sergy, Weathering of an Arctic oil spill over 20 years: The BIOS experiment revisited, *Marine Pollution Bulletin*, 2002, 44: 1236–1242.

Requejo, A. G. and H. I. Halpern, An unusual hopane biodegradation sequence in tar-sands from Point Arena (Monterey) formation, *Nature*, 1989, 342: 670–673.

Riva, A., P. Caccialanza, and F. Quagliaroli, Recognition of $18\alpha(H)$-oleanane in several crudes and Tertiary-Upper Cretaceous sediments, *Organic Geochemistry*, 1988, 13: 671–675.

Rogers, K. M., J. D. Collen, J. H. Johnston, and N. E. Elgar, A geological appraisal of oil seeps from the East Coast Basin, New Zealand, *Org. Geochem.*, 1999, 30: 593–605.

Rohmer, M., P. Bisseret, and S. Neunlist, The hopanoids, prokaryotic triterpenoids and precursors of ubiquitous molecular fossils, In: *Biological Markers in Sediments and Petroleum* (J. M. Moldowan, P. Albrecht, and R. P. Philip, eds), Prentice Hall, NJ, 1992, 1–17.

Seifert, W. K. and J. M. Moldowan, Application of steranes, terpanes, and monoaromatics to the maturation, migration, and source of crude oils, *Geochimica et Cosmochimica Acta*, 1978, 42: 77–95.

Seifert, W. K. and J. M. Moldowan, Paleoreconstruction by biological markers, *Geochimica et Cosmochimica Acta*, 1981, 45: 783–794.

Seifert, W. K., and J. M. Moldowan, Use of biological markers in petroleum exploration, In: *Methods in Geochemistry and Geophysics* (R. B. Johns, ed.), 1986, 24: 261–290.

Seifert, W. K., and J. M. Moldowan, and G. J. Demaison, Source correlation of biodegraded oils, *Org. Geochem.*, 1984, 6: 633–643.

Shen, J., Minimization of interferences from weathering effects and use of biomarkers in identification of spilled crude oils by gas chromatography/mass spectrometry, *Anal. Chem.*, 1984, 56: 214–217.

Sosrowidjojo, I. B., R. Alexander, and R. I. Kagi, The biomarker composition of some crude oils from Samatra, *Org. Geochem.*, 1994, 21: 303–312.

Stella A., M. T. Piccardo, R. Coradeghini, A. Redaelli, S. Lanteri, C. Armanino, and F. Valerio, Principal component analysis application in polycyclic aromatic hydrocarbons "mussel watch" analyses for source identification, *Anal. Chim. Acta*, 2002, 461: 201–213.

Stout S. A., A. D. Uhler, and K. J. McCarthy, A strategy and methodology for defensibly correlating spilled oil to source candidates, *Environ. Forensics*, 2001, 2: 87–98.

Stout, S. A., A. D. Uhler, K. J. McCarthy, and S. Emsbo-Mattingly, *Chapter 6: Chemical Fingerprinting of Hydrocarbons*, In: *Introduction to Environmental Forensics* (B. L. Murphy and R. D. Morrison, eds), Academic Press, London, 2002, 139–260.

Summons, R. E. and L. L. Jahnke, Hopenes and hopanes methylated in Ring A: Correlation of the hopanoids from extant methylotrophic bacteria with their fossil analogues, In: *Biological Markers in Sediments and Petroleum* (J. M. Moldowan, P. Albrecht, and R. P. Philp, eds), Prentice Hall, Englewood Cliffs, NJ, 1992, 182–194.

Summons, R. E., J. Thomas, J. R. Maxwell, and C. J. Boreham, Secular and environmental constraints on the occurrence of dinosterane in sediments, *Geochim Cosmochim Acta*, 1992, 56: 2437–2444.

Swannel, R. P. J., K. Lee, and M. McDonaph, Field evaluation of marine oil spill bioremediation, *Microbial Rev.*, 1996, 60: 342–365.

Telnaes, N. and B. Dahl, Oil-oil correlation using multivariate techniques, *Org. Geochem.*, 1986, 10: 425–432.

Tomasi, G., F. van den Berg, and C. Andersson, Correlation optimized warping and dynamic time warping as preprocessing methods for chromatographic data, *J. Chemometr.*, 2004, 18, 5: 231–241.

van Aarssen, B. G. K., H. C. Cox, P. Hoogendoorn, J. W. de Leeuw, A cadinene biopolymer present in fossil and extract Dammar resins as source for cadinanes and dicadinanes in crude oils from Southeast Asia, *Geochimica et Cosmochimica Acta*, 1990, 54: 3021–3031.

Volkman, J. K., P. Keaeney, and S. W. Jeffrey, A new source of 4-methyl and 5α(H)-stanols in sediments: Prymnesiophyte microalgae of the genus, *Pavlova*, *Organic Geochemistry*, 1990, 15: 489–497.

Volkman, J. K., D. G. Holdsworth, G. P. Neill, and Jr. H. J. Bavor, Identification of natural, anthropogenic and petroleum hydrocarbons in aquatic environments, *Sci. To. Environ.*, 1992, 112: 203–219.

Volkman, J. K., T. O'Leary, R. E. Summons, and M. R. Bendall, Biomarker composition of some asphaltic coastal bitumens from Tasmania, Australia, *Org. Geochem.*, 1992a, 18: 669–682.

Volkman, J. K., T. O'Leary, R. E. Summons, and M. R. Bendall, Biomarker composition of some asphaltic coastal bitumens from Tasmania, Australia, *Org. Geochem.*, 1992b, 18: 669–682.

Wang C. P. and T. L. Isenhour, Time-warping algorithm applied to chromatographic peak matching Gas-Chromatography Fourier-Transform Infrared Mass-Spectrometry, *Anal. Chem.*, 1987, 59: 649–654.

Wang, P., M. Li, and S. R. Larter, Extended hopanes beyond C_{40} in crude oils and source rock extracts from the Liaohe Basin, N. E. China, *Org. Geochem.*, 1996, 24: 547–551.

Wang, Z. D., M. Fingas, and G. Sergy, Study of 22-year-old Arrow oil samples using biomarker compounds by GC/MS, *Environ. Sci. Technol.*, 1994, 28: 1733–1746.

Wang, Z. D., M. Fingas, and G. Sergy, Chemical characterization of crude oil residues from an Arctic Beach by GC/MS and GC/FID, *Environ. Sci. Technol.*, 1995a, 29: 2622–2631.

Wang, Z. D. and M. Fingas, Study of the effects of weathering on the chemical composition of a light crude oil using GC/MS and GC/FID, *Journal of Microcolumn Separation*, 1995b, 7: 617–639.

Wang, Z. D., M. Fingas, M. Landriault, L. Sigouin, Y. Feng, and J. Mullin, Using systematic and comparative analytical data to identify the source of an unknown oil on contaminated birds, *Journal of Chromatography*, 1997, 775: 251–265.

Wang, Z. D., M. Fingas, S. Blenkinsopp, G. Sergy, M. Landriault, and L. Sigouin, Study of the 25-year-old Nipisi oil spill: Persistence of oil residues and comparisons between surface and subsurface sediments, *Environ. Sci. Technol.*, 1998a, 32: 2222–2232.

Wang, Z. D., M. Fingas, M. Landriault, L. Sigouin, B. Castel, D. Hostetter, D. Zhang, and B. Spencer, Identification and linkage of tarballs from the coasts of Vancouver Island and Northern California using GC/MS and isotopic techniques, *J. High Resolut. Chromatogr.*, 1998b, 21: 383–395.

Wang, Z. D., M. Fingas, S. Blenkinsopp, G. Sergy, M. Landriault, L. Sigouin, J. Foght, K. Semple, and D. W. S. Westlake, Oil composition changes due to biodegradation and differentiation between these changes to those due to weathering, *Journal of Chromatography A*, 1998c, 809: 89–107.

Wang, Z. D., M. Fingas, and D. Page, Oil spill identification, *J. Chromatogr.*, 1999a, 843: 369–411.

Wang, Z. D., M. Fingas, M. Landriault, L. Sigouin, S. Grenon, and D. Zhang, Source identification of an unknown spilled oil from Quebec (1998) by unique biomarker and diagnostic ratios of "source-specific marker" compounds, *Environmental Technology*, 1999b, 20: 851–862.

Wang, Z. D., M. Fingas, Y.Y. Shu, L. Sigouin, M. Landriault, and P. Lambert, Quantitative characterization of PAHs in burn residue and soot samples and differentiation of Pyrogenic PAHs from Petrogenic PAHs—the 1994 Mobile Burn Study, *Environ. Sci. Technol.*, 1999c, 33: 3100–3109.

Wang, Z. D., M. Fingas, and L. Sigouin, Characterization and Identification of a "mystery" Oil Spill from Quebec, (1999), *J. Chromatogr.*, 2001a, 909: 155–169.

Wang, Z. D., M. Fingas, E. H. Owens, L. Sigouin, and C. E. Brown, Long-term fate and persistence of the spilled Metula oil in a marine salt marsh environment: degradation of petroleum biomarkers, *J. Chromatogr.*, 2001b, 926: 275–190.

Wang, Z. D., M. Fingas, and L. Sigouin, Using multiple criteria for fingerprinting unknown oil samples having very similar chemical composition, *Environmental Forensics*, 2002, 3: 251–262.

Wang, Z. D., B. P. Hollebone, M. Fingas, B. Fieldhouse, and J. Weaver, Development of a physical and chemical property database for ten US EPA-selected oils, in *Proceedings of the 26th Arctic and Marine Oil Spill Program (AMOP) Technical Seminar*, Environment Canada, Ottawa, 2003, pp. 117–142.

Wang, Z. D., M. Fingas, and P. Lambert, Characterization and identification of Detroit River mystery oil spill (2002), *J. Chromatogr.*, 2004, 1038: 201–214.

Wold, S., C. Albano, W. J. Dunn, K. Esbensen, S. Hellberg, E. Johansson, W. Lindberg, and M. Sjostrom, Modelling data tables by principal components and PLS: Class patterns and quantitative predictive relations, *Analyst*, 1984, 12: 477–485.

Wold, S., K. Esbensen, and P. Geladi, Principal component analysis, *Chemometrics and Intelligent Laboratory Systems*, 1987, 2: 37–52.

Wolff, G. A., N. A. Lamb, and J. R. Maxwell, The origin and fate of 4-methyl steroid hydrocarbons, *Geochimica et Cosmochimica Acta*, 1986, 50: 335–342.

Zakaria, M. P., A. Horinouchi, S. Tsutsumi, H. Takada, S. Tanabe, and A. Ismail, Oil pollution in the Straits of Malacca, Malaysia: Application of molecular markers for source identification, *Environ. Sci. Technol.*, 2000, 34: 1189–1196.

Zakaria, M. P., T. Okuda, and H. Takada, PAHs and hopanes in stranded tar-balls on the coast of Peninsular Malaysia: Applications of biomarkers for identifying source of oil pollution, *Mar. Pollution. Bull.*, 2001, 12: 1357–1366.

Zhang, S., D. Liang, Z. Gong, K. Wu, M. Li, F. Song, Z. Song, D. Zhang, and P. Wang, Geochemistry of petroleum system in the eastern Pearl River Mouth Basin: Evidence for mixed oils, *Org. Geochem.*, 2003, 34: 971–991.

18 Automotive Gasoline

Scott A. Stout, Gregory S. Douglas, and Allen D. Uhler

Contents

18.1   INTRODUCTION

Environmental forensic investigations typically attempt to determine the criminal and civil liabilities associated with the impact of anthropogenic contamination on human health or the environment. Determining the liability associated with gasoline-derived contamination is one of the most common objectives of environmental forensics for the following reasons.

First, on a volumetric basis, gasoline perhaps is one of the most widely consumed man-made products. The volume of gasoline used in the United States has steadily increased for 50 years (Figure 18.1.1) such that in 2001 the United States consumed 360,000 gallons of gasoline per day, which represents 43% of the worldwide consumption (International Energy Annual, 2003). Second, on a geographic basis, gasoline is stored and dispensed from a large number of individual locations. For example, in 2002 there were more than 170,000 gasoline stations across the United States, many now appearing at supermarkets, mass merchandisers, and warehouse clubs. Third, gasoline has been demonstrated to pose measurable risks to human health and to the environment. In fact, if gasoline were a newly developed product being introduced to the market today, it would probably fail acceptance on the basis of the health and environmental risks it poses. Only by virtue of its historic acceptance and our individual reliance upon it has society accepted these risks.

This combination of facts – use in large volumes, widespread handling and storage, and human health and environmental risks – leave gasoline at the center of many environmental forensic investigations.

In this chapter, we address various issues relevant to environmental forensic investigations of gasoline. The chapter is organized to present the reader with the ability to find and read about specific issues or questions they may have, which does not necessarily require a front-to-back reading. The major topics covered include:

- Gasoline History
- Gasoline Refining
- Gasoline Additives
- Gasoline Fingerprinting Methods

- Gasoline Weathering
- Age-Dating Gasoline-Derived Contamination
- Specific Forensic Considerations

The last of these topics contains several specific sub-topics of relevance that are common or interesting issues faced in the course of environmental forensic investigations surrounding gasoline. These range from "gasoline distribution practices" to "indoor air issues".

18.2   HISTORY OF GASOLINE

The history of automotive gasoline has been thoroughly reviewed elsewhere (Jackman, 1985; Owen and Coley, 1990; Hamilton and Falkiner, 2003). From gasoline's first recorded use in a four-stroke engine in 1876 it has continually evolved into the product in use today. In this chapter we have divided this history into two eras – the *Performance Era* and the *Regulatory Era*. Some of the changes in gasoline during these eras are described in the following sections and key points are tabulated in Tables 18.2.1 and 18.2.2, respectively.

18.2.1   Performance Era (pre-1970)

The evolving requirements for gasoline's combustibility, antiknock quality (octane), chemical stability, volatility, gum content, and combustion emission yields, combined with the refiners' varying approaches to meet these performance requirements, has led to a steady change in the composition of gasoline over time. Consequently, generalizations regarding the chemical composition of "typical" automotive gasoline are often oversimplified. As a result, the forensic investigator can use this variability to distinguish different types and thereby sources of gasoline in the environment.

The first "gasolines" produced over 100 years ago bared little resemblance to what is now dispensed at the pump. These early fuels were essentially straight-run naphtha that was batch-distilled from crude oil, sometimes augmented with natural gas liquids (NGL) or town gas liquids. They had a carbon range from approximately C_4 to C_{13}, which

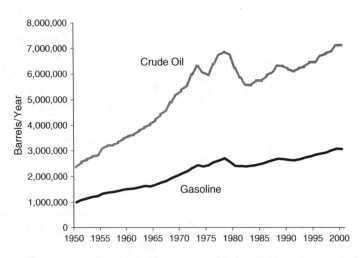

Figure 18.1.1 *Annual consumption of crude oil and motor gasoline in the US between 1950 and 2000 (Dept. of Energy, International Energy Annual, 2001). These data indicate that gasoline has represented between 38 and 44% of total crude oil consumed over the past 50 years.*

Table 18.2.1 Notable developments in the performance era (pre-1970) history of automotive gasoline

Year	Development
1913	Thermal cracking introduced (T ~450–540 °C)
1923	TEL introduced on limited basis by Standard Oil of Ohio
1924	Standard Oil of NJ (Esso) begins TEL production, Ethyl Corp. begins widespread marketing
1926	Dyes introduced (1–5 ppm), TEL widespread (max. 3.17 glpg)
1928	Lead scavengers introduced
1930	Antioxidants introduced to reduce oxidation of olefins (introduced by thermal cracking)
1933	MOR octane test developed
1936	Fixed-bed catalytic cracking introduced (T ~470–525 °C)
1937	ASTM 1st sets evaporation limits for summer vs winter gasolines
1938	H_2SO_4 Alkylation introduced (T <20 °C)
1939	Metal deactivators introduced to reduce oxidation catalyzed by trace metals (Cu); RON octane test developed
1940	Fluid catalytic cracking (FCC) and catalytic reformation introduced (T ~450–520 °C); ethanol-gasoline blend first introduced in Nebraska ("Argol")
1942	HF Alkylation introduced (T <20 °C)
1943	Isomerization introduced
1946	Oil-based corrosion inhibitors introduced to reduce rust (due to trace H_2O) in steel containers/pipelines
1947	ASTM sets protocol for research octane number measurement
1950	isopropyl alcohol first added to gasolines as a anti-icing agent
1953	tricresyl phosphate (TCP) introduced as a anti-fouling agent (added ~50% of Pb/gal)
1954	Carburetor detergents introduced to reduce engine deposits – 1st was "MPA" (marketed by Ethyl)
1956	Boron deposit modifiers introduced (used in Pb gasoline only)
1958	Hydrocracking introduced; FCC & alkylation widespread use
1959	TEL widespread (max. allowable increased to 4.23 glpg); MMT introduced by Ethyl Corp., marketed as "Motor Mix 33" 57.52% TEL: 6.97% MMT
1960	TML & reacted/physical mixes introduced; CA limits olefin content (Bromine Number <30)
1964	Nickel deposit modifiers introduced (used in Pb gasoline only)
1965	Polyglycol demulsifiers introduced to minimize hazing due to trace water
1966	Zinc deposit modifiers introduced (used in Pb gasoline only)
1968	Polybutene succinimide deposit control additives introduced
1969	TBA introduced (by Arco), typical use level of 2%

Table 18.2.2 Notable developments in the regulatory era (post-1970) history of automotive gasoline

Year	Development
1970	Lead content of gasoline maximizes; polybutene amine (~667 ppm) w/ carrier oil (~1600 ppm) deposit control additive introduced. Clean Air Act of 1970 harkens changes to come.
1971	Unleaded gasoline introduced required <0.07 glpg; Use of Nickel deposit modifiers stopped due to perceived health risks; CA set RVP limit at 9 psi in summer
1972	ASTM D 439 adopts maximum sulfur content of 0.10 wt% in gasoline
1973	Arab Oil Embargo initiated following 1973 Arab–Israeli War
1974	Unleaded gasoline required at all stations nationwide by July, lowered allowable lead to 0.05 glgp from 1971 level (and also <0.005 g P/gal to inhibit catalytic converter poisoning); phosphate detergents banned in all gasolines
1976	MMT in widespread use in leaded and unleaded gasoline (offsets loss of lead); CA sets sulfur limit of 500 ppm (max.)
1977	Clean Air Act Amendments set the Lead Phase down schedule; Dept. of Energy created to consolidate energy functions of US government
1978	Ethanol (re-)introduced (see 1940) on widespread scale due to embargo and United States Environmental Protection Agency waiver ~10% ("Gasohol"); TBA often used as a co-solvent (1:1) along with MeOH to improve handling of Gasohol; CA reduces sulfur limit to 400 ppm (max.); In Oct., MMT use banned in unleaded gasolines by CAA Amendment of 1977; MMT still allowed in leaded gasolines and is used inversely proportional to Pb; increase of maximum sulfur content in leaded gasoline to 0.15 wt% (ASTM D439)
1979	Methanol and MTBE introduced, TBA use becomes widespread after United States Environmental Protection Agency officially approves its use (even though many refiners had been using it). The typical TBA conc. ~1–5% (7% max. allowable by United States Environmental Protection Agency), largest producer of TBA is Oxirane (whose contract required them to sell 50% of TBA produced to Arco, and the other 50% to anyone else, of which Mobil was one customer using TBA in the Rocky Mt. region); MMT temporarily re-introduced to unleaded gasoline during Arab oil embargo; all gasolines avg. lead max. 0.8 glpg
1980	Federal Lead Phase down begins; MMT use again banned in unleaded gasoline (unless special waiver is granted) by Oct.; still allowed in leaded gasoline; mixtures of TBA:MeOH (1:1, 5.5% max.) used by some refiners, called "*Oxinol*™"; CA reduces sulfur limit to 300 ppm (max.)

(Continued)

Table 18.2.2 (Continued)

Year	Development
1981	Leaded premiums and the use of lead alkyl mixtures containing TML mostly have disappeared from market; MTBE used at up to 11 vol% (max); "Oxinol" (TBA:MeOH, 1:1) increased to 9 vol% (max.); 1st time in history where most gasoline sold is unleaded
1985	Lead content limited <1.1 (Jan 1) and 0.5 (July 1) glpg in leaded gasolines; unleaded gasoline max. lead is 0.05 glpg (actual levels much lower, <0.001 glpg); Ethyl Corp. ceases lead alkyl production leaving DuPont as the only supplier in the United States; 90–95% of lead used in 1985 was TEL (some TML mixtures still used, but minor)
1986	Lead content limited <0.1 glpg (Jan 1) but lead banking allows exceedences; most leaded gasolines still contain ~0.1 g Mn/gal (as MMT)
1987	MTBE production/use exceeds ethanol for the first time (nationwide)
1988	Winter oxygenates mandated in selected markets; lead banking ends, lead content of all gasolines <0.1 glpg; MTBE allowed up to 15 vol% (max.)
1989	Summer volatility (RVP) Federal Phase I implemented (max. RVP of 10.5 or 9.0 psi in different regions of the United States
1990	Clean Air Act Amendment calls for winter (Nov–Feb) oxygenate program (for CO non-attainments in 1992) and RFG program (for O_3 non-attainments in 1995)
1992	Summer volatility (RVP) Federal Phase II implemented (max. RVP 9.0 and 7.8 psi in different regions; 7.8 psi max. in CA); Winter oxygenated gasoline mandatory in 44 CO non-attainment areas (2.7 wt% O_2 min.), these 44 areas account for 27 vol% of the US gasoline market; CARB Phase I in CA requires O_2 1.8–2.2 wt% in Winter; all leaded gasoline banned in CA
1994	Clean Air Act, Tier 1 standards in effect
1995	Federal Phase I (Simple Model) RFG mandatory on Jan 1 in nine O_3 non-attainment areas that represent 22% of US gasoline market; many other areas "opt-in"; RFG benzene max. (1.0 wt%), O_2 min. (2.0 wt%), RVP max. (8.3/7.4 psi), sulfur max. (pool avg. of <338 ppm; single sample max. of 1000 ppm); MMT re-introduced into some unleaded gasolines after United States Environmental Protection Agency loses lawsuit with Ethyl Corp. (max. 0.031 gMMT/gal); MMT remains banned in CA
1996	All leaded gasolines banned nationwide for public use (e.g., racing gas still leaded); CARB Phase II in CA begins June 1, requires RVP max. (6.8 psi), sulfur max. (80 ppm), benzene max. (1.0 wt%), aromatics max. (22 vol%), olefins max. (4 vol%), O_2 max. (2.0 wt% summer)
1997	Oxygenated RFG in selected markets
1998	Federal Phase I (Complex Model) RFG mandatory after Jan 1
2000	Federal Phase II (Complex Model) RFG mandatory after Jan 1 in the (now) ten existing O_3 non-attainment areas, many (~20) other areas "opt-in" for 4-year minimum; new limits on emissions (Table 18.2.5); ~85% of RFG contains MTBE; ~8% of RFG contains ethanol
2003	California postpones ban on MTBE (CARB Phase III RFG), other states seek waiver of Federal Phase I requirement of 2 wt% O_2.
2004	California and New York MTBE after Jan 1; Clean Air Act, Tier 2 standards in effect; Federal sulfur phase down begins (max. allowable 378 ppm); excludes geographic phase in areas and small refiners
2005	Sulfur phase down continues (max. allowable 326 ppm)
2006	Sulfur phase down continues (max. allowable 95 ppm); average sulfur limit will be set at 30 ppm nationwide

essentially spans the same range as modern gasolines. A comparison of this feature is shown in Figure 18.2.1. This shows that while the boiling distribution in a gasoline from 1905 is unique, there is little difference between a 1920 and a 1995 gasoline, both spanning from approximately 150 °C to 232.2 °C. The low octane, early gasolines were subject to no particular specifications or testing, which led to engine knock and overall combustion inefficiency. This condition necessitated the search for a better quality gasoline to meet the growing demand for automotive performance.

Automotive gasoline's commercial success over the past 100+ years stems largely from the early and continuing cooperation between petroleum refiners and automobile manufacturers. The American Society for Testing and Materials (ASTM) provides a forum for this cooperation, and has ultimately led to gasoline specifications adopted by refiners, gasoline marketers, automakers, automotive engineers, and, more recently, regulatory agencies such as the United States Environmental Protection Agency (EPA) (Gibbs, 1998). Some of the earliest ASTM specifications for gasoline are still in use today; e.g., Reid Vapor Pressure (RVP) (ASTM D 323) and Petroleum Distillation (ASTM

D 86) were originally published in 1930 and 1921, respectively. Notably, ASTM D 323 is not appropriate for modern, alcohol-laden gasolines due to small amounts of water in the apparatus that can extract the alcohol (Furey et al., 1993).

ASTM's current standard for automotive gasoline, D 4814 – *Standard Specification for Automotive Spark-Ignition Engine Fuel*, refers to a variety of appropriate analytical methods and mandated limits for vapor pressure/volatility (seasonally and geographically variable), sulfur content, water tolerance, oxidation stability, existent gum content, lead content, and copper strip test. Additional regulatory limits (Federal and State) on properties not specifically addressed in ASTM D 4814, e.g., oxygen and phosphorous content, are described in an appendix (X3) to the standard. Notably, the octane of automotive gasolines is not specified by ASTM D 4814. This "performance-based" feature is left up to individual refiners and marketers.

Throughout the first 60 years of gasoline formulation the refining practices became steadily more advanced. As refining techniques became more sophisticated and more usable gasoline range hydrocarbons were extracted or produced from crude oil feedstocks, the overall composition of

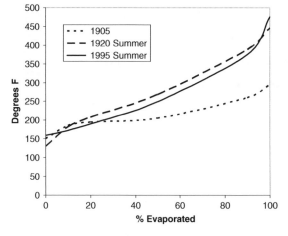

Figure 18.2.1 *Distillation curves for earlier automotive gasolines and a modern gasoline. (The 1905 and 1920 gasoline data are from Gibbs, 1990.)*

"gasoline" began to change. The refining practices such as "cracking" (thermal-, hydro-, and catalytic), reforming, and alkylation changed gasoline from a predominantly straight-run distillate or NGL product containing mostly naturally occurring paraffins and naphthenes into a blended product containing "man-made" aromatic and branched hydrocarbons. The effects of different refining processes on gasoline composition are described in greater detail later in this chapter.

In addition to improvements in the refining process, additives were being introduced into gasoline to continue to improve gasoline's performance, principally the antiknock properties or "octane rating." The most important additive in the early evolution of automotive gasoline was tetraethyl lead (TEL; $Pb(C_2H_5)_4$). Tetraethyl lead was first introduced into automotive gasolines in 1923 – in fact, gas-containing TEL produced the top three finishers at that year's Indianapolis 500 auto race (Needleman, 1998). After a few years of debate over its use due to lead's toxicity, the US Surgeon General decided to permit its widespread use in automotive gasoline at concentrations not to exceed 3.17 grams lead per gallon (glpg) in 1926. Overall use of TEL increased steadily in the following decades (except during World War II) as the demand for increased engine performance and fuel economy grew (Figure 18.2.2). The maximum allowable concentration of lead in gasoline was increased to 4.23 glpg in 1959.

Ultimately, the average levels of lead in the premium and regular gasoline pools in the United States reached a high of approximately 2.8 and 2.5 glpg, respectively, around 1970 (Figure 18.2.2). After 1970, improvements in refining practices and new Federally mandated restrictions led to a steady reduction in average lead levels in the US gasoline pool since 1970. This reduction culminated with a complete elimination of lead alkyls in automotive gasoline in the United States in 1996 (1992 in California). Other countries around the world have followed this same trend.

Figure 18.2.2 also shows the reduction in the average sulfur concentration in gasoline between 1949 and 1987. This reduction was not driven by regulatory changes but by improved refining techniques that eliminated naturally occurring sulfur from refinery feedstocks, mostly as a means to preserve the lifespan of expensive catalysts. In addition, refiners found that sulfur in leaded gasolines reduced the

octane gained from the lead alkyls due to a reaction(s) between sulfur and lead. Continued reduction in sulfur content of gasoline driven by environmental concerns is described in the next section.

18.2.2   Regulatory Era (1970–present)

As described in the previous section, the first 60 years or so of automotive gasoline formulation largely were driven by the need to increase performance (octane), while maximizing the yield from crude oil and minimizing production costs. By contrast, over the past 40 years the formulation (and reformulation) of gasoline has been driven by the introduction of various environmental regulations directed at improving automobile emissions and, more recently, at improving groundwater quality (Table 18.2.3). The regulatory history of gasoline, in particular the changes in chemical composition stemming from these regulations, is an important consideration for the forensic investigator (Dorn *et al.*, 1983; Gibbs, 1990, 1993, 1996; Frank, 1999; Morrison, 2000a).

The original Clean Air Act was passed in 1963 with emission standards to be set nationally starting in 1968 (1966 in California). Emission reductions were adequately achieved at this time through modifications to the carburetor and exhaust systems, with little if any affect on gasoline composition. In fact, as noted above (Figure 18.2.2), average lead concentrations reached their maximum after the 1968 emission standards were in place.

The Clean Air Act (CAA) of 1970 called for further reductions in emissions and in the use of lead, a known toxin. This Act promulgated the development of automobiles requiring exhaust systems equipped with oxidation catalysts in the early 1970s, which were subject to "poisoning" upon contact with organic lead. Phosphorous in gasoline had a similar effect. Thus, unleaded gasolines were necessarily developed by refiners while automobile makers developed suitable engine and exhaust modifications to maintain adequate performance. Unleaded gasolines containing a maximum of 0.05 glpg were required in July 1974, although some manufacturers introduced unleaded gasolines as early as 1971, which contained less than 0.07 glpg (Table 18.2.2). In addition, by 1973 most refiners produced a "low lead" gasoline, which (as per ASTM D 439) contained no more than 0.5 glpg. As such, most gasoline service stations in the 1970s had premium and regular leaded gasoline, low lead gasoline, and unleaded gasoline available for sale.

The CAA Amendments of 1977 dictated a systematic reduction in maximum allowable lead levels in the future US gasoline pool as shown in Table 18.2.4. The Code of Federal Regulations (40 CFR Subpart B, Part 80.20) required refiners and importers to provide the United States Environmental Protection Agency with the average concentration of lead in gasolines produced or imported quarterly. The specific CFR language was "a refiner shall not:

i) produce leaded gasoline whose average lead content during any calendar quarter ending prior to July 1, 1985, exceeds 1.10 grams of lead per gallon of leaded gasoline.

ii) produce leaded gasoline whose average lead content during any calendar quarter beginning on or after July 1, 1985, and ending prior to January 1, 1986, exceeds 0.50 grams of lead per gallon of leaded gasoline.

iii) produce leaded gasoline whose average lead content during any calendar quarter beginning on or after January 1, 1986, exceeds 0.10 grams of lead per gallon of leaded gasoline."

A

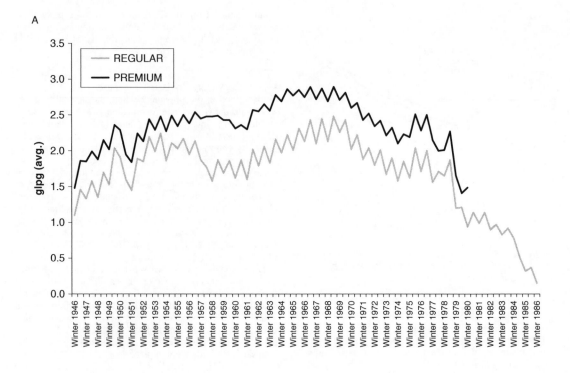

B

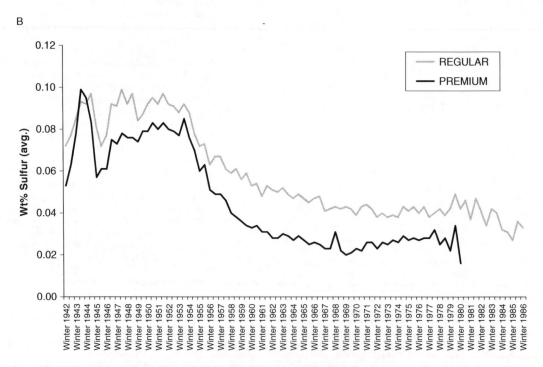

Figure 18.2.2 *Average concentrations of (A) total lead (grams lead per gallon) and (B) total sulfur (weight percent) in regular and premium gasolines from 1949 to 1987 (Shelton et al., 1982; Dickson et al., 1987). "Sawtooth" patterns for the lead concentration curves are due to the slightly higher concentrations typically used in summertime gasoline. Premium gasoline was no longer produced after 1980.*

The exact same language applied to importers of gasoline except the word "*produce*" was replaced with "*sell or offer for sale.*"

The 1977 CAA Amendments also introduced the "substantially similar" criteria which, for the first time, imposed restrictions on how refiners could modify future formulations of unleaded gasoline – unless granted a special waiver by the United States Environmental Protection Agency. The first United States Environmental Protection Agency waiver was granted in 1978 for the use of 10 vol% ethanol ("gasohol"), which subsequently led to other waivers for other oxygen-containing additives (e.g., *tert*-butyl alcohol (TBA) and MTBE; see Table 18.2.2). In 1981, the United States Environmental Protection Agency ruled that unleaded gasolines could contain up to 2 wt% oxygen and meet the "substantially similar" criteria of the 1977 Amendments. Thus, after this ruling many refiners began to use various oxygen-containing additives to boost octane in their unleaded gasolines.

Table 18.2.3 *Inventory of US Regulations between 1960 and 2005 affecting gasoline composition*

Year	Regulation
1963	Clean Air Act
1970	Clean Air Act
1974	Energy Policy Conservation Act
1977	Clean Air Act Amendments
1989	Summer Volatility (RVP) Regulation, Phase I
1990	Clean Air Act Amendments
1992	Summer Volatility (RVP) Regulation, Phase II
	Winter Oxygenated Gasoline
1993	California reformulated gasoline, Phase I
1994	Clean Air Act, Tier 1 standards in effect
1995	Federal RFG, Phase I
1996	California reformulated gasoline, Phase II
2000	Federal RFG, Phase II
2002	Mobile Source Air Toxics, Phase I
2003	California reformulated gasoline, Phase III
2004	Clean Air Act, Tier 2 low sulfur standards in effect

With the added octane achieved from these oxygen-containing additives, many refiners began to produce unleaded premium gasolines for the first time. Simultaneously, most refiners eliminated production of premium leaded gasolines (as per Figure 18.2.2). Throughout the 1980s the production of unleaded gasolines grew while leaded gasoline production declined. In 1980, unleaded gasolines comprised only 27% of the gasoline pool but had reached 98% by 1992.

The RVP requirements in gasoline were reduced in 1989 and 1992 by the Summer Volatility Regulation of 1989 (Table 18.2.3). This required that summertime gasolines meet stricter RVP requirements (see Table 18.2.2). Refiners met these reductions by reducing the amount of butanes used in gasoline blending, which necessitated the addition of aromatics and isoparaffins to offset the octane loss.

The CAA Amendments of 1990 culminated with a complete elimination of lead alkyls in automotive gasoline in the United States beginning on January 1, 1996 (1992 in California; Table 18.2.4). The 1990 Amendments also, *for the first time, required* the addition of oxygen-containing additives to some gasolines to be sold in certain markets. The two programs were introduced to reduce air emissions, particularly carbon monoxide (CO) and ozone (O_3). These were the "Oxyfuel" and Reformulated Gasoline (RFG) programs, respectively.

The Oxyfuel Program required that, starting in the Fall of 1992, gasoline sold in 39 CO non-attainment areas between November 1 and March 1 contain a minimum of 2.7 wt% oxygen (although 2.0 wt% minimum was adopted in California due to concern over NO emission increases at the higher oxygen level). The RFG Program required that, starting on January 1, 1995, RFG be used year-round in nine O_3 non-attainment areas (representing about 22% of the US gasoline market) – and its optional use in other areas. The primary emission and compositional requirements for the RFG program gasolines are shown in Table 18.2.5.

In order to meet the RFG requirements refiners had to reduce the proportion of aromatic-rich reformate blended into gasoline, or at least reduce the severity of the reformation process (a hydrogen-producing process), in order

Table 18.2.4 *Legal maximum lead content in US and Canadian gasoline. All concentration in grams lead per US gallon (after SAE Automotive Fuels Handbook)*

Date	Leaded	Unleaded	Regulation
United States			
1926	3.17		Surgeon General
1959	4.23		Surgeon General
Jul-74		0.05[b]	Federal Register 38(6), Part II, Jan. 10, 1973
Oct-82	1.1[a]		Federal Register, June 8, 1977
Jul-85	0.5[a]		
Jan-86	0.1		
Jan-92	banned in CA		
Jan-96	banned nationwide		Federal Register, 1990
Canada			
Jan-76	3.0	0.05	Clean Air Act, Section 22, Canada Gazette, Part II, 108(15), Aug. 14, 1974
Jan-87	1.1		
Dec-90	banned nationwide		
Dec-90	0.1[c]		

[a] Average quarterly leaded gasoline production
[b] Incidental lead in unleaded gasoline
[c] Only permitted in off-highway and marine use

Table 18.2.5 *Emission and compositional requirements for Oxyfuel and reformulated gasoline – Federal Phase I and II, California Phase II*

	Oxyfuel Federal	Phase I Federal RFG	Phase II Federal RFG	Phase II California RFG*
Implementation Dates	Nov 1, 1992	Jan 1, 1995	Jan 1, 2000	June 1, 1996
Emission Reductions (% versus 1990 levels)				
Volatile organic compounds	na	9	15	17
Nitrogen oxides	na	4	4	11
Carbon monoxide	na	11	11	11
Sulfur dioxide	na	0	0	80
Compositional Requirements				
RVP (psi) – max.	ASTM	7.0	6.7	6.8
Oxygen (wt%) – min.	2.7	2.0	2.0	2.0
Benzene (vol%) – max.	ASTM	1.0	1.0	1.0
Aromatics (vol%) – max.	ASTM	27	25	22
Olefins (vol%) – max.	ASTM	8.5	8.5	4
Sulfur (ppm) – max.	1000	1000	1000	80
Lead (glpg) – max.	0.05	0.05	0.05	0.05
Distillation temperatures (°F)				
T50 – max.	ASTM	210	207	200
T90 – max.	ASTM	329	321	290

*Satisfies Federal Phase II RFG requirements.
na – not applicable; ASTM – as per ASTM D4813

to meet the reduced total aromatic and benzene requirements in RFG. Simultaneously, hydrotreating (a hydrogen-consuming process) became more necessary in order to meet the lower sulfur requirement of RFG (Table 18.2.5) and the recent Tier 2 sulfur standards (Table 18.2.3). These two conditions created a "hydrogen deficit" for many refiners, which added costs to gasoline production.

Both the "Oxyfuel" and RFG Programs forced refiners to escalate the production and use of oxygenates in their gasolines and led to significant capital investments by many refiners. As a result, the production of MTBE and other oxygenates increased dramatically in the early 1990s (Figure 18.2.3). Because of the additional costs of refining Oxyfuel or RFG their use in CO and O_3 attainment areas was minimal. Thus, in most CO and O_3 attainment areas conventional modern gasoline is still used. The added costs of meeting the Oxyfuel and RFG requirements (along with other market pressures) led to the closure of nearly 50 United States refineries in the 1990s.

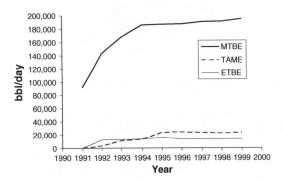

Figure 18.2.3 *US refinery production of ether oxygenates between 1991 and 1999. Data compiled from* Oil & Gas Journal *Annual Refining Reports.*

The environmental problems associated with MTBE following the release of MTBE-laden gasoline into the environment (United States Environmental Protection Agency Blue Ribbon Panel, 1999; Johnson *et al.*, 2000) has led to a movement to ban its use. Table 18.2.6 summarizes the state bans as of this writing. Although MTBE had been previously banned in some smaller (ethanol favoring) markets (e.g., Iowa), two major markets for RFG and oxygenated gasoline, namely California and New York, implemented MTBE bans beginning on January 1, 2004. Other states have already or are likely to follow. This will undoubtedly create a situation where the demand for MTBE will decline – while the demand for ethanol will rise.

The solution of most refiners is to replace MTBE with ethanol in order to meet the oxygen requirements of Oxyfuel and RFG (Table 18.2.5). The use of ethanol in Oxyfuel and RFG (in some markets; Chicago, California, New York) promulgated additional changes in gasoline composition by refiners. First, less ethanol is needed to meet the oxygen minimum requirements (Table 18.2.5). Second, replacing MTBE with ethanol increases the RVP of gasoline, thereby requiring refiners to offset this increase by reducing the amount of butanes and pentanes in ethanol-containing Oxyfuel and RFG. Third, because of concurrent changes in the T50 and T90 distillation temperatures, refiners must also reduce the proportion of heavier gasoline components (C_3 + aromatics) in order to meet these requirements. Overall these changes reduced the volume of gasoline able to be produced and increase costs (Stratco and Gertz, 2002).

The most recent regulatory changes to take effect, or scheduled to take effect, are related to the reduction of sulfur (Tables 18.2.2 and 18.2.3). As was depicted in Figure 18.2.2, the average sulfur content of US leaded gasolines had steadily decreased due to the deleterious effects of sulfur on the refining process, specifically expensive catalysts, and on its negative affect on the antiknock gain from lead. This resulted in average concentrations of sulfur on the order of 300–350 ppm in leaded gasoline in the 1980s. Unleaded gasolines also had remained at about this level through the 1980s and 1990s, although they are typically about 50% higher in the western United States, (Keesom

Table 18.2.6 *Inventory of State bans on the use of MTBE in RFG and oxygenated gasoline and average consumption (thousands of barrels per day) in those states in 2000. (Data from Dept. of Energy, Energy Information Administration website)*

State	MTBE Phaseout Date		MTBE Avg. Annual Consumption	State		MTBE Avg. Annual Consumption
MTBE bans enacted				No MTBE bans enacted		
California	Jan 1, 2004		102.4	Arizona	9	3.6
Connecticut	Jan 1, 2004		8.5	Delaware		3
Kentucky	Jan 1, 2006	1	2.2	Dist. Of Colombia		0.8
Missouri	Jul 1, 2005	1	3.3	Maine	10	0
New York	Jan 1, 2004		19.7	Maryland		11.7
Illinois	Jul 24, 2004	1,3	0	Massachusetts		16.5
Colorado	May 1, 2002		0	New Hampshire		2.9
Indiana	Jul 24, 2004	1	0	New Jersey		26.3
Iowa	May 11, 2000	1	0	North Carolina		0
Kansas	Jul 1, 2004	1,6	0	Pennsylvania		9.3
Michigan	Jun 1, 2003		0	Rhode Island		2.9
Minnesota	Jul 1, 2005	4	0	Texas		30.3
Nebraska	Jan 1, 2001	2	0	Utah		0
Nevada	Jan 1, 2004	5	0	Virginia		13.6
Ohio	Jul 1, 2005	6	0			
South Dakota	July 1, 2000	7	0			
Washington	Jan 1, 2004	8	0			

This table does not include MTBE that may be blended into conventional gasoline for octane gain only within attainment areas.
[1] Maximum 0.5 vol%/MTBE.
[2] Maximum 1.0 vol%/MTBE.
[3] MTBE banned in Chicago beginning Dec 2000.
[4] Year-round Sate-wide oxygenated gasoline requirement with ethers limited to 0.33 volume percent after July 1, 2000, and banned after July 1, 2005.
[5] This is not a State-wide ban. Washoe County (Reno) MTBE maximum limit of 0.3 volume percent effective from the same date as the California ban. Clark County (Las Vegas) adopted 10 volume percent ethanol requirement in 1999 for gasoline sold from October through March.
[6] This provision will take effect only if United States Environmental Protection Agency grants the State a waiver to control or prohibit MTBE in gasoline.
[7] MTBE limited to 2.0 volume percent beginning Feb 2000.
[8] Maximum 0.6 volume percent MTBE effective July 22, 2001
[9] Arizona Senate Bill 1504, approved by the Governor on April 28, 2000, states that it is the "policy" of the State that MTBE be phased out as soon as possible, but no later than 180 days after effective date of California ban if feasible.
[10] Maine Public Law Chapter 709 established a "goal" to eliminate MTBE in gasoline sold in the State by Jan 1, 2003.

et al., 1998), except in California where the maximum allowable was limited to 80 ppm after 1996 (Table 18.2.5).

In summary, the various changes in gasoline composition that occurred during the *Performance* and *Regulatory Eras* of gasoline's history are relevant to practitioners of environmental forensics. Specifically, the flexibility that refiners were given to achieve the performance standards up until the early 1970s and the changes in composition driven by regulatory standards after 1970 impart significant compositional heterogeneity in gasoline. In other words, *not all gasoline was (is) created equal.* As such, chemical differences among automotive gasolines released into the environment are anticipated to, at least in part (and subject to weathering effects), reflect differences in the general timeframe and manner by which they were manufactured. Toward this end, the gasoline-refining processes and use of additives are described in greater detail in the following sections.

18.3 GASOLINE REFINING & BLENDING

The modern petroleum refinery has evolved from a simple distillation facility whose primary goal was to produce fuel oils for heating and lighting (while discarding then uneconomical gasoline-range hydrocarbons), to modern operations of varying complexity focused on squeezing as much

automotive gasoline out of a barrel of crude oil; ~40% currently. Today, petroleum refineries are complex industrial plants where various crude oil feedstocks are utilized to produce not only gasoline, but other economically valuable distillate and residual products and chemical feedstocks.

18.3.1 Refining and Blending Practices

The sophistication of refineries varies, from "simple" to "complex." Simple refineries typically employ only a few processing steps (e.g., distillation and reformation) in the production of gasoline (Leffler, 2000). As a result, simple refineries are generally unable to produce the more complex gasolines such as Oxyfuels of RFG. Complex refineries have many more processing steps, the level of complexity is defined by the various types of equipment in use at the refinery (e.g., cracking, reforming, isomerization, alkylation, polymerization (Leffler, 2000)). From the environmental forensic investigator's standpoint, it is important to realize that virtually no two refineries are identically engineered or produce refined products that, at the molecular level, are identical. Armed with this knowledge, chemical fingerprinting of gasolines can be conducted and interpreted more fully.

Table 18.3.1 shows the variety of gasoline blending stocks available at a modern complex refinery and some of their important blending properties. The "recipe" for blending a

Table 18.3.1 *Inventory of gasoline blending stocks available from a single complex refinery producing oxygenated gasoline and RFG (using MTBE). (After Stout* et al.*, 2001)*

Blending Stock Name	Octane (RON)*	RVP** (psi)	Arom HC (vol%)	Aliph HC (vol%)	Benzene (vol%)	Olefins (vol%)
Straight-Run Gasoline	66.4	11.1	12.5	87.5	1.60	0
Alkylate	91.7	5.23	1.1	98.4	0.00	<1
Isomerate Unit A	89.0	12.67	0.6	99.4	0.00	<1
Isomerate Unit B	74.5	11.61	2.6	97.4	1.45	<1
MTBE	115.0	7.80	0.0	0.0	0.00	0.9
Light (FCC) Cat Gasoline	94.5	10.90	7.4	16.1	1.09	76.5
Heavy (FCC) Cat Gasoline	81.1	0.00	58.6	40.3	0.00	1.1
Light (Uni-) Crackate	80.7	9.01	3.4	96.6	1.69	<1
Light–Medium (Uni-) Crackate	65.0	1.29	4.9	94.7	0.07	<1
Mixed Pentanes	82.1	17.75	0.0	100.0	0.00	0.0
n-Butane	93.0	51.60	0.0	100.0	0.00	<1
Reformate Unit A	101.9	2.50	77.7	21.7	1.10	<1
Reformate Unit B	96.2	1.16	67.8	31.6	1.74	<1
Reformate Unit C	98.8	1.69	72.0	27.1	2.30	<1

* Research octane number
** RVP – Reid Vapor Pressure

specific gasoline grade or type at any given refinery depends upon several factors including: (1) the available inventories of the different gasoline blending stocks, (2) the operating status of the various refining units, (3) the specific regulatory requirements for the intended market, and, of course, (4) maximizing the profit margin. These factors, of course, change(d) over time.

Historic gasoline blending practices were far less complicated than they are today, due to the less stringent regulations governing gasoline. During the *Performance Era* (<1970; see above) many US refiners operated extremely simple, distillation-based refineries. Simple (distillation) refineries from this Era historically used a basic blending recipe to produce gasoline that relied largely upon straight-run gasoline, natural gas liquids, and lead alkyl antiknock (Beall *et al.*, 2002). Modern *Regulatory Era* refiners rely upon computer assistance in developing daily blending recipes from the available inventory of many different blending stocks (Table 18.3.1) that will meet the gasoline requirements. In-stream monitors provide continuous feedback on a product's properties so that blending can be automated and self-adjusting.

What follows is a brief description of the major refining units used to produce automotive gasoline. For an in-depth understanding of refinery operations, we recommend one of several excellent reference texts on the subject (Gary and Handwerk, 1984; Speight, 1991; Leffler, 2000).

18.3.2 Distillation
In distillation, a desalted and pre-heated crude oil feedstock (sometimes a blend of different crude oils) is fed into a distillation tower operated at atmospheric pressure (atmospheric distillation). Light gasses (C_1–C_4 hydrocarbons), light and heavy gasolines, various middle distillates, atmospheric gas oil, and atmospheric residuum are separated based upon volatility. More often than not in a modern refinery the atmospheric residuum feeds another distillation tower operated at reduced pressure (vacuum distillation), in which the atmospheric residuum is further fractionated into vacuum gas oil(s) and vacuum residuum. Thus, atmospheric and vacuum distillation produces a variety of "straight-run" intermediate products that can either act as feedstocks to other refinery units or, in some instances, as blending stocks in the production of finished products.

In the case of automotive gasoline blending, the most important distillation products available are light straight-run gasoline (LSRG), sometimes called light virgin naphtha, and heavy straight-run gasoline (HSRG). Both LSRG and HSRG generally boil in the C_5–C_{12} range (~100–400 °F) but have disparate mass distributions within this range. They will have octane values of between 60 and 70 (Gary and Handwerk, 1984), and therefore require enrichment in order to approach octane levels necessary for automobile gasoline.

Figure 18.3.1 shows the distributions of hydrocarbons in examples of LSRG and HSRG produced during distillation at a single refinery. The PIANO (Paraffins, isoparaffins, aromatics, naphthenes, and olefins) method of analysis is described later in this chapter. As can be seen both LSRG and HSRG contain similar types of hydrocarbons but with different boiling distributions. Both SRGs contain a relative abundance of *n*-alkanes and naphthenes, as is typical of crude oils and natural gasolines (i.e., gas condensates produced during natural gas production). This particular HSRG is highly enriched in methylcyclohexane (MCH).

Some chemical characteristics of the LSRG and HSRG are directly inherited from the parent crude oil feedstock, and as such they may reflect certain genetic features of the parent crude oil. Because different crude oils will exhibit different chemical characteristics, these differences may provide a means to distinguish gasolines derived from different crude oil feedstocks. This aspect of environmental forensics of gasoline is discussed later in this chapter – see *Genetic Influence of Crude Oil Feedstock on Gasoline Composition.*

18.3.3 Cracking Processes
As the demand for producing automotive gasoline exceeded that of kerosene, the need to "crack" larger hydrocarbons in crude oil into smaller ones arose. The history of refining has seen the development of several cracking processes. *Thermal cracking* was first introduced in 1913 (Speight, 1991). In this process hydrocarbon feedstocks were heated in a reactor so that C–C bonds were broken to yield smaller hydrocarbon fragments, and thereby increase the proportion of gasoline range hydrocarbons (as compared to the feedstock). As these bonds are broken, the fragments produced typically include unsaturated hydrocarbons, i.e., olefins (hydrocarbons containing at least one C=C bond). Mono-olefins (i.e., hydrocarbons with one C=C bond) are desirable components as they serve to increase the octane

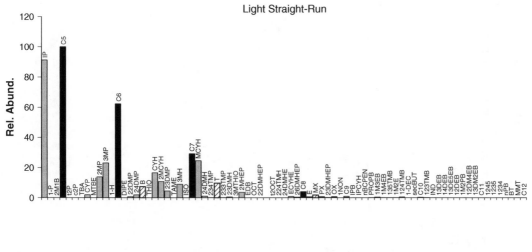

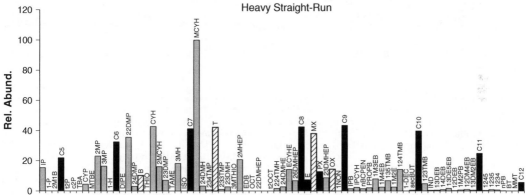

Figure 18.3.1 *PIANO histogram distributions of light and heavy straight-run gasolines produced from the distillation of crude oil. For compound identifications see Table 18.5.2 (Boiling point increases from left to right).*

rating of the cracked gasoline over that of the feedstock. Di-olefins (i.e., hydrocarbons with two C=C bonds) are unstable and readily oxidize, and therefore are undesirable components in gasoline.

The efficiency of hydrocarbon cracking was improved with the development of fixed bed *catalytic cracking* in 1936 and fluidized bed catalytic cracking (FCC) in 1940 (Table 18.2.1), processes in which an inert catalyst(s) was added to the reactor to further promote the cracking of hydrocarbons. Modern FCC still is the heart of gasoline production at most refineries. The FCC also produces mono-olefins, but generally in lower proportions relative to thermal cracking. In addition, the olefins produced during FCC typically contain three or more carbons, versus the one or two carbon fragments normally produced by thermal cracking. Also formed during FCC are iso-alkanes (branched hydrocarbons) and aromatic hydrocarbons, both of which further improve the octane rating of the cracked FCC gasoline. Catalyst and reactor improvements, along with the increasing demand for octane in the gasoline pool, has made thermal cracking obsolete in the modern refinery (except as it is used to reduce the viscosity of heavy feedstock, i.e., visbreaking).

The catalysts used in FCC prior to about 1965 were synthetic alumina-silicates. Since then catalysts have progressively shifted toward catalysts containing zeolites (Pines, 1981). This shift had reduced the proportion of olefins

formed (in favor of alkanes) and increased the rate of cracking and catalyst lifetime. (Notably, many *Performance Era* gasolines often contained more than 20 wt% olefins.) The reduction of olefins is a necessary property in the production of RFG and other modern gasolines (Table 18.2.5) but is also necessary to minimize gumming. Figure 18.3.2 shows the distributions of hydrocarbons in an FCC gasoline. The FCC gasoline is enriched in isopentane and hydrocarbons spanning the C5–C12 range. Although olefins are formed during the FCC cracking process, these typically comprise only a small weight percent of the FCC gasoline.

A third cracking process, *hydrocracking*, was introduced in 1958 (Table 18.2.1). During this process the breaking of C–C bonds is conducted in the presence of catalysts in a pressurized hydrogen atmosphere, a combination which immediately saturates any C=C bonds leading to the formation of an "olefin-free" cracked product. Thus, a gasoline blended using a hydrocracked product will not contain olefins (whereas an FCC gasoline will). This distinction can present the forensic investigator with an opportunity under certain circumstances. For example, since all *Regulatory Era* refiners employ some form of cracking process, an automotive gasoline that contains notable concentrations of olefins must have been blended using either a thermally cracked or FCC-cracked blending product. Nowadays, this is most likely an FCC; US refining charge capacity as of January 1, 2000, indicates that FCC capacity is about 4 times higher

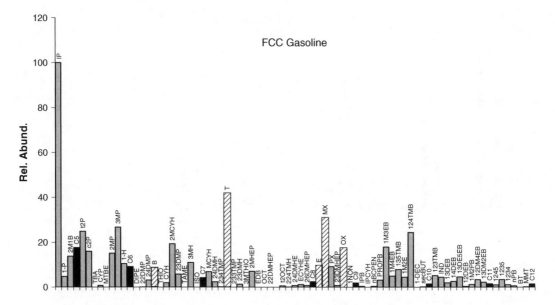

Figure 18.3.2 *PIANO histogram distributions of fluid cat-cracked (FCC) gasoline. For compound identifications see Table 18.5.2.*

than hydrocracking capacity (5.5 versus 1.4 million barrels/day – Radler, 1999). Oppositely, the absence of olefins argues for a gasoline source produced via hydrocracking. (Of course, weathering must be considered since olefins are subject to biodegradation and waterwashing; see below and Figure 18.6.8.)

18.3.4 Catalytic Reforming
The need for improved gasoline octane in the early 1900s also led to the development of thermal reforming processes in which re-arrangement ("reforming"), rather than cracking, reactions are dominant. This technology was replaced starting in around 1940 with the development of catalytic reforming (Table 18.2.1). In this process, reactions occur in the presence of a mixed-catalyst system and hydrogen, and promotes the re-arrangement of lower octane naphthenes into higher octane compounds without a significant reduction in carbon number (Speight, 1991). Because of the mixed catalysts used, some compounds tend to lose hydrogen (e.g., forming aromatics from naphthenes) while others tend to gain hydrogen (e.g., saturated alkanes from olefins). The product of catalytic reforming, *reformates*, are major blending stocks for most automotive gasolines (Table 18.3.1). There are often multiple catalytic reforming units operating at a modern refinery under slightly different conditions, thereby producing reformates with slightly different compositions (and octane ratings).

Dehydrogenation is the primary reaction sought during reformation since it yields the higher octane aromatics (and hydrogen that can be used elsewhere in the refinery, e.g., hydrocracking, desulfurization, or hydrofining). Consequently, most reformates are enriched in the monoaromatic (e.g., BTEX) compounds. Benzene concentrations in reformate will normally range from 5 to 10 wt%. Toluene occurs at higher concentration (20–25 wt%) and is often the most abundant compound, being formed primarily from the dehydrogenation of methylcyclohexane in the feedstock (e.g., note the abundance of methylcyclohexane in the HSRG; Figure 18.3.1). Figure 18.3.3 shows the distribution of hydrocarbons in two reformates produced under

varying conditions, which yields slightly different octane values. Both reformates are dominated by the BTEX compounds, particularly toluene. Among the C_3-alkylbenzenes, both reformates exhibit a predominance of 1-methyl-3-ethyl benzene (1M3EB) and 1,2,4-trimethylbenzene (124TMB), in a 1M3EB/124TMB ratio of ~0.7. These two isomers have the highest octane value among the C_3-alkylbenzenes (162 and 171, respectively; Pines, 1981) and appear preferentially formed during reformation. This ratio may change slightly under different reforming conditions yielding a slightly broader range observed in gasolines (~0.6–0.9).

Since reformates are typically blended directly into gasoline the proportions between various mono-aromatics can provide some diagnostic information potentially related to the reforming conditions (temperature, catalyst type and age, and feedstock). For example, conditions during reforming typically result in a predominance (~50% of total) of *m*-xylene among the C_2-benzene isomers. However, the relative proportions among these four isomers (ethylbenzene, *m*-, *p*-, and *o*-xylene) can vary depending upon reforming conditions (compare distributions in Figure 18.3.3). Substantially disparate distributions among these isomers, e.g., excessive *m*-xylene, can be caused by: (1) peculiar reforming conditions or (2) the use of *p*- or *o*-xylenes as chemical feedstocks to phthalic acid and anhydride production, respectively. The latter is more commonly observed in refineries that operate in connection with a petrochemical plant. Similarly, the percentage of ethyl-benzene to total C_2-aromatics is normally ~10–20% in reformates and fresh gasolines. A much lower percentage of ethyl-benzene versus total C_2-aromatics (<5%) could indicate ethyl-benzene's removal for specialty chemical production (e.g., styrene). Thus, relationships among BTEX compounds could prove useful in distinguishing gasolines containing reformate blends formed under different catalytic reforming conditions. However, and unfortunately, the BTEX compound distributions can be altered during environmental weathering, and therefore require caution when they are used to distinguish sources in environmental samples. On the other hand, the higher boiling aromatic components within reformate, e.g., C_3- and C_4-alkylbenzenes, naphthalenes, and

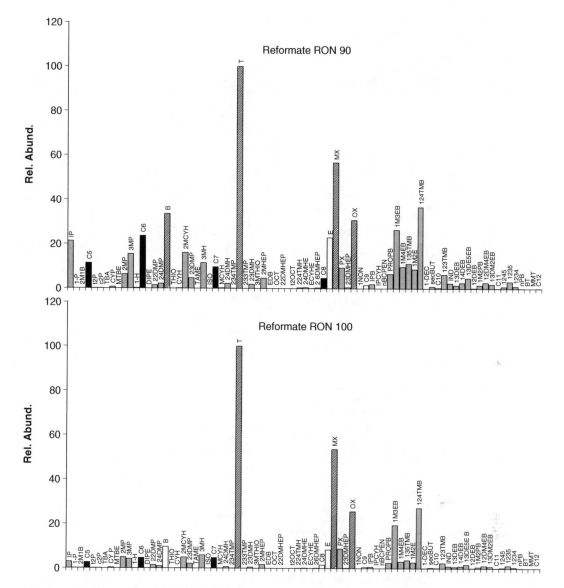

Figure 18.3.3 *PIANO histogram distributions of reformates of varying severity (RON – research octane number). For compound identifications see Table 18.5.2.*

C_1-alkylnaphthalences, are relatively stable and can also provide some basis to distinguish gasolines. Beall *et al.* (2002) describe the use of the ratio of naphthalene to n-C_{12} (a comparably boiling *n*-alkane) as a measure of the aromatic character of gasolines, which in turn is a reflection of how much reformate (a source of naphthalene) versus SRG (as source of n-C_{12}) was used in the blending the gasolines.

18.3.5 Isomerization

Isomerization of low boiling ($<C_6$) normal hydrocarbons to saturated branched hydrocarbons was another refining development triggered by the need for octane, that was introduced in 1943 (Speight, 1991). In this process various catalysts and reactor conditions are used to convert C_4 to C_6 feed streams (butane, pentane, and/or pentanehexane) into their various isomeric equivalents, i.e., isobutane, isopentane, 2,2-dimethylbutane, 2,3-dimethylbutane,

2-methylpentane, and 3-methylpentane. The specific reaction conditions during isomerization (type and age of catalyst, feed stream and rate, and temperature) will yield rather specific distributions of isomers, collectively known as *isomerate*.

Because isomerate is a common blending component in gasoline the distribution among these C_4–C_6 isomers can be used to distinguish among gasolines containing different isomerate blends. For example, it has been shown that the ratio between isopentane (i-C_5) and pentane (n-C_5) or between 2-methylpentane and 3-methylpentane can vary significantly depending upon the isomerization reaction conditions (Pines, 1981). For example, the i-C_5/(i-C_5 + n-C_5) ratio of LSRG ($\sim$0.40–0.50; see Figure 18.3.1) is significantly increased following isomerization, being greatest in the presence of pressurized hydrogen ($\sim$0.60–0.70; Pines, 1981; Gary and Handwerk, 1984). The degree of blending between

LSRG and isomerate will largely determine the $i\text{-}C_5/(i\text{-}C_5+n\text{-}C_5)$ ratio in the finished gasoline (Beall *et al.*, 2002). Therefore, this ratio can be useful in environmental forensic investigations that require distinction between fugitive gasolines in the environment. Of course, the high volatility of these $<C_6$ compounds warrants caution due to the potential effects of evaporation on environmental samples.

18.3.6 Alkylation

The acid-catalyzed reactions of olefins with normal hydrocarbons to yield higher boiling, and higher octane, gasoline range branched hydrocarbons (iso-alkanes) were first utilized in refining in 1938 (H_2SO_4) and 1943 (HF; Table 18.2.1). These two alkylation methods are still in use today at complex refineries that utilize alkylation. The product of the alkylation process is termed *alkylate* (Table 18.3.1).

Refineries without alkylation capabilities are becoming rarer and typically can only produce conventional gasoline, i.e., not Oxyfuels or RFG gasolines. The worldwide capacity for alkylate production by these two processes over the past 50 years has shown a decline in HF alkylation since $\sim$1990 (Beall *et al.*, 2002). This is primarily due to environmental concerns associated with the potential for an accidental release of HF from a failed alkylation unit. For example, an accidental release of HF vapor occurred in Texas City (Texas) in 1987, forcing an evacuation of the nearby community. Thus, some refiners have been converting HF units into H_2SO_4 units, further increasing the total H_2SO_4 alkylate capacity.

The acids promote the reaction of iso-butane and various C_3 or C_4 olefin streams (e.g., ethylene, propylene, butylenes). The type of olefin stream, combined with the reaction conditions (acid type, temperature, isobutane/olefin feed ratio, and olefin charge rate), will determine the mixture of iso-alkanes produced (Gary and Handwerk, 1984). Such differences have application in the environmental forensics of gasoline as described below.

The primary products of either H_2SO_4 or HF alkylation are trimethylpentanes (TMPs), e.g., 2,2,4-TMP (isooctane) or one of its three isomers (2,2,3-, 2,3,4- and 2,3,3-TMP). Because of its high octane and volatility, isooctane is the primary alkylation product sought by most refiners. Figure 18.3.4 shows the distributions of hydrocarbons in an alkylate and in an alkylate-dominated non-aqueous phase liquid (NAPL) collected from a refinery where alkylate had been spilled. Both alkylates were obtained from refineries operating an HF-type of alkylation unit. The predominance of the TMP isomers is evident in each, with isooctane comprising 60 and 66% of the total TMPs.

This observation is significant since the percentage of isooctane relative to the total of all four TMP isomers is expected to vary depending upon the alkylation conditions, e.g., H_2SO_4 versus HF-catalyzed reactions and the nature of the olefin feed material (e.g., Hutson and Hays, 1977; Pines, 1981; Gary and Handwerk, 1984).

The utility of this variation in forensic investigations was discussed by Beall *et al.* (2002). Based upon alkylate yield data it was demonstrated that the HF alkylation process produces a greater percentage of isooctane, versus the other trimethylpentane (TMP) isomers, than the H_2SO_4 alkylation process regardless of the olefin feed. Beall *et al.* (2002) used the data from Pines (1981) to calculate the percentage of isooctane relative to all four TMP isomers using the following formula:

$$2,2,4\text{-}TMP/(2,2,4\text{-}TMP+2,2,3$$
$$\text{-}TMP+2,3,4\text{-}TMP+2,3,3\text{-}TMP)$$

The results indicated that gasolines blended using an HF alkylate would be expected to contain a higher percentage of isooctane (54–73%) than gasolines blended using an H_2SO_4 alkylate (39–45%), regardless of the olefin feed. Note the HF alkylate data from Hutson and Hayes (1977) supported this contention with ratios between 60 and 70% isooctane for the various olefin feeds. Since the TMP isomers would, based upon their similar physico-chemical properties (Marchetti *et al.*, 1999), be anticipated to react similarly to environmental weathering, the percentage of isooctane (of the total TMP) could be useful in distinguishing gasolines blended using an HF or H_2SO_4 alkylation unit.

We have investigated this apparent relationship for 16 finished gasolines that were produced by refiners employing a specific alkylation unit type. More specific information regarding the particular catalysts, reaction conditions, or olefin feed types and rates were unknown. However, the nine gasolines containing an HF-type alkylate contained between 52 and 62% isooctane (avg. 58%; versus total TMP) and the seven gasolines containing an H_2SO_4-type alkylate contained between 37 and 47% isooctane (avg. 43%). These results are consistent with the ranges predicted by Beall *et al.* (2002). Figure 18.3.5(a) shows the population distribution for these 16 finished gasolines in which the two populations of alkylate type are evident.

Figure 18.3.5(b) also shows the percent isooctane in 75 NAPLs comprised of alkylated gasolines, but of an unknown alkylate type(s). The NAPL population contains between 37 and 69% isooctane. There is some indication of a bimodal character to the population that generally coincides with that of the two types of alkylate, with an apparent division between the two populations somewhere around 52%. The population with lower percentages of isooctane contained between 37 and 51% isooctane (avg. 44%) and the population with the higher percentages contained between 53 and 69% isooctane (avg. 59%). These ranges are also consistent with the predicted values for H_2SO_4 and HF-type alkylate, respectively. Of course, mixing of alkylates from different production methods could homogenize the percent of isooctane to total TMPs in the NAPL population, although this is not obvious. Weathering does not appear to have altered this ratio significantly, since the 75 NAPLs represent a wide range of weathering states. This is not unexpected since the TMPs share comparable physico-chemical properties that behave similarly in the subsurface (Marchetti *et al.*, 1999). Thus, the apparent differences among the percentages of isooctane in the NAPLs supports using the percent isooctane of the total TMPs to compare NAPLs to one another, and to determine the most likely alkylate type which they contain (i.e., HF versus H_2SO_4).

18.3.7 Butane Blending – Effect on normal- and iso-butane abundance

Historically, petroleum refiners have maximized the amount of butanes in gasoline due to their high octane and low cost. Because of the butanes' high vapor pressures this practice has been limited since the CAA Amendments of 1990. In spite of the restrictions, it is still beneficial (profitable) for modern refiners to blend some butane into gasoline and balance vapor pressure with other, higher boiling components.

Refiners have always preferred to add *n*-butane ($n\text{-}C_4$) over isobutane ($i\text{-}C_4$) because: (1) $n\text{-}C_4$ has lower vapor pressure (52 versus 71 psi) and (2) $i\text{-}C_4$ is a valuable process feed used in the alkylation process (see above). Beall *et al.* (2002) suggest that this practice can be monitored by measuring the ratio of:

$$n\text{-}C_4/(n\text{-}C_4 + i\text{-}C_4)$$

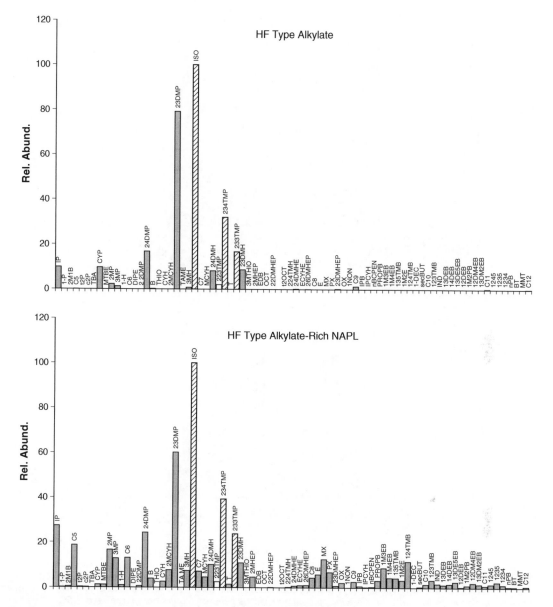

Figure 18.3.4 *PIANO histogram distributions in an alkylate and alkylate-rich NAPL. Both examples are from refineries operating HF-type alkylation units. For compound identifications see Table 18.5.2.*

A ratio of approximately 0.80 and higher has been and still is typical of automotive gasoline; i.e., 80% or more of the butane added to gasoline is *n*-C$_4$ (Beall *et al.* 2002). However, the butane ratio has tended to increase due to the even greater demand for isobutane as an alkylation feed in recent years. Of course, the potential affects of evaporation after a release to the environment must always be considered when comparing proportions of these highly volatile compounds. However, under some conditions where evaporation is retarded these compounds may prove diagnostic. The slight preferential evaporation of isobutane would tend to increase the ratio.

Notably, butanes are difficult to measure in environmental samples. Their analysis is restricted to NAPLs that can be analyzed by a direct-injection *Gas Chromatography/Mass*

Spectrometry (GC/MS) method. This method is described later in this chapter.

18.3.8 Modern Gasoline Blends

The gasoline blending stocks shown in Table 18.3.1 are combined to produce three primary types of gasoline in use today – namely Oxyfuels, Reformulated Gasolines (Federal and California), and Conventional Gasolines. The regulatory requirements of Oxyfuels and RFGs were listed in Table 18.2.5. The requirements of conventional gasolines are defined by ASTM D 4814. The specific chemical compositions of these three gasolines will vary slightly among refiners. Certainly, the type of oxygen-containing additives used in Oxyfuels and RFGs will vary by refiner and regionally. However, all refiners have the same basic "building

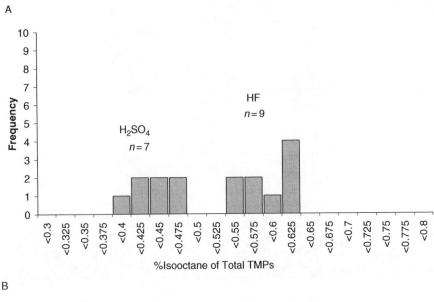

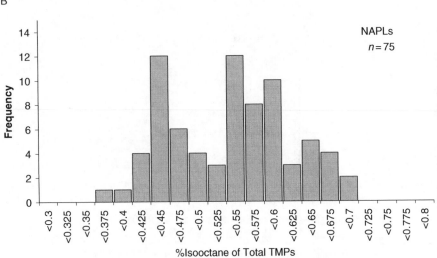

Figure 18.3.5 *Frequency-plot of the percent isooctane among the total trimethylpentanes in: (A) finished gasolines containing alkylate of a known type and (B) non-aqueous phase liquids (NAPLs; n = 75) containing an alkylate derived from unknown sources. The latter shows two apparent populations generally consistent with HF and H₂SO₄ alkylation.*

blocks" available to them to meet these regulatory and performance requirements – as described above.

Figure 18.3.6 shows the distribution of major hydrocarbons in regular and premium grades of dispensed gasolines sold as Oxyfuel, RFG (CARB Phase III), and Conventional gasoline in 2004. Each pair of gasolines was collected from a single retail outlet in Minnesota, California, and eastern Washington, respectively. Each pair represents the different blends/grades produced by a single refiner, but different refiners are represented at each location.

Comparison among the histograms reveals a predominance of isopentane, toluene, xylenes, and various other monoaromatics hydrocarbons in each of the gasolines. The prominence of isopentane indicates that all of the gasolines contain an isomerate blending stock. The concentration of isopentane varies only slightly among the six gasolines (~9–12 wt%). This compound, along with butanes and

ethanol (not shown; see below), is used as a means of regulating each gasoline's RVP.

The prominence of toluene and other monoaromatics indicates the presence of a reformate blending stock in each gasoline (as per Figure 18.3.3). Note the variations in the proportions of the BTEX compounds, particularly benzene, among the gasolines. The Oxyfuels each contains 2.2 and 2.6 wt% benzene, while the RFGs each contains less than 1 wt%, as is required in RFGs (Table 18.2.5). The modern (2004) conventional gasolines shown here each contains slightly more than 1 wt% benzene. However, today – as in the past – it is still possible to observe conventional gasolines with markedly higher benzene concentrations. For example, while a recent survey of 96 gasolines dispensed in 2004 determined that conventional gasoline contained between 0.5 and 3.0 wt%, the United States Environmental Protection

OXYFUEL

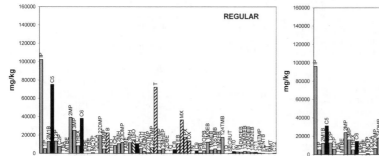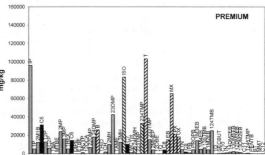

REFORMULATED

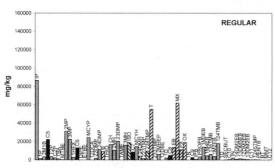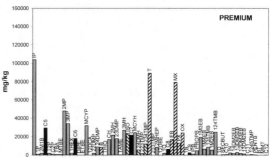

CONVENTIONAL

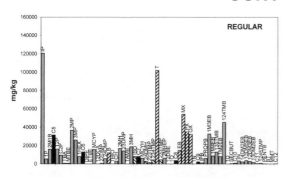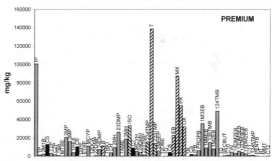

Figure 18.3.6 *PIANO histograms for regular (87 octane) and premium (91 octane) grades of Oxyfuel, Reformulated (RFG), and Conventional gasolines dispensed in 2004. Ethanol concentrations in Oxyfuels and RFGs are not shown – see text. See Table 18.5.2 for compound identifications.*

Agency-reported samples from 2002 and 2003 had contained up to 5.0 vol% (Weaver *et al.*, 2005).

The concentration of toluene is higher in each pair's premium gasoline than in the corresponding regular gasoline (Figure 18.3.6), likely indicating a greater proportion of reformate (and less straight-run gasoline) is used by each refiner in order to increase octane in their premium grades.

The conventional gasolines contain a greater concentration of C_3+ alkylbenzenes compared to the Oxyfuels or RFGs. This stems likely from the fact that there are no limits on the amount of total aromatics in conventional gasoline, thereby allowing a greater proportion of FCC gaso-

line (e.g., Figure 18.3.2) to be blended into these gasolines. The higher proportion of these higher boiling compounds requires that a greater proportion of low boiling compounds (e.g., butanes; not measured) also be added, in order to maintain some balance in vapor pressure of the gasoline.

Interestingly, the distributions among the four C_2-alkylbenzene isomers (EB, MX, PX, and OX) vary among the three refiner's gasolines, but are consistent between grades. As described above, this tends to indicate that the specific reaction conditions used during catalytic reforming varied among the three refiners.

Each of the refiners also have used varying amounts of alkylate (e.g., Figure 18.3.4) in blending their gasolines. As might be expected, each pair's premium grade contains a higher proportion of alkylate (isooctane and other TMPs) than the corresponding regular grade. The disparity is greatest in the Oxyfuels, in which the premium grade contains more than 5 times more isooctane than the regular grade (8.3 versus 1.5 wt%).

As noted above, the PIANO histograms shown in Figure 18.3.6 do not include ethanol, which was determined by a separate analysis of the Oxyfuels and RFGs. However, note that MTBE and other ether oxygenates were not present in any of these gasolines (Figure 18.3.6). This is not surprising given the markets from which they were obtained – Minnesota (ethanol "country", see below; California (which banned MTBE in 2004); and eastern Washington (where there is no oxygen minimum requirement in place). The concentrations of ethanol were measured by direct injection GC/MS in each of the Oxyfuels and RFGs, but not in the conventional gasoline. It is believed that ethanol was not in use in eastern Washington. The concentration of ethanol in the regular and premium Oxyfuels were 8.9 and 9.3 wt%, respectively. These ethanol concentrations satisfy the 2.7 wt% O_2 minimum of Oxyfuels (Table 18.2.5), which requires a minimum of 7.77 wt% ethanol. The concentrations of ethanol in the regular and premium RFGs were determined to be 6.0 and 6.5 wt%, respectively, which correspondingly meet the 2.0 wt% O_2 minimum in CARB RFG (Table 18.2.5), which requires 5.7 wt% ethanol.

The variation among these six modern dispensed gasolines demonstrates the utility in understanding how gasolines are blended by different refiners for different markets. These types of differences are at the heart of gasoline fingerprinting's application to environmental forensics. As stated previously, *not all gasolines are created equal*.

18.4   GASOLINE ADDITIVES

During the *Performance Era* (prior to 1970) fuel manufacturers had been striving to maximize gasoline's performance. During the *Regulatory Era* (since 1970) the challenge has been to maintain performance while meeting the regulatory guidelines introduced regularly (Tables 18.2.4–18.2.6) and the related automaker-driven changes (e.g., conversion from carburetors to fuel injectors in the mid-1980s). During both Eras of gasoline's history, performance has been maintained by blending various additives into gasoline so as to:

• increase power/reduce antiknock (i.e., increase octane),
• remove or prevent carburetor and spark plug deposits,
• prevent the formation of gums, peroxides, rust, and ice in engines,
• decrease tailpipe emissions, and
• distinguish different products or gasoline grades.

An inventory of the typical additives used in gasoline is shown in Table 18.4.1. Two of the more significant additives blended into gasoline, both which have forensic applications, are lead alkyls and the family of oxygen-containing compounds. The latter include both alcohols (e.g., ethanol) and ethers (e.g., methyl tertiary butyl ether, MTBE) – which because of the large volume at which they are often added (e.g., MTBE up to 15 vol%) might not be considered *additives* as much as *components* of the gasoline. Nonetheless, in this chapter we have used the term "additive" to describe these oxygen-containing compounds.

Reviews on the use and characteristics of various gasoline additives are available (Davis and Douthit, 1980; Tupa and Dorer, 1983; Gibbs, 1989, 1990). The reader is directed

Table 18.4.1 Inventory of the typical additives used in automotive gasoline. (After {Tupa & Dorer 1983 #366}; {Gibbs 1990 #889}; {Gibbs 1996 #925})

Purpose	Compound
Antiknock	Tetraethyllead (TEL)
	Tetramethyllead (TML)
	2-Methyl cyclopentadienyl manganese tricarbonyl (MMT)
Lead scavengers	1,2-Dibromoethane (EDB)
	1,2-Dichloroethane (EDC)
Oxygenates	Ethanol
	Methanol
	Methyl-*tert*-butyl ether (MTBE)
	tert-Butyl alcohol (TBA)
	di-isopropyl ether (DIPE)
	Ethyl-*tert*-butyl ether (ETBE)
	tert-Amyl methyl ether (TAME)
Detergents	Amino hydroxy amide
	Amines (e.g., polyisobutylene amine, PIBA)
	Alkyl ammonium dialkyl phosphate
	Imidazolines
	Succinimides
Antirust	Fatty acid amines (benzothiazoles)
	Sulfonates
	Amine/alkyl phosphates
	Alkyl carboxylates
Antioxdants	Hindered (*ortho*-alkylated) phenols [2-3 lb/1000 bbl]
	para-Phenylenediamine
	Aminophenols
	2,6-Di-*tert*-butyl-*para*-cresol
	ortho-Alkylated phenols combined with phenylenediamine
Dyes	Red: alkyl derivatives of azobenzene-4-azo-2-naphthol
	Orange: benzene-azo-2-naphthol
	Yellow: *para*-diethyl aminoazobenzene
	Blue: 1,4-diisopropylaminoanthraquinone
Anti-icing	Alcohols
	Amines/amines
	Organophosphate ammonium salts
	Glycols
Upper cylinder lubricants	Light mineral oils
	Cycloparaffins
Metal deactivators	N, N'-Disalicylidene-1,2-diaminopropane

to these for additional details. In this section, we offer discussions of the most common additives that are used in forensic investigation – namely lead alkyls and MTBE. Limited discussion on the use of TBA and fuel dyes are also presented.

18.4.1   Lead Alkyl Antiknocks

Among the most important additive in the early evolution of automotive gasoline were lead alkyls, specifically TEL ($Pb(C_2H_5)_4$). The history of its use was fully described by Robert (1983). An interesting collection of facts compiled from this history is given in Table 18.4.2.

Tetraethyl lead was first introduced into automotive gasolines in 1923 as an antiknock agent (Gibbs, 1990). After some initial controversy surrounding safety concerns

Table 18.4.2 Compilation of historical facts related to the development and demise of organic lead additives in US gasoline. Compiled from: Ethyl: A History of the Corporation and the People Who Made It by Robert (1983)

Year	Occurrence/Fact
Mar 1905	WWI research effort to boost engine compression results in development of "*Hecter*," a mixture of cyclohexane and benzene. After the war, the product was abandoned from large-scale, commercial development.
Dec 1921	GM/DuPont researchers in Dayton abandon "selenium oxychloride" for TEL.
Jun 1922	Bromide scavenger developed to reduce lead oxide buildup in engines; bromine availability was so limited that GM/DuPont builds a pilot plant in N. Carolina and develop a method to extract bromine from seawater
Sept 1922	GM/DuPont researchers (Midgley and Boyd) present paper at Am. Chem. Soc. Annual meeting touting TEL as an antiknock additive.
Feb 1923	First public sale of "Ethyl Gasoline" made at a Refiners Oil Company station in Dayton, OH. The gasoline was the product developed jointly by GM and DuPont which incorporated TEL and EDB (the latter produced by the ethyl bromide processing of seawater).
May 1923	GM Chemical Company executives convince some Indianapolis racers to use "Ethyl" gasoline. The first three finishers had used "Ethyl".
Sept 1923	GM enters into exclusive marketing contract with Standard Oil of Ohio for a 5-year period.
Aug 1924	GM (patent/process) and Standard Oil of New Jersey (manufacturing capacity) merge to create "Ethyl Gasoline Corporation" (renamed to Ethyl Corporation in 1942).
Jul 1924	workers at GM's Dayton laboratories die from TEL exposure.
Oct 1924	40 workers at Standard Oil of New Jersey's Bayway, NJ manufacturing facility become sick and 5 die. Newspapers label the new gasoline additive as "looney gas." Sales banned in NY and PA.
May 1, 1925	Surgeon General hosts a conference to discuss dangers of TEL. Ethyl's new president bans sale until after a scientific panel reviews issue.
Dec 1925	Scientific debate continues, Ethyl hires PR folks to help with marketing.
May 1926	Surgeon General's panel report is issued (Public Health Bulletin No. 163, "The Use of TetraEthyl Lead and Its Relation to Public Health") concluding "no health hazard to the public." Report set limit of TEL in gasoline at 3 ml TEL/gallon (3.17 glpg). Ethyl resumes sale of Ethyl gasoline on May 1, 1926, and experiences strong growth over the next few decades (in spite of a push by German company, Badische, to introduce iron carbonyl as an alternative to TEL. Iron carbonyl eventually proves problematic with spark plug fouling and disappears from even the European market; TEL remains "king").
Sept 1926	Ethyl researcher (Graham Edgar) develops the modern "octane scale" (0 – 100; 0 = *n*-heptane, 100 = isooctane) and publishes at the Am. Chem. Soc. Annual meeting. Isooctane was a newly developed compound.
Jan 1933	Ethyl first introduces TEL into "regular" gasoline. Up until this time it had only been used in premium gasoline (that had been known as "*Ethel*"). The new regular leaded gasoline was called "*Q Brand*." Embarrassingly, this new demand for TEL additive led to a shortage of EDB, which was offset by the addition of EDC with poor results.
Jun 1938	Ethyl builds a TEL production facility in Baton Rouge, LA. Up until this time they depended entirely upon production by DuPont from their Deepwater, NJ facility which was now maxed-out at 65 million pounds/year.
Jul 1943	German U-boat fires upon Ethyl's bromine extraction plant in N. Carolina – no damage and U-boat is sunk the next day. This emphasizes the importance of lead additive to the war effort.
Apr 1946	Ethyl files patent on scavenger mix of EDB and EDC, the mixture having been developed under WWII efforts to increase performance of gasolines.
Jan 1948	DuPont enters the market as an alternative supplier of TEL. Up until this time only Ethyl Corporation supplied TEL to refiners.
Jun 1957	Ethyl researchers develop and begin to temporarily market MMT, a straw-colored, stable antiknock compound. The market name was "AK-33X."
late 1950's	Ethyl researchers develop tetramethyl lead (TML) and various mixtures which are later marketed to refiners as alternatives to attain octane necessary for the newer car models.
Jan 1959	Permissible amount of TEL is increased from the 1926 level (3 ml/gal) to 4 ml/gal (4.23 glpg) by the Surgeon General's office. First wide marketing of MMT in "*Motor Mix 33*" (57.52%TEL, 6.97%MMT).
Dec 1970	Nixon signs the CAA and establishes United States Environmental Protection Agency as the regulator of fuel additives (despite Ethyl's lobbying efforts).
Feb 1972	United States Environmental Protection Agency announces plans to phase down lead content of leaded gasoline to 1.25 glpg by 1977 and to require production of unleaded gasoline by July 1, 1974. This announcement is met with much resistance by refiners and Ethyl.
Dec 1972	United States Environmental Protection Agency sticks to July 1, 1974, date for introduction of unleaded gasoline but revises schedule for lead phase down such that the mass average of all gasolines must be 1.25 glpg by 1978 (a 1-year delay from the original proposal).

(Continued)

Table 18.4.2 *(Continued)*

Year	Occurrence/Fact
1973–1974	Ethyl and United States Environmental Protection Agency wrangle in lawsuits, all the while Ethyl works to develop an engine that runs on leaded gasoline yet meets the emission requirements
Dec 1974	Ethyl wins Federal Appeals court ruling which delays the United States Environmental Protection Agency's phase-down schedule. No changed schedule is proposed. Leaded gasoline will continue for the foreseeable future.
Mar 1976	United States Environmental Protection Agency's appeal overturns the Dec 1974 decision. Ethyl, DuPont, PPG, and Nalco request hearing before US Supreme Court. Request is denied and United States Environmental Protection Agency is back in charge of lead phase down. The 1972 phase down schedule is now in place (1.25 glpg by 1978).
Jun 1976	Ethyl tries to offset loss of lead from gasoline with addition of MMT but is stopped by United States Environmental Protection Agency, which now wants to phase out MMT also. Ultimately, the United States Environmental Protection Agency's phase-down schedule for lead is delayed until 1980 (see Table 18.2.4).

(prompted by the deaths of five Standard Oil of New Jersey refinery workers in 1924 (Robert, 1983; Needleman, 1998), the use of TEL increased steadily in the following decades (except during World War II) as the demand for increased engine performance grew. Ultimately, the average levels of lead in the premium gasoline pool reached a high of approximately 3.0 glpg around 1970 (Figure 18.2.2); individual gasolines could legally contain up to 4.23 glpg (Table 18.2.4), though they rarely did. The reason that they rarely did is because the octane gained from the addition of lead alkyls to gasoline was non-linear, reaching a point above which the practical cost of the additive exceeded the added octane. The maximum allowable concentration of lead (4.23 glpg after 1959) was well above the practical limit, and therefore was seldom approached in practice, even during the "muscle car" era of the 1960s.

Tetraethyl lead was the lone lead alkyl compound added to automotive gasoline up until the early 1960s. Gibbs (1990) reports the first commercial use of the other lead alkyls in gasoline occurred in 1960. However, for years before it was known that other tetralead alkyl compounds (see below) that were more volatile than TEL could improve octane performance due to these compounds' greater thermal stability, which delayed their detonation compared to TEL, and thereby provided more sustained power during each engine cycle (Richardson *et al.*, 1961).

As a result, new antiknock packages were developed and marketed by the major lead producers in the 1960s (e.g,. Ethyl, DuPont, Nalco Chemical, and Houston Chemical). These new antiknock packages included both physical mixtures and reacted mixtures of TEL and TML. Reacted mixtures are the end products of a catalyzed TEL–TML reaction that results in the formation of five lead alkyls, namely

tetramethyl lead (TML)
trimethylethyl lead (TMEL)
diethyldimethyl lead (DEDML)
methyltriethyl lead (MTEL), and
tetraethyl lead (TEL)

in predictable proportion to one another.

Some properties of these individual lead alkyls and the resulting mixtures are shown in Table 18.4.3. The relative amounts of the five individual organo-lead compounds in any lead package marketed to refiners were dependent upon the molar ratios of the two starting materials (TEL and TML). A limited number (3) of reacted mixtures (RMs) were marketed by all of the lead manufacturers to all refiners (Gibbs, 1990). These typically had names based upon

the molar percent of TML in the mixture. Reacted mixtures containing 25, 50, and 75% TML were called RM25, RM50, and RM75, respectively (Table 18.4.3). These had market names such as *Tetarmix*-25, -50, or -75 (DuPont), *MLA*-250, -500, or -750 (Ethyl Corp.), *MAF*-25R, -50R, or -75R (Houston Chemical), and *Nalkyl ME*-25, -50, or -75 (Nalco Chemical.). Despite the nomenclature, each lead producer's reacted mixtures were compositionally similar (Table 18.4.3). Physical mixtures were also marketed. These were an unreacted combination of TEL and TML in known percentages. There were a greater variety of the physical mixtures marketed, but they ranged from TML to TEL proportions of 10:90 to 75:25. These had names such at PM10 or PM75, referring to the percentage of TML in the physical mixture (Table 18.4.3). In addition, additives containing TML only were available, but these were not as commonly used in gasoline as TEL, RMs, or PMs due to their higher cost and small octane gain over the mixtures available.

In our experience, it was more common for leaded gasolines to contain a single type of lead additive package rather than a combination of more than one lead packages. However, DuPont documents suggest that it was possible for refiners to use "*any combination of one or more of these lead antiknocks*" (DuPont marketing bulletin, undated). However, we are unaware of any document that indicates that this was done in practice. Therefore, when an NAPL is observed that contains more than one type of lead package it most likely represents a mixture of more than one leaded gasolines (Stout *et al.*, 2002).

Included in the commercially available lead packages for automotive gasolines were lead scavengers, 1,2-dibromoethane (EDB), and 1,2-dichloroethane (EDC), at constant weight percentages of the antiknock package (17.9 and 18.8 wt%, respectively; Table 18.4.3). This mixture results in a molar ratio of 1:2:1 for EDB/EDC/Lead since the mid-1940s (Table 18.4.3; Falta, 2004). On average, the concentration of EDB and EDC in leaded gasoline over the time period of 1950 through 1974 was 1.12 and 1.19 g/gallon, respectively (Falta, 2004). These scavengers had long been used to minimize the precipitation of non-volatile lead oxides in an automobile engine's combustion chambers through formation of volatile lead chloride and bromide salts.

During development of the TML mixtures, the added benefit of the other lead alkyls in gasoline was demonstrated to be greatest in gasolines containing higher amounts of aromatics (Figure 18.4.1; Richardson *et al.*, 1961). As a result, when various antiknock packages containing TML became available for use in gasoline, they were predominantly used

Table 18.4.3 *Selected properties of individual lead alkyl species, their equilibrium-predicted weight percentages in typical reacted and physical lead antiknock packages, and selected properties of the lead antiknock packages. (Actual weight percentages varied slightly from equilibrium values and between producers.)*

Organic Lead Species	Wt% Lead	Boil. Pt. (°C)	TEL only	TML only	Reacted Mixes			Physical Mixes		
					RM25	RM50	RM75	PM25	PM50	PM75
TEL	64.06	200[a]	100.0	0.0	28.8	4.8	0.1	75.0	50.0	25.0
MTEL	66.96	178[b]	0.0	0.0	49.5	25.6	3.6	0.0	0.0	0.0
DEDML	70.14	155[b]	0.0	0.0	18.6	42.4	20.5	0.0	0.0	0.0
TMEL	73.64	133[b]	0.0	0.0	3.0	23.4	49.6	0.0	0.0	0.0
TML	77.50	110	0.0	100.0	0.1	3.8	26.2	25.0	50.0	75.0
Lead Antiknock Package Compositions										
Wt% Alkyl Lead			61.48	50.82	58.82	56.15	53.49	58.82	56.15	53.49
Wt% EDB			17.86	17.86	17.86	17.86	17.86	17.86	17.86	17.86
Wt% EDC			18.81	18.81	18.81	18.81	18.81	18.81	18.81	18.81
Wt% Other[c]			1.79	12.50	4.45	7.12	9.78	4.45	7.12	9.78
Dye			0.06	0.06	0.06	0.06	0.06	0.06	0.06	0.06
Lead Antiknock Package Properties										
Wt.% Lead			39.39	39.39	39.39	39.39	39.39	39.39	39.39	39.39
Density (g/ml) 20 °C			1.586	1.583	1.587	1.585	1.854	1.583	1.583	1.583
Flash Point (°C)			118	36	82	48	NA	74	54	44
VP (mmHg 20 °C)			36	40	37	39	39	38	39	40

[a] decomposes
[b] interpolated
[c] solvent, inerts, and antioxidant

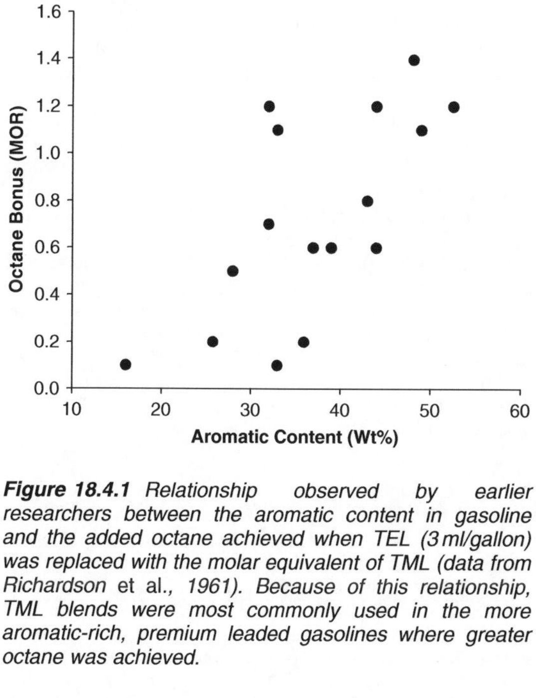

Figure 18.4.1 *Relationship observed by earlier researchers between the aromatic content in gasoline and the added octane achieved when TEL (3 ml/gallon) was replaced with the molar equivalent of TML (data from Richardson* et al., *1961). Because of this relationship, TML blends were most commonly used in the more aromatic-rich, premium leaded gasolines where greater octane was achieved.*

in higher octane (premium) gasolines that had contained higher aromatic contents, because those gasolines benefited most (Richardson *et al.*, 1961). Because of this added benefit, TML-bearing antiknocks cost more than TEL only, which further impacted their use pattern in different gasoline grades. As such, regular leaded gasoline more commonly contained TEL-only, which was capable of achieving research octane number (RON) values of up to 93. With the aid of TML-bearing antiknock mixtures, premium leaded

gasolines typically attained RON up to 96 by 1970 (Shelton *et al.*, 1982; Owens and Cowley, 1990).

The amount of a particular lead alkyl package that was used depended on the refiner's needs and economics. As stated above, to our knowledge, mixing of different lead packages in a single gasoline did not occur. However, very often different lead packages were added to different grades of gasoline produced at a single refinery. Corporate records of particular additive packages, which can become available in the course of a litigation matter's discovery process, can sometimes help unravel issues where it is important to determine specific antiknock packages in use in different grades at different times.

Related to the use of lead alkyls is the perception that after 1980 RMs and PMs antiknock packages were no longer in use and that leaded gasolines once again only contained TEL (e.g., Hurst *et al.*, 1996). However, many exceptions exist to defensibly utilize 1980 as a "cut-off" date after which lead alkyls other than TEL should not be present in leaded gasolines. However, it is likely that the use of TML-containing lead mixtures was in decline after 1980. For example, it is reported that by 1985, 90–95% of the lead used in gasoline was TEL-only (Wakim *et al.*, 1990).

18.4.2 MMT (2-Methylcyclopentadienyl manganese tricarbonyl)

Methylcyclopentadienyl manganese tricarbonyl (MMT; $C_9H_7MnO_3$) is another antiknock additive that was first commercially available in 1959, when it was blended with TEL and marketed in *"Motor Mix 33"* or *"AK-33X"* (57.5% TEL and 6.97% MMT – Gibbs, 1990). Its use was limited until the initial lead phase down in the early 1970s. At that time, MMT was first independently added to gasoline to offset the octane loss accompanying the removal of lead alkyls and was in widespread use in the United States in unleaded and leaded gasolines by 1976. The maximum allowable concentration at that time was 0.125 g Mn/gal. The CAA Amendments of 1977, however, banned the use of manganese additives in unleaded gasolines after October 1978, unless the

United States Environmental Production Agency granted a special waiver (Morrison, 2000b; Zayed *et al.*, 1999). The use of MMT continued in leaded gasoline and it was temporarily allowed in unleaded gasolines in the summer of 1979 (during the Arab oil embargo). In 1995, United States Environmental Protection Agency granted a waiver for MMT's use in conventional unleaded gasolines, but limited the maximum concentration to 0.031 g Mn/gal. Its use in the United States, though rare, in leaded gasolines ended in 1996 with the elimination of leaded gasoline. California regulations have continued to ban the use of manganese additives in unleaded gasoline. According to Kaplan (2003), MMT is only used by a few small refineries in the southern Rocky Mountain states. Its presence in NAPLs, however, may help to constrain the age of (at least) a component in a gasoline mixture.

MMT has been used much more widely in unleaded gasolines in Canada since 1976, when it was first introduced to offset octane lost with the removal of lead alkyls. Kaplan (2003) reports that MMT is still permitted in gasolines in Canada below 0.068 mg/L.

18.4.3 Oxygen-Containing Additives – General

As depicted in Tables 18.2.1 and 18.2.2, new classes of octane-boosting additives, with fewer (presumed) negative impacts than lead alkyl, were being developed and used in increasing frequency to augment gasoline performance. The additives included a variety of oxygen-containing additives such as alcohols and ethers.

Alcohols are a broad class of organic compounds containing a hydroxyl (–OH) functional group. Alcohols can be obtained from plant matter or synthetically from petroleum derivatives. Ethers are a class of organic compounds in which an oxygen atom is interposed between two carbon atoms: C–O–C. They both can be made from petroleum derivatives and are widely used as industrial solvents. Alcohols, predominantly TBA and ethanol, were introduced as a gasoline additive/supplement in the late-1960s and late-1970s, respectively, the latter being primarily in response to the Arab oil embargo. (Note, ethanol was actually first introduced unsuccessfully in Nebraska in the 1940s; Table 18.2.1) Throughout the 1970s a variety of alcohol blends and ether-based additives were developed which further "extended" the gasoline supply, and boosted octane (Guetens *et al.*, 1982). The most volumetrically important of the ethers, MTBE, was first added to commercially available gasolines in the United States in 1979 (Squillace *et al.*, 1995). An inventory of the more common oxygen-containing additives that has been used in gasoline is given in Table 18.4.4. Volumetrically, MTBE and ethanol are the most commonly used and are the focus of the remainder of this section. Some comments on TBA are also included since this compound has drawn some recent attention (Schaap, 2002; Rong and Kerfoot, 2003). A review of the other common ether oxygenates (TAME, ETBE) used in the United States was recently provided by Kaplan (2003).

MTBE has been the primary oxygen-containing additive used throughout most of the United States, except in the Midwest where tax advantages have favored the use of ethanol. Reviews of the geographic distribution of the types and concentrations of oxygen-containing additives are available in various gasoline databases (e.g., NIPER and United States Environmental Protection Agency). A review of these survey data has demonstrated that high concentrations (8–11 vol%) of ethanol were preferentially used in most metropolitan areas requiring Oxyfuels (Moran *et al.*, 2000). Similarly, there was good agreement between cities requiring RFG and the use of gasoline containing high concentrations of MTBE (averaging around 9–11 vol%).

One notable exception is Chicago in which RFG containing ethanol is used exclusively due to the city's ban on MTBE (Table 18.2.6). Conventional gasolines from areas not requiring Oxyfuel or RFG contained lower concentrations of MTBE <2 wt% (on average), most likely solely as an octane booster.

18.4.4 MTBE (Methyl Tertiary Butyl Ether)

In the early 1980s MTBE was used primarily as an octane booster within a minority of the gasoline pool, usually into the newly introduced premium unleaded blends. After 1988, and more so after 1990, MTBE's use expanded into Oxyfuels and RFG (as described earlier in this chapter) and production soared dramatically (Figure 18.2.3). By 1988 it was among the top 50 chemicals made in the United States, and by 1993 it was the second most produced chemical in the United States, essentially reaching a "commodity" status.

In the early 1980s, when MTBE was used primarily as an octane booster, it typically was added to unleaded gasoline at between 3 and 8 vol% (and mostly into premium unleaded grades). The use of MTBE at this level (3–8 vol%) persisted throughout the mid-1980s. After 1988, MTBE was being blended at up to 15 vol% (2.7 wt% oxygen) in order to produce Oxyfuel required in the CO non-attainment metropolitan areas. After 1995, MTBE was being blended at approximately 11 vol% (2.0 wt% oxygen) into RFG for use in the O_3 non-attainment areas and other "opt-in" markets. All the while MTBE was also being blended into conventional gasolines for octane enhancement, at lower concentrations (3–8 vol%). [The recent gasoline survey by Weaver *et al.* (2005) indicates conventional gasolines contained only 2–4 wt% MTBE.] Although figures are not yet available, it is anticipated that the use of MTBE has already begun to decline following the recent state bans (Table 18.2.6).

For the forensic investigator, the history of MTBE (or other oxygenate) use in gasoline would seemingly provide a gross method of age dating gasoline-derived contamination. However, the environmental behavior of this chemical compared to hydrocarbons (e.g., high aqueous solubility) and its convoluted use (e.g., winter only *versus* year-round) often makes its presence/absence or concentration difficult to interpret. The most useful forensic application of MTBE is in terms of its concentration gradients or spatial distribution in groundwater, which can reveal multiple sources/releases of the MTBE (e.g., Davidson and Creek, 1999). Recent attempts to use stable carbon isotopes to distinguish dissolved phase contaminants, such as MTBE, are discussed later in this chapter.

18.4.5 Ethanol

Ethanol blended into gasoline is produced from the fermentation of agricultural products, primarily corn and sugar cane. It was first introduced into gasoline in Nebraska in 1940 ("*Argol*") without much success (Table 18.2.1). It was reintroduced nationwide in the late 1970s when it was used to "stretch" the gasoline supply during the Arab oil embargo, and in doing so introduced "*gasohol*" (also called E-10) to the nation (Table 18.2.2). The concentration of ethanol in gasohol was (and still is) limited to 10 vol%. The National Energy Act of 1978 provided the tax advantages that promoted ethanol's quick introduction into the gasoline pool. In 1979 about 20 million gallons of ethanol were used in the production of gasohol (Wakim *et al.*, 1990).

Ethanol's use slowly grew throughout the 1980s, with the grain-producing states consuming most of the gasohol produced. In 1992, the tax waiver was renewed concurrent with initiation of the Oxyfuels program, which increased the demand and use of ethanol throughout the 1990s. Ethanol

Table 18.4.4 Inventory of oxygen-containing alcohols and ethers and their physical-chemical properties commonly used in gasoline

Chemical	Formula	Mol. Wt.	R+M/2	Blending RVP	Spec. Grav.	Boil. Pt.	Oxygen wt. %	Oxygen vol%	H$_2$O solubility mg/L @ 20°C
Alcohols									
Methanol	CH$_4$O	32.0	99.5	4.6 psi	0.792	65°C	50	3.7	miscible
Ethanol	C$_2$H$_6$O	46.0	113	17.6 psi	0.789	78°C	34.8	5.4	miscible
tert-butyl alcohol	C$_4$H$_{10}$O	74.1	100	6.8 psi	0.786	71°C	21.6	8.7	soluble
Ethers									
MTBE	C$_5$H$_{12}$O	88.2	109.5	7.8 psi	0.740	55°C	18.1	11	43000–54300
TAME	C$_6$H$_{14}$O	102.2	104	1.4 psi	0.770	71°C	15.7	12.2	5500–20000
ETBE	C$_6$H$_{14}$O	102.2	110	4.0 psi	0.742	72°C	15.7	12.7	7650–26000
DIPE	C$_6$H$_{14}$O	102.2	105	5.4 psi	0.726	68°C	15.7	13.0	9000–11250

is required at a concentration of 7.77 vol% in Oxyfuels in order to achieve 2.7 wt% oxygen minimum (Table 18.2.5).

Ethanol seems to be the oxygenate of choice as MTBE faces considerable pressures and state bans due to the environmental risks (Table 18.2.6; Powers *et al.*, 2001). States that have banned MTBE have replaced it with ethanol in RFG and Oxyfuels (e.g., California); however, numerous states are currently seeking waivers from the oxygen requirement of RFG in the hope of not requiring any oxygen-containing additive.

Because of ethanol's miscibility in water it cannot be blended into gasoline prior to transport through pipelines, trucks, or barges or it could quickly partition into small amounts of water present within the shipment infrastructure. As a result, ethanol must be batch-blended at the distribution terminals during transport to delivery trucks. This presents a logistical problem to the distribution of gasoline, and may have some implications for environmental forensic investigations in which "unfinished" gasolines (i.e., gasoline

to which ethanol has not yet been blended) or neat ethanol may become fugitive contaminants.

The forensic applications of ethanol are not well established. This derives, in part, from an overall lack of ethanol concentration data collected during most site assessments or other forensic studies. The analytical difficulties facing ethanol's analysis, particularly in groundwater, have only recently been achieved with direct aqueous injection (DAI) techniques (ASTM D 3695). Thus, the database on ethanol in the environment will undoubtedly grow.

It can be supposed that the utility of ethanol in forensic studies may be limited due to: (1) the extremely rapid rate at which it is lost from gasoline upon contact with water and (2) its susceptibility to rapid biodegradation under both aerobic and anaerobic conditions (Powers *et al.*, 2001). Combined, these two factors might remove ethanol from a spill site very rapidly. The rate of partitioning of ethanol from gasoline containing ethanol released into the environment has not yet been studied thoroughly. However, given its miscibility in water it can be presumed that ethanol will very rapidly enter the subsurface water phase – including both soil moisture and groundwater. Thus, the absence of detectable ethanol in NAPL, soils, or groundwater may not provide a sufficient basis to draw a conclusion surrounding its presence or absence in the parent gasoline. Ethanol's presence in the environment, on the other hand, would seem to implicate a relatively recent impact from ethanol-containing gasoline. How recent is likely to be function of the site-specific conditions that affect the rate of weathering – in fact, these conditions are relevant to the rate of weathering of all gasoline-derived chemicals (as is described later in this chapter).

The biodegradability of ethanol has raised concerns that it may slow the rate of degradation of other dissolved constituents (e.g., benzene), as various electron acceptors become consumed/depleted due to ethanol's potential for biodegradation.

One notable feature of ethanol's use in conventional gasoline, Oxyfuels, and RFG is that in a recent survey of gasolines from 2004, Weaver *et al.* (2005) showed that ethanol and MTBE did not co-occur in any gasolines surveyed nationwide. Their use was mutually exclusive in modern gasolines.

18.4.6 Tert-Butyl Alcohol

Tert-Butyl Alcohol (sometimes called gasoline-grade TBA or GTBA) has been used by some refiners since its introduction into the gasoline pool in about 1969 (Table 18.2.1). At that time some gasolines contained approximately 2 vol% TBA. Concentrations gradually increased to approximately 7 vol% by 1973. The use of TBA in gasoline became more widespread after the United States Environmental Protection Agency officially approved its use in 1979 even though many refiners had been using it already at concentrations up to 7 vol%. The typical TBA concentration in gasoline at that time was approximately 1–5 vol%, i.e., less than the maximum allowable by the United States Environmental Protection Agency waiver.

The largest producer of TBA at that time was Oxirane, a 50:50 joint venture between ARCO and Halcon. TBA was marketed alone or blended with methanol (1:1) to produce products called *Arconol* and *Oxinol*, respectively. These TBA-containing additives were blended into gasoline at a maximum of 5.5 vol% in 1979–1980. However, in November 1981 the United States Environmental Protection Agency issued a waiver for the use of *Arconol/Oxinol* at concentrations up to 9.5 vol% (Guetens *et al.*, 1982). A simple calculation suggests that gasoline blended after November 1981 could contain between 4.75 vol% and 9.5 vol% TBA.

The use of TBA was most likely a minor percentage of the oxygen-containing gasolines produced and sold in the United States throughout the 1980s as the use of TBA was generally supplanted by MTBE. Then ARCO abandoned TBA as they became a leading supplier of MTBE.

18.4.7 Fuel Dyes

Many finished automotive gasolines have been augmented with pigmented dyes to differentiate among fuel grades or to identify fuels for intended end use. For example, before being banned in 1996, leaded gasolines throughout the 1980s were required to be dyed to aid in distinguishing them from unleaded gasolines. Typical dye concentrations were on the order of 1–5 mg/kg (Gibbs, 1990).

The commercially used gasoline dyes were available in various colors—red (azobenzen-4-azo-2-napthol), orange (benzene-azo-2-napthol), yellow (*p*-diethyl aminoazobenzene), and blue (1,4-diisopropyl aminoanthraquinone). Reportedly, in NAPLs these dyes can be identified by thin layer chromatography (Kaplan and Galperin, 1996; Kaplan

et al., 1997). Identifying the presence, absence, or mixture of these dyes can be of great forensic utility in cases where one needs to differentiate between or among potentially similar fugitive petroleum products. One limitation of reliance on fuel dye data for forensic purposes is the reported susceptibility of these dye compounds to environmental weathering (Kaplan *et al.*, 1997). Thus, the absence of detectable dyes in a fugitive gasoline does not necessarily mean that the compound(s) was not initially present in the fresh fuel. Under some conditions (e.g., clay-rich soil and significant NAPL mass), however, dyes can persist in the subsurface environment for many years (see further discussion on dyes later).

18.5 GASOLINE FINGERPRINTING METHODS

In most environmental forensic investigations at sites contaminated with gasoline, it is important to identify the type or types of gasoline found at the site, to differentiate among different sourced gasolines and, to the extent possible, age-constrain the major product residues. Often, investigators must attempt to characterize gasoline or gasoline residues in various media, i.e., pure product, soil, water, and air/vapor. To achieve these goals, detailed chemical characterization of fugitive and often stored or dispensed gasoline samples is needed.

Certain bulk chemical and physical characteristics of gasoline that may be important in environmental forensics investigation—e.g., boiling point distributions, total sulfur, or total lead—can be achieved using standard methods of chemical analysis such as those based on ASTM protocols. Unfortunately, conventional methods of environmental sample chemical analysis for charactering the most important aspects of gasoline—namely the occurrence and concentrations of hydrocarbons and additives—prove inadequate for forensic chemistry applications (Stout *et al.*, 2002). These standard approaches, based on United States Environmental Protection Agency methods of analysis such as SW-846 (United States Environmental Protection Agency, 1996) techniques provide insufficient specificity for forensics investigations, largely because only a limited number of parameters and "forensic" chemicals of concern are measured with these techniques at appropriately low detection limits (Uhler 1998–1999).

In the case of gasoline analyses, the most common standard method for measurement of C_4–C_{15} gasoline-range hydrocarbons is United States Environmental Protection Agency Method 8260, *Volatile Organic Compounds by GC/MS*. This purge-and-trap GC/MS method, while theoretically capable of detecting hundreds of important gasoline-range hydrocarbons, is only used to determine the water soluble aromatic compounds benzene, toluene, ethylbenzene, and xylenes (BTEX), naphthalene and, in some implementations, MBTE and TBA. Alone, these few compounds are of limited use in forensic investigations requiring detailed fingerprinting of gasoline due to their ubiquity in many different types of gasoline.

In the last decade, there have been many useful analytical methods developed to support environmental forensic characterization of gasoline (Kaplan and Galperin, 1996; Morrison, 2000b; Philp, 2002; Stout *et al.*, 2002). The hallmarks of these methods are:

1) expansive target analyte lists that allow for thorough characterization and differentiation of gasoline,
2) sufficient sensitivity, to measure very low levels of hydrocarbons in environmental matrices,
3) selectivity, to minimize matrix interferences and false-positive biases in the measurements,
4) broad applicability across various matrices of possible interest.

Table 18.5.1 Methods of analysis for gasoline applicable in forensics investigations

Measurement	Applicability	Matrices	Determinative Method
Bulk Characterization			
Simulated distillation ASTM D2887	Gross boiling point distributions	Product	GC/FID {American Society for Testing and Materials, 1997 #557}
Total Sulfur ASTM D3227	Sulfur content	Product	GC/AED {American Society for Testing and Materials, 1997 #558}
Total Nitrogen ASTM D4629	Nitrogen content	Product	AS {American Society for Testing and Materials, 1997 #559}
Total Lead ASTM D3237	Lead content	Product	AS {American Society for Testing and Materials #560}
Fuel Dyes	Proprietarily marker	Product	LC {Kaplan, Galperin *et al.*, 1997 #815}
Bulk Isotopes	Source characteristic	Product	IRMS (Kaplan *et al.*, 1997)
Molecular Characterization			
Gas chromatography "fingerprint" United States Environmental Protection Agency 8015M	Composition, weathering	Product, Soil, Water, Air/vapor	GC/FID {United States Environmental Protection Agency, 1999 #556}
Paraffins, isoparaffins, aromatics naphthenes, olefins (PIANO)	Hydrocarbon composition	Product, Soil, Water	GC/MS {Uhler, Healey *et al.*, 2002 #1520}; {Kaplan and Galperin, 1996 #323}; {Zhu, Zhang *et al.*, 1999 #1275}
		Air/vapor	GC/MS {United States Environmental Protection Agency, 1999 #556}
Polycyclic aromatic hydrocarbons (PAH)	Hydrocarbon composition	Product, Soil, Water	GC/MS {Stout, Uhler *et al.*, 2002 #1203}
		Air/vapor	GC/MS {United States Environmental Protection Agency, 1999 #556}
Organic lead speciation	Lead additive formulation	Product	GC/MS or GC/ECD {Stout, Uhler *et al.*, 2002 #1203}
Oxygenates	Additive	Product, water, soil	GC/MS {Uhler, Healey *et al.*, 2002 #1520}; {Uhler, Stout *et al.*, 2001 #1497}
		Air/vapor	GC/MS {United States Environmental Protection Agency, 1999 #556}
Alcohols	Ethanol, C_1–C_4 alcohols	Product	GC/FID {American Society for Testing and Materials, 1999 #563}
		Water	GC/FID {United States Environmental Protection Agency, 1999 #556}
Nitrogen and Sulfur speciation	Non-Hydrocarbon composition	Product	GC/NCD ({Chawla, 1997 #1850}); GC/AED (Stumpf, Tolvay *et al.*, 1998 #1167) GC/MS (Coulombe, 1995 #379)
Chemical isotopic composition	Source characteristic	Product, soil, water	GCIRMS {Philp, 2002 #1482}

AS: Atomic Spectroscopy; GC/AED: gas chromatography/atomic emission detection; GC/FID: gas chromatography/flame ionization detection; GC/MS: gas chromatography/mass spectrometry; CIRMS: combustion isotope ratio mass spectrometry; GC/NCD: gas chromatography/nitrogen chemiluminescence detection; GCIRMS: gas chromatography-isotope ratio-mass spectrometry.

Table 18.5.1 lists examples of both conventional and advanced methods of chemical analysis suitable for environmental forensics investigations. The more useful methods used in environmental forensics investigations of gasoline are discussed below.

18.5.1 Simulated Distillation

Distillation analysis is a means to determine the boiling range distribution of gasolines and other petroleum products. Simulated distillation utilized GC as a convenient means to determine the boiling range of petroleum. In addition to providing the environmental forensics investigator

with information about the composition of petroleum products under study, simulated distillation curves are often the most convenient graphical means to convey differences in petroleum product composition to non-technical decision-makers (Figure 18.5.1).

The standard method for the determining the boiling range of petroleum product by GC is ASTM Method D 2887 (Table 18.5.1); however, other carefully calibrated GC methods can be used to reconstruct simulated distillation curves following the same guidelines presented in ASTM D 2887. For example, GC analysis following United States Environmental Protection Agency Method 8015 B *Nonhalogenated Organics Using* gas chromatography/flame ionization detec-

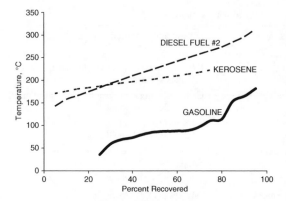

Figure 18.5.1 *Simulated distillation curves for gasoline, diesel fuel, and kerosene. The distinct boiling point distributions for each product can be clearly recognized. Unique boiling point features can be seen for the gasoline, reflecting the unique manufactured characteristics of this particular gasoline.*

tion (GC/FID) can be used to reconstruct simulated distillation curves. In order to use United States Environmental Protection Agency Method 8105 for simulated distillation, the analytical chemist must calibrate the gas chromatograph with an *n*-alkane solution containing hydrocarbons that cover the distillation range of interest. From this data, a calibration plot of boiling point versus retention time is constructed for the hydrocarbons contained in the calibration mixture. Authentic samples are analyzed under the same chromatographic conditions, and the area under the chromatographic curve between each calibration point is determined. The incremental areas are summed to determine a cumulative chromatographic area; fractional amounts of hydrocarbons from each boiling increment are thus calculated. Finally, a plot of cumulative evaporated hydrocarbon versus boiling point (determined from the initial calibration curve) is constructed. Reference standards (e.g., pure gasoline, diesel fuel, etc.) can be analyzed in conjunction with the field samples as a measure of method quality control.

18.5.2   Gas Chromatographic "Fingerprinting"

One of the most useful analyses for environmental samples suspected of containing gasoline (and/or other hydrocarbons products) is a GC/FID (United States Environmental Protection Agency Method 8015) or GC/MS (United States Environmental Protection Agency Method 8270) method. The results of such an analysis is a screening level "fingerprint," i.e., a chromatogram that shows both the boiling point range and the presence of major molecular constituents in the sample. Chromatographic "fingerprints" can be especially useful to the forensic investigator because mostly all hydrocarbon assemblages—crude oil, petroleum distillates, coal-derived liquids, and their combustion and waste products—have distinctive chromatographic signatures (Stout *et al.*, 2002). In the case of samples containing gasoline, differences in sample composition and in particular weathering (described later in this chapter) can be readily observed in the appearance of the gas chromatogram or "fingerprint." Such a "fingerprint" allows the forensic investigator to quickly map out the general nature of contamination, and to qualitatively identify important features of the gasoline residues in a study area. One of the benefits of GC

"fingerprinting" is that it is a powerful screening technique in the analysis of virtually all media—NAPL, soil, water, air, and vapor (Table 18.5.1).

The key to the successful application of this screening method is: (1) optimal operation of the GC inlet which will minimize mass discrimination, and (2) utilization of a very slow gas chromatograph oven temperature program, to facilitate optimal resolution of close-eluting compounds, and to allow subtle, but important, differences in chemical composition to emerge in the chromatogram. Total chromatographic run times of about 1 hour are typical for these high quality GC analyses. Obviously, for the analysis of gasoline, samples must be analyzed over an appropriate hydrocarbon range (C_5–C_{12}) to be useful.

18.5.2.1   PIANO and Ether Oxygenates

Hydrocarbons make up approximately 85–90 wt% of modern RFG and Oxyfuels, and 95–100 wt% of most conventional gasolines. The types, presence or absences, and relative distributions of hydrocarbons in gasoline reveal information about the feedstock, refining, weathering state, and vintage (see discussions elsewhere in this chapter).

The hydrocarbons that are present in gasoline occur within five compound classes: *P*araffins, *I*soalkanes, *A*romatics, *N*aphthenes, and *O*lefins—typically referred to by the acronym "PIANO." The major non-hydrocarbon classes in gasolines can include oxygen-containing ethers (e.g., MTBE, TAME, etc. – Gibbs, 1990), alcohols (e.g., ethanol, TBA, methanol, isopropyl alcohol – (Gibbs, 1990), sulfur- (e.g., mercaptans, thiophenes, disulfides, thiolanes, thianes – Coulombe, 1995; Keesom *et al.*, 1998; Stumpf *et al.*, 1998), and nitrogen-containing moieties (e.g., pyrroles, indoles, anilines, etc. – Chawla, 1996).

Gasoline is an extremely complex mixture of hundreds of compounds. For example, Whittmore, in 1979 resolved and identified 361 individual compounds in gasoline. However, it is neither practical nor necessary to measure this number of compounds for forensic chemistry purposes. Uhler *et al.* (2002) developed a PIANO target compound list for forensic analysis. In developing that target analyte list guidance was obtained from a review of the fuel (e.g., Adlard *et al.*, 1979; Whittmore, 1979; Shiomi *et al.*, 1991; Pauls, 1995) and environmental forensic literature (Kaplan and Galperin, 1996; Stout *et al.*, 1999a,b). The relative abundance of a compound, however, was not the only requirement for selecting target analytes because minor components (e.g., olefins and isoalkanes) are important in understanding the diagnostic features of gasoline (see *refining* discussion). The selection of target compounds also considered the approximate boiling range, with the intention being to have representatives of the hydrocarbon compound classes spanning as wide a boiling range as possible. This was particularly important if the method would be used to study weathered gasoline samples, in which the most volatile compounds might be absent.

Because of their utility in distinguishing gasoline types (e.g., Morrison, 1999), those gasoline additives that were amenable to study by the modified United States Environmental Production Agency 8260 Method were also considered as target analytes. These included the oxygenate additives (alcohols and ethers) used in oxygenated gasolines, lead scavengers (EDC, EDB), and MMT.

Given these considerations a list of more than 100 target analytes was proposed by Uhler *et al.* (2002). We have subsequently settled upon quantification of the 84 analytes listed in Table 18.5.2 as a means of achieving sufficient specificity among different gasoline types. This target analyte list includes representatives from the five PIANO classes, as well as selected oxygen- and sulfur-containing analytes. The

Table 18.5.2 *Inventory of typical PIANO analytes (reduced from Uhler* et al.*, 2002)*

Class*	Abbrev	Analytes	Class*	Abbrev	Analytes
I	IP	Isopentane	ADD	12DBE	1,2-Dibromoethane
O	1P	1-Pentene	A	EB	Ethylbenzene
O	2M1B	2-Methyl-1-butene	A	MPX	*p/m*-Xylene
P	C5	Pentane	O	1N	1-Nonene
O	T2P	2-Pentene (trans)	P	C9	Nonane
O	C2P	2-Pentene (cis)	A	STY	Styrene
OX	TBA	Tertiary butanol	A	OX	*o*-Xylene
N/I	CP	Cyclopentane/2,3-Dimethylbutane	A	IPB	Isopropylbenzene
I	2MP	2-Methylpentane	A	PROPB	*n*-Propylbenzene
OX	MTBE	MTBE	A	1M3EB	1-Methyl-3-ethylbenzene
I	3MP	3-Methylpentane	A	1M4EB	1-Methyl-4-ethylbenzene
O	1HEX	1-Hexene	A	135TMB	1,3,5-Trimethylbenzene
P	C6	Hexane	O	1D	1-Decene
OX	DIPE	Diisopropyl Ether (DIPE)	A	1M2EB	1-Methyl-2-ethylbenzene
OX	ETBE	Ethyl Tertiary Butyl Ether (ETBE)	P	C10	Decane
I	22DMP	2,2-Dimethylpentane	A	124TMB	1,2,4-Trimethylbenzene
N	MCYP	Methylcyclopentane	A	SECBUT	sec-Butylbenzene
I	24DMP	2,4-Dimethylpentane	A	1M3IPB	1-Methyl-3-isopropylbenzene
ADD	12DCA	1,2-Dichloroethane	A	1M4IPB	1-Methyl-4-isopropylbenzene
N	CH	Cyclohexane	A	1M2IPB	1-Methyl-2-isopropylbenzene
I	2MH	2-Methylhexane	A	IN	Indan
A	B	Benzene	A	1M3PB	1-Methyl-3-propylbenzene
I	23DMP	2,3-Dimethylpentane	A	1M4PB	1-Methyl-4-propylbenzene
S	THIO	Thiophene	A	BUTB	*n*-Butylbenzene
I	3MH	3-Methylhexane	A	12DM4EB	1,2-Dimethyl-4-ethylbenzene
OX	TAME	TAME	A	12DEB	1,2-Diethylbenzene
O/I	1HPT	1-Heptene/1,2-DMCP (trans)	A	1M2PB	1-Methyl-2-propylbenzene
I	ISO	Isooctane	A	14DM2EB	1,4-Dimethyl-2-ethylbenzene
P	C7	Heptane	P	C11	Undecane
N	MCYH	Methylcyclohexane	A	13DM4EB	1,3-Dimethyl-4-ethylbenzene
I	25DMH	2,5-Dimethylhexane	A	13DM5EB	1,3-Dimethyl-5-ethylbenzene
I	24DMH	2,4-Dimethylhexane	A	13DM2EB	1,3-Dimethyl-2-ethylbenzene
I	223TMP	2,2,3-Trimethylpentane	A	12DM3EB	1,2-Dimethyl-3-ethylbenzene
I	234TMP	2,3,4-Trimethylpentane	A	1245TMP	1,2,4,5-Tetramethylbenzene
I	233TMP	2,3,3-Trimethylpentane	A	PENTB	Pentylbenzene
I	3EH	3-Ethylhexane	P	C12	Dodecane
I	2MHEP	2-Methylheptane	A	N	Naphthalene
I	3MHEP	3-Methylheptane	S	BT	Benzothiophene
A	T	Toluene	ADD	MMT	MMT
O	1O	1-Octene	P	C13	Tridecane
P	C8	Octane	A	2MN	2-Methylnaphthalene
			A	1MN	1-Methylnaphthalene

*P – paraffin; A – aromatic; I – isoparaffin; N – naphthene; O – olefin; S – sulfur-containing; ADD – additive

oxygenated compounds include four ethers that have been used in gasoline (MTBE, TAME, DIPE, ETBE), as well as TBA. Methanol, ethanol, and isopropyl alcohol are notably absent from this list due to the difficulty of purging these compounds from water and other analytical constraints, particularly with water matrices. Sulfur compounds include various $C_0 – C_1$ thiophenes and benzothiophene. This target analyte list is not all inclusive or inflexible. Analytes can be added or removed from the list if sufficient reason exists to do so.

While this PIANO method has found favor in the environmental analysis of NAPLs, soils, and water in recent years, it is only of late that this same methodology has been applied to the characterization of hydrocarbons in subsurface vapor and ambient air. By utilizing the flexibility of United States Environmental Production Agency's advanced method for air sampling and analysis—TO-15—it is possible to develop sophisticated chemical profiles for volatile, air-borne PIANO hydrocarbons (United States Environmen-

tal Production Agency, 1999). Simply by extending the target analyte list of United States Environmental Production Agency TO-15 to include PIANO hydrocarbons of forensic utility (e.g., Table 18.5.2) as well as alcohols, it is possible to measure distinct chemical signatures and potentially distinguish among various gasoline and non-gasoline sources of hydrocarbon in air/vapor. Using properly prepared passive stainless steel canisters (e.g., 6-L SUMMA) for sample collection and sensitive GC/MS measurement techniques, it is possible to achieve detection limits of 0.5 parts per billion volume (ppbv) for individual PIANO compounds by United States Environmental Production Agency TO-15.

18.5.2.2 Alcohols

Alcohols—particularly ethanol—have been used as modern gasoline additives for more than 20 years. At the end of the 1970s, ethanol-gasoline blends were reintroduced to the US market when oil supplies from the Middle East were disrupted. In response to the CAA amendments of 1990, ethanol gained favor as a gasoline oxygenate additive in certain markets. As the petroleum industry phases out MTBE use in the early 2000s, ethanol is becoming the oxygenate of choice for many reformulated gasoline blends. Hence, the environmental forensics investigator will more frequently need to measure the presence/absence and concentration of ethanol at gasoline-impacted sites.

The analysis of ethanol and other alcohols in NAPL and dispensed gasoline can be achieved by direct injection of the product onto a gas chromatograph, followed by FID following adaptations of ASTM D 4815, *Standard Test Method for Determination of MTBE, ETBE, TAME, DIPE, tertiary-Amyl Alcohol and C_1 to C_4 Alcohols in Gasoline by Gas Chromatography*. Pre-concentration steps can achieve detection limits of approximately 200 ppb (Koester, 1999).

Ethanol, because of its significant water solubility, is difficult to analyze in water and soil samples because it is difficult to remove from water. For this reason, purge-and-trap GC methods such as United States Environmental Production Agency 8260 are poor choices for analysis of ethanol and other alcohols in water samples. While time-consuming and costly techniques such as azeotropic distillation (United States Environmental Production Agency Method 5031) are candidates techniques for measurement of ethanol and other alcohols in water, direct aqueous injection (DAI), followed by routine GC/FID can more readily and cost-effectively achieve detection limits of approximately 1 ppm (American Society for Testing and Materials, 1999) to as low as 20 μg/L (United States Environmental Production Agency, 1997). This same gas chromatographic approach, using a mass spectrometer detector, should give the investigator even more reliable data than the simple GC/FID approach. Emerging techniques, such as solid phase micro extraction followed by GC analysis, hold promise for even lower detection limits of ethanol and other alcohols (University of Wisconsin, 2000).

18.5.2.3 Organic Lead

In the United States, pre-1996 (pre-1992 in California) automotive gasoline could contain varying amounts of lead alkyls as octane enhancers (Gibbs, 1990). The measurement of these compounds can be used to determine the total lead concentration in a gasoline, potentially age-constrain its vintage (Johnson and Morrison, 1996; Kaplan et al., 1997; Stout et al., 1999b), or deduce the source of the gasoline based on the distribution of the individual lead alkyls (Kaplan and Galperin, 1996). The analysis of lead alkyls is currently limited to dispensed and NAPLs, since laboratory and field experiments have demonstrated that, in the absence of free phase gasoline, lead alkyls are strongly adsorbed to soils, precluding their efficient extraction and analysis (Mulroy and Ou, 1998).

Our own (unpublished) study conducted on sandy and clayey soils spiked with various concentrations of a hydrocarbon mixture containing all five lead alkyls, and allowed to sit undisturbed for 24 hours at room temperature, demonstrated recoveries following solvent extraction (with dichloromethane) on the order of only 0–40%. Recoveries generally increased proportional to the volume of the spiking solution. This study confirmed the claim of Mulroy and Ou (1998) and demonstrated the difficulty of quantitatively recovering organic lead from soils using conventional solvent extraction techniques. As such, a determination of the concentrations of lead alkyls in soil samples is not considered quantitative. However, such analyses might be useful in a qualitative sense, e.g., demonstrating the presence of extractable lead alkyl in soils would implicate the presence of a leaded gasoline component. Alternative solvents (other

than dichloromethane) and/or extraction techniques may prove more effective at removing organic lead from soils.

The five individual lead alkyls that were contained in antiknock additive packages (Table 18.4.3)—tetramethyllead (TML), trimethylethyllead (TMEL), diethyldimethyllead (DEDML), methyltriethyllead (MTEL), and tetraethyllead (TEL)—can be measured in NAPLs by GC/MS in the selected ion monitoring mode following adaptations of United States Environmental Production Agency Method 8270 (Healey et al., 2002). Detection limits as low as 5 μg/mL can be achieved. The results of such an analysis—typically reported in units of micrograms of individual lead alkyl per milliliter of gasoline—can be easily converted to units of glpg with the measurement or assumption of the product's density (needed to convert the measured lead concentration from a mass-to-volume basis). As mentioned above, soil extracts can be measured by the same GC/MS method to the qualitatively determine the presence or absence of lead alkyls in soils.

18.5.2.4 Stable Isotopes

Hydrocarbons have intrinsic ratios of the stable isotopes of the carbon and hydrogen atoms that comprise each molecule. As a bulk property, the stable isotope ratio for a given element in a hydrocarbon product or assemblage reflect the many different geochemical and biological processes to which the hydrocarbons were exposed during their formation, refining, or processing, and environmental degradation/fractionation (Wenger et al., 2001). These intrinsic isotopic features of petroleum have, in the case of gasoline, been shown to be useful in distinguishing between different source(s) of gasoline in free product and in groundwater (Kaplan et al., 1997).

Analytically, carbon and hydrogen isotopes are measured using an isotope ratio mass spectrometer. The isotope ratios measured for carbon and hydrogen are reported relative to international standards, in units of per mil (‰). A sub-sample of petroleum (pure or extracted from a sample) is combusted to produce CO_2 and H_2O. The carbon dioxide is purified, and the $^{13}C/^{12}C$ ratio is determined. Similarly, the water produced during the combustion reaction is converted to hydrogen gas with a catalyst. The deuterium-to-hydrogen ratio is measured for the resulting gas. Typically, isotope ratios for carbon and hydrogen can be measured with a precision of ±0.2‰ and ±2.0‰ respectively.

Gas chromatography-isotope ratio-mass spectrometry (GCIRMS) is a specialized application of isotope ratio analysis. In GCIRMS, the isotopic ratios of carbon (and theoretically other isotopes) for individual hydrocarbon compounds are measured. This method holds particular promise for environmental forensics investigations of gasoline, because isotopic differences in the chemicals that comprise gasoline blending stocks or in additives like MTBE or ethanol can be readily measured and compared among samples containing gasoline residues.

In this technique, individual hydrocarbons are first resolved from a complex sample using GC. The effluent from this separation is introduced into an isotope ratio mass spectrometer where isotope ratios are determined for individual compounds after combustion and catalytic conversion (Philp, 2002).

18.5.2.5 Gasoline Dyes

Many finished automotive gasolines have been amended by retailers with pigmented dyes to differentiate among fuel grades. As noted previously, in NAPLs these dyes can be quickly identified by thin layer chromatography (Kaplan and Galperin, 1996; Kaplan et al., 1997). More detailed fingerprinting of dyes using GC/MS (Youngless et al., 1985), thermospray HPLC/MS (Voyksner, 1985), or electrospray ionization mass spectrometry (Rostad, 2003) have been described but have not been widely used in forensic investigations. Identifying the presence, absence, or mixture of these dyes can be of potential benefit to the environmental forensic investigator in cases where it is necessary to differentiate between or among potentially similar fugitive petroleum products. One limitation of reliance on fuel dye data for forensic purposes is the fact that these dye compounds are environmentally unstable. Thus, the absence of dyes in a fugitive fuel does not necessarily mean that the compound(s) were not initially present in the fresh fuel.

18.6 GASOLINE WEATHERING

"Weathering" is the term commonly used to describe the influence of physical, chemical, and biological forces on the composition of gasoline hydrocarbons released into the environment (Kaplan et al., 1996; McCarthy et al., 1998). Depending on the site-specific conditions, environmental weathering often changes chemical composition of fugitive gasoline. Environmental forensics investigation of gasoline-derived contamination must consider the chemical changes due to weathering in order to reconcile the chemical characteristics of fugitive gasoline with those of candidate sources.

The three most recognized and well understood weathering processes that affect the chemical composition of fugitive gasoline are: (1) evaporation, (2) water-washing or solubilization, and (3) microbial degradation. However, other less studied processes, e.g., adsorption or hydrolysis, may also affect the composition of gasoline in the environment. For example, alkyl lead compounds (described above) may, under to certain conditions, both adsorb strongly to soil particles and/or hydrolyze to polar ionic alkyl leads (Ou et al., 1995a,b; Mulroy and Ou, 1998).

The extent of weathering on the composition of gasoline in the environment is simply a function of the rate(s) of weathering and the residence time in the environment. Unfortunately, the rate(s) of weathering is not simple, but is a complex function of combined effects of numerous, site-specific factors that will vary on many scales. These factors can include the nature of the release (e.g., surface versus subsurface, the volume, and rate of the release,), the mass and distribution in subsurface (which affects the surface area of exposure to air and water where weathering is greatest), the gasoline's initial product composition, soil properties (e.g., moisture and oxygen availability, temperature, chemical concentrations, etc.) and stratigraphy (continuity of strata), and the geohydrologic conditions. Since site-specific factors can change over time, naturally the rate of weathering can change over time.

The majority of environmental forensics investigations of gasoline involve studies of the impacts of gasoline released into the subsurface. The behavior of gasoline (or other petroleum products) in the subsurface have been fairly well established (Freeze and Cherry, 1979; Hinchee and Reisinger, 1987; Bruce, 1993 and refs therein). The brief review here is based upon these references. A figure depicting the components discussed is shown in Figure 18.6.1.

If a sufficient quantity of gasoline enters the soil, it can exceed the residual saturation of the soil. Under these conditions, the gasoline is mobile and will migrate downward through the forces of gravity as an NAPL with capillary forces causing some lateral spreading among the soil particles within the vadose zone. Downward migration stops when NAPL reaches the groundwater table or when the continuity of NAPL in soil pores can no longer facilitate downward movement/spreading. As the gasoline migrates through the soil column it begins to lose its more

volatile components into the soil gas between soil particles (volatilization) and its more water soluble compounds to the interstitial pore water (solubilization). Even upon reaching the water table some residual gasoline remains trapped within vadose zone soils due to the capillary attraction, i.e., the so-called residual saturation (Hoag and Marley, 1986).

During migration downward within the vadose zone gasoline can encounter various geological barriers (e.g., perched water table or clay lens) that alter its flow path and ultimately affect its chemistry – by affecting the rates of weathering. If sufficient gasoline is present to reach the water table, it will be begin to spread laterally as the soil pores become increasingly full with water, i.e., within the capillary fringe zone. In this zone the rate of solubilization of additional components within the gasoline will begin to increase as the surface area of the interfaces between gasoline and water increase. The so-called floating NAPL may accumulate in a mound above the groundwater table and spread laterally under its own weight (Figure 18.6.1). Spreading in the direction of groundwater flow is typically greatest. The lateral spreading of gasoline near the water table will progress until residual saturation conditions are achieved, at which point the NAPL movement stabilizes. Volatilization and solubilization continue to progress at the interfaces between gasoline and air, and gasoline and water, respectively. The affects of these weathering processes, and the affects of biodegradation, on gasoline are discussed in greater detail in the following sections.

18.6.1 Volatilization

When gasoline is exposed to air at the soil surface or in soil pore spaces, some of the gasoline will volatilize thereby altering its chemical signature. The fate of these volatile hydrocarbons is a function of the vapor pressure of the compounds, the surface area of exposure, their concentration in the gasoline, and the ambient temperature (Schwarzenbach et al., 1993), soil type, clay and organic matter content, and water content of the soil (Petersen et al., 1995). Since gasoline is composed of hundreds of compounds, the equilibrium vapor phase concentration of a specific compound in a complex mixture is significantly lower than its pure vapor phase equilibrium. This feature is described by Raoult's Law (Burris and MacIntyre, 1985), which states that the representative vapor concentration ($C_{eq,R}$) can be predicted by:

$$C_{eq,R} = m \times C_{eq,P}$$

where m is the mole fraction and $C_{ep,P}$ is the value for the pure-phase equilibrium vapor concentration. In general, highly volatile compounds will evaporate from the NAPL more rapidly and their relative mole fraction (within the NAPL) will decrease. The relative mole fractions of the constituents with lower vapor pressures (i.e., those compounds less prone to evaporation) concurrently increase, which results in increased and progressive volatilization of the compounds with increasingly lower vapor pressures. Table 18.6.1 lists the vapor pressures of common constituents of gasoline.

Gasoline is more susceptible to evaporative weathering than higher boiling petroleum products (e.g., diesel fuel #2) simply because gasoline in composed of a greater proportion of volatile hydrocarbons (Bruce, 1993). As a result, evaporation will have a greater effect on the composition of gasoline than on diesel fuel. This effect is reflected in a very pronounced reduction in the concentrations of the lower molecular weight hydrocarbons – i.e., those with the highest vapor pressures (Table 18.6.1).

The effect of evaporation on gasoline in the subsurface is likely to vary depending upon environment in which

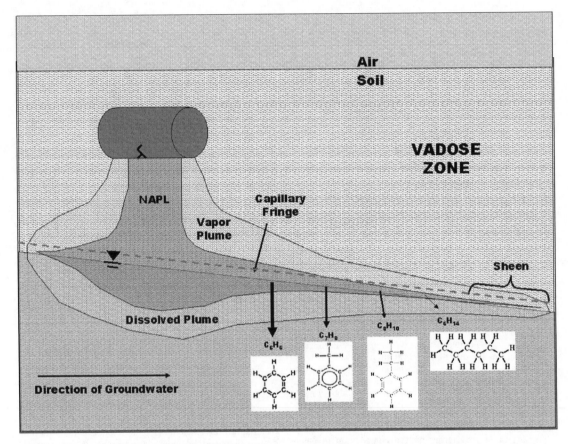

Figure 18.6.1 *Schematic showing migration and typical distribution of gasoline-derived NAPL in the subsurface.*

the gasoline occurs. For example, near the surface, where vapors are allowed to escape relatively freely (e.g., near-surface soils) via advection, the effect of evaporation is likely to be greater (other factors being equal) than might occur directly above a groundwater NAPL layer. In either case, however, the evaporation will follow Raoult's Law and a systematic and progressively increasing loss of compounds from gasoline down to the lowest hydrocarbons measured is anticipated (Stout *et al.*, 2002).

As a practical matter in forensic investigation of gasoline, the gasoline release circumstances and the site-specific conditions will have a strong influence on the degree of evaporation experienced by the gasoline over a given time period. Thus, it would be expected that gasoline found in near-surface soils will typically exhibit a higher rate and degree of evaporation than that found deeper in the subsurface when other factors are equal. Residual gasoline trapped in vadose zone soil pores generally will exhibit a higher degree of evaporation than free phase gasoline that exists in NAPL accumulations near the water table. To investigate these differences further, in the following sections we attempt to consider the effects of evaporation of gasoline under varying subsurface circumstances, namely in (near) surface soils saturated with gasoline, in sub-surface vadose zone soils with residual gasoline, and from NAPL from the capillary fringe.

18.6.2 Surface Soil Releases

As described above, gasoline released into near-surface soils might be anticipated to experience a greater degree of evap-

oration than gasoline released into subsurface soils. The extent to which gasoline released into surface soils will ultimately (100%) evaporate is an interesting question in certain forensic investigations. Does gasoline "disappear" following a release into near-surface soils? In an attempt to better understand the effects of evaporation of gasoline in surface soils (independent of other weathering effects), and the resulting chemical signatures, a simple laboratory study was conducted.

In this study, a surface soil was spiked with 10 wt% gasoline-derived NAPL. This concentration was approximately equal to the residual soil saturation of the soil, i.e., a soil containing "free phase" gasoline. Small aliquots of soil approximately 22.5 cm wide by 10 cm thick were exposed to extreme ambient temperatures (87°F) for 0 (T_0), 3 (T_3), and 30 (T_{30}) days in a convection oven. Composite soil samples from the piles were then analyzed by: (1) conventional PIANO analysis using GC/MS at T_0 and T_3 and (2) TPH (GC/FID, T_0, T_3, and T_{30} (analytical methods were described earlier in this chapter) in order to determine the mass loss and to characterize the chemical changes resulting from extensive levels of evaporation.

The results of this investigation documented a rapid loss (less than 3 days) of C_5–C_8 hydrocarbons resulting in only trace concentrations of these most volatile hydrocarbons (Figure 18.6.2). A slower though measurable loss of the C_9+ compounds were noted for the gasoline; e.g., C_3+ alkylbenzenes, naphthalenes, *n*-alkanes were reduced after 30 days. This group of higher boiling C_9+ compounds persisted in the soil after 30-day exposure at ambient tem-

Table 18.6.1 *Physical-Chemical Properties of selected compounds in gasoline*

Compound	Molecular Weight	Vapor Pressure (Atm)	Boiling Point (°C)	Solubility (mg/L)	Density
Benzene	78.11	1.25E-01	80.1	1780	0.8765
Toluene	92.13	3.75E-02	110.6	515	0.8669
Ethylbenzene	106.2	1.25E-02	136.2	152	0.867
m-Xylene	106.2	1.09E-02	139	160	0.8842
p-Xylene	106.2	1.15E-02	138	215	0.8611
o-Xylene	106.2	1.15E-02	144	220	0.8802
1,2,4-Trimethylbenzene	120.2	2.66E-03	169.4	57	0.8758
n-butylbenzene	134.22	1.35E-03	183	13.8	0.8601
Naphthalene	128.19	3.63E-04	218	31	1.03
2-Methylnaphthalene	142.2	1.11E-04	241.9	25	1.0058
1-Methylnaphthalene	142.2	8.72E-05	244.6	28	1.022
Iso-Butane	58.13	3.52E+00	−11.7	48.9	0.5571
Butane	58.13	2.40E+00	−0.5	61.4	0.5786
Iso-Pentane	72.15	9.04E-01	27.8	13.8	0.6193
Pentane	72.15	6.75E-01	36.07	38.5	0.6262
Methylcyclopentane	84.16	1.81E-02	71.8	42	0.7486
Hexane	86.17	1.99E-01	68.95	9.5	0.6593
Heptane	100.21	6.03E-02	98.42	2.93	0.6837
Methycyclohexane	98.19	6.10E-02	100.9	14	0.7694
Octane	114.23	1.78E-02	125.7	0.66	0.7027
Nonane	128.26	5.64E-03	150.8	0.22	0.7177
Decane	142.29	1.73E-03	174.1	0.052	0.7301
Undecane	156.32	5.15E-04	195.9	0.04	0.7402
Dodecane	170.33	1.55E-04	216.3	0.0037	0.7487
Tridecane	185.36	9.54E-05	235	0.013	0.755
2,2,4-Trimethypentane	114.23	6.47E-02	99.2	2.44	0.6919
MTBE	88.15	3.10E+00	55.2	51000	0.741
Ethanol	46.07	5.66E-02	78.3	Complete	0.79
EDB	187.9	1.45E-02	131.5	4300	2.17

Data obtained from Total Petroleum Criteria Working Group Series, Volume 2, Composition of Petroleum Mixtures, May 1998 Prepared by Thomas L. Potter and Kathleen E. Simmons, University of Massachusetts, Amherst, Massachusetts. pp. 33–45.

peratures (Figure 18.6.2). Overall greater than 97 wt% of the original hydrocarbons were lost from the samples due to evaporation after 30 days at 87 °F. Clearly, the impact on the chemical signature of the residual gasoline was dramatic. However, despite the excessive heating at 87 °F for 30 days, not all of the gasoline was evaporated (2910 mg/kg remained).

The residual gasoline was dominated by higher boiling constituents that represented a concentration of the highest boiling constituents present in the gasoline. The "fingerprint" for this excessively evaporated gasoline might be confused with a higher boiling petroleum product (e.g., kerosene). However, this is clearly not the case as it resembles other highly evaporated gasolines, as are typical of arson investigations. For example, Figure 18.6.3 compares the fingerprint of the residual gasoline-derived NAPL from this experiment (after 30 days at 87 °F) to that of a 98% evaporated gasoline reference standard (Newman *et al.*, 1998). Both samples exhibit a predominance of naphthalene and alkylated naphthalenes and selected C_3 + alkylbenzenes. A notable feature is the presence of a small alkane "tail" in both of these highly evaporated gasolines, which is common to gasolines weathered to this high degree.

As an important sidebar to this experiment, it should be noted that the PIANO and GC/FID analytical methods used to evaluate the samples detected thousands of parts per million of gasoline in soil even after extreme evaporation. Not all of the gasoline was evaporated. This speaks to the unlikelihood that natural evaporation processes in surface soil could result in the complete removal of all residual gasoline, even

in surface soils. The basis for this lies in the fact that the analytical methods available for detection of gasoline residues in soil can detect as low as 1 ppb – yet in this experiment approximately 3000 ppm residual gasoline remained. Therefore, even at much lower starting concentrations it is anticipated that some gasoline compounds in excess of 1 ppb would still remain after excessive evaporation. Additional evidence for the retention of detectable gasoline residues following extensive evaporation can be observed in environmental soil samples impacted by gasoline many years before. For example, Figure 18.6.4 shows a GC/FID chromatogram of residual gasoline in soil that was collected from a retail gasoline station site that had been closed more than 25 years before the sample was collected. The residue remaining in this vadose zone soil is dominated by C_3-alkylbenzene, naphthalenes, and C_{10}+ alkanes, similar to what remained in the laboratory-evaporated gasoline (Figure 18.6.2D).

Evaporated gasolines found in the environment often provide interesting and useful data to the forensic scientist. Evaporated gasoline often reveals the presence/absence of heavier hydrocarbons in the sample that might otherwise go unnoticed in less evaporated gasoline-derived products. For example, inspection of Figures 18.6.3 and 18.6.4 demonstrate the presence of the C_{13}+ hydrocarbons that may be present at low levels in gasoline. This so called "alkane tail" is not always observed but would be expected to be more common in gasolines containing a greater proportion of straight-run gasoline, e.g., as was common in *Performance Era* gasolines. In addition, some "alkane tails" might be attributable to minor amounts of heavier petroleum, such

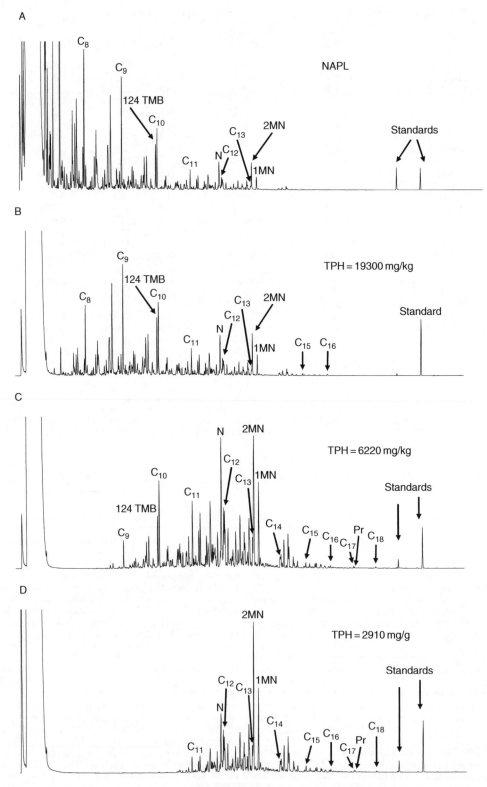

Figure 18.6.2 *The GC/FID chromatograms showing distribution of hydrocarbons in (A) starting gasoline-derived NAPL, (B) spiked soil with gasoline-derived NAPL (To), (C) spiked soil with gasoline-derived NAPL after 3 days at 87°F, (D) spiked soil with gasoline-derived NAPL after 30 days at 87°F; for compound identification see Table 18.5.2.*

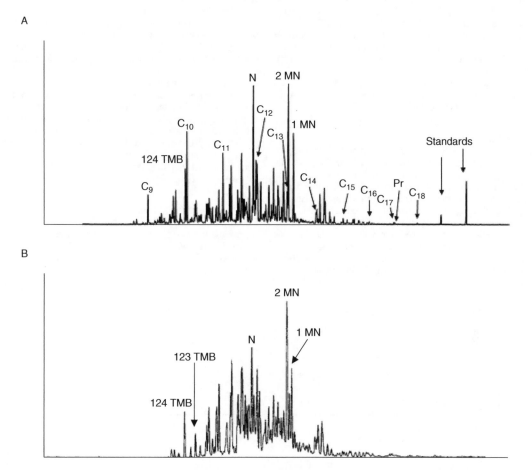

Figure 18.6.3 *The GC/FID chromatograms showing the distribution of hydrocarbons in (A) gasoline residues in laboratory-evaporated soil demonstrating a 97% weight loss and (B) dispensed gasoline laboratory-evaporated to 98 wt% (after Newman et al., 1998). For compound identification see Table 18.5.2.*

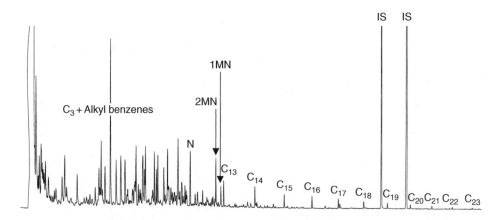

Figure 18.6.4 *The GC/FID chromatogram showing the character of residual gasoline in vadose zone soil after at least 25 years of evaporative weathering. Compare Figure 18.6.2D (IS- internal standard). For compound identification see Table 18.5.2.*

as distillate fuels. For example, Harvey in 1997 reported an alkane tail that comprise 3.2 wt% of a 75 wt% evaporated dispensed gasoline. She attributed this "alkane tail" to minor amounts of mixing of gasoline and middle distillate during pipeline transport of the products, rather than to the nature of the gasoline blending. Therefore, the forensic investigator that observes the "alkane tail" associated with gasoline-derived contamination needs to consider whether it may be attributed to the nature of the gasoline or to mixing with small amounts of distillate fuel(s).

18.6.3 Subsurface Releases

Gasoline deeper within the soil column may experience a different (slower) rate of evaporation than surface soils. Nonetheless, as a general rule, the loss of volatile compounds deeper within vadose zone soil will still be governed by the same processes as evaporation in surface soils. In practice, the degree to which subsurface gasoline is evaporated in vadose zone soils will rarely alter the composition of product to the degree that will confound forensic interpretations. Stout *et al.* (2002) have demonstrated the progressive losses of hydrocarbons and alteration of chemical signatures with increased evaporation in subsurface

vadose zone soils. Figure 18.6.5 shows a series of PIANO (GC/MS) fingerprints for vadose zone soils from a single site that has been impacted by automotive gasoline. The series demonstrates the variable degrees to which evaporation can impact the distribution of hydrocarbons due overwhelmingly to evaporation. Of course, in field samples it is difficult to demonstrate evaporative weathering independent of other weathering processes. For example, the gasoline hydrocarbons shown in Figure 18.6.5 may have been subject to a certain degree of solubilization due to infiltrating water. Nonetheless, most of the changes noted in this figure are reasonably attributed to evaporation, which progressively has removed the more volatile compounds, leaving a residual gasoline product dominated by C_3 + alkyl benzenes, naphthalene, alkylnaphthalenes-and high molecular weight alkanes (Figure 18.6.5).

18.6.4 Non-aqueous Phase Liquid

Gasoline-derived NAPL recovered from monitoring wells represents the free phase gasoline typically residing in soils at or near the water table. In general, these may appear less impacted by evaporation than surface and subsurface soil samples. The basis for this is that NAPLs are usually

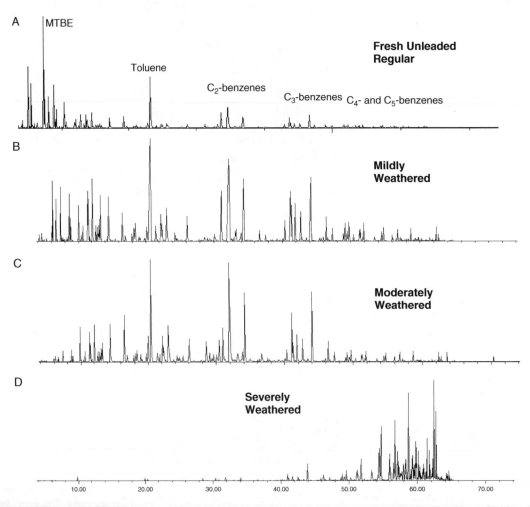

Figure 18.6.5 *The GC/MS total ion chromatograms showing the progressive evaporation of gasoline in vadose zone soils (after Stout et al., 2002).*

perched on impervious soil, or upon the top of a water table, where pore volumes are often filled with water or NAPL. As a result, the interface between gasoline and soil vapor is significantly reduced, thereby retarding the effects of evaporation.

Figure 18.6.6 shows the GC/MS total ion chromatograms (PIANO) of dispensed regular unleaded gasoline and three NAPL samples that exhibit increasing degrees of evaporative weathering. The hydrocarbon distributions in these samples show a progressive loss of the C_5–C_8 hydrocarbons followed by a relative enrichment of the less volatile C_9–C_{12} hydrocarbons (e.g., C_3 + alkylbenzenes, naphthalene, alkanes). It is uncommon that evaporation from mobile NAPL (i.e., those recovered from monitoring/recovery wells) will progress to the same degree observed in vadose zone soil samples (e.g., Figure 18.6.5d). However, under conditions in which there is a rising or falling water table the effect of evaporation on NAPLs may be greater than in static water table conditions. In addition, in locations where the mass of NAPL is low enough so that only sheens are observed in monitoring/recovery wells, the effects of evaporation may be higher than in wells containing measurable thickness of NAPL.

18.6.5 Soil Gas

Evaporation of gasoline in the subsurface produces a vapor phase that can also be beneficial in environmental forensic investigations. Soil gas surveying techniques can be used to rapidly identify the presence/absence of gasoline-derived contamination in the subsurface (Jones, 1995). In addition, the carbon isotopic composition of the soil gas may be used to distinguish between naturally occurring hydrocarbons and anthropogenic sources (Kaplan, 1994; Kaplan et al., 1997; Lundegard et al., 1999). In some cases, the soil gas may also provide additional insight into the source of the volatile hydrocarbons if sufficiently diagnostic data are collected. The use of PIANO-type fingerprinting on soil gas vapors is discussed later in this chapter – as it is relevant to forensic investigation of subsurface gasoline impacts on indoor air. Some value in "fingerprinting" soil vapors may lie in the proportions of different C_4 isomers. For example, some refiners may remove valuable isobutane from crude oil or straight-run gasoline for use as a chemical feedstock (Beall et al., 2002) and add n-butane to gasoline to control vapor pressure of the finished gasoline (see *Butane Blending* section within the discussion of refining, above). As such, the ratio of n-butane/(n-butane + isobutane) in soil gas (and NAPL) may reflect different gasoline sources, and further aid in distinguishing gasoline-derived vapors from other sources (e.g., natural gas). The rate of evaporative loss of these C_4 isomers will depend upon multiple site-specific conditions. This is discussed further later in this chapter (see *Rate of Weathering with respect to Age-Dating*, in Section 18.7).

18.6.6 Water Washing

Solubilization or "water washing" of gasoline occurs when there is a partitioning of gasoline constituents from NAPL into the water phase. This process will predictably alter the chemical fingerprint of the fugitive gasoline over time and is therefore relevant to forensic investigations involving chemical fingerprinting. Water washing can occur (1) during percolation of recharge water through vadose zone soil contaminated with residual gasoline or (2) at the interface between NAPL and groundwater within the capillary fringe zone. The impact of water washing on gasoline is a function of three factors, namely: (1) the composition of the

gasoline and its additive (e.g., reformate-based gasoline versus alkylate-based gasoline or Oxyfuel/RFG versus conventional gasoline), (2) the molar concentration of a given compound in the product (mole fraction or concentration), and (3) the relative solubility of a given compound in water versus its solubility in the hydrocarbon mixture. The fuel–water partition coefficient (K_{fw}) for each compound describes its tendency to migrate from the fuel to the water and is estimated by the equation:

$$Log\ K_{fw} = 6.099 - 1.15\ log\ S$$

where S is the equilibrium aqueous solubility for the compound (Bruce, 1993).

Table 18.6.1 contains the equilibrium water solubilities of the important gasoline compounds. The rate of partitioning of individual compounds from fugitive gasoline into water will be highest for those with the highest solubilities. Several general observations can be made concerning the solubility of gasoline compounds, e.g.,

1) Aromatic hydrocarbons are generally more soluble than aliphatic hydrocarbons.
2) Increasing degrees of alkylation generally reduces the solubility of structurally similar hydrocarbons.
3) Lower molecular weight hydrocarbons within a compound class are generally more soluble than higher molecular weight hydrocarbons in that class.
4) Polar compounds such as ethers (e.g., MTBE) and alcohols (e.g., ethanol) are generally more soluble than hydrocarbons.

In addition the rate of partitioning due to water washing is a function of the degree of contact between NAPL and water. A large mass of NAPL in the surface will have a smaller surface-area-to-mass ratio than a small mass of NAPL. Since water washing will occur at the interface between NAPL and water, the rate of partitioning (and the effect on the remaining NAPL fingerprint) will be proportional to the mass of the NAPL – under most conditions a larger mass of NAPL will weather more slowly.

As might be predicted by these general "rules," gasoline is more susceptible to water washing than higher boiling petroleum products. Because water washing will preferentially remove compounds that are often dominant constituents in gasoline (i.e., oxygen-containing additives and monoaromatics), the gas chromatographic character of water-washed gasolines are markedly distinct from dispensed gasolines. Solubilization primarily affects gasoline that is in direct contact with groundwater. However, given the right surface and soil conditions and rainfall/infiltration rates, water washing of gasoline within the vadose zone can influence the composition sufficiently to affect the fingerprinting of gasoline residues within the vadose zone. In many situations (and as was mentioned above with respect to evaporation), gasoline in contact with the water table may be subject to a high rate of water washing due to fluctuations in the water table – e.g., infiltration/recharge fluctuations or tidal "pumping" (API, 2004).

Demonstrating the effect of water washing on gasoline independently of other forms of weathering is difficult in field settings since other weathering processes occur contemporaneously (Stout et al., 2002). Therefore, laboratory studies provide a means to determine the specific effects of water washing on gasoline and other petroleum products independent of evaporation and biodegradation (e.g., American Petroleum Institute, 1985; Burris and MacIntyre, 1985). Most of these studies focused on the composition of the water in contact with gasoline while only a few studies have addressed the compositional changes to the petroleum

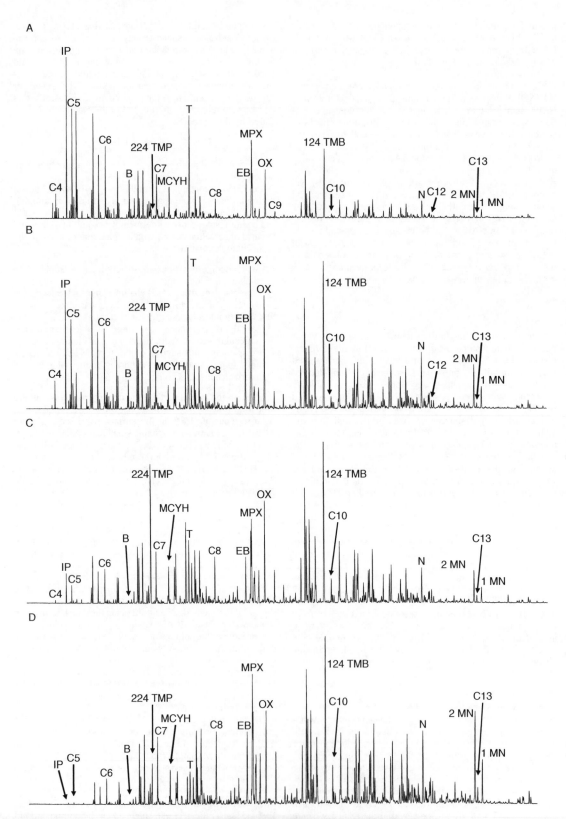

Figure 18.6.6 *The GC/MS total ion chromatograms showing effect of evaporation on (A) a dispensed gasoline and (B–D) a series of NAPLs recovered from monitoring wells. Compound identification see Table 18.5.2.*

(Larafgue and Thiez, 1996). However, understanding the compositional changes of water washing on gasoline are particularly important to environmental forensic investigations involving gasoline-derived NAPL.

Laboratory studies can provide additional information that may be more difficult to recognize in field studies due to the simultaneous effects of other weathering processes. However, under field condition when NAPL evaporation is minimized (e.g., clayey soils), water washing may be the predominant weathering pathway. Thus, we designed and conducted a series of laboratory water-washing experiments in which the effects on the gasoline "fingerprint" were closely monitored using conventional PIANO fingerprinting methods. In our experiments, glass dissolution vessels (4 L) were designed to allow for a small volume of gasoline to be in contact with gently swirling water that was constantly flowing through the vessel at a slow rate for 6 months (~1 ml/minute) and then "batch" mode (replaced weekly) thereafter (Figure 18.6.7). Except for the input/output ports for the water flow, and a sampling port for the floating gasoline, the chamber was sealed thereby permitting minimal or no evaporation. Photolysis was inhibited by keeping the chambers covered and biodegradation was inhibited through the addition of mercuric chloride to the water. The experiment was performed by adding (150 mL; <1 cm NAPL thickness) different types of gasoline, including reformate- and alkylate-dominated gasolines, MTBE- and ethanol-bearing gasoline, and leaded gasoline. The experiments were conducted for a period of 20 months. Samples of floating gasoline were collected periodically using gastight syringes in the course of the experiments and analyzed using conventional PIANO techniques (described elsewhere in this chapter).

Figure 18.6.8 shows the PIANO histograms for an MTBE-laden (10 wt%) gasoline (T_0) that had been artificially water washed in this system for six (T_6) and 20 months (T_{20}). Comparison of the C_4 and C_5 hydrocarbon isomers, particularly the presence of *n*-butane in the T_{20} NAPL sample indicates that minimal evaporation had occurred. The results demonstrate the impact of water washing on the gasoline signature due to the partitioning of soluble aromatic hydrocarbons (e.g., BTEX) and MTBE to the underlying water. Benzene and MTBE were fully removed from the NAPL within the first 6 months of the study. On the other hand, the less soluble hydrocarbons such as normal, iso-, and cylco-alkanes were minimally affected by water washing.

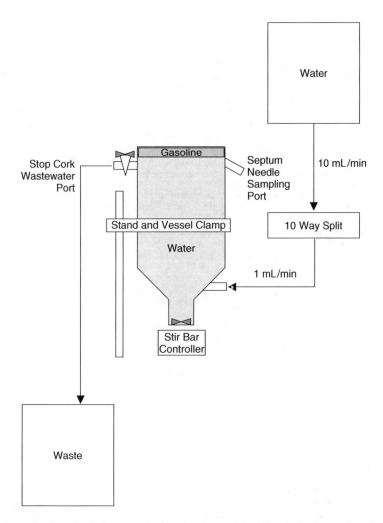

Figure 18.6.7 *Schematic diagram of apparatus designed to evaluate the effects of water washing (independent of other weathering phenomena) on gasoline.*

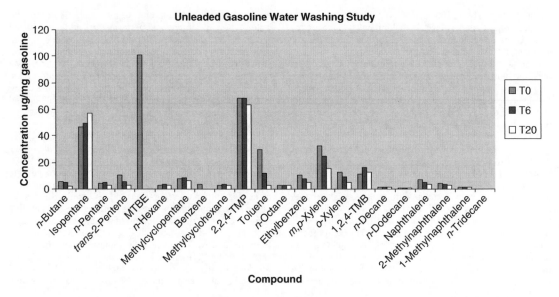

Figure 18.6.8 *The PIANO histogram showing the distribution of major constituents in an unleaded MTBE-laden gasoline* (T_0) *and artificially water-washed for 6 months* (T_6) *and 20 months* (T_{20}).

This argues for the potential for retaining important diagnostic feature of a gasoline despite the impact of water washing. For example, less soluble compounds such as 2,2,4-trimethypentane (isooctane), methycyclohexane, and decane (Table 18.6.1) remained virtually unchanged over the 20-month study. Similarly, the proporations among various C_3 + alkylbenzenes remained relatively constant. Thus, forensic evaluations requiring comparison among water-washed gasolines benefit from comparisons of the distributions among the less soluble compounds.

The results of our laboratory experiments indicate that over time, the aromatic hydrocarbons and olefines will become substantially depleted in the NAPL, and the relative concentrations of the less water soluble compounds (e.g., methylcyclohexane) will correspondingly increase (on an oil weight basis). On this basis it might be predicted that the chemical signature of reformate-enriched gasolines will be more highly influenced by water washing than those containing predominantly straight-run gasoline or alkylate. For example, Figure 18.6.9 shows the GC/MS total ion chromatograms for an unweathered reformate-enriched gasoline and two reformate-enriched NAPLs that have experienced water washing. Each of the NAPLs demonstrates a significant reduction in the relative abundance of the most soluble monoaromatics (BTEX) and a corresponding enrichment in less soluble compounds, including methylcyclohexane and *n*-alkanes.

18.6.7   Biodegradation

Biodegradation should be considered a more complicated process than physical weathering (evaporation and solubilization) as it requires a specific suite of conditions in order to degrade substantial amounts of gasoline in the environment. For example, the rate and effects of biodegradation of gasoline depend not only on the properties of the gasoline and but also upon parameters that affect the ability for gasoline-degrading microbial communities to flourish (Prince, 1998; Solano-Serena *et al.*, 1999), e.g., availability of (1) oxygen or other suitable electron acceptor, (2) nutrients, and (3) moisture and the toxicity of NAPL to the microbes.

These are of particular interest in forensic investigations involving hydrocarbon fingerprinting as they can impart certain spatial heterogeneities in the degree of biodegradation within an NAPL accumulation. For example, the presence of continuous phase NAPL in soils precludes or minimizes the available oxygen or moisture in those pores. The high concentration of hydrocarbons within those pores creates an environment that is toxic to microorganisms, thereby prohibiting or significantly retarding biological activity. As such, any biodegradation effects on gasoline-derived NAPL would be anticipated to be greatest in areas where the soil pores contain a greater proportion of water, such as within the capillary fringe where water and NAPL co-exist. Greater degree of biodegradation was observed in the capillary fringe zone in an NAPL accumulation of diesel fuel #2, whereas NAPL above this zone was undegraded (Stout and Lundegard, 1998) and the same disparity is expected in gasoline-derived NAPL accumulations. In addition, gasoline-derived hydrocarbons can be biodegraded at a higher rate/extent within the upgradient edge of the NAPL accumulation, i.e., in the vicinity where dissolved oxygen is constantly replenished with fresh groundwater.

Given the complex number of factors required for biodegradation to proceed, physical weathering processes (evaporation and water washing) initially (at least) have the largest impact on the chemical signature of the gasoline. It is likely that biodegradation's greatest impact on gasoline-derived contamination is upon those components that have already evaporated (Roggemans *et al.*, 2001) or been water-washed (Brauner and Killingstad, 1996; Larafgue and Thiez, 1996). One difficulty in assessing the impact of biodegradation on gasoline in the subsurface is the concurrent (and dominating) effects of evaporation and water washing. However, laboratory studies on gasoline in soil (Zhou and Crawford, 1995) and in water (Yerushalmi and Guiot, 1998), in which the effects of physical weathering have been minimized, have demonstrated that gasoline is intrinsically biodegradable under aerobic conditions. Other studies have examined the behavior of discrete gasoline compounds (e.g., BTEX) and have concluded that many are highly biodegradable (Paje *et al.*, 1997; Alvarez and Vogel, 1991).

A

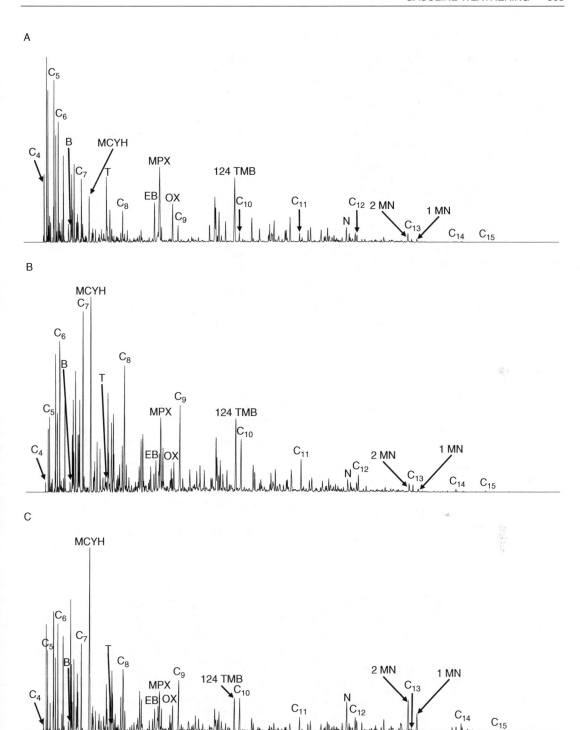

Figure 18.6.9 *The GC/MS total ion chromatograms for (A) unweathered reformate-enriched gasoline, (B) and (C) increasingly water-washed NAPLs derived from reformate-based gasoline. For compound identification see Table 18.5.2.*

Solano-Serena *et al.* (1999) examined the biodegradation of whole gasoline and its constituents. In their study 94 wt% of the gasoline was biodegraded under optimal growth conditions using microbial inoculums derived from activated sludge. Respirometry analysis indicated that the gasoline hydrocarbons were being converted to CO_2 and biomass. All of the hydrocarbon PIANO classes were extensively degraded, however, each class degraded at a different rate. Aromatic hydrocarbons (e.g., BTEX) were most rapidly degraded, followed by olefins.

The biodegradation rate of paraffins and naphthenes appeared to be slower than the aromatic hydrocarbons, with the isoparaffins being the most recalcitrant. Specifically, the 2,2,4-trimethylpentane (isooctane) was exceptionally recalcitrant most likely due to the steric hindrance induced by the quaternary carbon/consecutive carbon side chains on the molecule. Marchal *et al.* (2003) reported similar results with 96 wt% of the whole gasoline biodegraded in 28-day tests. Of the 95 major compounds measured, 72 were completely degraded and 23 were partially consumed. The aromatic, olefins, and paraffin compound classes were effectively degraded; however, only 51% of the isoparaffins were degraded. The recalcitrant isoparaffins included 2,2-dimethylbutane, 2,2,4-trimethylpentane, 2,2,5-trimethylhexane, 2,3,4-trimethypentane, and 2,3,3-trimethypentane. This is consistent with an API study (American Petroleum Institute, 2001) that evaluated the biodegradability of three gasoline blending streams: light alkylate naphtha, light catalytically cracked naphtha, and light catalytic reformed naphtha. They observed 40–60 wt% biodegradation depending on the blending stream, with the light alkylate naphtha exhibiting the lowest biodegradation potential (40 wt%). The more refractory nature of the isoparaffins under ideal biodegradation conditions and their relatively low water solubility (Table 18.6.1) makes them ideal candidates for gasoline "fingerprinting" indices in forensic studies (e.g., see trimethylpentane ratio discussed earlier in this chapter).

Most work on gasoline biodegradation referenced above has been performed in aerobic systems. Field observations of anaerobic groundwater systems suggest that gasoline-range hydrocarbons can also be biodegraded within these environments (Alvarez *et al.*, 1998). Small, water soluble aromatic compounds, such as benzene and toluene, have been shown to undergo biodegradation under sulfate-reducing, nitrate-reducing, perchlorate-reducing, ferric ion reducing, humic acid–reducing and methanogenic conditions (Widdel and Rabus, 2001; Chakraborty and Coates, 2004). Less soluble compounds, such as the alkanes have also been shown to be degraded under sulfate-reducing, nitrate-reducing, and methanogenic conditions (Widdel and Rabus, 2001). Cyclic alkanes have also been shown to be readily degraded under sulfate-reducing and methanogenic conditions (Townsend *et al.*, 2004).

18.6.8 Weathering Indices
Based upon anticipated and demonstrated effects of evaporation, water washing, and biodegradation described above, a set of diagnostic ratios that may reflect these effects in gasoline-derived NAPL was developed (Table 18.6.2). Several ratios of compounds within the same hydrocarbon class but with widely varying vapor pressures are reasonably expected to reflect variations in the degree of evaporation among gasoline-derived NAPLs (i.e., *n*-pentane/*n*-heptane). Similarly, ratios between compounds with comparable vapor pressures and disparate aqueous solubilities will reflect variation in the degree of water washing in NAPLs (e.g., toluene/methylcyclohexane). Finally, ratios between susceptible and less susceptible to biodegradation will reflect variation in the degree of biodegradation (e.g., methylcyclohexane/*n*-heptane). In practice, it may be difficult to attribute changes in these ratios exclusively to any one of the weathering processes. Nonetheless, these ratios (or others that may be similarly developed) can be used in an attempt to quantitatively compare the degree of weathering among NAPLs from a particular forensic investigation. This type of quantitative comparison can be used to establish biodegradation trends within a site which

Table 18.6.2 *Inventory of parameters that can reflect different degrees and types of weathering in gasoline-derived NAPL (modified from Kaplan et al., 1996)*

Ratios
Evaporation
n-pentane/*n*-heptane
2-methylpentane/2-methylheptane
isopentane/*n*-heptane
n-heptane/*n*-decane
Water washing
MTBE/2methylpentane
benzene/cyclohexane
toluene/methylcyclohexane
aromatics/total alkanes (*n* + iso + cyc)
aromatics/naphthenes
benzene/toluene
toluene/xylenes
toluene/*n*-octane
1,2,4-trimethylbenzene/*n*-decane
naphthalene/*n*-dodecane
Biodegradation
C4–C8 (*n* + iso) alkanes/C4–C8 olefins
3-methylhexane/*n*-heptane
methylcyclohexane/*n*-heptane
sum trimethylpentanes/*n*-octane

in turn may be useful in unraveling sources for the contamination. Notably, under some circumstances (e.g., minimal weathering of any sort) some of these ratios may reflect differences in the source of gasoline-derived NAPL, rather than weathering effects (e.g., naphthalene/*n*-dodecane).

18.7 AGE-DATING GASOLINE-DERIVED CONTAMINATION

This topic is one of the most common aspects of environmental forensics of gasoline investigations. There are three principal methods by which an environmental forensic investigation can determine the age of contamination, i.e., the time of its original release into the environment. These are: (1) chemical fingerprinting, (2) site-specific and regulatory history, and (3) fate and transport (reverse and confirmation) modeling. The users of these techniques must be aware of the assumptions, caveats, and oversimplifications that may influence any conclusions drawn from any one approach. Reliance upon any one of these methods can yield over-precise or erroneous conclusions. Thus, considerable caution is necessary to avoid extending any conclusions beyond defensible scientific constraints. In the best situation, i.e., when multiple approaches are found that constrain the age of a gasoline within a comparable range, the technical credibility and defensibility of the conclusion is increased.

The focus of this section rests primarily with the first of the aforementioned approach, namely chemical fingerprinting. There are several approaches to age-dating gasoline that have been purported in the literature and/or in the course of litigation that we will describe herein. These are:

- Rate of Weathering
- Temporal Trends in Refining/Blending
- Gasoline Additives
- Stable Lead Isotopes
- Groundwater BTEX Ratios.

How elegantly simple it would be to measure a single chemical fingerprinting parameter, or even a set of parameters, for any site where there is gasoline contamination, and compare that value(s) against some simple empirical relationship to establish *"years since release"* or *"year of production"*? If this could be done to the scientific satisfaction of everyone, litigation costs surrounding liability for that contamination would be drastically reduced. Unfortunately, few tasks in science are this simple and the age-dating of gasoline-derived contamination is no exception. Under most scenarios, there are far too many unknowns to rely upon simple empirical or scientific relationships between simple parameters and the age of gasoline contamination. Each objective and site needs to be evaluated independently and within the constraints of scientific and historical facts.

18.7.1 Age-Dating Based on Weathering

Gasoline released into the environment will weather at widely varying rates depending upon the site- or incident-specific conditions. This was described in considerable detail earlier in this chapter. It should be clear from this earlier discussion that the degree of weathering is not a scientifically reliable basis to establish age of contamination.

The difficulties of this approach can be understood considering two simple scenarios. Consider a small volume, surface spill of gasoline on a warm day versus a large volume, subsurface release due to catastrophic underground storage tank (UST) failure. In the first instance a significant mass of the gasoline is evaporated over a period of hours whereas in the second instant the gasoline would be more slowly altered by evaporation, solubilization, and biodegradation over a period of many years. These two simple scenarios demonstrate the range of weathering effects and rates that must be woven into any interpretation regarding the age of a gasoline impact based upon the degree of weathering.

Senn and Johnson (1987) suggested that the relative degree of weathering might be used to represent the *relative* age of gasoline-derived NAPL from a particular site – more weathered NAPLs samples being older than less weathered NAPLs. However, this simplistic approach would seem to be easily confounded if weathering heterogeneities that may exist within a single NAPL accumulation have not been evaluated. For example, the NAPL accumulation may be more highly weathered at the locations where the ratio of surface area to mass was greatest and less weathered at the locations where the ratio of surface area to mass was lowest (see weathering discussion).

As a result, we caution against the use of the degree of weathering alone to establish an age of gasoline-derived contamination, in either a *relative* or an *absolute* sense at a given site in the absence of a defensible basis to quantify the rate(s) of weathering (e.g., biodegradation of n-alkanes) at that particular site. Recall from the discussions above that the rate of weathering depends primarily upon the:

1) spill/release conditions (rate/volume/composition),
2) depth (surface, subsurface, depth to groundwater), and
3) soil type and conditions (moisture, O_2, microbial populations, stratigraphy).

This summary does not even consider the difficulties in using the degree of weathering to establish age at sites subject to multiple release events and/or mixtures. Thus, while it is reasonable to examine a soil containing a moderately weathered gasoline and conclude that it was not spilled yesterday or last week, perhaps not even last year, establishing an age with greater precision is scientifically indefensible.

Contrary to this discussion, the detection of butenes or n-butane in gasoline-derived NAPLs has (in the course of

litigation, not the literature) been purported to indicate the presence of a gasoline released within about two years. The basis for this claim is that these compounds represent the most volatile compounds in gasoline – which is true. However, as discussed earlier, the rate of evaporation of these or any other compounds from gasoline will depend upon the surface-area-to-mass ratio of the NAPL accumulation, as well as the soil conditions (e.g., the presence of clayey soils might retard evaporation) that affect rate of diffusion away from the NAPL. Therefore, there is no reliable means to claim that certain highly volatile compounds (e.g., butenes or n-butane) will not be present after a prescribed amount of time. This point is demonstrated by the retention of n-butane in gasoline-derived NAPL from clay-rich soils that is approximately 10 years old or more (e.g., Figure 18.6.6). The NAPL in Figure 18.6.9b was collected directly below a tank pit from a gasoline station that was closed in 1980.

Thus, the use of the degree of weathering to establish an *absolute* age for gasoline in the environment is not defensible without site-specific data to defensibly determine the rate(s) of weathering for that site. Determining a *relative* age of gasoline-derived contamination requires thorough consideration of all available site-specific information and consideration of the appropriate degree of error.

18.7.2 Age-Dating Based on Temporal Changes in Refining and Blending

The evolution of the refining capabilities over time (Tables 18.2.1 and 18.2.2) may provide a basis to grossly determine the age of gasoline-derived contamination. However, since most refining processes (e.g., alkylation, reformation, hydrocracking, etc.) have been available for more than 50 years, the application of such gross compositional features is of little practical use in environmental forensic investigations. However, as described above, the overall blending characteristics of gasoline has changed due to the regulatory-driven changes described previously in this chapter. These changes were primarily driven by the need to maintain octane (performance) while reducing lead alkyl additives.

Specifically, *Regulatory Era* gasolines many contain as many as ten different gasoline components – including oxygen-containing octane boosters or oxygenates – blended together to meet finished gasoline specifications (Table 18.3.1; Stout *et al.*, 2001). Thus, the blending of gasoline at a modern refinery can impart a significant amount of character and diagnostic information to finished gasoline produced there. Understanding these processes, and their implications on the character of the gasoline, can be valuable to the forensic investigation of modern, *Regulatory Era* fugitive gasolines (Beall *et al.*, 2002). This is particularly true when the gasoline supply history for a site includes a defined change from one supplier to another.

Oppositely, gasoline blending practices during the *Performance Era* were somewhat less complicated than they are today due to the less stringent regulatory specifications for gasoline. Prior to the stricter regulatory control on gasoline composition that began in the 1970s, many gasoline blenders operated "simple" distillation-based refineries (in contrast to modern complex refineries). Simple (distillation) refineries historically used a basic blending recipe to produce a gasoline that performed well, namely:

- Blending butanes: ~5 volume %
- Straight-run gasoline: ~95 volume %
- TEL: ~3 grams per gallon (Beall *et al.*, 2002).

Other early "complex" refineries, although having the capability to produce reformate and alkylate blending stocks, used these higher octane blending stocks sparingly in lieu of straight-run gasoline and lead alkyl. Thus, *Performance Era* gasolines were generally different from *Regulatory Era* gasolines. The features of these different eras are more thoroughly described later in this chapter (see Figure 18.7.5 and corresponding text in *Genetic Influence of Crude Oil Feedstock on Gasoline Composition*). This overall blending characteristic of gasoline approach to age-dating is not a precise method as it (alone) is unable to yield any particular year or set of years. However, in combination with other age-dating approaches it may prove useful in certain forensic investigations.

Schmidt *et al.* (2002a) described the use of the ratio of toluene to *n*-octane ($T/n\text{-}C_8$) as a proxy for the temporal change in blending practices. An increase in the ratio reflects a greater proportion of reformate (a source of toluene) and lesser proportion of straight-run gasoline (a source of $n\text{-}C_8$) used during the conversion of leaded (*Performance Era*) gasolines to unleaded (*Regulatory Era*) gasolines. (This approach is comparable to that of Beall *et al.* (2002) in which the ratio of naphthalene/$n\text{-}C_{12}$ was used.) The $T/n\text{-}C_8$ ratio demonstrated an empirical relationship with the age of more than 100 regular and mid-grade gasolines spanning a 29-year period (1973–2001). This relationship is shown in Figure 18.7.1.

Premium grades were excluded since they often contained excess toluene (reformate) relative to lower grades of gasolines. Schmidt *et al.* (2002a) contend that the $T/n\text{-}C_8$ ratio can distinguish the approximate age of regular and mid-grade gasolines produced between: (1) 1973 and 1983, (2) 1984 and 1993, and (3) 1994 and 2001. The data clearly show a distinction of gasolines before and after 1994 (Figure 18.7.1); however, a strong statistical basis for distinguishing the first two groups from one another is not as obvious. Nonetheless, at a minimum, this approach may prove useful in constraining the age(s) of gasoline before and after 1994 – once weathering or mixing effects are considered.

For example, through laboratory evaporation studies Schmidt *et al.* (2002a) determined that the $T/n\text{-}C_8$ ratio remains diagnostic of the original gasoline up to a 50% mass loss due to evaporation. These authors acknowledge that site-specific considerations warrant caution in applying this age-dating approach, particularly in instances of multiple

releases or excessive weathering. For example, weathering due to water washing would artificially decrease the ratio due to the preferential partitioning of toluene from the NAPL. Some added caution may be appropriate since it may be difficult to recognize an environmental impact due of premium (toluene-augmented) gasolines *a priori*.

Another approach to age-dating that should be mentioned herein is the reliability of comparing gasoline-derived contamination to any one particular "reference" gasoline of a known or presumed age. One example is the comparison of gasoline contamination to the reference gasoline provided to researchers by the American Petroleum Institute, known as "PS-6." The history and chemical composition of PS-6 was recently reviewed by Wade (2003). Wade critiques the inference that the PS-6 gasoline represents an unleaded gasoline produced in 1984/1985, and that any gasoline-derived contamination that is chemically similar to PS-6, must be from this time period. After a thorough review of the historical documents associated with the production of PS-6, Wade (2003) concluded that the PS-6 gasoline represents an artificial blend of 6500 gallons of gasoline produced in late 1977 using four individual blending gasoline components – light and FCC naphthas, reformate, and alkylate – whose production date back to the early 1970s. The specific origins of each blending component are unclear and therefore Wade (2003) concludes that PS-6 should not be considered representative of any particular age of gasoline.

Wade (2003) further argues that no single gasoline should be considered a representative formulation for any particular era. Although we would agree that no single parameter (e.g., relative abundance of *n*-propylbenzene) is likely to be reflective of any particular vintage gasoline, we disagree that the overall refining/blending characteristics cannot be attributed to a certain era. For example, an NAPL containing a significant straight-run gasoline component, little or no alkylate component, and with an elevated lead alkyl concentration (>2 glpg) are characteristic features of *Performance Era* gasoline (prior to mid-1970s). Oppositely, NAPLs comprised of a reformate-alkylate-based gasoline containing significant MTBE (>5 wt%) are consistent features of *Regulatory Era* gasolines. Examples of these are shown in Figures 8.6.8 and 8.7.5 and are described more thoroughly later. Thus, we contend that certain eras may

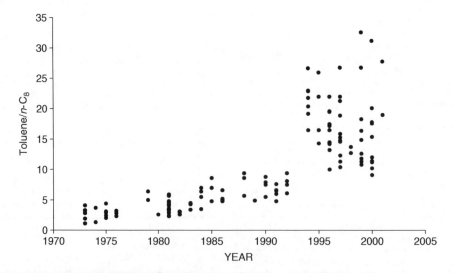

Figure 18.7.1 *Cross-plot showing the ratio of toluene to n-octane for 133 regular and mid-grade gasolines versus the year of production. Data from Schmidt et al. (2002a).*

in some instances be represented by the collective features of gasoline-derived contamination. However, we agree with Wade (2003) that because of the heterogeneity among refiners at any one time, assigning a specific age on the basis of any specific feature is unreliable.

18.7.3 Age-Dating Based on Lead Alkyl Additives

As described earlier in this chapter, gasoline has experienced numerous changes in the additives that have been used, abandoned, permitted, or banned. Certain time periods or constraints can be developed based upon the commercial availability or use for most of these additives (Morrison, 1999). The historical summaries provided in Tables 18.2.1 and 18.2.2, as well as the regulatory history described (e.g., Tables 18.2.4–18.2.6), can be combined to aid in constraining the age of some gasoline-derived contamination. However, the precision to which this can be achieved needs to be carefully considered.

In the case of lead alkyls, the presence of organic lead in gasoline-derived NAPLs does indicate the presence of an organic lead component in that NAPL. (The measured concentration should exceed 0.05 glpg, which was/is the maximum concentration permitted in unleaded gasolines.) However, the mass of the leaded component, versus any other number of components contributing to the NAPL, could range from the highest regulated concentrations to only trace levels. Additional features of the NAPL, heterogeneities among multiple NAPLs at a site (used to evaluate proportional mixing), or other facts might help determine the most likely proportion of leaded gasoline present.

With respect to age-dating, at the most basic level, the presence of organic lead indicates that (at least) there is a pre-1996 component present – since December 31, 1995, represented the last date leaded gasoline was legally available for on-road use in the United States (Table 18.2.4). In California, the presence of organic lead in NAPL would dictate the presence of a pre-1992 component, since California banned its sale earlier than the Federal government. Because most individual stations stopped carrying leaded gasoline before these final dates (often years prior), greater restraints on the latest date that leaded gasoline might have been available at any particular location might be obtained from site-specific (bill of ladings) or region-specific (NIPER survey) documents. Armed with this historical information one might be able to determine the last time leaded gasolines was available for sale at a particular location. Notably, because TEL is still used in some aviation gasolines today, its presence alone in the environment at any location where avgas might be present would not necessarily implicate the presence of leaded automotive gasoline. (The presence of avgas should be easily confirmed or refuted with conventional PIANO fingerprinting.)

Further constraint of the age of a leaded gasoline component in an NAPL might be achieved from careful consideration of the lead rollback schedule (Table 18.2.4). This approach can sometimes be useful to the forensic investigator for bracketing or constraining the age of spilled gasoline in the subsurface (Johnson and Morrison, 1996; Kaplan *et al.*, 1997). However, this approach requires that a significant number of variables that can affect gasoline lead levels in terms of both pre-spill and post-spill conditions be considered. Careful consideration of these conditions are necessary before the legal threshold of *"more likely than not"* can be confidently crossed. Simply and blindly comparing a measured lead concentration in an NAPL against the rollback schedule severely oversimplifies a very complex topic.

In a pre-spill sense, the rollback originally involved the reporting of *quarterly averages* for each refiner (see the actual 40 CFR language provided earlier in this chapter). This meant that individual leaded gasolines in the early and mid-1980s contained lead concentrations above and below the rollback schedule limits. Because of this, the "cut-off" dates for the rollback schedule are not hard-and-fast dates after which no leaded gasolines exceeding the limits were present in the market. In addition, a system of waivers permitted refiners to build up lead credits that could be "banked" for future use during the rollback. The lead banking program allowed some refiners to exceed the pool average standards; the lead banking program expired in January 1988 where after the 0.1 glpg maximum (Table 18.2.4), which had been in place since 1986, was strictly enforced. Finally, not all leaded gasolines produced throughout the *Performance Era* contained high concentrations of lead. For example, although the average lead concentrations for regular and premium leaded gasolines exceed 1.5 glpg for decades (Figure 18.2.2), some "low lead" gasolines (<0.5 glpg as per ASTM D 439) were available long before the lead rollback started.

In a post-spill sense, the concentration of organic lead may have been altered due to weathering – as is described later in this chapter (see *The Fate of Organic Lead*) – which requires careful assessment of the weathering-related features of the NAPL before comparing the concentration of total lead against the rollback schedule. Another post-spill consideration is the potential for mixing of multiple gasolines with different lead concentrations. The potential for mixing might be evident in the distributions of the individual lead alkyls (e.g., if, after considering the effects of weathering, they do not "match" a particular antiknock package; Kaplan *et al.*, 1996) or other chemical features of the NAPL (e.g., the presence of oxygenates).

Thus, considering these pre- and post-spill factors, under most circumstances the most defensible position that the concentration of total organic lead in an NAPL sample may allow is to restrict the most recent possible date for (at least one component in) the spilled gasoline. For example, if an NAPL is determined to contain >0.1 glpg it may (after considering the weathering characteristics of the NAPL) be reasonable to conclude that it (at least) contains a leaded gasoline component produced prior to January 1988 (after all banking and credits were exhausted). Further constraining the age requires thorough consideration of all of the pre- and post-spill factors discussed above. If, however, the concentration measured in NAPL is >0.5 glpg (after considering the weathering characteristics of the NAPL) it is reasonable to conclude the NAPL contains a leaded gasoline component produced sometime prior to the 1986. The basis for this is that it is unlikely that any gasoline produced after January 1, 1986, could have contained >0.5 glpg and that the refiner still met the 0.1 glpg quarterly maximum. Thus, in both these examples the total organic lead concentrations could be useful in determining (albeit conservatively) the latest possible date for a leaded gasoline contribution. However, in both these examples it is much more tenuous to defensibly define the earliest possible date due to the presence of lower lead gasolines in the market long before the rollback.

Some aid in constraining the earliest possible date for a leaded gasoline component in NAPL lies in the presence of multiple lead alkyls in an NAPL. Since TML and the various reacted and physical mixes created with TML were not commercially available until 1960 (Gibbs, 1990), their presence argues for a post-1960 leaded gasoline component in the NAPL. However, Gibbs (1990) reports that the first commercial use of the TML-containing lead antiknocks in gasoline occurred in 1960 by Standard Oil of California (now Chevron). How soon and where other refiners introduced these alternatives to TEL is unclear, although there may have been a lag period in which TEL continued to predominate in certain areas or by certain refiners. This seems likely since the reported production of TML was much lower than TEL (18 versus 494 million pounds) in 1962 (Wakim *et al.*, 1990). This disparity suggests that TML-derived antiknock mixtures were not rapidly introduced nationwide. Thus, for age-dating purposes, the presence of lead alkyls other than TEL in NAPL conservatively indicates the presence of post-1960, though perhaps slightly later, leaded gasoline component.

On the "back end" of this approach is the perception that after 1980 RM and PM lead packages were no longer in use and that leaded gasolines once again, only contained TEL (e.g., Hurst *et al.*, 1996). However, in our experience, which included access to confidential records in the course of litigation, too many exceptions exist to defensibly utilize 1980 as a "cut-off" after which lead alkyls other than TEL were not present in leaded gasolines. However, it is likely that the use of TML-containing lead mixtures was in decline after 1980. For example, it is reported that by 1985 90–95% of the lead used in US gasoline was TEL-only (Wakim *et al.*, 1990). Thus, the presence of lead alkyls other than TEL in NAPL *suggests* the presence of a leaded gasoline component older than the mid-1980s. More specific information (records) for a refiner or region might help to constrain this age more defensibly. Because TEL-only antiknocks have been used for decades, and right up until the 1996 ban on leaded gasoline, the presence of TEL-only in NAPL provides no basis for constraining the age of the NAPL.

Finally, the trend in the national average concentration of lead in regular and premium gasoline (Figure 18.2.2) provides a tempting basis for estimating the age of NAPL. However, this approach is unlikely to be a defensible basis to determine the specific age of gasoline-derived NAPL. At a minimum, the data provided in the (Dickson *et al.*, 1987) or (Shelton *et al.*, 1982) reports are national averages and no measures of each year's population statistics are provided therein. Without more detailed data it is difficult to evaluate the variability represented by the national average trend (Figure 18.2.2). However, more detailed regional and annual data are contained within annual reports by the Department of Energy's Bartlesville Energy Technology Center. For example, Shelton (1980a,b) provides data that indicates leaded gasoline sold in the New York district in 1979 contained between 0.51 and 3.40 glpg. This wide range of lead concentrations in one region in one year demonstrates the inappropriateness of using national average lead concentrations (as in Figure 18.2.2) to draw conclusions regarding the specific age of any given NAPL. At a minimum, any attempt at this approach requires that the regional data for an area of interest be evaluated using appropriate population statistics to provide some measure of the variability in lead concentrations within that region over time. With this more detailed information in hand it may be possible to conservatively restrict the age of an NAPL on the basis of total organic lead concentration.

18.7.4 Age-Dating Based on MTBE

The detection of MTBE in environmental samples (at concentrations in excess of non-point sources, e.g., >10 µg/kg in water; Pankow *et al.*, 1996) would similarly indicate the presence of an MTBE-laden gasoline component. While it is true that MTBE was first used in gasoline in 1979 (Table 18.2.2), its use was not widespread and did not spread nationally until the mid-to-late 1980s. For example, MTBE's first documented use on the East Coast reportedly was in 1982 whereas it first use in California was not until 1986 (Squillace *et al.*, 1996; Davidson and Creek, 1999). Other reports indicate MTBE's first appearance in East Coast gasoline was earlier than 1982 (e.g., Garrett *et al.*, 1986 in Morrison, 2000a). Thus, precise dates of use are not well established in published documents. Litigation in specific regions may generate specific company records that might reveal whether and when MTBE was being added to a specific gasoline grade in a specific market. With respect to grade, it was more common that MTBE first appeared as an octane booster in premium unleaded gasolines that were being introduced by various refiners in the early 1980s— rather than in other grades of unleaded gasoline.

Oil and Gas Journal annually published refining records can provide some insight as to the "*who, when, and where*" of MTBE production. This information might serve as a proxy for whose gasoline probably contained MTBE. In the course of litigation, company records (purchase records, bills of lading, etc.) might better constrain when MTBE was first used in certain markets. The same due diligence approach is necessary for other ether oxygenates (e.g., TAME)— some of which are more easily traced to particular refiners since there were fewer of them manufacturing TAME than MTBE. With the ban of MTBE in many markets on certain dates (Table 18.2.6), the presence of MTBE in NAPL, soil, or groundwater might be constrained to a certain period of time. For example, the presence of MTBE in states with MTBE bans would dictate the presence of a pre-ban component. Careful research, however, is necessary to cautiously constrain the age of gasoline contamination when specific records are absent due to the complexities of the gasoline distribution network.

In addition to lead alkyls and oxygenates, the additives that were used at only trace levels in gasolines may present another opportunity for age-dating. These additives include various antioxidants, detergents, dyes, etc. (Table 18.4.1). The current theories surrounding the use of these additives in environmental forensic investigations is that they are inherently unstable in the environment due to the presence of *N*- and *O*-containing moieties and/or conjugated C=C bonds (Kaplan *et al.*, 1997). As such they are not anticipated to persist in the environment. To our knowledge there are no published accounts concerning the rate(s) of deterioration of gasoline antioxidants, detergents, or dyes. Certainly contributing to the problem of "additive fingerprinting" is the inherent difficulty of analyzing for compounds of unknown compositions or molecular weights. This results from the proprietary nature of many of these additives.

Youngless *et al.* (1985) indicated that some gasoline additives were prone to sorption in soils, thereby confounding analysis of the polar fraction of soil extracts by GC/MS. However, more sensitive analytical techniques have become available (see *Gasoline Dye Analysis*, earlier in this chapter) and some reconsideration of the utility of additives in forensic studies, particularly age-dating, may be warranted. It is likely, in our opinion, that the fate/rate of weathering of trace level additives is a function of the site-specific conditions—as described above for other components in gasoline. Why should additives be any different? For example, the preservation of a fuel dye recently was demonstrated to exist in NAPLs that were reportedly more than 10-years old (Rostad, 2002). Additional research on the potential of additives in environmental forensics of gasoline warrants further consideration.

18.7.5 Age-Dating Based on Stable Lead Isotopes—ALAS Model

The stable lead isotope ratios of organo-lead compounds have been used to determine the age of leaded gasoline-derived contamination (Hurst, *et al.*, 1996; Hurst, 2000,

2002). The underlying basis for age-dating leaded gasoline in the environment using the anthropogenic lead archeostratigraphic (ALAS) model lies in the increase in lead isotope ratios (in particular, ^{206}Pb$/^{207}$Pb) caused by a systematic change in the geologic source of lead used by the manufacturers of lead alkyl antiknocks over time (Hurst *et al.*, 1996; Hurst, 2000), which reportedly affected the lead isotope ratios of the lead alkyls in the antiknock packages.

To demonstrate this change over time, Hurst (2000) had assembled a suite of calibration samples that were used to develop the ALAS model. Unfortunately the numerical lead isotope data for the calibration samples have not been published. The data have been presented in figures in which only a subset of the reported calibration samples (e.g., 41 of 76 calibration points) can be resolved (see Figure 1 from Hurst (2000), which we have re-created in Figure 18.7.2 using planimetry. As can be seen there is a remarkably precise "sigmoidal" shift in the isotopic composition over time conveyed by the calibration samples.

Hurst (2000) state that the calibration samples' absolute ages (x-axis in Figure 18.7.2) are known to be accurate to within <0.5–1 year. However, 51 of the calibration samples were NAPLs collected from contaminated sites, the ages of which were determined via "*environmental releases documented via legal records*" that included fire department records, site inventory records, and repair records (see the appendix to Hurst, 2000). Our experience has demonstrated that the use of such records in determining the absolute ages of releases at UST sites rarely would permit accuracy to a <0.5- to 1-year timeframe. The reason for this is that most UST releases are long-term occurrences and that the "legal records" will reflect when the release was discovered and/or reported to authorities—not the time or duration over which the release actually occurred.

In addition, Philp (2002) provides some relevant comments on the issue of lead ore supply as a "driver" of the ALAS model's isotopic shift. He notes that the gradual, annual shift in lead isotope ratios in the ALAS model implies Mississippi Valley (MV) lead was being systematically mixed with lead from other ore bodies. However, Philp (2002) suggests that lead ores from multiple ore bodies were probably being purchased on the spot market thereby introducing monthly variations in supply, which he suggests makes such a systematic annual change in the lead isotope ratio for all gasoline produced in a given year unlikely.

Support for this contention lies in lead isotopic data produced for gasolines in the 1960s and 1970s by Chow *et al.* (1975). These authors noted "*lead additives in various brands of gasoline sold in one region do not have identical lead isotope ratios. This reflects the fact that petroleum refineries obtain the additives from different lead alkyl manufacturers which utilize lead from different mining districts.*" They go on to state "*the same brand of gasoline sold in various regions of a country may also vary in lead isotope ratios. For example, our 1964 data show that a brand of gasoline sold on the Atlantic coast of the USA contained lead which was more radiogenic than that of the same brand of the Pacific coast.*" The Chow *et al.* (1975) data indicate that isotopic variations clearly existed for gasoline collected at the same time from different suppliers and different locations.

We have compiled the published lead isotopic data representing urban aerosols attributable to leaded gasoline combustion in California (Rabinowitz and Wetherill, 1972; Hirao and Patterson, 1974; Chow *et al.*, 1975; Smith *et al.*, 1992), Texas (Manton, 1977; Manton, 1985), Rhode Island (Sturges and Barrie, 1987), Missouri (Rabinowitz and Wetherill, 1972), and the upper Midwest (Sturges and Barrie, 1987; Graney *et al.*, 1995), which are shown in Figure 18.7.3. The

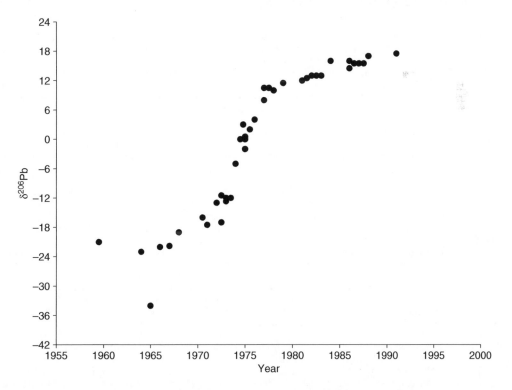

Figure 18.7.2 *Reproduction of the ALAS Model's calibration data for 41 samples from Figure 1 in Hurst (2000) and Hurst (2002). Data points determined by planimetry.* δ^{206}Pb $=[1000 \times^{206}$Pb$/^{207}$Pb$)]/1.1966$ *after Hurst (2000).*

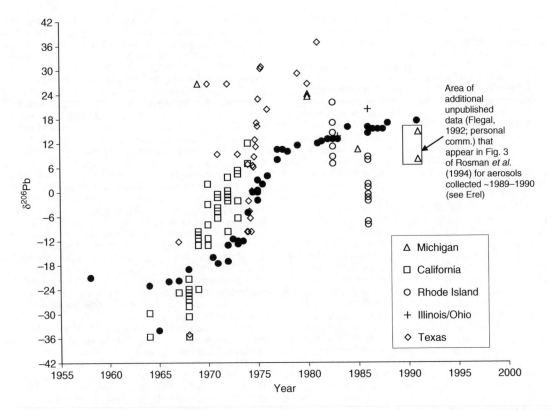

Figure 18.7.3 *Comparison of stable lead isotope ratios (as $\delta^{206}Pb$) in published urban aerosols versus the ALAS model curve (solid black).* $\delta^{206}Pb = [1000\times \ (^{206}Pb/^{207}Pb)]/1.1966.$

original ALAS calibration curve (as per Figure 18.7.2) is included for comparison. The published aerosol data show that: (1) there is a recognized shift in the $\delta^{206}Pb$ between 1965 and 1980, (2) however, the relationship between age of the gasoline reflected in the urban aerosol data and $\delta^{206}Pb$ is not nearly as precise as conveyed in the ALAS model, (3) distinct regional variations appear to exist in different parts of the United States, and (4) the overall "sigmoidal" shape of the ALAS model is incorrect. In particular regard to the latter, data compiled for urban aerosols after 1980 indicate a reduction in the $\delta^{206}Pb$ of atmospheric lead attributable to gasoline combustion clearly occurred (Veron *et al.*, 1992; Erel and Patterson, 1994; Rosman *et al.*, 1994). This would seem to confound the ability to age-date gasoline-derived contamination over most of the ALAS model's applicable range 1965–1990 (Hurst, 2000). For example, an NAPL with a $\delta^{206}Pb$ of ~6 would indicate an age of both ~1975 (ALAS) and 1985 (aerosol decline after 1980).

Hurst (2000) argues that such comparisons of historic $^{206}Pb/^{207}Pb$ data to ALAS model are inappropriate because the historic data suffer from lower analytical precision and lack of correction versus a lead standard. However, even considering that some error may exist, to dismiss the wide variation in the data and the overall trends they depict, particularly the decline in $\delta^{206}Pb$ after 1980 and the regional variations, is unwarranted.

Published lead isotope ratio data also exists for dispensed gasolines from the 1960s and 1970s (Chow and Johnstone, 1965; Chow, 1970). After converting to $\delta^{206}Pb$ these data are plotted in Figure 18.7.4, along with the ALAS calibration for comparison. These data clearly show considerable variation in $\delta^{206}Pb$ existed in dispensed leaded gasolines: (1) within

a given year, (2) within the same geographic region (e.g., California samples in Figure 18.7.4), and (3) between different geographic regions within the same year. All three of these conclusions are contrary to the ALAS model. In fact, data for leaded gasoline from early 1990s exhibit variable $\delta^{206}Pb$ values (Hurst *et al.*, 1996) that would appear contrary to the model.

It could be claimed that these historic gasoline data suffer from analytical scatter inherent in the measurements. However, this explanation would not account for the observed variability within any given sample set. For example, the data reported by Chow and Johnstone (1965) for six San Diego (California) gasolines collected in May 1964 demonstrate considerable $\delta^{206}Pb$ variability (Figure 18.7.4), yet all of these data were produced in the same laboratory under the same conditions (therefore introducing the same level of analytical scatter). Thus, the relative differences observed among these gasolines' lead isotope ratios cannot be reasonably dismissed as attributable to analytical scatter alone. Hurst (2000) does not acknowledge these historic data nor explain the variations they depict.

Notably, one of the historic gasolines analyzed by Chow *et al.* (1975) was later provided to Hurst (2000) for inclusion in this ALAS calibration sample suite. A 1965 leaded gasoline provided by Dr. Chow reportedly had a $\delta^{206}Pb$ value of ~−34 when analyzed by Hurst (2000). Chow *et al.* (1975) reported a $^{206}Pb/^{207}Pb$ ratio of 1.151, which corresponds to a $\delta^{206}Pb$ of −38.11. The original Chow *et al.* value is only about 12% lower than the value reported by Hurst (2000). Thus, if only ~12% error is attributable to the historic nature of the published data (Chow and Johnstone, 1965; Chow, 1970; Chow *et al.*, 1975), then the scatter observed in the pub-

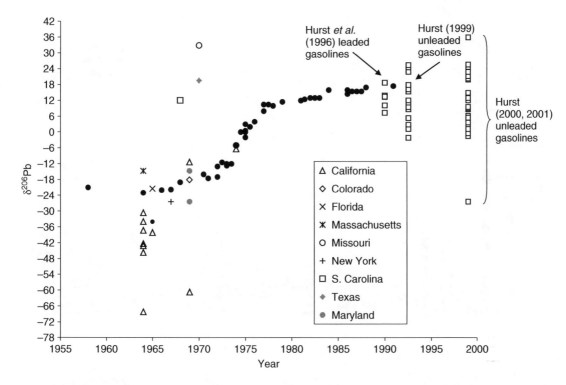

Figure 18.7.4 *Comparison of stable lead isotope ratios (as $\delta^{206}Pb$) in published dispensed gasoline versus the ALAS model curve (solid black).* $\delta^{206}Pb = [1000 \times (^{206}Pb/^{207}Pb)]/1.1966$.

lished data for dispensed gasolines in the 1960s far exceeds that conveyed in the ALAS calibration (Figure 18.7.4). This raises a significant question with respect to the accuracy of the ALAS model's claim of consistency among leaded gasolines.

Philp (2002) notes a potential problem of the ALAS model that arises due to the potential for mixing of different vintages of leaded gasoline at a site, thereby distorting the age estimated based on the ALAS model. Hurst (2000) argues that commingling of free phase gasoline of different ages in the subsurface occurs too slowly to produce homogenized lead isotope ratios in a NAPL accumulation. However, Hurst (2000) also argues against the use of lead concentrations for estimating the age of NAPL because the measured concentrations may vary, at least in part, "due to mixing." These two positions—that mixing of younger and older leaded gasoline-derived NAPLs can alter the commingled NAPL's lead concentrations but not affect its lead isotope ratios—seem contradictory.

Finally, Kaplan (2003) has provided a detailed independent assessment of the ALAS model. After considering data specific to southern California, he concluded that there is a predictable relationship between the $\delta^{206}Pb$ ratio and the age of gasoline used in the southern California market between 1964 and 1983. However, he raises numerous questions with respect to the applicability of the ALAS model nationwide.

18.7.6 Age-Dating Based on Groundwater BTEX Ratios

Kaplan *et al.* (1997) and Kaplan (2003) described a method for age-dating gasoline-derived contamination on the basis of the ratio among BTEX compounds in groundwater. The ratio of:

$$R_b = (B + T)/(E + X)$$

was purported to vary at a predictable and exponential rate in groundwater near a source due to benzene's and toluene's greater aqueous solubilities compared to ethylbenzene and xylenes. The basis for this empirical relationship includes field and laboratory data that produced a best-fit equation:

$$R_b = 6.0 \ \exp(-0.308T),$$

where T is equal to the time since release in years (Kaplan *et al.*, 1997). Although the authors warn of the effects of site-specific conditions and the benefit of site-specific temporal data in improving this approach, the published "equation" is highly tempting to an unwary investigator with BTEX data.

This age-dating approach had been critically evaluated by Alvarez *et al.* (1998), who concluded that it was an oversimplified approach that was valid in only rare, well-constrained situations. One factor that will affect this ratio, which we have discussed herein, is the varying proportions of the BTEX constituents in the parent gasoline—e.g., the prominence of toluene in higher octane reformate-based gasolines (see Figure 18.7.5). The Kaplan *et al.* (1997) equation dictates a starting $(B + T)/(E + X)$ ratio of 6.0 for all gasolines, which is clearly not the case. Furthermore, the multitude of site-specific factors that affect the rate of weathering for any given site (see *Gasoline Weathering* above) would seem to preclude the universal application of any one empirical relationship at all sites. Alvarez *et al.* (1998) demonstrated this through the use of a gasoline–water partitioning model that

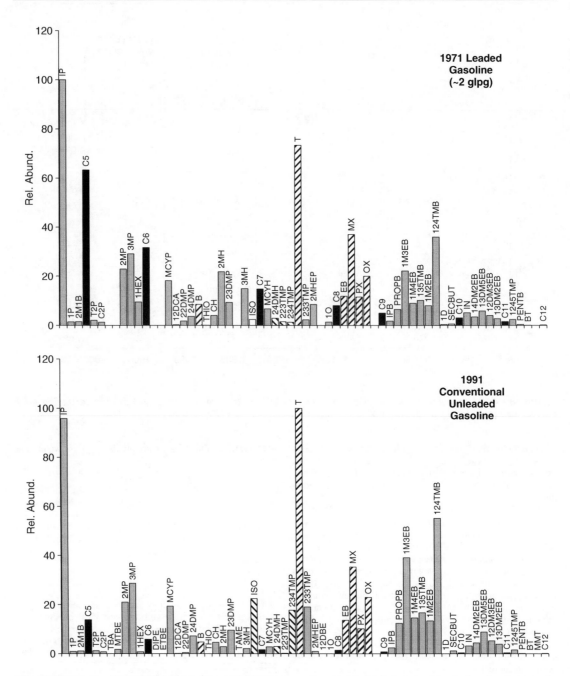

Figure 18.7.5 *The PIANO histograms for a 1971 leaded gasoline and a 1991 (non-oxygenated) unleaded gasoline showing the greater abundance of straight-run gasoline in the 1971 gasoline, as evidenced by the higher relative abundance of n-alkanes (black bars). See Table 18.5.2 for compound identifications.*

extended laboratory-measured BTEX ratios vary over time. They further discuss the confounding effects other weathering phenomena (e.g., volatilization, biodegradation, sorption, and retardation) can have on BTEX ratios in groundwater, in both temporal and spatial sense. In conclusion, Alvarez *et al.* (1998) state that "one should not use the relative concentrations of individual BTEX compounds in ground water to determine the age of a petroleum product release reliably."

18.8 SPECIFIC FORENSIC CONSIDERATIONS

In the following sections, several specific and (we think) interesting issues are discussed that can have relevance in environmental forensic investigations of gasoline impacts.

18.8.1 Genetic Influence of Crude Oil Feedstock on Gasoline Composition

Stout *et al.* (2002) described the three *controlling factors* that influence the interpretation of hydrocarbon fingerprinting data—primary, secondary, and tertiary. The primary controls are those features of hydrocarbon fingerprinting data that are inherited directly from the parent form of the contamination, i.e., these features might be considered "genetic" to the contamination. For instance, the chemical compositions of crude oils around the world are highly variable due to the conditions under which they have formed, migrated, and accumulated prior to their discovery and production (Tissot and Welte, 1984). The secondary controls are those features imparted in the course of anthropogenic processing of the crude product; e.g., petroleum refining in the case of gasoline. Some of the effects of the different refining processes were described previously in this chapter. The tertiary controls are those features imparted after a contaminant has been released into the environment, i.e., weathering, which was also previously discussed. In this section, we describe the first controlling factor's affect on gasoline composition.

As was described earlier in this chapter, gasoline is a blend of refinery streams that have been produced and altered in various ways from the raw crude oil feedstock. The degree to which any of the "genetic" features of the crude oil feedstock are retained during this process is a function of the severity of the alteration. The least alteration of crude oil's genetic features occurs during the distillation step. As described previously, distillation simply separated the components of the feedstock into different fractions based upon the boiling points.

The most important distillate fraction used in the production of automotive gasoline are straight-run gasolines (SRG; e.g., Figure 18.3.1). In some regions natural gas production also yields natural gas liquids (NGL), sometimes called gas condensate or "natural gasoline" that can be used in blending gasoline. The amount of SRG or NGL used in the production of automotive gasoline has changed over time. Prior to the 1970s (i.e., during the Performance Era; see above) gasoline blending was less complex than during the Regulatory Era. That is to say that there was less need

for complex blending of many different blending stocks in order to achieve the desired performance—i.e., refiners could more easily and cost-effectively adjust octane by the addition of lead antiknocks. In fact, many "simple" refineries existed in the Performance Era but were forced to close during the Regulatory Era due to their inability to produce suitable gasoline without the use of lead additives (Beall et al., 2002). As such, Performance Era gasolines typically contained a higher proportion of SRG or NGL than later gasolines. This difference is exemplified in Figure 18.7.5 which shows the PIANO distributions of two gasolines representing each "era." The 1971 gasoline (obtained from a re-discovered 1971 UST that still contained gasoline) contains a greater relative abundance of straight-run gasoline, as evidenced by the greater relative abundance of n-alkanes. Octane gain in the 1971 gasoline was significantly supplemented using organic lead at about 2 glpg. The 1991 gasoline, which contained only a small amount of MTBE as an octane booster, achieved an 89 octane by blending proportions of reformate and some alkylate—and less straight-run gasoline.

Simple gasolines containing a higher proportion of straight-run or natural gasoline would contain a higher proportion of n-alkanes and naphthenes (e.g., methylcyclohexane; see Figure 18.3.1) than more complex gasolines blended using multiple refinery streams.

Another consequence of using a greater proportion of straight-run or natural gasoline is the "sharpness" of the gasoline end-points. Variations in the specific end-points employed during distillation or gas production will produce straight-run or natural gasoline blending stocks with slightly different boiling distributions, which alter the shape and extent of the n-alkane envelope. In gasoline, this can affect the proportion of $C_{13}+$ compounds present in gasoline, i.e., the "tail." An example of this feature is shown in Figure 18.8.1.

The prominence of straight-run gasoline or natural gasoline in some gasolines is also important to forensic investi-

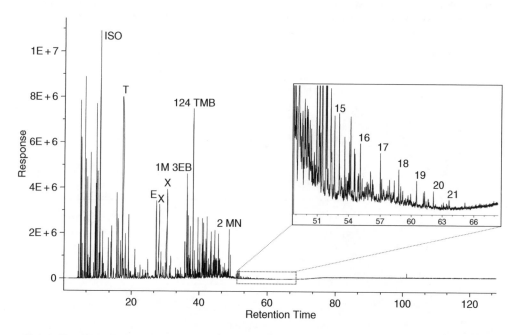

Figure 18.8.1 The GC/FID chromatogram of a NAPL comprised of a weathered gasoline with a $C_{13}+$ "tail" that extends into the diesel range. This feature is typical of gasolines containing a significant straight-run gasoline component, which was more common in performance era gasolines.

gations because these resulting gasolines will retain certain "genetic" features of the parent crude oil feedstock or natural gasoline liquids. Since, for geologic reasons, these can exhibit different genetic features, certain of these features can be transferred to the gasoline.

For example, Stout and Douglas (2004) described the use of diamondoids in gasoline fingerprinting. Diamondoids are a class of naturally occurring, saturated hydrocarbons in petroleum that consist of three or more fused cyclohexane rings, which results in a "diamond-like" structure (Fort and Schleyer, 1964). The diamondoids that can be found in natural gasoline, straight-run gasolines, and finished petroleum products (e.g., automotive gasoline) include adamantane (boiling point ~190°C) and diamantane (boiling point ~272°C) and their various substituted equivalents (Stout and Douglas, 2004). These naturally occurring compounds are extremely resistant to weathering, are unaffected by the distillation process, and as such their distribution and relative abundance in environmental samples can be useful in the chemical fingerprinting of gasoline, particularly in gasolines containing a significant straight-run or natural gasoline component (Stout and Douglas, 2004).

Another "genetic" feature that can be inherited from the parent crude oil feedstock is the distribution of C_7 isoparaffins. Mango (1987) described the remarkable invariance in the distributions of selected isoheptanes (i.e., methylhexanes and dimethylpentanes) in a large number (~2000) of crude oils due to a steady-state catalytic process during oil formation. The so-called "K1" ratio, defined as:

$$\frac{(2\text{-methylhexane} + 2, 3\text{-dimethylpentane})}{(3\text{-methylhexane} + 2, 4\text{-dimethylpentane})}$$

was shown to remain near equity in crude oils and gas condensates (~1; Mango, 1987). It would be anticipated that gasolines containing a significant straight-run or natural gasoline component would be expected to approach a value of 1.0 whereas gasolines containing a significant alkylate fraction (in which additional isoheptanes were created and added to the gasoline) might have K1 values that vary from equity. Figure 18.8.2 shows the distribution of K1 values

for 98 gasolines and gasoline-derived NAPLs. Several racing gasolines exhibit K1 ratios that are markedly higher than typical gasolines. The high values in these racing gasolines is readily attributed to the addition of a significant alkylate component (which is necessary to achieve their 100+ octane values). However, most gasolines exhibit only subtle (<2-fold) increases in the K1 from equity. This increase suggests that most gasoline in this sample suite probably contains some amount of alkylate, which has increased the values above the value expected in straight-run gasoline (~1). Alkylate is typically enriched in 2,3-dimethylpentane (relative to 2,4-dimethylpentane; see Figure 18.3.4). Thus, the addition of alkylate to gasoline has resulted in an increase in the K1 ratio above equity in most gasolines and gasoline-derived NAPL (see Figure 18.8.2). Based upon this result, the use of K1 parameter may have application in distinguishing gasolines containing variable (including no) alkylate components.

18.8.2 Gasoline Distribution Practices

In 2002 there were approximately 170,000 gasoline retailers in the United States. This number represents a reduction from the 1960s when nearly 220,000 service stations existed. This reduction has been accompanied by a general change in the marketing strategy in which low-profit service stations have been closed while new gasoline retail outlets associated with other businesses are developed (e.g., grocery stores, warehouse clubs). Getting the gasoline to all these retail outlets from the refineries represents a significant logistical effort. Domestic transport of gasoline through pipelines, on barges, and by railcar and trucks has been used for decades. The national product distribution system's infrastructure was mostly developed when far fewer formulations of gasoline were being used. Gasoline distribution through this system became much more complex in the 1990s as many more types of gasoline (Oxyfuels, RFG, and conventional unleaded gasolines) destined for distinct markets at specific times required transport. Some understanding of this complex distribution process can be useful in approaching gasoline forensic issues in which the supply

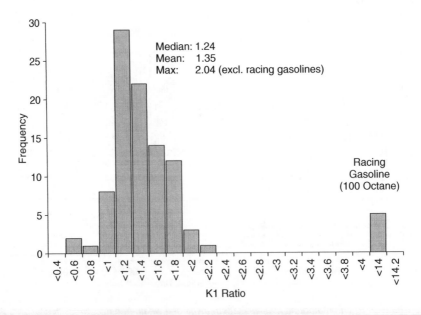

Figure 18.8.2 *Population statistics for 98 gasolines and gasoline-derived NAPLs for the K1 ratio (2 MH + 23 DMP)/ (3 MH + 24 DMP) as per Mango (1987).*

of gasoline to a region or individual retail outlet is relevant. In this section, this process is briefly reviewed.

Table 18.8.1 summarizes some gasoline production statistics for different Petroleum Administration for Defense Districts (PADDs) within the United States. These data were compiled from the Department of Energy's annual publication of petroleum use (www.eia.doe.gov/oil_gas/petroleum/data). Some interesting features of these data are the fact that PADD I (the Mid-Atlantic and New England) requires that 66% of the gasoline consumed in that district be brought in from other parts of the country or from outside the US Although PADD I receives about 94% of all US gasoline foreign imports, most of this PADD I's imported gasoline comes from other PADDs, particularly PADD III (Gulf Coast). Transport of gasoline from the Gulf Coast to the East Coast and Northeast is accomplished through multiple, interstate product pipelines that carry gasoline and other products. The foreign import of gasoline to the East Coast is notable as it represents about 15% of the gasoline consumed in PADD I.

The Mid-West also runs a gasoline "deficit" and imports about 23% of the gasoline consumed there (almost exclusively) from other PADDs—again, mostly from PADD III (Table 18.8.1). Similar to the situation in the Mid-Atlantic and Northeast, product pipelines carry gasoline up from the Gulf Coast toward the Mid-West's population centers. Further inspection of Table 18.8.1 shows that the Rocky Mountain and Western districts require much smaller amounts of gasoline produced outside of the districts.

Overall, it is estimated that about 60% of all US gasoline is shipped to the market by product pipeline (Gibbs, 1989). Some pipelines are company-owned while others are common carriers. The point of origin of gasoline carried through company-owned pipelines is more easily unraveled (with the right documents) than the sources of gasoline transported through common carrier pipelines.

The transport of gasoline through pipelines occurs in "batches." These batches of distinct products may or may not be physically separated from one another. Physical separation can be attained through "spheres" inserted between batches. In successive batches without spheres, the difference in density between the two batches maintains the separation (under pressure and turbulent flow; Kennedy, 1993). Notably, since the late 1940s pipeline companies often add carboxylic acids to transported fuels as corrosion inhibitors and drag reducing agents (Gibbs, 1989).

Some of the gasoline transported by common carrier pipeline is considered "fungible," meaning that it is interchangeable with the next "batch." However, the degree to which gasolines are fungible is lower than in the past when fewer gasoline types were manufactured for larger markets. Thus, nowadays there is a better chance to determining the original refinery source of any particular gasoline transported through common carrier pipelines than in the past. However, the temporal complexities of the market can render specific determination of the source refinery tenuous.

Pipelines typically carry gasoline to petroleum terminals where the products are stored prior to local distribution by trucks. Some terminals ("jobbers") supply gasoline to multiple "unbranded" retailers while others (oil company-owned terminals) are devoted to a single gasoline retail brand. Jobbers typically obtain gasoline from multiple sources. In either case, before gasolines are distributed to service stations they will receive their final additive package upon transfer to the delivery trucks (Gibbs, 1989). This will include detergents, deposit control additives, and demulsifiers that are specific to different grades and manufacturers. This final step imparts the "brand" of the particular gasoline.

Table 18.8.1 *Inventory of US gasoline refining and transport by PADD for 2003 (from Energy Information Administration, 2003). All units in 1000 bbl*

	PADD					US Total
	I—East Coast	II—Mid West	III—Gulf Coast	IV—Rocky Mtn.	V—West	
Operating Refineries	14	26	54	16	36	146
Total Gasoline Production	388,787	655,450	1,307,805	104,109	534,798	2,990,949
Reformulated Gasoline	242,676	128,641	233,416	0	386,118	990,851
Oxygenated Gasoline	14,581	192,311	5,015	13,149	35,922	260,978
Conventional Gasoline	131,530	334,498	1,069,374	90,960	112,758	1,739,120
Gasoline Export to other PADDs	76,650	25,614	816,908	14,790	785	934,747
Gasoline Import from other PADDs	652,791	224,543	11,351	17,243	28,819	934,747
Foreign Imports	177,930	681	2,353	195	7,869	189,028
Foreign Exports	1131	378	41,678	2	2580	45,769
Net Consumption	1,141,727	854,682	462,923	106,755	568,121	3,134,208
Gasoline Deficit/Surplus[a]	−752,940	−199,232	844,882	−2,646	−33,323	−143,259
%Gasoline consumed produced outside of the PADD (or US)	66	23	na	2	6	5

[a] Total Production − Consumption.

Sometimes even unbranded gasolines, i.e., gasolines purchased by retailers that are not obligated to sell any particular brand of gasoline, will contain a "branded" additive package. However, this is not always the case, particularly when unbranded stations are purchasing gasoline through jobbers. The complexity of "whose additives are present in which gasolines" is a hurdle for the forensic investigator that requires due diligence of records—usually only available during litigation, if at all. Furthermore, as described above, the chemical analysis of the brand-specific additives (detergents, deposit control additives, and demulsifiers) is a prospective research area.

Notably, shipment of gasolines containing ethanol through pipelines is problematic due to ethanol's susceptibility to partition into water within the pipeline system. Thus, ethanol (and other alcohols) often is transported separately (i.e., as its own "batch") from the unfinished gasoline and then blended at terminals prior to distribution via trucks. On the other hand, even though ethers are partially soluble in water they are added at the refinery and transported via pipeline, barge, and truck.

Gasoline without final additive packages (typically proprietary to each major oil company) is exchanged or traded among producers to meet contract and demand requirements as well as to improve transportation logistics. As a result of exchange agreements, a branded retailer may sell a competitor's gasoline that contains different additives than the producer would add at its own refinery. Thus, an individual oil company's recorded use of particular additives may not be indicative of all of the additives present in fuel at that company's service stations.

Determining what types of gasolines were used at a particular service station can only be achieved through careful records research. Retailers typically retain delivery records (bill of lading) that provide information regarding the supplier and some basic compositional information, such as: "Reformulated gasoline: This product contains no ethanol; contains MTBE or other ethers" or "Oxygenated gasoline: contains 15–17 vol% MTBE." This information might prove useful in some circumstances in trying to determine the homogeneity or heterogeneity of gasolines dispensed over a specific time period. Establishing the volumes of different gasoline types used at a particular service station over a certain time through delivery records research *may* yield a sufficient basis to determine certain compositional features and volumes of different gasolines that were dispensed.

18.8.3 Gasoline Releases at Retail Sites

Forensic investigations at retail sites require some understanding of where and how gasoline releases may have occurred. In this section, some considerations of the nature of releases are described. This information can provide a basis for developing a sampling strategy for a forensic investigation.

It is likely that most gasoline releases at retail sites are caused by customers. We suspect that accidents (drive offs), negligence (spillage), and problems with the hanging hardware are common occurrences. Most such releases, however, usually only permit small volumes of gasoline to be released above ground. These circumstances would not appear to have significant environmental impact. However, considering accidental drips and drops of gasoline may go on for decades, it is conceivable that some impact may result, particularly in shallow soils near dispenser islands.

The most notorious impacts from gasoline releases at retail sites are those which occur underground and go unnoticed for long periods of time. Such releases are the result of compromised equipment, either USTs or the associated piping, including valves, elbows, joints, flanges, and flexible connectors. A survey of gasoline releases in the late 1980s indicated that, contrary to perceptions at the time, most releases at service stations occurred from piping—and not from leaking USTs (United States Environmental Protection Agency, 1987).

Historically, USTs were subject to spills from overfilling of the tanks and from leaks caused by corrosion. The propensity for these spills and leaks has been greatly reduced with improvements in technology and practice fostered by implementation of the 1998 deadline for compliance with new Federal UST standards. However, even modern leak detection systems cannot detect leaks less than 0.1 gallon per hour.

Steel USTs were the common form of gasoline storage at retail sites for decades. These suffered from corrosion problems due to water that accumulated in tank pits or within the tanks themselves, which over time corroded the tank wall leading to "pinhole" leaks. The survey conducted in the late 1980s had revealed that most (72%) of the more than 24,000 steel tanks that failed did so after 10–20 years of service (United States Environmental Production Agency, 1987). Notably, Morrison (2000a) provides a review of the various UST corrosion models that are used to predict corrosion rates of steel USTs. Steel tanks in use today are typically protected from corrosion by cathodic protection systems, as per the API Recommended Practice 1632, "Cathodic Protection of Underground Petroleum Storage Tanks and Piping Systems." This functions when a small electrical charge is applied to the tank, which helps prevent the steel from dissolving.

Underground storage tanks made of fiberglass were first introduced in the late 1970s. These were quickly shown to be problematic because they lacked metal striker plates. As a result, during manual stick-gauging holes were commonly and sometimes unknowingly poked in the tanks leading to "catastrophic" releases. Around 1990, double-walled fiberglass tanks were introduced. These USTs were equipped with a conducting brine between the walls that functioned as part of a sophisticated leak detection system. Most USTs in use today are double-walled fiberglass. However, some older tanks (steel and single-wall fiberglass tanks) are still in use.

Underground piping has also improved. Historically, underground piping at retail sites were also constructed of steel, and thus were prone to corrosion. More importantly, however, these systems had many pipe joints/elbows as they snaked around the property, each joint being a potential leak point. Newer piping systems are often constructed using flexible piping that reduces the number of joints (i.e., potential leak sites).

Although catastrophic losses do occur, most releases of gasoline from USTs or the associated piping systems at retail stations more commonly occur at low rates for long periods of time. Periodic testing of USTs ("tank tightness") or piping tests are designed to detect pressure losses or volume reduction during the course of the test. These tests are temperature sensitive, due to the effect of temperature on pressure and volume, which can affect the precision of these tests. As such, these tests are designed to detect leaks larger than 0.1 gallons per hour.

18.8.4 Chemical Fingerprinting of Dissolved Phase Contamination

Water-soluble gasoline hydrocarbons—particularly benzene, toluene, ethylbenzene, and o-, m-, and p-xylenes (BTEX) and oxygen-containing blending agents (Table 18.4.4)—are common groundwater contaminants, often found proximal to automotive gasoline service stations, bulk petroleum storage and distribution facilities, and petroleum pipelines. Virtually all States have established maximum contamination levels for these chemicals in groundwater that are extremely low (e.g., $<10\,\mu g/L$). Because of the high solubility and mobility of BTEX and oxygen-containing compounds in groundwater and the relatively high concentration of these compounds in gasoline, these action levels are trigger forensic investigations as to the source(s) of these chemicals.

The detailed chemical analysis of free phase gasoline (NAPLs) and impacted soils for the purpose of distinguishing between or among potential sources of contamination is challenging but can be readily accomplished with conventional fingerprinting data (Table 18.5.1). When impacted groundwater within monitoring wells contains sheens, small quantities of entrained oil droplets, or emulsions, conventional fingerprinting also yields detailed fingerprints that can be readily compared to NAPLs and soils. However, when a forensic investigation is faced with fingerprinting truly dissolved phase contamination only, additional issues are introduced due to the effects of partitioning on the chemical fingerprints.

The two particular issues faced in forensic investigations of dissolved phase contaminants are: (1) there is the need to acquire chemical fingerprinting data of suitable detail and spatial (aerial) extent to permit recognition of chemical trends that may have source implications (e.g., "hotspots" or other anomalies) and (2) there is the need to compare dissolved phase contamination to NAPL or residual gasoline in soil. Each of these issues is discussed in the following paragraphs. This section concludes with some comments on the sampling strategies for forensic investigation of gasoline-impacted groundwater.

Chemical fingerprinting of dissolved phase contamination derived from gasoline can be achieved using conventional PIANO fingerprinting methods and emerging methods. Conventional methods rely upon purge-and-trap GC/MS analysis of an expansive PIANO analyte list (Table 18.5.2) at low reporting limits (e.g., $<1\,\mu g/L$ in water; Uhler et al., 2003). Emerging methods include the use of stable isotopes to characterize and differentiate individual compounds dissolved in groundwater, such as MTBE or benzene, using GCIRMS (Table 18.5.1; Kolhatkar et al., 2002; Smallwood et al., 2001) or methods focused on the highly soluble polar constituents contained in gasoline, such as nitrogen-containing anilines and pyridines (Schmidt et al., 2001, 2002b).

An example of the chemical detail achievable using conventional PIANO data is shown in Figure 18.3.6. Figure 18.8.3 shows the PIANO histogram for a water sample that had been in contract with a fresh, MTBE-laden gasoline for 28 days (as per the laboratory weathering experiments described earlier in this chapter). Although the water sample's PIANO fingerprint is dominated by the highly soluble constituents (e.g., MTBE and TBA; 212 and 14.3 mg/L), there is a full suite of hydrocarbons dissolved into the water at concentrations exceeding $1\,\mu g/L$. The wide range of concentrations of dissolved volatiles in groundwater samples often results in laboratories diluting samples a priori in order to keep the higher concentration analytes within calibration. However, in doing so, minor constituents are "diluted out." Therefore, care is necessary in conducting analyses of groundwater samples so as to avoid diluting out low-level analytes with forensic value.

Collection of high-quality PIANO data on groundwater permits detailed comparisons among groundwater samples across a study area, thereby allowing recognition of compositional trends. In our experience, however, groundwaters containing extremely low concentrations of total petroleum hydrocarbons (e.g., $<500\,\mu g/L$ TPH_{gas} or

$<100\,\mu g/L$ BTEX) may not provide sufficient chemical details using conventional PIANO methodology to permit comparisons among samples. However, in such cases the detection of individual compounds at very low levels (e.g., MTBE) can still provide important forensic information.

A real-world example of conventional PIANO fingerprinting of groundwater is shown in Figure 18.8.4. This shows the PIANO fingerprints for two groundwater samples collected from wells located on adjacent service stations. The groundwater contained approximately 160 and 6.3 mg/L of total PIANO analytes. Both samples contained a full suite of detectable PIANO analytes which provided an opportunity to compare them. It was evident that the parent gasolines for the two samples were not identical on the basis of the absence (Figure 18.8.4A) or prominence (Figure 18.8.4B) of various trimethylpentane isomers—i.e., components indicating the absence and prominence of an alkylate blending component in the parent gasolines (e.g., see Figure 18.3.4). In addition, groundwater from one well (Figure 18.8.4B) contained a higher proportion of C_3+ alkylbenzenes which were likely attributable to the presence of a greater abundance of FCC gasoline in the parent gasoline at this location or the presence of a more highly weathered, additional gasoline component at this location. These two wells were part of a widespread groundwater survey that revealed comparable differences among the other groundwater samples in the area. Spatial reconciliation of these differences allowed for the distinction between the groundwater impacts attributable to the two parent gasoline types/sources.

As noted above, one of the emerging techniques for groundwater fingerprinting relies upon the stable carbon isotopic composition of individual compounds dissolved in groundwater. These studies have relied upon the GCIRMS (Table 18.5.1) of individual compounds extracted or purged from groundwater. A recent review of the abilities and limitation of this application was provided by Slater (2003), and therefore will not be discussed in detail herein. The GCIRMS studies have been directed at BTEX (Kelly et al., 1997) and MTBE (Slater et al., 1998; Hunkeler et al., 2001; Kolhatkar et al., 2002; Kuder et al., 2003) dissolved in groundwater. Many other GCIRMS studies have been directed at monitoring the degradation of these compounds, which produces a shift in the stable carbon isotopic composition due to preferential biodegradation of ^{12}C over ^{13}C (e.g., Wang et al., 2004 and refs. therein). This shift can complicate forensic interpretations to the same degree that weathering can complicate any of the fingerprinting methods. In the end, GCIRMS may hold promise as a forensic tool so long as it is used in conjunction with other fingerprinting and forensic techniques.

Of course, it is possible that PIANO fingerprints across an area will exhibit differences that are unrelated to the source of the contamination—but are instead attributable to various fractionation processes. For example, (1) partitioning due to differential evaporation (as estimated from Henry's Law constant), (2) sorption (as estimated from soil-water sorption coefficient (K_s) or organic-carbon partitioning coefficient (K_{oc}), or (3) degradation of some compounds during dissolved phase transport are expected to occur. In other words, the groundwater "fingerprints" are expected to change with distance from their source(s). In such instances, a numerical- or modeling-based reconciliation of these changes may be necessary to interpret the fingerprint differences to either source differences or fractionation processes. Thus, the use of conventional fingerprinting data in groundwater studies requires careful interpretation and should focus on those features least affected by various fractionation processes. This leads to the second issue facing groundwater fingerprinting studies.

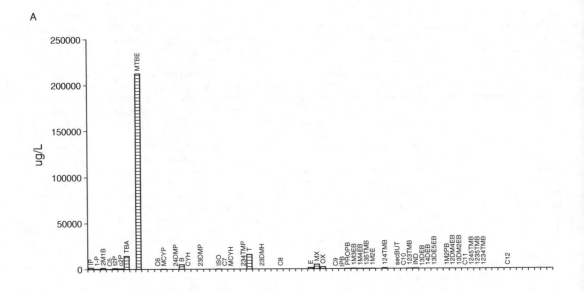

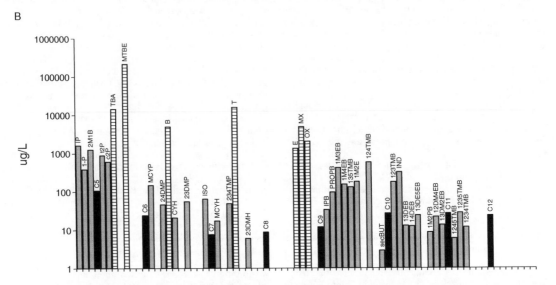

Figure 18.8.3 *The PIANO histograms showing the concentration of dissolved compounds in water after 28 days of contact with MTBE-laden gasoline. (A) normal scale and (B) logarithmic scale. The latter shows the full suite of PIANO compounds detected in the water above 1 μg/L. Detailed analysis like this permits greater "fingerprinting" of groundwater samples than conventional EPA Method 8260 or 8020 data.*

The second issue facing groundwater fingerprinting studies is the difficulty of comparing groundwater fingerprints to those from other matrices—NAPL and soils—due to partitioning affects that will alter the "fingerprints" of compounds derived from the same gasoline within the different matrices. Partitioning of individual compounds from NAPL into groundwater is a function of the compound's aqueous solubility and the mole fraction of the compound in the gasoline mixture—Raoult's Law (Burris and MacIntyre, 1985). The mole fraction of any component in a gasoline-derived NAPL is difficult to determine because the composition contains hundreds of individual compounds. Therefore, a proxy for this measure is the fuel–water partitioning coefficient (K_{fw}) defined as:

$$K_{fw} = C_f/C_w,$$

where C_f anc C_w are the equilibrium concentrations of a compound in gasoline and water, respectively (e.g., Cline *et al.*, 1991). In the absence of experimentally determined C_f anc C_w data Bruce (1993) suggests that the K_{fw} can be estimated from the aqueous solubility (S) according to the following equation:

$$\log K_{fw} = 6.099 - 1.15 \log(S)$$

Compounds with a low K_{fw} will be efficiently extracted from gasoline and dissolve into groundwater. Because of the relationship between aqueous solubility (S) and K_{fw}, aqueous solubility can generally serve as a proxy for the partitioning behavior of gasoline constituents between NAPL and groundwater.

Sauer and Costa (2003) utilized this relationship to develop a solubility-based approach to comparing NAPL

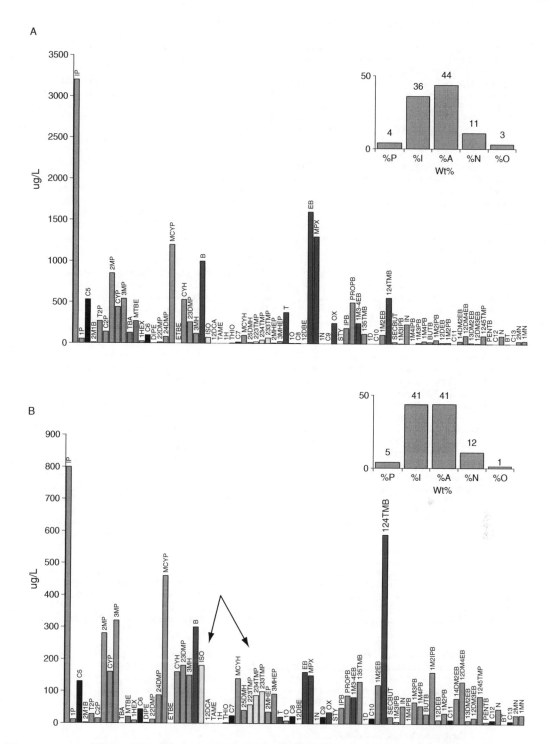

Figure 18.8.4 *The PIANO histograms for two groundwater samples from monitoring wells located on adjacent service stations. Detailed comparison reveals that the parent gasolines for (A) contained no alkylate component and (B) contained significant alkylate component as indicated by the near absence and prominence of trimethylpentanes (arrows) in the groundwater. Note the reduced amount of toluene relative to other BTEX compounds, which might be due to preferential degradation of toluene under anaerobic conditions. For compound identifications see Table 18.5.2.*

and groundwater fingerprints. These authors developed a suite of ratios among monoaromatics compounds with comparable aqueous solubilities. Solubility ratios approaching equity were anticipated to remain relatively constant upon the dissolution from the NAPL into groundwater. This consistency of various solubility-based ratios was demonstrated for a gasoline-derived NAPL, the underlying groundwater (from the same well), and groundwater from a downgradient well. These authors point out the importance of using multiple solubility ratios before drawing conclusions since the effects of biodegradation or sorption during migration are not accounted for in these ratios.

This solubility-ratio approach of Sauer and Costa (2003) might also aid in groundwater studies in which there is no NAPL, but detailed comparisons between groundwater samples is necessary. Data demonstrating this potential application at a site without NAPL are shown in Table 18.8.2. Six solubility-based ratios with values near 1 were used to compare the compositions in six groundwater samples from a study area where gasoline-impacted groundwater extended across multiple potential sources. One sample (A) demonstrated multiple disparate ratios that suggest it had been impacted by a different source of gasoline than the other wells. Although exhibiting a generally similar PIANO fingerprint (Figure 18.8.5), the unique nature of sample A was further substantiated by the presence of small amounts of oxygen-containing compounds that were not detected in the other groundwaters.

A different approach to the problem of comparing different environmental matrices was developed by Reisinger and Burris (2004). They used the commonly measured BTEX concentrations in NAPL, soil, or groundwater to calculate the BTEX concentrations in a "hypothetical" parent NAPL, thereby converting each matrix to a common frame of reference. This approach assumes the "hypothetical" parent NAPL contains only BTEX (rather than hundreds of compounds), which permits calculation of the mole fraction of each of the four BTEX components. This approach also assumes a minimal extent of biodegradation, which would require evaluation on a site-specific basis. An example is given in Table 18.8.3. The mole fraction fingerprints of NAPL and soil can be determined by dividing each component's concentration (g/kg) by its molecular weight (g/mol), and then dividing by the total BTEX concentration (mol/kg) to get mole fraction of each component (i.e., normalizing to the 4-component BTEX mixture). For groundwater, the measured concentrations (mg/L) are divided by the pure compound aqueous solubility (mg/L) to obtain the relative mole fraction of each of the BTEX components (i.e., application of the effective solubility relationship) and then divided by the total BTEX relative mole fraction to obtain the

mole fraction in the "hypothetical" parent NAPL (i.e., normalizing to the 4-component BTEX mixture). This approach allows for the comparison among BTEX distributions in multiple matrices, thereby allowing for determination of a common or disparate source of BTEX to each. In the example data shown in Table 18.8.3, the fingerprints obtained from the BTEX mole fractions in a hypothetical parent NAPL are highly comparable among all three matrices, suggesting they share a common source. This commonality was not evident in comparing the measured concentrations of BTEX in each matrix (Table 18.8.3). Notably, this "hypothetical mole fraction" approach is not limited to BTEX components but could rely upon any number of gasoline-derived constituents.

Finally, any type of groundwater forensic study can benefit from a sampling design that provides good spatial (sometimes temporal) coverage of the area. At a minimum, in most studies the groundwater from all of the accessible monitoring and recovery wells should be collected and analyzed. This type of "survey" provides an opportunity to recognize aerial trends after considering any well screen depth or geologic controls (e.g., clastic versus fracture bedrock) on groundwater composition that can reveal hotspots, anomalies, or other trends across the study area (or over time). An example of the value of a thorough groundwater survey, in combination with conventional PIANO fingerprinting, is given in the case study below.

In this study, a large number of monitoring wells had been used to determine the extent of groundwater contamination that led to a lawsuit filed by a downgradient property owner against two upgradient service stations. Station A had a known release of gasoline more than 10 years prior, whereas Station B had no reported releases. Groundwater flow was, in part, controlled by a paleochannel that crossed all three properties (dashed line in Figure 18.8.6A). The PIANO analysis of the area's groundwater revealed concentration (Figure 18.8.6B) and compositional (Figure 18.8.6C) differences in the nature of the impacted groundwater along the axis of the paleochannel. Of particular note was the appearance of markedly higher levels of TAME into groundwater at the location of Station B's UST field. This result implicated a release(s) of TAME-bearing gasoline had occurred from Station B's tank field. Faced with this evidence, Station B cooperated with Station A in settling the case prior to trial. Only through the combination of high-quality data (e.g., including low level analysis for TAME) and good spatial coverage was the obvious spatial trend revealed.

Another important consideration in a sampling strategy for impacted groundwater is the collection of soil samples. Vadose zone and capillary fringe zone soils in the vicinity of known of suspected sources (e.g., UST fields, piping runs,

Table 18.8.2 Data demonstrating the solubility-based ratio approach to fingerprinting (as per Sauer and Costa, 2003) for six groundwater samples (A through F) from monitoring wells impacted with gasoline-derived dissolved constituents. Various disparate ratios for "A" (boldface) suggest it has been impacted from a different source of gasoline than the other five wells. Representative PIANO histograms for "A" and "B" are shown in Figure 18.8.5

Compound Ratios	Solubility Ratio	A	B	C	D	E	F
135TMB/124TMB	0.9	0.29	0.33	0.35	0.26	0.30	0.23
1M3PB/1M4PB	0.9	0.80	2.71	2.67	2.32	2.78	2.50
14M2EB/12M4EB	1.2	0.86	0.91	0.94	0.94	0.96	1.00
1M2PB/12M4EB	1.0	0.63	0.18	0.28	0.27	0.22	0.22
1M2EB/1M3EB	0.9	0.26	0.28	0.48	0.29	0.33	0.24
2MN/1MN	0.9	0.19	1.79	1.62	0.70	1.83	1.82

See Table 18.5.2 for compound abbreviations.

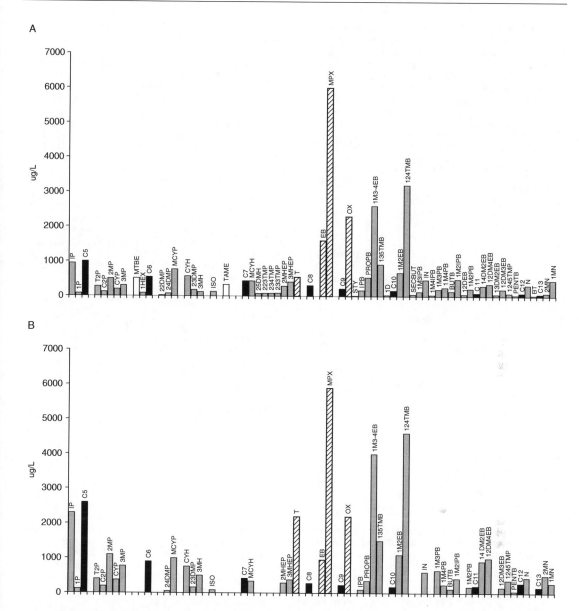

Figure 18.8.5 *The PIANO histograms for two groundwater samples, "A" and "B" per Table 18.5.2, exhibited generally comparable fingerprints. However, multiple solubility-based ratios (Table 18.5.2) demonstrate considerable disparity exists. Minor fingerprinting differences (e.g., presence of MTBE and TAME in "A" support a difference source. See Table 18.5.2 for compound identifications).*

or dispensers) should be considered for sampling and forensic analysis. These soils may contain residual free phase gasoline that is trapped in pores. Concentrations of total VOC on the order of 500 mg/kg in soil indicate the presence of sufficient residual gasoline adsorbed or trapped by the soil to obtain suitable PIANO fingerprints for comparison. In addition, because of the nature of gasoline releases (see above section), the identification and characterization of gasoline-impacted vadose zone soils, which are stratigraphically above any historic ground water highs, provide strong evidence that a gasoline release(s) occurred in that area. The PIANO fingerprinting of these impacted soils will provide the chemical characterization of that release(s) and

allow for comparison to impacted groundwater in the area. Thus, groundwater studies often can benefit from the collection of vadose zone and capillary fringe zone soils in suspected source areas.

18.8.5 Fate of Organic Lead in the Environment

The environmental fate of lead alkyls derived from gasoline is an important consideration to the forensic chemist. As discussed earlier, the distribution and concentration of lead alkyls in NAPL are often used to identify gasoline contamination derived from different types of lead antiknocks and to constrain the time of release to groundwater. Little attention has been given to the potential affects of weathering on

Table 18.8.3 *Data demonstrating the conversion of measured BTEX concentrations (C$_i$) in groundwater, NAPL, and soil into moles, and then into a hypothetical mole fraction for a hypothetical parent gasoline (after Reisinger and Burris, 2004). Note the three matrices exhibit comparable distributions of BTEX in the hypothetical parent gasoline suggesting they are all impacted by the same gasoline type, despite measured concentration differences*

$C_{i(aq)}$ (mg/L)		Relative Mole Fraction[a]	Hypothetical NAPL Mole Fraction[c]
GroundWater			
B	6.4	0.0036	0.12
T	3.2	0.0061	0.20
E	0.7	0.0041	0.14
X	2.8	0.0160	0.54

$C_{i(napl)}$ (g/kg)	$C_{i(napl)}^{b}$ (mol/kg)	Hypothetical NAPL Mole Fraction[c]
NAPL		
B 10	0.13	0.14
T 18	0.20	0.22
E 12	0.11	0.13
X 50	0.47	0.52

$C_{i(soil)}$ (mg/kg)	$C_{i(soil)}^{b}$ (mol/kg)	Hypothetical NAPL Mole Fraction[c]
Soil		
B 60	0.77	0.15
T 130	1.40	0.27
E 53	0.50	0.09
X 280	2.60	0.50

[a] divide by pure aqueous solubility (mg/L).
[b] divide by molecular weight (g/mol).
[c] normalize to moles of total BTEX.

these distributions or concentrations. Like all compounds contained within gasoline, the alkyl leads each exhibit a range of physical properties that present an opportunity for preferential alteration by different weathering processes.

For example, the boiling points of the different lead alkyls (Table 18.4.3) range from 200 °C (TEL) to 110 °C (TML). Thus, there is the potential for evaporation to impact the proportion of TML to a greater extent than TEL. This will change both the distribution among the five lead alkyls and the overall concentration. Of course, evaporation will simultaneously influence hydrocarbon that may offset to some degree any increase or decrease in concentration. Therefore, the potential effect of evaporation needs to be evaluated before alkyl lead distributions or concentrations are used to draw forensic conclusions. In addition to the potential effects of evaporation, within the past decade research has been performed to more fully understand the environmental fate of alkyl leads in the environment by lesser-known weathering processes—e.g., hydrolysis.

Laboratory studies by Mansell *et al.* (1995) examined the fate and transport of TEL and leaded gasoline in loam and sand soil columns. They concluded that pure TEL was unstable in soil and rapidly undergoes dealkylation to the more water-soluble triethyllead (TREL), diethyllead (DEL), and Pb^{+2}. The rapid breakdown of TEL, however, was not observed when TEL was added to soil columns as leaded gasoline. Based upon experiments of immiscible flow of

leaded gasoline through soil columns of loam and sand, Mansell *et al.* (1995) observed that TEL was co-transported through the soil with gasoline. Tetraethyl degradation products TREL and DEL, however, were also detected in the effluent soil indicating that some of the TEL was unstable during the gasoline transport through soil. Additional testing of TREL in water-saturated soil indicated that TREL was converted to Pb^{+2} which was subsequently adsorbed by the soils. This data and the general reactivity of TEL with water (Feldhake and Stevens, 1963; Mansell *et al.*, 1995) and the formation of ionic degradation products (TREL and DEL) indicate that the chemical degradation is a result of hydrolysis.

Ou *et al.* (1995a,b) further examined the mechanisms by which TEL degrades in subsurface soils. Using ^{14}C-labeled TEL in sterile and non-sterile soils, they determined that TEL degradation could occur by both biological and chemical processes, and that chemical degradation of TEL to ionic ethyllead compounds (TREL and DEL) was the major process. (Mulroy and Ou, 1998) took the next step and examined the fate of TEL in leaded gasoline contaminated soils. As noted above, soils contaminated with pure TEL rapidly degraded and disappeared within 14 days (Mansell *et al.*, 1995). In soil samples spiked with leaded gasoline at 1000 µg/g soil and 5000 µg/g soil, Mulroy and Ou (1998) observed an initial rapid loss of TEL followed by decreasing rates of TEL degradation. This decrease appeared to be related to soil gasoline loading with TEL degradation occurring more slowly in the samples with higher concentrations of gasoline. They concluded that gasoline hydrocarbons in contaminated soils partially protect TEL from rapid chemical degradation.

The main degradation pathway for lead alkyls in gasoline is reportedly sorption onto soils or hydrolysis at the NAPL groundwater interface (Morrison, 2000a; Kaplan *et al.*, 1997) where the NAPL mass/surface area exposure is considered to be a critical factor (Morrison, 2000a). For example, a NAPL with an apparent thickness of only 1 cm will have a greater surface area in contact with groundwater than a NAPL with an apparent thickness of 5 m. Therefore, as is the case in other weathering processes affecting gasoline, the potential effect of chemical degradation of TEL in leaded gasoline is most likely greatest in intervals of lower concentrations/thinner accumulations of gasoline.

Kaplan *et al.* (1997) has suggested that lead alkyls are strongly absorbed to soil and hydrolyzed by water. Similarly, Morrison (2000a) suggests that lead alkyls may even be completely removed (sorbed or hydrolyzed) from the subsurface and that the lead scavengers (EDB and EDC) may be the only evidence of leaded gasoline's impact.

In order to verify if TEL does indeed hydrolyze at the NAPL/water interface a laboratory NAPL water washing study was performed. Samples of leaded gasoline-derived NAPL were analyzed as part of the water washing studies described earlier in this chapter using the apparatus shown in Figure 18.6.7. Two NAPL samples were selected with initial TEL concentrations of 1.7 glpg (NAPL 1) and 0.6 glpg (NAPL2), respectively. The study was performed in duplicate for the NAPL 1 sample. The experiment was carried out for a period of 20 months after which the T_0 and T_{20} samples were analyzed for lead alkyls by GC/MS (as described earlier in this chapter).

Analysis of the hydrocarbon distributions in the floating NAPLs (T_0 versus T_f) clearly indicated that aromatic hydrocarbon water washing had occurred and that there was only a small loss of more highly volatile alkanes suggesting that evaporation was minor. There was no evidence of hydrocarbon biodegradation in the samples. Between 77 and 94 wt% of the original TEL in the NAPL samples was removed over

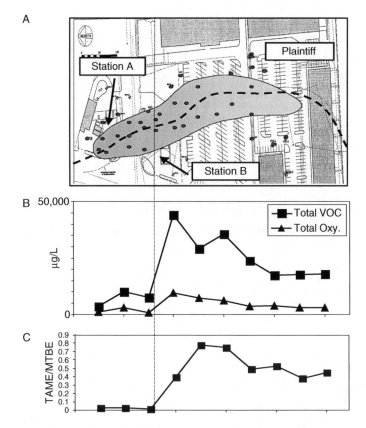

Figure 18.8.6 *Groundwater fingerprinting case study: (A) map showing locations of two service stations (co-defendants) and the impacted groundwater plume that extended downgradient (south) onto the plaintiff's property. (B) Dissolved phase concentration of total VOC and total ether oxygenates increased markedly in wells located near Station B's tank field (indicated by vertical line through the figure). (C) The character of the oxygenates also changed at this location as evidenced by the increased the TAME/MTBE ratio. This finding implicated a contribution by Station B and led to joint settlement of the case by both stations.*

Table 18.8.4 *Initial and final concentration of TEL in NAPLs following a 20-month water washing experiment demonstrating the percent of TEL reduction due to chemical degradation. See text for details*

Sample ID	Initial TEL glpg[a]	Final TEL glpg	Wt% TEL Degraded
NAPL 1	1.7	0.39	77
NAPL 1 Duplicate	1.7	0.11	94
NAPL 3	0.6	0.04	94

[a] glpg = grams lead per gallon.

the 20-month period (Table 18.8.4). This data supports the contention of Kaplan *et al.* (1997) and Morrison (2000a) and results of other laboratory studies (e.g., Mulroy and Ou, 1998), who had concluded that TEL in gasoline can be chemically degraded in the environment.

A recent field investigation lends additional support for the chemical degradation of TEL in the environment. Two NAPL samples were collected near a retail gasoline station and analyzed for hydrocarbon composition and organic

lead content (Figure 18.8.9). Non-aqueous phase liquid1 is determined to be a leaded gasoline-derived NAPL containing 2.0 glpg as TEL. It was collected from a monitoring well located directly downgradient from an area where leaded gasoline was stored and released in the past—the presumed source. Non-aqueous phase liquid2 was collected from a monitoring well downgradient from NAPL1 and contained markedly lower concentration of TEL (0.1 glpg). A clay layer existed above the NAPL preventing extensive evaporation of the two samples.

The hydrocarbon signatures of the two NAPLs indicate both are consistent with a water-washed leaded gasoline, as evidenced by the relative depletion of soluble constituents (BTEX; Figure 18.8.7). Water washing has altered the fingerprint of the NAPLs so that they are enriched in less soluble components (e.g., methylcyclohexanes and normal paraffins). As such, each NAPL's fingerprint is now dominated by a significant straight run naphtha component.

The remarkable similarity in the hydrocarbon distributions in the two NAPLs, along with the locations relative to the likely source, strongly argues that the NAPLs were derived from the same parent gasoline. Thus, the only chemical difference between the NAPLs rests in the concentration of the TEL, being approximately 20 times lower in the downgradient NAPL (Figure 18.8.7B). This disparity is strong

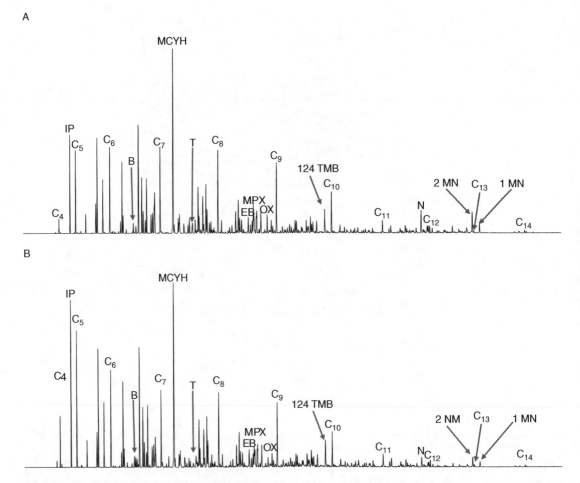

Figure 18.8.7 *The GC/MS total ion chromatograms for (A) NAPL1 containing 2.0 glpg (as TEL) and (B) NAPL2 containing 0.1 glpg (as TEL). See text for description. Compound identifications as per Table 18.5.2*

evidence that chemical degradation has decreased the concentration of TEL in the downgradient NAPL sample.

Thus, based upon laboratory and field observations described above, environmental forensic investigations that rely upon alkyl lead concentrations and distributions in NAPL require some caution. Investigators must, at least, consider the potential confounding effect of weathering processes (e.g., evaporation or chemical degradation) on lead concentration and distribution in NAPL.

18.8.6 Effect of Soil Vapor Extraction on Vadose Zone Soils Impacted with Gasoline

The effect of evaporation on residual gasoline in the vadose zone occurs in a predictable manner that is largely a function of the vapor pressure of each compound and the temperature (see *Gasoline Weathering* section). In this short section, the effects of anthropogenic-enhanced evaporation on residual gasoline in soils is described—namely soil vapor extraction (SVE).

Soil vapor extraction is an *in situ* remedial technology used at many gasoline-impacted sites as a means of recovering vapors from unsaturated soils. It operates when a vacuum is applied to the soil matrix through an extraction well(s), usually placed near a known source area. The vacuum creates a negative pressure gradient that facilitates the

evaporation and movement of the most volatile constituents toward these wells. This process artificially evaporates the residual gasoline in soil, thereby changing its fingerprint.

At two service station locations we have observed the impact of SVE on gasoline-impacted soils. In each case the effects were markedly different from those imparted by natural evaporation—which we have observed at many locations. The difference between SVE and natural evaporation are demonstrated in Figure 18.8.8. This figure shows the PIANO histograms for vadose zone soils from a site that had been impacted with gasoline and that employed SVE technology for nearly 1 year. Soils collected nearest to the SVE extraction wells had experienced the near complete removal of all compounds with vapor pressures higher than that of the C_2-alkylbenzenes (i.e., all compounds to the left of ethylbenzene), with only minor relative reduction of ethylbenzene and *m-/p*-xylene (Figure 18.8.8A). Even *o*-xylene does not appear reduced due to SVE relative to less-volatile C_3-alkylbenzenes. The more remote soil (Figure 18.8.8B) was not particularly affected by SVE and experienced a more gradual and relative loss of compounds that was directly proportional to their vapor pressures.

The C_2-alkylbenzenes have vapor pressures in the range of 1.09–1.25 atm (TPH Criteria Working Group, 1997). Based upon our observations, SVE will have little effect on the hydro-

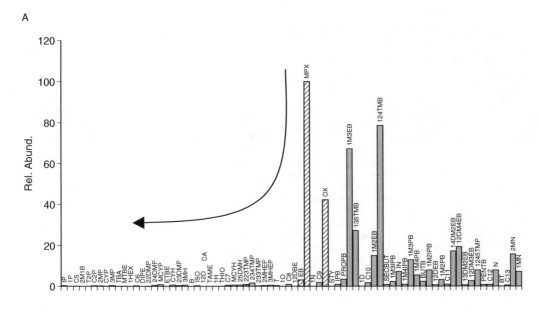

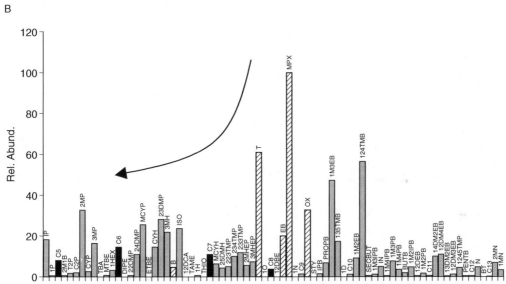

Figure 18.8.8 *Histograms showing the varying affects of volatilization of residual gasoline from vadose zone soils at a site due to: (A) soil vapor extraction (SVE) and (B) natural evaporation. Soil shown in (B) was obtained from a location outside the area of influence of the SVE system. For compound identifications see Table 18.5.2.*

carbons with vapor pressures above some particular level. This level is likely to be a function of the magnitude of the vacuum drawn, proximity to the extraction point, and on soil properties. Remarkably, at both the service stations we observed this phenomenon the affect was drastic and affected only those compounds with vapor pressures higher than the C_2-alkylbenzenes. Thus, fingerprinting studies involving soils from SVE sites may need to necessarily focus on the higher boiling, least volatile components.

18.8.7 Gasoline-Derived Soil Gas and Indoor Air
Vapor transport of gasoline-derived contamination has been evaluated for more than 10 years (Johnson and Ettinger, 1991). However, only in the last five years, there has been

significant interest in this potential transport of subsurface vapors of gasoline and other volatile hydrocarbon chemicals emanating from contaminated water or soil, and the intrusion of these vapors into homes and other buildings. This concern prompted United States Environmental Protection Agency to issue draft guidance for evaluating the sources, potential pathways, and mechanisms for migration of volatile hydrocarbons from subsurface soil and water (United States Environmental Protection Agency, 2002). Central to this concern is the recognition that volatile organic compounds such as those that comprise gasoline can partition from water, soil, or NAPL into the subsurface gas phase and then migrate to the ground surface and into buildings.

Investigators at suspected vapor intrusion sites typically develop conceptual screening models to determine if subsurface hydrocarbons have the potential to migrate into homes or other buildings. The most common conceptual model, advocated by United States Environmental Protection Agency, relies upon the Johnson and Ettinger model for calculating the ratio of indoor air vapor concentrations to soil vapor concentrations at a specific depth (Johnson and Ettinger, 1991). This model incorporates the so-called "attenuation factors" of the subsurface vapors as they migrate through the soil column, across building slabs, and mix with indoor air. The attenuation factors included in this model incorporate a number of simplified assumptions regarding contaminant distribution and occurrence, subsurface characteristics, transport mechanisms, and building construction. These assumptions do not account for important physical and biological degradation of the subsurface hydrocarbon vapors (United States Environmental Protection Agency, 2004). Because of these assumptions, the model is inherently conservative in its prediction of potential migration of subsurface vapors into indoor air. As such, in cases where modeling suggests the possibility of indoor air intrusion by subsurface hydrocarbons, it is incumbent upon investigators to confirm such hypotheses with actual field measurements. The challenge is developing sufficiently convincing monitoring data to confirm that any indoor air pollutants indeed have their source as subsurface and subslab contaminants.

One of the principal confounding factors in confirming vapor intrusion investigations is the ubiquitous presence of anthropogenic volatile hydrocarbons found inside buildings. Such chemicals are unrelated to volatile hydrocarbons that might intrude into the building from the subsurface. Sources of indoor air-borne chemicals include solvents, paints, paint strippers, and other solvents; wood preservatives; aerosol sprays; cleansers and disinfectants; moth repellents and air fresheners; stored fuels and automotive products; hobby supplies; and dry-cleaned clothing (USEPA and the United States Consumer Product Safety Commission Office of Radiation and Indoor Air, 1995). Recent studies have suggested that the preponderance of indoor volatile hydrocarbons at sites situated above contaminated soil and groundwater are actually the result of these anthropogenic indoor sources—not the subsurface contamination (McHugh *et al.*, 2004).

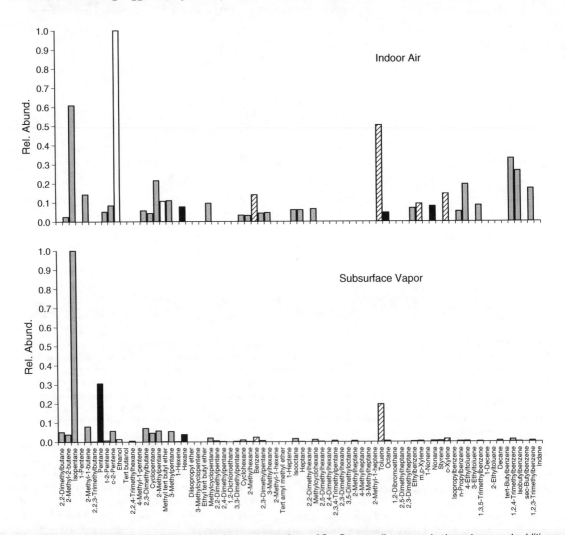

Figure 18.8.9 *The PIANO histograms of relative concentrations of* C_4–C_{10} *gasoline range hydrocarbons and additives in subsurface vapor and indoor air of a residential building situated above gasoline-contaminated soil and groundwater.*

In the cases where subsurface gasoline (either NAPL or dissolved phase) is suspected as the source of indoor air contamination, the findings of gasoline-related chemicals such as BTEX or MTBE does not necessarily mean that the subsurface is the source of these chemicals. More detailed chemical fingerprinting is warranted if investigators wish to link subsurface gasoline contamination with hydrocarbons found in indoor air.

Standard implementations of United States Environmental Protection Agency-approved methods for indoor and ambient air sampling and analysis such as United States Environmental Protection Agency TO-15 measure an impressive list of potential indoor air pollutants (United States Environmental Protection Agency, 1999). However, the standard target analyte list falls for this and related methods short of that necessary to "fingerprint," i.e., compare chemical signatures, of indoor air-borne hydrocarbons versus potential subsurface gasoline, or other indoor or outdoor anthropogenic sources of hydrocarbons. Assessment of potential indoor intrusion of subsurface gasoline can best be done by analyzing samples using a target analyte list such as that in Table 18.5.2. Using this target compound list, direct comparisons of subsurface and subslab vapor compositions with those found in indoor air are possible. Forensic data analysis techniques, including simple graphical analysis as well as more advanced statistical methods such as principal components analysis (Johnson *et al.*, 2002) can then be used to determine if there is a chemical linkage between indoor air-borne hydrocarbons and subsurface gasoline.

For an example of the use of appropriately tailored forensic chemical measurements at a suspected gasoline vapor intrusion site, consider data compiled in Figure 18.8.9. Figure 18.8.9 shows the PIANO histograms of the relative concentrations of 73 C_4–C_{10} gasoline hydrocarbons measured using a modified United States Environmental Protection Agency TO-15 method for subsurface vapor (lower panel) and indoor air (upper panel) of a residence situated above groundwater and soil contaminated with gasoline. The data reveal that while there are some common chemicals found in both samples (e.g., varying concentrations of BTEX) there are also significant differences in the chemical compositions of the subsurface vapor and the indoor air. Notably, the indoor air contained elevated concentrations of ethanol and MTBE, benzene, and toluene relative to the subsurface vapor. Also evident is the significant amounts of C_7 – C_{10} hydrocarbons—particularly alkylated benzenes— in the indoor air that was largely absent in the subsurface vapor. Such distinct chemical differences must be reconciled before a clear connection between subsurface contamination and indoor air-borne hydrocarbons can be made.

18.9 CONCLUSION

The environmental forensics of gasoline poses multiple challenges. An understanding of the effects the history, refining, additive use, and environmental weathering of gasoline, along with specific "fingerprinting" data, can provide the investigator with a means to reliably address the objectives surrounding source and age of gasoline-derived contamination in the environment.

REFERENCES

Adlard, ER, AW Bowen, and DG Salmon. Automatic system for the high-resolution gas chromatographic analysis of gasoline-range hydrocarbon mixtures. *Journal of Chromatography*. 186: 207–218. 1979.

Alvarez, PJJ and TM Vogel. Substrate interactions of benzene, toluene, and para-xylene during microbial degradation by pure cultures and mixed culture aquifer slurries. *Appl. Environ. Microbiol.* 57: 2981–2985. 1991.

Alvarez, PJJ, RC Heathcote, and SE Powers. Caution against interpreting gasoline release dates based on BTEX ratios in ground water. Ground Water Monitoring and Remediation. (Fall), 69–76. 1998.

American Petroleum Institute. *Laboratory study on solubilities of petroleum hydrocarbons in groundwater*. API Publ. No. 4395, August 1985, 21. 1985.

American Petroleum Institute. Gasoline blending streams test plan. API Report to US Environmental Protection Agency, dated December 12, 2001, p. 38, 2001.

American Society for Testing and Materials. ASTM D4815, *Standard Test Method for Determination of MTBE, ETBE, TAME, DIPE, tertiary-Amyl Alcohol and C_1 to C_4 Alcohols in Gasoline by Gas Chromatography*. West Conshohocken, PA. 1999.

American Petroleum Institute. *Chemical fate and impact of oxygenates in groundwater: Solubility of BTEX form gasoline-oxygenate compounds: American Petroleum Institute*, Publ. No. 4531. 1999.

American Petroleum Institute 2004. API interactive LNAPL guide. http://groundwater.api.org/inaplguide.

Beall, PW, SA Stout, GS Douglas, and AD Uhler, 2002. On the role of process forensics in the characterization of fugitive gasoline. *Env. Claims J.*, 14, 487–506.

Brauner, JS and M Killingstad. In *situ Bioredeation of Petroleum Aromatic Hydrocarbons*. Groundwater Pollution Primer CE 4594, 1996.

Bruce, LG. *Refined gasoline in the subsurface*. The American Association of Petroleum Geologists Bulletin, 77 (2): 212–224. 1993.

Burris, DR and WG MacIntyre. Water solubility behavior of hydrocarbon mixtures—Environ. Tox. Chem. 4: 371–377. 1985.

Chakraborty, R and JD Coates. Anaerobic degradation of monoaromatic hydrocarbons. *Applied Microbiology and Biotechnology*. 64: 437–446. 2004.

Chawla, B. Paraffin, olefin, naphthene, and aromatic determination of gasoline and JP-4 jet fuel with supercritical fluid chromatography. *Journal of Chromatographic Science*. 35: 97–103. 1996.

Chow, TJ. Lead accumulation in roadside soil and grass. *Nature*. 225: 295–296. 1970.

Chow, TJ and MS Johnstone. Lead isotopes in gasoline and aerosols of Los Angeles Basin, California. *Science*. 147: 5002–503. 1965.

Chow, TJ, CB Snyder, and JL Earl. Isotope ratios of lead as pollutant source indicators. In: *Proc, United Nations FAO Int'l. At. Ener. Assoc. Symp.* IAEA-SM-191/4, 95–108. 1975.

Cline, PV, JJ Delfino, and PSC Rao. Partitioning of aromatic constituents into water from gasoline and other complex solvent mixtures. *Environmental Science and Technology*. 25 (5): 914–920. 1991.

Coulombe, R. Chemical markers in weathered gasoline. *Journal of Forensic Sciences*. 40 (5): 867–873. 1995.

Davidson, JM and DN Creek. Using the gasoline additive MTBE in forensic environmental investigations. *International Journal of Environmental Forensics*. 1 (1): 57–67. 1999.

Davis, BC and WH Douthit. The use of alcohol mixtures as gasoline additives. Presented at the 1980 NPRA Annual Meeting. 1980. Washington, DC, National Petroleum Refiners Association. March 23–25, 1980.

Dickson, CL, PW Woodward, and PL Bjugstad. *Trends of Petroleum Fuels, 1987*. Bartlesville, OK, National Institute for Petroleum and Energy Research. 1987.

Dorn, P, AM Mourao, and S Herbstman. The properties and performance of modern automotive fuels. Society of Automotive Engineers, Paper No. 861178, 51–67. 1983.

Erel, Y and CC Patterson. Leakage of industrial lead into the hydrocycle. *Geochimica et Cosmochimica Acta*. 58 (15): 3289–3296. 1994.

Falta, R. The potential for ground water contamination by the gasoline scavengers ethylene dibromide and 1,2-dichloroethane. *Ground Water Monitoring and Remediation*. 24 (3): 76–87. 2004.

Feldhake, CJ and CD Stevens. The solubility of tetraethyllead in water. *J. Chem. Eng.* Data 8, 196–197. 1963.

Fort, RC and PVR Schleyer, 1964, Adamantane: Consequences of the diamondoid structure. *Chemical Reviews*, pp. 277–300.

Frank, JL. New mandates present fuel challenges for US refiners. *Oil & Gas Journal*, December 13 edition, 118–121. 1999.

Freeze, RA and JA Cherry. *Groundwater*. Prentice Hall, Englewood Cliffs, NJ , 604. 1979.

Furey, RL, AM Horowitz and NJ Schroeder. Automotive Gasolines. Dyroff, G.V. *ASTM Manual on Significance of Tests for Petroleum Products*, 6th edition, 24–33. 1993. West Conshohocken, PA, American Society for Testing and Materials. 1993.

Gary, JH and GE Handwerk. *Petroleum Refining*. New York, NY, Marcel Dekker, Inc. 1984.

Gibbs, LM. Additives boost gasoline quality. *Oil & Gas Journal*, April 29 edition, 60–63. 1989.

Gibbs, LM. *Gasoline Additives—When and Why*. Warrendale, Pennsylvania, Society of Automotive Engineers. SAE Technical Paper Series No. 902104. 1990.

Gibbs, LM. How gasoline has changed. SAE International Fuels and Lubricants Meeting and Exposition. S01007-S010025. 1993. Warrendale, PA, Society of Automotive Engineers, SAE Technical Paper Series 932828. 1993.

Gibbs, LM. How gasoline has changed II—The impact of air pollution regulations. Warrendale, PA, Society of Automotive Engineers. SAE Technical Paper Series 961950. 1996.

Gibbs, L. The ASTM gasoline specification. ASTM Standardization News April, 40–46. 1998.

Graney, JR, AN Halliday, GJ Keeler, JO Nriagu, JA Robbins, and SA Norton. Isotopic record of lead pollution in lake sediments from the northeastern United States. Geochim. Cosmochim. Acta 59, 1715–1728. 1995.

Guetens, EG, Jr., JM DeJovine, and GJ Yogis. TBA aids methanol/fuel mix. Hydrocarbon Processing May 1982, 113–117. 1982.

Hamilton, B and RJ Falkiner. Motor gasoline. In: *Fuels and Lubricants Handbook: Technology, Properties, Performance, and Testing (G.E. Totten, ed.) ASTM International*, W. Conshohocken, PA, 61–88. 2003.

Harvey, E. Interpretative considerations for pattern matching refined petroleum products. Proc. of Hydrocarbon Pattern Recognition and Dating. 17. 1997. University of Wisconsin. Program No. 7675.

Healey, E, SA Smith, KJ McCarthy, SA Stout, RM Uhler, AD Uhler, SD Emsbo-Mattingly, and GS Douglas. Fingerprinting organic lead species in automotive gasolines and free products using direct injection GC/MS. *Int'l. Conf. Contaminated Soils, Sediments and Water, 18th Annual Meeting* Amherst, MA. 2002.

Hincher, R.E. and H.J. Reisinger. A practical application of multiphase transport theory to ground water contamination problems: Ground Water Monitoring Review, Winter Issue, pp. 84–92, 1987.

Hirao, Y and CC Patterson. Lead aerosol pollution in the high Sierra override natural mechanisms which exclude lead from a food chain. *Science*. 184: 989–992. 1974.

Hoag, GE and MC Marley. Gasoline residual saturation in unsaturated uniform aquifer materials. *Journal of Environmental Engineering*. 112 (3): 586–604. 1986.

Hunkeler, D, BJ Butler, R Aravena, and JF Barker. Monitoring biodegradation of methyl *tert*-butyl ether (MTBE) using compound-specific carbon isotope. *Environmental Science and Technology*. 35 (4): 676–681. 2001.

Hurst, RW. Applications of anthropogenic lead archaeostratigraphy (ALAS Model) to hydrocarbon remediation. *Environmental Forensics*. 1 (1): 11–23. 2000.

Hurst, RW. Lead isotopes as age-sensitive, genetic markers in hydrocarbons: 3. Leaded gasoline, 1923–1990 (ALAS Model). *Environmental Geosciences*. 9 (2): 43–50. 2002.

Hurst, RW, TE Davis, and BD Chinn. The lead fingerprints of gasoline contamination. *Environmental Science and Technology*. 30 (7): 304A–307A. 1996.

Hutson, T and GE Hays, 1977, Reaction mechanisms for hydrofluoric acid alkylation: In: *Industrial and Laboratory Alkylations* (L.F. Albright and A.R. Goldsby, eds.), *Am. Chem. Soc.* Symposium Series No. 55, p. 27–55.

International Energy Annual, US Department of Energy, 2003; http://www.eia.doc.gov/emeu/iea.

Jackman, FA. The history of gasoline. In: *Technology of Gasoline* (E.G. Hancock, ed.) *Soc. Chemical Industry*, 2–19. 1985.

Johnson, R Ehrlich, and W Full. Principal components analysis and receptor models in environmental forensics. In: Introduction to Environmental Forensics, Murphy, Brian L. and Morrison, Robert D. eds. San Diego CA. Academic Press. 2002.

Johnson, MD and RD Morrison. Petroleum fingerprinting: Dating a gasoline release. *Environmental Protection*. 37–39. 1996.

Johnson, PC and RA Ettinger. Heuristic model for the intrusion rate of contaminant vapors into buildings. *Environmental Science and Technology*. 25 (8): 1445–1452. 1991.

Johnson, R, J Pankow, D Bender, C Price, and J Zogorski. MTBE. To what extent will past releases contaminate community water supply wells? *Environmental Science and Technology*/News May 1, 2000, 210A–217A. 2000.

Jones, VT. Characterization and mapping of underground product contamination using soil vapor techniques. National Institute of Hydrocarbon Fingerprinting. Santa Fe, NM, March 2–3, 1995.

Kaplan, IR. Identification of formation process and source of biogenic gas seeps. *Isr. J. Earth Sci.* 43: 297–308. 1994.

Kaplan, IR. Age dating of environmental organic residues. *Environmental Forensics*. 4: 95–141. 2003.

Kaplan, IR and Y Galperin. How to recognize a hydrocarbon fuel in the environment and estimate its age of release. *Groundwater and Soil Contamination: Technical Preparation and Litigation Management*. 1996. New York, John Wiley & Sons, Inc.

Kaplan, IR, Y Galperin, H Alimi, RP Lee, and S-T Lu. Patterns of chemical changes during environmental alteration of hydrocarbon fuels. *Ground Water Monitoring and Remediation*, 113–124. 1996.

Kaplan, IR, Y Galperin, S-T Lu, and R-P Lee. Forensic environmental geochemistry: Differentiation of fuel-types, their sources and release time. *Organic Geochemistry*. 27 (5–6): 289–317. 1997.

Keesom, W, J Gieseman, and B Wood. Commercial strategies for gasoline sulfur reduction. Presented at the 1998 NPRA Environmental Conference. 1998. Washington, DC, National Petrochemical & Refiners Association. November 3–5, 1998.

Kelly, WR, JS Herman, and AL Mills. The geochemical effects of benzene, toluene, and xylene (BTX) biodegradation. *Applied Geochemistry*. 12: 291–303. 1997.

Kennedy, JL. *Oil and Gas Pipeline Fundamentals*, 2nd edition Tulsa, OK, Pennwell Publ. 1993.

Koester, C. Evaluation of analytical methods for the detection of ethanol in ground and surface water. Volume 4: Potential Ground and Surface Water Impacts. Health and Environmental Assessment of the use of Ethanol as a Fuel Oxygenate. Report to the California Environmental Policy Council in Response to Execuitive Order D-5-99. 1999.

Kolhatkar, R, T Kuder, P Philp, J Allen, and JT Wilson. Use of compound-specific stable carbon isotope analyses to demonstrate anaerobic biodegradation of MTBE in groundwater at a gasoline release site. *Environmental Science and Technology*. 36 (23): 5139. 2002.

Kuder, T, P Philp, J Alen, R Kolhatkar, J Wilson, and J Landmeyer. Compound-specific stable isotope analysis to demonstrate in situ MTBE biotransformation. In situ and On-site bioremediation. 2003.

Larafgue, E and PL Thiez. Effect of waterwashing on light ends compositional heterogeneity. *Org. Geochem*. 24 (12): 1141–1150. 1996.

Leffler, WL. *Petroleum Refining*. 3rd edition. 2000. Tulsa, OK, PennWell Corp.

Lundegard, PD, RE Sweeney, and GT Ririe. Soil gas methane at petroleum contaminated sites; forensic determination of origin and source. *International Journal of Environmental Forensics*. 1 (1). 1999.

Mango, F. An invariance in the isoheptanes of petroleum. *Science*. 237: 514–517. 1987.

Mansell, RS, L Ou, RD Rhue, and Y Ouyang, 1995, The fate and behavior of lead alkyls in the subsurface environment: Final Tech. Rpt. to Armstrong Laboratory, Tyndall, Air Force Base, Battelle Memorial Institute, p. 102.

Manton, WI. Sources of lead in blood. *Arch. Environ. Health*. 42: 168–172. 1977.

Manton, WI. Total contribution of ariborne lead to blood lead. *Br. J. Ind. Med*. 42: 168–172. 1985.

Marchal, R, S Penet, F Solano-Serena, and JP Vandecasteele. Gasoline and Diesel Oil Biodegradation. *Oil & Gas Science and Technology*. 58(4), 441–448. 2003.

Marchetti, A, J Daniels, and D Layton. Chapter 6: Environmental transport and fate of fuel hydrocarbon alkylates. *Potential Ground and surface water impacts*. 4: 1–20. 1999.

McCarthy, KJ, AD Uhler, and SA Stout. Weathering affects petroleum ID. *Soil & Groundwater Cleanup*. 1998.

McHugh, TE, JA Connov, and F Ahmad. An empirical analysis of the groundwater-to-indoor-air exposure pathway: The role of background concentrations in indoor air. *Environmental Forensics*. 5 (1) 33–44. 2004.

Moran, MJ, RM Clawges, and JS Zogorski. Identifying the usage patterns of methyl tert-butyl ether (MTBE) and other oxygenates in gasoline using gasoline surveys. Exploring the Environmental Issues of Mobile, *Recalcitrant Compounds in Gasoline*. 40 (1): 209–212. 2000. American Chemical Society, Division of Environmental Chemistry. March 26–30, 2000.

Morrison, RD. Use of proprietary additives to date petroleum hydrocarbons. *Environmental Claims Journal*. 11 (3): 81–90. 1999.

Morrison, RD. *Critical review of environmental forensic techniques: Part II. Environmental Forensics* 1, 175–195. 2000a.

Morrison, RD. *Environmental Forensics*. 2000b. Boca Raton, CRC Press.

Mulroy, PT and L-T Ou. Degradation of tetraethyllead during the degradation of leaded gasoline hydrocarbons in soil. *Environmental Toxicology and Chemistry*. 17 (5): 777–782. 1998.

Needleman, HL. Review. Clair Patterson and Robert Kehoe: Two views of lead toxicity. *Environmental Research*, Section A 78, 79–85. 1998.

Newman, R, M Gilbert, and K Lothridge. *GC-MS Guide to Ignitable Liquids*. CRC Press, New York, p. 750. 1998.

Ou, L-T, W Jing, and JE Thomas. Biological and chemical degradation of ionic ethyllead compounds in soil. *Environmental Toxicology and chemistry*. 14 (4): 545–551. 1995a.

Ou, L-T, W Jing, and JE Thomas. Degradation and metabolism of tetraethyllead in soils. *Journal of Industrial Microbiology*. 14: 312–318. 1995b.

Owen, K and T Coley. A history of gasoline and diesel fuel development. *Automotive Fuels Handbook SAE International*, 5–23. 1990.

Paje, MLF, BA Neilan, and I Couperwhite. A Rhodococcus species that thrives on medium saturated with liquid benzene. *Microbiology*. 143: 2975–2981. 1997.

Pankow, JF, RE Rathbun, and JS Zogorski. Calculated volatilization rates of fuel oxygenate compounds and other gasoline-related compounds from rivers and streams. *Chemosphere*. 33 (5): 921–937. 1996.

Pauls, RE. Chromatographic characterization of gasolines. *Advances in Chromatography*. 35: 259–335. 1995.

Petersen, LW, P Moldrup, YH El-Farhan, OH Jacobsen, T Yamaguchi, and DE Rolston. The effect of moisture and soil texture on the adsorption of organic vapors. *Journal of Environmental Quality*. 24: 752–759. 1995.

Philp, RP. Application of Stable Isotopes and Radioisotopes in Environmental Forensics. Murphy, Brian L. and Morrison, Robert D. *Introduction to Environmental Forensics*. 99–136. 2002. San Diego, CA, Academic Press.

Pines, H. *The Chemistry of Catalytic Hydrocarbon Conversions*. 1981. New York, Academic Press.

Powers, SE, D Rice, B Dooher, and PJJ Alvarez. Will ethanol-blended gasoline affect groundwater quality? *Environmental Science and Technology*. 35(1), 24A–30A, 2001.

Prince, RC. Bioremediation. In Encyclopedia of Chemical Technology, Supplement to 4th edition Wiley, New York, 48–89. 1998.

Rabinowitz, MB, and GW Wetherill. Identifying sources of lead contamination by stable isotope techniques. *Environmental Science and Technology*. 6(8), 705–709. 1972.

Radler, M. Worldwide Refining Survey. *Oil & Gas Journal*. December 20, 1999, 45–89. 1999.

Reisinger, HJ and DR Burris. A Novel Environmental Forensics Tool to Relate NAPL, Soil and Groundwater Data by Conversion to a Common Frame of Reference. *Environmental Forensics: Theory, Applications, and Case Studies*, November 9–10, 2004, North Charleston, South Carolina. 2004.

Richardson, WL, MR Barusch, GJ Kautsky, and RE Steinke. Tetramethyllead–an improved antidetonant. *Journal of Chemical and Engineering Data*. 6 (2): 305–309. 1961.

Robert, JC. *Ethyl: A History of the Corporation and the People Who Made It*. 1983. Charlottesville, VA, University of Virginia Press.

Roggemans, S, CL Bruce, PC Johnson, and RL Johnson. Vadose Zone Natural Attenuation of Hydrocarbon Vapors: An Empirical Assessment of Soil Gas Vertical Profile Data. A Summary of Research Results from API's Soil and Groundwater Technical Task Force 15. 2001.

Rong, Y and H Kerfoot. Much ado: How big is the problem of hydrolysis of methyl tertiary butyl ether (MtBE) to Form tertiary butyl alcohol (tBA)? *Environmental Forensics*. 4: 239–243. 2003.

Rosman, KJR, W Chisholm, CF Boutron, JP Candelone, and S Hong. Isotopic evidence to account for changes in the concentration of lead in Greenland snow between 1960 and 1988. *Geochimica et Cosmochimica Acta*. 59 (15): 3265–3269. 1994.

Rostad, C. Analysis for solvent dyes in light non-aqueous phase liquid samples by electrospray ionization mass spectrometry. *Petrol. Chem. Div.* 47, 1. 2002.

Rostad, C. Analysis of solvent dyes in petroleum products by electrospray ionization mass spectrometry. *Anal. Chem.* 2003.

Sauer, T and H Costa, Fingerprinting of gasoline and coal tar NAPL volatile hydrocarbons dissolved in groundwater. *Environmental Forensics.* 4: 319–329, 2003.

Schaap, B. Occurrence of T-butanol (*tert*-butyl-alcohol, TBA, etc.) in gasoline samples from the NIPER survey. 2002. US Geological Survey.

Schmidt, GW, DD Beckmann, and BE Torkelson. A Technique for Estimating the Age of Regular/Mid-grade Gasolines Released to the Subsurface Since the Early 1970's. *Environmental Forensics.* 3: 145–162. 2002a.

Schmidt, TC, P Kleinert, C Stengel, KU Gross, and SB Haderlein. Polar fuel constituents: Compound identification and equilibrium partitioning between non-aqueous phase liquids and water. *Environmental Science and Technology.* 36 (19): 4074–4080. 2002b.

Schmidt, TC, P Kleinert, C Stengel, and SB Haderlein. Polar fuel constituents: Compounds identification and partitioning between nonaqueous-phase liquids and water. In: *Oxygenates in Gasoline—Environmental Aspects*, (A.F. Diaz and D.L. Drogos, eds.) *Am. Chem. Soc. Symposium Series 799*, 281–287. 2001.

Schwarzenbach, RP, PM Gschwend, and DM Imboden. *Environmental Organic Chemistry.* 1993. New York, John Wiley & Sons, Inc.

Senn, RB and MS Johnson, 1987, Interpretation of gas chromatographic data in subsurface hydrocarbon investigations. *GWMR*, v. Winter, pp. 58–63.

Shelton, EM. Motor gasoline, Winter 1979–1980. Depart. of Energy, Bartlesville Energy Technology Center DOE/BETC/PPS-80/3, Report No. 115. 1980a.

Shelton, EM. Motor gasolines, Summer 1979. Dept. of Energy, Bartlesville Energy Technology Center DOE/BETC/PPS-80/3, Report No. 115. 1980b.

Shelton, EM, ML Whisman, and PW Woodward. Trends in Motor Gasolines: 1942–1981. 1982. United States Department of Energy.

Shiomi, K, H Shimono, H Arimoto, and S Takahashi. High resolution capillary gas chromatographic hydrocarbon-type analysis of naphtha and gasoline. *Journal of High Resolution Chromatography.* 14, 729–737. 1991.

Slater, G, 2003, Stable isotope forensics—when isotopes work. *Environmental Forensics*, 4: 13–23.

Slater, GF, HD Dempster, B Sherwood Lollar, J Spivack, M Brennan, and P Mackenzie. Isotopic tracers of degradation of dissolved chlorinated solvents. Wickramanayake, Godage B. and Hinchee, Robert E. Natural Attenuation: Chlorinated and Recalcitrant Compounds (C1-3). *Proceedings of the First International Conference on Remediation of Chlorinated and Recalcitrant Compounds.* 133–138. 1998. Columbus, Ohio, Battelle Press. May 18–21, 1998.

Smallwood, BJ, RP Philp, TW Burgoyne, and JD Allen. The use of stable isotopes to differentiate specific source markers for MTBE. *Environmental Forensics.* 2 (3): 215–221. 2001.

Smith, DR, S Niemeyer, and AR Flegal. Lead sources to California sea otters: Industrial inputs circumvent natural lead biodepletion mechanisms. *Environ. Res.* 57: 163–175. 1992.

Solano-Serena, F, R Marchal, M Ropars, JM Lebeault, and JP Vandecasteele. Biodegradation of Gasoline: kinetics, mass balance and fate of individual hydrocarbons. *J. App. Microbio.* 86: 1008–1016. 1999.

Speight, JG. *The Chemistry and Technology of Petroleum.* 1991. New York, NY, Marcel Dekker, Inc.

Squillace, PJ, DA Pope, and CV Price. Occurrence of the gasoline additive MTBE in shallow ground water in urban and agricultural areas. 1995. Rapid City, SD, United States Geological Survey, National Water Quality Assessment Program. US Geological Survey Fact Sheet.

Squillace, PJ, JS Zogorski, WG Wilber, and CV Price. Preliminary assessment of the occurrence and possible sources of MTBE in groundwater in the United States, 1993–1994. *Environmental Science and Technology.* 30 (5): 1721–1730. 1996.

Stout, SA, JM Davidson, KJ McCarthy, and AD Uhler. Gasoline Additives—Usage of lead and MTBE. *Soil & Groundwater Cleanup* [Feb/Mar]. 1999b.

Stout, SA and, GS Douglas. Diamondoid hydrocarbons Application in the chemical fingerprinting of natural gas condensate and gasoline. *Environmental Forensics.* 5 (4): 225–235. 2004.

Stout, SA, AD Uhler, and KJ McCarthy. Fingerprinting of Gasolines. *Soil & Groundwater Cleanup.* Dec/Jan. 1999a.

Stout, SA, AD Uhler, KJ McCarthy, and S Emsbo-Mattingly. Chemical fingerprinting of hydrocarbons. Murphy, B.L. and Morrison, R.D. Introduction to Environmental Forensics. 137–260. 2002. Boston, Academic Press.

Stout, SA and PD Lundegard. Intrinsic biodegradation of diesel fuel in an interval of separate phase hydrocarbons. *Applied Geochemistry.* 13 (7), 851–859. 1998.

Stout, SA, AD Uhler, KJ McCarthy, S Emsbo-Mattingly, and GS Douglas. The influences of refining on petroleum fingerprinting. Part 2. Gasoline blending practices. *Contaminated Soil Sediment and Water* [December]. 2001.

Stratco, P and A Gertz. Refining options for MTBE-free gasoline. *NPRA Annual Mtg.*, Publ. No. AM-00-53, San Antonio, TX. 2002.

Stumpf, A, K Tolvay, and M Juhasz. Detailed analysis of sulfur compounds in gasoline range petroleum products with high-resolution gas chromatography-atomic emission detection using group-selective chemical treatment. *Journal of Chromatography.* A 819: 67–74. 1998.

Sturges, WT and LA Barrie. Lead 206/207 isotope ratios in the atmosphere of North America as tracers of the US and Canadian emissions. *Nature.* 329, 144–146. 1987.

Tissot, BP and DH Welte. *Petroleum Formation and Occurrence.* 1984. Berlin, Springer-Verlag.

Townsend, GT, RC Prince, and JM Suflita. Anaerobic biodegradation of alicyclic constituents of gasoline and natural gas condensate by bacteria from an anoxic aquifer. *FEMS Microbiology Ecology.* 49: 129–135. 2004.

TPH Criteria Working Group. *Selection of representative TPH fractions based on fate and transport considerations.* 3: 102. 1997.

Tupa, RC, and CJ Dorer. Gasoline and diesel fuel additives for performance/distribution quality—II. 69–97. 1983.

United States Environmental Protection Agency. Causes of release from UST systems: (and) attachments. Final Report to US Environmental Protection Agency, Office of Underground Storage Tanks. 1987. Washington, DC, US Environmental Protection Agency, Solid Waste and Emergency Response.

United States Environmental Protection Agency. Test methods for evaluating solid waste, Physical/Chemical Methods (SW-846). 3rd Edition, Office of Solid Waste and Emergency Response. 1996. Washington, DC, US Environmental Protection Agency.

United States Environmental Protection Agency. Compendium Method TO-15. Compendium of methods for the determination of toxic organic compounds in ambient air. Second edition. United States Environmental Protection Agency/625/R-96/010b. Center for Environmental Research Information. 1999. Washington, DC, US Environmental Protection Agency.

United States Environmental Protection Agency. Draft guidance for evaluating the vapor intrusion to indoor air pathway from groundwater and soil (Subsurface Vapor Intrusion Guidance),. *Federal Register* 67, 2002. Washington, DC, US Environmental Protection Agency.

US Environmental Protection Agency Blue Ribbon Panel. Achieving clean air and clean water. The Report of the Blue Ribbon Panel on Oxygenates in Gasoline. 116 pp. 1999. Washington, DC, US Environmental Protection Agency.

United States Environmental Protection Agency. Office of Emergeny and Remedial Response. Users guide for evaluating subsurface vapor intrusion into buildings. United States Environmental Protection Agency Contract Number: 68-W-02-33. 2004. Washington, DC, US Environmental Protection Agency.

United States Environmental Protection Agency. The Inside Story: A Guide to Indoor Air Quality. United States Environmental Protection Agency Document # 402-K-93-007. 1995. Washington, DC, US Environmental Protection Agency.

Uhler, AD, SA Stout, and KJ McCarthy. Increase success of assessments at petroleum sites in 5 steps. *Soil & Groundwater Cleanup.* 13–19. 1998–1999.

Uhler, RM, EM Healey, KJ McCarthy, AD Uhler, and SA Stout. Molecular fingerprinting of gasoline by a modified United States Environmental Protection Agency 8260 gas chromatography/mass spectrometry method. *Int. J. Environ.* Anal. Chem. 83 (1): 1–20. 2002.

United States Environmental Protection Agency and the United States Consumer Product Safety commission Office of Radiation and Indoor Air. The Inside Story: A Guide to Indoor Air Quality. Environmental Protection Agency Document #402-K-93-007. 1995.

University of Wisconsin. Ethanol detection methods. *Underground Tank Technology Update.* 14 (4): 12–14. 2000.

Veron, A, TM Church, CC Patterson, Y Erel, and JT Merrill. Continental origin and industrial sources of trace metals in the northwest Atlantic troposphere. *Journal of Atmospheric Chemistry.* 14: 339–351. 1992.

Voyksner, R. Characterization of dyes in environmental samples by thermospray high-performance liquid chromatography/mass spectrometry. *Anal. Chem.* 57: 2600–2605. 1985.

Wade, M. History and composition of PS-6 gasoline – Implications to age-dating gasoline contamination. *Environmental Forensics.* 4, 89–92. 2003.

Wakim, JM, H Schwendener, and J Shimosato. Gasoline octane improvers. CEH Marketing Research Report. 1990. Menlo Park, CA, SRI International.

Wang, Y, Y Huang, J Huckins, and J Petty. Compound-specific carbon and hydrogen isotope analysis of sub-parts per billion level waterborne petroleum hydrocarbons. *Environmental Science Technology.* 38: 3689–3697. 2004.

Weaver, JW, L Jordan, and DB Hall. Predicted ground water, soil, and soil gas impacts from US gasolines, 2004—First Analysis of the Autumnal Data. US Environmental Protection Agency Publication 600/R-05/032, February 2005, p. 89. 2005.

Wenger, LM, CL Davis, and GH Isaksen. Multiple controls on petroleum biodegradation and impact on oil quality. 2001. Richardson, TX, Society of Petroleum Engineers, Inc.

Whittmore, IM, KH, Altegelt, and Gouw, TH. *Chromatography in Petroleum Analysis.* pp. 50–70. 1979. New York, Marcel Dekker.

Widdel, F and R Rabus. Anaerobic biodegradation of saturated and aromatic hydrocarbons. *Current Opinions in Biotechnology.* 12: 259–276. 2001.

Yerushalmi, L and SR Guiot. Kinetics of biodegradation of gasoline and its hydrocarbon constituents. *Appl. Microbiol.* 49: 475–481. 1998.

Youngless, T, J Swansiger, D Danner, and M Greco. Mass Spectral characterization of petroleum dyes, tracers, and additives. *Anal. Chem.* 57: 1894–1902. 1985.

Zayed, J, B Hong, and G L'Esperance. Characterization of manganese-containing particles collected from the exhaust emissions of automobiles running with MMT additive. *Environmental Science and Technology.* 33: 3341–3346. 1999.

Zhou, E and RL Crawford. Effects of oxygen, nitrogen and temperature on gasoline biodegradation. *Biodegradation.* 6: 127–140. 1995.

Subject Index

Page numbers in *italics* refer to figures and tables, margin notes by suffix 'n' (e.g. '404n')

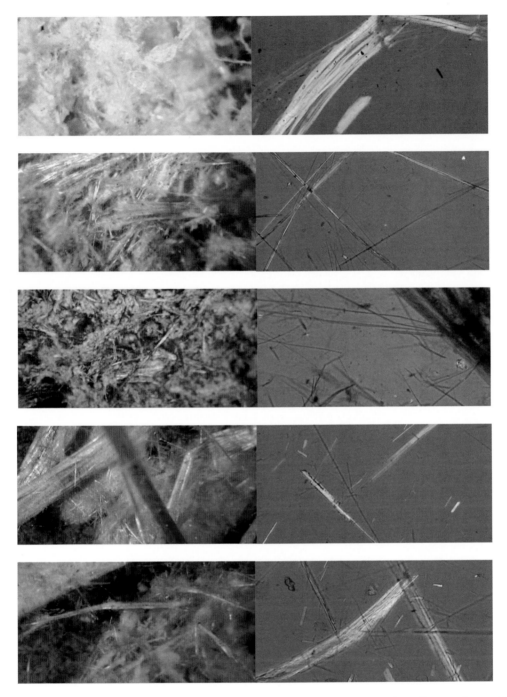

Figure 2.2.1 Photographs of the regulated asbestos minerals. From top to bottom: chrysotile, amosite, crocidolite, anthophyllite, and tremolite/actinolite. The photographs on the left are taken at a magnification of 3×. The photographs on the right are taken in a polarized light microscope at a magnification of approximately 100×, with crossed polars, inserted wave retardation plate, and the mineral immersed in a matching refractive index oil.

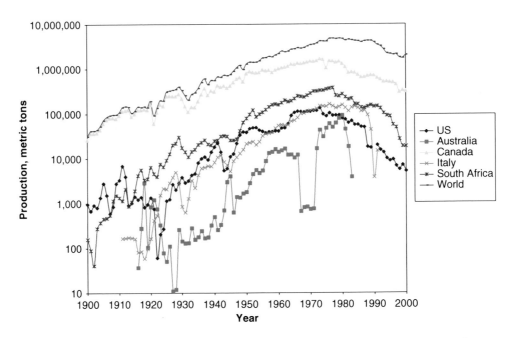

Figure 2.2.5 Historical production of asbestos for selected countries, compiled from U.S. Geological Survey records (Virta, 2003).

CARBOFURAN
ESTIMATED ANNUAL AGRICULTURAL USE

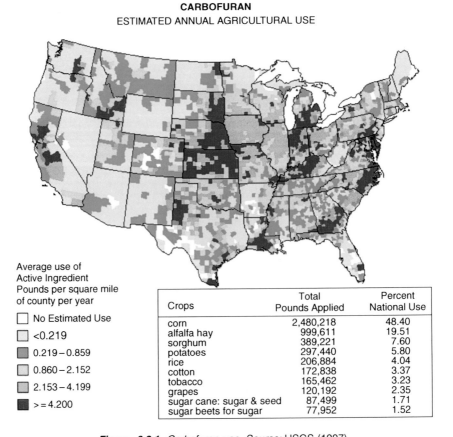

Average use of
Active Ingredient
Pounds per square mile
of county per year

☐ No Estimated Use
☐ <0.219
▨ 0.219 – 0.859
☐ 0.860 – 2.152
☐ 2.153 – 4.199
■ >= 4.200

Crops	Total Pounds Applied	Percent National Use
corn	2,480,218	48.40
alfalfa hay	999,611	19.51
sorghum	389,221	7.60
potatoes	297,440	5.80
rice	206,884	4.04
cotton	172,838	3.37
tobacco	165,462	3.23
grapes	120,192	2.35
sugar cane: sugar & seed	87,499	1.71
sugar beets for sugar	77,952	1.52

Figure 8.2.1 Carbofuran use. Source: USGS (1997).

Figure 9.3.1 Soil sample containing perchlorate from Mission Valley Formation, California.

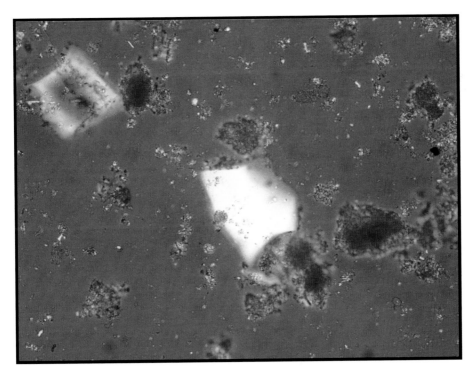

Figure 9.3.3 Polarized light microscope image of perchlorate-enriched soil from Mission Valley Formation, San Diego County, California.

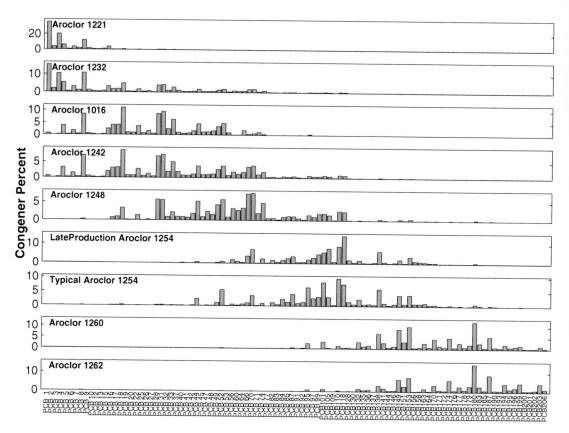

Figure 10.1.2 *Congener-specific compositions of Aroclor formulations (Frame et al., 1996). Frame reports data for all 209 congeners. Only the 100 most abundant congeners are shown in this figure.*

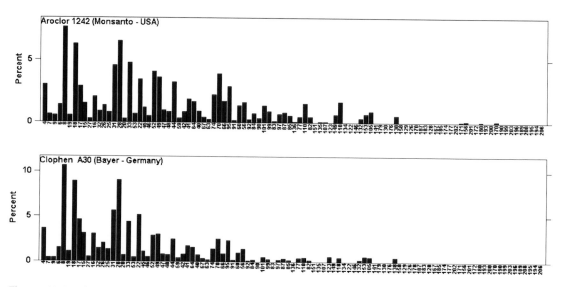

Figure 10.1.3 *Comparison of congener profiles of Aroclor 1242 (Monsanto–USA) and Clophen A30 (Bayer–Germany). The similarity between these patterns is visually evident, and the cos-θ similarity metric calculated between these two patterns is 0.96 (data from Shulz et al., 1989).*

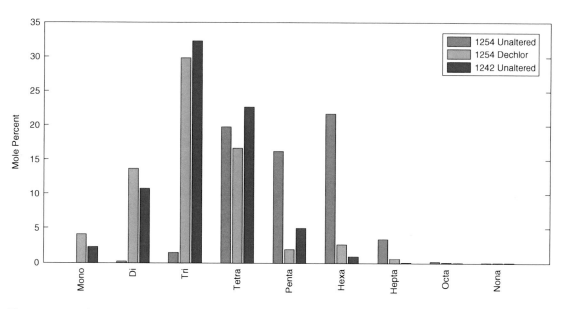

Figure 10.2.2 *Comparison of homolog data for unaltered Aroclor 1254; dechlorinated Aroclor 1254, and unaltered Aroclor 1242. The dechlorinated Aroclor 1254 is more similar to 1242 than it is to the original 1254 from which it was derived. Data from Quensen et al., (1990).*

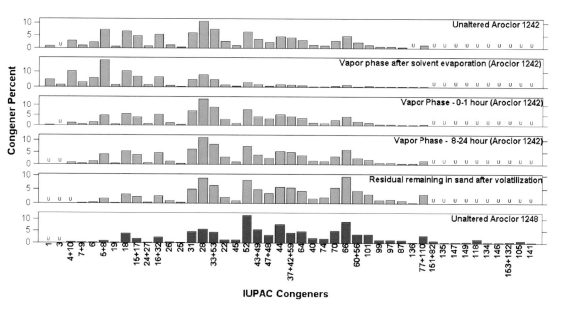

Figure 10.3.1 *Congener patterns observed in Aroclor 1242 volatilization experiments (Chiarenzelli et al., 1997). An unaltered Aroclor 1248 pattern is shown in the bottom graph for comparison to the Aroclor 1242 residual pattern. These data are included in Appendix, along with data from Chiarenzelli's other Aroclor volatilization experiments.*

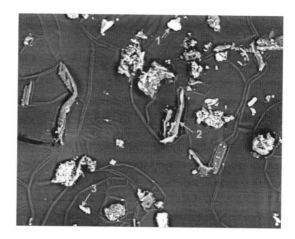

Figure 13.3.1 *Particles identified on the basis of semiquantitative EDS analysis of arsenic and other elements within a 240 × SEM image. Particle 1 – clay; Particle 2 – particle from arsenic source material; Particle 3 – quartz. Figure reproduced courtesy of Battelle.*

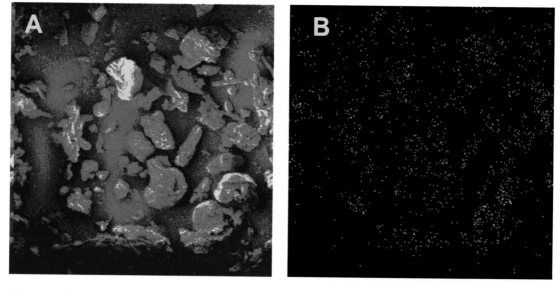

Figure 13.3.2 *TOF SIMS image of a 300 μm × 300 μm area containing mixed particles collected from outdoor surfaces where an arsenic source is present. (A) shows a total ion map and (B) shows an arsenic map. Figure reproduced courtesy of Battelle.*

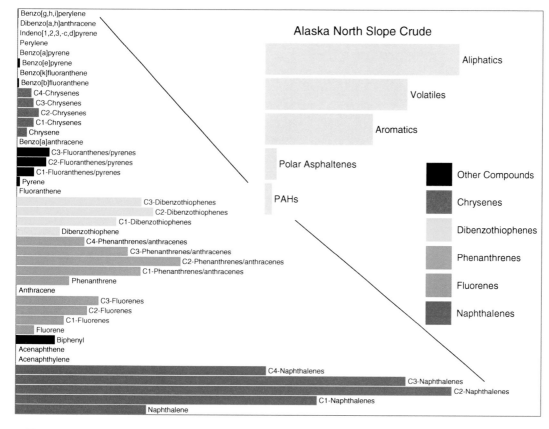

Figure 15.2.2 *Polycyclic aromatic hydrocarbons in Alaska North Slope Crude Oil (from Boehm et al., 2001a).*

- Petrogenic
  - ➤ Alkyl > Parent
  - ➤ Little 4 to 6 Ring

- Pyrogenic – Type 1
  - ➤ Parent > Alkyl
  - ➤ High 2 and 3 Ring

- Pyrogenic – Type 2
  - ➤ Parent > Alkyl
  - ➤ High 4 to 6 Ring

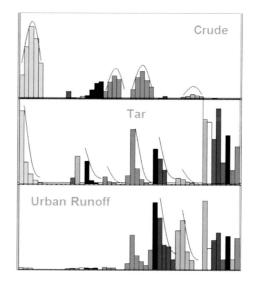

Figure 15.4.2 *Typical Characteristics of PAH Assemblages for Petrogenic and Pyrogenic Sources.*

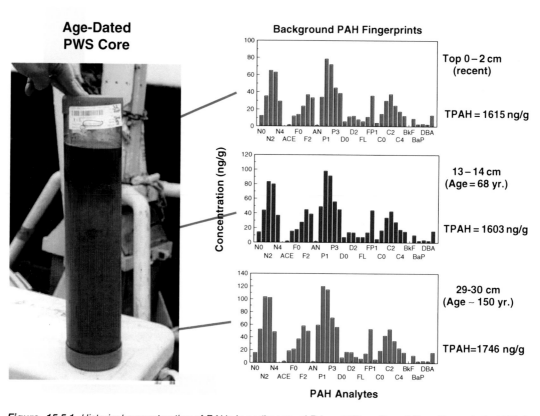

Figure 15.5.1 *Historical reconstruction of PAHs in sediments of Prince William Sound (from Page et al., 1995a).*